PLEASE STAMP DATE DUE

Geology
GC84 .N65 2001
Non-volcanic rifting of
continental margins : a
comparison of evidence from
land and sea

AF479059

Non-Volcanic Rifting of Continental Margins: A Comparison of Evidence from Land and Sea

Geological Society Special Publications
Series Book Editors
A. J. FLEET (CHIEF EDITOR)
P. DOYLE
F. J. GREGORY
J. S. GRIFFITHS
A. J. HARTLEY
R. E. HOLDSWORTH
A. C. MORTON
N. S. ROBINS
M. S. STOKER
J. P. TURNER

Special Publication reviewing procedures

The Society makes every effort to ensure that the scientific and production quality of its books matches that of its journals. Since 1997, all book proposals have been refereed by specialist reviewers as well as by the Society's Publications Committee. If the referees identify weaknesses in the proposal, these must be addressed before the proposal is accepted.

Once the book is accepted, the Society has a team of series editors (listed above) who ensure that the volume editors follow strict guidelines on refereeing and quality control. We insist that individual papers can only be accepted after satisfactory review by two independent referees. The questions on the review forms are similar to those for *Journal of the Geological Society*. The referees' forms and comments must be available to the Society's series editors on request.

Although many of the books result from meetings, the editors are expected to commission papers that were not presented at the meeting to ensure that the book provides a balanced coverage of the subject. Being accepted for presentation at the meeting does not guarantee inclusion in the book.

Geological Society Special Publications are included in the ISI Science Citation Index, but they do not have an impact factor, the latter being applicable only to journals.

More information about submitting a proposal and producing a Special Publication can be found on the Society's web site: www.geolsoc.org.uk.

GEOLOGICAL SOCIETY SPECIAL PUBLICATION No. 187

Non-Volcanic Rifting of Continental Margins: A Comparison of Evidence from Land and Sea

EDITED BY

R.C.L. WILSON
The Open University, UK

R.B. WHITMARSH
Southampton Oceanography Centre, UK

B. TAYLOR
University of Hawaii, Hawaii

N. FROITZHEIM
University of Bonn, Germany

2001
Published by
The Geological Society
London

THE GEOLOGICAL SOCIETY

The Geological Society of London (GSL) was founded in 1807. It is the oldest national geological society in the world and the largest in Europe. It was incorporated under Royal Charter in 1825 and is Registered Charity 210161.

The Society is the UK national learned and professional society for geology with a worldwide Fellowship (FGS) of 9000. The Society has the power to confer Chartered status on suitably qualified Fellows, and about 2000 of the Fellowship carry the title (CGeol). Chartered Geologists may also obtain the equivalent European title, European Geologist (EurGeol). One fifth of the Society's fellowship resides outside the UK. To find out more about the Society, log on to www.geolsoc.org.uk.

The Geological Society Publishing House (Bath, UK) produces the Society's international journals and books, and acts as European distributor for selected publications of the American Association of Petroleum Geologists (AAPG), the American Geological Institute (AGI), the Indonesian Petroleum Association (IPA), the Geological Society of America (GSA), the Society for Sedimentary Geology (SEPM) and the Geologists' Association (GA). Joint marketing agreements ensure that GSL Fellows may purchase these societies' publications at a discount. The Society's online bookshop (accessible from www.geolsoc.org.uk) offers secure book purchasing with your credit or debit card.

To find out about joining the Society and benefiting from substantial discounts on publications of GSL and other societies world-wide, consult www.geolsoc.org.uk, or contact the Fellowship Department at: The Geological Society, Burlington House, Piccadilly, London W1J 0BG: Tel. +44 (0)20 7434 9944; Fax +44 (0)20 7439 8975; Email: enquiries@geolsoc.org.uk.

Published by The Geological Society from:
The Geological Society Publishing House
Unit 7, Brassmill Enterprise Centre
Brassmill Lane
Bath BA1 3JN, UK
(*Orders*: Tel. +44 (0)1225 445046
Fax +44 (0)1225 442836)
Online bookshop: http://bookshop.geolsoc.org.uk

The publishers make no representation, express or implied, with regard to the accuracy of the information contained in this book and cannot accept any legal responsibility for any errors or omissions that may be made.

© The Geological Society of London 2001. All rights reserved. No reproduction, copy or transmission of this publication may be made without written permission. No paragraph of this publication may be reproduced, copied or transmitted save with the provisions of the Copyright Licensing Agency, 90 Tottenham Court Road, London W1P 9HE. Users registered with the Copyright Clearance Center, 27 Congress Street, Salem, MA 01970, USA: the item-fee code for this publication is 0305-8719/01/$15.00.

British Library Cataloguing in Publication Data
A catalogue record for this book is available from the British Library.

ISBN 1-86239-091-6
ISSN 0305-8719

Typeset by Alden Bookset
Printed by The Alden Press, Oxford, UK.

Distributors

USA
  AAPG Bookstore
  PO Box 979
  Tulsa
  OK 74101-0979
  USA
  Orders: Tel. +1 918 584-2555
      Fax +1 918 560-2652
      E-mail *bookstore@aapg.org*

Australia
  Australian Mineral Foundation Bookshop
  63 Conyngham Street
  Glenside
  South Australia 5065
  Australia
  Orders: Tel. +61 88 379-0444
      Fax +61 88 379-4634
      E-mail *bookshop@amf.com.au*

India
  Affiliated East-West Press PVT Ltd
  G-1/16 Ansari Road, Daryaganj,
  New Delhi 110 002
  India
  Orders: Tel. +91 11 327-9113
      Fax +91 11 326-0538
      E-mail *affiliat@nda.vsnl.net.in*

Japan
  Kanda Book Trading Co.
  Cityhouse Tama 204
  Tsurumaki 1-3-10
  Tama-shi
  Tokyo 206-0034
  Japan
  Orders: Tel. +81 (0)423 57-7650
      Fax +81 (0)423 57-7651

Contents

It is recommended that reference to all or part of this book should be made in one of the following ways:

WILSON, R.C.L., WHITMARSH, R.B., TAYLOR, B. & FROITZHEIM, N. 2001. *Non-Volcanic Rifting of Continental Margins: A Comparison of Evidence from Land and Sea.* Geological Society, London, Special Publications, **187**.

CHALMERS, J.A. & PULVERTAFT, T.C.R. 2001. Development of the continental margins of the Labrador Sea—a review. *In:* WILSON, R.C.L., WHITMARSH, R.B., TAYLOR, B. & FROITZHEIM, N. 2001. *Non-Volcanic Rifting of Continental Margins: A Comparison of Evidence from Land and Sea.* Geological Society, London, Special Publications, **187**, 77–105.

Introduction: the land and sea approach

R. C. L. WILSON[1], R. B. WHITMARSH[2], N. FROITZHEIM[3] & B. TAYLOR[4]

[1]*Department of Earth Sciences, The Open University, Milton Keynes MK7 6AA, UK*
(e-mail: r.c.l.wilson@open.ac.uk)
[2]*Southampton Oceanography Centre, European Way, Southampton SO14 3ZH, UK*
(e-mail: bob.whitmarsh@soc.soton.ac.uk)
[3]*Geologisches Institut, Nussallee 8, Bonn D-53115, Germany*
(e-mail: froitzheim@ubaclu.unibas.ch)
[4]*SOEST, University of Hawaii, 2525 Correa Road, Honolulu HI-96822, USA*
(e-mail: taylor@soest.hawaii.edu)

The idea for a special publication on non-volcanic margins arose during Ocean Drilling Program (ODP) Leg 173 off West Iberia, prompted by ODP's decision to cease publishing the scientific results volumes as hard copy. The Shipboard Scientific Party favoured an open scientific meeting and associated publication. But they did not want to produce a book that was a scientific results volume by another name, but rather contribute to a publication that had a much broader scope than just reporting results obtained off West Iberia. These thoughts, and many scientific discussions during the Leg, were influenced by the presence on board of scientists who also work on Alpine geology: hence the 'evidence from land and sea' approach that underlies the content of this publication.

However, when planning the meeting, we were very conscious of the fact that the West Iberia and Alpine examples might not be typical of other non-volcanic margins. We were keen, therefore, to ensure that margins in other parts of the world were discussed, including a margin that is active today, and that was visited by the *JOIDES Resolution* not long after Leg 173 took place (Leg 180: Woodlark Basin). We caution, therefore, that it may be premature to use models based on the Iberia and Tethyan margins as the paradigm for all non-volcanic margins.

The first paper in this book, by **Boillot & Froitzheim**, reviews the synergies that have occurred between investigations of the eastern Atlantic non-volcanic margins and remnants of similar Mesozoic margins preserved in the Alps. The key points they make about the way oceanic evidence has influenced Alpine work, and vice versa, are as follows:

Sea to land: (1) tilted fault blocks first recognized beneath the Armorican margin were later interpreted to be present in fragments of Mesozoic margins preserved in the Alps; (2) identification of the serpentinite ridge off Galicia led to the interpretation of basalt-deficient ophiolites (the Steinmann Trinity of peridotite, pillow lavas and radiolarites) in the Alps as pieces of the ocean–continent transition (OCT).

Land to sea: (1) the Basin and Range simple shear detachment model was used to explain the 'unroofing' of mantle rocks off Iberia, the nature of the S-reflector off Armorica and Galicia, and the significance of low-angle detachment faults in rifted margins in general; (2) field studies of ancient detachment faults and shear zones in the Alps aided the interpretation of the structure of the Iberian margin.

In addition to demonstrating the benefits of integrating offshore and onshore studies, we also wished to illustrate the different kinds of data that are obtained at different scales when comparing evidence from land and sea. This is not straightforward, for different datasets collected across a range of spatial scales have to be evaluated: thin sections, cores, outcrops, seismic reflection profiles and other geophysical data (see **Wilson et al.**, fig. 1). The outcrop scale is crucial because it allows the spatial gulf to be bridged between Deep Sea Drilling Project (DSDP) and ODP cores, and marine seismic data. There is also the problem that basins on land and beneath the sea inevitably have had different post-rift histories resulting in their contrasting present-day elevation. In mountain belts, portions of continental margins and oceanic crust are superbly exposed, but dismembered by subsequent compressional tectonics. Off present-day passive margins, extensional features

From: WILSON, R.C.L., WHITMARSH, R.B., TAYLOR, B. & FROITZHEIM, N. 2001. *Non-Volcanic Rifting of Continental Margins: A Comparison of Evidence from Land and Sea.* Geological Society, London, Special Publications, **187**, 1–8. 0305-8719/01/$15.00 © The Geological Society of London 2001.

have been only slightly deformed, if at all, by compressional movements, but are buried beneath significant thicknesses of post-rift sediments and so can only be sampled by ocean drilling at a small number of points.

Four sections follow the introductory paper by **Boillot & Froitzheim**: Margin overviews; Exhumed crust and mantle; Tectonics and stratigraphy; Numerical models of extension and magmatism.

Margin overviews

The first four papers in this section present the results of marine geophysical studies of fossil non-volcanic margins; the fifth paper examines the history of a rifted margin exposed in the Alps. Key issues addressed in the papers are highlighted below.

Rift propagation, timing, distribution and crustal models

Huchon *et al.* discuss a propagating rift system that was active beneath the South China Sea before *c.* 15 Ma. They suggest that strain localization occurs at the tip of the propagating rift, where continental crust is stretched by a factor of four. Stretching was significantly less in regions further from the tip, falling to a factor of two some 40 km ahead of it. This 3D view of rifting is important, as the process is so often considered in a simple 2D framework, a point highlighted by **Whitmarsh *et al.*** Three of the margins described show evidence for rifting leading to continental break-up being preceded by earlier rift phases. Two such phases occurred in the central Great Australian Bight (**Sayers *et al.***) and one in both the Labrador Sea (**Chalmers & Pulvertaft**) and the Southern Alps (**Bertotti**). The three stages of rifting in the central Great Australian Bight span 55 Ma, and are, according to **Sayers *et al.***, linked to pure shear occurring at different levels within a four-layer system (brittle upper crust and upper mantle, ductile lower crust and lower lithospheric mantle). The ductile lower crust is interpreted to have acted as a décollement zone between the brittle upper crust and mantle, so that by the time extension had ceased, brittle upper crust and upper mantle were juxtaposed. The model used by **Sayers *et al.*** was first developed by Brun & Beslier (see reference in their paper), and explains many of the features of the West Iberia margin. At the lithospheric scale it is a pure shear model, but involves the development of asymmetric structures produced by simple shear.

Bertotti describes two stages of rifting in the Southern Alps, in Norian–early Liassic and Toarcian–Mid-Jurassic time, that led to break-up. These were preceded by a major thermal event in Mid-Triassic (Ladinian) time. The latter event resulted in cooling of the lithosphere during rifting. This led to higher subsidence rates during the early rifting episode than predicted by crustal stretching models, and a shift in the locus of the second episode of rifting to a weaker less extended region relative to the strengthened lithosphere under the early rift. The **Bertotti** model, like that of **Sayers *et al.***, assumes that lower crust is ductile rather than brittle. It should be noted, however, that in Part 3 of this book field results reported by **Müntener & Hermann** indicate boudinage of a strong lower crust during extension in another part of the Alps. Bertotti suggests that the lack of a thermal anomaly at the end of rifting explains the lack of major post-rift subsidence.

Origin of the ocean–continent transition (OCT) zone

Discussions of the OCT are frequently dogged by semantic problems; here we favour the definition that the OCT is that region of uncertain affinity lying between fault blocks of thinned continental crust and crust that has the unequivocally normal geophysical characteristics of oceanic crust.

Sayers *et al.* briefly review the three main hypotheses that have been advanced for the origin of the OCT: (1) highly extended continental lithosphere characterized by crustal-scale detachment faults; (2) oceanic lithosphere formed by ultra-slow sea-floor spreading; (3) tectonically and magmatically disrupted continental crust with underplated gabbros. **Sayers *et al.*** conclude that the OCT in the Great Australian Bight is composed of highly deformed and intruded continental crust. They suggest that a significant amount of lava may have erupted at the time of break-up and flowed in a landward direction across the transition zone. A basement ridge complex on the inboard edge of the transition zone is interpreted as a combination of serpentinized peridotite and gabbros derived from mantle upwelling that produced a magnetic anomaly unrelated to sea-floor spreading.

Chalmers & Pulvertaft review the features and origin of continental margins bordering the Labrador Sea, and conclude that south of 62°N the margins are non-volcanic. These authors are equivocal about the nature of the OCT: it may consist of very thin continental crust overlying serpentinized peridotites, or entirely of serpentinized peridotites, or thin basalt overlying serpentinized peridotites.

Whitmarsh *et al.* review magnetic and seismic data that indicate that the 30–170 km wide OCT off West Iberia is distinct from adjacent oceanic and continental crusts. Their interpretation of these data suggests that only small amounts of synrift melt occur within the OCT. They coin the term 'octagenesis' for the processes that produced the OCT in this area. This involved tectonic exhumation of mantle, and the emplacement of 'isolated, margin-parallel, intrusive melt bodies'. The bodies are scattered in the landward part of the OCT and occur well below the top of acoustic basement. On the oceanward (western) side of the OCT, the intrusions reached higher levels as extension proceeded, so that in the westernmost 30 km a continuous belt of intrusions occurred that is, magnetically, indistinguishable from crust produced by sea-floor spreading. The process of octagenesis proposed by **Whitmarsh *et al.*** is, therefore, a fourth hypothesis for the origin of the OCT to be added to the three alternatives reviewed by **Sayers *et al.*** **Whitmarsh *et al.*** suggest that octagenesis has occurred to form the OCT along other non-volcanic margins that show low basement relief, weak magnetic anomalies parallel to the margin, and a similar velocity structure to that observed off West Iberia.

The examination of the nature and evolution of the OCT is continued in the next section.

Evolution during exhumation

Four papers in this section report analyses of ODP core material from the West Iberia margin, and two examine the evolution of a Jurassic passive margin, now exposed in the Eastern Alps, that was once situated on the eastern margin of the Ligurian–Piemontese ocean.

West Iberia

A trio of papers (**Abe**, **Hébert *et al.*** and **Gardien *et al.***) present petrological, geochemical and isotopic age data obtained from serpentinized peridotites, gabbros and amphibolites drilled during ODP Leg 173. The fourth paper (**Zhao**) uses rock magnetic studies to attempt to constrain the age of peridotite emplacement. During this Leg, and during Leg 149, there was considerable debate as to whether the cored mafic rocks were related to Cretaceous rifting

processes, or were derived from older subcontinental lower crust.

On the basis of her geochemical data (Cr/(Cr + Al) atomic ratios, Na_2O content and REE patterns), **Abe** concludes that the peridotites encountered in Holes 1067A and 1070A are intermediate between abyssal and subcontinental peridotites. In particular, the trace elements present in clinopyroxenes are similar to peridotite xenoliths from arc settings, suggesting they originated as pre-rift, possibly Precambrian Ibero-Armorican arc material. **Hébert *et al.*** present the first PGE (platinum group elements) analyses on peridotites and gabbros obtained from passive margins. Like **Abe**, they conclude that the peridotites are of subcontinental origin. **Gardien *et al.*** document three stages in the exhumation of amphibolites and acidic gneisses recovered from Hole 1067A. The first, amphibolite-facies, stage indicates a temperature of 670 $\pm$ 40 °C and a pressure of 7 $\pm$ 1 kbar. This was followed by conditions in the range of 550 $\pm$ 50 °C and 5.5 $\pm$ 1 kbar (low-grade amphibolite facies) after which temperatures fell below 500 °C and pressures below 3 kbar (greenschist facies), with the final metamorphic evolution ending under ocean-floor conditions. Oxygen isotope and fluid inclusion studies indicate that magmatic water-rich fluids occurred during the first stage, but that low-temperature water-rich fluids occurred in the two later stages, probably containing a proportion of sea water. Apatite fission-track dating reveals two thermal excursions below 120 °C. That between 113 and 100 Ma is probably related to cooling after Cretaceous rifting, but the later episode at 75–55 Ma may be linked to uplift following the Pyrenean orogeny.

The radiometric ages obtained from the amphibolites sampled by ODP Legs 149 and 173 reported here by **Gardien *et al.*** and the more recently acquired dates reported by **Manatschal *et al.*** are related to Cretaceous rifting save for one enigmatic Jurassic (161 Ma) date. However, the formation of the gabbro protolith occurred much earlier, during Late Palaeozoic time. **Hébert *et al.*** conclude that the Site 1070 gabbros they studied were derived from E-MORB magma, probably emplaced in a very slow-spreading setting. **Manatschal *et al.*** report hornblende and plagioclase cooling ages of 119 Ma and 110 Ma from a hornblende gabbro pegmatite at this site, and suggest that the gabbros at Site 1070 'document the end of rifting or the onset of sea-floor spreading'. **Hébert *et al.*** document deformation and retrograde metamorphism of peridotites and gabbros that is 'consistent with exhumation in a rift environment postdating the 120 Ma magmatic stage'.

The paper by **Zhao** reports the results of palaeomagnetic and rock magnetic studies of serpentinized peridotites from Sites 897, 899 and 1070 in the Iberia Abyssal Plain. At all three Sites two zones occur in the serpentinized peridotites: an upper zone 'pervasively veined and altered brown-coloured' (comparable with the Alpine ophicalcites discussed by **Desmurs et al.**) and a lower zone of fresher rocks. In the upper, more altered zone the remanent magnetization is dominated by a single stable component of normal polarity, but in the lower zone reversed polarity dominates. In addition, the inclinations in both zones are much shallower than that of the present-day magnetic field and are similar to those that would have characterized the mid-Cretaceous long normal Superchron at the latitude of Iberia. The reversed polarity exhibited by the fresher peridotites could, therefore, indicate emplacement during a short reversed polarity event in mid-Cretaceous time (probably Late Barremian–Aptian time, *c.* 121 Ma). However, given the suggestion by **Wilson et al.** in Section 4 that rifting may have occurred during late Berriasian and Early Valanginian time, emplacement during several earlier reversed polarity events is also possible.

Eastern Alps

Desmurs et al. describe new evidence concerning the 'Steinmann Trinity' of serpentinites, basalts and radiolarites, the characteristic stratigraphy of which corresponds broadly to the stratigraphy of the OCT along some non-volcanic margins except for the scarcity of basalts in the West Iberia OCT. They describe the stages of mantle exhumation and magmatism inferred from field observations made in the Platta nappe of eastern Switzerland. They conclude that the serpentinized peridotites are subcontinental mantle rocks that were exhumed along low-angle detachment faults. These rocks show evidence that deformation occurred at falling temperatures, culminating in exposure at the sea floor and the formation of tectonosedimentary breccias (ophicalcites). Gabbros were intruded into the peridotites at shallow depths, and subsequently suffered low-pressure hydration as hot oceanic water permeated them, followed by static alteration immediately beneath the sea floor. Pillow lavas overlie the mafic rocks and the tectonosedimentary breccias, but as their geochemical character is similar to that of the gabbros, they 'may be the products of a steady process that combined extensional deformation with magma generation and emplacement'. **Des-murs et al.** suggest that the lavas record the onset of slow sea-floor spreading.

Müntener & Hermann focus on the lower continental crust and upper mantle, concluding that the peridotites exposed in the Eastern Alps preserve a high-temperature evolution unrelated to Jurassic rifting and break-up. Petrological and microstructural data indicate a two-stage history for the peridotites and gabbros. Pre-rift high temperature (>650 °C) shearing and annealing of microstructures during Permian time were followed by early rift-related localized mylonitic shear zones cutting earlier structures and developing 'during nearly isothermal decompression' (<600 °C) followed by cooling and hydration. The mylonitic zones formed during exhumation of the lower crust and upper mantle to depths between 10 and 15 km. The large shear zones excised 10–20 km of intermediate and lower crust and were linked in space and time to the formation of rift basins. **Müntener & Herman** suggest that the lower crust was boudinaged during this stage, and that this process may explain the absence or scarcity of lower crust in some passive margins.

Tectonics and stratigraphy

Some papers in this section examine the development of both high- and low-angle normal faults. They include the results of both numerical modelling studies and the examination of samples recovered from fault zones in the Southern Alps, and from faults drilled by ODP in the Woodlark Basin and off West Iberia. Other papers deal with the stratigraphy of both these latter areas, and the Red Sea, and discuss how the timing of rifting and break-up may be determined. The last paper examines the role of transfer zones in partitioning sub-basins within the Perth Basin on the Western Australia margin.

Low-angle faults

The first paper, by **Buck & Lavier**, uses modelling studies to search for a description of fault formation that can explain the formation of large-offset (up to 50 km), low-angle (<30°) normal faults *and* small-offset (<10 km), high-angle (>30°) normal faults. Conventional fault mechanics precludes the occurrence of brittle slip along low-angle normal faults. Yet such structures are common along some non-volcanic margins and in metamorphic core complexes. This contradiction can be explained if high-angle faults were subsequently rotated by later normal faults or developed as rolling-hinge faults. The paper by **Buck & Lavier** models the latter scenario. In their models, the developments of large offset, low-angle normal faults depends on the magnitude and rate of strain softening (modelled as cohesion reduction) and the thickness of the brittle layer. Such faults are possible if this layer is thin, the magnitude of cohesion reduction high, and the rate of cohesion reduction moderate. **Abers** draws attention to recent seismic observations that show that movement does occur along low- to moderate-angle ($<30°$) normal faults, focusing on examples from the Woodlark Basin and Gulf of Corinth. His observations suggest that processes associated with high extension rate on individual fault systems may produce the conditions needed for low-angle faulting. For example, well-worn faults should achieve effective frictional coefficients near 0.5. Such coefficients do not require very unusual materials but may be expected in faults that achieve large offsets in short time spans and so can develop mature, uncemented gouge zones.

Roller et al. describe ODP samples from the recently active low-angle detachment in the Woodlark Basin buried beneath Plio-Pleistocene synrift sediments, and from faults in its hanging wall and footwall. Ductile deformation dominates the detachment zone, but rocks above and below it show dominantly brittle behaviour. Ductile creep in the detachment occurred under at least lower greenschist facies grade metamorphism, so up to 10 km of tectonic unroofing may have occurred, as the submarine setting of the fault precludes unroofing by erosion. **Roller et al.** observe transformation of basement rocks into clay-rich gouges indicating a reduction of frictional coefficient, and fibrous veins indicating high fluid pressure. Both conditions favour low-angle faulting, as also suggested by **Abers**.

Stratigraphy of the Woodlark Basin

The next three papers examine the stratigraphy of the Woodlark Basin. **Robertson et al.** provide a review of the regional setting of the basin and a synthesis of the stratigraphy based on ODP Leg 180 results and some commercial wells. **Lackschewitz et al.** provide added precision to the ages of the sediments recovered during Leg 173 based on their ^{40}Ar/^{39}Ar results. These show that there was a transition from shallow (<150 m) to deeper ($150-500$ m) water conditions at 3 Ma. This corresponds to the commencement of submergence that formed the Woodlark Basin in latest Miocene time, as reported by **Robertson et al.**, who note that more focused subsidence, and uplift of the

southern margin of the basin, occurred during Pleistocene time. This change may be associated with the progressive westward advance of the zone of sea-floor spreading at *c.* 2 Ma. Tectonically active basins such as the Woodlark Basin are prone to stratigraphic problems caused by slumping, which emplaces older sediments into younger successions. **Resig et al.** use a combination of palaeomagnetic data, first and last appearances of Foraminifera, and coiling changes in one foraminiferal species to identify a slump of Pliocene sediments emplaced between 1.10 and 0.65 Ma.

Detachments and rolling-hinge faulting off Iberia

Two papers (**Manatschal et al.**, **Wilson et al.**) discuss structural and stratigraphic aspects of rifting beneath the Iberia Abyssal Plain off West Iberia and present descriptions and interpretations of breccias comparable with the Alpine ophicalcites discussed by **Desmurs et al.** **Manatschal et al.** identify three detachment faults. They conclude from a kinematic inversion of their structural interpretation of a seismic section that the faults became active only when the crust had already thinned to 12 km. They show that the style of deformation changed in a seaward direction. Shallowing downward (listric) faulting with relatively low amounts of extension and associated asymmetric basins was followed by a high extension phase during which developed a steepening downward fault along which the footwall was pulled from beneath the hanging wall (rolling-hinge fault). **Wilson et al.** show that this final stage of faulting produces stratal relationships that are completely different from those widely recognized in classical relatively low extension rift basins. They also suggest that published identifications of synrift intervals in basins that developed close to the sites of future sea-floor spreading off West Iberia (Galicia Margin and Iberia Abyssal Plain) and in the Eastern Alps have overestimated the duration of rifting by as much as 20 Ma. This is because previous studies did not use objective criteria, including divergence of stratal units towards faults and onlapping relations to demonstrate fault block rotation, and also failed to correctly identify break-up unconformities. **Wilson et al.** conclude that off West Iberia, rifting that preceded sea-floor spreading may have lasted for <5 Ma probably from Late Berriasian to Early Valanginian time, much shorter than the durations used in the modelling studies by **Bowling & Harry** and **Minshull et al.** in the last section of this book.

Red Sea and Western Australia

In contrast to the Woodlark Basin, West Iberia margin and Alpine examples described in this volume, **Khalil & McClay** and **Song** *et al.* show that no low-angle normal faults developed during extension of the non-volcanic NW Red Sea and Gulf of Suez rift system and the Perth Basin, Western Australia. In the Red Sea and Gulf of Suez system planar domino-style faults dominate. **Khalil & McClay** show that Precambrian basement fabrics exerted a strong control on fault distribution, and that initially individual faults were strongly segmented and offset across 'soft-linked' relay structures. As extension increased, the relay structures were replaced by 'hard-linked' transfer faults. The authors suggest that the progressive focusing of rifting towards the centre of the rift systems favours a crustal-necking model. **Song** *et al.* discuss the role of transfer faults in the Perth Basin, which formed during rifting phases during Permian and Late Jurassic–Early Cretaceous time, the latter phase preceding continental break-up. The later NW-trending transfer zones are manifested by changes in the trends of normal faults, or by their terminations, and are probably contiguous with ocean transform faults.

Numerical models of extension and magmatism

The papers in this section focus on numerical modelling at the crustal and upper-mantle scale, seeking explanations for the lack of significant magmatism along non-volcanic margins. The first paper, by **Clift** *et al.*, is a combination of a description of features of the South China margin, and forward modelling based on the observational data using a flexural-cantilever model. This model assumes that strain occurs in brittle upper crust by simple shear accommodated by steep normal faults, and by pure shear in ductile lower crust and mantle. As the total subsidence across the margin is in excess of that predicted using this model, and that indicated by the degree of faulting, the authors suggest that preferential extension of the lower crust occurred. The presence of probable volcanic features observed on seismic lines landward of the OCT indicates that the margin is intermediate between the non-volcanic and volcanic end members observed in the Atlantic Ocean. Such magmatism is not predicted by simple adiabatic decompression melting models and so **Clift** *et al.* suggest that magmatism resulted from a pre-rift thermal perturbation and the presence of

water in the mantle lithosphere. They do not, however, explain why the presence of such water did not result in serpentinization, but **Pérez-Gussinyé** *et al.* suggest that this would not occur if the mantle was hot.

The lack of magmatism along non-volcanic margins and its abundance in volcanic margins has been explained by different mantle temperatures (cold and hot margins), or contrasting structural styles (simple or pure shear, with the former involving crustal-scale detachments), or differences in the intensity and scale of mantle convection beneath break-up rifts. **Bowling & Harry** propose an alternative hypothesis involving a rheologically homogeneous crust and moderately rapid extension rate. In their model, the rise of the asthenosphere to shallow depths at which it begins to melt is delayed until after significant lithosphere necking has occurred. Decompression melting is therefore confined to the very last stages of rifting, after which break-up and sea-floor spreading follow rapidly. The duration of synrift magmatism and its areal extent, are, therefore, minimized. However, this hypothesis does not explain the small volumes of melt inferred by **Whitmarsh** *et al.* to exist in the (post-break-up and pre-sea-floor spreading) OCT off West Iberia. Using detailed analysis of the basement structure of two geophysical profiles across the formerly conjugate margins of the Iberia Abyssal Plain and Newfoundland, **Minshull** *et al.* offer an alternative explanation for the lack of synrift magmatism beneath the conjugate margins. They suggest that the duration of rifting was relatively short (*c.* 10 Ma, rather than the widely accepted 25 Ma but not the even shorter period of 5 Ma proposed by **Wilson** *et al.*). They propose, therefore, that melt generation predicted by models of pure shear continental rifting was reduced by the combination of effects of lateral heat conduction, depth-dependent stretching and reduced mantle temperature. They also suggest that post break-up melt generation was limited to *c.* 17 Ma during the time the OCT 'was formed as the thermal structure evolved to that of a spreading ridge axis and asthenosphere flow became increasingly focused at the site of the new ridge axis'.

Finally, the modelling studies reported by **Pérez-Gussinyé** *et al.* examine the effects of initial lithospheric structure on the structural evolution and magma generation (or the lack of it) along non-volcanic margins. The authors use three types of initial lithospheric models: a craton (e.g. SW Greenland and Labrador), an old orogen (e.g. West Iberia) and a young orogen (e.g. Woodlark Basin). They show that for the cratonic and older orogen models (in which the crust is thinner: *c.* 30 km instead of 50 km, and cooler than a young orogen), the entire crust becomes brittle when β values reach 3–4. This results in faults that cut the entire crust, permitting water to reach the mantle and serpentinize it. This leads to the development of décollements at the crust–mantle boundary, and exhumation of the mantle along the OCT. The absence of both serpentinites and a broad OCT along the margins of the Woodlark Basin is explained by the fact the crust is thicker and hotter. The weak lithosphere was faulted and break-up occurred when β reached 4–5, long before the crust became brittle.

Concluding comments

We conclude by offering some comments on some key issues that are covered in several sections.

The record of rifting and break-up

The difficulties in interpreting the tectonic setting of sedimentary successions that accumulated on continental margins is examined by **Wilson *et al.***, who highlight the problems of identifying synrift intervals. They suggest that objective criteria were not always rigorously applied in previous published studies of the Iberian and Alpine examples they discuss. The same problem may apply to interpretations of other margins, including some of those described in this book. Defining more accurately when continental rifting started and stopped, and when sea-floor spreading began, is crucial information needed to refine models of margin development. We should not assume that the end of rifting coincides with the onset of sea-floor spreading. In the Woodlark Basin and the South China Sea (**Clift *et al.***) it continues after spreading begins. Radiometric or fission-track dating cannot date rifting events precisely, as what is recorded is the time at which rocks cooled through key temperatures as a result of cooling after thermal events, and/or by exhumation by tectonic and/or erosional processes.

The nature and scale of faulting

Were subhorizontal extensional faults like the Galicia margin S-reflector (**Boillot & Froitzheim**, fig. 2) active in their present-day shallow orientation or were they originally steeper and have been rotated to their shallow attitude after ceasing to be active (rolling-hinge model)?

There are a few examples of important 30° normal faults that are at present seismically active, such as the Moresby seamount fault in the Woodlark Basin (**Abers, Roller *et al.***). But what about the really low-angled (<10°) structures, such as the Galicia S-reflector? Development of such shallowly dipping structures by the rolling-hinge mechanism is successfully modelled numerically by **Buck & Lavier**. Retrodeformation of a profile across the Iberia Abyssal Plain (**Manatschal *et al.***) also requires moderate instead of very low dip angles for the normal faults during their activity. On the other hand, thermo-mechanical modelling by **Pérez-Gussinyé *et al.***, field observations from the Alps (**Müntener & Hermann, Desmurs *et al.***) and geophysical observations and modelling (**Clift *et al.***) suggest that normal faults may cut across the entire continental crust but only after some significant amount of stretching. In some cases, this may have resulted in mantle serpentinization and the development of décollements along this weaker material.

Are deep crustal and mantle rocks exposed in OCTs related to pre-final phase of rifting?

Radiometric dating both in the Alps and the Atlantic (**Gardien *et al.*, Manatschal *et al.*, Müntener & Hermann**) shows that the formation and high-temperature mylonitization of gabbros from the OCT are not necessarily rift related but may substantially predate the final rifting process. Such rocks are exposed in the OCT not necessarily because they are genetically related to the formation of the OCT, but only because the OCT allows sampling of the crust–mantle transition, which is otherwise seldom possible. These observations highlight the importance of accurate radiometric dating of both crystallization and cooling ages for the understanding of continental break-up.

How does the lower crust behave during rifting?

Rocks of the continental lower crust are only occasionally observed along non-volcanic rifted margins. This may be explained either by ductile flow from the rift axis outward, which requires the lower crust to be weak, or by boudinage of the lower crust, which requires the opposite. Modelling studies yield satisfactory results for the Alps assuming the lower crust to behave in a ductile manner (**Bertotti**), but field observations in another part of the Alps (**Mün-**

tener & Hermann) favour boudinage of a strong lower crust.

Are non-volcanic margins amagmatic?

A conundrum concerning non-volcanic margins is how the onset of sea-floor spreading, which is principally a magmatic process, develops out of the amagmatic rifting of continental crust. The numerical models of **Bowling & Harry** and **Minshull** *et al.* propose different solutions for what is clearly a complex and changing thermal and rheological situation. The geophysical observations and interpretations of **Whitmarsh** *et al.* lead to a plausible concept of steadily increasing magmatism towards the ocean, in time and space, but one that clearly needs to be verified by further drilling off Iberia, and by comparable geophysical studies along other margins.

The importance of studies of conjugate margins

The majority of papers in this book are 'one-sided' in their approach, in that they examine the evolution of only one of a pair of formerly conjugate margins. Even for a well-studied margin such as West Iberia (three ODP Legs, numerous diving and dredging investigations, and many geophysical campaigns), the issue as to whether symmetric (pure shear) or asymmetric (simple shear) rifting operated has still not been resolved. Comparable data are not available for the conjugate Newfoundland margin and are crucial evidence needed to resolve this issue.

We express our sincere appreciation to Marie-Odile Beslier for her work as co-convenor with us of the meeting at the Geological Society of London in September 1999, from which this book arose, and for her help in identifying reviewers. She was to have played a major role as editor, but unfortunately ill health prevented her from doing so. We thank the Geological Society Conference Office for its help in arranging the meeting, and its Publishing House staff for their support in the publication process. A. Hills and A. Morton provided advice at review stage, and D. Swan ensured that the editing, page design and proof stages proceeded smoothly. J. Matela of the Open University provided valuable support to C. Wilson in keeping track of submissions, and correspondence between authors, academic editors and reviewers. We wish to thank the following for contributing to the review process: S. Arai, G. Axen, J. Beard, D. Bernoulli, J. Brown, R. Buck, P. Burgess, R. Caby, J. Chalmers, F. Chalot-Prat, P. Clift, M. Coffin, K. Crook, M. Davis, J. Escartin, B. Evans, D. Forsyth, P. Fryer, E. Hailwood, M. Handy, D. Harry, J. Hermann, R. Hey, D. Hindle, J. Hopper, J. Jackson, J. Jones, G. Karner, Y. Lagabrielle, M. Leckie, K. Louden, G. Manatschal, G. McIntosh, F. Martinez, T. A. Minshull, G. Molli, C. Monnier, O. Müntener, S. Prosser, T. Reston, D. G. Roberts, A. Robertson, M. Rubenach, D. Sawyer, M. Seyler, I. Sharp, G. Stampfli, R. Steel, P. Symonds, B. Turrin, R. S. White, S. Wise.

Non-volcanic rifted margins, continental break-up and the onset of sea-floor spreading: some outstanding questions

G. BOILLOT[1] & N. FROITZHEIM[2]

[1]*Observatoire Océanologique, Géosciences Azur, B.P. 48, 06235 Villefranche-sur-Mer, France (e-mail: boillot@obs-vlfr.fr)*
[2]*Geologisches Institut, Nussallee 8, D-53115 Bonn, Germany*

Abstract: During the last 20 years, regional studies on the West Iberia margin and on the former margins of the Tethys have considerably advanced the understanding of processes related to continental break-up and the onset of sea-floor spreading. However, some questions remain outstanding. To tentatively answer these, a coherent interpretation of available data is proposed, based on the detachment fault concept applied to the continental as well as the oceanic lithosphere, and on the hypothesis of a multi-staged rifting process. The interpretation addresses the nature of the lower crust beneath non-volcanic passive margins, the origin of ophicalcites, the probable time gap between syn- or post-rift crystallization of gabbros and extrusion of basalts on the sea floor, and the significance of dipping reflectors within oceanic lithosphere adjacent to non-volcanic passive margins. The interpretation also considers the symmetry v. asymmetry of continental rifting and break-up, the location of the ocean–continent boundary, and the possible association of magnetic quiet zones with ultramafic sea floor (serpentinized peridotite) bordering non-volcanic passive margins.

Twenty years after the discovery by dredging of the peridotite ridge bounding the Galicia margin (Boillot *et al.* 1980; Sibuet *et al.* 1987), and after three Ocean Drilling Program (ODP) Legs (103, 149, 173; Fig. 1), three French diving cruises, and several British, German, US and French geophysical surveys on the West Iberia margin, the advance in the understanding of processes controlling continental break-up and the onset of sea-floor spreading appears to be spectacular, and many of the major problems seem to be solved. It is of interest, however, to consider also some outstanding questions, which necessitate both further research and acquisition of new data: (1) What are the respective components of serpentinized peridotite, syn-rift gabbro and pre-rift lower continental crust in the lower crust of passive margins? (2) Does a time gap exist between crystallization of gabbros and extrusion of basalts at slow-spreading oceanic ridges, as predicted by the model of unroofing of peridotite and gabbros by detachment faulting? (3) What is the geodynamic significance of ophicalcites covering ultramafic sea floor in the present ocean and in Alpine ophiolites? What are the respective roles of, among other possible processes, hydraulic fracturing and extensional detachment faulting in the genesis of these rocks? (4) What is the origin of dipping reflectors (continentward or

not) occurring in slow-spreading oceanic lithosphere, especially in the vicinity of passive margins? Are they seismic images of detachments? (5) Are the fault systems accommodating rifting and continental break-up symmetric or asymmetric (with respect to the rift axis) on a lithospheric scale? What are the evolutionary stages of the rift from the initial lithosphere stretching to continental break-up? (6) Where is the actual ocean–continent boundary (OCB)? Is it located at the lithospheric or at the crustal OCB? (7) What is the significance of magnetic quiet zones adjacent to rifted margins, in so far as they do not correspond to periods of stability of the global magnetic field? Do they represent serpentinitic sea floor resulting from tectonic unroofing of mantle rocks and their hydrothermal alteration? (8) Where are the best study areas, at sea and on land, for answering these questions?

This paper does not attempt a review of the scientific results recorded by the international community from the regional study of the West Iberia margin or of Tethyan margins in the Alps, nor does it present new data. The data mentioned in the text and their interpretations are, or will soon be, published and accessible. Instead, we will focus on some remaining scientific problems, to point out possible new research objectives for the near future. In

From: WILSON, R.C.L., WHITMARSH, R.B., TAYLOR, B. & FROITZHEIM, N. 2001. *Non-Volcanic Rifting of Continental Margins: A Comparison of Evidence from Land and Sea.* Geological Society, London, Special Publications, **187**, 9–30. 0305-8719/01/$15.00 © The Geological Society of London 2001.

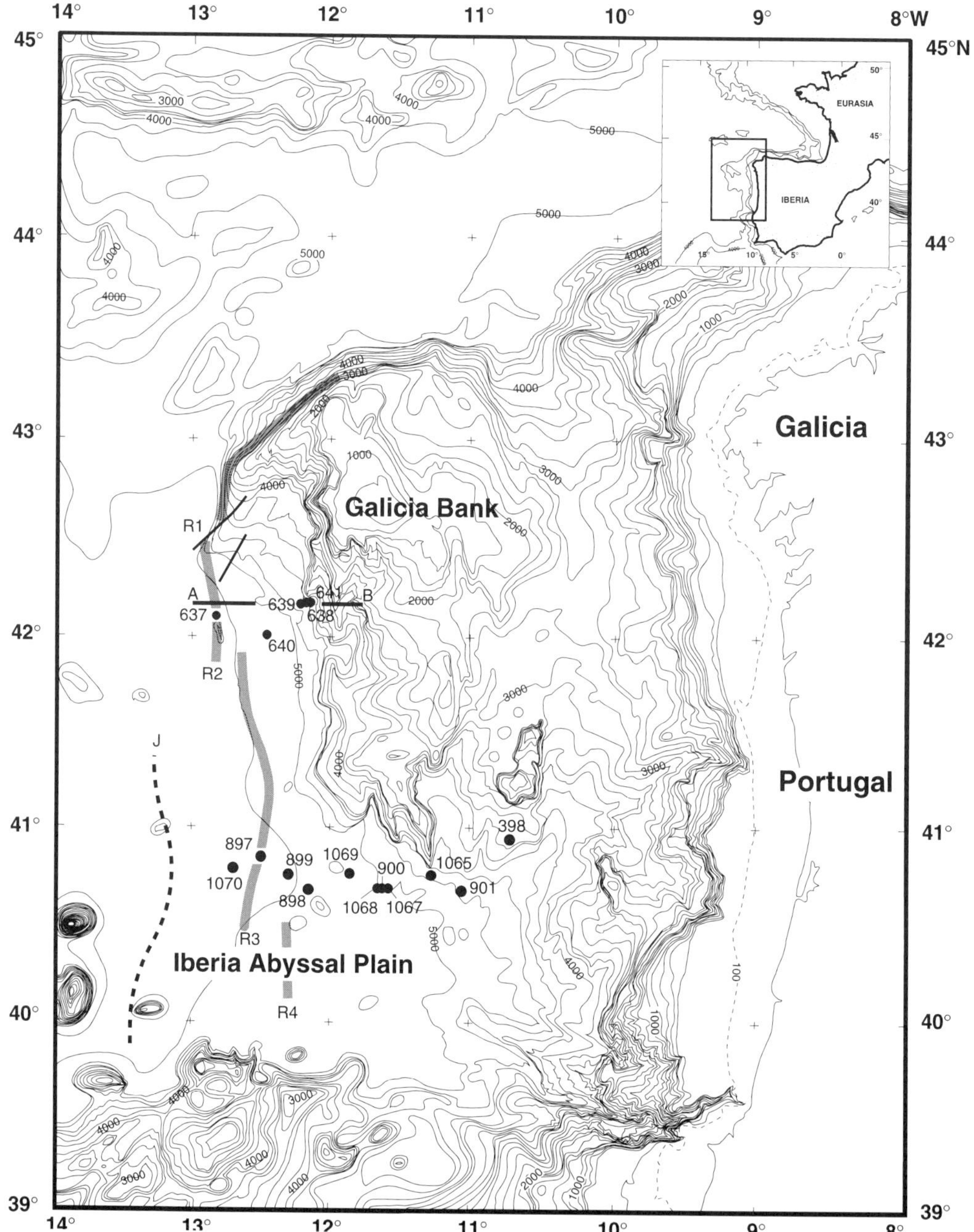

Fig. 1. The West Iberia margin to the north of 39°N. Numbers refer to drilling sites of ODP Legs 103, 149 and 173. Drilling data were completed by diving data on the Galicia margin and adjacent sea floor (about 300 samples recovered from pre-rift sediments and oceanic or continental basement; Boillot *et al.* 1988*a*, 1995*b*). R1–R4, segments of the peridotite ridge (after Beslier *et al.* 1993). A–B, location of the cross-section depicted in Figure 2.

addition, we will discuss the process of slow sea-floor spreading, which immediately follows continental break-up. It is only at fast-spreading ridges, where the lithosphere is extremely thin, that the partial melting of peridotite as a result of adiabatic decompression is efficient enough to completely accommodate plate divergence. At slow-spreading ridges, as in continental rifts, extensional tectonics play a very important role, exclusive in some cases, in balancing plate divergence. The data collected from the West Iberia margin and adjacent oceanic area strongly constrain the models accounting for these tectonic processes.

In this paper, we consider as firmly established the following results of previous studies.

(1) The passive margin of Iberia, and probably many other non-volcanic passive margins in the world, is bordered by a continuous belt of serpentinized peridotite and, locally, gabbro. This ultramafic sea floor culminates along the peridotite ridge, a structural high of the basement, partly buried by post-rift sediment on its eastern side, 10–12 km wide and 2–3 km high (Mauffret & Montadert 1987; Boillot *et al.* 1987*b*, 1988*a,b,c*; Beslier *et al.* 1990, 1993). However, the ultramafic sea floor is wider than the peridotite ridge (>100 km wide in the Iberia Abyssal Plain; Beslier *et al.* 1996; Whitmarsh & Sawyer 1996; see also Manatschal *et al.* 2001; Pérez-Gussinyé *et al.* 2001).

(2) The ultramafic sea floor results from syn-rift tectonic unroofing of subcontinental peridotite intruded by gabbro. Simple shear along lithospheric detachments may have been the main process in stretching and thinning of the lithosphere and in the unroofing of subcrustal mantle (Boillot *et al.* 1987*a*, 1989*a*, 1995*a*; Beslier & Brun 1991; Brun & Beslier 1996).

(3) The geophysical signatures of the main lithospheric faults and shear zones responsible for mantle unroofing are strong seismic reflectors, the most famous of them being the S reflector of the Armorican and Galicia margins (De Charpal *et al.* 1978; Le Pichon & Barbier 1987; Hoffmann & Reston 1992; Sibuet 1992; Krawczyk & Reston 1995; Reston *et al.* 1996), named H in the Iberia Abyssal Plain (Beslier *et al.* 1995; Krawczyk *et al.* 1996; Manatschal *et al.* 2001). The geological signature of the tectonic shearing and exhumation of subcrustal mantle is the foliation and brecciation of the peridotite and intruded gabbro, caused by deformation at falling temperature and decreasing pressure (Girardeau *et al.* 1988, 1998, 1999; Kornprobst & Tabit 1988; Beslier *et al.* 1990, 1996; Cornen *et al.* 1996*a,b*).

(4) Beneath the most distal part of the passive margin, the detachments imaged by seismic reflectors result in the tectonic contact between the thinned continental crust of the margin and the underlying serpentinized peridotites. Consequently, the oceanward boundary of the continental crust coincides with the trace of the detachment on the sea floor (Etheridge *et al.* 1989).

(5) The serpentinization of peridotites and the retrometamorphism of gabbros in the greenschist facies result from syn-, and possibly post-rift hydrothermal activity at shallow levels of the lithosphere (Agrinier *et al.* 1988, 1996; Schärer *et al.* 1995). The seismic Moho is the boundary at depth between fresh and serpentinized peridotite, i.e. the hydrothermal front (lower limit of hydrothermal circulation, or, more probably, thermal barrier for serpentinization; see Pérez-Gussinyé *et al.* 2001). Moreover, the serpentinite layer extends continentward beneath the most thinned continental crust of the margin and oceanward beneath a thin, post-rift basalt layer (Boillot *et al.* 1989*b*, 1991; Recq *et al.* 1996; Whitmarsh *et al.* 1996*c*; Dean *et al.* 2000).

(6) Oceanic basalts locally cover serpentinized peridotite and gabbro to the west of the peridotite ridge (Kornprobst *et al.* 1988; Malod *et al.* 1993; Charpentier *et al.* 1998) and thicken oceanward (Sibuet *et al.* 1995; Whitmarsh *et al.* 1996*c*). Gabbros, however, appear to be older than the continental break-up (and thus, older than post-rift oceanic basalts) when radiometric ages are available (Féraud *et al.* 1988, 1996; Schärer *et al.* 1995, 2001).

Figures 2 and 3 are schematic depictions of the crustal structure of the OCB off the Galicia margin (Fig. 2) and of the African (Apulian) margin of the Mesozoic Tethys (Fig. 3) as reconstructed from field data in the Swiss–Italian Alps by Trommsdorff *et al.* (1993). Figure 4a and b shows simplified diagrams of continental break-up inspired by the analogue model experiments of Brun & Beslier (1996).

The lower crust in non-volcanic passive margins

To correctly address the problem of the nature, behaviour and fate of the pre-rift continental crust during rifting and continental break-up, it is necessary to distinguish between three kinds of lower crust beneath passive margins, as follows.

(1) The seismic lower crust made of serpentinized peridotite, of syn-rift (and possibly post-

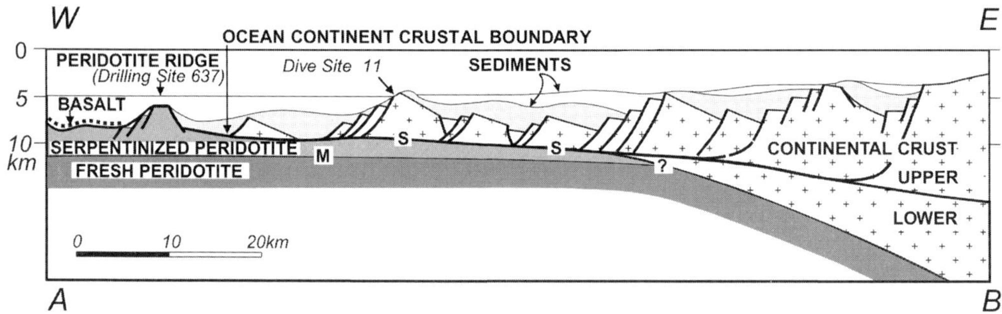

Fig. 2. Synthetic cross-section of the Galicia passive margin; location shown in Figure 1. Beneath the peridotite ridge, the Moho is the boundary between 'fresh' and partly serpentinized peridotite, i.e. a limit of mineralogical phase change corresponding to a palaeo-hydrothermal front. Oceanward, to the west of the peridotite ridge, the sea floor is made also of serpentinized peridotite, covered by a thin basaltic layer emplaced after the continental break-up (see Fig. 5). On the eastern side of the ridge, the deepest blocks of the margin, made of upper continental crust, rest directly on serpentinized peridotite through a tectonic contact (a detachment), the image of which is the S seismic reflector. Normal faults bounding the crustal blocks of the margin are steep (60–70°), and root at depth in the subhorizontal detachment located at the bottom of the blocks (after Boillot *et al.* 1995c).

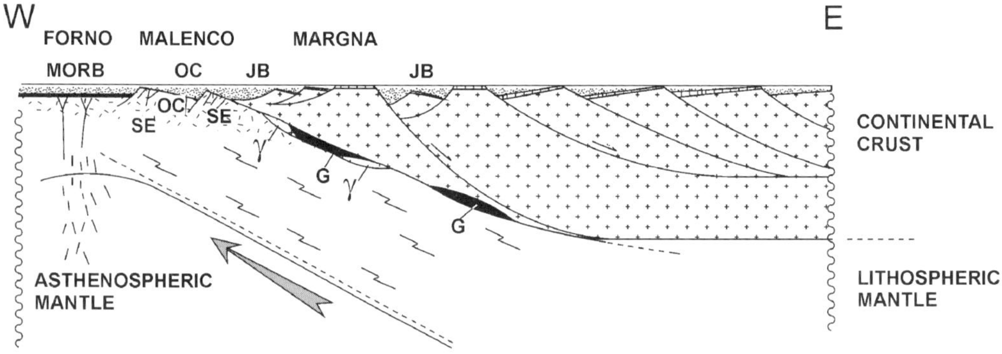

Fig. 3. Ocean–continent transition in the Tethyan Ocean (Ligurian basin) from which the Forno–Malenco and Margna nappes are derived. G, gabbro; γ, pieces of lower continental crust, metamorphosed in the granulite facies; SE, serpentinite; OC, ophicalcite resting on the serpentinite sea floor; JB, Jurassic breccias; MORB, mid-ocean ridge basalts (after Trommsdorff *et al.* 1993).

rift) origin. This is necessarily located in the most distal part of the margin, where hydrothermal circulation and metamorphism of mantle rocks are possible, owing to the thinning and fracturing of the upper continental crust (Boillot *et al.* 1989b). In this case, the definition as lower crust is based on the seismic character (P-wave velocities ranging between 6.5 and 7.9 km s^{-1}; see Recq *et al.* 1996; Whitmarsh *et al.* 1996c; Chian *et al.* 1999; Dean *et al.* 2000) and not on its geological nature. Actually, 'undercrusting' (crustal thickening by addition of serpentinite at the bottom of the crust) results in this case from hydrothermal transformation of fresh peridotite into serpentinized peridotite of lower density and seismic velocity.

(2) The lower crust made of underplated gabbro, of syn-rift origin. Non-volcanic passive margins are not entirely amagmatic. For instance, peridotites sampled along the West Iberia margin underwent about 8–10% of magma extraction by syn-rift partial melting (Girardeau *et al.* 1988; Cornen *et al.* 1996a,b). The resulting tholeiitic magma, of which no superficial volcanic equivalent is known on the margin, was necessarily underplated as gabbro, at least partly, beneath the thinned crust of the rift (Fig. 4a and b). On the Galicia margin, such gabbros, initially crystallized at high temperature, were sheared at falling temperature and finally retrometamorphosed in the greenschist facies. They were recovered at the top of the 'lower plate', close to the tectonic contact sep-

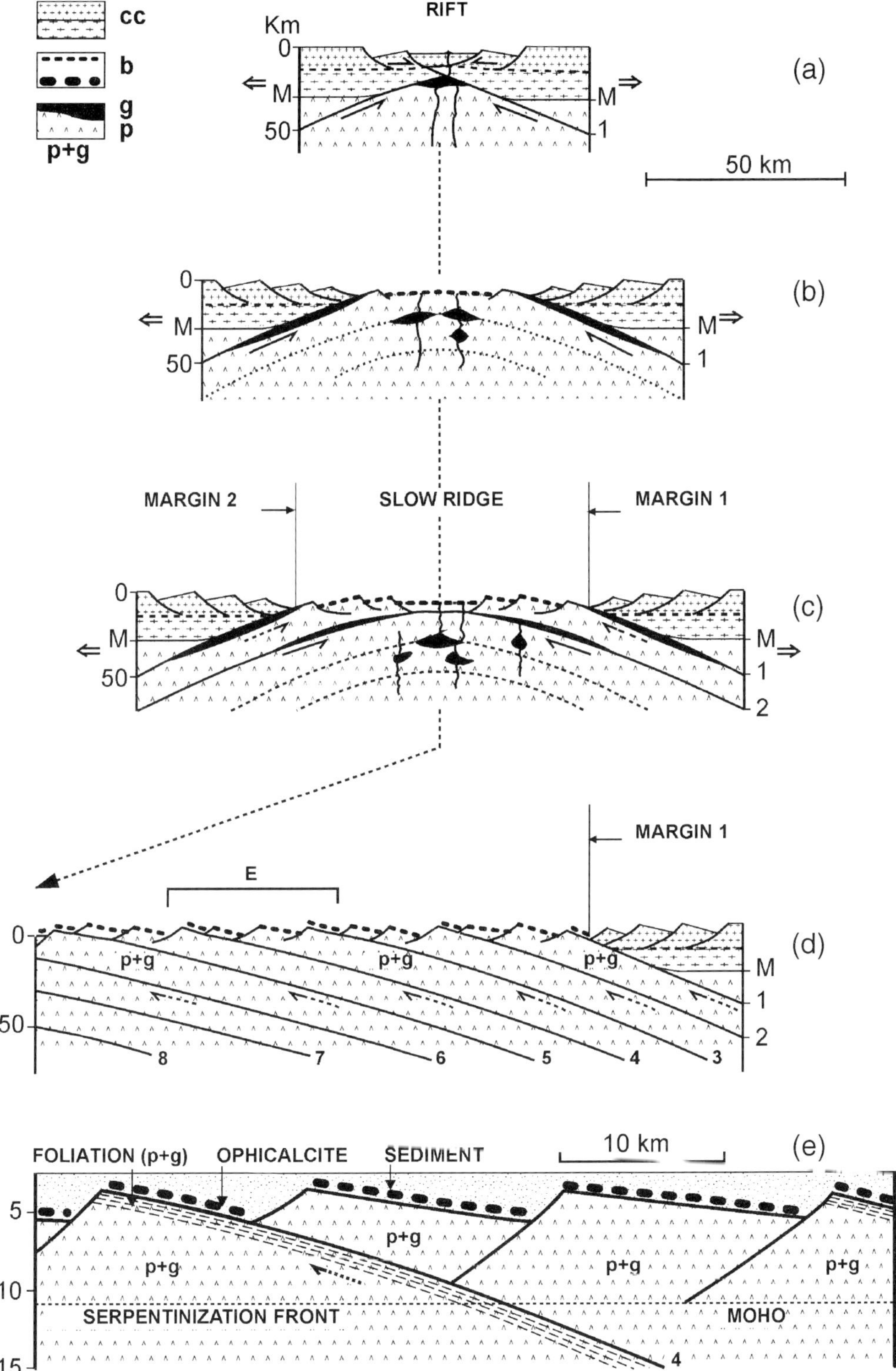

Fig. 4. Conceptual model of continental break-up and onset of sea-floor spreading at non-volcanic passive margins. cc, continental crust; b, basalt; g, gabbro; p, peridotite (serpentinized near the surface). The asthenosphere–lithosphere boundary is not represented (after Boillot & Coulon 1998).

arating the upper continental crust from serpentinized peridotite (chlorite-bearing schist mylonite in Fig. 5). The U–Pb age of zircons included in the rock (122.1 ± 0.3 Ma; Schärer *et al.* 1995, 2001) clearly indicates that the gabbros crystallized during the rifting stage, which lasted from 140 to 115 Ma in this part of the margin (Boillot *et al.* 1987*a*, 1988*b*). On the other hand, the current position of metamorphosed gabbros at the top of mantle rocks suggests that they were underplated beneath the continental crust of the rift, before they underwent shearing, uplift and tectonic denudation in the latest stage of the rifting (Schärer *et al.* 1995, 2001).

(3) The pre-rift lower continental crust. Close to the OCB in the Iberia Abyssal Plain, metagabbro was cored during ODP Leg 149 (Sawyer *et al.* 1994; Cornen *et al.* 1996*b*; Whitmarsh *et al.* 1996*b*). The rock was tentatively interpreted as a syn-rift intrusion, based on a plagioclase $^{40}Ar/^{39}Ar$ cooling age of 136 Ma (Féraud *et al.* 1996). During Leg 173 (Ocean Drilling Program Leg 173 Shipboard Scientific Party 1998; Whitmarsh *et al.* 1998), metagabbro and amphibolite were cored nearby, and yielded a similar plagioclase $^{40}Ar/^{39}Ar$ age of 137 Ma and a hornblende $^{40}Ar/^{39}Ar$ age of 161 Ma (Turrin, pers. comm.; see Manatschal *et al.*

2001). However, U/Pb SHRIMP analysis of zircons from the metagabbro yielded 270 Ma (Rubenach, pers. comm.). These results show that the Leg 173 metagabbros–amphibolites are not syn-rift, but belong to Variscan (pre-rift) lower continental crust captured in the OCB.

A similar but clearer situation is recorded in former Tethys margins (Malenco and Tasna areas; Fig. 3) where sheared gabbros intercalated between upper continental crust of the margin and exhumed mantle peridotite have yielded pre-rift, Permian U/Pb zircon ages (Müntener & Hermann 1996; Hermann *et al.* 1997; Froitzheim & Rubatto 1998). These gabbros demonstrate a widespread thermal and tectonic event that affected the Hercynian belt at *c.* 270 Ma. Their exhumation is due to Mesozoic rifting, and not their formation (see Desmurs *et al.* 1999; Müntener & Hermann 2001).

These findings indicate that pre-rift lower crust exists along the OCB of passive margins, although it is rare and scattered. In fact, it seems from available sampling that the pre-rift lower crust is thin and dismembered at the rift axis when continental break-up occurs, and that it can apparently be completely missing at the OCB (this is the case for the Galicia margin; see Fuegenschuh *et al.* 1998.

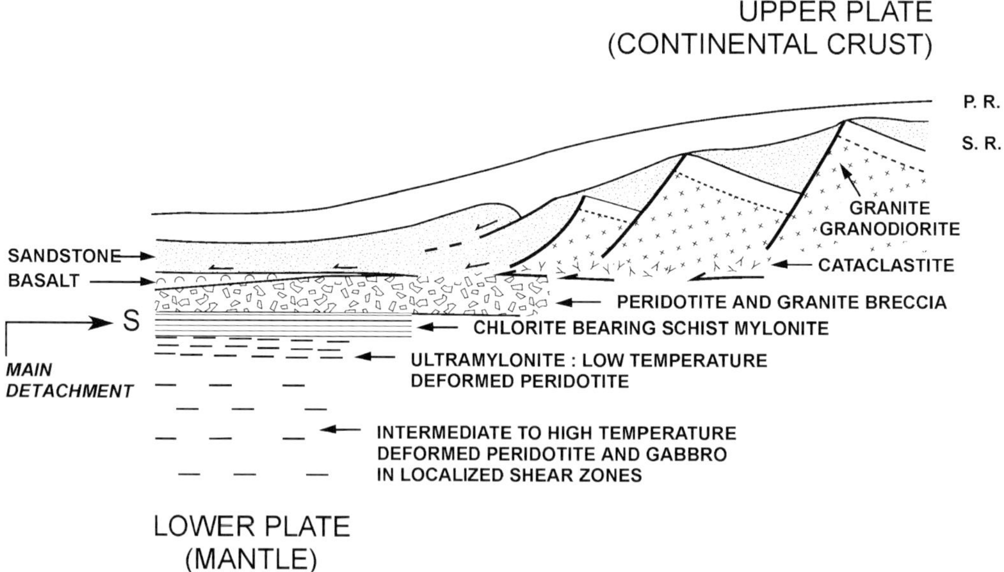

Fig. 5. Schematic cross-section showing the tectonic contact between the upper continental crust and serpentinized peridotites at the edge of the Galicia passive margin, from diving data (after Boillot *et al.* 1995*b*). P.R., post-rift sediments; S.R., syn-rift sediments. Not to scale.

Time gap between gabbro crystallization and superficial basalt flooding

In the previous section, we mentioned some Late Palaeozoic ages of gabbros recovered close to the OCB in the Iberia Abyssal Plain or associated with Alpine ophiolites. In contrast, gabbros at the OCB of the Galicia margin, where continental break-up occurred at 115 Ma, are 122.1 ± 0.3 Ma old (Schärer et al. 1995) whereas the oldest oceanic basalt sampled several kilometres off the margin is c. 100 Ma old (Féraud, in Malod et al. 1993). Therefore, it is necessary to consider, for oceanic crust of the current oceans as well as for ophiolites, the probability of significant time gaps between the crystallization of gabbros and the emplacement of overlying basalts. This point is of importance as it concerns the assumed unroofing process of gabbro, first crystallized at depth, then sheared and brought up by extensional detachment faulting to the sea floor. How much time does this process take, from the formation of the gabbro to its final exposure on the sea floor?

Published data that would allow comparison of the ages of gabbro and basalt at the same locality in present oceanic areas formed by slow or ultra-slow spreading have not been found. With the notable exception of Permian gabbros (see above), modern radiometric ages for gabbros in Alpine ophiolites from the Ligurian basin range between 185 and 158 Ma, very close to the age of the first sediment (radiolarite) deposited on the mafic basement, generally of Callovian–Oxfordian age (e.g. Ohnenstetter et al. 1981; Bill et al. 1997; Rubatto 1998; Rubatto et al. 1998). However, Pinet et al. (1989); Caby (1995); Costa & Caby (1997), from their studies of the Queyras and Mont-Genèvre ophiolites, concluded that intrusion and intra-oceanic deformation of gabbros clearly predate extrusion of the overlying pillow basalts. The occurrence of radiolarite beneath basalt (Bortolotti et al. 1991) also confirms that basalts postdate gabbros in many cases.

The question posed above concerning the time taken for gabbro formation and subsequent unroofing remains open because of the lack of high-precision radiometric ages for gabbros and basalts from the same locality, in the current ocean as well as in ophiolites. However, determination of the actual time gap, if any, between gabbro crystallization and basalt extrusion is a major objective in constraining the tectonic processes at slow-spreading, or very slow-spreading oceanic ridges (for further discussion see Desmurs et al. 2001).

Ophicalcite and lithospheric shear zones

Ophicalcites (Figs 6 and 7) are serpentinite and gabbro breccias covering ultramafic–gabbroic basement in many ophiolite sheets from the Tethyan ocean (e.g. Bernoulli & Weissert 1985;

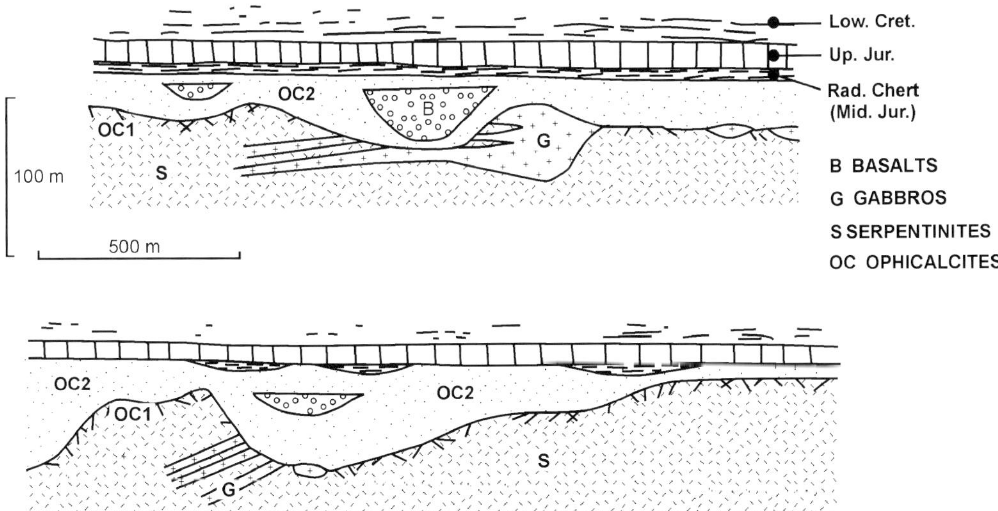

Fig. 6. Sea floor of the Ligurian ocean in late Jurassic time. Post-rift sediments directly overlie serpentinized peridotites, gabbros or basalts, depending on location. Ophicalcite 1 (OC1) is a tectonic breccia, whereas ophicalcite 2 (OC2) is a sedimentary formation resulting from the reworking of OC1 on the sea floor. Rad. Chert: Radiolarian chert (after Lagabrielle & Lemoine 1997).

Fig. 7. Photograph of ophicalcite in the French Queyras ophiolite (Western Alps). (Compare with Fig. 8)

Treves & Harper 1994; Lagabrielle & Lemoine 1997). The matrix of the breccia is either carbonate (mainly calcite) or second-generation serpentinite. The 'clasts' are variable in size and shape (Fig. 7). According to the fabric of the rock, two kinds of ophicalcites can be defined (Tricart & Lemoine 1989). Ophicalcites 1 (OC1) are tectonic breccias resting directly on deeply fractured basement. Consisting of serpentinite and gabbro cross-cut by calcite, dolomite or serpentinite veins, OC1 show a gradual transition into fractured ultramafic basement with which they share the same petrological composition. Ophicalcites 2 (OC2) are detrital sediments generally devoid of fossils. OC2 result from submarine reworking and short-distance transport of underlying OC1.

Ophicalcites 1 and 2 are locally covered by oceanic basalt (pillow basalt and pillow breccia), which formed the actual sea floor for sediments (Lagabrielle *et al.* 1984; Lagabrielle & Lemoine 1997; Manatschal & Nievergelt 1997). This stratigraphic setting implies that sedimentation (OC2) occurred before basalt extrusion, at least in some places. This conclusion is confirmed by the local occurrence of radiolarite between the ophicalcites and basalt (Bortolotti *et al.* 1991), which implies a time gap between the crystallization of gabbro and the emplacement of oceanic basalts as mentioned in the previous section.

Serpentinite and gabbro breccias cemented by calcite, similar to ophicalcite, have been found at the top of ultramafic basement in the

Iberia Abyssal Plain (Whitmarsh *et al.* 1996*b*, 1998; Wilson *et al.* 2001). Cataclasis, at least in some cases, appears to have occurred relatively late, possibly after continental break-up, as the calcite cement crystallized at low temperatures from sea water (<80°C; Agrinier *et al.* 1988). Another example was discovered on the Galicia margin in the form of a tectonic breccia located at the top (the brittle level) of the lithospheric shear zone separating the peridotite ridge and the upper continental crust of the margin (Boillot *et al.* 1995*b*; Figs 5 and 8). This tectonic breccia includes clasts from both the upper and lower plates, i.e. of continental and mantle origin (granite and metamorphic sandstone, serpentinite and gabbro, respectively). This mix of fragments originating from rocks in tectonic contact is strong evidence in support of the relationship of the breccia to the shear zone separating the two kinds of rocks. Continental and oceanic clasts have also been found interbedded in some Alpine ophicalcites, but are of detrital character (e.g. Polino & Lemoine 1984; Bernoulli & Weissert 1985).

The resemblance in lithological character and structural position between the Galicia margin breccia (Figs 5 and 8) and Alpine ophicalcite (Figs 6 and 7) suggests a similar origin for

both. Alpine ophicalcite OC1 may then be interpreted as a brittle fault rock formed along extensional detachment faults (Florineth & Froitzheim 1994). In the case of the Galicia margin, the detachment was clearly syn-rift and associated with the continental break-up process; in the Alps, part of the ophicalcites are syn-rift (e.g. in the Tasna nappe of the Eastern Alps; Florineth & Froitzheim 1994); others may be post-rift and formed at slow-spreading oceanic ridges (e.g. in the Western Alps, Lagabrielle & Lemoine 1997; see below for further discussion). The possible setting of ophicalcites formed during slow spreading is indicated in Figure 4e.

The problem of the origin of ophicalcite concerns directly the process of sea-floor formation. Is the fracturing of serpentinite a hydraulic process related to the oceanic hydrothermal circulation (Früh-Green *et al.* 1990; Treves & Harper 1994), a result of brittle deformation along transform faults (Lemoine 1980; Weissert & Bernoulli 1985), or a result of cataclasis along extensional faults or detachments, as proposed in this paper? In the case of the Galicia margin, the detachment fault interpretation (which does not exclude the hydraulic fracturing hypothesis, as fluids are active in faulted zones) is sup-

Fig. 8. Submarine photograph showing the tectonic breccia ('peridotite and granite breccia' in Fig. 5) located at the top of the mantle terrane, on the northwestern slope of the Galicia Bank. (Compare with Fig. 7.)

ported by the structural setting of the breccia. In the case of Alpine ophicalcites, more field studies are necessary to reconstruct the geometry of ophicalcite-bearing faults and to distinguish between rift-related, transform-related, and slow-spreading-related ophicalcites. The challenge is important: if it can be established that ophicalcites are the geological signature of uppermost parts of extensional shear zones, then ultramafic sea floors covered by the breccia have to be considered as structural surfaces (unroofed detachment fault planes). Beyond, the question concerns the process of tectonic sea-floor spreading at slow ridges, and the possible role of dipping reflectors in the sea-floor emplacement of ultramafic rocks and foliated gabbros (see the next two sections).

Dipping reflectors and shear zones in oceanic lithosphere

Dipping reflectors are not restricted to volcanic margins. Sloping mostly continentward, sometimes oceanward, they appear also on many seismic lines recorded on oceanic lithosphere adjacent to non-volcanic passive margins. For example, clear images of such reflectors were recorded in the Iberia Abyssal Plain off the Portugal passive margin (Pickup *et al.* 1996; Fig. 9), and in the Biscay Abyssal Plain bounding the Goban Spur (Masson *et al.* 1985). Both these areas comprise oceanic lithosphere emplaced immediately after the rifting of the margin, when one can assume that sea-floor spreading was particularly slow. Similar reflectors have also been observed in the oceanic lithosphere bordering the Eastern Canada margin (Keen & De Voogd 1988). These features cannot be interpreted as images of volcanic piles, as there is no regional evidence of extensive volcanism. Moreover, they dip preferentially toward the continent, whereas along volcanic margins the reflectors dip toward the ocean.

The continentward dipping reflectors imaged in the oceanic lithosphere bordering passive margins resemble the S reflector of the Galicia and Armorican margin, or the H reflector in the Iberia Abyssal Plain. It is probable that they also represent palaeo-shear zones within the crust and mantle terranes. The frequent association of reflectors dipping both toward the continent and toward the ocean suggests the occurrence of conjugate shear zones that have acted simultaneously during tectonic extension of the lithosphere at slow ridges (Cannat *et al.* 1997; Lagabrielle *et al.* 1998). On the other hand, the common occurrence of flaser gabbros and foliated serpentinized peridotites in ophiolites and in the oceanic lithosphere indicates deformation under elevated temperature and pressure, i.e. in shear zones at depth, before uplift and unroofing on the sea floor. Moreover, extensional detachment faults and related core complexes have also been identified along the present Mid-Atlantic Ridge (e.g. Tucholke *et al.* 1997, 1998; Blackman *et al.* 1998).

It is concluded that shear zones probably play an important role in the sea-floor spreading process following continental break-up, and, more generally, at slow ridges, where the magmatism is episodic. The next section of the paper is an attempt to visualize this process.

A conceptual model for the onset of sea-floor spreading

The detachment hypothesis, first applied to continental rifts, has also been tested by analogue modelling of tectonic accretion at slow ridges (Schemenda & Grocholsky 1994). The experiments showed that sea-floor spreading may be accommodated by upward motion of mantle rocks along detachments, dipping continentward on each side of the ridge. The following conceptual model (Fig. 4) accounts both for these experiments and for the available seismic and geological data from passive margins and the adjacent oceanic lithosphere. It addresses the process of ultra-slow sea-floor spreading immediately following rifting and continental break-up. The model is a simplification in that it does not account for the 3D structure of slow ridges, where detachments seem to be confined to the extremities of ridge segments or to non-transform offset zones of the ridges (e.g. Tucholke & Lin 1994; Tucholke *et al.* 1997, 1998; Blackman *et al.* 1998; Ranero & Reston 1999; Gracia *et al.* 2000).

In stage A, the lithosphere is thinned as a result of motion along conjugate shear zones (Brun & Beslier 1996), and gabbros are intruded and underplated beneath the rift at the top of the mantle (see Schärer *et al.* 1995, 2001).

In stage B, the continental crust is broken up at the rift axis. Mantle peridotite and syn-rift intruded or underplated gabbros are unroofed and metamorphosed as a result of hydrothermal processes (including serpentinization). Simultaneously, unroofed serpentinized peridotite experiences ultimate tectonic extension, indicated by normal faulting of the serpentinite at every scale, especially on both sides of the peri-

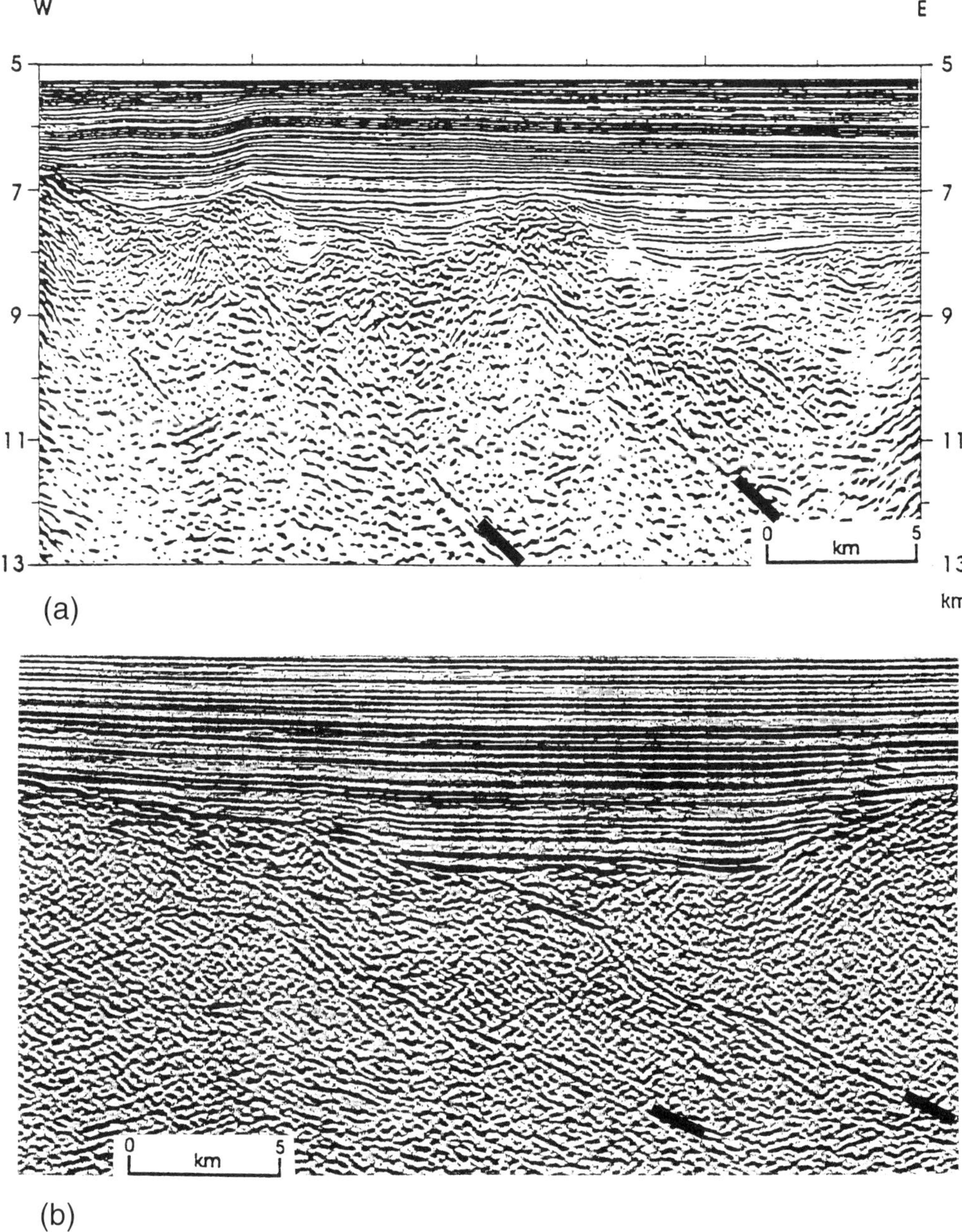

Fig. 9. Continentward dipping reflectors (indicated by bold lines) in the Iberia Abyssal Plain basement (off Portugal). (**a**) Vertical exaggeration ×2; (**b**) another section of the same line, without vertical exaggeration (after Pickup *et al.* 1996).

dotite ridge (Fig. 2). As plate divergence accelerates, the asthenosphere rises more rapidly and undergoes more partial melting because of faster decompression. New basaltic magma intrudes the thin lithosphere remaining between the two conjugate margins, erupts onto the sea floor and covers the previously tectonically denuded peridotite, gabbro and ophicalcite. As a result, there is a time gap between gabbro and basalt formation.

In stage C, new sheets of peridotite and gabbro are emplaced. The plate divergence is not rapid enough to allow the asthenosphere to reach the surface. As beneath continental rifts (stage A), mantle rocks are exhumed by low-angle normal shear zones or faults now imaged as continentward dipping reflectors. Successive shear zones are localized, or initiated, at the level of superficial magma chambers, where the strength of the lithosphere is lowered by the presence of magma. For example, at Gorringe Bank, off South Portugal, a gabbro laccolith is included within serpentinized peridotite. The top and bottom boundaries of the laccolith were intensely deformed under high temperatures along normal shear zones dipping shallowly towards the continent, before peridotite and gabbro were exposed on the sea floor at the onset of sea-floor spreading (Girardeau *et al.* 1998, 1999). The kinematic evolution and structure are in agreement with the above interpretation of continentward dipping reflectors beneath the Iberia Abyssal Plain.

Finally, as a result of increasing rate of plate divergence and a correlative increase in partial melting within the mantle, the basaltic layer covering the mantle rocks becomes thicker and thicker oceanward and the crust thus gradually becomes typically oceanic.

According to the model, the role of tectonics is greater than that of magmatism in the initial stages of sea-floor spreading, and, more generally, in the case of slow- or ultra-slow-spreading ridges. The model accounts in a coherent way for field data, recovered at sea and in ophiolites on land: (1) in many cases, peridotites and gabbros are foliated as a result of deformation in shear zones at depth, under high temperatures. Overlying basalts are undeformed, except for local superficial faulting. (2) Where age constraints are available, gabbros often appear to be older than the overlying basalts. The time gap is variable, from several tens (to hundreds) of million years at the foot of passive margins where pre-rift gabbros crop out in some places, to a few million years and probably less at slow oceanic ridges. The delay between gabbro crystallization and basalt flooding is also supported by the frequent intercalation of a sedimentary event (OC2 and/or radiolarite) between intrusive and extrusive igneous rocks. (3) A progressive transition appears to exist between the areas where the peridotites exposed at the sea floor are of subcontinental (lithospheric) origin, and areas where such rocks are of oceanic (asthenospheric) origin. (4) Continentward dipping reflectors occur, at least in some places, within the ultramafic–mafic basement bordering the

passive margins and within the oceanic lithosphere resulting from slow spreading. These reflectors can be interpreted as seismic signatures of low-angle normal faults (detachments).

However, considerable uncertainties remain about the transition in space and time between continental rifting and sea-floor spreading *sensu stricto*. In particular, the lithospheric shear zones, depicted with a uniform, continentward dip in Figure 4, may form conjugate sets accounting for the subordinate occurrence of oceanward dipping reflectors in addition to the continentward dipping ones.

Rift evolution and the geometry of extensional fault systems

A major contribution towards resolution of whether the fault and shear zone systems accommodating rifting are symmetric or asymmetric, with respect to the axis of the rift, on a lithospheric scale has been made by analogue experiments of continental rifting (Beslier & Brun 1991; Brun & Beslier 1996). These experiments, which used sand as analogue for brittle layers and silicone for ductile layers of the lithosphere, resulted in boudinage of brittle layers, representing the upper crust and the uppermost mantle, accommodated by shear zones located (1) in the lower crust and (2) below the base of the uppermost mantle. The model developed from these experiments is fundamentally symmetric with respect to the rift axis, and envisages the continuous activity of the shear zones mentioned above, from the early stages of rifting to the continental break-up. Although this model represents a good first-order approximation of the rifting process, three sets of observations necessitate some modifications: (1) the model does not take into account the rheological and thermal changes in the lithosphere during the rifting and continental break-up processes, and especially the presence of magma chambers at the top of the mantle. (2) Field observations in the Alpine former margins (Froitzheim & Manatschal 1996; Hermann & Müntener 1996; see also Handy *et al.* 1999) suggest a discontinuous, two-stage evolution of rifting, with a fundamental reorganization of fault patterns between the two stages. In the Alps, the second stage appears to be governed by asymmetric, unidirectionally dipping detachment faults (Fig. 10). A similar two-stage evolution is also proposed for the Iberia–Newfoundland conjugate margins (Manatschal & Bernoulli 1999). (3) Drilling in the Iberia Abyssal Plain during Leg 173 revealed the presence of probable pre-rift lower crust (Rubenach *et al.*, pers. comm.; see the first section of this paper). The lower-crustal terrane is made up of metagabbro to amphibolite. The scarcity of quartz within these rocks casts doubt on the common inference that the rheology of the lower crust is generally governed by quartz and therefore that the lower crust behaves as a weak layer during rifting. If the drilled rocks are representative of the pre-rift lower crust of the

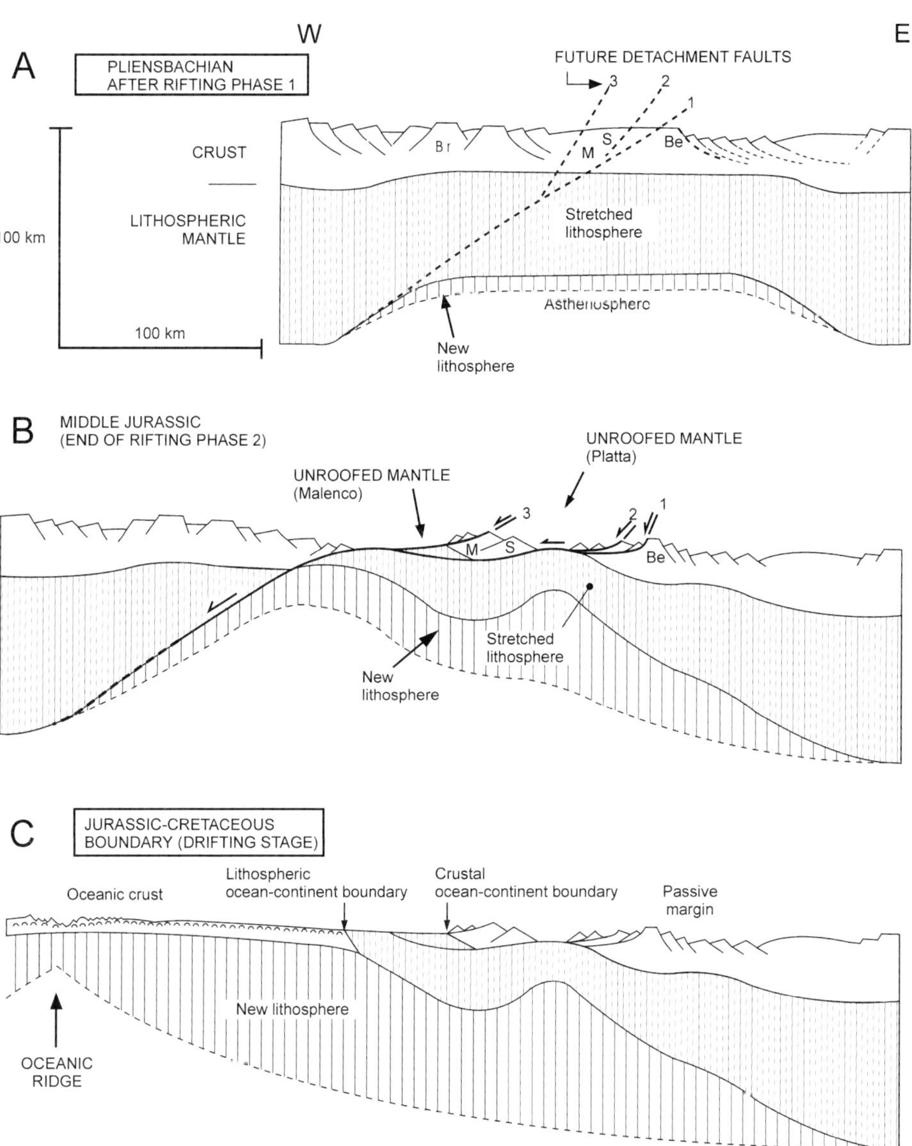

Fig. 10. Hypothetical kinematic evolution of the Tethyan ocean (Ligurian basin) and its passive margins in Mesozoic times. Pre-, syn- and post-rift sedimentary series are neglected. 'Stretched lithosphere' is pre-rift continental lithosphere; 'new lithosphere' is syn- or post-rift lithosphere derived from cooling asthenosphere. Be, Bernina; Br, Briançonnais; M, Margna; S, Sella (after Froitzheim & Manatschal 1996).

margin, the rheology of the lower crust is governed by plagioclase rather than quartz, and a six-layer rheological profile needs to be envisaged instead of the four-layer profile assumed by Brun & Beslier (1996). The same conclusion may be drawn from the field study of a lower-crustal terrane in the Malenco area (Italian Alps) where quartz is a minor constituent and did not control the rheology of the rocks (O. Müntener, pers. comm.)

Accordingly, it is proposed that the initial lithosphere configuration comprises six layers: a brittle upper crust, ductile middle crust (hot enough for viscous behaviour of quartz), brittle lower crust, ductile lowermost crust (hot enough for viscous behaviour of plagioclase), brittle uppermost mantle, and ductile lithospheric mantle overlying asthenosphere.

From this initial rheological structure, it is anticipated that the rifting evolves in two stages: (1) the boudinage stage, controlled by boudinage of brittle layers, including the uppermost crust, part of the lower crust and the uppermost layer of the mantle; (2) the detachment fault stage, preceding and allowing continental break-up and tectonic unroofing of lower crust and mantle rocks. In a sense, the first stage exhibits an overall pure-shear geometry, whereas the second stage is typically governed by simple shear.

The proposed scenario accounts for the fact that detachment faults are generally late structures in the evolution of the margin (e.g. Manatschal & Nievergelt 1997), and neither coincide nor are necessarily linked kinematically with the earlier high-angle faults bounding tilted blocks. It is also supported by thermo-mechanical modelling (see Pérez-Gussinyé *et al.* 2001).

Another question is whether (or why) the continental break-up is a symmetrical or asymmetrical process. From their analogue experiment, Brun & Beslier (1996) came to the conclusion that conjugate margins are probably symmetrical at the lithospheric scale, whereas other workers have proposed they are fundamentally asymmetric (Wernicke 1985; Boillot *et al.* 1987*a*, 1989*a*; Lemoine *et al.* 1987; Etheridge *et al.* 1989; Wernicke & Tilke 1989; Froitzheim & Manatschal 1996, Fig. 10). In fact, both possibilities may exist, depending on the partition of motion between the two possible sets of faults during the detachment fault stage of the rifting. The final stretching of the lithosphere can be accommodated either by subequal distribution of motion between conjugate structures (subsymmetrical margins; Fig. 4), or by motion of one of the two sets that becomes dominant (asymmetric margins; Fig. 10). However, this

question remains open, mainly because of the difficulty of obtaining comparable datasets from conjugate margins (see below).

Thus, we suggest that rifting starts as an overall symmetric boudinage process, and continues with an asymmetric or symmetric detachment fault stage. It is difficult, if not impossible, to test this hypothesis using data from the Iberia margin alone. New numerical or analogue modelling is necessary. Kinematic balancing of whole-crust profiles of conjugate margins (e.g. the Iberia and Newfoundland margins) is also necessary. Another promising target is to restudy the (unfortunately strongly metamorphosed and deformed) former margins that once were conjugate with the well-preserved Tethys margins NW of Apulia (Froitzheim & Manatschal 1996; Hermann & Müntener 1996; Manatschal & Nievergelt 1997; Desmurs *et al.* 1999, 2001) and NW of the Briançonnais terrane (Florineth & Froitzheim 1994).

Location of the ocean–continent boundary, and significance of some magnetic quiet zones bordering passive margins

If the lithosphere is considered as a whole, the peridotite ridge, derived from terranes initially located beneath a continental rift, belongs to the continent and not to the ocean. The limit between the continental lithosphere and the 'true' oceanic lithosphere derived from asthenosphere (convective mantle) is farther offshore. The OCB is then located between the 'stretched lithosphere' delineated in Figure 10 (i.e. the unroofed continental mantle) and the 'new lithosphere' (i.e. the oceanic lithosphere derived from cooling asthenosphere). In other words, the 'transition zone' (also referred to as the ocean–continent transition (OCT) by many workers) composed of subcontinental mantle is inseparable from the margin. Moreover, the seismic velocity structure in the OCT is different from that in typical oceanic crust (Dean *et al.* 2000).

It is, however, difficult to accurately locate the actual OCB according to this definition. How can an unambiguous distinction be made between mantle rocks derived from subcontinental lithosphere and those derived from the asthenosphere rising beneath a slow-spreading ridge at the beginning of sea-floor spreading? To do so would require many rock samples to be recovered from the basement of the margin and the adjacent ocean and analysed geochemically. Both conditions are rarely fulfilled. Moreover, because of ubiquitous serpentinization, the geochemical signature of the original peridotite is difficult to determine, and the subcontinental v. suboceanic origin of the rocks remains somewhat ambiguous and controversial (e.g. Evans & Girardeau 1988; Kornprobst & Tabit 1988; Cornen *et al.* 1996*a*). The origin of mafic rocks (gabbro, basalt, etc.) is more easily determined using isotope data (e.g. Schärer *et al.* 1995; Cornen *et al.* 1996*b*; Seifert *et al.* 1997; Charpentier *et al.* 1998). However, the genesis of these mafic rocks, as previously discussed, is not necessarily related to the emplacement of the surrounding peridotite at the sea floor. Wide-angle seismic data can be a criterion for separating the transition zone from the oceanic crust (Dean *et al.* 2000). Unfortunately, the method is difficult to apply everywhere in the global ocean, and does not permit a precise location of the oceanic crust boundary.

For these reasons it is probably more realistic to locate the OCB with reference to the continental crust boundary and not to the transition zone–oceanic crust boundary. The continental crust of the margin can be identified without ambiguity using its geophysical and geological characters. On the other hand, oceanic domains can be defined as areas where no continental crust occurs, the oceanic crust being made of mafic or ultramafic rocks derived either from subcontinental lithosphere or from asthenospheric mantle. In the case where mantle windows and extensional allochthons of continental crust exist along the margin (Figs 2 and 10), the edge of the most oceanward located continental terrane can represent the OCB.

According to this pragmatic definition, the peridotite ridge bordering the West Iberia margin is part of the ocean, like the 'normal' oceanic crust and lithosphere, even if it originates from subcontinental mantle. The OCB is defined as the trace on the sea floor of the tectonic contact (the detachment fault) separating the continental crust of the margin from serpentinized peridotites and gabbros exhumed in the final stage of the rifting (Fig. 2). The boundary corresponds to a major tectonic structure active during extensional tectonics and passive later. It is not a more or less blurred transition zone but a linear boundary that is relatively easy to delineate. And finally, the definition proposed here is in agreement with the continental break-up concept, which concerns the continental crust rather than the continental lithosphere.

This discussion is not only semantic, as it concerns the validity of plate kinematic reconstructions. To reconstruct the pre-break-up locations of continents, it is appropriate to fit the boundaries of continental crust on both sides of the ocean together, and not the boundaries of exhumed subcontinental mantle. Before fitting the continents together, the amount of syn-rift stretching of the margins has to be accounted for, which is possible for continental crust but would be very difficult, if not impossible, for subcontinental mantle.

Another linked question concerns the origin of some magnetic quiet zones bordering passive margins, when these quiet zones do not result from sea-floor spreading during periods of stability of the global magnetic field. Do these correspond, at least in some cases, to a zone of ultramafic sea floor?

With few local exceptions, serpentinized peridotites and associated gabbros bordering the West Iberia margin do not give rise to significant magnetic anomalies (Sibuet *et al.* 1995; Whitmarsh & Miles 1995; Whitmarsh *et al.* 1996*a*) and so they constitute a magnetic quiet zone. Typical oceanic anomalies are recognized only to the west of the peridotite ridge where basalts occur. Then, the question is: are magnetic anomalies missing there because of the lack or scarcity of basalt on the serpentinized peridotite, or because tectonic unroofing and serpentinization of mantle rocks occurred during the Cretaceous magnetic quiet period? To answer the question unambiguously it is necessary to study passive margin and magnetic quiet zones younger than the West Iberia ones. (The reason why serpentinized peridotite does not give rise to significant magnetic anomalies is the subject of another debate.)

Summary and conclusions

(1) The detachment–shear-zone models for evolved rifts, as depicted in Figure 4, partly derived from the analogue experiments of Schemenda & Grocholsky (1994) and of Brun & Beslier (1996), account for the available geological and geophysical data and processes recorded and interpreted on the West Iberia margin, especially: (a) the syn-rift tectonic unroofing of gabbro and mantle peridotite (transformed into serpentinite as a result of hydrothermal activity) in the uppermost level of the lithosphere; (b) the age discrepancy between pre- or syn-rift gabbros and overlying post-rift basalts cropping out at the edge of the margin; (c) the structural contrast between undeformed basalts and peridotite or gabbro, both sheared at depth and at falling temperature before unroofing; (d) the rare occurrence of pre-rift, lower continental crust at the OCB; (e) the presence of strong reflectors (S on the Galicia and

Armorican margins; H in the Iberia Abyssal Plain), which mark the tectonic contact between thinned continental crust and underlying syn-rift lower seismic crust (serpentinized peridotite and gabbro); (f) the location of the seismic Moho at the boundary between fresh and serpentinized peridotite, i.e. at the hydrothermal front.

(2) To account for the geological and geophysical characters of the oceanic lithosphere bordering the current margins and of some Alpine ophiolites as well, the detachment model can be extended to slow or ultra-slow oceanic ridges (Fig. 4). Specifically, the model explains: (a) the frequent age discrepancy and unconformity between gabbros and basalts within oceanic lithosphere resulting from slow spreading; (b) the ubiquitous occurrence of tectonic breccia (ophicalcite) at the top of serpentinite basement, interpreted as a result of brittle (possibly hydraulic) deformation in the uppermost level of former shear zones (detachments); (c) the seismic images of the upper level of the lithosphere, where strong low-angle reflectors dip continentward (most frequently) or oceanward (less frequently). Although they are located within oceanic lithosphere, it is suggested that these reflectors are the seismic signatures of former lithospheric, conjugate shear zones, similar to the S or H reflectors imaged beneath the deep margin. More generally, the model accounts for the nature of sea-floor spreading along slow ocean ridges, where thin lithosphere, continuously collapsed and extended, covers the asthenosphere at the boundary between the two divergent plates.

(3) However, several questions remain open, including the following: (a) Is the evolution of the rift actually a two-stage process as proposed in this paper (pure shear in the boudinage stage, simple shear in the detachment fault stage)? (b) Where is the actual OCB? Pragmatically, a location is proposed at the limit of the continental crust (the margin *sensu stricto*). The definition implies considering the new lithospheric surface created by unroofing of subcontinental mantle terranes (the peridotite ridge and adjacent areas) as part of the oceanic domain. (c) What is the actual time gap between the formation of gabbro and basalt now cropping out together on the sea floor, and formed by slow spreading, and what are the relations between that time gap and the spreading rate at slow ridges? (d) Where and how does the pre-rift, old lower crust of the continent give way to the syn-rift, new seismic lower crust of the deepest part of the margin formed by undercrusted serpentinite or underplated gabbro? So far, available geophysical data do not permit an unambiguous distinction between these different kinds of lower crust, which have practically the same seismic velocities and densities. (e) Where are the best places on land to test the proposed model of tectonic sea-floor spreading by direct observations of lithospheric shear zones in ophiolites?

(4) The last question introduces a final remark on the synergy between offshore and inland studies of passive margins and oceanic lithosphere. In the late 1970s, marine geologists imaged the first tilted crustal blocks of the Armorican passive margin (e.g. Montadert *et al.* 1979). Soon after, field geologists recognized similar tilted blocks, in some cases inverted into thrust nappes by Cenozoic tectonics, for instance in the French Alpine crystalline massifs (e.g. Lemoine *et al.* 1981; Lombardo & Pognante 1982). The approach was then typically uniformitarian, with the model of a modern rifted margin (Armorican margin) being used to interpret palaeo-rifted margins (e.g. French Western Alps). Later, the knowledge of the structure of the tilted blocks as imaged by seismic data in passive margins or in extensional basins profited considerably from the field observations, allowing interpretation of the seismic images recorded at sea in a realistic way. This is a clear example of the synergy between research work conducted at sea and on land.

In the 1980s, the discovery of the serpentinized peridotite ridge bounding the Galicia margin provided a new uniformitarian model to interpret the ultramafic ophiolites of the internal Franco-Italian Alps (e.g. Lemoine *et al.* 1987). As a result of the field-based observations of the ophiolites, the relations between peridotite, gabbro and ophicalcite in the oceanic crust and ultramafic sea floor near the OCB were clarified. This is a second example of fruitful transfer of knowledge and concept from ocean to mountain belt, and vice versa.

The detachment model for unroofing of metamorphic core complexes was based on studies of the Basin and Range province (e.g. Lister & Davis 1989). Wernicke (1985) applied the model to the whole lithosphere, to account for the unroofing of lower continental crust and even the uppermost mantle. Later the concept was used to explain: (1) the occurrence of ultramafic sea floor at the foot of the Iberia margin, and the shearing of mantle rocks at falling temperature and decreasing pressure; (2) the possible age discrepancy between gabbro and basalt ophiolites derived from the Ligurian ocean; (3) the occurrence of detachment faults marked by the S reflector within the seismic crust of the deepest part of the Armorican and Galicia margins. In this case, the insight came first from field studies on the continent, to later illuminate the geodynamics of the passive margin and of the ocean. More recently the interpretation of the West Iberia OCB as a result of low-angle normal shearing of the lithosphere was successfully applied to the palaeo-OCB of the Valais and Ligurian oceans preserved in the Alpine belt (Florineth & Froitzheim 1994; Froitzheim & Manatschal 1996; Hermann & Müntener 1996; Manatschal & Nievergelt 1997; Manatschal & Bernoulli 1999). In turn, Alpine studies have aided interpretation of the Iberia margin, mainly because of the relatively easy availability of structural kinematic data (shear-sense determination, overprinting relations between faults and shear zones, etc.).

These three examples illustrate the benefits of integrated marine and onshore studics for the advance of geosciences. The first success of this strategy was the understanding of the oceanic crust by comparison with ophiolitic complexes. The next and not less important success was the transfer of experience and concepts concerning passive margins and adjacent ultramafic sea floor. The two communities (onshore and marine geologists) do not work at the same scale nor with the same tools, but they have the same scientific objectives.

We thank M. O. Beslier for useful scientific discussion, and D. Bernoulli, J. Girardeau, Y. Lagabrielle, M. Lemoine and anonymous reviewers for comments and suggestions on the first version of this paper. This paper is Contribution 325 of the UMR Géosciences Azur (CNRS, UPMC, UNSA, IRD).

References

AGRINIER, P., CORNEN, G. & BESLIER, M.O. 1996. Mineralogical and oxygen isotopic features of serpentinites recovered from the ocean–continent transition in the Iberia Abyssal Plain. *In*: WHITMARSH, R.B., SAWYER, D.S., KLAUS, A. *et al.* (eds) *Proceedings of the Ocean Drilling Program, Scientific Results, 149*. Ocean Drilling Program, College Station, TX, 541–551.

AGRINIER, P., MEVEL, C. & GIRARDEAU, J. 1988. Hydrothermal alternation of the peridotites cored at the ocean/continent boundary of the Iberian margin: petrologic and stable isotope evidences. *In*: BOILLOT, G., WINTERER, E.L., MEYER, A.W. *et al.* (eds) *Proceedings of the Ocean Drilling Program, Scientific Results (Part B), 103*. Ocean Drilling Program, College Station, TX, 225–234.

BERNOULLI, D. & WEISSERT, H. 1985. Sedimentary fabrics in Alpine ophicalcites, South Pennine Arosa zone, Switzerland. *Geology*, **13**, 755–758.

BESLIER, M.O. & BRUN, J.-P. 1991. Boudinage de la lithosphère et formation des marges passives. *Comptes Rendus de l'Académie des Sciences, Série II*, **313**, 951–958.

BESLIER, M.O., ASK, M. & BOILLOT, G. 1993. Ocean–continent boundary in the Iberia Abyssal Plain from multichannel seismic data. *Tectonophysics*, **218**, 383–393.

BESLIER, M.O., BITRI, A. & BOILLOT, G. 1995. Structure de la transition continent–océan d'une marge passive: sismique réflexion multitrace dans la plaine abyssale ibérique (Portugal). *Comptes Rendus de l'Académie des Sciences*, **320**, 969–976.

BESLIER, M.O., CORNEN, G., GIRARDEAU, J. 1996. Tectono-metamorphic evolution of peridotites from the ocean/continent transition of the Iberia Abyssal Plain margin. *In*: WHITMARSH, R.B., SAWYER, D.S., KLAUS, A. *et al.* (eds) *Proceedings of the Ocean Drilling Program, Scientific Results, 149*. Ocean Drilling Program, College Station, TX, 397–412.

BESLIER, M.O., GIRARDEAU, J. & BOILLOT, G. 1990. Kinematics of peritotite emplacement during North Atlantic continental rifting, Galicia, NW Spain. *Tectonophysics*, **184**, 321–343.

BILL, M., BUSSY, F., COSCA, M., MASSON, N. & HUNZIKER, J. 1997. High-precision U–Pb and ^{40}Ar/^{39}Ar dating of an Alpine ophiolite (Gets nappe, French Alps). *Eclogae Geologicae Helvetiae*, **90**, 43–54.

BLACKMAN, D.K., CANN, J.R., JANSSEN, B. & SMITH, D.K. 1998. Origin of extensional core complexes: evidence from the Mid-Atlantic Ridge at Atlantic fracture zone. *Journal of Geophysical Research*, B9, **103**, 21315–21333.

BOILLOT, G. & COULON, C. *La Déchirure continentale et l'Ouverture océanique: Géologie des Marges passives*. Gordon and Breach, Paris.

BOILLOT, G., AGRINIER, P., BESLIER, M.O. & 9 OTHERS 1995*b*. A lithospheric syn-rift shear zone at the ocean–continent transition: preliminary results of the GALINAUTE II cruise (Nautile dives on the Galicia Bank, Spain). *Comptes Rendus de l'Académie des Sciences, Série IIa*, **322**, 1171–1178.

BOILLOT, G., BESLIER, M.O. & COMAS, M.C. 1991. Seismic image of undercrusted serpentinite beneath a rifted margin. *Terra Nova*, **4**, 25–33.

BOILLOT, G., BESLIER, M.O. & GIRARDEAU, J. 1995*c*. Nature, structure and evolution of the ocean–continent boundary: the lesson of the West Galicia Margin (Spain). *In*: BANDA, E., TORNÉ, M. & TALWANI, M. (eds) *Rifted Ocean–Continent Boundaries*. Kluwer, Dordrecht, 219–229.

BOILLOT, G., BESLIER, M.O., KRAWCZYK, C.M., RAPPIN, D. & RESTON, T.J. 1995*a*. The formation of passive margins: constraints from the crustal structure and segmentation of the deep Galicia margin, Spain. *In*: SCRUTTON, R.A., STOKER, M., SCHIMMIELD, G.B. & TUDHOPE, A.W. (eds) *The Tectonics, Sedimentation and Palaeooceanography of the North Atlantic*

Region. Geological Society, London, Special Publications, **90**, 71–91.

BOILLOT, G., COMAS, M.C., GIRARDEAU, J. & 5 OTHERS 1988a. Preliminary results of the Galinaute cruise (dives of the submersible 'Nautile' on the West Galicia margin, Spain). *In*: BOILLOT, G., WINTERER, E.L. & MEYER, A.W. *et al.* (eds) *Proceedings of the Ocean Drilling Program, Scientific Results, 103*. Ocean Drilling Program, College Station, TX, 733–740.

BOILLOT, G., FÉRAUD, G., RECQ, M. & GIRARDEAU, J. 1989b. 'Undercrusting' by serpentinite beneath rifted margins. *Nature*, **341**, 523–525.

BOILLOT, G., GIRARDEAU, J. & KORNPROBST, J. 1988b. Rifting of the Galicia margin: crustal thinning and emplacement of mantle rocks on the seafloor. *In*: BOILLOT, G., WINTERER, E.L., MEYER, A.W. *et al.* (eds) *Proceedings of the Ocean Drilling Program, Scientific Results, 103*. Ocean Drilling Program, College Station, TX, 741–756.

BOILLOT, G., GRIMAUD, S., MAUFFRET, A., MOUGENOT, D., KORNPROBST, J., MERGOIL-DANIEL, J. & TORRENT, G. 1980. Ocean–continent boundary of the Iberian margin: a serpentinite diapir west of the Galicia Bank. *Earth and Planetary Science Letters*, **48**, 23–34.

BOILLOT, G., MOUGENOT, D., GIRARDEAU, J. & WINTERER, E.L. 1989a. Rifting processes on the West Galicia Margin, Spain. *In*: TANKARD, A.J. & BALKWILL, H.R. (eds) *Extensional Tectonics and Stratigraphy of the North Atlantic Margins*. American Association of Petroleum Geologists, Memoirs, **46**, 363–377.

BOILLOT, G., RECQ, M., WINTERER, E.L. & 21 OTHERS 1987a. Tectonic denudation of the upper mantle along passive margin: a mode based on drilling results (Ocean Drilling Program Leg 103, Western Galicia Margin, Spain). *Tectonophysics*, **132**, 335–342.

BOILLOT, G., WINTERER, E.L., MEYER, A.W. *et al.* (eds) 1987b. *Proceedings of the Ocean Drilling Program, Initial Reports (Part A), 103*. Ocean Drilling Program, College Station, TX.

BOILLOT, G., WINTERER, E.L., MEYER, A.W. *et al.* (eds) 1988c. *Proceedings of the Ocean Drilling Program, Scientific Results (Part B), 103*. Ocean Drilling Program, College Station, TX.

BORTOLOTTI, V., CELLAI, D., VAGGELLI, G. & VILLA, I.M. 1991. $^{40}Ar/^{39}Ar$ dating of Apenninic ophiolites: 2. Basalts from the Aiola sequence, Southern Tuscany, Italy. *Ofiolitti*, 1, **16**, 37–42.

BRUN, J.P. & BESLIER, M.O. 1996. Mantle exhumation at passive margin. *Earth and Planetary Science Letters*, **142**, 161–173.

CABY, R. 1995. Plastic deformation of gabbros in a slow-spreading Mesozoic ridge: example of the Montgenèvre ophiolite, Western Alps. *In*: VISSERS, R.L.M. & NICOLAS, A. (eds) *Mantle and Lower Crust Exposed in Oceanic Ridges and in Ophiolites*. 123–145. Kluwer, Dordrecht.

CANNAT, M., LAGABRIELLE, Y., BOUGAULT, T.H., CASEY, J., DE COUTURE, N., DIMITRIEV, L. & FOUQUET, Y. 1997. Ultramafic and gaboric

exposures at the Mid-Atlantic Ridge: geological mapping in the 15°N region. *Tectonophysics*, **279**, 193–213.

CHARPENTIER, S., KORNPROBST, J., CHAZOT, G., CORNEN, G. & BOILLOT, G. 1998. Interaction entre lithosphère et asthénosphère au cours de l'ouverture océanique: données isotopiques préliminaires sur la marge passive de Galice (Atlantique-Nord). *Comptes Rendus de l'Académie des Sciences, Sciences de la Terre et des Planètes*, **326**, 757–762.

CHIAN, D., LOUDEN, K., MINSHULL, T.A. & WHITMARSH, R.B. 1999. Deep structure of the ocean–continent transition in the southern Iberia Abyssal Plain from seismic refraction profiles: Ocean Drilling Program (Legs 149 and 173) transect. *Journal of Geophysical Research*, **104**, 7443–7462.

CORNEN, G., BESLIER, M.O. & GIRARDEAU, J. 1996a. Petrologic characteristics of the ultramafic rocks from the ocean–continent transition in the Iberia Abyssal Plain. *In*: WHITMARSH, R.B., SAWYER, D.S., KLAUS, A. *et al.* (eds) *Proceedings of the Ocean Drilling Program, Scientific Results, 149*. Ocean Drilling Program, College Station, TX, 377–395.

CORNEN, G., BESLIER, M.O. & GIRARDEAU, J. 1996b. Petrology of the mafic rocks cored in the Iberia Abyssal Plain. *In*: WHITMARSH, R.B., SAWYER, D.S., KLAUS, A. *et al.* (eds) *Proceedings of the Ocean Drilling Program, Scientific Results, 149*. Ocean Drilling Program, College Station, TX, 449–470.

COSTA, S. & CABY, R. 1997. Evolution of the Ligurian Tethys in the Western Alps: Sm–Nd and U–Pb geochronology and REE study of the Mont-Genèvre ophiolite (Southeast of France). *In*: *EUG 9th Meeting Abstracts*. Cambridge Publications, Strasbourg, 510.

DEAN, S.M., MINSHULL, T.A., WHITMARSH, R.B. & LOUDEN, K.E. 2000. Deep structure of the ocean–continent transition in the southern Iberia Abyssal Plain from seismic refraction profiles: the IAM-9 transect at 40°20′N. *Journal of Geophysical Research*, **105**, 5859–5885.

DE CHARPAL, O., GUENNOC, P., MONTADERT, L. & ROBERTS, D.G. 1978. Rifting, crustal attenuation and subsidence in the Bay of Biscay. *Nature*, 5682, **275**, 706–711.

DESMURS, L., SCHALTEGGER, U., MANATSCHAL, G. & BERNOULLI, D. 1999. Geodynamic significance of gabbros along ancient ocean–continent transitions: Tasna and Platta nappes, Eastern Alps. *Journal of Conference Abstracts 4, 1(EUG)*, **10**, 379.

DESMURS, L., MANATSCHAL, G. & BERNOULLI, D. 2001. The Steinmann Trinity revisited: mantle exhumation and magmatism along an ocean–continent transition: the Platta nappe, eastern Switzerland. *In*: WILSON, R.C.L., WHITMARSH, R.B., TAYLOR, B. & FROITZHEIM, N. (eds) *Nonvolcanic Rifting of Continental Margins: a Comparison of Evidence from Land and Sea*. Geological Society, London, Special Publications, **187**, 235–266.

ETHERIDGE, M.A., SYMONDS, P.A. & LISTER, G.S. 1989. Application of the detachment model to reconstruction of conjuguate passive margins. *In*: TANKARD, A.J. & BALKWILL, H.R. (eds) *Extensional Tectonics and Stratigraphy of the North Atlantic Margins*. American Association of Petroleum Geologists, Memoirs, **46**, 23–40.

EVANS, C.A. & GIRARDEAU, J. 1988. Galicia margin peridotites: underplated abyssal peridotites from the North Atlantic. *In*: BOILLOT, G., WINTERER, E.L., MEYER, A.W. *et al.* (eds) *Proceedings of the Ocean Drilling Program, Scientific Results, 103*. Ocean Drilling Program, College Station, TX, 195–208.

FÉRAUD, G., BESLIER, M.O. & CORNEN, G. 1996. ^{40}Ar/^{39}Ar dating of gabbros from the ocean/continent transition of the western Iberia Margin: preliminary results. *In*: WHITMARSH, R.B., SAWYER, D.S., KLAUS, A. *et al.* (eds) *Proceedings of the Ocean Drilling Program, Scientific Results, 149*. Ocean Drilling Program, College Station, TX, 489–498.

FÉRAUD, G., GIRARDEAU, J., BESLIER, M.O. & BOILLOT, G. 1988. Datation ^{39}Ar–^{40}Ar de la mise en place des péridotites bordant la marge de Galice (Espagne). *Comptes Rendus des Séances de l'Académie des Sciences, Série III*, **307**, 49–55.

FLORINETH, D. & FROITZHEIM, N. 1994. Transition from continental to oceanic basement in the Tasna nappe (Engadine window, Graubünden, Switzerland): evidence for early Cretaceous opening of the Valais ocean. *Schweizerische Mineralogische und Petrographische Mitteilungen*, **74**, 437–448.

FROITZHEIM, N. & MANATSCHAL, G. 1996. Kinematics of Jurassic rifting, mantle exhumation, and passive-margin formation in the Austroalpine and Penninic nappes (eastern Switzerland). *Geological Society of America Bulletin*, **108**, 1120–1133.

FROITZHEIM, N. & RUBATTO, D. 1998. Continental break-up by detachment faulting: field evidence and geochronological constraints (Tasna nappe, Switzerland). *Terra Nova*, **10**, 171–176.

FRÜH-GREEN, G.L., WEISSERT, H. & BERNOUILLI, D. 1990. A multiple fluid history recorded in Alpine ophiolites. *Journal of the Geological Society, London*, **147**, 959–970.

FUEGENSCHUH, B., FROITZHEIM, N. & BOILLOT, G. 1998. Cooling history of granulite samples from the ocean–continent transition of the Galicia margin: implications for rifting. *Terra Nova*, **10**, 96–100.

GIRARDEAU, J., CORNEN, G., AGRINIER, P. & 7 OTHERS 1998. Preliminary results of nautile dives on the Gorringe Bank (West Portugal). *Comptes Rendus de l'Académie des Sciences*, **326**, 247–254.

GIRARDEAU, J., CORNEN, G., BESLIER, M.O. & 7 OTHERS 1999. Extensional tectonics in the Gorringe Bank rocks, Eastern Atlantic ocean: evi-

dence of an oceanic ultra-slow mantellic accreting centre. *Terra Nova*, **10**, 330–336.

GIRARDEAU, J., EVANS, C.A., BESLIER, M.O. *et al.* 1988. Structural analysis of plagioclase-bearing peridotite emplaced at the end of continental rifting: Hole 637A, Ocean Drilling Program Leg 103 on the Galicia margin. *In*: BOILLOT, G., WINTERER, E.L., MEYER, A.W. *et al.* (eds) *Proceedings of the Ocean Drilling Program, Scientific Results, 103*. Ocean Drilling Program, College Station, TX, 209–223.

GRACIA, E., CHARLOU, J.L., RADFORD-KNOERY, J. & PARSON, L.M. 2000. Non-transform offsets along the Mid-Atlantic Ridge south of the Azores (38°N–34°N): ultramafic exposures and hosting of hydrothermal vents. *Earth and Planetary Science Letters*, **177**, 89–103.

HANDY, M.R., FRANZ, L., HELLER, F., JANOTT, B. & ZURBRIGGEN, R. 1999. Multistage accretion and exhumation of the continental crust (Ivrea crustal section, Italy and Switzerland). *Tectonics*, **18**(6), 1154–1177.

HERMANN, J. & MÜNTENER, O. 1996. Extension-related structures in the Malenco–Margna System: implications for paleogeography and consequences for rifting and Alpine tectonics. *Schweizerische Mineralogische und Petrographisches Mitteilungen*, **76**, 501–519.

HERMANN, J., MÜNTENER, O., TROMMSDORFF, V., HANSMANN, W. & PICCARDO, G.B. 1997. Fossil crust-to-mantle transition, Val Malenco (Italian Alps). *Journal of Geophysical Research*, B9, **102**, 20123–20132.

HOFFMANN, H.J. & RESTON, T.J. 1992. Nature of the S reflector beneath the Galicia Banks rifted margin: preliminary results from pre-stack depth migration. *Geology*, **20**, 1091–1094.

KEEN, C.E. & DE VOOGD, B. 1988. The continent–ocean boundary at the rifted margin off Eastern Canada: new results from deep seismic reflection studies. *Tectonics*, **7**, 107–124.

KORNPROBST, J. & TABIT, A. 1988. Plagioclase bearing ultramafic tectonites from the Galicia margin (Leg 103, Site 637): comparison of their origin and evolution with low pressure ultramafic bodies in Western Europe. *In*: BOILLOT, G., WINTERER, E.L., MEYER, A.W. *et al.* (eds) *Proceedings of the Ocean Drilling Program, Scientific Results, 103*. Ocean Drilling Program, College Station, TX, 253–268.

KORNPROBST, J., VIDAL, Ph. & MALOD, J. 1988. Les basaltes de la marge de Galice (NO de la Péninsule ibérique): hétérogénéité des spectres de terres rares à la transition continent–océan. Données géochimiques préliminaires. *Comptes Rendus de l'Académie des Sciences, Série II*, **306**, 1359–1364.

KRAWCZYK, C.M. & RESTON, T.J. 1995. Detachment faulting and continental break-up: the S reflector offshore Galicia. *In*: BANDA, E., TORNÉ, M. & TALWANI, M. (eds) *Rifted Ocean–Continent Boundaries*. Kluwer, Dordrecht, 231–246.

KRAWCZYK, C.M., RESTON, T.J., BESLIER, M.O. & BOILLOT, G. 1996. Evidence for detachment tectonics on the Iberia Abyssal Plain rifted margin. *In*: WHITMARSH, R.B., SAWYER, D.S., KLAUS, A. *et al.* (eds) *Proceedings of the Ocean Drilling Program, Scientific Results, 149.* Ocean Drilling Program, College Station, TX, 603–615.

LAGABRIELLE, Y. & LEMOINE, M. 1997. Alpine, Corsican and Apennine ophiolites: the slow spreading ridge model. *Comptes Rendus de l'Académie des Sciences, Sciences de la Terre et des Planètes*, **325**, 909–920.

LAGABRIELLE, Y., POLINO, R., AUZENDE, J.-M. & 8 OTHERS 1984. Les témoins d'une tectonique intra-océanique dans le domaine téthysien: analyse des rapports entre les ophiolites et leur couverture méta-sédimentaire dans la zone piémontaise des Alpes franco-italiennes. OFIOLITTI, **9**(1), 67–88.

LAGABRIELLE, Y., BIDEAU, D., CANNAT, M., KARSON, J.A. & MEVEL, C. 1998. Ultramafic–mafic plutonic rocks suites exposed along the mid-Atlantic ridge (10°N–30°N). Symmetrical–asymmetrical distribution and implications for seafloor spreading processes. *Faulting and Magmatism at Mid-Ocean Ridges.* Geophysical Monograph, American Geophysical Union, **106**, 153–176.

LEMOINE, M. 1980. Serpeninites, gabbros and ophicalcite in the Piemont–Ligurian domain of the Western Alps: possible indicators of oceanic fracture zones and of associated serpentinite protrusions in the Jurassic–Cretaceous Tethys. *Archives des Sciences, Genève*, **33**, 103–115.

LEMOINE, M., GIDON, M. & BARFETY, J.C. 1981. Les massifs cristallins externes des Alpes occidentales: d'anciens blocs basculés nés au Lias lors du rifting téthysien. *Comptes Rendus de l'Académie des Sciences, Série II*, **292**, 917–920.

LEMOINE, M., TRICART, P. & BOILLOT, G. 1987. Ultramafic and gabbroic ocean floor of the Ligurian Tethys (Alps, Corsica, Apennines): in search of a genetic model. *Geology*, **15**, 622–625.

LE PICHON, X. & BARBIER, F. 1987. Passive margin formation by low-angle faulting within the upper crust: the northern Bay of Biscay margin. *Tectonics*, 2, **6**, 133–150.

LISTER, G.S. & DAVIS, G.A. 1989. The origin of metamorphic core-complexes and detachment faults formed during Tertiary continental extension in the northern Colorado Riover region, USA. *Journal of Structural Geology*, **11**, 65–94.

LOMBARDO, B. & POGNANTE, V. 1982. Tectonic implication in the evolution of the western Alps ophiolite metagabbros. *Ofiolitti*, **7**, 371–395.

MALOD, J.A., MURILLAS, J., KORNPROBST, J. & BOILLOT, G. 1993. Oceanic lithosphere at the edge of a Cenozoic active continental margin (northwest slope of the Galicia Bank, Spain). *Tectonophysics*, **221**, 195–206.

MANATSCHAL, G. & BERNOULLI, D. 1999. Architecture and tectonic evolution of non-volcanic margins: present-day Galicia and ancient Adria. *Tectonics*, **18**, 1099–1119.

MANATSCHAL, G. & NIEVERGELT, P. 1997. A continent–ocean transition recorded in the Err and Platta nappes (Eastern Switzerland). *Eclogae Geologicae Helvetiae*, **90**, 3–27.

MANATSCHAL, G., FROITZHEIM, N., RUBENACH, M. & TURRIN, B.D. 2001. The role of detachment faulting in the formation of an ocean–continent transition: insights from the Iberia Abyssal Plain. *In*: WILSON, R.C.L., WHITMARSH, R.B., TAYLOR, B. & FROITZHEIM, N. (eds) *Non-volcanic Rifting of Continental Margins: a Comparison of Evidence from Land and Sea.* Geological Society, London, Special Publications, **187**, 405–428.

MASSON, D.G., MONTADERT, L. & SCRUTTON, R.A. 1985. Regional geology of the Goban Spur continental margin. *In*: DE GRACIANSKY, P.C., POAG, C.W. *et al.* (eds) *Initial Reports, Deep Sea Drilling Project, 80.* US Government Printing Office, Washington, DC, 1115–1139.

MAUFFRET, A. & MONTADERT, L. 1987. Rift tectonics on the passive continental margin off Galicia (Spain). *Marine and Petroleum Geology*, **4**, 49–70.

MONTADERT, L., DE CHARPAL, O., ROBERTS, D., GUENNOC, P. & SIBUET, J.C. 1979. Deep drilling results in the Atlantic Ocean: continental margins and palaeoenvironment. *In*: TALWANI, M., HAY, W. & RYAN, B.F. (eds) *Northeast Atlantic Passive Continental Margins: Rifting and Subsidence Processes.* American Geophysical Union, Maurice Ewing Series, **3**, 154–186.

MÜNTENER, O. & HERMANN, J. 1996. The Val Malenco lower crust–upper mantle complex and its field relations (Italian Alps). *Schweizerische Mineralogisches und Petrographishes Mitteilungen*, **76**, 475–500.

MÜNTENER, O. & HERMANN, J. 2001. The role of lower crust and continental upper mantle during formation of non-volcanic passive margins: evidence from the Alps. *In*: WILSON, R.C.L., WHITMARSH, R.B., TAYLOR, B. & FROITZHEIM, N. (eds) *Non-volcanic Rifting of Continental Margins: a Comparison of Evidence from Land and Sea.* Geological Society, London, Special Publications, **187**, 267–288.

Ocean Drilling Program Leg 173 Shipboard Scientific Party 1998. Drilling reveals transition from continental break-up to early magmatic crust. *EOS Transactions, American Geophysical Union*, **79**, 180–181.

OHNENSTETTER, M., OHNENSTETTER, D., VIDAL, Ph., CORNICHET, J., HERMITTE, D. & MACE, J. 1981. Crystallization and age of zircon from Corsican ophiolitic albitite: consequences for oceanic expansion in Jurassic times. *Earth and Planetary Science Letters*, **54**, 397–408.

PÉREZ-GUSSINYÉ, M., RESTON, T.J. & PHIPPS MORGAN, J. 2001. Serpentinization and magmatism during extension at non-volcanic margins: the effect of initial lithospheric structure. *In*: WILSON, R.C.L., WHITMARSH, R.B., TAYLOR, B. & FROITZHEIM, N. (eds) *Non-volcanic Rifting of Continental Margins: a Comparison of Evidence from Land and Sea.* Geological Society, London, Special Publications, **187**, 551–576.

PICKUP, S.L.B., WHITMARSH, R.B., FOWER, C.M.R. & RESTON, T.J. 1996. Insight into the nature of the ocean–continent transition off the West Iberia from a deep multichannel seismic reflexion profile. *Geology*, **24**, 1079–1082.

PINET, N., LAGABRIELLE, Y. & WHITECHURCH, H. 1989. Le complexe du Pic des Lauzes (Haut Queyras, Alpes occidentales, France): structure alpines et océaniques dans un massif ophiolitique de type Liguro-Piémontais. *Bulletin de la Société Géologique de France*, **V**, 317–328.

POLINO, R. & LEMOINE, M. 1984. Détritisme mixte d'origine continentale et océanique dans les sédiments jurassico-crétacé supra ophiolitiques de la Téthys Ligure: la serre du Lago Nero (Alpes occidentales franco-italiennes). *Comptes Rendus de l'Académie des Sciences, Série II*, **298**, 359–364.

RANERO, C.R. & RESTON, T.J. 1999. Detachment faulting at ocean core complexes. *Geology*, **27**, 983–986.

RECQ, M., WHITMARSH, R.B., SIBUET, J.C., WHITE, R.S. & LYNESS, D. 1996. Structure sismique de la ride de péridotite à l'ouest du Banc de Galice (Ouest-Ibérie). *Comptes Rendus de l'Académie des Sciences, Série IIa*, **322**, 571–578.

RESTON, T.J., KRAWCZYK, C.M. & KLAESCHEN, D. 1996. The S reflector west of Galicia (Spain): evidence from pre-stack depth migration for detachment faulting during continental break-up. *Journal of Geophysical Research*, B4, **101**, 8075–8091.

RUBATTO, D. 1998. *Dating of pre-Alpine magmatism, Jurassic ophiolites and Alpine subductions in the Western Alps*. PhD thesis, ETH Zurich.

RUBATTO, D., GEBAUER, D. & FANNING, M. 1998. Jurassic formation and Eocene subduction of the Zermatt–Saas–Fee ophiolites: implications for the geodynamic evolution of the Central and Western Alps. *Contributions to Mineralogy and Petrology*, **132**, 269–287.

SAWYER, D.S., WHITMARSH, R.B., KLAUS, A. *et al.* (eds) 1994. *Proceedings of the Ocean Drilling Program, Initial Reports, 149*. Ocean Drilling Program, College Station, TX.

SCHÄRER, U., GIRARDEAU, J., CORNEN, G. & BOILLOT, G. 2001. 138–121 Ma asthenospheric magmatism prior to continental break-up in the North Atlantic and geodynamic implications. *Earth and Planetary Science Letters*, **181**, 555–572.

SCHÄRER, U., KORNPROBST, J., BESLIER, M.O., BOILLOT, G. & GIRARDEAU, J. 1995. Gabbro and related rock emplacement beneath rifting continental crust: U–Pb geochronological and geochemical constraints for the Galicia passive margin (Spain). *Earth and Planetary Science Letters*, **130**, 187–200.

SCHEMENDA, A.I. & GROCHOLSKY, A.L. 1994. Physical modeling of slow seafloor spreading. *Journal of Geophysical Research*, B5, **990**, 9137–9153.

SEIFERT, K.E., CHENG-WEN, C. & BRUNOTTE, D.A. 1997. Evidence from Ocean Drilling Program Leg 149 mafic igneous rocks for oceanic crust in the Iberia Abyssal Plain ocean–continent transition zone. *Journal of Geophysical Research*, B4, **102**, 7915–7928.

SIBUET, J.C. 1992. New constraints on the formation of non-volcanic continental Galicia–Flemish Cap conjugate margins. *Journal of the Geological Society, London*, **149**, 829–840.

SIBUET, J.C., LOUVEL, V., WHITMARSH, R.B. & 5 OTHERS 1995. Constraints on rifting processes from refraction and deep-tow magnetic data: the example of the Galicia Continental margin. *In*: BANDA, E., TORNÉ, M. & TALWANI, M. (eds) *Rifted Ocean–Continent Boundaries*. Kluwer, Dordrecht, 197–217.

SIBUET, J.C., MAZE, J.P., AMORTILLA, Ph. & LE PICHON, X. 1987. Physiography and structure of the western Iberian continental margin off Galicia from seabeam and seismic data. *In*: BOILLOT, G., WINTERER, E.L., MEYER, A.W. *et al.* (eds) *Proceedings of the Ocean Drilling Program, Initial Reports (Part A), 103*. Ocean Drilling Program, College Station, TX, 77–97.

TREVES, B.E. & HARPER, G.D. 1994. Exposure of serpentinite on the ocean floor: sequence of faulting and hydrofracturing in the northern Apennine ophicalcites. *Ofiolitti*, **196**, 435–466.

TRICART, P. & LEMOINE, M. 1989. The Queyras ophiolite west of Monte Viso (Western Alps): indicator of a peculiar ocean floor in the Mesozoic Tethys. *Journal Geodynamics*, **13**, 163–181.

TROMMSDORFF, V., PICCARDO, G.P. & MONTRASIO, A. 1993. From magmatism through metamorphism to seafloor emplacement of subcontinental Adria lithosphere during pre-Alpine rifting (Malenco, Italy. *Schweizerische Mineralogisches und Petrographishes Mitteilungen*, **73**, 191–203.

TUCHOLKE, B.E. & LIN, J. 1994. A geological model for the structure of ridge segments in slow spreading ocean crust. *Journal of Geophysical Research*, **99**, 11937–11958.

TUCHOLKE, B.E., LIN, J. & KLEINROCK, M.C. 1998. Megamullions and mullion structure defining oceanic metamorphic core complexes on the Mid-Atlantic Ridge. *Journal of Geophysical Research*, **103**, 9857–9866.

TUCHOLKE, B.E., LIN, J., KLEINROCK, M.C., TIVEY, M.A., REED, T.B., GOFF, J. & JAROSLOW, G. 1997. Segmentation and crustal structure of the western Mid-Atlantic Ridge flank, 25°25′–27°10′N and 0–29 m.y. *Journal of Geophysical Research*, B5, **102**, 10203–10223.

WEISSERT, H.J. & BERNOULLI, D. 1985. A transform margin in the Mesozoic Tethys: evidence from the Swiss Alps. *Geologische Rundschau*, **74**, 665–679.

WERNICKE, B. 1985. Uniform-sense normal simple shear of the continental lithosphere. *Canadian Journal of Earth Sciences*, **22**, 108–125.

WERNICKE, B. & TILKE, P.G. 1989. Extensional tectonic framework of the U.S. Central Atlantic passive margin. *In*: TANKARD, A.J. & BALK-WILL, H.R. (eds) *Extensional Tectonics and Stratigraphy of the North Atlantic Margins.* American Association of Petroleum Geologists, Memoirs, **46**, 7–21.

WHITMARSH, R.B. & MILES, P.R. 1995. Models of the development of the West Iberia rifted continental margin at 40°30′N deduced from surface and deep-tow magnetic anomalies. *Journal of Geophysical Research*, B3, **100**, 3789–3806.

WHITMARSH, R.B. & SAWYER, D.S. 1996. The ocean/continent transition beneath the Iberia Abyssal Plain and continental rifting to seafloor-spreading processes. *In*: WHITMARSH, R.B., SAWYER, D.S., KLAUS, A. *et al.* (eds) *Proceedings of the Ocean Drilling Program, Scientific Results, 149.* Ocean Drilling Program, College Station, TX, 713–733.

WHITMARSH, R.B., BESLIER, M.O., WALLACE, P.J. *et al.* (eds) 1998. *Proceedings of the Ocean Drilling Program, Initial Reports, 173.* Ocean Drilling Program, College Station, TX.

WHITMARSH, R.B., MILES, P.R., SIBUET, J.C., LOUVEL, V. *et al.* 1996*a*. Geological and geophysical implications of deep-tow magnetometer observations near Sites 897, 898, 899, 900 and 901 on the West Iberia Continental margin. *In*: WHITMARSH, R.B., SAWYER, D.S., KLAUS, A. *et al.* (eds) *Proceedings of the Ocean Drilling Program, Scientific Results, 149.* Ocean Drilling Program, College Station, TX, 665–674.

WHITMARSH, R.B., SAWYER, D.S., KLAUS, A. *et al.* (eds) 1996*b*. *Proceedings of the Ocean Drilling Program, Scientific Results, 149.* Ocean Drilling Program, College Station, TX.

WHITMARSH, R.B., WHITE, R.S., HORSEFIELD, S.J., SIBUET, J.C., RECQ, M. & LOUVEL, V. 1996*c*. The ocean–continent boundary off the western continental margin of Iberia: crustal structure west of Galicia Bank. *Journal of Geophysical Research, 101(B)*, **12**, 28291–28314.

WILSON, R.C.L., MANATSCHAL, G. & WISE, S. 2001. Rifting along non-volcanic passive margins: stratigraphic and seismic evidence from the Mesozoic successions of the Alps and western Iberia. *In*: WILSON, R.C.L., WHITMARSH, R.B., TAYLOR, B. & FROITZHEIM, N. (eds) *Non-volcanic Rifting of Continental Margins: a Comparison of Evidence from Land and Sea.* Geological Society, London, Special Publications, **187**, 429–452.

Propagation of continental break-up in the southwestern South China Sea

P. HUCHON[1,2], T. N. H. NGUYEN[1] & N. CHAMOT-ROOKE[1]

[1]*Laboratoire de Géologie, Ecole Normale supérieure & CNRS, UMR 8538, 24 rue Lhomond, 75231 Paris Cedex 05, France*

[2]*Present address: Géosciences Azur, Pierre & Marie Curie University & CNRS, UMR 6526, Observatoire océanologique de Villefranche, La Darse, BP 48, 06235 Villefranche-sur-Mer, France (e-mail: huchon@obs-vlfr.fr)*

Abstract: We present new bathymetric, seismic and gravity data on the southwestern tip of the South China Sea oceanic basin, where propagation of continental break-up occurred before *c.* 15 Ma. The oceanic domain has a V-shape typical of oceanic propagating rifts. The tectonic fabric of its margins shows that the main stretching direction was slightly oblique to that of the rift axis. A 2D gravity anomaly inversion, corrected for the thermal effect, is used to estimate the crustal structure. At the continent−ocean boundary, the continental crust is stretched by a factor of about four, rapidly decreasing to about two over a few tens of kilometres, a distance corresponding to just over 1 Ma of break-up propagation. Thus, strain localization occurs at the tip of the propagating oceanic crust just before break-up. The along-axis variation in continental crustal stretching is in good agreement with the kinematics of the oceanic crust derived from magnetic anomalies. This analysis suggests that break-up propagates toward the pole of relative rotation and is primarily controlled by the amount of stretching of the continental crust before oceanization.

Recent studies of continental and oceanic rifts suggest that extension does not occur synchronously along strike. Several models of rift propagation have been proposed (Hey *et al.* 1980; Courtillot 1982; Vink 1982; Bosworth 1985). Hey's concept of oceanic rift propagation differs from models dealing with continental break-up propagation as it implies extension beyond a transform zone of an oceanic rift (spreading centre) into undeformed oceanic lithosphere. In contrast, what we call break-up propagation corresponds to strain localization within a stretched continental lithosphere, i.e. the start of sea-floor spreading. In this regard, the process of strain localization is a direct consequence of the relative motion between two lithospheric plates (Martin 1984). As stretching increases with the distance to the pole of rotation, break-up should initiate away from the pole and propagate towards it, into the stretched continental lithosphere (Vink 1982). Courtillot (1982) presented a different concept of propagation, in which the continental rift is characterized by the presence of 'locked zones' unevenly distributed along the rift axis. Spreading then nucleates between the locked zones, in a way similar to what Bonatti (1985) called punctiform initiation of sea-floor spreading, then propagates into the locked zones. However, McKenzie (1986) pointed out that this mechanism corresponds to strain localization rather than propagation, because the distributed strain in the continental locked zones is progressively replaced by localized strain at the oceanic spreading axis.

In this paper, we attempt to quantify the amount of stretching that occurred during break-up propagation at the tip of the South China Sea basin, one of the best examples of an oceanic basin with a propagating ridge geometry, leading to a typical V-shape. We base our analysis on bathymetry, gravity and seismic data acquired in a 3° × 3° area located at the southwestern tip of the South China Sea (Fig. 1). Through a structural and gravimetric analysis, we estimate the direction and amount of stretching that occurred before break-up, to compare it with the overall kinematics of the opening and to test the various models of propagation.

Kinematic framework

The survey area is located at the southwestern tip of the South China Sea basin (Fig. 1). Early

From: WILSON, R.C.L., WHITMARSH, R.B., TAYLOR, B. & FROITZHEIM, N. 2001. *Non-Volcanic Rifting of Continental Margins: A Comparison of Evidence from Land and Sea.* Geological Society, London, Special Publications, **187**, 31–50. 0305-8719/01/$15.00 © The Geological Society of London 2001.

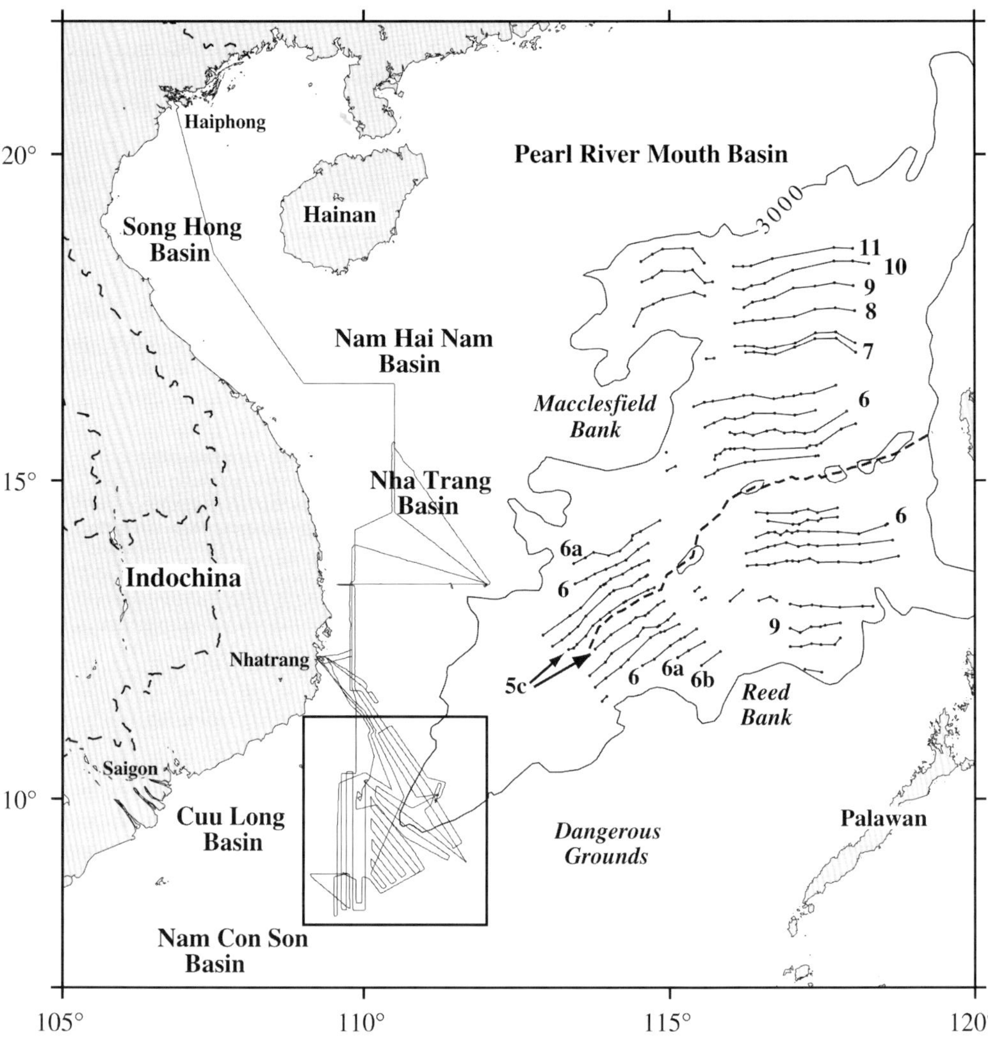

Fig. 1. Location map of the South China Sea. Magnetic anomalies are from Briais *et al.* (1993). Tracks of detailed surveys during the Ponaga cruise are shown. Inset shows location of Figure 2.

work on this basin recognized a typical oceanic basin, based on bathymetry (Chase & Menard 1969), seismic reflection (Emery & Ben-Avraham 1972) and refraction profiles (Ludwig *et al.* 1979), and magnetic profiles (Ben Avraham & Uyeda 1973). The V-shape of the basin mentioned above is well illustrated by the 3000 m isobath (Fig. 1). Taylor & Hayes (1980, 1983) proposed the first comprehensive kinematic model for the opening. They recognized E–W-trending magnetic anomalies 11 to 5D in the eastern part of the basin, corresponding to ages (32–17 Ma, using the time scale of Harland *et al.* (1990)) consistent with the heat flow measurements of Watanabe *et al.* (1977). They also identified the southwestern rift axis with

gravity data. This analysis was further extended and refined by Lu *et al.* (1987); Briais *et al.* (1993), who confirmed the age of the onset of spreading at 32 Ma. Furthermore, Briais *et al.* showed that spreading continued until 15 Ma (between anomalies 5B and 5C), after a reorientation of the direction of spreading from N–S to NW–SE shortly after anomaly 7 (23 Ma) (Fig. 1). Pautot *et al.* (1986) and Briais *et al.* (1989) used multibeam bathymetric data to support this reorientation, to which the general trend of the southwestern basin conforms. Magnetic anomalies are not well identified in the southwestern part of the South China Sea. However, the continuity of bathymetric features and the gravimetric signature of the spreading axis suggests that anomaly 5C should be present at the southwestern tip of the basin.

Although the kinematics of opening of the South China Sea is moderately constrained, the mechanism is much debated. Marginal basins in the western Pacific appear to open (and close) in a large variety of tectonic contexts and have been related either to the effect of subduction (Mariana, Okinawa, Shikoku and Parece Vela Basins) or to the collision of India with Eurasia (Japan Sea, South China Sea) (Jolivet *et al.* (1989) and references therein). The South China Sea is both one of the largest marginal basins and the closest to the collision zone. Analogue experiments (Tapponnier *et al.* 1982) suggest that the basin could have opened at the termination of the left-lateral Red River Fault, as a consequence of the extrusion of the Indochina block. But Taylor & Hayes (1980) suggested that subduction of the proto-South China Sea (south of Reed Bank and Dangerous Grounds) below Palawan and Borneo may have induced its opening by the slab pull effect. These two mechanisms have opposite consequences for the relation between the South China Sea and the Indochina continental margin, which would be a left-lateral strike-slip margin in the extrusion model, whereas the slab pull model predicts a right-lateral margin. Field studies (Rangin *et al.* 1995) have shown that, following a pervasive NW–SE left-lateral strike-slip faulting, the central and southern Vietnam has been affected by large N–S- to N160°E-trending right-lateral strike-slip faulting. To the east of Central Vietnam, analysis of academic and industrial data has led to the conclusion that the N–S-trending margin was active as a dextral transform fault between about 28 and 20 Ma (Marquis *et al.* 1997; Roques *et al.* 1997*a,b*). In the basins offshore from southern Vietnam, a kinematic analysis based on crustal structure derived from gravity data led Huchon *et al.* (1998) to recognize dextral decoupling between the opening of the South China Sea and the stretching on the continental margin, also supporting the hypothesis of formation of the South China Sea by southward subduction of the proto-South China Sea.

Geophysical data acquisition and processing

The new data presented here have been acquired during the Ponaga cruise on R.V. *L'Atalante* of 6–30 May 1993 (Fig. 1). They consist of swath bathymetry, six-channel reflection seismic profiles, and 3.5 kHz echo-sounder, gravity and magnetic data. We also dredged at locations where the 3.5 kHz records indicated that basement rocks crop out, mostly on volcanic structures. The Simrad EM 12 dual multibeam bathymetry system enabled us to obtain data in a corridor up to 20 km wide at great depths (>4000 m) and hence to map with full coverage the area shown in Figure 2. The seismic data went through a standard processing sequence: trace editing, common mid-point (CMP) gathering, pulse-shaping and predictive deconvolution, band-pass filtering, stack and f–k (frequency–wavenumber) migration. Processed profiles have been interpreted to produce a structural map, as well as sediment thickness and basement depth maps. The gravity measurements have been corrected for navigational effects and converted to free-air anomalies (Fig. 3). Comparison with the satellite-derived free-air anomaly data (Sandwell & Smith 1992) shows a good agreement, except for the shortest wavelengths (less than 15 km) where the ship data provide more information. The total magnetic field measured using a proton precession magnetometer towed 300 m behind the ship was reduced to magnetic anomalies using the IGRF90 reference field. Comparison of cross tracks, however, showed large discrepancies and we were consequently unable to map the anomalies. As the survey area is located close to the magnetic equator, we tried to apply a correction for the diurnal variation of the magnetic field. Data recorded in the Dalat observatory operated by the Institute of Geophysics of NCNST, Vietnam (11°55'N, 108°25'E) and in a temporary station in Nam Yet island (10°11'N, 114°21'E) show peak-to-peak diurnal variation for quiet days often reaching 150 nT. However, the two sets of data were not consistent and we could not obtain any reliable correction. A plot of the raw data is available in the study by Nguyen (1997).

In the following, we first describe the general structure of the oceanic tip and its margins and then discuss the direction as well as the timing of rifting that will be used as an input for modelling the gravity data in terms of crustal structure.

Shape of the oceanic basin and structure of the margins

The detailed bathymetric map (Fig. 2) shows a typical V-shaped domain whose tip is located near 9°N, 110°E. This domain is limited by a N55°E bathymetric trend to the southeast and a N30°E trend to the northwest. Water depths increase from *c.* 2700 m at the tip to >4000 m

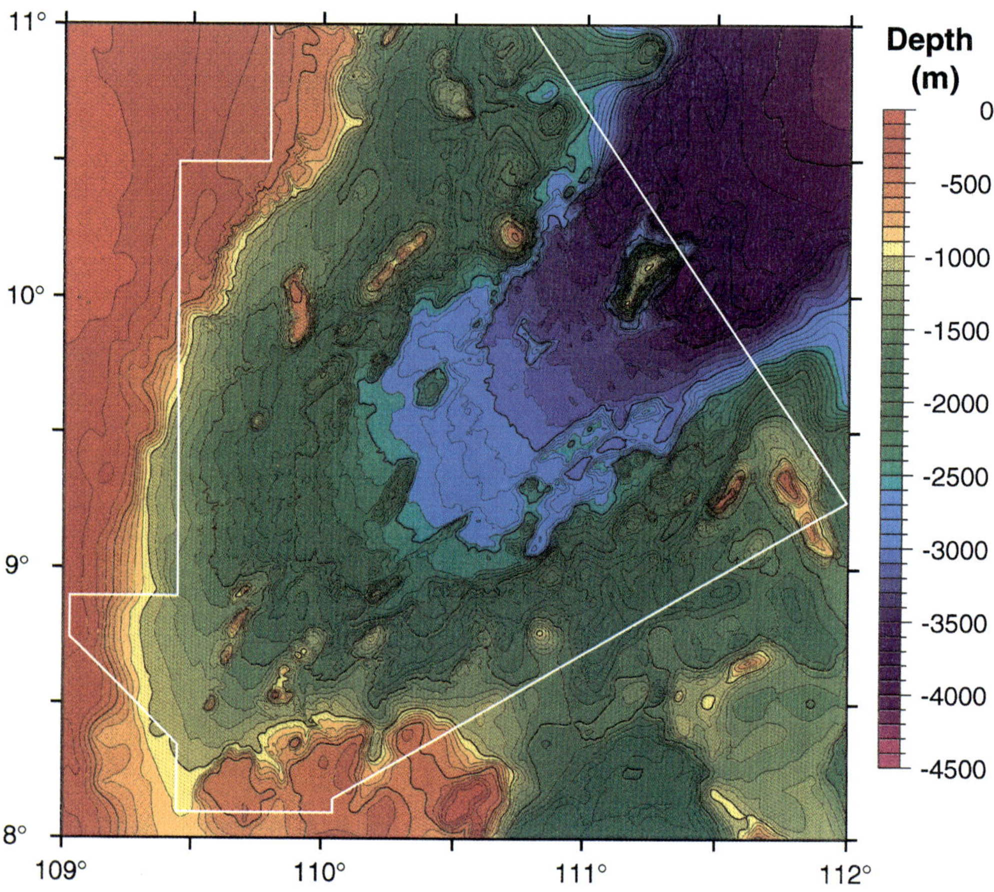

Fig. 2. Detailed bathymetric map (EM12 multibeam data) of the propagating tip of the South China Sea, superimposed on ETOPO5 data. The EM12 survey area is delineated by a white line.

toward the northeast. A large 'butterfly-shaped' volcano occupies the middle of the V-shaped domain near 10°N, 111°15'E. Fresh basalts were dredged from the flank of the volcano, from which a K/Ar age of 4 Ma was obtained (Bellon *et al.* 1994). This volcano, however, is not related to the spreading phase but is part of a large alkalic volcanic event affecting Indochina during Plio-Quaternary time (Flower *et al.* 1996).

Although the area is heavily sedimented, the bathymetry reveals horst and graben structures that parallel the axis of the V-shaped domain, in addition to a few more northerly oriented structures. In Figure 4 we locate the profile Ponaga 9, which shows that the block-faulted margins are clearly separated by an area of highly reflective crust, *c.* 30 km wide (between 90 km and 120 km on the profile, Fig. 5a) and covered by 0.5–1.0 s TWTT (two-way travel time) of post-rift sediment. On the free-air anomaly profile (Fig. 5b), this central area shows low-amplitude peak-to-peak anomalies whereas the stretched continental crust displays higher amplitude, with gravity lows associated with deep basins. We tentatively interpret the deep, central area as underlain by oceanic crust, a hypothesis we shall later test using gravity data modelling. By contrast, the magnetic data show no obvious relationship to the nature of the crust, probably because of the small width of the oceanic domain and the effect of post-spreading volcanism. On the contrary, except over the rift axis, the magnetic anomalies anticorrelate with the free-air anomaly, suggesting normally magnetized continental basement.

The structural map superimposed on the depth to basement (Fig. 6) has been established using the seismic profiles obtained during the cruise (Fig. 4) and complemented by a few industrial profiles. It shows that the deepest basement area (>4500 m) does not strictly

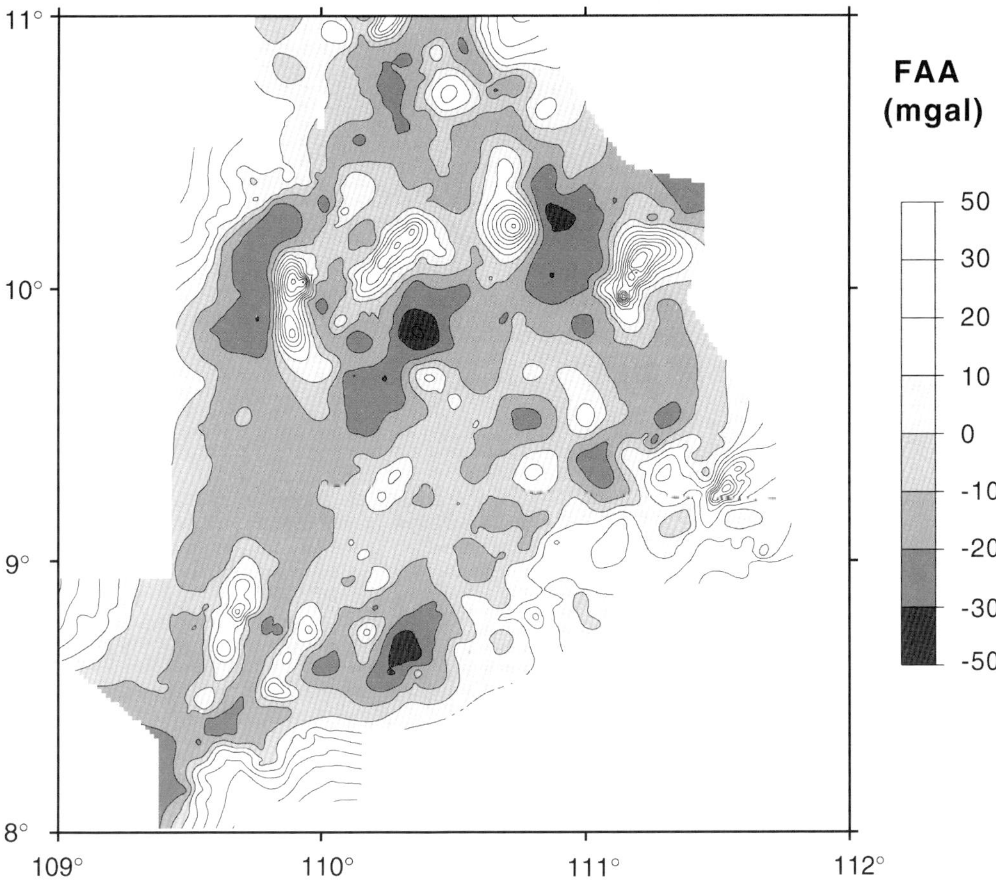

Fig. 3. Free-air gravity anomaly (FAA) map of the propagating tip of the South China Sea.

coincide with the V-shaped domain highlighted by the bathymetry, but also includes a significant part of the lower part of both margins. The northern border of the V-shaped axial domain consists of N30°E–N45°E-trending normal faults, whereas the southern border corresponds to a succession of N45°E–N70°E-trending, right-handed en echelon stepping faults. The tip of the inferred oceanic domain continues into a narrow N45°E-trending depression limited by NW-facing faults, forming a half-graben. At the scale of the whole area, the faulting pattern is dominated by N45°E-trending faults parallel to the axis of the V-shaped domain. This is the main difference between our observations and the scenario devised by Whitmarsh & Miles (1995) for the west Iberia margin, because most of the normal faults on the South China Sea margin do not parallel the continent–ocean boundary, but instead the axis of the oceanic V-shaped domain. However, the fault directions are somewhat scattered, with many N60–70°E-trending faults to the south and more northerly oriented faults (N20–30°E) to the north.

In addition to along-strike variations in the faulting pattern, the structural map (Fig. 6) also reveals an obvious asymmetry. Whereas the southern margin displays mostly NW-facing normal faults, thus facing the rift axis, the northern one shows not only SE-facing faults, but also numerous NW-facing ones, giving rise to an overall asymmetry. This is particularly well expressed at the tip of the V-shaped domain, in the southwest corner of the map, where NW-facing faults delineate a half-graben system. This asymmetry may suggest the occurrence of deep, listric normal faults, which were, however, not imaged on our seismic profiles because of the limited penetration. Listric faults were not imaged on the northern margin of the South China Sea (Pearl River Mouth Basin) by Hayes *et al.* (1995) although they used deep penetration, multi-channel seismic profiles. However, Nissen *et al.* (1995) concluded that

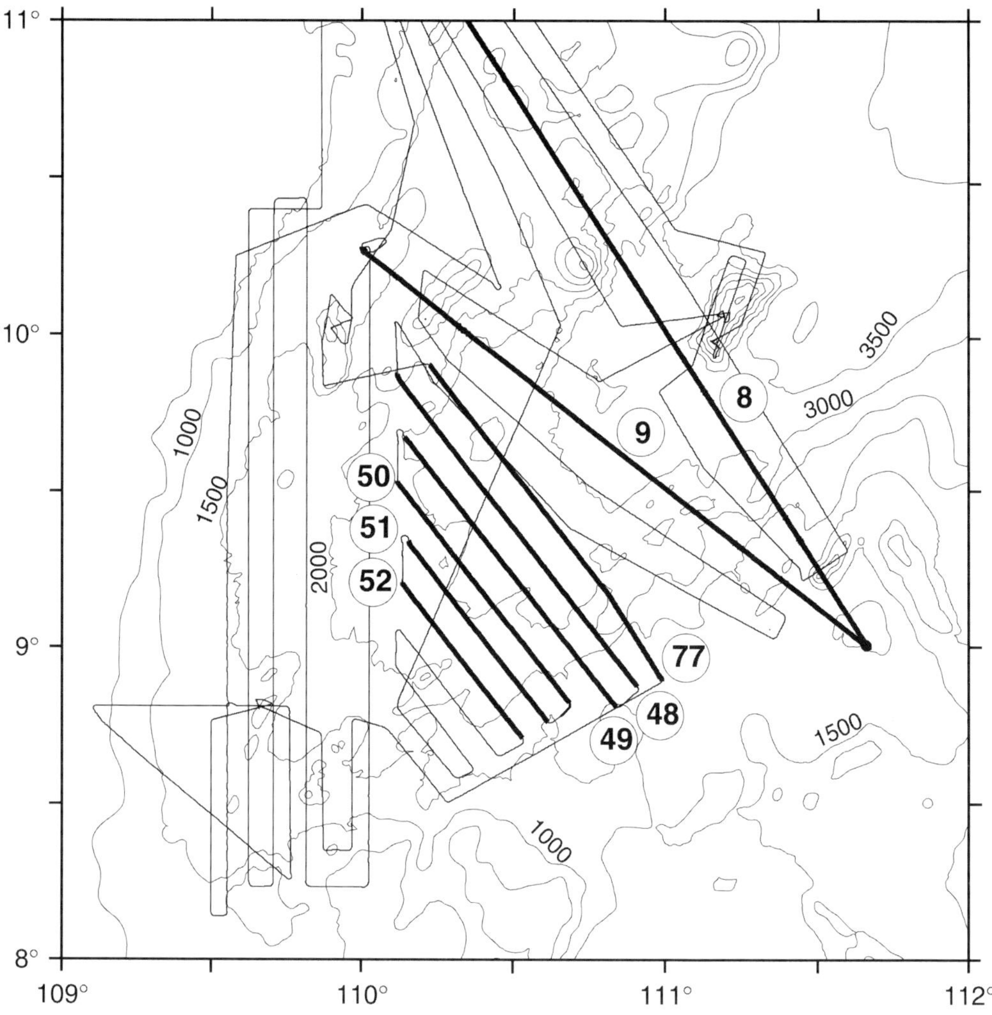

Fig. 4. Location of seismic profiles (bold lines with circled numbers) discussed in the text, superimposed on bathymetry (contour interval 500 m).

matching the observed subsidence and heat flow probably requires a combination of both pure and simple shear in the crust.

Rifting and opening directions

In the studied area, the direction of spreading is not constrained by magnetic anomalies or by transform faults. Hence, it can be only inferred from indirect evidence. In this regard, the N45°E-trending axis of the V-shaped oceanic domain is probably not coincidentally perpendicular to the N136°E direction of motion predicted by Briais *et al.* (1993). This estimate was derived from rotation parameters for the last phase of spreading, from anomaly 5D (17.8 Ma)

to the end of spreading at 15.6 Ma. However, the average direction of the normal faults in the stretched continental crust is not strictly parallel to the axis of the propagating oceanic tip, but deviates from the N45°E direction by about 5° (Fig. 7a). In detail, the frequency histogram of Figure 7a shows a slight asymmetry with peaks at N40°E (especially for the area located north of the axis), N50°E and N70°E. This faulting pattern is typical of oblique rifting. Using small-scale models, Tron & Brun (1991) have established experimental histograms of fault directions for various obliquities of rifting. They summarized these experiments in a diagram showing the distribution of the angle between the faults and the rift axis as a function

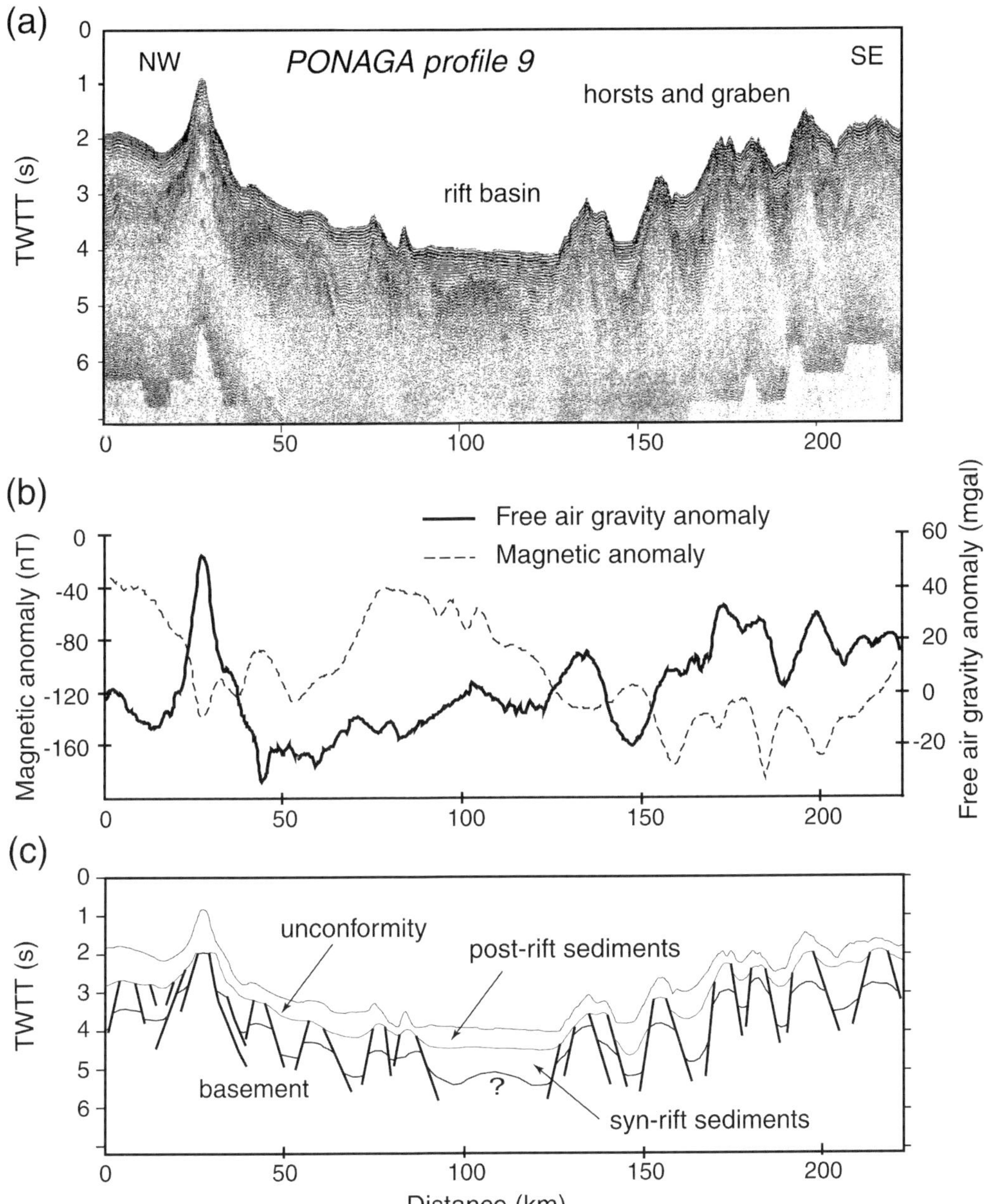

Fig. 5. (**a**) Single-channel seismic profile Ponaga 9 across the propagating tip of the South China Sea. Location indicated in Figure 4. (**b**) Free-air gravity and magnetic anomalies along the same profile. (**c**) Schematic interpretation of the seismic profile. TWTT, two-way travel time.

of the obliquity (Fig. 7b). Comparison of the faulting azimuthal distribution around the tip of the South China Sea (Fig. 7a) with the diagram of Tron & Brun (1991) (Fig. 7b) predicts an obliquity of 25°, and thus a N160°E direction of rifting (Fig. 7c). This is in good agreement with Briais *et al.*'s reconstruction pole for anomaly 6 (20.5 Ma), which predicts a N161°E oriented spreading direction. If one considers that rifting and spreading directions are coaxial, the faulting pattern would then reflect the finite kinematics since 20.5 Ma, and not that of the

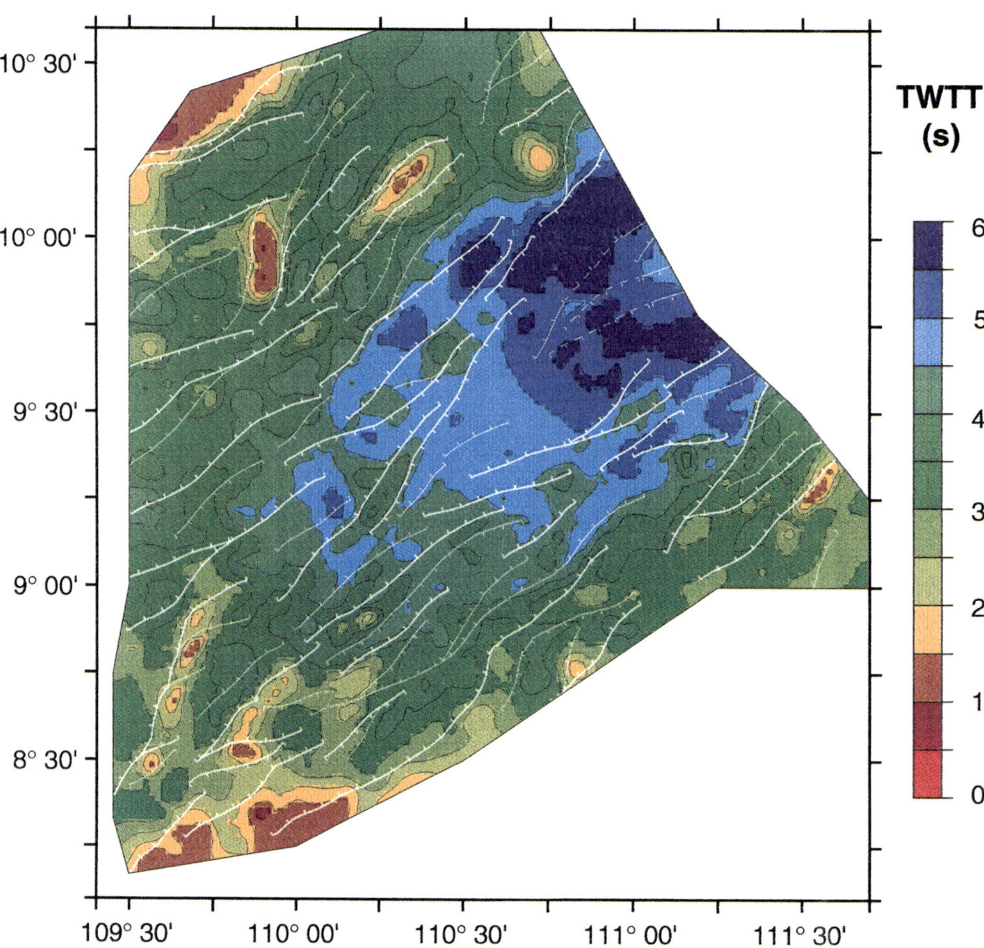

Fig. 6. Structural map of the propagating tip of the South China Sea, superimposed on the depth to basement map (in seconds of two-way travel time; TWTT). Major (>100 m of vertical throw) and minor (<100 m) normal faults are shown by bold and fine fault lines (dashed where unclear), respectively, with ticks on the footwall side.

very last phases of spreading from 17.8 to 15.6 Ma, with a N136°E direction of motion. A N160°E stretching direction implies that the N15°E–N30°E faults (Figs 6 and 7) must have a significant component of dextral strike-slip.

Age of rifting and timing of propagation

The age of the oceanic crust is not constrained in the survey area because the oceanic domain is very narrow and the magnetic anomalies are not well expressed. The magnetic anomaly 5C (16.6 Ma) is the youngest anomaly, identified *c.* 300 km to the NE of the survey area (Fig. 1), and the axis of the extinct oceanic ridge has been estimated to be 15.6 Ma in age (Briais *et al.* 1993). The width of the oceanic crust measured along the N147°E-trending profile Ponaga 8 is *c.* 90 km (Figs 4 and 8). Here, Briais *et al.*'s model predicts a 3.2 cm a^{-1} rate

Fig. 7. (a) Cumulated fault length v. azimuth of faults shown in Figure 6. (b) Experimental relationship between the obliquity of rifting and the angle between faults and rift axis (redrawn from Tron & Brun 1991). The shaded area is the projection of the fault azimuth histogram: its left and right boundaries are shown by bold lines whereas the two peaks of the histogram are shown by fine lines. The fit of our data to Tron & Brun's diagram is depicted by bold dashed lines. (c) Sketch showing the definition of obliquity.

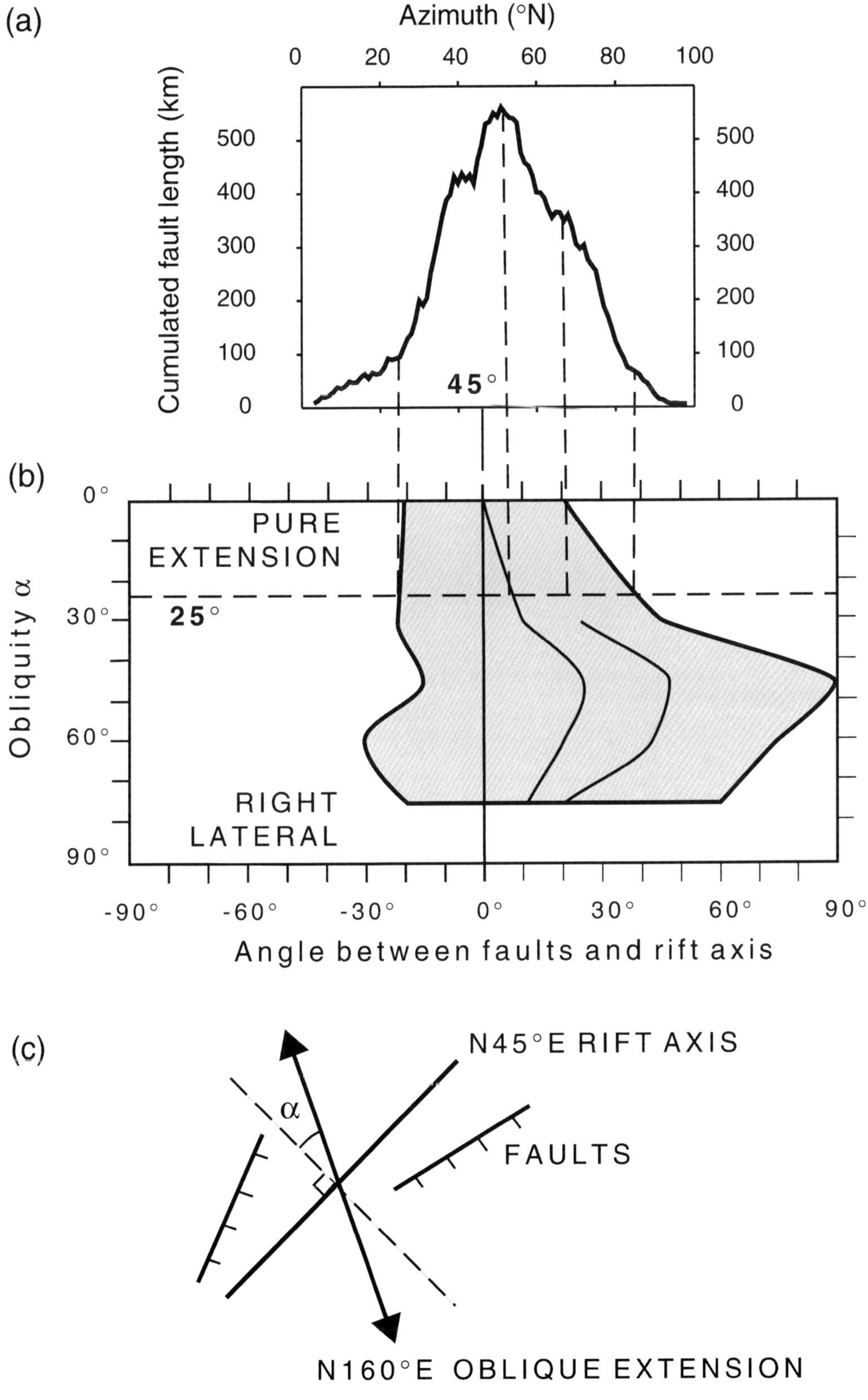

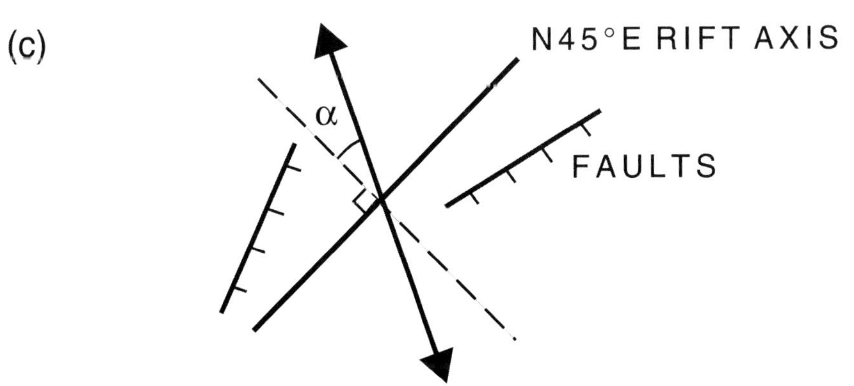

(a)
Azimuth (°N)
0 20 40 60 80 100
Cumulated fault length (km)
500 500
400 400
300 300
200 200
100 100
0 0
45°
(b)
Obliquity α
0°
PURE EXTENSION
25°
30°
60°
90°
RIGHT LATERAL
-90° -60° -30° 0° 30° 60° 90°
Angle between faults and rift axis
(c)
α
N45°E RIFT AXIS
FAULTS
N160°E OBLIQUE EXTENSION

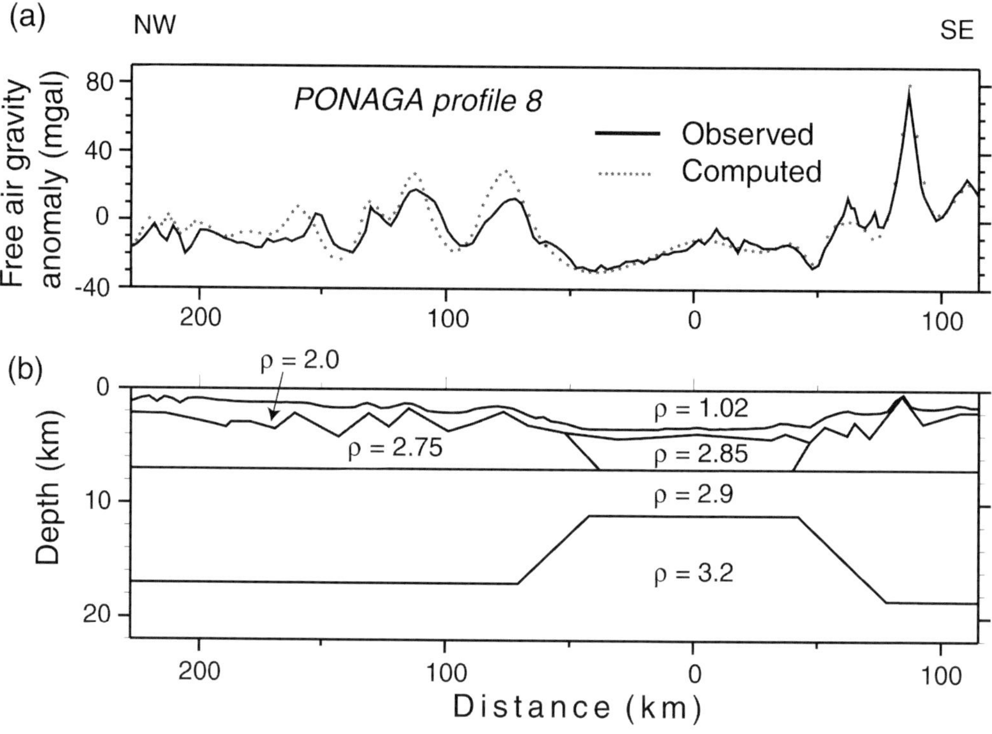

Fig. 8. (**a**) Observed free-air gravity anomaly along profile Ponaga 8 (location indicated in Fig. 4) compared with the anomaly computed using a simple 2D forward modelling technique. (**b**) Geometry and densities (ρ) of layers used for the forward model.

and a N136°E direction of opening during the last phase of spreading, since anomaly 5D (17.8–15.6 Ma). Therefore, the 90 km of oceanic crust seen along profile Ponaga 8 could have been formed in 2.8 Ma, and thus between 18.4 and 15.6 Ma, assuming that spreading ceased at the same time (15.6 Ma) as further to the northeast. This age range is fairly consistent with the depth of the oceanic crust. Unloaded for sediments assuming local isostatic compensation, the average depth of *c.* 4000 m of the oceanic crust corresponds to 18 Ma using the age–depth relationship of Parsons & Sclater (1977). At the scale of the surveyed area, the break-up would thus be diachronous by 3 Ma over a distance of *c.* 100 km. This implies a rate of break-up propagation of *c.* 30 km Ma^{-1}, smaller than the 90 km Ma^{-1} rate computed from the shape of the whole South China Sea (Fig. 1) and the average opening rate since anomaly 7 (*c.* 40 km Ma^{-1}).

The age of the onset of rifting is more difficult to assess. Although rift subsidence in the basins located south of Vietnam is inferred to have started in Late Oligocene time (Matthews

et al. 1997) or even earlier (during Paleocene–Eocene time) regionally, the major rifting event is marked in the Early Miocene period, with thickening of deposits onto N–S- and NE–SW-oriented normal faults (Matthews *et al.* 1997). Most subsidence curves from industrial wells in the Nam Con Son basin (location shown in Fig. 1) indeed show that subsidence started between 20 and 23 Ma (see, for example, Anonymous 1989). To the north of the survey area, in the Nha Trang Basin (Fig. 1), Marquis *et al.* (1997) suggested that rifting could have started at *c.* 28 Ma, which allows the onset of rifting at the present-day tip of the oceanic crust to be bracketed between 28 and 21 Ma. This would imply a short duration for the rifting phase, between 3 and 10 Ma. Whereas rifting is often considered to have occurred for tens of million years in some basins, it is worth noting that the upper estimate (10 Ma) is that often considered for basins such as the Gulf of Aden (Cochran 1981). Even the 3 Ma lower estimate may not be unreasonable inasmuch as such a short duration of rifting before oceanization has been proved in the case of the active Woodlark Basin (Taylor *et al.* 1995).

Gravimetric modelling

In the following section, we perform a two-step analysis of the gravity data, to estimate the crustal structure as well as the location of zone of maximum stretching (continent–ocean transition). The first step consists of a simple forward modelling and in the second step we introduce a correction for the thermal structure and perform a 2D inversion of the gravimetric data.

Simple 2D gravity models

We used the values of water depth and sediment thickness (measured on seismic profiles, assuming a $2\,km\,s^{-1}$ velocity in sediments) to make 2D forward models of the free-air gravity data for profile Ponaga 8, which is nearly perpendicular to the rift axis (Fig. 8). The adopted densities are $1.03\,g\,cm^{-3}$ for water, $2.85\,g\,cm^{-3}$ for the upper oceanic crust, $2.75\,g\,cm^{-3}$ for the upper continental crust, $2.9\,g\,cm^{-3}$ for the lower crust and $3.2\,g\,cm^{-3}$ for the mantle. The best-fitting model is consistent with the presence of $7\,km$ thick, probably oceanic crust in the centre of the basin and of thinned continental crust on the edges of the basin. This crude model shows an average thinning factor (in the sense of McKenzie (1978)) of about two over both margins. It also confirms the asymmetry, the continental crust being slightly thinner (by $1.5\,km$) to the north than to the south.

However, the age of opening of the southwestern tip of the South China Sea is sufficiently recent (*c.* 16 Ma) to still have a significant residual thermal anomaly (e.g. Chamot-Rooke *et al.* 1999). This thermal effect cannot be neglected, because it results in less dense mantle below the rift compared with the flanks, a lateral variation in density that is not taken into account in the simple gravity model.

The 2D inversion of gravity data with thermal correction

The 2D rifting model of Alvarez *et al.* (1984) provides a way to compute the thermal anomaly resulting from the rifting, the spreading and the subsequent cooling phase, taking into account lateral heat conduction and radiogenic heat production in the continental crust. In the studied area, we assumed that the rifting started *c.* 30 Ma ago, which is an upper bound (see discussion above). As discussed above, the estimated $90\,km$ width of oceanic crust corresponds to 2.8 Ma of spreading (from 18.4 to 15.6 Ma). Consequently, the end of rifting, coinciding with the beginning of spreading, was taken at 20 Ma, to account for possible magmatic injection before true oceanic crust emplacement. The end of spreading (beginning of cooling phase) was rounded to 16 Ma. The computed thermal structure for profile Ponaga 8 at 20 Ma, 16 Ma and at the present time is shown in Figure 9. The present-day thermal anomaly results in a gravimetric effect of the order of $-100\,mgal$ at the rift axis, decreasing back to nearly zero over a distance of $300\,km$ on both sides.

To perform the inversion of the gravity data, we use the 2D gravity model of Talwani & Ewing (1960), taking the continental crust as the reference layer with a density $\rho_c = 2.75\,g\,cm^{-3}$. The mantle Bouguer anomaly g_b is then the sum of the observed free-air anomaly g_{obs} and the gravimetric contributions of the various layers: the anomaly g_w due to water of density $\rho_w = 1.03\,g\,cm^{-3}$; the anomaly g_s due to sediments of average density $\rho_s = 2.0\,g\,cm^{-3}$; the anomaly g_{co} due to the presence of oceanic crust with a density $\rho_{co} = 2.85\,g\,cm^{-3}$, instead of continental crust (the boundary of which is adjusted iteratively to fit the observed anomaly); the anomaly g_T caused by the thermal anomaly T with a density $\rho_m = \rho_{mo}(1 - \alpha T)$, where $\rho_{mo} = 3.3\,g\,cm^{-3}$ and α is the thermal expansion coefficient.

To compute the gravimetric Moho, we used a 2D inversion by fast Fourier transform (FFT). The gravity effect of the Moho interface with variation $h(x)$, density contrast $\Delta\rho_m = \rho_m - \rho_c$ at a mean depth d, provided that $d \gg h(x)$ (Parker 1972) can be approximated by

$$g(x) = 2\pi\Delta\rho G h(x) e^{-kd}$$

where G is the universal constant of gravitation and k is the wavenumber. In the Fourier domain we can write

$$G(k) = 2\pi\Delta\rho G H(k) e^{-kd}$$

Here, $G(k) = TF[g(x)]$ and $H(k) = TF[h(x)]$, where TF is the direct Fourier transform. Hence, we can determine undulations $h(x)$ of the gravimetric Moho by the inverse Fourier transform (TF^{-1}) of $H(k)$:

$$h(x) = TF^{-1}[H(k)] = TF^{-1}\left[\frac{G(k)}{2\pi G \Delta\rho e^{-kd}}\right]$$

$$= TF^{-1}\frac{TF[g(x)]}{2\pi G \Delta\rho e^{-kd}}$$

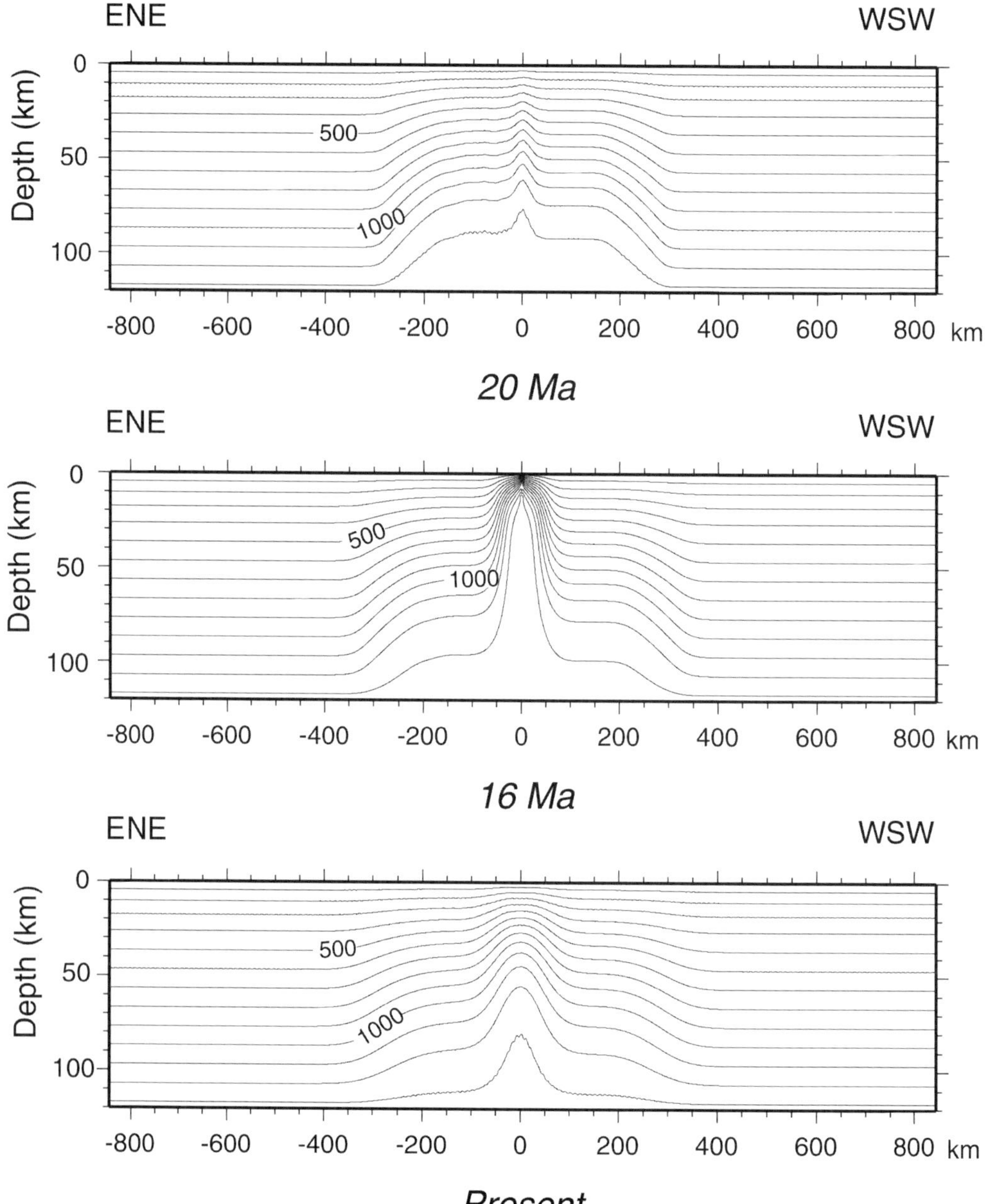

Fig. 9. Isotherms (in °C) computed for profile Ponaga 8 using the 2D rifting model of Alvarez *et al.* (1984), shown at 20 Ma (beginning of spreading), 16 Ma (end of spreading) and at present.

Figure 10 shows the result of the inversion for the profile Ponaga 8. The calculated thickness of what we interpret as oceanic crust is *c.* 6 km, whereas that of the continental crust varies between 8 km near the continent–ocean boundary and 20 km at 150 km to the northwest. The shape of the Moho is obviously more realistic than with the simple forward modelling, which provided only an estimate of the mean Moho depth. For comparison, we also computed the depth to the isostatic Moho, which is obtained by summing the water depth h_w, the sediment

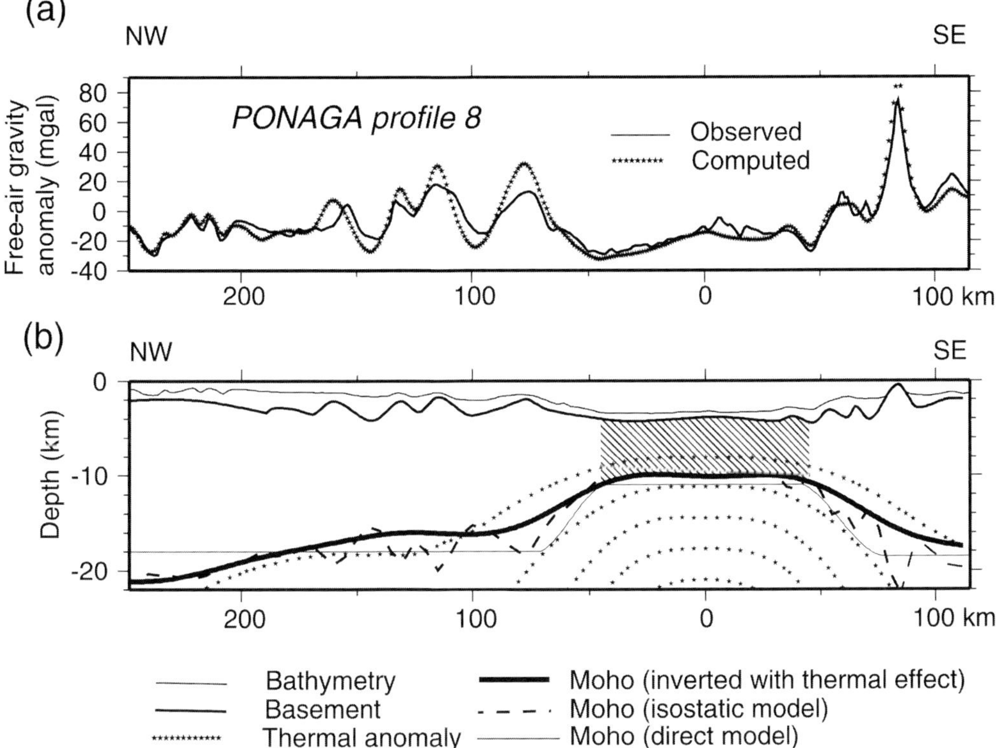

Fig. 10. (a) Observed free-air gravity anomaly along profile Ponaga 8 compared with the anomaly computed from the model shown in (b) with the Moho derived from the 2D inversion of gravity data, including the thermal effects of rifting and spreading as discussed in the text. (b) Comparison of the depth to Moho inverted with thermal effect with the depth obtained by forward modeling as in Fig. 8 and by a local isotatic model. The bathymetry to top of basement and the thermal anomaly (contoured interval 50°C) derived from Figure 9 are also shown.

thickness h_s and the crustal thickness h_c obtained by isostatically unloading the sediments:

$$h_c = H - [h_w(\rho_c - \rho_w) + h_s(\rho_c - \rho_s)]/(\rho_m - \rho_c)$$

In this equation, H is the initial crustal thickness (before stretching), taken as 30 km to be consistent with the observed crustal thickness along the coast of Vietnam (Bui 1993). The Moho obtained by inversion is smoother than the isostatic Moho because of the filtering by FFT, but also slightly higher as it takes into account the lateral variation of density.

The crustal structures obtained for the eight profiles identified in Figure 4 have several common features (Fig. 11): (1) a 6–7 km thick oceanic crust in the centre of the basin (profiles 8, 9, 77 and 48), which narrows towards the southwest (90 km on profile 8, 30 km on profile 48); (2) a narrow (c. 10 km) transition zone between oceanic crust and continental crust; (3)

a thinned continental crust on the basin edges, with Moho depths ranging from 17 to 21 km in the north and slightly deeper on the southern margin; this confirms that the continental crust is asymmetrically thinned.

In the following sections, we first explore the consequences of these results in terms of the geometry of the ocean–continent boundary and the nature of continental crustal extension before break-up. We then use these data to derive the kinematics of rifting.

Continent–ocean boundary and continental crust stretching

A primary result of the gravimetric modelling is the crustal thickness map shown in Figure 12. Outside of the area where the crustal thickness is constrained by our gravity modelling, we used the regional crustal thickness map of Huchon *et al.* (1998), obtained from a 3D inversion of

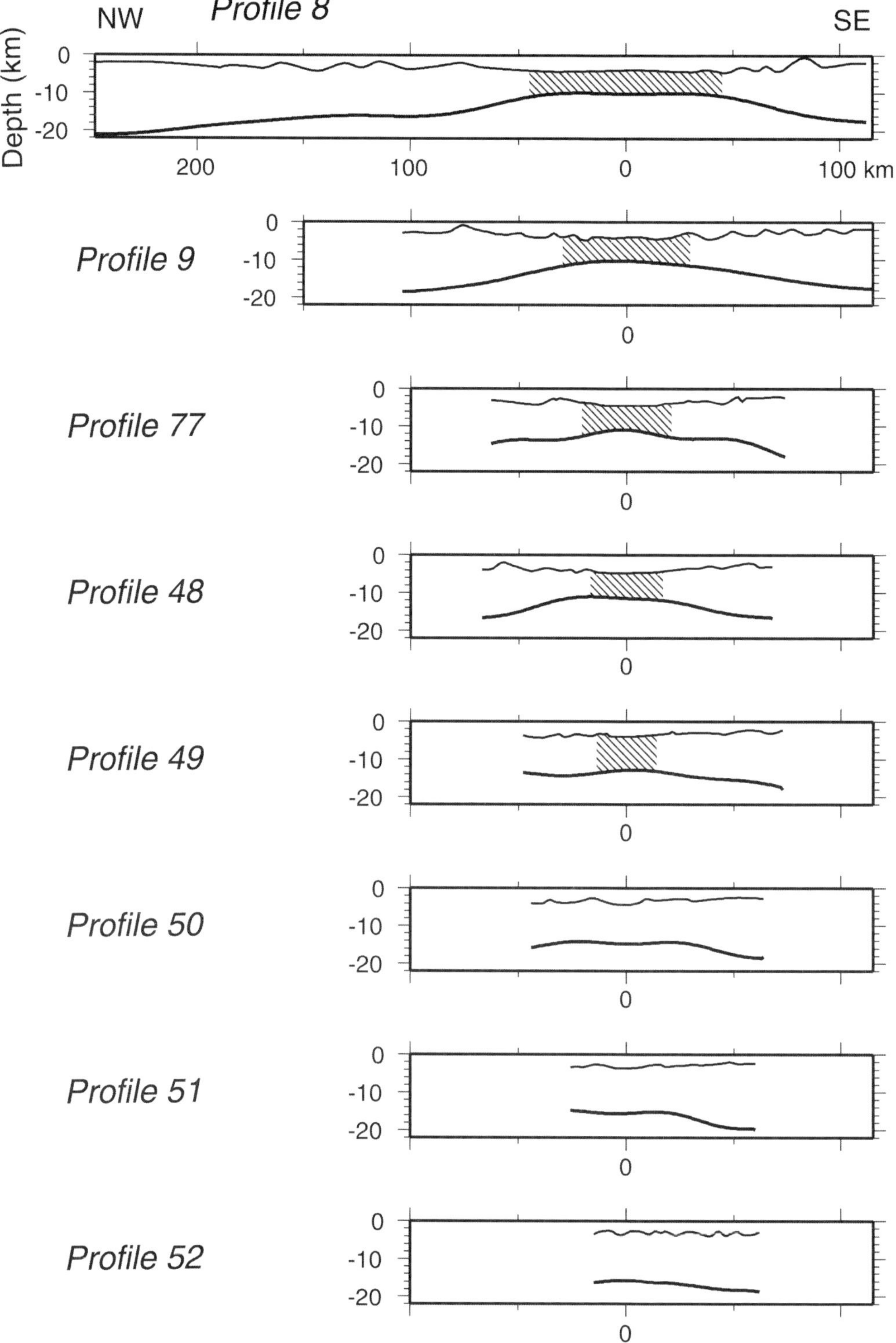

NW
Profile 8
SE
Depth (km)
0
-10
-20
200
100
0
100 km
Profile 9
0
-10
-20
0
Profile 77
0
-10
-20
0
Profile 48
0
-10
-20
0
Profile 49
0
-10
-20
0
Profile 50
0
-10
-20
0
Profile 51
0
-10
-20
0
Profile 52
0
-10
-20
0

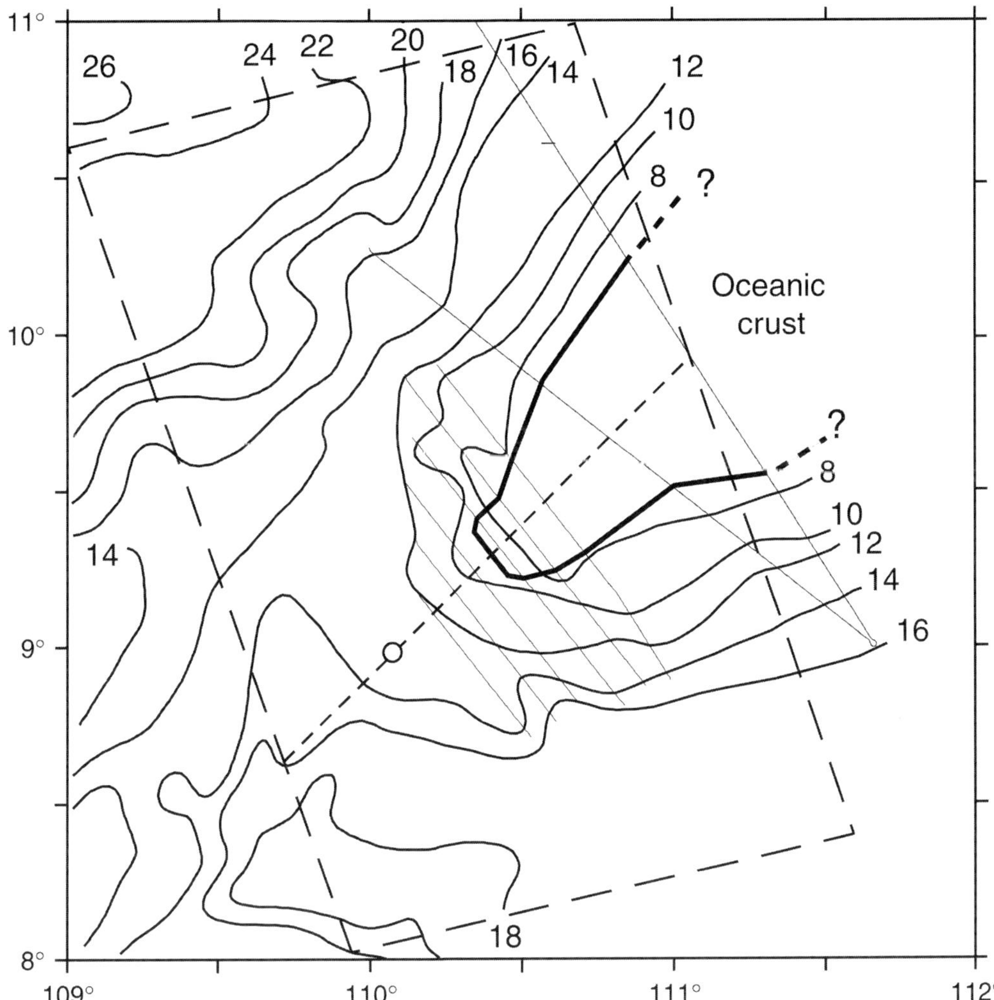

Fig. 12. Continental crustal thickness map (in km) obtained using the eight profiles of Figure 10 (shown by fine lines) complemented with a regional estimate of the crustal thickness (Huchon *et al.* 1998). Bold line: continent–ocean boundary; short-dashed line, *x*-axis of Figures 12 and 13; long-dashed line delineates the area of computation for Figure 14.

gravity data over the South Vietnam shelf, without correction for thermal effect. In areas where the two datasets overlap, the agreement is fair, with <2 km offsets. On the same map, we also plot the location of the continent–ocean boundary obtained from the gravity modelling. The corresponding width of the oceanic crust is plotted in Figure 13a. Figure 13b shows the crustal thickness measured along the axis of the oceanic propagating crust (shown as a short-dashed line in Fig. 12). Whereas on profiles 8, 9, 77 and 48 the crustal thickness is nearly constant and typical of oceanic crust (6.5–7 km), it rapidly increases toward the southwest on profiles 49, 50, 51 and 52 (Fig. 13b). The nature of the crust on profile 49, where the crustal thickness is *c.* 9 km, is questionable. The gravity data on this profile are best modelled with a denser

Fig. 11. Crustal structure obtained by 2D inversion of gravity data along eight profiles across the propagating tip of the South China Sea (location shown in Fig. 4). Shaded area is inferred oceanic crust.

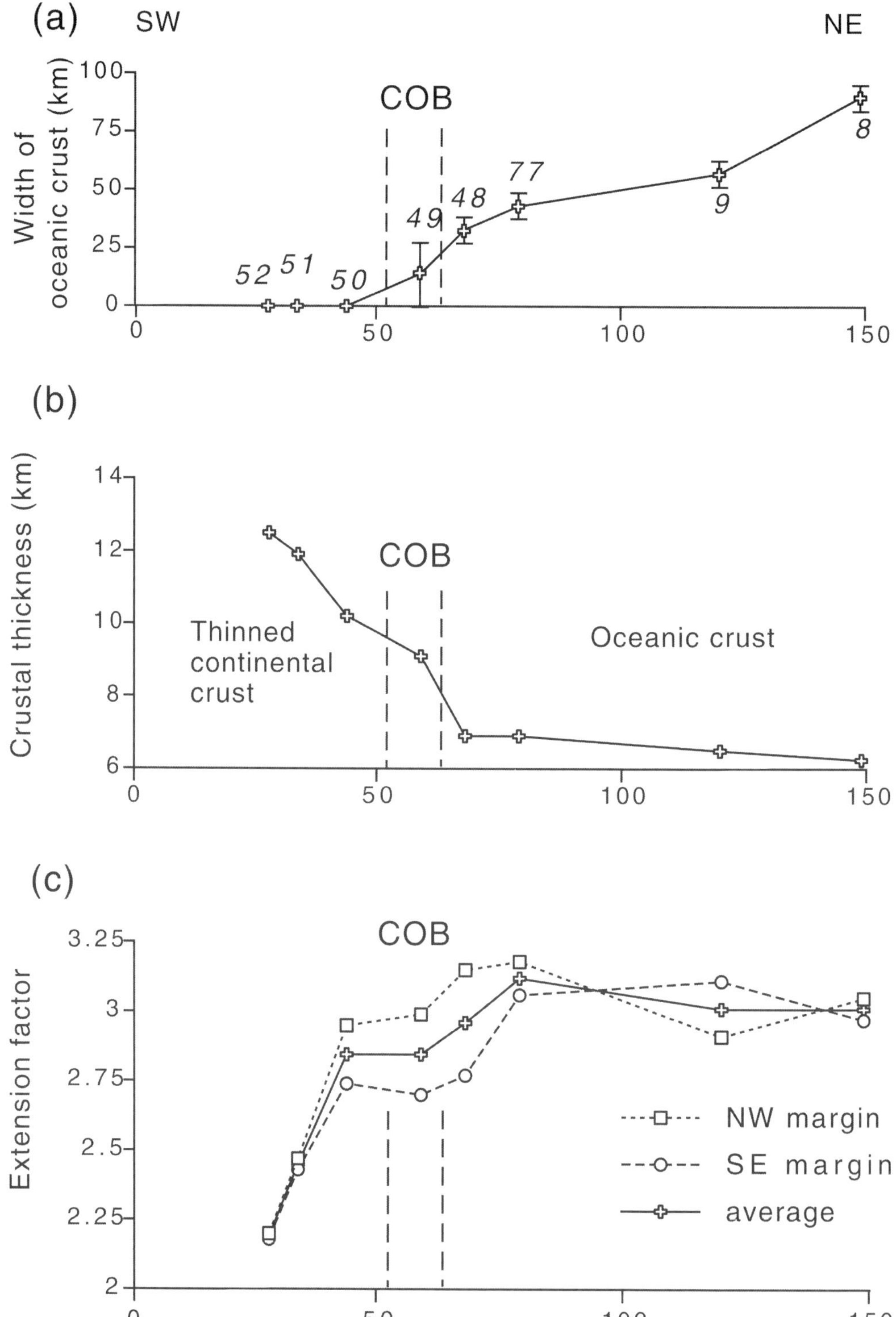
(a)
SW
NE
Width of oceanic crust (km)
100
75
50
25
0
COB
52 51 50
49
48
77
9
8
0 50 100 150

(b)
Crustal thickness (km)
14
12
10
8
6
COB
Thinned continental crust
Oceanic crust
0 50 100 150

(c)
Extension factor
3.25
3
2.75
2.5
2.25
2
COB
NW margin
SE margin
average
0 50 100 150
Distance (km)

upper crust in the middle part of the profile (see Fig. 11), suggesting the presence of oceanic crust.

However, there is an obvious trade-off between the shape of the Moho and the density of the crustal layer, so that gravity alone cannot solve simultaneously for Moho shape and width of the oceanic crust. Based on the thermal models of generation of the oceanic crust, we see no rational way to increase the oceanic crust thickness at the tip of the propagator, at least in a permanent regime of propagation. Larger oceanic crust thickness at the propagator tip may thus be a transient effect related to the dying rift. The obvious alternative is that the crust is continental on profile 9. We conclude that the total width of the oceanic crust on this profile is between 0 and 25 km (Fig. 13a).

The similarity between the highly thinned area ahead of the tip of the propagating oceanic rift and the outer continental margin on both sides of the V-shaped domain is illustrated in Figure 13c. The mean stretching factor in a strip 50 km wide, running parallel to the inferred continent–ocean boundary on both sides of the oceanic crust is about three. It slightly decreases on profiles 49 and 50 to a value of 2.85 then suddenly decreases to 2.5 (profile 51) and 2.2 (profile 52). This suggests that oceanic propagation occurs when the continental crust in front of the propagating tip is stretched enough (<10 km) to allow oceanic spreading to start. On the various profiles analysed, the crustal thickness at the continent–ocean boundary ranges from 7 to 10 km. The corresponding threshold stretching factor of three to four for break-up is that generally considered for mechanical reasons (see, for example, Le Pichon & Sibuet 1981).

In more detail, the crustal thickness mapped in Figure 12 and the stretching factors plotted in Figure 13c reveal the asymmetry of rifting already pointed out. Although the situation is reversed on profile 9, the average stretching factor within 50 km of the continent–ocean boundary is larger to the northwest, compared with the southeast (Fig. 13c), an observation consistent with both the depth to basement map (Fig. 6) and the free-air gravity anomaly map (Fig. 3), which shows that basins are deeper to the northern side of the propagating tip compared with the southern side. This can be explained by the obliquity of rifting, as the direction of propagation (toward azimuth N225°E) is directed more southeastward than the mean trend of normal faults (N50°E, or N230°E), leaving more extended crust to the northwest.

Break-up propagation and oceanic kinematics

In the model we envision the propagation of sea-floor spreading follows the model of Martin (1984), in which the sum of oceanic spreading and continental stretching should coincide with the amount of separation between the plates, and thus be described by the rules of plate kinematics. Taken in the simplest way, extension should decrease towards the pole of rotation. To test this idea, and thus to compare the amount of extension corresponding to the crustal thickness map of Figure 12 with the prediction of oceanic kinematics, we measured the continental crustal thickness along N160°E-trending sections on the map of Figure 12. We then converted the crustal thickness profiles into horizontal stretching, assuming an initial 30 km thick continental crust, equal to the present-day crustal thickness in the unstretched coastal area of South Vietnam (Bui 1993). These are, of course, lower estimates, as they ignore the stretching that occurred on the inner (or upper part) of the margin, to the north and to the south of the studied area. However, the stretching factor in this area is small (c. 1.2) and, given an approximate width of 150 km, would result in an additional 30 km of stretching at most, compared with c. 130 km for the outer (or lower) part of the margin (Fig. 14). We also measured the width of the oceanic crust along the same profiles. Whereas the amount of sea-floor spreading decreases toward the southwest, the amount of continental stretching slightly increases (Fig. 14). This is only an apparent paradox because the continental crust located ahead of the propagating oceanic crust could be subjected to stretching for a longer period of

Fig. 13. (a) Width of the oceanic crust measured along the eight profiles where gravity data have been modelled (location shown in Fig. 4) and plotted along the axis of the propagating tip of the South China Sea (location shown in Fig. 12). The origin is taken at the tip of the V-shaped domain (9°N, 110°E). COB, continent–ocean boundary. The larger error bar for profile 49 should be noted (see discussion in text). (b) Along-axis variation in crustal thickness; x-axis as in (a). (c) Extension factor of the continental crust measured in a 50 km wide strip parallel to the continent–ocean boundary; x-axis as in (a). The northwestern and southeastern margins are plotted separately to show the asymmetry.

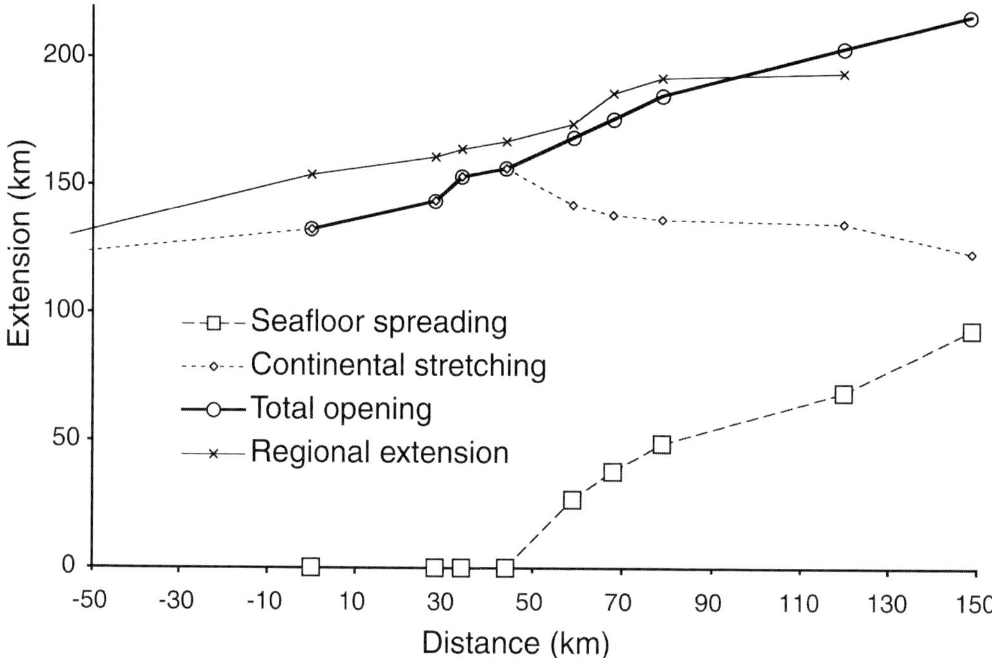

Fig. 14. Variation in the amount of oceanic spreading and continental stretching along the axis of the propagating tip of the South China Sea. Origin and *x*-axis as in Figure 13 (see Fig. 12 for location). The values have been computed along N160°E-trending, 300 km long profiles in the dashed box shown in Figure 12. The total opening (spreading plus stretching) shows a decrease toward the southwest that predicts a pole of rotation located *c.* 285 km to the southwest (see text). The regional extension computed by Huchon *et al.* (1998) is shown for comparison.

time than that on both sides of the oceanic crust. In our case, the amount of oceanic spreading decreases southwestward from 90 km to zero whereas the amount of continental stretching increases from 125 km to 155 km. The sum is then decreasing toward the southwest (Fig. 14). It is noteworthy that the same observation of increasing continental extension toward the pole of relative rotation has been made in the Woodlark Basin (see Taylor *et al.* 1999, fig. 5).

For comparison, we also show in Figure 14 the amount of extension estimated by Huchon *et al.* (1998) from a 3D gravity model of the South Vietnam basins. The agreement is fair, except to the northeast where our estimate is larger: this is because Huchon *et al.* did not take into account the amount of oceanic spreading, leading to an underestimation of the extension to the northeast. The profiles used by Huchon *et al.* were slightly longer (350 km instead of 300 km here); however, the difference is only about 10 km, showing that the stretching of the inner margin is almost negligible with respect to that of the most stretched area around the oceanic propagating tip.

The least-squares fit of the total opening curve in Figure 14 shows that the corresponding pole of rotation must be located 285 km to the southwest of the oceanic tip (taken at profile 49), not far from the pole of finite rotation for anomaly 5D (5°N, 105.5°E) of Briais *et al.* (1993). Again, we should stress that we do not take into account the stretching of the inner margin, outside the surveyed area. This would result in a slightly more distant pole of rotation, which incidentally would better fit with the pole of finite rotation determined from magnetic anomalies. Such an agreement between two totally independent data sets therefore suggests that the Martin (1984) model of strain localization can be considered as basically valid in the case of the South China Sea.

Conclusion

In spite of a complex geodynamic setting, the propagation of break-up in the southwestern South China Sea appears to have occurred in a rather simple way, with spreading occurring as soon as thinning of the continental crust reaches a factor of about four. Ahead of the propagating tip, the stretching factor drops from this value to a value of two over a distance of c. 40 km. The continental crust thus thins from c. 15 km to <10 km over a distance corresponding to just over 1 Ma of break-up propagation. Compared with the average thinning during the total rifting period, with an estimated duration of 3–10 Ma, this observation suggests that strain localization, as expected, does occur at the tip of the propagating oceanic crust just before break-up. This has been best demonstrated in the case of the active Woodlark Basin, where the local strain rate immediately preceding break-up is often one or two orders of magnitude larger than the average strain rate across the whole margin (Taylor *et al.* 1999).

We warmly thank the captain, officers and crew of the R.V. *L'Atalante*, as well as the GENAVIR technical team. IFREMER and CNRS–INSU supported the cost for operations and travel, respectively. Additional funding for data processing was provided by CNRS–INSU through a 'Géosciences marines' grant. Nguyen Thi Kim Thoa kindly provided us with magnetic data from Vietnamese observatories. Maps were prepared using the GMT freeware (Wessel & Smith 1991). The paper benefited from the careful reviews of R. Hey, B. Taylor and an anonymous referee. This paper is UMR Géosciences Azur Contribution 336.

References

ALVAREZ, F., VIRIEUX, J. & LE PICHON, X. 1984. Thermal consequences of lithosphere extension over continental margins: the initial stretching phase. *Geophysical Journal of the Royal Astronomical Society*, **78**, 389–411.

Anonymous *Report on Comprehensive Investigation and Research of Nansha Islands and Adjacent Maritime Regions*. Academia Sinica Publishing House, Beijing.

BELLON, H., RANGIN, C. & HUCHON, P. 1994. Late Miocene to Quaternary volcanism of Vietnam: time and space geochemical variations (abstract). *Cenozoic Evolution of the Indochina Peninsula, Hanoi and Do Son, Vietnam, 25–29 April 1994*, 47–48.

BEN AVRAHAM, Z.B. & UYEDA, S. 1973. The evolution of the China basin and the Mesozoic paleogeography of Borneo. *Earth and Planetary Science Letters*, **18**, 365–376.

BONATTI, E. 1985. Punctioform initiation of seafloor spreading in the Red Sea during transition from a continental to an oceanic rift. *Nature*, **316**, 33–37.

BOSWORTH, W. 1985. Geometry of propagating continental rifts. *Nature*, **316**, 625–627.

BRIAIS, A., PATRIAT, P. & TAPPONNIER, P. 1993. Updated interpretation of magnetic anomalies and reconstructions of the South China basin: implications for the Tertiary evolution of Southeast Asia. *Journal of Geophysical Research*, **98**, 6299–6328.

BRIAIS, A., TAPPONNIER, P. & PAUTOT, G. 1989. Constraints of SeaBeam data on crustal fabrics and seafloor spreading in the South China Sea. *Earth and Planetary Science Letters*, **95**, 307–320.

BUI, C.Q. 1993. Some characteristics of the deep crustal structure and the geodynamics in the territory of Viet Nam and neighbouring sea area. *Journal of Geology, Geological Society of Vietnam*, **1-2**, 39–50.

CHAMOT-ROOKE, N., GAULIER, J.M. & JESTIN, F. 1999. Constraints on Moho depth and crustal thickness in the Liguro-Provençal Basin from a 3D gravity inversion: geodynamic implications. *In*: DURAND, B., JOLIVET, L., HORVÀTH, F. & SÉRANNE, M. (eds) *The Mediterranean Basins: Tertiary Extension within the Alpine Orogen*. Geological Society, London, Special Publications, **156**, 37–61.

CHASE, T.E. & MENARD, H.W. *Bathymetric Atlas of the Northwestern Pacific*. US Naval Oceanographic Office, Washington, DC.

COCHRAN, J.R. 1981. The Gulf of Aden: structure and evolution of a young ocean basin and continental margin. *Journal of Geophysical Research*, **86**, 263–287.

COURTILLOT, V. 1982. Propagating rifts and continental breakup. *Tectonics*, **1**, 239–250.

EMERY, K.O. & BEN-AVRAHAM, Z. 1972. Structure and stratigraphy of China Basin. *AAPG Bulletin*, 5, **56**, 839–859.

FLOWER, M., NGUYEN, H., NGUYEN, X.B. & NGUYEN, T.Y. 1996. Implications of basalt major element compositions for melting beneath Indochina: response to reorganised spreading and thermally-anomalous asthenosphere. *Bulletin de la Société Géologique de France*, 6, **167**, 773–784.

HARLAND, W.B., ARMSTRONG, R.L., COX, A.V., CRAIG, L.E. & SMITH, A.G. *A Geologic Time Scale 1989*. Cambridge University Press, Cambridge.

HAYES, D.E., NISSEN, S.S., BUHL, P. & DIEBOLD, J. 1995. Throughgoing crustal faults along the northern margin of the South China Sea and their role in crustal extension. *Journal of Geophysical Research*, **100**, 22435–22446.

HEY, R.N., DUENNEBIER, F.K. & MORGAN, W.J. 1980. Propagating rifts on mid-ocean ridges. *Journal of Geophysical Research*, **85**, 3647–3658.

HUCHON, P., NGUYEN, T.N.H. & CHAMOT-ROOKE, N. 1998. Finite extension across the South Vietnam basins from 3D gravimetric modeling: relation to South China Sea kinematics. *Marine and Petroleum Geology*, **15**, 619–634.

JOLIVET, L., HUCHON, P. & RANGIN, C. 1989. Tectonic setting of western Pacific marginal basins. *Tectonophysics*, **160**, 23–47.

LE PICHON, X. & SIBUET, S. 1981. Passive margins: a model of formation. *Journal of Geophysical Research*, B5, **86**, 3708–3720.

LU, W., KE, C., WU, J., LIU, J. & LIN, C. 1987. Characteristics of magnetic lineations and tectonic evolution of the South China Sea basin. *Acta Oceanologica Sinica*, **6**, 577–588.

LUDWIG, W.J., KUMAR, N. & HOUTZ, R.E. 1979. Profiler-Sonobuoy measurements in the South China Sea basin. *Journal of Geophysical Research*, **84**, 3505–3518.

MARQUIS, G., ROQUES, D., HUCHON, P., COULON, O., CHAMOT-ROOKE, N., RANGIN, C. & LE PICHON, X. 1997. Amount and timing of extension along the continental margin off central Vietnam. *Bulletin de la Société Géologique de France*, 6, **168**, 15–24.

MARTIN, A.K. 1984. Propagating rifts: crustal extension during continental rifting. *Tectonics*, **3**, 611–617.

MATTHEWS, S.J., FRASER, A.J., LOWE, S., TODD, S.P. & PEEL, F.J. 1997. Structure, stratigraphy and petroleum geology of the south east Nam Con Son basin, offshore Vietnam. *In*: FRASER, A.J., MATTHEWS, S.J. & MURPHY, R.W. (eds) *Petroleum Geology of South East Asia*. Geological Society, London, Special Publications, **126**, 89–106.

MCKENZIE, D.P. 1978. Some remarks on the development of sedimentary basins. *Earth and Planetary Science Letters*, **40**, 25–32.

MCKENZIE, D.P. 1986. The geometry of propagating rifts. *Earth and Planetary Science Letters*, **77**, 176–186.

NISSEN, S.S., HAYES, D.E., YAO, B., ZENG, W., CHEN, Y. & NU, X. 1995. Gravity, heat flow, and seismic constraints on the process of crustal extension: northern margin of the South China Sea. *Journal of Geophysical Research*, **100**, 22447–22483.

NGUYEN, T.N.H. 1997. Structure et cinématique de l'extrémité de la Mer de l'Est (Mer de Chine méridionale) et des bassins du sud Vietnam. PhD thesis, University of Paris 6.

PARKER, R.L. 1972. The rapid computation of potential anomalies. *Geophysical Journal of the Royal Astronomical Society*, **31**, 447–455.

PARSONS, B. & SCLATER, J.G. 1977. An analysis of the variation of ocean floor bathymetry and heat flow with age. *Journal of Geophysical Research*, **82**, 803–827.

PAUTOT, G., RANGIN, C., BRIAIS, A. & 9 OTHERS 1986. Spreading direction in the central South China Sea. *Nature*, **321**, 150–154..

RANGIN, C., HUCHON, P., BELLON, H., NGUYEN, D.H., LE PICHON, X., PHAN, V.Q. & ROQUES, D. 1995. Cenozoic deformation of central and south Vietnam: evidences for superposed tectonic regimes. *Tectonophysics*, 1-4, **251**, 179–196.

ROQUES, D., MATTHEWS, S.J. & RANGIN, C. 1997*a*. Constraints on strike-slip motion from seismic and gravity data along the Vietnam margin offshore Danang: implications for hydrocarbon prospectivity and opening of the East Vietnam Sea. *In*: FRASER, A.J., MATTHEWS, S.J. & MURPHY, R.W. (eds) *Petroleum Geology of South East Asia*. Geological Society, London, Special Publications, **126**, 342–353.

ROQUES, D., RANGIN, C. & HUCHON, P. 1997*b*. Geometry and sense of motion along the Vietnam continental margin: onshore/offshore Da Nang area. *Bulletin de la Société Géologique de France*, 4, **168**, 413–422.

SANDWELL, D.T. & SMITH, W.H.F. 1992. Global marine gravity from ERS-1, Geosat and Seasat reveals new tectonic fabric. *EOS Transactions, American Geophysical Union*, 43, **73**, 133.

TALWANI, M. & EWING, M. 1960. Rapid computation of gravitational attraction of three-dimensional bodies of arbitrary shape. *Geophysics*, 1, **25**, 203–225.

TAPPONNIER, P., PELTZER, G., LE DAIN, A.Y., ARMIJO, R. & COBBOLD, P. 1982. Propagating extrusion tectonics in Asia: new insights from simple experiments with plasticine. *Geology*, **10**, 611–616.

TAYLOR, B. & HAYES, D.E. 1980. The tectonic evolution of the South China basin. *In*: HAYES, D.E. (ed.) *The Tectonic and Geologic Evolution of Southern Asian Seas and Islands. Part 1*. Geophysical Monographs, American Geophysical Union, **23**, 89–104.

TAYLOR, B. & HAYES, D.E. 1983. Origin and history of the South China Basin. *In*: HAYES, D.E. (ed.) *The Tectonic and Geologic Evolution of Southern Asian Seas and Islands. Part 2*. Geophysical Monographs, American Geophysical Union, **27**, 23–56.

TAYLOR, B., GOODLIFFE, A. & MARTINEZ, F. 1999. How continents break up: insights from Papua New Guinea. *Journal of Geophysical Research*, **104**, 7497–7512.

TAYLOR, B., GOODLIFFE, A., MARTINEZ, F. & HEY, R. 1995. Continental rifting and initial sea-floor spreading in the Woodlark basin. *Nature*, **374**, 534–537.

TRON, V. & BRUN, J.-P. 1991. Experiments on oblique rifting in brittle–ductile systems. *Tectonophysics*, **188**, 71–84.

VINK, G.E. 1982. Continental rifting and the implications for plate tectonic reconstructions. *Journal of Geophysical Research*, **87**, 10677–10688.

WATANABE, T., LANGSETH, M.G. & ANDERSON, N.R. 1977. Heat flow in back-arc basins of the Western Pacific. *In*: TALWANI, M. & PITMAN, C.W. (eds) *Island Arcs, Deep Sea Trenches and Back-Arc Basins*. Maurice Ewing Series, American Geophysical Union, **1**, 137–161.

WESSEL, P. & SMITH, W.H.F. 1991. Free software helps map and display data. *EOS Transactions, American Geophysical Union*, 441, **72**, 445–446.

WHITMARSH, R.B. & MILES, P.R. 1995. Models of the development of the West Iberia rifted continental margin at 40°30′N deduced from surface and deep-tow magnetic anomalies. *Journal of Geophysical Research*, **100**, 3789–3806.

Nature of the continent–ocean transition on the non-volcanic rifted margin of the central Great Australian Bight

JACQUES SAYERS, PHILIP A. SYMONDS, NICHOLAS G. DIREEN & GEORGE BERNARDEL

Australian Geological Survey Organisation, GPO Box 378, Canberra, ACT 2601, Australia
(e-mail: Jacques.Sayers@agso.gov.au)

Abstract: A region of 50–120 km width defines the continent–ocean transition (COT) in the central Great Australian Bight. It is characterized by a thin apron of post-break-up sediments overlying complexly deformed sediments and intruded crust bounded landward by a basement ridge complex and oceanward by rough oceanic basement. Recently acquired deep reflection and refraction seismic data have significantly enhanced understanding of the COT and basement ridge. Modelled gravity and magnetic data, and features interpreted from seismic data, are consistent with aspects of extensional and break-up models proposed for the West Iberia margin. Many of the features and relationships observed beneath the outer margin of the central Great Australian Bight can be explained by extension within a lithosphere-scale 'pure-shear' environment involving four layers: brittle upper crust and upper mantle, and ductile lower crust and lower lithospheric mantle. The COT is interpreted to be underlain by extended continental lithosphere. Thus, the continent–ocean boundary is unequivocally defined between oceanic crust and the COT and appears to be associated with sea-floor spreading magnetic anomaly 33, indicating that break-up and sea-floor spreading did not commence until *c.* 83 Ma (early Campanian time), later than the currently accepted 95 Ma age. The major part of the basement ridge complex is probably a combination of serpentinized peridotites and mafic intrusions or extrusions derived by mantle upwelling and limited partial melting. The magmatic products of this process probably cooled during chron 34 producing a distinctive magnetic anomaly, but one that does not relate to break-up and sea-floor spreading.

The mode of formation of non-volcanic rifted margins, and the relationship between extension, magmatism and the transition to sea-floor spreading during break-up, have been the focus of intensive recent study on the relatively sediment-starved central segment of the West Iberia margin of the eastern North Atlantic. These studies utilized seismic reflection and refraction and potential field techniques, as well as sampling by submersibles and cores from the Ocean Drilling Program (ODP Legs 149 and 173) (e.g. Whitmarsh & Miles 1995; Whitmarsh & Sawyer 1996; Whitmarsh *et al.* 1996, 1998; Krawczyk *et al.* 1996). Other non-volcanic margins around the Atlantic are less well understood, as their synrift and early post-break-up sections are generally buried beneath up to 10–15 km of post-rift sediment (Poag & de Graciansky 1992). Previous work on non-volcanic margins has focused largely on testing and application of pure shear and simple shear rifting models, and various combinations of these, to explain the various features formed during the rift, break-up and post-break-up phases of

margin evolution (Royden & Keen 1980; Le Pichon & Sibuet 1981; Wernicke 1985; Lister *et al.* 1986, 1991; Le Pichon & Barbier 1987; Etheridge *et al.* 1989). Until recently, a lack of high-quality seismic data over continent–ocean transition (COT) zones has hampered understanding of the late rift and break-up phases.

Despite the number and range of investigations over the West Iberia non-volcanic margin, the nature of its COT is still controversial. None of the three main hypotheses (Whitmarsh & Miles 1995; Whitmarsh & Sawyer 1996; Krawczyk *et al.* 1996) proposed so far accounts for the full range of geophysical and geological observations (Whitmarsh & Sawyer 1996; Whitmarsh *et al.* 2001). The three hypotheses are as follows:

(1) The COT is highly extended continental lithosphere. The model put forward by Krawczyk *et al.* (1996) proposes extension of the upper lithosphere above a large-scale detachment fault system. Those workers have suggested that deep lithospheric levels were progressively tectonically unroofed as a result

From: WILSON, R.C.L., WHITMARSH, R.B., TAYLOR, B. & FROITZHEIM, N. 2001. *Non-Volcanic Rifting of Continental Margins: A Comparison of Evidence from Land and Sea.* Geological Society, London, Special Publications, **187**, 51–76. 0305-8719/01/$15.00 © The Geological Society of London 2001.

of conjugate, lithospheric shear zone activity during rifting.

(2) The COT is oceanic lithosphere with a crust formed by ultraslow sea-floor spreading (Whitmarsh & Sawyer 1996), with peridotite and gabbro exposed at the sea-floor by extensive faulting (e.g. Cannat 1993).

(3) The COT is tectonically and magmatically disrupted continental crust with gabbroic material either underplated beneath thinned continental crust, or associated with areas of aborted sea-floor spreading (Whitmarsh & Miles 1995; Whitmarsh & Sawyer 1996).

Various detachment models have been proposed to explain the development of the COT of the Iberia Abyssal Plain (e.g. Krawczyk *et al.* 1996). These models envisage that the detachment system exposed a cross-section through the upper lithosphere from upper crust in the east, through the lower crust, to upper mantle in the west, adjacent to oceanic crust. An objective of Ocean Drilling Program (ODP) Leg 173 was to drill through the detachment, the so-called 'H' reflector. However, the results did not provide unequivocal evidence for a major low-angle detachment fault, although a major ductile shear zone was penetrated (Whitmarsh *et al.* 1998). Brun & Beslier (1996) proposed a composite four-layer (brittle upper crust, ductile lower crust, brittle upper mantle, ductile lower lithosphere–mantle) lithosphere model that appears to explain many of the features observed on the West Iberia margin, and is likely to have wider application on other non-volcanic margins. The model involves bulk pure shear at the lithospheric scale, but with asymmetric, simple shear structures developing internally as a result of heterogeneous boudinage and/or faulting of brittle layers.

Non-volcanic margins are now considered to be the least common form of rifted margins, and are thought to constitute only about 30% of the margins of the Atlantic Ocean (Hinz, pers. comm.). The central Great Australian Bight (GAB) margin off southern Australia is one of the few segments of Australia's rifted margins that can be categorized as non-volcanic (Symonds *et al.* 1998*b*). On this margin and its conjugate Antarctic margin, the oldest magnetic anomaly, originally identified as anomaly 22 (49 Ma) by Weissel & Hayes (1972), was reinterpreted as anomaly 34y (83 Ma) by Cande & Mutter (1982). This reinterpretation invoked an initial episode of very slow spreading from anomaly 34 to 21 at a half-rate of *c.* 4.5 mm a^{-1}. Veevers (1986) used the Cande & Mutter (1982) model to extrapolate an age of 95 $\pm$ 5 Ma for the magnetic anomaly at the seaward

edge of the Magnetic Quiet Zone (MQZ), and they interpreted this anomaly as a magnetic edge effect associated with the continent–ocean boundary (COB). Veevers *et al.* (1990) and Tikku & Cande (1999) further refined anomaly picks and spreading rates, but the timing of break-up and start of sea-floor spreading remained dated at Late Cretaceous time (*c.* 95 Ma; Cenomanian time on the Australian Geological Survey Organisation (AGSO) timescale used throughout this paper; Young & Laurie 1996).

This paper builds on marine magnetic, gravity, seismic reflection and refraction data that have been acquired in the GAB over the past 30 years, largely as a result of reconnaissance surveys by the Lamont–Doherty Geological Observatory, AGSO (formerly the Bureau of Mineral Resources), and petroleum exploration companies. In particular, we have utilized public domain satellite-derived free-air gravity data (Sandwell & Smith 1997) and their integration with marine free-air gravity data (Petkovic *et al.* 1999). These datasets were combined with newly available marine magnetic, gravity, seismic reflection and refraction data acquired during a 1997 survey in the deep-water part of the GAB by AGSO's research vessel *Rig Seismic* (Symonds *et al.* 1998*a*). We present the results of a 'first look' at the new data, which consist of 11 seismic reflection lines, totalling 3448 line-km, with concurrent shipboard gravity and towed magnetometer observations (Fig. 1). Bathymetric data have been integrated from all available sources, including conventional single-beam bathymetric profiling, a small amount of multi-beam swath-mapping and, where ship data are too sparse, satellite-derived predicted bathymetry (Smith & Sandwell 1997). Twenty-six sonobuoys were deployed during the course of the survey, and data from 15 of the 17 sonobuoys that recorded successfully (Fig. 1) were analysed to provide velocity and density constraints on both seismic reflection and gravity interpretations.

This paper attempts to better constrain models for Mesozoic–Cenozoic rifting, break-up and initiation of sea-floor spreading between Australia and Antarctica, in particular the nature and evolution of the Australian part of the COT. The new data provide a basis with which to evaluate previously proposed extensional and break-up models for the GAB margin (e.g. Etheridge *et al.* 1989; Lister *et al.* 1991; Tikku & Cande 1999), as well as the new four-layer lithosphere extensional model (Brun & Beslier 1996) proposed for the West Iberia margin. We also briefly examine the implications of our results for the general definition of COTs, COBs, and identification of the initiation of sea-floor spreading on non-volcanic margins where only potential field data may be available.

Geological and physiographic framework

The area referred to in this paper covers over 0.5 × 10^6 km^2 and encompasses part of the abyssal plain, continental rise and slope of the central GAB (Fig. 2). Physiographic elements beyond the continental shelf include the Eyre and Ceduna Terraces (Fig. 2) overlying the Eyre and Ceduna Sub-basins (Fig. 1). West of the large Ceduna Terrace, a relatively narrow lower continental slope (20–50 km), merges with a broad (up to 200 km) continental rise. The Eyre and Ceduna sub-basins, together with the deep-

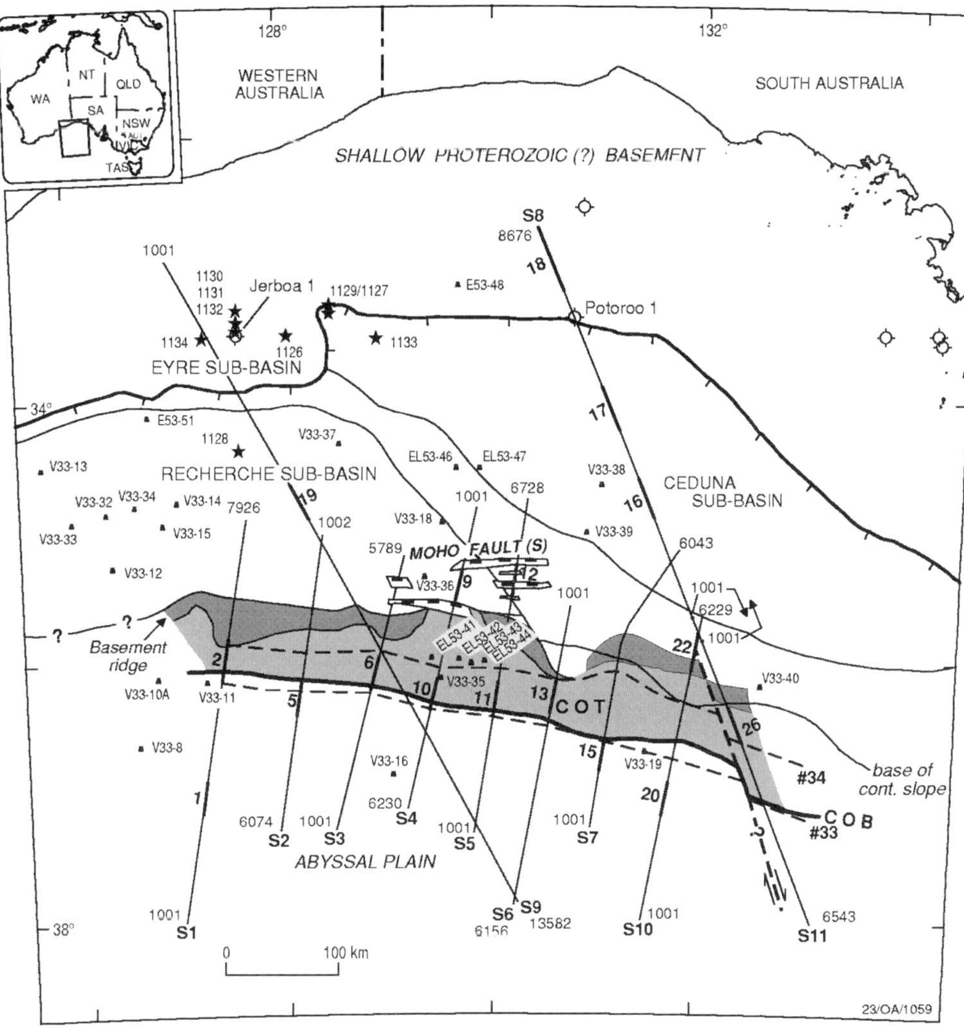

Fig. 1. Location map of the central Great Australian Bight. Straight lines, seismic lines S1–S11 from the 1997 AGSO survey; black box symbols, sonobuoy sites from Talwani *et al.* (1978); bold numbers in the range 1–26 indicate sonobuoy sites from the 1997 survey. Black stars and well symbols indicate the location of ODP Leg 182 drill sites (Shipboard Scientific Party 2000), and petroleum exploration wells, respectively. Line with ticks marks edge of shelf and major bounding fault for the Recherche and Ceduna sub-basins; dashed lines show location of magnetic anomalies 33 and 34; COT, continent–ocean transition; COB, continent–ocean boundary. Grey shading shows both the extent of the COT and basement ridge (dark grey), the latter at the level of the pre-break-up sediments.

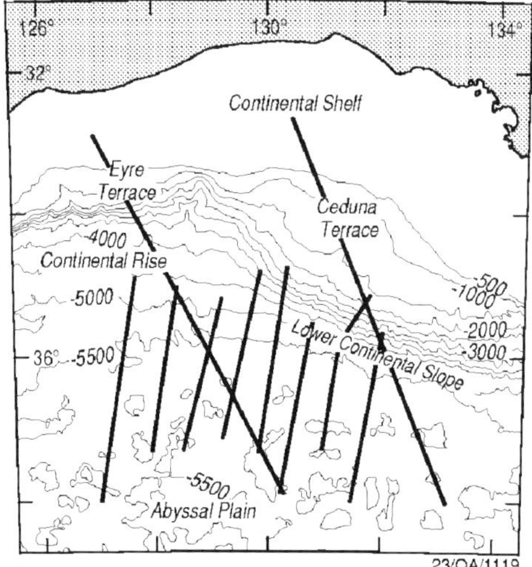

Fig. 2. Physiography of the GAB. Bathymetric contour interval is 500 m. Bold lines show the locations of seismic reflection profiles from the 1997 AGSO survey interpreted in this study.

water Recherche Sub-basin, constitute the Bight Basin (Stagg *et al.* 1990). The most substantial of these depocentres, the Ceduna Sub-basin, contains up to 15 km of mostly Mesozoic sediments (Totterdell *et al.* 2000). To the west another Mesozoic depocentre, the Recherche Sub-basin, underlies the continental rise to the north of an east–west-trending basement ridge (Fig. 1) that coincides with a positive, linear free-air gravity anomaly (Fig. 3a). Pre-break-up synrift and pre-rift sediments within both the Recherche and Ceduna sub-basins are interpreted to extend into the COT zone beyond the basement ridge (Fig. 4). The general absence of volcanogenic features (seaward-dipping reflector sequences, volcanic constructions and large intrusions), and the low apparent volume of landward-flowing flood basalts around break-up time, characterize this margin as a 'cold, non-volcanic' margin (Symonds *et al.* 1998*b*).

Recent swath-mapping, geophysical measurements or observations and dredging have provided important insights into the nature of the 200 km wide COT zone off southwestern Australia, which includes the MQZ and the Diamantina Zone, lying about 1000–1500 km west of the central GAB (Munschy 1998). The dredge samples are particularly important as they provide insights into the petrology of rocks that are likely to be present in the equivalent tectonic position in the central GAB. Dredging in the Diamantina Zone recovered peridotites that are thought to represent continental mantle, and basalts resulting from limited partial melting; the latter returned a 93 Ma ^{40}Ar/^{39}Ar age (Chatin *et al.* 1998). Some samples exhibit intense ductile shearing compatible with mantle exhumation associated with extension (Chatin *et al.* 1998).

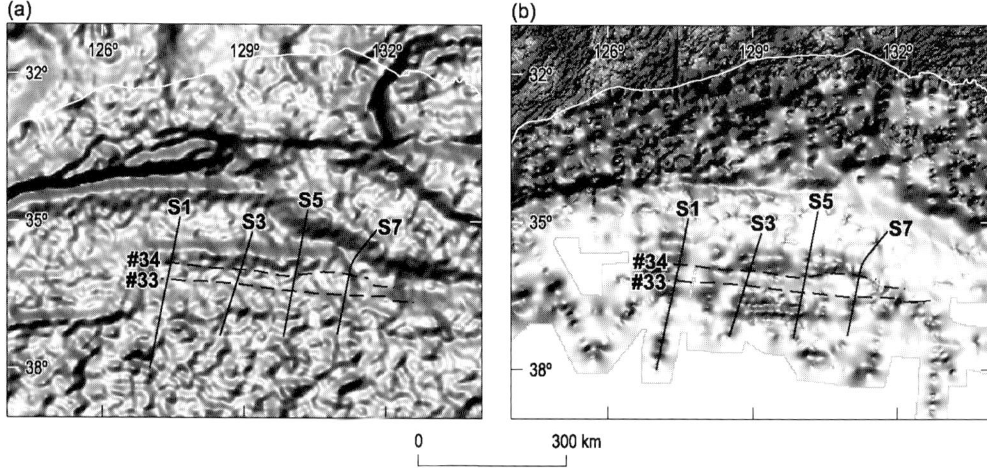

Fig. 3. (**a**) Grey-scale free-air satellite gravity anomaly image of the central GAB in the marine areas derived from satellite and ship survey data and Bouguer gravity anomaly for onshore areas. (**b**) Grey-scale total magnetic anomaly image. Dashed lines indicate the locations of magnetic anomalies 33o and 34y from Tikku & Cande (1999); straight lines are locations of seismic profiles S1, S3, S5 and S7.

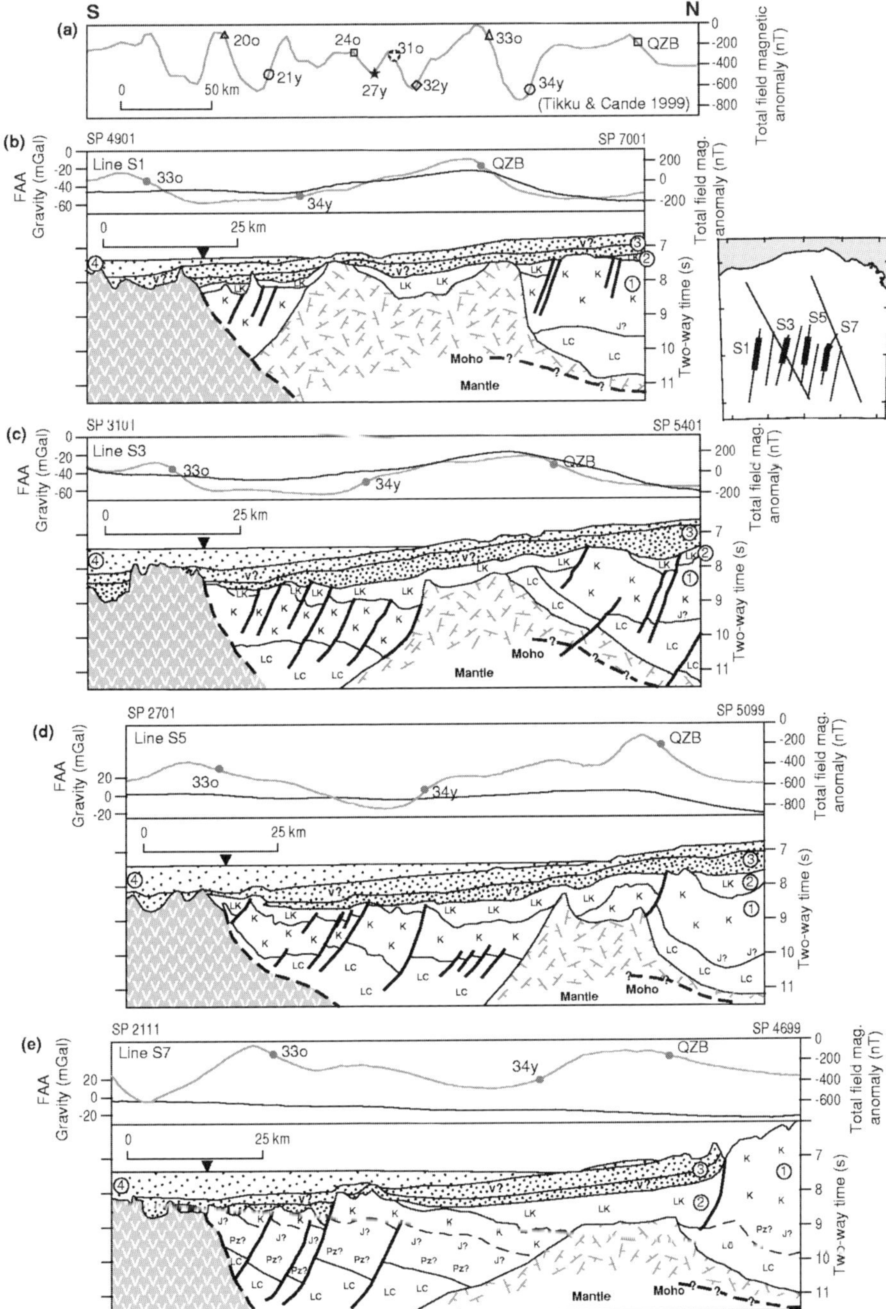

Fig. 4. (a) Preferred spreading rate model for the southern Australian continental margin after Tikku & Cande (1999). (b–e) Line drawings of seismic profiles showing observed gravity and magnetic data over the COT, with superimposed magnetic anomaly identifications based on the model of Tikku & Cande (1999) above (locations shown in Fig. 1). LC, lower crust; Pz, Palaeozoic; J, Jurassic; K, Cretaceous (Megasequence 1); LK, Late Cretaceous (late-rift Megasequence 2); heavy and intermediate stipples represent Megasequence 3 and light stipple Megasequence 4; 'v' pattern, oceanic crust; hinge pattern, basement ridge (altered mantle, serpentinized mantle, associated igneous rocks, altered sediments); filled triangle symbol denotes the COB; v?, possible volcanic flows.

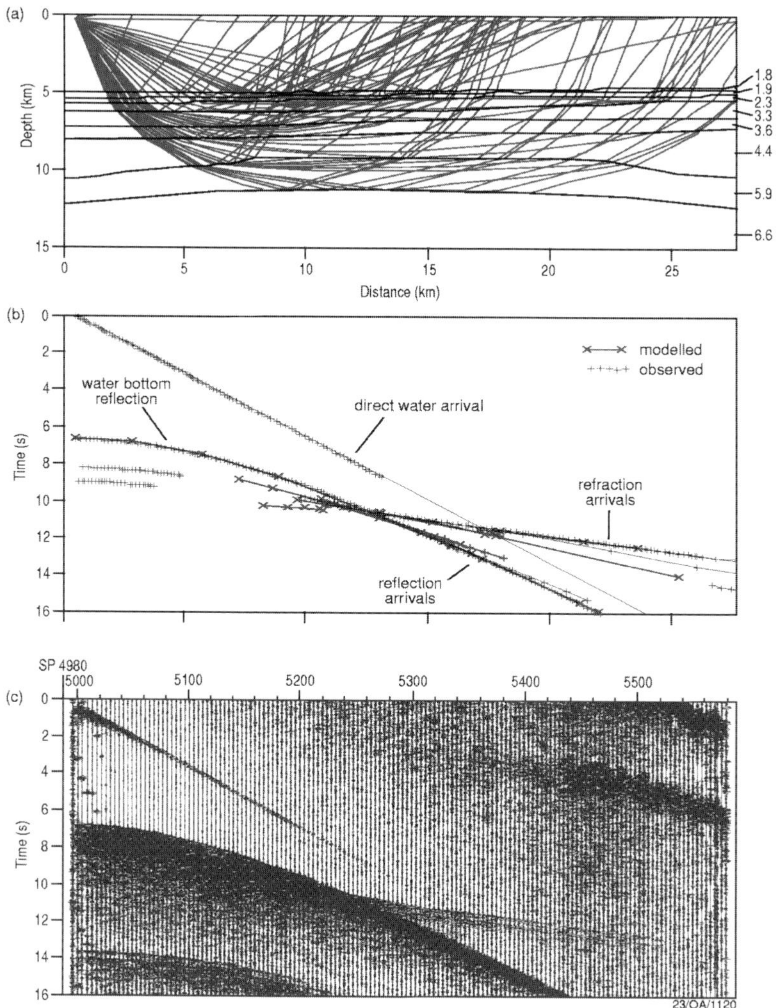

Fig. 5. (**a**) Ray-tracing of depth-converted seismic reflection profile S5 at sonobuoy 12 (location shown in Figs 1 and 7a). Final modelled layer velocities (km s^{-1}) are shown. The direct water arrival and water bottom reflection, and wide-angle reflections and refractions for all layers, are shown. (**b**) Observed and modelled picks for sonobuoy 12. The direct water arrival, water bottom reflection, and identified wide-angle reflections and refractions are shown. (**c**) Processed seismic record from sonobuoy 12 showing every fourth trace. A 16 s automatic gain control (AGC) and a band-pass filter of 6–30 Hz were used.

Interpretation of seismic refraction and reflection results

Seismic refraction data and modelling

Refraction modelling was carried out on records acquired using sonobuoys during the 1997 AGSO GAB deep-seismic survey. Fifteen of the 26 sonobuoys that successfully recorded data during the survey (Fig. 1) were interpreted and subsequently modelled. Modelling results were used to complement the

interpretation of the seismic reflection data, which are the fundamental dataset underpinning this study. An unreversed refraction profile was acquired at each sonobuoy location, with an average maximum receiver–source offset of 30 km. The nominal shotpoint (SP) interval for the refraction profiles was 50 m.

Paper displays of the sonobuoy profiles (Figs 5c and 6c) were initially interpreted and then all significant events (the direct arrival, the pre-critical water bottom reflection, other reflections, refraction arrivals and deeper pre-critical

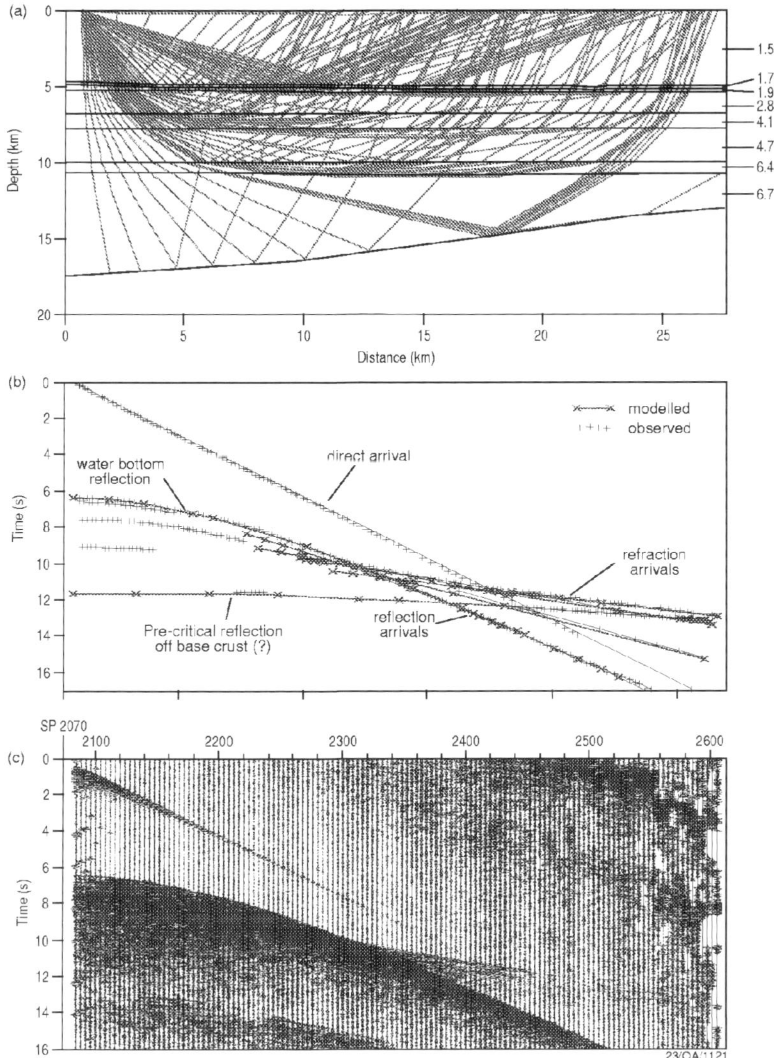

Fig. 6. (**a**) Ray-tracing of depth-converted seismic reflection profile S4 at sonobuoy 9 (location shown in Fig. 1). Final modelled layer velocities (km s^{-1}) are shown. The direct water arrival and water bottom reflection, and wide-angle reflections and refractions for all layers, are shown. Pre-critical reflections are shown for the deepest layer. (**b**) Observed and modelled picks for sonobuoy 9. The direct water arrival, water bottom reflection, wide-angle reflections, refractions and pre-critical reflection for the deepest layer are shown. (**c**) Processed seismic record from sonobuoy 9 showing every fourth trace. A 16 s AGC and a band-pass filter of 6–30 Hz were used.

arrivals where present) were digitized (Figs 5b and 6b). The interpreted events were exported into a forward modelling package to assess the drift of the sonobuoy with respect to the seismic survey line caused by current and wind. The component of drift along the seismic line was accounted for by adjusting the constant SP interval so that the observed direct arrivals had a velocity equal to the speed of sound in the surface water (1500 m s^{-1}). This approach cannot specifically correct for non-linear or lateral drift, but incorporates any constant velocity drift components into the along-line correction described above. The digitized interpreted events from the refraction profile were then scaled using the new drift-corrected SP interval in readiness for forward modelling (Figs 5b and 6b).

Preliminary depth models for ray-tracing were derived by depth-converting interpreted seismic reflection data coincident with each sonobuoy refraction profile. The thicknesses of the various layers in the seismic reflection interpretation were determined using average interval velocities derived from stacking velocities computed while processing the seismic reflection data. These initial models were then iteratively ray-traced (Figs 5a and 6a) until a good fit was obtained between the modelled and observed picks (Figs 5b and 6b).

The final velocity–depth models are subject to a variety of errors that are difficult to quantify, related to inadequately corrected sonobuoy drift, use of apparent velocities as a result of unreversed refraction profiles, and use of averaged interval velocities derived from stacking velocities to depth convert the shallow sedimentary layers. These layers do not produce refracted first arrivals because of the deep water of the survey area (in excess of 4 km; Figs 5c and 6c). Despite the inherent limitations, all 15 sonobuoy profiles give relatively consistent model results, and thus provide a reasonable first-pass estimate of crustal velocity structure. These estimates are compared with those determined by Konig & Talwani (1977) and Talwani *et al.* (1978) in later sections.

Seismic reflection data and interpretation

All seismic reflection profiles presented in this paper have been processed to 16 s two-way-time (TWT) by Robertson Research Australia Pty Ltd, and interpretations were based on migrated stacked displays. The interpretation of the seismic reflection and refraction data is best described in terms of four main provinces, each characterized by a distinct tectonic style: from south to north these are oceanic crust south of the COB; transitional crust of the COT to the north of the COB; an E–W-trending basement ridge complex; and the outer Recherche and Ceduna sub-basins to the north of the basement ridge (Figs 1 and 4).

The nature and age of seismic sequences mapped throughout the study area, particularly within the COT and adjacent Recherche and Ceduna sub-basins, have been deduced by tying the post-middle Eocene succession to ODP Leg 182 Site 1128 (Shipboard Scientific Party 2000); and by linking into a new sequence stratigraphic framework (Totterdell *et al.* 2000) established throughout the sub-basins in the shallower parts of the margin. Totterdell *et al.* (2000) identified ten supersequences (related to second-order transgressive–regressive cycles)

for the region. To simplify the discussion of the sedimentary succession in this paper we have generally grouped these supersequences into four megasequences (Fig. 4). These megasequences are related to major phases of basin and margin evolution.

Oceanic crust

The oceanic crust is readily identified by its relatively shallow basement with an absence of coherent intra-basement reflectors compared with the adjacent COT zone (Fig. 7a). The very slow-spreading oceanic crust of the central GAB has a generally rough basement surface created by submarine volcanic build-ups and tilt blocks. The term 'build-up' is used here to define volcanic constructions that formed (Fig. 8a) during sea-floor spreading. The volcanic build-ups are irregular in shape, vary in width from <1 km to >10 km, and protrude through the sedimentary cover, in places to as much as 1500 m above the surrounding abyssal plain. It has not been possible to ascertain whether these build-ups form ridges, principally because of the large (average of 90 km) seismic line spacing. However, ridge formation is likely as some volcanic build-ups appear to correspond to episodes of extremely slow spreading (e.g. half-rate of 1.5 mm a^{-1} between anomalies 24o and 31o; Tikku & Cande 1999), and could have formed elongate features during periods of minimal movement between Australia and Antarctica.

The second type of oceanic basement high occurs beneath the sedimentary cover. These highs tend to be asymmetric and appear to be fault blocks (Fig. 8b). The longer limbs of the blocks dip landward, and are generally overlain by a thin, seismically transparent layer (up to 0.3 km thick) that commonly thickens slightly towards the adjacent block. The transparent layer is onlapped by strongly reflecting sediments, and the nature of the intervening sequence boundary may imply a continuation of faulting and block rotation (amagmatic extension) of oceanic crust for some time after its initial formation. Fault planes associated with the rotated oceanic basement highs are generally poorly imaged at depth.

Scismic velocities for oceanic crust derived from modelling the 1997 sonobuoy data can be broadly grouped into three main layers: velocities range from 3.5–4.6 km s^{-1} in an upper layer just below the sediment veneer to 5.1–6.3 km s^{-1} in a second layer, to 6.3–7.3 km s^{-1} in the deepest layer, which

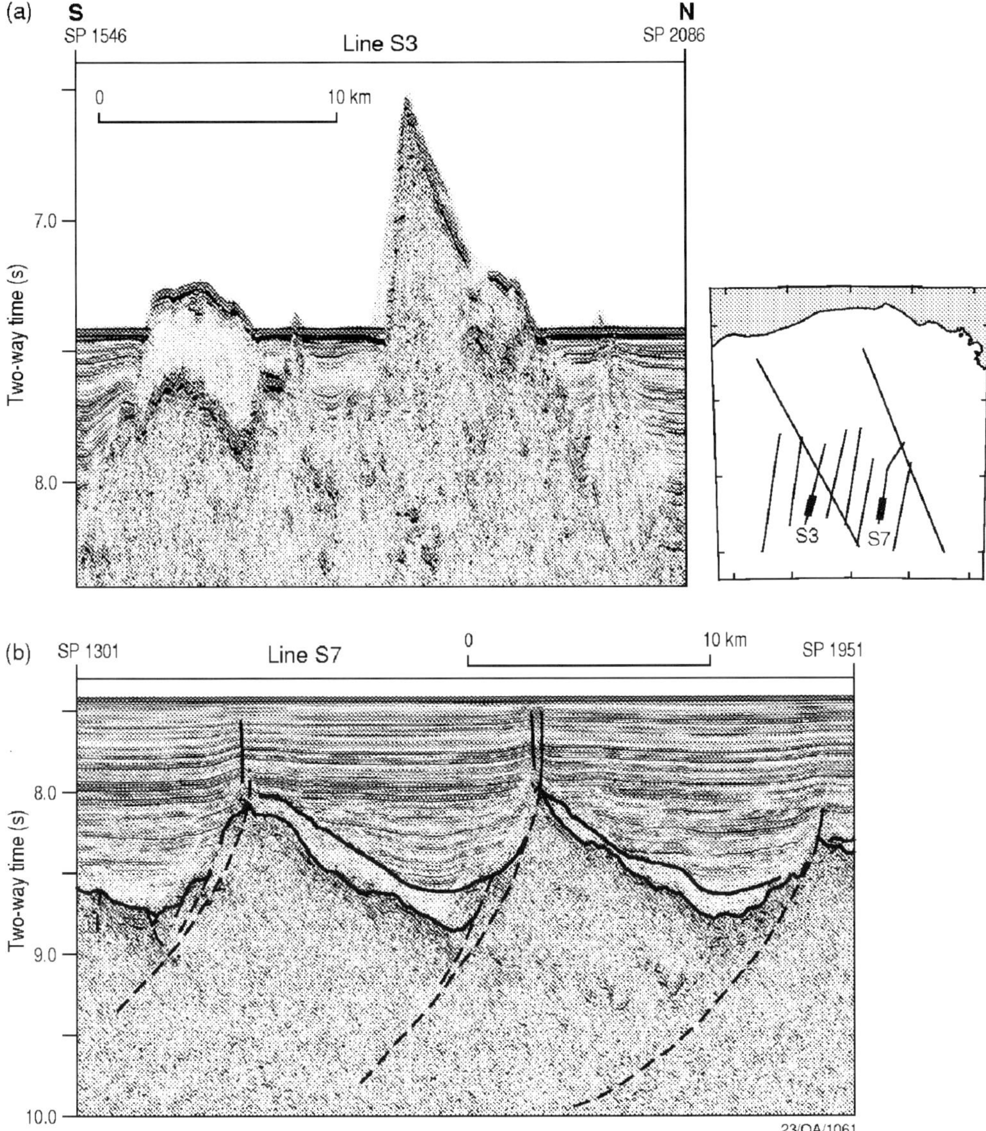

Fig. 8. Seismic sections showing: (**a**) typical volcanic build-ups on oceanic crust that may be related to episodes of extremely slow spreading; (**b**) faulted (dashed) blocks of oceanic basement that are characteristic of slow-spreading crust. The basal transparent sequence is conformable with basement and is onlapped by the overlying sequence. Rotation of basement and the transparent sequence may indicate that faulting (extension) of oceanic crust continued for some time after its initial formation.

is poorly defined in thickness. The oceanic crustal layers of the potential field model (Fig. 7b) are based on the above results. Weak 'first breaks' observed at far offsets on some refraction profiles may define the base of the third layer, the probable base of oceanic crust. Refraction modelling using these 'first breaks' indicates that igneous oceanic crust could be up to 9 km thick in places, and averages about 6 km in thickness. This average thickness agrees with earlier estimates by Talwani *et al.* (1978) of a 5–6 km thick oceanic crust. The variation in oceanic crustal thickness derived from the sonobuoy refraction profiles may reflect modelling difficulties, data deficiencies and/or a real variability of the crust. A degree of caution needs to accompany use of velocities or thicknesses derived from the 1997 refraction data as the estimates were derived from unreversed sonobuoy profiles, as noted previously. However, potential field modelling (see below) also supports an oceanic crust of variable thickness, particularly adjacent to the COB.

A significant change in seismic reflection character occurs at the location of anomaly 33o (83 Ma, early Campanian time) from the typical very slow-spreading oceanic crust to the south, to the relatively thick, faulted, folded and probably intruded crust in the COT to the north (Figs 4 and 7a). The change in seismic character is interpreted to represent a clear COB. This interpretation implies that break-up and sea-floor spreading in the central GAB commenced at *c.* 83 Ma (early Campanian time), and not 95 ± 5 Ma (Cenomanian time) as previously suggested by Cande & Mutter (1982), Veevers (1986) and Tikku & Cande (1999).

Transitional crust

Transitional crust refers to crust that exhibits neither typically oceanic nor continental seismic reflection characteristics. This crust lies in the COT between the very slow-spreading oceanic crust to the south and the basement ridge complex to the north, which in turn separates it from the extended continental crust of the outer Recherche and Ceduna sub-basins (Figs 1, 4 and 7). Magnetic anomaly 34, as mapped by Cande & Mutter (1982), coincides with the southern flank of the basement ridge (Figs 3 and 4), which was previously interpreted to lie close to the COB (Veevers 1986). The basal section within the COT has seismic characteristics that are more like continental crust rather than oceanic crust. This section consists of relatively thick sedimentary sequences that exhibit several phases and styles of deformation and intrusion (Figs 4, 7a and 9a), and have similar seismic characteristics to sequences occurring in the outer Ceduna Sub-basin to the north of the basement ridge complex (Figs 7a and 9a).

Seismic sequences within the COT include a pre-break-up (pre-rift) megasequence (of Late Jurassic to Cenomanian age, Megasequence 1), characterized by a rough upper surface that is onlapped by the overlying 'late-rift' sequence (Megasequence 2; Figs 4, 7a and 9a). The seismic character within Megasequence 1 is variable: it exhibits characteristics and dips similar to those of Megasequence 1 north of the basement ridge (Fig. 9a), but also includes non-reflective zones with discordant, high-amplitude events that probably represent sills and dykes (Figs 7a and 9a).

A seismically transparent layer underlying Megasequence 1 within the COT has a rough upper surface, possibly resulting from localized block faulting. The seismic characteristics of this layer are similar to those of the interpreted lower continental crust north of the basement ridge complex (Figs 4 and 7a, 'LC'). We use the term lower crust to refer to crystalline crust found in the lower 15 km of 'normal' thickness continental crust. Both the upper and lower continental crust are present beneath the inner part of the continental margin of the central GAB; however, the upper crust has been largely removed by extension beneath the outer margin, and may form a significant component of the conjugate Antarctic margin (Etheridge *et al.* 1989). Within the COT, the upper crystalline crust appears to be absent, and the lower crust is highly extended and directly overlain by probable Mesozoic sediments. Refraction-derived velocities have also helped in establishing the presence of crystalline continental crust within the COT (velocity of 5.9–6.7 km s^{-1}), by comparison with the area north of the basement ridge (velocity of 5.9–6.8 km s^{-1}) where the crust is better imaged by seismic reflection data (Fig. 7a). In this northern location, the lower crust thins dramatically to the south and has been uplifted on the northern flank of the basement ridge, together with uplift and deformation of Megasequence 1. The similarity of the seismic character of the crystalline crust and the basal sedimentary section to the north and south of the basement ridge (Fig. 7a) provides strong support for the COT being a zone of highly extended and intruded continental crust, rather than deformed oceanic crust from an early episode of sea-floor spreading.

Synrift Megasequence 2 (of Turonian to Santonian age) onlaps Megasequence 1, and its upper bounding unconformity (Figs 4 and 7a) correlates with the time of break-up and the start of sea-floor spreading. Faulting does not generally extend through Megasequence 2, implying that upper-crustal extension and emplacement of the basement ridge was largely complete before its deposition (i.e. during Cenomanian time).

Sediments deposited during the post-break-up thermal sag phase constitute Megasequence 3 (of early Campanian to Mid-Eocene age), which downlaps onto Megasequence 2 (Figs 4 and 7a). High-amplitude, chaotic seismic facies

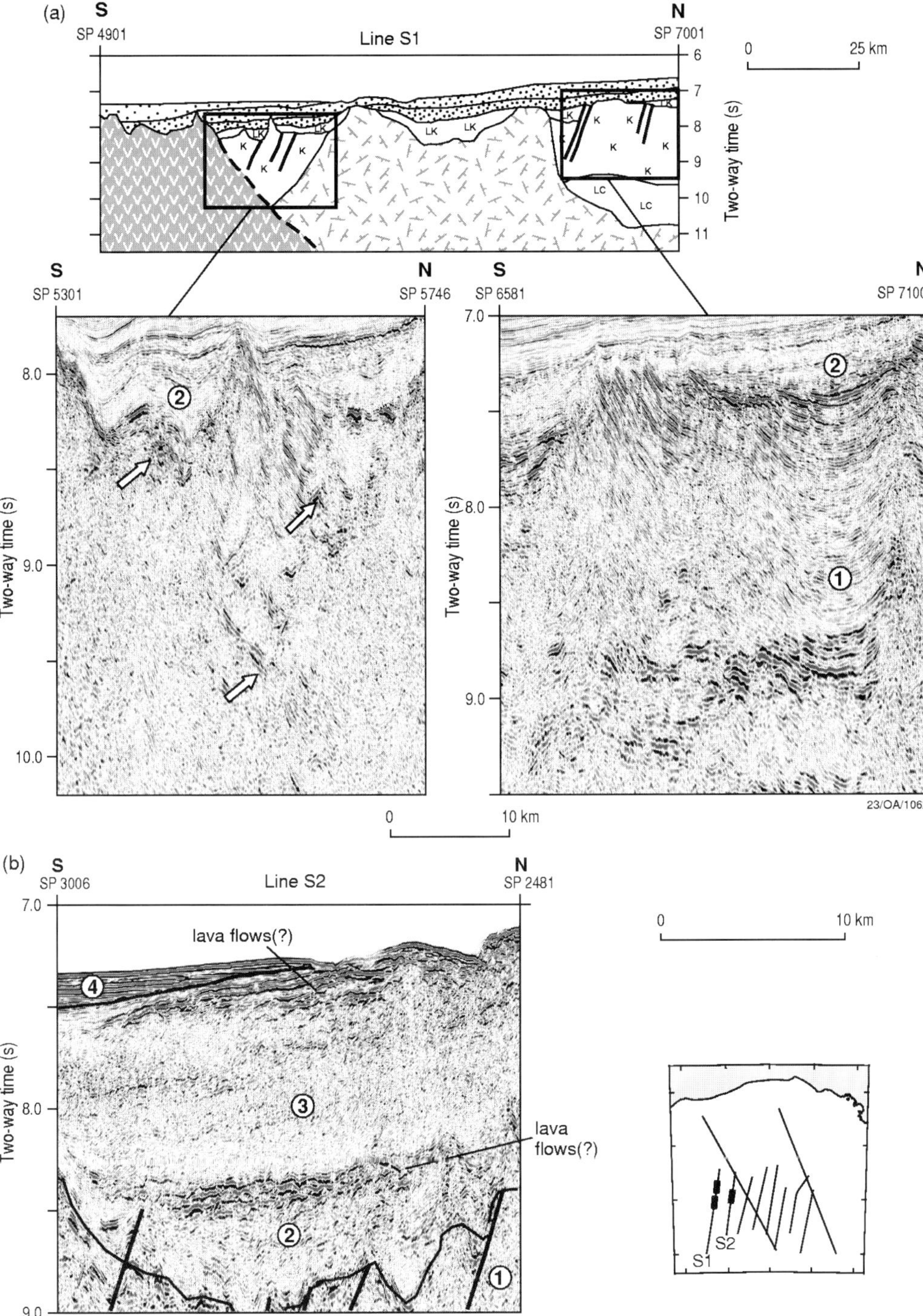

Fig. 9. (a) Line drawing of seismic profile S1 across the COT and associated seismic section details. K, Cretaceous; LK, Late Cretaceous; LC, lower crust; 1–2, Megasequences 1 and 2; arrows indicate possible sills and dykes. 'v' pattern, oceanic crust; hinge pattern, basement ridge; stippled sequences, post-break-up sediments. **(b)** Possible volcanic flows within the COT near the boundary of Megasequences 2 and 3, and near the top of Megasequence 3.

within this sequence, particularly near its base (Fig. 9b), are interpreted as lava flows and/or volcaniclastic deposits resulting from restricted and episodic volcanic activity during Campanian to early Cenozoic time. This interpretation is supported by magnetic field modelling, which requires a moderately negative susceptibility layer within this sequence to match the observed magnetic anomaly profile (see below).

Megasequence 4 (of Mid-Eocene to Recent age) is a southward-thickening wedge of onlapping sediments that forms part of the Australia–Antarctic Basin. This sequence reflects a new depositional regime corresponding to a change from very slow-spreading (<10 mm a^{-1} half-rate) to slow-spreading ($10–70$ mm a^{-1} half-rate) oceanic crust at chron 18 (Tikku & Cande 1999), leading to a changed oceanic circulation system related to the opening of the Southern Ocean gateway.

Basement ridge complex

The east–west-trending basement ridge is located landward of anomaly 34 (Figs 1 and 4). It extends along strike for at least 500 km, and averages 20 km in width where it protrudes through sediments of Megasequence 1 and broadens to about 60 km at depth. Figure 1 only shows the extent of the ridge at about Megasequence 1 level, and highlights its irregular and variable geometry at this upper level. Although the exact shape of the basement ridge is difficult to map at depth, its general form and continuity, as illustrated in Figure 4, are supported by the dimensions of the associated gravity and magnetic anomaly highs (Fig. 3), at least as far east as line S6 (Fig. 1). In detail the basement ridge is a complex feature, in width, relief, seismic character and probably composition. It appears to be segmented (Fig. 1), and in places may consist of two or more coalescing ridges or peaks (Fig. 4, line S1). The general structural position of the basement ridge between the extended crust of the outer Recherche and Ceduna sub-basins to the north and the highly deformed and intruded crust and sediments of the COT to the south is maintained along its full strike-length (Fig. 4); however, other characteristics vary, particularly to the east. In the west, the basement ridge is associated with a shallowing of mantle, uplift and deformation of Megasequence 1 above its northern flank (Figs 4, 7a and 9a), and more localized culminations at the crest (Fig. 7a). These features begin to disappear between lines S6 and S7, and to the east the ridge appears to be underlain by a very poorly defined Moho. The

Moho in the east is interpreted to have a consistent northerly dip based on the geometry of the overlying crust. This northerly dip may be offset by seaward-dipping faults. To the east the ridge has a more transparent seismic character typical of the lower continental crust to the north, and its top has a characteristic strongly reflecting, chaotic appearance (Fig. 10). East of the apparent dextral offset in the COB (Fig. 1, line S11) at the edge of our study area, the basement ridge complex is difficult to map, and its character, and that of the COT, changes significantly. The upper surface of the basement ridge retains its chaotic reflection character, but here may be underpinned by thicker, transparent lower crust. The ridge, and possible large tilt blocks within and adjacent to the COT, appear to be bounded by southward-dipping, low-angle normal faults.

The change in seismic character and thickness of the top of the highly attenuated lower crust and overlying Jurassic to Cenomanian Megasequence 1, across the basement ridge (Figs 4 and 7a), may imply the involvement of a major southward-dipping upper-crustal fault during the early rift phase and before the formation of the ridge. Such a fault, which may or may not have linked with a major upper-mantle fault (see below), could have influenced the location of the basement ridge. The fault would no longer be imaged on seismic data because of overprinting by the deformation and magmatism associated with the development of the ridge. Such faults could be similar to those described above to the east of the dextral offset in the COB.

Seismic velocities modelled for the basement ridge in the east (Fig. 1; profile S10, sonobuoy 22) indicate that here it is a three-layer body composed of *c.* 1 km of 5.2 km s^{-1}, 1 km of 6.2 km s^{-1} and an underlying layer with a velocity of 7.2 km s^{-1} (Fig. 10). Refraction ray-tracing through the southern flank of the feature further west (Fig. 1; sonobuoy 13) indicates a deep layer of 7.4–7.6 km s^{-1} velocity. Velocities of 7.2–7.6 km s^{-1} contrast significantly with velocities of 6.5–6.8 km s^{-1} modelled for the 'normal' lower crust (Figs 1 and 7a; sonobuoys 9–12). Similar velocities of 7.6 km s^{-1} underlying the COT of the Iberia margin (Whitmarsh *et al.* 1993), and 7.2–7.6 km s^{-1} for the conjugate Newfoundland margin (Reid 1994), have been associated with altered upper mantle. Whitmarsh *et al.* (1993, 1998) proposed two alternatives to explain the relatively low upper-mantle velocities: (1) underplated material emplaced at time of rifting and break-up; (2) serpentinized upper-mantle peridotite

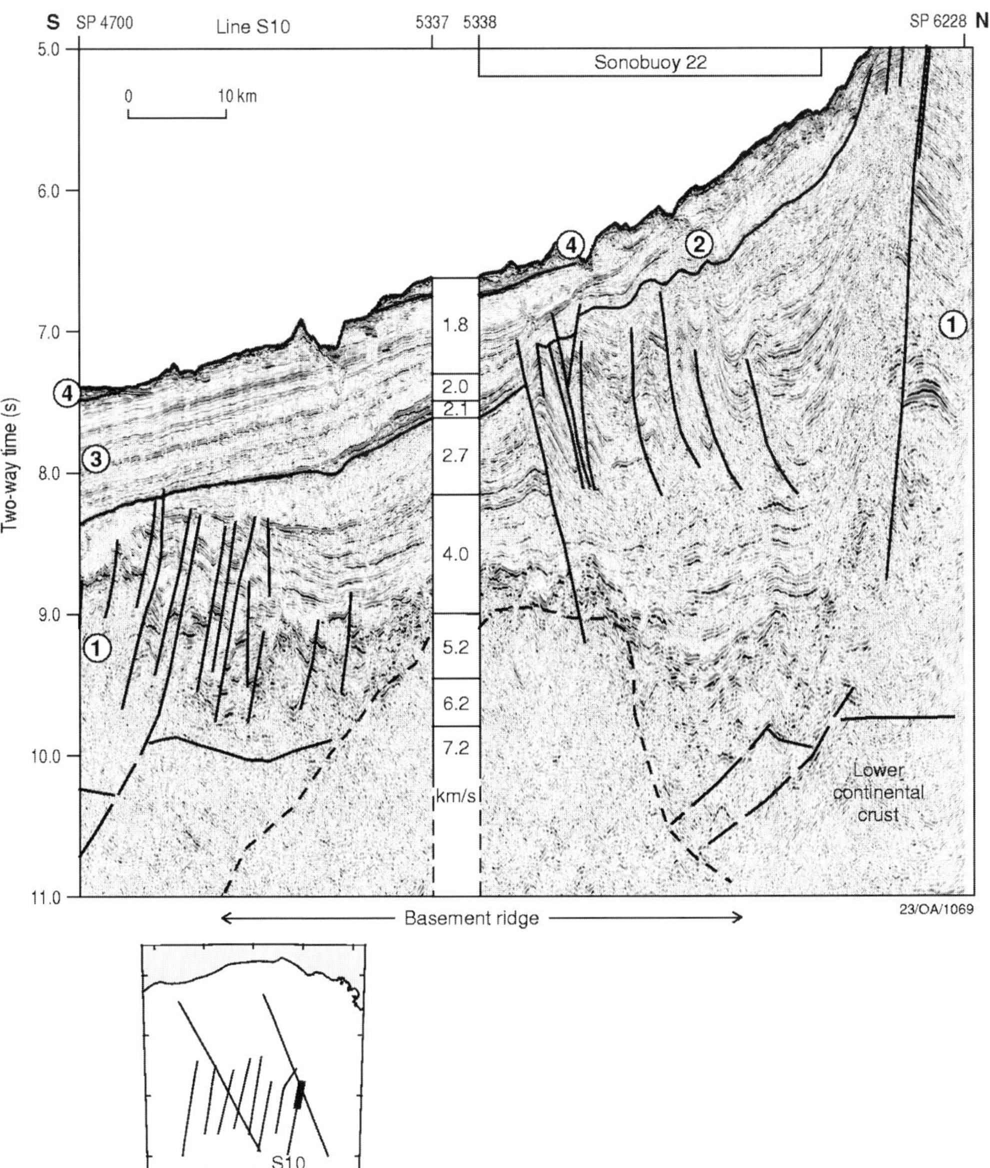

Fig. 10. Refraction velocity model at sonobuoy site 22 over the basement ridge, shown as a vertical strip positioned at one end of the shot range for display purposes (location shown in Fig. 1). 1–4, Megasequences 1–4; dashed line represents approximate outline of basement ridge; subvertical lines are faults (dashed where uncertain).

emplaced in the late stages of rifting during the onset of sea-floor spreading. Several recent studies (Discovery 215 Working Group 1998; Chian *et al.* 1999; Dean *et al.* 2000) reconfirmed velocities between 7.0 and 7.9 km s^{-1} at the base of crust that support the hypothesis of exhumed serpentinized upper mantle. Sampling and ODP drilling on the West Iberia margin (Beslier *et al.* 1994; Whitmarsh *et al.* 1996, 1998) have recovered both peridotite and serpentinized peridotite, although the timing of the serpentinization is difficult to ascertain. Beslier *et al.* (1994) suggested that serpentinization post-dates exhumation, although recent studies

(Skelton & Valley 2000) suggest that serpentinization occurred before exhumation. We realize that in the absence of other evidence velocities in the range of 7.2–7.6 km s^{-1} can be interpreted as intruded lower continental crust, underplating beneath the crust or altered, possibly serpentinized, mantle. In the central GAB, we suggest that the integration of the seismic reflection and refraction data supports a variable composition for the basement ridge: altered, possibly serpentinized, mantle in the west; altered lower continental crust in the east; and both overlain to a variable degree by lower-velocity (5.2–6.2 km s^{-1}) bodies, that could represent less dense igneous suites (Christensen & Mooney 1995), possibly fractionated melts, and/or intruded sediments.

Definition of Moho

The northern flank of the basement ridge on lines S3, S4, and S5 merges with high-amplitude coherent reflectors at about 10–11 s TWT (Fig. 7a). These reflectors, which deepen northward to as much as 12 s TWT (Fig. 11), show offsets ranging from a few hundred metres to as much as 6 km (Fig. 7b). The offsets dip southwards at between 15 and 30° (Fig. 1). The reflectors (Figs 7a and 11) are interpreted as faulted, altered mantle based on their high reflectivity, velocity structure and depth (TWT *c.* 11–12 s). The tectonic significance of the faulted mantle is elaborated on in the discussion below.

A pre-critical reflection from the base of lower crust (Fig. 6; layer with a velocity of 6.7 km s^{-1}) was modelled from the 1997 refraction data (Fig. 1, sonobuoy 9) but refraction first breaks from a deeper event were absent, thus providing no conclusive evidence for a velocity characteristic of upper mantle at this location (see Moho faults, Fig. 1). Talwani *et al.* (1978) did, however, model refraction first breaks from unaltered mantle (velocities >8.0 km s^{-1}) at six locations (Fig. 1; V33-6, -8, -11, -12, -15, -40) and interpreted ambiguous unaltered mantle at four other locations (Fig. 1; V33-13, -21, -27, -36). Based on these data, depth to Moho, within the COT and immediately to the north of the basement ridge, is in the range of 12–16 km (TWT 9.9–11.5 s) and 13–17 km (TWT 9.7–11.5 s), respectively. The above TWTs agree well with our results that estimate a depth to Moho (unaltered mantle) of 10–12 s TWT, marginally below the deep, high-amplitude coherent reflectors (Figs 7 and 11).

The top of a thin (2–5 km) 7.4 km s^{-1} layer overlying unaltered mantle (>8.0 km s^{-1}) as modelled by Talwani *et al.* (1978; site V33-36, Fig. 1) on the northern limb of the basement ridge, correlates well with the deep, high-amplitude coherent reflectors mapped on lines S3, S4 and S5 (Fig. 1). We believe that, on a balanced consideration of the integrated dataset, this thin layer is likely to represent altered mantle rather than altered lower continental crust. Further, the absence of similar high-amplitude coherent reflectors elsewhere may imply an absence of alteration of the upper mantle in these areas.

Ceduna, Eyre and Recherche sub-basins

Unequivocal evidence of a Jurassic rifting event is documented in the Eyre Sub-basin, where the geometry of faulted basement tilt blocks indicates that extension took place in a generally NW–SE direction (Etheridge *et al.* 1989). The same ?Precambrian- to Jurassic-aged sediments are interpreted to be highly attenuated in the Recherche Sub-basin, and under the COT where remnant blocks are thought to be present (Fig. 4, line S7). It is probable that faulting of any crystalline upper crust and pre-rift or rift sediments during an episode of Cenomanian–Turonian extension leading to break-up involved reactivation of Jurassic rift faults.

A major décollement surface within the middle Albian succession of the Ceduna Sub-basin (Totterdell *et al.* 2000) extends under the continental slope where gravity-driven listric growth faults, and associated toe-thrusts, sole onto it (Fig. 12). The décollement, growth faults and associated deformation are interpreted to have been active during Cenomanian time following deposition of Albian–Cenomanian marine shales (Totterdell *et al.* 2000).

A significant reduction in thickness of the lower crust (i.e. 7 km to <2 km under the lower continental slope north of the basement ridge) and the relative absence of penetrative faulting (Fig. 7a) indicate that the lower crust may have behaved in a ductile manner. An initial brittle–ductile transition at the top of the lower crust (now at base of Megasequence 1) may have acted as a second, deeper shear zone, or possibly extensional detachment beneath the outer margins to the north of the basement ridge. We suggest that movement along both the gravity-driven décollements and possible extensional detachment, and associated faulting and deformation, peaked in late Cenomanian time (near the end of deposition of Megasequence 1, *c.* 10 Ma before break-up; Fig. 13a). A second phase of extension from Turonian to Santonian time leading to break-up (Fig. 13b), was charac-

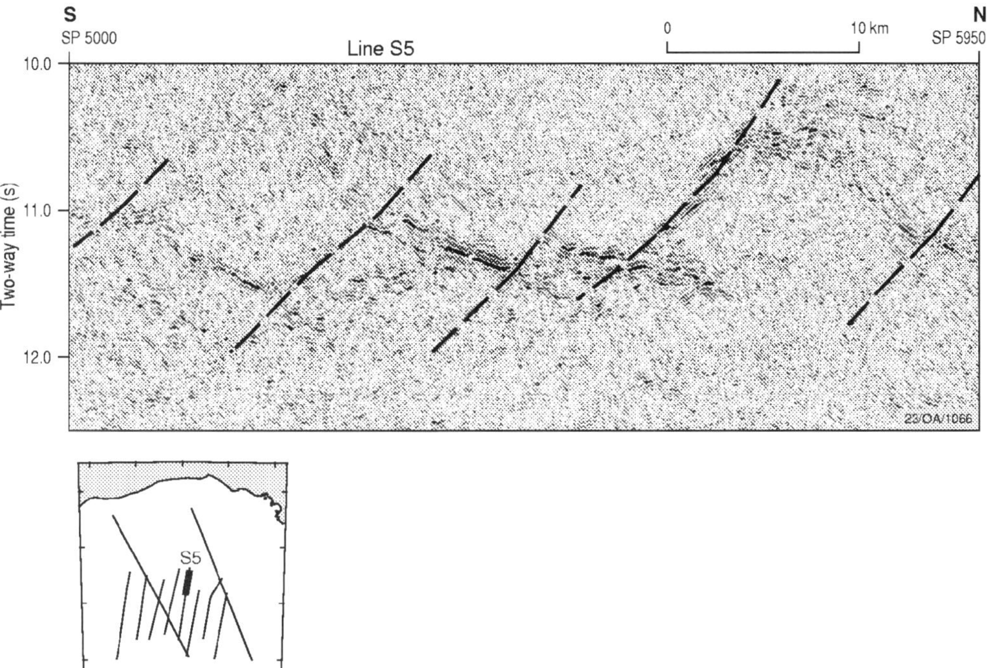

Fig. 11. Seismic section detail showing brittle faulting of the upper mantle. Dashed lines are probable faults; strong coherent reflectors represent top of altered mantle. Independent support for this interpretation is provided by $7.2-7.6 \, \text{km s}^{-1}$ refraction velocities.

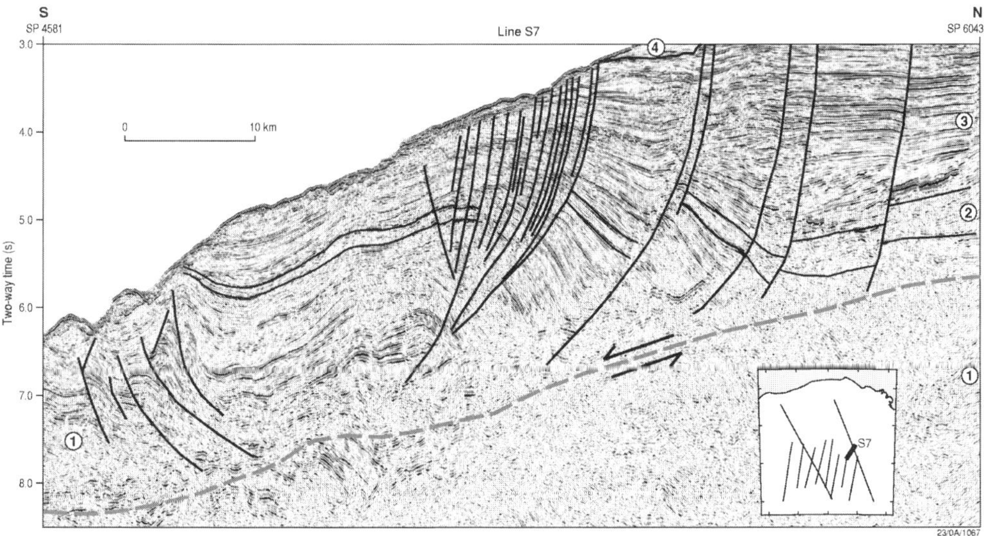

Fig. 12. Décollement (subhorizontal dashed line) near base of middle Albian sediments underlying the continental slope, outboard of the Ceduna Terrace. The transparent zone under the décollement merges into a more reflective section beneath the central Ceduna Sub-basin. Subvertical lines are late Cenomanian faults that were reactivated during Tertiary time. Note toe-thrusts associated with the décollement at bottom left-hand side of figure. 1–4, Megasequences 1–4.

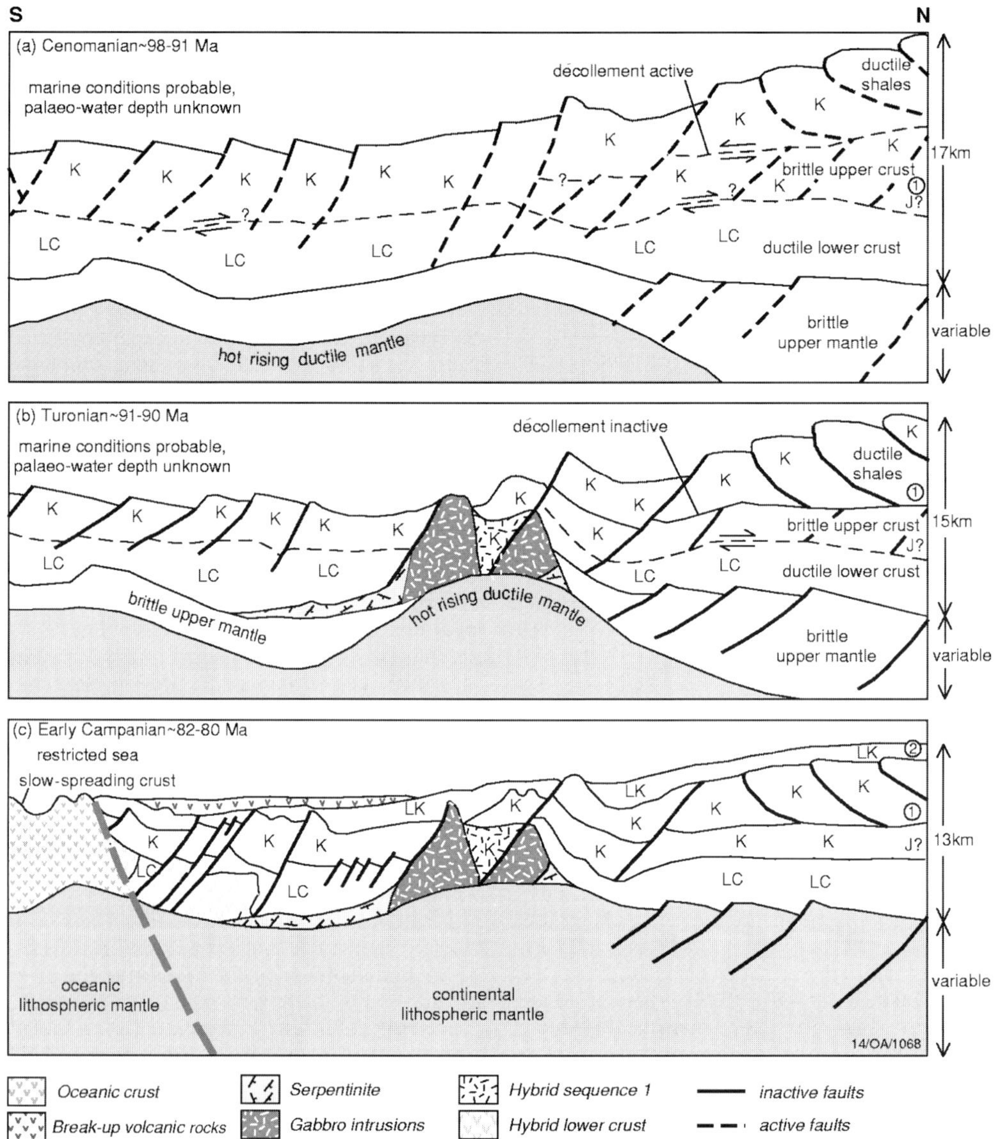

Fig. 13. Schematic conceptual model showing the development of the COT based on a four-layer lithosphere (i.e. ductile and brittle crust and mantle, similar to the model of Brun & Beslier (1996)). (**a**) Upper- and lower-crustal extension active, and a décollement developed in mobile shales of the Ceduna Sub-basin. (**b**) Lower-crust–upper-mantle extension, hot rising ductile mantle, gabbroic intrusions from decompression melting localized at the present-day basement ridge, and mobilization of serpentinized peridotite in the vicinity of the ridge. (**c**) Progressive thinning leads to break-up, and formation of oceanic lithospheric mantle and slow-spreading oceanic crust, accompanied by intrusions into adjacent lower crust of COT. The subhorizontal dashed lines indicate levels of décollement (upper) and possible 'detachment' (lower). Arrows on the 'detachment' zone only indicate that relative motion has occurred along it and not necessarily the sense of motion. Dashed line in (**c**) denotes conceptual boundary. LK, Late Cretaceous; K, Cretaceous; J, Jurassic; LC, lower crust; 1–2, Megasequences 1 and 2. (See text for detailed discussion.)

terized by further ductile deformation and thinning of the lower crust.

Potential field analysis

Gravity field

Satellite-derived free-air gravity anomalies (FAA, Sandwell & Smith 1997), combined with gridded ship-track data (Petkovic *et al.* 1999; Fig. 3a), show a sharp, slightly arcuate bipolar anomaly at the shelf edge running east–west along the margin (*c.* 34°S); a large lobate anomaly (*c.* 34–36°S) associated with thick deltaic sediments on the Ceduna Terrace (Willcox & Stagg 1990); and an east–west positive FAA trend (*c.* 36°S), possibly offset at *c.* 127°E. A large proportion of the area to the south of the shelf-break anomaly is characterized by high-frequency random processing and acquisition noise.

The east–west positive FAA trend (*c.* 36°S) of about 30 mGal relative amplitude is paralleled to the north by a gravity low of about −40 mGal relative amplitude (Fig. 4). The gradient between this pair of features is about 1.5 mGal km^{-1}. The positive FAA trend dies out between lines S6 and S7 (Figs 1 and 3a). Seaward of this positive FAA trend, the anomalies rise to progressively higher values, reaching around −20 mGal over known oceanic crust, where superimposed relative anomalies of 10 mGal and wavelengths of about 40 km are present.

Magnetic field

Shiptrack magnetic data have had the International Geomagnetic Reference Field (1990 epoch) removed, but have not been corrected for quiet-day diurnal variation or reduced to the pole (Petkovic *et al.* 1999). The resultant anomalies (Fig. 3b) show considerably more complexity than the gravity anomalies. The main features of the magnetic anomaly have been comprehensively reviewed by Tikku & Cande (1999). Of principal interest in the study area is a composite magnetic high of 230 nT amplitude that is approximately spatially coincident with the prominent east–west gravity trend (*c.* 36°S; Fig. 4), but extends further east beyond line S7 (Figs 1 and 3). Landward of this anomaly is the broad negative anomaly termed the Magnetic Quiet Zone (MQZ), first recognized by Weissel & Hayes (1972). Seaward of the magnetic high is a deep magnetic trough of −200 nT. Further south lie anomalies of between 30 and 150 nT that have been inter-

preted as sea-floor spreading anomalies 33 to 22 (Tikku & Cande 1999).

Modelling

Two and a half dimensional forward modelling was used to test the validity of the seismic refraction and reflection interpretations on several lines crossing the COT. In particular, modelling of gravity and magnetic fields was used to assess the range of plausible density and susceptibility values for possible source bodies associated with the basement ridge (*c.* 36°S, Fig. 7b) interpreted from seismic data to lie near the northern margin of the COT.

Megasequence boundaries interpreted from the seismic reflection data were depth converted using velocities derived from both sonobuoy refraction data and seismic reflection processing. Voluminous depth data for these horizons were decimated and subsequently imported into the ModelVision modelling package (Encom Technology 1998). Lines S3, S5 and S10 were then modelled together to ensure a consistency of solutions along the length of the ridge. Free-air gravity and magnetic data were modelled simultaneously for all lines. However, because of space constraints, only the model for line S5 is presented and discussed here. Table 1 describes body parameters used for line S5 (Fig. 7b).

The fact that the modelled lines straddle the COT, from a thicker section of continental crust to a thinner oceanic crust with strikingly different physical properties, has led to the adoption of a number of assumptions. As the bulk of the models lie on the landward side of the COB as defined from the seismic reflection data, these assumptions are consistent with onshore modelling methods, rather than the approach normally adopted in purely oceanic environments. Thus, the background density for calculating the free-air gravity is assumed to be 2.67 g cm^{-3}, reflecting continental basement. The assumed magnetic field equals the Earth's present field strength of 60000 nT inclined at −70° with a 2°E declination, as used by Cande & Mutter (1982). In modelling continental crust, the effect of magnetic remanence is usually complex and tends to be much less than induced magnetization; anomaly modelling is therefore usually assumed to reflect susceptibility contrasts alone. In this model, domains of remanently magnetized crust have been represented with bulk apparent susceptibilities. This approximation attempts to account for both induced effects and remanent components of magnetization parallel to the overall inducing field vector, and ignores the component perpen-

Table 1 *Petrophysical properties of layers in potential field models*

Layer	Density ($g\ cm^{-3}$)	Susceptibility (c.g.s.)
Sea water	1.027	0
Megasequence 4	1.10	0
Megasequences 2–3	2.12	0
Megasequence 3 lavas	2.75	−0.0030
Upper Megasequence 1	2.50	0
Lower Megasequence 1	2.50	−0.01
Lower crust	2.93	0
Modified lower crust	2.72	0
Hybrid lower crust	2.90−3.03	−0.0018 to −0.005
Hybrid lower Megasequence 1	2.90	−0.006
Gabbroic intrusions	2.90	0.001−0.006
Serpentinite	2.67	0.0100
Oceanic crust layer 3	3.00	0
Oceanic crust layer 2	2.75	0
Oceanic crust layer 2	2.50	variable
Mantle	3.20	0
Altered mantle	3.07	0

dicular to the field. It must be noted that oceanic crust in the model probably acquired remanent magnetizations at latitudes much farther south than their current positions. The orientation of these magnetizations has been modelled by other workers as $I = 74°$, $D = 4°E$ (Cande & Mutter 1982). As noted above, the present field from all sources, both geocentric axial dipole and non-dipole, has an orientation that very nearly parallels this, indicating that the magnitude of the vector component perpendicular to the field must be very minor, and can thus be ignored with little impact on the model. Thus, any bodies modelled with negative apparent bulk susceptibility are expected to have a major remanent component of magnetization opposed to the current inducing field.

In general, support for the inclusion of model bodies has been derived from analysis of changes in seismic reflection characteristics. Where the seismic reflection data are ambiguous, or carry little information on anomaly sources, for example in the oceanic crust, bodies are only included where the signal in the potential field anomalies demands a contrast. For example, no information is available from the seismic reflections on transitions between normal and reverse remanently magnetized portions of layer 2 in the oceanic crust; these bodies are thus inferred from changes in the magnetic signal alone, in accord with standard practice.

The coincident gravity and magnetic highs associated with the basement ridge complex were modelled by a combination of effects. The large gravity effect derives principally from a zone of elevated mantle underlying this feature, rising to depths of about 11 km (Fig. 7b). Above this elevated mantle is a lower-density layer, thought to represent magnetite-bearing serpentinized peridotite on the grounds of its extremely high magnetic susceptibility and transparent seismic reflection character. This layer contributes substantially to the broad magnetic high. Also modelled are two igneous bodies with high densities and moderate to high susceptibilities producing a higher-frequency magnetic response. These bodies are thought to represent gabbroic plugs, and possible associated extrusive rocks, sourced from the decompressed elevated mantle. The densities and magnetic susceptibilities required to produce the coincident geophysical features associated with the ridge preclude it from being a sedimentary feature such as a salt diapir. The variety of modelled bodies is consistent with the seismic data, and the modelled solution provides no support for the possibility that the ridge feature at this location is a metamorphic core complex (Stagg & Willcox 1992). At the eastern end of the seismically defined basement ridge, where there is no associated gravity high, and a poorly defined magnetic high, a different combination of effects will be required to model the potential field data. The absence of a significant gravity high implies that the mantle is less elevated to the east, which is supported by seismic reflection data.

The mantle shows considerable relief in the model consistent with brittle faulting. On the northern end of the line, this interpretation is supported by 6–8 km offsets of high-amplitude reflectors, thought to represent the top of altered mantle (Fig. 7a). At the southern end of the line, there is no clear altered mantle (or Moho) reflection, possibly as a result of masking by the high seismic velocity contrast between ocean-floor basalts and the overlying sediment veneer. Nevertheless, longer-wavelength features in the gravity suggest deep-seated density variations, which have also been modelled using mantle topography. This result is tentatively supported by variable depth-to-Moho determinations in Talwani *et al.* (1978) and minimum crystalline oceanic crustal thickness estimates from this study. Depression of the mantle beneath oceanic crust at the southern end of the model appears to coincide with substantial basement highs and volcanic build-ups on the ocean floor. This coincidence may imply the operation of Airy-type isostasy to support this basement topography, which rises to 1500 m above the surrounding basement level.

The very large gravity low on the landward side of the basement ridge is modelled with a lower-density zone in the upper mantle, as suggested by refraction data (Fig. 7b, hachured area). Alternatively, it is possible to model this zone with a lithological heterogeneity (e.g. felsic granulites inside the dashed lines, Fig. 7b) in the lower crust.

On the seaward side of the basement ridge, the modelled lower crust requires very different magnetic properties: three zones of reversely magnetized, dense lower crust suggest massive intrusion of the transitional lower crust by mafic dykes and sills during the magnetic reversal between anomalies 34y and 33o. A large number of irregular high-amplitude reflectors in this area provide supporting seismic evidence for such features (Figs 7a and 9a).

The layer overlying the lower crust (Megasequence 1) is modelled with a significant reversely magnetized component. Megasequence 1 is a composite body, representing deposition of a variety of sedimentary facies from at least Late Jurassic to Cenomanian time, and thus integrates petrophysical properties from a variety of magnetic sources. The basal part of the megasequence is assumed to represent the integration of numerous periods of reversal during latest Jurassic to earliest Cretaceous time, and is coincident with a zone of more transparent seismic character above the lower crust (Fig. 7a).

The use of negative susceptibility in sediments to explain the uniform negative anomaly over the MQZ needs more robust testing along the length of the margin. A reasonable analogue for the modelled section might be found in the equivalent Upper Jurassic–Lower Cretaceous rift sequence (the Casterton Formation) of the Otway Basin some 1000 km to the east. The

Casterton Formation can be up to 500 m thick, and comprises a clastic lacustrine synrift sequence overlying and intercalated with volcanic rocks and basalt flows. K–Ar radiometric dating of the volcanic rocks (Mitchell *et al.* 1997) gave ages ranging from Late Jurassic to Early Cretaceous time, but at least as old as 153 Ma in places. An equivalent Upper Jurassic to Lower Cretaceous basaltic sequence is likely to be present throughout the Southern Australian rift system, providing support for magnetized rocks of Jurassic age within the central GAB.

Also, the long, linear flood basaltic belt within Gondwana that stretched from southern Africa (Karoo province) through Antarctica (Ferrar province) to southern Australia (Tasman province) (Elliot 1992) is represented by the 175 Ma Tasmanian dolerites (Hergt *et al.* 1989) and equivalent age basaltic rocks on Kangaroo Island, just east of the Ceduna Terrace. This phase of Mid-Jurassic magmatism, which probably represents the initial stages of Gondwana break-up in the region (Elliot 1992) may also be represented in the central GAB, providing another source of magnetized Jurassic rocks.

Megasequence 1 is intruded at the crest of the basement ridge, forming a dense hybrid domain (Fig. 7b) which also retains a reverse magnetization. The upper part of Megasequence 1, mainly represents the thick mid-Cretaceous (Albian–Cenomanian) package, that was deposited during a long period of normal polarity. Its lack of apparent susceptibility probably reflects a lack of volcanic activity at this time.

The base of the next layer (a combination of Megasequences 2 and 3), is represented at the seaward end by a series of bands of high-amplitude reflectors (Fig. 9b). These are more common towards the base of the lower Campanian to Middle Eocene Megasequence 3 and have been modelled by a single, moderately dense and reversely magnetized layer, which may represent basaltic lava flows formed early in chron 33. Break-up appears to have occurred at about the start of deposition of this layer, with the first slab of oceanic crust being reversely magnetized (Fig. 7b). In this scenario, the initial oceanic tholeitiic lavas could have flowed landward into the depression adjacent to the elevated area along the locus of break-up, in a similar manner to the early stage landward flows of volcanic margins (Symonds *et al.* 1998*b*). In general, the top kilometre of the oceanic crust (>7000 m below sea level (mbsl)) has been given a lower density (2.50 g cm^{-3}) than normal (2.85 g cm^{-3}), to represent fractured, very slow-spreading oceanic crust (Mutter & Karson 1992; Tikku & Cande 1999). The deeper layers are overlain by an upper onlapping wedge of low-density, non-magnetic sediment that represents the Mid-Eocene to Recent GAB Basin deposition (Megasequence 4).

The crustal model (Fig. 7b), whose geometry is constrained by seismic reflection interpretation and densities obtained from sonobuoy velocities (Fig. 7a), depicts a complex zone of transitional crust between the thinned continental crust landward of the basement ridge, and very slow-spreading oceanic crust seaward of the COB. Key features include indications of a highly extended, normally faulted, brittle mantle with significant throws on some faults (up to 6000 m); a COT composed of highly extended continental crust that is increasingly intruded by mafic dykes and sills towards the COB; possible operation of Airy isostasy to support oceanic basement loads within the oceanic lithosphere; evidence for mantle diapirism and upwelling in the form of a mantle bulge, with attendant melting and pluton formation; and significant modification of the lower crust by ductile thinning, fracturing and massive intrusion. Supporting evidence is also found for a significant pulse of basaltic magmatism around the time of break-up, in the form of stacked, reversely magnetized lava flows that have flowed landward over the extended continental crust and sediments of the COT.

Evolutionary model

The schematic evolutionary model in Figure 13 outlines the proposed evolution of the central GAB and development of the COT by progressive extension from Late Jurassic to Campanian time, when break-up and sea-floor spreading commenced.

(1) Initial Late Jurassic–Berriasian, roughly NW–SE-directed upper-crustal extension occurred throughout the region, probably including parts of the COT and adjacent seaward portions of the Recherche and Ceduna sub-basins. Extension was accompanied by thinning of the lower crust–upper mantle in all areas seaward of the Eyre Sub-basin. Thinning of the deep lithosphere produced subsidence and resulted in the accumulation of widespread, relatively thick Early Cretaceous to Cenomanian mudstone-dominated sediments. This phase of extension is not illustrated in Figure 13.

(2) Enhanced mid-Albian–Cenomanian extension involved faulting in the remaining brittle upper crust and pre- and rift-phase sediments, ductile deformation of the lower crust and faulting and/or boudinage of the brittle upper mantle. This second phase of mainly deep (lower-crust–upper-mantle) thinning caused accelerated subsidence and the deposition of marine shales (Fig. 13a). The rapid Cenomanian deposition resulted in gravity-induced deformation and the development of a series of listric growth faults and toe-thrusts that sole out on a mid-Albian décollement (Fig. 13a). Totterdell *et al.* (2000) have suggested that movement on the décollement was purely gravity driven and unrelated to tectonism. However, ductile thinning of the lower crust and the formation of pinch-and-swell features probably increased gradients beneath the outer margin and contributed to the triggering of the growth faulting and associated deformation. This phase of development may have been accompanied by some extension within the sediments above a 'detachment' at the top of the deforming lower crust particularly beneath the outermost part of the margin.

(3) A final Turonian pulse of extension in the lower crust–upper mantle led to break-up beneath the outer part of the margin. This extension event allowed the ductile mantle to rise in places, particularly beneath the western part of the central GAB, perhaps controlled by major faulting and thinning in the overlying brittle and ductile crust (Fig. 13b). The extension was accompanied by decompression melting and intrusion of gabbroic melts above the rising mantle 'diapir', mobilization of serpentinized peridotite, and uplift and deformation of the overlying crust and sedimentary sequences (Fig. 13b). This phase of extension produced the basement ridge that marks the landward edge of the COT and probably caused a final episode of anomalously high subsidence just before break-up. During this extreme thinning, the ductile lower crust acted as a 'décollement' zone between the brittle upper crust and mantle (Brun & Beslier 1996). By the end of thinning, ductile mantle was juxtaposed against the remnants of the ductile lower crust.

(4) The progressive localization of thinning, intrusion into the lower crust, decompression melting and dyke injection within the developing COT resulted in complete continental break-up in early Campanian time and the creation of new oceanic crust by very slow spreading (Fig. 13c). Even in the relatively cold, very slow-spreading setting of the GAB, break-up was associated with a significant outpouring of lava that flowed away from the locus of break-up and over the adjacent COT. As very slow spreading continued, it was associated with faulting and rotation of the newly formed oceanic crust and its overlying sediments.

Discussion

Basement ridge models

It is suggested that in the GAB the major part of the basement ridge complex that forms the inboard margin of the COT, where it is associated with the prominent east–west FAA trend, may be the result of upwelling and alteration of ductile mantle. We also suggest that the upwelled mantle is associated with irregular high-relief bodies (velocities of 5.2, 6.2 km s^{-1}) that form the upper part of the basement ridge (Figs 7b and 10). These bodies may have formed by the intrusion and extrusion of melts derived from decompression melting in the zone of mantle upwelling during a final Turonian episode of crustal extension leading to break-up. An earlier hypothesis, that the ridge may be a metamorphic core complex (Stagg & Willcox 1992), is considered unlikely because of: a lack of seismic evidence for detachment faults bounding the ridge complex; the observation that its deep flanks merge with a reflection marking the top on altered mantle; and its high refraction velocities. These, together with the results of the potential field modelling, lead us to favour a composite igneous-mantle origin for the formation of the main part of the ridge.

The recognition of a magnetic anomaly, previously described as sea-floor spreading anomaly 34, along nearly the entire length of the GAB and conjugate Antarctica margin (Tikku & Cande 1999), and its association with the basement in the central GAB, has implications for any interpreted magnetic sources. Considering all available data, we favour serpentinization of mantle peridotites on a large scale to explain the broad form of this anomaly, rather than local alteration of the lower crust, or local emplacement of a metamorphic core complex. Shorter-wavelength magnetic features superimposed on this broad anomaly appear to be related to the intrusion and extrusion of magma formed by decompression melting in the zone of mantle upwelling. The location of the ridge was possibly controlled by a major, pre-existing upper-crustal fault that may have been responsible for significant relief within the original extended terrane.

Further east, where the more poorly defined basement ridge complex is not associated with a gravity high, mantle upwelling appears to have played no significant role in the formation of the ridge. In this area, it is likely that the ridge complex is composed largely of altered and deformed lower continental crust, with magmatic products present in places. On the edge of the study area to the east of the COB offset (Fig. 1) low-angle, southward-dipping normal faults bounding tilt blocks could be interpreted in terms of a detachment-controlled extensional model. However, although the structural geometries indicate a distinctly different tectonic regime to that further west, they do not provide convincing evidence for a classic core complex origin. Rather, they may indicate exposure of deeper levels of the crust by detachment-type faults in a similar manner to that suggested for the West Iberia margin (Krawczyk et al. 1996). However, these low-angle faults could also be readily explained as brittle fractures connecting to shear zones in the ductile lower crust as proposed in the four-layer lithospheric model of Brun & Beslier (1996). Regardless of which model is favoured, this style of extensional tectonism may have preceded and controlled the location of the mantle upwelling, as well as contributed to the rotation and deformation of blocks within the COT.

Synopsis of rifting and break-up models

The results of this study are broadly consistent with the hypothesis that the central GAB formed in a similar way to the North Atlantic Iberian non-volcanic rifted margin; that is, via a largely pure-shear mechanism at the lithosphere scale (Brun & Beslier 1996). The spatially coincident gravity and magnetic high associated with the basement ridge complex, crustal structures constrained by modelled gravity, magnetic and seismic refraction data, and features within the seismic reflection dataset all support the interpretation. This outcome contrasts with earlier extensional models proposed for this margin, which inferred the operation of upper-crustal detachment systems linked to zones exhibiting various degrees of sub-detachment pure shear (Etheridge et al. 1989; Lister et al. 1991). We find no substantial evidence for major upper-crustal detachment systems, and no evidence at all for the existence of large lithosphere-penetrating shear zones demanded by some earlier models (e.g. Wernicke 1985) beneath the seaward part of the GAB extensional terrane.

Instead, the data indicate that extension and rifting have resulted in a partitioning of deformation within both the crust and mantle. The upper crust, before break-up (Megasequence 1 and underlying pre-rift rocks), shows some brittle response, with intense normal and reverse faulting on closely spaced planar to listric faults (Figs 7a and 13a). The brittle behaviour of the upper crust is not confined to the continental crust. It is also characteristic of the adjacent pre-Early Eocene (chron 22) very slow-spreading oceanic crust (Cande & Mutter 1982; Tikku & Cande 1999), where extensional faulting produced rotated blocks of basaltic crust and synrift relationships in the overlying sedimentary cover. After that time, increased magma production, possibly related to a hotter asthenosphere, created more typical, largely unstructured, oceanic basement. A link between deformation of the extended continental crust and oceanic crust is also implied by the apparent offset of the basement ridge, the COT and linear magnetic anomalies at *c.* 132°E (Fig. 1). This offset may represent a transfer or accommodation zone in continental crust linked to a transform fault in oceanic crust.

The brittle rheological behaviour described above stands in sharp contrast to that of the lower continental crust, which contains very few penetrative faults. North of the basement ridge, the gross geometry of the interpreted lower crust has a pinch-and-swell character, suggestive of extension involving a ductile rheology. A ductile rheology is also consistent with the interpretation of near-complete rupture of lower crust above the basement ridge. Beneath the ductile lower crust under the lower slope, the uppermost mantle has responded to extension through brittle faulting producing large offsets of the upper mantle (Figs 7 and 11). In places, some of the upper-mantle faults may have penetrated into the more ductile lower crust and influenced the geometry of the pinch-and-swell structures (Fig. 7a). These pinch-and-swell structures have in turn influenced the style and location of late-stage deformation in the overlying Jurassic to Cenomanian rift section.

In addition, the behaviour of the brittle mantle beneath oceanic crust appears consistent with the operation of Airy isostasy to support upper-crustal loads. This process has not been sufficiently investigated in many models, which have tended to focus more on thermal effects, and may provide grounds for further study.

Beneath the inner part of much of the COT, a different mantle rheology appears to have operated, with a mantle bulge rising to shallow depths. The ascent and depressurization of this hotter material under the extended and thinned continental crust resulted in melt formation and the subsequent intrusion of gabbroic bodies under the COT, as well as possible associated extrusive units (Figs 7b and 13b). Geochemical and petrological results provide evidence for a similar model to explain the exposure of mantle rocks and associated partial melting products in the Diamantina Zone about 1000–1500 km to

the west (Chatin *et al.* 1998). In places, particularly near the COB of the central GAB, a significant episode of intrusion has led to modification of the geophysical character of the ?Proterozoic to Mesozoic pre- and synrift sediments and the lower crust. Focusing of extension within this zone, together with intense faulting, resulted in partial unroofing and alteration of the mantle and the lateral squeezing of highly magnetic serpentinized peridotite out along the flattened planes of listric faults at the crust–mantle boundary (Fig. 13b). This combination of processes, and the associated production of normally magnetized rocks during Cenomanian time, has given rise to a magnetic anomaly at chron 34 time within extended continental crust that is indistinguishable from a normal sea-floor spreading magnetic anomaly.

The combined evidence provides strong support for a four-layer lithosphere model of the type postulated to explain the extensional style and mantle exhumation processes of the Iberia non-volcanic margin by Brun & Beslier (1996). To our knowledge, the evidence from our new deep-seismic data supporting ductile thinning of the lower crust and faulting of the brittle upper mantle, has not been available on other rifted non-volcanic margins. Brun & Beslier (1996) suggested that this type of model, which gives rise to complex strain patterns, layer-parallel shearing of the ductile lower crust, boudinage of the brittle upper mantle and the formation of asymmetric fault patterns, and diapiric upwelling of the deeper ductile mantle is most likely to occur in an environment of pure shear operating at a lithosphere scale.

Thus, the GAB–Antarctic conjugate margins may more appropriately be considered as a product of lithospheric-scale pure shear rather than of a composite model involving simple-shear extension of the upper crust on detachment faults above zones of distributed pure shear (Etheridge *et al.* 1989; Lister *et al.* 1991). Our study has found no evidence of either lithosphere-penetrating detachments or major upper-crustal detachments beneath the COT. It could be reasonably argued that such detachments would be very difficult to image seismically in the complex COT zone. In Figure 13a we schematically illustrate a possible shear zone at the top of the lower crust that could be interpreted as an extensional detachment in the sense of Lister *et al.* (1991), and in the easternmost part of the study area detachment-type faults may be present. It should be pointed out that the mid-crustal detachment models previously proposed to explain the extensional development of the southern Australian margin (Etheridge *et al.*

1989) would produce brittle upper-crustal faulting and deep, pure-shear thinning of the lower crust and upper mantle not that dissimilar in broad terms to the Brun & Beslier (1996) four-layer model, except for the lack of faulting in the brittle upper mantle.

The removal and thinning of a large portion of the upper crust beneath the outer GAB margin during Late Jurassic–Early Cretaceous and mid-Cretaceous extensional episodes is difficult to explain without resorting to mid-crustal detachment models. It seems likely that different styles of extension may have operated at various stages during the development of the non-volcanic central GAB margin: mid-crustal detachment faulting with some sub-detachment pure shear during the earlier intra-continental rift phase (in Late Jurassic–Early Cretaceous time); followed by lithosphere-scale pure-shear extension (in mid-Cretaceous time), involving substantial deep (lower-crust–upper-mantle) thinning leading to continental break-up. This change in extensional style may reflect progressive heating of the lithosphere leading to break-up.

Conclusions

This study indicates that previously identified sea-floor spreading magnetic anomaly 34 (Cande & Mutter 1982; Tikku & Cande 1999) lies over extended continental crust, an important outcome having both global and regional implications. It indicates that the use of magnetic data alone to define the onset of sea-floor spreading and the location of the COB may not always be a simple matter and needs to be approached with caution. Linear magnetic anomalies of appropriate form and age can be generated by a variety of processes that commonly occur in and adjacent to the COT. Without supporting data on the nature of the crust being investigated, the COB can be misidentified.

The interpretation that anomaly 34 previously identified in the central GAB is not the result of sea-floor spreading requires a re-evaluation of the break-up history of the region. The result has important implications for the thermal and subsidence history, and palaeogeography of the margin. Early chron 33 (*c.* 83 Ma) now corresponds to the first unambiguous emplacement of oceanic crust by sea-floor spreading, and defines true continental break-up as occurring in early Campanian time (much the same time as the commencement of sea-floor spreading in the Tasman Basin to the east) and some 15 Ma later than previously proposed. This result also

throws some doubt on the identification of magnetic anomaly 34 on the conjugate Antarctic margin (Tikku & Cande 1999).

Evidence from the GAB margin indicates that, even on relatively cold non-volcanic margins, a significant volume of lava can erupt at the time of break-up and flow landward over the adjacent COT. This is important for understanding and imaging COTs, because it is plausible that with slightly elevated asthenospheric temperatures and increased melt production, the volume of lavas generated could mask extended continental crust within the COT. That is, on some margins, highly extended continental basement blocks and associated synrift sediments could lie beneath basaltic lavas, complicating the identification of sea-floor spreading magnetic anomalies and the COB, definition of the COT and the interpretation of potential field and seismic data.

The main conclusions of this paper are underpinned by deductions made from a seismic reflection dataset that clearly images the COT in the central GAB and thus allows an understanding of events immediately preceding break-up. In this area, a clearly defined COB lies at the seaward edge of a COT that is composed of highly deformed and intruded extended continental crust. The data and interpretations are well explained by a four-layer extensional model, similar to that of Brun & Beslier (1996), that includes a brittle upper crust and upper mantle and a ductile lower crust and lower lithospheric mantle. Understanding the process of break-up using this model involves a consideration of the significance of extension in each layer during the various extensional events leading up to break-up, as well as the contribution from a hot rising ductile mantle and associated melts.

We thank AGSO's Marine Operations Group and the AMSA crew of R.V. *Rig Seismic* for their professionalism in carrying out the surveys on which this study is based. The authors also wish to thank L. Emerton and B. Hack for helping with the illustrations; and members of AGSO's Southern Margins Frontiers project, in particular J. Totterdell, H. Struckmeyer and D. Scott, and members of AGSO's Law of the Sea project, for many valuable discussions. D. Scott, in particular, provided an important balance in considering alternative models for the evolution of the region. We thank R. Parums for assisting with the seismic refraction modelling. We are grateful to B. Willcox and P. McFadden for constructive and expeditious reviewing of early drafts of the manuscript and for enhancing its quality. We particularly thank R. B. Whitmarsh, T. Minshull and S. Cande for useful reviews, and editors of the volume for their patience and forbearance in finalizing and submitting the manuscript. This paper is published with the permission of the Chief Executive Officer, Australian Geological Survey Organisation, Canberra, ACT, Australia.

References

BESLIER, M.-O., CORNEN, G. & GIRARDEAU, J. 1994. Petro-structural evolution of basement rocks drilled at the ocean–continent transition of the Iberia abyssal plain passive margin during the ODP Leg 149. *EOS Transactions, American Geophysical Union*, 44, **75**, 318.

BRUN, J.P. & BESLIER, M.-O. 1996. Mantle exhumation at passive margins. *Earth and Planetary Science Letters*, **142**, 161–173.

CANDE, S.C. & MUTTER, J.C. 1982. A revised identification of the oldest sea-floor spreading anomalies between Australia and Antarctica. *Earth and Planetary Science Letters*, **58**, 151–160.

CANNAT, M. 1993. Emplacement of mantle rocks in the seafloor at mid-ocean ridges. *Journal of Geophysical Research*, **98**, 4163–4172.

CHATIN, F., ROBERT, U., MONTIGNY, R. & WHITE-CHURCH, H. 1998. The Diamantina rift zone (eastern Indian Ocean): petrological and geochemical approach. *Comptes Rendus de l'Académie des Sciences, Science de la Terre et des Planètes*, **326**, 839–845.

CHIAN, D., LOUDEN, K.E., MINSHULL, T.A. & WHITMARSH, R.B. 1999. Deep structure of the ocean–continent transition in the southern Iberia Abyssal Plain from seismic refraction profiles: Ocean Drilling Program (Legs 149 and 173) transect. *Journal of Geophysical Research*, B4, **104**, 7443–7462.

CHRISTENSEN, N.I. & MOONEY, W.D. 1995. Seismic velocity structure and composition of the continental crust: a global view. *Journal of Geophysical Research*, B7, **100**, 9761–9788.

DEAN, S.M., MINSHULL, T.A., WHITMARSH, R.B. & LOUDEN, K.E. 2000. Deep structure of the ocean–continent transition in the southern Iberia Abyssal Plain from seismic refraction profiles: the IAM-9 transect at 40°20′N. *Journal of Geophysical Research*, B3, **105**, 5859–5885.

Discovery 215 Working Group 1998. Deep structure in the vicinity of the ocean–continent transition zone under the southern Iberia Abyssal Plain. *Geology*, **26**, 8, 743–746.

Encom Technology *ModelVision LE/SE/AutoMag: the 3d Workbench for Magnetics and Gravity Interpretation*. Encom Technology, Sydney, NSW.

ELLIOT, D.H. 1992. Jurassic magmatism and tectonism associated with Gondwanaland break-up: an Antarctic perspective. *In*: STOREY, B.C., ALABASTER, T. & PARKHURST, R.J. (eds) *Magmatism and the Causes of Continental Break-up*. Geological Society, London, Special Publications, **68**, 165–184.

ETHERIDGE, M.A., LISTER, G.S. & SYMONDS, P.A. 1989. Application of the detachment model to reconstruction of conjugate passive margins. *In*: TANKARD, A.J. & BALKWILL, H.R. (eds) *Extensional Tectonics and Stratigraphy of the North American Margins*. Memoirs, American Association of Petroleum Geologists, **46**, 23–40.

HERGT, J.M., CHAPPELL, B.W., FAURE, G. & MENSING, T.M. 1989. The geochemistry of Jurassic dolerites from Portal Peak Antarctica. *Contributions to Mineralogy and Petrology*, **102**, 298–305.

KONIG, M. & TALWANI, M. 1977. A geophysical study of the southern continental margin of Australia: Great Australian Bight and western sections. *Geological Society of America Bulletin*, **88**, 1000–1014.

KRAWCZYK, C.M., RESTON, T.J., BESLIER, M.-O. & BOILLOT, G. 1996. Evidence for detachment tectonics on the Iberia Abyssal Plain rifted margin. *In*: WHITMARSH, R.B., SAWYER, D.S., KLAUS, A. & MASSON, D.G. (eds) *Proceedings of the Ocean Drilling Program, Scientific Results, 149*. 603–615. Ocean Drilling Program, College Station, TX.

LE PICHON, X. & BARBIER, F. 1987. Passive margin formation by low-angle faulting within the upper crust: the northern Bay of Biscay margin. *Tectonics*, **6**, 133–150.

LE PICHON, X. & SIBUET, J.C. 1981. Passive margins: a model of formation. *Journal of Geophysical Research*, B5, **86**, 3708–3729.

LISTER, G.S., ETHERIDGE, M.A. & SYMONDS, P.A. 1986. Detachment faulting and the evolution of passive continental margins. *Geology*, **14**, 246–250.

LISTER, G.S., ETHERIDGE, M.A. & SYMONDS, P.A. 1991. Detachment models for the formation of passive continental margins. *Tectonics*, 5, **10**, 1038–1064.

MITCHELL, M.M., DUDDY, I.R. & O'SULLIVAN, P.B. 1997. Reappraisal of the age and origin of the Casterton Formation, western Otway Basin, Victoria. *Australian Journal of Earth Sciences*, **44**, 819–830.

MUNSCHY, M. 1998. The Diamantina zone as the result of rifting between Australia and Antarctica: geophysical constraints. *Comptes Rendus de l'Académie des Sciences, Science de la Terre et des Planètes*, **327**, 533–540.

MUTTER, J.C. & KARSON, J.A. 1992. Structural processes at slow-spreading ridges. *Science*, **257**, 627–634.

PETKOVIC, P., BRETT, J., MORSE, M.P., HATCH, L., WEBSTER, M.A. & ROCHE, P. *Gravity, Magnetic and Bathymetry Grids from Levelled Data for Southwest Australia–Great Australian Bight*. Australian Geological Survey Organisation, Record, **1999/48**.

POAG, W.C. & DE GRACIANSKY, P.C. 1992. Geologic evolution of continental rises: a circum-Atlantic perspective. *In*: POAGEE, W.C. & DE GRACIANSKY, P.C. (eds) *Geologic Evolution of*

Atlantic Continental Rises. 347–368. Van Nostrand Reinhold, New York.

REID, I. 1994. Crustal structure of a nonvolcanic rifted margin east of Newfoundland. *Journal of Geophysical Research*, **99**, 15161–15180.

ROYDEN, L. & KEEN, C.E. 1980. Rifting process and thermal evolution of the continental margin of eastern Canada determined from subsidence curves. *Earth and Planetary Science Letters*, **51**, 343–361.

SANDWELL, D.T. & SMITH, W.H.F. 1997. Marine gravity from Geosat and ERS-1 altimetry. *Journal of Geophysical Research*, **102**, 10039–10054.

Shipboard Scientific Party, 2000. Leg 182 summary. *In*: FEARY, D.A., HINE, A.C., MALONE, M.J., *et al.* (eds) *Proceedings of the Ocean Drilling Program, Initial Reports, 182*: College Station, TX, Ocean Drilling Program, 1–58.

SKELTON, A.D.L. & VALLEY, J.W. 2000. The relative timing of serpentinization and mantle exhumation at the ocean–continent transition, Iberia: constraints from oxygen isotopes. *Earth and Planetary Science Letters*, 3–4, **178**, 327–338.

SMITH, W.H.F. & SANDWELL, D.T. 1997. Global seafloor topography from satellite altimetry and ship depth soundings. *Science*, **277**, 1956–1962.

STAGG, H.M.J. & WILLCOX, J.B. 1992. A case for Australia–Antarctica separation in the Neocomian (*ca* 125 Ma). *Tectonophysics*, **210**, 21–32.

STAGG, H.M.J., WILLCOX, J.B., NEEDHAM, D.J.L. & 5 OTHERS 1990. Basins of the Great Australian Bight region: geology and petroleum potential. *Continental Margins Program Folio, 5*. Bureau of Mineral Resources, Geology & Geophysics and Department of Mines & Energy, South Australia, Canberra, A.C.T., 1–143.

SYMONDS, P.A., MURPHY, B., RAMSAY, D.C., LOCKWOOD, K.L. & BORISSOVA, I. 1998*a*. The outer limits of Australia's resource jurisdiction off Western Australia. *In*: PURCELL, P.G. & PURCELL, R.R. (eds) *The Sedimentary Basins of Western Australia 2. Proceedings of Petroleum Exploration Society of Australia Symposium, Perth, WA, 1998.* 3–19.

SYMONDS, P.A., PLANKE, S., FREY, O. & SKOGSEID, J. 1998*b*. Volcanic evolution of the western Australian continental margin and its implications for basin development. *In*: PURCELL, P.G. & PURCELL, R.R. (eds) *The Sedimentary Basins of Western Australia 2. Proceedings of Petroleum Exploration Society of Australia Symposium, Perth, WA, 1998*, 33–54.

TALWANI, M., MUTTER, M., HOUTZ, R. & KONIG, M. 1978. The crustal structure and evolution of the area underlying the magnetic quiet zone on the margin south of Australia. *In*: WATKINS, J., MONTADERT, L. & DICKENSON, P. (eds) *Geological and Geophysical Investigations of Continental Margins*. Memoirs, American Association of Petroleum Geologists, **29**, 151–175.

TIKKU, A.A. & CANDE, S.C. 1999. The oldest magnetic anomalies in the Australian–Antarctic Basin: Are they isochrons? *Journal of Geophysical Research*, B1, **104**, 661–677.

TOTTERDELL, J.M., BLEVIN, J.E., STRUCKMEYER, H.I.M., BRADSHAW, B.E., COLWELL, J.B. & KENNARD, J.M. 2000. A new sequence framework for the Great Australian Bight: starting with a clean slate. *Australian Petroleum Production and Exploration Association Journal*, **40**, Part I, 95–117.

VEEVERS, J.J. 1986. Breakup of Australia and Antarctica estimated as mid-Cretaceous (95 ± 5 Ma) from magnetic and seismic data at the continental margin. *Earth and Planetary Science Letters*, **77**, 91–99.

VEEVERS, J.J., STAGG, H.M.J., WILLCOX, J.B. & DAVIES, H.L. 1990. Pattern of slow seafloor spreading (<4 mm/year) from breakup (96 Ma) to A20 (44.5 Ma) off the southern margin of Australia. *Bureau of Mineral Resources Journal of Australian Geology & Geophysics*, **11**, 499–507.

WEISSEL, J.K. & HAYES, D.E. 1972. Magnetic anomalies in the southeast Indian Ocean. *In*: HAYES, D.E. (ed.) *Antarctic Oceanology II. The Australian–New Zealand Sector*. Antarctic Research Series, American Geophysical Union, **19**, 165–196.

WERNICKE, B. 1985. Uniform-sense normal simple shear of the continental lithosphere. *Canadian Journal of Earth Sciences*, **22**, 108–125.

WHITMARSH, R.B. & MILES, P.R. 1995. Models of the development of the West Iberia rifted continental margin at 40°30′N deduced from surface and deep-tow magnetic anomalies. *Journal of Geophysical Research*, B3, **100**, 3789–3806.

WHITMARSH, R.B. & SAWYER, D.S. 1996. The ocean/continent transition beneath the Iberia Abyssal Plain and continental-rifting to seafloor-spreading processes. *In*: WHITMARSH, R.B., SAWYER, D.S., KLAUS, A. & MASSON, D.G. (eds) *Proceedings of the Ocean Drilling Program, Scientific Results, 149*. 713–733. Ocean Drilling Program, College Station, TX.

WHITMARSH, R.B., BESLIER, M.-O., WALLACE, P.J. *et al.* (eds) 1998. *Proceedings of the Ocean Drilling Program, Initial Reports, 173*. Ocean Drilling Program, College Station, TX.

WHITMARSH, R.B., MINSHULL, T.A., RUSSELL, S.M., DEAN, S.M., LOUDEN, K.E. & CHIAN, D. 2001. The role of syn-rift magmatism in the rift-to-drift evolution of the West Iberia continental margin: geophysical observations. *In*: WILSON, R.C.L., WHITMARSH, R.B., TAYLOR, B. & FROITZHEIM, N. (eds) *Non-volcanic Rifting of Continental Margins: a Comparison of Evidence from Land and Sea*. Geological Society, London, Special Publications, **187**, 107–124.

WHITMARSH, R.B., PINHEIRO, L.M., MILES, P.R., RECQ, M. & SIBUET, J.C. 1993. Thin crust at the western Iberia ocean–continent transition and ophiolites. *Tectonics*, **12**, 1230–1239.

WHITMARSH, R.B., SAWYER, D.S., KLAUS, A. & MASSON, D.G. (eds) 1996. *Proceedings of the Ocean Drilling Program, Scientific Results, 149.* Ocean Drilling Program, College Station, TX.

WILLCOX, J.B. & STAGG, H.M.J. 1990. Australia's southern margin: a product of oblique extension. *Tectonophysics*, **173**, 269–281.

YOUNG, G.C. & LAURIE, J.R. (eds) 1996. *An Australian Phanerozoic Timescale.* Oxford University Press, Melbourne.

Development of the continental margins of the Labrador Sea: a review

JAMES A. CHALMERS & T.C.R. PULVERTAFT

Geological Survey of Denmark and Greenland, Thoravej 8, DK-2400 Copenhagen NV, Denmark (e-mail:jac@geus.dk)

Abstract: The Labrador Sea is a small oceanic basin that developed when the North American and Greenland plates separated. An initial period of stretching in Early Cretaceous time formed sedimentary basins now preserved under the continental shelves and around the margins of the oceanic crust. The basins subsided thermally during Late Cretaceous time and a second episode of tectonism took place during latest Cretaceous and early Paleocene time, before the onset of sea-floor spreading in mid-Paleocene time. Around the northern Labrador Sea, Davis Strait and in southern Baffin Bay, voluminous picrites and basalts were erupted at and shortly after the commencement of sea-floor spreading. Volcanism occurred again in early Eocene time at the same time as sea-floor spreading commenced in the northern North Atlantic. Farther southeast, along the Labrador and southern West Greenland margins, oceanic crust is separated from continental crust by highly stretched but non-magmatic transition zones which developed before sea-floor spreading. A complex transform zone, which developed during sea-floor spreading in late Paleocene and early Eocene time, separates continental and oceanic crust along the Baffin Island margin. The Greenland and Labrador ocean–continent transitions are asymmetric across the only available conjugate cross-sections. However, a cross-section through the Labrador margin farther north resembles the Greenland cross-section in the conjugate pair more than it does the Labrador cross-section of this pair. Consideration of the geological history of the area suggests that the non-magmatic transition zones may have formed by slow extension of a few millimetres per year through a period of 53 Ma during Cretaceous and early Paleocene time.

The Labrador Sea is a small oceanic basin about 900 km wide that opens to the southeast into the North Atlantic Ocean. To the north it shallows and passes into the Davis Strait, a seaway of 300 km width leading into Baffin Bay (Fig. 1). The Labrador Sea is flanked by typical Arctic continental shelves with banks less than 200 m deep separated by glacially eroded channels: the Labrador shelf to the southwest, the southeast Baffin Island shelf to the northwest and the southern West Greenland shelf to the northeast.

The geological history of the area is best known within the sedimentary basins under and adjacent to the continental shelves, where exploration for hydrocarbons took place off southern West Greenland during the 1970s and 1990s and offshore Labrador during the 1970s and 1980s. Over the Labrador shelf, extensive seismic surveys were carried out and 25 wells were completed in the period 1971–1983. Six discoveries of gas were made, but none of them has been exploited. Three wells, one of which recovered condensate, were drilled on the southeast Baffin Island shelf during the early 1980s. Balkwill (1987) summarized the knowledge of the geology of the Labrador and southeast Baffin Island shelves that resulted from that phase of exploration, and a summary of the maps of Balkwill (1987) is included in Figure 2.

Off Greenland some 37 000 km of multi-channel seismic data were acquired between 1970 and 1978 and five wells were drilled, all of which were declared dry. The early results from the Greenland shelf were described by Manderscheid (1980) and Henderson *et al.* (1981), who concluded that the sedimentary basins off Greenland consist of uniformly westward-dipping sediments bordered by basement ridges along what Srivastava (1978) interpreted as the continent–ocean boundary (COB). The hydrocarbon potential of these ridges was tested by two of the five wells, and the other three drilled into either Paleogene volcanic rocks or relatively shallow continental basement (Rolle 1985).

From: WILSON, R.C.L., WHITMARSH, R.B., TAYLOR, B. & FROITZHEIM, N. 2001. *Non-Volcanic Rifting of Continental Margins: A Comparison of Evidence from Land and Sea*. Geological Society, London, Special Publications, **187**, 77–105. 0305-8719/01/$15.00 © The Geological Society of London 2001.

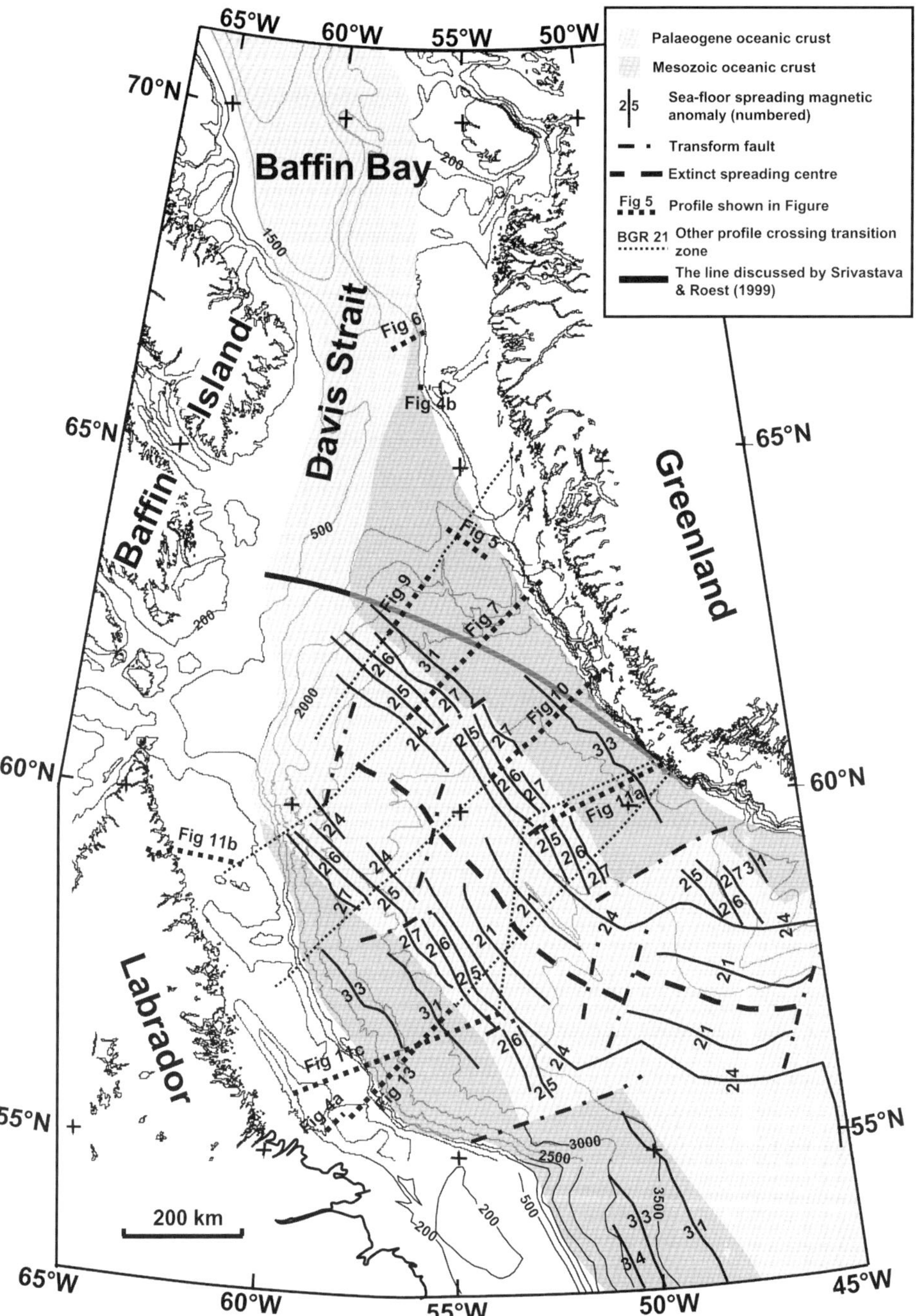

Fig. 1. Bathymetry of the Labrador Sea and the extent of Mesozoic and Cenozoic oceanic crust according to Wheeler *et al.* (1996). Sea-floor spreading anomalies are taken from Roest & Srivastava (1989) and were also published by Bell (1989). The locations of profiles shown in other figures are marked.

A reassessment of the hydrocarbon potential of southern West Greenland was carried out by Chalmers & Pulvertaft (1993), who concluded that all of the prospects drilled on the Greenland shelf were in some way defective and that the basin was more structured than previously realized (Chalmers 1989). These results prompted the Danish and Greenland authorities to fund acquisition of new multichannel seismic data to reinvestigate the area's hydrocarbon potential. The Geological Survey of Greenland (GGU) and commercial seismic companies have acquired multichannel seismic data in nine of the 10 years since 1990. Exploration for hydro-

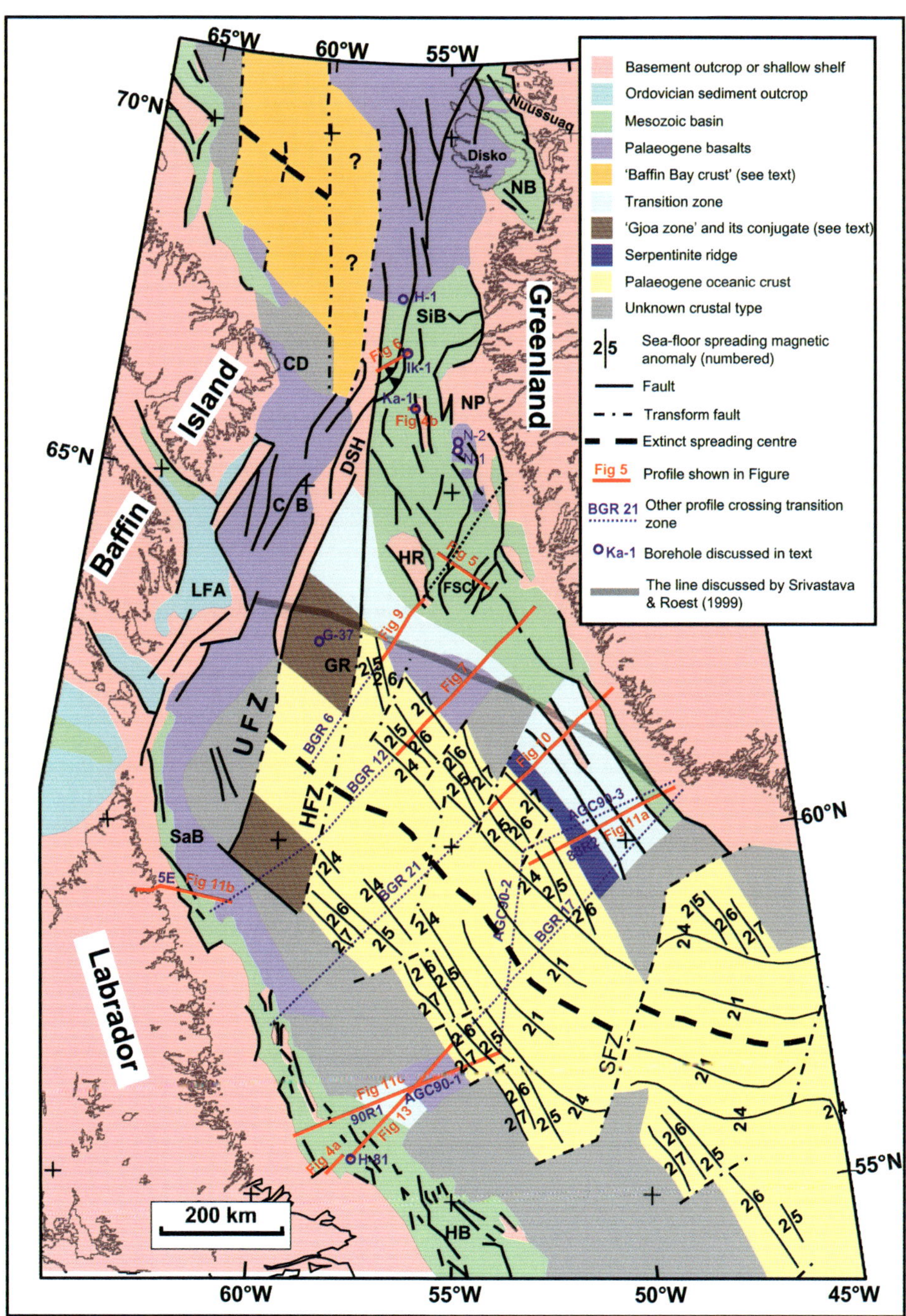

Fig. 2 caption on next page.

carbons, including the drilling of several shallow wells and one deep exploration well, also took place in the Nuussuaq Basin, onshore central West Greenland, during the 1990s. Summaries of the regional geology of the southern West Greenland basins have been presented by Chalmers *et al.* (1993, 1995, 1999), and the most promising new hydrocarbon prospects that have been identified from the new work have been described by Bate *et al.* (1994), Aram (1999) and Isaacson & Neff (1999). A summary geological map showing the currently published interpretations of the sedimentary basins around the Labrador Sea is shown in Figure 2.

The early history of the development of the oceanic crust in the Labrador Sea is a subject of controversy. Srivastava (1978) interpreted the earliest sea-floor spreading magnetic anomaly as 31 (68 Ma, Mesozoic time scale from Gradstein *et al.* (1995)) and oceanic crust extending to the foot of the bathymetric continental slopes. Srivastava (1978) also pointed out that a reorientation of the spreading axis took place during magnetochron 24r (55.9–53.3 Ma, Cenozoic time scale from Berggren *et al.* (1995)), at the same time as sea-floor spreading started between Europe and Greenland. Roest & Srivastava (1989) modified slightly the linear magnetic anomaly map to identify the oldest magnetic anomaly as 33 (79.7–74.5 Ma), and this interpretation is the one that is commonly used in the literature for plate reconstructions of Greenland relative to North America. Srivastava & Roest (1999) further modified their maps by suggesting that sea-floor spreading in the northern Labrador Sea took place via a series of ridge jumps.

Chalmers (1991) and Chalmers & Laursen (1995) tested the Roest & Srivastava (1989) identifications of magnetic anomalies and were unable to model sea-floor spreading anomalies older than 27n. Chalmers (1991) and Chalmers & Laursen (1995) concluded that (1) sea-floor spreading began in the Labrador Sea during magnetochron 27 (61.3–60.9 Ma), and (2) continental crust extends under deep water and is

separated from oceanic crust by some kind of transition zone whose character and width are different in different parts of the Labrador Sea. There is general agreement that the original map of Srivastava (1978) is on the whole correct for anomalies 27n and younger.

The transition zone has been investigated by deep reflection seismic lines (Keen *et al.* 1994*b*) and by wide-angle reflection–refraction lines (Chian & Louden 1994; Chian *et al.* 1995*b*), which indicate the presence of serpentinized mantle peridotites at shallow depth, either directly under sediments or below continental crust 1–3 km thick, or (Srivastava & Roest 1995) under basaltic oceanic crust 1 km thick. Whatever its nature, the thin crust of the transition zone passes northwestwards into a volcanic margin where sea-floor spreading during Paleocene time appears to have been subaerial (Chalmers & Laursen 1995).

The present paper is a short review of the current understanding of the development of the area. An attempt is made to present observations critical to the discussion so that they are available from a single source. Only a little of what is published here has not been published before.

Geology of the sedimentary basins

Onshore outcrops

Cretaceous and Paleogene sediments and Paleogene volcanic rocks (Fig. 3) are exposed on the island of Disko and the peninsulas of Nuussuaq and Svartenhuk (69°–72°N) in West Greenland (the Nuussuaq Basin) and near Cape Dyer (67°N) in southeast Baffin Island (Fig. 2) (Burden & Langille 1990; Chalmers *et al.* 1999).

The Cretaceous sediments in the Nuussuaq Basin were deposited in a delta system that fanned out to the west and northwest from a point east of Disko island. Southeast and east of the outcrop area these consist of fluvial–deltaic sandstones, mudstones and coal seams of Albian–early Campanian age, which constitute the Atane Formation. The fluvio-deltaic sedi-

---

Fig. 2. Geological map over the same area as for Figure 1, but showing only Cenozoic ocean crust after Chalmers (1991); Chalmers & Laursen (1995). Other data from Balkwill (1987), Jackson *et al.* (1992), Chalmers *et al.* (1993, 1995, 1999), Wheeler *et al.* (1996), Whittaker *et al.* (1997) and Skaarup *et al.* (2000). The locations of profiles shown in other figures are marked in red and the locations of other seismic reflection lines that cross from continental to oceanic crust are shown in blue. CB, Cumberland Basin; CD, Cape Dyer; DSH, Davis Strait High; FSC, Fylla Structural Complex; GR, Gjoa Rise; G-37, Gjoa G-37 borehole; HB, Hopedale Basin; HFZ, Hudson Fracture Zone; HR, Hecla Rise; H-1, Hellefisk-1 borehole; Ik-1, Ikermiut-1 borehole; Ka-1, Kangâmiut-1 borehole; N-1 and N-2, Nukik-1 and Nukik-2 boreholes; LFA, Lady Franklin Arch; NP, Nukik Platform; NB, Nuussuaq Basin; SaB, Saglek Basin; SFZ, Snorri Fracture Zone; SiB, Sisimiut Basin; UFZ, Ungava Fault Zone.

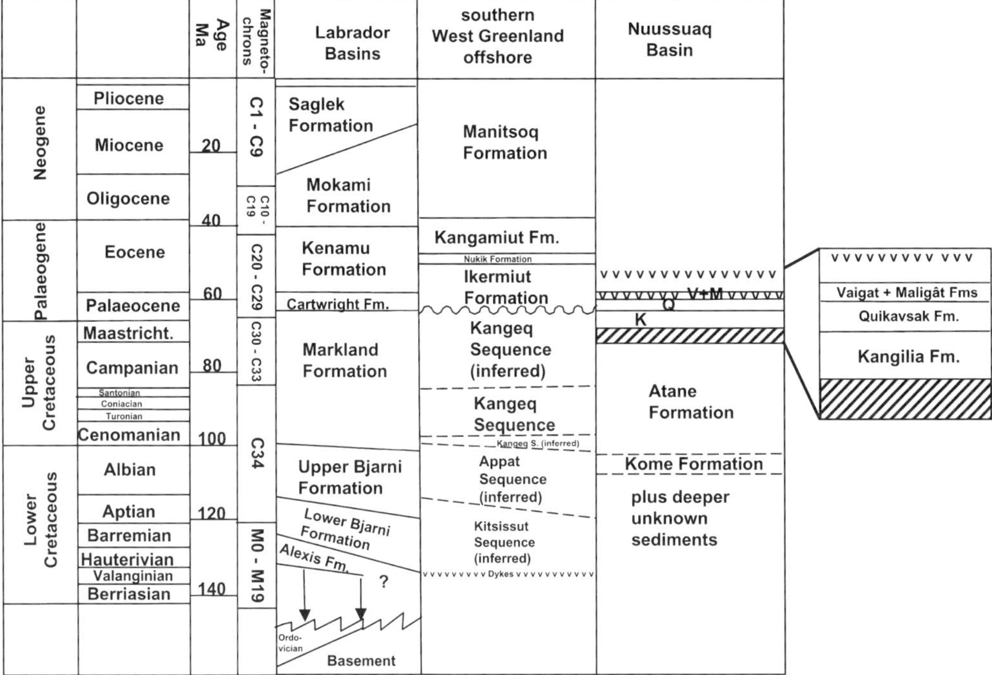

Fig. 3. Stratigraphic columns from the Labrador (Hopedale and Saglek) Basins, southern West Greenland basins offshore and the Nuussuaq Basin onshore central West Greenland (see Fig. 2). The time scales are taken from Berggren *et al.* (1995) for the Cenozoic units and from Gradstein *et al.* (1995) for the Mesozoic units. Stratigraphy redrawn from Balkwill (1987) for the Labrador Basins, Rolle (1985) for the Cenozoic units and Chalmers *et al.* (1993) for the Mesozoic units of the southern West Greenland basins offshore, and Dam *et al.* (1999) for the Nuussuaq Basin.

ments pass northwestwards into marine mudstones alternating with turbidite channel sandstones of up to 50 m thickness that were deposited on a submarine slope, the Itilli succession (Dam & Sønderholm 1994). The present-day eastern boundary of Cretaceous outcrops is a system of major faults. In northern Upernivik Ø (an island just north of the northern limit of Fig. 2), the occurrence of very coarse conglomerate in Cenomanian sediments adjacent to one of the faults indicates active faulting at that time (Rosenkrantz & Pulvertaft 1969), but farther south, Upper Albian–Cenomanian fluvial sands and shales are cut off by the boundary fault without any evidence of the proximity of synsedimentary fault scarps, suggesting that fault movement here was of post-Cenomanian age (Pulvertaft 1979, 1989). The oldest Cretaceous sediments are exposed on the north side of Nuussuaq near the boundary fault system. These are of Late Albian age (Nøhr-Hansen, pers. comm.) and consist of conglomerates, sandstones, heteroliths and mudstones of fluvio-deltaic origin that are overlain by inner shelf mudstones; together these sediments constitute the Kome Formation (Midtgaard 1996). Midtgaard (1996) suggested that these sediments were deposited in an active half-graben with its bounding fault to the west. A low-angle unconformity separates these sediments from the overlying sediments of the Atane Formation (Midtgaard 1996).

Important faulting started in early-Campanian time and was resumed in mid-Maastrichtian time (Dam & Sønderholm 1994; Dam *et al.* 2000). These movements resulted in both rotation and uplift of the Atane Formation sediments in fault blocks (Chalmers *et al.* 1999), and were followed by the incision of a series of large, deep channels which were filled by transgressive successions of coarse conglomerates and turbidites or by fluvial sandstones (Dam & Sønderholm 1994, 1998; Dam *et al.* 1998). An angular unconformity separates mid-Campanian–Paleocene sediments from the underlying Atane Formation in many parts of the area.

Paleogene volcanism began in the Nuussuaq Basin with the eruption of the high-temperature, plume-related picrites of the Vaigat Formation (Clarke & Pedersen 1976; see also Gill *et al.* 1992; Holm *et al.* 1993; Graham *et al.* 1998), which is generally <2 km thick (Chalmers *et al.* 1999). This episode was followed by the eruption of the feldspar-phyric tholeiites of the Maligât Formation (Hald & Pedersen 1975), at least 2 km thick. Six ^{40}Ar/^{39}Ar age determinations of samples from the Vaigat and Maligât formations yielded good plateau ages between 60.4 ± 0.5 and 59.4 ± 0.5 Ma (Storey *et al.* 1998). These ages are within magnetochron 26r (Berggren *et al.* 1995), but the lowermost lavas of the Vaigat Formation show normal magnetization and must therefore have been erupted during chron 27n (Riisager & Abrahamsen 1999). Conspicuous dolerite sheets intruded into the boundary fault and surrounding rocks (the Tartunaq intrusions) belong to a group of intrusions that have been dated at 54.8 ± 0.4 Ma, which is within magnetochron 24r. At the northwest extremity of Nuussuaq peninsula there are *c.* 2 km of lower Eocene basalts. A comendite tuff within these basalts has yielded an age of 52.5 ± 0.2 Ma corresponding to the transition between magnetochrons 24n and 23r (Berggren *et al.* 1995). There is evidence of even younger volcanism in the area in the form of dykes and necks as young as 27.4 ± 0.6 Ma (Storey *et al.* 1998).

Near Cape Dyer, on eastern Baffin Island, there are fluvial sediments of Late Aptian–Early Cenomanian age that appear to have been deposited during active faulting (Burden & Langille 1990). The sediments are overlain unconformably by lower Paleocene mudstones, arkoses and conglomerates also deposited during an episode of active faulting. Here also sedimentation ended when picritic basalts were erupted (Clarke & Upton 1971).

A swarm of dolerite dykes that run parallel to the coast of southern West Greenland (Watt 1969) have recently been redated at 138–133 Ma (Valanginian, magnetochron M-12) by the ^{40}Ar/^{39}Ar technique (Larsen *et al.* 1999).

Labrador shelf basins

The sediments encountered under the Labrador shelf are predominantly from Early Cretaceous to Recent in age; locally these are underlain by palaeo-erosional remnants of Lower Palaeozoic sediments. Ordovician sediments underlie the Mesozoic–Cenozoic succession over an extensive area off SE Baffin Island (Balkwill 1987) (Fig. 2). The stratigraphy of the sediments is shown in Figure 3 (McWhae *et al.* 1980). They

were described by Balkwill (1987), and the following is a précis of that study.

The oldest Mesozoic rocks in the Labrador basins are more or less altered basalts referred to as the Alexis Formation (Umpleby 1979); K–Ar whole-rock ages determinations on these rocks have yielded ages between 131 and 104 Ma (Late Valanginian–middle Albian time). Basalts of the Alexis Formation were penetrated in seven wells in the southern Hopedale Basin (Balkwill 1987). In the adjacent coastal region of Labrador, the only reported evidence of coeval mafic volcanic activity is a possible mid-Mesozoic lamprophyre dyke in Precambrian rocks of the Makkovik sub-province (King & McMillan 1975). Keen *et al.* (1994*a*) and Williamson *et al.* (1994) showed that Alexis Formation melts were consistent with adiabatic stretching of the lithosphere, but over mantle with the higher than 'normal' temperature of 1400 °C.

Syn-rift sedimentation offshore Labrador began with deposition of the Bjarni Formation, which occupies coast-parallel grabens and half-grabens. The formation is divided into two members: the Lower Bjarni Member consisting of non-marine (probably fluvial) sandstones and fluvio-lacustrine shales, and the Upper Bjarni Member consisting of sandy, clayey and carbonaceous, predominantly non-marine siltstone and shale, with intercalated beds of sandstone, that were deposited in a deltaic to shallow marine environment. The oldest palynomorph assemblages recovered from the Bjarni Formation indicate a Barremian age, but the unsampled lowest beds in the deepest basins may well be older than this. The youngest sediments of the Bjarni Formation are of Late Albian or perhaps Cenomanian age. The unconformity between the lower and upper members (Fig. 4a) is probably of mid-Albian age.

The Markland Formation is a very widespread succession of marine shelf shales of Cenomanian–Danian age. It abruptly but conformably overlies the Bjarni Formation in the central parts of structural depressions, but oversteps this formation onto Precambrian basement, Palaeozoic strata, or Alexis Formation basalts towards the margins of basins. Most of the Markland Formation probably represents sedimentation in a thermally subsiding basin; however, along the inner (SW) side of the Upper Cretaceous basins there are syn-rift, half-graben-confined wedges of intercalated shale, siltstone and sandstone, that thin seawards and grade into the Markland Formation.

The Markland Formation is overlain by the late Paleocene and Eocene Cartwright and

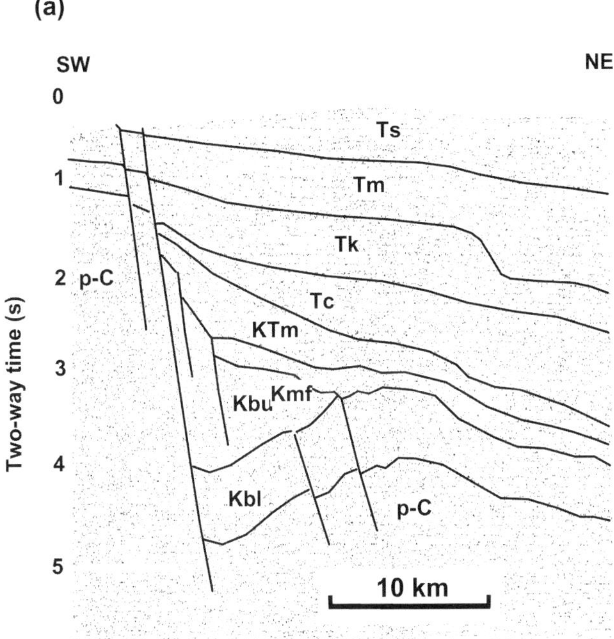

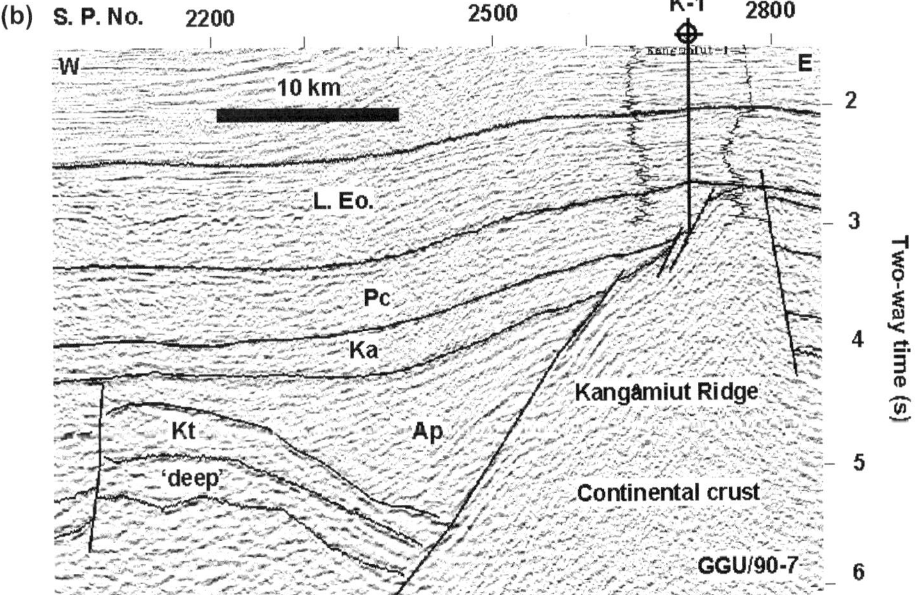

Fig. 4. Caption on next page.

Kenamu formations, both of which are dominated by shales. Along the southwest flank of the basin there are wedges of shallow marine sands that are proximal equivalents of the Cartwright Formation. The Kenamu Formation is in turn overlain unconfomably by the Neogene Mokami and Saglek Formations and glacial deposits.

Offshore southern and central West Greenland

Knowledge of the sedimentary succession offshore southern West Greenland is derived mainly from the five exploration wells in the area and from interpretation of a grid of multichannel seismic data of all vintages over the entire shelf and slope within Greenland territorial waters. None of the wells penetrated the deepest sediment sequences visible on the seismic sections, and the dating of these is provisional and based on analogies drawn with the Labrador shelf and to a lesser extent with the onshore outcrops in West Greenland and eastern Baffin Island. Rolle (1985) reported on the lithostratigraphy of the wells, and Nøhr-Hansen (1998) has revised the biostratigraphy of four of them.

Of the five wells drilled offshore West Greenland (Rolle 1985), three penetrated only Cenozoic sediments before being terminated in Paleocene basalts (Hellefisk-1 and Nukik-2) or Precambrian basement (Nukik-1). The Kangâmiut-1 well is situated on the western flank of a fault-bounded basement ridge (Fig. 4b). The well drilled through an Eocene and younger, sand-dominated succession, then lower Eocene (L. Eo.) and Paleocene (Pc) mudstones, below which the well penetrated a coarse arkosic sand interleaved with mudstone, before being terminated in Precambrian basement. Bate (1997) interpreted the coarse arkosic and interleaved mudstone section as a syntectonic fan of Paleocene age that developed on the hanging wall of an active fault. It is possible that the sands are a hydrocarbon reservoir that was not tested properly because of technical difficulties during drilling of the well. The interpretation by Bate (1997) of the age of this section is controversial and the unit could be of Santonian age (Nøhr-Hansen 1998). Only one well, Ikermiut-1, drilled a significant section of pre-Cenozoic sediments, sampling an 850 m section of Turonian–Santonian (Nøhr-Hansen 1998) mudstone before drilling was stopped.

Mesozoic succession. Four seismic sequences have been recognized below the Cenozoic succession offshore southern West Greenland (Chalmers *et al.* 1993, 1995). Only the uppermost of these has been penetrated, and that by only the Ikermiut-1 well. Dating of the remaining pre-Cenozoic sequences is therefore a problem.

The Paleocene mudstones (Pc) encountered in the Kangâmiut-1 well can be traced to the west on the seismic data and their base is marked by a continuous reflection (Fig. 4b). Below this there is another transparent sequence (Ka) about 1 km thick, which was not reached by the Kangâmiut-1 well. This sequence was named the Kangeq Sequence by Chalmers *et al.* (1993). Below the Kangeq Sequence is another sequence, named the Appat Sequence by Chalmers *et al.* (1993) (Ap in Fig. 4b), which is over 2 km thick against the fault on the west side of the Kangâmiut Ridge. Two additional sequences can be identified lying between the Appat Sequence and the basement; the upper of the two was called the Kitsissut Sequence by Chalmers *et al.* (1993) (Kt) and the lower was referred to as 'the deep sequence' by Chalmers *et al.* (1995) ('deep').

The Kangeq Sequence can be traced to the north, though with difficulty because of tectonic complications, to the Ikermiut-1 well where the upper part of the sequence ties with the lowermost 850 m of Turonian–Santonian mudstones encountered in this well. Thus the Kangeq Sequence is established as being of Upper Cretaceous age, equivalent to the older part of the Markland Formation, and at the Ikermiut-1 well it is separated from the overlying upper Paleo-

Fig. 4. Seismic lines from (**a**) the Hopedale Basin (redrawn from Balkwill 1987, fig. 10) and (**b**) southern West Greenland to show the similarity between the cross-sections of the Upper and Lower Bjarni Formation in (**a**) and the Appat and Kitsissut Sequences in (**b**). (See Fig. 2 for locations of these lines.) The comparison shown here suggests that the Appat Sequence may be the same age as the Upper Bjarni Member (Kbu) (Aptian–Albian) and the Kitsissut Sequence the same age as the Lower Bjarni Member (Kbl) (Barremian–Aptian). The borehole (K-1) shown in (**b**) is Kangâmiut-1 and the electric logs are the gamma log (on the left) and sonic log (on the right). Contractions in (**a**) p-C, Precambrian basement; Kbl, Lower Bjarni Member; Kbu, Upper Bjarni Member; Kmf, Freydis Member; KTm, Markland Formation; Tc, Cartwright Formation; Tk, Kenamu Formation; Tm, Mokami Formation; Ts, Saglek Formation. Contractions in (**b**) 'deep', deep sequence (see text); Kt, Kitsissut Sequence; Ap, Appat Sequence; Ka, Kangeq Sequence; Pc, Paleocene sediments; L.Eo., lower Eocene sediments.

cene sediments by a substantial unconformity. The other three sequences must be older than Turonian–Santonian age, but just how old can only be judged by analogies to the Labrador shelf, using relationships to tectonic events as a further guide.

Figure 4 shows seismic lines from both the Labrador and Greenland shelves. The sequences shown on the Labrador shelf line have been penetrated by wells and dated. It has already been shown that the Markland Formation (KTm in Fig. 4a) on the Labrador shelf is approximately the same age as the Kangeq Sequence (Ka) on the Greenland shelf, i.e. of Late Cretaceous (and possibly as young as Danian) age (100–62 Ma). Below these mudstone-dominated units, both the Upper Bjarni Formation (Kbu) of the Labrador shelf and the Appat Sequence (Ap) of the Greenland shelf show a similar geometry that is obviously related to faults (Fig. 4). The sequences show typical wedge shapes in cross-section, being thickest in the footwall close to fault planes and thinning away from this towards the crests of the rotated fault blocks. The Upper Bjarni Formation (Kbu) is in deltaic facies, in contrast to the marine mudstones of the Markland Formation (KTm) and Kangeq Sequence (Ka). The internal reflection characteristics of the Appat Sequence (Ap) also indicate a more heterogeneous lithology than the Kangeq Sequence (Ka), possibly indicating the presence of graben-fill sands. The Upper Bjarni Formation (Kbu) is of Aptian–Albian age (115–100 Ma), so this is possibly also the age of the Appat Sequence (Ap).

The Kitsissut (Kt) and 'deep' sequences have restricted distributions. They appear to have been carried down and rotated with the fault blocks. The Kitsissut Sequence is much more reflective than the Appat Sequence and tends to show downlap to the west, possibly indicating a prograding delta front. Chalmers *et al.* (1993) suggested that the Kitsissut Sequence (Kt) could be correlated with the Lower Bjarni Formation (Kbl). If this correlation is valid, it implies that the Kitsissut sequence is of Barremian–Aptian age (*c.* 125–115 Ma).

The oldest sequence (called 'the deep sequence' by Chalmers *et al.* (1995)) is interpreted on seismic sections only in the Sisimiut Basin (Fig. 2), between 66° and 68°N. This sequence may consist of Ordovician limestones equivalent to those exposed in eastern Canada (Bell & Howie 1990), encountered in four of the wells on the Labrador shelf, and also occurring in a fault breccia in West Greenland (Stouge & Peel 1979). Other possibilities are Lower Cretaceous volcanic rocks equivalent to the Alexis Formation on the Labrador shelf (Balkwill 1987) or hitherto completely unknown Mesozoic sediments.

Cenozoic succession. The base of the Cenozoic succession was penetrated in both the Kangâmiut-1 and Ikermiut-1 wells. In the Ikermiut-1 well there is a hiatus at the Cretaceous–Cenozoic boundary where upper Paleocene mudstones lie on Santonian mudstones (Nøhr-Hansen 1998). At Kangâmiut-1 there may be a hiatus between upper Paleocene mudstones and the possibly Santonian syntectonic fan developed at the foot of a fault scarp that has already been mentioned, although the possibility that the fan sediments are of early Paleocene age must be kept open.

As discussed above, the fault blocks in the Nuussuaq Basin were produced during an early Campanian to early Paleocene extensional tectonic episode. Additional evidence for the extent of such an episode is the syntectonic fan penetrated by the Kangâmiut-1 well (assuming that it is of Paleocene (Bate 1997) and not Santonian age (Nøhr-Hansen 1998)), the observations that some of the volcanic rocks near the Nukik-2 well appear to have been erupted into half-grabens (Chalmers *et al.* 1993) and the tectonic episode that produced the large fault blocks of the Fylla structural complex (Fig. 5) (Bate *et al.* 1994; Aram 1999) that appears to have taken place during Palaeogene time. This tectonic episode does not appear to have affected the Labrador Basins, where Markland Formation sedimentation appears to have continued without major interruption from Maastrichtian into early Paleocene time.

Paleocene (and possibly Eocene) igneous rocks are present around the Nukik-2 well (Rolle 1985). The Nuussuaq Basin basalts can be traced offshore to south of 68°N, where they were penetrated by the Hellefisk-1 well (Rolle 1985) (Fig. 2), and as far north as 73°N (Whittaker *et al.* 1997), a total area of around 75 000 km^2. Palaeogene volcanism affected many areas around the northern Labrador Sea (Fig. 2) but the ages of the rocks are poorly constrained, reported ages being mostly from unreliable K–Ar measurements.

Later tectonic activity can be seen along the Ikermiut Ridge, where the Ikermiut-1 well appears to have penetrated the easternmost fold of a flower structure (Fig. 6) which has all the characteristics of having been produced by a step-over on a strike-slip fault (compare Fig. 6 with, for example, fig. 8 of Dooley *et al.* (1999)). The unconformity that terminates the deformed sediments above is between early and mid-Eocene age (Nøhr-Hansen 1998).

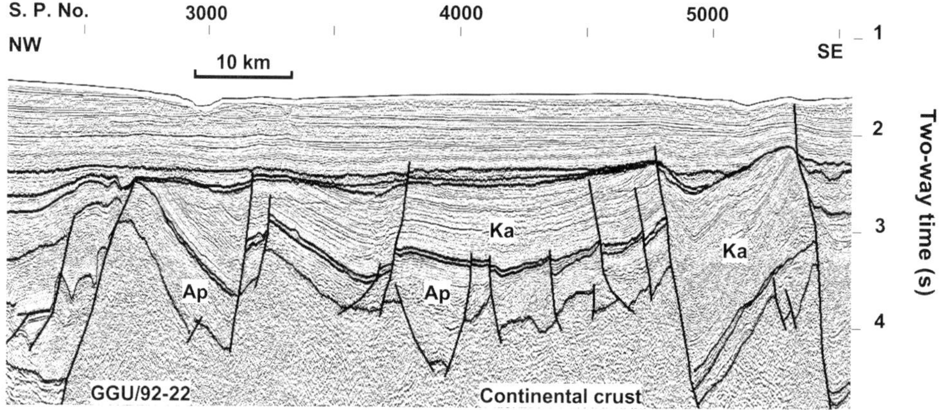

Fig. 5. Seismic line through the so-called Fylla Structural Complex offshore southern West Greenland. (See Fig. 2 for location.) The line runs across the northern extension of the 'propagating rift' hypothesized by Srivastava & Roest (1999, fig. 4). The inferred presence of Appat Sequence (Ap, of inferred Aptian–Albian age) and Kangeq (Ka, of known Cenomanian–Santonian and probably younger age) is inconsistent with the hypothesis of Srivastava & Roest (1999) that this seismic line runs over oceanic crust of Campanian or younger age.

From mid-Eocene time, the West Greenland Basin subsided, without further obvious evidence of tectonism, until late Neogene time. However, sedimentation patterns are complex, caused presumably by the interaction of varying shelf-margin currents, sea level and sediment supply. During Neogene time, uplift of the eastern parts of the basin by 2–3 km (Chalmers 2000) exposed the area known as the Nuussuaq Basin.

Summary. The sedimentary–tectonic history of the basins around the Labrador Sea is as follows: (1) pre- to early syn-rift deposition of the Kitsissut (and possibly 'deep') sequence of the basin off southern West Greenland, and the Alexis Formation and Lower Bjarni Member of the basins off Labrador in Valanginian to Aptian time (*c.* 130–115 Ma); (2) faulting and fault-block rotation and deposition of the Appat Sequence and Upper Bjarni Member onto the hanging walls of the fault blocks during Aptian and Albian time (115–100 Ma); (3) post-rift deposition of the Kangeq Sequence and Atane and Markland Formations during a phase of thermal subsidence from Cenomanian to mid-Campanian or even Danian time (100 Ma to 80 Ma or even 62 Ma); (4) extension and fault-block rotation in the Nuussuaq Basin onshore and Fylla area offshore and possibly elsewhere in the West Greenland basins in early Campanian to early Paleocene time (80–63 Ma); uplift and erosion caused by impact of the Iceland plume slightly later in the Paleocene period (63–62 Ma) produced major channel systems in the Nuussuaq Basin and the combined effect of these episodes probably caused the hiatus encountered in the Ikermiut-1 well between Santonian and late Paleocene sediments; (5) volcanism during Paleocene time at 61–59 Ma, and again in early Eocene time at 54.8–52.5 Ma; (6) strike-slip tectonism along the Ikermiut Ridge that ended at the end of early Eocene time (*c.* 49 Ma); (7) regional subsidence and sedimentation from mid-Eocene time to some time during the Neogene period; (8) uplift of basin margins during Neogene time to expose the Nuussuaq Basin.

Sea-floor spreading in the Labrador Sea

The early sea-floor spreading history of the Labrador Sea and the distribution of continental and oceanic crust in the region are poorly constrained and a subject of controversy. The starting point for modern interpretations of the structure and sea-floor spreading history is the study by Srivastava (1978), in which he identified and numbered linear magnetic anomalies in the Labrador Sea. Srivastava (1978) suggested that sea-floor spreading started during magnetochron 31 (68.7–67.8 Ma) and that oceanic crust extended to the foot of the bathymetric continental slopes. Srivastava (1978) also pointed out that a reorientation of the spreading axis

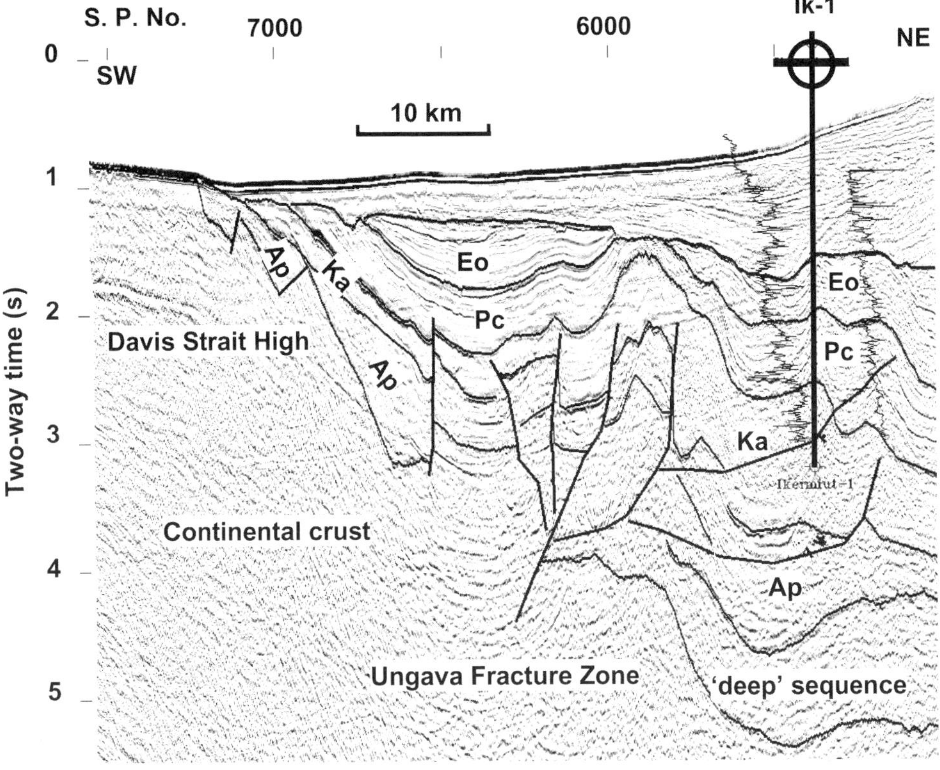

Fig. 6. Seismic section from the Sisimiut Basin through the Ikermiut-1 borehole and onto the Davis Strait High. (See Fig. 2 for location.) The flower structure is the Ikermiut Ridge, part of the strike-slip Ungava Fault Zone (UFZ, Fig. 2). The Ikermiut-1 well (Rolle 1985) drilled through Cenozoic sediments and into Upper Cretaceous sediments of the Kangeq Sequence (Ka). The redating by Nøhr-Hansen (1998) of this well shows upper Paleocene sediments lying unconformably on lower Santonian–Turonian (85–93.5 Ma) sediments of the Kangeq Sequence. The borehole's maximum depth was in sediments possibly as old as Cenomanian age (93.5–99 Ma). The seismically transparent Kangeq Sequence can be traced across folds in the UFZ and southwestwards above the eastern flank of the Davis Strait High. Below the Kangeq Sequence, two older sequences can be seen, identified as the Appat (Ap) and 'deep' Sequences. Both of these must be older than the oldest sediments identified in Ikermiut-1, of Turonian–early Santonian age. The Appat Sequence can be traced through the faults of the flower structure and onto the flank of the Davis Strait High. These observations mean that the basement of the Davis Strait High must be older than Turonian (89–93.5 Ma) and probably older than Aptian age (112.2–121 Ma). Since the start of sea-floor spreading in the Labrador Sea is dated to either Campanian (magnetochron 33, *c.* 75–80 Ma) (Roest & Srivastava 1989) or mid-Paleocene time (magnetochron 27, 61 Ma) (Chalmers & Laursen 1995) this must mean that the crust of the Davis Strait High cannot be oceanic. We interpret the Davis Strait High to be a ridge of continental basement within the Ungava Fracture Zone that has been uplifted to sea bed, probably by local transpressional forces.

took place during magnetochron 24r (55.9–53.3 Ma), at the same time as sea-floor spreading started between Europe and Greenland. As a result of the reorientation of the spreading axis there is a discordance of about 13° between anomaly 25n, which trends *c.* N27°W, and anomaly 24n, which trends *c.* N40°W in the central part of the Labrador Sea. According to Srivastava (1978), the youngest clearly identifiable magnetic anomaly is anomaly 20 (43.8–42.5 Ma), and spreading had died out entirely

by magnetochron 13 (33 Ma). Roest & Srivastava (1989) slightly modified the linear magnetic anomaly map to identify the oldest magnetic anomaly as 33 (79.7–74.5 Ma), but otherwise the original map of Srivastava (1978) has proved on the whole to be correct, at least regarding anomalies as far back as 27n.

Chalmers (1991) used reprocessed multichannel seismic and magnetic data, originally acquired by Bundesanstalt für Geowissenschaften und Rohstoffe (BGR) in 1977 (Hinz *et al.*

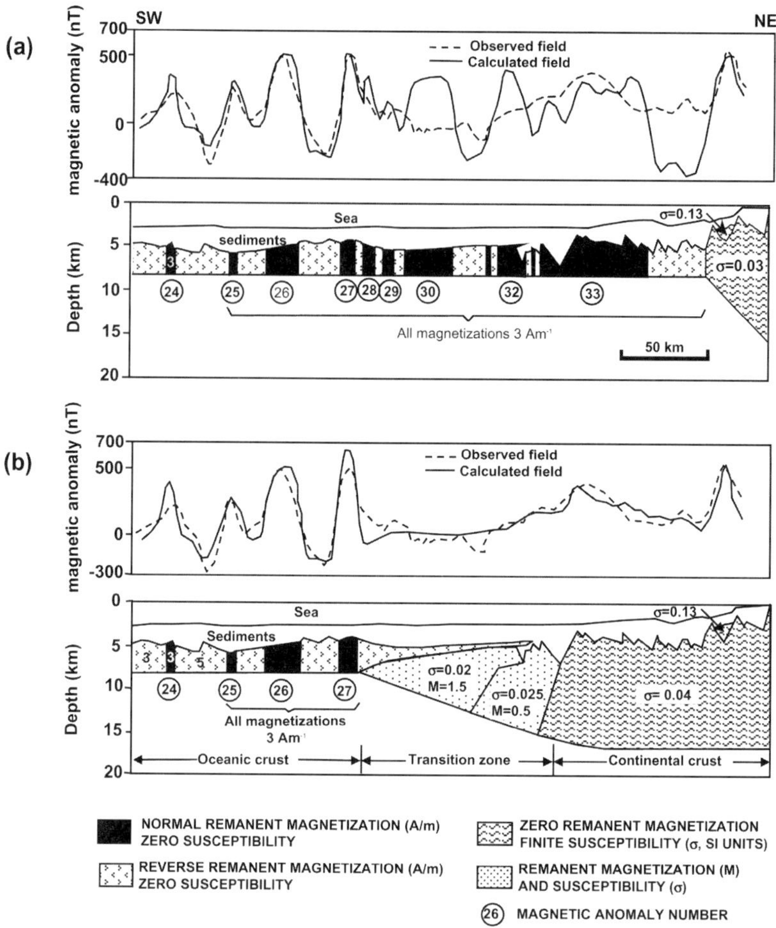

Fig. 7. Magnetic modelling along seismic line BGR 12 and comparison of the calculated and observed residual magnetic fields that result. (See Fig. 2 for location.) (**a**) 'Oceanic crust' model in which the model profile was generated from the interpretations by Roest & Srivastava (1989) for the locations of sea-floor spreading anomalies (as shown in Fig. 1). (**b**) 'Continental crust' model in which oceanic crust is interpreted at the south-western end of the profile, continental crust at the northeastern end and a transition zone between. The transition zone is modelled as continental crust containing reversely magnetized intrusions and overlain by reversely magnetized flood basalts (of presumably magnetochron 27r age). These interpretations were first published by Chalmers (1991, figs. 4 and 5) and again by Chalmers & Laursen (1995, fig. 3), and those studies should be consulted for technical details of the interpretation.

1979) to test the interpretation by Roest & Srivastava (1989), initially on a single traverse across the continent–ocean boundary off southern West Greenland. The magnetic modelling (Fig. 7), constrained by interpretations of depth to basement on the seismic line, showed that the Roest & Srivastava (1989) hypothesis fitted the data seaward of anomaly 27 (mid-Paleocene time, 61 Ma), but landward of this a model of block-faulted continental crust separated from oceanic crust by a reversely magnetized transition zone fitted the observed magnetic data better. Chalmers & Laursen

(1995) extended the tests to other reprocessed BGR lines on both the Greenland and Canadian sides of the Labrador Sea and concluded that (1) sea-floor spreading began in the Labrador Sea during Chron 27 (61.3–60.9 Ma), not Chron 33 (79.7–74.5 Ma) as proposed by Roest & Srivastava (1989), and (2) there is no continent–ocean boundary (COB) situated at the foot of the bathymetric continental slope as previously proposed by Srivastava (1978); Roest & Srivastava (1989). Instead, continental crust extends under deep water and is separated from oceanic crust by some kind of transition zone

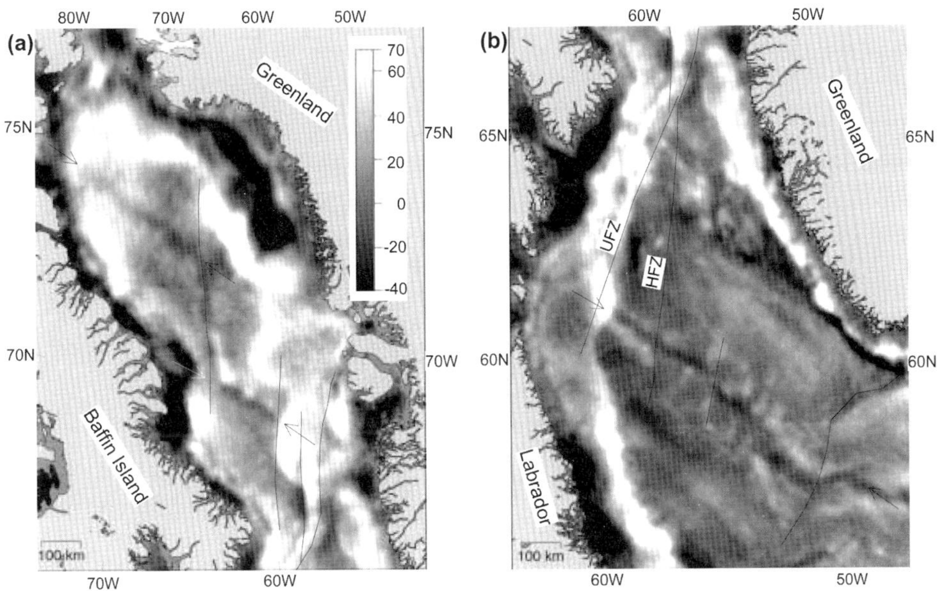

Fig. 8. Free-air gravity over (**a**) Baffin Bay and the Davis Strait, and (**b**) the Labrador Sea and the Davis Strait. Map (**a**) is north of and overlaps map (**b**). A linear gravity low ((**b**), arrows) marks the location of the extinct spreading centre in the Labrador Sea, which continues NW of the Hudson Fracture Zone (HFZ) to terminate at the Ungava Fault Zone (UFZ). We interpret the linear gravity lows ((**a**), arrows) in Baffin Bay to mark the presence of extinct spreading centres there too. The gravity data are from Sandwell & Smith (1992) south of 72°N and from Laxon & McAdoo (1998) north of 72°N.

whose character and width is different in different parts of the Labrador Sea (Chalmers 1997). Figure 2 shows the distribution of oceanic crust based on the interpretation of Chalmers & Laursen (1995), the extent of continental crust and of the transitional area between.

As Srivastava (1978) pointed out, the position of the extinct mid-Labrador Sea spreading axis coincides with a pronounced linear gravity low (Fig. 8b). In Figure 2 the extinct spreading axis has been drawn in the position of the gravity low, which was originally interpreted on gravity maps from Geological Survey of Canada (1988). This position differs only slightly from the position shown by Srivastava (1978), except that, in contrast to the maps of Srivastava (1978) and Roest & Srivastava (1989), the extinct spreading axis has been shown in Figure 2 as continuing northwest of the Hudson Fracture Zone (HFZ) as far as the Ungava Fracture Zone (UFZ), with a slight offset at the HFZ. The reason for drawing the axis thus is that the negative gravity anomaly in the middle of the Labrador Sea is no less distinct northwest of the HFZ than it is to the southeast (Fig. 8b).

The distribution of oceanic crust northwest of the HFZ is open to discussion. Roest & Srivastava (1989), Srivastava & Keen (1995) and Srivastava & Roest (1995) all showed magnetic anomalies 25–27 continuing northwest of the line of the HFZ into the area surrounding the Gjoa G-37 well. This, however, conflicts with the published results from this well, which terminated 3 km below sea level in mudstones that are overlain by Paleocene basalts. Klose *et al.* (1982) stated that biostratigraphic evidence indicates that these mudstones are of Maastrichtian age, but no details concerning the nature of this evidence have been released. Assuming that the 'Maastrichtian' fossils are diagnostic of *in situ* Maastrichtian sediments and that they are not reworked, the crust underlying the Gjoa G-37 well site must be of greater than Maastrichtian age and therefore cannot be a continuation of the Paleocene oceanic crust from southeast of the HFZ. Otherwise the nature of the crust in this area and its conjugate on the other side of the spreading axis is unknown.

In Figure 2 a narrow area between the HFZ and UFZ is shown as underlain by oceanic crust of uncertain age. From the orientation of the extinct spreading axis, which appears to be a continuation of the Eocene-age (magnetochron 24–13) spreading axis to the southeast, one would conclude that this crust is of Eocene age. The absence of detectable linear magnetic anomalies is this area could be due to oblique spreading; an explanation of how oblique spreading can result in magnetic quiet zones has been provided by Roots & Srivastava (1984), and their study should be consulted for further details. However, an alternative interpretation of the age of oceanic crust between the HFZ and the UFZ cannot at present be discounted. According to this interpretation, this area of oceanic crust was formed during Paleocene time between magnetochrons 27n and 24r, and Eocene sea-floor spreading took place only southeast of the HFZ. In other words, during Paleocene spreading, the UFZ was the main transform structure in the northern Labrador Sea and Davis Strait, but during Eocene spreading, movement between the North American and Greenland plates was transferred from the Labrador Sea to Baffin Bay along the HFZ.

The UFZ marks the NW termination of oceanic crust in the Labrador Sea. Its position is taken to be along the east–southeast side of a line of positive gravity anomalies and weak positive magnetic anomalies that runs from east of the southern tip of Baffin Island to the northern part of Davis Strait (Fig. 8b). Whatever the origin of the anomalies themselves, it is obvious from their strength and alignment that they must reflect a major structure in the lithosphere. As early as 1967 Kerr (1967) interpreted a transform fault here and named this the Ungava transform fault (Kerr 1967, p. 286). Later workers, e.g. Klose *et al.* (1982), Menzies (1982) and Rice & Shade (1982), interpreted this structural zone as part of the transform system linking sea-floor spreading in the Labrador Sea with spreading in Baffin Bay, and this interpretation is adopted here. In Davis Strait the UFZ meets the HFZ, and where Davis Strait is narrowest the combined fracture zones become a zone of transpression, with eastward-directed reverse faults and thrusts developed on the Greenland side (Fig. 6).

Baffin Bay

Baffin Bay crust

Baffin Bay is a major sedimentary basin, with sediment thicknesses up to 12 km in the northwest part of the bay, diminishing to <3 km just north of Davis Strait (Keen *et al.* 1974; Reid & Jackson 1997). The nature of the crystalline rocks underlying these sediments has not yet been unequivocally demonstrated, and there is no obvious pattern of linear magnetic anomalies in this region. For this reason the neutral term 'Baffin Bay crust' has been adopted for the substratum to the thick sediments in central Baffin Bay.

Evidence indicating that Baffin Bay is oceanic in the geological and not merely maritime sense has been published by Barrett *et al.* (1971), Keen & Barrett (1972), Hood & Bower (1973) and Keen *et al.* (1974). The principal evidence for this interpretation has been provided by seismic refraction lines, which show that the crust underlying the deeper part of Baffin Bay is very thin, the M discontinuity (Moho) lying at a depth of 8–9 km below sea bed in the deeper parts of the Bay at latitude 72°N, where the average thickness of sediments is *c.* 4 km (Keen & Barrett 1972; Keen *et al.* 1974). In this area Keen & Barrett (1972) identified a layer with velocities of $5.0-6.3\,\mathrm{km\ s^{-1}}$, which they interpreted as oceanic layer 2, and a layer with velocities of $6.5-6.9\,\mathrm{km\ s^{-1}}$ interpreted as oceanic layer 3. Later Srivastava *et al.* (1981) and Balkwill *et al.* (1990) grouped velocities between 5.7 and $6.8\,\mathrm{km\ s^{-1}}$ in central Baffin Bay into a single layer interpreted as problematic oceanic layers 2 and 3. However, in a more recent paper, Reid & Jackson (1997) reported on the results of detailed seismic refraction traverses in northernmost Baffin Bay, one of which crosses the margin of continental crust. Southeast of the continental crust the material forming the subsurface to the sedimentary section has a velocity of $6.8\,\mathrm{km\ s^{-1}}$, but there is no evidence in the data for the presence of any oceanic layer 2 with velocity around 5 km $\mathrm{s^{-1}}$. The $6.8\,\mathrm{km\ s^{-1}}$ material might be interpreted as oceanic layer 3, were it not for the fact that it has a high velocity gradient that appears to provide a smooth transition to mantle. For this reason Reid & Jackson (1997) suggest that spreading here was amagmatic and that the $6.8\,\mathrm{km\ s^{-1}}$ material is serpentinized mantle, a situation known to occur off western Iberia (ODP Leg 173 Shipboard Scientific Party 1998) and that may also occur off SW Greenland (Chian & Louden 1994; Chalmers 1997) (see below).

An explanation of the difference between central and northwest Baffin Bay with regard to interpretation of crustal structure could be that the situation in Baffin Bay is analogous to that in the Eurasia Basin. Here the spreading rate along the Gakkel (= Nansen) Ridge becomes slower passing from the northern end of the Fram Strait to the Laptev Sea in the east, so that approaching the Laptev shelf the spreading rate is as low as $6\,\mathrm{mm\ a^{-1}}$. At the same time the crust becomes vanishingly thin, indicating that the Gakkel Ridge is virtually amagmatic in the east and the sediment subsurface here is serpentinized mantle (Coakley & Cochran 1998).

Transform fracture zones; extinct spreading axis

In the absence of any agreed pattern of linear magnetic anomalies in Baffin Bay, only the position of an extinct spreading axis and possible transform faults in the basin can be determined from gravity and magnetic anomaly maps.

In gravity anomaly maps (Geological Survey of Canada 1988; Sandwell & Smith 1992; Laxon & McAdoo 1998) there is a distinct negative gravity anomaly running approximately north–south between 73°N, 65°W and 71°N, 64°W (Fig. 8a). Seismic data show that there is a deepening of the sediment–basement interface in this position (Jackson *et al.* 1979; Rice & Shade 1982). Both Jackson *et al.* (1979) and Rice & Shade (1982) interpreted this feature as the site of the extinct spreading axis, although Jackson *et al.* (1979) also considered the possibility that it is the site of a fracture zone. Whittaker *et al.* (1997) pointed out that the feature is parallel to the Hudson Fracture Zone and other late fracture zones in the Labrador Sea, and that it is therefore most probably the expression of a fracture zone, the interpretation favoured by Roest & Srivastava (1989) and Wheeler *et al.* (1996). For convenience this fracture zone is referred to in the following as the 64°W fracture zone. However, the position of the 64°W fracture zone shown by Wheeler *et al.* (1996) does not exactly coincide with that shown in Figure 2, which is based entirely on the negative gravity anomaly.

The gravity anomaly maps also show a NW–SE-trending anomaly that closely resembles the negative anomaly in the position of the extinct spreading axis in the Labrador Sea (compare Fig. 8a with Fig. 8b). We interpret this anomaly as arising from the extinct spreading axis in Baffin Bay, an explanation first proposed by Whittaker *et al.* (1997).

One segment of this NW–SE-trending anomaly terminates to the southeast at the 64°W fracture zone. Here it is offset about 300 km right-laterally, so that it reappears on the east side of the fracture zone at 70°N (Figs 2 and 8a). From here it continues to the SE with minor right-lateral offsets until it terminates at longitude 60°W, where another *c.* north–south fracture zone is shown in Figures 2 and 8a. The nature of the crust to the east is a matter of pure conjecture, hence the question marks in Figure 2.

Whatever the nature of the basement in Baffin Bay, the above interpretation of the position of fracture zones and the extinct spreading axis in Baffin Bay implies that spreading in this direction took place in an approximately north–south direction, and that spreading was strongly oblique to the spreading axis. As Roots & Srivastava (1984) have explained, oblique spreading can account for the lack of distinct linear magnetic anomalies in areas such as Baffin Bay where the underlying crust is oceanic. Another factor that could diminish the strength of linear magnetic anomalies in Baffin Bay is the blanketing effect of the cover of sedimentary rocks of 4–12 km thickness.

The nature of crust in the Davis Strait

One of the most important differences between the maps in Figures 1 and 2 concerns nature of the crust in the Davis Strait. Seismic refraction measurements in the central part of the Davis Strait (Keen & Barrett 1972) are consistent with continental crust underlying the strait. The depth to the Moho here is 22 km. The lowest crustal layer is 18 km thick and has a velocity of 6.2 km s^{-1}, which is consistent with continental crust. However, seismically, Davis Strait appears similar not only to continental fragments such as the Seychelles or Rockall Plateau but also to Iceland (Keen & Barrett 1972; Keen *et al.* 1974). The similarity to Iceland and the occurrence of plume-derived volcanic rocks both at Cape Dyer (Clarke 1970) and in West Greenland led Keen *et al.* (1974) to suggest that a mixture of continental crust and plume-related volcanic rocks might underlie Davis Strait. There is, however, no indication in the seismic reflection data of the presence of volcanic rocks penetrating sediments west of the Ikermiut-1 well, although shallow drill cores of basalt have been recovered from the Davis Strait High to the west (Srivastava *et al.* 1982).

The structure of Davis Strait was discussed by Srivastava *et al.* (1982) and Srivastava (1983). In these papers it was argued that, whereas the Davis Strait High, an elevated basement feature in the central part of the strait, is probably continental, the crust underlying the sedimentary basins that flank the high to the east and west is oceanic. This is not so as far as crust between the high and the Ikermiut-1 and Kangâmiut-1 wells on the Greenland shelf is concerned. The lower part of the Ikermiut-1 well penetrated Turonian and possibly Cenomanian mudstones of the Kangeq Sequence (Nøhr-Hansen 1998). The Kangeq Sequence and two seismic sequences underlying it can be traced westwards on seismic lines to where they overlie the eastern flank of the Davis Strait High (Chalmers *et al.* 1995) (Fig. 6). The Kangeq Sequence has also been traced to the median line in the area west of the Kangâmiut-1 well, and it is underlain by sequences that resemble the Lower Cretaceous Bjarni Formation in the Labrador basins. It is therefore concluded that both the Davis Strait basement high and the crust underlying the sedimentary basin to the east must be formed of Early Cretaceous or older rocks. As no-one has proposed that there is Early Cretaceous or older oceanic crust in this region, this crust must be continental. Extensive Paleogene basalts can be seen on the seismic data from the basins to the west of the Davis Strait High; such rocks are known to mask reflections from deeper rocks. Until decisive evidence to the contrary is presented, it is assumed that the crust to the west of the Davis Strait High is also continental.

The structure of the continent–ocean transitions in the Labrador Sea

The oceanic crust of the Labrador Sea is bordered by transitions to continental crust on three sides only; to the southeast, it merges with the oceanic crust of the North Atlantic via a ridge–ridge–ridge triple junction that existed in Eocene time. To the northwest, Labrador Sea oceanic crust is separated from continental crust on Baffin Island by a transform margin along the UFZ and HFZ, as has already been discussed. The detailed geology around the UFZ and HFZ is, however, poorly known and requires substantial work.

Extensional margins border the oceanic crust along both the Labrador margin to the southwest and the Greenland margin to the northeast. The nature of the transition from continental to oceanic crust across these two margins has been the subject of debate during recent years.

The transition zone off southern West Greenland

Geophysical evidence across the Greenland transition zone consists of five migrated multichannel seismic reflection lines (Fig. 2) (Hinz *et al.* 1979; Chalmers 1991; Keen *et al.* 1994*b*; Chalmers & Laursen 1995) and one refraction–wide-angle reflection line (Chian & Louden 1994; Louden & Chian 1999) in addition to magnetic and gravity data. The appearance of the Greenland transition zone is very different on the two northwesternmost multichannel lines from what can be seen on those farther southeast, and Chalmers (1997) discussed the significance of this along-strike variation.

Modern understanding of the nature of these transition zones dates back only to the early 1990s. Chalmers (1991) first showed that block-faulted continental crust west of Greenland is separated from oceanic crust by some kind of transition zone. On the one seismic and magnetic line (BGR 12, Fig. 7) that Chalmers (1991) examined, this transition zone seems to consist of reversely magnetized flood basalts underlain by material that Chalmers (1991) modelled to be reversely magnetized. Chalmers (1991) suggested that it consists of thinned continental crust and sediments intruded by dykes and sills. Later, Chalmers & Laursen (1995) published interpretations of the upper-crustal structure along all four BGR seismic lines that cross the Greenland transition zone (Fig. 2).

The northernmost line, BGR 6, crosses a volcanic margin where seaward-dipping reflectors (SDRs) and a landward-facing escarpment can be seen (Fig. 9) in an area known as the Gjoa Rise (Tucholke & Fry 1985). The Gjoa Rise has been drilled by the commercial well Gjoa H-37 (Klose *et al.* 1982), which reached final depth in sediments, reported by Klose *et al.* (1982) to be of Maastrichtian age, below Paleocene basalts. Landward of the transition zone is the Hecla Rise, which was formerly thought to be an oceanic rise (Tucholke & Fry 1985), but was interpreted by Chalmers & Laursen (1995) to be an area of shallow continental basement.

Lines BGR 21, BGR 17 (Hinz *et al.* 1979; Chalmers & Laursen 1995) and AGC/90-3 (Keen *et al.* 1994*b*) are all multichannel seismic lines that show similar features (Fig. 10). The margin of the continental shelf is steep and formed by one or more faults that are either exposed at the sea bed or covered only by thin sediments. Seaward of the fault zone is the transition zone. An irregular 'basement' surface can be seen between 4.5 and 10 km below sea level, broken into grabens and half-grabens containing sediments and possibly sills, and covered by 2–3 km of flat-lying sediment. Chalmers & Laursen (1995) interpreted features that resemble the peridotite ridges off western Iberia (Sawyer *et al.* 1994 and references therein; ODP Leg 173 Shipboard Scientific Party 1998), and Louden & Chian (1999) also discussed the similarities between these areas. Thermal subsidence, indicated by the deposition of flat-lying sediments above the rifted blocks, appears to be confined to the period after sea-floor spreading started. Seaward of this transition zone, a much less irregular surface indicates undisputed oceanic crust, where sea-floor spreading anomalies 27N and younger can be interpreted (Chalmers & Laursen 1995; Srivastava & Roest 1995).

A refraction–wide-angle reflection line, 88R2 (Fig. 11a) was described by Chian & Louden

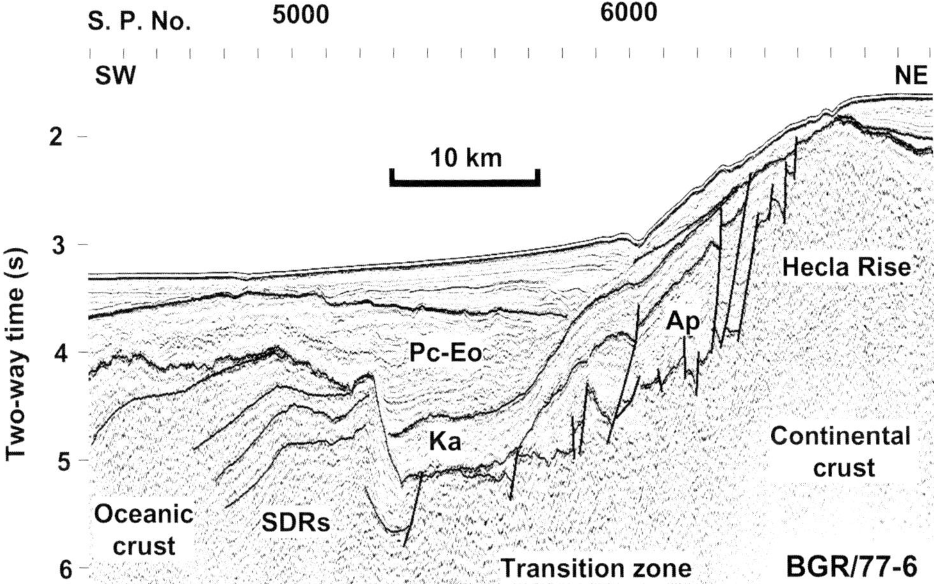

Fig. 9. Part of seismic line BGR 6 (Fig. 2) extending from the shallow basement of the Hecla Rise (Tucholke & Fry 1985), which we interpret to consist of continental crust, across the transition zone to the seaward-dipping reflectors (SDRs) of the Gjoa Rise oceanic crust. Ap, Appat Sequence sediments; Ka, Kangeq Sequence sediments; Pc-Eo, sediments of Paleocene to Eocene age.

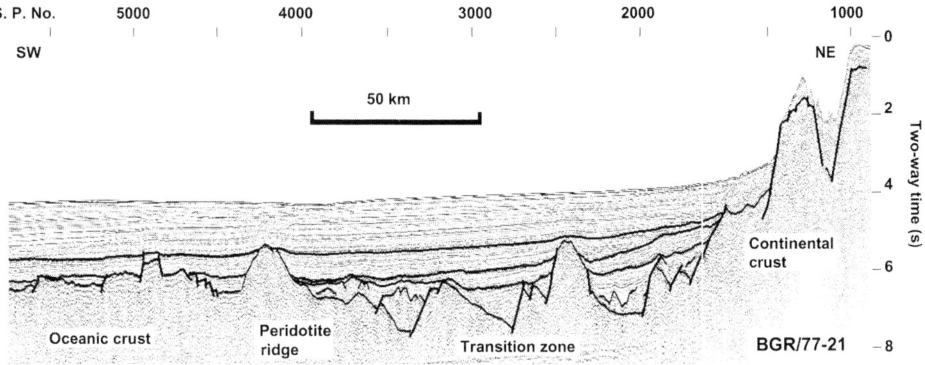

Fig. 10. Seismic line BGR 21 across the transition zone offshore southern West Greenland (Fig. 2). Oceanic crust with magnetic anomalies 27 (61 Ma) and younger (Chalmers & Laursen 1995) is found southwest of SP 4500, and continental crust is found at the northeast end of the line. Half-grabens and a ridge that resembles the peridotite ridges west of Iberia (Pickup *et al.* 1996) can be interpreted within the transition zone and the infilling sediments are probably older that the onset of true sea-floor spreading. However, all the immediately succeeding post-rift sediments can be interpreted also to lie over oceanic crust, suggesting that rifting in the transition zone continued until immediately before the onset of sea-floor spreading.

(1994). It shows 1.5–2 km of sediments overlying 3–3.5 km of material with a P-wave velocity between 5.6 and 6 km s^{-1} (Fig. 11a). This layer is interrupted by a zone of 40 km width of lower-velocity (4.4–4.7 km s^{-1}) material, which may indicate high-velocity sediments, sediments intruded by igneous material or anomalously low-velocity basement. Below this there is a 3 km layer of material with velocities between 6.8 and 7.6 km s^{-1}. This deeper layer lies on mantle material with a velocity of 8 km s^{-1} at a depth of 12 km. The boundary between the two crustal layers corresponds to the depth of a strong intracrustal reflection on line AGC90-3 (Keen *et al.* 1994*b*; Chian *et al.* 1995*a*), which may be similar to the S-reflector observed on

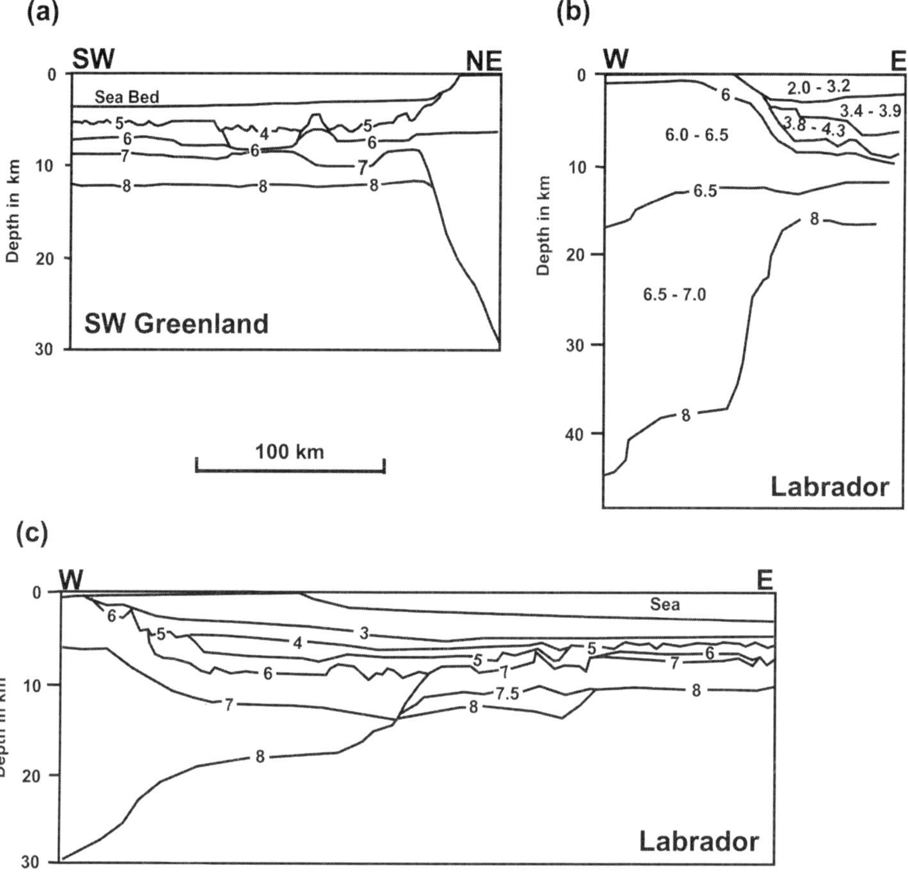

Fig. 11. Velocity structure (P-wave, in km s^{-1}) across (**a**) the southern West Greenland transition zone, along line 88R2, (**b**) in northern Labrador along line 5E, redrawn from Funck & Louden (1999) and (**c**) across the continent to ocean transition off central Labrador along line 90R1. The velocity boundaries in (**a**) and (**c**) were redrawn from the gridded velocities shown by Louden & Chian (1999, figs. 9 and 10). (See Fig. 2 for locations.)

the Iberia margin (e.g. Hoffman & Reston 1992; Reston *et al.* 1996).

The geophysical evidence in the transition zone off southern West Greenland between 59°N and 62°N has been interpreted as indicating oceanic crust (Roest & Srivastava 1989), highly extended continental crust containing synrift sediments and possibly igneous intrusions (Keen *et al.* 1994*b*; Chalmers & Laursen 1995), highly extended continental crust overlying serpentinized mantle (Chian & Louden 1994) or a layer of 1 km thickness of basaltic 'oceanic' crust overlying serpentinized mantle (Srivastava & Roest 1995).

There is a distinct resemblance between the transition zone off southern West Greenland between 59°N and 62°N and the transition zone west of Iberia (Whitmarsh *et al.* 1993; Pickup

et al. 1996; Discovery 215 Working Group 1998; Chian *et al.* 1999; Dean *et al.* 2000). There, too, the crust is thin, around 3 km, and a layer of abnormal velocities in the range 7–7.5 km s^{-1} separates crustal material from mantle material. On the Iberian margin, ridges of serpentinized peridotite have been sampled by submersible (Boillot *et al.* 1988), by dredging (Beslier *et al.* 1990) and by ODP drilling (Boillot *et al.* 1987, 1988; Sawyer *et al.* 1994; ODP Leg 173 Shipboard Scientific Party 1998).

The nature of the uppermost 3 km of crust in both of these areas is still unclear, despite ODP drilling on the Iberian margin. The Iberian margin data indicate that only small amounts of melt are present, and mantle material that has either never been melted or suffered only low partial melting lies at the sea bed or at most

3 km under the sediment cover. An S-reflector has been observed under both the Iberian (Hoffman & Reston 1992) and southern West Greenland margins (Chian *et al.* 1995*a*), and Reston *et al.* (1996) have shown evidence that the S-reflector under the Galician margin may be a detachment fault, either at the base of continental crust (Reston *et al.* 1996) or in places within the continental crust (Whitmarsh *et al.* 1996). It is likely therefore that the 3 km of material with velocities of 5.6–6 km s^{-1} consists of continental crust.

Chalmers (1997) published schematic representations of possible cross-sections across the Greenland transition zone, reproduced here as Figure 12a–d. The shallow structure is based on the four BGR multichannel seismic lines (Chalmers & Laursen 1995) and the deep structure is based on the wide-angle reflection–refraction line 88R2 (Fig. 11a) (Chian & Louden 1994; Louden & Chian 1999).

The transition zone off Labrador

There are two migrated (BGR 17 and AGC90-1) and two unmigrated (BGR 21 and BGR 12) multichannel seismic reflection lines, and one complete (90R1) and one part (5E) refraction–wide-angle reflection lines across the Labrador transition zone (Fig. 2).

One of the best images of the upper-crustal structure across this margin is BGR 17, shown in Figure 13. At the southwest end of this line is the Bjarni H-81 borehole, which drilled through *c.* 2250 m of Cenozoic, Upper Cretaceous and Lower Cretaceous clastic sediments and was terminated within the basalts of the Alexis Formation after drilling through 250 m of these Lower Cretaceous rocks (Bell 1989). Plate 60 of Bell (1989) shows a detailed interpretation of a seismic line close and parallel to line BGR/77-17.

The top of the Alexis Formation volcanic rocks acts as an acoustic basement and the base of the volcanic rocks cannot be followed on the seismic sections. However, the Cenozoic and Cretaceous sediments can be interpreted. The Lower Cretaceous Bjarni Formation can be seen to occupy half-graben structures, which tend to be faulted down to the southwest. The half-grabens appear at about 2 s TWT (two-way time) at the southwestern end of the line and can be followed tentatively to about 8 s TWT near SP 2300, despite interference from noise from multiples northeast of about SP 800 (Fig. 13). Near SP 2000, fault-block structures appear to be present but have been over-migrated.

Northeast of approximately SP 2850, a strong, irregular reflection exists at about 7 s TWT, and may exist farther southwest in the noise generated by the sea-bed multiple. The topography of this reflector is rugged northeast of about SP 3700 and it also gradually shallows to the northeast, to about 6.5 s at SP 5000. The reflection from the top of the Markland Formation (Danian) can be followed from the Bjarni H-81 well (at about 2 s TWT). Again, despite interference from multiples, the top Markland Formation reflection can tentatively be followed to and below the irregular reflection as far as SP 3950. The landward limit of the irregular reflection corresponds to the 'inboard edge of seismically perceptible basalt' of Balkwill (1987). The overlying Cenozoic sediments are generally flat-lying and undisturbed, except along the continental rise between SPs 900 and 1600, where major slumping has taken place along fault planes that dip seaward at shallow angles.

The only other reflection line across the Labrador margin on which much structural detail can be seen is AGC90-1, which crosses BGR 17 at an acute angle within the transition zone (Fig. 2). The reflection patterns visible on AGC90-1 are similar to those on BGR 17. Processing of the other two lines, BGR 21 and BGR 12, is relatively elementary. They have never been migrated and specialized algorithms to reduce the intensity of multiples have never been applied to them. The Cenozoic section visible on these two lines is very thick, and no details of underlying structure can be seen because of strong multiple energy.

A wide-angle reflection–refraction profile, 90R1 (Fig. 11c), exists along the same track as AGC90-1 (Fig. 2) (Chian & Louden 1994; Louden & Chian 1999), which, combined with the reflection data, permits construction of a schematic cross-section of this part of the Labrador margin (Fig. 12e). Between continental and definite oceanic crust is a transition zone similar to that on the Greenland margin with a 5 km thick, high-velocity, high-gradient (6.4–7.7 km s^{-1}) layer overlain by a thin (<2 km), low-velocity (4–5 km s^{-1}) layer. In contrast to line AGC90-3 over the Greenland margin, no S reflection is observed on line AGC90-1 across the Labrador transition zone (Louden & Chian 1999).

A second wide-angle reflection–refraction line (5E), which images part of the transition zone farther north, has been published by Funck & Louden (1999). This line is the eastern extension of a survey designed to investigate the structure of the Proterozoic rocks farther west and does not extend as far as oceanic crust (Fig. 2). None the less, the line (Fig. 11b) shows that

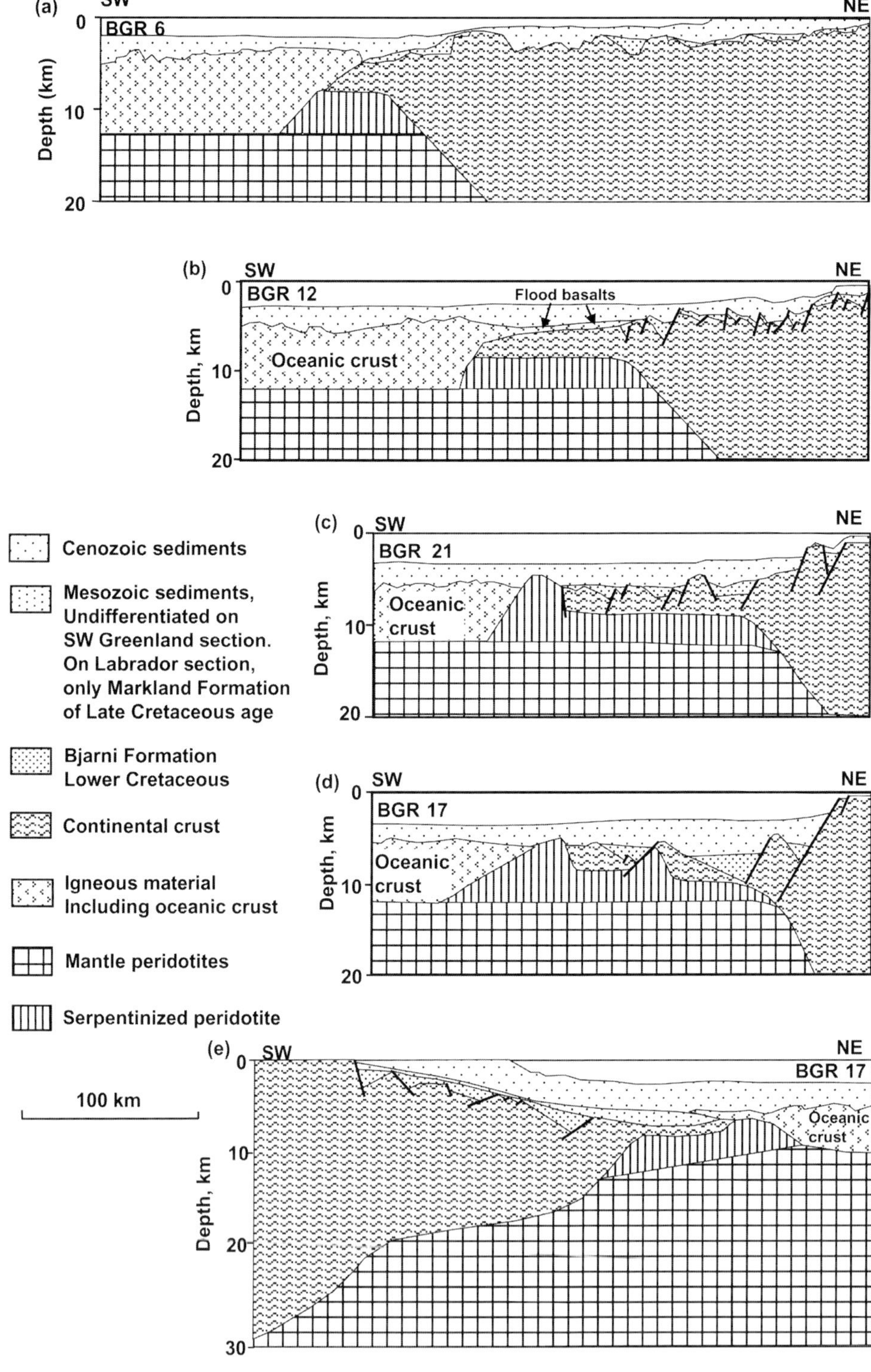
(a)
SW
NE
BGR 6
Depth (km)
0
10
20
(b)
SW
NE
BGR 12
Flood basalts
Oceanic crust
Depth, km
0
10
20
(c)
SW
NE
BGR 21
Oceanic crust
Depth, km
0
10
20
(d)
SW
NE
BGR 17
Oceanic crust
Depth, km
0
10
20
(e)
SW
NE
BGR 17
Oceanic crust
Depth, km
0
10
20
30
Cenozoic sediments
Mesozoic sediments, Undifferentiated on SW Greenland section. On Labrador section, only Markland Formation of Late Cretaceous age
Bjarni Formation Lower Cretaceous
Continental crust
Igneous material Including oceanic crust
Mantle peridotites
Serpentinized peridotite
100 km

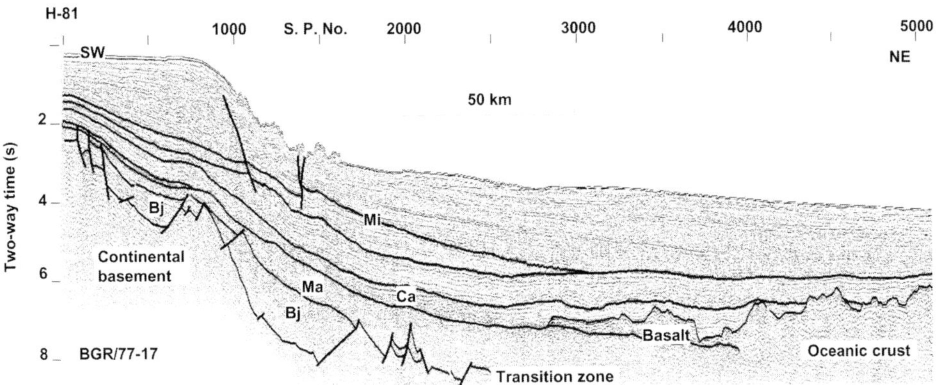

Fig. 13. Seismic line BGR 17 extending from the Bjarni H-81 borehole (Bell 1989) in the Hopedale Basin to the SW, across the Labrador transition zone to oceanic crust with magnetic anomalies 27 (61 Ma) and younger (Chalmers & Laursen 1995) to the NE. Half-grabens containing Bjarni Formation (Bj) sediments of Barremian–Albian age (127–99 Ma) can be followed to the edge of the transition zone, and sediments of the Markland Formation (Ma), of Cenomanian–Danian age (99–61 Ma) appear to continue across the transition zone to disappear under lower Paleogene basalts just landwards of the oceanic crust. This interpretation is tentative, and justification for it can be seen more clearly on the larger-scale version of this figure published by Chalmers & Laursen (1995, foldout 5). Ca, Cartwright Formation; Mi, the Hinz *et al.* (1979) reflector U, of Miocene age.

crustal thinning across this part of the Labrador margin takes place over a much narrower zone than farther southeast. On 90R1, the depth of the Moho rises from 30 to 20 km at a dip angle of 9°, whereas on 5E it rises from 38 km to 18 km at a dip of 30°, the same dip angle at which the Moho rises from 30 km to 8 km depth on 88R2 in the Greenland transition zone (Fig. 11a).

Formation of the Labrador and Greenland transition zones

Structural development

Figure 14b shows a cross-section of the transition zones during Chron 27, just before the start of sea-floor spreading. This cross-section has been redrawn from Louden & Chian (1999) and a similar cross-section has been published by Chian *et al.* (1995a). The cross-section, and particularly thinning of the crust, is asymmetric, with the location of break-up offset towards the Greenland margin and a broad zone of extended crust beneath the outer Labrador margin. On the

Greenland margin, the crust thins from 30 to 3 km over a distance of only 45 km, whereas the equivalent thinning on the Labrador margin takes place over a zone 175 km wide. There appears to be a zone where continental crust had already broken and serpentinized peridotites had reached either the sea floor or the base of the sedimentary layer before the onset of true sea-floor spreading. The first oceanic crust appeared within the zone of serpentinite diapirs. If extension and thinning of the continental crust developed by pure shear alone, then it did so to produce the asymmetric profile shown in Figure 14a. However, the asymmetry shown in Figure 14a suggests that simple shear may have been involved in the extension process. On the other hand, the early extension to about $\beta = 2$ seems to have been symmetric (Fig. 14b) and may have taken place by pure shear.

Understanding of how the transition zones developed is further complicated by the observation that the cross-section of the Labrador margin varies along strike. The section along line 5E (Fig. 11b) resembles that along 88R2

Fig. 12. Schematic cross-sections across the southern West Greenland transition zone (**a**)– (**d**) and the Labrador transition zone (**e**). The interpretations in (**a**)– (d) were published by Chalmers (1997) and are along multichannel seismic lines BGR 6, BGR 12, BGR 21 and BGR 17, respectively (see also Chalmers & Laursen 1995). The deep structure is based on the wide-angle reflection–refraction line 88R2 published by Chian & Louden (1994) and shown in Figure 11a. The interpretation in (**e**) is along multichannel seismic line BGR 17 as shown by Chalmers & Laursen (1995). The deep structure is based on the wide-angle reflection–refraction line 90R1 published by Louden & Chian (1999) and shown in Figure 11c.

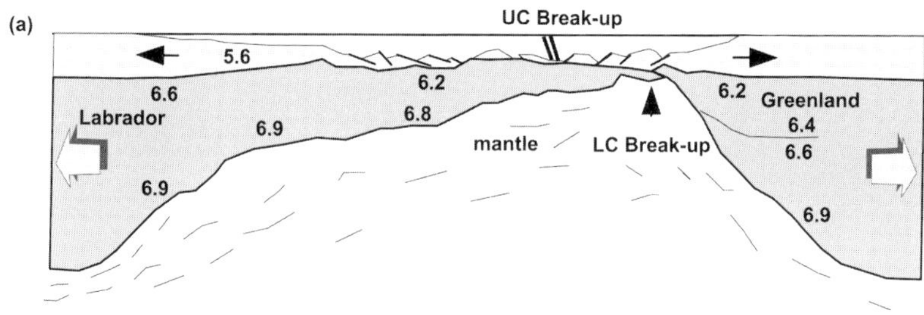

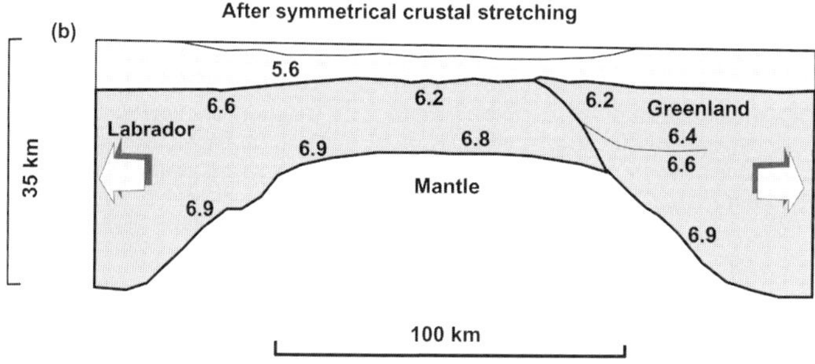

Fig. 14. Reconstructed schematic cross-sections originally published by Louden & Chian (1999, fig. 11). (**a**) is based on cross-sections similar to those shown in Figure 11c–e and is reconstructed to the situation at magnetochron 27 (61 Ma, mid-Paleocene time) at the start of sea-floor spreading. If the transition zones formed by pure shear, with no upper or lower plate, it is not clear how the observed asymmetric cross-sections observed could develop. (**b**) Reconstruction to an earlier period than shown in (**a**) assuming simple shear as discussed in the text. (**b**) was constructed assuming that the Labrador transition zone formed the lower plate and the Greenland margin formed the upper plate. Louden & Chian (1999) postulated that Greenland formed the upper plate because they interpret an S-reflector at the crust–mantle boundary on the Greenland side, but not on the Labrador side and (Louden, pers. com.) because of the velocity structure of the crust and location of continental fault blocks. A similar reconstruction using the opposite polarity can also be made because evidence for sediments (Markland Formation) deposited during an episode of Late Cretaceous thermal subsidence can be seen in the Labrador transition zone (Fig. 13) but not in the Greenland transition zone (Fig. 10).

(Fig. 11a) across the Greenland margin more than it does that along 90R1 (Fig. 11c). Line 5E is located west of the limit of oceanic crust shown in Figure 2 and continues right into the coast, and there are no deep seismic data that image the deep crust and upper mantle in the equivalent area of the Greenland margin. The significance of the along-strike change in structure of the Labrador margin is not understood.

Chian *et al.* (1995*a*) suggested two hypotheses for the origin of the non-magmatic transition zone based on mass-balanced reconstructions of crustal cross-sections across both margins. Both models incorporated the interpretation that the lower 5 km thick, high-velocity, high-velocity-gradient (6.4–7.7 km s^{-1}) layer consists of serpentinized mantle peridotite. The upper thin (<2 km), low-velocity (4–5 km s^{-1}) layer could be either oceanic (basaltic) crust or thin continental crust. Chian

et al. (1995*a*) argued that such thin crust could result from very slow (*c.* 6 mm a^{-1}) sea-floor spreading, as suggested by Srivastava & Roest (1995). The thinner than normal crustal thickness could be related to this very slow rate of spreading, which restricts the volume of melt produced (Reid & Jackson 1981; Bown & White 1995).

Chian *et al.* (1995*a*) also considered the possibility that the transition zone could be formed by simple shear continental rifting where the transition zone crust on the West Greenland margin represents block-faulted and thinned continental upper crust, the underlying S-reflector images the detachment surface and the pre-rift lower-plate is now preserved on the Labrador side of the rift zone. This asymmetric break-up was followed by continued necking of lithosphere and serpentinization of upper mantle, which formed the observed high-velocity layer in the transition zone. Louden & Chian (1999) reconstructed the transition zone through this process (Fig. 14) and found that, at its start, the continental crust had already been thinned to half thickness ($\beta \approx 2$) and this early rift was symmetric. Louden & Chian (1999) concluded that the early rifting process might therefore have been pure shear.

The arguments of Chian *et al.* (1995*a*) and Louden & Chian (1999) in favour of a simple-shear origin of the transition zones are based on the observation of an S-reflector under the Greenland margin on reflection seismic line AGC90-3 and the absence of one on the Labrador margin on line AGC90-1, the velocity structure of the crust and the attitude of the continental fault blocks. No S-reflector can be observed on either line BGR 21 or BGR 17 on the Greenland side, and inspection of the original migrated version of line AGC/90-3 shows that any S-reflector is much more poorly imaged on that line than the well-known S-reflector off western Iberia (Reston *et al.* 1996).

Chalmers (1997) observed that there is no post-rift, pre-drift sedimentary infill in the Greenland transition zone older than the onset of sea-floor spreading (see Fig. 10). The irregular 'basement' surface between 4.5 and 10 km below sea level is broken into grabens and half-grabens containing sediments and possibly sills, and covered by 2–3 km of flat-lying sediment (Keen *et al.* 1994*a*; Chalmers & Laursen 1995, foldouts 3 and 4). The deepest of the flat-lying sedimentary layers can be followed onto undoubted oceanic crust, indicating that there was no period of thermal subsidence above the rifted blocks before the onset of sea-floor spreading. The absence of pre-drift, thermal subsidence phase sediments suggests that the transition zone was still extending until immediately before the onset of sea-floor spreading in mid-Paleocene time (chron 27N, 62 Ma). On the other hand, if the interpretation shown in Figures 12c and 13 is correct, it does indicate the existence of a thermal subsidence basin within the Labrador transition zone as the Upper Cretaceous Markland Formation may extend into the transition zone and possibly be capped by Paleogene basalts in the outer part of the transition zone. These observations are tentative and are based on only one poor-quality seismic line (BGR 17, Fig. 13) uncalibrated by borehole data. However, it may be possible that the orientation of simple shear during the final stages of continental break-up was in the opposite direction to that argued by Chian *et al.* (1995*a*) and Louden & Chian (1999). The Markland Formation, of Late Cretaceous–

Danian age, could have survived in the transition zone only if the upper plate was on the Labrador side. In that case, residual continental crust on the Greenland side constitutes the lower plate, and no overlying thermal subsidence basin would be found there.

Thermal restrictions on extension rates in the transition zone

It is extremely difficult to stretch the lithosphere by the amounts implied in the interpretations of the transition zones of the Labrador Sea without melting the upper mantle (see, e.g. discussion after Louden & Chian 1999). When the lithosphere extends, it must also thin to conserve volume (in a 2D model, cross-section area would be conserved) and the upper asthenosphere must rise to maintain isostatic balance. Rapid extension and asthenosphere uplift take place, by definition, so that heat is not lost from the asthenosphere during its ascent; that is, it rises adiabatically. To produce crust only 3 km thick by pure shear implies that the extension parameter β is in excess of eight and perhaps as much as 12. To extend and thin the lithosphere by this amount during a rapid adiabatic event such as described by McKenzie (1978) would result in the generation of several kilometres of melt (McKenzie & Bickle 1988) (Fig. 15, curve A), some of which would be erupted at the contemporaneous surface, and the result would be difficult to distinguish geophysically from oceanic crust. To extend and thin continental crust by a factor of eight or more without producing melt requires that heat be transported out of the mantle, and this is most easily accomplished if the spreading is sufficiently slow that the excess heat can be lost and hence the temperature–depth curve nowhere crosses the solidus for dry peridotite (Fig. 15, curve S). All quantitative models published to date deal only with a conduction model. A more accurate model would take account of heat transport by hydrothermal circulation in the crust in the zone undergoing serpentinization, and of heat generated during the exothermic serpentinization reaction. None the less, a purely conduction model should give a reasonable first approximation, as both of the other processes are confined to the uppermost few kilometres of an initially nearly 100 km thick lithospheric mantle.

Calculations about the effect of finite extension rates on melt generation have been carried out by Pedersen & Ro (1992) and Bown & White (1995). The two methods led to rather different conclusions, and Chalmers (1997) dis-

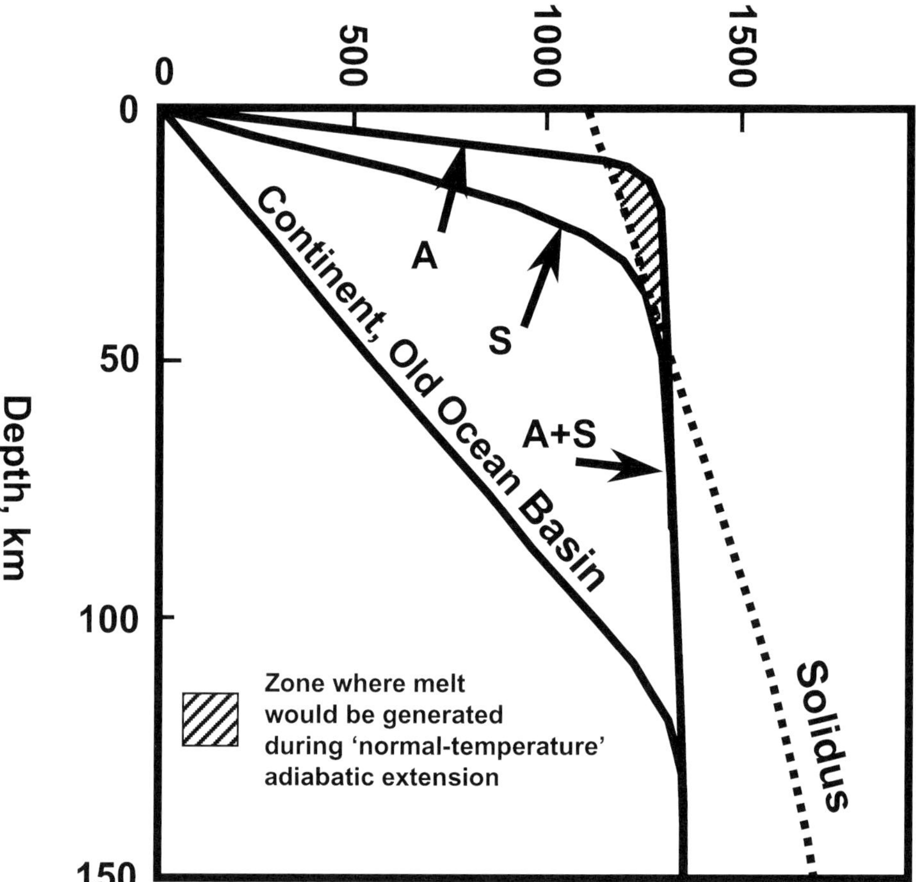

Fig. 15. Temperature–depth profiles for various scenarios discussed in the text. The temperature variation through lithosphere that has been stable for a long time is shown by the 'continent, old ocean basin' curve. Curve A shows an example of temperature variation with depth that can result if continental crust is extended adiabatically (McKenzie 1978) to a high beta factor, and melt will be produced in the region where the curve is shallower than solidus for mantle peridotites. To avoid melting, heat must be lost, either by conduction or by a combination of conduction in the deeper lithosphere and hydrothermal circulation in the shallower lithosphere, so that the temperature–depth curve S nowhere crosses the solidus for dry peridotite. If that is accomplished, a 'thin-spot' (Thompson & Gibson 1991) is formed. Redrawn from McKenzie & Bickle (1988).

cussed models with specific reference to the Labrador Sea margins.

In the following, it is assumed that the transitional crust was formed by stretching continental crust originally 35 km thick. Regardless of whether the stretching was by pure or simple shear, the thinnest crust (Fig. 14) is 3 km thick, which implies that effectively $\beta = 12$ at that location. Using the calculations of Pedersen & Ro (1992), it can be shown (Chalmers 1997) that rifting can proceed to $\beta = 12$ after a mini-

mum of 55 Ma, during which the rate of extension, assuming constant pure shear would have been 5 mm a^{-1}. The model of Bown & White (1995, fig. 13) shows that no melt would be produced for $\beta = 12$ if extension took 32 Ma or more, during which time the extension rate assuming uniform pure shear was 8.7 mm a^{-1}.

If extension and delamination of the transition zone started at *c.* 125 Ma and lasted for 30–35 Ma, as suggested by the Bown & White (1995) model, there should have been a thermal cooling phase that lasted about 30 Ma before the start of sea-floor spreading. Sediments deposited during such a phase should form a layer 1–2 km thick, which should be visible on the seismic lines in both the Greenland and Labrador transition zones. As discussed above, the Markland Formation in the Labrador basins and the Kangeq seismic sequence in the southern West Greenland basins were probably deposited into such a thermally subsiding basin. The Markland Formation can be traced across the Labrador transition zone (Fig. 13) but no such sedimentary interval can be seen on the seismic sections through the transition zone off Greenland (Fig. 10; see also Keen *et al.* 1994*b*; Chalmers & Laursen 1995). This observation suggests that slow extension–delamination of the continental crust in the transition zone continued until just before the onset of sea-floor spreading.

The Bown & White (1995) model forecasts that extension should have lasted at least 32 Ma. If their calculations are valid, extension began at or before 95 Ma. The main phase of extension in the Labrador basin took place during deposition of the Upper Bjarni Member, from 115 to 100 Ma (Fig. 3), which is earlier than Bown & White's latest date, but within the bounds of their model. The model of Pedersen & Ro (1992), in contrast, forecasts that 55 Ma is necessary for the extension phase, which is also consistent with extension having continued from 115 Ma to the onset of sea-floor spreading at 62 Ma. During those 53 Ma the average extension rate was 5 mm a^{-1}.

The above arguments are clearly uncertain and simplified. In particular, they take no account of the effects of varying rates of extension and, as discussed above, extension and crustal thinning during Early Cretaceous time appears to have been fast enough to have caused the partial melting that produced the Alexis Formation basalts in the Labrador shelf basins (Keen *et al.* 1994*a*), and possibly dykes of 135 Ma age (Larsen *et al.* 1999) in southern West Greenland. However, as long as extension rates during the later stages of rifting were low, the conclusions reached above are little affected.

Conclusions

(1) The geological history of the sedimentary basins offshore Labrador and southern West Greenland is becoming clear, and forthcoming work in connection with hydrocarbon exploration off Greenland will continue to improve our knowledge.

(2) The geology of the area southeast of Baffin Island is poorly known other than that major strike-slip movements are interpreted to have taken place during sea-floor spreading in the Labrador Sea and Baffin Bay. Existing open-file reflection seismic data, originally acquired by the oil industry in the 1970s and early 1980s, could be used to improve understanding of this area.

(3) True sea-floor spreading probably started in the Labrador Sea during Paleocene time (magnetochron 27N, 61 Ma).

(4) Continental crust and oceanic crust in the Labrador Sea are separated by some form of transition zone. Off southern West Greenland, the nature of the transition zone is different in different places. Between 60°N and 62°N, the transition zone may consist of very thin continental crust overlying serpentinized peridotites, or entirely of serpentinized peridotites, or of thin basalt overlying serpentinized peridotites. Near 63°N, continental and oceanic crust are separated by a volcanic continental margin containing seaward-dipping reflectors.

(5) Velocity profiles show that conjugate margins off Labrador and Greenland are asymmetric, and also that the Labrador margin changes significantly along strike. Only one deep profile is available off southern West Greenland, so nothing is known about any lateral change of crustal structure here.

(6) The structure of the lateral transition from the volcanic to non-volcanic margins off southern West Greenland is not understood.

(7) Solving some of the problems about the nature and development of the transition zones requires substantial amounts of new reflection seismic data, a number of carefully sited wide-angle reflection–refraction lines, a number of high-resolution magnetic profiles from deep-towed magnetometers, and sampling of the sub-surface, which can only be done by drilling.

We wish to thank T. Dahl-Jensen for help in producing Figure 8. We also wish to thank K. Louden, R. B. Whitmarsh and an anonymous referee for constructive comments that improved the paper. The paper is published with permission from the Geological Survey of Denmark and Greenland.

References

ARAM, R.B. 1999. West Greenland versus Vøring Basin: comparison of two deepwater frontier plays. *In:* FLEET, A.J. & BOLDY, S.A.R. (eds) *Petroleum Geology of Northwest Europe: Proceedings of the 5th Conference.* Geological Society, London, 315–324.

BALKWILL, H.R. 1987. Labrador Basin: structural and stratigraphic style. *In:* BEAUMONT, C. & TANKARD, A.J. (eds) *Sedimentary Basins and*

Basin-forming Mechanisms. Canadian Society of Petroleum Geologists, Memoirs, **12**, 17–43.

BALKWILL, H.R., MCMILLAN, N.J., MACLEAN, B., WILLIAMS, G.L. & SRIVASTAVA, S.P. 1990. Geology of the Labrador Shelf, Baffin Bay and Davis Strait. *In*: KEEN, M.J. & WILLIAMS, G.L. (eds) *Geology of the Continental Margins of Eastern Canada.* Geological Survey of Canada, Geology of Canada, **2**, 293–348.

BARRETT, D.L., KEEN, C.E., MANCHESTER, K.S. & ROSS, D.I. 1971. Baffin Bay—an ocean. *Nature*, **229**, 551–553.

BATE, K.J. 1997. Interpretation of the basal section of well Kangâmiut-1, offshore southern West Greenland. *Danmarks og Grønlands Geologiske Undersøgelse Rapport*, 1997/76.

BATE, K.J., WHITTAKER, R.C., CHALMERS, J.A. & DAHL-JENSEN, T. 1994. Fylla complex—possible very large gas reserves off S.W Greenland. *Oil and Gas Journal*, **92**, 79–82.

BELL, J. S. (coordinator) 1989. *East Coast Basin Atlas Series, Labrador Sea.* Atlantic Geoscience Centre, Dartmouth, N.S.

BELL, J.S. & HOWIE, R.D. 1990. Paleozoic geology. *In*: KEEN, M.J. & WILLIAMS, G.L. (eds) *Geology of the Continental Margin of Eastern Canada.* Geological Survey of Canada, Geology of Canada, **2**, 141–165.

BERGGREN, W.A., KENT, D.V., SWISHER, C.C. III & AUBRY, M.-P. 1995. A revised Cenozoic geochronology and chronostratigraphy. *In*: BERGGREN, W.A., KENT, D.V., AUBRY, M.-P. & HARDENBOL, J. (eds) *Geochronology, Time Scales and Global Stratigraphic Correlation.* Society of Econonomic Paleontologists and Mineralogists, Special Publications, **54**, 129–212.

BESLIER, M.-O., GIRARDEAU, J. & BOILLOT, G. 1990. Kinematics of peridotite emplacement during North Atlantic continental rifting, Galicia, northwest Spain. *Tectonophysics*, **184**, 321–343.

BOILLOT, G., WINTERER, E.L. *et al.* 1988. *Proceedings of the Ocean Drilling Program, 103.* Ocean Drilling Program, College Station, TX, 37–51.

BOILLOT, G., WINTERER, E. L., MEYER, A. W. *et al.* (eds) 1987. *Proceedings of the Ocean Drilling Program, Initial Reports 103.* Ocean Drilling Program, College Station, TX.

BOWN, J.W. & WHITE, R.S. 1995. Effect of finite extension rate on melt generation at rifted continental margins. *Journal of Geophysical Research*, B9, **100**, 18011–18027.

BURDEN, E.T. & LANGILLE, A.B. 1990. Stratigraphy and sedimentology of Cretaceous and Paleocene strata in half-grabens on the southeast coast of Baffin Island, Northwest Territories. *Bulletin of Canadian Petroleum Geology*, **38**, 185–195.

CHALMERS, J.A. 1989. A pilot seismo-stratigraphic study on the West Greenland continental shelf. *Rapport Grønlands Geologiske Undersøgelse*, 142.

CHALMERS, J.A. 1991. New evidence on the structure of the Labrador Sea/Greenland continental margin. *Journal of the Geological Society, London*, **148**, 899–908.

CHALMERS, J.A. 1997. The continental margin off southern Greenland: along-strike transition from an amagmatic to a volcanic margin. *Journal of the Geological Society, London*, **154**, 571–576.

CHALMERS, J.A. 2000. Offshore evidence for Neogene uplift in central West Greenland. *Global and Planetary Change*, **24**, 311–318.

CHALMERS, J.A. & LAURSEN, K.H. 1995. Labrador Sea: the extent of continental crust and the timing of the start of seafloor spreading. *Marine and Petroleum Geology*, **12**, 205–217.

CHALMERS, J.A. & PULVERTAFT, T.C.R. 1993. The southern West Greenland continental shelf—was petroleum exploration abandoned prematurely? *In*: VORREN, T.O., BERGSAGER, E., DAHL-STAMNES, O.A., HOLTER, E., JOHANSEN, B., LIE, E. & LUND, T.B. (eds) *Arctic Geology and Petroleum Potential.* Norwegian Petroleum Society, Special Publications, **2**, 55–66.

CHALMERS, J.A., DAHL-JENSEN, T., BATE, K.J. & WHITTAKER, R.C. 1995. Geology and petroleum prospectivity of the region offshore southern West Greenland—a summary. *Rapport Grønlands Geologiske Undersøgelse*, **165**, 13–21.

CHALMERS, J.A., PULVERTAFT, T.C.R., CHRISTIANSEN, F.G., LARSEN, H.C., LAURSEN, K.H. & OTTESEN, T.G. 1993. The southern West Greenland continental margin: rifting history, basin development, and petroleum potential. *In*: PARKER, J.R. (ed.) *Petroleum Geology of Northwest Europe: Proceedings of the 4th Conference.* Geological Society, London, 915–931.

CHALMERS, J.A., PULVERTAFT, T.C.R., MARCUSSEN, C. & PEDERSEN, A.K. 1999. New insight into the structure of the Nuussuaq Basin, central West Greenland. *Marine and Petroleum Geology*, **16**, 197–224.

CHIAN, D. & LOUDEN, K.E. 1994. The continent–ocean crustal transition across the southwest Greenland margin. *Journal of Geophysical Research*, **99**, 9117–9135.

CHIAN, D., KEEN, C., REID, I. & LOUDEN, K.E. 1995a. Evolution of nonvolcanic rifted margins: new results from the conjugate margins of the Labrador Sea. *Geology*, **23**, 589–592.

CHIAN, D., LOUDEN, K.E., MINSHULL, T.A. & WHITMARSH, R.B. 1999. Deep structure of the ocean–continent transition in the southern Iberia Abyssal Plain from seismic refraction profiles: Ocean Drilling Program (Legs 149 and 173) transect. *Journal of Geophysical Research*, **104**, 7443–7462.

CHIAN, D., LOUDEN, K.E. & REID, I. 1995b. Crustal structure of the Labrador Sea conjugate margin and implications for the formation of nonvolcanic continental margins. *Journal of Geophysical Research*, **100**, 24239–24253.

CLARKE, D.B. 1970. Tertiary basalts of Baffin Bay: possible primary magma from the mantle. *Contributions to Mineralogy and Petrology*, **25**, 203–224.

CLARKE, D.B. & PEDERSEN, A.K. 1976. Tertiary volcanic province of West Greenland. *In*: ESCHER, A. & WATT, W.S. (eds) *Geology of Greenland*. Geological Survey of Greenland, Copenhagen, 365–385.

CLARKE, D.B. & UPTON, B.G.J. 1971. Tertiary basalts of Baffin Island: field relations and tectonic setting. *Canadian Journal of Earth Sciences*, **8**, 248–258.

COAKLEY, B.J. & COCHRAN, J.R. 1998. Gravity evidence of very thin crust at the Gakkel Ridge (Arctic Ocean). *Earth and Planetary Science Letters*, **162**, 81–95.

DAM, G. & SØNDERHOLM, M. 1994. Lowstand slope channels of the Itilli succession (Maastrichtian–Lower Paleocene), Nuussuaq, West Greenland. *Sedimentary Geology*, **94**, 49–71.

DAM, G. & SØNDERHOLM, M. 1998. Sedimentological evolution of a fault-controlled Early Paleocene incised-valley system, Nuussuaq Basin, West Greenland. *In*: SHANLEY, K.W. & MCCABE, P.J. (eds) *Relative Role of Eustasy, Climate, and Tectonism in Continental Rocks*. Society of Economic Paleontologists and Mineralogists, Special Publications, **59**, 109–121.

DAM, G., LARSEN, M., NØHR-HANSEN, H. & PULVERTAFT, T.C.R. 1999. Discussion on the erosional and uplift history of NE Atlantic passive margins: constraints on a passing plume; reply by Clift. P. D., Carter, A. & Hurford, A. J. *Journal of the Geological Society, London*, **156**, 653–656.

DAM, G., LARSEN, M. & SØNDERHOLM, M. 1998. Sedimentary response to mantle plumes: implications from Paleocene onshore successions, West and East Greenland. *Geology*, **26**, 207–210.

DAM, G., NØHR-HANSEN, H., PEDERSEN, G.K. & SØNDERHOLM, M. 2000. Sedimentary and structural evidence of a new early Campanian rift phase in the Nuussuaq Basin, West Greenland. *Cretaceous Research*, **21**, 127–154.

DEAN, S.M., MINSHULL, T.A., WHITMARSH, R.B. & LOUDEN, K.E. 2000. Deep structure of the ocean–continent transition in the southern Iberia Abyssal Plain from seismic refraction profiles: the IAM-9 transect at 40°20′N. *Journal of Geophysical Research*, B3, **105**, 5859–5885.

DISCOVERY 215 WORKING GROUP 1998. Deep structure in the vicinity of the ocean–continent transition zone under the southern Iberia Abyssal Plain. *Geology*, **26**, 743–746.

DOOLEY, T., MCCLAY, K. & BONORA, M. 1999. 4D evolution of segmented strike-slip fault systems: applications to NW Europe. *In*: FLEET, A.J. & BOLDY, S.A.R. (eds) *Petroleum Geology of Northwest Europe: Proceedings of the 5th Conference*. Geological Society, London, 215–225.

FUNCK, T. & LOUDEN, K.E. 1999. Wide-angle seismic transect across the Torngat Orogen, northern Labrador: evidence for a Proterozoic crustal root. *Journal of Geophysical Research*, B4, **104**, 7463–7480.

GEOLOGICAL SURVEY OF CANADA. *Gravity Anomaly Map of the Continental Margin of Eastern Canada, 1:5 000 000*. Geological Survey of Canada Map, **1708A**.

GILL, R.C.O., PEDERSEN, A.K. & LARSEN, J.G. 1992. Tertiary picrites in West Greenland: melting at the periphery of a plume? *In*: STOREY, B.C., ALABASTER, T. & PANKHURST, R.J. (eds) *Magmatism and the Causes of Continental Break-up*. Geological Society, London, Special Publications, **68**, 335–348.

GRADSTEIN, F.M., AGTERBERG, F.P., OGG, J.G., HARDENBOL, J., VAN VEEN, P., THIERRY, J. & HUANG, Z. 1995. A Triassic, Jurassic and Cretaceous time scale. *In*: BERGGREN, W.A., KENT, D.V., AUBRY, M.-P. & HARDENBOL, J. (eds) *Geochronology, Time Scales and Global Stratigraphic Correlation*. Society of Economic Paleontologists and Mineralogists, Special Publications, **54**, 95–126.

GRAHAM, D.W., LARSEN, L.M., HANAN, B.B., STOREY, M., PEDERSEN, A.K. & LUPTON, J.E. 1998. Helium isotope composition of the early Iceland mantle plume inferred from the Tertiary picrites of West Greenland. *Earth and Planetary Science Letters*, **160**, 241–255.

HALD, N. & PEDERSEN, A.K. 1975. Lithostratigraphy of the Early Tertiary volcanic rocks of central West Greenland. *Rapport Grønland Geologiske Undersøgelse*, **69**, 17–24.

HENDERSON, G., SCHEINER, E.J., RISUM, J.B., CROXTON, C.A. & ANDERSEN, B.B. 1981. The West Greenland Basin. *In*: KERR, J.W. & FERGUSSON, A.J. (eds) *Geology of the North Atlantic Borderlands*. Canadian Society of Petroleum Geologists, Memoirs, **7**, 399–428.

HINZ, K., SCHLUTER, H.-U., GRANT, A.C., SRIVASTAVA, S.P., UMPLEBY, D. & WOODSIDE, J. 1979. Geophysical transects of the Labrador Sea: Labrador to southwest Greenland. *Tectonophysics*, **59**, 151–183.

HOFFMAN, H.J. & RESTON, T.J. 1992. Nature of the S reflector beneath the Galicia Banks rifted margin; preliminary results from prestack depth migration. *Geology*, **20**, 1091–1094.

HOLM, P.M., GILL, R.C.O., PEDERSEN, A.K., LARSEN, J.G., HALD, N., NIELSEN, T.F.D. & THIRLWALL, M.F. 1993. The Tertiary picrites of West Greenland; contributions from 'Icelandic' and other sources. *Earth and Planetary Science Letters*, **115**, 227–244.

HOOD, P.J. & BOWER, M.E. 1973. Low-level aeromagnetic surveys of the continental shelves bordering Baffin Bay and the Labrador Sea. *In*: HOOD, P.J. (ed.) *Earth Science Symposium on Offshore Eastern Canada*. Geological Survey of Canada, Paper, **71-23**, 573–598.

ISAACSON, E.S. & NEFF, D.B. 1999. A, B AVO cross-plotting and its application to Greenland and the Barents Sea. *In*: FLEET, A.J. & BOLDY, S.A.R. (eds) *Petroleum Geology of Northwest Europe: Proceedings of the 5th Conference*. Geological Society, London, 1289–1298.

JACKSON, H.R., DICKIE, K. & MARILLIER, F. 1992. A seismic reflection study of northern Baffin Bay: implication for tectonic evolution. *Canadian Journal of Earth Sciences*, **29**, 2353–2369.

JACKSON, H.R., KEEN, C.E., FALCONER, R.K.H. & APPLETON, K.P. 1979. New geophysical evidence for seafloor spreading in central Baffin Bay. *Canadian Journal of Earth Sciences*, **16**, 2122–2135.

KEEN, C.E. & BARRETT, D.L. 1972. Seismic refraction studies in Baffin Bay: an example of a developing ocean basin. *Geophysical Journal of the Royal Astronomical Society*, **30**, 253–271.

KEEN, C.E., COURTNEY, R.C., DEHLER, S.A. & WILLIAMSON, M.-C. 1994a. Decompression melting at rifted margins: comparison of model predictions with the distribution of igneous rocks on the eastern Canadian margin. *Earth and Planetary Science Letters*, **121**, 403–416.

KEEN, C.E., KEEN, M.J., ROSS, D.I. & LACK, M. 1974. Baffin Bay: small ocean basin formed by sea-floor spreading. *AAPG Bulletin*, **58**, 1089–1108.

KEEN, C.E., POTTER, P. & SRIVASTAVA, S.P. 1994b. Deep seismic reflection data across the conjugate margins of the Labrador Sea. *Canadian Journal of Earth Sciences*, **31**, 192–205.

KERR, J.W. 1967. A submerged continental remnant beneath the Labrador Sea. *Earth and Planetary Science Letters*, **2**, 283–289.

KING, A.F. & MCMILLAN, N.J. 1975. A mid-Mesozoic breccia from the coast of Labrador. *Canadian Journal of Earth Sciences*, **12**, 44–51.

KLOSE, G.W., MALTERRE, E., MCMILLAN, N.J. & ZINKAN, C.G. 1982. Petroleum exploration offshore southern Baffin Island, northern Labrador Sea, Canada. *In*: EMBRY, A.F. & BALKWILL, H. (eds) *Arctic Geology and Geophysics*. Canadian Society of Petroleum Geologists, Memoirs, **8**, 233–244.

LARSEN, L.M., REX, D.C., WATT, W.S. & GUISE, P.G. 1999. ^{40}Ar–^{39}Ar dating of alkali basaltic dykes along the south-west coast of Greenland: Cretaceous and Tertiary igneous activity along the eastern margin of the Labrador Sea. *Geology of Greenland Survey Bulletin*, **184**, 19–29.

LAXON, S. & MCADOO, D. 1998. Satellites provide new insights into polar geophysics. *EOS Transactions, American Geophysical Union*, **79** (69), 72–73.

LOUDEN, K. & CHIAN, D. 1999. The deep structure of non-volcanic rifted continental margins. *Philosophical Transactions of the Royal Society of London*, **357**, 767–804.

MANDERSCHEID, G. 1980. The geology of the offshore sedimentary basin of West Greenland. *In*: MIALL, A.D. (ed.) *Facts and Principles of World Oil Occurrence*. Canadian Society of Petroleum Geologists, Memoirs, **6**, 951–973.

MCKENZIE, D.P. 1978. Some remarks on the development of sedimentary basins. *Earth and Planetary Science Letters*, **40**, 25–32.

MCKENZIE, D.P. & BICKLE, M.J. 1988. The volume and composition of melt generated by extension of the lithosphere. *Journal of Petrology*, **29**, 625–679.

MCWHAE, J.R.H., ELIE, R., LAUGHTON, K.C. & GUNTHER, P.R. 1980. Stratigraphy and petroleum prospects of the Labrador Shelf. *Bulletin of Canadian Petroleum Geology*, **28**, 460–488.

MENZIES, A.W. 1982. Crustal history and basin development of Baffin Bay. *In*: DAWES, P.R. & KERR, J.W. (eds) *Nares Strait and the Drift of Greenland: a Conflict in Plate Tectonics*. Meddelelser om Grønland, Geoscience, **8**, 295–312.

MIDTGAARD, H.H. 1996. Inner-shelf to lower-shoreface hummocky sandstone bodies with evidence for geostrophic influenced combined flow, Lower Cretaceous, West Greenland. *Journal of Sedimentary Research*, **66**, 343–353.

NØHR-HANSEN, H. 1998. Dinoflagellate cyst stratigraphy of the Upper Cretaceous to Paleogene strata from the Hellefisk-1. *Ikermiut-1, Kangâmiut-1 and Nukik-1*, **1998/54**.

ODP Leg 173 Shipboard Scientific Party 1998. Drilling reveals tranistion from continental breakup to early magmatic crust. *EOS Transactions, American Geophysical Union*, **79**, 173, 180–181.

PEDERSEN, T. & RO, H.E. 1992. Finite duration extension and decompression melting. *Earth and Planetary Science Letters*, **113**, 15–22.

PICKUP, S.L.B., WHITMARSH, R.B., FOWLER, C.M.R. & RESTON, T.J. 1996. Insight into the nature of the ocean–continent transition off West Iberia from a deep multichannel seismic reflection profile. *Geology*, **24**, 1079–1082.

PULVERTAFT, T.C.R. 1979. Lower Cretaceous fluvial–deltaic sediments at Kûk. Nûgssuaq, West Greenland. *Geological Society of Denmark, Bulletin*, **28**, 57–72.

PULVERTAFT, T.C.R. 1989. The geology of Sarqaqdalen, *West Greenland, with special reference to the Cretaceous boundary fault system*. Grønlands Geologiske Undersøgelse Open File Series, **89/5**.

REID, I. & JACKSON, H.R. 1981. Oceanic spreading rate and crustal thickness. *Marine Geophysical Researches*, **5**, 165–172.

REID, I. & JACKSON, H.R. 1997. Crustal structure of northern Baffin Bay: seismic refraction results and tectonic implications. *Journal of Geophysical Research*, B1, **102**, 523–542.

RESTON, T.J., KRAWCZYK, C.M. & KLAESCHEN, D. 1996. The S-reflector west of Galicia (Spain). Evidence for detachment faulting during continental breakup from pre-stack depth migration. *Journal of Geophysical Research*, **101**, 8075–8091.

RICE, P.D. & SHADE, B.D. 1982. Reflection seismic interpretation and seafloor spreading history of Baffin Bay. *In*: EMBRY, A.F. & BALKWILL, H.R. (eds) *Arctic Geology and Geophysics*. Canadian Society of Petroleum Geologists, Memoirs, **8**, 245–265.

RIISAGER, P. & ABRAHAMSEN, N. 1999. Magnetostratigraphy of Paleocene basalts from the Vaigat

Formation of West Greenland. *Geophysical Journal International*, **137**, 774–782.

ROEST, W.R. & SRIVASTAVA, S.P. 1989. Sea-floor spreading in the Labrador Sea: a new reconstruction. *Geology*, **17**, 1000–1003.

ROLLE, F. 1985. Late Cretaceous–Tertiary sediments offshore central West Greenland: lithostratigraphy, sedimentary evolution, and petroleum potential. *Canadian Journal of Earth Sciences*, **22**, 1001–1019.

ROOTS, W.D. & SRIVASTAVA, S.P. 1984. Origin of the marine magnetic quiet zones in the Labrador and Greenland Seas. *Marine Geophysical Researches*, **6**, 395–408.

ROSENKRANTZ, A. & PULVERTAFT, T.C.R. 1969. Cretaceous–Tertiary stratigraphy and tectonics in northern West Greenland. *In*: KAY, M. (ed.) *North Atlantic—Geology and Continental Drift*. American Association of Petroleum Geologists, Memoirs, **12**, 883–898.

SANDWELL, D.T. & SMITH, W.H.F. 1992. *Global Marine Gravity Field from ERS-1, Geosat and Seasat*. Scripps Institute of Oceanography, La Jolla, CA.

SAWYER, D. S., WHITMARSH, R. B., KLAUS, A. *et al.* (eds) 1994. *Proceedings of the Ocean Drilling Program, Initial Reports, 149*. Ocean Drilling Program, College Station, TX.

SKAARUP, N., CHALMERS, J.A. & WHITE, D. 2000. An AVO study of a possible new hydrocarbon play, offshore central West Greenland. *AAPG Bulletin*, **84**, 174–182.

SRIVASTAVA, S.P. 1978. Evolution of the Labrador Sea and its bearing on the early evolution of the North Atlantic. *Geophysical Journal of the Royal Astronomical Society*, **52**, 313–357.

SRIVASTAVA, S.P. 1983. Davis Strait: structures, origin and evolution. *In*: BOTT, M.H.P., SAXOV, S., TALWANI, M. & THIEDE, J. (eds) *Structure and Development of the Greenland–Scotland Ridge: New Methods and Concepts*. Plenum, New York, 159–189.

SRIVASTAVA, S.P. & KEEN, C.E. 1995. A deep seismic reflection profile across the extinct mid-Labrador Sea spreading center. *Tectonics*, **14**, 372–389.

SRIVASTAVA, S.P. & ROEST, W.R. 1995. Nature of thin crust across the southwest Greenland margin and its bearing on the location of the ocean–continent boundary. *In*: BANDA, E., TORNÉ, M. & TALWANI, M. (eds) *Rifted Ocean–Continent Boundaries*. Kluwer, Dordrecht, 95–120.

SRIVASTAVA, S.P. & ROEST, W.R. 1999. Extent of oceanic crust in the Labrador Sea. *Marine and Petroleum Geology*, **16**, 65–84.

SRIVASTAVA, S.P., FALCONER, R.K.H. & MACLEAN, B. 1981. Labrador Sea, Davis Strait, Baffin Bay: geology and geophysics—a review. *In*: KERR, J.W. & FERGUSSON, A.J. (eds) *Geology of the North Atlantic Borderlands*. Canadian Society of Petroleum Geologists, Memoirs, **7**, 333–398.

SRIVASTAVA, S.P., MACLEAN, B., MACNAB, R.F. & JACKSON, H.R. 1982. Davis Strait: structure and evolution as obtained from a systematic geophysical survey. *In*: EMBRY, A.F. & BALKWILL, H.R. (eds) *Arctic Geology and Geophysics*. Canadian Society of Petroleum Geologists, Memoir, **8**, 267–278.

STOREY, M., DUNCAN, R.A., PEDERSEN, A.K., LARSEN, L.M. & LARSEN, H.C. 1998. ^{40}Ar/^{39}Ar geochronology of the West Greenland Tertiary volcanic province. *Earth and Planetary Science Letters*, **160**, 569–586.

STOUGE, S. & PEEL, J.S. 1979. Ordovician conodonts from the Precambrian Shield of southern West Greenland. *Rapport Grønlands Geologiske Undersøgelse*, **91**, 105–109.

THOMPSON, R.N. & GIBSON, S.A. 1991. Subcontinental mantle plumes, hotspots and pre-existing thinspots. *Journal of the Geological Society, London*, **148**, 973–977.

TUCHOLKE, B.E. & FRY, V.A. 1985. Basement structure and sediment distribution in Northwest Atlantic Ocean. *AAPG Bulletin*, **69**, 2077–2097.

UMPLEBY, D.C. *Geology of the Labrador Shelf*. Geological Survey of Canada Paper, **79-13**.

WATT, W.S. 1969. The coast-parallel dike swarm of southwest Greenland in relation to the opening of the Labrador Sea. *Canadian Journal of Earth Sciences*, **6**, 1320–1321.

WHEELER, J. O., HOFFMAN, P. F., CARD, K. D., DAVIDSON, A., SANFORD, B. V., OKULITCH, A. V. & ROEST, W. R. (compilers) 1996. *Geological Map of Canada, scale 1:5 000 000*. Geological Survey of Canada, Map **1860A**.

WHITMARSH, R.B., PINHEIRO, L.M., MILES, P.R., RECQ, M. & SIBUET, J.C. 1993. Thin crust at the western Iberia ocean–continent transition and ophiolites. *Tectonics*, **12**, 1230–1239.

WHITMARSH, R.B., WHITE, R.S., HORSEFIELD, S.J., SIBUET, J.-C., RECQ, M. & LOUVEL, V. 1996. The ocean–continent boundary off the western continental margin of Iberia: crustal structure west of Galicia Bank. *Journal of Geophysical Research*, **101**, 28291–28314.

WHITTAKER, R.C., HAMANN, N.E. & PULVERTAFT, T.C.R. 1997. A new frontier province offshore northern West Greenland: structure, basin development and petroleum potential of the Melville Bay area. *AAPG Bulletin*, **81**, 979–998.

WILLIAMSON, M.-C., COURTNEY, R.C., KEEN, C.E. & DEHLER, S.A. 1994. Relationship between crustal deformation and magmatism in rift zones: modelling approach and applications to the eastern Canadian margin. *Geological Survey of Canada, Current Research*, **1994**, 251–258.

The role of syn-rift magmatism in the rift-to-drift evolution of the West Iberia continental margin: geophysical observations

R.B. WHITMARSH[1], T.A. MINSHULL[2], S.M. RUSSELL[3], S.M. DEAN[2], K.E. LOUDEN[4] & D. CHIAN[5]

[1]*Southampton Oceanography Centre, European Way, Southampton SO14 3ZH, UK (e-mail: rbw@soc.soton.ac.uk)*
[2]*Southampton Oceanography Centre, European Way, Southampton SO14 3ZH, UK*
[3]*Totalfinaelf SA, 2 Place de la Coupole, La Defense 6, 92400 Courbevoie, France*
[4]*Department of Oceanography, Dalhousie University, Halifax, N.S. B3H 4J1, Canada*
[5]*Geological Survey of Canada—Atlantic Region, Bedford Institute of Oceanography, P.O. Box 1006, Dartmouth, N.S. B2Y 4A2, Canada*

Abstract: The presence of a well-defined ocean–continent transition (OCT) and the absence of large volumes of extrusive or intrusive rocks on the West Iberia margin make it a good place to investigate how the largely amagmatic rifting and break-up of continental lithosphere evolves into oceanic crust produced by magmatic sea-floor spreading. In the southern Iberia Abyssal Plain there is a broad OCT with a characteristic seismic and magnetic character, distinct from both thinned continental crust and normal oceanic crust, which supports the notion that it consists predominantly of exhumed and serpentinized mantle. Interpretations of magnetic and seismic data indicate that on average only small amounts of syn-rift melt exist within the OCT. Isolated, probably margin-parallel, intrusive melt bodies are scattered within the eastern part of the OCT well beneath the top of acoustic basement. Within the western part of the OCT, closer to unambiguous sea-floor spreading magnetic anomalies, such bodies were later(?) emplaced at higher levels and more closely together in the basement until eventually sea-floor spreading began. The evidence does not support the hypothesis that ultraslow sea-floor spreading can explain the magnetic anomalies observed within the wider parts of the West Iberia OCT, where the OCT evolution is best resolved.

It is widely accepted that the basement structure of rifted continental margins presents two contrasting styles, volcanic and non-volcanic. Whereas considerable attention has been paid to explaining and predicting the volume of syn-rift melt produced by adiabatic decompression melting of the asthenosphere at volcanic margins (Mutter *et al.* 1988; Keen *et al.* 1994; Pedersen 1994; Bown & White 1995*a,b*) much less attention has been devoted to explaining the apparent absence of syn-rift melt products at non-volcanic margins (Harry & Bowling 1999; Bowling & Harry 2001; Minshull *et al.* 2001). An aspect of the problem, which is addressed here, is to explain how the largely amagmatic rifting and break-up of continental lithosphere at such margins evolves, in space and time, into oceanic crust that is produced largely or entirely by the magmatism that accompanies sea-floor spreading. This leads in turn to more philosophical

questions of how to define the onset of sea-floor spreading and the formation of oceanic crust. Here, we define sea-floor spreading as that essentially axisymmetric process active at an accretionary plate boundary that leads, within the resolution of current magnetochronological time scales, to the continuous formation of new oceanic crust. Oceanic crust is recognizable always by its characteristic two-layer velocity–depth structure including the presence of layer 3 (White *et al.* 1992) and often by its magnetic anomaly pattern (caused by alternating normally and reversely magnetized blocks) and by a layered sequence of volcanic extrusive rocks, dykes and gabbros; it is seismically distinct from the underlying mantle. This definition holds even for the unusually thin crusts formed at ultraslow (<10 mm a^{-1} half-rate) spreading ridges, which exhibit a normal volcanic layer 2 of 1.5–2.0 km thickness over a thinner than

From: WILSON, R.C.L., WHITMARSH, R.B., TAYLOR, B. & FROITZHEIM, N. 2001. *Non-Volcanic Rifting of Continental Margins: A Comparison of Evidence from Land and Sea.* Geological Society, London, Special Publications, **187**, 107–124. 0305-8719/01/$15.00 © The Geological Society of London 2001.

usual layer 3 (Muller *et al.* 1997, 1999; Klingel-hoefer *et al.* 2000).

Although rifting is often considered as a relatively simple 2D process one may expect that this is an oversimplification. The local mechanical conditions along a profile normal to a margin will alter with time during rifting; for example, as temperature changes cause the rheology, and probably the strain rate, to change. Further, many margins show clear evidence of the systematic propagation of the onset of sea-floor spreading, and hence of the break-up of thinned continental crust, along the margin (e.g. Laughton *et al.* 1970; Taylor *et al.* 1999). Thus, important changes in the third dimension may also be expected during the development of rifting. Taken to its limit therefore one is faced with a 4D problem very different from the almost 2D steady-state (on geological time scales) generation of oceanic crust at a mid-ocean ridge.

The West Iberia margin has been extensively studied for over two decades and a wide range of geophysical data, including seismic refraction, seismic reflection, surface and deep-towed magnetic and surface gravity profiles have been collected. The margin clearly has a non-volcanic style (although, as we shall see, it is not entirely amagmatic). For example, multichannel seismic reflection profiles frequently exhibit tilted fault blocks and seaward-dipping volcanic reflectors are never seen (Groupe Galice 1979; Krawczyk *et al.* 1996; Pickup *et al.* 1996), and seismic velocity models lack a clear underplated layer at the base of the crust (Pinheiro *et al.* 1992; Whitmarsh *et al.* 1996*b*; Chian *et al.* 1999; Dean *et al.* 2000). Evidence of significant syn-rift magmatism is also absent on land (Pinheiro *et al.* 1996) and in samples of the offshore acoustic basement obtained by dredging, manned submersibles and Ocean Drilling Program (ODP) coring. All these results can contribute to a discussion of the problem elicited above. Here the geophysical evidence for slight syn-rift magmatism on this margin is summarized and a hypothesis of how the margin evolves from amagmatic rifting to the onset of sea-floor spreading is proposed. This necessarily involves a discussion of the origin of the region that lies seaward of thinned continental crust and landward of the region with geophysical characteristics of oceanic crust, which we define as the ocean–continent transition (OCT) zone. We argue that the OCT is not the result of any form of sea-floor spreading. A companion paper (Minshull *et al.* 2001) discusses the same problem from the perspective of numerical modelling of asthenospheric melting.

Background

The western continental margin of Iberia extends from Cape Finisterre in the north to Cape Saint Vincent in the south (Fig. 1). The continental margin has a straight shelf, only a few tens of kilometres wide, and a steep continental slope. South of 40°N the slope is cut by numerous canyons. This simple picture is complicated by several offshore bathymetric features. The largest feature is Galicia Bank, an area of 200 km x 150 km within which the sea floor shoals to about 600 m water depth. Galicia Bank is characterized by a series of isolated seamounts on its southern edge (Vasco da Gama and Vigo; Fig. 2), and is separated from NW Iberia to the east by the Interior Basin, a broad submarine valley. At 39°N, the Estremadura Spur extends over 100 km offshore and, with Tore Seamount, forms a barrier between the Iberia and Tagus Abyssal Plains. Lastly, the ENE-trending Gorringe Bank forms the southern boundary of the Tagus Abyssal Plain and marks the surface expression of the seismically active Eurasia–Africa plate boundary. Final break-up between Iberia and the Grand Banks off Newfoundland occurred in Early Cretaceous time, just after anomaly M0 time, after a prolonged history of rifting beginning in the Triassic period (Pinheiro *et al.* 1996). Break-up and the onset of sea-floor spreading proceeded from south to north over some 16 Ma (Pinheiro *et al.* 1992).

A magnetic anomaly chart of the West Iberia margin is shown in Figure 1 (Miles *et al.* 1996). The 015°-trending sea-floor spreading anomaly 34 is clearly seen along the west edge of the chart. To the east of anomaly 34 a region of irregular anomalies corresponds to the Cretaceous constant polarity interval. This region is bounded between 37° and 41.5°N by the high-amplitude J anomaly, which lies parallel to anomaly 34 around 13–13.5°W. This enigmatic anomaly, unrelated to any pair of magnetic reversals, can be modelled throughout the North Atlantic by a zone of anomalous magnetization in the interval M0–M1 or even slightly later (Pitman & Talwani 1972; Rabinowitz *et al.* 1978; Tucholke & Ludwig 1982). East of J the anomaly amplitudes are greatly reduced except for an east–west chain of anomalies caused by Tore Seamount and the Estremadura Spur and in the relatively shallow region over, and south of, Galicia Bank. South of Galicia Bank in this eastern region some of the low-amplitude anomalies are elongated for many tens of kilometres parallel to the margin (Miles *et al.* 1996).

A large part of the deep margin of West Iberia is concealed beneath 1–3 km of post-rift

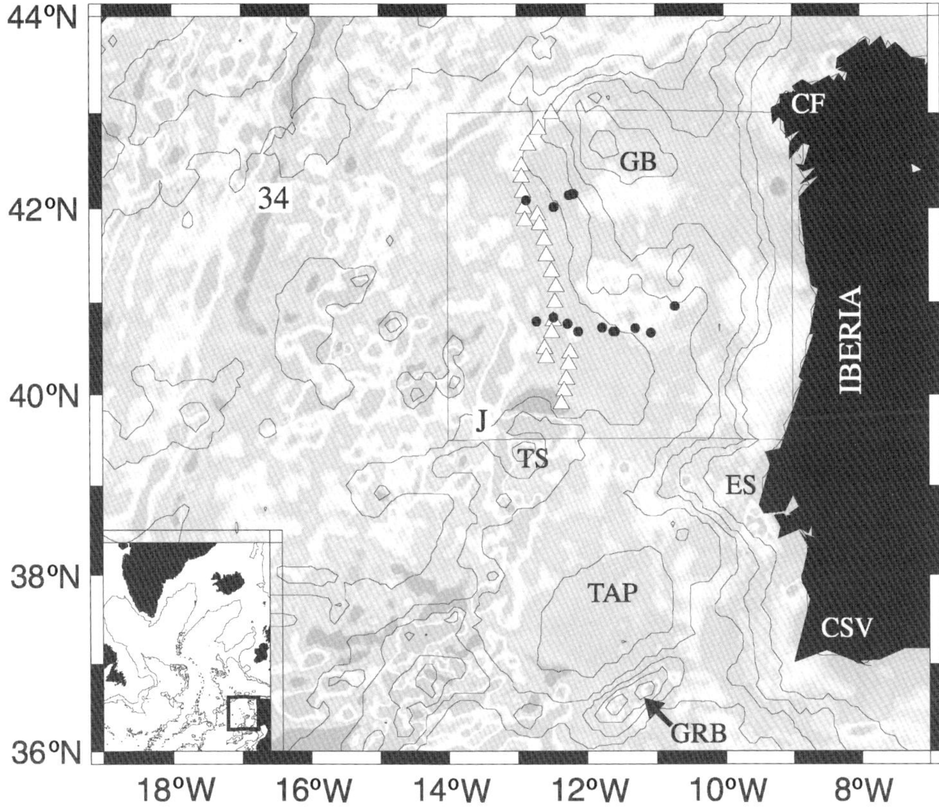

Fig. 1. Magnetic anomaly chart off West Iberia (from Miles *et al.* 1996) overlain by 1000 m bathymetric contours. Box denotes area of Figure 2. △, peridotite ridge; ●, ODP or DSDP drill sites on the margin. Anomalies J and 34 are labelled on the chart. Inset relates the West Iberia margin to the rest of the North Atlantic Ocean. CF, Finisterre; CSV, Cape St. Vincent; GB, Galicia Bank; TS, Tore Seamount; ES, Estremadura Spur; TAP, Tagus Abyssal Plain; GRB, Gorringe Bank.

sediments. However, the coverage of multi-channel seismic reflection profiles in part of the southern Iberia Abyssal Plain is sufficient to construct a contoured chart of the depth to acoustic basement (Fig. 3). This chart can be divided into four regions of contrasting relief, which will be used as a basis for later discussion in this paper. In a western region A, the basement surface consists of elongated ridges and troughs that parallel the sea-floor spreading isochrons further west. Modelling of magnetic anomalies and seismic velocities indicate that this region has the geophysical characteristics of oceanic crust, albeit perhaps slightly thinner than normal; magnetic modelling suggests that it was formed from the time of anomaly M3 (Whitmarsh & Miles 1995; Whitmarsh *et al.* 1996*a*; Dean *et al.* 2000). Tholeiitic pillow basalts have been dredged from the crest of a basement ridge only 170 km west of Site 1070 (Matthews 1962). Region B is relatively shallow.

It is flanked to the west by a basement ridge, which drilling has shown to consist of serpentinized peridotite and which extends along the West Iberia margin for *c.* 350 km in three or four en echelon, but otherwise continuous, segments (Figs. 1–3; Beslier *et al.* 1993; ODP Leg 149 Shipboard Party 1993. This ridge may be offset eastwards around 40°25′N. Peridotite breccia was also encountered in the basement of region B at ODP Site 899. The peridotite ridge is an enigmatic feature that west of Galicia Bank corresponds to an abrupt boundary across which basement magnetization falls by an order of magnitude going from west to east (Sibuet *et al.* 1995). Region B, described here as the region of offset peridotite ridges, is also associated with post-rift sediments that were folded into two NNE-trending monoclines during Mid-Miocene compression (North, pers. comm.; Masson *et al.* 1994). Seismic observations (see below) indicate that the Region C basement is

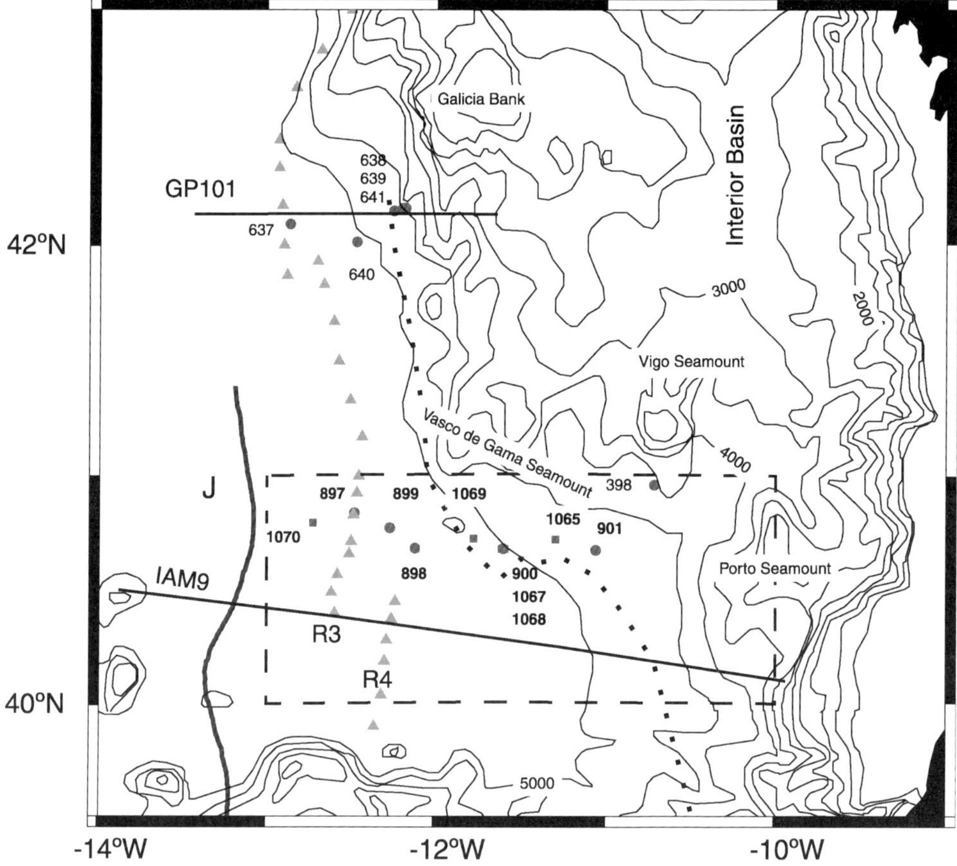

Fig. 2. Bathymetry of the Galicia Bank and southern Iberia Abyssal Plain margins of West Iberia (contours in metres). Box denotes area of Figure 3. Bold lines denote the seismic transects referred to in the text across Galicia Bank (Whitmarsh *et al.* 1996*b*) and within the southern Iberia Abyssal Plain (gridded lines, Chian *et al.* 1999; IAM9, Dean *et al.* 2000). DSDP and ODP sites appear as numbered dots and squares. Thick line denotes the trend of the J magnetic anomaly. Grey triangles denote the peridotite ridge; segments R3 and R4 are indicated. Small squares denote the estimated seaward edge of thinned continental crust.

also largely serpentinized mantle but, in contrast to B, it is deeper and the basement has much less relief (Pickup *et al.* 1996; Discovery 215 Working Group 1998; Dean *et al.* 2000). Region D contains the southern flank of Galicia Bank and seismic observations (see below), dredging and drilling strongly suggest that it consists of thinned continental crust (Capdevila & Mougenot 1988; ODP Leg 173 Shipboard Scientific Party 1998; Chian *et al.* 1999). Regions B and C correspond to the OCT.

Seismic observations

The deep structure of the West Iberia continental margin west of Galicia Bank and in the southern Iberia Abyssal Plain has been investigated along three transects of coincident multichannel seis-mic reflection and wide-angle OBS profiles and, in the Tagus Abyssal Plain, by more limited observations not presented here (Purdy 1975; Mauffret *et al.* 1989; Pinheiro *et al.* 1992).

West of Galicia Bank, Whitmarsh *et al.* (1996*b*) presented a seismic velocity model con-strained by, and coincident with, multichannel seismic reflection profile GP101 (Fig. 2; Groupe Galice 1979). They found that the continental crust thins westward from 17 km at 11.6°W to 2 km immediately east of the peridotite ridge. Immediately west of the peridotite ridge oceanic crust is only 2.5–3.5 km thick but 15 km further west (oceanward) it is 7 km thick. The peridotite ridge represents the apex of a triangular-shaped serpentinized mantle body of *c.* 60 km width (mostly 7.2–7.6 km s^{-1}) underlying both thinned continental and thin oceanic crust. On this transect any OCT zone, which has yet to be distinguished from adjacent crust, must be relatively narrow because Palaeozoic sediments were dredged *c.* 30 km east of the peridotite ridge (Mamet *et al.* 1991). Simple 1D numerical models of melting, assuming pure shear (Bown & White 1995*a*), and borehole subsidence information were used by Whitmarsh *et al.* (1996*b*) to constrain a rifting model. The easternmost continental crust experienced a total stretching factor of 4.3 (probably in two stages). Extension probably occurred over *c.* 25 Ma, with the highest rate of stretching at the beginning of the main earlier rift phase (in Valanginian time; 141–135 Ma). Modelling the extension of the 3 km thick continental crust requires a stretching factor (β) of >11 and a rift duration of >25 Ma. However, Whitmarsh *et al.* (1996*b*) could not exclude the presence of some melt products in the serpentinized peridotite layer; if this layer is included in the 'crust' then $\beta \geq 7$ and rifting lasted at least 13 Ma. The atypically thin oceanic crust immediately west of the peridotite ridge was explained by conductive cooling of the mantle during the long pre-break-up stretching phase, which temporarily caused reduced melting immediately after break-up.

In the southern Iberia Abyssal Plain two transects across the margin both produced rather similar structures of the OCT and thinned continental crust, although the width of the OCT was different (Fig. 2). The first transect includes the deep multichannel seismic reflection profile IAM9, which extends from the continental shelf to unequivocal oceanic crust, and a velocity model along a coincident wide-angle OBS line (Pickup *et al.* 1996; Dean *et al.* 2000). The velocity model of Dean *et al.* clearly shows the four-fold division referred to above (Fig. 4). Within 50 km west of the region of overlapping peridotite ridges a crust of 6.5–7.0 km thickness with a typical Atlantic oceanic velocity structure is seen (region A; Fig. 5b). No abrupt lateral change in upper basement velocities is observed between oceanic crust and the overlapping peridotite ridges (region B). In the easternmost part of the profile (region D) continental crust is defined by velocities in the range 5.5–6.8 km s^{-1} associated with tilted fault blocks and a reflection Moho on IAM9. Across the lower continental slope and rise the crust thins from 28 to 7 km over a distance of 80 km. In detail, the crust appears to be as thin as 4 km immediately adjacent to the OCT (Pickup *et al.* 1996). Coincidentally there is an abrupt lateral reduction in the velocity at the top

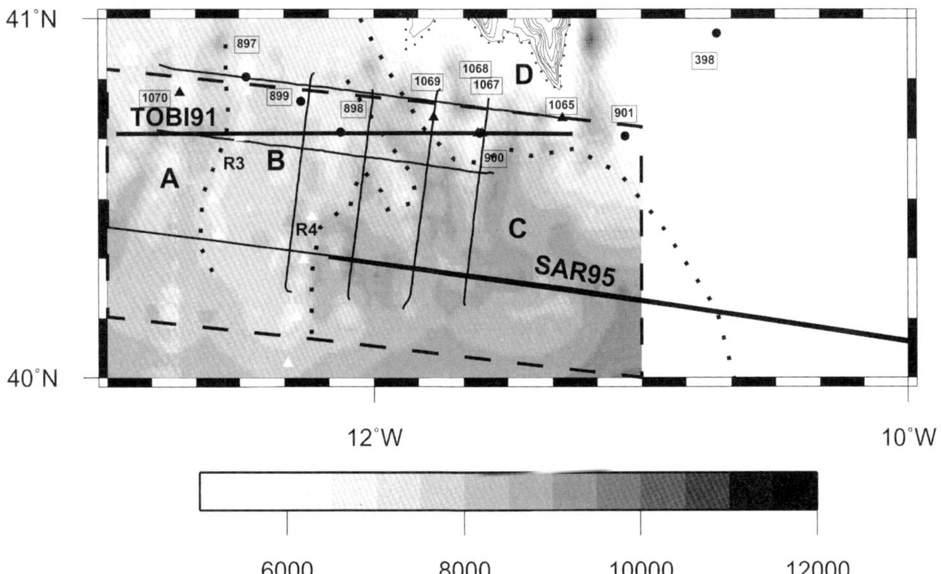

Fig. 3. Depth-to-basement chart (from Cole, pers. comm., after Discovery 215 Working Group (1998); scale in metres below sea level, see Figure 6 for a coloured version) showing regions A–D bounded by dotted lines (see text; from Russell (1999)). Bold lines denote SAR and TOBI deep-towed magnetometer tracks discussed in the text; other lines are wide-angle seismic lines of Chian *et al.* (1999); Dean *et al.* (2000). White triangles, peridotite ridge; black triangles and dots, ODP or DSDP drill sites on basement highs. Box bounded by dashed lines represents the area within which 29 regularly spaced, 010°-trending, synthetic basement (and magnetic) profiles were computed (see text).

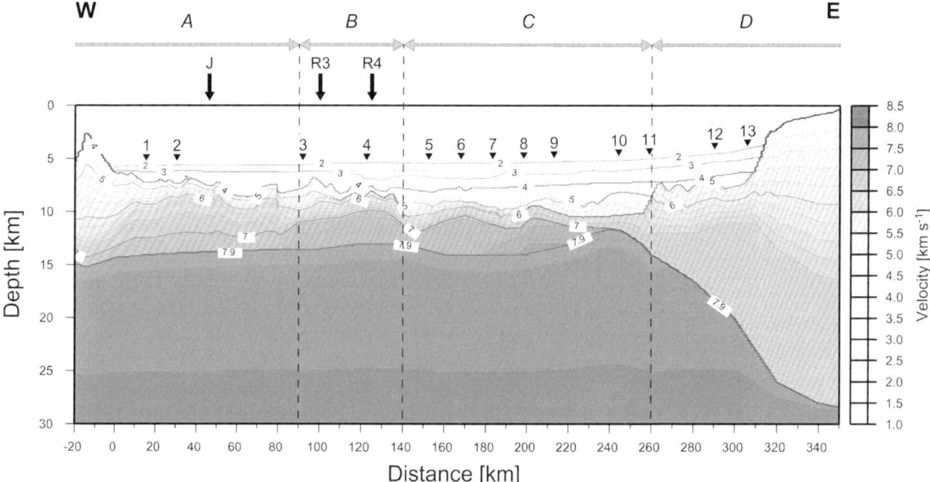

Fig. 4. Velocity model along profile IAM9 from Dean *et al.* (2000) based on observations at 13 ocean-bottom hydrophones and seismographs (▼). Velocity contours every $0.2\,\text{km s}^{-1}$. The four regions discussed in the text are labelled A–D. R3, R4 and J correspond to the two offset peridotite ridges and the J magnetic anomaly, respectively (see text).

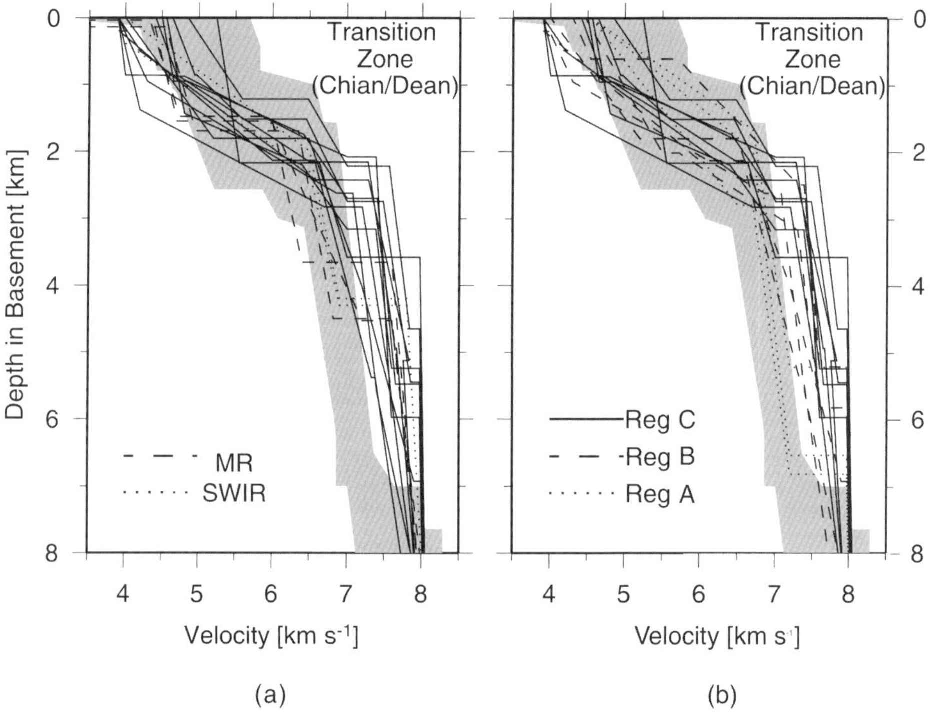

Fig. 5. Velocity–depth profiles below the top of acoustic basement at OBS–OBH locations in the ocean–continent transition zone (Regions B and C) and in region A from Chian *et al.* (1999) and Dean *et al.* (2000). For comparison the bounds for 59–142 Ma Atlantic Ocean normal oceanic crust are shown in grey (White *et al.* 1992). (a) comparison of OCT (regions B and C combined), Southwest Indian Ocean Ridge (SWIR; from Muller *et al.* (1997)) and Mohns Ridge profiles 2, 3 and 4 (MR; from Klingelhoefer *et al.* (2000)). The ridge profiles were selected away from the ridge axes and are over crust at least 9 Ma old. (b) A comparison of the region A, B and C profiles.

of acoustic basement and a decrease in basement relief. On this profile the OCT, including 50 km of region B, is 170 km wide. The OCT basement is characterized everywhere by a velocity structure consisting of a layer of high-velocity gradient (c. $1.25\,\text{s}^{-1}$) and of 2–3.5 km thickness, which corresponds to the seismically unreflective layer of Pickup *et al.* (1996), over a 7.3–7.9 km s^{-1} lower layer up to 4 km thick (Fig. 5). There is a gradual transition downwards from the lower layer to a normal mantle velocity of 8 km s^{-1}. Therefore, only weak wide-angle Moho reflections are observed; Moho reflections are absent within the OCT along profile IAM9. This structure resembles neither oceanic crust nor thinned continental crust. Instead, a basement of serpentinized mantle has been proposed to explain these and similar observations in the OCT (Whitmarsh *et al.* 1993; Pickup *et al.* 1996; Dean *et al.* 2000). The high-gradient upper layer represents the result of vigorous hydrothermal circulation, which has led to 25–100% serpentinization, whereas the lower layer, being deeper and perhaps less generally fractured, has a bulk serpentinization of <25%.

Dean *et al.* (2000) discussed the possibility that the OCT basement could contain some proportion of syn-rift melt products. They discussed an extreme argument, based on their velocity model, that all the 3 km thick upper layer represents extrusive rocks and that the lower layer represents some *ad hoc* 'underplated' mixture of 8 km s^{-1} mantle and <7.3 km s^{-1} intrusive rocks equivalent to a total of 4.5 km melt in all. First, they commented that the 1D model of Bown & White (1995*a*), assuming a β factor of at least 50, as implied by the absence of continental crust within the OCT, indicates minimum melt thicknesses of 3–5 km, depending on the duration of rifting. Thus it appears that either the effective β factor is ≤50 or the model is inappropriate. Second, there is also a problem in explaining how relatively dense extrusive rocks can rise buoyantly through lower-density serpentinized mantle. Last, the above hypothesis is not consistent with the possibly complete absence of mafic rocks with a syn-rift intrusion age in basement cores recovered at ODP Sites 897, 899, 900, 1067 and 1068 (mafic cores from Sites 900, 1067 and 1068 no longer provide support for Early Cretaceous syn-rift melt products there because a metagabbro from Site 1067 now appears to be much older (270 ± 3 Ma; Manatschal *et al.* 2001); however, the Site 899 basalt clasts remain undated). Dean *et al.*'s argument does not explain the ubiquitous c. $1.25\,\text{s}^{-1}$ velocity gradient in the upper layer because extrusive-dominated layer 2 in oceanic crust, when

59–142 Ma old, typically has a mean gradient of only $0.7\,\text{s}^{-1}$ (Fig. 5). Further, if 4.5 km melt was intruded and extruded, significant magnetic anomalies (and even sea-floor spreading anomalies) would be expected and these are not observed within region C. Dean *et al.* (2000) concluded that either no melt is present or that it is very much less than 4.5 km thick and distributed in such a way that it was neither sampled by drilling (at sites limited to basement highs) nor detected by seismic wavelengths of c. 1 km. We shall return later to a discussion of magnetic anomalies and the possible distribution of melt.

Chian *et al.* (1999) described the seismic velocity structure along a rectangular grid of seismic lines within a 40 km x 80 km box that includes most of the ODP Leg 149 and 173 drill sites and encompasses the southern flank of Galicia Bank (region D), the adjacent part of region C and the northern part of region B (Fig. 2). Their results are entirely consistent with the above classification of Dean *et al.* (2000). The thinned upper continental crust of region D, of 2–5 km thickness, has a velocity of 5.0–6.6 km s^{-1} and is limited in extent to a number of blocks bounded by probably north–south-striking, seaward-dipping normal faults; some of these blocks have been inferred to be continental crust by results from ODP Sites 900, 901, 1065, 1067, 1068 and 1069 (Sawyer *et al.* 1994; Whitmarsh *et al.* 1998). This continental crust is underlain everywhere by a 7.3–7.9 km s^{-1} layer that extends southwards into the OCT (region C) where it underlies an upper basement layer of c. 2 km thickness in which velocity increases rapidly with depth from c. 4 km s^{-1} at the top of basement. The velocity–depth profiles of Chian *et al.* (1999) emphasize the previously noted simple two-layer structure of the OCT and the lack of a clear velocity contrast at the Moho (Fig. 5). They also point out that the velocity of the lower basement layer is well in excess of normal oceanic layer 3. Further west, in region B of the overlapping peridotite ridges, velocities are reduced both within the upper (3.5–6.0 km s^{-1}) and lower (6.4–7.5 km s^{-1}) basement layers. Dean *et al.* (2000) similarly observed a lower top basement velocity in region B than region C along IAM9.

Chian *et al.* (1999) proposed a serpentinized peridotite origin for the basement in regions B and C for similar reasons to Dean *et al.* (2000). An interesting implication of their interpretation is that serpentinized peridotite underlies thinned continental crust up to 5 km thick. They also concluded that their velocity models, although not being able to distinguish the presence of small amounts of melt, imply that very little syn-rift melt was produced. Chian *et al.* (1999) suggested that factors that may have reduced the volume of melt include partial extension during earlier episodes of rifting, cooling by 2D lateral conduction and advective cooling by penetration of water into the basement.

Figure 5 emphasizes the typical two-layer OCT velocity structure, its difference from normal 59–142 Ma Atlantic oceanic crust (White *et al.* 1992) and the small to zero velocity contrast at the Moho. A feature of the combined OCT structures in Figure 5 is that there appears to be a systematic reduction in velocity, below 2 km sub-basement, going from region C to region B to region A (oceanic crust). The velocity decreases towards values typical of oceanic layer 3 and the Moho deepens. It is interesting to examine whether this can be explained by a steadily increasing proportion of gabbroic melt in the lower basement in the direction of oceanic crust. Although the introduction of *c.* 1 km of $6.9–7.0 \, \text{km s}^{-1}$ melt into the lower basement can explain the slightly different structure in region B it is not a viable explanation for the structure in region A because there the equivalent velocity is on average about $7.0 \, \text{km s}^{-1}$ too. Similarly, the structures in the upper 2 km of regions B and C are similar (with a mean gradient of $1.25 \, \text{s}^{-1}$) but they differ significantly from that of region A, which has a smaller gradient. Therefore, it appears that a fundamental change in structure occurs in the vicinity of the boundary between regions B and A.

Magnetic observations

Deep-towed three-component fluxgate magnetometer profiles as well as a contoured magnetic anomaly chart of surface observations (Miles *et al.* 1996) are available to study the deep margin off West Iberia. A variety of magnetic inversion and forward modelling techniques have recently been applied to these data (Russell 1999) and some of the results are reported here. These results provide conclusive evidence that some, probably syn-rift, melt was emplaced within the OCT on this margin.

The first step in this study was to consider to what extent magnetic anomalies over the margin are caused by relief of the top basement surface where there is a potentially strong contrast in magnetization between weakly magnetized sediment and the underlying igneous rocks. This will allow us to estimate what contribution, if any, is provided by magnetization contrasts within the basement such as might be caused by intrusive (or extrusive) magmatic bodies with uniform magnetization. Because the basement is deeply buried it was possible to compute the 3D contribution of basement relief only in the area covered by the basement chart (Figs 2 and 3).

A potential field can be described as the convolution of a source distribution with a function dependent on the assumptions of potential theory (such as the potential field satisfying Euler's equation). Given the function then the source distribution can be determined by a form of deconvolution (Thompson 1982; Reid *et al.* 1990). Thus, 3D Euler deconvolution solutions were computed from the magnetic anomaly chart for a structural index of 0.5 (equivalent to a step in the basement surface) and their depths compared with the depth of underlying acoustic basement (Fig. 6). It is estimated that the areal distribution and trends of only about 25% of the most reliable solutions, which lay within 1 km beneath the top of basement, correlate with significant features of the basement topography.

Another more quantitative approach was to compare observed anomaly profiles with profiles computed from models in which the only magnetization contrast existed at the top of basement (and at the base 14 km below sea level). Twenty-nine synthetic magnetic anomaly and basement profiles were constructed from the anomaly and basement charts, respectively. The position and orientation (100°) of the profiles were chosen to be normal to the dominant trend of linear basement features and observed magnetic anomalies in region A (Fig. 3). A constant magnetization of $\pm 1 \, \text{A m}^{-1}$ and several different magnetization directions, corresponding to two likely remanent directions and to the present-day field (induced magnetization), were applied to the models. The models were assumed to be 2D. Pairs of computed and observed anomalies were compared statistically in the space and wavenumber domains. In the space domain, only three profiles had a correlation coefficient, regardless of sign, greater than 0.5. Differences between pairs of profiles were also used to estimate the degree of fit by region. Not surprisingly, the worst fit was in region A (where oceanic crust is known to contain strong intrabasement magnetization contrasts). The best fit was in region C but, as this region has low relief anyway and is on average *c.* 2 km deeper than the other regions, this result must be treated with caution. In the wavenumber domain, squared coherency spectra between pairs of profiles (wavelengths of 5–30 km) were computed. Significant (>0.5) coherence was found on only a single line at a wavelength of *c.* 15 km.

It is concluded, therefore, that the observed magnetic anomalies in the southern Iberia Abyssal Plain must be attributed largely to variations

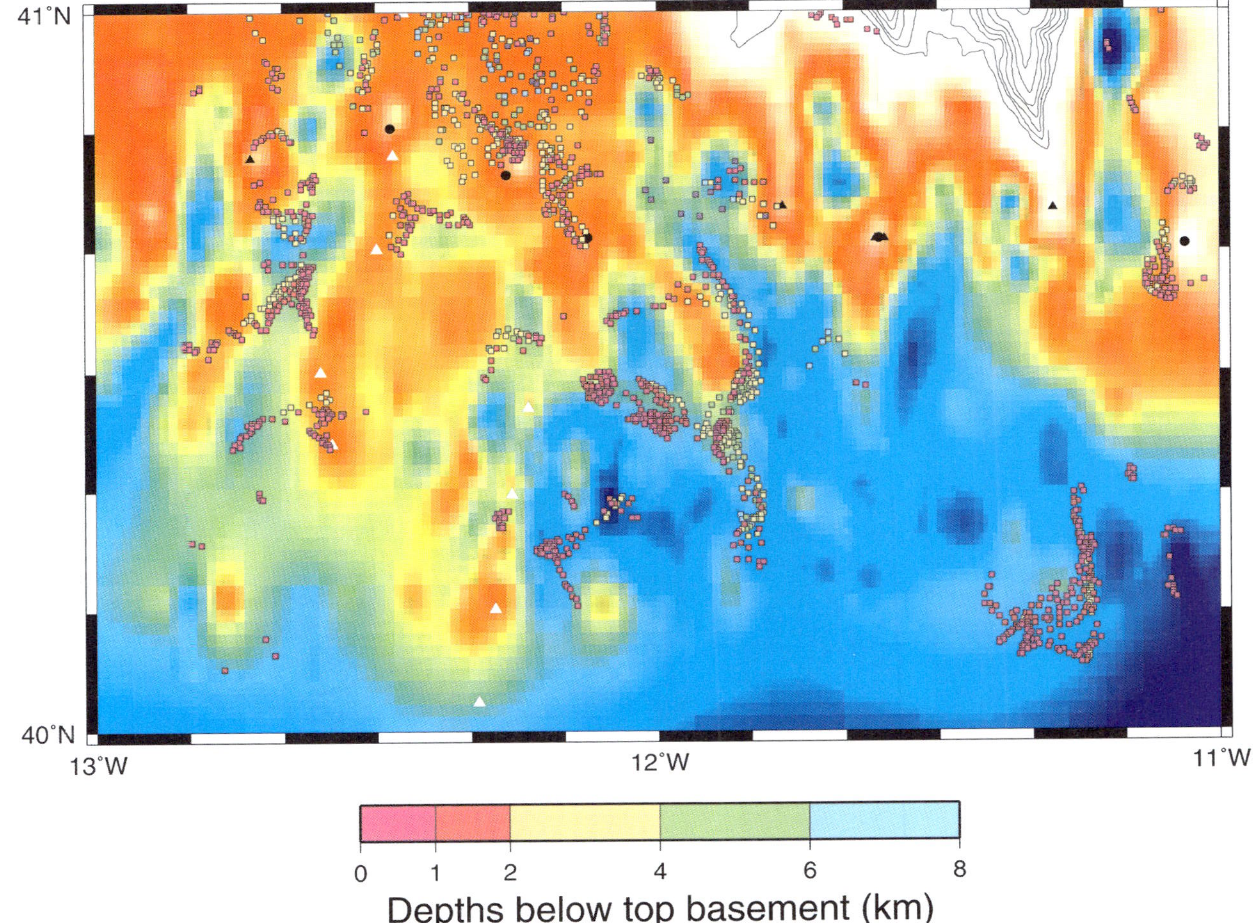

Fig. 6. Euler deconvolution solutions for a structural index of 0.5 (colour coded circles), that have depth uncertainties of $<1\%$ of the estimated depth, presented as depth below top of basement (km) overprinted on a depth-to-basement chart (from Cole, pers. comm., after Discovery 215 Working Group (1998); highs are orange and deeps are blue). The 500 m bathymetric contours show southern tip of Vasco da Gama seamount.

in magnetic character of the acoustic basement other than its surface relief, although the latter should not be neglected in further modelling.

Given that basement relief is not the only source of the anomalies it is possible to assess regional differences in the bulk magnetization of basement. If certain assumptions are made the 3D inverse technique of Parker & Huestis (1974) can be applied to the anomaly and depth-to-basement charts whereby the areal distribution of basement magnetization can be determined (see also Discovery 215 Working Group (1998)). A constant source layer thickness of 4 km was assumed. The inversion was conducted so as to include magnetization variations with wavelengths of >5 km, this limit being imposed by the average depth of the top of basement. The resulting magnetization distributions exhibit a marked difference in character either side of an approximately NNE–SSW boundary approximately coincident with peridotite ridge segment R3 and the west edge of region B (Fig. 7). West of this boundary magnetizations (regardless of sign) commonly exceed 0.5 A m^{-1}, even 1.5 A m^{-1}, and are strongly aligned NNE–SSW. Within region B there is a NNE–SSW alignment

too but the magnetizations are weaker. The area east of the boundary is characterized by less strongly developed linear trends and by magnetizations (regardless of sign) less than 0.5 A m^{-1}. Exceptions are a few isolated highs, four of which exceed 1 A m^{-1}. The high H1 at 40°40'N, 12°10'W coincides with a pair of basement highs that include Site 899, at which 1.5 A m^{-1} serpentinized peridotite breccia basement was cored. The other highs lie in regions where depth-to-basement control is poor or absent and so magnetization is poorly constrained. Qualitatively, the same sort of results were obtained when a source layer of 7 km thickness was assumed.

A similar, but 2D, approach was taken with two deep-towed total field profiles acquired near the sea bed, which allow greater along-track resolution in the inverted magnetizations (Figs 3 and 8). Profile TOBI91 traverses regions B, C and D. A strong contrast in magnetization occurs at *c.* 75 km near the B–C boundary. Peak–trough amplitudes to the west are 3–3.5 A m^{-1} and to the east 1 A m^{-1} or less. Profile SAR95 exhibits a similar contrast, around 20 km near the western edge of region C, but in addition there

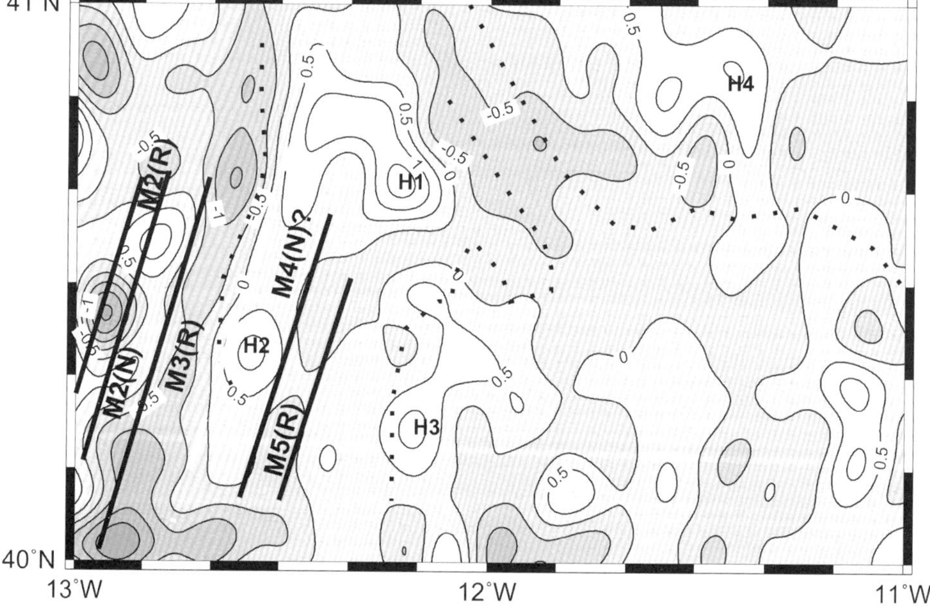

Fig. 7. Magnetization solution (0.5 A m^{-1} contours) from the 2D inversion of the sea-surface magnetic anomaly chart by the method of Parker & Huestis (1974). A source layer thickness of 4 km and a source magnetization of $I = 46°$, $D = 0°$ were assumed. The upper surface of the magnetic layer was assumed to coincide with the top of basement. Dotted lines bound regions A–D (see Fig. 3). Thick black lines mark sharp magnetization contrasts, which can be modelled as the edges of sea-floor spreading blocks M2(R)–M5(R) and correspond to a spreading rate of 10 mm a^{-1} if block M3(R) is arbitrarily narrowed by 50%. Dashed white line marks a boundary between a region of higher (to the west) and lower (to the east) magnetization. H1–H4 are mentioned in the text.

are two peaks with peak–trough amplitudes of c. $1.5\,\text{A m}^{-1}$ between 70 and 120 km. Initially magnetization variations with wavelengths >5 km were sought but when shorter wavelength (2.5 km) magnetization variations were included in the model better fits were obtained between observed profiles and profiles computed by forward modelling of the inverted magnetizations. Further, it appeared that the magnetization structure beneath SAR95 (dominantly over region C) occurs on length scales of 5 km or more whereas on TOBI91 (dominantly over regions B and D) it happens on scales of c. 2.5 km.

Source depths were already estimated by Euler deconvolution from the surface anomaly chart but better resolution is expected from the deep-tow profiles. The spectral character of each of three profiles (e.g. Spector & Grant 1971) reveals consistently that one generic source ensemble lies at the top of acoustic basement. Profile SAR95, however, reveals a source ensemble at a depth of c. 6.5 km below the basement surface as well. Two-dimensional Euler deconvolution along SAR95 (Discovery 215 Working Group 1998) suggests that this deeper source ensemble is best modelled by a structural index of 1.0 corresponding to linear dyke-like source bodies. Such bodies are inferred, however, only at three locations along the profile (at 78, 96 and 112 km).

In summary, therefore, various interpretation techniques applied to the magnetic anomalies consistently suggest the following. Region A has the magnetic characteristics of oceanic crust. Sea-floor spreading at c. $10\,\text{mm a}^{-1}$ can explain magnetization boundaries in region A and part of region B (Fig. 7) but none of the magnetization boundaries on a deep-towed magnetometer profile across region C. There is a significant reduction in basement magnetization going from west to east between regions A and C. Although relief of the basement surface contributes to the observed anomalies, a greater contribution is from a few dyke-like magnetization contrasts that extend well below the top of basement, even as deep as c. 6.5 km (just below the Moho) in region C.

Discussion

In the light of the above, several points require discussion. What can the seismic and magnetic data tell us about the quantity and distribution of syn-rift melt products within the OCT off West Iberia and, assuming some melt is present, by what process was it emplaced? Finally, to what extent can any model for West Iberia be considered to be more generally applicable?

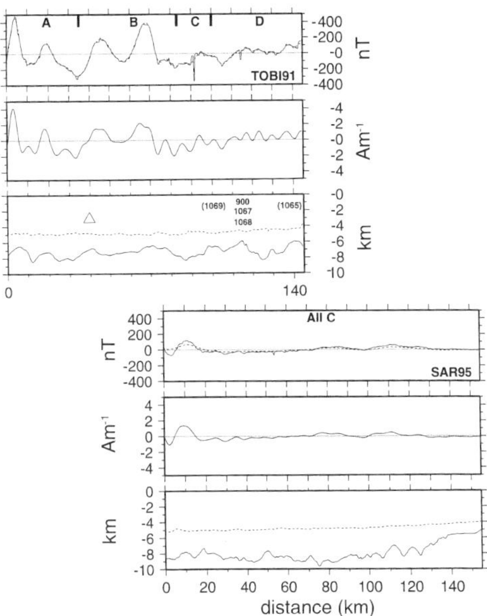

Fig. 8. Result of inverting deep-tow magnetometer profiles TOBI91 (with peridotite ridge shown as a triangle and ODP sites) and SAR95 (see Fig. 3 for locations) with a pass filter of 5–200 km cosine tapered to 2.5 km. The region(s) traversed by each profile are shown within each top panel. Top panel, observed (thin line) and computed (dashed line) profiles. Middle panel, inverted magnetization values. Bottom panel, acoustic basement surface (thin line) and magnetometer trajectory (dashed line).

The magnetic and seismic evidence for the presence of significant amounts of melt in the OCT is constrained by the resolution of the methods employed. The typical 8 Hz seismic signal, generated by the airguns every c. 100 m along-track, has a wavelength of c. 900 m in the lower basement and therefore is unlikely to resolve seismic inhomogeneities with horizontal dimensions of less than 1–2 km and vertical dimensions less than c. 450 m. The resolution of the deep-towed magnetometer profiles sampled every few metres (which must be higher than that of the surface observations) is determined by the depth, orientation and magnetization vector, as well as the lateral extent, of the sources in the basement and is much harder to quantify. However, given the probable depth and magnetic characteristics of melt bodies these profiles are likely to provide better along-track resolution than the seismic observations. It is also important to note that a discrete seismic body cannot necessarily be equated with a discrete magnetic body, and vice versa.

The seismic evidence for the presence of melt is weak and the arguments are tenuous. In the absence of seaward-dipping volcanic reflector sequences in the upper basement on multi-channel reflection profiles only the seismic velocity can be used as a guide to estimating the amount of melt. There is no clear argument that the upper basement velocity structure represents extrusive rocks; the average *c.* $1.25 \, s^{-1}$ velocity gradient is significantly greater than oceanic layer 2 of the same age. Likewise, the magnetic data do not suggest that a significant extrusive layer exists within the OCT. Neither Chian *et al.* (1999) nor Dean *et al.* (2000) detected any indication of an underplated layer at the top of the mantle but they discussed the possibility that an *ad hoc* mixture of serpentinized mantle and melt could lead to the observed velocities. Dean *et al.* (2000), when considering the structure along profile IAM9, concluded that either no melt is present or it is very much less than 4.5 km thick and distributed in such a way that it was not detected seismically as a discrete layer. Chian *et al.* (1999), when considering the OCT and adjacent edge of thinned continental crust, also concluded that their velocity models, although not being able to distinguish the presence of 'small' melt bodies, i.e. with dimensions less than one or two seismic wavelengths, imply that very little syn-rift melt was produced. The only seismic indication on a regional scale that melt may have contributed to the velocity structure in the OCT is the fact that the lower basement velocity under region B can be considered numerically as equivalent to that of region C but modified by the uniform admixture of *c.* 1 km of melt (Fig. 5).

The magnetic evidence for the presence of melt within the OCT is much stronger and less ambiguous. The already mentioned space- and wavenumber-domain properties of the deep-tow magnetic anomaly profiles provide strong evidence of source bodies, located both near the surface of the OCT basement and at a depth of *c.* 6.5 km, with length scales normal to the margin of ≥ 5 km. Such source bodies could represent magnetization contrasts within the largely serpentinized peridotite basement caused either by peridotite inhomogeneities or by intrusive magmatic bodies. Serpentinization can lead to an increase in peridotite magnetization (Coleman 1971), in which case it is expected that the strongest magnetization contrasts are associated with either shallow basement or basement fault zones. However, the along-margin extent for many tens of kilometres, the weak correlation with basement relief of some anomalies and the depth distribution as deep as 6.5 km of several OCT source bodies do not support the shallow basement hypothesis, except locally in the vicinity of Sites 898 and 899 where there is an anomaly that is anomalously strong for the OCT (Whitmarsh *et al.* 1996a). The basement fault zone hypothesis is not favoured because it is hard to correlate even the most obvious faults over *c.* 10 km, between adjacent seismic reflection profiles, let alone over many tens of kilometres (we consider that any serpentinized fault zone exhibiting a magnetic contrast should also exhibit an acoustic impedance contrast). Instead, it is more plausible that anomalies are caused by intrusive magmatic bodies that are continuous parallel to the margin for many tens of kilometres. However, within region C, along the single deep-tow profile SAR95, such bodies are infrequent and discontinuous. Within region B, however, they may have merged into an effectively continuous layer because here it is possible to set up sea-floor spreading models to explain the observed anomalies (Fig. 7).

It is concluded therefore that generally, except for the west side of region B, the OCT basement has a much weaker bulk magnetization than in region A west (oceanward) of the peridotite ridge. However, both magnetic and, to a lesser extent, seismic interpretations of the data suggest that magnetic source bodies, which it is suggested are magmatic in origin, become more common within region C, going from east to west, and eventually form an effectively continuous layer in the 30 km wide western part of region B. These sources lie in the top 6 km of basement, which includes the uppermost (unserpentinized) mantle. A scheme expressing this concept is shown in Figure 9.

Assessing the wider applicability of the model in Figure 9 to other non-volcanic margins is difficult because of the limited number of such margins that have been investigated to a sufficient level of detail to determine the presence or absence of an OCT as defined here. The velocity structures adjacent to several other non-volcanic margins appear similar to that within the West Iberia OCT (Minshull *et al.* 2001). The sea-floor spreading half-rate immediately west of the peridotite ridge has been estimated to be about $10 \, mm \, a^{-1}$ (Whitmarsh *et al.* 1990, 1996a; Whitmarsh & Miles 1994) although subsequently the mean rate over the 39 Ma interval from anomaly M0 until anomaly C33R is $8.5 \, mm \, a^{-1}$. Therefore the earliest oceanic crust formed off West Iberia is characterized by slow (or even ultraslow) sea-floor spreading. Such rates characterize the earliest-formed oceanic crust at other rifted non-volcanic margins where a broad ocean–continent transition has been proposed, e.g. SW Greenland, Labrador and western Jan Mayen Ridge (Chian *et al.* 1995; Chalmers 1997; Kodaira *et al.* 1998). Elsewhere, where

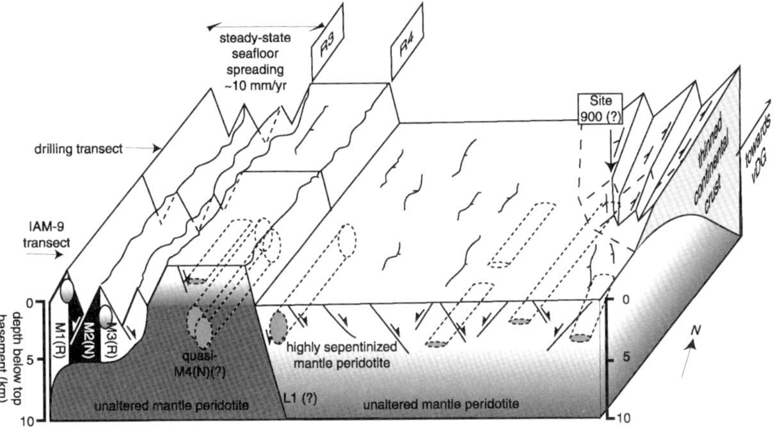

Fig. 9. Schematic 3D model of the proposed geological structure beneath the southern Iberia Abyssal Plain. During prolonged extension of the lithosphere continental crust 'breaks' and mantle peridotite is exhumed at the sea floor. Linear 350°-trending anomalies result from synrift intrusion of dyke- or sill-like gabbroic bodies (horizontal rod-like bodies outlined by dashed lines) into the exhumed and sub-continental mantle under rifting related stress conditions. Serpentinization, and hence reduced density, of uppermost mantle basement prevents melt rising above sub-basement depths of *c.* 3 km. As extension proceeds increased focusing of melt oceanwards (westwards) creates basement that, in the region of the offset peridotite ridges R3 and R4, is magnetically indistinguishable from oceanic crust.

initial sea-floor spreading was at an intermediate rate (25–50 mm a^{-1}) margins tend to be more volcanic in character and the OCT, if not described as an abrupt ocean–continent boundary, is estimated to be <50 km wide, e.g. Lofoten Islands and Exmouth Plateau (Mutter *et al.* 1989; Mjelde *et al.* 1996). In the well-surveyed Woodlark Basin, where an OCT has yet to be detected, the initial half-rate was 20–35 mm a^{-1}. Here a sharp (<5 km wide) transition is proposed, on the basis of swath bathymetry, and magnetic and single-channel seismic reflection data, from stretched continental or 'transitional' (partly intruded) crust to oceanic crust, but at present no published seismic refraction or basement samples are available to constrain this interpretation (Taylor *et al.* 1999). Thus, the question of whether our model fits just those margins that appear non-volcanic and that are also characterized by slow initial sea-floor spreading (and even those that have only a broad OCT) or is applicable to a wider class of margins will have to await further observations.

Weak magnetic anomalies that parallel rifted margins and lie adjacent to thinned continental crust have been observed on several other non-volcanic rifted margins, e.g. on both margins of the Labrador Sea (Roest & Srivastava 1989; Srivastava & Roest 1995, 1999) and off Newfoundland (Srivastava *et al.* 1998) and southern Australia (Koenig & Talwani 1977) and in the Laxmi Basin, Arabian Sea (Bhattacharya *et al.*

1994). Some of these anomalies have been explained by sea-floor spreading, often involving a spreading half-rate of less than 10 mm a^{-1} (i.e. ultraslow) (Bhattacharya *et al.* 1994; Srivastava & Roest 1999). However, such hypotheses overlook three important factors. First, to fit these anomalies abrupt and often substantial (factors of 2–3) temporal changes in spreading rate and, in some cases, magnetization have to be proposed, which we regard as implausible. Second, the magnetic anomalies are always noticeably weaker in amplitude than adjacent unambiguous sea-floor spreading anomalies identified further seaward, and the hypotheses explain this observation only in general terms of episodic crustal accretion at ultraslow spreading rates resulting in fragmentation that will give rise to inhomogeneities in magnetic source rocks (Srivastava & Roest 1995, p. 110). Third, no account is taken of the significance of the anomalies being located close to the break-up of continental crust. The last factor implies a non-steady-state thermal regime immediately after break-up as colder continental lithosphere drifts away from an upwelling asthenospheric high, which eventually provides a focused source of sea-floor spreading just below the sea bed. The cooling effect of lateral heat conduction has been proposed as one explanation of the abnormally thin oldest oceanic crust off West Iberia (Bown & White 1995*a*; Whitmarsh *et al.* 1996*b*). This situation is therefore very different from the

Table 1 *Examples of ultraslow sea-floor spreading ridges*

Accretionary ridge	Age range (Ma)	Half-rate (mm a^{-1})	Identified magnetic anomalies	Bathymetric relief	Crustal thickness (km)	Source
Nansen (Gakkel) Ridge	0–45	3–6	5 and 6			Vogt *et al.* (1979)
				continuous axial rift valley, blocky ridge flank	0–4, estimated from gravity	Coakley & Cochran (1998)
				very rough topography	6 on flank (unreversed sonobuoy)	Kristoffersen *et al.* (1982); Jokat *et al.* (1995)
Mohns Ridge	0–22	8	0–7	strong ridge-parallel fabric, *c.* 1000 m relief across median valley	highly variable, averages 4.0±0.5	Geli (1994); Klingelhoefer *et al.* (2000)
Kolbeinsey Ridge	13–26	7.5	5B–6B (east flank)			Vogt *et al.* (1980)
Cayman Trough	0–30	7.5 ± 2.5	1–5C	extensive relief, >1000 m peak–trough		Rosencrantz & Sclater (1986)
SW Indian Ridge	0 to >5.9	8		2200 ± 600 m across axial median valley; ridges and fault scarps continuous over several tens to *c.* 100 km		Grindlay *et al.* (1998)
Atlantis II Fracture Zone (SWIR)	0–30	5–11	1–10		4±1	Dick *et al.* (1991); Muller *et al.* (1997)

steady-state generation of normal oceanic crust at a mid-ocean ridge.

Several ultraslow spreading accretionary plate boundaries have been studied (Table 1) and have very different characteristics from the low basement relief and weak magnetic anomalies of region C. The ultraslow ridges are characterized by lineated basement with substantial (up to a few thousand metres) relief and by lineated magnetic anomalies, which can be related directly to continuous sea-floor spreading models with rates that are not required to change abruptly. Published velocity structures of non-fossil ultraslow ridges are rare, but over the Southwest Indian Ridge (Muller *et al.* 1997, 1999) and the Mohns Ridge (Klingelhoefer *et al.* 2000) the seismic structure is recognizably oceanic, with a clear layer 3 (Fig. 5a), and, unlike region C (Dean *et al.* 2000), high-amplitude post-critical P_mP (Moho) reflections are seen. Off West Iberia we are convinced that slow or ultraslow sea-floor spreading, originally proposed by Sawyer (1994) and recently by Srivastava *et al.* (2001), is a poor hypothesis to explain the observed magnetic anomalies within the whole OCT. Within region C the basement relief is subdued and not clearly lineated parallel to the margin (Fig. 3). Both seismic layer 3 and an abrupt velocity contrast at the Moho are absent from the many velocity–depth profiles that have been measured (Fig. 5). Basement within the OCT clearly has a significantly weaker bulk magnetization, compared with the oceanic crust west of the peridotite ridge (Fig. 7), which has not been quantitatively explained by any hypothesis involving ultraslow sea-floor spreading. Whereas slow ($10 \, \text{mm a}^{-1}$) sea-floor spreading can explain anomalies back to M4(N) or even M5(R), within region A and the west part of B, neither sea-surface nor deep-towed magnetometer profiles across region C fit computed slow or ultraslow sea-floor spreading profiles whether the profiles are forward modelled empirically or objectively (from magnetization inversions). Inversion of the deep-tow profile SAR95 across region C indicates that magnetization contrasts are at least $5–10 \, \text{km}$ apart whereas pre-M4 sea-floor spreading at $8 \, \text{mm}$ a^{-1} or less should involve many blocks with widths of $5 \, \text{km}$ or less. Inversion also suggests that some of the source bodies are as deep as $6.5 \, \text{km}$ below the top of basement; sea-floor spreading west of the peridotite ridge, and generally in the oceans, is associated with sources that reach to the basement surface. Finally, it is well documented that one of the $010°$-trending magnetic anomalies within region C that parallels the margin continues over the thinned continental crust of region D (Russell 1999). This strongly suggests that the process giving rise to this anomaly cannot have been sea-floor spreading. Instead, it could be the result of the more diffuse pre-sea-floor spreading magmatism that has been outlined above. The difference between these two processes is subtle but appears, on the still limited evidence available, to hinge on the flux and continuity of the supply of melt from the asthenosphere. We conclude that off West Iberia ultraslow sea-floor spreading does not match the geophysical observations within the OCT, which currently are better explained by the process of diffuse, but systematically evolving, magmatism already described and for which the name of octagenesis (genesis of OCT) is suggested.

Conclusions

(1) The $30–170 \, \text{km}$ wide ocean–continent transition (OCT) zone off West Iberia has a characteristic seismic and magnetic character, distinct from both thinned continental crust and normal oceanic crust, which supports the notion that it consists predominantly of exhumed and serpentinized mantle.

(2) Interpretations of magnetic and, to a lesser extent, seismic data indicate that on average (at lateral and vertical scales greater than *c.* 2.5 and $1 \, \text{km}$, respectively) only small amounts of syn-rift melt exist within the OCT.

(3) A sequence of events is proposed to describe the origin of the OCT (octagenesis) within the general rift-to-drift evolution of the West Iberia margin. Isolated, margin-parallel, intrusive melt bodies are scattered within the eastern (landward) part of the OCT well beneath the top of acoustic basement. As extension, and melt production, proceed these bodies are emplaced further west (oceanward) at higher levels and more closely together in the basement. Eventually, in the westernmost $30 \, \text{km}$ of the OCT, an effectively continuous layer is formed, which is magnetically indistinguishable from crust generated by the earliest sea-floor spreading immediately to the west of the adjacent peridotite ridge. This process has been identified in the southern Iberia Abyssal Plain where the OCT is at least $70 \, \text{km}$ wide; where the OCT is much narrower, for example west of Galicia Bank, the process may be harder to resolve.

(4) The available evidence does not support the hypothesis that ultraslow sea-floor spreading can explain the magnetic anomalies observed within the wider parts of the OCT off West Iberia, where the evolution of the OCT is better resolved. We suspect that this conclusion may

be applicable to other non-volcanic rifted margins, such as SW Greenland and Labrador (Louden & Chian 1999), which also exhibit an OCT with the same characteristic velocity structure, low basement relief and weak margin-parallel magnetic anomalies.

We thank B. Taylor and F. Martinez for reviews that enabled us to improve the manuscript. NERC Research Grant GR3/9354, awarded to T.A.M. and R.B.W., made possible the collection of many of the data reported here. S.M.R. and S.M.D. were supported by NERC Research Studentships. T.A.M. is supported by a Royal Society University Research Fellowship. K.E.L. and D.C. acknowledge support from NSERC.

References

BESLIER, M.-O., ASK, M. & BOILLOT, G. 1993. Ocean–continent boundary in the Iberia Abyssal Plain from multichannel seismic data. *Tectonophysics*, **218**, 383–393.

BHATTACHARYA, G.C., CHAUBEY, A.K., MURTY, G.P.S., SRINIVAS, K., SRAMA, K.V.L.N.S., SUBRAHMANYAM, V., KRISHNA, K.S. Evidence for seafloor spreading in the Laxmi Basin, northeastern Arabian Sea. *Earth and Planetary Science Letters*, **125**, 211–220.

BOWLING, J.C. & HARRY, D.L. 2001. Geodynamic models of continental extension and the formatin of non-volcanic rifted continental margins. *In:* WILSON, R.C.L., WHITMARSH, R.B., TAYLOR, B. & FROITZHEIM, N. (eds) *Non-volcanic Rifting of Continental Margins: a Comparison of Evidence from Land and Sea.* Geological Society, London, Special Publications, **187**, 511–536.

BOWN, J.W. & WHITE, R.S. 1995*a*. Effect of finite extension rate on melt generation at continental rifts. *Journal of Geophysical Research*, **100**, 18011–18030.

BOWN, J.W. & WHITE, R.S. 1995*b*. Finite duration rifting, melting and subsidence at continental margins. *In:* BANDA, E., TORNÉ, M. & TALWANI, M. (eds) *Rifted Ocean–Continent Boundaries.* Kluwer, Dordrecht, 31–54.

CAPDEVILA, R. & MOUGENOT, D. 1988. Pre-Mesozoic basement of the western Iberian continental margin and its place in the Variscan Belt. *In:* BOILLOT, G. & WINTERER, E.L. (eds) *Proceedings of the Ocean Drilling Program, Scientific Results, 103.* Ocean Drilling Program, College Station, TX, 3–12.

CHALMERS, J.A. 1997. The continental margin off southern Greenland: along-strike transition from an amagmatic to a volcanic margin. *Journal of the Geological Society, London*, **154**, 571–576.

CHIAN, D., KEEN, C.E., REID, I. & LOUDEN, K.E. 1995. Evolution of non-volcanic rifted margins: new results from the conjugate margins of the Labrador Sea. *Geology*, **23**, 589–592.

CHIAN, D., LOUDEN, K.E., MINSHULL, T.A. & WHITMARSH, R.B. 1999. Deep structure of the ocean–continent transition in the southern Iberia Abyssal Plain from seismic refraction profiles: Ocean Drilling Program Legs 149 and 173) transect. *Journal of Geophysical Research*, **104**, 7443–7462.

COAKLEY, B.J. & COCHRAN, J.R. 1998. Gravity evidence of very thin crust at the Gakkel Ridge (Arctic Ocean). *Earth and Planetary Science Letters*, **162**, 81–95.

COLEMAN, R.G. 1971. Petrologic and geophysical nature of peridotite. *Geological Society of America Bulletin*, **82**, 897–917.

DEAN, S.M., MINSHULL, T.A., WHITMARSH, R.B. & LOUDEN, K. 2000. Deep structure of the ocean–continent transition in the southern Iberia Abyssal Plain from seismic refraction profiles: II The IAM-9 transect at 40°20′N. *Journal of Geophysical Research*, **105**, 5859–5886.

DICK, H., SCHOUTEN, H., MEYER, P.S. & 7 OTHERS 1991. 21. Tectonic evolution of the Atlantis II Fracture Zone. *In:* STEWART, S.K. (ed.) *Proceedings of the Ocean Drilling Program, Scientific Results, 118.* Ocean Drilling Program, College Station, TX.

DISCOVERY 215 WORKING GROUP 1998. Deep structure in the vicinity of the ocean–continent transition zone under the southern Iberia Abyssal Plain. *Geology*, **26**, 743–746.

GELI, L. 1994. Ocean crust formation processes at very slow spreading centres: a model for the Mohns Ridge, near 72°N, based on magnetic, gravity and seismic data. *Journal of Geophysical Research*, **99**, 2995–3013.

GRINDLAY, N.R., MADSEN, J.A., ROMMEVAUX, J.C. & SCLATER, J. 1998. A different pattern of ridge segmentation and mantle Bouguer gravity anomalies along the ultra-slow spreading Southwest Indian Ridge (15°30′E to 25°E). *Earth and Planetary Science Letters*, **161**, 243–253.

GROUPE GALICE 1979. The continental margin of Galicia and Portugal, acoustic stratigraphy, dredge stratigraphy and structural evolution. *In:* RYAN, W.B.F. & SIBUET, J.-C. (eds) *Proceedings of the Deep Sea Drilling Project, Leg 47.* US Government Printing Office, Washington, DC, 633–662.

HARRY, D.L. & BOWLING, J.C. 1999. Inhibiting magmatism on nonvolcanic rifted margins. *Geology*, **27**, 895–898.

JOKAT, W., WEIGELT, E., KRISTOFFERSEN, Y., RASMUSSEN, T. & SCHONE, T. 1995. New geophysical results from the south-western Eurasian Basin (Morris Jesup Rise, Gakkel Ridge, Yermak Plateau) and the Fram Strait. *Geophysical Journal International*, **123**, 601–610.

KEEN, C.E., COURTNEY, R.C., DEHLER, S.A. & WILLIAMSON, M.-C. 1994. Decompression melting at rifted margins: comparison of model predictions with the distribution of igneous rocks on the eastern Canadian margin. *Earth and Planetary Science Letters*, **121**, 403–416.

KLINGELHOEFER, F., GELI, L., MATIAS, L., STEINSLAND, N. & MOHR, J. 2000. Crustal structure of a super-slow spreading centre: a seismic refraction study of Mohns Ridge, 72°N. *Geophysical Journal International*, **141**, 509–526.

KODAIRA, S., MJELDE, R., GUNNARSSON, K., SHIOBARA, H. & SHIMAMURA, H. 1998. Structure of the Jan Mayen micro-continent and implications for its evolution. *Geophysical Journal International*, **132**, 383–400.

KOENIG, M. & TALWANI, M. 1977. A geophysical study of the southern continental margin of Australia: Great Australian Bight and western sections. *Geological Society of America Bulletin*, **88**, 1000–1014.

KRAWCZYK, C.M., RESTON, T.J., BESLIER, M.-O. & BOILLOT, G. 1996. Evidence for detachment tectonics on the Iberia Abyssal Plain rifted margin. *In*: WHITMARSH, R.B., SAWYER, D.S., KLAUS, A. & MASSON, D.G. (eds) *Proceedings of the Ocean Drilling Program, Scientific Results, 149*. Ocean Drilling Program, College Station, TX, 603–616.

KRISTOFFERSEN, Y., HUSEBYE, E.S., BUNGUM, H. & GREGERSEN, S. 1982. Seismic investigations of the Nansen Ridge during the FRAM I experiment. *Tectonophysics*, **82**, 57–68.

LAUGHTON, A.S., WHITMARSH, R.B. & JONES, M.T. 1970. The evolution of the Gulf of Aden. *Philosophical Transactions of the Royal Society of London, Series A*, **267**, 227–266.

LOUDEN, K.E. & CHIAN, D. 1999. The deep structure of non-volcanic rifted continental margins. *In*: WHITE, R.S., HARDMAN, R.F.P., WATTS, A.B. & WHITMARSH, R.B. (eds) *Response of the Earth's Lithosphere to Extension*. Philosophical Transactions of the Royal Society Series A, **357**, 767–799.

MAMET, B., COMAS, M.C. & BOILLOT, G. 1991. Late Paleozoic basin on the West Galicia Atlantic margin. *Geology*, **19**, 738–741.

MANATSCHAL, G., FROITZHEIM, N., RUBENACH, M. & TURRIN, B.D. 2001. The role of detachment faulting in the formation of an ocean–continent transition: insights from the Iberia Abyssal Plain. *In*: WILSON, R.C.L., WHITMARSH, R.B., TAYLOR, B. & FROITZHEIM, N. (eds) *Non-volcanic Rifting of Continental Margins: a Comparison of Evidence from Land and Sea*. Geological Society, London, Special Publications, **187**, 405–428.

MASSON, D.G., CARTWRIGHT, J.A., PINHEIRO, L.M., WHITMARSH, R.B., BESLIER, M.-O. & ROESER, H. 1994. Compressional deformation at the ocean–continent transition in the NE Atlantic. *Journal of the Geological Society, London*, **151**, 607–614.

MATTHEWS, D.H. 1962. Altered lavas from the floor of the eastern North Atlantic. *Nature*, **194**, 368–369.

MAUFFRET, A., MOUGENOT, D., MILES, P.R. & MALOD, J.A. 1989. Cenozoic deformation and Mesozoic abandoned spreading centre in the Tagus Abyssal Plain (west of Portugal): results of a multichannel seismic survey. *Canadian Journal of Earth Sciences*, **26**, 1101–1123.

MILES, P.R., VERHOEF, J. & MACNAB, R. 1996. Compilation of magnetic anomaly chart west of Iberia. *In*: WHITMARSH, R.B., SAWYER, D.S., KLAUS, A. & MASSON, D.G. (eds) *Proceedings of the Ocean Drilling Program, Scientific Results 149*. Ocean Drilling Program, College Station, TX, 659–664.

MINSHULL, T.A., DEAN, S.M., WHITE, R.S. & WHITMARSH, R.B. 2001. Anomalous melt production after continental break-up in the southern Iberia Abyssal Plain. *In*: WILSON, R.C.L., WHITMARSH, R.B., TAYLOR, B. & FROITZHEIM, N. (eds) *Non-volcanic Rifting of Continental Margins: a Comparison of Evidence from Land and Sea*. Geological Society, London, Special Publications, **187**, 537–550.

MJELDE, R., KODAIRA, S., HASSAN, R.K. & 7 OTHERS 1996. The continent/ocean transition of the Lofoten volcanic margin, N. Norway. *Journal of Geodynamics*, **22**, 189–206.

MULLER, M.R., MINSHULL, T.A. & WHITE, R.S. 1999. Segmentation and melt supply at the Southwest Indian Ridge. *Geology*, **27**, 867–870.

MULLER, M.R., ROBINSON, C.J., MINSHULL, T.A., WHITE, R.S. & BICKLE, M.J. 1997. Thin crust beneath Ocean Drilling Program borehole 735B at the Southwest Indian Ridge? *Earth and Planetary Science Letters*, **148**, 93–107.

MUTTER, J.C., BUCK, W.R. & ZEHNDER, C.M. 1988. Convective partial melting 1, A model for the formation of thick basaltic sequences during the initiation of spreading. *Journal of Geophysical Research*, **93**, 1031–1048.

MUTTER, J.C., LARSON, R.L. & NORTHWEST AUSTRALIA STUDY GROUP 1989. Extension of the Exmouth Plateau, offshore northwestern Australia: deep seismic reflection/refraction evidence for simple and pure shear mechanisms. *Geology*, **17**, 15–18.

ODP LEG 149 SHIPBOARD SCIENTIFIC PARTY 1993. ODP drills the West Iberia rifted margin. *EOS Transactions, American Geophysical Union*, **74**, 454–455.

ODP LEG 173 SHIPBOARD SCIENTIFIC PARTY 1998. Drilling reveals transition from continental break-up to early magmatic crust. *EOS Transactions, American Geophysical Union*, **79**, 173, 180–181.

PARKER, R.L. & HUESTIS, S.P. 1974. The inversion of magnetic anomalies in the presence of topography. *Journal of Geophysical Research*, **79**, 1587–1593.

PEDERSEN, T. 1994. Some remarks on lithospheric forces and decompression magmatism. *In*: CLOETINGH, S., ELDHOLM, O., LARSEN, B.T., GABRIELSEN, R. & SASSI, W. (eds) *Dynamics of Extensional Basin Formation and Inversion*. Elsevier, Amsterdam, 11–19.

PICKUP, S.L.B., WHITMARSH, R.B., FOWLER, C.M.R. & RESTON, T.J. 1996. Insight into the nature of the ocean–continent transition off West Iberia from a deep multichannel seismic reflection profile. *Geology*, **24**, 1079–1082.

PINHEIRO, L.M., WHITMARSH, R.B. & MILES, P.R. 1992. The ocean–continent boundary off the western continental margin of Iberia II. Crustal structure in the Tagus Abyssal Plain. *Geophysical Journal International*, **109**, 106–124.

PINHEIRO, L.M., WILSON, R.C.L., PENA DOS REIS, R., WHITMARSH, R.B. & RIBEIRO, A. 1996. The western Iberia margin: a geophysical and geological overview. *In*: WHITMARSH, R.B., SAWYER, D.S., KLAUS, A. & MASSON, D.G. (eds) *Proceedings of the Ocean Drilling Pro-*

gram, Scientific Results 149. Ocean Drilling Program, College Station, TX, 1–26.

PITMAN, W.C. & TALWANI, M. 1972. Sea-floor spreading in the North Atlantic. *Geological Society of America Bulletin*, **83**, 619–646.

PURDY, G.M. 1975. The eastern end of the Azores–Gibraltar plate boundary. *Geophysical Journal of the Royal Astronomical Society*, **43**, 973–1000.

RABINOWITZ, P.D., CANDE, S.C. & HAYES, D.E. 1978. Grand Banks and J-Anomaly Ridge. *Science*, **202**, 71–73.

REID, A.B., ALLSOP, J.M., GRANSER, H., MILLETT, A.J. & SOMERTON, I.W. 1990. Magnetic interpretation in three dimesnions using Euler deconvolution. *Geophysics*, **55**, 80–91.

ROEST, W.R. & SRIVASTAVA, S.P. 1989. Seafloor spreading in the Labrador Sea: a new reconstruction. *Geology*, **17**, 1000–1003.

ROSENCRANTZ, E. & SCLATER, J.G. 1986. Depth and age in the Cayman Trough. *Earth and Planetary Science Letters*, **79**, 133–144.

RUSSELL, S.M. 1999. *A magnetic study of the West Iberia and conjugate rifted continental margins: constraints on rift-to-drift processes.* PhD thesis, University of Durham.

SAWYER, D.S. 1994. The case for slow-spreading oceanic crust landward of the peridotite ridge in the Iberia Abyssal Plain (abstract). *EOS Transactions, American Geophysical Union*, **75**, 616.

SAWYER, D.S., WHITMARSH, R.B., KLAUS, A. & SHIPBOARD SCIENTIFIC PARTY (eds) 1994. *Proceedings of the Ocean Drilling Program, Initial Reports, 149.* Ocean Drilling Program, College Station, TX.

SIBUET, J.-C., LOUVEL, V., WHITMARSH, R.B. & 5 OTHERS 1995. Constraints on rifting processes from refraction and deep-tow magnetic data: the example of the Galicia continental margin (West Iberia), *In:* BANDA, E., TORNÉ, M. & TALWANI, M. (eds) *Rifted Ocean–Continent Boundaries.* Kluwer, Dordrecht, 197-218.

SPECTOR, A. & GRANT, F.S. 1971. Statistical models for interpreting aeromagnetic data. *Geophysics*, **35**, 293–302.

SRIVASTAVA, S.P. & ROEST, W.R. 1995. Nature of thin crust across the southwest Greenland margin and its bearing on the location of the ocean–continent boundary. *In:* BANDA, M.T., TORNÉ, M. & TALWANI, M. (eds) *Rifted Ocean–Continent Boundaries.* Kluwer, Dordrecht, 95–120.

SRIVASTAVA, S.P. & ROEST, W.R. 1999. Extent of oceanic crust in the Labrador Sea. *Marine and Petroleum Geology*, **16**, 65–84.

SRIVASTAVA, S.P., SIBUET, J.C. & CANDE, S. 1998. Origin of thin crust across the Iberia and Newfoundland passive continental margins: a result of ultraslow seafloor spreading (abstract). *EOS Transactions, American Geophysical Union*, F854.

SRIVASTAVA, S.P., SIBUET, J.C., CANDE, S., ROEST, W.R. & REID, I.R. 2001. Magnetic evidence for slow seafloor spreading during the formation of the Newfoundland and Iberian margins. *Earth and Planetary Science Letters*, **182**, 61–76.

TAYLOR, B., GOODLIFFE, A.M. & MARTINEZ, F. 1999. How continents break up: insights from Papua New Guinea. *Journal of Geophysical Research*, **104**, 7497–7512.

THOMPSON, D.T. 1982. EULDPH—a new technique for making computer-assisted depth estimates from magnetic data. *Geophysics*, **47**, 31–37.

TUCHOLKE, B.E., LUDWIG, W.J. Structure and origin of the J-anomaly Ridge, western North Atlantic Ocean. *Journal of Geophysical Research*, **87**, 9389–9407.

VOGT, P.R., JOHNSON, G.L. & KRISTJANSSON, L. 1980. Morphology and magnetic anomalies north of Iceland. *In:* JACOBY, W., BJORNSSON, A. & MOLLER, D. (eds) *Iceland; Evolution, Active Tectonics, and Structure.* Springer, Berline, 67–80.

VOGT, P.R., TAYLOR, P.R., KOVACS, L.C. & JOHNSON, G.L. 1979. Detailed aeromagnetic investigations of the Arctic Basin. *Journal of Geophysical Research*, **84**, 1071–1089.

WHITE, R.S., MCKENZIE, D. & O'NIONS, K. 1992. Oceanic crustal thickness from seismic measurements and rare earth element inversions. *Journal of Geophysical Research*, **97**, 19683–19715.

WHITMARSH, R.B. & MILES, P.R. 1994. Propagating rift model for the West Iberia rifted margin based on magnetic anomalies and ODP drilling (abstract). *EOS Transactions, American Geophysical Union*, **75**, 616.

WHITMARSH, R.B. & MILES, P.R. 1995. Models of the development of the west Iberia rifted continental margin at 40°30'N deduced from surface and deep-tow magnetic anomalies. *Journal of Geophysical Research*, **100**, 3789–3806.

WHITMARSH, R.B., BESLIER, M.-O. & WALLACE, P.J. (eds) *Proceedings of the Ocean Drilling Program, Initial Reports, 173.* Ocean Drilling Program, College Station, TX.

WHITMARSH, R.B., MILES, P.R. & MAUFFRET, A. 1990. The ocean–continent boundary off the western continental margin of Iberia I. Crustal structure at 40°30'N. *Geophysical Journal International*, **103**, 509–531.

WHITMARSH, R.B., MILES, P.R., SIBUET, J.-C. & LOUVEL, V. 1996a. Geological and geophysical implications of deep-tow magnetometer observations near Sites 897, 898, 899, 900 and 901 on the west Iberia continental margin. *In:* WHITMARSH, R.B., SAWYER, D.S., KLAUS, A. & MASSON, D.G. (eds) *Proceedings of the Ocean Drilling Program, Scientific Results 149.* Ocean Drilling Program, College Station, TX, 665–674.

WHITMARSH, R.B., PINHEIRO, L.M., MILES, P.R., RECQ, M. & SIBUET, J.-C. 1993. Thin crust at the western Iberia ocean–continent transition and ophiolites. *Tectonics*, **12**, 1230–1239.

WHITMARSH, R.B., WHITE, R.S., HORSEFIELD, S.J., SIBUET, J.-C., RECQ, M. & LOUVEL, V. 1996b. The ocean–continent boundary off the western continental margin of Iberia: crustal structure west of Galicia Bank. *Journal of Geophysical Research*, **101**, 28291–28314.

Subsidence, deformation, thermal and mechanical evolution of the Mesozoic South Alpine rifted margin: an analogue for Atlantic-type margins

Giovanni Bertotti

Department of Tectonics, Faculty of Earth Sciences, Vrije Universiteit, De Boelelaan 1085, 1081 HV Amsterdam, Netherlands (e-mail: bert@geo.vu.nl)

Abstract: The geological record of the currently exposed South Alpine transect of the Mesozoic passive continental margin provides information on the evolution of Atlantic-type margins. Before the onset of rifting, in Mid-Triassic to Carnian time, strong subsidence affected the central parts of the Southern Alps, in the Lombardian basin. No major fault is documented for this time span. Thermal conditions were strongly perturbed. Continental rifting began in Norian times, and until the beginning of the Liassic period was characterized by overall subsidence of the Lombardian basin with rates up to 200 mm ka^{-1}. The strong subsidence is a result of continuing extension and of generalized crustal cooling. Subsidence rates were still important in Liassic times, although lower than in Late Triassic time. At the end of the Liassic period, the site of extension shifted towards the west, where crustal break-up eventually took place in Mid-Jurassic times. Previously poorly documented features such as the very strong subsidence in the initial rifting stages, the changing geometry and mechanics of normal faults are here associated with the thermal interactions between pre-existing thermal anomalies and rifting.

A >300 km long transect of the Mesozoic passive continental margin of the Adriatic plate (or microplate) is exposed in the Southern Alps of North Italy and Switzerland (Fig. 1). The margin formed during the Late Triassic to Mid-Jurassic break-up of Pangaea, ending, in the Alpine domain, with the opening of the Liguria–Piedmont ocean. The South Alpine domain formed at that time a roughly east–west-trending transect across the western Adriatic passive continental margin. Beginning from Late(?) Cretaceous times, convergent movements between Adria and Europe caused north–south-directed shortening with consequent folding and thrusting (e.g. Dewey *et al.* 1989).

The Southern Alps, defined as the units found south of the Insubric Line (Fig. 1), provide a unique set of information on rifted margin formation and development mainly for two reasons: (1) because of the unusually large angle between the Mesozoic extension (roughly east–west in present-day coordinates) and Tertiary north–south contraction, rift-related normal faults escaped major reactivation and lateral relations are substantially preserved along most of the margin; (2) because of their upper-plate position in the Alpine subduction zone, rocks exposed at present in the Southern

Alps did not experience significant Alpine metamorphism. Consequently, the South Alpine basement still preserves the record of Palaeozoic and, more importantly for the purpose of this study, of Mesozoic thermal and deformational events.

In a very schematic fashion, the Southern Alps consist at present of a limited amount of mainly south-vergent, both thin- and thick-skinned thrust sheets (e.g. Schumacher *et al.* 1997). The rock units composing them become roughly younger from north to south. In the north, the pre-Mesozoic crystalline basement crops out, replaced to the south by Permian and Mesozoic pre-rift, syn-rift and post-rift sediments and eventually by shortening-related Cretaceous and Tertiary clastic rocks at the transition with the Po Plain. On the whole, the Southern Alps can be looked at as a vertical section through the Adriatic continental margin, deformed, tilted and exposed at the surface.

The Mesozoic sediments of the Southern Alps have been studied for more than 100 years and no major uncertainties exist on their stratigraphy and thickness. Indeed, abrupt thickness and/or facies changes have been recognized for several decades and attributed to Mesozoic normal faulting (von Bistram 1903; Vonderschmitt

From: Wilson, R.C.L., Whitmarsh, R.B., Taylor, B. & Froitzheim, N. 2001. *Non-Volcanic Rifting of Continental Margins: A Comparison of Evidence from Land and Sea.* Geological Society, London, Special Publications, **187**, 125–141. 0305-8719/01/$15.00 © The Geological Society of London 2001.

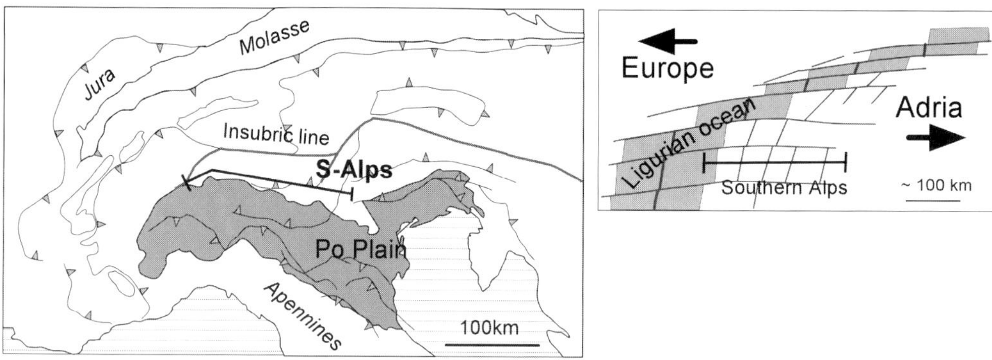

Fig. 1. Schematic map of the Southern Alps and of their position in the Late Jurassic plate tectonic configuration (from Weissert & Bernoulli 1985). The thick line marks the approximate position of the profile shown in Figure 4.

1940). With the appearance of plate tectonics, these observations were integrated into a systematic frame providing important constraints on the geology of passive continental margins (Bernoulli *et al.* 1979). In general, however, sediments were studied to obtain information on the Mesozoic to Tertiary sedimentation and/ or constraints on Alpine shortening. Studies on sediments were also decoupled from those being performed in the crystalline basement, which were largely aimed at the reconstruction of Variscan metamorphism. Mesozoic tectonics received much less attention. It was at the beginning of the 1980s that a bridge was created between the domains. From the observation that the sediments in the Lake Como area were steeply south dipping, Bally *et al.* (1981) concluded that the deeper continuation of the Mesozoic normal faults documented in the sedimentary cover had to be looked for in the metamorphic basement cropping out north of those sediments. Consequently, geologists began to look at Mesozoic sediments and at the metamorphic basement of the Southern Alps with a new perspective, namely that of Mesozoic rifting. This approach was further encouraged by structural and radiometric studies carried out in the western part of the Southern Alps, which suggested the existence of important Permian to Jurassic vertical movements controlled by normal faulting (e.g. Hodges & Fountain 1984; Handy 1987). Particularly illustrative is the situation in the Lake Como area where a section of *c.* 15 km thickness through the Mesozoic crust is at present exposed, which includes the deep continuation of a major listric normal fault (Fig. 2) (Bertotti 1991).

Recently, the available database has been significantly expanded, resulting in a fairly detailed picture of the kinematic and thermal evolution of the South Alpine margin during and after rifting (Handy & Zingg 1991; Bertotti *et al.* 1993*a*, Bertotti *et al.* 1997). The wealth of data has eventually allowed for the use of numerical modelling techniques to quantitatively integrate the observations.

In this paper I present for the first time a complete analysis of the subsidence patterns of the margin and propose an integrated reconstruction of its evolution taking into consideration all significant aspects, from vertical movements, to deformation mechanisms, overall thermal evolution and mechanical changes. The reconstruction is limited to the western and central Southern Alps and does not take into account the easternmost part of the profile including large-scale tectonic features such as the Belluno trough and the Friuli Platform (e.g. Winterer & Bosellini 1981). In the first part of the paper the techniques adopted and the main results obtained with each method will be briefly reviewed. These will be the 'bricks' used to develop the reconstruction presented in the second part of the study. Finally, some implications for the general understanding of passive continental margins will be discussed.

The picture that emerges is a remarkably complete one providing an integrated reconstruction of kinematic, thermal and mechanical patterns through rifting. The importance of these studies is not limited to the Alpine system. In fact, various tectonic characteristics of the South Alpine rift such as dimensions, duration of rifting, and amount and rate of defor-

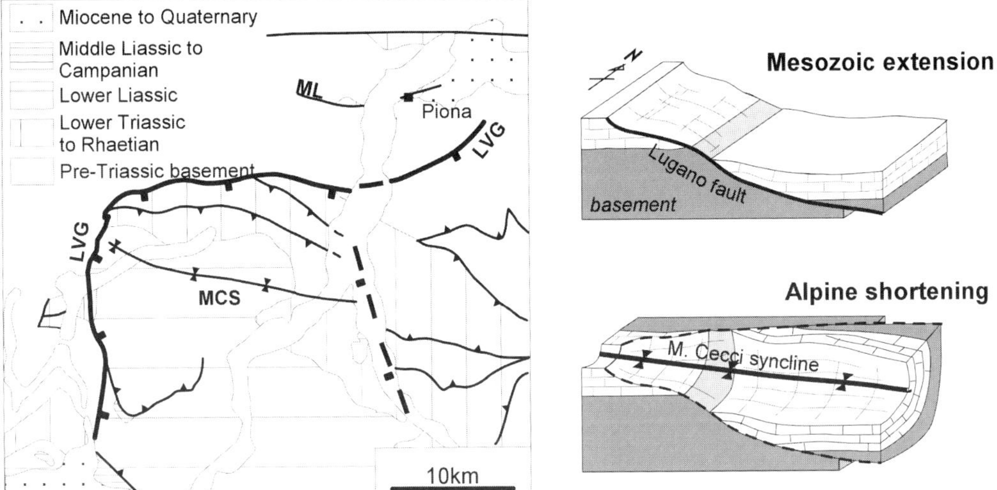

Fig. 2. Schematic geological map of the Lake Como area (left) and scheme of the effects of north–south Alpine shortening overprinting Mesozoic east–west extension (to the right). The scheme illustrates how pre-Alpine rocks and structures have been exhumed in Alpine times and are at present accessible. The map-view of the Lake Como area is a slightly deformed Mesozoic vertical section. LVG, Lugano–Val Grande Fault; ML, Musso line; MCS, Monte Cecci syncline. In the scheme, syn- and post-rift sediments have been omitted for clarity.

mation are comparable with those of other Atlantic-type passive continental margins (e.g. Manatschal & Bernoulli 1999). By contrast, they are of more difficult application to margins formed in back-arc settings.

Constraints on the evolution of the rifted margin

The reconstruction of the South Alpine evolution presented here is based on a large amount of data. In this section, the most important techniques and methods adopted are briefly mentioned. The goal in doing this is not to provide a methodological discussion but rather to present the steps performed and the data used for the reconstruction.

Geometries and kinematics

Any complete quantitative reconstruction of the evolution of a rifted margin is impossible in the absence of a well-constrained kinematic scheme. It is indeed surprising how few such reconstructions exist in the literature. The first step to reconstruct the kinematics of the South Alpine margin is the definition of the architecture of the margin before Alpine shortening. In the case of the Southern Alps, this would correspond, for instance, to the Late Jurassic profile. To obtain such a profile, data on the geometry

of sedimentary bodies and on their stratigraphy have been collected through detailed mapping and measurement of stratigraphic columns. Removing Alpine deformation, these data are then assembled to form an upper-crustal profile across the Late Jurassic margin (Fig. 3d). This reconstruction is fairly reliable, with the exception of the westernmost, distal, parts, where Alpine deformation has been very intense. More specifically, depicted fault geometries are well constrained to depths of 5–10 km.

To reconstruct pre-break-up kinematics, the Late Jurassic profile has then been stepwise retrodeformed by removing sedimentary packages, correcting the remaining sediments for compaction and retrodeforming movements of fault blocks. As a result, a complete kinematic description is obtained not only for the rifting period as a whole but also for some intervals of it and for specific areas of the South Alpine margin (Fig. 3a–d). Assuming an initial crustal thickness of 30 km and applying thinning factors derived from the previously obtained stretching factors, a complete crustal profile across the South Alpine margin is obtained (Fig. 4).

For the specific purpose of the analysis presented in this paper, the availability of a well-constrained kinematic reconstruction means that the position of rock masses along the profile is known in each time step. More specifi-

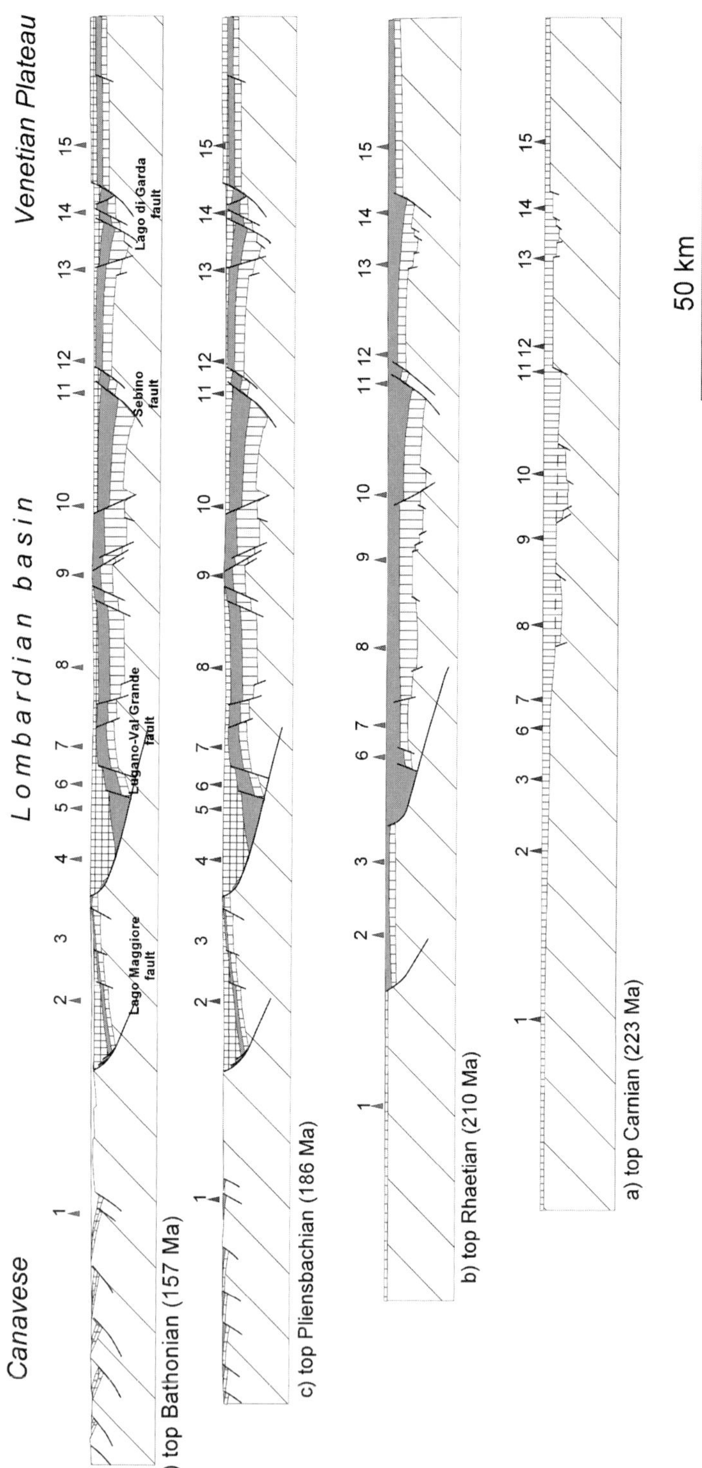

Fig. 3. Upper-crustal profiles through the South Alpine rifted margin (from Bertotti *et al.* 1993*a*). Grey triangles indicate the profiles for which subsidence curves have been constructed (Fig. 5).

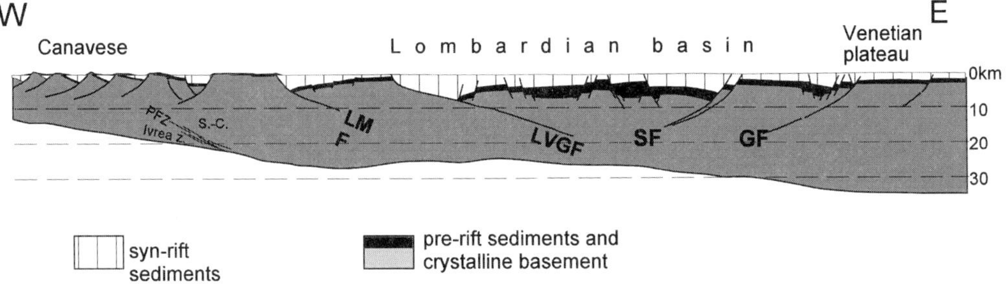

Fig. 4. Crustal profile across the Southern Alps at the end of rifting (Mid-Jurassic time). Upper-crustal geometries are constrained by field observations. An initial crustal thickness of 30 km has been assumed (see text). Crustal thinning corresponds to stretching factor reconstructed from the upper-crustal record. LMF, Lago Maggiore fault; LVGF, Lugano–Val Grande Fault; SF, Sebino Fault; GF, Garda Fault.

cally, if a geographical point at present at the surface can be positioned in the Late Jurassic profile then its position can be tracked backwards in time. This has been carried out only on geometrical arguments, that is, without invoking any assumption on geothermal gradients, etc. (see below).

Subsidence analysis

To constrain the history of vertical movements along the South Alpine rifted margin, subsidence curves have been constructed for various localities along the section (Figs 3 and 5). Stratigraphic columns have been measured in the field and, when needed, compiled from the literature (for details, see Bertotti *et al.* 1993*a*). They have then been de-lithified using standard procedures (e.g. Bond & Kominz 1984). Water depths are well constrained until earliest Liassic time. After this time, pelagic conditions prevailed over large parts of the margin. Water depths <1500 m have been assumed for the subsidence curve. This is the value estimated for a continental crust of initially 30 km thickness thinned to 20 km (e.g. Bertotti 1991; Bertotti *et al.* 1993*a*). The bathymetry is sedimentologically reasonable. Subsidence curves are also used to obtain rates of subsidence.

Geochronology

Until a few years ago, the number of geochronological data available in the South Alpine basement was fairly limited (Hunziker *et al.* 1992, and references therein) and mainly concentrated in the western regions where some post-Variscan overprint had been detected. During the last decade, there has been a substantial increase both in the quantity and in the quality of available data. These include, among others, high-resolution ^{40}Ar/^{39}Ar data but also fission-track measurements on zircons and apatites. A significant portion of these data were produced for the central parts of the Southern Alps, thereby providing a wider and more complete coverage (Bertotti *et al.* 1999). These data have generated substantially new and partly unexpected constraints on the thermal evolution of the South Alpine margin.

In this study, geochronological data are taken as indicators of temperatures using the concept of closing or annealing temperatures. According to this concept, the age determined corresponds to the moment when the rock was cooled below the closing temperature of a given system. This implies that the rock was at closing temperature at the moment yielded by the measurement. The correlation of a dated rock and thus of a temperature estimate with a specific crustal position is usually made difficult by our inability to determine independently the depth of the sampled rock at the moment corresponding to the measured absolute age. In general, assumptions are made on the geotherms. Beside the difficulty in estimating thermal gradients for the past, especially in tectonically active areas, it is clear that the obtained data cannot be used to reconstruct the thermal field.

In the central Southern Alps, the problem is solved by the possibility of locating sample points in the pre-Alpine crustal section and, therefore, follow their movement during rifting with the available kinematic reconstruction. The diagrams presented hereafter are thus constructed according to the following scheme: (1) the position of the samples in the Late Jurassic section is determined by measuring present-day distances from pre-Alpine features such as the basement–sedimentary cover contact and Mesozoic normal faults; (2) a value correspond-

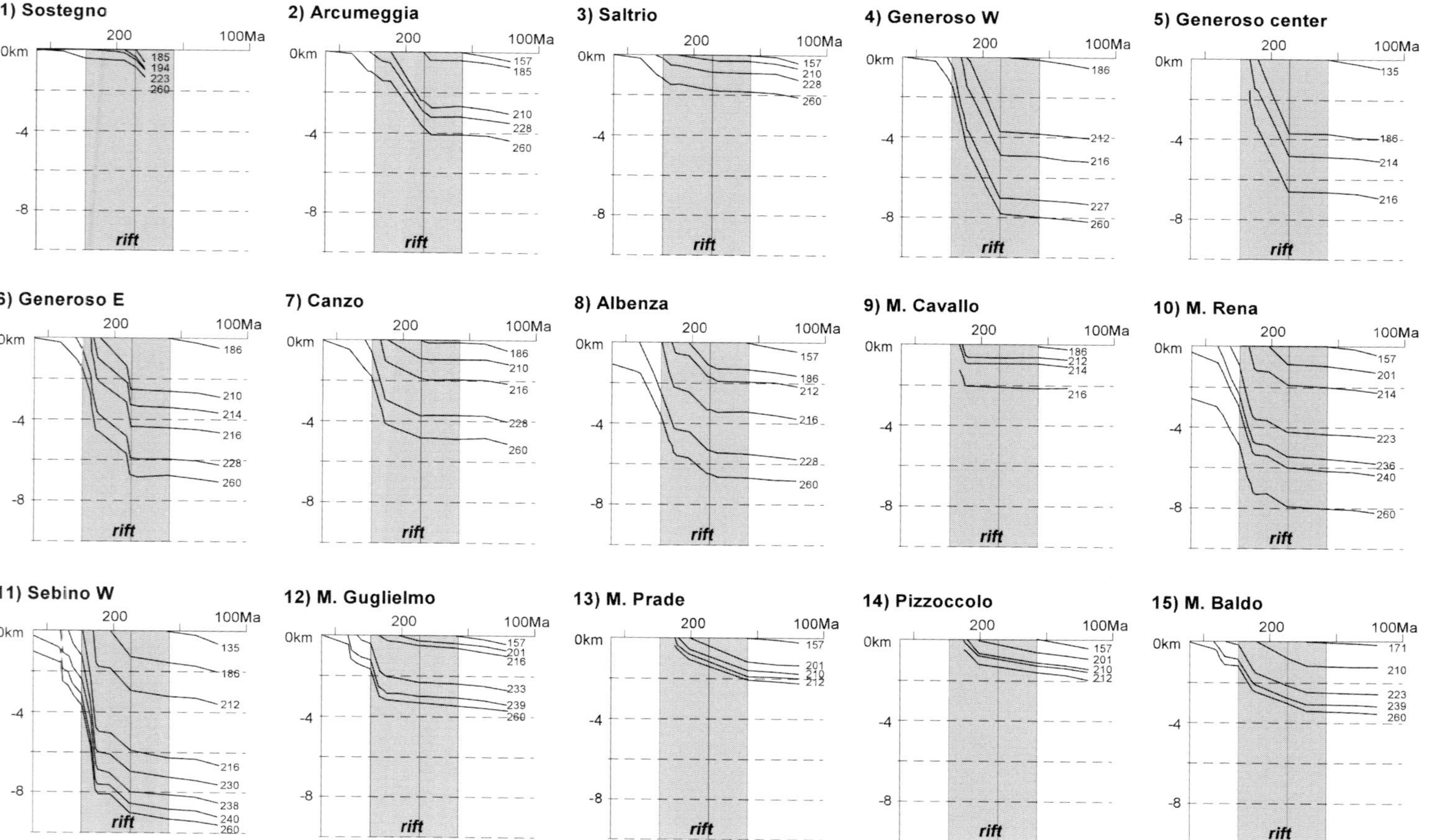

Fig. 5. Subsidence curves for the profiles shown in Figure 3. The rifting period marked by the grey bar is defined on the basis of onset and end of normal faulting. It should be noted that all diagrams have the same scale.

ing to the closing or annealing temperature of a given system is assigned to the sample point at the age provided by dating; before this moment, the same sample must have been at higher temperatures, and following that moment it was cooler; (3) for each time step, temperature values for different points have been contoured showing the pattern of isotherms; comparison of profiles drawn for different time steps provides an illustration of temporal changes. The situation in the western part of the section is less constrained because of the difficulty of independently determining the depth of rocks of the Strona–Ceneri and Ivrea zones in Jurassic time. However, we have made use of the very extensive petrological work that has been performed on these areas, which provides fairly reliable estimates of palaeobarometry (e.g. Handy & Zingg 1991, and references therein). Data used in our diagrams are shown in Table 1.

Deformation along fault zones

The simple recognition that normal faults defined in the sedimentary succession necessarily had a continuation into the exposed basement was a breakthrough in South Alpine studies. This offered the possibility to acquire fundamental data on events taking place within the crust during rifting, and opened unprecedented possibilities to analyse deformation mechanism along rift-related normal faults. This potential could be exploited along the Pogallo fault separating the Ivrea zone from the Strona–Ceneri zone (Handy 1987) and along the Lugano–Val Grande fault, which is continuously exposed from the Jurassic surface to a former depth of 12–15 km (Bertotti 1991). In both cases, fault rocks have been studied and their distribution has been mapped, producing significant information on how these faults operated. It is very likely that other, hitherto unrecognized, rift-related shear zones exist in the South Alpine basement.

Numerical modelling

The use of numerical models in the Southern Alps (and indeed in the Alps in general) is fairly young. The results of numerical studies that have been performed on the Mesozoic evolution of the South Alpine rifted margin are presented here. A code developed by ter Voorde (1996) was used. The model correctly describes the kinematics and the thermal evolution of the rifted margin (Fig. 6). It is composed of an upper layer where extension is accommodated by discrete faults, the number and kinematics of which can be varied at will, and of a lower layer where stretching is distributed. The depth of transition between the two layers can be fixed by the operator. A thermal grid is superimposed on the model to describe thermal changes during and after stretching.

Table 1. *Sources of geochronological data*

Point no.	Notes	References
1	Representative of the Strona–Ceneri zone along the Val d'Ossola section	1, 2
2	Samples from northern Lake Maggiore (Brissago area)	1, 2
3	Samples from the Val Colla Passo S. Iorio area (north of Lake Lugano)	2, 3, 4
4	Points representative of the hanging wall of the Lugano–Val Grande normal fault (east side of Lake Como)	5
5	Points representative of the footwall of the Lugano–Val Grande normal fault (east side of Lake Como, northern part)	5

Specification of data used to reconstruct the thermal evolution of the Southern Alps shown in Fig. 5. References: 1, Handy & Zingg 1991; 2, Hunziker *et al.* 1992; 3, Internal reports Swiss NFP 20; 4, Hunziker *et al.* 1997; 5, Bertotti *et al.* 1999.

Rheology

Rheology profiles have been constructed for representative sections across the South Alpine margin for different moments to derive lithospheric strength estimates and obtain information on the way this is supported by the different lithospheric layers. Standard techniques have been adopted for rheological modelling (Ranalli & Murphy 1985); rheological profiles have been then translated into integrated strength and effective elastic thickness values following procedures developed by Burov & Diament (1995).

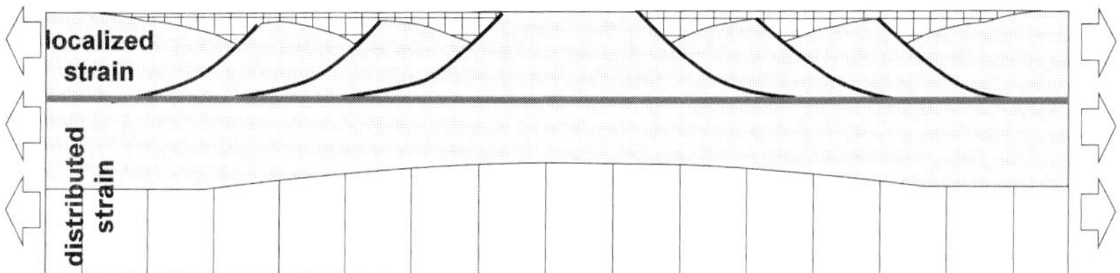

Fig. 6. The set-up of the numerical model adopted to describe the kinematic and thermal evolution of the South Alpine rifted margin (from ter Voorde 1996). In the upper part of the model, stretching is accommodated by localized shearing, in the lower one by distributed deformation. The kinematics of the various faults (geometry and deformation) can be varied at will.

The pre-rift stage: from Late Permian to Carnian time

Late Permian to Early Triassic time

Following the end of Variscan orogeny and magmatism, and the vertical and horizontal movements that took place during its final stages, the South Alpine crust was gradually stabilized. Relief had been practically eliminated in Late Permian time, and the Upper Permian to Lower Triassic continental to transitional clastic sediments were deposited over large plains close to sea level (Assereto *et al.* 1973). These observations suggest that the crust had regained 'normal' thickness, probably close to 30 km. Upper Permian to Lower Triassic sediments gradually thicken from west to east and reach values up to several hundred metres in the central and eastern parts of the section roughly coinciding with the depocentres of Upper Carboniferous to Lower Permian clastic sediments (De Sitter & De Sitter-Koomans 1949). Lateral thickness changes seem to be gradual over the entire South Alpine domain and only marginally, namely in the east, controlled by faults (Assereto *et al.* 1973) (Fig. 3a). A significant part of Late Permian subsidence was probably controlled by differential compaction of underlying sediments.

Thermal conditions in the Permian to Early Triassic crust are difficult to assess mainly because very high temperatures in Mid-Triassic time obliterated the record of previous situations. High thermal gradients during Permian time are suggested by widespread volcanic rocks in the Lugano and Bozen regions and by geological reconstructions from the Ivrea zone (e.g. Quick *et al.* 1994).

Mid-Triassic to Carnian time

Sedimentation and subsidence. Beginning from Anisian times, marine conditions pre-vailed across the future South Alpine margin and a complex but fairly well-known system of carbonate platforms and intervening basins developed (Brack & Rieber 1993, and references therein). Volcanic and volcanoclastic layers are commonly intercalated in the Ladinian platform and basin sediments.

In Carnian time, the carbonate domains of the Southern Alps were 'polluted' by the arrival of terrigenous conglomerates to silts from the south. As a result, most of the carbonate platforms died. Volcanic clasts are very common and testify the erosion and destruction of a volcanic belt south of the present-day Southern Alps (Garzanti 1985).

Subsidence in Mid-Triassic to Early Carnian times was extremely high. The largest thicknesses are recorded in the central parts of the future margin, in what was to become the Lombardian basin (Figs 3 and 5). Subsidence rates of up to $180\,\text{mm ka}^{-1}$ are obtained for these regions (Fig. 7). Domains with thick successions roughly coincide with the Late Permian to Early Triassic depocentres. Very detailed compilation and mapping (Brack & Rieber 1993) have excluded the presence of major faults associated with the observed thickness changes. Indeed, the origin of the observed subsidence during the considered time span, that is, before the onset of rifting leading to the opening of the Liguria–Piedmont ocean, is enigmatic.

Thermal evolution. Common volcanoclastic intercalations and volcanic clasts in Middle Triassic to Carnian conglomerates and siltstones document the activity and subsequent cessation of magmatism in the South Alpine domain. The tectonic significance of this magmatism has been debated and related either to continental subduction or to incipient rifting (Castellarin *et al.* 1988). Without entering the geochemical discussion, what is relevant for this paper is the physical side of the problem,

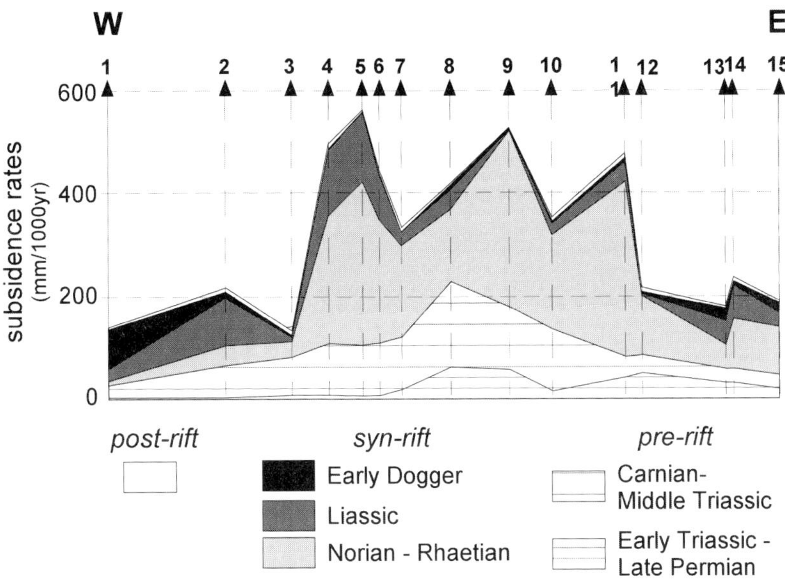

Fig. 7. Subsidence rates for representative localities along the South Alpine rifted margin. The location of the analysed profiles is given in Figure 3. Values have been corrected for compaction.

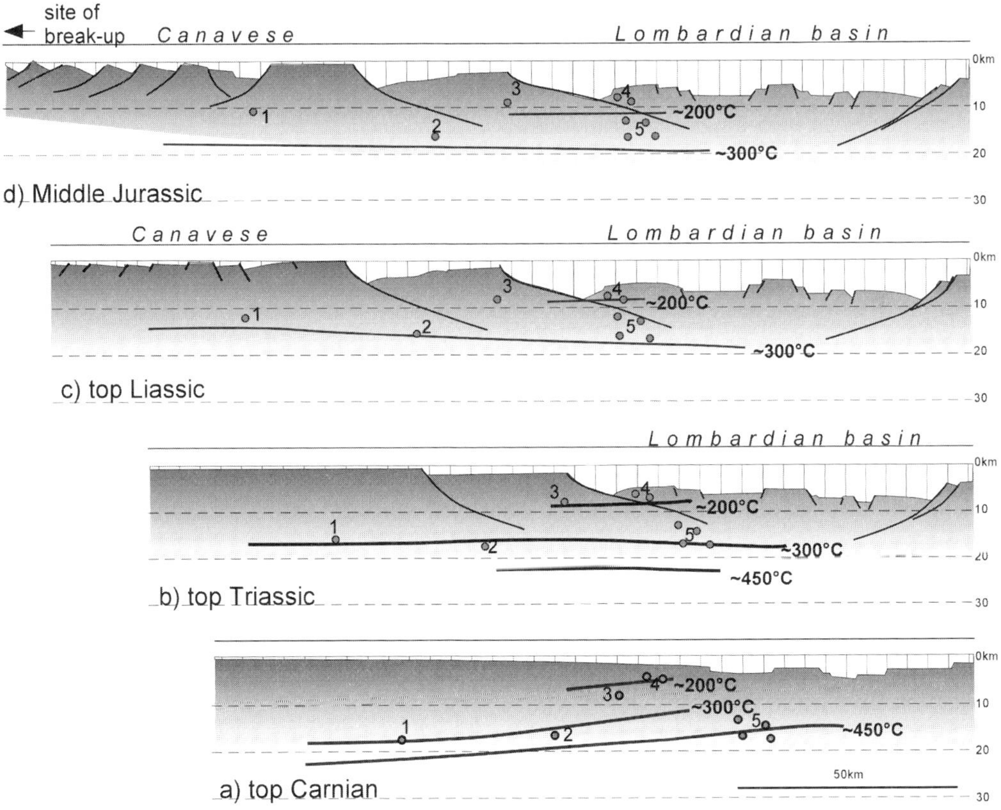

Fig. 8. Reconstruction of the thermal evolution of the South Alpine rifted margin based on observed geometries of extension and on geochronological data (see text for explanation).

namely the presence in Ladinian time of a large thermal anomaly able to produce melts.

Recent geochronological work has demonstrated that the Carnian crust was very hot. In the Lake Como area, schists of the Variscan basement lying in Carnian times at a depth of 12–15 km underwent static high-temperature metamorphism characterized by the formation of sillimanite + biotite aggregates pseudomorphically replacing garnet porphyroblasts and by fibrolite growth across older parageneses (Sanders *et al.* 1996). Pegmatites were emplaced during this time span. *P–T* estimates yielded temperatures of >600 °C (Diella *et al.* 1992). Combined with the geological reconstruction, this suggests thermal gradients in excess of 60 °C km^{-1}. These estimates are probably representative of a large domain in the Lombardian area (Fig. 8a) and possibly also of domains further to the north in parts of the future Austroalpine zone (Bertotti *et al.* 1993*b*). Interestingly, Late Triassic isobaric cooling has been detected in Lower Austroalpine units (Müntener *et al.* 1999).

Thermal conditions in the western part of the section are less well constrained but there are indications of close-to-normal gradients. Rocks of the Strona–Ceneri zone, located at this time at depths of 12–15 km, were already below the Rb/Sr and K/Ar closing temperatures for biotite (Handy & Zingg 1991).

To estimate the characteristics of the heat source needed to produce the documented thermal anomaly, we have performed a numerical modelling study (Fig. 9). Parameters and other modelling specifics have been given by Bertotti *et al.* (1997). Because the geological record excludes the presence of relevant intrusive rocks in the upper crust above 15 km, the anomaly was introduced in the lower crust. The lower crust of the Lombardian basin is not exposed and, thus, the dimensions and position of the intrusion are unconstrained. Our modelling provides best-fit curves assuming a body of 100 km width and 10 km thickness emplaced

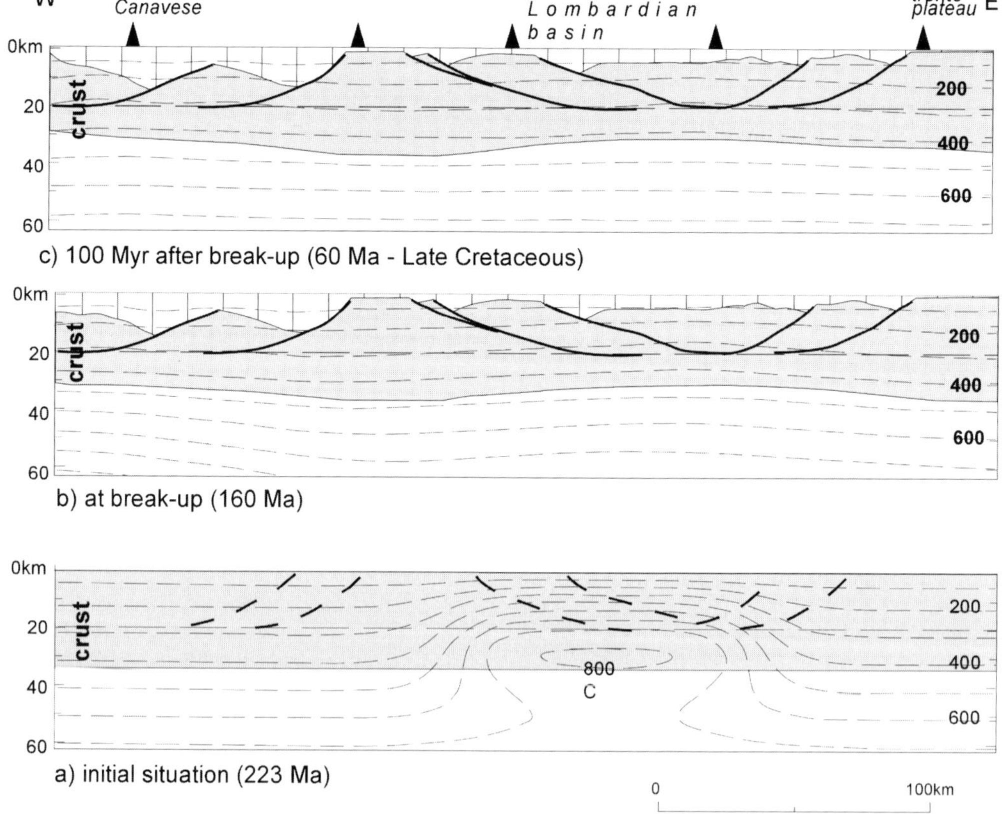

Fig. 9. Results of thermal modelling of the South Alpine rifted margin.

immediately above the crust–mantle boundary. Such dimensions are similar to those observed, for instance, along the Vøring margin (Planke *et al.* 1991; Skogseid 1994; Richardson *et al.* 1998). Although the precise dimensions and position of the intrusive body are irrelevant for the purpose of this study, modelling is positive in requiring an intrusive body of several tens of kilometres size to adequately disturb the upper-crustal thermal field.

The rift stage: from Norian to Mid-Jurassic time

Norian to Rhaetian time

Tectonics and sedimentation. The Norian time represents a fundamental step in the evolution of the Southern Alps (and possibly of the entire Alpine domain) because it marks the onset of the rifting that was eventually to lead to crustal separation. Norian to Rhaetian subsidence was very high and allowed for the deposition of >2000 m of sediments. A large number of normal faults were active in Norian time during the deposition of the regional Hauptdolomit carbonate platform (Bertotti *et al.* 1993*a*). Normal faults were typically kilometres long and had a fairly scattered orientation although a NE–SW trend seems to prevail. The lozenge shape of these basins (Bertotti *et al.* 1993*a*) could point to a strike-slip component in their formation. A significant proportion of these faults did not cause any morphological depression at the sea floor and rocks deposited on the footwall and on the hanging wall are sedimentologically similar. Only in places was a submarine relief created, in which case the faults became also facies boundaries. Norian basins are typically <20 km across and host very characteristic well-laminated and bituminous fine-grained carbonate mudstones. Breccias are areally limited and found only close to the normal faults.

A substantial change in the sedimentological pattern took place in Rhaetian time, when large quantities of terrigenous clay were abruptly introduced into the South Alpine domain, causing the extinction of large parts of the Hauptdolomit platform especially in the Lombardian Basin. Here, basinal areas developed with thick sequences and water depths up to a few hundred metres connected by gentle ramps to the surviving platforms. Terrigenous input decreased towards the end of the Rhaetian period and shallow-water carbonates spread over the entire domain. To the west of the Lombardian basin, Norian to Rhaetian deposits are basically lacking (e.g. Elter *et al.* 1966). East of the Lombardian basin, Norian to Rhaetian sediments are shallow-water carbonates, a few hundred metres thick (Fig. 3).

The Rhaetian period is a tectonically interesting time because most of the Norian normal faults were deactivated and deformation was progressively localized along a few (four) major normal faults (Fig. 3). This concentration is partly a direct consequence of the lateral growth of normal faults (Gupta *et al.* 1998) and partly the signal of a progressive strengthening of the South Alpine lithosphere (see below).

The subsidence pattern. The Norian to Rhaetian interval is characterized by an area of very strong subsidence, the Late Triassic Lombardian basin, surrounded by domains with little to no subsidence (Figs 3 and 5). Norian to Rhaetian sediments in the Lombardian basin east of the Lugano fault are up to 3000 m thick. Very thick successions are found across most the Lombardian basin and not only in the proximity of large faults. This is demonstrated, for instance, by profiles 9 (M. Cavallo) and 10 (M. Rena) (Figs 3 and 5), which are tens of kilometres away from important normal faults. Subsidence rates (Fig. 7) along these profiles are very high (e.g. Schlager 1999) and of the order of 200–300 mm ka^{-1}. The reconstructed magnitude of subsidence is far larger than that predicted from the observed amount of extension. This implies the need for a deeper-seated motor to explain the extra subsidence and is compatible with the overall pattern of Norian subsidence.

Areas west of the Lombardian basin, namely west of the Monte Generoso basin, were fairly stable. This is compatible with the absence of documented normal faults in the area. The subsidence observed to the east of the basin is, in contrast, important and is in apparent contradiction with the absence of substantial extension in this very internal part of the margin (>200 km away from the site of continental break-up). The proximity of the Vardar and/or Meliata (e.g. Froitzheim *et al.* 1996) ocean might provide a partial explanation for such behaviour.

The thermal evolution. Following years of intense structural, petrographic and geochronological work, it is now demonstrated that the South Alpine crust underwent cooling during Norian to Rhaetian times, that is, during the initial stages of continental rifting (see also Hermann 1999). This is documented by a large body of Rb–Sr, K–Ar, ^{40}Ar/^{39}Ar and fission-track ages from the Lake Como–Lake Lugano area and from other regions of the Southern

Alps (Fig. 8) (Bertotti *et al.* 1999). Thermal gradients reconstructed for the Lake Como area were >60 °C km^{-1} at the beginning of Norian time and <30 °C km^{-1} at the end of Triassic time. Crustal thermal gradients west and east of the Lombardian basin were already fairly normal and cooling, therefore, was only limited. In the west, rocks of the Strona–Ceneri zone were at temperatures lower than 300 °C (Handy & Zingg 1991; Hunziker *et al.* 1992). Cooling of upper-crustal rocks during extension is obviously incompatible with the consequences of rifting and can only be explained by the deactivation of a strong pre-existing thermal anomaly.

Numerical modelling demonstrates the physical feasibility of such dramatic changes. According to our results, a time span of *c.* 15 Ma (from Late Carnian to Rhaetian time) is sufficient to let most of the thermal effects disappear which were formed during the activity of a deep-seated thermal anomaly. The cooling pattern dominant during the initial stages of rifting also suggests that rift-related thermal perturbations were only minor.

Early to Mid-Liassic time

Tectonics and sedimentation. During the Liassic period and especially in its early part, the carbonate platforms of the Lombardian basin were drowned and thick successions of deep-water carbonate turbidites were deposited (Bernoulli *et al.* 1990). West of the Lombardian basin only little subsidence took place. Here, Liassic sediments are absent or represented by very condensed carbonate successions deposited on submarine highs (e.g. Elter *et al.* 1966). To the east of the basin, shallow-water conditions persisted throughout Liassic time with the deposition of up to 600 m of platform carbonates (Winterer & Bosellini 1981).

Liassic rifting took place following the pattern acquired towards the end of Triassic time. Crustal extension is basically limited to the Lombardian basin and accommodated by four major structures: the eastward-dipping Lago Maggiore and Lugano–Val Grande faults, and the westward-dipping Sebino and Lago di Garda faults (Fig. 3c). The displacements accommodated on these two groups of faults are roughly comparable, thereby providing a fairly symmetric aspect to the Lombardian basin. A listric shape for these faults is determined on the basis of cut-off relations and the geometry of syn-extensional sedimentary bodies.

Subsidence analysis. Liassic, especially Lower and Middle Liassic, sediments are up to several kilometres thick in the Lombardian basin and much thinner outside the basin (Fig. 3c). In the Lombardian basin, the largest thicknesses are found close to the major normal faults and are thus associated with the downward movements of their hanging walls. Areas on the footwalls and away from the fault show much less subsidence. This suggests that, in contrast to the Late Triassic conditions, accommodation space was mainly created by the downward movement of fault blocks rather than with isostasy-related crustal movements. To the west, areally restricted deposits are thin limestones deposited in relatively shallow-water but subphotic conditions.

Quantitatively correct subsidence curves in times of deeper-water sedimentation are crucially dependent on the knowledge of water depths. In the Lombardian basin, these cannot be sedimentologically and/or palaeontologically defined and are thus poorly constrained. Water depths of the order of several hundred metres to 1.5 km are predicted estimating the amount of crustal thinning associated with the 50% extension reconstructed by kinematic studies (Bertotti *et al.* 1993*a*). Substantially higher values would require a much thinner crust, which is not substantiated by observations. Sedimentation rates (Fig. 7) show fairly high values in Liassic time decreasing towards the end of the Liassic period.

Thermal evolution. Absolute ages from the South Alpine crust demonstrate persisting, although slower cooling during Liassic time (Fig. 8). Fission-track ages on zircons from the Lake Como area are scattered between 180 and 150 Ma (Bertotti *et al.* 1999). In the west, even lower-crustal rocks of the Ivrea Zone were at this stage cooler than 300 °C (Handy & Zingg 1991).

The thermal model (Fig. 9) demonstrates the overall cooling pattern. The thermal anomaly that was present in Carnian time had practically disappeared and geothermal gradients of the order of 20–25 °C km^{-1} are estimated.

Toarcian to Mid-Jurassic time

The Toarcian (Late Liassic) period marks a new, fundamental, step in the evolution of the South Alpine rift. During this time interval, the four major faults that had accommodated the regional extension and formed the Lombardian basin were progressively deactivated and sealed by deep-water turbidites or basinal and seamount pelagic deposits (Fig. 3). The Lombardian basin at this time became starved and only thin pelagic successions were deposited. East of the Lombardian basin, shallow-water conditions prevailed but subsidence was still significant. This led to the Mid-Jurassic drowning of the Trento Plateau.

To the west of the Lombardian basin a wide area with little subsidence, called the Gozzano swell, was present. Further to the west (Canavese region), however, Toarcian to Middle Jurassic shales are found, indicative of deeper-water conditions. The onset of deep-water sedimentation and the presence of olistholiths and breccias (Levone Breccia) intercalated in the shaly successions suggest extensional faulting in this region. Normal faulting is also necessary, to allow for the crustal separation that took place in the Mid-Jurassic time. This reconstruction would thus indicate that the site of crustal extension had shifted from the Lombardian basin to the west into the region of break-up (Bernoulli *et al.* 1990; Bertotti *et al.* 1993*a*).

Thermal data for the central Southern Alps indicate no major changes during Mid-Jurassic time, suggesting that (1) the Triassic thermal anomaly had practically disappeared and (2) rifting had caused little warming. In the Lake Como area, fission-track data on zircons constrain the position of the annealing temperature isotherm at *c.* 12 km. Also, the model (Fig. 9) shows that the thermal effects of rifting had practically disappeared from the crust underneath the Lombardian basin. In the western part of the profile, a thermal anomaly developed as a response of strong extension, which eventually led to crustal separation.

The post-rift stage

Following crustal separation in Mid-Jurassic time, thin successions of deep-water sediments (chert and calcareous ooze) were deposited over most of the margin (Winterer & Bosellini 1981). Thicknesses are of the order of several tens of metres and fairly constant across the Lombardian segment of the margin. Tectonic instabilities are suggested by the widespread occurrence of slumps in the Lower Cretaceous limestones. At the eastern boundary of the Lombardian basin, activity along the Lago di Garda fault is testified by Triassic to Jurassic blocks within Lower Cretaceous sediments close to the fault (Castellarin 1972).

The only thermal change affecting the South Alpine lithosphere was the waning of the thermal anomaly developed at the western termination of the section in association with the final rifting stages (Fig. 9). Thermal conditions over the rest of the margin had already relaxed in Mid-Jurassic time and remained so until the Late Cretaceous onset of Alpine shortening.

Relevance of the South Alpine case for rifted margin studies

Thermal anomalies and the onset of rifting

It has been postulated in this study that an important thermal anomaly was present in the South Alpine crust when rifting began. The age of its emplacement is basically unconstrained because of the inability of geochronological methods to detect heating phases. Modelling studies underline the physical difficulty of maintaining strongly perturbed upper-crustal geotherms for periods of tens of millions of years. I therefore favour the hypothesis of an emplacement fairly close to the onset of rifting, probably in Anisian or Ladinian times. It is unlikely that the intrusion of such an underplated body caused major uplift, as densities of accreted basic magmas do not substantially differ from those of the lower crust. On cooling, in contrast, densities increased, which could explain the observed subsidence before and during rifting.

Although it is difficult to say when and why the intrusion was emplaced, there is an apparent coincidence between the location of the thermal anomaly and the site where extension began. Modelling studies have shown that 'normal' tectonic forces are unable to cause extension unless the lithosphere is thinner than normal, that is, if heat flows are $< 60\,\text{mW m}^{-2}$ (e.g. Kusznir & Park 1987). It is therefore here suggested that the Mid-Triassic thermal anomaly was instrumental in the onset of extension in the South Alpine margin and in its localization in the Lombardian basin.

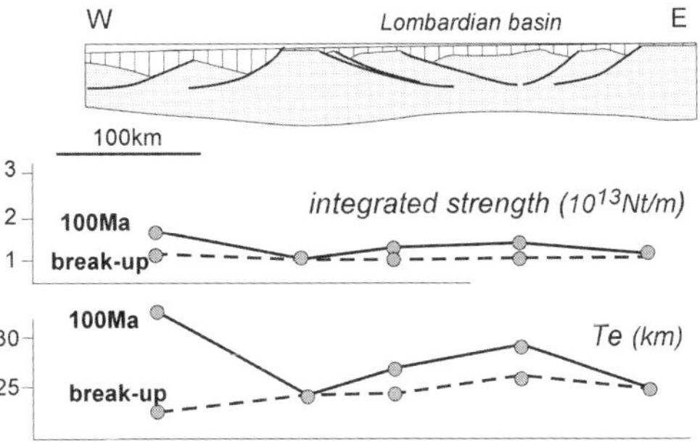

Fig. 10. Integrated strength and effective elastic thickness values along the South Alpine rifted margin at break-up and 100 Ma later, at the onset of Alpine shortening.

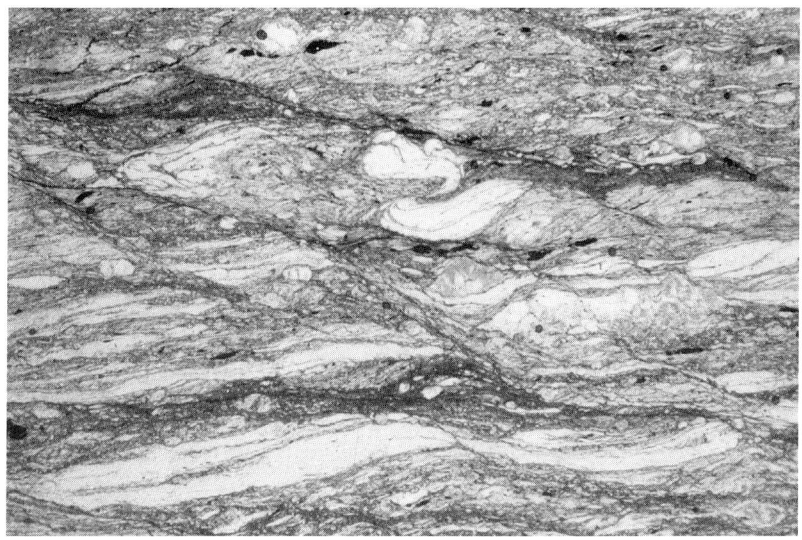

Fig. 11. Photomicrographs of fault rocks along the deep segment of the Lugano–Val Grande normal fault. (**a**) Shearing along quartz ribbons in quartz mylonites (light) is progressively replaced by ultramylonites (dark and finer grained). Microphotograph is 6 cm across; parallel nicols. (**b**) Mylonitic quartz bands are overprinted by brittle deformation. Micrograph is 10 cm across; crossed nicols.

Cooling during rifting

Geochronological data and related thermal modelling support the conclusion that the South Alpine crust was cooling during rifting. This is supported by the observation that the high-T–low-P metamorphic paragenesis and textures in the Lake Como area are perfectly preserved and undeformed (Sanders *et al.* 1996). Cooling rates were particularly high during the first sev-

eral millions of years of rifting and decreased thereafter. The process controlling this pattern was the deactivation of the Mid-Triassic thermal anomaly. General cooling during rifting had very important consequences for the tectonic and sedimentary evolution of the margin.

(1) Subsidence during the initial rifting stages was higher than predicted from crustal stretching. This is evident from the overall subsidence pattern and from specific subsidence

curves. Late Triassic subsidence affected the entire Lombardian basin even in areas where little normal faulting occurred. This is in contrast to the situation in Liassic time, when accommodation space was mainly created by downward movement of hanging walls. Cooling during the initial stages of rifting is also evident from the subsidence curves, which show very high rates in the first 10–20 Ma in the Lombardian basin. Rates are comparable with or even higher than those estimated for the later time intervals when extension was at least as important.

(2) Cooling generally causes a strengthening of the lithosphere despite continuing stretching. Areas such as the Lombardian basin, which underwent both cooling and thinning, became stronger than surrounding domains, eventually resulting in a shift of the site of extension towards other, less extended areas. Syn-rift strengthening of the lithosphere as first suggested by England (1983) was in the South Alpine margin amplified by the decay of a thermal anomaly. Lithospheric strength and effective elastic thickness estimates derived for the moment of break-up and for 100 Ma later show clearly this effect (Fig. 10). Abandoned extensional structures such as the Lombardian basin are common in most passive continental margins and are well exemplified by the Norwegian margin (Skogseid 1994). Syn-rift cooling also causes the thickening of the brittle layer of the crust, which, in turn, imposes an increase in the spacing between faults. This partly explains the change in faulting pattern from Late Triassic to Early Jurassic time, when deformation was progressively localized along a small number of major faults. This phenomenon is also a consequence of the growth of faults (Gupta *et al.* 1998).

(3) A further, hitherto unexplored, consequence of cooling concerns the change of deformation mechanisms along crustal-scale normal faults. Deepening of the isotherms during normal faulting implies that the segment of the fault deforming in a brittle manner increases in length during rifting, thereby changing the rheological behaviour of the lithosphere. In the field, the fall in ambient temperatures is documented by cataclasis overprinting mylonitic deformation as observed, for instance, along the Lugano–Val Grande fault zone (Fig. 11). Differently from core complex settings, such a pattern cannot be explained by the exhumation of fault rocks together with the fault foot-wall and necessarily requires a downward movement of isotherms.

Thermal effects of lithospheric extension

The observation that the thermal evolution of the South Alpine rift was dominated by waning of a pre-existing anomaly, that is, by processes not directly related to rifting, implies that stretching of the lithosphere in itself was unable to produce substantial thermal perturbations. Our numerical models confirm this and support the notion that rifts >100–200 km width and operating over time intervals >20–30 Ma produce negligible thermal perturbations.

The absence of a major thermal anomaly at the end of rifting means that only little thermal subsidence could affect the margin during drifting even in regions where large extension had taken place. In the specific case of the South Alpine margin, this suggests that very characteristic deposits such as the Upper Jurassic–Lower Cretaceous cherts and calcareous ooze have been deposited at water depths not significantly different from those of the late syn-rift sediments. Our modelling results indicate for these sediments water depths <1500 m.

A second, more intriguing consequence of the limited thermal perturbation associated with rifting is the difficulty of creating thermal anomalies large enough to cause melting. Although our modelling was not designed to investigate melting conditions and is, therefore, somewhat inadequate for this purpose, the difficulties of creating large thermal disturbances in the crust and/or in the uppermost mantle are apparent. This is indeed in accordance with the observation that a large proportion of 'Atlantic-type' passive continental margins are 'non-volcanic'. Following this line of reasoning, volcanism in 'volcanic' margins should not be directly correlated with rifting but rather associated with other phenomena such as, for instance, hotspots. This is the case for the Norwegian margin, which, according to some workers (White 1988), became 'volcanic' when it interacted with the Icelandic hotspot.

It is a pleasure to thank D. Bernoulli for sharing his ideas and competence throughout these years. Discussions with V. Picotti were always a precious source of information and thoughts. N. Froitzheim and D. Bernoulli are thanked for their careful review and editorial work. This is publication 20001001 of the Netherlands Research School of Sedimentary Geology (NSG).

References

ASSERETO, R., BOSELLINI, A, FANTINI-SESTINI, N. & SWEET, W.C. 1973. The Permian–Triassic boundary in the Southern Alps. *In*: LOGAN, A.

& HILLS, L.V. (eds) *The Permian and Triassic Systems and their Mutual Boundaries*. Canadian Society of Petroleum Geologists, Memoir, **2**, 176–199.

BALLY, A.W., BERNOULLI, D., DAVIS, G.A. & MONTADERT, L. 1981. *Listric normal faults*, Oceanologica Acta, 26th International Congress, Paris. 87–101.

BERNOULLI, D., BERTOTTI, G. & FROITZHEIM, N. 1990. Mesozoic faults and associated sediments in the Austroalpine–South Alpine passive continental margin. *Memorie della Società Geologica Italiana*, **45**, 25–38.

BERNOULLI, D., CARON, C., HOMEWOOD, P., KÄLIN, O. & VAN STUIJVENBERG, J. 1979. Evolution of continental margins in the Alps. *Schweizerische Mineralogische und Petrographische Mitteilungen*, **59**, 165–170.

BERTOTTI, G. 1991. Early Mesozoic extension and Alpine shortening in the western Southern Alps: the geology of the area between Lugano and Menaggio (Lombardy, Northern Italy). *Memorie di Scienze Geologiche (Padova)*, **43**, 17–123.

BERTOTTI, G., PICOTTI, V., BERNOULLI, D. & CASTELLARIN, A. 1993*a*. From rifting to drifting: tectonic evolution of the South-Alpine upper crust from the Triassic to the Early Cretaceous. *Sedimentary Geology*, **86**, 53–76.

BERTOTTI, G., SILETTO, G.-B. & SPALLA, M.-I. 1993*b*. Deformation and metamorphism associated with crustal rifting: the Permian to Liassic evolution of the Lake Lugano–Lake Como area (Southern Alps). *Tectonophysics*, **226**, 271–284.

BERTOTTI, G., TER VOORDE, M., CLOETINGH, S. & PICOTTI, V. 1997. Thermomechanical evolution of the South-Alpine rifted margin (North Italy): constraints on the strength of passive continental margins. *Earth and Planetary Science Letters*, **146**, 181–193.

BERTOTTI, G., SEWARD, D., WIJBRANS, J., TER VOORDE, M. & HURFORD, A.J. 1999. Crustal thermal regime prior to, during and after rifting: a geochronological and modeling study of the Mesozoic South Alpine rifted margin. *Tectonics*, **18**, 185–200.

BOND, G.C. & KOMINZ, M.A. 1984. Construction of tectonic subsidence curves for the Early Paleozoic miogeocline, southern Canadian Rocky Mounains: implications for subsidence mechanisms, age of break-up and crustal thinning. *Geological Society of America Bulletin*, **95**, 155–173.

BRACK, P. & RIEBER, H. 1993. Towards a better definition of the Anisian/Ladinian boundary: new biostratigraphic data and correlations of boundary sections from the Southern Alps. *Eclogae Geologicae Helvetiae*, **86**, 415–527.

BUROV, E. & DIAMENT, M. 1995. The effective elastic thickness (Te) of continental lithosphere. What does it really mean? *Journal of Geophysical Research*, **100**, 3905–3928.

CASTELLARIN, A. 1972. Evoluzione paleotettonica sinsedimentaria del limite tra la 'Piattaforma Veneta' e 'Bacino Lombardo' a Nord di Riva del Garda. *Giornale di Geologia*, **38**, 11–212.

CASTELLARIN, A., LUCCHINI, F., ROSSI, P.L., SELLI, L. & SIMBOLI, G. 1988. The Middle Triassic magmatic–tectonic arc development in the Southern Alps. *Tectonophysics*, **146**, 79–89.

DE SITTER, L.U. & DE SITTER-KOOMANS, C.M. 1949. The geology of the Bergamasc Alps. *Leidse Geologische Mededelingen*, **14B**, 1–257.

DEWEY, J.F., HELMAN, M.L., TURCO, E., HUTTON, D.H.W. & KNOTT, S.D. 1989. Kinematics of the western Mediterranean. In: COWARD, M.D., DIETRICH, D. & PARK, R.G. (eds) *Alpine Tectonics*. Geological Society, London, Special Publications, **45**, 265–283.

DIELLA, V., SPALLA, M.-I. & TUNESI, A. 1992. Contrasting thermomechanical evolutions of the South-Alpine metamorphic basement of the Orobic Alps (Central Alps, Italy). *Journal of Metamorphic Geology*, **10**, 203–219.

ELTER, G., ELTER, P., STURANI, C. & WEIDMANN, M. 1966. Sur la prolongation du domaine ligure dans le Montferrat et les Alpes et sur l'origine de la nappe de la Simme s.l. des Préalpes romandes et chablaisiennes. *Archives des Sciences Genève*, **19**, 279–375.

ENGLAND, P. 1983. Constraints on extension of lithosphere. *Journal of Geophysical Research*, **88**, 1145–1152.

FROITZHEIM, N., SCHMID, S.M. & FREY, M. 1996. Mesozoic paleogeography and the timing of eclogite facies metamorphism in the Alps: a working hypothesis. *Eclogae Geologicae Helvetiae*, **89**, 81–110.

GARZANTI, E. 1985. The sandstone memory of the evolution of a Triassic volcanic arc in the Southern Alps, Italy. *Sedimentology*, **32**, 423–433.

GUPTA, S., COWIE, P.A., DAWERS, N.H. & UNDERHILL, J.R. 1998. A mechanism to explain rift-basin subsidence and stratigraphic patterns through fault-array evolution. *Geology*, **26**, 595–598.

HANDY, M.R. 1987. The structure, age and kinematics of the Pogallo fault zone: southern Alps, northwestern Italy. *Eclogae Geologicae Helvetiae*, **80**, 593–632.

HANDY, M. & ZINGG, A. 1991. The tectonic and rheological evolution of an attenuated cross section of the continental crust; Ivrea crustal section, southern Alps, northwestern Italy and southern Switzerland. *Geological Society of America Bulletin*, **103**, 236–253.

HERMANN, J. 1999. *The Braccia Gabbro (Malenco, Alps): Permian intrusion at the crust to mantle interface and Jurassic exhumation during rifting*. PhD thesis, ETH Zurich.

HODGES, K.V. & FOUNTAIN, D.M. 1984. Pogallo Line, Southern Alps, Northern Italy: an intermediate crustal level, low-angle normal fault? *Geology*, **12**, 151–155.

HUNZIKER, J.C., DESMONS, J. & HURFORD, A.J. 1992. Thirty-two years of geochronological work in the Central and Western Alps: a review on seven maps. *Mémoires Géologie (Lausanne)*, **13**, 1–59.

HUNZIKER, J.-C., HURFORD, A.J. & CALMBACH, L. 1997. Alpine cooling and uplift. *In*: PFIFFNER, O., LEHNER, P., HEITZMANN, P., MUELLER, S. & STECK, A. (eds) *Deep Structure of the Swiss Alps*. Birkhäuser, Basel, 260–265.

KUSZNIR, N.J. & PARK, R.G. 1987. The extensional strength of the continental lithosphere: its dependence on geothermal gradient, and crustal composition and thickness. *In*: COWARD, M.P., DEWEY, J.F. & HANCOCK, P.L. (eds) *Continental Extensional Tectonics*. Geological Society, London, Special Publications, **28**, 35–52.

MANATSCHAL, G. & BERNOULLI, D. 1999. Architecture and tectonic evolution of nonvolcanic margins: present-day Galicia and ancient Adria. *Tectonics*, **18**, 1099–1119.

MÜNTENER, O., HERMANN, J. & TROMMSDORFF, V. 1999. Cooling history and exhumation of lower-crustal granulites and upper mantle (Malenco, Eastern Central Alps). *Journal of Petrology*, **41**, 175–200.

PLANKE, S., SKOGSEID, J. & ELDHOLM, O. 1991. Crustal structure off Norway, 62 degrees to 70 degrees north. *Tectonophysics*, **189**, 91–107.

QUICK, J.E., SINIGOI, S. & MAYER, A. 1994. Emplacement dynamics of a large mafic intrusion in the lower crust, Ivrea–Verbano zone, northern Italy. *Journal of Geophysical Research*, **99**, 21559–21573.

RANALLI, G. & MURPHY, D.C. 1985. Rheological stratification of the lithosphere. *Tectonophysics*, **132**, 281–295.

RICHARDSON, K.R., SMALLWOOD, J.R., WHITE, R.S., SNYDER, D.B. & MAGUIRE, P.K.H. 1998. Crustal structure beneath the Faroe Islands and the Faroe–Iceland Ridge. *Tectonophysics*, **300**, 159–180.

SANDERS, C.A.E., BERTOTTI, G., TOMMASINI, S., DAVIES, G.R. & WIJBRANS, J.R. 1996. Triassic pegmatites in the Mesozoic middle crust of the Southern Alps: fluid inclusions, radiometric dating and tectonic implications. *Eclogae Geologicae Helvetiae*, **89**, 505–525.

SCHLAGER, W. 1999. Scaling of sedimentation rates and drowning of reefs and carbonate platforms. *Geology*, **27**, 183–186.

SCHUMACHER, M.E., SCHÖNBORN, G., BERNOULLI, D. & LAUBSCHER, H.P. 1997. Rifting and collision in the Southern Alps. *In*: PFIFFNER, O., LEHNER, P., HEITZMANN, P., MUELLER, S. & STECK, A. (eds) *Deep Structure of the Swiss Alps*. Birkhäuser, Basel, 186–204.

SKOGSEID, J. 1994. Dimensions of the Late Cretaceous–Paleocene Northeast Atlantic Rift derived from Cenozoic subsidence. *Tectonophysics*, **240**, 225–247.

TER VOORDE, M. 1996. *Tectonic modelling of lithospheric extension along faults; implications for thermal and mechanical structure and basin stratigraphy*. PhD thesis, Vrije Universiteit Amsterdam..

VON BISTRAM, A. 1903. Das Dolomitgebiet der Luganeralpen. *Berichte der Naturforschenden Gesellschaft Freiburg in Breisgau*, **14**, 1–84.

VONDERSCHMITT, L. 1940. Bericht über die Exkursion der Schweizerischen Geologischen Gesellschaft in den Süd Tessin. *Eclogae Geologicae Helvetiae*, **33**, 205–219.

WEISSERT, H.J. & BERNOULLI, D. 1985. A transform margin in the Mesozoic Tethys: evidence from the Swiss Alps. *Geologische Rundschau*, **74**, 665–679.

WHITE, R.S. 1988. A hot-spot model for early Tertiary volcanism in the N Atlantic. *In*: MORTON, A.C. & PARSON, L.M. (eds) *Early Tertiary Volcanism and the Opening of the NE Atlantic*. Geological Society, London, Special Publications, **39**, 49–56.

WINTERER, E.L. & BOSELLINI, A. 1981. Subsidence and sedimentation in Jurassic passive continental margin, Southern Alps, Italy. *AAPG Bulletin*, **65**, 394–421.

Petrochemistry of serpentinized peridotite from the Iberia Abyssal Plain (ODP Leg 173): its character intermediate between sub-oceanic and sub-continental upper mantle

NATSUE ABE[1,2]

[1]*Department of Earth and Planetary Sciences, Tokyo Institute of Technology, 2-12-1 Ookayama, Meguro-ku, Tokyo 152-8551, Japan*
[2]*Present address: GEMOC ARC National Key Centre, Department of Earth and Planetary Sciences, Macquarie University, Sydney, N.S.W. 2109, Australia*
(e-mail: nabe@laurel.ocs.mq.edu.au)

Abstract: Highly serpentinized peridotites derived from the upper mantle beneath a non-volcanic continental margin were sampled from the southern Iberia Abyssal Plain (Ocean Drilling Program Leg 173), and the primary mantle minerals were analysed for major and trace elements to petrologically characterize the upper mantle. The *Cr*-number (= Cr/(Cr + Al) atomic ratio) of the peridotite spinels varies widely from 0.1 to 0.6. The Na_2O content of clinopyroxene is rather constant, 0.5–0.8 wt %. The chondrite-normalized rare earth element (REE) patterns have light REE (LREE) depleted convex-upward shapes. The LREE/HREE (heavy REE) ratio in clinopyroxene is constant irrespective of the degree of melt extraction of the sample as measured by the *Cr*-number of spinel. The trend of the peridotites' mineral chemistry is different from both the simple melt extraction and the general mantle metasomatic trends. The geochemical character of the Iberia Abyssal Plain peridotite is intermediate between those of abyssal peridotites and peridotite xenoliths from continental regions. These geochemical features, especially for the trace elements in clinopyroxene, are rather similar to those in peridotite xenoliths from arcs. This chemical trend is probably the result of 'open-system melting', which involves melting simultaneously with enrichment of trace elements by the influx agent.

In recent years, the progress of *in situ* analytical techniques has allowed us to obtain a wealth of information about trace-element mineral chemistry of mantle peridotite, despite the very low concentrations of trace elements in mantle minerals. The *in situ* technique is represented by secondary ion mass spectrometry (SIMS) and laser ablation inductively coupled plasma mass spectrometry (LA-ICPMS). There are many papers on trace-element mineral chemistry by *in situ* analysis of the peridotite xenoliths, especially from continental areas (e.g. Salters & Shimizu 1988; Rivalenti *et al.* 1996), and of abyssal peridotites from oceanic fracture zones (e.g. Johnson *et al.* 1990; Seyler & Bonatti 1997).

On the other hand, there are few studies of mantle geochemistry beneath convergent plate margins (arcs). Only for the western Pacific area have some localities with mantle xenoliths from arcs representative of the sub-arc mantle in a narrow sense been described by Abe

(1997), Abe *et al.* (1998) and Schiano *et al.* (1995) among others. Fore-arc peridotite was drilled from the Izu–Bonin–Mariana arc during Ocean Drilling Program (ODP) Leg 125, and was described by Ishii *et al.* (1992), Parkinson *et al.* (1992) and Parkinson & Pearce (1998).

Data from upper-mantle peridotites beneath ocean–continent transition (OCT) zones are far fewer than those from arcs. Kornprobst & Tabit (1988) and Cornen *et al.* (1996) reported petrological and mineral chemical characteristics of peridotites from the Galicia margin and the Iberia Abyssal Plain, respectively, in detail, but their studies were only for major and minor elements. Seifert & Brunotte (1996) reported the geochemistry of Iberia Abyssal Plain OCT mantle peridotite including trace-element analysis for bulk rocks and mineral separates.

In this study, the OCT mantle peridotites sampled during ODP Leg 173, Iberia Abyssal Plain, were analysed for their trace-element chemistry of clinopyroxene by SIMS at Tokyo

From: WILSON, R.C.L., WHITMARSH, R.B., TAYLOR, B. & FROITZHEIM, N. 2001. *Non-Volcanic Rifting of Continental Margins: A Comparison of Evidence from Land and Sea*. Geological Society, London, Special Publications, **187**, 143–159. 0305-8719/01/$15.00 © The Geological Society of London 2001.

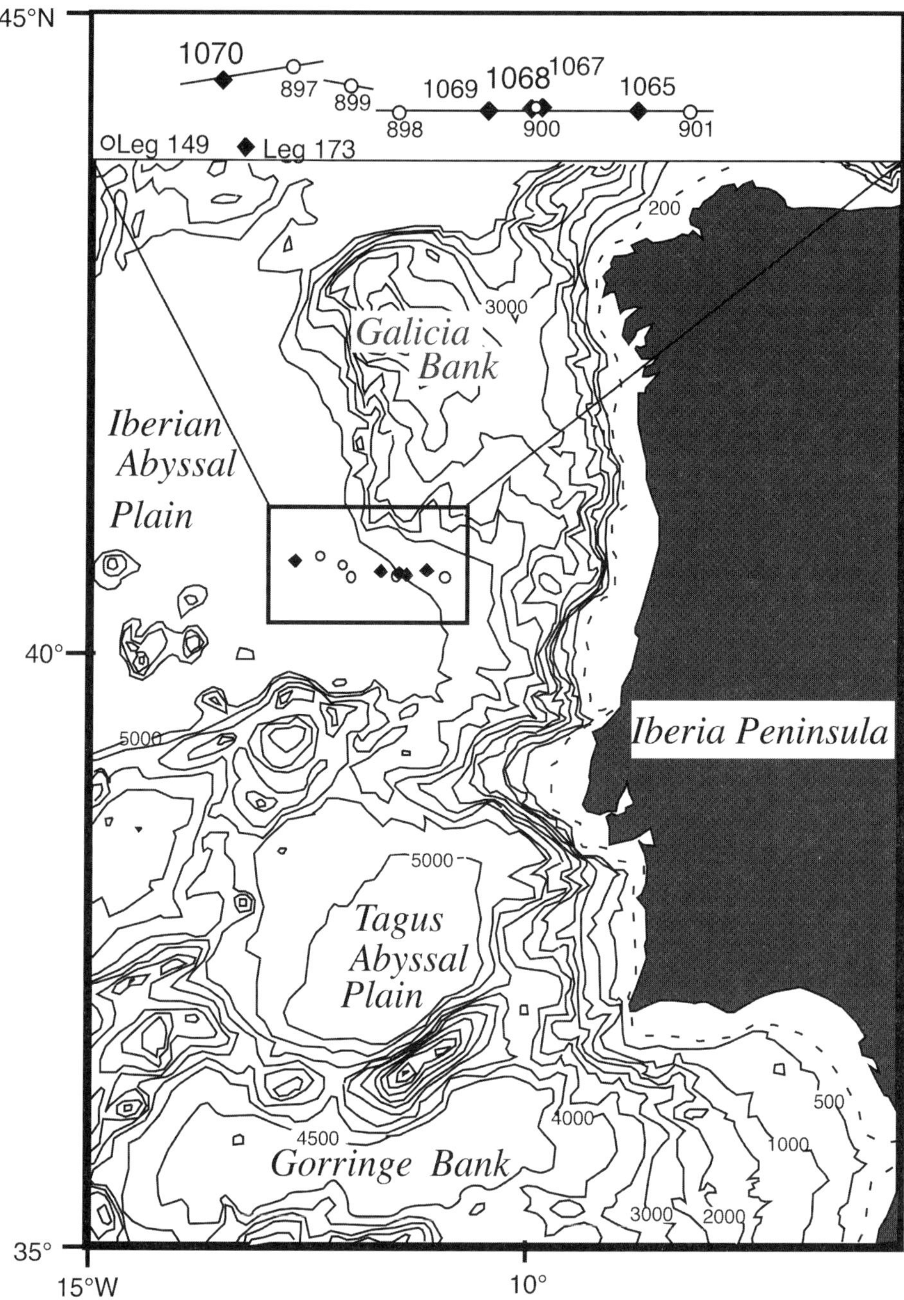

Fig. 1. Bathymetric map of West Iberia margin located in the northeastern part of the Atlantic Ocean (500 m contour intervals). Sites 897–901 were occupied during ODP Leg 149, and Sites 1065–1070 were occupied during Leg 173.

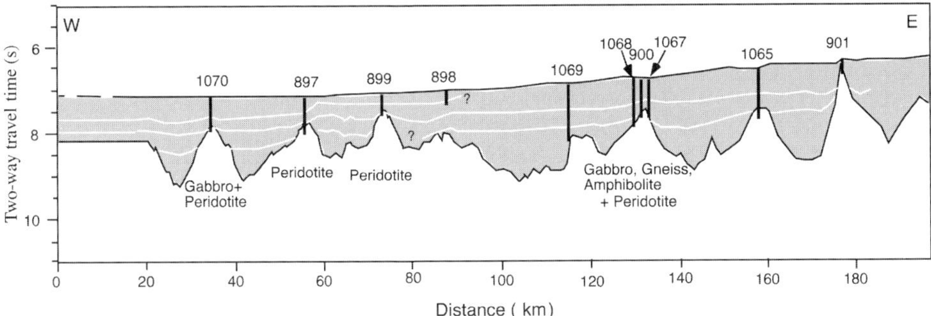

Fig. 2. Synthetic cross-section of the Iberia Abyssal Plain along 41°N. Depths are given as two-way travel times (s).

Institute of Technology. It is difficult to specify the processes responsible for the peridotite genesis by major-element mineral chemistry for highly altered samples. The *in situ* trace-element analysis on clinopyroxene is highly useful because of the relatively high resistance of clinopyroxene to serpentinization. The geochemical data of the OCT mantle peridotite will provide us with important constraints on the processes of the break-up of continental lithosphere, whole-mantle evolution and mantle convection.

Geological and tectonic settings

The Iberia Abyssal Plain is one of the segments of the West Iberia margin, which is located in the northeastern part of the Atlantic Ocean (Fig. 1). It is one of the ocean–continent transition (OCT) zones of a non-volcanic rifted margin. A detailed description of its tectonic setting has been given by Whitmarsh *et al.* (1998). Research on the West Iberia margin has been concentrated in three separate segments of the margin: the west Galicia Bank segment

(drilled during Leg 103), the southern Iberia Abyssal Plain segment (drilled during Legs 47B, 149 and 173), and the Tagus Abyssal Plain segment (Fig. 1).

These investigations significantly contributed to our understanding of lithospheric extensional structures in the crust and mantle (shear zones, detachment faults, block faulting), emplacement and sea-bed exposure of mantle rocks, synrift magmatism, onset of sea-floor spreading, and characterization of the OCT.

Serpentinized peridotites were obtained from two holes, 1068A and 1070A, during ODP Leg 173 (Fig. 2). Hole 1068A lies near the southern edge of the Iberia Abyssal Plain, which is about 200 km west of the coast of the Iberian Peninsula. Hole 1070A lies over an elongated basement ridge about 15 km east of the crest of the J magnetic anomaly and about 100 km west of Hole 1068A (Fig. 2).

In Hole 1068A, completely serpentinized peridotite breccia overlies massive serpentinized peridotite (>62.7 m thick). In Hole 1070A serpentinite breccia (17.8 m thick) and massive serpentinized peridotite (>39 m thick) were

Table 1. *Samples in this study*

Hole	Core	Section	Piece	Interval (cm)	Rock type	Primary minerals
1068A	27R	3	5	42–44	Pl peridotite	Sp, Cpx
	28R	1	1	1–5	Pl peridotite	Sp, Cpx
	28R	4	9	101–106	Peridotite	Sp, Ol
	29R	3	1C	67–68	Peridotite	Sp, Cpx
1070A	9R	3	7	60–65	Serpentinite breccia	Sp, Cpx, Opx
	10R	1	2A	14–17	Peridotite	Sp, Cpx, Opx
	10R	2	5A	43–47	Websterite	Sp, Cpx, Opx

Cpx, clinopyroxene; Sp, chromian spinel; Opx, orthopyroxene; Ol, olivine; Pl, Cr-free and/or Cr-bearing chlorite as thin rim surrounding chromian spinel. Primary minerals: 1068A-28R-4, 101–106 cm sample includes tremolite; 1070A-10R-1, 14–17 cm sample includes hornblende and pargasite.

obtained with intervening pegmatitic gabbro (3.5 m thick).

The peridotites are highly serpentinized: serpentinization is complete in the breccia and the upper part of the massive serpentinized peridotite units. Only the bottom part of the drilled serpentinized peridotite units has a small amount (<10 vol. %) of primary minerals. Primary pyroxenes in the pegmatitic gabbro and the pyroxenite vein in Hole 1070A are mostly preserved as relicts. Drilling at both Holes 1068A and 1070A was eventually abandoned because it appeared unlikely that less serpentinized peridotites would be encountered.

Sample descriptions

A total of 46 samples (20 samples from Hole 1068A and 26 samples from Hole 1070A) were examined, and six of these serpentinized peridotite samples and one pyroxenite sample are described in detail in this paper (Table 1). They

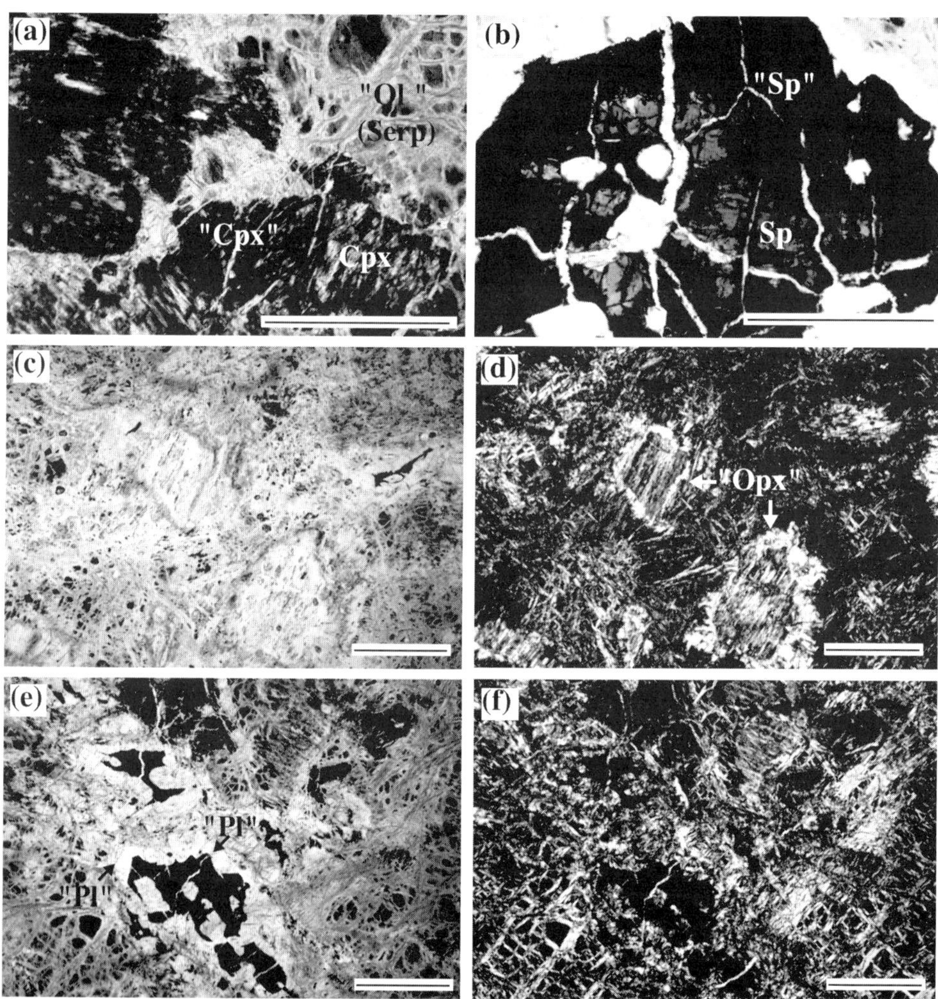

Fig. 3. (a) Clinopyroxene and its pseudomorph, set within mesh-textured serpentinite matrix (1068A-27R-3, 42–44 cm). Plane-polarized light. (b) Spinel relict with magnetite rim (1068A-28R-4, 101–106 cm). Plane-polarized light. (c) 'Bastite' as a pseudomorph after pyroxene, mainly orthopyroxene. Plane-polarized light. (d) The same 'bastite' taken with cross-polarized light. (e) Holly-leaf textured spinels, surrounded by pale chlorite, set within a mesh-textured serpentinite matrix (1068A-29R-3, 62–66 cm). Plane-polarized light. (f) Spinel mantled with plagioclase pseudomorph taken with cross-polarized light. 'Opx', pseudomorph after orthopyroxene (bastite); Serp, serpentine; 'Ol', serpentine after olivine with mesh texture; 'Pl', pseudomorph after plagioclase; Cpx, clinopyroxene; 'Cpx', pseudomorph after clinopyroxene. All scale bars represent 0.5 mm.

Table 2a. *Representative electron microprobe analysis of primary minerals*

1068A:	27R-3, 42–44 cm					28R-1, 1–5 cm				28R-4, 101–106 cm					29R-4, 67–68 cm			
	Cpx		Sp			Cpx		Sp		Sp		Ol		Tr	Cpx		Sp	
No.:	1	2	1	2	3	1	2	1	2	1	2	3	3	4	1	2	1	2
SiO$_2$	51.53	52.79	–	–	–	53.09	52.70	–	–	–	–	40.87	40.54	55.08	50.35	51.64		
TiO$_2$	0.49	0.45	0.17	0.20	0.16	0.08	0.04	–	–	0.04	–	–	–	0.03	0.74	0.61	0.07	0.08
Al$_2$O$_3$	5.14	3.63	31.75	35.57	18.89	4.58	4.08	46.27	46.27	41.21	38.29	0.01	0.02	3.04	6.21	5.59	47.02	36.07
Cr$_2$O$_3$	1.09	1.22	33.48	29.65	44.72	1.10	1.04	19.09	19.40	25.53	28.46	–	–	0.27	0.84	1.00	20.39	30.24
FeO*	2.17	2.45	20.87	21.61	27.28	2.55	2.23	14.19	13.68	15.61	18.57	8.53	9.06	1.83	2.33	2.43	16.24	19.44
MnO	0.07	0.08	0.33	0.30	0.52	0.07	0.08	0.21	0.18	0.20	0.22	0.14	0.16	–	0.07	0.12	0.24	0.28
MgO	15.05	16.15	11.67	10.31	6.67	16.30	15.96	18.02	18.52	15.19	12.41	50.02	49.25	22.35	14.40	15.50	15.83	12.29
CaO	23.52	22.17	–	–	–	21.44	22.75	–	–	–	–	0.03	0.01	13.25	22.60	23.33	–	–
Na$_2$O	0.76	0.54	–	–	–	0.82	0.77	–	–	–	–	–	–	0.69	0.69	0.72	–	–
K$_2$O	–	–	–	–	–	–	–	–	–	–	–	–	–	–	–	–	–	–
NiO	–	–	0.09	0.11	0.05	–	–	0.20	0.30	0.14	0.10	0.38	0.32	0.10	–	–	0.22	0.06
Total	99.82	99.48	98.36	97.76	98.30	100.03	99.65	97.98	98.35	97.92	98.06	100.01	99.36	96.68	98.23	100.93	100.01	98.47
mg–no.	0.925	0.922	0.530	0.468	0.331	0.919	0.927	0.744	0.760	0.651	0.546	0.913	0.906	0.956	0.917	0.919	0.652	0.545
cr–no.	0.125	0.184	0.414	0.359	0.614	0.138	0.146	0.217	0.220	0.294	0.333			0.056	0.083	0.107	0.225	0.360
Fe^{3+}			0.027	0.005	0.041			0.036	0.038	0.012	0.002						0.012	0.012
Al^{IV}	0.120	0.076				0.081	0.081								0.137	0.136		
Al^{VI}	0.102	0.080				0.115	0.094								0.134	0.102		
Al^{VI}/Al^{IV}	0.849	1.062				1.423	1.167								0.979	0.752		

Oxide values are in weight per cent. Amphiboles in 1068A-28R-4, 101–106 cm, and 1070A-10R-1, 14–17 cm, are also shown in this table. Cpx, clinopyroxene; Sp, chromian spinel; Opx, orthopyroxene; Ol, olivine; Tr, tremolite; Am, hornblende and pargasite. *mg*-number = Mg/(Mg + Fe^{2+}) ratio. Fe^{2+} and Fe^{3+} in spinel were calculated assuming spinel stoichiometry after subtraction all Ti as ulvöspinel molecule (Fe$_2$TiO$_4$). The same number (No.) in the sample means the grains exist adjacent to each other. Al^{VI} and Al^{IV} are aluminium cationic proportions in six- and four-coordination site of clinopyroxene calculated on the basis of six oxygens.

Table 2b. *Representative electron microprobe analysis of primary minerals*

1070A:	9R-3, 60-65 cm						10R-1, 14-17 cm								10R-2, 43-47 cm					
	Cpx		Opx		Sp		Cpx		Opx		Sp		Am		Cpx		Opx		Sp	
No.:	1	2	1	2	1	2	1	3	3	4	1	3	1	2	1	2	1	2	1	2
SiO_2	50.71	49.94	52.63	53.37	–	–	51.24	52.44	53.63	54.15	–	–	51.17	43.27	51.31	50.93	54.54	54.23	–	–
TiO_2	0.31	0.25	0.17	0.10	0.06	0.32	0.05	–	–	–	0.03	–	0.02	0.23	0.23	0.23	0.02	0.08	0.01	–
Al_2O_3	5.56	5.31	4.32	4.94	45.63	28.54	5.71	3.82	5.03	4.92	53.92	51.49	7.06	14.93	6.17	5.93	5.06	4.87	52.23	50.27
Cr_2O_3	1.18	1.18	0.71	0.75	16.58	35.03	1.15	0.53	0.70	0.71	13.31	15.29	0.78	1.79	1.02	0.97	0.62	0.65	14.12	15.72
FeO	2.66	2.67	7.23	7.25	17.55	19.20	2.27	2.55	5.50	5.52	11.47	12.14	2.42	3.35	2.76	3.76	6.93	6.53	12.93	12.01
MnO	0.07	0.10	0.13	0.10	0.18	0.33	0.07	0.08	0.11	0.13	0.16	0.11	0.06	0.09	0.14	0.13	0.17	0.15	0.13	0.17
MgO	14.48	15.47	27.07	28.85	17.28	14.17	15.77	16.79	32.82	32.36	19.32	18.69	20.72	17.61	14.70	17.32	30.76	30.09	18.34	19.43
CaO	22.91	23.12	5.86	3.12	–	–	22.20	22.71	0.62	0.56	0.01	–	12.17	12.00	22.78	18.83	0.85	2.52	–	0.02
Na_2O	0.58	0.58	0.11	0.03	–	0.02	0.48	0.46	0.02	0.01	0.04	0.02	1.63	3.17	0.57	0.46	0.03	0.05	–	0.03
K_2O	–	–	–	–	–	–	–	–	–	–	–	–	0.04	–	–	–	–	–	–	–
NiO	–	–	0.06	0.08	0.30	0.18	–	–	–	0.15	0.25	0.25	0.09	0.21	–	–	–	–	0.18	0.24
Total	98.47	98.52	98.28	98.60	97.58	97.78	98.96	99.38	98.44	98.50	98.50	97.99	96.16	96.67	99.69	98.56	98.97	99.16	97.94	97.89
mg–no.	0.907	0.912	0.870	0.877	0.721	0.647	0.925	0.921	0.914	0.913	0.769	0.755	0.938	0.903	0.905	0.892	0.888	0.891	0.740	0.785
cr–no.	0.124	0.129	0.099	0.092	0.196	0.452	0.119	0.085	0.085	0.088	0.142	0.166	0.069	0.074	0.100	0.099	0.076	0.082	0.154	0.173
Fe^{3+}					0.064	0.062					0.012	0.015							0.016	0.029
Al^{IV}	0.123	0.147					0.123	0.085							0.127	0.131				
Al^{VI}	0.120	0.085					0.124	0.080							0.138	0.125				
Al^{VI}/Al^{IV}	0.973	0.574					1.012	0.944							1.085	0.953				
$T(°C)$	804.5						943.0								875.3					

T, equilibrium temperature calculated by Wells (1977) two-pyroxene geothermometer.

were carefully selected because of their relative freshness; they preserve relics of primary minerals, clinopyroxene, spinel, and/or orthopyroxene, and/or olivine.

The primary mantle mineralogy of the peridotite samples can hardly be described because all of them are severely serpentinized, sometimes even without spinel relicts. The spinel grain with thick magnetite rim shown in Figure 3a is a good example. Primary lithologies of the serpentinized peridotites are roughly recognized by the modal amounts of the relict pyroxene and pseudo-

morphs (black hazy aggregate after clinopyroxene, Fig. 3b; bastite after orthopyroxene, Fig. 3c and d). It is very difficult to estimate the primary clinopyroxene modal volume in these highly altered peridotites, but it was estimated from the black hazy aggregate as 2–10 vol. % and 4–6 vol. % in Hole 1068A and 1070A serpentinized peridotites, respectively. The most serpentinized peridotites of Hole 1068A have chromian spinel mantled by a thin leucocratic rim, which is composed of a fine aggregate of Cr-free and Cr-bearing chlorite and is prob-

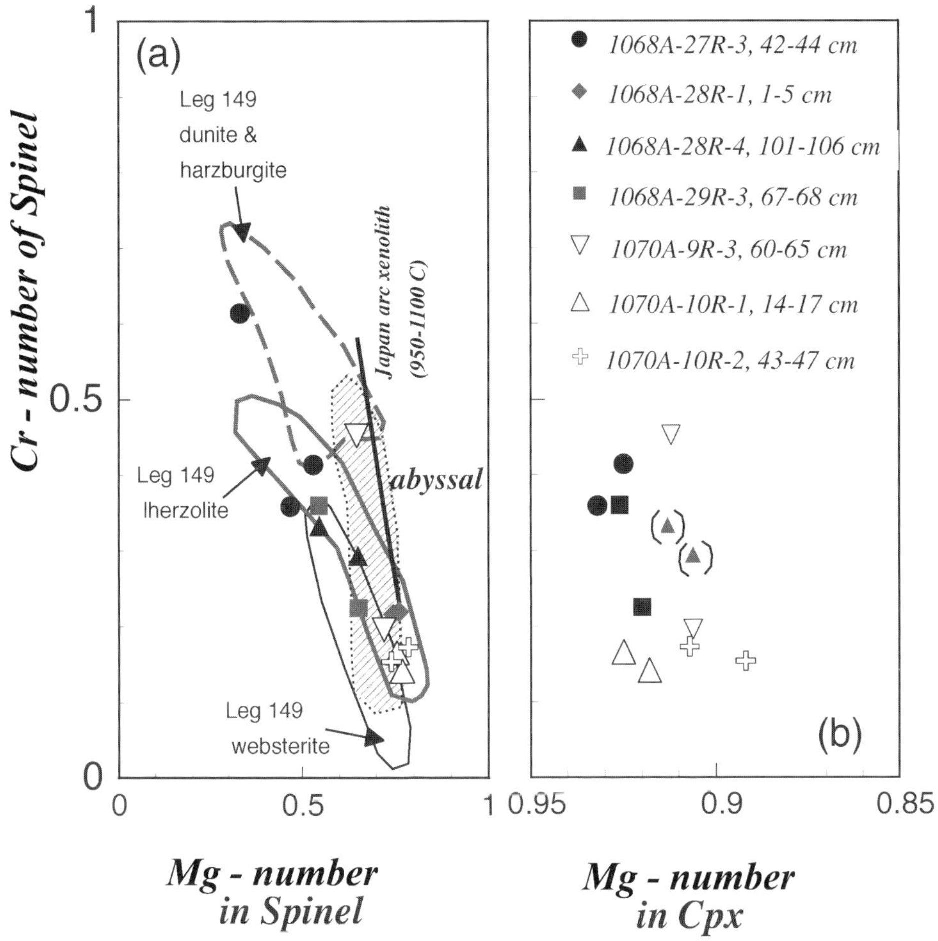

Fig. 4. (a) Compositions of Iberia Abyssal Plain mantle spinels plotted as *Cr*-number (= Cr/(Cr + Al) atomic ratio) v. *Mg*-number (Mg/(Mg + Fe^{2+}) atomic ratio). Composition of abyssal peridotite from Dick & Bullen (1984); Leg 149 dunite and harzburgite, lherzolite and websterite from Cornen *et al.* (1996); and the data of Japan arc xenoliths after Abe (1997). (b) Relationships between *Cr*-number of spinel and *Mg*-number in coexisting clinopyroxene. Olivine *Mg*-number is shown for sample 1068A-28R-4, 101–106 cm, instead of the clinopyroxene *Mg*-number.

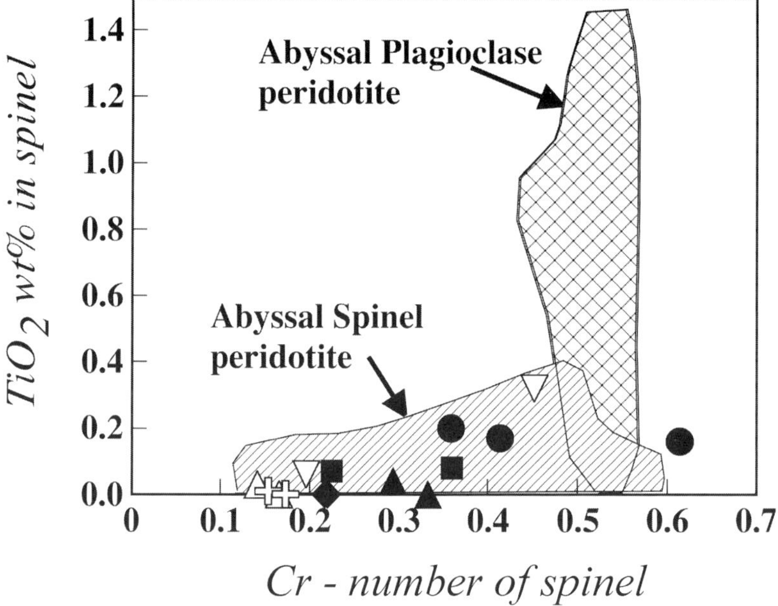

Fig. 5. Weight per cent TiO_2 plotted against the *Cr*-number of chromian spinel in Iberia Abyssal Plain peridotites. The ranges of abyssal plagioclase peridotite and abyssal spinel peridotite (after Dick & Bullen 1984) are also shown for comparison. Symbols as in Figure 4.

ably a pseudomorph after plagioclase (Fig. 3e and f). Plagioclase pseudomorph exists only around chromian spinel, except for sample 1068A-28R-4, 101–106 cm, which also has a few volume per cent of large (up to 0.5 mm) discrete plagioclase pseudomorph grains. Some portions of the cores exhibit alternations of chromian spinel-rich, almost bastite-free and bastite-rich parts, which possibly correspond originally to dunite and pyroxene-rich peridotite (lherzolite), respectively.

Orthopyroxene in the Hole 1070A samples remains as small relicts in large pseudomorphs (bastite). Amphiboles were found in two samples: 1068A-28R-4, 101–106 cm, and 1070A-10R-1, 14–17 cm. The amphibole in the former sample is tremolite, which is very small and appears to be a pseudomorph after primary clinopyroxene. On the other hand, amphibole in the latter sample is rather large (up to 0.5 mm across) and occurs in pyroxene-rich portions as interstitial grains within pyroxenes.

The foliation is marked by elongate spinel grains in the Hole 1068A serpentinite. Detailed core and thin-section descriptions of serpenti-nized peridotites from Holes 1068A and 1070A have been given by Whitmarsh *et al.* (1998).

Major-element mineral chemistry

Major-element analyses of olivine, clinopyroxene and chromian spinel were made with a four-spectrometer JEOL 8800 superprobe at Tokyo Institute of Technology, using a focused beam, with an accelerating voltage of 15 keV and a beam current of 12 nA for all phases.

Selected analyses are listed in Table 2. The cation ratios of spinel were calculated assuming spinel stoichiometry after subtracting all Ti as an ulvöspinel molecule (Fe_2TiO_4). All Fe is assumed to be Fe^{2+} in silicates.

Spinel

The *Cr*-number of chromian spinel in the Hole 1068A samples varies widely from 0.217 to 0.614 (Fig. 1). It covers almost the whole range of abyssal peridotites (Dick & Bullen 1984; Arai 1994). The highest *Cr*-number of spinel is, however, for a small euhedral inclusion in the rim of the pseudomorph after orthopyroxene. Except for that sample, the *Cr*-number of most

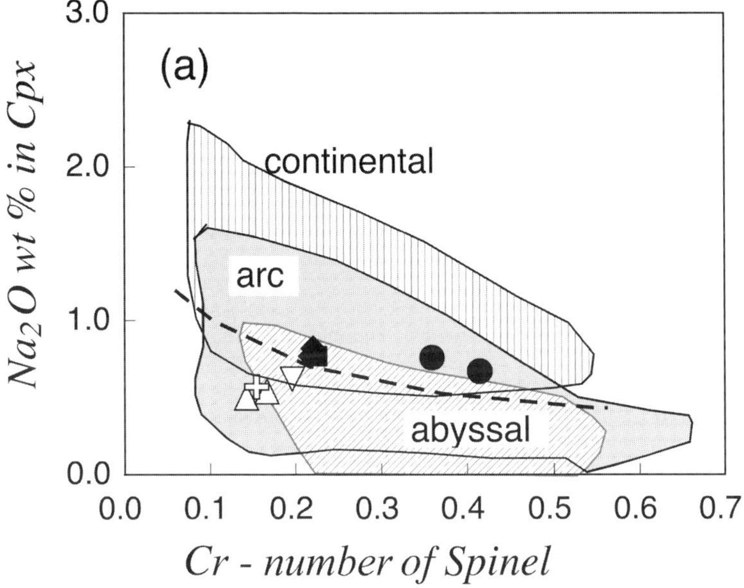

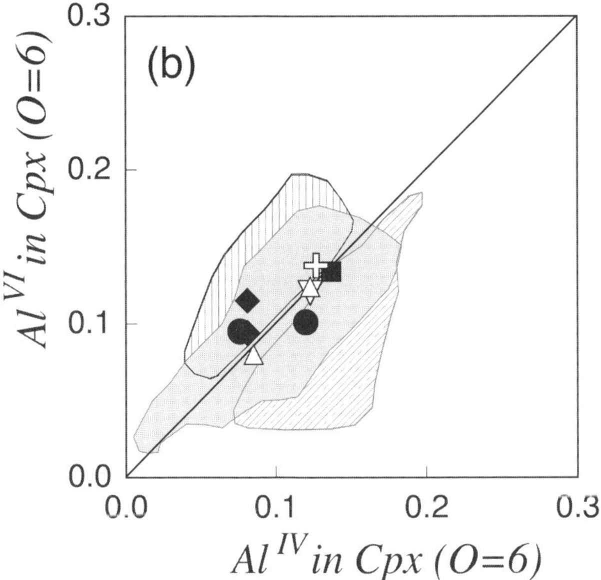

Fig. 6. (**a**) Na$_2$O content in clinopyroxene v. *Cr*-number of coexisting spinel. Continental and abyssal areas are after Arai (1991) compilations; the arc area shows the data of peridotite xenoliths from Japanese arcs after Abe (1997). (**b**) Cationic proportions of Al^{VI} v. Al^{IV} in clinopyroxenes. Continental and abyssal area after Seyler & Bonatti (1994). The data from arc areas after Abe (1997). Symbols as in Figure 4.

spinels is <0.4 (Fig. 4). The TiO_2 content in all chromian spinels is low, except for a euhedral small inclusion in 1068A-27R-3, 42–44 cm. The correlation between *Cr*-number and TiO_2 content in chromian spinels (Fig. 5) is consistent with the abyssal spinel peridotite region shown by Dick & Bullen (1984).

Clinopyroxene

The *Mg*-number (= $Mg/(Mg + Fe^{2+})$ atomic ratio) of clinopyroxene is relatively low in pyroxenite (1070A-10R-2, 43–47 cm) and in the Hole 1070A peridotites (1070A-9R-3, 60–65 cm, and 10R-1, 14–17 cm) (Fig. 4b). In Figure 4b, the olivine *Mg*-number is shown for sample 1068A-29R-3, 67–78 cm, which is free from relict clinopyroxene. The Na_2O content of clinopyroxene is slightly higher in the Hole 1068A samples than in those from Hole 1070A (Fig. 6a; Table 2). The clinopyroxene has around 23 wt % of CaO (Table 2). The Al^{VI}/Al^{IV} atomic ratios are almost unity (Fig. 6b).

Olivine

Olivine is preserved in only one sample (1068A-28R-4, 101–106 cm). Its *Mg*-number (0.91) is lower than that of the clinopyroxenes in other samples (Fig. 4b).

Orthopyroxene

Three samples from Hole 1070A have orthopyroxene, which is associated with clinopyroxene (Tables 1 and 2). The orthopyroxene contains appreciable amounts of Cr_2O_3 (0.6–0.8 wt %) and Al_2O_3 (4.3–5.0 wt %). The apparent CaO content is variable, possibly as a result of contamination by thin clinopyroxene lamellae to various extents.

Amphiboles

The amphibole in sample 1068A-28R-4, 101–106 cm, is tremolite with a high *Mg*-number (0.956, Table 2). The amphibole is hornblende to pargasitic hornblende with a very low Na content in sample 1070A-10R-1, 14–17 cm (Table 2).

Trace-element chemistry of clinopyroxene

In situ trace-element analyses of clinopyroxenes on polished thin sections were carried out by SIMS with a Cameca IMS-3f instrument at Tokyo Institute of Technology. The primary ion beam was mass filtered, $^{16}O^-$-accelerated

to 12.5 keV, adjusted for a beam current of about 20 nA and focused to a spot diameter of 20 μm. Kinetic energy filtering was achieved by offsetting the sample accelerating voltage (−60 V for REE and −100 V for Ti, Zr, Sr and Y) while keeping the setting of the electrostatic analyser voltage and the width and position of the energy slit constant. Other analytical and instrumental conditions were similar to those employed by Yurimoto *et al.* (1989) and Wang & Yurimoto (1993). Analytical techniques were the same as those used by Abe *et al.* (1998) for mantle xenoliths from the Japanese island arcs.

The concentrations of REE, Sr, Zr, Y and Ti in clinopyroxenes are shown in Table 3. Their chondrite-normalized REE patterns are light rare earth element (LREE) depleted convex-upward, except for clinopyroxene in sample 1068A-29R-3, 67–68 cm, which has a slightly LREE-depleted flat pattern (Fig. 7). Ce and Yb contents vary widely from 1.2 to 7.7 and 5.6 to 17.6 times the chondrite value (Anders & Grevesse 1989), respectively (Fig. 8a). Sr contents also vary from 0.8 to 3.5 times (Fig. 8b) the chondrite value (Anders & Grevesse 1989). The peridotites from the Iberia Abyssal Plain are intermediate in terms of the concentrations of trace elements between ordinary abyssal peridotites and peridotite xenoliths from continental rifts (Fig. 8). It is noteworthy that their concentrations in clinopyroxene are not so well correlated with the degree of melt extraction measured by the *Cr*-number of coexisting spinel (Fig. 9). The Ti/Zr atomic ratios of all clinopyroxenes are remarkably constant, around 100 (Fig. 9f).

Discussion

Equilibrium temperature and pressure

The equilibrium temperatures of the Hole 1070A peridotites and pyroxenite were calculated to be 800–940 °C by the Wells (1977) two-pyroxene geothermometer using the compositions of relict orthopyroxene and clinopyroxene in the samples (Table 2). There is no relict orthopyroxene in the Hole 1068A samples, but the equilibrium temperatures of their protolith are possibly the same as those for Hole 1070A samples. The CaO and Al_2O_3 contents of clinopyroxene, which are sensitive to the equilibrium temperature in the spinel peridotite field (e.g. Gasparik 1984), are the same for the Hole 1070A and Hole 1068A samples (Table 2). A possible primary coexistence of Mg-rich olivine and plagioclase indicates low equilibrium pressures (<1 GPa) and

Table 3. *Rare earth and other trace element abundances (in ppm) in clinopyroxenes*

	La	Ce	Nd	Sm	Eu	Dy	Er	Yb	Ti	Sr	Y	Zr	(Ce/Yb)$_N$	Ti/Zr
1068A-27R-3, 42–44 cm	0.47	2.33	3.18	1.79	0.55	3.70	2.26	1.80	1971	19.59	20.22	27.42	0.34	71.9
1068A-28R-1, 1–5 cm	0.67	1.42	0.93	0.40	0.25	1.24	0.98	0.91	742	19.17	11.05	7.13	0.40	104.1
1068A-28R-4, 101–106 cm	0.94	4.62	5.30	2.94	0.85	4.73	3.55	2.63	3888	7.47	35.80	47.83	0.45	81.3
1070A-9R-3, 60–65 cm	0.07	0.95	4.28	2.24	0.36	4.33	3.26	2.86	2078	1.08	33.86	9.42	0.09	220.6
1070A-10R-1, 14–17 cm	0.09	0.72	2.61	1.05	0.30	1.90	1.58	1.27	615	6.30	13.46	6.79	0.15	90.5
1070A-10R-2, 43–47 cm	0.07	1.29	4.28	1.65	0.44	2.54	1.73	1.55	1920	26.11	18.55	9.93	0.21	193.4

Subscript N indicates chondrite-normalized value.

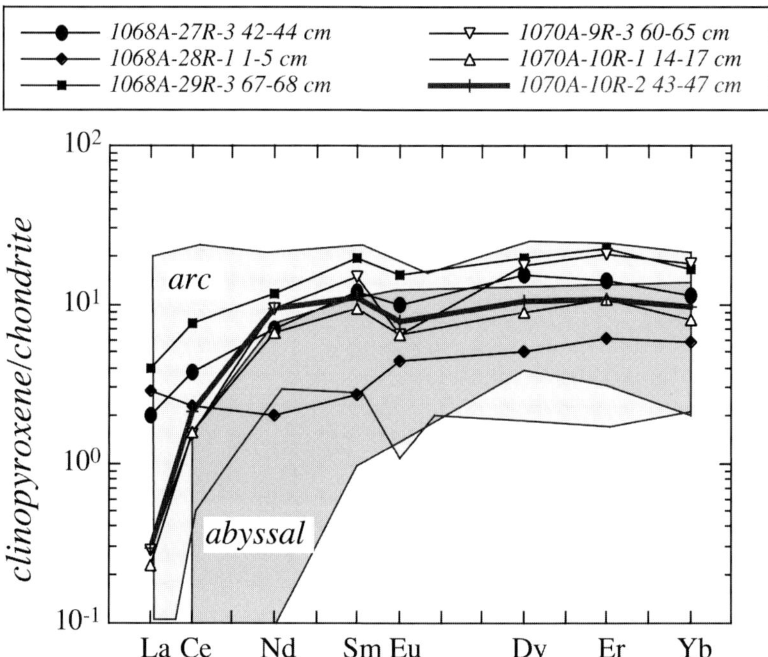

Fig. 7. Chondrite-normalized REE patterns of clinopyroxene. The plotted abyssal area is after Johnson *et al.* (1990), the arc area after Abe (1997). The REE normalization values are from Anders & Grevesse (1989). Symbols as in Figure 4.

high equilibrium temperature (>900 °C) (e.g. Kushiro & Yoder 1966; Green & Hibberson 1970). The existence of plagioclase pseudomorphs as the rim around spinel in some massif peridotites (e.g. External Liguride units, Northern Apennines, Rampone *et al.* 1993) suggests that they are formed by subsolidus reactions during decompression from the spinel to plagioclase stability fields. On the other hand, spinel associated with plagioclase formed by injected melt in abyssal peridotites has high TiO_2 content and high *Cr*-number (Fig. 5; Dick & Bullen 1984). TiO_2 content in spinel increases slightly by subsolidus reaction during the spinel- to plagioclase-facies transition (Rampone *et al.* 1993), but this increase is much less than when melt impregnation produces plagioclase peridotite (Fig. 5). The low TiO_2 content (<0.4 wt %) and rather low *Cr*-number of chromian spinel in the Iberia Abyssal Plain peridotites does not support the hypothesis that their plagioclase was formed by melt impregnation.

Comparison with peridotites from other tectonic settings

The Iberia Abyssal Plain peridotites are obviously distinguished from ordinary abyssal peridotites produced beneath mid-ocean ridges (e.g. Johnson *et al.* 1990) in their trace-element chemistry (Figs 6, 8, 9 and 10). Peridotite xenoliths from young continental lithosphere, which have been modified by metasomatic enrichment, also show different geochemical trends in Figures 6, 8 and 9. The Ti/Zr ratio in the Iberia Abyssal Plain clinopyroxenes is mostly around 100 (71.9–220.6; Table 3; Figs 9f and 10b). Residual peridotites (or their clinopyroxenes) should have higher Ti/Zr ratios than their source because Zr is more incompatible than Ti (Hart & Dunn 1993); the Ti/Zr ratio of clinopyroxene from abyssal peridotite increases up to 4430 according to the extent of melt extraction (Johnson *et al.* 1990). The clinopyroxenes in the Iberia Abyssal Plain peridotites are

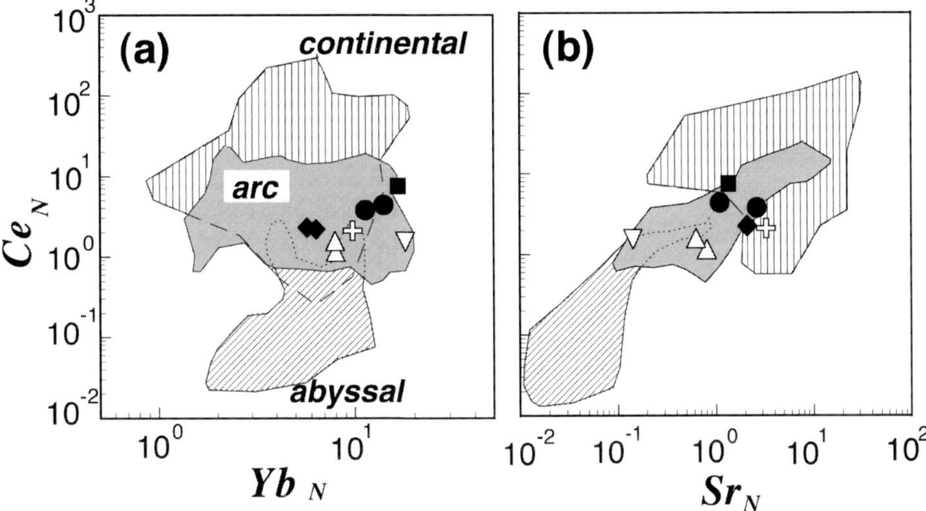

Fig. 8. Relationships between (**a**) Ce_N v. Yb_N and (**b**) Ce_N v. Sr_N in clinopyroxenes in mantle-derived peridotites from various tectonic settings. Subscript N indicates normalization by chondrite values. It should be noted that the Iberia Abyssal Plain clinopyroxenes plot between abyssal peridotite (Johnson *et al.* 1990) and peridotite xenoliths from young continental lithosphere. For continental areas (young continental lithosphere) the data are from Menzies *et al.* (1985), Stosch & Lugmair (1986), Salters & Shimizu (1988), Witt & Seck (1989), Witt-Eickschen *et al.* (1993), Alibert (1994), Blusztaijn & Shimizu (1994), Witt-Eickschen & Harte (1994) and Rivalenti *et al.* (1996). The arc areas after Abe (1997). Symbols as in Figure 4.

enriched in incompatible elements such as Sr, Zr and Ce relative to abyssal peridotites (Johnson *et al.* 1990), which form a simple residual suite. Addition of a metasomatic component of continental type, which is highly enriched in alkalis, Ti, Zr and CO_2, commonly leads to a decrease of the Ti/Zr ratio in clinopyroxenes (Blusztaijn & Shimizu 1994). If locally the metasomatic melt or fluid has percolated chromatographically into the peridotite, the Ti/Zr ratio may be rather variable spatially because of fractionation as in the continental samples. These tendencies in clinopyroxene indicate that the Iberia Abyssal Plain clinopyroxenes can be neither a series of simple restites nor a series of simple metasomatites. It is most probable that the peridotites with these clinopyroxenes are the residual product of 'open-system melting' (e.g. Ozawa & Shimizu 1995); that is, partial melting assisted by, or associated with, addition of fluid or melt with a constant Ti/Zr ratio (Fig. 9f). That is, the clinopyroxene Ti/Zr trend for the OCT peridotites can be formed by abyssal-type melt extraction combined with addition of melt or fluid with a constant and low Ti/Zr ratio. Alternatively, a metasomatic agent has

been continuously and constantly fluxed into the upper mantle until the Ti/Zr ratio in clinopyroxene equilibrated at *c.* 100.

The Leg 173 trace-element chemistry is similar to that of supra-subduction zone mantle, which has an intermediate geochemical character between abyssal peridotite and peridotite xenoliths from young continental lithosphere (Abe 1997; Abe *et al.* 1998). Supra-subduction zone mantle, however, shows a much clearer correlation between melt extraction and metasomatic trend.

There are two possibilities for the origin of the Iberia Abyssal Plain mantle. One of them is that it was made by open-system melting with a flux agent that was probably from the plume causing the rifting. The other is that it is a mantle fragment, which experienced the supra-subduction zone processes of an active margin before rifting. Silva *et al.* (2000) have suggested that an ancient arc, the Precambrian Ibero-Armorican Arc, exists adjacent to the Iberia, northwestern France and the Canadian Grand Banks margins, on the basis of a compilation of magnetic data from the Iberia and Grand Banks margins. The geochemical data in

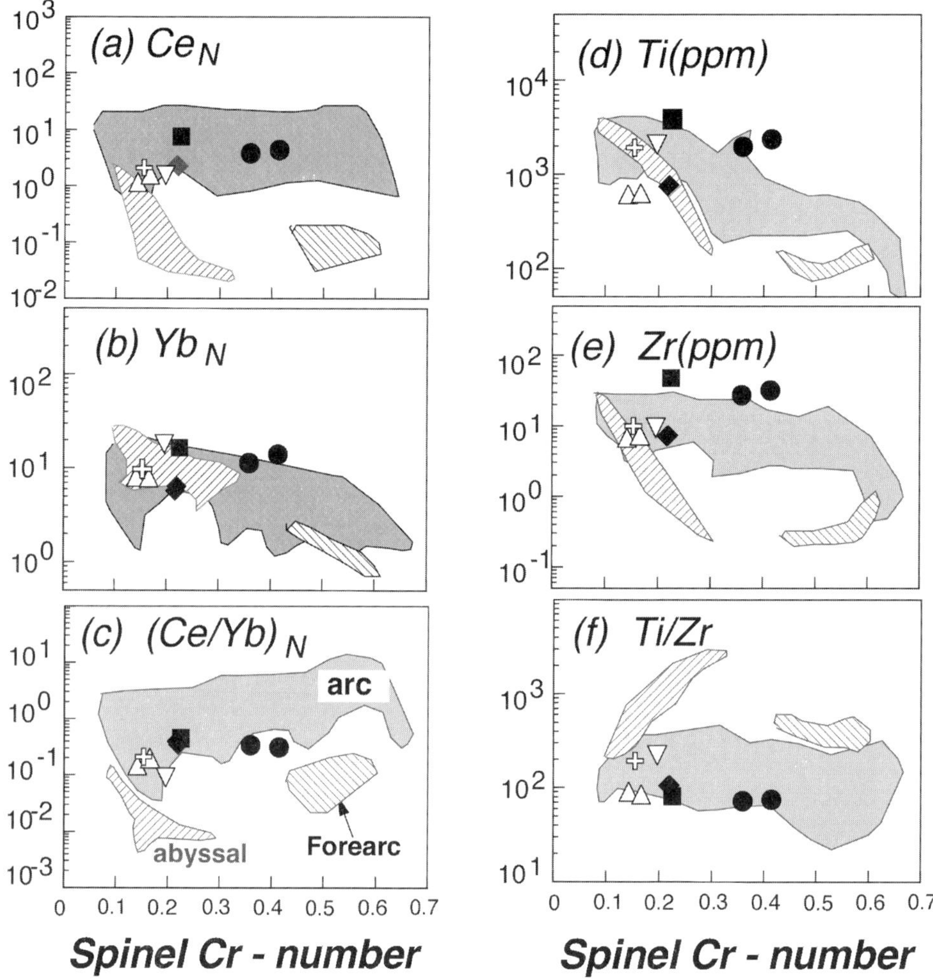

Fig. 9. Spinel *Cr*-number v. (**a**) Ce_N abundance, (**b**) Yb_N abundance, (**c**) $(Ce/Yb)_N$ ratio, (**d**) Ti content (ppm), (**e**) Zr content (ppm) and (**f**) Ti/Zr weight ratio. For comparison, ranges for abyssal peridotites (Dick 1989; Johnson *et al.* 1990), fore-arc peridotite (Parkinson *et al.* 1992) and arc peridotite xenoliths (Abe 1997) are also shown. Symbols as in Figure 4.

this paper are not sufficient to judge which is correct. Isotope measurements on the samples will be necessary, although it will be very difficult to obtain valid isotope compositions for these highly serpentinized mantle peridotites at this stage.

Summary and Conclusions

(1) The mantle peridotites beneath the OCT zone west of Iberia are harzburgite to lherzolite with subordinate dunite layers. Plagioclase coexists with chromian spinel in most cases.

(2) The trace-element chemistry shows that the mantle materials were formed neither by simple melt extraction from primitive mantle nor by simple mantle metasomatism. They most probably represent a series of residues after open-system melting, that is, melting assisted by an influx enriched with incompatible components.

(3) The geochemical features of the OCT mantle are similar to those of a sub-arc mantle.

H. Yurimoto provided enormous help in running the secondary ion mass spectrometer at Tokyo Institute

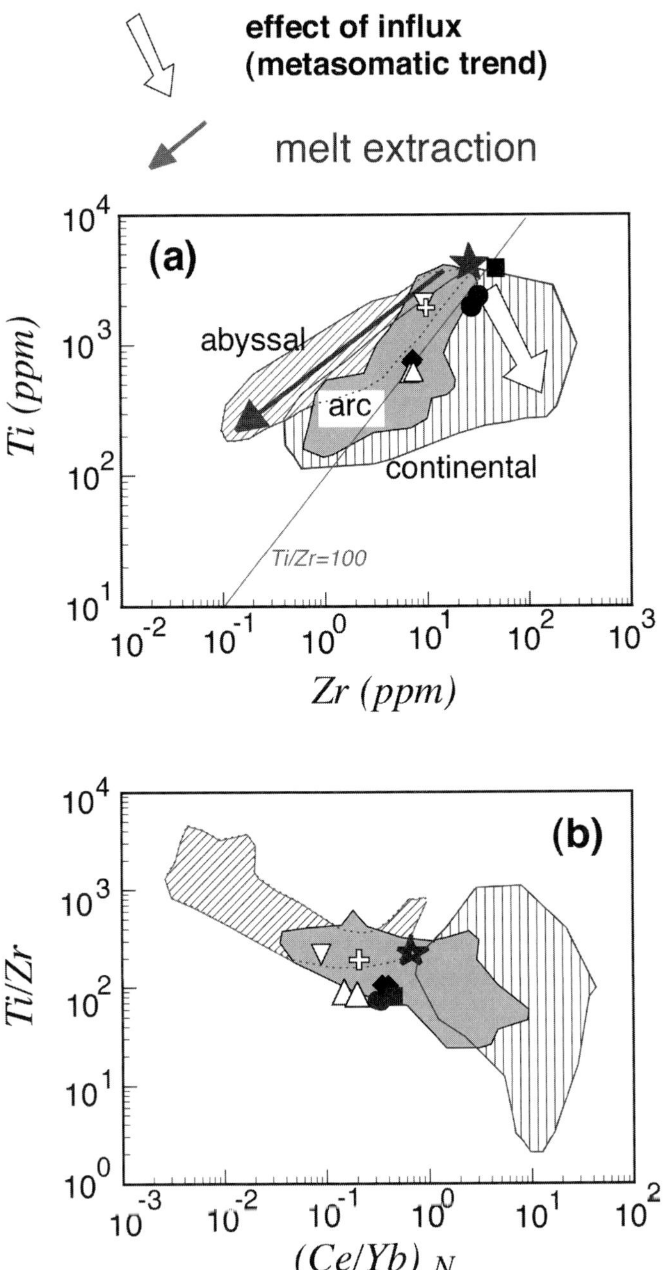

Fig. 10. Relationships between (**a**) Ti and Zr contents and (**b**) Ti/Zr and $(Ce/Yb)_N$ in clinopyroxenes in the Iberia Abyssal Plain peridotites. The trends can be explained by a melt extraction process combined with contamination by melt or fluid. The simple melt extraction trend may be simulated by the trend of abyssal peridotites (Johnson *et al.* 1990). Estimated primitive mantle (solid star) is after Sun & McDonough (1989). The continental data (young continental mantle) are from Witt & Seck (1989), Witt-Eickschen *et al.* (1993), Alibert (1994), Blusztaijn & Shimizu (1994), Witt-Eickschen & Harte (1994), Rivalenti *et al.* (1996) and Vanucci *et al.* (1994). The arc areas after Abe (1997). Symbols as in Figure 4.

of Technology. The author is grateful to the Ocean Drilling Program for permitting her to participate in Leg 173. She is greatly indebted to J. Beard, L. Hopkinson and A. Skelton for their encouragement and help during the cruise. The author wishes to express her deep thanks to M.-O. Beslier, R. B. Whitmarsh, P. Wallace and all shipboard scientists, all marine technicians and specialists, and the crew of JOIDES *Resolution* for their collaboration, kindness, and friendship during Leg 173.

References

ABE, N. 1997. *Petrology of mantle xenoliths from the arcs: implications for the petrochemical evolution of the wedge mantle.* PhD thesis, Kanazawa University.

ABE, N., ARAI, S. & YURIMOTO, H. 1998. Geochemical characteristics of the uppermost mantle beneath the Japanese island arcs: implications for upper mantle evolution. *Physics of the Earth and Planetary Interiors*, **107**, 233–248.

ANDERS, E. & GREVESSE, N. 1989. Abundances of the elements: meteoritic and solar. *Geochimica et Cosmochimica Acta*, **53**, 197–214.

ALIBERT, C. 1994. Peridotite xenoliths from western Grand Canyon and the Thumb: a probe into the subcontinental mantle of the Colorado Plateau. *Journal of Geophysical Research*, **99**, 21605–21620.

ARAI, S. 1991. Petrological characteristics of the upper mantle peridotites beneath the Japan island arcs—petrogenesis of spinel peridotites. *Soviet Geology and Geophysics*, **32**, 8–25.

ARAI, S. 1994. Characterization of spinel peridotites by olivine–spinel compositional relationships: review and interpretation. *Chemical Geology*, **113**, 191–204.

BLUSZTAIJN, J. & SHIMIZU, N. 1994. The trace-element variations in clinopyroxenes from spinel thermobarometers, and practical assessment of existing thermobarometers. *Journal of Petrology*, **31**, 1353–1378.

CORNEN, G., BESLIER, M.-O. & GIRARDEAU, J. 1996. 21. Petrologic characteristics of the ultramafic rocks from the ocean/continent transition in the Iberia Abyssal Plain. *In*: Whitmarsh, R.B., Sawyer, D.A., Masson, D.G. *et al.* (eds) *Proceedings of the Ocean Drilling Program, Scientific Results, 149*. Ocean Drilling Program, College Station, TX, 377–395.

DICK, H.J.B. 1989. Abyssal peridotites very slow spreading ridges and ocean ridge magmatism. *In*: SAUNDERS, A.D. & NORRY, M.J. (eds) *Magmatism in the Ocean Basins*. Geological Society, London, Special Publications, **42**, 597–628.

DICK, H.J.B. & BULLEN, T. 1984. Chromian spinel as a petrogenetic indicator in abyssal and alpine-type peridotites and spatially associated lavas. *Contributions to Mineralogy and Petrology*, **86**, 54–76.

GASPARIK, T. 1984. Two-pyroxene thermobarometry with new experimental data in the system CaO–MgO–Al$_2$O$_3$–SiO$_2$. *Contributions to Mineralogy and Petrology*, **87**, 87–97.

GREEN, D.H. & HIBBERSON, W. 1970. The instability of plagioclase in peridotite at high pressure. *Lithos*, **3**, 209–221.

HART, S.R. & DUNN, T. 1993. Experimental cpx/melt partitioning of 24 elements. *Contributions to Mineralogy and Petrology*, **113**, 1–8.

ISHII, T., ROBINSON, P.T., MAEKAWA, H. & FISK, R. 1992. Petrological studies of peridotites from diapiric serpentinite seamounts in the Izu–Bonin–Mariana forearc, Leg 125. *In*: Fryer, P., Pearce, J.A., Stokking, L.B. *et al.* (eds) *Proceedings of the Ocean Drilling Program, Scientific Results 125*. Ocean Drilling Program, College Station, TX, 445–486.

JOHNSON, K.T., DICK, H.J.B. & SHIMIZU, N. 1990. Melting in the oceanic upper mantle: ion microprobe study of diopsides in abyssal peridotites. *Journal of Geophysical Research*, **95**, 2661–2678.

KORNPROBST, J. & TABIT, A. 1988. 17. Plagioclase-bearing ultramafic tectonites from the Galicia margin (Leg 103, Site 637): comparison of their origin and evolution with low-pressure ultramafic bodies in western Europe. *In*: Boillot, G., Winterer, E.L. *et al.* (eds) *Proceedings of the Ocean Drilling Program, Scientific Results 103*. Ocean Drilling Program, College Station, TX, 253–263.

KUSHIRO, I. & YODER, H.S. Jr 1966. Anorthite–forsterite and anorthite–enstatite reactions and their bearing on the basalt–eclogite transformation. *Journal of Petrology*, **7**, 337–362.

MENZIES, M.A., KEMPTON, P. & DUNGAN, M. 1985. Interaction of continental lithosphere and asthenospheric melts below the Geronimo Volcanic Field, Arizona, U.S.A. *Journal of Petrology*, **26**, 663–693.

OZAWA, K. & SHIMIZU, N. 1995. Open-system melting in the upper mantle: Constraints from the Hayachine–Miyamori ophiolite, northeastern Japan. *Journal of Geophysical Research*, **100**, 22315–22335.

PARKINSON, I.J. & PEARCE, J.A. 1998. Peridotites from the Izu–Bonin–Mariana forearc (ODP Leg 125): evidence for mantle melting and melt–mantle interaction in a supra-subduction zone setting. *Journal of Petrology*, **39**, 1577–1618.

PARKINSON, I.J., PEARCE, J.A., THIRLWALL, M.F., JOHNSON, K.T.M. & INGRAM, G. 1992. Trace element geochemistry of peridotites from the Izu–Bonin–Mariana forearc, Leg 125. *In*: Fryer, P., Pearce, J.A., Stokking, L.B. *et al.* (eds) *Proceedings of the Ocean Drilling Program, Scientific Results 125*. Ocean Drilling Program, College Station, TX, 487–506.

RAMPONE, N., PICCARDO, G.B., BANNUCCI, R., BOTTAZZI, P. & OTTOLINI, L. 1993. Subsolidus reactions monitored by trace element partitioning: the spinel to plagioclase-facies transition in mantle peridotites. *Contributions to Mineralogy and Petrology*, **115**, 1–17.

RIVALENTI, G., VANNUCCI, R., RAMPONE, E. & 5 OTHERS 1996. Peridotite clinopyroxene chemistry reflects mantle processes rather than continental versus oceanic settings. *Earth and Planetary Science Letters*, **139**, 423–437.

SALTERS, V.J.M. & SHIMIZU, N. 1988. World-wide occurrence of HFSE-depleted mantle. *Geochimica et Cosmochimica Acta*, **52**, 2177–2182.

SCHIANO, P., CLOCCHIATTI, R., SHIMIZU, N., MAURY, R.C., JOCHUM, K.P. & HOFMANN, W. 1995. Hydrous, silica-rich melts in the sub-arc mantle and their relationship with erupted arc lavas. *Nature*, **377**, 595–600.

SEIFERT, K. & BRUNOTTE, D. 1996. 23. Geochemistry of serpentinized mantle peridotite from Site 897 in the Iberia Abyssal Plain. *In*: Whitmarsh, R.B., Sawyer, D.A. & Masson, D.G. *et al.* (eds) *Proceedings of the Ocean Drilling Program, Scientific Results 149*. Ocean Drilling Program, College Station, TX, 413–424.

SEYLER, M. & BONATTI, E. 1994. Na, Al^{IV} and Al^{VI} in clinopyroxenes of subcontinental and suboceanic ridge peridotites: a clue to different melting processes in the mantle? *Earth and Planetary Science Letters*, **122**, 281–289.

SEYLER, M. & BONATTI, E. 1997. Regional-scale melt–rock interaction in lherzolitic mantle in the Romanche Fracture Zone (Atlantic Ocean). *Earth and Planetary Science Letters*, **146**, 273–287.

SILVA, E.A., MIRANDA, J.M., LUIS, J.F. & GALDEANO, A. 2000. Correlation between the Palaeozoic structures from West Iberian and Grand Banks margins using inversion of magnetic anomalies. *Tectonophysics*, **321**, 57–71.

STOSCH, H.-G. & LUGMAIR, G.W. 1986. Trace element and Sr and Nd isotope geochemistry of peridotite xenoliths from the Eifel (West Germany) and their bearing on the evolution of the subcontinental lithosphere. *Earth and Planetary Science Letters*, **80**, 281–298.

SUN, S.-S. & MCDONOUGH, W.F. 1989. Chemical and isotopic systematics of oceanic basalts: implications for mantle composition and processes. *In*: SAUNDERS, A.D. & NORRY, M.J. (eds) *Magmatism in Ocean Basins*. Geological Society, London, Special Publications, **42**, 313–345.

VANUCCI, R., OTTOLINI, L., BOTTAZZI, P., DOWNES, H. & DUPY, C. 1994. INAA, IDMS, SIMS comparative investigations of clinopyroxenes from mantle xenoliths with different textures. *Chemical Geology*, **118**, 85–108.

WANG, W. & YURIMOTO, H. 1993. Analysis of rare earth elements in garnet by SIMS. *Annual Report of the Institute of Geoscience, University of Tsukuba*, **19**, 87–91.

WELLS, P.R.A. 1977. Pyroxene thermometry in simple and complex systems. *Contributions to Mineralogy and Petrology*, **106**, 431–439.

WHITMARSH, R.B., BESLIER, M.-O., WALLACE, P.J. et al. (eds) 1998. *Proceedings the Ocean Drilling Program, Initial Reports, 173*. College Station, TX, Ocean Drilling Program.

WITT, G. & SECK, H.A. 1989. Origin of amphibole in recrystallized and porphyroclastic mantle xenoliths from the Rhenish Massif: implications for the nature of mantle metasomatism. *Earth and Planetary Science Letters*, **91**, 327–340.

WITT-EICKSCHEN, G. & HARTE, B. 1994. Distribution of trace elements between amphibole and clinopyroxene from mantle peridotites of the Eifel (western Germany): an ion-microprobe study. *Chemical Geology*, **117**, 235–250.

WITT-EICKSCHEN, G., SECK, H.A. & REYS, C.H. 1993. Multiple enrichment processes and their relationships in the subcrustal lithosphere beneath the Eifel (Germany). *Journal of Petrology*, **34**, 1–22.

YURIMOTO, H., YAMASHITA, A., NISHIDA, N. & SUENO, S. 1989. Quantitative SIMS analysis of GSJ rock reference samples. *Geochemical Journal*, **23**, 215–235.

Petrology and geochemistry of exhumed peridotites and gabbros at non-volcanic margins: ODP Leg 173 West Iberia ocean–continent transition zone

RÉJEAN HÉBERT[1], K. GUEDDARI[2], M.R. LAFLÈCHE[2], M.-O. BESLIER[3] & V. GARDIEN[4]

[1]*Département de Géologie et de Génie géologique, Pavillon Adrien-Pouliot, Université Laval, Ste-Foy, Qc., Canada, G1K 7P4 (e-mail: hebert@ggl.ulaval.ca)*

[2]*INRS-Géoressoures, Centre géoscientifique de Québec, 880 Chemin, Ste-Foy, Québec, Canada, G1V 4C7*

[3]*Géosciences Azur, Observatoire océanologique, B.P. 48, 06235 Villefranche-sur-Mer Cedex, France*

[4]*Université Claude Bernard Lyon I, 43 Boulevard du 11 Novembre 1918, Laboratoire de Pétrologie et Tectonique, UMR 5570, Bât. 402, 69622 Villeurbanne Cédex, France*

Abstract: Ultramafic and mafic rocks recovered at Holes 1068A and 1070A were drilled during Leg 173 of the Ocean Drilling Program (ODP) in the ocean–continent transition zone of the Iberia Abyssal Plain. Peridotites show contrasting petrographic characteristics. Hole 1068A peridotites are fine grained and show a well-defined high-temperature foliation marked by elongated pyroxene as well as aligned spinels. Hole 1068A peridotites are strongly serpentinized. Hole 1070A peridotites are coarse grained and show little evidence of high-temperature foliation. The degree of serpentinization is lower and relicts of silicate minerals are preserved. In both sets of recovered material, spinels show a wide range of composition and suggest a complex magmatic evolution. Gabbros dykes, which are found only in Hole 1070A, are very coarse grained and are locally sheared and/or crushed. Magmatic amphiboles are kaersutites and Ti-rich tschermakites that are partially replaced by hornblende and actinolite, and are associated with plagioclase of intermediate composition. Peridotites and pyroxenite have low TiO_2, Al_2O_3 and CaO contents in carbonate-free samples. Ni and Cr contents fall into the upper-mantle array. On the other hand, gabbros have relatively high TiO_2 and V contents reflecting modal ilmenite, and suggesting that they are relatively differentiated. This paper presents the very first geochemical data on platinum group elements (PGE) of peridotites and gabbros from passive margins. Peridotites and gabbros show low PGE (25.83 ppb and 1.44 ppb), Pd (2.75 ppb and 0.15 ppb), Pd/Ir ratios (1.45 and 1.3) and mafic index. Pyroxenite has the highest PGE (27.97 ppb), Pd/Ir (19.87) and Pt/Ir (10.25). Interelemental correlation and observation of PGE-bearing sulphide phases suggest that the PGE are hosted by single sulphide phases. From a PGE point of view, extraction of magmas involved very PGE-depleted liquids similar to gabbroic veins cutting the peridotites at Hole 1070A. Partial melting is interpreted as occurring just before oceanic accretion. Geochemical attributes suggest that the peridotites belong to the Ronda and Beni Bousera peridotitic depleted end-member clan. Thus they are believed to be of subcontinental origin. Deformation and retrograde metamorphism of peridotites and gabbros are consistent with exhumation in a rift environment post-dating the 120 Ma magmatic stage.

Ocean Drilling Program (ODP) Leg 173 (April–June 1997) was devoted to better constrain the processes and timing of continental rifting and associated exhumation of deep lithospheric levels by determining the nature and evolution of crystalline basement rocks in the ocean–continent transition zone (OCT) outlined by seismic lines Lusigal 12, Resolution 3 and Sonne 16 (Beslier *et al.* 1993, 1995, 1996; Krawczyk *et al.* 1996; Pickup *et al.* 1996). Leg 173 complements the drilling transect initiated during Leg 149 (ODP Leg 149 Shipboard Scientific Party 1993; Whitmarsh *et al.* 1996). Two sites, some 100 km apart, recovered strongly to partly serpentinized peridotites and partly metamorphosed gabbros (Holes 1068A

From: WILSON, R.C.L., WHITMARSH, R.B., TAYLOR, B. & FROITZHEIM, N. 2001. *Non-Volcanic Rifting of Continental Margins: A Comparison of Evidence from Land and Sea.* Geological Society, London, Special Publications, **187**, 161–189. 0305-8719/01/$15.00 © The Geological Society of London 2001.

and 1070A, Fig. 1) and provide an excellent opportunity to document the composition and evolution of upper mantle exhumed in the OCT (ODP Leg 173 Shipboard Scientific Party 1998; Fig. 2). This paper presents results from a petrological and geochemical study of the recovered peridotites and gabbros and outlines implications for understanding the process of continental rifting.

Previous work

The West Iberia margin, comprising the Galicia Bank, is among the best studied passive margins in the world. It is a sediment-starved non-volcanic margin formed during the North Atlantic opening. Several oceanographic cruises focused on critical geological and geophysical data so as to understand the processes involved during continental rifting and subsequent break-up in Early Cretaceous time (Srivastava *et al.* 1990) including exhumation of deep lithosphere, and establishment of sea-floor spreading in the pre-Late Cretaceous constant polarity interval (Whitmarsh & Miles 1995).

The existence of peridotites from the Galicia Bank was first established by Boillot *et al.* (1980). The presence of plagioclase peridotites was confirmed by many other cruises deploying drilling or submersible diving (Agrinier *et al.* 1988; Boillot *et al.* 1988, 1989, 1995*a,b*; Girardeau *et al.* 1988; Kornprobst & Tabit 1988; Beslier *et al.* 1990; Charpentier *et al.* 1998; Fig. 1). These findings helped to define the position of a peridotite ridge that trends north–south along the west edge of continental crust of Galicia Bank. The petro-structural evolution of the peridotites is compatible with uplift and exhumation at the rift axis during the rifting (Evans & Girardeau 1988; Girardeau *et al.* 1988; Kornprobst & Tabit 1988; Boillot *et al.* 1989; Beslier *et al.* 1990). The evolution is controlled by limited partial melting and a continuum of deformation at progressive decreasing temperature. Gabbroic masses intruded into the upper mantle are observed on both sides of the ridge and exhumed along synrift Mesozoic ductile to Cenozoic compression-mode brittle faults associated with the Pyrenean Orogeny (Scharer *et al.* 1995). Gabbros have high-temperature assemblages that are partly retrograded to low-temperature parageneses.

Massive or brecciated ultramafic rocks were also drilled in the Iberia Abyssal Plain during Leg 149 at Holes 897C, 897D and 899B (Beslier *et al.* 1996; Cornen *et al.* 1996*a*; Seifert & Brunotte 1996; Figs 1 and 2). These ultramafic rocks are websterites, dunites and harzburgites, with some minor plagioclase-rich lherzolites. The ultramafic rocks underwent several stages of partial melting, recrystallization and impregnation under both plastic and subsolidus conditions. The geochemical signature suggests metasomatic enrichment by addition of mid-ocean ridge basalt (MORB)-type basaltic magma (Seifert & Brunotte 1996). The magmatic signature of the Leg 149 ultramafic rocks, their petrological and tectonometamorphic evolution, and their similarities with the peridotites recovered at the western edge of the Galicia Bank support the hypothesis that they originated as a piece of asthenospheric mantle that has been accreted to the lithosphere in a rift system during stretching of the continental lithosphere (Cornen *et al.* 1996*b*).

Mafic rocks were recovered at Holes 899A and 900A (Cornen *et al.* 1996*b*). At Hole 899A they are clasts of low-grade chlorite-bearing schists and metamorphosed to fresh lavas, microgabbros and leucogabbros or plagiogranites (Cornen *et al.* 1996*b*). The magmatic affinities are connected with E-MORB to alkaline series. At Hole 900A, the flaser gabbro clasts

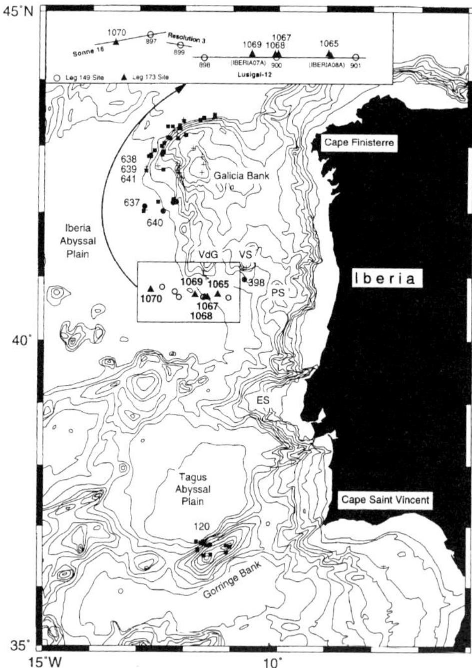

Fig. 1. Location map of the study area. Legs 13, 149 and 173 sites as well as several dredging sites and seismic lines Sonne 16, Resolution 3 and Lusigal 12 are indicated. After ODP Leg 173 Shipboard Scientific Party (1998).

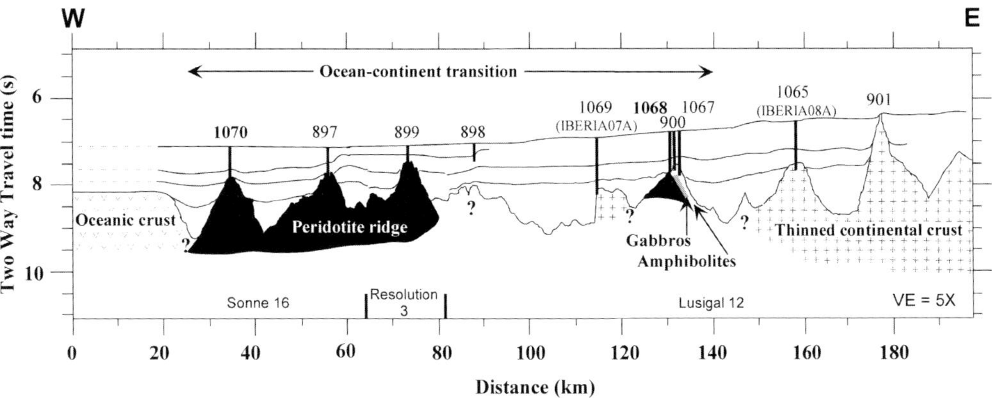

Fig. 2. Summary of the basement cores obtained along the Leg 173 drilling transect. After ODP Leg 173 Shipboard Scientific Party (1998).

are believed to derive from cumulative tholeiitic to transitional magmas (Cornen *et al.* 1996*b*; Seifert *et al.* 1996, 1997). The mineralogy, chemistry and tectonometamorphic evolution of these mafic rocks are consistent with an origin during early stages of continental rifting. The late-stage event that affected the gabbros is dated to 136 ± 0.3 Ma (Féraud *et al.* 1996). This age was interpreted as the cooling age of the gabbro.

Leg 173 occurrences

Serpentinized peridotites were recovered from two sites during 1997 ODP Leg 173 (ODP Leg 173 Shipboard Scientific Party 1998; Whitmarsh *et al.* 1998). In both cases basement peridotites were immediately preceded by tectonic or sedimentary breccia.

Hole 1068A peridotites were recovered from 904 to 956 m below seafloor (mbsf) a few hundred metres west of Leg 149 Hole 900 (Sawyer *et al.* 1994). They are overlain by <20 m of tectonic breccia made of strongly serpentinized peridotite fragments set in black serpentinite matrix. The tectonic breccia is overlain by <30 m of a sedimentary breccia characterized by polygenic angular clasts set in carbonate-dominated matrix. The peridotites are fine grained and, based on visual macroscopic estimations, contain 5–20% of light-coloured chlorite occurring as haloes around spinels. The degree of serpentinization is high and initial modal mineralogy is difficult to ascertain from sample description.

In thin sections most original minerals except chromiferous spinels are replaced by serpentine. However, the amount of relict orthopyroxene and clinopyroxene increases towards the bottom of the hole. No fresh olivine was detected. Foliation is ubiquitous and marked by aligned elongated lobate spinels and aligned lobate clinopyroxene. The orthopyroxene and pseudomorphs of orthopyroxene also show preferred orientation and few signs of high-temperature deformation such as kinks and ductile stretching. They are relatively fine grained and maximum observed size is 5 mm. Granular clinopyroxene is <1 mm in diameter and almost undeformed. It occurs as small lobate grains. Spinel estimated mode accounts for 1–3 vol. % of the peridotites. Spinels showing worm-like to lobate shapes are partially replaced by ferritchromite and surrounded by elongate green chlorite haloes. Chlorite also occurs as veinlets.

At Hole 1070A, basement rocks were recovered at 658.5 mbsf and consist of matrix-supported serpentinite tectonic breccia overlying pegmatitic gabbro and strongly to partly serpentinized peridotites (Whitmarsh *et al.* 1998). Syn-rift sediments capping the acoustic basement are Aptian in age. The breccia is largely matrix supported, the matrix representing 40–85 vol. % of the breccia and composed of calcite. Only few relicts of clinopyroxene are observed in the serpentinite clasts, whereas olivine, orthopyroxene and spinel are altered to a mixture of serpentine and magnetite. A few altered gabbroic to leucogabbroic clasts were also recovered. A fault gouge made of very fine-grained altered serpentinite and chlorite mixture was also observed (Whitmarsh *et al.* 1998).

Hole 1070A coherent serpentinized to partly serpentinized peridotites were first encountered between 659 and 719 mbsf. In addition to peridotite some intrusive pyroxenite (e.g. 173-1070A-10R2, 60–65 cm), dunite and pegmatitic gabbro were also sampled. The peridotites are coarse grained (5 mm to 1 cm). They contain very few chlorite pseudomorphs surrounding spinels. The 1070A peridotites are relatively better preserved than 1068A serpentinized peridotites. Relicts of orthopyroxene, clinopyroxene and spinel were observed in the upper part of the cores, and olivine is common from Core 11R downhole. Most of the recovered peridotites are lherzolite and harzburgite. Foliation is generally weak to absent, but is obvious in Core 13R-3 and marked by aligned coarse-grained spinel. Foliation is also locally roughly defined by preferred orientation of pyroxene or pyroxene pseudomorphs. Orthopyroxene shows clinopyroxene exsolution lamellae. Granular clinopyroxene is finer grained than orthopyroxene and is generally attached to the latter. Most spinels are relatively coarse grained and do not usually show preferred orientation. Some spinels have worm-like to lobate irregular shapes. Spinel estimated mode accounts for <1 vol. % of the peridotites. Fresh olivine shows kink bands whereas pyroxenes are mostly undeformed. They show irregular shapes and occur as interstitial phases. Pyroxenite layers, up to 4 cm in thickness, are more easily studied

where serpentinization is not extensively developed. Late carbonate and serpentine veinlets cut the serpentinized peridotites.

Pegmatitic gabbros occur in several intervals of the hole. The main occurrence, a pegmatitic gabbro of 2.7 m thickness, is observed in Core 9R (Fig. 3), and has a tectonized contact with the top of the mantle rocks. The recovered contact between gabbro and peridotite is made of sheared retrograded gabbro in contact with totally serpentinized and chloritized peridotite.

Other veins or dykes occur in Cores 11R, 13R and 14R (Whitmarsh *et al.* 1998) and show clear igneous contacts. The width of the veins or dykes ranges from 1 to 10 cm. Turrin (1999) dated the magmatic amphiboles at 120.8 Ma and partly recrystallized (?) plagioclase at 108.6 Ma using the $^{40}Ar/^{39}Ar$ method. He suggested that the age obtained on amphiboles corresponds to the onset of mid-ocean ridge magmatism. On the other hand, the age obtained on plagioclase is explained in terms of exhumation of the upper mantle.

The primary mineralogy of the gabbro in Core 9R includes plagioclase, red–brown kaersutite, clinopyroxene, ilmenite and red biotite. This corresponds to an evolved ferrogabbro mineral assemblage. Textural relationships suggest the crystallization order plagioclase, clinopyroxene, amphibole then ilmenite. Amphibole occurs as large grains and as a pseudomorph phase within clinopyroxene. Pegmatitic gabbros record a continuous deformation at progressively lower temperatures. Locally, a strong foliation involving elongation of plagioclase, ilmenite and red amphibole porphyroclasts was probably developed in plastic to brittle conditions at temperatures exceeding 500 °C. Amphibole porphyroclasts show subgrain boundaries and wavy extinction, and are locally oriented parallel to the foliation. Locally amphibole shows lower-temperature cataclastic deformation and is recrystallized into fine grains. Secondary minerals develop as coronitic or pseudomorphic replacement or as vein infilling. For instance, amphibole is rimmed and partly replaced by deep green edenite and magnesio-hornblende and then by light green actinolitic hornblende and actinolite. These gabbros are strongly to moderately altered and zeolite minerals are ubiquitous, implying very low-grade metamorphism. Typically, plagioclase is replaced by albite, zeolite, chlorite and some amphibole, red–brown amphibole is rimmed by green to blue–green amphibole, clinopyroxene is rimmed by amphibole, ilmenite is partly converted into magnetite and sphene, whereas red biotite is partly replaced

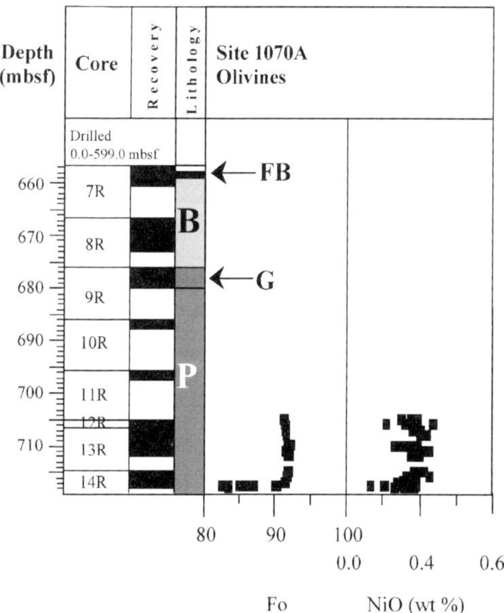

Fig. 3. Downhole Fo and NiO contents of olivine in Hole 1070A peridotites. B, breccia and fault gouge; FB, nannofossil chalk breccia; G, gabbro; P, peridotites.

by green biotite and magnetite. Series of low-grade hydrothermal minerals are observed in late veins and include vesuvianite, zeolite, prehnite and chlorite. Most of these low-grade minerals show syntectonic growth. These veins are cut by late calcite veins. These features are compatible with progressive exhumation of the gabbros.

Mineral chemistry

In this section we present results from Cameca SX-100 microprobe spot analyses performed on minerals from peridotites of both holes and gabbros from Hole 1070A. Single grains were checked for heterogeneity and both cores and margins were analysed. On average, five spot analyses were carried out on each grain. It was found that the grains are unzoned and, accordingly, no distinction will be made between cores and margins in diagrams. Total variation as well as averages are shown in diagrams dealing with log-variations. Analytical conditions were set to accelerating potential 15 kV, counting time 10 s and sample current 20 nA. A complete set of microprobe data can be obtained from the Society Library or the British Library Document Supply Centre, Boston Spa, Wetherby, West Yorkshire LS23 7BQ, UK as Supplementary Publication No. SUP 18168.

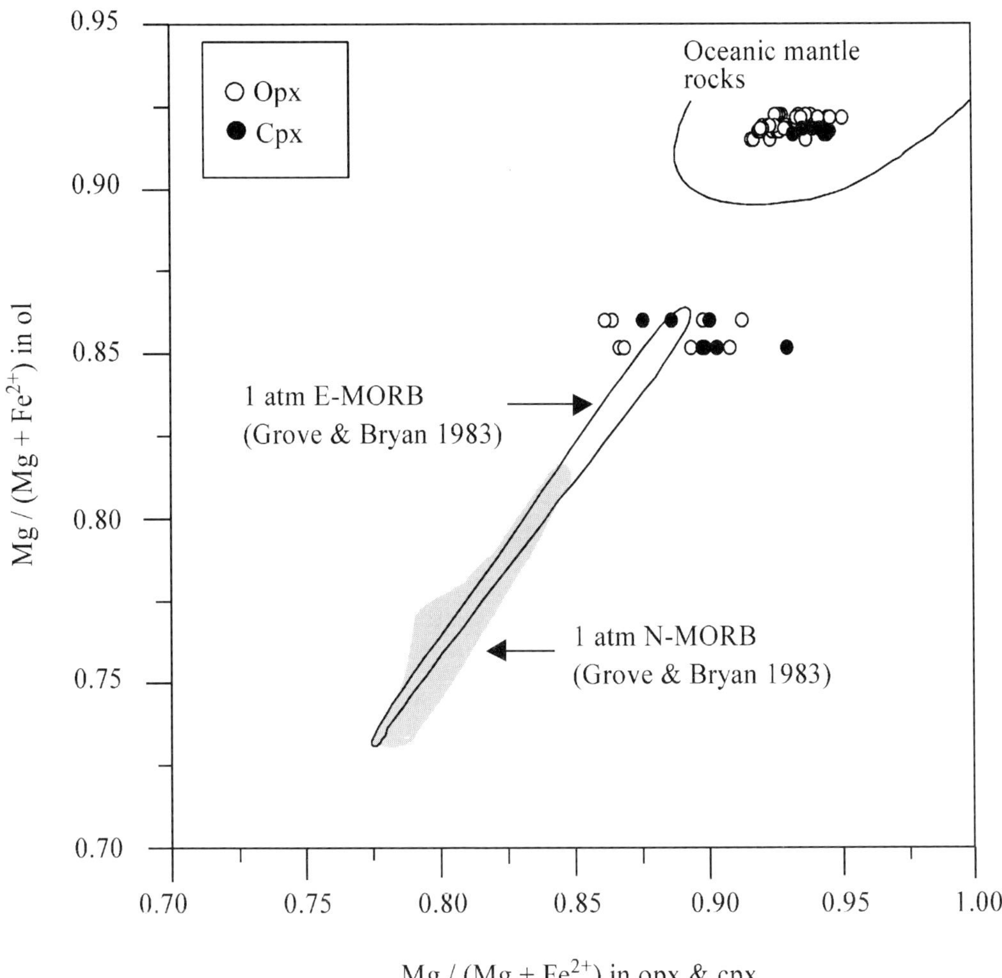

Fig. 4. Mg/(Mg + Fe^{2+}) covariation of olivine, clinopyroxene and orthopyroxene in Hole 1070A peridotites for two gabbroic and three peridotitic samples.

Olivine

Olivine is preserved only in Hole 1070A. Figure 3 shows the variation of Fo mol % and NiO contents against stratigraphic position. Fresh olivine relicts are common towards the bottom of the hole, from sections 11R down to 14R. Most olivine falls in the range Fo_{91-92} whereas NiO contents are more variable (0.48–0.22 wt %). Departing from this distribution are olivine analyses from sample 173-1070A-14R-3, 80–85 cm, piece 6. This sample, adjacent to a gabbro vein, shows iron-rich olivine (Fo_{82-87}) and NiO contents 0.38 down to 0.13 wt %. This sample plots away from the usual mantle olivine composition and is more like compositions found in E-MORB type ultramafic crustal rocks (see Fig. 4). Samples 173-1070A-13R3, 73–77 cm, piece 6B, and 173-1070A-14R3, 110–116 cm, piece 7, also show low NiO contents but the Fo content remains high. Core 13R also contains a gabbro vein (Whitmarsh *et al.* 1998). CaO contents tend to be higher (up to 0.07 wt %) in olivine grains showing low NiO content whereas the average mantle value is 0.04 wt %.

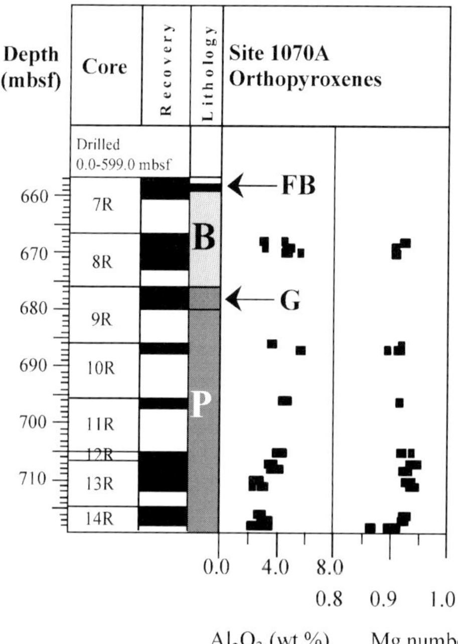

Fig. 5. Downhole *Mg*-number and Al_2O_3 wt % in orthopyroxene from 1070A peridotites. TB, tectonic breccia. Other symbols as in Figure 3.

Orthopyroxene

Orthopyroxene is observed in both Holes 1068A and 1070A but fresh relicts are more abundant in Hole 1070A. The *Mg*-number is generally between 0.90 and 0.95, and Al_2O_3 ranges between 3 and 6 wt % (Fig. 5). A few exceptions to this scheme are observed in samples 173-1070A-14R2, 30–35 cm, piece 2B; 173-1070A-13R3, 73–77 cm, piece 6B; 173-1070A-14R3, 80–85 cm, piece 6; and 173-1070A-13R4, 103–108 cm, piece 4. In these samples orthopyroxene *Mg*-number is as low as 0.86 and Al_2O_3 as low as 2.2 wt %. Orthopyroxene from Leg 173 has Al_2O_3 content similar to abyssal peridotite but higher *Mg*-number (Fig. 6). There is a general decrease in *Mg*-number along with Al_2O_3 increase (Fig. 6). Maximum Al_2O_3 values are observed for *Mg*-number 0.89. Cr_2O_3 content is generally between 0.60 and 1.00 wt %. Low values (0.45–0.55 wt %) are observed in samples 173-1070A-10R2, 60–65 cm, piece 7, and 173-1070A-8R3, 39–45 cm, piece 2. TiO_2 values are generally <0.14 wt %. The lowest values are observed for the highest *Mg*-number. Temperatures inferred by using the quadrilateral diagram calibrated by Lindsley (1983) show a large span of variation from 1300 °C to <500 °C (Fig. 7). The highest temperatures are observed for samples 1730-1068A-25R2, 29–35 cm, piece 3B; 173-1070A-8R3, 39–45 cm, piece 2; and 173-1070A-12R1, 27–32 cm, piece 2; and probably reflect close to magmatic temperatures.

Clinopyroxene

Clinopyroxene is the best preserved silicate phase. Relicts of fresh clinopyroxene occur in Cores 25R to 28R at Hole 1068A and it is ubiquitous at Hole 1070A (Fig. 8). *Mg*-number is variable; it ranges from 0.78 to 0.95 at Hole 1068A and from 0.77 to 0.94 at Hole 1070A. At both sites a general increase of *Mg*-number maximum values with depth is observed. However Al_2O_3 values show considerable variations. At Hole 1068A, Al_2O_3 content ranges from 3.5 to 6 wt %. No covariation is observed against depth. At Hole 1070A, Al_2O_3 is even more scattered (from 2.2 to 7.1 wt %). However, there is a general decrease in Al_2O_3 from Cores 10R down to 14R. Al_2O_3 contents show a slight downhole negative correlation with average *Mg*-number. Highest Al_2O_3 contents are observed in samples 173-1070A-8R2, 93–98 cm, piece 8A; 173-1070A-8R3, 39–45 cm, piece 2; and 173-1070A-10R2, 60–65 cm,

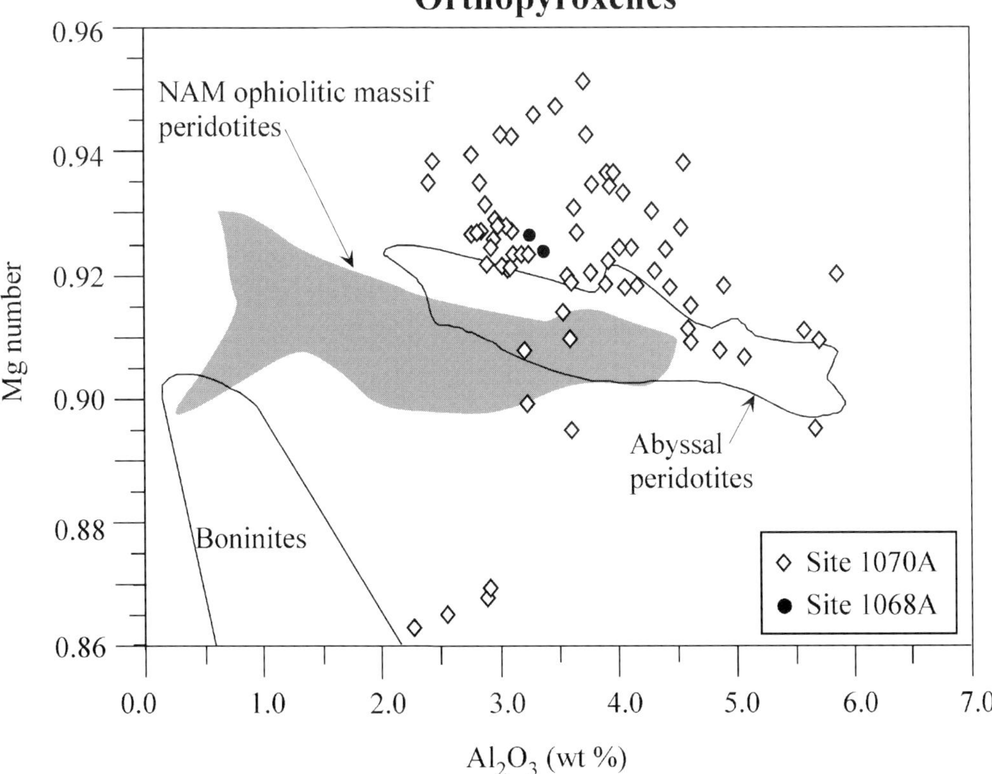

Fig. 6. The *Mg*-number and Al$_2$O$_3$ wt % variation diagram in orthopyroxene from Hole 1068A and 1070A peridotites. Fields of abyssal peridotites, boninites and North Arm Mountain (NAM) massif are shown for comparison. Modified after Varfalvy *et al.* (1997).

piece 7 (6.0–7.1 wt %; *Mg*-number varies from 0.84 to 0.91). Most values fall into the field defined by *Mg*-number and Al$_2$O$_3$ boundaries 0.90–0.94 and 3.7–6.6 wt %, respectively. TiO$_2$ content in clinopyroxene at both sites is largely overlapping from low values of 0.05 wt % to 0.55 wt % (Fig. 9). Sample 173-1068A-26R2, 36–41 cm, piece 1B, shows significant higher TiO$_2$ values, from 0.54 to 0.89 wt %.

Clinopyroxene from Hole 1070A pegmatitic gabbro shows more restricted *Mg*-number values in the range 0.80–0.85. Clinopyroxene from gabbro shows Al$_2$O$_3$ content varying from 1.2 to 5.8 wt % in sample 173-1070A-9R2, 82–87 cm, piece 10. At Hole 1070A, TiO$_2$ contents of clinopyroxene in gabbro are highly variable, as low as 0.1 and up to 2.2 wt %, whereas the Cr$_2$O$_3$ content is relatively high, in the range 0.75–1.75 wt %. The Cr$_2$O$_3$ content of clinopyroxene in gabbro is below detection limit. When plotted into Ti–Al^{IV} and Al^{VI}–Al^{VI} spaces (not shown here) clinopyroxene from

gabbro falls into metamorphic and Southwest Indian Ridge fields, respectively. It is not as aluminous as clinopyroxene reported from Zabargad Island, Red Sea (Piccardo *et al.* 1988; Seyler & Bonatti 1994). In Ti–Na space (not shown here) clinopyroxene data from both Holes 1068A and 1070A are almost exactly overlapping. Most of the Na contents of clinopyroxene are higher than those in South Pacific peridotites and pyroxenite (Constantin *et al.* 1995) for the same Ti compositional range. Clinopyroxene from Hole 1068A peridotites plots into the diopside-rich part of the quadrilateral of Lindsley (1983) whereas Hole 1070A clinopyroxene shows a widespread compositional distribution. The temperature record ranges from 900 °C down to 500 °C. These temperatures are subsolidus to metamorphic in origin. The most enriched Fs component (9–10 mol %) clinopyroxene is observed in gabbro (sample 173-1070A-9R2, 82–87 cm, piece 10) and peridotite samples: 173-1070A-8R3, 39–

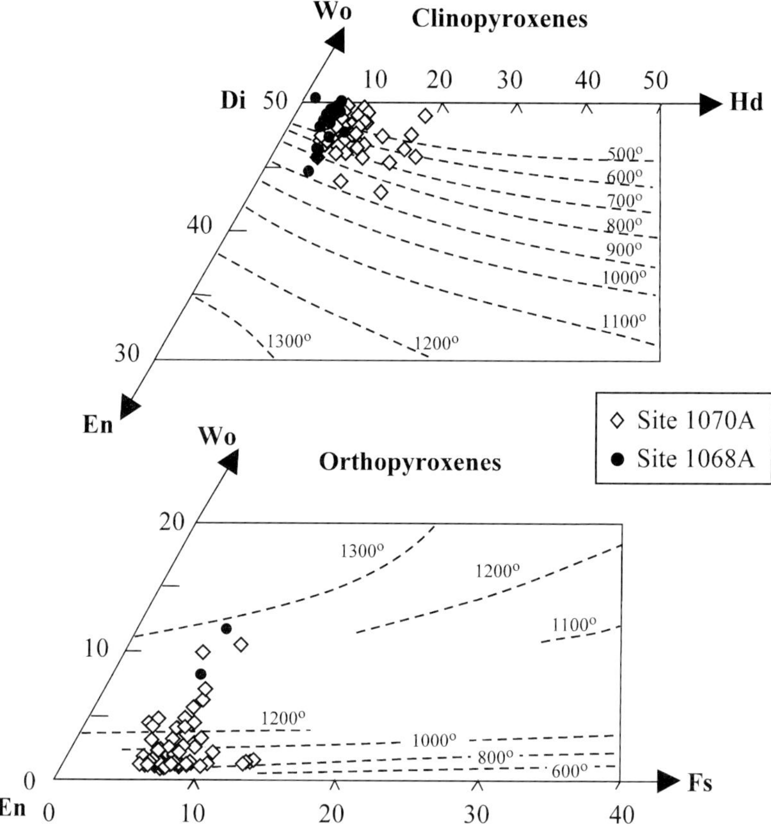

Fig. 7. Quadrilateral projection for orthopyroxene and clinopyroxene from Hole 1068A and 1070A peridotites. After Lindsley (1983).

45 cm, piece 2 (4.6–8.7 mol %); 173-1070A-10R2, 60–65 cm, piece 7 (4.6–6.0 mol %); and 173-1070A-14R3, 80–85 cm, piece 6 (3.5–5.2 mol %).

Spinel

Spinel is the best preserved oxide primary phase in Leg 173 peridotites (Fig. 10). The *Cr*-number in Hole 1068A spinels is mostly restricted to values between 0.10 and 0.40, averaging 0.25. Some spinels showing higher *Cr*-number are observed in Cores 22R, 25R and 29R. The *Mg*-number average value for Hole 1068A spinels is *c.* 0.65. The *Mg*-number, at fixed *Cr*-number, is low for 1068A spinels. The peak TiO_2 values for 1068A spinels are observed in Cores 14R, and 22R, 25R, 28R and 29R respectively (e.g. 173-1068A-28R2, 36–41 cm, piece 1B; 173-1068A-29R2, 8–13 cm, piece 1B) but does not exceed 0.60 wt %. NiO

content ranges from 0.05 to 0.35 wt %. However, at fixed *Cr*-number, 1068A peridotites have spinel containing relatively high NiO (up to 0.35 wt %). NiO content is higher for low *Cr*-number. Fe^{3+}-number tends to be very low (<0.02) in 1068A spinels peridotites. Highest Fe^{3+}-number occurs in 1068A Cores 22R and 29R (up to 0.24; 173-1068A-22R1, 54–59 cm, piece 7) and 29R (up to 0.15; 173-1068A-29R2, 8–13 cm, piece 1B), respectively.

In Hole 1070A, *Cr*-number of spinel is variable but a cluster of data is observed around 0.50. In Core 8R a large compositonal span is observed (0.10–0.60), whereas in Cores 12R, 13R and 14R values are 0.50–0.65. In *Cr*-number–*Mg*-number space the data for 1070A spinels plot towards the *Cr*-number-rich side (Fig. 11). Very few data overlap. TiO_2 and NiO contents show a very large scatter. The TiO_2 content in spinel is several orders of magnitude higher than those of residual peridotites (Hébert

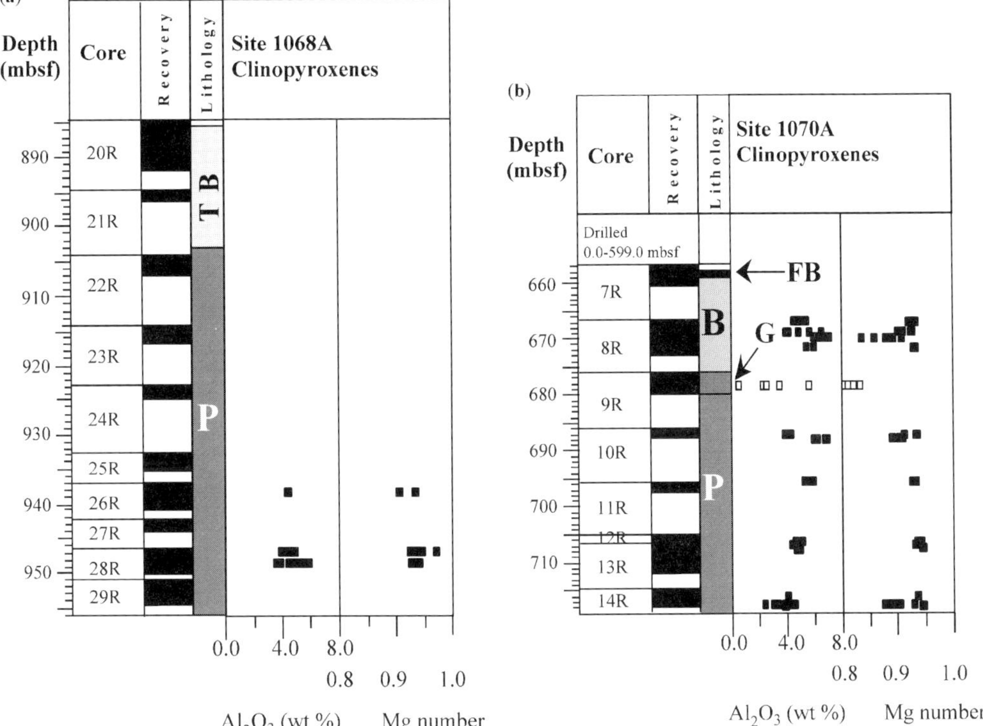

Fig. 8. Downhole *Mg*-number and Al_2O_3 wt % variations in clinopyroxene from (**a**) Hole 1068A and (**b**) Hole 1070A peridotites and gabbros. Symbols as in Figs. 3 and 5.

et al. 1989), which is generally <0.1 wt %. The highest TiO_2 content is observed in 1070A spinels (173-1070A-8R2, 93–98 cm, piece 8A, and 173-1070A-14R3, 80–85 cm, piece 6, content up to 2.2 wt %). There is no apparent systematic variation with increasing depth. The Fe^{3+}-number v. *Cr*-number shows a gradual sympathetic increase of both ratios. Highest Fe^{3+}-number ratios are observed in 1070A Cores 9R and 14R (up to 0.15; 173-1070A-14R3, 80–85 cm, piece 6). Significant ZnO content was detected in samples 173-1070A-12R2, 32–37 cm, piece 1A (up to 0.35 wt %); 173-1070A-13R1, 29–35 cm, piece 1B (up to 0.41 wt %); and 173-1070A-14R3, 80–85 cm, piece 6 (up to 0.49 wt %).

Plagioclase

Fresh plagioclase was found in amphibole gabbros only (173-1070A-9R1 and -9R2). The anorthite content of plagioclase ranges from 42.5 to 52.4 mol %. FeO content is up to 0.22 wt % and is higher for less anorthitic com-

positions. K_2O content ranges from 0.19 to 0.37 wt %.

Amphibole

Amphibole is observed mainly in gabbroic samples and grouped into red–brown Ti-rich tchermakite, kaersutite (Ti >0.3, Al^{IV} >1.5) of probable primary origin and low-Ti green to blue–green coronitic edenite, magnesio-hornblende, actinolitic hornblende and actinolite (Ti <0.3, Al^{IV} <1.5) of metamorphic origin (Fig. 12). The amphibole show a pargasitic type substitution pattern. The most titaniferous amphiboles are found in samples 173-1070A-9R1, 49–55 cm, piece 5A, and 173-1070A-9R2, 82–87 cm, piece 10. Similar compositional variations were shown by Hébert & Constantin (1991). Cl content is generally low but is as high as 0.22 wt % in magnesio-hornblende of sample 173-1070A-9R2, 51–56 cm, piece 8. Amphiboles also show high $(Na + K)_A$ content between 0.75 and 0.86. Most amphiboles have low Al^{VI} content (<0.40), and at high Al^{IV}

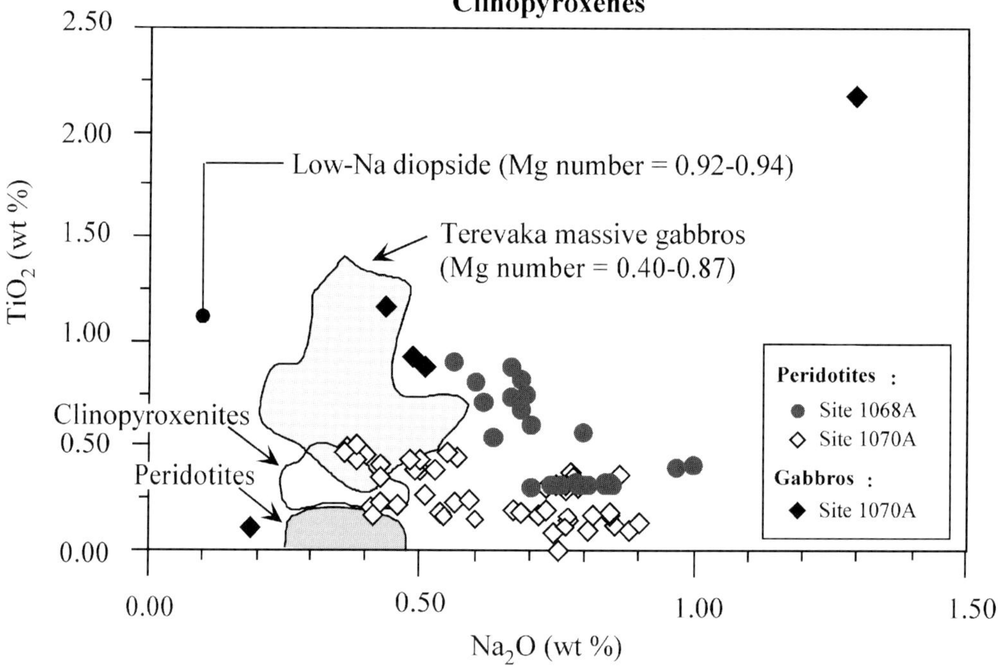

Fig. 9. TiO$_2$ wt % and Na$_2$O wt % covariations in clinopyroxene from Hole 1068A and 1070A peridotites. Fields of oceanic upper-mantle peridotites, clinopyroxenite and Terevaka gabbros are shown for comparison. Modified from Constantin *et al.* (1995).

levels there is no correlation observed with Al^{VI}. However, there is a positive correlation for low-Ti metamorphic amphiboles between Al^{IV} and Al^{VI}. K$_2$O content is medium for low-Ti amphiboles (<0.40 wt %) to high for high-Ti amphiboles (0.60–0.95 wt %) when compared with Atlantic amphiboles reported by Cannat & Casey (1995). There is a positive correlation between K$_2$O and Na + K for low-Ti amphiboles. High-K$_2$O amphiboles have a restricted composition typical of magmatic amphiboles (Fig. 11). Ti-rich amphiboles have *Mg*-number restricted to 0.65–0.75, whereas Ti-poor amphiboles show a wider interval of variation from 0.62 to 0.88. We have compared amphibole compositions with Leg 149 Holes 899 and 900 amphiboles. Ti-rich amphiboles have similar composition to Leg 149 outlined fields for high-temperature amphiboles of probable magmatic origin (Hébert & Constantin 1991). Ti-poor amphiboles of metamorphic origin are similar to the low-temperature field as defined for Leg 149 amphiboles (Cornen *et al.* 1996*b*).

Serpentine

Serpentine pseudomorphs in Hole 1070A peridotites are relatively poor in Al$_2$O$_3$ (<4.6 wt %) and Cr$_2$O$_3$ rich (up to 3 wt %). Serpentine analysed in Hole 1068A, in contrast, shows low Cr$_2$O$_3$ contents (<1.2 wt %) and relatively high aluminium content (up to 8.87 wt %). Relatively high aluminium content of serpentine reflects the aluminous attribute of the replaced primary phases (e.g. spinel, plagioclase?). NiO content is below detection limit in 1068A serpentine but reaches 0.43 wt % in 1070A peridotites.

Chlorite

Chlorite is also found as a replacement product of an aluminous phase and occurs as haloes around spinels. These haloes are observed almost exclusively in 1068A peridotites. In sample 173-1068A-22R-1, 54–59 cm, piece 7, chlorite has a low Ti content (<0.19 wt %), is aluminous (up to 11.18 wt %), and is depleted in NiO and MnO (<0.02 wt %). Cr$_2$O$_3$ content averages 0.95 wt %. Chlorite analysed in

1070A peridotites can be also aluminous (up to 13 wt % in sample 173-1070A-9R2, 51–56 cm, piece 7) but Al_2O_3 content is generally <9.9 wt %. Significant TiO_2 (up to 0.46 wt %), MnO (up to 0.18 wt %) and NiO (up to 0.28 wt %) are reported. Cr_2O_3 in 1070A chlorite content can be as high as 1.19 wt %. In Hole 1068A peridotites, plagioclase was probably a significant reactant with metamorphic fluid to produce a mixture of serpentine and chlorite. In 1070A peridotites, the chemistry of chlorites suggests that spinel was the major aluminous reactant. The data presented here are in agreement with descriptions of similar spinel-related haloes in fresh Leg 149 plagioclase peridotites (Cornen *et al.* 1996*a*).

Ilmenite

Large centimetre-size ilmenite crystals occur only in the pegmatitic gabbros. They are spatially associated with kaersutite and tchermakites, and contain inclusions of idiomorphic apatite grains. Numerous exsolutions of Ti-magnetite are also observed on back-scattered microprobe images.

PGE hosting minerals

Preliminary scanning electron microscope (SEM) survey reveals that rare grains of Cu–Pt–Pd and Cu– (Pd–Pt) are spatially associated with Ni–Fe sulphide (e.g. sample 173-1068A-24R, 51–57 cm; Fig. 13). The PGE grains are small irregular aggregates 20 μm across attached to pentlandite grains. Exsolution of a Cu-rich PGE compound is observed.

Discussion of the petrographic and mineral chemistry results

Hole 1068A and 1070A peridotites bear many common mineral chemistry features but also differ in some aspects, as is discussed below. Some primitive characteristics were probably

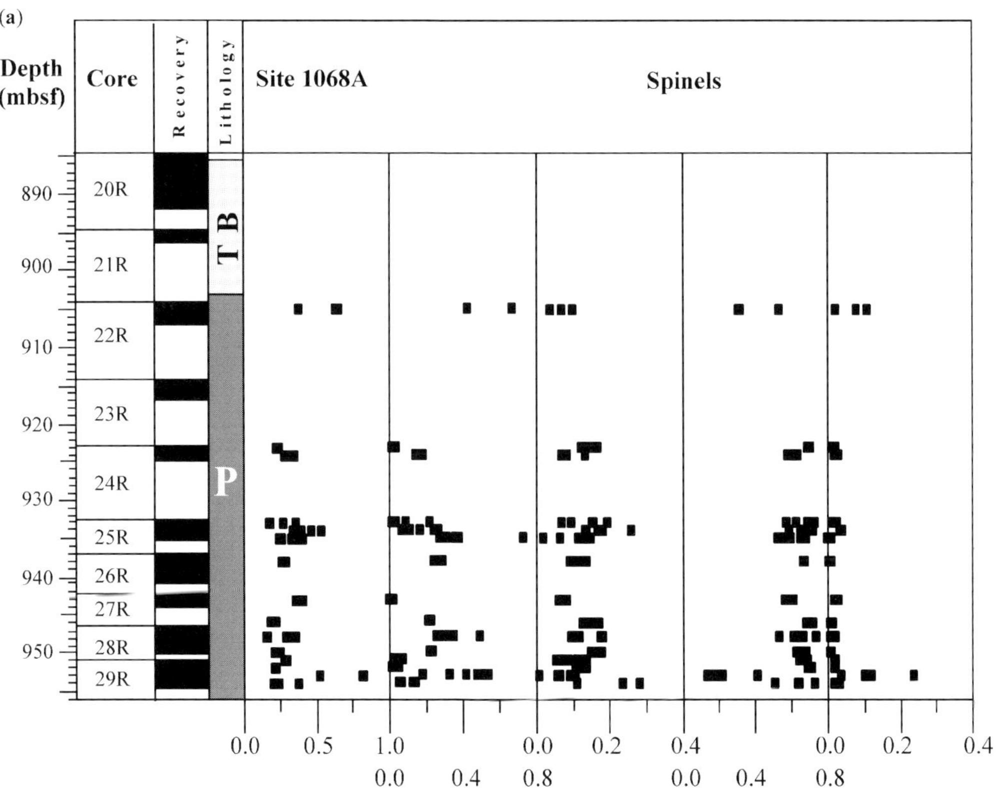

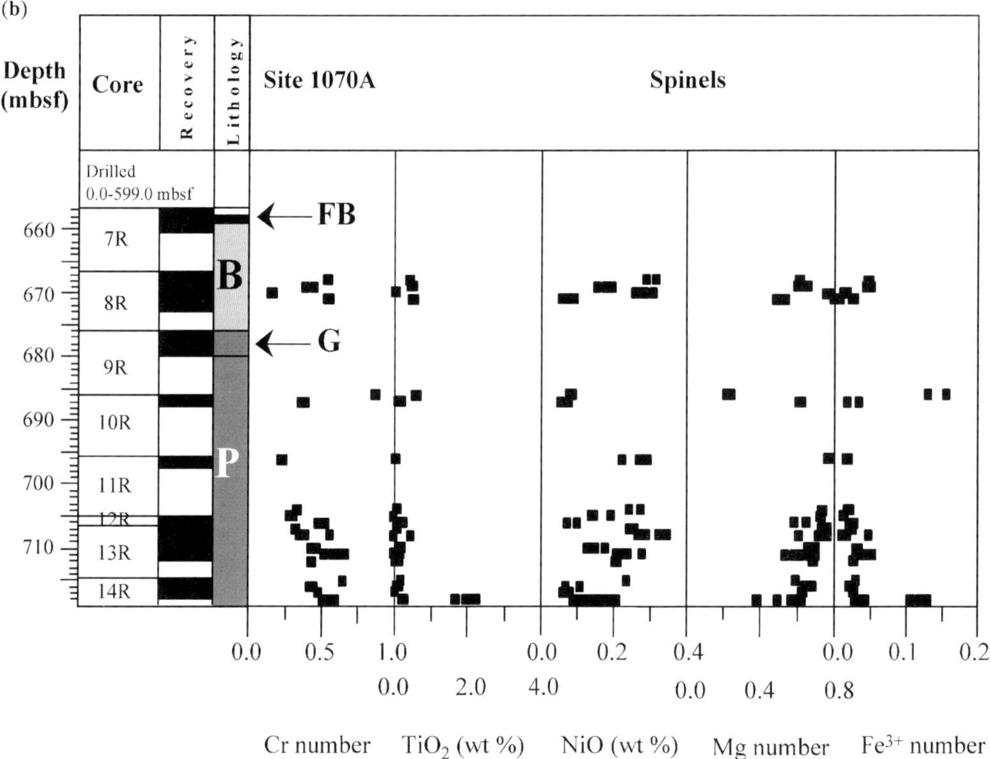

Fig. 10. Downhole *Cr*-number, TiO_2 wt %, NiO wt %, *Mg*-number and *Fe* $^{3+}$-number variations in chromite from (**a**) Hole 1068A and (**b**) Hole 1070A peridotites. Symbols as in Figures 3 and 5.

obliterated by strong serpentinization in Hole 1068A peridotites.

From spinel composition we cannot distinguish the peridotitic series on the basis of *Cr*-number. Spinels from both sites show a very large interval of variation, with some 1068A peridotites containing the most aluminous spinel. It is interesting to note that the most titaniferous spinels are found in cores or close to cores exhibiting gabbroic veins. Several workers have established a link between incorporation of Ti in spinel structure by magma circulation in peridotite (e.g. Cannat *et al.* 1990). The slightly lower *Mg*-number of some spinels in 1068A peridotites could also be explained in terms of magmatic interaction and subsolidus re-equilibration (Hébert *et al.* 1989). However, they show distinctive *Mg*-number at constant *Cr*-number. Hole 1068A peridotites show the lowest *Mg*-number. Compositions of both orthopyroxene and clinopyroxene show that aluminous and aluminium-poor varieties coexist at both sites. Na content of clinopyroxene is relatively high when compared with

oceanic depleted peridotites (Kornprobst *et al.* 1981). These characteristics suggest that the peridotites' initial compositions from both sites (as observed for Leg 149 Hole 897) covered a large spectrum from aluminous lherzolite to more aluminium-poor harzburgite. Hole 1070A peridotites contain the most fayalitic olivine and ferrosilite-rich clinopyroxene pairs.

Gabbroic intrusions were observed only within 1070A peridotites. They are moderately differentiated kaersutite (tschermakite) gabbros, as plagioclase falls into the intermediate range of An_{42-52} and titaniferous clinopyroxene is still diopsidic with up to 10 mol % ferrosilite. The high K_2O content of primary amphibole as well as occurrence of primary biotite suggest that initial magma was relatively K rich and water bearing. These mineral characteristics are typical of E-MORB generated at a mid-ocean ridge. This conclusion put a major constraint on the origin of 1070A peridotites and by extension on 1068A peridotites, situated a further 100 km towards the actual continental margin.

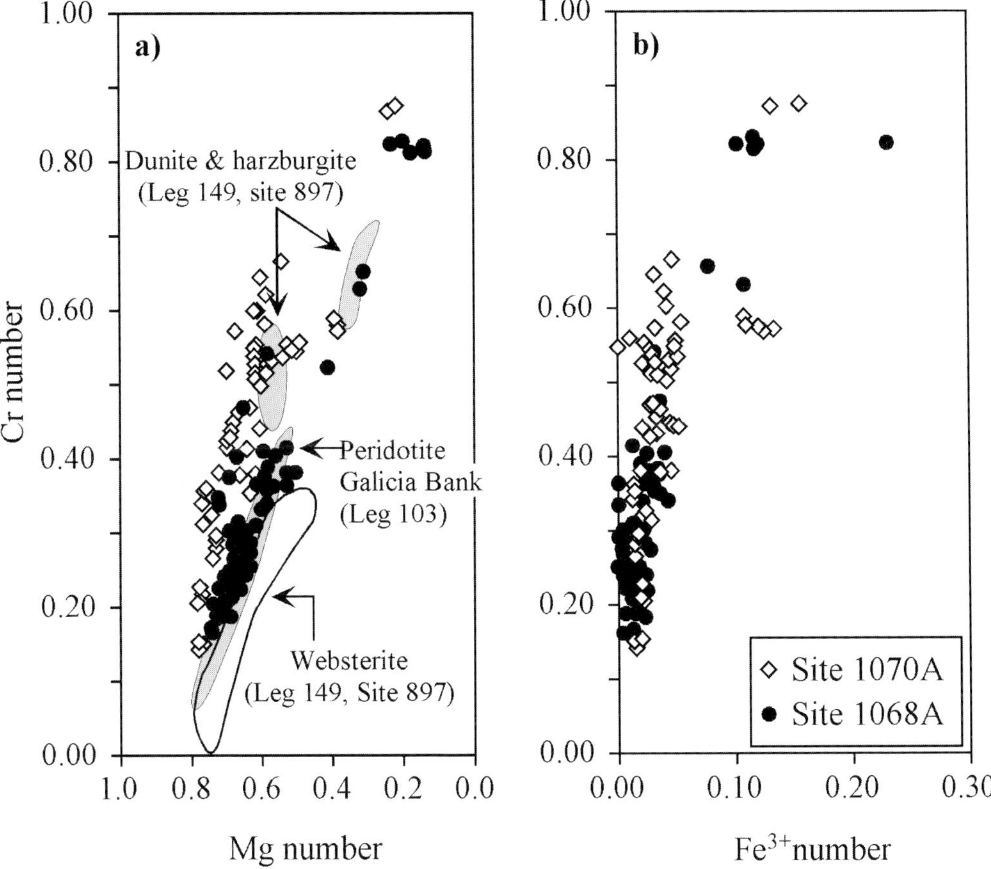

Fig. 11. (a) *Cr*-number and *Mg*-number diagram of chromite from Hole 1068A and 1070A peridotites. Fields for Galicia Bank peridotites, Leg 149, Site 897 (websterite), and Leg 149, Hole 897 (dunite and harzburgite) from Cornen *et al.* (1996a) are shown for comparison. (**b**) *Cr*-number and *Fe*$^{3+}$-number diagram for chromite from Hole 1068A and 1070A peridotites.

Orthopyroxene and clinopyroxene plotted in the quadrilateral diagram of Lindsley (1983) show re-equilibrated compositions (Fig. 7). Temperature recorded ranges from 1300 °C to <500 °C. Molecular wollastonite increase is attributed to subsolidus re-equilibration and to metamorphic recrystallization (Hébert *et al.* 1989).

Geochemistry

Major, minor and trace elements, as well as PGE analyses for 1068A and 1070A peridotites and 1070A gabbros are shown in Tables 1 and 2. Major and trace elements were determined using X-ray fluorescence and inductively coupled plasma atomic emission spectrometry at INRS-Géoressources Laboratories, Sainte-Foy (Québec). PGE were extracted according to the method described by Gueddari *et al.* (1996) and determined using inductively coupled plasma mass spectrometry.

Peridotites and pyroxenite

The geochemistry of peridotites shows remarkable homogeneity regarding distinctive oxide and trace element contents. Peridotites from both Holes 1068A and 1070A show high loss on ignition, reflecting modal hydrous minerals such as serpentine, chlorite and locally carbonate. TiO$_2$ content is low and averages 0.05 wt % (0.01–0.12 wt %), the maximum value being observed in sample 173-1070A-10R2, 60–65 cm, piece 7. CaO is low, averaging 0.72 and 3.97 wt % in 1068A and 1070A peridotites, respectively (total variation 0.37–1.54 wt % and 0.12–16.01 wt %, respectively). The high

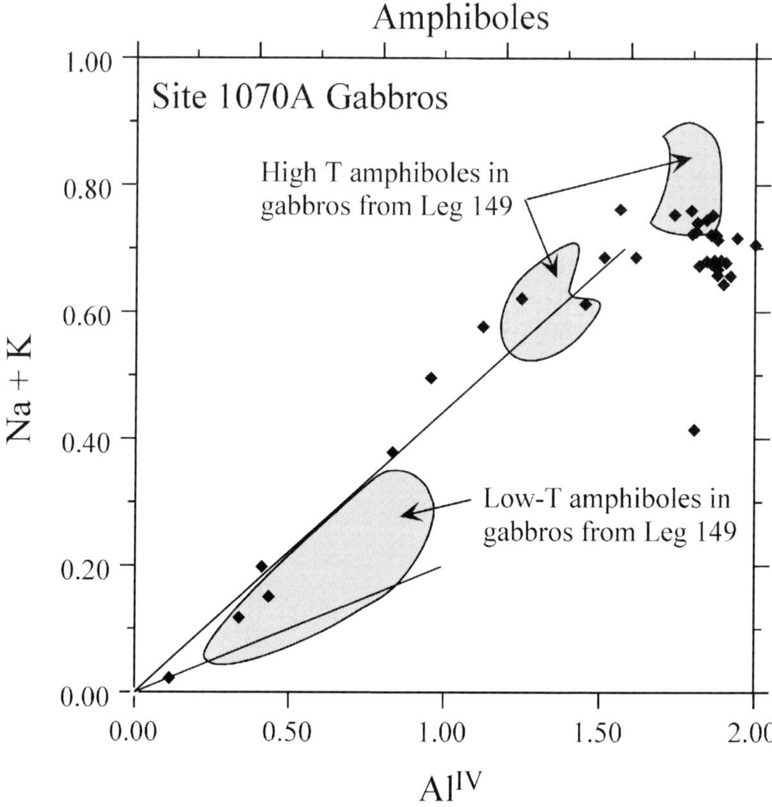

Fig. 12. Na + K v. Al^{iv} in Leg 173 Hole 1070A gabbros. Fields for Leg 149 amphiboles are from Cornen *et al.* (1996*a*).

CaO content in 1070A peridotites is correlated with modal carbonate observed in thin sections (samples 173-1070A-8R1, 57–63 cm, piece 2D; 173-1070A-8R2, 93–98 cm, piece 8A; and 173-1070A-8R3, 39–45 cm, piece 2; see petrography section). Average Al$_2$O$_3$ content is almost the same at both sites (1.64 v. 1.70 wt %). The maximum values are observed in samples 173-1070A-10R, 60–65 cm, piece 7; 173-1070A-8R, 57–62 cm, piece 2D; 173-1068A-29R, 56–61 cm, piece 1C; and 173-1070A-8R, 39–45 cm, piece 2. Excluding samples containing carbonate veins, there is a sympathetic variation between CaO and Al$_2$O$_3$ contents (see PGE sections). In low-Ca samples, MgO content is very homogeneous in 1068A peridotites, averaging 43.96 wt % (42.89–45.12 wt %). MgO content is more variable in 1070A peridotites, averaging 41.35 wt % (30.59–46.62 wt %). The lowest MgO contents are observed for samples showing higher LOI (up to 19.62 wt % including carbonate). Both 1068A and 1070A peridotites show high Cr, Ni and Co contents as

expected from upper-mantle ultramafic rocks. The respective averages are: 2395 vs 2353 ppm (1937–2595 ppm and 1242–4014 ppm), 2091 vs 1936 ppm, (1837–2326 ppm and 1542–2241 ppm) and 59 vs 55 ppm (49–71 ppm and 39–69 ppm). Most (27 out 31) of the samples fall within a relatively restricted compositional interval between 2000 and 3000 ppm Cr and 1500 and 2400 ppm Ni. Average Sc values are very low in peridotites (1068A: 8.6 ppm; 1070A: 10.4 ppm) suggesting very low initial clinopyroxene content. V content is slightly higher in 1070A peridotites (51.7 ppm) than in Site 1068A samples (38 ppm). This would indicate a slightly higher modal spinel content for the first group of peridotites. Sr values are surprisingly very low given the high water content of the peridotites (5 ppm for 1068A; 16.5 ppm for 1070A). The same remark is valid for Ba (22.4 ppm v. 21.7 ppm). Cu content is highly variable in 1068A rocks (average: 18.05 ppm; range: from not detected to 89.01 ppm) and in 1070A peridotites (average: 33.37 ppm; range:

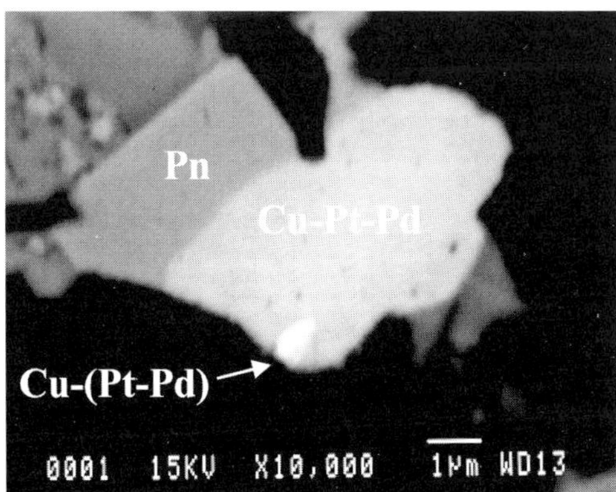

Fig. 13. Back-scattered electron photomicrograph shows textural relationships between pentlandite crystal (Pn) and Cu–Pt–Pd alloys. The bright spot is a Cu-rich Pt–Pd alloy. Sample 173-1068A-24R2, 51–57 cm, piece 4.

from not detected to 245.18 ppm). These variations do not show systematic relationship with any other trace elements.

Gabbros

Five samples of gabbro from Hole 1070A Core 9R were analysed. The high TiO_2 content (3.67–10.16 wt %) of the gabbro reflects modal ilmenite, Ti-rich kaersutite and Ti-diopside. V content is relatively high, averaging 480 ppm, and correlates with TiO_2 content.

MgO content is highly variable (5.30–24.44 wt %). Although variable, Al_2O_3 shows relatively high values from 11.57 to 20.38 wt % (average 15.22 wt %). This is related to initial plagioclase content now partially replaced by Al-bearing phase such as albite, zeolite and vesuvianite. Average CaO content of 7.84 wt % (5.74–9.04 wt %) is not balanced with Na_2O content (average 1.97 wt %; 0.46–4.51 wt %) suggesting that hydrothermal alteration significantly modified these parameters. K_2O content averages 0.24 wt % (0.12–0.35 wt %), in agreement with low modal K-bearing phases (biotite, amphibole) and low-K plagioclase. Compatible elements such as Cr and Ni are reduced to low average values (62 and 50 ppm, respectively) suggesting that the gabbros experienced significant differentiation. These elements are not correlated with MgO content. Sc is five times higher in gabbros than in the peridotites and is related to higher modal clinopyroxene and amphibole. Sr and Ba contents are respectively 10–15 times and 3–4 times higher in gabbbros than in peridotites (averages: Sr, 352 ppm; Ba,

75 ppm). This suggests that sea water circulated into the gabbros, which acted as natural sinks, and lost part of its Sr and Ba content to Ca-rich low-temperature minerals. However, there is no obvious correlation between LOI and Sr and Ba contents. S content is very low and several times below detection limits.

Platinum group elements

The PGE data presented here are the first for peridotites recovered from a modern passive margin. Hole 1068A and 1070A peridotite PGE contents along with data taken from the literature are reported in Table 2. The peridotites have approximately the same total PGE ranges, 9.19–25.83 and 5.44–19 ppb, respectively. Carbonate-rich peridotites do not show different PGE content but are not considered in the discussion. The highest total PGE contents are observed in samples 173-1068A-28R, 107–112 cm, piece 9 (25.83 ppb); 173-1068A-24R, 51–57 cm, piece 4A (24.43 ppb); 173-1070A-13R1, 98–104 cm (19 ppb); 173-1068A-27R, 28–33 cm, piece 3 (17.49 ppb); and 173-1068A-24R, 95–100 cm, piece 7A (17.38 ppb). Pd/Ir ratios vary from 0.17 to 1.45. The most fractionated ratio is observed in sample 173-1068A-24R, 51–57 cm, piece 4A. Rh/Pd and Pt/Ir ratios vary from 0.31 to 1.88 and 0.35 to 4.88, respectively. On a mantle-normalized diagram (Fig. 14) the profiles show a significant Pt and Pd depletion with respect to other PGE. These data result in negative profiles and, except for two samples, Pd/Ir ratios lower than unity. These patterns are similar to those of Ronda and Beni Bousera depleted peridotites (Gueddari *et al.* 1996; Table 3). The PGE values are also generally lower than the theoretical fertile mantle of Barnes *et al.* (1988). Only sample 173-1068A-26R1, 107–111 cm, piece 4B, plots away from the general total PGE–Pt trend. The (Pt/Pd)/(Ir + R) ratio vary from 0.45 to 1.33. This type of variation is similar to S- and As-bearing mineralizations compiled by Ohnenstetter (1996). The highest Pt/Pd ratios are found in samples 173-1070A-13R1, 98–104 cm (10.38); 173-1068A-26R4, 1–7 cm, piece 1A (8.33); 173-1068A-28R4, 107–112 cm, piece 9 (5.01); and 173-1068A-25R2, 29–35 cm, piece 3B (3.19). Samples showing maximum Cr and Ni also have the maximum Pt, Ir and total PGE contents. The lowest Pt/Pd ratio is found in sample 173-1068A-24R2, 51–57 cm, piece 4 (1.26). Interestingly, there is a strong correlation among all PGE (Ir–Ru, Ir–Rh, Ir–Pt, Ru–Rh, Ru–Pt, Rh–Pt) except Pd (Fig. 15). These correlations

Table 1a. *Major and trace elements, Leg 173, Sites 1068A and 1070A*

Site:	1068	1068	1068	1068	1068	1068	1068	1068	1068	1068	1068	1068	1068	1068	1068	1070A	1070A	1070A
Core:	22R	24R	24R	25R	25R	25R	26R	26R	27R	28R	28R	28R	29R	29R	29R	8R	8R	8R
Interval (cm):	54–59	95–100	51–57	128–133	29–35	111–116	107–111	1–7	28–33	5–10	36–41	107–112	74–79	8–13	56–61	57–62	93–98	39–45
Rock type:	P	P	P	P	P	P	P	P	P	P	P	P	P	P	P	Pc	Pc	Pc
Major elements (wt %)																		
SiO_2	36.04	35.74	35.76	37.13	36.73	38.15	36.55	37.09	34.80	36.34	36.36	37.48	35.79	35.27	36.02	32.40	32.82	32.92
TiO_2	0.06	0.03	0.05	0.02	0.05	0.04	0.02	0.01	0.02	0.02	0.05	0.03	0.04	0.05	0.08	0.06	0.07	0.06
Al_2O_3	1.20	1.05	1.61	1.36	1.45	0.99	0.66	0.33	1.37	1.35	1.96	1.29	1.45	1.90	2.34	2.62	2.07	2.18
Fe_2O_3	6.88	7.80	7.89	7.34	7.67	7.31	7.10	6.94	7.04	7.15	7.56	7.09	7.17	7.83	7.03	5.75	7.59	7.02
MgO	34.93	37.03	35.83	36.76	36.26	37.75	37.07	36.91	36.70	37.40	36.06	37.57	36.63	35.42	34.06	25.25	25.13	24.74
MnO	0.09	0.11	0.11	0.10	0.10	0.10	0.10	0.09	0.10	0.10	0.11	0.10	0.11	0.13	0.10	0.11	0.09	0.09
CaO	0.42	0.42	0.64	0.43	0.63	0.53	0.42	0.30	0.38	0.70	0.89	0.59	0.63	0.71	1.25	12.75	11.84	13.05
Na_2O	0.09	nd	nd	nd	nd	nd	nd	nd	0.08	0.06	0.08	0.07	0.07	0.11	0.07	nd	nd	0.11
K_2O	0.01	nd	0.02	nd	0.00	nd	nd	nd	nd	nd	nd	nd	nd	nd	0.01	0.12	0.08	0.09
P_2O_5	nd	nd	nd	nd	0.01	0.01	nd	nd	0.01	0.01	0.02	0.01	0.01	0.01	0.01	0.02	0.02	0.01
LOI	18.25	16.13	16.02	15.74	15.87	16.22	16.78	16.24	16.94	16.41	16.32	16.52	16.33	16.25	17.07	19.62	19.24	18.96
Total	98.39	98.86	98.46	99.38	99.27	101.56	99.16	98.33	97.93	100.00	99.89	101.24	98.70	98.20	98.47	99.14	99.37	99.71
Trace elements (ppm)																		
Cr	2253	2592	2754	2433	2572	2312	2168	1937	2515	2279	2414	2595	2372	2584	2140	2421	2140	2753
Co	59.4	71.3	70.6	64.3	69.2	63.2	62.7	61.0	52.0	50.6	52.2	49.1	51.2	56.1	49.4	53.6	43.1	54.3
Ni	1837	2326	2266	2133	2233	2069	2048	2001	2127	2020	2058	2038	2057	2248	1899	1542	1598	1595
Cu	14.5	57.6	nd	1.3	nd	13.1	3.3	4.2	89.0	13.8	10.8	9.7	8.6	4.5	4.4	37.1	4.8	6.1
As	23.8	nd	nd	nd	nd	nd	nd	nd	16.8	10.5	nd	51.0	61.2	nd	18.8	nd	nd	nd
Ba	40.3	28.0	27.2	26.2	27.3	26.5	26.6	26.0	17.9	16.6	14.8	14.9	15.0	14.5	13.5	29.1	26.5	26.1
Pb	59.9	266.4	63.3	76.9	69.2	70.8	72.2	81.5	89.1	49.4	70.7	63.3	76.2	84.6	64.4	82.6	72.7	72.7
Sc	9.4	9.3	12.1	10.4	10.4	8.1	7.0	6.5	7.1	7.7	8.1	6.4	7.6	9.9	9.7	11.0	13.1	12.5
Sr	6.0	6.4	8.4	6.8	8.1	9.3	7.4	7.0	1.1	1.8	2.1	3.7	2.2	1.8	3.2	58.4	42.5	52.3
V	39.8	34.2	50.6	37.5	42.2	32.1	27.8	20.7	34.3	38.9	39.6	37.0	38.7	49.1	47.8	46.3	57.8	51.1
Y	2.9	2.5	2.6	2.1	2.5	2.2	2.1	1.7	1.9	1.7	2.3	1.4	1.9	2.4	2.6	4.0	3.9	3.3
Zn	36.5	113.1	46.4	23.2	34.1	37.3	36.5	35.7	36.1	40.1	42.0	39.7	35.1	33.9	35.4	60.7	56.0	54.7
Zr	5.2	nd	nd	4.0	2.1	1.7	0.9	nd	nd	nd	nd	nd	nd	nd	1.2	nd	nd	nd
Cd	0.2	0.2	0.5	0.5	0.5	0.6	0.4	0.4	0.6	0.4	0.8	0.7	0.3	0.4	0.8	0.1	0.3	0.3

LOI, loss on ignition; nd, not detected. P, peridotite; Pc, peridotite with carbonate veins; Px, pyroxenite; G, gabbro.

Table 1b. *Major and trace elements, Leg 173, Sites 1068A and 1070A (continued)*

Site:	1070A	1070A	1070A	1070A	1070A	1070A	1070A	1070A	1070A	1070A	1070A	1070A	1070A	1070A	1070A	1070A	1070A
Core:	8R	10R	11R	12R	12R	13R	13R	13R	14R	14R	14R	10R	9R	9R	9R	9R	9R
Interval (cm):	75–81	128–133	13–18	27–32	32–37	29–35	98–104	26–32	36–42	30–35	110–116	60–65	49–55	96–101	110–115	51–56	82–87
Rock type:	P?	F	P	P	P	P	P	P	P	P	P	Px	G	G	G	G	G
Major elements (wt %)																	
SiO$_2$	36.80	37.93	37.76	36.24	37.65	37.90	37.29	37.41	38.21	36.12	36.36	39.54	42.30	35.22	32.85	43.83	39.71
TiO$_2$	0.04	0.04	0.03	0.02	0.02	0.02	0.01	0.02	0.02	0.03	0.02	0.11	3.73	3.25	9.45	4.08	4.73
Al$_2$O$_3$	1.29	0.81	0.98	1.14	0.94	0.25	nd	0.56	0.31	0.75	0.57	5.37	11.19	15.68	13.78	18.37	10.77
Fe$_2$O$_3$	7.61	6.63	6.91	6.75	5.84	8.06	7.15	6.45	7.07	5.70	6.85	5.93	10.41	7.00	13.69	6.46	10.52
MgO	33.57	34.31	36.99	38.48	38.39	37.28	38.10	38.11	38.32	37.77	38.97	29.25	18.36	21.63	14.53	4.78	17.54
MnO	0.07	0.07	0.09	0.06	0.06	0.08	0.06	0.06	0.07	0.06	0.07	0.11	0.13	0.13	0.22	0.10	0.17
CaO	1.94	1.19	0.11	0.24	0.28	0.10	0.04	0.28	0.17	0.14	0.14	6.14	7.91	5.08	6.71	8.15	8.37
Na$_2$O	0.17	nd	nd	0.12	nd	nd	nd	nd	nd	0.12	0.12	nd	1.95	0.41	1.64	4.07	1.01
K$_2$O	0.01	nd	nd	nd	nd	nd	nd	nd	nd	nd	nd	nd	0.34	0.11	0.16	0.29	0.20
P$_2$O$_5$	0.01	nd	0.01	0.01	nd	nd	nd	nd	nd	0.01	0.01	nd	0.02	0.01	0.01	0.01	0.01
LOI	15.32	14.72	15.43	16.35	16.61	15.28	15.79	16.82	16.83	16.90	16.67	13.68	5.38	10.78	7.49	2.36	6.64
Total	97.27	96.18	98.75	99.93	100.30	99.34	98.79	100.16	101.50	98.06	100.16	100.79	101.88	99.38	100.74	92.62	99.82
Trace elements (ppm)																	
Cr	2243	2472	2251	2764	3017	1242	1034	2461	2617	2460	1408	4014	85	46	56	32	90
Co	38.9	59.2	58.9	49.6	60.7	68.8	68.9	57.8	65.4	48.3	52.2	47.9	49.8	36.1	76.8	33.9	55.0
Ni	1918	1907	1934	2016	1959	2227	2241	1925	2163	1995	2156	1863	184	112	208	123	174
Cu	6.5	7.7	17.2	22.1	12.3	nd	nd	4.8	nd	18.5	18.2	245.2	nd	nd	nd	49.5	nd
As	26.9	nd	nd	52.7	nd	nd	nd	23.8	nd	63.0	nd	nd	nd	45.1	113.5	51.8	27.5
Ba	14.4	24.9	25.6	15.2	27.1	24.0	20.4	22.2	19.6	14.6	15.7	19.9	89.2	52.1	60.0	107.3	64.6
Pb	65.7	88.2	92.9	57.5	64.7	80.8	76.3	60.7	70.4	54.4	87.0	70.0	55.9	84.4	59.5	56.4	46.7
Sc	8.8	9.2	11.0	6.6	8.7	5.2	4.2	7.6	5.9	5.6	5.7	41.1	76.3	29.0	43.5	22.7	69.1
Sr	5.5	6.4	0.5	nd	0.4	0.5	nd	0.9	0.9	nd	nd	13.5	119.7	143.2	669.7	481.2	348.1
V	39.4	39.9	44.8	36.6	39.8	20.9	14.5	32.6	25.8	28.0	26.6	271.2	701.0	314.0	617.1	258.6	507.4
Y	1.5	2.3	2.8	2.1	2.6	1.3	1.0	1.7	0.9	2.1	1.9	5.6	63.0	24.2	25.3	9.9	28.4
Zn	34.9	39.1	40.0	35.6	41.1	35.7	36.2	34.8	42.1	28.3	31.4	31.8	64.6	24.8	167.8	19.6	49.9
Zr	nd	1.3	1.8	nd	nd	nd	nd	1.0	0.8	nd	nd	1.7	81.1	39.3	99.2	37.5	61.0
Cd	0.8	0.6	0.4	0.3	0.4	0.7	0.9	0.7	1.1	0.5	0.6	1.2	1.4	nd	0.7	1.5	0.8

LOI, loss on ignition; nd, not detected. P, peridotite; Pc, peridotite with carbonate veins; Px, pyroxenite; G, gabbro.

Table 2a. *PGE Leg 173, Sites 1068A and 1070A*

Site:	1068A	1068A	1068A	1068A	1068A	1068A	1068A	1068A	1068A	1068A	1068A	1068A	1068A	1068A	1068A	1070A	1070A
Core:	22R	24R	24R	25R	25R	25R	26R	26R	27R	28R	28R	28R	29R	29R	29R	8R	8R
Interval (cm):	54–59	95–100	51–57	128–133	29–35	111–116	107–111	1–7	28–33	5–10	36–41	107–112	74–79	8–13	56–61	57–62	93–98
Rock type:	P	P	P	P	P	P	P	P	P	P	P	P	P	P	P	Pc	Pc
Ir	2.21	3.38	3.59	2.58	2.49	3.38	2.97	3.34	2.98	3.03	2.29	3.86	3.07	2.29	1.49	3.95	1.24
Ru	4.53	7.03	7.36	4.66	4.73	5.98	4.94	6.09	6.21	5.38	4.25	6.71	5.51	4.17	2.98	5.79	2.22
Rh	0.83	1.27	1.69	0.86	0.86	1.21	0.79	1.04	1.11	0.99	0.76	1.19	1.05	0.89	0.57	0.94	0.29
Pt	1.74	3.28	6.58	3.58	2.35	3.94	2.75	4.61	4.56	4.01	2.77	11.73	3.26	3.51	2.57	6.34	1.00
Pd	1.27	2.44	5.21	2.69	0.74	2.74	1.72	0.55	2.64	3.15	2.00	2.34	2.51	2.63	1.57	1.63	0.70
Sum PGE	10.59	17.39	24.43	14.38	11.17	17.25	13.17	15.63	17.49	16.55	12.07	25.83	15.39	13.49	9.19	18.65	5.44
Pd/Ir	0.58	0.72	1.45	1.04	0.30	0.81	0.58	0.17	0.89	1.04	0.87	0.61	0.82	1.15	1.06	0.41	0.56
Pt/Ir	0.79	0.97	1.83	1.39	0.95	1.17	0.93	1.38	1.53	1.32	1.21	3.04	1.06	1.53	1.73	1.61	0.81
Pt/Pd	1.37	1.34	1.26	1.33	3.19	1.44	1.61	8.33	1.72	1.27	1.39	5.01	1.30	1.33	1.63	3.88	1.43
Ru/Ir	2.05	2.08	2.05	1.80	1.90	1.77	1.66	1.82	2.08	1.77	1.86	1.74	1.79	1.82	2.00	1.47	1.79
Rh/Ir	0.38	0.38	0.47	0.33	0.35	0.36	0.27	0.31	0.37	0.33	0.33	0.31	0.34	0.39	0.39	0.24	0.23
Rh/Pd	0.66	0.52	0.32	0.32	1.17	0.44	0.46	1.88	0.42	0.31	0.38	0.51	0.42	0.34	0.36	0.58	0.41
Rh/Pt	5.43	5.54	4.36	5.42	5.50	4.92	6.23	5.85	5.61	5.43	5.58	5.65	5.27	4.66	5.20	6.15	7.65

P, peridotite; Pc, peridotite with carbonate veins; Px, pyroxenite; G, gabbro.

Table 2b. *PGE Leg 173, Sites 1068A and 1070A (continued)*

Site:	1070A	1070A	1070A	1070A	1070A	1070A	1070A	1070A	1070A	1070A	1070A	1070A	1070A	1070A	1070A	1070A	1070A
Core:	8R	8R	10R	11R	12R	12R	13R	13R	13R	14R	14R	14R	10R	9R	9R	9R	9R
Interval (cm):	39–45	75–81	128–133	13–18	27–32	32–37	29–35	98–104	26–32	36–42	30–35	110–116	60–65	49–55	110–115	51–56	82–87
Rock type:	Pc	P?	P	P	P	P	P	P	P	P	P	P	Px	G	G	G	G
Ir	1.88	2.49	3.74	2.95	2.05	2.00	1.42	2.21	2.57	2.31	2.36	3.19	0.52	0.12	0.10	0.05	0.06
Ru	2.96	4.21	4.93	5.10	3.02	3.13	2.59	4.37	4.02	3.98	4.20	5.30	1.31	0.04	0.12	0.09	0.03
Rh	0.40	0.64	0.93	0.80	0.59	0.58	0.38	0.57	0.67	0.57	0.67	0.78	0.47	0.07	0.07	0.05	0.01
Pt								4.61	4.56	4.01	2.77	11.73	3.26	3.51	2.57	6.34	1.00
	2.17	2.64	5.27	3.62	1.89	2.72	1.32	10.80	3.69	0.82	1.78	1.13	5.33	0.01	0.30	0.05	1.33
Pd	1.24	1.13	2.75	2.38	1.37	1.19	0.60	1.04	1.44	0.59	1.07	0.83	10.33	0.08	0.13	0.15	0.01
Sum PGE	8.66	11.10	17.61	14.84	8.92	9.62	6.30	19.00	12.39	8.27	10.08	11.23	17.95	0.32	0.73	0.39	1.44
Pd/Ir	0.66	0.45	0.74	0.81	0.67	0.60	0.42	0.47	0.56	0.26	0.46	0.26	19.96	0.68	1.23	2.84	0.17
Pt/Ir	1.16	1.06	1.41	1.23	0.92	1.36	0.93	4.88	1.44	0.36	0.75	0.35	10.30	0.08	2.89	1.01	22.51
Pt/Pd	1.75	2.34	1.92	1.52	1.38	2.28	2.20	10.38	2.56	1.39	1.65	1.36	0.52	0.12	2.34	0.36	132.91
Ru/Ir	1.58	1.69	1.32	1.73	1.48	1.57	1.82	1.97	1.57	1.73	1.78	1.66	2.52	0.30	1.19	1.81	0.49
Rh/Ir	0.21	0.26	0.25	0.27	0.29	0.29	0.27	0.26	0.26	0.25	0.28	0.25	0.90	0.57	0.70	0.87	0.17
Rh/Pd	0.32	0.56	0.34	0.34	0.43	0.49	0.63	0.55	0.46	0.97	0.62	0.94	0.05	0.84	0.56	0.31	1.00
Rh/Pt	0.18	0.24	0.18	0.22	0.31	0.21	0.29	0.05	0.18	0.70	0.37	0.69	0.09	6.86	0.24	0.86	0.01

P, peridotite; Pc, peridotite with carbonate veins; Px, pyroxenite; G, gabbro.

suggest that Platinum-like platinum group elements (Rh, Pt) and Iridium-like platinum group elements (Ir, Ru) have similar behaviour and are hosted in the same phases, in agreement with Platinum group minerals observations. General positive covariations are observed between Al_2O_3 and CaO concentrations (Fig. 16) and PI index (primitiveness index, i.e. $(SiO_2 + Al_2O_3 + CaO + FeO)/(FeO + MgO)$ molar; Garuti *et al.* 1997) and Pd/Ir ratios.

Pyroxenite and gabbros

The only pyroxenite sample is distinctive in many aspects. It shows the highest total PGE (27.97 ppb), the highest Pd/Ir (19.87) and Pt/Ir (10.25) ratios, and the lowest Rh/Pd ratios of all analysed samples. The gabbro samples show very low PGE contents (0.32–1.44 ppb), low Pd/Ir ratios (1.3), and low Pd values (0.15 ppb) that are close to the detection limits. On a mantle-normalized diagram for samples the gabbros show very low PGE content and a U-shaped pattern (Fig. 14). Pyroxenite shows 10–100 times enrichment when compared with gabbros, and is particularly enriched in Pd and Cu. These results show that magmatic pyroxene and accumulation process are the major controls on geochemistry of peridotites.

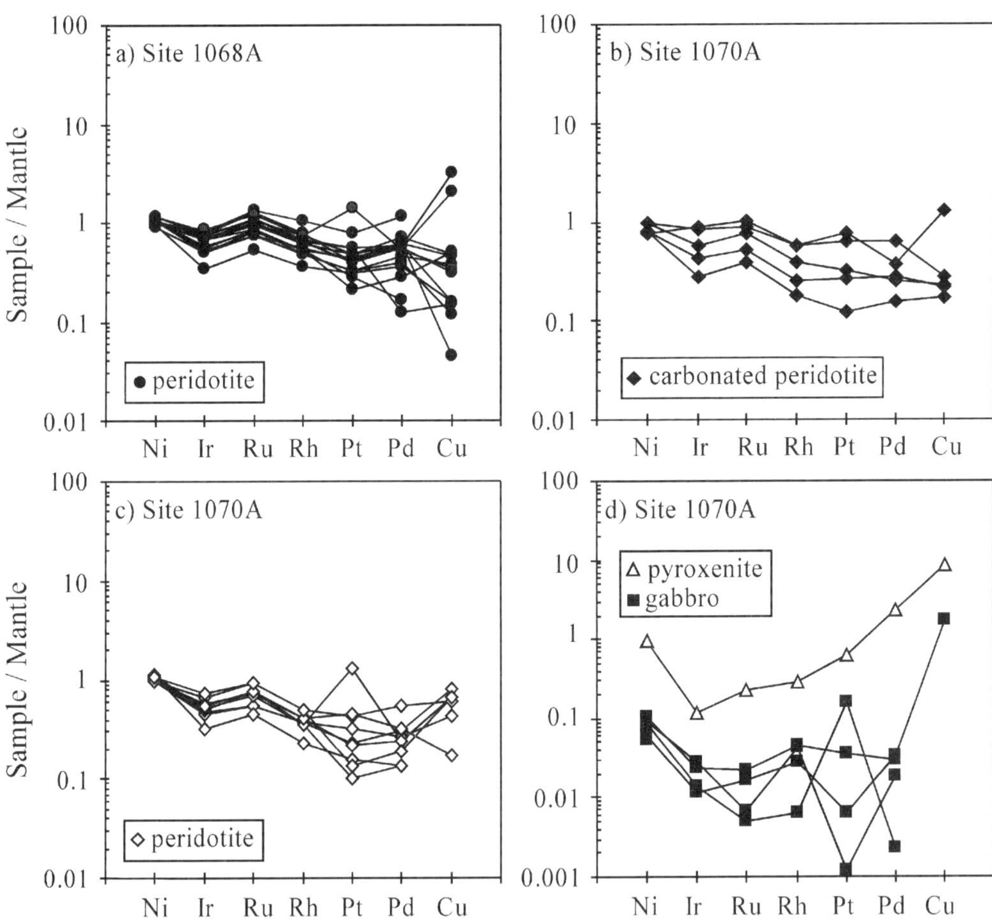

Fig. 14. Mantle-normalized Ni–PGE–Cu patterns for all samples of Hole 1068A and 1070A peridotites. (**a**) Hole 1068A; (**b**) carbonate-rich peridotites from Hole 1070A; (**c**) carbonate-free peridotites from Hole 1070A; (**d**) pyroxenite (open symbol) and gabbros (closed symbols).

Table 3. *Pd and Pd/Ir mean values, Leg 173, Sites 1068A and 1070A, and other massifs*

		n	Minimum	Maximum	Average	References
1. Peridotites from Sites 1068A and 1070A (Leg 173)	Pd	29	0.55	5.21	1.8	This work
	Pd/Ir	29	0.17	1.45	0.66	
2. Peridotites from Site 1068A (Leg 173)	Pd	15	0.55	5.21	2.28	This work
	Pd/Ir	15	0.17	1.45	0.8	
3. Peridotites from Site 1070A (Leg 173)	Pd	14	0.59	2.75	1.28	This work
Pd/Ir		14	0.26	0.81	0.52	
4. Peridotites (Ronda, Spain)	Pd	17	0.66	5.93	3.76	Gueddari *et al.* (1996)
Pd/Ir		17	0.33	2.39	1.37	
5. Peridotites (Beni Bousera, Morocco)	Pd	11	0.5	6.7	3.6	Gueddari *et al.* (1996)
Pd/Ir		11	0.23	3.37	1.65	
6. Cpx-poor lherzolites and harzburgites (Mont Albert, Québec)	Pd	17	1.82	7.71	4.63	Gueddari *et al.* (1999)
	Pd/Ir	17	0.52	2.16	1.25	
7. Peridotites (Baldissero, Balmuccia and Finero, Italy)	Pd	23	0.5	13.5	5.4	Garuti *et al.* (1984)
8. Peridotites (Baldissero, Italy)	Pd	3	6.7	9.5	8.5	Garuti *et al.* (1997)
	Pd/Ir	3	1.89	2.17	2.03	
9. Lherzolites (Balmuccia, Italy)	Pd	4	9.12	13.4	11.53	Garuti *et al.* (1997)
	Pd/Ir	4	2.74	5.85	4.59	
10. Harzburgites–dunites (Finero, Italy)	Pd	6	0.88	9.46	4.44	Garuti *et al.* (1997)
	Pd/Ir	6	0.26	5.11	1.81	
11. Lherzolites (Lanzo, Italy)	Pd	24	3.5	33.3	12.59	Lorand *et al.* (1993)
	Pd/Ir	24	1.9	16.05	4.8	
12. Lherzolites (Pyrenean massifs, France)	Pd	14	4	6.9	5.38	Pattou *et al.* (1996)
	Pd/Ir	14	1.48	2.14	1.8	
13. Massive peridotites (Pyrenean massifs, France)	Pd	8	8	19	12.75	Lorand (1989)
	Pd/Ir	8	2.35	3.87	2.51	
14. Spinel peridotite xenoliths (French Massif Central, France)	Pd	38	0.21	7.55	2.85	Gueddari (1996)
	Pd/Ir	38	0.26	3.81	1.02	
15. Harzburgites (Thetford Mine ophiolite, Québec)	Pd	7	0.45	10	3.8	Oshin & Crocket (1982)
	Pd/Ir	7	0.21	3.03	1.17	
16. Harzburgites (Lewis Hill massif, Bay of Islands, Newfoundland)	Pd	2	3.87	8.16	6.01	Edwards (1990)
	Pd/Ir	2	1.24	1.71	1.47	
17 Harzburgites, North Arm Mountain massif (Bay of Islands, Newfoundland)	Pd	12	2.3	20.4	7.03	Pagé *et al.* (1998); Pagé (unpublished)
	Pd/Ir	12	0.97	10.73	4.3	
18. Harzburgites–dunites, Garrett Transform Fault, East Pacific Rise	Pd	6	2.8	10.4	5.53	Pagé *et al.* (1998); Pagé (unpublished)
	Pd/Ir	6	1.36	17.62	5.69	

n, number of samples.

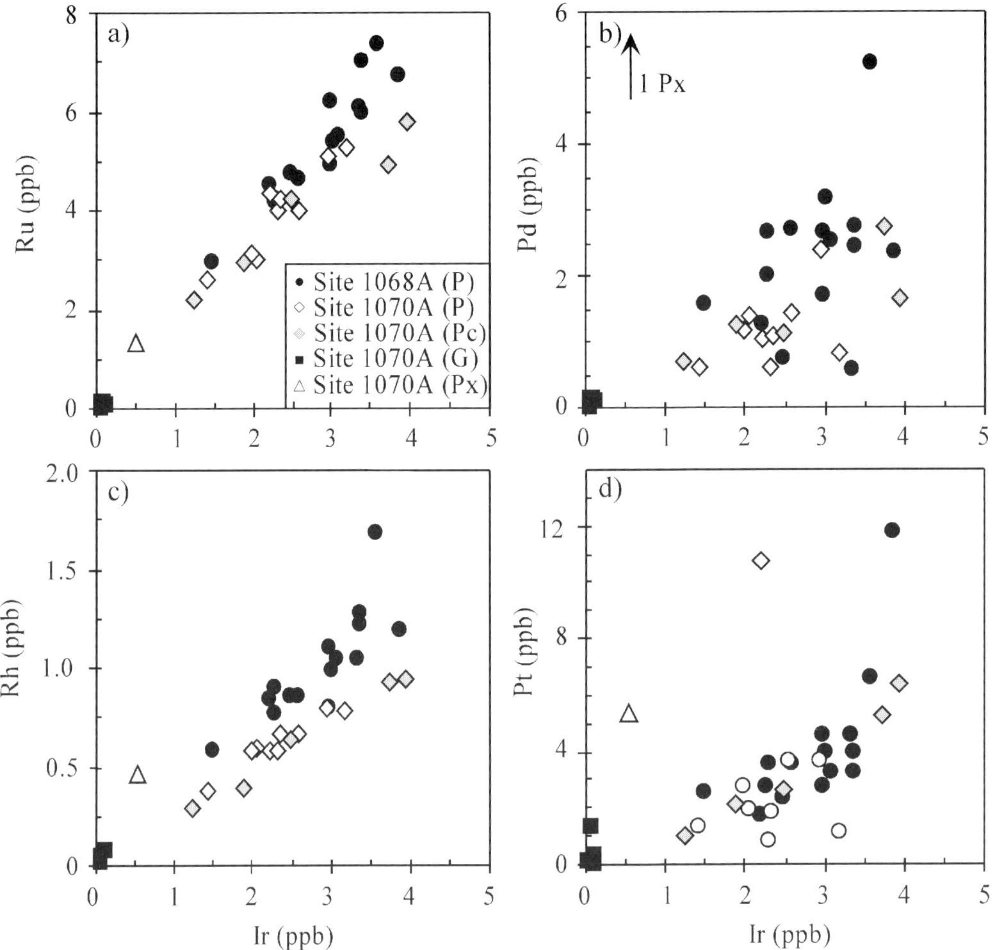

Fig. 15. PGE interelemental covariations (in ppb) for Leg 173 peridotites, pyroxenite and gabbros. (**a**) Ru v. Ir; (**b**) Pd v. Ir; (**c**) Rh v. Ir; (**d**) Pt v. Ir. G, gabbro; P, peridotite: Pc, carbonate-bearing peridotite; Px, pyroxenite.

Discussion

The relatively low TiO_2, Al_2O_3 and CaO (excluding carbonatized samples) contents are indicative of residual nature of Leg 173 peridotites.

PGE contents of mantle peridotites have been reported from several locations including the lherzolite type-locality (Bégou *et al.* 1989; Lorand 1989), ophiolitic massifs (Monte Maggiore, Ohnenstetter 1992; Bay of Islands, Edwards 1990; Pagé *et al.* 1998; Pagé unpublished; Thetford Mines, Oshin & Crocket 1982; Tanguay *et al.* 1990; Leka, Pedersen *et al.* 1993); Pindos, Economou-Eliopoulos & Via-

condios 1995; Kempirsai, Melcher *et al.* 1997), orogenic massifs (Finero massif, Garuti *et al.* 1995; Ronda and Beni Bousera, Leblanc *et al.* 1990; Gueddari *et al.* 1998), and in numerous publications on ore-grade massifs such the Bushveld, Stillwater and Great Dyke; see Naldrett (1989) for a review), and a few from the oceanic domain, namely, Leg 147 (Hess Deep, Pritchard *et al.* 1996), Leg 37 (Crocket & Teruta 1977) and the Garrett Transform Fault (Pagé *et al.* 1998; Pagé unpublished).

Hole 1068A and 1070A peridotites are characterized by relatively low Pd content, low Pd/Ir and relatively low total PGE contents when compared with peridotites from both con-

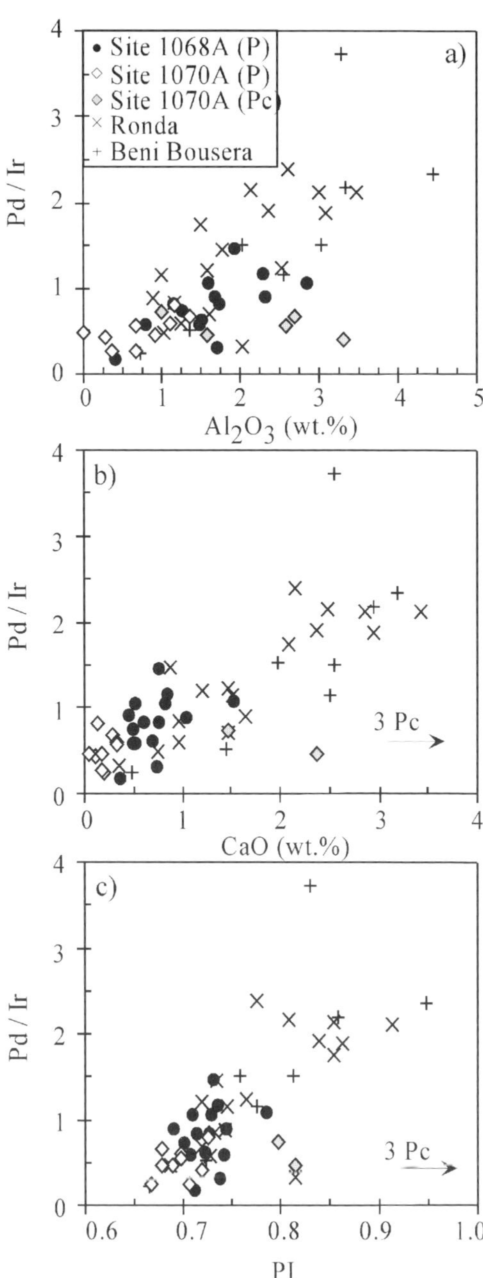

Fig. 16. Variation diagrams of Pd/Ir ratios as a function of (**a**) Al₂O₃ wt %, (**b**) CaO wt % and (**c**) PI (primitiveness index) in Hole 1068A and 1070A peridotites (P). Pc, carbonate-rich peridotites from Hole 1070A. Data from Ronda and Beni Bousera peridotites after Gueddari *et al.* (1996).

tinental and oceanic environments (Fig. 17; Tables 2 and 3). The PGE abundances, Pd/Ir and total PGE are generally comparable with values found in Ronda and Beni Bousera highly depleted peridotites, i.e. having $<1.7\%$ CaO (Gueddari *et al.* 1996). This covariation is suggestive of partial melting being a major control on PGE distribution. Thus regarding CaO, Al₂O₃, PI and PGE contents, the Leg 173 rocks are considered to be severely depleted upper-mantle peridotites. PGE contents are known to be little affected by alteration processes, as suggested by experimental studies of Wood *et al.* (1989); Lorand *et al.* (1993). Thus the analytical PGE data reflect magmatic characteristics. The PGE content shown in Table 2 and Figure 17 is similar to that of other occurrences such as Monte Maggiore, Corsica (Ohnenstetter 1992), Ronda (moderate to very low PGE contents, Gueddari *et al.* 1998), the Bay of Islands ophiolite (Edwards 1990; Pagé unpublished), and the Shetland ophiolite (Pritchard *et al.* 1986). The high PGE contents found in pyroxenite are similar to results from the Shetland ophiolite (Pritchard *et al.* 1986). In the Garrett Transform Fault the Pd/Ir ratio averages 5.69 (1.36–17.62) and total PGE sums to 7–27.6 ppb in mantle peridotites and 3 ppb in gabbros (Pagé *et al.* 1998; Pagé unpublished). These results are slightly higher than those for Leg 173 peridotites (see also Table 3). PGE depletion is less pronounced at least for fast-spreading generated peridotites. The 1070A pyroxenite, which has a relatively high Pd content and Pd/Ir ratio, is unlikely to be representative of a circulating magma and is probably the result of accumulation of pyroxene, which could explain the high Pd/Ir ratio. Even though Leg 173 peridotites have very low sulphide content, the PGM should be mostly associated with S-bearing compounds as noted in nearby similar upper-mantle material from Lherz (Bégou 1989), Ronda and Beni Bousera (Gueddari *et al.* 1996) or from xenoliths enclosed in alkalic basalt (Mitchell & Keays 1981; Gueddari 1996). Recently, Scharer *et al.* (1999); Charpentier *et al.* (1999) suggested respectively a lithospheric origin for Galicia Bank peridotites and an asthenospheric origin for Galicia Bank and Gorringe Bank gabbros. Geochemical data, such as PGE content, for the Leg 173 peridotites suggest a complex evolution for the depleted subcontinental lithospheric mantle. This interpretation is consistent with the geochronological results of Turrin (1999) on 1070A gabbros.

Interelemental correlation also suggests that the PGE are enclosed in the same phases. This

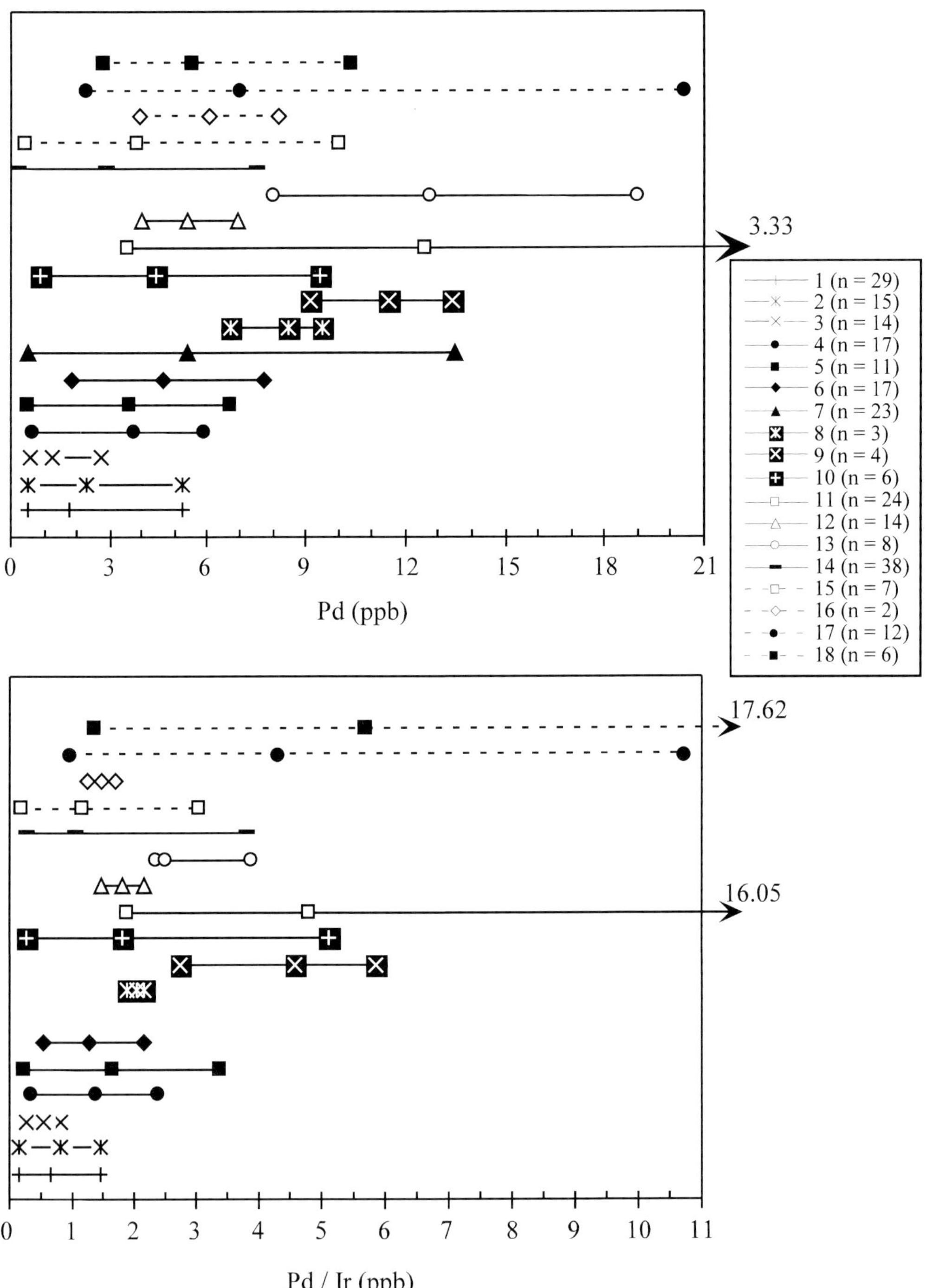

Fig. 17. Mean values and total range for Pd contents and Pd/Ir ratios for Hole 1068A and 1070A peridotites and for ophiolitic and subcontinental peridotites. *n*, number of samples. 1, Peridotites from Holes 1068A and 1070A (this work); 2, peridotites from Hole 1068 (this work); 3, peridotites from Hole 1070A (this work); 4 and 5, peridotites from Ronda and Beni Bousera, respectively (Gueddari *et al.* 1996); 6, Cpx-poor lherzolites and harzburgites from Mont Albert massif, Québec (Gueddari *et al.* 1999); 7, peridotites from Baldiserro, Bal-

study has shown that PGE are unlikely to be associated with silicates and spinel. This is in agreement with our preliminary SEM results showing that Cu–Pt–Pd and Cu– (Pd–Pt) grains are associated with a Ni–Fe sulphide phase. On the other hand, the correlations between PGE, Ni and Cr are not fully understood. This non-ideal behaviour of PGE in the partial melting scheme outlined above reflects complex extraction and redistribution of PGE during subcontinental and subsequent exhumation at the onset of rifting. Modelled ϵNd values suggest depleted to moderately depleted N-MORB sources for Galicia Bank and Gorringe Bank gabbros (Seifert & Brunotte 1996). Our results suggest that Leg 173 gabbros are derived from E-MORB magma because of the inferred water and K (occurrence of magmatic amphibole) contents and high Na content of clinopyroxene, and low total PGE content. The magmatic ages range from 140 Ma (Gorringe Bank) to 122 Ma (Galicia Bank) and 120 Ma (Leg 173 gabbros). Thus it is likely that a depleted mantle source was involved during production of the magma batches through time. However, we believe that the oceanic asthenospheric mantle was the main reservoir that was variably melted and percolated before oceanic drifting producing pyroxenite and gabbro intrusive rocks. According to geochronological constraints, melting and percolation occurred soon before oceanic accretion. These conditions are likely to be realized in a very slow spreading environment.

Contrasting Sr and Ba contents were outlined between highly hydrated peridotites and metamorphosed gabbros. We believe that this behaviour is related to the particular mineral composition of gabbros, containing high modal plagioclase, now partly replaced by aluminous low-grade metamorphic phases. This interpretation makes it unnecessary that the peridotites and the gabbros have been altered by different fluids. The fluids circulated at very low temperatures during the exhumation period, allowing sea water to penetrate the rising upper mantle.

Conclusion

Hole 1068A and 1070A peridotites show contrasting petrographic characteristics. Mineral chemistry, especially the very large spectrum of spinel composition, occurrence of pyroxenite, and gabbroic veins in 1070A peridotites, as well as interstitial undeformed clinopyroxene, suggest percolation of magmas. The averaged spinel composition is Cr-number 0.25 for 1068A peridotites, whereas a Cr-number of 0.50 is observed in 1070A peridotites. TiO_2 content up to 2.2 wt % is higher close to gabbro intrusions. Hole 1068A and 1070A peridotites, pyroxenite and gabbros are products of partial melting of continental-type mantle that was percolated by K-rich, Na-rich and water-rich E-MORB-type magmas having lower Fo content in olivine, lower Mg-number in pyroxenes, very low PGE, Pd and S contents, and very low Pd/Ir ratios. The Leg 173 peridotites have geochemical characteristics similar to depleted Ronda and Beni Bousera peridotite end-members. The E-MORB magmas were generated in a rifting environment and percolated into the asthenospheric mantle just before the oceanic accretion. The retrograde metamorphic evolution of the gabbroic veins is compatible with progressive cooling and associated exhumation (Beslier *et al.* 1993, 1995; Kornprobst *et al.* 1981; Naldrett & Duke 1980; Pickup *et al.* 1996; Scharer *et al.* 1995; Seifert *et al.* 1997).

We thank the Captain and crew of the Joides Resolution for their help during Leg 173 drilling operations. R. Hébert benefited from NSERC grant 1253. P. Pagé, V. Varfalvy, F.-D. Cloutier and E. Giguère are acknowledged for discussions, preparation of figures and sharing of unpublished pertinent data. This paper is dedicated to the first author's mother Aurore and brother Edgar, who passed away on 31 October and 6 November 2000, respectively, during the period of final polishing of this paper.

References

AGRINIER, P., MÉVEL, C. & GIRARDEAU, L. 1988. Hydrothermal alteration of the peridotite cored

---

muccia and Finero (Garuti *et al.* 1984); 8, peridotites from Baldissero (Garuti *et al.* 1997); 9, lherzolites from Balmuccia (Garuti *et al.* 1997); 10, harzburgites– dunites from Finero (Garuti *et al.* 1997); 11, lherzolites from Lanzo (Lorand *et al.* 1993); 12, lherzolites from Pyrenean massifs (Pattou *et al.* 1996); 13, massive peridotites from Pyrenean massifs (Lorand 1989); 14, peridotite xenoliths from alkaline basalts, French Massif Central (Gueddari 1996); 15, harzburgites from Thetford Mines (Oshin & Crocket 1982); 16, harzburgites from Lewis Hill massif, Bay of Islands ophiolite (Edwards 1990); 17, harzburgites from North Arm Mountain massif, Bay of Islands ophiolite (Pagé *et al.* 1998; Pagé unpublished); 18, harzburgites–dunites from Garrett Transform Fault (Pagé *et al.* 1998; Pagé unpublished).

at the ocean–continent boundary of the Iberian margin: petrologic and stable isotope evidence. *In*: BOILLOT, G., WINTERER, E.L. & MEYER, P.S. (eds) *Proceedings of the Ocean Drilling Program, Scientific Results, 103*. Ocean Drilling Program, College Station, TX, 225–234.

BARNES, S.-J., BOYD, R., KORNELIUSSEN, A., NILSSON, L.P., OFTEN, M., PEDERSEN, R.-B. & ROBINS, B. 1988. The use of mantle normalization and metal ratios in discriminating between effects of partial melting, crystal fractionation and sulphide segregation of platinum-group elements, gold, nickel and copper: examples from Norway. *In*: PRITCHARD, H.M., POTTS, P.J., BOWLES, J.F.W. & CRIBB, S.J. (eds) *Geoplatinum 87*. Elsevier, Amsterdam, 113–143.

BÉGOU, P. 1989. *Distribution des éléments du groupe du platine et de l'or dans les roches basiques et ultrabasiques. Approche de leur comportement géochimique orthomagmatique pendant les phénomènes de fusion partielle et de cristallisation fractionnée*. PhD thesis, Joseph Fourier University, Grenoble.

BÉGOU, P., AMOSSÉ, J., FISHER, W. & PIBOULE, M. 1989. Distribution des éléments du groupe du platine (PGE) dans les péridotites massives à spinelle de l'étang de Lherz (Ariège, France): résultats préliminaires. *Comptes Rendus de l'Académie des Sciences, Série II*, **309**, 1177–1183.

BESLIER, M.-O., ASK, M. & BOILLOT, G. 1993. Ocean–continent boundary in the Iberia abyssal plain from multichannel seismic data. *Tectonophysics*, **218**, 383–393.

BESLIER, M.-O., BITRI, A. & BOILLOT, G. 1995. Structure de la transition continent–océan d'une marge passive: sismique réflection multitrace dans la Plaine Abyssale Ibérique (Portugal). *Comptes Rendus de l'Académie des Sciences, Série IIa*, **320**, 969–976.

BESLIER, M.-O., CORNEN, G. & GIRARDEAU, J. 1996. Tectono-metamorphic evolution of the peridotites from the ocean/continent transition of the Iberia Abyssal Plain margin. *In*: WHITMARSH, R.B., SAWYER, D.S., KLAUS, A. & MASSON, D.G. (eds) *Proceeding of the Ocean Drilling Program, Scientific Results, 149*. Ocean Drilling Program, College Station, TX, 397–412.

BESLIER, M.-O., GIRARDEAU, J. & BOILLOT, G. 1990. Kinematics of peridotite emplacement during North Atlantic continental rifting, Galicia, northwestern Spain. *Tectonophysics*, **184**, 321–343.

BOILLOT, G., AGRINIER, P., BESLIER, M.-O., CORNEN, G. *et al.* 1995a. A lithospheric syn-rift shear zone at the ocean–continent transition: preliminary results of the Galinaute II cruise (Nautile dives on the Galicia Bank, Spain). *Comptes rendus de l'Académie des Sciences*, **321**, 1171–1178.

BOILLOT, G., BESLIER, M.-O., KRAWCZYK, C.M., RAPPIN, D. & RESTON, T.J. 1995b. The formation of passive margins: constraints from the crustal structure and segmentation of the deep Galicia margin. *In*: SCRUTTON, R.A., STOKER, M.S., SHIMMIELD, G.B. & TUDHOPE, A.W. (eds) *The Tectonics, Sedimentation and Palaeoceanography of the North-Atlantic Region*. Geological Society, London, Special Publications, **90**, 71–90.

BOILLOT, G., FÉRAUD, G., RECQ, M. & GIRARDEAU, J. 1989. 'Undercrusting' by serpentinite beneath rifted margins. *Nature*, **341**, 523–525.

BOILLOT, G., GRIMAUD, S., MAUFFRET, A., MOUGENOT, D., KORNPROBST, J., MERGOIL-DANIEL, J. & TORRENT, G. 1980. Ocean–continent boundary off the Iberian margin: a serpentinite diapir west of the Galicia Bank. *Earth and Planetary Science Letters*, **48**, 23–34.

BOILLOT, G., WINTERER, E.L. & MEYER, P.S. *Proceedings of the Ocean Drilling Program, Scientific Results, 103*. Ocean Drilling Program, College Station, TX.

CANNAT, M. & CASEY, J. 1995. An ultramafic lift at the Mid-Atlantic Ridge: successive stages of magmatism in serpentinized peridotites from the 15°N region. *In*: VISSERS, R.L.M. & NICOLAS, A. (eds) *Mantle and Lower Crust Exposed in Oceanic Ridges and in Ophiolites*. Kluwer, Dordrecht, 5–34.

CANNAT, M., BIDEAU, D. & HÉBERT, R. 1990. Plastic deformation and magmatic impregnation in serpentinized ultramafic rocks from the Garrett Transform Fault (East Pacific Rise). *Earth and Planetary Science Letters*, **101**, 216–232.

CHARPENTIER, S.J., KORNPROBST, J., CHAZOT, G., BOTTAZZI, P., VANNUCCI, R. & LUAIS, B. 1999. *Dénudation du manteau lithosphérique sous-continental au cours de l'océanisation: exemple de la Marge passive de Galice (Espagne)*. Société Géologique de France, Paris.

CHARPENTIER, S.J., KORNPROBST, G., CHAZOT, G., CORNEN, G. & BOILLOT, G. 1998. Interaction entre lithosphère et asthénosphère au cours de l'ouverture océanique: données isotopiques préliminaires sur la marge de Galice (Atlantique Nord). *Comptes Rendus de l'Académie des Sciences*, **326**, 757–762.

CONSTANTIN, M., HEKINIAN, R., BIDEAU, D. & HÉBERT, R. 1995. Construction of the oceanic lithosphere by magmatic intrusions: petrological evidence from plutonic rocks formed along fast-spreading East Pacific Rise. *Geology*, **24**, 731–734.

CORNEN, G., BESLIER, M.-O. & GIRARDEAU, J. 1996a. Petrologic characteristics of the ultramafic rocks from the ocean/continent transition in the Iberia Abyssal Plain. *In*: WHITMARSH, R.B., SAWYER, D.S., KLAUS, A. & MASSON, D.G. (eds) *Proceedings of the Ocean Drilling Program, Scientific Results, 149*. Ocean Drilling Program, College Station, TX, 377–387.

CORNEN, G., BESLIER, M.-O. & GIRARDEAU, J. 1996b. Petrology of the mafic rocks from the ocean/continent transition in the Iberia Abyssal Plain. *In*: WHITMARSH, R.B., SAWYER, D.S., KLAUS, A. & MASSON, D.G. (eds) *Proceedings of the Ocean Drilling Program, Scientific*

Results, 149. Ocean Drilling Program, College Station, TX, 449–469.

CROCKET, J.H. & TERUTA, Y. 1977. Determination of palladium, iridium, and gold in mafic and ultramafic rocks drilled from the Mid-Atlantic Ridge, DSDP Leg 37. *In*: AUMENTO, F., MELSON, W.G. & ROBINSON, P.T. (eds) *Deep Sea Drilling Project, 37.* US Government Printing Office, Washington, DC, 577–579.

ECONOMOU-ELIOPOULOS, M. & VIACONDIOS, I. 1995. Geochemistry of chromitites and host rocks from the Pindos ophiolite complex, northwestern Greece. *Chemical Geology*, **122**, 99–108.

EDWARDS, S.J. 1990. Harzburgites and refractory melts in the Lewis Hills massif, Bay of Islands ophiolite complex: the base-metals and precious metals story. *Canadian Mineralogist*, **28**, 537–552.

EVANS, C.A. & GIRARDEAU, J. 1988. Galicia margin peridotites: undepleted abyssal peridotites from the North Atlantic. *In*: BOILLOT, G., WINTERER, E.L. & MEYER, P.S. (eds) *Proceedings of the Ocean Drilling Program, Scientific Results, 103.* Ocean Drilling Program, College Station, TX, 209–223.

FÉRAUD, G., BESLIER, M.-O. & CORNEN, G. 1996. $^{40}Ar/^{39}Ar$ dating of gabbros from the ocean/ continent transition of the western Iberia margin: preliminary results. *In*: WHITMARSH, R.B., SAWYER, D.S., KLAUS, A. & MASSON, D.G. (eds) *Proceedings of the Ocean Drilling Program, Scientific Results, 149.* Ocean Drilling Program, College Station, TX, 489–495.

GARUTI, G., GAZZOTTI, M. & TORRES-RUIZ, J. 1995. Iridium, rhodium, and platinum sulphides in chromitites from ultramafic massifs of Finero, Italy, and Ojén, Spain. *Canadian Mineralogist*, **33**, 509–520.

GARUTI, G., GORGONI, C. & SIGHINOLFI, G.P. 1984. Sulfide mineralogy and chalcophile and siderophile elements abundances in the Ivrea-Verbano mantle peridotites (Western Italian Alps). *Earth and Planetary Science Letters*, **70**, 69–87.

GARUTI, G., ODDONE, M. & TORREZ-RUIZ, J. 1997. Platinum-group element distribution in subcontinental mantle: evidence from the Ivrea zone (Italy) and the Betic–Rifean cordillera (Spain and Morocco). *Canadian Journal of Earth Sciences*, **34**, 444–463.

GIRARDEAU, J., EVANS, C.A. & BESLIER, M.-O. 1988. Structural analysis of plagioclase-bearing peridotites emplaced at the end of continental rifting: Hole 637A, ODP Leg 103 on the Galicia margin. *In*: BOILLOT, G., WINTERER, E.L. & MEYER, P.S. (eds) *Proceedings of the Ocean Drilling Program, Scientific Results, 103.* Ocean Drilling Program, College Station, TX, 209–223.

GROVE, T.L. & BYRAN, W.B. 1983. Fractionation of pyroxene phyric MORB at low pressure: an experimental study. *Contributions to Mineralogy and Petrology*, **84**, 293–309.

GUEDDARI, K. 1996. *Approche géochimique et physico-chimique de la différenciation des éléments du groupe du platine (PGE) et de l'or dans le manteau supérieur bético-rifain et dans les xénolites de péridotites sous-continentales.* PhD thesis, Joseph Fourier University, Grenoble.

GUEDDARI, K., LAFLÈCHE, M.R. & PIBOULE, M. 1998. Evidence of two contrasting garnet pyroxenite group in Ronda ultramafic Massif (Betic Range, Spain). *Geological Association of Canada, Mineralogical Association of Canada, Association profesionnelle des géologues et géophysiciens du Québec, International Association of Hydrogeologists, Canadian Geophysical Union, Annual Meeting, Québec City, Abstracts Volume*, **23**, 69.

GUEDDARI, K., LA FLÈCHE, M.R. & CAMIRÉ, G. 1999. Premières donées sur la géochimie des éléments du groupe du platine (EGP) dans les péridotites du Mont Albert (Québec). *Comptes-Rendus Académie des Sciences de Paris*, **329**, 479–486.

GUEDDARI, K., PIBOULE, M. & AMOSSÉ, J. 1996. Differentiation of platinum-group elements (PGE) and of gold during partial melting of peridotites in the lherzolitic massifs of the Betico-Rifean range (Ronda and Beni Bousera). *Chemical Geology*, **134**, 181–197.

HÉBERT, R. & CONSTANTIN, M. 1991. Petrology of hydrothermal metamorphism of oceanic Layer 3: implications for sulfide parageneses and redistribution. *Economic Geology*, **86**, 472–485.

HÉBERT, R., SERRI, G. & HEKINIAN, R. 1989. Mineral chemistry of ultramafic tectonites and ultramafic to gabbroic cumulates from the major oceanic basins and Northern Apennine ophiolites (Italy): a comparison. *Chemical Geology*, **77**, 183–207.

KORNPROBST, J. & TABIT, A. 1988. Plagioclase-bearing ultramafic tectonites from the Galicia margin (Leg 103, Site 637): comparison of their origin and evolution with low-pressure ultramafic bodies in Western Europe. *In*: BOILLOT, G., WINTERER, E.L. & MEYER, P.S. (eds) *Proceedings of the Ocean Drilling Program, Scientific Results, 103.* Ocean Drilling Program, College Station, TX, 253–263.

KORNPROBST, J., OHNENSTETTER, D. & OHNENSTETTER, M. 1981. Na and Cr contents in clinopyroxenes from peridotites: a possible discriminant between sub-continental and sub-oceanic mantle. *Earth and Planetary Science Letters*, **53**, 241–254.

KRAWCZYK, C.M., RESTON, T.J., BESLIER, M.-O. & BOILLOT, G. 1996. Evidence for detachment tectonics on the Iberia Abyssal Plain rifted margin. *In*: WHITMARSH, R.B., SAWYER, D.S., KLAUS, A. & MASSON, D.G. (eds) *Proceedings of the Ocean Drilling Program, Scientific Results, 149.* Ocean Drilling Program, College Station, TX, 603–615.

LEBLANC, M., GERVILLA, F. & JEBWAB, J. 1990. Noble metals segregation and fractionation in magmatic ores from Ronda and Beni Bousera lherzolite massifs (Spain, Morocco). *Mineralogy and Petrology*, **42**, 233–248.

LINDSLEY, D.H. 1983. Pyroxene thermometry. *American Mineralogist*, **68**, 477–493.

LORAND, J.P. 1989. Abundance and distribution of Cu–Fe–Ni sulfides, sulfur, copper, and platinum-group elements in orogenic-type spinel lherzolite massifs of Ariège (Northeastern Pyrenees, France). *Earth and Planetary Science Letters*, **93**, 50–64.

LORAND, J.P., KEAYS, R.R. & BODINIER, J.L. 1993. Copper and noble metal enrichments across the lithosphere–asthenosphere boundary of mantle diapirs: evidence from the Lanzo lherzolite masif. *Journal of Petrology*, **34**, 1111–1140.

MELCHER, F., GRUM, W., SIMON, G., THALHAMMER, T.V. & STUMPFL, E. 1997. Petrogenesis of the giant chromite deposits of Kempirsai, Kazakstan: a study of solid and fluid inclusions in chromite. *Journal of Petrology*, **38**, 1419–1458.

MITCHELL, R.H. & KEAYS, R.R. 1981. Abundance and distribution of gold, palladium and iridium in some spinel and garnet lherzolites: implications for the nature and origin of precious metal-rich intergranular componenets in the upper mantle. *Geochimica et Cosmochimica Acta*, **45**, 2425–2442.

NALDRETT, A.J. 1989. *Magmatic sulfide deposits.* Clarendon Press, Oxford.

NALDRETT, A.J. & DUKE, J.M. 1980. Platinum metals in magmatic sulfide ores: the occurrence of these metals is discussed in relation to the formation and importance of these ores. *Science*, **208**, 1417–1424.

ODP Leg 149 Shipboard Scientific Party 1993. ODP drills the West Iberia rifted margin. *EOS Transactions, American Geophysical Union*, **74**, 454–455.

ODP Leg 173 Shipboard Scientific Party 1998. Drilling reveals transition from continental breakup to early magmatic crust. *EOS Transactions, American Geophysical Union*, **79**, 173–181.

OHNENSTETTER, M. 1992. Platinum Group Element enrichment in the upper mantle peridotites of the Monte Maggiore ophiolitic massif (Corsica, France): mineralogical evidence for ore-fluid metasomatism. *Mineralogy and Petrology*, **46**, 85–107.

OHNENSTETTER, M. 1996. Diversity of PGE deposits in basic–ultrabasic intrusives—single model of formation. *In*: DEMAIFFE, D. (ed.) *Petrology and Geochemistry of Magmatic Suites of Rocks in the Continental and Oceanic Crusts. A volume dedicated to Professor Jean Michot.* Université libre de Bruxelles, Royal Museum for Central Africa, Tervuren, 337–354.

OSHIN, I.O. & CROCKET, J.H. 1982. Noble metals in Thetford Mines ophiolite, Quebec, Canada. Part I: Distribution of gold, iridium, platinum, and palladium in the ultramafic and gabbroic rocks. *Economic Geology*, **77**, 1556–1570.

PAGÉ, P., HÉBERT, R., GUEDDARI, K. & LAFLÈCHE, M.R. 1998. Géochimie des EGP dans les roches mafiques et ultramafiques: Massif de North Arm Mountain (Ophiolite de Bay of Islands, Terre Neuve) et la Faille Transformante Garrett (Pacifique sud). *Geological Association of Canada, Mineralogical Association of Canada, Association des géologues et géophysiciens du Québec, International Association of Hydrogeologists, Canadian Geophysical Union, Annual Meeting, Québec City, Abstracts Volume*, **23**, 140.

PATTOU, L., LORAND, J.-P. & GROS, M. 1996. Non-chrondritic platinum-group element ratios in the Earth's mantle. *Nature*, **379**, 712–715.

PEDERSEN, R.B., JOHANNESEN, G.M. & BOYD, R. 1993. Stratiform platinum-group element mineralizations in the ultramafic cumulates of the Leka ophiolite complex, central Norway. *Economic Geology*, **88**, 782–803.

PICCARDO, G.B., MESSIGA, B. & VANNUCCI, R. 1988. The Zabargad peridotite–pyroxenite association: petrological constraints on its evolution. *Tectonophysics*, **150**, 209–227.

PICKUP, S.L.B., WHITMARSH, R.B., FOWLER, M.R. & RESTON, T.J. 1996. Insight into the nature of the ocean–continent transition off West Iberia from a deep multichannel seismic reflection profile. *Geology*, **24**, 1079–1082.

PRITCHARD, H.M., NEARY, C.R. & POTTS, P.J. 1986. Platinum group minerals in the Shetland ophiolite. *In*: GALLAGHER, M.J., IXER, R.A., NEARY, C.R. & PRITCHARD, H.M. (eds) *Metallogeny of Basic and Ultrabasic Rocks.* Institution of Mining and Metallurgy, London, 395–414.

PRITCHARD, H.M., PUCHELT, H., ECKHARDT, J.-D. & FISHER, P.C. 1996. Platinum-group elements concentrations in mafic and ultramafic lithologies drilled from Hess Deep. *In*: MÉVEL, C., GILLIS, K.M., ALLAN, J.F. & MEYER, P.S. (eds) *Proceedings of the Ocean Drilling Program, Scientific Results, 147.* Ocean Drilling Program, College Station, TX, 77–90.

SAWYER, D., WHITMARSH, R., KLAUS, A., et al. (eds) 1994. *Proceedings of the Ocean Drilling Program, Initial Reports, 149.* Ocean Drilling Program, College Station, TX.

SCHARER, U., GIRARDEAU, J., CORNEN, G. & BOILLOT, G. 1999. *Èvidences géochimiques et géochronologiques pour une origine asthénosphérique du magmatisme antérieur à l'accrétion océanique atlantique.* Société Géologique de France, Paris.

SCHARER, U., KORNPROBST, J., BESLIER, M.-O., BOILLOT, G. & GIRARDEAU, J. 1995. Gabbro and related rock emplacement beneath rifting continental crust: U–Pb geochronological and geochemical constraints for the Galicia passive margin (Spain). *Earth and Planetary Science Letters*, **130**, 197–200.

SEIFERT, K. & BRUNOTTE, D. 1996. Geochemistry of serpentinized mantle peridotite from Site 897 in the Iberia Abyssal Plain. *In*: WHITMARSH, R.B., SAWYER, D.S., KLAUS, A. & MASSON, D.G. (eds) *Proceedings of the Ocean Drilling Program, Scientific Results, 149.* Ocean Drilling Program, College Station, TX, 413–424.

SEIFERT, K.E., CHANG, C.-W. & BRUNOTTE, D.A. 1997. Evidence from Ocean Drilling Program

Leg 149 mafic igneous rocks for oceanic crust in the Iberia Abyssal Plain ocean–continent transition zone. *Journal of Geophysical Research*, **102**, 7915–7928.

SEIFERT, K., GIBSON, I., WEIS, D. & BRUNOTTE, D. 1996. Geochemistry of the metamorphosed cumulate gabbros from Hole 900A, Iberia Abyssal Plain. *In*: WHITMARSH, R.B., SAWYER, D.S., KLAUS, A. & MASSON, D.G. (eds) *Proceedings of the Ocean Drilling Program, Scientific Results, 149*. Ocean Drilling Program, College Station, TX, 471–488.

SEYLER, M. & BONATTI, E. 1994. Na, Al^{iv} and Al^{vi} in clinopyroxenes of subcontinental and suboceanic ridge peridotites: a clue to different melting processes in the mantle? *Earth and Planetary Science Letters*, **122**, 281–289.

SRIVASTAVA, S.P., ROEST, W.R., KOVACS, L.C., OAKEY, S., LEVESQUE, J., VERHOEF, J. & MCNAB, R. 1990. Motion of Iberia since the Late Jurassic: results from detailed aeromagnetic measurements in the Newfoundland basin. *Tectonophysics*, **184**, 229–260.

TANGUAY, S., HÉBERT, R. & BERGERON, M. 1990. Distribution of PGE in pyroxene-bearing ultramafic cumulates in the Thetford Mines ophiolitic complex, Quebec. *Canadian Mineralogist*, **28**, 597–605.

TURRIN, B.D. 1999. $^{40}Ar/^{39}Ar$ ages from the ocean–continental transition zone, Iberia Abyssal Plain, ODP Leg 173: chronologic and thermochronologic constraints on the kinematics of lithospheric extension and continental breakup.

In: Non-volcanic Rifting of Continental Margins: a Comparison of Evidence from Land and Sea, Abstracts of Talks and Posters. Geological Society, London.

VARFALVY, V., HÉBERT, R., BÉDARD, J.H. & LA FLÉCHE, M.R. 1997. Petrology and geochemistry of pyroxenite dykes in upper mantle peridotites of the North Arm Mountain massif, Bay of Islands ophiolites, Newfoundland: implications for the genesis of boninitic and related magmas. *The Canadian Mineralogy and Petrology*, **84**, 293–309.

WHITMARSH, R.B., BESLIER, M.-O., WALLACE, P. J. *et al.* (eds) 1998. *Proceedings of the Ocean Drilling Program, Initial Reports, 173*. Ocean Drilling Program, College Station, TX.

WHITMARSH, R.B. & MILES, P.R. 1995. Models of the development of the west Iberia rifted continental margin at 40°30′ deduced from surface and deep-tow magnetic anamalies. *Journal of Geophysical Research*, **100**, 3789–3806.

WHITMARSH, R.B., SAWYER, D.S., KLAUS, A. & ODP SHIPBOARD SCIENTIFIC PARTY, LEG 149 1996. *Proceedings of the Ocean Drilling Program, Scientific Results 149*. Ocean Drilling Program, College Station, TX.

WOOD, S.A., MOUNTAIN, B.W. & FENLON, B. 1989. Thermodynamic constraint on the solubility of platinum and palladium in hydrothermal solutions: reassessment of hydroxide, bisulphide and ammonia complexing. *Econonic Geology*, **84**, 2020–2028.

The evolution of amphibolites from Site 1067, ODP Leg 173 (Iberia Abyssal Plain): Jurassic rifting to the Pyrenean compression

V. GARDIEN[1], G. POUPEAU[2], B. MUCEKU[2], R. HÉBERT[3], G. BEAUDOIN[3] & E. LABRIN[2]

[1]1- UMR 5570, CNRS and UCB-Lyon1, 43 Bd 11 Novembre 1918, 69622 Villeurbanne Cedex, France

[2]2- UPRES-A 5025, CNRS-University Joseph Fourrier, Institut Dolomieu, 15 rue Maurice Gignoux, 38031 Grenoble, France

[3]3- Département de Géologie et de Génie géologique, Pavillon Adrien-Pouliot, Université Laval, Ste-Foy, Québec G1K 7P4, Canada

Abstract: During ODP Leg 173 (April–May 1997) five new sites (1065, 1067–1070) were drilled at the ocean–continent transition (OCT) zone off the West Iberia margin in the Iberia Abyssal Plain. At Site 1067 the 92 m of cored amphibolites were subdivided into three units based on textural criteria. Unit 1 consists of highly foliated and folded amphibolites and acidic gneiss concordant to and folded along with the foliation of the amphibolite. Unit 2, in the middle part of the section, consists of brecciated amphibolites. Unit 3, at the bottom of the hole, is a weakly deformed zone where magmatic textures are observed in the amphibolites and in associated anorthosites. The amphibolites contain tschermakitic to magnesio-hornblende amphibole, plagioclase, zircon, apatite $\pm$ titanite $\pm$ Fe-oxide $\pm$ quartz. The acidic gneiss consists of garnet, plagioclase, alkali-feldspar, quartz, biotite and zircon. Chlorite, sericite and ilmenite occur as secondary phases. The metamorphic evolution of the amphibolite and acidic gneiss started under amphibolite-facies conditions (Stage I: 670 $\pm$ 40 °C and 7 $\pm$ 1 kbar). Further exhumation took place through low-grade amphibolite-facies (Stage II: 550 $\pm$ 50 °C and 5.5 $\pm$ 1 kbar) to green-schist-facies (Stage III: < 500 °C and < 3 kbar) conditions contemporaneous with the development of ductile structures. The late metamorphic evolution of the amphibolite ended under ocean-floor conditions. Oxygen isotope ratios and studies of fluid inclusion indicate that a magmatic water-rich fluid in equilibrium with the igneous protolith predominated during Stage I. During Stages II and IIII low-temperature water-rich fluids of metamorphic origin predominated, with a probable contribution of sea water. Apatite fission-track dating indicates that the amphibolites record two thermal excursions below 120 °C. The first took place at *c.* 113–100 Ma and could be related to Cretaceous rifting. The second was brief, and so did not anneal older tracks; it occurred between 75 and 55 Ma and could be related to the Pyrenean orogeny.

Studies of fossil and present-day ocean–continent transition (OCT) zones are of great importance for the understanding of mechanisms responsible for the thinning and break-up of continental lithosphere leading to the formation of passive margins. Some fossil passive margins, now tectonically integrated into mountain belts, are very often readily accessible and offer large exposures of outcrops giving the opportunity for rigorous structural and petrological study. However, the obduction of passive margins onto continents can cause tectonic and metamorphic events that obliterate the tectonic and metamorphic evolution related to the rift-ing. Present-day passive margins that are submerged thousands of metres below sea level are difficult to reach and usually or mostly covered by thick post-rift sediments. Only starved non-volcanic margins can be sampled by diving or drilling, resulting in very small samples that display the structures and the metamorphic evolution related to the rifting. The West Iberia margin is a good example of a starved non-volcanic margin comparable in many ways with the Jurassic margin of the Alpine Tethys (Lemoine *et al.* 1987; Trommsdorff *et al.* 1993; Manatschal & Bernoulli 1999; Müntener *et al.* 2000). This work focuses on the study of the

From: WILSON, R.C.L., WHITMARSH, R.B., TAYLOR, B. & FROITZHEIM, N. 2001. *Non-Volcanic Rifting of Continental Margins: A Comparison of Evidence from Land and Sea.* Geological Society, London, Special Publications, **187**, 191–208. 0305-8719/01/$15.00 © The Geological Society of London 2001.

amphibolites at Ocean Drilling Program (ODP) Site 1067 (Iberia Abyssal Plain), presenting the *P–T* evolution of the rocks together with apatite fission-track dating to document the processes of continental rifting. Preliminary $\delta^{18}O$ and fluid inclusion analyses are presented to constrain the origin and the role of fluid circulation during the metamorphic evolution of the rocks. The petrological and geochronological data collected on amphibolites from the Iberia Abyssal Plain are compared with other results obtained from the Jurassic margin of the Tethys.

Site descriptions

During ODP Leg 173 (April–May 1997) five Sites (1065 and 1067–1070) were drilled at the OCT zone of the West Iberia margin in the Iberia Abyssal Plain (Fig. 1). Three of these Sites (1067, 1068 and 1070) reached the crystalline basement. At Sites 1068 and 1070 the basement section is almost completely ultramafic. In Hole 1070 a dyke of pegmatitic amphibolite–gabbro occurs within the ultramafic rocks. At Site 1067 variably foliated and brecciated amphibolites were recovered between 763.8 and 855.6 m below seafloor (mbsf) (Fig. 2). This site is located 800 m east of Site 900, where gabbro was drilled during ODP Leg 149 (Whitmarsh *et al.* 1994). The amphibolites were recovered below 5021 m of water and 763.8 m of middle Eocene to upper Paleocene sedimentary rocks (Whitmarsh *et al.* 1998).

Three units were defined at Site 1067 on the basis of structural characteristics and lithological and mineralogical variations (Whitmarsh *et al.* 1998).

Unit 1 occurs between cores 173-1067A-14R-1 (763.8 mbsf) and 173-1067A-17R (801.9 mbsf); however, recovery was very poor (9.6%) in this section of the hole as only 3.7 m of rocks were cored. The main lithology of this unit is a strongly foliated amphibolite associated with foliated quartz layers and acidic gneiss. The latter were found only on top of Unit 1 as isolated core fragments or as a layer associated with the amphibolites; in this case, the contact between amphibolite and acidic gneiss is parallel to the foliation of the amphibolite. The foliation plane is folded around tight to open folds. Some samples show acidic layers folded along with the foliation of the amphibolites. Shallowly inclined epidote and chlorite-rich veins crosscut the ductile struc-

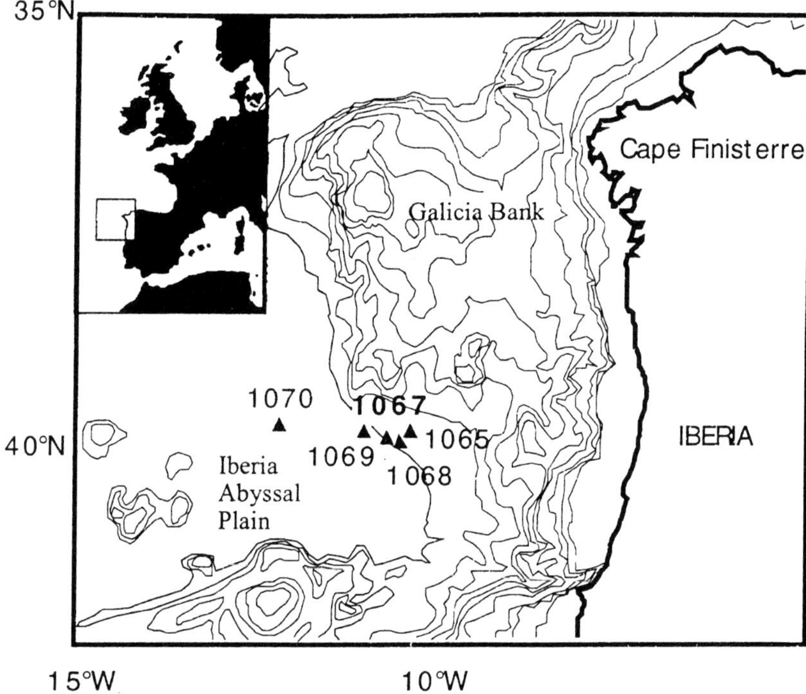

Fig. 1. Bathymetric map of the West Iberia margin showing the location of sites drilled during ODP Leg 173.

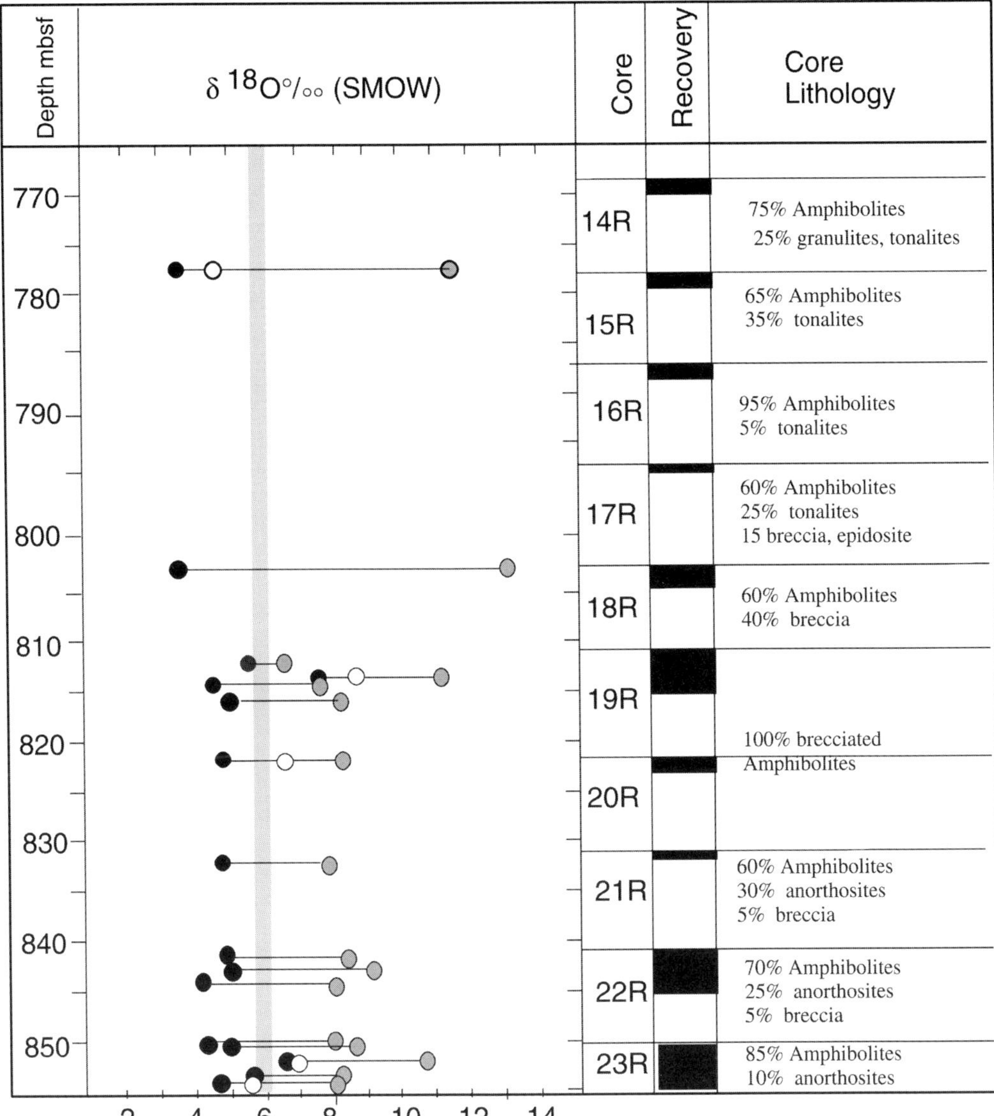

Fig. 2. Distribution of whole-rock and mineral $\delta^{18}O$ values with depth (metres below sea floor) at ODP Site 1067. The grey area covers the oxygen isotope compositions of fresh magmatic rocks derived from an oceanic mantle source. Numbers in column refer to core sections. Open circles, whole rock; grey circles, amphibole; filled circles, plagioclase.

tures (foliation and folds) and are cut by later vertical calcite-rich veins.

Unit 2 occurs between cores 173-1067A-18R (801.9 mbsf) and 173-1067A-20R (822.6 mbsf) and is characterized by the presence of strongly brecciated rocks. The recovery of this unit was 26% (7.5 m of core), consisting of 75% brecciated amphibolite and acidic gneiss, 20%

matrix-supported breccia and 5% indurated fault gouge. The brecciated amphibolite and acidic layer consists of centimetre-scale angular fragments included in a network of 1–5 mm veins crosscutting the rocks.

Unit 3 ranges from the bottom of Core 173-1067A-20R (822.6 mbsf) to core 23R (855.6 mbsf) and in this section the recovery

Table 1. *Oxygen isotope analyses for whole-rock composition, amphibole and plagioclase of the Site 1067 amphibolites*

1067A Core, cm, piece	Hbl mg	$\delta^{18}O$	Plg mg	$\delta^{18}O$	WR $\delta^{18}O$
15R-1W, 42–47, 8B	15.4	3.9	6.1	11.5	
18R-2W, 68–73, 7A	11.0	3.9	8.0	13.0	4.5
18R-2W, 114–118, 9	9.3	3.7	65.8	5.8	
19R-1W, 34–40, 5B	22.8	6.5	16.4	8.8	
19R-2W, 102–107, 6B	6.8	7.8	10.1	8.0	11.2
19R-3W, 30–34, 5	12.6	4.8	9.0	8.5	
19R-3W, 59–64, 7	8.0	5.9	8.6	8.4	
20R-1W, 36–40, 4	8.9	5.1	21.5	7.9	6.4
21R-1W, 1–7, 1	20.3	5.2	8.5	8.3	
22R-1W, 90–94, 8B	8.6	5.3	16.4	9.2	
22R-2W, 132–136, 9A	5.8	5.6	8.1	8.1	
22R-3W, 137–140, 11B	11.8	4.6	8.0	7.9	
23R-3W, 95–100, 3B	8.5	4.6	8.0	8.4	
23R-1W, 88–94, 2E	7.5	5.2	10.3	8.0	
23R-1W, 127–132, 2F	9.2	6.2	8.0	8.1	
23R-2W, 49–54, 1B	32.6	6.6	8.2	7.7	6.6
23R-2W, 84–88, 1D	10.4	5.7	24.5	8.2	
23R-4W, 65–69, 7	8.6	4.7	6.1	6.3	
23R-5W, 36–41, 4	19.2	4.8	5.3	7.1	5.5

rose to 39%. The material consists mainly of amphibolite associated with minor anorthosite. Most of the amphibolite is weakly foliated, preserving euhedral plagioclase included in poikilitic amphibole. This texture is interpreted as relicts of magmatic texture inherited from a probable gabbroic protolith; however, a foliation plane locally overprints the magmatic texture.

Age determination using SHRIMP on zircon was performed on both amphibolite and acidic gneiss. It gave an age of 270 ± 3 Ma for the amphibolite, which is interpreted as representing the emplacement of the mafic protolith (Manatschal *et al.* 2001). One acidic gneiss (sample 1067A, 15R-1, piece 2) associated with the amphibolites gave inherited Cadomian (595 Ma) and Hercynian (340 Ma) ages from the cores of zircons, and a Permian age (290 Ma) from their outer parts (Rubenach & Wycsoczanski 1998). $^{40}Ar/^{39}Ar$ dating on amphibole and plagioclase from the amphibolite (the researchers did not specify the number of the sample) gave respectively 161 and 138 Ma (Manatschal *et al.* 2001). $^{40}Ar/^{39}Ar$ analyses performed on two bulk plagioclase samples extracted from flaser gabbro cored at Site 900, located on the same structural high as Site 1067, yielded an age of 136.4 ± 0.3 Ma (Feraud *et al.* 1996). These ages are interpreted as cooling stages related to the exhumation of the rocks during the Mesozoic continental rift-

ing (Feraud *et al.* 1996; Manatschal *et al.* 2001).

Oxygen isotopes

Samples representative of the three units defined above were selected for geochemical studies. We obtained $\delta^{18}O$ for whole-rock and mineral separates (amphibole and plagioclase) and the values are reported in Table 1. Purity of mineral separates was checked by X-ray diffraction on homogenized powder. Oxygen was extracted from rock powders using the BrF_5 method (Clayton & Mayeda 1963) and analysed as CO_2 gas on a VG SIRA 10 mass spectrometer at the Laval University, Ste-Foy, Quebec. Isotopic compositions are quoted in the standard δ notation relative to Standard Mean Ocean Water (SMOW). Results for the NBS28 standard gave $\delta^{18}O = +9.5 \pm 0.2$‰ ($2\sigma m$).

Five whole rocks and 38 mineral separates were analysed for oxygen isotope compositions (Table 1). In Figure 2 whole-rock and mineral $\delta^{18}O$ values are plotted against depth (mbsf). The distribution of $\delta^{18}O$ values does not show any regular trend of isotopic enrichment or depletion with depth. From Unit 3, two whole-rock samples were analysed. One sample (1067A 23R-5W, 36–41 cm, piece 4) of amphibolite at the bottom of the hole yielded a $\delta^{18}O$ value similar to the mantle reference of 5.7 ± 0.2 (Pineau *et al.* 1976) whereas the other samples have $\delta^{18}O$ slightly higher (6.2–6.4) or slightly lower (4.5) than the mantle reference value.

The $\delta^{18}O$ values for amphibole and plagioclase are distinct because of specific exchange rates. Plagioclase has a wide range of $\delta^{18}O$ values (8.0–13.0) higher than the mantle reference, with only one analysis of plagioclase giving values of 5.8. Only three analyses of amphibole gave $\delta^{18}O$ values of 5.6, 5.7 and 5.9; the other samples have lower (3.7–4.8) or higher (6.2–6.6) values when compared with the mantle reference. Plagioclase ($5.8 < \delta^{18}O < 13.0$) is principally responsible for the increase of whole-rock $\delta^{18}O$ values obtained for most of the analysed samples. Oxygen isotope exchange is fast and extensive between plagioclase and fluid whereas it is slower and limited between amphibole and fluid. The $\delta^{18}O$ depletion for amphibole can be explained by isotope exchange with a high-temperature (c. 450 °C) aqueous fluid phase. The very low Cl content of amphibole (Table 2) precludes a significant contribution from sea water, and so suggests the intervention of a magmatic water-rich fluid in equilibrium with the igneous protolith. The ^{18}O enrichment (11.5–13.0) for plagioclase indicates isotope exchange with a low-temperature (c. 200 °C) aqueous fluid phase under greenschist- to zeolite-facies metamorphism.

Petrological description of the samples

Amphibolites

Amphibolites from Site 1067 were derived from a gabbroic protolith. The paragenesis observed in the amphibolites, therefore, represents a retrograde assemblage. However, to simplify the description of the rocks, we consider in the following discussion whether the amphibolite-facies paragenesis represents the 'starting material' of the samples studied.

Amphibolites consist of an assemblage of amphibole, plagioclase, quartz, titanite, apatite and zircon. Epidote, chlorite and albite are secondary minerals. Pyrite and calcite occur within the matrix of the breccias and veins. A modal increase in quartz, sphene and apatite is observed towards the top of the hole whereas zircons and iron oxide are more abundant towards the bottom of the hole.

Amphibole grains occur as brown–green porphyroclast (defined as Hbl I) up to 10 mm in size and show a prismatic shape. The porphyroclastic amphiboles (Fig. 3a) are transformed into green lenticular-shaped hornblende (defined as Hbl II) progressively oriented parallel to the foliation plane (Fig. 3b). With increasing deformation the green hornblende recrystallized into blue–green neoblasts (defined as Hbl III) elongated in the foliation plane (Fig. 3c). Plagioclase grains occur as euhedral porphyroclasts up to 5 mm in size (defined as Plg I) in association with the brown–green amphiboles (Fig. 3a). Within the foliated amphibolites the plagioclases occur as recrystallized neoblasts (defined as Plg II) forming aggregates elongated in the foliation plane in association with the blue–green lenticular amphibole (Fig. 3b). A late plagioclase with an albitic composition (Plg III) occurs associated with the green amphibole (Fig. 3c). Thus, the modification of amphibole and plagioclase compared with the 'starting material' comprises a change of shape and an important grain-size reduction. Moreover, the modification of amphibole grains implies a variation of colour from brown–green to green and to blue–green.

Within the foliated amphibolites, titanite and apatite grains elongated parallel to the foliation plane, and quartz forming layers of 0.5–1 cm width, are folded along with the foliation of the amphibolite. Epidote and sericite occur in minor amounts, principally as an alteration product of plagioclase. Amphibole is replaced by chlorite, and the latter can also develop as elongated flakes parallel to the foliation or as unoriented grains.

Acidic gneiss

This rock type consists of a plagioclase, quartz, alkali-feldspar, garnet, biotite, chlorite and sericite assemblage plus apatite and zircon as accessory minerals. Garnet occurs either as equant grains containing quartz and chlorite inclusions and surrounded by biotite corona, or as elongated grains parallel to the foliation. Oligoclase and orthoclase are present as lenticular porphyroclasts elongated in the foliation plane. Plagioclase shows orthoclase exsolution defining an antiperthitic texture (Fig. 4). Albite developed after oligoclase, chlorite occurs as an alteration product of biotite and garnet, and sericite results from the destabilization of plagioclase and alkali feldspar.

Mineral chemistry

Phase compositions were obtained using the CAMECA SX100 microprobe (Université de Clermont-Ferrand, France) using silicate, vanadate and Fe-oxide standards for calibration. The

Table 2. *Selected analyses of the Site 1067 amphibolites*

Amphibole

	Core 14				Core 15			Core 19				Core 22			Core 23	
	Hbl 1	Hbl 1	Hbl 1	Hbl 1	Hbl 2	Hbl 3	Hbl 1	Hbl 2	Hbl 3	Hbl 3	Hbl 2	Hbl 3	Hbl 3	Hbl 1	Hbl 1	Hbl 2
SiO_2	41.98	41.77	42.36	42.58	42.37	45.61	43.05	45.117	46.61	47.06	43.78	46.23	45.83	41.96	41.93	42.18
TiO_2	0.61	0.87	0.70	1.04	0.88	0.93	1.11	1.20	0.73	1.06	1.40	1.33	1.233	1.34	1.23	1.20
Al_2O_3	12.64	12.46	12.67	12.78	11.94	8.93	12.66	10.04	8.78	7.63	11.48	8.89	8.34	12.20	12.26	11.44
FeO	19.03	19.88	17.39	16.82	17.11	16.38	16.99	16.64	16.04	16.69	17.01	15.90	16.13	18.87	18.58	18.73
MnO	0.34	0.38	0.30	0.32	0.30	0.31	0.31	0.31	0.32	0.27	0.31	0.27	0.3	0.36	0.35	0.37
MgO	8.97	9.08	9.76	9.95	10.03	11.41	10.28	11.52	11.71	11.59	10.43	11.96	12.15	9.36	9.46	9.64
CaO	11.68	11.33	11.76	11.51	11.75	11.64	11.20	11.00	11.65	11.50	10.98	11.52	11.11	10.62	10.69	10.56
Na_2O	1.53	1.6	1.52	1.58	1.26	1.14	1.58	1.52	1.07	1.06	1.86	1.411	1.41	2.01	1.93	1.92
K_2O	0.55	0.58	0.69	0.73	0.72	0.52	0.54	0.40	0.34	0.357	0.46	0.35	0.32	0.68	0.70	0.64
F	0	0.23	0	0	0	0	0	0.05	0.07	0	0	0	0	0	0	0
Cl	0.01	0.008	0.01	0.01	0.001	0.01	0.01	0.01	0	0	0	0.01	0.01	0.01	0	0.01
H_2O	1.99	2	1.98	1.98	1.97	1.20	1.98	1.98	2	2	2.00	2.02	2.00	1.97	1.97	1.96
Total	99.33	99.95	99.22	99.07	98.35	99.00	99.87	99.87	99.49	99.31	99.93	99.91	98.92	99.39	99.19	98.68
Si	6.32	6.25	6.40	6.42	6.45	6.39	6.43	6.71	6.92	7.01	6.54	6.85	6.86	6.38	6.38	6.45
Ti	0.07	0.098	0.08	0.09	0.10	0.10	0.12	0.13	0.08	0.12	0.16	0.13	0.14	0.15	0.14	0.14
Al	2.24	2.20	2.26	2.27	2.14	1.58	2.23	1.76	1.54	1.34	2.02	1.55	1.47	2.19	2.20	2.06
Fe	2.32	2.38	2.70	2.19	2.04	2.20	2.12	2.07	1.99	2.08	2.12	1.97	2.02	2.40	2.36	2.40
Mn	0.04	0.05	0.04	0.04	0.04	0.04	0.04	0.04	0.04	0.03	0.04	0.03	0.04	0.05	0.05	0.05
Mg	2.01	2.02	1.85	2.30	2.56	2.20	2.29	2.55	2.59	2.58	2.32	2.64	2.71	2.12	2.15	2.20
Ca	1.88	1.82	1.92	1.88	1.88	1.90	1.79	1.75	1.85	1.84	1.76	1.83	1.78	1.73	1.74	1.73
Na	0.45	0.46	0.40	0.42	0.35	0.45	0.46	0.44	0.31	0.31	0.54	0.40	0.41	0.59	0.57	0.57
K	0.10	0.11	0.10	0.14	0.10	0.13	0.10	0.07	0.06	0.07	0.09	0.06	0.06	0.13	0.13	0.13
Feldspar		Core 15R		Plg 1		Core 23R				Core 15R		Plg 2			Core 23R	
SiO_2	59.17	58.29	60.68	59.09	59.28	59.09	59.45	58.67	59.44	62.64	61.21	63.82	59.68	60.96	61.54	66.41
Al_2O_3	24.55	24.79	24.18	24.73	24.5	24.21	24.5	24.19	24.94	23.04	23.14	22.84	24.09	23.13	22.6	20.37
MgO	0.03	0	0	0	0.02	0.01	0.01	0	0	0.02	0.03	0	0.01	0	0	0
FeO	0.17	0.26	0.01	0.36	0.07	0.22	0.11	0.17	0.17	0.16	0.16	0.11	0.18	0.01	0.01	0.16
MnO	0.01	0	0.05	0	0.03	0.01	0	0.01	0.01	0	0	0	0	0.02	0	0.02
TiO_2	0	0.01	0	0	0	0	0	0	0	0	0.01	0.01	0.02	0	0	0
CaO	7.83	8.57	7.92	7.77	8	8.03	7.18	7.47	7.54	5.25	5.77	4.59	6.03	4.2	3.29	2.47
Na_2O	7.39	7.1	7.77	7.32	7.31	7.35	7.86	7.37	7.56	8.91	8.57	9.34	8.46	10.39	10.58	10.48
K_2O	0.12	0.08	0.08	0.07	0.15	0.12	0.06	0.14	0.07	0.15	0.1	0.25	0.11	0.06	0.06	0.05
Total	99.26	99.1	100.69	99.33	99.36	99.04	99.16	98.02	99.73	100.17	98.99	100.96	98.57	98.77	98.07	99.96
Si	2.69	2.66	2.73	2.69	2.69	2.67	2.69	2.69	2.68	2.79	2.77	2.81	2.71	2.77	2.79	2.94
Al	1.31	1.33	1.28	1.32	1.31	1.33	1.31	1.31	1.32	1.21	1.23	1.19	1.29	1.24	1.21	1.06
Mg	0	0.01	0	0	0	0	0	0	0	0	0	0	0	0	0	0
Fe	0.01	0	0	0.01	0	0.01	0	0	0	0	0.01	0	0.01	0	0	0.01
Mn	0	0	0	0	0	0	0	0	0	0	0	0	0	0	0	0
Ca	0.38	0.42	0.38	0.38	0.39	0.40	0.35	0.37	0.37	0.25	0.28	0.22	0.29	0.10	0.06	0.12
Na	0.65	0.63	0.68	0.64	0.64	0.66	0.69	0.66	0.66	0.77	0.75	0.80	0.74	0.94	0.95	0.99
K	0.01	0	0	0	0.01	0.01	0	0.01	0	0.01	0.01	0.01	0.01	0	0	0
% Ab	62.64	59.73	63.71	62.77	61.8	61.92	66.25	63.57	64.22	74.81	72.46	77.58	71.33	80.76	84.34	88.21
% An	36.69	39.82	35.88	36.86	37.38	37.39	33.45	35.63	35.39	24.36	26.97	21.04	28.06	18.88	15.32	11.49
% Or	0.67	0.46	0.4	0.37	0.81	0.68	0.31	0.81	0.4	0.83	0.57	1.38	0.61	0.36	0.34	0.3

Formula units for amphiboles based on 23 oxygen and two (OH, Cl).

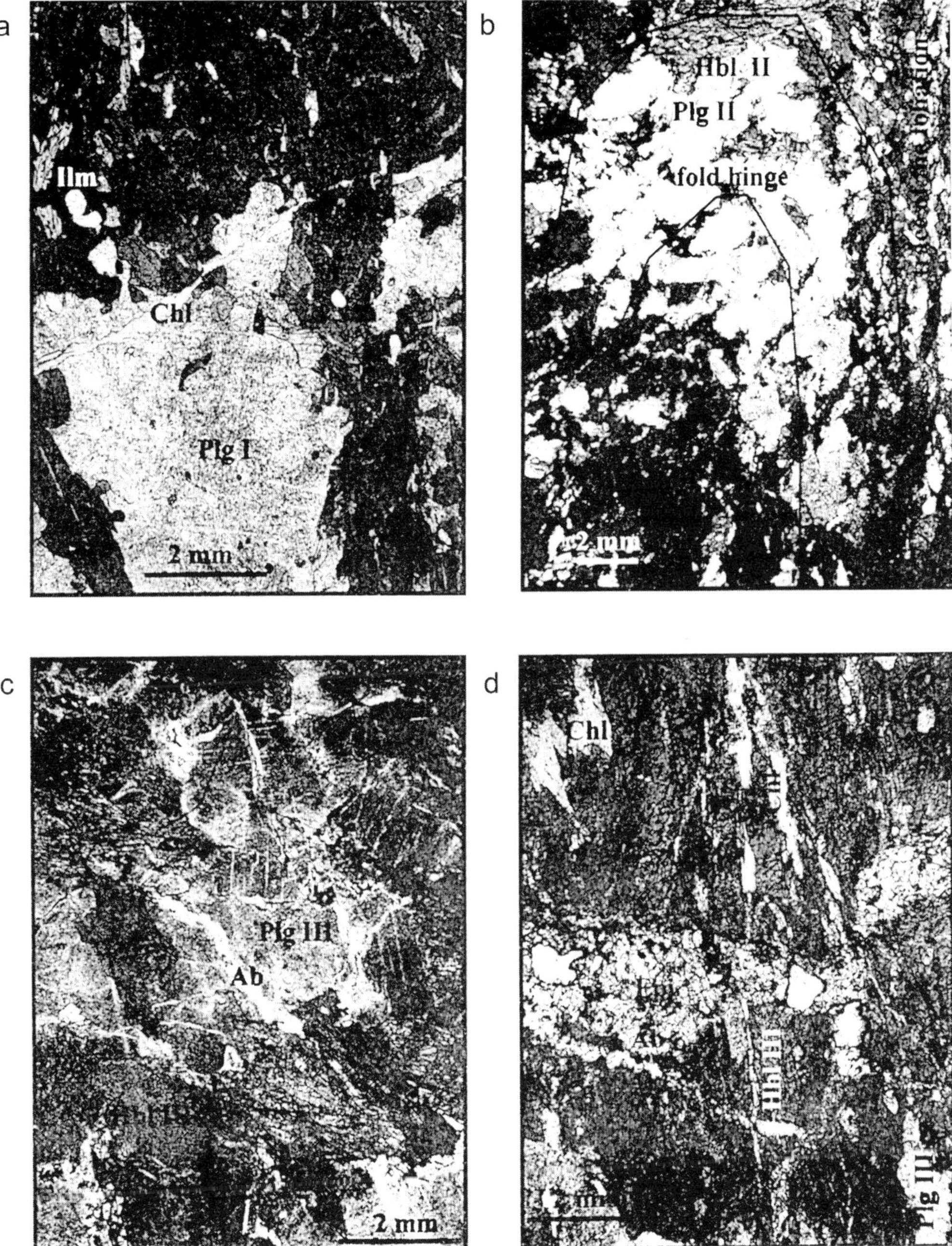

Fig. 3. Photomicrographs of Site 1067 amphibolites; plane-polarized light. (**a**) Poikilitic amphibole (Hbl I) containing euhedral plagioclase (Plg I) preserving a magmatic texture (sample 1067A-23R2-75-78). (**b**) Recrystallized amphibole (Hbl I) and plagioclase (Plg II) elongated in the foliation plane (sample 1067A-16R, 109–111 cm). (**c**) Blue amphibole (Hbl II) transformed into chlorite and associated with the albitic plagioclase (Plg III). The foliation plane is crosscut by vein filled with epidote and albite (sample 1067A-18R-2W, 28–30 cm). (**d**) Breccia composed of amphibole (Hbl II) + plagioclase (Plg III) fragments separated by albite-rich veins (sample 1067A-20R2, 68–73 cm).

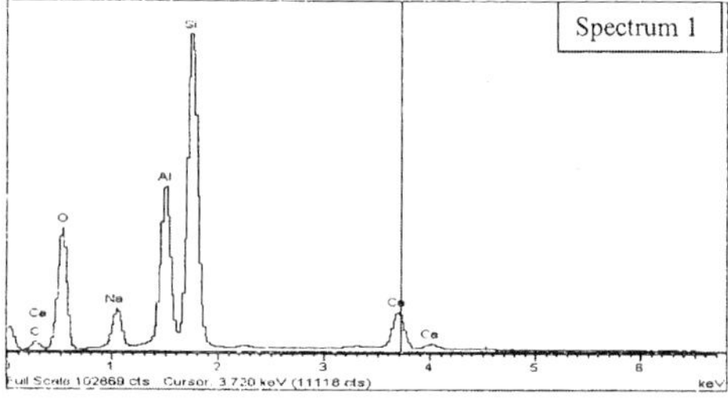

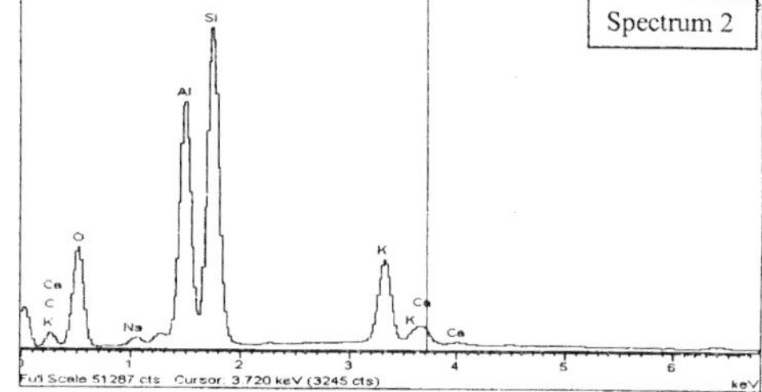

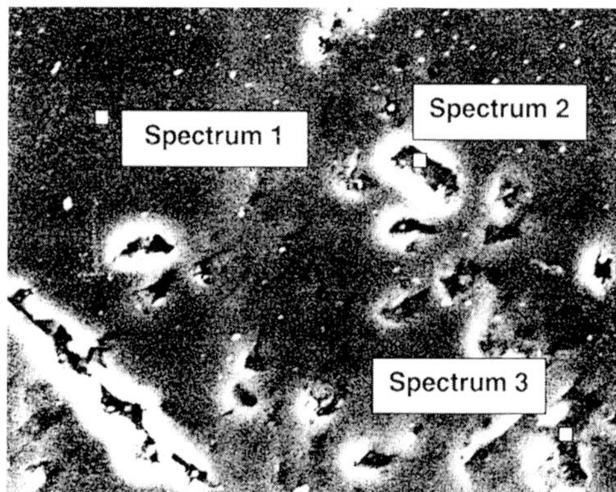

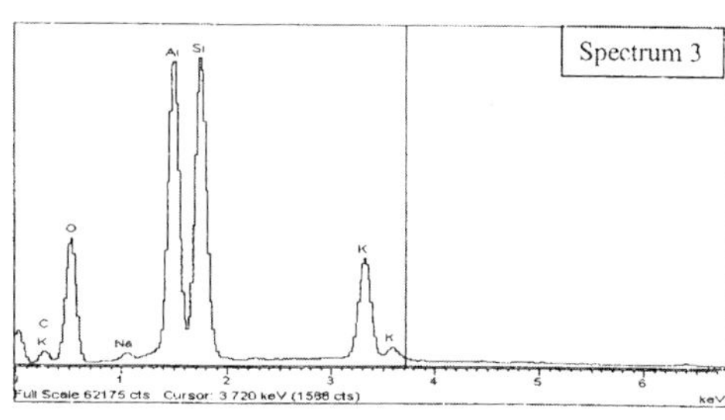

Fig. 4. The antipertithic texture (electron microscope image) is shown by the presence of alkali feldspar exsolution (spectra 2 and 3) within the plagioclase matrix (spectrum 1), sample 1067A-15R1, interval 36–46 cm, piece 8.

accelerating voltage was 15 kV, the beam current 20 nA and the counting time for each element was 10 s.

Amphibolite phase composition

The amphibole composition varies from a ferro-tschermakite, for Hbl I, or tschermakitic hornblende, for Hbl II, to a magnesio-hornblende composition, for Hbl III (Table 2, nomenclature after Leake 1978). Prismatic amphiboles occur in the brecciated amphibolites (Unit 2), and in those in Unit 3 have a magnesio-hornblende composition similar to the composition of Hbl III. They are associated with plagioclase having an albitic composition (Fig. 3d). Differences in amphibole composition are shown in Fig. 5, which is independent of the normalization scheme of amphibole. The Na/(Na + Ca) and Al/(Al + Si) ratios (Laird & Albee 1981) are high in Hbl I and II (close to the garnet zone) and very low in Hbl III (close to the biotite zone). Two types of feldspar were analysed within a single specimen (Table 2): they are either oligoclase ($Or_{0.5-1}Ab_{75}An_{25}$) or andesine

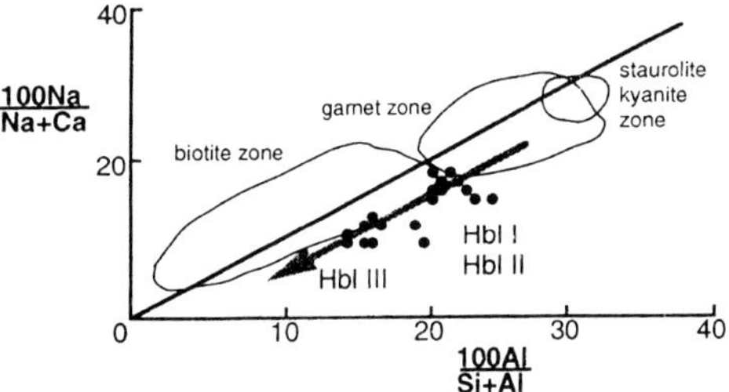

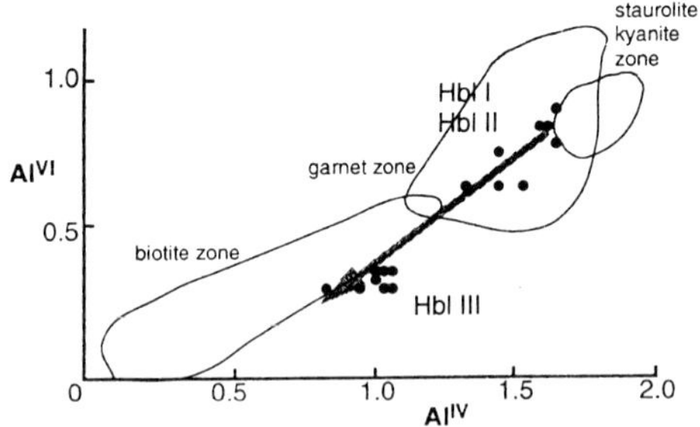

Fig. 5. Plots to show the correlation between 100Na/(Na + Ca) v. 100Al/(Al + Si) and Al^{VI} v. AL^{IV} of amphiboles (Laird & Albee 1981).

Table 3. *Selected analyses of garnet, alkali feldspar and plagioclase in the acidic gneiss cored at Site 1067*

Sample 15R-1W, piece 8				Garnet						Plg				K-fels	
	1	2	3	4	5	6	7	8		1	2	3	1	2	3
SiO_2	36.39	37.28	36.86	37.24	36.14	36.33	37.55	37.06	SiO_2	60.48	61.5	61.51	54.64	54.84	55.42
TiO_2	0.03	0.11	0.06	0.13	0.15	0	0.06	0.05	TiO_2	0	0	0	0	0	0
Al_2O_3	20.85	20.55	20.02	20.06	20.494	20.44	20.44	20.66	Al_2O_3	22.74	22.2	22.78	30.27	33.23	31
FeO	25.42	25.23	25.25	25.38	25.84	24.25	25.73	25.62	FeO	0.06	0.07	0.04	0.17	1.35	0.13
MnO	5.74	5.59	5.58	5.72	5.51	5.47	5.46	5.42	MnO	0.04	0.02	0.03	0	0.05	0.05
MgO	1.74	1.56	1.57	1.6	1.59	1.58	1.58	1.62	MgO	0.05	0	0	0.06	0.18	0.18
CaO	9.06	9.24	9.41	9.71	9.44	9.32	9.32	9.23	CaO	6.06	4.89	5.54	0.5	0.03	0.36
Na_2O	0.05	0.02	0.1	0.05	0.11	0.04	0.04	0.05	Na_2O	8.24	7.92	8.42	0.74	0.11	0.22
K_2O	0	0	0	0.01	0.01	0	0	0.05	K_2O	0.25	1.57	0.36	9.04	10.46	9.49
Total	99.27	99.31	98.86	99.89	99.27	98.65	100.18	99.74	Total	97.86	98.17	98.7	97.46	100.25	99.84
24 Ox									24 Ox						
Si	5.85	6.00	5.96	5.96	5.93	5.95	6.09	5.93	Si	2.77	2.81	2.78	2.45	2.99	2.41
Ti	0	0.01	0.01	0.02	0.02	0	0	0	Ti	0	0	0	0	0	0
Al	3.95	3.85	3.81	3.78	3.78	3.85	3.90	3.90	Al	1.23	1.19	1.21	1.55	1.78	1.59
Fe	3.42	3.40	3.41	3.40	3.41	3.18	3.4	3.43	Fe	0	0	0	0.01	0.05	0
Mn	0.78	0.76	0.76	0.77	0.77	1.08	0.73	0.74	Mn	0	0.01	0	0	0	0
Mg	0.42	0.37	0.38	0.38	0.39	0.44	0.37	0.39	Mg	0	0	0	0	0.01	0.01
Ca	1.56	1.59	1.63	1.67	1.66	1.50	1.58	1.58	Ca	0.30	0.24	0.27	0.02	0	0.02
Na	0.01	0.01	0.03	0.01	0.03	0.01	0.01	0.01	Na	0.73	0.70	0.74	0.32	0	0.27
K	0	0	0	0	0	0	0	0.01	K	0.01	0.09	0.02	0.40	0.60	0.41
Alm	55.34	55.43	55.2	54.63	54.78	51.37	55.9	55.91	% Ab	70.09	67.97	71.85	1.38	1.58	0.57
Gro	25.27	26.01	26.34	26.79	26.67	24.17	25.95	25.79	% An	28.5	23.16	26.12	0.25	0.2	2.35
Pyr	6.73	6.12	6.11	6.12	6.24	7.08	6.13	6.31	% Or	1.42	8.87	2.03	98.37	98.22	97.08
Spe	12.66	12.44	12.35	12.46	12.31	17.38	12.03	11.99							

Table 4. *Ice melting and homogenization temperatures for the first and the second generation of fluid inclusions at Site 1067*

Inc I	T_m ice	TH L-V	Wt % NaCl	Density
1	−2.1	159.4	3.44	0.94
2	−2.4	159.6	3.92	0.94
3	−2	186.2	3.28	0.91
4	−1.8	163.9	2.96	0.93
5	−1.8	141.5	2.96	0.95
6	−1.8	148.6	2.96	0.94
7	−2.6	153.1	4.23	0.95
8	−2.6	169.7	4.32	0.93
9	−2.6	183.2	4.23	0.92
10	−2.5	175.3	4.1	0.92
11	−2.3	180.1	3.76	0.92
12	−2.3	171.1	3.76	0.93
13	−2.5	195	4.07	0.90
14	−2.3	163.5	3.76	0.93
15	−2.3	163.5	3.76	0.93
16	−1.7	147.3	2.80	0.94
17	−1.8	142.7	2.96	0.95
18	−2.3	158.9	3.76	0.94
19	−2.3	145.2	3.76	0.95
20	−2.1	149.7	3.44	0.94
21	−2.5	145.5	4.07	0.95
22	−2.4	156.8	3.92	0.98
1	−3.2	221.8	5.17	0.88
2	−1.9	196.2	3.12	0.88
3	−1.8	215.9	2.96	0.87
4	−1.9	268.1	3.12	0.79
5	−2.4	280.2	3.92	0.78

($Or_{0.5-1}Ab_{60-65}An_{35-40}$). The difference in composition between porphyroclasts (Plg I) and neoblasts (Plg II) is not clear. Indeed, the An content in neoblasts is always low (An_{25}) but it varies from An_{25} to An_{40} in porphyroclasts. Oligoclase and andesine are generally transformed into albite ($Or_{0.2-6.8}Ab_{90-97}An_{2-4}$) coexisting with the blue–green amphibole (Hbl III) and chlorite.

Gneiss phase composition

All of the garnet grains analysed are mostly almandine rich (55%) with grossular, pyrope and spessartine end-members of respectively 26%, 7% and 12% (Table 3). Zoning profiles of equant garnet grains shows insignificant variations in Fe, Ca, Mn and Mg contents from the core to the rim of garnet (see analyses 1–6 in Table 4) and the Fe/(Fe + Mg) values range from 0.928 to 0.940. Three types of feldspars were analysed (Table 4); they are oligoclase ($Or_{1.5-2}Ab_{70-78}An_{20-28}$) albite ($Or_{0.7}Ab_{90-95}An_{5-;8}$) or orthoclase ($Or_{98}Ab_{0.5-1.5}An_{0.2-2.5}$).

Analysis of fluid inclusions

Fluid inclusions from the amphibolites were analysed to determine the pressure conditions during the metamorphic evolution of the amphibolites and to specify the nature of the fluids percolating during the metamorphic evolution. Fluid inclusions observed in quartz grains occur in two distinctive settings: larger ($10-15\,\mu$m) isolated inclusions and smaller ($5-8\,\mu$m) inclusions that are distributed as planar trails. Both isolated and trail inclusions consist of two-phase systems (liquid and vapour) at room temperature. The presence of primary inclusions in metamorphic rocks is difficult to interpret, as inclusions of different generations can be redistributed during the annealing of the host mineral. However, isolated inclusions possess higher densities (lower homogenization temperatures) than the inclusions occurring as planar trails corresponding unequivocally to a second generation of inclusions. This suggests that the isolated inclusions, termed first-generation inclusions, were trapped at higher pressure conditions than the planar trail inclusions during metamorphic evolution. Microthermo-

Frequency

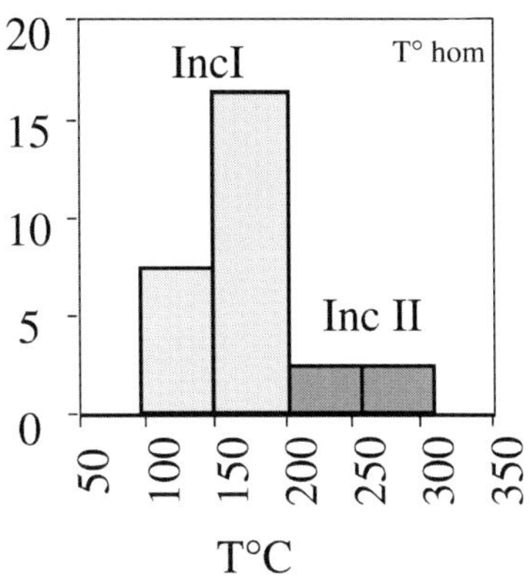

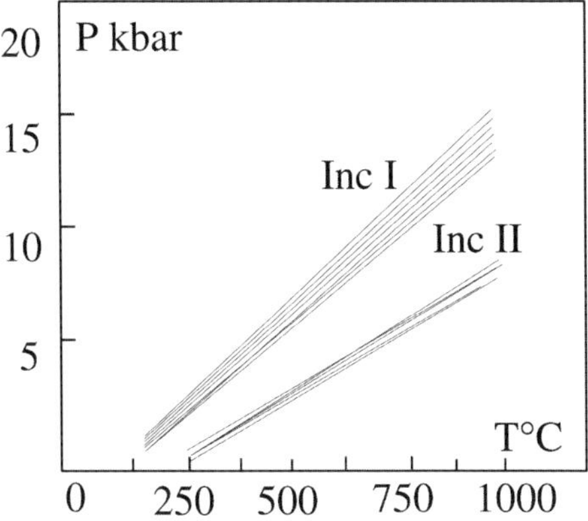

Fig. 6. Calculated isochores for the primary and secondary inclusions of quartz grains from the amphibolites at ODP Site 1067.

metric measurements were made using a Linkam TMS 92 temperature control unit and microscope stage calibrated using a pure CO_2 inclusion at $-56.6\,°C$, distilled water at $0\,°C$ and a range of pure solids (including acetanilide at $114\,°C$ and benzanilide at $163\,°C$). Measurements were made of final ice melting temperature ($T_{m, ice}$) and homogenization temperature (T_h). The final ice melting temperature for the first generation of inclusions (Inc I in Table 4) showing equilibrium melting ranged from -1.7 to $-2\,°C$, corresponding to a salinity range of 2.8–4.3 wt % NaCl equivalent (calculated using the equation of Potter *et al.* 1978). Because of their small size very few measurements were made on secondary inclusions (Inc II in Table 4). However, some inclusions could be analysed and their final ice melting temperature for inclusions showing equilibrium melting

ranged from -2.4 to $-3.2\,°C$ corresponding to a salinity range of 3.9–5.2 wt % NaCl equivalent. In all the studied samples, fluid inclusion salinities are close to or slightly higher than that of sea water (modern sea-water salinity *c.* 3.3 wt % NaCl equivalent) leading us to suggest that the metamorphic fluid had a marine origin with little modification of its salinity during equilibration with the host rock. However, the very low Cl content of amphiboles indicates that the introduction of sea-water-rich fluid into the system occurred after amphibole development. All the quartz inclusions studied show a restricted range in Th between 143 and 280 °C. Isochores for the modal Th and the mean fluid inclusion salinity for inclusions showing equilibrium ice melting were calculated using the data of Zhang & Frantz (1987) and the FLINCOR program of Brown & Hagemann (1994). Figure 6 shows the calculated isochores for the first and the second generation of inclusions.

Metamorphic evolution

Amphibolites

It is very difficult to calculate $P–T$ conditions of equilibration from metamorphic amphibolites especially when garnet is not present. However, it is also well known that the Al tends to replace Si in tetrahedral coordination in Ca-amphibole with rising temperature, whereas Al substitutes for Mg + Fe in the M2 octahedral site to progressively greater extents with increasing pressure (Raase 1974; Hawthorne 1981; Gilbert *et al.* 1982; Robinson *et al.* 1982; Anderson & Smith 1985). Accordingly, it seems likely that the Al_2O_3 content of Ca-amphiboles increases as a function of both pressure and temperature (Moody *et al.* 1983). In Ca-amphiboles the Ti concentration evolves positively with temperature (Raase 1974; Ernst & Liou 1998) but negatively with pressure (Ernst & Liou 1998). The plot of the amphibole compositions in Figure 5 shows the transition from the garnet zone to the biotite zone during the crystallization of Hbl II and Hbl III. The evolution of the amphibole composition is in agreement with the observed amphibolite-facies paragenesis, defined by the assemblage hornblende + plagioclase ± quartz ± ilmenite (Spear 1980), overprinted by greenschist-facies paragenesis as defined by the assemblage chlorite + epidote + albite + calcite +quartz (Liou *et al.* 1974, 1985; Apted & Liou 1983). Thus, to a first approximation, the general trend observed in the Ca-amphiboles composition from Site

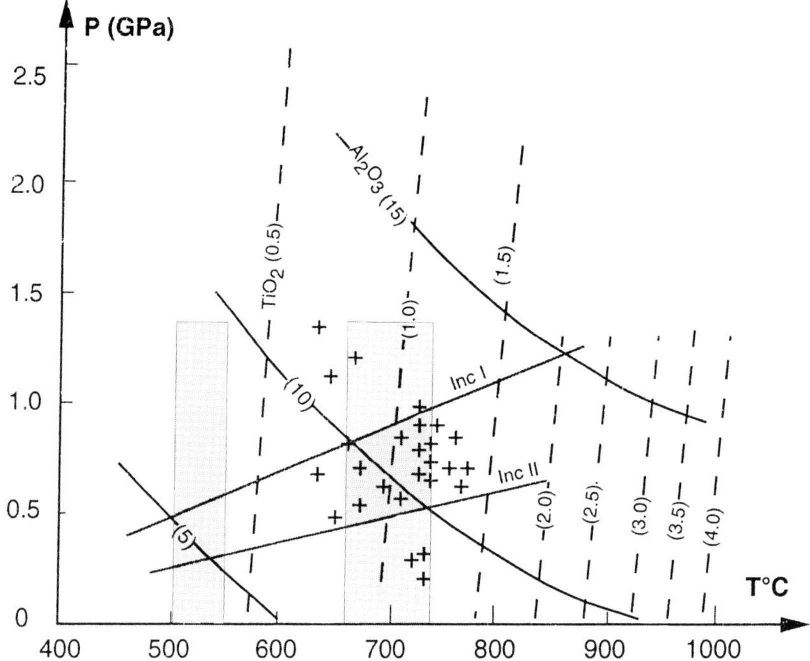

Fig. 7. Petrogenetic grid showing the isopleths of Al$_2$O$_3$ and TiO$_2$ in weight per cent of Ca-amphiboles. The Al$_2$O$_3$ and TiO$_2$ contents of the amphiboles from ODP Site 1067 indicate P between 10 and 5 kbar for temperature of c. 700 ± 50 °C.

1067 amphibolites is consistent with cooling and decompression during the crystallization of the successive generations of hornblende.

The semiquantitative thermobarometer of Ernst & Liou (1998) calibrated for metamorphic basaltic rocks consists of a petrogenetic grid that allows the simultaneous Al$_2$O$_3$ and TiO$_2$ contents of Ca-amphiboles to be used to obtain P–T of crystallization (Fig. 7). Amphibole composition plotted on this grid indicates pressures of equilibration that range between 10 and 5 kbar for temperatures between 650 and 750 °C. Assuming that quartz layer deformation was synchronous with fluid inclusion entrapment, the fluid present in inclusions is representative of the transition between ductile and brittle deformation of the rocks, under respectively low amphibolite-facies and greenschist-facies conditions. Pressures calculated using isochores coupled with the hornblende–plagioclase thermometer range between 8 and 4 kbar (Fig. 6) consistent with the pressure obtained using the petrogenetic grid of Ernst & Liou (1998).

The presence of hornblende and plagioclase in the samples permits the use of the thermometer of Blundy & Holland (1990) and Holland & Blundy (1994) based on the reaction Ed + 4 Qtz = Tr + Ab. This thermometer can also be used for samples without quartz, as the authors specified that the silica activity does not influence the calibration (Blundy & Holland 1992). For the assemblage Hbl I + Plg I the thermometer gives a temperature of c. 670 ± 40 °C. For the amphibole and the plagioclase elongated in the foliation plane (Hbl II and Plg II) the thermometer gives a temperature of c. 500 ± 50 °C.

Acidic gneiss

The retrograde overprint is developed strongly in the acidic gneiss. Examples include the development of chlorite after biotite and garnet, or the development of sericite after plagioclase. Thus it was not possible to use classical thermobarometers to estimate the pressure and the temperature of equilibration of the rock. However, the granulite-facies condition defined by the assemblage garnet + K-feldspar + H$_2$O = quartz + muscovite + biotite suggests a temperature range of 680 °C (for XH$_2$O of 0.4) to 750 °C (for XH$_2$O of unity) for a pressure of c. 7–9 kbar (Vielzeuf & Boivin 1984). The granulitic facies is also defined by the presence of antiperthitic feldspar (Fig. 4) in the sample

(Waard 1965). The greenschist retrogression marked by the development of chlorite after biotite and garnet according to the reaction Fe chlorite + quartz = almandine + H_2O (for XH_2O of unity) indicates a pressure below 5 kbar for a temperature of 550 °C (Hsü 1968.

To summarize, the thermobarometric calculations supported by microstructural analysis indicate that amphibolites and acidic gneiss underwent a metamorphic evolution characterized by an early high P–T stage (7 ± 1 kbar, 670 ± 40 °C) and a subsequent retrogression through low-grade amphibolite- and greenschist-facies conditions (5 ± 1 kbar and 550 ± 60 °C) contemporaneous with the development of the pervasive foliation and folds. The tectonic and metamorphic evolution of the rocks ended under ocean-floor conditions during which occurred brittle deformation characterized by the formation of veins filled with epidote and albite (Fig. 8).

Apatite fission-track dating

Background

The fission-track analysis of metamorphic or intrusive basement is based on the determination of (1) an apparent sample fission-track age and (2) the distribution of the total etchable track lengths. The meaning of this age, for a given phase, depends on its cooling history. During a cooling process, fission tracks become progressively more stable through the 120–60 °C temperature range, the so-called partial annealing zone (PAZ), where their total etchable length increases from zero to its maximum value of around 16 μm.

Total etchable lengths are determined by the measurement of confined tracks, i.e. tracks that are totally included in the grains. In the case of a fast (<5 Ma) cooling through the PAZ, confined tracks will have an average length value around or higher than 14 μm and a standard deviation lower than 1.5 μm. In the case of a slow but constant cooling through the PAZ, the length distribution will be biased towards shorter values, resulting in a shorter average length and a larger standard deviation. More complex t–T histories including mild reheating (<120 °C) event(s) might result in variously deformed length distributions with average values down to <12 μm and standard deviations up to >3 μm.

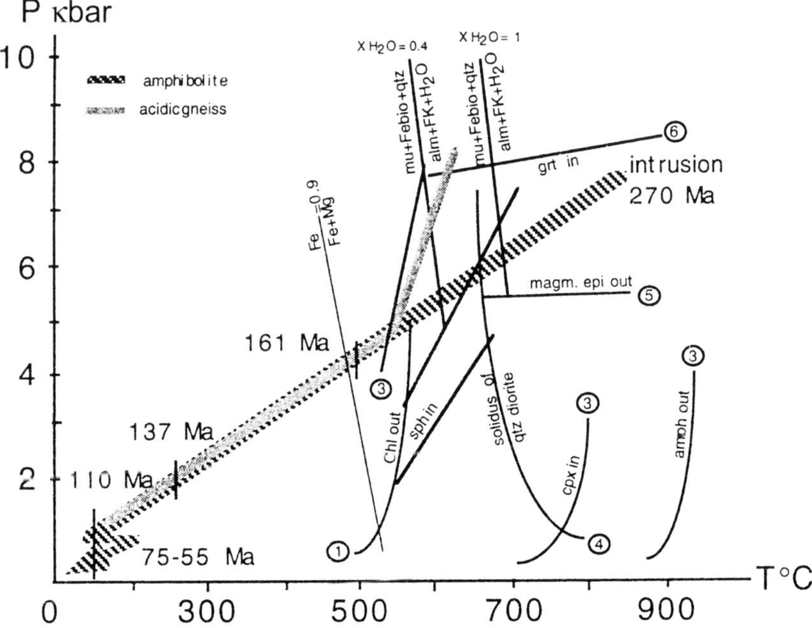

Fig. 8. The P–T–t evolution of the amphibolites and acidic gneisses from ODP Site 1067. Numbers refer to references cited: 1, Liou *et al.* (1974); 2, Apted & Liou (1983); 3, Spear (1981); 4, Piwinskii (1973); 5, Schmidt (1992); 6, Ernst & Liou (1998). Reactions mu + Febio + qtz = alm + FK + H_2O (for XH_2O of 0.4 and unity) from Vielzeuf & Boivin (1984). Fe/(Fe + Mg) is 0.9 in garnet, from Spear (1981).

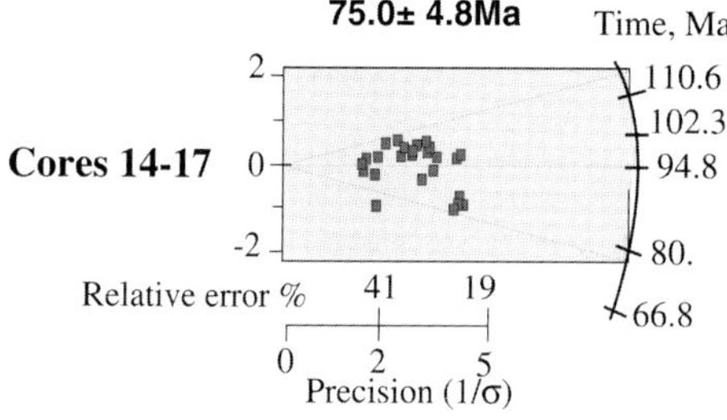

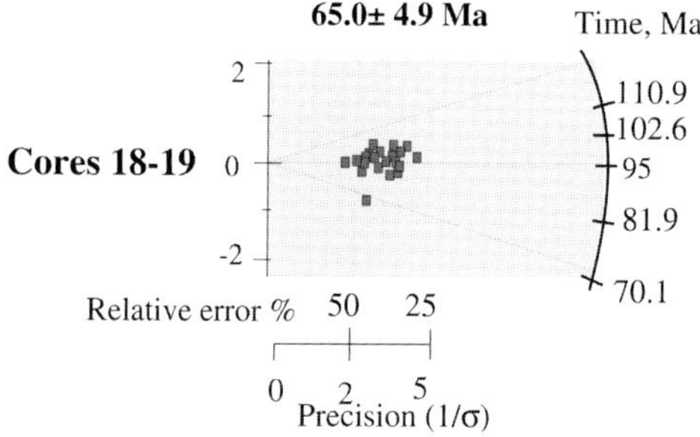

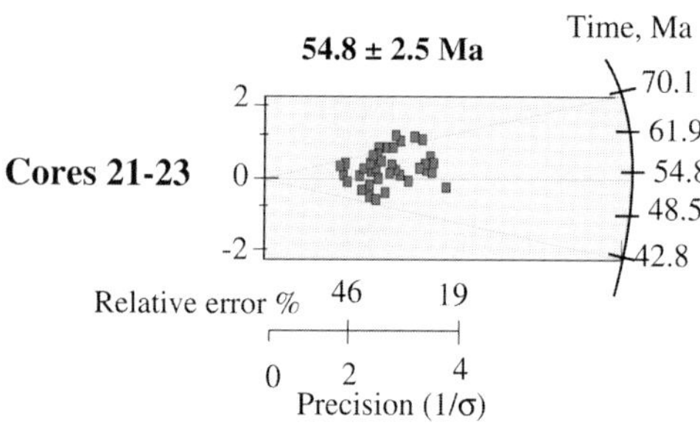

Fig. 9. Apatite fission tracks from the selected apatites from Units 1 (cores 14–17), 2 (cores 18–19) and 3 (cores 21–23) of the ODP Site 1067 amphibolites.

Sampling and results

The apatites from one sample of each of the amphibolite units were analysed. At least 20 grains per sample were dated individually. The CO_2 test of Galbraith *et al.* (1993) shows that in each sample only one age population is present, as illustrated by the radial plots (Fig. 9), where both the age of each grain and its precision are given. The central ages (Galbraith *et al.* 1993) of the samples are respectively 75 ± 4.8 and 65 ± 4.9 Ma for Units 1 and 2 amphibolites, and 54.8 ± 2.5 Ma for the weakly deformed amphibolites of Unit 3.

Fluorine and chlorine contents of the analysed apatite grains are shown in Table 5. Apa-

Table 5. *Selected analyses of the apatite grains from Units 1, 2 and 3 of the Site 1067 amphibolites*

Hole 1067A cores	F (wt %)	Cl (wt %)
Unit I		
14R	2.03	0.038
14R	1.927	0.03
14R	1.551	0.022
14R	1.678	0.022
14R	1.32	0.024
15R	1.664	0.018
15R	1.622	0.013
15R	1.596	0.14
16R	2.095	0.07
16R	1.838	0.016
16R	2.326	0.009
17R	1.964	0.025
17R	1.885	0.011
17R	1.808	0.01
17R	2.296	0.016
17R	1.913	0.006
17R	1.919	0.005
17R	1.653	0.016
17R	1.672	0.013
17R	1.268	0.018
Unit II		
18R	2.283	0.005
18R	2.891	0.003
18R	2.678	0.016
18R	2.541	0.021
19R	2.272	0.015
19R	2.428	0.014
19R	2.011	0.058
19R	2.511	0.017
19R	2.763	0.061
20R	2.727	0.012
20R	2.944	0.018
20R	3.557	0.062
20R	4.111	0.041
20R	2.72	0.01
20R	3.192	0.017
Unit III		
21R	2.143	0.017
21R	1.944	0.013
21R	1.751	0.002
22R	1.488	0.001
22R	1.614	0.004
22R	2.121	0.023
22R	1.752	0.013
23R	2.051	0.02
23R	1.776	0.063
23R	1.607	0.02

tites from all units have a chloride content range from 0.001 to 0.14 wt %. Fluoride contents are in the range 1.27–2.33 wt % in apatites from Units 1 and 3 and 2.68–4.11 wt % in apatite from Unit 2. Despite the highest fluoride contents of apatites occurring in the brecciated amphibolites (Unit 2), all grains are non-zoned pure fluorapatite.

Interpretation

These fission-track analysis cooling ages are younger than the Cretaceous rifting (122–114 Ma) and give an unexpected large age spectrum (25 Ma) through the 92 m of the cored amphibolites. The Monte Trax algorithm (Gallagher *et al.* 1995) suggests that the oldest recorded tracks in apatites are of about 100–110 Ma in Units 1 and 2 and 75–69 Ma in Unit 3 (Fig. 10). Differences between Units 1–2 and 3 are, however, most probably not attributable to differences in apatite chemical composition, as all belong to the fluorapatite pole. We suggest that the age distribution might result from local hydrothermal effects and/or a late tectonic event. Unimodal track length distributions suggest constant cooling in Units 1 and 3, whereas bimodal distributions in Unit 2 imply two thermal excursions below 120 °C. Monte Trax $t-T$ calculations indicate that this second excursion would have occurred some 65 Ma ago (Fig. 8). Thus it appears that the amphibolites record two thermal peaks below 120 °C. The first occurred before *c.* 100–110 Ma and would be post-dated by the oldest tracks preserved in the top and medium parts of the hole (Units 1 and 2), whereas the second was probably associated with fluid circulation and the development of fractures that occurred at *c.* 65 Ma. The presence of pyrite, epidote and albite filling the fractures indicates that the temperature of the fluid was above 150 °C. Such a temperature was high enough to totally erase the oldest post-rift tracks in Unit 3, and to partly erase them in Units 1 and 2.

Discussion

The oxygen isotope results indicate that the amphibolites have preserved a strong magmatic signature (whole-rock and relatively constant mineral $\delta^{18}O$ values) during their metamorphic evolution, suggesting that after the magmatic stage any high-temperature metamorphic event biased the $\delta^{18}O$ values. According to the magmatic ages, Site 1067 amphibolites derived from a gabbroic source underplated at the base of the thinned continental crust during Permian time (270 Ma). The acidic gneiss was previously interpreted as tonalite dyke of Hercynian age (340 Ma), which intruded the mafic rock (Manatschal *et al.* 2001). However, the younger magmatic age obtained from the amphibolites clearly shows that the acidic

gneisses are older than the amphibolites. We believe that the acidic layer represents the base of the Hercynian crust under which the gabbroic rock was underplated during Permian time. The intrusion of the gabbro melted the lower crust, causing growth of the outer rim of the zircons that are dated at 290 Ma in the acidic gneisses. The melt intruded through solidified gabbro, after which the gabbro and the acidic gneiss were exhumed during the Jurassic rifting (between 160 and 140 Ma). These results indicate that there was no continuous extension between the Permian intrusion and the Cretaceous rifting. Similar assemblage and evolution are preserved in the Malenco–Forno and Margna nappes in the eastern central Alps (Müntener *et al.* 2000). In this area, upper-mantle and lower-crust associations are exposed, showing that a gabbroic intrusion occurred during Permian time, inducing local partial melting and the granulitic-facies metamorphism in the lower crust (Hermann *et al.* 1997; Müntener *et al.* 2000). The gabbroic intrusion remained at 30 km depth for at least 50 Ma. Age determination using Ar/Ar analyses of amphibolite from the retrogressed gabbro indicate that the exhumation of the lower crust took place between 225 and 190 Ma (Müntener *et al.* 2000). Sm–Nd ages gave an age of 274 ± 11 Ma for a gabbroic dyke intruded into the Balmuccia peridotite from the Ivrea–Verbano Zone (Mayer *et al.* 2000). The same method gave ages of 267 ± 21 Ma and 244 ± 4 Ma for respectively a sample of amphibole gabbro and a sample of granulitic paragneiss from the Sesia and Sessera Valleys (Mayer *et al.* 2000).

Apatite fission-track dating also records the final uplift recorded by the Site 1067 amphibolites, The Monte Trax $t-T$ calculations indicate that in the Iberia Abyssal Plain the unroofing of the lower crust occurred until 100–110 Ma. Similar ages were obtained with different methods all along the West Iberia margin. At Gorringe Bank recrystallized plagioclase and amphibole from a flaser gabbro gave an age of 110 Ma (Feraud *et al.* 1986). At Galicia Bank, recrystallized plagioclase from a diorite gave an age of 118 Ma (Feraud *et al.* 1996). These ages are interpreted as the last step of unroofing of the mantle and underplated rocks (Cornen *et al.* 1999). Our results also demonstrate that this tectonic exhumation took place all along the West Iberia margin.

The very young ages that range from 75 to 55 Ma indicate that the Pyrenean compression overprinted earlier events. Until now, the compressional effect related to the Europe–Africa convergence has been observed in the northern

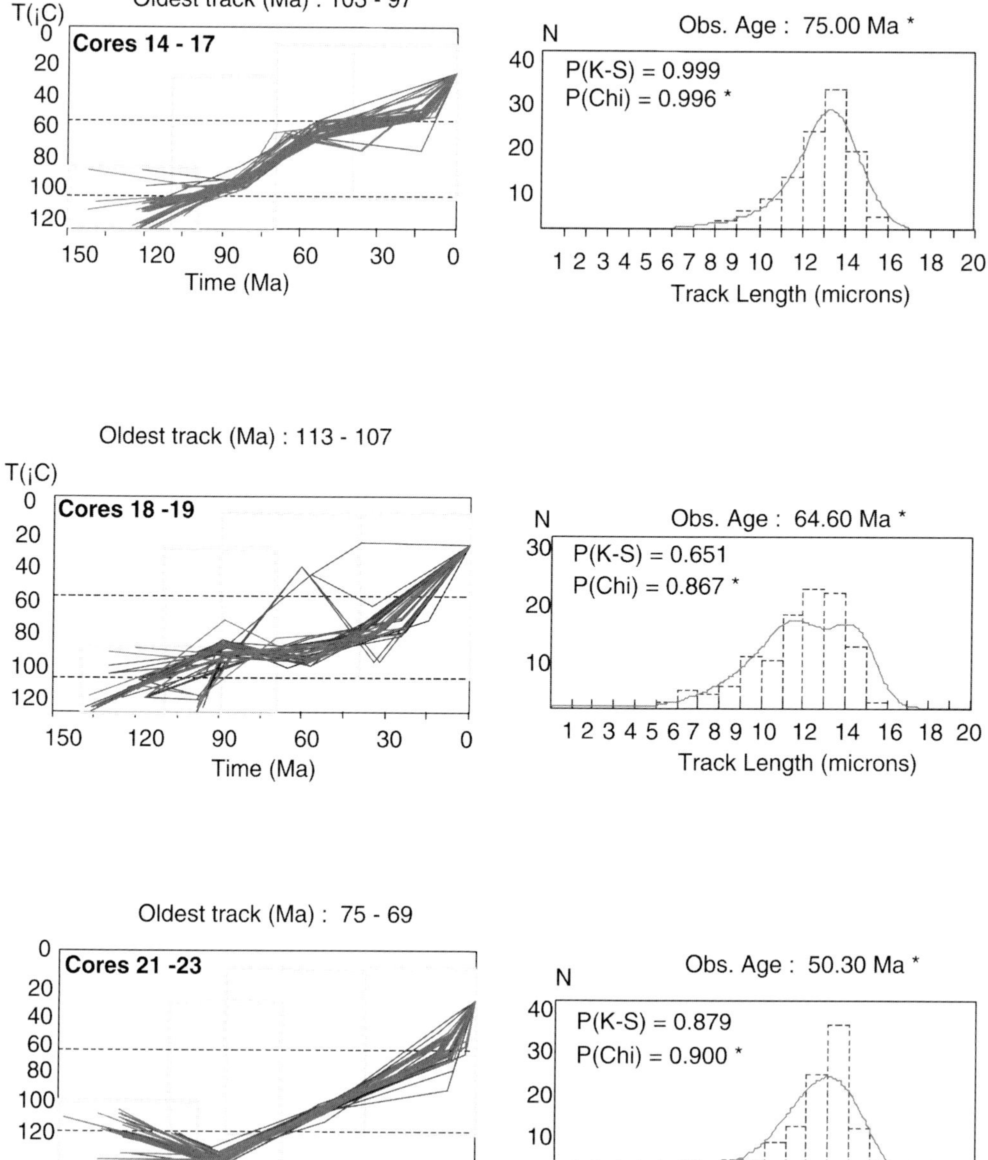

Fig. 10. Track length (in microns) of apatites from Units 1, 2 and 3, showing the bimodal distribution of the track length in Unit 2. Optimization of the data with the Monte Trax model of Gallagher *et al.* (1995) Gallagher (1995) showing the best thermal history for apatites with respect to the track length.

(Galicia Bank) and southern (Gorringe Bank) segments of the Iberia margin (Masson *et al.* 1994; Girardeau *et al.* 1999). The apatite fission-track dating on amphibolites from the Iber-ia Abyssal Plain indicates that the compression affects most of the margin. This late tectonic event must be associated with relatively hot (>120 °C) fluid circulation, which partly reset the apatite fission-track chronometer, annealing most of the old tracks related to Cretaceous rifting.

Conclusions

The amphibolite and the acidic gneisses from the Iberia Abyssal Plain (Site 1067) record similar metamorphic evolution characterized by an early high-T and -P stage ($670 \pm 40\,°C$, $7 \pm 1\,kbar$) followed by retrogression under medium-T and -P ($550\,°C$, $5.5 \pm 1\,kbar$) greenschist-facies conditions. The dynamic recrystallization responsible for the development of ductile structures (foliation plane, folds) occurred under amphibolite- to greenschist-facies conditions. The metamorphic retrogression was dominated by hydration reactions in the gabbroic rocks and occurred at temperatures between $500\,°C$ and $200\,°C$. Fluid inclusion analysis indicates that penetration of sea-water-dominated fluid occurred at a late stage (under greenschist-facies conditions) during the exhumation of the amphibolites. The fluid circulation is also responsible for the opening of veins and fractures filled with quartz, epidote and pyrite. Apatite fission-track dating gives central ages of $75 \pm 4.8\,Ma$ for the amphibolites in Unit 1, $65 \pm 4.9\,Ma$ for the amphibolites in Unit 2, and $54.8 \pm 2.5\,Ma$ for the weakly deformed amphibolites at the bottom of the hole (Unit 3). These cooling ages are clearly younger than the Cretaceous rifting and are probably related to the Pyrenean orogeny. Optimization of the data with the Monte Trax model of Gallagher *et al.* (1995) indicates that the amphibolites record a previous thermal excursion below $120\,°C$ at *c.* $113–100\,Ma$ related to Cretaceous rifting.

References

ANDERSON, L.P. & SMITH, D.R. 1985. The effects of temperature and fO_2 on the Al-in-hornblende barometer. *American Mineralogist*, **80**, 549–559.

APTED, M. & LIOU, J.G. 1983. Phase relations among greenschist, epidote-amphibolite, and amphibolite in a basaltic system. *American Journal of Science*, **283A**, 328–354.

BLUNDY, J.D. & HOLLAND, T.B. 1990. Calcic amphibole equilibrium and a new amphibole–plagioclase geothermometer. *Contributions to Mineralogy and Petrology*, **104**, 208–224.

BLUNDY, J.D. & HOLLAND, T.B. 1992. Calcic amphibole equilibrium and a new amphibole–plagioclase geothermometer: Reply to comments of Hammarstom and Zen, and Rutherford and

Johnson. *Contributions to Mineralogy and Petrology*, **111**, 269–272.

BROWN, P.E. & HAGEMANN, S.G. 1994. MacFlincor: a computer program for fluid inclusion data reduction and manipulation. *In*: DE VIVO, B. & FREZZOTTI, M.L. (eds) *Fluid Inclusion in Minerals: Methods and Applications*. VPI Press, 231–250.

CLAYTON, R.N. & MAYEDA, T.K. 1963. The use of bromine pentafluoride in the extraction of oxygen from oxides and silicates for isotopic analyses. *Geochimica et Cosmochimica Acta*, **27**, 43–52.

CORNEN, G., GIRARDEAU, J. & MONNIER, C. 1999. Basalts, underplated gabbros and pyroxenites record the rifting process of the west Iberian margin. *Mineralogy and Petrology*, **67**, 111–142.

ERNST, W.G. & LIOU, J. 1998. Experimental phase equilibrium study of Al- and Ti-contents of calcic amphibole in MORB—a semiquantitative thermometer. *American Mineralogist*, **83**, 952–969.

FERAUD, G., BESLIER, M.O. & CORNEN, G. 1996. $^{40}Ar/^{39}Ar$ dating of gabbros from the ocean/continent transition of the western Iberia Margin: preliminary results. *In*: WHITMARSH, R.B., SAWYER, D.S., KLAUS, A. & MASSON, D.G. (eds) *Proceedings of the Ocean Drilling Program, Scientific Results, 149*. Ocean Drilling Program, College Station, TX, 489–495.

FERAUD, G., YORK, D., MEVEL, C. & AUZENDE, J.M. 1986. Additional $^{40}Ar–^{39}Ar$ dating of the basement and alkaline volcanism of the Gorringe Bank (Atlantic Ocean). *Earth and Planetary Science Letters*, **79**, 255–269.

GALBRAITH, R.F., DUBBY, I.R. & LASLETT, G.M. 1993. Statistical models for mixed fission track ages. *Nuclear Tracks and Radioactive Measurement*, **21**, 459–470.

GALLAGHER, K., HAWKESWORTH, C.J. & MANTOVANI, M. 1995. Denudation, fission track analysis and the long-term evolution of passive margin topography: application to the southeast Brazilian margin. *Journal of South American Earth Science*, **8**, 65–77.

GILBERT, M.C., HELZ, R.T., POPP, R.K. & SPEAR, F.S. 1982. Experimental studies of amphibole stability. *In*: VEBLEN, D.R. & RIBBE, P.H. (eds) *Amphiboles: Petrology and Experimental Phase Relations*. Mineralogical Society of America, Reviews in Mineralogy, **9B**, 229–353.

GIRARDEAU, J., CORNEN, G., BESLIER, M.O. & 7 OTHERS 1999. Extensional tectonics in the Gorringe Bank gabbros, eastern Atlantic Ocean: witness of an ultra-slow accreting center. *Terra Nova*, **8**, 312–324.

HAWTHORNE, F.C. 1981. Crystal chemistry of the amphiboles. *In*: RIBBE, P.H. (ed.) *Amphiboles: Petrology and Experimental Phase Relations*. Mineralogical Society of America, Reviews in Mineralogy, **9B**, 1–102.

HERMANN, J., MÜNTENER, O., TROMMSDORFF, V. & HANSMANN, W. 1997. Fossil crust to mantle transition, Val Malenco (Italian Alps). *Journal of Geophysical Research*, **102**, 20123–20132.

HOLLAND, T. & BLUNDY, J. 1994. Non-ideal interactions in calcic amphiboles and their bearing on amphibole–plagioclase thermometry. *Contributions to Mineralogy and Petrology*, **116**, 433–447.

HSÜ, L.C. 1968. Selected phase relationships in the system Al–Mn–Fe–Si–O–H: a model for garnet equilibria. *Journal of Petrology*, **9**, 49–83.

LAIRD, J. & ALBEE, A.L. 1981. High-pressure metamorphism in mafic schist: its application to reconstructing the polymetamorphic history of Vermont. *American Journal of Science*, **281**, 127–175.

LEAKE, B.E. 1978. Nomenclature of amphiboles. *Mineralogical Magazine*, **42**, 533–563.

LEMOINE, M., TRICART, P. & BOILLOT, G. 1987. Ultramafic and gabbroic ocean floor of the Ligurian Tethys (Alps, Corsica, Apennines): in search of a genetic model. *Geology*, **15**, 622–625.

LIOU, J.G., KUNIYOSHI, S. & ITO, K. 1974. Experimental studies of the phase relations between greenschist and amphibolite in basaltic system. *American Journal of Science*, **274**, 613–632.

LIOU, J.G., MARUYAMA, S. & CHO, M. 1985. Phase equilibria and mineral paragenesis of metabasites in low-grade metamorphism. *Mineralogical Magazine*, **49**, 321–333.

MANATSCHAL, G. & BERNOULLI, D. 1999. Architecture and tectonic evolution of nonvolcanic margins: present day Galicia and ancient Adria. *Tectonics*, **18**, 1099–1119.

MANATSCHAL, G., FROITZHEIM, N., RUBENACH, M. & TURRIN, B.D. 2001. The role of detachment faulting in the formation of an ocean–continent transition: insights from the Iberia Abyssal Plain. *In*: WILSON, R.C.L., WHITMARSH, R.B., TAYLOR, B. & FROITZHEIM, N. (eds) *Non-volcanic Rifting of Continental Margins: a Comparison of Evidence from Land and Sea*. Geological Society, London, Special Publications, **187**, 405–428.

MASSON, D.G., CARTWRIGHT, J.A., PINHEIRO, L.M., WHITMARSH, R.B., BESLIER, M.O. & ROESER, H. 1994. Compressional deformation at the ocean–continent transition in the NE Atlantic. *Journal of the Geological Society, London*, **154**, 607–614.

MAYER, A., MEZGER, K. & SINIGOI, S. 2000. New Sm–Nd ages for the Ivrea–Verbano Zone, Sesia and Sassera Valleys (Northern Italy). *Journal of Geology*, **30**, 147–166.

MOODY, J.B., MEYER, D. & JENKINS, J.E. 1983. Experimental characterization of the greenschist/amphibolite boundary in mafic systems. *American Journal of Science*, **283**, 48–92.

MÜNTENER, O., HERMANN, J. & TROMMSDORFF, V. 2000. Cooling history and exhumation of lower-crustal granulite and upper mantle (Malenco, eastern central Alps). *Journal of Petrology*, **41**, 175–200.

PINEAU, F., JAVOY, M., HAWKINS, J.W. & GRAIG, H. 1976. Oxygen isotope variations in marginal basin and ocean ridge basalts. *Earth and Planetary Science Letters*, **28**, 299–307.

PIWINSKII, A.J. 1973. Experimental studies bearing on the origin of the central and southern coast range granitoids, California. *Tschermaks Mineralogische und Petrographische Mitteilungen*, **20**, 107–130.

POTTER, R.W., CLYNNE, M.A. & BROWN, D.L. 1978. Freezing point depression of aqueous sodium chloride solution. *Economic Geology*, **73**, 283–285.

RAASE, P. 1974. Al and Ti contents of hornblende, indicators of pressure and temperature of regional metamorphism. *Contributions to Mineralogy and Petrology*, **45**, 231–236.

ROBINSON, P., SPEAR, F.S., SCHUMACHER, J.C., LAIRD, J., KLEIN, C., EVANS, B.W. & DOOLAN, B.L. 1982. Phase relations of metamorphic amphiboles: natural occurrence and theory. *In*: VEBLEN, D.R. & RIBBE, P.H. (eds) *Amphiboles: Petrology and Experimental Phase Relations*. Mineralogical Society of America, Reviews in Mineralogy, **9B**, 1–227.

RUBENACH, M. & WYCSOCZANSKI, R., 1998. Metagabbros and metatonalites from the ocean–continent transition, Iberia Abyssal Plain: synrift or Hercynian basement? *Geological Society of Australia, 14th Australian Convention, Abstracts*, 386..

SPEAR, F.S. 1980. NaSi–CaAl exchange equilibrium between plagioclase and amphibole. *Contributions to Mineralogy and Petrology*, **72**, 33–41.

SPEAR, F.S. 1981. An experimental study of hornblende stability and compositional variability in amphibole. *American Journal of Science*, **281**, 697–734.

TROMMSDORFF, V., PICCARDO, G. & MONTRASIO, A. 1993. From magmatism through metamorphism to sea floor emplacement of subcontinental Adria lithosphere during pre-Alpine rifting (Malenco, Italy). *Schweizerische Mineralogische und Petrographische Mitteilungen*, **72**, 191–203.

VIELZEUF, D. & BOIVIN, P. 1984. An algorithm for the construction of petrogenetic grids: application to some equilibria in granulitic paragneisses. *American Journal of Science*, **284**, 760–791.

WAARD, D. 1965. A proposed subdivision of the granulites facies. *American Journal of Science*, **263**, 455–461.

WHITMARSH, R.B., SAWYER, D.S., KLAUS, A. & MASSON, D.G. (eds) *Proceedings of the Ocean Drilling Program, Initial Reports, 149*. Ocean Drilling Program, College Station, TX.

WHITMARSH, R.B., BESLIER, M.O., WALLACE, P.J. *et al.* 1998. *Proceedings of the Ocean Drilling Program, Initial Reports, 173*. College Station, TX, Ocean Drilling Program.

ZHANG, Y.G. & FRANTZ, J.D. 1987. Determination of the homogenization temperature and density of supercritical fluids inclusions. *Chemical Geology*, **64**, 335–340.

Palaeomagnetic and rock magnetic results from serpentinized peridotites beneath the Iberia Abyssal Plain

XIXI ZHAO

Center for Studies of Imaging and Dynamics of the Earth, Institute of Geophysics and Planetary Physics, University of California, Santa Cruz, CA 95064, USA (e-mail: xzhao@es.ucsc.edu)

Abstract: A palaeomagnetic and rock magnetic study has been performed on serpentinized peridotite rocks recovered at Ocean Drilling Program (ODP) Leg 149 and 173 sites, off the west coast of Portugal (Sites 897, 899 and 1070). Stable components of magnetization are revealed after detailed thermal and alternating field demagnetization. Results from a series of rock magnetic measurements with advanced instrumentation all corroborate the demagnetization behaviour and show that (titano)magnetites and maghemite are present in the peridotites. At Leg 149 and 173 sites, the inclinations of characteristic magnetization in the 'fresher' lower part of the serpentinized peridotite section show a predominantly reversed polarity in a depth zone of $c.$ 25 m, which is compatible with a reversed event (probably correlated with marine Anomaly M0 at $c.$ 121 Ma) before the Cretaceous Long Normal Superchron (83–120 Ma) and is incompatible with a Holocene field direction. In contrast, the inclinations of samples from the more 'altered' upper part are predominantly normal. In view of the polarity of magnetization identified from the overlying Cretaceous and Tertiary sediments and the newest interpretation of the drilling results for the Iberia margin, our results suggest that the Iberian peridotites were emplaced during Aptian–Barremian time and recorded the middle Cretaceous geomagnetic field during Anomaly M0 time ($c.$ 121 Ma). Our data are consistent with new radiometric dates for the Iberian peridotites at Leg 173 drill sites.

The west Iberia margin is a non-volcanic rifted margin characterized by an apparent lack of syn-rift volcanism. Quantitatively rigorous fits of North America to Europe repeatedly show that, once the Bay of Biscay itself has been closed by a clockwise rotation of Iberia against Europe, the SE Grand Banks margin can be matched with the west Iberia margin (Bullard *et al.* 1965; Le Pichon *et al.* 1977; Courtillot 1982; Masson & Miles 1984; Klitgord & Schouten 1986; Srivastava & Tapscott 1986; Srivastava & Verhoef 1992; Sibuet & Srivastava 1994). Detailed studies of the magnetic anomalies of the west Iberia margin have underpinned many such reconstructions and suggested that Iberia separated from the Newfoundland margin of the Grand Banks in Early Cretaceous time (Whitmarsh *et al.* 1990, 1996*a,b*; Pinheiro *et al.* 1992; Whitmarsh & Miles 1995).

As shown in Figure 1, the western continental margin of Iberia encompasses a north–south-trending narrow shelf and a steep continental slope. Sea-floor spreading between the Iberia margin and its conjugate Newfoundland margin propagated south to north in three stages, which led to the development of three segments with different structural histories (see the review by Pinheiro *et al.* (1996)). Broadly, these segments of margin can be divided from south to north into the Tagus Abyssal Plain segment (earliest sea-floor spreading at 136 Ma), the southern Iberia Abyssal Plain segment (130 Ma), and the Galicia Bank segment (120 Ma). The precise dating and duration of each stage of rifting are arguable (Fuegenschuh *et al.* 1998), spanning a period of up to 20 Ma (Féraud *et al.* 1982, 1986, 1988, 1996; Schärer *et al.* 1995, 2000).

Research on the west Iberia margin has been concentrated on these three separate segments of the margin. Sampling of the basement has shown that a north–south basement ridge, which marks the oceanward edge of the ocean–continent transition zone (OCT) and lies along two of the segments (open diamonds in Fig. 1), is made of serpentinized peridotite and intrusive gabbros (Boillot *et al.* 1980, 1987*a,b*, 1988, 1995; Whitmarsh *et al.* 1996*a*, 1998; Discovery 215 Working Group 1998; Girardeau *et al.*

From: WILSON, R.C.L., WHITMARSH, R.B., TAYLOR, B. & FROITZHEIM, N. 2001. *Non-Volcanic Rifting of Continental Margins: A Comparison of Evidence from Land and Sea*. Geological Society, London, Special Publications, **187**, 209–234. 0305-8719/01/$15.00 © The Geological Society of London 2001.

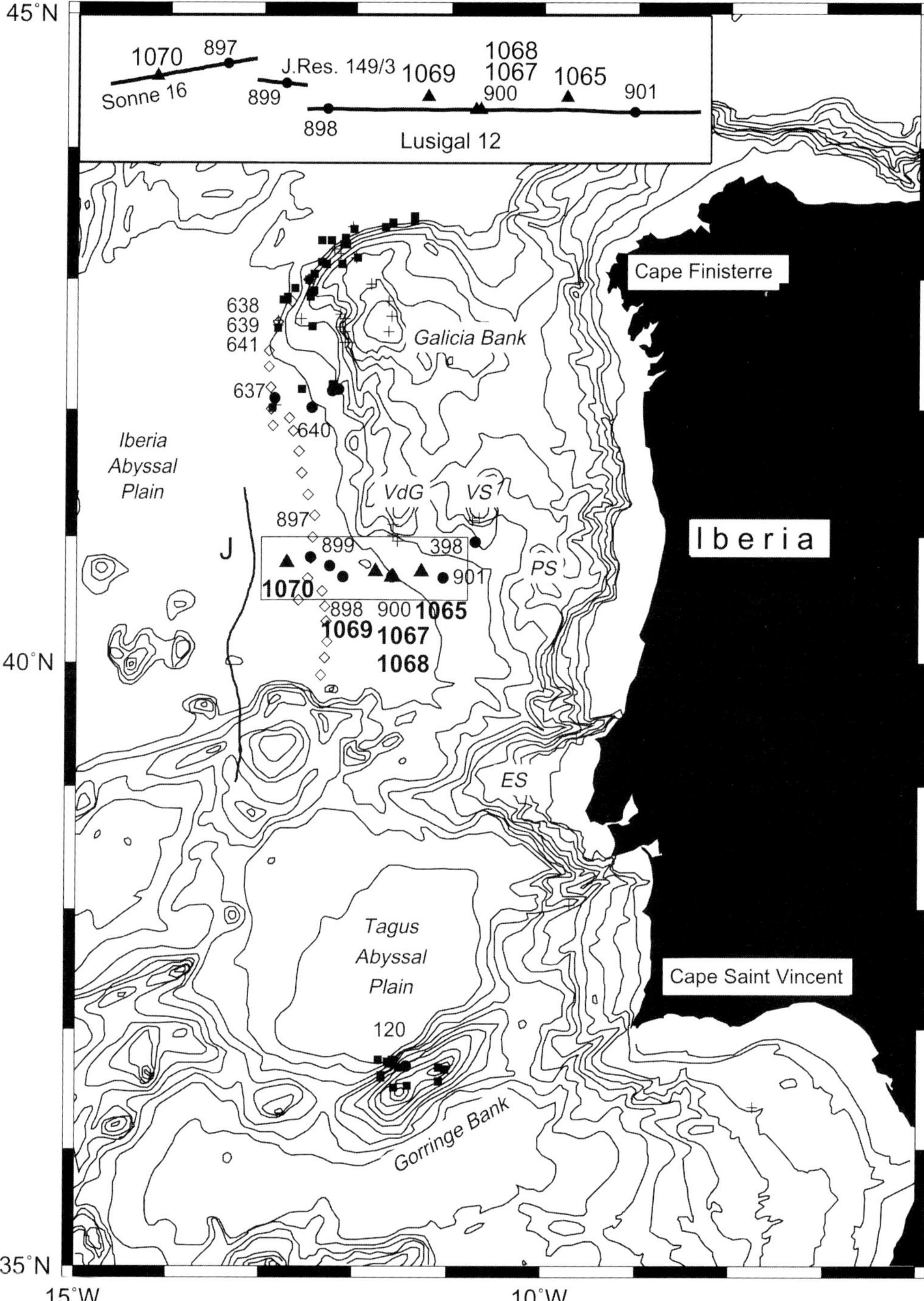

Fig. 1. Contoured regional bathymetric chart of the west Iberia margin. Contours at 500 m intervals. Rectangular inset in the centre of the chart shows locations of Leg 149 (●) and Leg 173 sites (▲ with bold numbers). J, J magnetic anomaly; PS, VS and VdG, Porto, Vigo and Vasco da Gama Seamounts, respectively; ES, Estremadura Spur. ● outside the inset, sites drilled on earlier legs. ■, +, rock samples obtained by submersible and dredge, respectively. ◇, location of the crest of the peridotite ridge. The top inset is a composite of seismic tracks showing the relative locations of Leg 149 and Leg 173 drill sites.

1999). Two en echelon ridges in the southern Iberia Abyssal Plain have also been proposed to consist of serpentinized peridotite (Beslier *et al.* 1993; Sawyer *et al.* 1994), which overlap horizontally by >25 km. No evidence exists to suggest that faulting activity in the region produced the offset ridges. Such a distribution of peridotite makes the peridotite ridges in the western Iberia margin one of the most enigmatic elements in the geological history of the region (Whitmarsh *et al.* 1996*b*, 1998). In addition, these ridges are parallel to the sea-floor spreading isochrons to the west and possess a dominantly remanent magnetization with a Cretaceous declination (Discovery 215 Working Group 1998), suggesting that they may have been formed by sea-floor spreading. The intimate coexistence between the marine magnetic anomaly and the margin-parallel peridotite ridge implies an integrated dynamic process and may contain clues to our understanding of the rifting process.

Palaeomagnetism continues to play a pivotal role as a technique for dating and correlation in Ocean Drilling Program (ODP) studies. Palaeomagnetic polarity dating is based on the facts that the Earth's magnetic field occasionally reverses polarity and that many rocks retain a magnetic imprint of the field at the time they were formed or altered. In favourable circumstances, palaeomagnetic dating can furnish highly resolved numerical ages by identifying the polarity patterns and fitting the polarity patterns into biostratigraphically identified zones, the geomagnetic polarity time scale, or other geochronological frameworks. Today, the general pattern of geomagnetic polarity reversals during Cenozoic and late Mesozoic time is well documented (Cande & Kent 1995). A number of short-period reversed events have been identified within the Cretaceous interval of dominantly normal polarity (Tarduno *et al.* 1991). The best documented of these is the reversed polarity event, of possible *c.* 1 Ma duration, situated close to the Barremian–Aptian boundary (Helsey & Steiner 1969; McElhinny & Burek 1971; Pechersky & Khramov 1973; Channell *et al.* 1987). This reversal has been correlated with marine Anomaly M0. The M0 used in the Gradstein *et al.* (1995) time scale is placed at 121 Ma, which immediately precedes the Cretaceous Long Normal Superchron (83–120 Ma).

Recently, measurement of rock magnetic parameters has also been demonstrated to be useful for studying various rock-forming and rock-altering geological processes (Banerjee 1992; Halgedahl 1992; Jackson *et al.* 1993,

1998). Rock magnetic properties can be used to identify the magnetic mineralogy and particle sizes, to subdivide or to correlate different sequences of the deposits, and to help assess the origin and stability of remanent magnetization. The combination of palaeomagnetic and rock magnetic studies has importance in understanding the relationship of magnetic properties to the geological processes in the Iberian region and may help answer key questions such as why the peridotite ridge is >40 km wide at the western edge of the OCT.

The purpose of this paper is to present the results of a palaeomagnetic and rock magnetic investigation carried out on cores recovered from Leg 149 and 173 sites. We will first briefly introduce the background information about the study, and then describe the palaeomagnetic and rock magnetic results from our own work to date on Leg 149 and 173 cores, focusing on the cores that provided the most readily interpretable data. We then summarize the data and explore their implications for understanding the rifting to sea-floor spreading history of the west Iberia margin.

Background information and site setting

In 1993, ODP Leg 149 drilled a west-to-east transect of five sites west of Portugal (see Fig. 1, inset). At three of these sites (Sites 897, 899 and 900) acoustic basement was reached, and at Sites 897 and 899 a sequence of serpentinized peridotite and associated mantle rocks was recovered from the basement. Site 897 was drilled on the crest of a narrow elongate basement ridge that had been linked to a peridotite ridge drilled west of Galicia Bank 140 km to the north (see Fig. 1). Site 899 also lay within the OCT and is *c.* 20 km east of Site 897. At both sites, the upper part of the peridotite section is pervasively veined and altered to a brown serpentinized peridotite. The finer-grained matrix within the upper part is dominantly dark yellowish brown (10YR 4/2) to moderate yellowish brown (10YR 5/4). The lower part of the peridotite section consists of 'fresher' (compared with the upper part of the section) serpentinized peridotites and no carbonate, and is generally dark green (5GY 4/1) or greenish black (5G 2/1). At both holes, upper Cretaceous sediments were also penetrated immediately above the basement, and lower Cretaceous (late Hauterivian) claystone and siltstone were found in mass-flow deposits or breccias dominated by basement clasts that appeared to merge downwards into basement.

Leg 173 was a sequel to Leg 149. A total of five sites were drilled during Leg 173 in 1997 (Fig. 1). No rocks characteristic of the upper oceanic crust were found at these sites. Basement rocks at two of the five sites (Sites 1068 and 1070) included a complex exposure of crust and mantle rocks including serpentinized peridotite, alternating tectonic breccia, and amphibolite-grade metagabbro. The enigmatic J anomaly (bold line in Fig. 1), which has been shown in the Central Atlantic to be 120–125 Ma old (Tucholke & Ludwig 1982), is identified from just west of Site 1070 (only 30 km). Similar to basement at Sites 897 and 899, the upper part of the basement at Site 1070 consists of a carbonate–serpentine zone containing abundant carbonate veins. These rocks are yellowish to brown and exhibit the chemical effects of alteration by cold sea water. Below this upper zone rocks lack carbonates and comprise green, grey and black serpentinized peridotites. It appears that this colour variation and calcite veining are the two most obvious visible features in the Iberian peridotites. For convenience, the terms 'altered' and 'fresher' peridotites are used in this paper to denote the upper and lower zones, respectively, even thought the Iberian peridotites were all serpentinized.

Laboratory and analytical methods

Palaeomagnetic sampling

All the cores at Holes 897C, 897D, 899B and 1070A were drilled using a rotary core barrel (RCB). Therefore, these cores were not oriented with respect to north. Samples were all taken from long and continuous pieces of core sections so as to exclude those that had flipped over inside the drilling pipe, thus ensuring the inclinations of the samples were not disturbed. These 2.5 cm cylindrical samples were drilled using a water-cooled nonmagnetic drill bit attached to a standard drill press. The depth, freshness, colour and general appearance of each sample were carefully recorded. This sampling procedure facilitated the process of comparing the palaeomagnetic results of individual samples with the corresponding core photo and assessing the validity of the palaeomagnetic results in a reliable manner. All samples were kept in a field-free room to prevent viscous remanence acquisition.

Magnetic measurement procedure

The palaeomagnetic and rock magnetic data presented in this paper are of two types: those obtained using the shipboard pass-through cryogenic magnetometer and those derived from analysis of discrete samples both on ship and onshore. In the shipboard pass-through system, magnetic measurements were performed by passing continuous archive half-core sections through a 2 G cryogenic magnetometer, and were taken at intervals of 5 or 10 cm along the core, and after alternating field (AF) demagnetization at 10 and 15 mT. Whole-core magnetic susceptibility was measured at 3 cm intervals on selected sections using the Bartington Instrument susceptibility meter mounted on the multisensor track. During thermal demagnetization, the initial susceptibility was monitored between each temperature step as a means of assessing any irreversible mineralogical changes associated with heating. A vector diagram was used for each sample to identify the components of magnetization that were present. Magnetic components were determined by fitting least-squares lines to segments of the vector demagnetization plots or 'the principal component analysis' (Zijderveld 1967; Kirschvink 1980) that were linear in 3D space.

Rock magnetic data of discrete samples in shore-based studies were obtained at the palaeomagnetism laboratories at the University of California at Santa Cruz and at the Institute for Rock Magnetism of the University of Minnesota. For rock magnetic characterization, samples were subjected to a wide range of magnetic measurements. These included: (1) the initial natural remanent magnetization (NRM) measurement followed by 2 weeks of zero-field storage; (2) remeasurement of NRM; (3) initial susceptibility; (4) Königsberger ratio; (5) hysteresis loop parameters: saturation magnetization (Js), saturation remanence (Jrs), coercivity (Hc) and remanent coercivity (Hcr); (6) thermomagnetic measurements including low-temperature (5 and 10 K) thermal demagnetization of saturation isothermal remanent magnetization (SIRM) and high-temperature (up to 700 °C) dependence of the magnetic moment and Curie temperature determination; (7) alternating current (a.c.) susceptibility measurements as a function of field amplitude and of frequency.

Magnetic results

The results of measurements (1)–(5) (see preceding paragraph) are given in Tables 1 and 2.

Table 1. *Remanent magnetic properties of minicore samples from Leg 149 and 173 sites*

Core, section, interval (cm)	Depth (mbsf)	NRM ($\times 10^{-1}$ A m^{-1})	Incl. (°)	χ (10^{-6} SI)	Q ratio
Site 897					
149-897C-					
64R-3, 64–66	660.74				
64R-5, 47–49	662.72				
67R-2, 98–100	689.58	3.22	−34.9	4660	1.93
69R-1, 51–53	706.71	2.51	−14.4	10400	0.67
70R-1, 58–60	710.78	0.83	−40.5	4690	0.49
72R-1, 49–51	726.09	5.91	2.8	14050	1.17
72R-1, 91–93	726.51	1.90	7.6	19680	0.27
72R-2, 108–110	728.13	1.54	22.6	11550	0.37
73R-3, 42–44	737.50	0.98	−61.2	11470	0.24
149-897D-					
11R-1, 69–71	694.49	3.12	14.5	15550	0.56
11R-2, 10–12	695.37	5.62	−69.9	19960	0.79
12R-1, 44–46	703.94	1.85	−22.1	15290	0.34
12R-4, 88–90	708.56	0.877	44.3	16580	0.15
13R-1, 62–64	713.62	5.68	20.6	14940	1.06
13R-5, 65–67	718.72	3.36	−23.9	12420	0.75
14R-3, 117–119	726.71	2.57	18.1	30750	0.23
15R-2, 86–88	734.50	5.04	44.4	24560	0.57
16R-2, 72–74	743.15	1.80	−52.1	26770	0.19
16R-3, 30–32	743.96	2.06	−31.9	12120	0.47
16R-3, 42–44	744.08	3.15	−35.0	7171	1.23
16R-4, 63–65	745.78	2.58	−2.3	20100	0.34
16R-5, 61–63	747.20	2.12	49.6	20920	0.28
17R-3, 26–28	754.86	3.81	−32.3	25360	0.42
17R-4, 14–16	755.81	0.29	−34.8	1254	0.65
17R-4, 118–120	756.85	0.32	−16.4	3123	0.27
17R-5, 7–9	757.10	0.27	−18.4	1746	0.43
17R-6, 15–17	757.93	0.78	−24.1	8652	0.25
18R-1, 10–12	761.36	0.21	−38.5	32390	0.02
18R-1, 16–18	763.24	0.14	−40.4	31960	0.01
19R-2, 58–60	772.69	1.1	33.4	14280	0.22
20R-1, 118–120	781.78	9.24	23.2	38080	0.68
20R-2, 135–137	783.42	7.39	−37.8	36690	0.56
21R-2, 16–18	791.10	4.61	10.0	23910	0.54
23R-3, 35–37	812.07	3.60	−10.5	25500	0.39
23R-4, 79–81	813.90	3.87	2.0	28890	0.37
23R-5, 31–33	814.84	7.56	5.4	30440	0.69
23R-5, 78–80	815.31	4.38	−26.3	37030	0.33
23R-6, 44–46	816.34	3.66	1.7	25150	0.41
24R-1, 88–90	819.48	1.05	−11.5	5805	0.51
24R-3, 57–59	821.29	2.17	38.5	22560	0.27
24R-4, 63–65	822.81	1.23	18.3	43860	0.08
25R-1, 20–22	828.40	1.34	−4.7	8304	0.45
Site 899					
149-899B-					
16R-2, 13–15	371.25	12.06	12.2	24590	1.37
16R-3, 57–59	372.96	8.24	20.4	17970	1.28
17R-1, 86–88	380.26	12.0	16.4	23460	1.43
17R-2, 4–6	380.87	7.38	12.4	21220	0.97
18R-1, 66–68	389.76	21.15	61.7	47060	1.25
18R-3, 57–59	392.22	8.76	−7.6	20060	1.22
18R-4, 15–17	393.28	8.2	10.5	23030	0.99
18R-5, 8–10	394.32	16.14	20.2	27410	1.64
19R-1, 24–26	398.51	21.66	−16.7	39570	1.53
19R-3, 70–72	401.85	22.45	57.5	73770	0.85
19R-4, 1–3	402.25	15.27	9.7	35700	1.19

Table 1. *Continued*

Core, section, interval (cm)	Depth (mbsf)	NRM ($\times 10^{-1}$ A m^{-1})	Incl. (°)	χ (10^{-6} SI)	Q ratio
20R-2, 112–114	410.32	11.29	17.2	8081	3.90
21R-1, 64–66	417.74	19.92	21.3	30290	1.84
21R-2, 75–77	418.99	7.29	26.7	7078	2.88
21R-3, 57–59	420.05	3.09	3.7	1509	5.72
21R-4, 97–99	421.93	8.89	50.7	1048	23.68
21R-5, 43–45	422.85	15.54	−6.8	27450	1.58
22R-1, 134–136	427.64	11.53	12.8	27970	1.15
22R-2, 27–29	428.03	21.76	−44.0	62650	0.97
23R-1, 4–6	435.64	11.56	29.6	27420	1.18
23R-1, 12–14	435.72	14.08	−23.1	32410	1.21
23R-4, 16–18	440.14	6.71	−26.9	20450	0.92
24R-1, 38–40	445.68	3.31	−33.8	16310	0.57
24R-2, 33–35	447.13	14.2	9.5	26390	1.50
24R-3, 121–123	449.47	12.88	−13.0	31770	1.13
25R-1, 67–69	455.37	6.16	13.2	23270	0.74
25R-3, 29–31	457.76	14.36	−17.8	36320	1.10
25R-3, 34–36	457.81	14.23	−23.3	40500	0.98
25R-1, 26–28	464.46	24.55	16.2	40850	1.68
26R-1, 60–62	464.8	13.71	12.6	33930	1.13
27R-2, 2–4	475.1	7.89	−25.3	34070	0.65
28R-1, 106–108	483.96	2.9	−22.0	34960	0.23
28R-2, 0–2	484.35	6.06	24.7	13360	1.27
29R-1, 118–120	493.28	22.17	48.8	40710	1.54
30R-1, 99–101	502.99	19.55	−79.7	42330	1.29
30R-2, 21–23	503	15.6	67.6	50230	0.87
Site 1070					
173-1070A-					
8R-1, 33–35	666.83	20.7	60.3	12146	4.76
8R-1, 107–109	667.57	10.8	61.4	40790	1.00
8R-3, 33–35	669.74	9.16	−7.3	30100	0.57
8R-4, 72–74	671.51	10.9	−15.7	53760	0.56
9R-1, 130–132	677.50	1.49	−51.6	2010	2.07
10R-1, 25–27	685.15	11.1	−24.5	52920	0.58
10R-1, 72–74	686.62	20.9	51.4	397	0.80
10R-1, 91–93	686.81	15.2	21.9	48010	0.88
10R-1, 95–97	686.85	23.3	53.2	3509	18.5
10R-2, 18–20	687.50	10.0	−5.3	73210	0.38
11R-1, 71–73	696.21	14.9	40.9	2733	15.2
11R-1, 81–83	696.31	17.0	−41.2	72400	0.66
11R-2, 4–6	696.64	5.12	21.9	91260	0.12
11R-2, 29–31	696.89	7.68	64.4	8292	2.60
11R-2, 80–82	697.40	5.95	−5.0	6300	2.63
11R-2, 99–101	697.59	53.8	−30.4	5510	27.3
12R-1, 21–23	705.31	8.76	1.6	23254	1.05
12R-1, 54–56	705.64	14.2	46.6	5496	7.20
12R-1, 78–80	705.88	10.2	−2.9	63300	0.45
12R-2, 5–7	706.01	11.6	−0.1	77310	0.42
12R-2, 40–42	706.36	58.5	−29.3	853	191.2
12R-2, 83–85	706.79	12.2	−20.0	4748	7.2
12R-2, 109–110	707.05	8.29	−4.8	37540	0.62
13R-1, 11–13	706.71	3.39	47.0	12635	0.75
13R-1, 44–46	707.04	4.35	−11.4	12630	0.96
13R-1, 50–52	707.10	60.8	62.5	7783	21.8
13R-1, 89–91	707.49	5.39	−10.8	46122	0.33
13R-2, 51–53	708.44	3.03	−3.1	15613	0.54
13R-2, 131–133	709.24	1.31	−14.9	5011	0.73
13R-3, 55–57	709.88	1.35	−0.8	7247	0.52
13R-4, 13–15	710.70	6.44E-2	−10.7	2336	0.77
13R-4, 68–70	711.25	9.39	41.3	4802	5.50

Table 1. *Continued*

Core, section, interval (cm)	Depth (mbsf)	NRM ($\times 10^{-1}$ A m^{-1})	Incl. (°)	χ (10^{-6} SI)	Q ratio
13R-4, 95–97	711.52	22.3	−20.4	12955	4.81
14R-1, 62–64	715.12	12.2	−15.1	56985	0.59
14R-1, 95–97	715.45	9.86	−1.0	19832	1.39
14R-2, 21–23	715.89	2.50	−10.2	54158	0.13
14R-3, 16–18	716.96	7.56	8.1	60903	0.35
14R-3, 62–64	717.72	10.7	−3.0	28901	1.03
14R-3, 102–104	717.82	8.74	−2.0	45392	0.54

NRM, NRM intensity; Inc., characteristic or stable remanent magnetization inclination after demagnetization; χ, low-field magnetic susceptibility; Q-ratio, Königsberger ratio.

Measurements (6) and (7) will be discussed separately. The most common features of the palaeomagnetic results of the peridotites are outlined below.

NRM and magnetic susceptibility

The intensities of NRM of the peridotites (NRM intensity in Table 1) are in the range of 14–924 mA m^{-1} in Hole 897D, 290–2455 mA m^{-1} in Hole 899B, and 64–6080 mA m^{-1} for samples from Hole 1070A. Although the intensity in Hole 899B is scattered (Fig. 2a), the peaks and lows in the downhole profile appear to coincide with petrographical boundaries in the section (Zhao 1996). In Hole 897D, the remanence successively shows a step-like decrease at first, and then increases from top to bottom. In detail it followed three trends: (1) from 694.49 to 754.86 m below sea floor (mbsf), it is scattered with a mean of 296 mA m^{-1}; (2) from 755.81 to 763.24 mbsf, the intensities are much weaker (mean 33 mA m^{-1}); (3) from 772.69 to 828.40 mbsf, the intensities are stronger (mean 394 mA m^{-1}) and more scattered (Fig. 2b). In Hole 1070A, the intensities are scattered (Fig. 2c) and appear stronger in the lower part of the hole. It is interesting to note that the NRM intensity of peridotites that were recovered at Site 637A by Leg 103 from the northernmost part of the peridotite ridge appears to increase with depth as well (Ogg 1988).

Volume magnetic susceptibility of natural materials in a weak magnetic field depends on the abundance and grain size of ferromagnetic minerals as well as on grain size and other factors such as stress. At all Leg 149 and 173 sites, variations in magnetic susceptibility of serpentinized peridotite generally parallel the

Table 2. *Hysteresis properties of minicore samples from the Leg 149 and 173 sites*

Core, section, interval (cm)	Depth (mbsf)	Hc (mT)	Hcr (mT)	Hcr/Hc	Js (mAm^2 kg^{-1})	Jrs (mAm^2 kg^{-1})	Jrs/Js
Site 897							
897C-67R-2, 98–100[1]	689.58	13.1	26.7	2.04	1170	176	0.150
897C-69R-1, 51–53[1]	706.71	4.94	25.4	5.14	731	32.7	0.045
897D-14R-3, 117–119[1]	726.71	4.8	17.0	3.54	273	18.6	0.068
897D-16R-2, 72–74	743.15	8.75	23.1	2.64	3820	367	0.096
Site 899							
899B-16R-3, 57–59[1]	372.96	9.55	26.7	2.80	521.3	66.2	0.127
899B-20R-2, 112–114[1]	410.32	14.6	32.4	2.22	8000	1240	0.156
899B-29R-1, 118–120	493.28	19.0	32.0	1.68	25.7	6.25	0.243
Site 1070							
1070A-8R-1, 107–109[1]	667.57	13.5	25.1	1.86	496	95	0.192
1070A-10R-1, 25–27	685.15	12.8	24.2	1.89	1910	333	0.175
1070A-11R-1, 81–83	696.31	9.6	20.7	2.16	1940	227	0.117
1070A-13R-1, 44–46	707.04	7.61	19.8	2.60	6860	473	0.069
1070A-13R-4, 95–97	711.52	8.35	27.0	3.23	9.53	1.02	0.107
1070A-14R-2, 21–23	715.89	10.4	20.9	2.01	466	56.9	0.122

[1] Samples from the 'altered' zone. Other symbols are defined in the text.

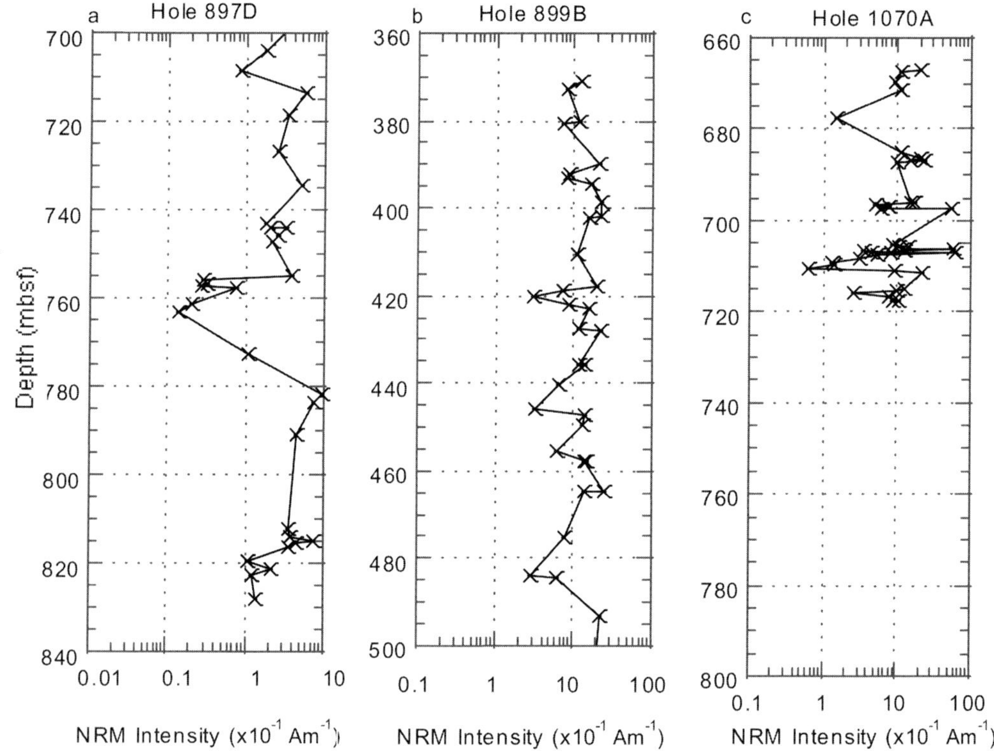

Fig. 2. Intensity of NRM (Table 1) plotted against depth in (**a**) Hole 897D, (**b**) Hole 899B, and (**c**) Hole 1070A.

variations in NRM intensity, although the scatter in susceptibility is less than that of the NRM intensities (Fig. 3). The susceptibility of serpentinized peridotite samples from Sites 897, 899 and 1070 is much higher than that of the overlying sediments, as documented in appropriate site chapters (Sawyer *et al.* 1994; Whitmarsh *et al.* 1998). The magnetic susceptibility of peridotites in Hole 899B (mean 2.959 × 10^{-2} SI units) is *c.* 1.4 times higher than those of Hole 897D (mean 2.067 × 10^{-2} SI units). Hole 1070A peridotites have a mean susceptibility of 2.902 × 10^{-2} SI units, similar to those of Holes 897 and 899.

Demagnetization behaviour

One of the major experimental requirements in palaeomagnetic research is to isolate the characteristic remanent magnetization by selective removal of secondary magnetization. To investigate the nature of the remanent magnetization of the peridotites, discrete samples were stepwise AF or thermally demagnetized. The NRM intensity or direction of the minicore samples did not change significantly (<5%) after zero-field storage for two weeks. As mentioned above, we used magnetic susceptibility to monitor the production of new magnetic materials during thermal demagnetization that might have altered the remanence. Apart from small insignificant fluctuations, the susceptibility of minicore samples generally did not change until after they were heated to 350 °C. Above this temperature, many samples showed a decrease or increase in susceptibility. The maghemite to hematite transition at about 300 °C produces a decrease in susceptibility, which is a useful means of identifying the presence of this mineral (Opdyke & Channell 1996). In addition, progressive thermal demagnetization on several minicores of serpentinized peridotite revealed that a component of viscous demagnetization parallel to the present-day magnetic field has been recorded in these rocks. The measured viscous remanent magnetization has been demonstrated to be useful for orienting cores for structural studies (see Shipboard Scientific Party 1998*a,b,c,d*; chapters for Sites 1067, 1068, 1069 and 1070, respectively).

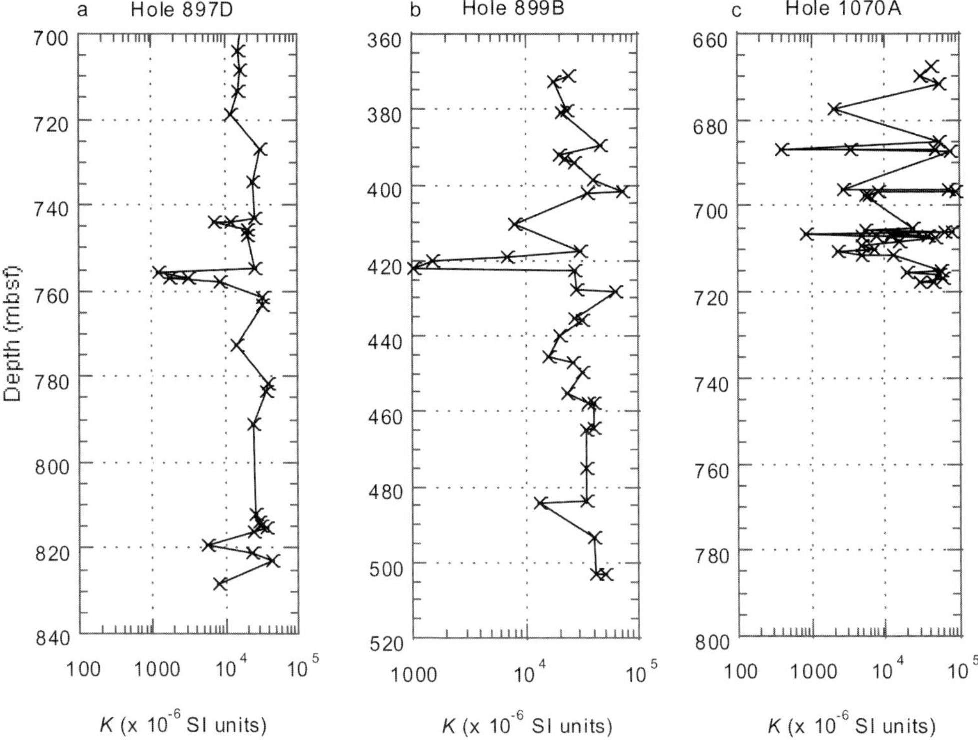

Fig. 3. Magnetic susceptibility (Table 1) plotted as a function of depth from (**a**) Hole 897D, (**b**) Hole 899B, and (**c**) Hole 1070A.

As shown in Figure 4a and b, thermal demagnetization on Samples 149-899B-23R-1, 12–14 cm, and 149-897D-16R-2, 72–74 cm (both from the 'fresher' lower part of the section) removed a 'soft' component, probably of viscous origin, at low to intermediate temperatures (200–400 °C). Demagnetization up to 585–620 °C revealed the stable component of magnetization. In the 'altered' upper part, however, most samples show a single component of magnetization during demagnetization (Fig. 4e–h). In addition, it appears that there is no significant difference in demagnetization behaviour of peridotite samples throughout this part of the section whether samples are from matrix, clast or even veins. Examples of this behaviour are shown in Figure 4g: Sample 149-899B-20R-2, 112–114 cm, which was specially chosen from a heavily veined piece of core, exhibited identical magnetic properties to a veinless sample (Sample 149-899B-21R-2, 75–77 cm, Fig. 4h).

The magnetically cleaned inclinations from all three holes are systematically shallower than the inclination expected today (59°) at the drill sites, but are statistically indistinguishable from the Jurassic–Cretaceous inclinations for Iberia (33.4–45.2°, α_{95} c. 10°; see Van der Voo 1969; Galdeano *et al.* 1989). Assuming the inclination represents the primary remanence at the time when these rocks were formed, the similarity between the observed and the expected inclinations is consistent with the notion that the drill sites were part of the Iberia plate at the time of acquisition of the magnetization. Alternatively, the shallow inclination could indicate that these sections have been tilted after the acquisition of the magnetization.

Königsberger ratio

The Königsberger ratio Q is defined as the ratio in a rock of remanent magnetization to the induced magnetization in the Earth's field. In general, the Königsberger ratio is used as measure of stability to indicate a rock's capability of maintaining a stable remanence. The International Geomagnetic Reference Field (IGRF) value at the Leg 149 and 173 sites (45 000 nT = 35.83 A m^{-1}, IAGA Division V,

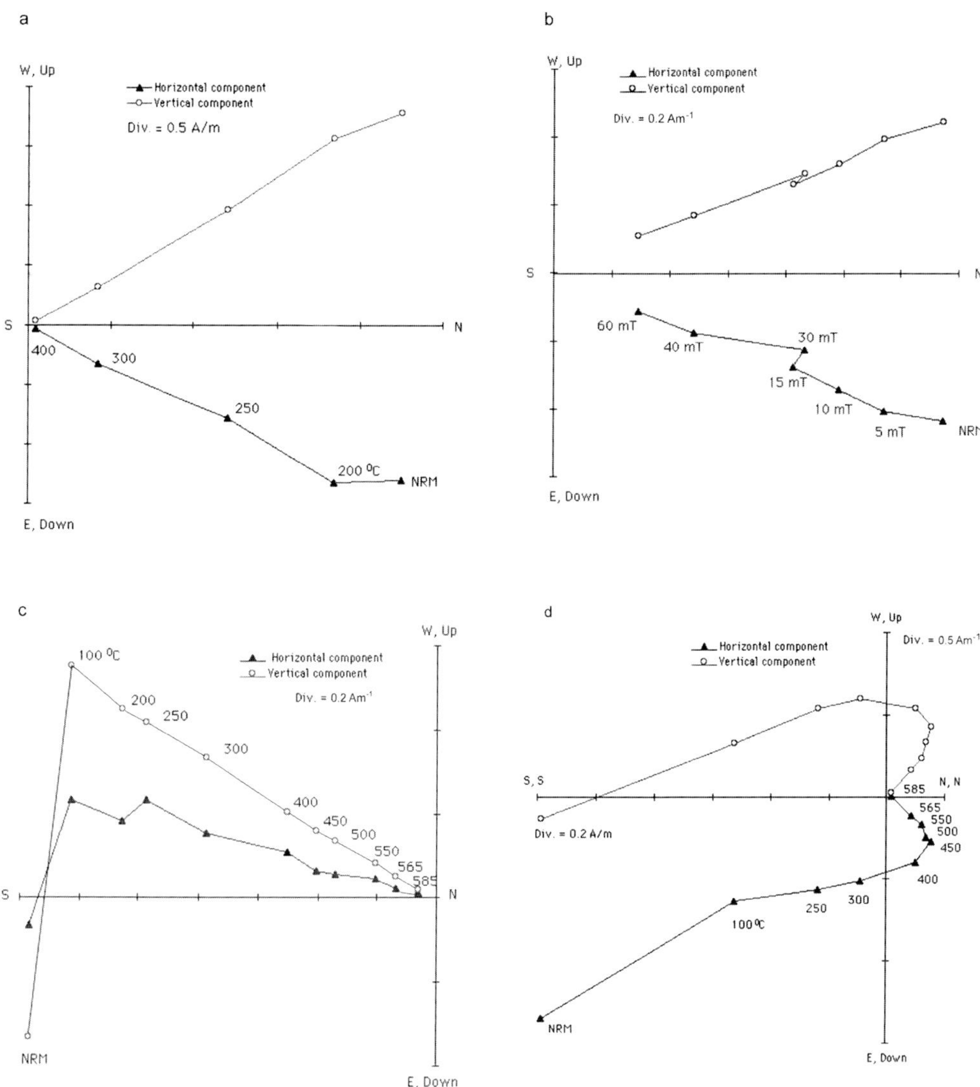

Fig. 4. Representative vector endpoint diagrams showing the results of thermal demagnetization for samples from the peridotite section. Samples from the 'fresher' lower part of the peridotite section at both Holes 897D and 899B are displayed in (**a**)– (**d**). Reversed polarity of magnetization is revealed in all samples. (**a**) Sample 149-897D-16R-3, 42–44 cm. (**b**) Sample 149-897D-16R-3, 30–32 cm. (**c**) Sample 149-899B-23R-1, 12–14 cm. (**d**) Sample 149-897D-16R-2, 72–74 cm. It should be noted that Samples 149-897D-16R-3, 30–32 cm (**a**) and 149-897D-16R-3, 42–44 cm (**b**) were taken from same piece of continuous core section. Samples from the 'altered' upper part of peridotite section are shown in (**e**)– (**h**). Normal polarity of magnetization is revealed in all samples. (**e**) Sample 149-899B-16R-3, 57–59 cm. (**f**) Sample 149-899B-21R-4, 77–79 cm. (**g**) Sample 149-899B-20R-2, 112–114 cm. (**h**) Sample 149-899B-21R-2, 75–77 cm. Sample 149-899B-20R-2, 112–114 cm (**g**) was specially chosen from a heavily veined core section, whereas Samples 149-899B-16R-3, 57–59 cm (**e**) and 149-899B-21R-2, 75–77 cm (**h**) were from relatively 'vein-free' pieces.

Working Group 8 1995) was used for calculating Q, where $Q = \text{NRM}$ (A m^{-1})/(k (SI) $\times$ H (A m^{-1})), and H is the local geomagnetic field.

The distribution of Q values in both Leg 149 and 173 holes (Table 1, Fig. 5) in general resembles that of the NRM. Although the mean susceptibility values of peridotites from Holes

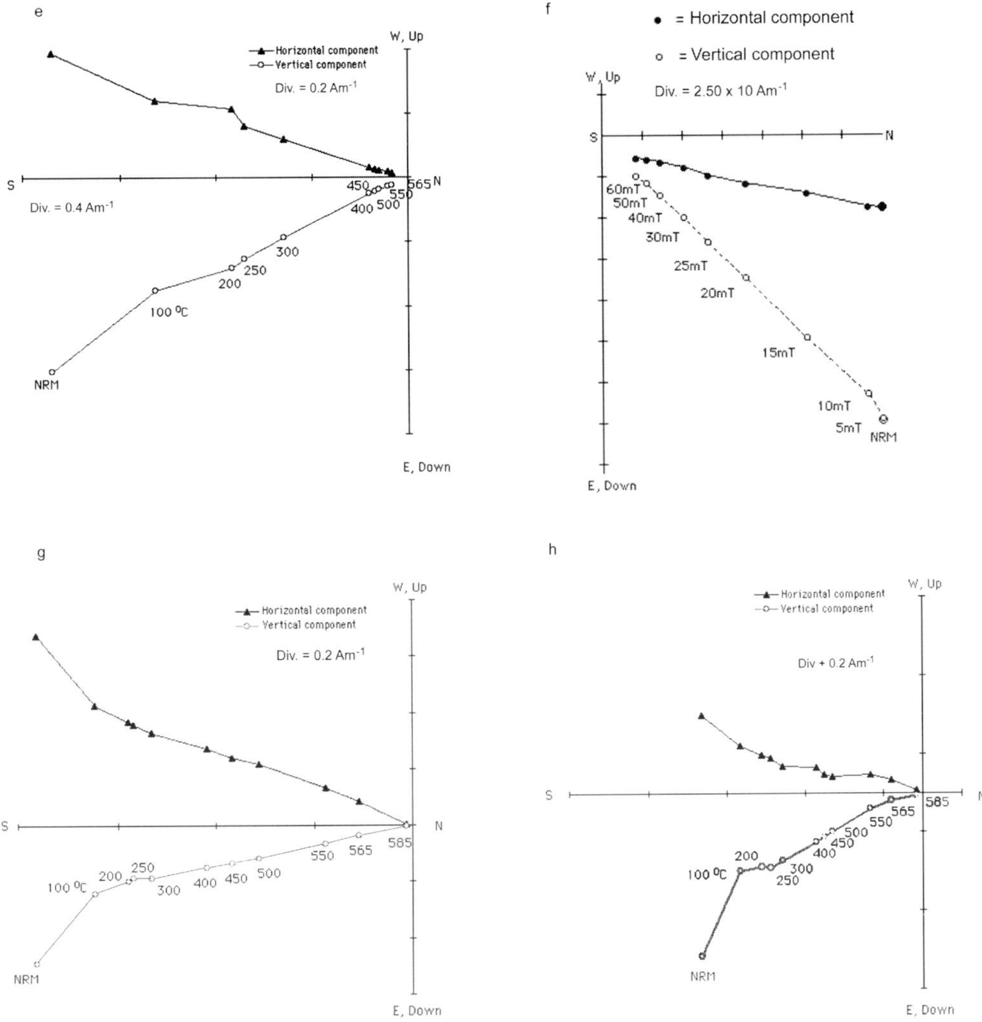

1070A and 899B are comparable, those from Hole 1070A have a much higher Königsberger ratio (mean 8.549) than those of Hole 899B (mean 2.049). Hole 899B has a higher mean intensity of the remanence, and consequently, the Q value of Hole 897D (mean 0.423) is lower than that of Hole 899B. Within the altered upper section in Hole 899B, a sharp increase in Q value occurs at 421.93 m depth (Sample 149-899B-21R-4, 97–99 cm). A similar spike is also present in Hole 897D (Sample 149-897D-13R-1, 62–64 cm). The Königsberger ratio peaks found in both holes may serve as stratigraphic markers for these two sites. The Q ratio is <1.0 in all but two of the samples in Hole 897D, whereas in Hole 899B it is nor-

mally close to or slightly greater than 1.0 (with only four exceptions). The low values of Q demonstrate that induced magnetization would be more or less comparable with that of remanent magnetization if present in an external magnetic field. Samples with low Q values often display multicomponent magnetization.

Magnetic polarity

In this study, polarity sense can be assigned with reasonable assurance on the basis of the inclination determined from discrete samples. The unexpected result generated from our preliminary study is the identification of what appears to be a reversed magnetic polarity zone

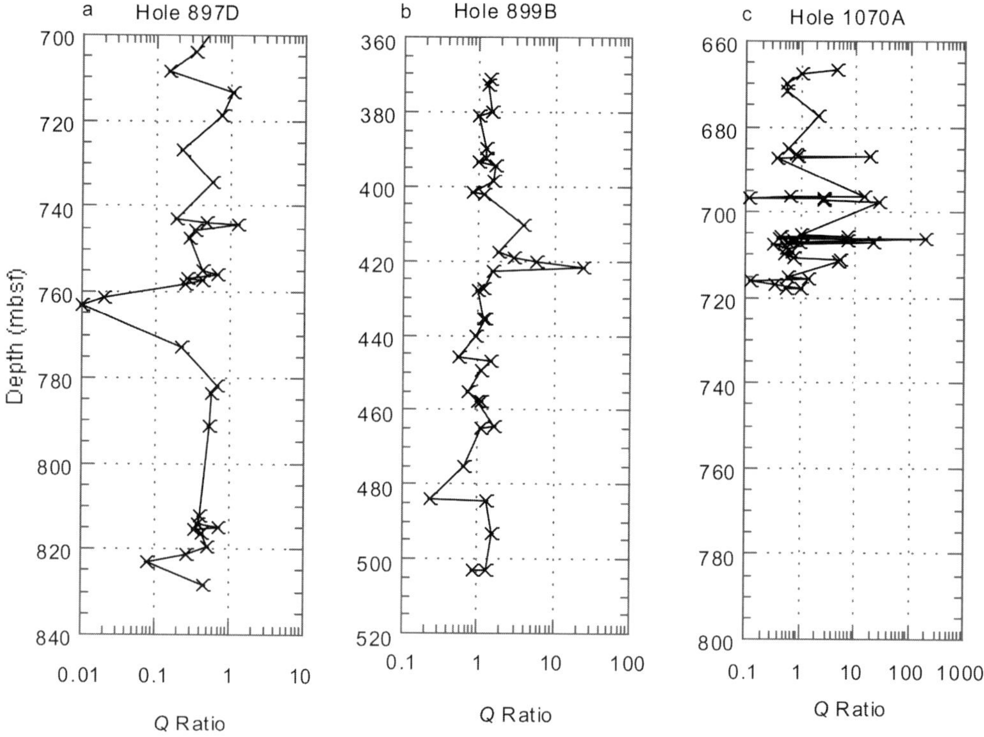

Fig. 5. Königsberger ratio of peridotite samples plotted against depth in (**a**) Hole 897D, (**b**) Hole 899B, and (**c**) Hole 1070A.

in the serpentinized peridotites (Fig. 6). At Leg 149 and 173 sites, the inclinations of characteristic magnetization in the 'fresher' lower part of the peridotite section show a predominantly reversed polarity in a depth zone of c. 20 m. In contrast, the inclinations of samples from the more 'altered' upper part are predominantly normal. One diagnostic feature is a pink-coloured gabbro in Hole 1070A (Sample 173-1070A-9R-1, 130–132 cm), which was dated at 119 ± 0.7 Ma based on $^{40}Ar/^{39}Ar$ step-heating ages for hornblende (Manatschal *et al.* 2001). This rock intruded the 'altered' upper part and has a reversed magnetization after removal of a normal component during the initial demagnetization steps (Shipboard Scientific Party 1998*d*). The reversed magnetization is at odds with the age of the rock, but we must keep in mind that this sample itself cannot ensure averaging out of secular variation. Interestingly, two samples both above and below this gabbro sample also revealed negative inclination (Fig. 6). This observation not only suggests that the gabbro (and its reversed magnetization) may have also remagnetized these two samples, but also

implies that the magnetization of the peridotites in the 'altered' upper part predates 119 Ma, at least for Hole 1070A.

Although colour of a peridotite itself does not carry much weight in overall palaeomagnetic interpretation, the onset depth of colour changes (see the Munsell colour index in Fig. 6) coincides with that of the reversed magnetic polarity zone. In Hole 1070A, this zone ranges from 697.59 mbsf (Sample 173-1070A-11R-2, 99–101 cm) to 717.82 mbsf (Sample 173-1070A-14R-3, 102–104 cm), which coincides with the intensity increase and first appearance of 'fresher' peridotites. In Hole 897D, this zone starts at 743.15 mbsf (Sample 149-897D-16R-2, 72–74 cm), which corresponds to the onset depth of the 'fresher' lower part of the section, and ends at 763.24 mbsf (Sample 149-897D-18R-1, 16–18 cm). In Hole 899B, it ranges from 435.72 mbsf (Sample 149-899B-23R-1, 12–14 cm) to 457.81 mbsf (Sample 149-899B-25R-3, 34–36 cm) and also coincides with the first appearance of greenish, 'fresher' peridotites in the lower part of the section.

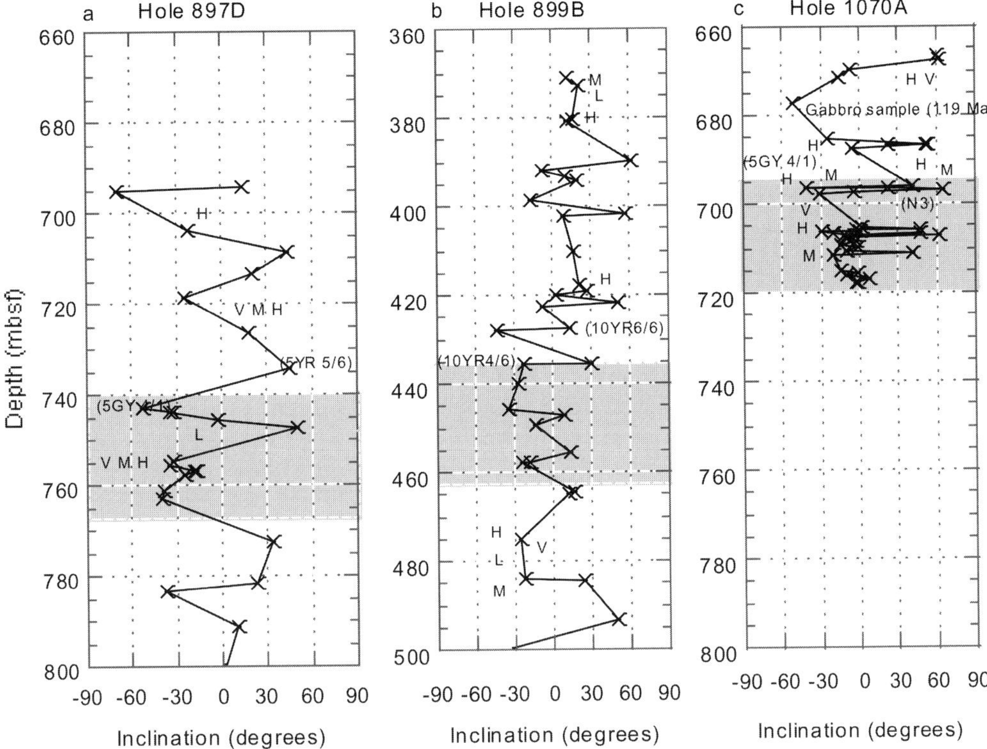

Fig. 6. Summary of downhole magnetic properties and inclinations observed within the peridotite sections of Leg 149 and Leg 173 holes. Shaded area is the inferred reversed magnetic polarity zone within the serpentinized peridotites. Rock magnetic characterization of discrete samples, taken from both the 'altered' and 'fresher' zones, is shown by bold letters: M, low-temperature cycling carried out on MPMS; V, Curie temperature determined on VSM; L, temperature dependence of in-phase susceptibility measurement determined on Lakeshore susceptometer; H, hysteresis parameters measured on Micromag. Munsell colour index: 5YR 5/6, light brown; 5GY 4/1, dark greenish grey; 10YR 6/6, yellowish brown; 10YR 4/6, reddish brown; N3, dark grey. Magnetic mineral in the 'altered' zone is mainly maghemite, and in the 'fresher' zone is magnetite.

Two observations in palaeomagnetic data from Site 897 are worth further mention. First, several samples from the very top of the 'altered' upper part display negative inclinations (hence, reversed polarity?), suggesting that the normal polarity magnetization in the 'altered' upper part was probably not a Holocene overprint. The most diagnostic examples are shown in Figure 7: a light-coloured clayey limestone sample (Sample 149-897C-66R-4, 15–17 cm), which was assigned a Late Cretaceous age based on the preliminary shipboard biostratigraphic ages, as well as a peridotite sample from Hole 897D (Sample 149-897D-11R-4, 71–73 cm), are both reversely magnetized. Similarly, indications of this reversed signal are present in two other peridotite

samples in Table 1 (Samples 149-897D-11R-2, 10–12 cm, and 149-897D-12R-1, 44–46 cm, at 695.37 and 703.94 mbsf, respectively). Second, there is a possible narrow normal-polarity interval at 747.2 mbsf in Hole 897D, in which no obvious physical disturbance is present. It appears that this polarity transition is represented simultaneously in both inclination and declination, although declination shows some additional shifts as a result of the lack of internal orientation of these cores upon coring. It is also interesting to note that below this transition cores have the weakest NRM intensities within the basement section (see Table 1). These features increase our confidence in the polarity determinations for both the upper and lower parts of the peridotite sections.

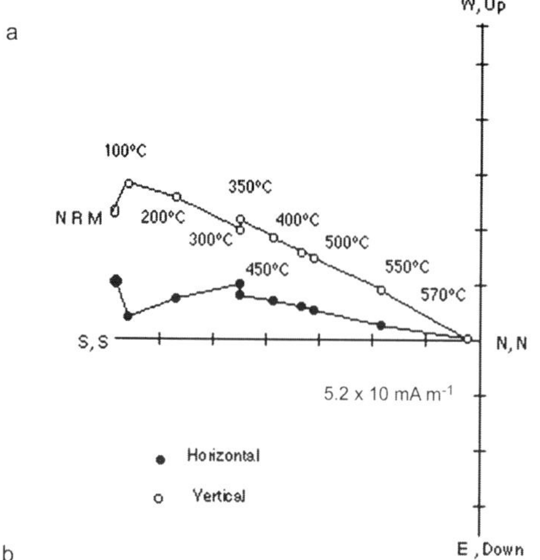

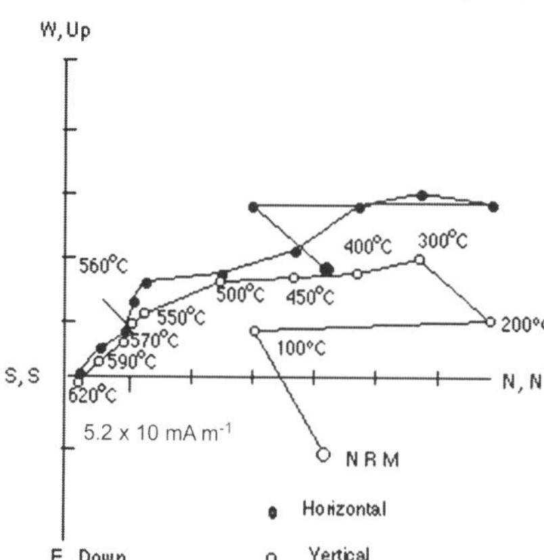

Fig. 7. Representative vector end-point diagrams showing thermal demagnetization results from cores overlying the 'altered' upper section of peridotite at Site 897, showing that the stable component of magnetization is probably of reversed polarity. (**a**) Sample 149-897C-66R-4, 15–17 cm. (**b**) Sample 149-897D-11R-4, 71–73 cm. Conventions as in Figure 4.

Rock magnetic properties

Hysteresis loop parameters

Saturation magnetization (Js), saturation remanence (Jrs), coercivity (Hc), and remanent coercivity (Hcr, determined from backfield demagnetization experiments) are parameters that can be determined from a hysteresis loop. Hysteresis loop parameters are useful in characterizing the intrinsic magnetic behaviour of rocks. Thus, they are helpful in studying the origin of remanence. In this study, hysteresis loops and the associated parameters Jrs/Js, Hc

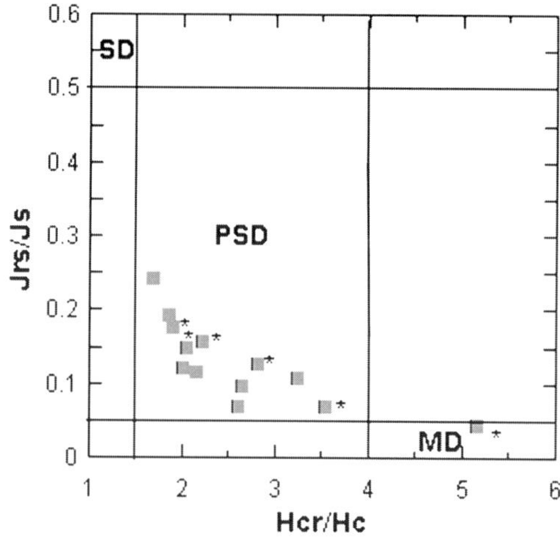

Fig. 8. The hysteresis ratios plotted on a Day *et al.* (1977) diagram suggest that the bulk magnetic grain size of the Iberian peridotite is mainly in the PSD region. *Js*, saturation magnetization; *Jrs*, saturation remanent magnetization; *Hc*, coercivity; *Hcr*, remanent coercivity. *Samples from the 'altered' upper part of peridotite sections (see Table 2 for details).

and Hcr were obtained using an alternating gradient magnetometer (AGFM; Princeton Measurements Corporation) capable of resolving magnetic moments as small as 5×10^{-8} e.m.u. (Flanders 1988). Hysteresis parameters determined from 13 representative samples from Leg 149 and 173 sites are presented in Table 2. Saturation magnetization (Js) is a measure of the total amount of magnetic mineral in the sample. The coercivity (Hc) is a measure of magnetic stability. The two ratios, Jr/Js and Hcr/Hc, are commonly used as indicators of domain states and, indirectly, grain size. For magnetite, high values of Jr/Js (>0.5) indicate small (<0.1 μm or so) single-domain (SD) grains, and low values (<0.1) are characteristic of large (>15–20 μm) multidomain (MD) grains. The intermediate regions are usually referred to as pseudo-single domain (PSD). Hcr/Hc is a much less reliable parameter, but conventionally SD grains have a value close to 1.1, and MD grains should have values >3–4 (Day *et al.* 1977). Figure 8 displays the ratios of the hysteresis parameters plotted on a Day *et al.* (1977) diagram. Such a representation provides qualitative information on the magnetic grain sizes from SD to PSD to large MD. The samples analysed in this study indicate that the magnetic grain sizes of samples from both the 'fresher' and 'altered' parts of the peridotite sections are in the PSD

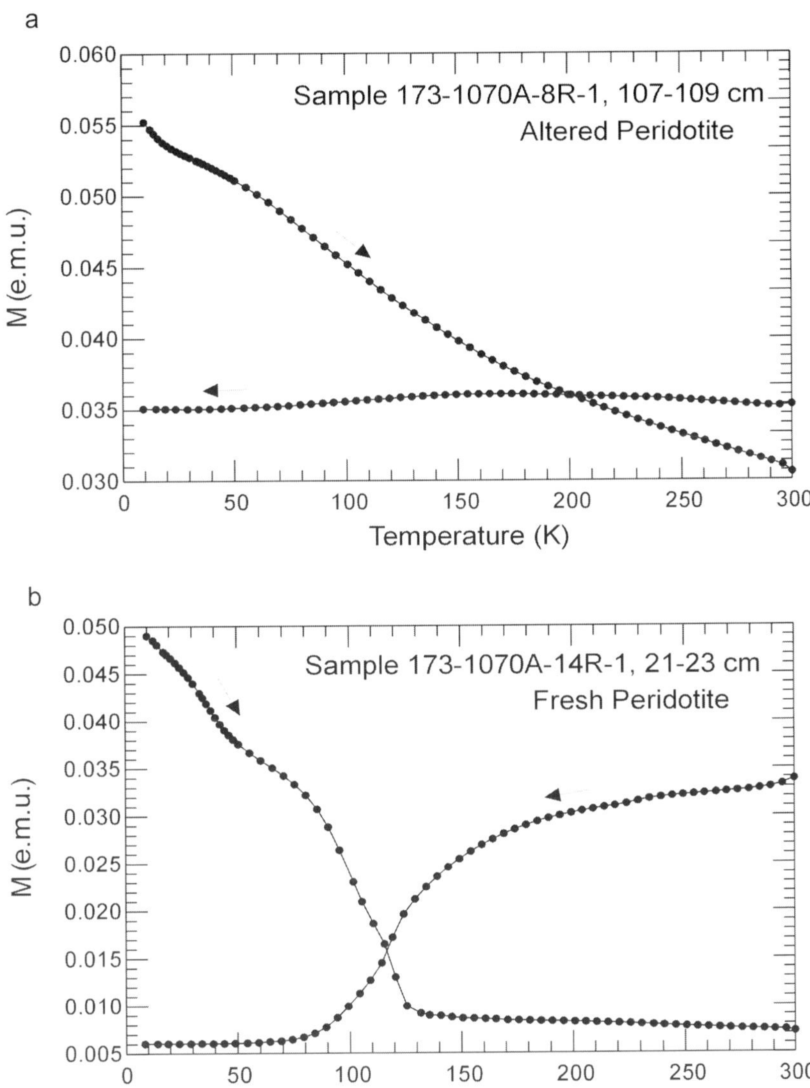

Fig. 9. Low-temperature heating curves of saturation remanence normalized to 5 K ($SIRM_5$) for representative samples in Hole 1070A. (**a**) Altered peridotite (Sample 173-1070A-8R-1, 107–109 cm). (**b**) Fresher peridotite (Sample 173-1070A-14R-1, 21–23 cm) from Site 1070.

range, but one 'altered' sample falls in the MD area (Fig. 8).

Low-temperature properties

Transitions in the magnetic properties of magnetite, pyrrhotite and hematite occur at low temperatures and potentially they provide a means of magnetic mineral identification. Magnetite exhibits a crystallographic phase tran-sition from cubic to monoclinic at 110–120 K. Associated with this transition, the anisotropy constant goes through zero as the easy axis of magnetization changes from [100] to [111] (Nagata *et al.* 1964). Low-temperature mea-surements were made on 16 representative samples from both the 'altered' and 'fresher' sections to help characterize the magnetic min-erals and understand their rock magnetic prop-erties. These measurements were designed to

                                    XIXI ZHAO

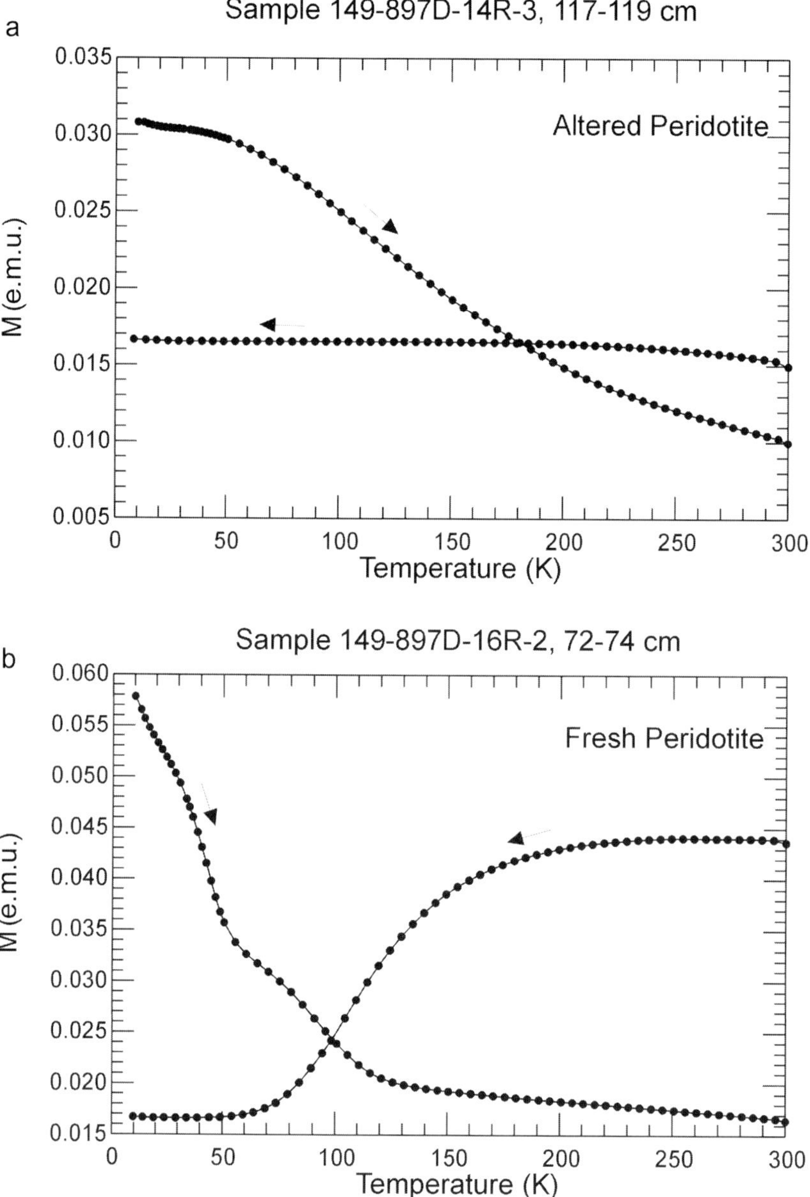

Fig. 10. Low-temperature heating curves of saturation remanence normalized to 5 K (SIRM$_5$) for representative samples in Hole 897D. (**a**) Altered peridotite (Sample 149-897D-14R-3, 117–119 cm). (**b**) Fresher peridotite (Sample 149-897D-16R-2, 72–74 cm).

determine the Néel temperature and other critical temperatures of a magnetic substance, and were made from 10 K to room temperature on 100–300 mg subsamples in a Quantum Design Magnetic Property Measurement System (MPMS) at the University of Minnesota. By definition, a ferrimagnetic mineral grain is superparamagnetic if its volume is smaller than the critical value $25kT/K$ (where k is Boltzmann's constant, T is 300 K in this study, and K is the magnetic anisotropy constant per unit volume), so that its net remanence over 100 s is zero (Cullity 1972). However, as T approaches 0 K, the thermal energy kT decreases and K

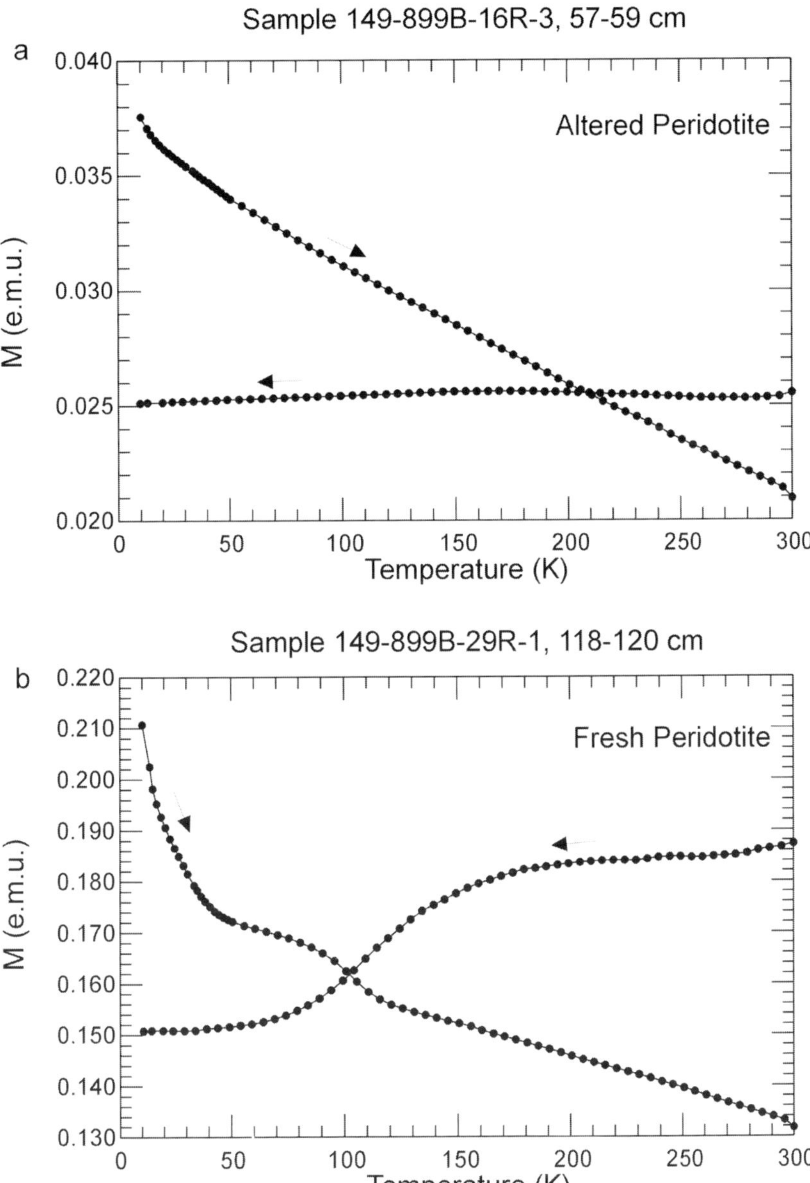

Fig. 11. Low-temperature heating curves of saturation remanence normalized to 5 K ($SIRM_5$) for several representative samples in Hole 899B. (**a**) Altered peridotite (Sample 149-899B-16R-3, 57–59 cm). (**b**) Fresher peridotite (Sample 149-899B-29R-1, 118–120 cm).

increases (both serving to aid magnetic stability), so that all grains that were superparamagnetic at 300 K will be able to retain thermally stable remanent magnetization near 0 K (Banerjee 1992). The so-called Verwey transition in magnetite can be observed by a decrease in intensity of an isothermal remanent magnetization (IRM) at low temperature as a sample is allowed to warm through or cool through the transition temperature. In this study, these experiments included (1) cooling the sample from room temperature (300 K) to 10 K (in some cases to 5 K) in a steady magnetic field of 2.5 T and measuring the remanence at 5 K inter-

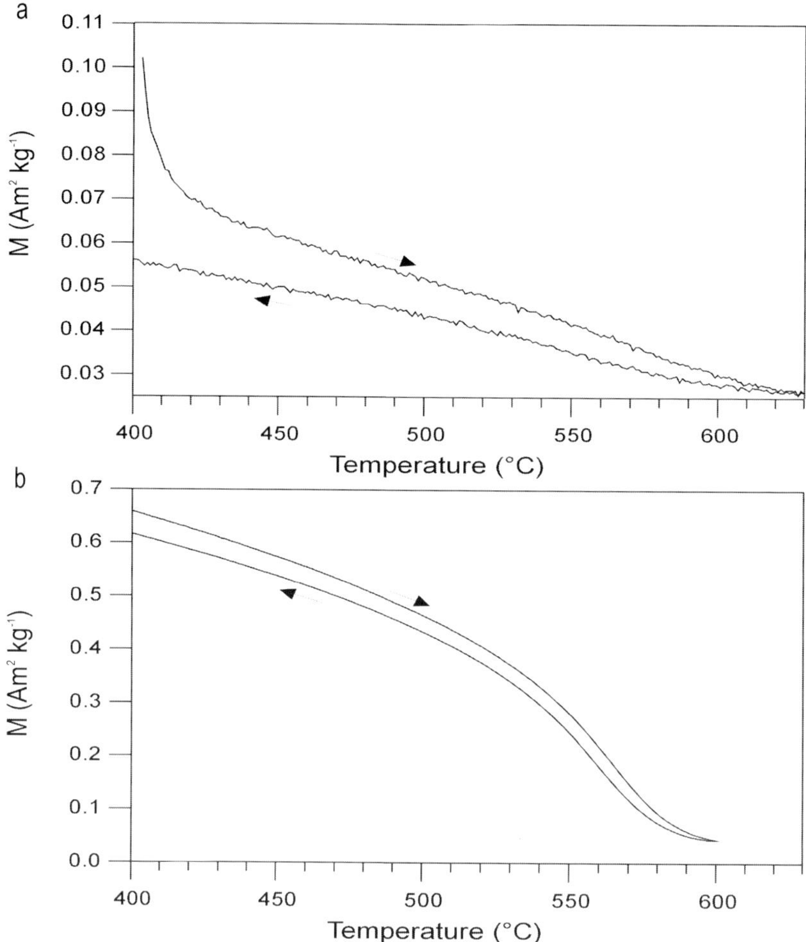

Fig. 12. Typical thermomagnetic curves for Leg 173 peridotites. (a) Altered peridotite sample (Sample 173-1070A-8R-1, 107–109 cm) from Site 1070. (b) Fresher peridotite sample (Sample 173-1070A-11R-1, 81–83 cm) from Site 1070.

vals; (2) measuring the saturation isothermal remanence at 5 K and 10 K ($SIRM_5$ and $SIRM_{10}$, respectively) and then warming the sample to 300 K in zero field while measuring the remanence value every 5 K.

As shown in Figures 9–11, the low-temperature curves of SIRM both in zero field warming and in a 2.5 T field cooling display a variety of features. These include an unblocking temperature in the vicinity of 40–60 K, and a decrease in remanence in the 100–120 K range, probably caused by the magnetocrystalline anisotropy constant, k_1, of magnetite going to zero in this temperature range. As shown in Figure 9b, titanomagnetites are clearly present in the 'fresher' peridotites, with a distinct Verwey transition in the vicinity of 110 K. The magnitude of the Verwey transition suggests that non-superparamagnetic magnetite is the dominant magnetic carrier in the sample. On the other hand, samples from the 'altered' peridotites do not display any Verwey transition (Fig. 9a). The same features as shown in Figure 9 are seen in the other seven pairs of samples (e.g. Figs 10 and 11). The low-temperature data are one of the major lines of evidence for the presence of magnetites in the 'fresher' peridotites, and maghemites in the 'altered' part of the peridotite section.

Curie temperature determinations

Curie temperature is the temperature below which a magnetic mineral is magnetically

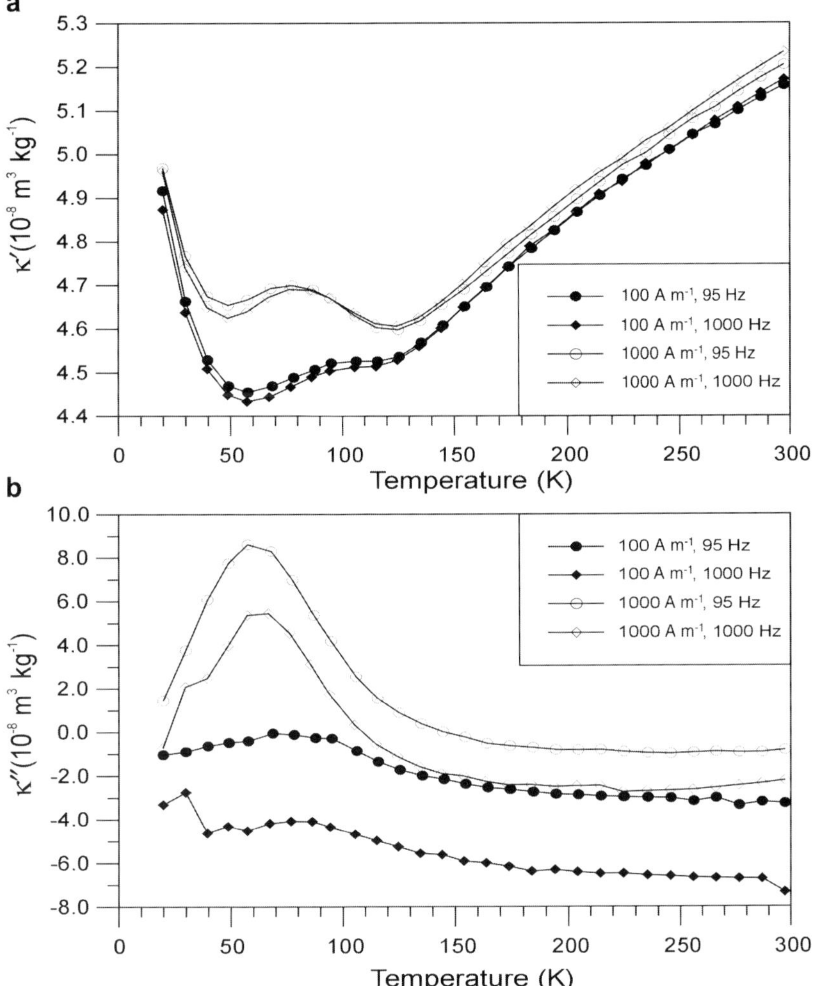

Fig. 13. Representative diagram showing the temperature dependence of mass susceptibility for altered peridotite sample (Sample 173-1070A-8R-1, 107–109 cm) from Site 1070 during warming from 15 K to 300 K. (**a**)
In-phase (κ') susceptibility measured at two frequencies (95 and 1000 Hz) with two field amplitudes (100 and
1000 A m^{-1}). (**b**) Quadrature (κ'') susceptibility measured at two frequencies (95 and 1000 Hz) with two field
amplitudes (100 and 1000 A m^{-1}).

ordered. Because this value is a sensitive indicator of composition, it is useful in understanding the magnetic mineralogy. In this study,
Curie temperature was determined by measurement of magnetic moment v. temperature
(using the Princeton MicroMag Vibrating
Sample Magnetometer (VSM) at the University
of Minnesota), where the magnetic moment
drops to zero at about the Curie temperature.
We conducted thermomagnetic analyses in an
inert atmosphere on 16 samples chosen to be
representative of the Leg 149 and 173 cores.

Figure 12 shows high-temperature magnetic
moment runs of a representative 'altered'
(Fig. 12a) and a 'fresher' (Fig. 12b) peridotite
sample from Site 1070. The heating and cooling curves for the 'altered' sample display a
significant drop of magnetic moment around
420 °C (Fig. 12a). This drop may be indicative
of a fraction of maghemite, which could be
responsible for the observed remanent magnetization. For the 'fresher' peridotite sample, the
results show Curie temperatures between 550
and 580 °C, indicative of the presence of titanomagnetite (Fig. 12b).

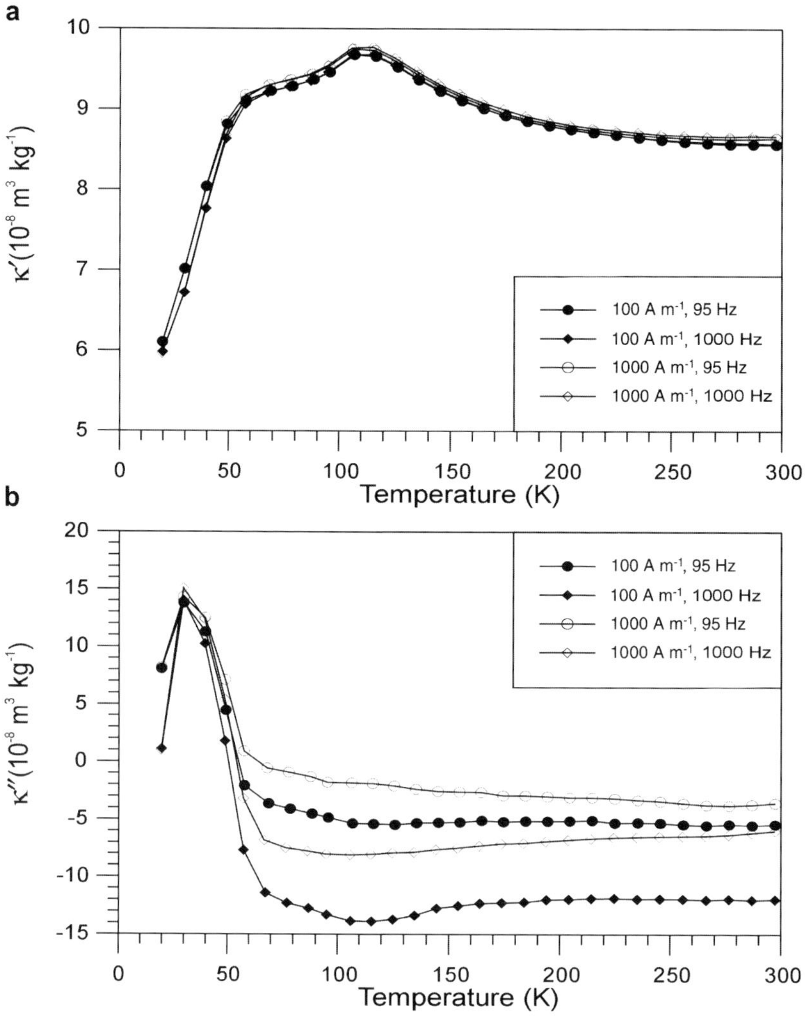

Fig. 14. Representative diagram showing the temperature dependence of mass susceptibility for fresh peridotite sample (Sample 149-899B-29R-1, 118–120 cm) from Site 899 during warming from 15 to 300 K. (**a**) In-phase (κ') susceptibility measured at two frequencies (95 and 1000 Hz) with two field amplitudes (100 and 1000 A m⁻¹). (**b**) Quadrature (κ'') susceptibility measured at two frequencies (95 and 1000 Hz) with two field amplitudes (100 and 1000 A m⁻¹).

Low-temperature a.c. susceptibility measurements

In a non-linear and sinusoidal-varying applied field, the magnetic response is determined by factors including field dependence, time or frequency dependence (i.e. viscosity), and electromagnetic induction (in sufficiently conductive material). If the applied field is not a linear function, then the magnetic responses will not be a pure sine function but will also contain higher (mainly odd) harmonics. In this case, there is a phase lag between the magnetic response and the applied field and the magnetic response can be resolved into in-phase and out-of-phase (also known as quadrature or imaginary) components (Jackson *et al.* 1998). Previous work by Worm *et al.* (1993) and Jackson *et al.* (1998) has suggested that ferrimagnetic pyrrhotite and titanomagnetites exhibit strong field dependence of a.c. susceptibility in a large applied field. To investigate the field- and frequency-dependent susceptibility of the Iberian peridotite samples, a.c. susceptibility measurements were made on selected samples with a LakeShore Model 7130 a.c. susceptometer at the University of Minnesota. The temperature dependence of the in-phase (κ') susceptibility between 15 and 300 K was measured at two frequencies (95 and 1000 Hz), with a.c. field amplitudes between 100 and 1000 A m^{-1}.

The field- and frequency-dependent susceptibility curves for the 'fresher' peridotites from Site 1070 are similar to those of synthetic TM55 (Jackson *et al.* 1998), with some field dependence and no frequency dependence. For the 'altered' peridotite, the field- and frequency-dependent relationships are directly opposite to those in the 'fresher' zone, with some frequency dependence but no field dependence. Figures 13 and 14 show the temperature dependence of magnetic susceptibility between 15 and 300 K for an 'altered' and 'fresher' peridotite sample, respectively.

Discussion

The rock magnetic properties of Legs 149 and 173 peridotites are generally characterized by relative strong NRM intensities. Variations of the Königsberger ratio and the initial magnetic susceptibilities generally resemble those of NRM values. A recent magnetic survey revealed that there is a magnetic anomaly high (about 0.3 A m^{-1}) in the vicinity of Site 899, which was *c.* 40 km eastward and was previously thought to result from a relatively strongly magnetized non-oceanic crust (Whitmarsh *et al.* 1996*a*; Discovery 215 Working Group 1998). The much stronger magnetization intensity of Site 899 samples is apparently in excellent agreement with the observed magnetic anomaly high. In addition, several samples from Hole 899B retained nearly half of their NRM intensity after 400 °C demagnetization, suggesting that the remanence of the serpentinized peridotite body under Site 899 can contribute significantly to the magnetic anomaly

The serpentinized peridotites that were recovered at Site 637 by Leg 103 have not been subjected to detailed palaeomagnetic and rock magnetic study. Nine oriented small slabs (thin-section billets or petrographic samples) of the serpentinized peridotite were treated by AF demagnetization to 100 Oe only. All nine pieces yielded normal polarity. Although not documented in detail, the NRM intensity appears to increase with depth. The entire section of peridotite is altered, but the alteration is more pervasive in the upper 60 cm (Boillot *et*

al. 1987*b*). The rock has a leached appearance and the alteration involves serpentinization and veining by serpentine and calcite. The peridotite is generally yellow to pale green, but local sections are black (e.g. Samples 103-637A-26R-4, 60 cm; 103-637A-27R-2, 15 cm; 103-637A-27R-4, 45–120 cm; 103-637A-29R-1, 31–125 cm). These features fit what we saw in Leg 149 and 173 cores, and warrant future studies to compare and ascertain whether or not the magnetic histories of the three ODP Sites are related.

Results from low-temperature measurements in the MPMS (high-field, 2.5 T) and Lakeshore (low-field, 100–1000 A m^{-1}) susceptometers show that (titano)magnetites are present in the 'fresher' peridotites, with a strong Verwey transition in the vicinity of 110 K, and with field- and frequency-dependent susceptibility curves that resemble those of synthetic TM55. These results are in good agreement with the thermomagnetic characteristics of the 'fresher' peridotites in which titanomagnetites, with Curie temperatures around 580 °C, were identified. The hysteresis ratios suggest that the bulk magnetic grain size is in the PSD field (e.g. with lower *Hcr/Hc* values). In contrast to the magnetic properties observed from the 'fresher' peridotites, the low-temperature curves for the 'altered' peridotites did not show any Verwey transition. Thermomagnetic analysis using the high-temperature VSM also failed to show evidence for titanomagnetites. The remanent magnetization is carried by a thermally unstable mineral that breaks down at *c.* 420 °C, probably maghemite. The field- and frequency-dependent relationships are also directly opposite to those in the reversed magnetic polarity zone, with no signs of titanomagnetite characteristics. Although the hysteresis ratios still fall in the PSD region, the cluster is centred towards the MD region (with higher *Hcr/Hc* values). Altogether, these rock magnetic data seem to be sensitive indicators of alteration and support the contention that maghemite is responsible for the magnetic signatures displayed in the altered peridotites of the upper part of the peridotite section. The rock magnetic properties are useful in evaluating the fidelity of the natural magnetic memory in upper-mantle rocks and understanding the alteration processes through time.

Stable components of magnetization are revealed in the results of demagnetization on Leg 149 and 173 samples (Table 1). Because the remanence in the 'altered' upper part of the peridotite section is dominated by a single stable component of normal polarity with nearly identical demagnetization behaviour regardless of lithology and depth, the palaeomagnetic data of the 'altered' upper part seem best interpreted as indicating that the magnetization is an early overprint. The mean inclination of this overprinted magnetization is consistently shallower (*c.* 30°) than that of a Holocene field (*c.* 60°), suggesting that the overprint was unlikely to have been acquired in Holocene time. In view of the polarity of magnetization identified from the overlying Cretaceous and Tertiary sediments, this overprint is probably older than the magnetization of the overlying Tertiary sediments. The estimated palaeolatitudes from the mean inclinations are in agreement with the expected Cretaceous palaeolatitudes for Iberia. Recent radiometric ages have indicated that the Iberian peridotites at Leg 173 Site 1070 were emplaced between 121 and 110 Ma (Manatschal *et al.* 2001), which puts direct age constraints on the magnetization. Thus, the normal polarity is consistent with the Cretaceous Normal Superchron and the age of the rocks. Micropalaeontological dating of cores from Hole 899B and dynamic modelling revealed that the emplacement of these peridotites probably lasted only *c.* 1–2 Ma (de Kaenel & Bergen 1996; Bowling & Harry 2001). Therefore, the overprinted normal magnetization signal could have been imposed within the Cretaceous Long Normal Superchron, probably soon after M0 time (Zhao 1999).

In the 'fresher' lower part, although the remanence is still dominated by a single component of magnetization in most cases, several samples display a multicomponent nature, with a characteristic component isolated after removal of a secondary component of opposite polarity. The characteristic component is carried by magnetite, which is known to be capable of preserving an early remanent magnetization over a long time. These samples maintain inclinations close to the theoretically predicted values for the palaeolatitude of drill sites, indicating they may represent the primary magnetization. Because of the demagnetization behaviour of these samples, we have no reason not to believe that the magnetization directions derived from these fresher peridotites are primary. Moreover, the reversed magnetization zone in the beginning of the fresher lower part is correlative between Sites 897 and 899 (Zhao 1996), which is perhaps the most compelling argument for the presence of primary magnetization because it would be difficult for a later remagnetization to produce such a preferential polarity pattern at depth. As Chron M0 is the only reversed polarity zone during this time span (Aptian–Barremian time), it is most likely that the reversed magnetization in the 'fresher' lower part was imposed at M0 time. Our recently completed post-cruise research on magnetostratigraphy of the overlying sediments also suggests that the reversal signal in the peridotite section is older than Cenozoic time. Several lines of geological observations agree with the above age assignments to the magnetic polarity zones:

(1) Biostratigraphic arguments suggest that the oldest units of these peridotites are of Early Cretaceous age (Hauterivian, about 135–132 Ma), and the first sediment deposited on top of the peridotite section was found to be of Late Cretaceous age. Therefore, the entire peridotite sections should have been emplaced during Cretaceous time.

(2) Magnetic Anomaly M0 (*c.* 121 Ma) is tentatively identified from the west side of the peridotite sites (<150 km away), suggesting the peridotites are of approximately the same age as Anomaly M0. This has been confirmed by recent available radiometric ages (121–110 Ma) for Leg 173 peridotites (Manatschal *et al.* 2001), as mentioned above.

(3) Numerous major unconformities (at *c.* 120 Ma) in many sedimentary basins surrounding the peridotite sites are concurrent with sea-floor spreading at M0 time (Tankard *et al.* 1989; Wilson *et al.* 1989; Murillas *et al.* 1990). These unconformities not only provide a direct linkage between the evolution of these basins and separation of Iberia from North America, but also suggest that the emplacement and alteration events in these peridotites should have taken place at about M0 time.

(4) Shipboard observations suggest that the alteration of the peridotites recovered during Legs 149 and 173 took place soon after the peridotites were placed at or near the sea-floor surface and that the fluid responsible for serpentinization was sea water. Therefore, both the palaeomagnetic and geological data are compatible with the suggestion that alteration by fluid circulation was associated with the first stages of accretion of the oceanic crust, during mid-Cretaceous time.

Assuming that the origins of the identified magnetic components of the peridotites are correctly established, constraints from palaeomagnetic data can be placed on the mode of emplacement and the late-stage alteration of the Iberian peridotites. The peridotite ridge sites are locations where the crust is missing and where the mantle rocks once cropped out directly on the sea floor. This would imply that progressive uplift and final emplacement of peridotites at the Earth's surface seem to be related to the thinning of the continental crust beneath the rift. Because uplift of the peridotite by some kind of diapiric mechanism (e.g. Bonatti 1987; Nicolas *et al.* 1987) is needed and detachment faulting accounts well for the crustal thinning and stretching (e.g. Boillot *et al.* 1987*a,b*; Beslier *et al.* 1996; Krawczyk *et al.* 1996), we suggest a combination of these two models for the process of emplacement of the Iberian peridotites. Uplift by diapirism could occur before the main stage of stretching. At the end of the stretching phase, when the crust is particularly thinned, these peridotites would be serpentinized by sea-water circulation and hence become less dense than the surrounding crustal material. The serpentinized peridotites finally rose to the sea floor, recorded the mid-Cretaceous (probably at M0 time, 121 Ma) field, and cropped out at the ocean–continent boundary just before 'true' sea-floor spreading started between the Iberian and Newfoundland margins (i.e. the opening of the North Atlantic). If we assume that the average rifting rate (*c.* 4 cm a^{-1}) and the time span of Anomaly M0 (*c.* 1 Ma) are both correctly established, our suggestion would explain the existence of wide exposure of the peridotite ridge, and accommodate the observed magnetic anomalies in the region.

Conclusions

Preliminary palaeomagnetic study has revealed important information about the origin of remanence and on the magnetic minerals present in the Leg 149 and 173 cores, which in turn provide physical insight into the geodynamic evolution of the Iberian margin (from continental rifting to sea-floor spreading). The Cretaceous (shallow) inclinations, magnetic polarity patterns and rock magnetic properties suggest that the emplacement of the Iberian peridotites probably took place during the mid-Cretaceous period around M0 time (*c.* 121 Ma), which agrees very well with recent radiometric ages (121–110 Ma) for the Iberian peridotites at Leg 173 drill sites. The magnetization of the altered zone was also imposed during Cretaceous time, probably soon after Anomaly M0 during the opening of the North Atlantic. These peridotites were probably brought up by a combination of mantle upwelling and lithospheric stretching, and reached the last stage of their evolution in Late Cretaceous time (110 Ma). A detailed rock magnetic and isotopic study of peridotites recovered from Legs 103, 149 and 173 is still

needed to be able to interpret the remarkable magnetic signatures of the Iberian peridotites.

I thank R.B. Whitmarsh of the book editorial board, G. McIntosh and E. Hailwood for insightful reviews and constructive suggestions on the original manuscript, and U. Draeger for her help with graphs. I also wish to express my appreciation to Leg 149 and Leg 173 shipboard scientists, crews of the *JOIDES Resolution*, the ODP staff, the SEDCO crew, and the Catermar staff for their help and company during Legs 149 and 173. My special thanks are extended to the palaeomagnetism group of the Institute for Rock Magnetism at the University of Minnesota for support and fruitful discussions, and to R. Coe, who encouraged me to participate in ODP research and provided guidance during the study. Financial support for various parts of this research was provided by two grants from the US Science Support Program of the Joint Oceanographic Institution, Inc., and NSF grants EAR 97-06439 and EAR 9805444. This manuscript is Contribution 423 of the Institute of Tectonics and Palaeomagnetism Laboratory at the University of California, Santa Cruz.

References

BANERJEE, S.K. 1992. Applied rock magnetism in the 1990s: potential breakthrough in a new user-driven science. *EOS Transactions, American Geophysical Union*, **73**, 142–143.

BESLIER, M.-O., ASK, M. & BOILLOT, G. 1993. Ocean–continent boundary in the Iberia Abyssal Plain from multichannel seismic data. *Tectonophysics*, **218**, 383–393.

BESLIER, M.-O., CORNEN, F. & GIRARDEAU, J. 1996. Tectono-metamorphic evolution of peridotites from the ocean/continent transition of the Iberia Abyssal Plain margin. *In*: WHITMARSH, R.B., SAWYER, D.S., KLAUS, A. & MASSON, D.G. (eds) *Proceedings of the Ocean Drilling Program, Scientific Results, 149*. Ocean Drilling Program, College Station, TX, 397–412.

BOILLOT, G., AGRINIER, P., BESLIER, M.-O. & 9 OTHERS 1995. A lithospheric syn-rift shear zone at the ocean–continent transition: preliminary results of the GALINAUTE II cruise (Nautile dives on the Galicia Bank, Spain). *Comptes Rendus de l'Académie des Sciences, Série 2a*, **321**, 1171–1178.

BOILLOT, G., GRIMAUD, S., MAUFFRET, A., MOUGENOT, D., KORNPROBST, J., MERGOIL-DANIEL, J. & TORRENT, G. 1980. Ocean–continent boundary off the Iberian margin: a serpentinite diapir west of the Galicia Bank. *Earth and Planetary Science Letters*, **48**, 23–34.

BOILLOT, G., RECQ, M., WINTERER, E.L. & 21 OTHERS 1987*a*. Tectonic denudation of the upper mantle along passive margins: a model based on drilling results (ODP Leg 103, western Galicia margin, Spain). *Tectonophysics*, **132**, 335–342.

BOILLOT, G., WINTERER, E.L., *et al.* (eds) 1988*b*. *Proceedings of the Ocean Drilling Program, Scientific Results, 103*. Ocean Drilling Program, College Station, TX.

BOILLOT, G., WINTERER, E.L., MEYER, A.W., *et al.* (eds) 1987*b*. *Proceedings of the Ocean Drilling Program, Initial Reports, 103*. Ocean Drilling Program, College Station, TX.

BONATTI, E. 1987. The rifting of continents. *Scientific American*, **258**, 96–103.

BOWLING, J.C. & HARRY, D.L. 2001. Geodynamic models of continental extension and the formation of non-volcanic rifted continental margins. *In*: WILSON, R.C.L., WHITMARSH, R.B., TAYLOR, B. & FROITZHEIM, N. (eds) *Non-volcanic Rifting of Continental Margins: a Comparison of Evidence from Land and Sea*. Geological Society, London, Special Publications, **187**, 511–536.

BULLARD, E., EVERETT, J.E. & SMITH, A.G. 1965. The fit of the continents around the Atlantic. *Philosophical Transactions of the Royal Society of London, Series A*, **258**, 41–75.

CANDE, S.C. & KENT, D.V. 1995. Revised calibration of the geomagnetic polarity time-scale for the Late Cretaceous and Cenozoic. *Journal of Geophysical Research*, **100**, 6093–6095.

CHANNELL, J.E.T., BRALOWER, T.J. & GRANDESSO, P. 1987. Biostratigraphic correlation of Mesozoic polarity chrons CM1 and CM23 at Capriolo and Xausa (Southern Alps, Italy). *Earth and Planetary Science Letters*, **85**, 203–221.

COURTILLOT, V. 1982. Propagating rifts and continental breakup. *Tectonics*, **1**, 239–250.

CULLITY, B.D. *Introduction to Magnetic Minerals*. Addison Wesley, Reading, MA.

DAY, R., FULLER, M. & SCHMIDT, V.A. 1977. Hysteresis properties of titanomagnetites: grain-size and compositional dependence. *Physics of the Earth and Planetary Interiors*, **13**, 260–266.

DE KAENEL, E. & BERGEN, J.A. 1996. Mesozoic calcareous nanofossil biostratigraphy from Sites 897, 899, and 901, Iberia Abyssal Plain: new biostratigraphic evidence. *In*: WHITMARSH, R.B., SAWYER, D.S., KLAUS, A. & MASSON, D.G. (eds) *Proceedings of the Ocean Drilling Program, Scientific Results, 149*. Ocean Drilling Program, College Station, TX, 79–146.

DISCOVERY 215 WORKING GROUP 1998. Deep structure in the vicinity of the ocean–continent transition zone under the southern Iberia Abyssal Plain. *Geology*, **8**, 743–746.

FÉRAUD, G., BESLIER, M.-O. & CORNEN, G. 1996. $^{40}Ar-^{39}Ar$ dating of gabbros fromthe ocean/continent transition of the West Iberia margin: preliminary results. *In*: WHITMARSH, R.B., SAWYER, D.S., KLAUS, A. & MASSON, D.G. (eds) *Proceedings of the Ocean Drilling Program, Scientific Results, 149*. Ocean Drilling Program, College Station, TX, 489–495.

FÉRAUD, G., GASTAUD, J., AUZENDE, J.-M., OLIVET, J.L. & CORNEN, G. 1982. $^{40}Ar-^{39}Ar$ ages for the alkaline volcanism and basement of the Gorringe Bank, North Atlantic Ocean. *Earth and Planetary Science Letters*, **57**, 211–226.

FÉRAUD, G., GIRARDEAU, J., BESLIER, M.-O. & BOILLOT, G. 1988. Datation $^{39}Ar-^{40}Ar$ de la mise en place des peridotites bordant la marge de Galice (Espagne). *Comptes Rendus de l'Académie des Sciences*, **307**, 49–55.

FÉRAUD, G., YORK, D., MÉVEL, C., CORNEN, G. & AUZENDE, J.-M. 1986. Additional $^{40}Ar-^{39}Ar$ dating of the basement and alkaline volcanism of the Gorringe Bank (Atlantic Ocean). *Earth and Planetary Science Letters*, **79**, 255–269.

FLANDERS, P.J. 1988. An alternating-gradient magnetometer. *Journal of Applied Physics*, **63**, 3940–3945.

FUEGENSCHUH, B., FROITZHEIM, N. & BOILLOT, G. 1998. Cooling history of granulite samples from the ocean–continent transition of the Galicia margin: implications for rifting. *Terra Nova*, **10**, 96–100.

GALDEANO, A., MOREAU, M.-G., POZZI, J.-P., BERTHOU, P.-Y. & MALOD, J.-A. 1989. New palaeomagnetic results from Cretaceous sediments near Lisbon (Portugal) and implications for the rotation of Iberia. *Earth and Planetary Science Letters*, **94**, 95–106.

GIRARDEAU, J., CORNEN, G., SCHÄRER, U., BESLIER, M-O., BOILLOT, G. & CHARPENTIER, S. 1999. Petrological and structural characteristics of the Gorringe Bank basement rocks: comparison with IAP and Galicia, geodynamic implications. *Abstract Volume, Non-volcanic Rifting of Continental Margins: a Comparison of Evidence from Land and Sea*. Geological Society, London **48**.

GRADSTEIN, F.M., AGTERBERG, F.P., OGG, J.G., HARDENBOL, J., VAN VEEN, P., THIERRY, J. & HUANG, Z. 1995. A Mesozoic timescale. *Journal of Geophysical Research*, **99**, 24051–24074.

HALGEDAHL, S.L. 1992. Future research topics in rock magnetism. *EOS Transactions, American Geophysical Union*, **73**, 142.

HELSEY, C.E. & STEINER, M.D. 1969. Evidence for long intervals of normal polarity during the Cretaceous period. *Earth and Planetary Science Letters*, **5**, 325–332.

IAGA Division V, Working Group 8 1995. International geomagnetic reference field, 1995 revision. *Journal of Geomagnetism and Geoelectricity*, **47**, 1257–1261.

JACKSON, M., ROCHETTE, P., FILLION, G., BANERJEE, S. & MARVIN, J. 1993. Rock magnetism of remagnetized Palaeozoic carbonates: low-temperature behavior and susceptibility characteristics. *Journal of Geophysical Research*, **98**, 6217–6225.

JACKSON, M.J., MOSKOWITZ, B.M., ROSENBAUM, J. & KISSEL, C. 1998. Field-dependence of AC susceptibility in titanomagnetites. *Earth and Planetary Science Letters*, **157**, 129–139.

KIRSCHVINK, J.L. 1980. The least-squares line and plane and the analysis of palaeomagnetic data. *Geophysical Journal of the Royal Astronomical Society*, **62**, 699–718.

KLITGORD, K.D. & SCHOUTEN, H. 1986. Plate kinematics of the central Atlantic. *In*: VOGT, P.R. & TUCHOLKE, B.E. (eds) *The Geology of North America (Vol. M): The Western North Atlantic Region*. Geological Society of America, Boulder, CO, 351–378.

KRAWCZYK, C.M., RESTON, T.J., BESLIER, M.-O. & BOILLOT, G. 1996. Evidence for detachment tectonics on the Iberia Abyssal Plain rifted margin. *In*: WHITMARSH, R.B., SAWYER, D.S., KLAUS, A. & MASSON, D.G. (eds) *Proceedings of the Ocean Drilling Program, Scientific Results, 149*. Ocean Drilling Program, College Station, TX, 603–615.

LE PICHON, X., SIBUET, J.-C. & FRANCHETEAU, J. 1977. The fit of the continents around the North Atlantic Ocean. *Tectonophysics*, **38**, 169–209.

MANATSCHAL, G., FROITZHEIM, N., RUBENACH, M. & TURRIN, B.D. 2001. The role of detachment faulting in the formation of an ocean–continent transition: insights from the Iberia Abyssal Plain. *In*: WILSON, R.C.L., WHITMARSH, R.B., TAYLOR, B. & FROITZHEIM, N. (eds) *Non-volcanic Rifting of Continental Margins: a Comparison of Evidence from Land and Sea*. Geological Society, London, Special Publications, **187**, 405–428.

MASSON, D.G. & MILES, P.R. 1984. Mesozoic sea-floor spreading between Iberia, Europe and North America. *Marine Geology*, **56**, 279–287.

MCELHINNY, M.W. & BUREK, P.J. 1971. Mesozoic palaeomagnetic stratigraphy. *Nature*, **232**, 98–102.

MURILLAS, J., MOUGENOT, D., BOILLOT, G., COMAS, M.C., BANDA, E. & MAUFFRET, A. 1990. Structure and evolution of the Galicia interior basin (Atlantic western Iberian continental margin). *Tectonophysics*, **184**, 297–319.

NAGATA, T., KOBAYASHI, K. & FULLER, M.D. 1964. Identification of magnetite and hematite in rocks by magnetic observation at low temperature. *Journal of Geophysical Research*, **69**, 2111–2121.

NICOLAS, A., BOUDIER, F. & MONTIGNY, R. 1987. Structure of Zabargad Island and early rifting of the Red Sea. *Journal of Geophysical Research*, **92**, 461–474.

OGG, J.G. 1988. Early Cretaceous and Tithonian magnetostratigraphy of the Galicia margin. *In*: BOILLOT, G., WINTERER, E.L., *et al.* (eds) *Proceedings of the Ocean Drilling Program, Scientific Results, 103*. Ocean Drilling Program, College Station, TX, 659–682.

OPDYKE, N.D. & CHANNELL, J.E.T. *Magnetic Stratigraphy*. International Geophysics Series, **64**. Academic Press, New York.

PECHERSKY, D.M. & KHRAMOV, A.N. 1973. Mesozoic palaeomagnetic scale of the USSR. *Nature*, **244**, 499–501.

PINHEIRO, L.M., WHITMARSH, R.B. & MILES, P.R. 1992. The ocean–continent boundary off the western continental margin of Iberia, II. Crustal structure in the Tagus Abyssal Plain. *Geophysical Journal*, **109**, 106–124.

PINHEIRO, L.M., WILSON, R.C.L., PENA DOS REIS, R., WHITMARSH, R.B. & RIBEIRO, A. 1996. The western Iberia Margin: a geophysical and geological overview. *In*: WHITMARSH, R.B., SAWYER, D.S., KLAUS, A. & MASSON, D.G. (eds) *Proceedings of the Ocean Drilling Program, Scientific Results, 149*. Ocean Drilling Program, College Station, TX, 3–23.

SAWYER, D.S., WHITMARSH, R.B., KLAUS A., *et al.* (eds) 1994. *Proceedings of the Ocean Drilling Program, Initial Reports, 149*. Ocean Drilling Program, College Station, TX.

SCHÄRER, U., GIRARDEAU, J., CORNEN, G. & BOILLOT, G. 2000. 138–121 Ma asthenospheric magmatism pre-dating oceanic crust formation in the North Atlantic. *Earth and Planetary Science Letters*, **181**, 555–572.

SCHÄRER, U., KORNPROBST, J., BESLIER, M.-O., BOILLOT, G. & GIRARDEAU, J. 1995. Gabbro and related rock emplacement beneath rifting continental crust: U–Pb geochronological and geochemical constraints for the Galicia passive margin (Spain). *Earth and Planetary Science Letters*, **130**, 187–200.

SHIPBOARD SCIENTIFIC PARTY, 1998*a*. Site 1067. *In*: WHITMARSH, R.B., BESLIER, M.-O., WALLACE, P.J., *et al.* (eds) *Proceedings of the Ocean Drilling Program, Initial Reports, 173*. Ocean Drilling Program, College Station, TX, 107–162.

SHIPBOARD SCIENTIFIC PARTY, 1998*b*. Site 1068. *In*: WHITMARSH, R.B., BESLIER, M.-O., WALLACE, P.J., *et al.* (eds) *Proceedings of the Ocean Drilling Program, Initial Reports, 173*. Ocean Drilling Program, College Station, TX, 163–218.

SHIPBOARD SCIENTIFIC PARTY, 1998*c*. Site 1069. *In*: WHITMARSH, R.B., BESLIER, M.-O., WALLACE, P.J., *et al.* (eds) *Proceedings of the Ocean Drilling Program, Initial Reports, 173*. College Station, TX, Ocean Drilling Program, 219–264.

SHIPBOARD SCIENTIFIC PARTY, 1998*d*. Site 1070. *In*: WHITMARSH, R.B., BESLIER, M.-O., WALLACE, P.J., *et al.* (eds) *Proceedings of the Ocean Drilling Program, Initial Reports, 173*. Ocean Drilling Program, College Station, TX, 265–294.

SIBUET, J.-C. & SRIVASTAVA, S. 1994. Rifting consequences of three plate separation. *Geophysical Research Letters*, **21**, 521–524.

SRIVASTAVA, S.P. & TAPSCOTT, C.R. 1986. Plate kinematics of the North Atlantic. *In*: VOGT, P.R. & TUCHOLKE, B.E. (eds) *The Geology of North America, Vol. M. The Western North Atlantic Region*. Geological Society of America, Boulder, CO, 379–404.

SRIVASTAVA, S.P. & VERHOEF, J. 1992. Evolution of Mesozoic sedimentary basins around the North Central Atlantic: a preliminary plate kinematic solution. *In*: PARNELL, J. (ed.) *Basins of the Atlantic Seaboard: Petroleum Geology, Sedimentology and Basin Evolution*. Geological Society, London, Special Publications, **62**, 397–420.

TANKARD, A.J., WELSINK, H.J. & JENKINS, W.A.M. 1989. Structure styles and stratigraphy of the Jeanne d'Arc basin, Grand Banks of Newfoundland. *In*: TANKARD, A.J. & BALKWILL, H.R. (eds) *Extensional Tectonics and Stratigraphy of the North Atlantic Margins*. American Association of Petroleum Geologists, Memoirs, **46**, 265–282.

TARDUNO, J.A., LOWRIE, W., SLITER, W.V., BRALOWER, T.J. & HELLER, F. 1991. Reversed polarity characteristic magnetizations in the Albian Contessa Section (Umbrian Apennines, Italy): implications for the existence of a Mid-Cretaceous mixed polarity interval. *Journal of Geophysical Research*, **97**, 241–271.

TUCHOLKE, B.E. & LUDWIG, W.J. 1982. Structure and origin of the J-anomaly Ridge, western North Atlantic Ocean. *Journal of Geophysical Research*, **87**, 9389–9407.

VAN DER VOO, R. 1969. Palaeomagnetic evidence for the rotation of the Iberia peninsula. *Tectonophysics*, **7**, 5–56.

WHITMARSH, R.B. & MILES, P.R. 1995. Models of the development of the West Iberia rifted continental margin at 40°30'N deduced from surface and deep-tow magnetic anomalies. *Journal of Geophysical Research*, **100**, 3789–3806.

WHITMARSH, R.B., BESLIER, M.-O., WALLACE, P.J., *et al*. 1998. *Proceedings of the Ocean Drilling Program, Initial Reports, 173*. Ocean Drilling Program, College Station, TX.

WHITMARSH, R.B., MILES, P.R. & MAUFFRET, A. 1990. The ocean–continent boundary off the western continental margin of Iberia—I, Crustal structure at 41°30'N. *Geophysical Journal International*, **103**, 509–531.

WHITMARSH, R.B., MILES, P.R., SIBUET, J.-C. & LOUVEL, V. 1996a. Geological and geophysical implications of deep-tow magnetometer observations near Sites 897, 898, 899, 900, and 901 on the west Iberia continental margin. *In*: WHITMARSH, R.B., SAWYER, D.S., KLAUS, A. & MASSON, D.G. (eds) *Proceedings of the Ocean Drilling Program, Scientific Results, 149*. Ocean Drilling Program, College Station, TX, 665–674.

WHITMARSH, R.B., SAWYER, D.S., KLAUS, A. & MASSON, D.G. (eds) 1996. *Proceedings of the Ocean Drilling Program, Scientific Results, 149*. Ocean Drilling Program, College Station, TX.

WILSON, R.C.L., HISCOTT, R.N., WILLIS, M.G. & GRADSTEIN, F.M. 1989. The Lusitanian Basin of west–central Portugal: Mesozoic and Tertiary tectonic, stratigraphic and subsidence history. *In*: TANKARD, A.J. & BALKWILL, H.R. (eds) *Extensional Tectonics and Stratigraphy of the North Atlantic Margins*. American Association of Petroleum Geologists, Memoirs, **46**, 341–362.

WORM, H.-U., CLARK, D. & DEKKERS, M.J. 1993. Magnetic susceptibility of pyrrhotite: grain size, field and frequency dependence. *Geophysical Journal International*, **114**, 127–137.

ZHAO, X. 1996. Magnetic signatures of peridotite rocks from Sites 897 and 899 and their implications. *In*: SAWYER, D.S., WHITMARSH, R.B., KLAUS, A. & MASSON, D.G. (eds) *Proceedings of the Ocean Drilling Program, Scientific Results, 149*. Ocean Drilling Program, College Station, TX, 431–446.

ZHAO, X. 1999. Palaeomagnetic and rock magnetic results from the Iberian Plateau: a tale from peridotite rocks recovered in three ODP legs? *American Geophysical Union 1999 Fall Meeting Abstract Volume. Supplement, EOS Transactions, American Geophysical Union*, **80**, 46, F509.

ZIJDERVELD, J.D.A. 1967. A.C. demagnetization of rocks: analysis of results. *In*: COLLISON, D.W., CREER, K.M. & RUNCORN, S.K. (eds) *Methods in Palaeomagnetism*. Elsevier, New York, 254–286.

The Steinmann Trinity revisited: mantle exhumation and magmatism along an ocean–continent transition: the Platta nappe, eastern Switzerland

Laurent Desmurs[1], Gianreto Manatschal[1,2] & Daniel Bernoulli[1]

[1]Geology Institute, ETH-Zentrum, CH-8092 Zürich, Switzerland
(e-mail: desmurs@erdw.ethz.ch)

[2]EOST-UMR 7517 CNRS, Université Louis Pasteur, F-67084 Strasbourg, France

Abstract: The close association of serpentinites, basalts and radiolarites, later known as the Steinmann Trinity, was clearly described by Steinmann from the south Pennine Arosa zone and its southern prolongation, the Platta nappe of the eastern Swiss Alps. This classical 'ophiolite' is distinctly different from typical fast-spreading ridge associations and can be compared with the transitional crust occurring along non-volcanic passive continental margins in present-day oceans. It includes serpentinized peridotites that we interpret as subcontinental mantle rocks, which were exhumed along low-angle detachment faults and locally overlain by extensional allochthons of continental crust, minor gabbroic intrusions, tholeiitic pillow lavas and flows and a succession of oceanic sediments. The serpentinized peridotites record deformation at falling temperatures during extension leading to final exposure of the mantle rocks at the sea floor and their inclusion in tectono-sedimentary breccias (ophicalcites). Field relationships, and mineral-chemical and radiometric data show that the gabbros intruded already serpentinized mantle rocks at shallow depth 161 Ma ago. They are Mg gabbros, Fe gabbros and Fe–Ti gabbros, cut by dioritic pegmatoid veins and albitite dykes, which originated by differentiation from the same parental magma. All gabbros show the same metamorphic evolution, i.e. intrusion at relatively low pressure, oceanic hydration at elevated temperature and a subsequent static oceanic alteration. The pillow lavas stratigraphically overlie the exhumed mantle rocks and the tectono-sedimentary breccias related to the exhumation of both mantle rocks and gabbros. However, both gabbros and pillow basalts are characterized by εNd values typical for an asthenospheric mid-ocean ridge-type source of the melts. They may be the products of a steady process that combined extensional deformation with magma generation and emplacement. They appear to document the onset of sea-floor spreading across an exhumed subcontinental mantle during the earliest phases of a slow-spreading ridge.

'*Mapping the ophiolite group as a unit has often obscured critical relationships of its various members to the tectonic cycle*' (H.H. Hess 1955).

In 1905, Gustav Steinmann noted the close association of serpentinites, 'diabase' and radiolarite and considered this 'green rock' or ophiolitic association as characteristic for the axial part of the 'geosyncline' and the deep ocean floor. Although Steinmann considered diabase, spilite and variolite (variolitic pillow lavas) as intrusive rocks distinctly younger than the associated sediments, he stressed their characteristic association with deep-sea sediments, notably radiolarian cherts but also deep-water limestones of Maiolica facies (Steinmann 1925, 1927). In his view, the 'consanguineous' association of ultramafic and mafic material was characteristic for suboceanic environments from where these magmas ascended during folding of the oceanic sediments. Eventually, the importance of Steinmann's discovery was recognized and the association of serpentinites, pillow lavas and radiolarites became known as the Steinmann Trinity: 'Steinmann's realization of this trinity as a world-wide, age-long characteristic of geosynclinal deposition ranks among the most exciting achievements of geological research' (Bailey & McCallien 1950). With time, the Steinmann Trinity became a synonym for ophiolitic associations in the sense of the 1972 Penrose Conference on ophiolites (Anonymous 1972). As a consequence, the examples described by Steinmann from the Alps and the Apennines were considered as tectonically dismembered ophiolites, incomplete equivalents of

From: Wilson, R.C.L., Whitmarsh, R.B., Taylor, B. & Froitzheim, N. 2001. *Non-Volcanic Rifting of Continental Margins: A Comparison of Evidence from Land and Sea*. Geological Society, London, Special Publications, **187**, 235–266. 0305-8719/01/$15.00 © The Geological Society of London 2001.

ancient oceanic lithosphere (e.g. Laubscher 1969; Dewey & Bird 1970).

In the 1970s, detailed fieldwork showed that the stratigraphy of the ophiolites of the Alps and the Apennines did not match the classical ophiolite stratigraphy as defined by the Penrose Conference in 1972 and observed in large ophiolitic complexes such as Troodos (Moores & Vine 1971) or Semail (Glennie *et al.* 1974). No substantial relics of a sheeted dyke complex were found, and gabbros appeared to play a subordinate role. Instead, oceanic sediments, radiolarian cherts and pelagic limestones were observed to stratigraphically overlie (1) serpentinized mantle rocks and tectono-sedimentary breccias ('ophicalcites') derived from them, and (2) pillow lavas that were extruded onto the serpentinites previously exposed on the ocean floor (Fig. 1; Decandia & Elter 1972; Lagabrielle *et al.* 1984; Bernoulli & Weissert 1985; Weissert & Bernoulli 1985). At an early stage, exposure of mantle rocks at the sea floor was postulated to be associated with oceanic transform faults in analogy with serpentinite occurrences along such faults in present-day oceans (Lemoine 1980; Bernoulli & Weissert 1985; Weissert & Bernoulli 1985); however, the data now available from the south Pennine–Austroalpine margin suggest that in this area subcontinental mantle was exhumed along a system of low-angle extensional detachment faults (Lemoine *et al.* 1987; Froitzheim & Manatschal 1996; Manatschal & Nievergelt 1997). It appears today that the trinity of serpentinites, pillow lavas and radiolarites Steinmann (1905, 1927) described from the Alps and the Apennines corresponds, by and large, to the stratigraphy of transitional crust, as observed along 'non-volcanic' continental margins (Manatschal & Bernoulli 1999) laterally grading into that of oceanic crust generated along a slow-spreading ridge (Lagabrielle & Lemoine 1997). Indeed, exhumation of mantle rocks and their exposure at the sea floor is now well documented along present-day margins where it occurred during the final stage of rifting and break-up of the continental lithosphere (Boillot *et al.* 1987, 1988). However, the origin and significance of the magmatic rocks associated with exhumation are still discussed.

In this paper, we shall describe in detail the stages of mantle exhumation and associated magmatic activity along the south Pennine–Austroalpine margin of the Adria plate, the type area of the Steinmann Trinity. In an earlier paper, two of us described the extensional tectonic structures associated with rifting (Manatschal & Bernoulli 1999), whereas here we deal with the retrograde metamorphism and the submarine alterations associated with mantle uplift and the magmatic processes associated with the latest phases of rifting at the onset of sea-floor spreading. Our argumentation is based on field relationships, and petrological, geochemical and geochronological data, and will focus on: (1) the evolution of the mantle rocks during their exhumation to the sea floor; (2) the $P–T$ conditions during the intrusion of the gabbros; (3) the genetic and time relationships between the gabbros and the overlying pillow basalts and their relation to the rifting processes. From our data it will become clear that, in our study area, peridotites and mafic rocks are not a con-

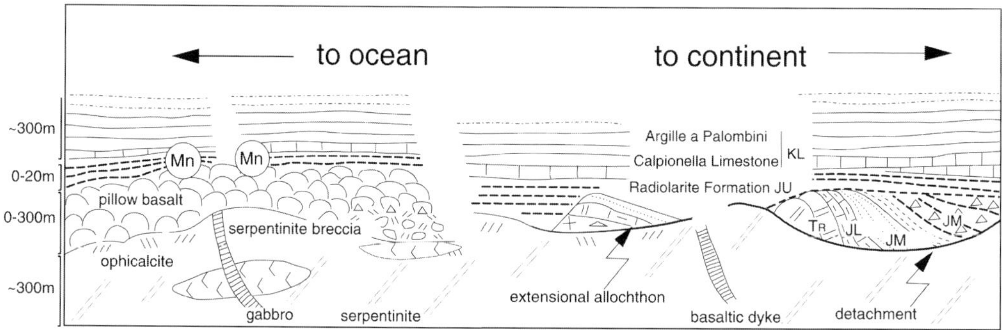

Fig. 1. Generalized stratigraphy of transitional crust (ocean–continent transition) in the Platta nappe, eastern Switzerland. TR, Triassic; JL, Lower Jurassic; JM, Middle Jurassic; JU, Upper Jurassic; KL, Lower Cretaceous; Mn, manganese deposits. After Manatschal & Nievergelt (1997).

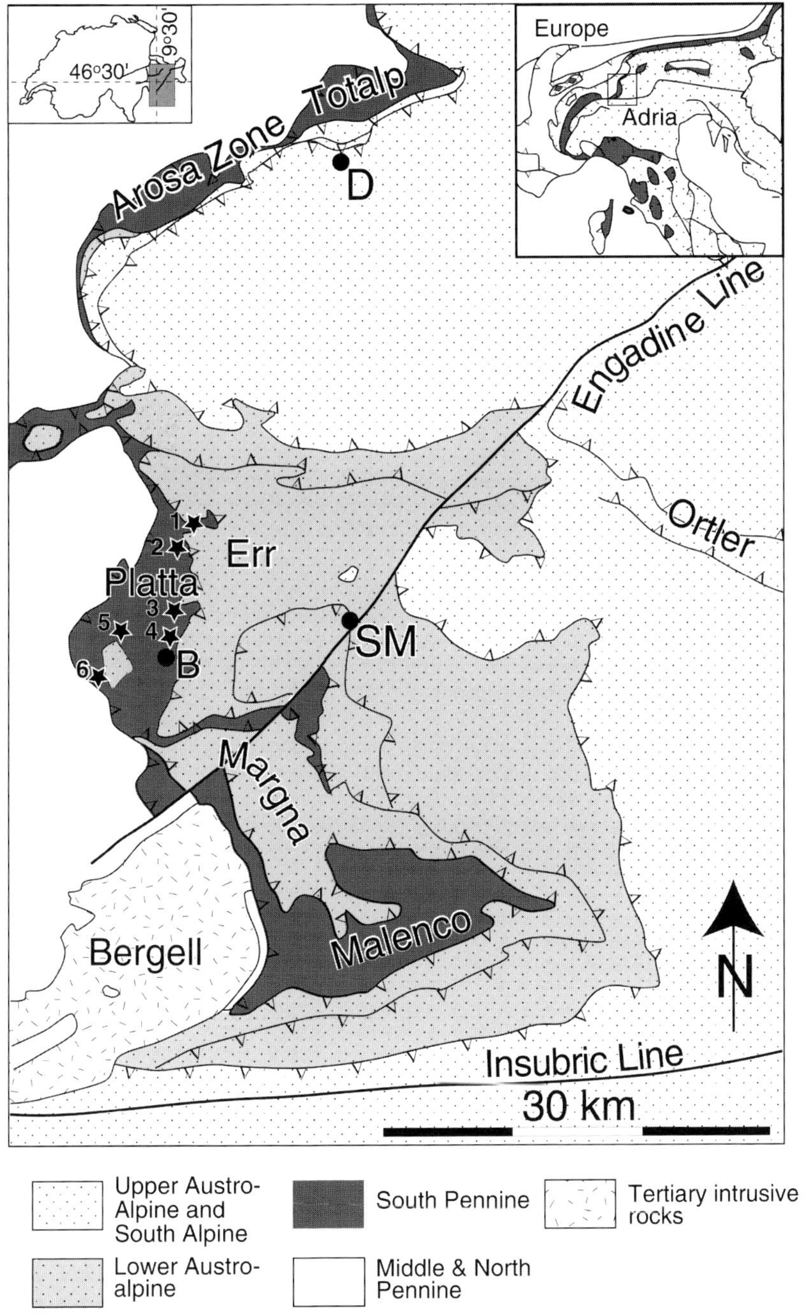

Fig. 2. Simplified geological map and location of the study area. B, Bivio; D, Davos; SM, St. Moritz; 1, Val d'Err; 2, Falotta; 3, Val Savriez; 4, Val da Natons; 5, Starschagns; 6, Fuorcla da Faller; modified after Froitzheim *et al.* (1994).

sanguineous association as originally conceived by Steinmann.

General situation

Steinmann (1905) recognized that in the Alps his ophiolitic trinity always occurred in the same tectonic unit, which he called the Rhetic nappe, a term that did not survive but included the ophiolite-bearing south Pennine units of eastern Switzerland and the adjacent eastern Alps as well as the south Pennine relics in the pre-Alpine nappes of western Switzerland (nappe des Gets). Remnants of this oceanic suture can be traced across the Alpine system from Corsica and the Apennines along the south Pennine nappes to the eastern end of the Alps and possibly beyond. In eastern Switzerland and western Austria, the relics of this ophiolite zone can be followed below the Austroalpine nappes from Val Malenco (Malenco complex) across the Engadin valley to the Sursés valley (Platta nappe), and across a number of imbricates (Totalp slice) and local occurrences in tectonic melanges (Arosa zone) to the northern margin of the Northern Calcareous Alps (Fig. 2). It is, however, only in the area of the Platta and Err nappes of Sursés that the ocean–continent transition including the distal continental margin can be reconstructed with some confidence (Froitzheim & Manatschal 1996; Manatschal & Nievergelt 1997; Manatschal & Bernoulli 1999). In this area, the ophiolites were always part of the upper plate of the Alpine system and were never subducted to great depth. Here the Alpine nappe movements can be kinematically inverted and the pre-Alpine geometrical and age relationships

between the exhumed mantle rocks, the gabbroic intrusions and the tholeiitic pillow lavas can be studied in detail. In many places, the primary contacts between plutonic, volcanic and sedimentary rocks are well exposed. Locally the pre-Alpine mineral parageneses are excellently preserved, and a complete sedimentary record of the late syn- and post-rift history is available. As temperatures, in our area, never exceeded *c.* 250 °C during Alpine orogeny, the pre-Alpine history of mantle deformation and hydration during exhumation is well documented.

The Platta nappe

Together with the lower Austroalpine Err nappe, the Platta nappe preserves the remnants of the southeastern margin of the Liguria–Piedmont segment of the Alpine Tethys (Figs 2 and 3). The Err nappe includes the remnants of the distal Adriatic margin and was thrust onto the Platta nappe during late Cretaceous time. It consists of a thinned continental crust, overlain by a number of extensional allochthons, slivers of continental crust and pre-rift sediments, which were tectonically emplaced during mid-Jurassic time along an oceanward-dipping low-angle detachment system (Fig. 3; Froitzheim & Eberli 1990; Froitzheim & Manatschal 1996; Manatschal & Nievergelt 1997). The Platta nappe, originally situated oceanward of the Err nappe, represents the ocean–continent transition *sensu stricto* and exposes the Steinmann Trinity, i.e. serpentinized peridotites, basaltic volcanic rocks and oceanic sediments. Far-travelled extensional allochthons, derived from the distal continental margin, were emplaced on

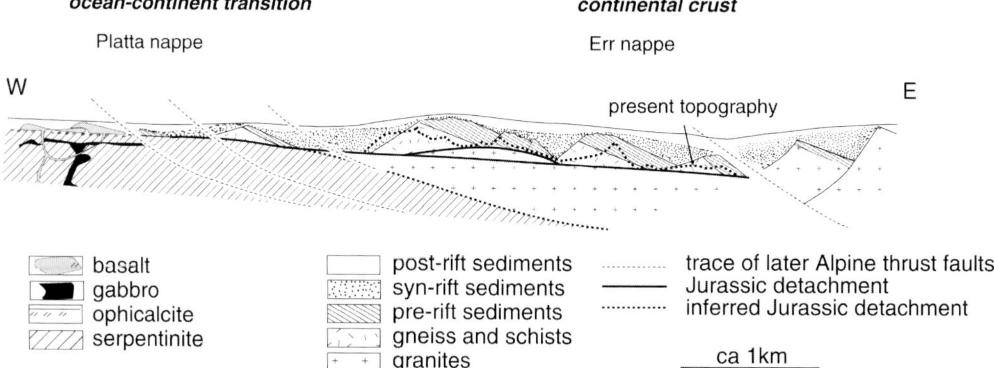

Fig. 3. Palinspastic reconstruction of the Err–Platta ocean–continent transition, after Manatschal & Nievergelt (1997).

the exhumed mantle rocks before the deposition of the basalts and sediments (Figs 1 and 3). The Err and Platta nappes are now part of a late Cretaceous west-directed thrust wedge, which was probably associated with subduction along the eastern border of Adria (Froitzheim *et al.* 1996). Subduction of the Liguria–Pied-mont ocean initiated later during late Cretaceous time and continued during Tertiary time as indicated by the early Tertiary ages of high-pressure metamorphism within the ophiolites of the western Alps (Rubatto *et al.* 1998). In contrast to these ophiolites, the south Pennine ophiolites of the Arosa zone, the Platta nappe

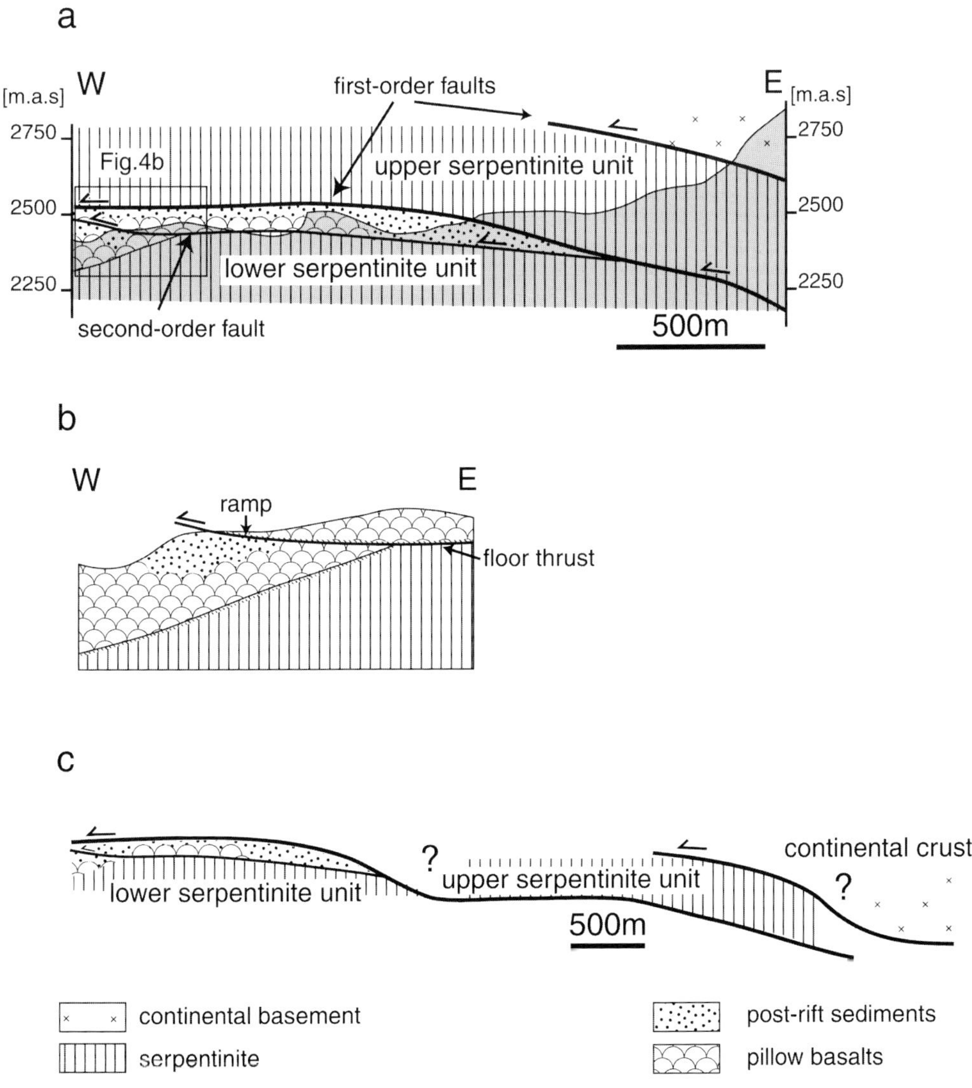

Fig. 4. Qualitative palinspastic reconstruction of the Platta nappe in the Falotta area. (**a**) Alpine structure. The shaded lower part of the section is observed, the white upper part is a lateral projection of geometries mapped in the surroundings of the section. (**b**) Detail of (**a**) showing a second-order fault ramping across the basalts into the overlying sediments. (**c**) Kinematic inversion of section (**a**). The question marks indicate the uncertainties in the estimate of the amounts of displacement (pre-thrust distance between the first-order tectonic units).

and the Malenco Complex lack a high-pressure metamorphic overprint, and show only a general increase of Alpine metamorphism from deep burial diagenesis in the north (Arosa zone) to epidote–amphibolite facies conditions in the south (Malenco complex, Ferreiro Mählmann 1995, and references therein). The serpentine mineral phases are lizardite–chrysotile north of Bivio and antigorite–lizardite–chrysotile south of this location, confirming the increase in Alpine metamorphism from north to south (Trommsdorff & Evans 1974; Trommsdorff 1983). Obviously, these nappes were part of the hanging wall of the south-directed subduction zone, which apparently first developed within the ocean (Lagabrielle 1987; Avigad *et al.* 1993).

Alpine deformation

The occurrence of 'middle' Cretaceous (Albian to (?) Cenomanian; Dietrich 1970) hemipelagic and turbiditic sediments in the Platta nappe provides a lower age bracket for the Alpine deformation (Froitzheim *et al.* 1994), which is consistent with radiometric age data from the overlying Austroalpine nappes (Handy *et al.* 1993). The oldest structures related to Alpine convergence are localized thrust faults of late Cretaceous age with a top-to-the-west sense of shear. The occurrence of sediments and/or basalts stratigraphically overlying the exhumed mantle rocks suggests that décollement was near-surface and collected only the uppermost crust. The geometry of the thrust system is locally complicated by lateral thickness variations of the basalts, pre-existing basement topography, the occurrence of extensional allochthons of continental provenance overlying the serpentinites, and later, Tertiary, deformation. The resulting complex geometries probably led Ring *et al.* (1988; 1989) and Dürr (1992) to the conclusion that parts of the Platta ophiolites formed a tectonic mélange generated during subduction of the upper Penninic below the Austroalpine units, an interpretation that is not supported by our field observations.

By and large, the large-scale geometry of the Platta nappe is defined by two serpentinite thrust sheets, which can be traced across the entire nappe and are interleaved with imbricated cover sequences. This geometry resembles that of a large-scale duplex structure and is particularly well illustrated in the area of Falotta (for location see Fig. 2). In Falotta, the major tectonic units consist of massive sheets, of up to 300 m thickness, of serpentinized peri-

dotites (upper and lower serpentinites in Figure 4a), which are floored by basal décollement horizons and are interleaved with basalts and post-rift sediments. Along the basement–cover contacts, i.e. between serpentinites and massive pillow basalts, thrust faults occur, typically along ophicalcites, which eventually ramp across the basalts into the sediments and lead to local repetitions in the cover sequence (Fig. 4b). Locally, isoclinal folds preserving preferentially the inverted limb are associated with these faults. Thus, the simplified overall geometry of the thrust stack is that of a duplex structure with first-order thrust faults lying along the base of the large serpentinite sheets, and second-order thrust faults along the basement–cover contact and ramping into basalts and overlying sediments (compare roof and floor thrusts in figure 19 of Boyer & Elliott (1982)).

The amount of displacement accommodated by the first-order faults cannot be determined directly. A minimum value, however, can be estimated by measuring the distance in transport direction between the westernmost occurrence of a serpentinite sheet and the easternmost outcrop of sediments in the underlying sheet, which results in a minimum displacement of 9 km between the two serpentinite units. The displacement along the second-order thrust faults is variable but is an order of magnitude less than that of the first-order faults, i.e. of the order of 500 m or less.

The most prominent post-stack structures are metre- to kilometre-scale folds that deform the thrust-stack including the thrust contact between the Err and the Platta nappe. This folding phase is associated with extensional faults (Manatschal & Nievergelt 1997), which, however, are not easily visible in the Platta nappe. The fold axes of the large-scale folds plunge slightly to the east and their axial planes are subhorizontal. Micas crystallizing within the fold axial-plane cleavage of these folds yielded ages of 72.5 ± 6.4 Ma (Handy *et al.* 1993). Structures postdating this folding event become more pronounced towards the south (south of Bivio) and towards the base of the Platta nappe.

A qualitative kinematic inversion of the Platta nappe is possible only in the northern part of the Platta nappe; for instance, in the area of Falotta, where the architecture of the late Cretaceous thrust stack is still preserved (Fig. 4a). The kinematic inversion is achieved by aligning the stacked serpentinite sheets in a horizontal order in which the highest sheet is placed furthest to the east (opposite to the transport direction) (Fig. 4c). The imbricate structures between two serpentinite sheets can be reconstructed using the same approach, assuming that the imbricates lying between two serpentinite sheets were the cover of the underlying sheet. This qualitative kinematic inversion of the thrust stack leads to a palinspastic reconstruction for the Platta domain, which allows us to understand the relative position of individual outcrops and the spatial distribution of specific rock associations relative to the ocean–continent transition (Fig. 3).

Pre-Alpine situation and stratigraphy

Kinematic inversion of the movements that assembled the imbricates of the Platta nappe resulted in a zone of 20–30 km width of transitional to incipient oceanic crust. The serpentinites of the individual imbricates of the Platta nappe are about 300 m thick and therefore document a crustal section about 300 m deep (Fig. 1). Deeper levels of the crust are not observed in the Platta nappe.

The base of the imbricates of the Platta nappe is invariably formed by serpentinized peridotites derived from spinel lherzolites and harzburgites into which gabbros and basaltic dykes intruded. The top of the serpentinites is usually formed by ophicalcites, tectono-sedimentary breccias (Bernoulli & Weissert 1985; Lemoine *et al.* 1987) overlain by basalts toward the former ocean. Toward the former continent, the extensional allochthons mentioned above overlie the exhumed mantle rocks (Fig. 1; Froitzheim & Manatschal 1996; Manatschal & Nievergelt 1997). Clasts of continental basement rocks and pre-rift sediments occur within tectonic and sedimentary breccias associated with the extensional allochthons.

The gabbros occur as planar, sill-like bodies within the serpentinized peridotites. They are relatively rare, <10% of the total observed serpentinite volume. Massive basalts, pillow lava, pillow breccias and hyaloclastites occur in patches of variable thickness and size, and their abundance appears to increase oceanward. They stratigraphically overlie the serpentinites, ophicalcites, gabbros and associated breccias, and their extrusion therefore postdates emplacement of the mantle rocks and gabbros at the sea floor. The original thickness of the basalts is of the order of 150 m, but was overestimated in the past: basalt bodies appearing to be several 100 m thick are either folded or duplicated by thrust faults as indicated by relics of post-rift sediments or slices of serpentinite occurring between different basalt sequences.

Exhumed mantle rocks, ophicalcites, pillow lavas, extensional allochthons and syn-rift sediments are all overlain by post-rift deep-water sediments (Fig. 1). Post-rift sediments also seal the faults associated with the emplacement of the extensional allochthons. These oldest sediments, found on both continental and oceanic basement rocks, define the base of the post-rift sedimentary sequence; they are thin-bedded cherts, siliceous shales, and turbidites of the upper Middle to Upper Jurassic Radiolarite Formation (Baumgartner 1987). This formation is often only a few metres thick. It is overlain by (1) well-bedded, light-coloured micritic limestones with shale intercalations (Calpionella Limestone or Aptychus Limestone) of Berriasian age; (2) dark siliceous shales and calcarenites alternating with dark grey micritic limestones (Argille a Palombini, of approximately Valanginian to Barremian age); (3) hemipelagic marls with interbedded sandstone turbidites, similar to the Scisti di Val Lavagna of the northern Apennines, and dated by planktonic Foraminifera (Aptian–Albian– (?) Cenomanian age, Dietrich 1970). In the radiolarites, the pelagic limestones of the Calpionella Limestone Formation, and the Argille a Palombini, intercalations of pebbly mudstones, clast-supported breccias and turbidites yield locally derived clasts of serpentinite and basalt and of crystalline basement rocks and pre-rift sediment derived from the distal continental margin and the extensional allochthons. These mass-flow deposits document a long-lived submarine topography.

Mantle exhumation

Serpentinized lherzolite is the major rock type throughout the Platta nappe. The peridotites of the upper serpentinite unit (Fig. 4), which are derived from the originally continentward part of the transitional crust, preserve a spinel foliation and contain pyroxenite dykes parallel to the foliation, indicating that they equilibrated in the spinel stability field. The peridotites of the lower serpentinite unit (Fig. 4), which originally were nearer to the ocean, are usually less deformed and generally free of pyroxenites. Because all these rocks are stratigraphically overlain by deep-marine sediments they must have been uplifted from mantle depth to the sea floor. Serpentinization generally masks the earlier stages of retrograde evolution during exhumation; however, at places the relationships between deformation and retrograde evolution can be recognized.

Petrology and microstructures of the lower serpentinized peridotite

The serpentinized lherzolites of the lower serpentinite unit show a porphyroclastic structure with holly-leaf spinel. They consist of large porphyroclastic Cr–diopside, orthopyroxene, Cr–Al spinel and olivine, which is always replaced by serpentine minerals. In a few samples, disseminated grains of phlogopite occur within the porphyroclastic assemblage. This assemblage partly re-equilibrated in the spinel stability field as shown by the recrystallization of granoblastic diopside, Cr–Al spinel and Ti–pargasite between the large porphyroclastic pyroxenes.

The evidence for retrograde high-temperature hydration in this unit is the appearance of Mg–

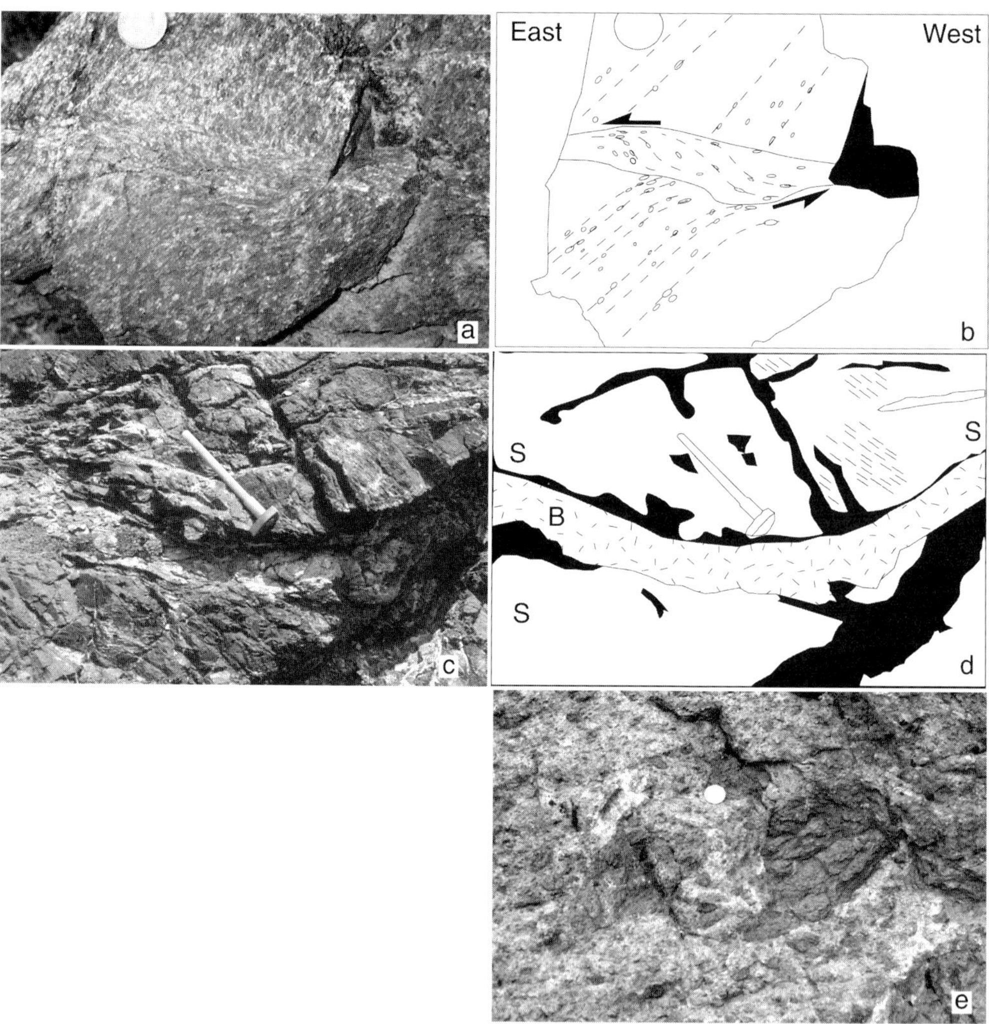

Fig. 5. (**a**) and (**b**) High-temperature top-to-the-east shear zone deforming the pre-existing spinel foliation of the peridotite (diameter of coin is 2.3 cm). Val d'Err (coordinates of Swiss Topographic Map: 772'750/ 159'750). (**c**) and (**d**) Serpentinite mylonite (S), deformed under low-temperature conditions and cut by undeformed basaltic dyke (B) indicating a pre-Alpine age of deformation (hammer is 50 cm long). Near Pt. 2594, east of Falotta (coordinates of Swiss Topographic Map: 771'250/156'930). (**e**) Tectonic breccia with clasts of mylonitic serpentinite embedded in a serpentinite matrix (diameter of coin is 2 cm). Falotta (coordinates of Swiss Topographic Map: 770'330/157'350).

hornblende in coronae around clinopyroxene and along its cleavage planes. This was followed by a more widespread hydration at lower temperature characterized by the crystallization of tremolite at the expense of clinopyroxene and of Mg–hornblende. Olivine was transformed into a serpentine mesh with magnetite, whereas the spinel was altered to Cr–chlorite and magnetite. Finally the serpentine minerals were partially replaced by the assemblage calcite + talc.

Petrology and deformation of the upper serpentinized peridotite

In the upper serpentinite unit (Falotta and Val d'Err areas), several mylonitic shear zones occur within the peridotite, which affected the previous high-temperature spinel foliation (Fig. 5a,b). The microstructures associated with these shear zones are still visible within small pyroxenite dykes (olivine websterite) present within

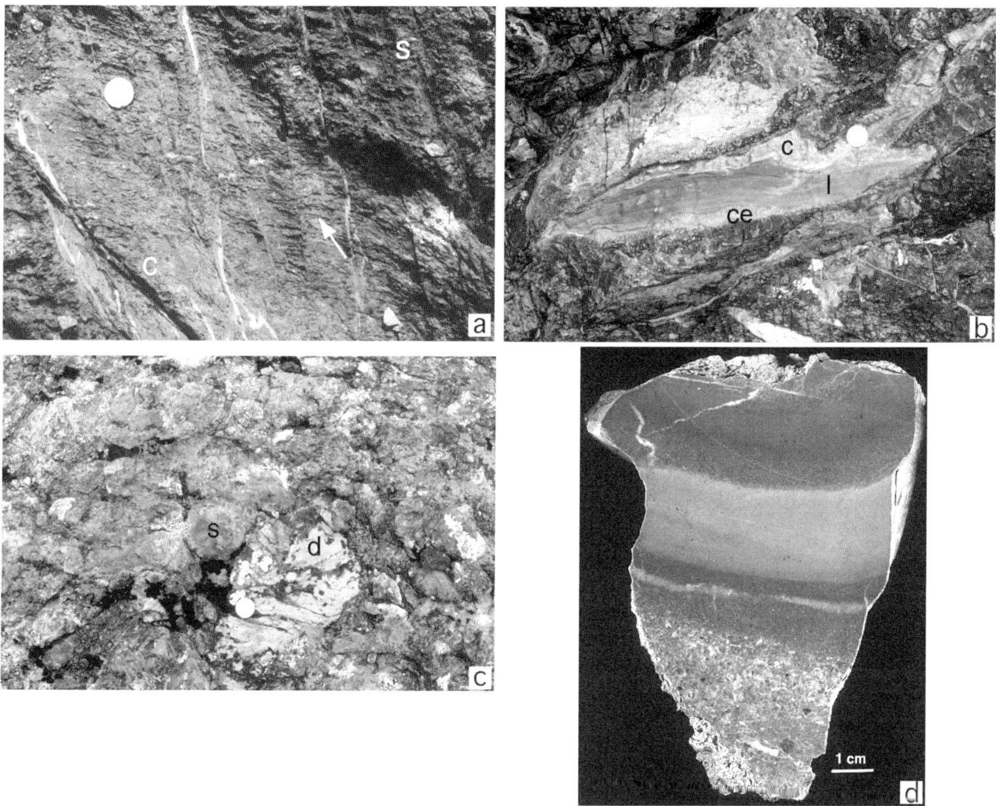

Fig. 6. (a) Calcite (c) partially replacing serpentinite (s). The replacement process is recorded by the perfectly preserved high-temperature spinel foliation (indicated by the arrow) floating in the newly formed calcite matrix (diameter of coin 2 cm). Falotta (coordinates of Swiss Topographic Map: 769′550/157′350). **(b)** Geopetal structure in sediment-filled pocket in serpentinite (ophicalcite I). A first generation of cement (ce) fringes the void and is overlain by grey lime mudstone (l) filling the lower part of the pocket. Clear sparry calcite (c) cements the remaining pore space above (diameter of coin 2 cm). Near Pt. 2594, east of Falotta (coordinates of Swiss Topographic Map: 771′250/156′930). **(c)** Sedimintary breccia with clasts of serpentinite (s) and of pre-rift Triassic dolomite (d) embedded in a serpentinite matrix. This breccia stratigraphically overlies the exposed mantle rocks; the dolomite clasts are derived from an extensional allochthon (diameter of coin 2 cm). Plang Tguils (coordinates of Swiss Topographic Map: 767′500/147′400). **(d)** Graded lithoclastic sandstone composed of serpentinite and ophicalcite clasts intercalated in the lowermost Radiolarite Formation. Polished surface, Arosa Zone, Obersasställi, Totalp (coordinates of Swiss Topographic Map: 780′650/191′150).

Table 1. *Representative chemical compositions of the minerals characterizing the metamorphic evolution of the lherzolites and the pyroxenites*

	Lherzolite of the lower serpentinite unit													
	Porphyroclast							granoblast				oceanic alteration		
	MSP		SUP2			NAP99-6		MSP		SUP2		MSP		NAP99-7-2	
	cpx	opx	cpx	opx	spi	cpx	phl	spi	cpx	cpx	Ti-par	serp	mt	chl	serp
SiO$_2$	51.03	55.28	50.37	56.08	0.00	52.70	38.38	0.00	52.10	49.67	43.40	44.43	0.49	27.97	36.94
TiO$_2$	0.56	0.23	0.81	0.34	0.65	0.32	1.84	0.63	0.49	0.96	3.75	0.03	0.17	0.01	0.03
Cr$_2$O$_3$	1.04	0.79	1.04	0.64	29.16	0.86	0.88	33.03	0.94	1.04	1.84	0.00	0.00	0.18	0.00
Al$_2$O$_3$	5.74	3.43	4.83	3.06	33.39	4.96	17.48	29.81	3.30	4.26	12.26	0.47	0.01	20.39	0.45
Fe$_2$O$_3$	0.23	0.14	2.28	0.00	6.09	0.00	0.00	4.28	0.59	2.59	0.00	0.00	67.88	0.00	0.00
FeO	2.99	6.65	1.01	6.56	14.88	2.46	3.65	18.44	2.03	0.65	4.36	1.56	31.09	12.15	9.55
MnO	0.11	0.19	0.13	0.19	0.29	0.08	0.00	0.14	0.06	0.09	0.05	0.09	0.09	0.19	0.02
MgO	15.73	31.80	15.98	32.37	14.48	15.94	22.75	11.59	16.48	16.30	16.82	39.09	0.20	25.92	35.28
CaO	22.18	1.97	22.83	1.15	0.05	21.34	0.00	0.05	23.11	22.30	12.34	1.25	0.04	0.15	0.09
Na$_2$O	0.43	0.05	0.67	0.04	0.00	0.93	0.43	0.04	0.41	0.65	3.38	0.02	0.05	0.00	0.02
K$_2$O	0.00	0.00	0.00	0.00	0.00	0.00	8.74	0.00	0.00	0.00	0.05	0.02	0.00	0.00	0.01
H$_2$O	–	–	–	–	–	–	4.21	–	–	–	2.10	12.92	–	12.11	11.61
Total	100.11	100.61	99.97	100.53	99.19	99.61	98.55	98.20	99.54	98.56	100.53	100.00	100.14	99.07	94.12
Si	1.86	1.91	1.85	1.93	0.00	1.91	2.74	0.00	1.91	1.85	6.20	2.06	0.02	2.77	3.82
Ti	0.02	0.01	0.02	0.01	0.01	0.01	0.10	0.01	0.01	0.03	0.40	0.00	0.00	0.00	0.00
Cr	0.03	0.02	0.03	0.02	0.68	0.02	0.05	0.80	0.03	0.03	0.21	0.00	0.00	0.01	0.00
Al	0.25	0.14	0.21	0.12	1.16	0.21	1.47	1.08	0.14	0.19	2.06	0.03	0.00	2.38	0.05
Fe^{3+}	0.01	0.00	0.06	0.00	0.13	0.00	0.00	0.10	0.02	0.07	0.00	0.00	1.96	0.00	0.00
Fe^{2+}	0.09	0.19	0.03	0.19	0.37	0.07	0.22	0.47	0.06	0.02	0.52	0.06	1.00	1.01	0.82
Mn	0.00	0.01	0.00	0.01	0.01	0.00	0.00	0.00	0.00	0.00	0.01	0.00	0.00	0.02	0.00
Mg	0.85	1.64	0.88	1.66	0.64	0.86	2.42	0.53	0.90	0.91	3.58	2.70	0.01	3.83	5.43
Ca	0.87	0.07	0.90	0.04	0.00	0.83	0.00	0.00	0.91	0.89	1.89	0.06	0.00	0.02	0.01
Na	0.03	0.00	0.05	0.00	0.00	0.07	0.06	0.00	0.03	0.05	0.94	0.00	0.00	0.00	0.00
K	0.00	0.00	0.00	0.00	0.00	0.00	0.79	0.00	0.00	0.00	0.01	0.00	0.00	0.00	0.00
OH	–	–	–	–	–	–	2.00	–	–	–	2.00	8.00	–	8.00	8.00
Mg#	0.90	0.89	0.90	0.90	0.56	0.92	0.92	0.48	0.92	0.91	0.87	0.98	0.00	0.79	0.87

Ions calculated on the basis of six oxygens (pyroxene), three cations (spinel, magnetite), 23 oxygens and $\sum$(cat)–Ca–Na–K $= 13$ (amphibole), 18 oxygens (chlorite, serpentine).

Table 1. *continued*

	Lherzolite of the upper serpentinite unit			Pyroxenite of the upper serpentinite unit										
	porphyroclast		neoblast	porphyroclast			neoblast	static hydration		oceanic alteration				
	FAP4	VEP8	VEP8	VEP3	FAP4	CUP1	CUP1	VEP3		VEP3		CUP1		
	cpx	cpx	cpx	cpx	cpx	cpx	cpx	Mg-hbl	tr	chl	Cr-chl	Cr-chl	serp	mt
SiO_2	51.17	51.87	52.51	51.87	51.74	52.12	51.84	48.45	54.67	34.38	22.39	24.39	42.00	0.42
TiO_2	0.80	0.53	0.44	0.47	0.81	0.60	0.48	0.14	0.24	0.04	0.10	0.10	0.03	0.59
Cr_2O_3	0.58	0.95	0.43	0.71	0.35	0.51	0.47	0.35	0.97	0.62	13.51	4.93	0.24	0.38
Al_2O_3	6.42	6.19	4.29	6.91	7.42	6.89	3.93	10.80	3.85	12.83	11.21	21.81	4.95	0.04
Fe_2O_3	1.75	1.30	1.70	1.51	0.00	0.42	1.54	4.12	2.75	0.00	0.00	0.00	0.00	65.28
FeO	1.60	1.32	0.74	1.20	3.20	2.84	1.28	0.00	0.00	3.70	17.56	7.38	5.16	30.95
MnO	0.09	0.11	0.09	0.04	0.10	0.10	0.12	0.11	0.06	0.03	0.82	0.27	0.19	0.15
MgO	15.15	15.18	16.30	14.43	13.67	15.96	16.57	19.93	22.57	34.36	23.55	27.46	34.46	0.21
CaO	21.41	21.21	22.67	21.34	21.27	19.28	22.11	12.35	12.46	0.03	0.10	0.26	0.71	0.14
Na_2O	1.41	1.46	1.07	1.86	1.58	1.47	0.82	2.20	0.86	0.00	0.18	0.01	0.01	0.00
K_2O	0.00	0.01	0.02	0.03	0.03	0.02	0.00	0.02	0.04	0.00	0.03	0.02	0.02	0.00
H_2O	–	–	–	–	–	–	–	2.18	2.21	12.56	11.29	12.09	12.83	–
Total	100.41	100.19	100.30	100.48	100.26	100.32	99.21	100.66	100.77	98.56	101.07	98.92	100.62	98.16
Si	1.86	1.88	1.90	1.88	1.87	1.88	1.90	6.67	7.41	3.28	2.38	2.42	3.93	0.02
Ti	0.02	0.01	0.01	0.01	0.02	0.02	0.01	0.01	0.02	0.00	0.01	0.01	0.00	0.02
Cr	0.02	0.03	0.01	0.02	0.01	0.01	0.01	0.04	0.10	0.05	1.13	0.39	0.02	0.01
Al	0.27	0.26	0.18	0.29	0.32	0.29	0.17	1.75	0.61	1.44	1.40	2.55	0.55	0.00
Fe^{3+}	0.05	0.04	0.05	0.04	0.00	0.01	0.04	0.43	0.28	0.00	0.00	0.00	0.00	1.92
Fe^{2+}	0.05	0.04	0.02	0.04	0.10	0.09	0.04	0.00	0.00	0.30	1.56	0.61	0.40	1.01
Mn	0.00	0.00	0.00	0.00	0.00	0.00	0.00	0.01	0.01	0.00	0.07	0.02	0.02	0.01
Mg	0.82	0.82	0.88	0.78	0.74	0.86	0.91	4.09	4.56	4.89	3.73	4.06	4.80	0.01
Ca	0.83	0.82	0.88	0.83	0.82	0.74	0.87	1.82	1.81	0.00	0.01	0.03	0.07	0.01
Na	0.10	0.10	0.08	0.13	0.11	0.10	0.06	0.59	0.22	0.00	0.04	0.00	0.00	0.00
K	0.00	0.00	0.00	0.00	0.00	0.00	0.00	0.00	0.01	0.00	0.00	0.00	0.00	0.00
OH	–	–	–	–	–	–	–	2.00	2.00	8.00	8.00	8.00	8.00	–
Mg#	0.90	0.92	0.93	0.91	0.88	0.90	0.92	0.91	0.94	0.94	0.70	0.87	0.92	0.00

the lherzolite host. They consist of rotated or kinked porphyroclastic ortho- and clinopyroxene showing undulatory extinction and of neoblastic minerals growing in the plane of the new foliation or recrystallizing in a mosaic fabric. The asymmetric clasts associated with this event of high-temperature shearing show a top-to-the-east, i.e. top-to-the-continent, sense of shear. The neoblastic assemblage consists of secondary Al-diopside, orthopyroxene, olivine and spinel, which are altered to chlorite, serpentine minerals and magnetite. We interpret this association as formed under spinel peridotite facies conditions with a later low-grade overprint. As no pressure constraints are available for this deformation event, we cannot relate it with certainty to rifting. Only the late, hydrous serpentinite mylonites indicate defor-

mation of the mantle rocks during their exhumation to the sea floor.

Deformation under greenschist-facies conditions produced a foliation defined by the assemblage chlorite, serpentine and rare talc. The associated shear zones show a top-to-the-west, i.e. oceanward, sense of shear, and are cut by undeformed basaltic dykes, demonstrating their pre-Alpine age (Fig. 5c,d).

Deformation under low-temperature conditions is recorded by tectonic breccias. In these, clasts from the greenschist-facies shear zones are embedded in a serpentinite matrix, indicating that deformation extended into the brittle field during final exhumation of the mantle rocks (Fig. 5e). The clasts are rounded by frictional wear, suggesting rotational deformation. Up-section the breccia is cut by chrysotile and calcite veins.

Replacement of serpentine minerals under static conditions by calcite occurred at still lower temperatures. This is illustrated by serpentinites that, in a calcitized matrix, preserve the relics of the high-temperature foliation (Fig. 6a). These calcitized serpentinites are transitional to tectono-sedimentary breccias (ophicalcite type I of Lemoine *et al.* (1987)). Typically, the serpentinites grade rapidly through a narrow zone of host rock, cut by different generations of neptunic dykes, into complex breccias dominated by a carbonate matrix and occasionally preserving a jigsaw-puzzle structure documenting *in situ* fragmentation of the host rock. Crevasses and dykes are filled by red or grey microsparitic limestone and/or white sparry calcite preserving typical cement fabrics (Fig. 6b). Geopetal infill of internal pelagic and/or diagenetic sediment into crevasses and pockets of the serpentinites (Fig. 6b) indicates proximity to the sea floor (Bernoulli & Weissert 1985), which is also evidenced by their stratigraphic position underlying pillow basalts or radiolarian cherts. Further evidence of the occurrence of the mantle rock on the sea floor is the presence of mass-flow breccias (ophicalcite type II) with clasts of serpentinite and Triassic dolomites within a serpentine arenite or calcite matrix (Fig. 6c) and the occurrence of serpentinite arenites within the post-rift sediments (Fig. 6d).

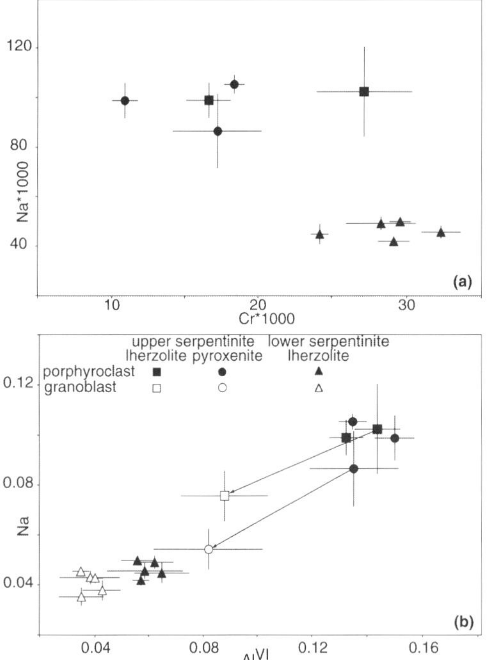

Fig. 7. Chemical evolution of the clinopyroxenes of the peridotites. Each point represents the average composition of clinopyroxenes from one sample. The error bars represent the standard deviation. (**a**) Na v. Cr diagram showing two groups of clinopyroxenes: the pyroxenes from the lherzolites and the pyroxenites of the upper serpentinite unit show high Na and low Cr contents compared with the clinopyroxenes from the lherzolites of the lower serpentinite unit. (**b**) Na v. Al^{VI} diagram showing the chemical evolution of the clinopyroxenes from porphyroclastic to neoblastic minerals. (See text for discussion.)

Mineral chemistry

Clinopyroxene (Table 1, Fig. 7). The porphyroclastic clinopyroxenes of the lower serpentinized lherzolite are always Cr-diopsides with Cr_2O_3 contents of up to 1.22 wt %. They are characterized by high Al_2O_3 (4.24–5.38 wt %) and TiO_2 (up to 1 wt %) contents. The granoblasts are depleted in Na and Al with respect to the porphyroclasts (Fig. 7b).

The clinopyroxenes of the upper serpentinized lherzolite and of its pyroxenite dykes are both systematically enriched in Al_2O_3 (6.19–6.99 wt %) and Na_2O (1.23–1.46 wt %) in comparison with the clinopyroxenes of the pyroxenite-free lower serpentinite unit (Fig. 7a). The clinopyroxenes of the pyroxenites are depleted in Cr in comparison with those of the lherzolite (Fig. 7a). In the high-temperature shear zones, a difference in composition exists between the granoblasts underlining the mylonitic foliation and the porphyroclasts; the granoblasts show lower Al, Na and Cr contents (Fig. 7b); however, the granoblastic clinopyroxenes show still relatively high Al and Na values indicating that deformation occurred under spinel peridotite metamorphic conditions (see Müntener *et al.* 2000).

Orthopyroxene. Orthopyroxene is rarely preserved and fresh minerals have been found only within the lower serpentinite unit. The orthopyroxene shows an enstatite-rich composition with a constant *Mg*-number of 0.89. A zonation can be shown, with Al, Cr and Ca decreasing from core to rim.

Spinel. Fresh spinel compositions are present only in the lower serpentinite unit. The large porphyroclasts and the small granoblasts occurring along the interface between clino- and orthopyroxenes show the same composition. They are Cr-spinels with >30 wt % Cr_2O_3. The rim is systematically altered to magnetite.

Amphibole. First-generation amphiboles are Ti-pargasites occurring as small granoblasts between the large porphyroclastic clino- and orthopyroxenes of the lower serpentinite unit. These amphiboles are rich in Al_2O_3 (12.1–12.9 wt %), TiO_2 (2.9–3.8 wt %) and Cr_2O_3 (1.7–1.9 wt %) but poor in K_2O (<0.07 wt %). As these high-temperature amphiboles are stable within a mantle assemblage, they do not necessarily record hydration of the mantle during rifting but rather indicate re-equilibration at falling temperature ($T < 1100\,^{\circ}C$) and under lithospheric mantle conditions. In contrast, the amphiboles present in coronae statically overgrowing the clinopyroxenes record hydration of the mantle rocks during rifting. The first generation of these retrograde amphiboles is a Mg–hornblende with an Al_2O_3 content of up to 10 wt %. These hornblendes are further retrogressed to tremolitic hornblende characterized by lower Al_2O_3 and Na_2O contents.

Chlorite. Chlorites display different types of composition depending on their occurrence. The chlorites overgrowing the spinel in the mylonitic foliation of the deformed rocks are all chromian clinochlores with up to 13 wt % of Cr_3O_3. They are characterized by a low Si value, high Al and Cr contents and a highly variable *Mg*-number between 0.69 and 0.86. The chlorites overgrowing the orthopyroxenes contain no Cr, have higher Si and lower Al contents, and display higher *Mg*-number values between 0.92 and 0.95.

Retrograde metamorphism of the mantle rocks

Thermobarometric calculations have been applied only to the porphyroclastic assemblages as they represent the only microstructures in which fresh orthopyroxene has been found. The single-orthopyroxene thermometers (Witt-Eickschen & Seck 1991) and the Al net-transfer thermometry (Carrol-Webb & Wood 1986) yield temperatures of $1000 \pm 50\,^{\circ}C$ for pressures between 10 and 15 kbar. The two-pyroxene thermometer (Brey & Kohler 1990) gives similar temperatures of $1010 \pm 100\,^{\circ}C$ for one of these samples and $810 \pm 120\,^{\circ}C$ for another. The coexistence of spinel, Ti-pargasite and Al-clinopyroxene in the recrystallized assemblage indicates spinel peridotite facies conditions (Evans 1977, 1982), suggesting that recrystalli-

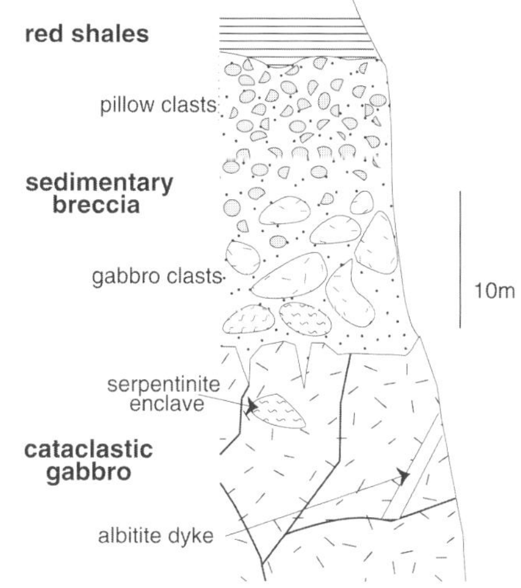

Fig. 8. Stratigraphic succession from exhumed gabbro to overlying sediments. Val da Natons (coordinates of Swiss Topographic Map: 770'690 to '850/152'200 to '300).

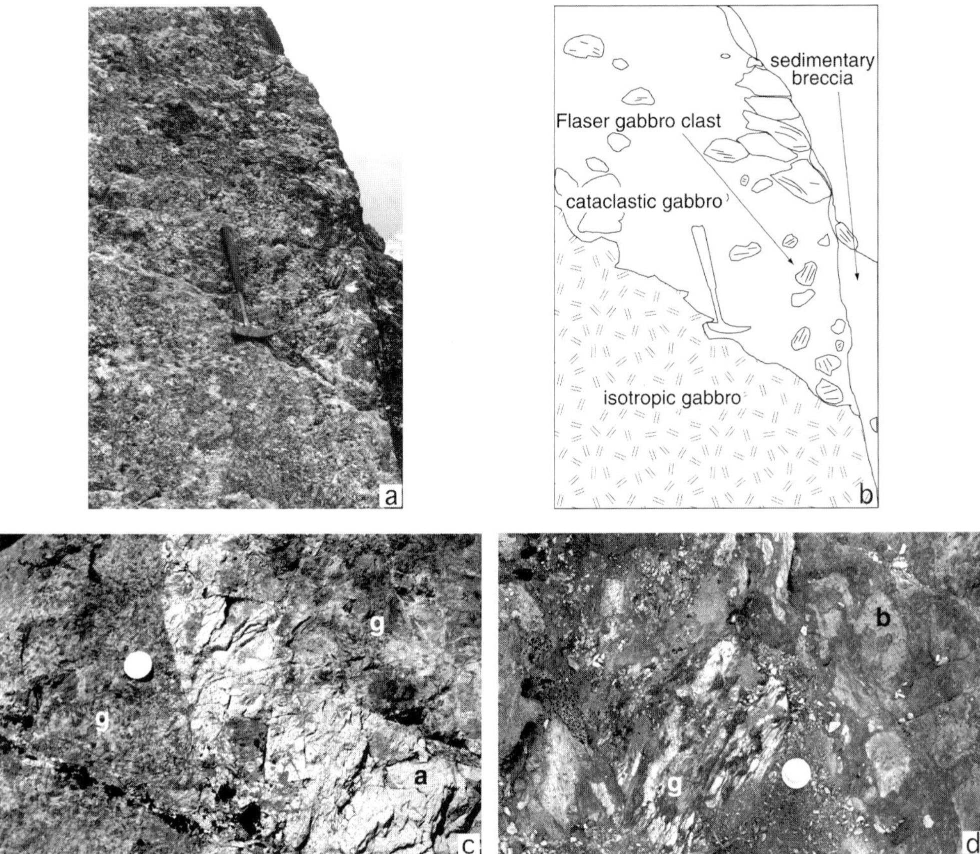

Fig. 9. (**a**) and (**b**) coarse-grained isotropic Mg gabbro (below) with a zone of flaser gabbro overprinted and disrupted by cataclastic deformation (above). The unconformably overlying sedimentary breccia to the right contains clasts of both flaser and undeformed gabbro (hammer is 33 cm long). Val da Natons (coordinates of Swiss Topographic Map: 770'690/152'200). (**c**) Albitite dyke (a) within a Mg gabbro (g) (diameter of coin 2.3 cm). Val da Natons (coordinates of Swiss Topographic Map: 770'830/152'250). (**d**) Pillow (b) breccia with fragments of flaser gabbro (g) (diameter of coin 2 cm). Val Savriez (coordinates of Swiss Topographic Map: 771'725/153'660).

zation occurred in the upper mantle. We interpret these values as defining the $P-T$ conditions present in the mantle before the onset of rifting.

The crystallization of coronitic Mg- and tremolitic hornblende around the clinopyroxene documents high-temperature hydration of the mantle rocks during rifting. Such amphiboles are often found to be stable in the presence of orthopyroxene and spinel (e.g. Müntener *et al.* 2000) and therefore may have formed at high temperatures in the ultrabasic rocks (Evans 1982). We suggest that hydration of the mantle rocks started at upper amphibolite facies conditions, i.e. at temperatures between 600 and 700 °C. Later, pervasive alteration of the mantle rocks below and near the sea floor is finally

indicated by the crystallization of chlorite, tremolite and serpentine minerals replacing the previous assemblages.

Gabbroic intrusions

Field relationships

The mantle rocks of the lower serpentinite unit of the Platta nappe are intruded by gabbros and locally by rodingitized mafic dykes. Magmatic intrusions within the upper serpentinite unit are limited to few dolerite dykes and a single small gabbro sill. In Val da Natons, the exhumation history of a gabbro from shallow intrusion into serpentinized peridotite to exposure on the sea floor can be documented (Fig. 8). A cataclastic gabbro that is cut by small dykes of albitite (Fig. 9c) includes enclaves of serpentinite and fractures filled by ophicalcite. It is stratigraphically overlain by a sedimentary megabreccia including first metre-sized blocks of serpentinite, then blocks of gabbro set in a matrix of serpentine arenite. A few gabbro blocks, several metres across, show localized bands of flaser gabbro fragmented and overprinted by cataclastic deformation (Fig. 9a,b). An overlying breccia contains clasts of these flaser gabbros together with serpentine and abundant pillow-basalt fragments embedded in a matrix of serpentine arenite. The breccias are overlain by pillow breccias covered in turn by red shales. The absence of a pre-Alpine foliation and the random arrangement of the unsorted polymictic blocks suggest that the gabbro breccias have a sedimentary origin. Together, these observations show that the gabbros were exhumed at the sea floor along discrete shear zones active at declining temperatures. Clasts of flaser gabbro and albitite occur also in pillow breccias

overlying the exposed mantle rocks in Val Savriez (Fig. 9d).

In the Fuorcla da Faller area, a gabbro body shows a great diversity in composition and mineralogy suggesting a complex intrusion history. The main part of this body (90%) consists of coarse-grained Mg gabbro (clinopyroxene–plagioclase) with localized high-temperature shear zones. In places, the gabbro grades into a microgabbro of the same mineralogy (Fig. 10a), locally it is cut by the latter (Fig. 10b). The microgabbro crystallized simultaneously with a Fe gabbro (clinopyroxene–hastingsitic hornblende–Fe-oxide–plagioclase, 5% of the outcrop; Fig. 10a) and is cut by a Fe–Ti–P gabbro (clinopyroxene–pargasitic hornblende–ilmenite–plagioclase–apatite–zircon). Pegmatitic diorite (pargasitic hornblende–plagioclase–clinopyroxene–quartz; Fig. 10c) occurs as patches or dykes, which can form boudins or folds within the Mg gabbro. The entire gabbro body is cut by basaltic dykes with chilled margins indicating that emplacement of the basalts took place after cooling of the gabbro.

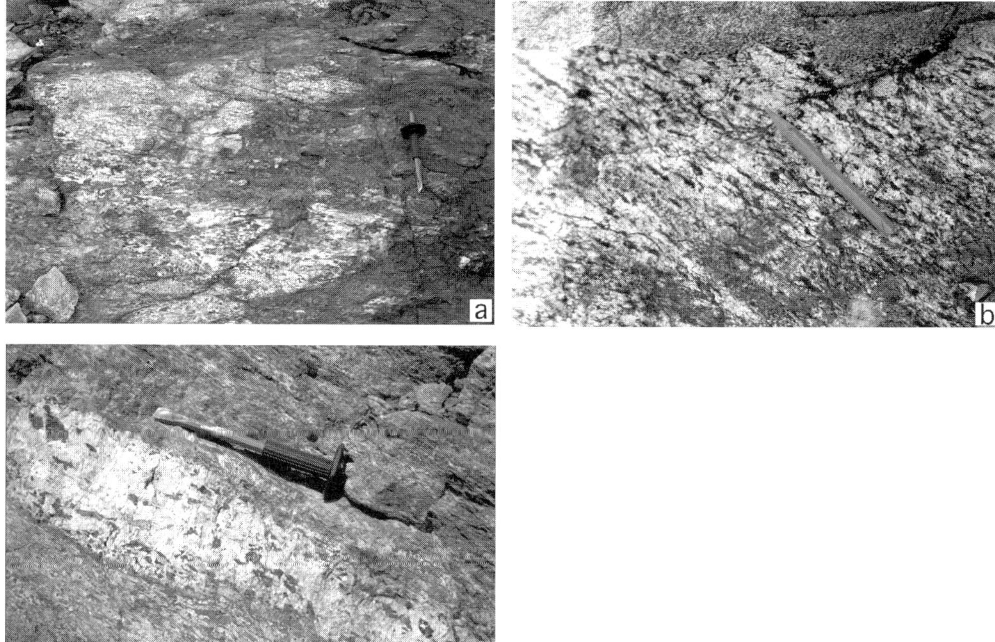

Fig. 10. (a) Mg gabbro grading into a darker Fe gabbro (chisel is 28 cm long). Fuorcla da Faller (coordinates of Swiss Topographic Map: 765′150/177′650). (b) Mg flaser gabbro cut by Mg microgabbro (pencil is 9.5 cm long). Fuorcla da Faller (coordinates of Swiss Topographic Map: 765′150/177′650). (c) Pegmatitic diorite dyke cutting across a Mg gabbro (chisel is 28 cm long). Fuorcla da Faller (coordinates of Swiss Topographic Map: 765′150/177′650).

Contact with mantle rocks

Above Alp da Starschagns, the serpentinites show evidence of contact metamorphism along the contact with the gabbro sills intruding them. These intrusions are characterized by chilled margins and intense veining is observed within the serpentinite near the contact with the gabbros. Thin-sections show the veins to cut across the pre-existing serpentine mesh; they consist of chlorite minerals growing parallel to the edges of the veins and of tremolite growing

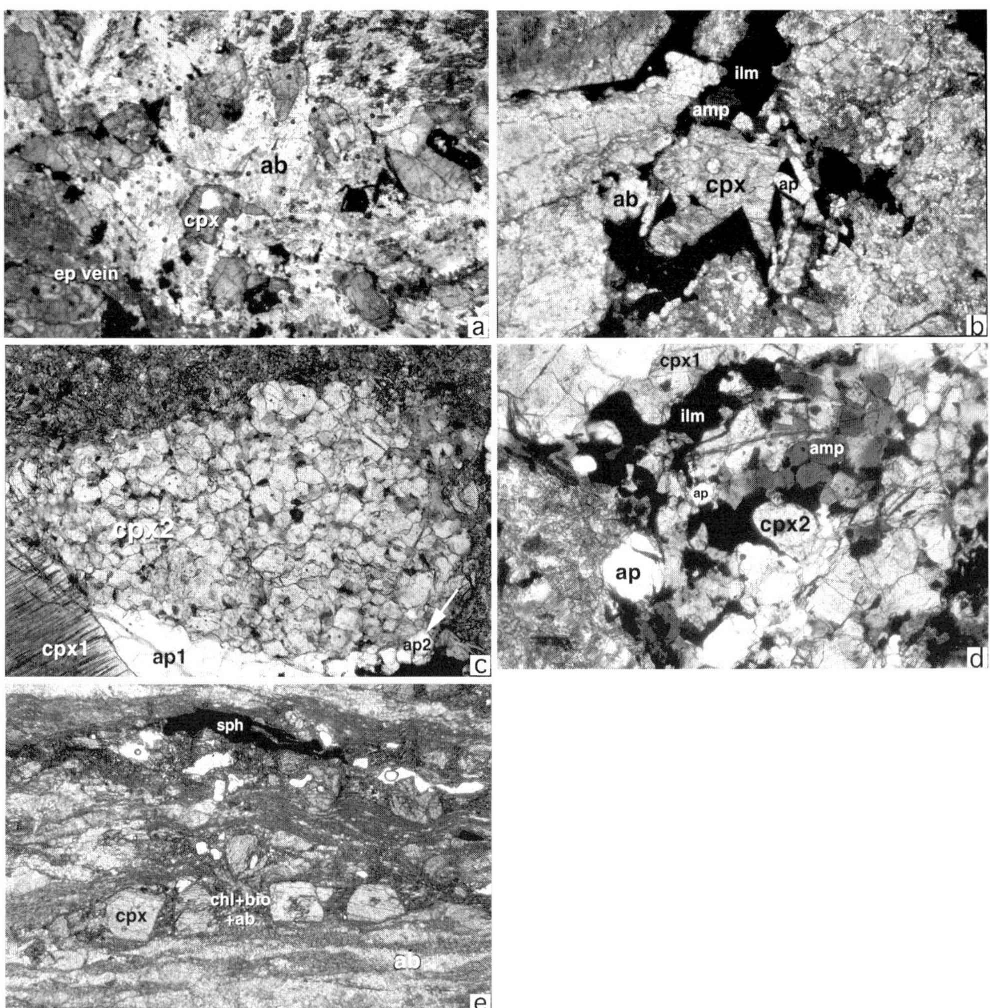

Fig. 11. (**a**) Magmatic assemblage of Mg gabbro with euhedral plagioclase (now albite, ab) and interstitial clinopyroxene (cpx). (Note the late epidote (ep) vein cutting across the magmatic texture.) Fuorcla da Faller. Field of view 6.3 mm wide. (**b**) Magmatic assemblage of Fe–Ti–P gabbro with numerous grains of apatite (ap), pargasite (amp) and interstitial ilmenite (ilm). Fuorcla da Faller. Field of view 6.3 mm wide. (**c**) Neoblastic clinopyroxene (cpx2) within a high-temperature shear zone in a Fe–Ti–P gabbro. (Note the large porphyroclasts of clinopyroxene (cpx1) and apatite (ap1) and the interstitial brown magmatic amphiboles between the clinopyroxene neoblasts.) Val da Natons. Field of view 1.2 mm wide. (**d**) Neoblastic pargasite and apatite with interstitial magmatic ilmenite within the same shear zone. Val da Natons. Field of view 1.2 mm wide. (**e**) Low-temperature shear zone within a Mg gabbro. The clinopyroxenes are rotated or show brittle stretching whereas the albite and sphene (sph) were ductilely deformed; the space between the clinopyroxene clasts is filled by the assemblage chlorite (chl) + biotite (bio) + albite (ab). Val da Natons. Field of view 6.3 mm wide. Plane-polarized light for all pictures.

perpendicular to them. Inside the veins, equigranular grains of diopside and (?)antigorite overgrow tremolite and serpentine minerals. This mineral assemblage occurs also in patches within the serpentinites, suggesting that these were metamorphosed by hot fluids during the intrusion of the dyke. Where the contact was not overprinted by later deformation, it appears that the intrusion of the gabbroic sills led to prograde metamorphism of the already serpentinized mantle rocks.

Age

One Fe–Ti–P gabbro from the Fuorcla da Faller area, an associated diorite, a pegmatitic Mg gabbro from Val da Natons and an albitite clast from a pillow breccia in Val Savriez were dated by U–Pb on zircon (Desmurs *et al.* 1999). These four rocks gave an identical intrusion age of 161 ± 1 Ma (i.e. of mid–late Callovian age in the time scale of Gradstein *et al.* (1995)) and belong to the same very short-lived magmatic event.

Magmatic and deformational textures

Most of the gabbros of the Platta nappe show a well-preserved magmatic texture, whereas the magmatic mineral assemblage was partially hydrated. This is shown by the growth of amphibole at the expense of pyroxene. Hydration was followed by a more widespread low-temperature alteration recorded by the systematic breakdown of plagioclase to albite, chlorite and epidote.

At Fuorcla da Faller, the Mg gabbros (coarse grained and microgabbro) show euhedral plagioclase, sub- to anhedral clinopyroxene and interstitial Ti-pargasite (Fig. 11a). In many places clinopyroxene and plagioclase show graphic intergrowth indicating cotectic crystallization of the two phases. The plagioclase was altered to a fine-grained mixture of albite, chlorite and epidote and later to prehnite and pumpellyite; the pyroxenes show a rim of hornblende. The Fe gabbros consist of anhedral to subhedral clinopyroxene and plagioclase with interstitial iron oxide. Hastingsitic hornblende developed between the iron oxide and the clinopyroxene. The Fe–Ti–P gabbro is made of euhedral plagioclase and apatite, subhedral clinopyroxene, and interstitial pargasite and ilmenite (Fig. 11b); euhedral zircon occurs as an accessory mineral. As in the Fe gabbro, pargasitic hornblende developed between ilmenite and clinopyroxene and the plagioclase shows the same alteration as in the other gabbro types.

A younger rim of actinolite is present around the pyroxene and the pargasite. It is interpreted to result from Alpine overprint. The pegmatitic diorite is made of large euhedral plagioclase crystals, large sub- to anhedral magnesian hornblende, rare and strongly altered clinopyroxene showing symplectitic texture with an unidentified phase, quartz and zircon. Clinopyroxenes with a similar texture occur also in the Mg gabbro at the contact with the diorite, suggesting that the clinopyroxenes within the diorite are xenocrysts from the Mg gabbro.

In Val da Natons, all gabbros types (Mg and Fe–Ti–P gabbro) are cut by discrete mylonitic shear zones. Deformation occurred with falling temperatures. The highest temperature deformation event is recorded by a high-temperature foliation defined by small clinopyroxene, pargasite and apatite granoblasts showing typical triple-junction grain boundaries. Pargasite and all the ilmenite occur also as undeformed interstitial grains between the neoblasts, indicating that they are of magmatic origin (Fig. 11c,d). These observations suggest that the earliest deformation was synmagmatic because granoblastic and interstitial magmatic grains coexist within the foliation. As no evidence for the deformation of ilmenite was found, deformation probably ceased when the temperature fell below the solidus.

In other shear zones, still in Val da Natons, the clinopyroxene porphyroclasts were statically retrogressed to brown Mg-hornblende and later to actinolite, and the mylonitic foliation is defined by small neoblasts of actinolite and albite. In a few shear zones, albite, elongated chlorite and biotite crystallized between clasts of stretched pyroxene aligned parallel to the foliation (Fig. 11e). The recrystallization of albite and actinolite and the deformation of chlorite and biotite suggest that these shear zones were active at lower amphibolite- to greenschist-facies conditions. These zones of deformation were later affected by cataclastic deformation and intense veining during which epidote, chlorite and albite crystallized.

Mineral chemistry

Clinopyroxene (Table 2). The augitic clinopyroxenes of the gabbros of the Platta nappe are characterized by their low Al_2O_3 (<3.4 wt %), Na_2O (0.6 wt %) and Cr_2O_3 (<0.07 wt %) contents, which are similar for the various gabbro types. However, the *mg*-number of these minerals continuously decreases from the Mg gabbro and microgabbro (0.74–0.80) to the Fe gabbro (0.69–0.71) and the highly differentiated Fe–Ti–P gabbro (0.61–0.69). This decrease in *Mg*-number correlates with an increase of MnO and a decrease of TiO_2 and Al^{IV} (Fig. 12a–c). According to Ulmer (1986), this decrease could be related to a fall in the temperature of crystallization, however, the clinopyroxenes of the highly differentiated pegmatitic diorite show the highest *Mg*-number (0.83–0.89) of the entire gabbro suite. These clinopyroxenes could therefore represent xenocrysts from the Mg gabbro host-rock preserved within the diorite. This interpretation is favoured by their composition being similar to that of the clinopyroxenes of the Mg gabbro.

In the high-temperature mylonitic Fe–Ti–P gabbro, the neoblasts are only slightly different from the porphyroclasts. They show only a decrease in Al_2O_3 from 1.1–1.8 wt % for the porphyroclasts to 0.7–1.1 wt % for the neoblasts and in Na_2O from 0.6–0.95 to 0.42–0.53 wt %, respectively.

Plagioclase. All plagioclases in the gabbros of the Platta nappe are now pure albite and no relics of a more anorthitic composition have been found.

Amphibole. Crystallization of magmatic brown amphibole was observed in all gabbro types but is more frequent in the differentiated gabbros (Fe gabbro, Fe–Ti–P gabbro and diorite). These magmatic amphiboles are Ti-rich calcic amphiboles and plot in the Ti–pargasite to edenitic hornblende fields according to the classification of Leake (1978). Only the amphiboles of the Fe gabbro show a hastingsitic composition. They are characterized by rather high TiO_2 (4.3–3.1 wt %) and Al_2O_3 (up to 13 wt %) contents and a low Na^{M4} content (0.13–0.29 p.f.u.). The neoblastic amphiboles of the mylonitic shear zones show a composition similar to that of the magmatic amphiboles, suggesting that deformation started at high temperature (synmagmatic) compatible with the textural observations.

The retrograde coronitic amphiboles in the diorites and the Fe–Ti–P gabbros are magnesian hornblendes whereas those of the Mg gabbros plot in the edenite field. They show lower Al_2O_3 (4.6–9.1 wt %) and TiO_2 (1.5–2.1 wt %) contents than the first generation of amphiboles. Again, their Na^{M4} (0.07–0.2 p.f.u) is rather low. They show systematically higher *mg*-number than the high-temperature amphiboles in the same rock type.

A last generation of pre-Alpine amphiboles consists of actinolites forming rims around the pargasitic hornblende and the clinopyroxene or underlining, together with albite neoblasts, a mylonitic foliation. They are characterized by low Al_2O_3 (0.62–2.53 wt %) and TiO_2 (0.02–0.09 wt %) contents, and differ from the Alpine actinolite by their low Na^{M4} (0.02–0.09 p.f.u. v. 0.49–1.02 p.f.u. for the Alpine actinolite).

The evolution of the composition of the amphiboles is shown in Figure 13a, where two trends can be observed, one related to magmatic, the other to metamorphic processes. The first trend shows a decrease of the *Mg*-number with Ti for the high-temperature amphiboles from the Mg gabbro to the Fe–Ti–P gabbro. This suggests that the Ti-rich amphiboles record magmatic differentiation (reflected by the *Mg*-number) at falling crystallization temperatures (reflected by the Ti content; Heltz 1973; Otten 1984). The second trend shows increasing *Mg*-number with decreasing Ti from the high-temperature to the retrograde amphiboles, suggesting continuing hydration of the gabbro at falling temperatures. Figure 13b–d shows a general decrease of Al^{IV} and Na + K^A together with Na^{M4} and Al^{VI} + Fe^{3+} + Cr. The decrease in Al^{IV} and Na + K^A is thought to be temperature sensitive whereas Na^{M4} and Al^{VI} reflect the pressure at a given Al^{IV} and Na + K^A (Laird 1982, and references therein). Thus, these diagrams allow us to qualitatively evaluate the $P–T$ conditions of crystallization of the various generations of amphiboles. The magmatic amphiboles are characterized by high Al^{IV} and Na + K^A values, indicating high-temperature crystallization, and rather low Na^{M4} and Al^{VI}, suggesting a low-pressure environment. Part of the coronitic amphiboles plot within the same field, suggesting high-temperature hydration of the gabbros. A second group of coronitic and neoblastic amphiboles is characterized by lower Al^{IV} and Na + K^A and very low Na^{M4} and Al^{VI} + Fe + Cr^{3+}, indicating that they formed when the gabbro was at relatively shallow depth. A third group can be distinguished by their low Al^{IV} and Na + K^A contents, and high Na^{M4} and Al^{VI} + Fe^{3+} + Cr. These low-temperature, high-pressure amphiboles most probably crystallized during Alpine metamorphism.

Conditions of intrusion

The conditions of the intrusion of the gabbros are determined by the crystallization sequence and by the chemical composition of the clinopyroxenes. All the gabbros except the Fe–Ti–P gabbro show idiomorphic plagioclase and interstitial clinopyroxene, suggesting that the plagioclase was the first phase to crystallize. This crystallization sequence suggests an emplacement at a pressure <8 kbar (Green & Ringwood

1967; Bender *et al.* 1978; Elthon & Scarfe 1984). This is supported by the low Al_2O_3 and Na_2O contents of the augite in all the gabbro types. Figure 14 shows that the clinopyroxenes have a composition close to that of clinopyroxenes from oceanic gabbros and different from that of the pyroxenes of high-pressure gabbros such as the Permian Braccia Gabbro of Val Malenco (Hermann 1997). As magmatic amphiboles are observed in all gabbro types, the temperature of crystallization of the gabbros can be constrained to below 1100 °C, which is the upper thermal stability for amphibole in basaltic systems (Allen *et al.* 1975). The clinopyroxene geobarometer (Nimis & Ulmer 1998) gives pressures below 5 kbar at temperatures between 1000 and 1100 °C for all gabbros. The thermometer of Otten (1984), based on the Ti content of hornblende, gives temperatures between 950 and 1025 °C for the magmatic amphiboles of the diorite, the Fe gabbro and the Fe–Ti–P gabbro. The observation that the gabbros intruded already serpentinized peridotites places the intrusion of the gabbros above the serpentinization front, i.e. around 6–8 km or around 2 kbar according to the interpretation of the change of seismic velocities west of Iberia (Boillot *et al.* 1992).

High-temperature deformation and hydration

The microstructural data suggest synmagmatic deformation of the gabbros. This is supported by the similar composition of the magmatic and neoblastic amphiboles and clinopyroxenes in the high-temperature shear zones. The Otten (1984) thermometer applied to the neoblastic amphiboles gives temperatures between 900 and 975 °C, similar to those obtained on the magmatic amphiboles. This thermometer is less reliable for coronitic amphiboles as the Ti content of the amphiboles depends on the bulk-rock chemistry, and the temperatures obtained should be considered as minimum temperatures. However, the coronitic amphiboles yield temperatures between 900 and 600 °C, suggesting high-temperature static hydration of the gabbros during cooling. This high temperature of crystallization is in accordance with the high Al^{IV} content of the amphiboles, which increases with temperature in both metamorphic and magmatic amphiboles (Heltz 1982; Laird 1982; Anderson & Smith 1995), and with the experimental data of Spear (1981) indicating that the assemblage clinopyroxene + hornblende + ilmenite is stable at temperature above 750 °C.

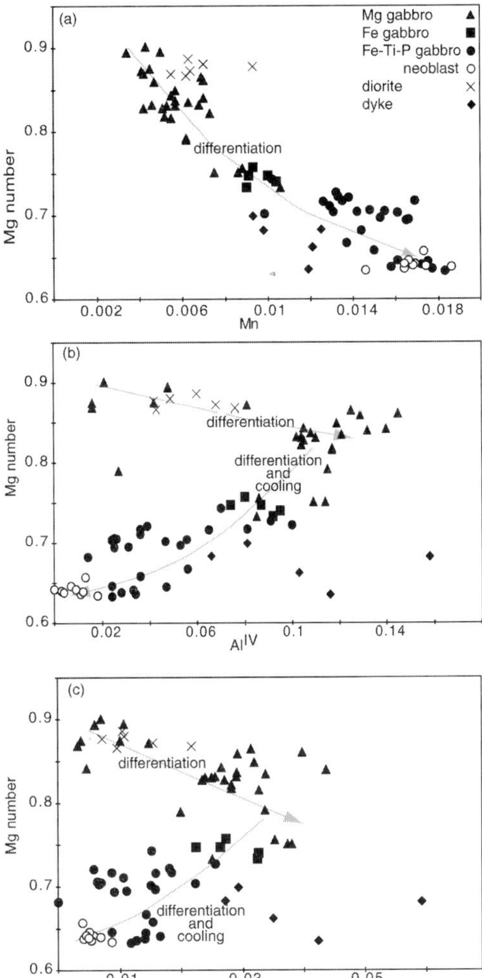

Fig. 12. Chemical evolution of the clinopyroxenes of the various gabbros. (**a**) *Mg*-number (Mg/(Mg + Fe^{2+})) v. Mn diagram showing an increase of Mn from Mg gabbro to Fe–Ti–P gabbro and decreasing *Mg*-number reflecting the magmatic differentiation. (**b**) *Mg*-number v. Al^{IV} showing first an increase of Al^{IV} during magmatic differentiation and then a decrease of Al^{IV} during further differentiation because of lower crystallization temperatures of the more differentiated gabbros. (**c**) *Mg*-number v. Ti diagram showing Ti first increasing with magmatic differentiation and then decreasing with continuing differentiation because of lower crystallization temperatures of the more differentiated gabbros.

Low-temperature deformation

The gabbros were deformed again at lower amphibolite to greenschist temperature conditions and finally in the brittle field before

Table 2. *Representative chemical compositions of the magmatic and retrograde minerals of the gabbros*

	Magmatic assemblages																	
	Mg gabbro				Fe gabbro		Fe–Ti–P gabbro									Diorite	
	MS8		MS9	NAG3	NAG9	MS7		MS11			NAG5			NAG7			MS12	
	cpx	Hbl	cpx	cpx	cpx	cpx	Hbl	cpx	Hbl	ilm	cpx	hbl	ilm	cpx	Hbl	ilm	cpx	Hbl
SiO$_2$	50.40	42.52	50.87	50.88	51.93	50.96	42.05	51.51	42.50	0.04	52.30	42.50	0.45	51.70	43.08	0.01	53.03	44.76
TiO$_2$	1.43	4.31	1.21	0.85	0.53	0.93	4.37	0.56	3.11	47.98	0.39	3.13	49.86	0.54	2.15	49.18	0.25	3.36
Cr$_2$O$_3$	0.03	0.06	0.07	0.13	0.16	0.00	0.02	0.01	0.00	0.03	0.03	0.02	0.02	0.00	0.00	0.01	0.05	0.02
Al$_2$O$_3$	3.41	12.00	3.44	2.74	3.35	2.55	10.85	1.89	11.26	0.39	1.20	9.86	0.05	1.38	8.99	0.00	1.37	8.12
Fe$_2$O$_3$	3.69	2.11	2.16	0.00	1.57	1.82	2.40	2.35	1.03	6.35	1.00	3.06	3.17	1.09	4.23	4.28	1.69	0.00
FeO	4.30	8.49	5.18	5.70	4.19	8.63	11.93	9.32	13.30	39.91	10.26	13.73	41.86	11.12	13.62	39.37	3.99	12.89
MnO	0.22	0.12	0.20	0.18	0.13	0.31	0.13	0.40	0.24	2.97	0.52	0.32	2.20	0.47	0.30	4.07	0.30	0.59
MgO	14.99	14.60	14.73	15.71	15.94	14.28	12.46	13.19	11.86	0.07	13.12	11.26	0.31	12.01	10.52	0.06	16.00	15.00
CaO	21.66	11.21	21.78	22.61	20.77	19.87	10.79	19.93	11.11	0.01	20.63	10.34	0.15	20.75	10.88	0.31	22.75	10.64
Na$_2$O	0.54	3.17	0.51	0.41	0.83	0.58	2.95	0.71	3.33	0.03	0.59	3.14	0.10	0.68	2.72	0.00	0.52	3.07
K$_2$O	0.02	0.14	0.00	0.01	0.04	0.01	0.29	0.01	0.30	0.00	0.00	0.16	0.01	0.02	0.28	0.00	0.01	0.23
H$_2$O	–	2.08	–	–	–	–	2.03	–	2.03	–	–	2.00	–	–	1.98	–	–	2.05
Total	100.70	100.81	100.15	99.27	99.43	99.95	100.27	99.88	100.08	97.77	100.04	99.57	98.23	99.75	98.82	97.53	99.97	100.74
Si	1.85	6.14	1.88	1.90	1.92	1.91	6.21	1.94	6.30	0.00	1.97	6.37	0.01	1.96	6.52	0.00	1.96	6.55
Ti	0.04	0.47	0.03	0.02	0.01	0.03	0.49	0.02	0.35	0.93	0.01	0.35	0.96	0.02	0.24	0.96	0.01	0.37
Cr	0.00	0.01	0.00	0.00	0.00	0.00	0.00	0.00	0.00	0.00	0.00	0.00	0.00	0.00	0.00	0.00	0.00	0.00
Al	0.15	2.04	0.15	0.12	0.15	0.11	1.89	0.08	1.97	0.01	0.05	1.74	0.00	0.06	1.60	0.00	0.06	1.40
Fe^{3+}	0.10	0.23	0.06	0.00	0.04	0.05	0.27	0.07	0.12	0.12	0.03	0.34	0.06	0.03	0.48	0.08	0.05	0.00
Fe^{2+}	0.13	1.03	0.16	0.18	0.13	0.27	1.47	0.29	1.68	0.86	0.32	1.72	0.90	0.35	1.73	0.85	0.12	1.58
Mn	0.01	0.01	0.01	0.01	0.00	0.01	0.02	0.01	0.03	0.07	0.02	0.04	0.05	0.02	0.04	0.09	0.01	0.07
Mg	0.82	3.14	0.81	0.87	0.88	0.80	2.74	0.74	2.62	0.00	0.74	2.52	0.01	0.68	2.37	0.00	0.88	3.27
Ca	0.85	1.73	0.86	0.90	0.82	0.80	1.71	0.80	1.77	0.00	0.83	1.66	0.00	0.84	1.76	0.01	0.90	1.67
Na	0.04	0.89	0.04	0.03	0.06	0.04	0.84	0.05	0.96	0.00	0.04	0.91	0.00	0.05	0.80	0.00	0.04	0.87
K	0.00	0.03	0.00	0.00	0.00	0.00	0.05	0.00	0.06	0.00	0.00	0.03	0.00	0.00	0.05	0.00	0.00	0.04
OH	–	2.00	–	–	–	–	2.00	–	2.00	–	–	2.00	2.00	–	2.00	–	–	2.00
Mg#	0.78	0.71	0.79	0.83	0.84	0.71	0.61	0.67	0.59	0.00	0.68	0.55	0.01	0.64	0.52	0.00	0.84	0.67

Calculations as in Table 1; two cations (ilmenite), 12 oxygens (epidote) and three cations (sphene).

Table 2. *continued*

	HT deformation	Static HT hydration			LT deformation				Static oceanic alteration						
	Fe–Ti–P gabbro	Mg-	Fe–Ti–P	diorite	Mg gabbro				Mg gabbro			Fe gabbro		Fe–Ti–P gabbro		
	NAG7	NAG9	NAG5	MS12	NAG9	NAG3			MSG			MS7		NAG7		
	cpx	hbl	hbl	hbl	hbl	tr	biot	chl	ab	chl	epi	ab	sph	act	sph	act	chl
SiO_2	52.53	43.35	47.95	43.64	47.66	54.90	37.29	29.25	69.13	26.78	37.01	67.23	30.53	56.55	31.17	54.05	27.03
TiO_2	0.17	2.77	0.86	1.24	1.82	0.46	0.23	0.09	0.00	0.16	0.00	0.00	39.67	0.05	34.89	0.05	0.00
Cr_2O_3	0.00	0.03	0.21	0.00	0.00	0.09	0.02	0.04	0.00	0.01	0.03	0.01	0.00	0.01	0.01	0.00	0.00
Al_2O_3	0.82	8.94	8.11	9.89	6.53	2.51	15.94	16.73	20.39	20.61	21.18	20.88	0.43	0.62	1.51	0.72	16.49
Fe_2O_3	0.46	4.18	0.00	8.63	1.75	1.09	0.00	0.00	0.00	0.00	15.88	0.00	0.79	0.31	1.59	0.86	0.00
FeO	12.03	12.41	8.22	9.47	12.65	4.63	18.04	22.47	0.19	24.60	0.00	0.12	0.00	7.18	0.00	15.86	26.80
MnO	0.54	0.29	0.09	0.52	0.50	0.10	0.23	0.41	0.02	0.45	0.08	0.03	0.00	0.19	0.00	0.25	0.31
MgO	11.96	11.41	16.55	12.49	13.32	20.54	11.72	13.90	0.02	13.48	0.05	0.10	0.01	19.77	0.14	13.18	13.81
CaO	21.08	10.55	13.04	8.59	11.23	12.79	0.46	0.18	0.06	0.05	23.16	0.99	28.77	12.80	29.10	11.76	0.23
Na_2O	0.49	3.08	2.00	2.92	2.01	0.76	0.09	0.04	10.38	0.02	0.06	9.71	0.00	0.55	0.01	0.81	0.01
K_2O	0.00	0.26	0.09	0.10	0.33	0.06	8.08	1.96	0.05	0.01	0.01	0.03	0.00	0.08	0.00	0.03	0.02
H_2O	–	2.00	2.08	2.03	2.04	2.16	3.82	15.54	–	15.76	1.85	–	0.17	2.15	0.45	2.05	15.20
Total	100.09	99.28	99.26	99.52	99.86	100.08	96.00	100.62	100.28	101.97	99.40	99.10	100.38	100.25	98.88	99.62	99.93
Si	1.99	6.49	6.92	6.46	7.00	7.64	2.93	4.51	3.04	4.08	2.99	3.00	0.99	7.90	1.02	7.92	4.26
Ti	0.00	0.31	0.09	0.14	0.20	0.05	0.01	0.01	0.00	0.02	0.00	0.00	0.97	0.01	0.86	0.01	0.00
Cr	0.00	0.00	0.02	0.00	0.00	0.01	0.00	0.00	0.00	0.00	0.00	0.00	0.00	0.00	0.00	0.00	0.00
Al	0.04	1.58	1.38	1.72	1.13	0.41	1.48	3.04	1.06	3.70	2.02	1.10	0.02	0.10	0.06	0.12	3.07
Fe^{3+}	0.01	0.47	0.00	0.96	0.19	0.11	0.00	0.00	0.00	0.00	0.97	0.00	0.02	0.03	0.04	0.09	0.00
Fe^{2+}	0.38	1.55	0.99	1.17	1.55	0.54	1.19	2.90	0.01	3.13	0.00	0.00	0.00	0.84	0.00	1.94	3.54
Mn	0.02	0.04	0.01	0.07	0.06	0.01	0.02	0.05	0.00	0.06	0.01	0.00	0.00	0.02	0.00	0.03	0.04
Mg	0.67	2.55	3.56	2.75	2.92	4.26	1.37	3.20	0.00	3.06	0.01	0.01	0.00	4.12	0.01	2.88	3.25
Ca	0.85	1.69	2.02	1.36	1.77	1.91	0.04	0.03	0.00	0.01	2.01	0.05	1.00	1.92	1.02	1.85	0.04
Na	0.04	0.89	0.56	0.84	0.57	0.20	0.01	0.01	0.89	0.01	0.01	0.84	0.00	0.15	0.00	0.23	0.00
K	0.00	0.05	0.02	0.02	0.06	0.01	0.81	0.39	0.00	0.00	0.00	0.00	0.00	0.01	0.00	0.01	0.00
OH	–	2.00	2.00	2.00	2.00	2.00	2.00	16.00	–	16.00	1.00	–	0.04	2.00	0.10	2.00	16.00
$Mg\#$	0.63	0.56	0.78	0.56	0.63	0.87	0.54	0.52	0.16	0.49	0.01	0.60	0.03	0.83	0.15	0.59	0.48

Calculations as in Table 1; two cations (ilmenite), 12 oxygens (epidote) and three cations (sphene).

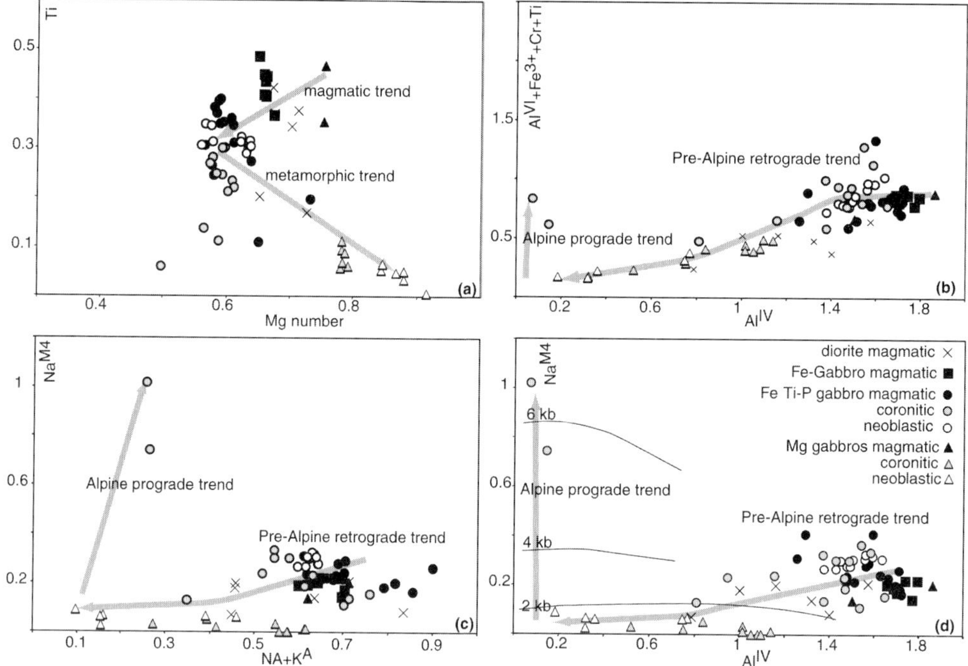

Fig. 13. Chemical evolution of the amphiboles of the various gabbros. (**a**) Ti v. *Mg*-number diagram showing two trends. The first is a magmatic trend showing decreasing Ti contents and decreasing *Mg*-number in the high-temperature amphiboles from the Mg gabbro to the Fe–Ti–P gabbro. The second, metamorphic trend shows increasing *Mg*-number and decreasing Ti contents in all gabbro types indicating hydration at falling temperatures. (**b**) $Al^{IV} + Fe^{3+} + Cr + Ti$ v. Al^{IV} diagram. (**c**) Na^{M4} v. $Na + K^A$ diagram and (**d**) Na^{M4} v. Al^{IV} from Brown (1977). The pressure estimates apply only to the low Al(IV), i.e. the actinolitic amphiboles. (See text for discussion.)

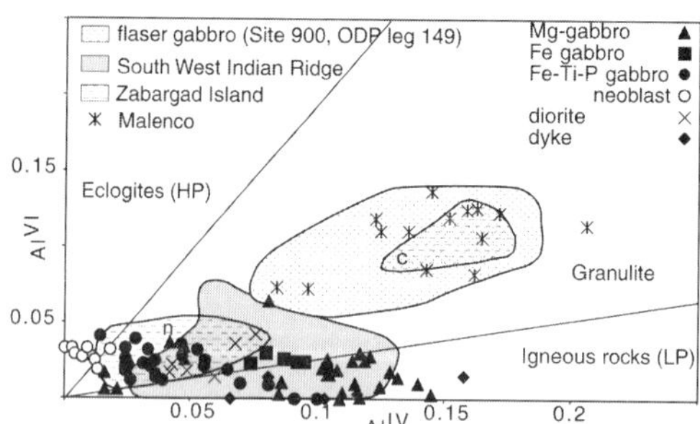

Fig. 14. Al^{VI} v. Al^{IV} diagram for the clinopyroxenes of the various gabbros. All clinopyroxenes plot in the low-pressure field in contrast to the clinopyroxenes of the high-pressure Permian Braccia Gabbro from Val Malenco. Areas for high, low and intermediate pressures from Aoki & Shiba (1973), clinopyroxenes from the South West Indian Ridge from Hébert *et al.* (1991); Stakes *et al.* (1991), clinopyroxenes from Zabargad Island (c, porphyroclasts; n, neoblasts) from Bonatti & Seyler (1987); Boudier *et al.* (1988), clinopyroxenes from ODP Site 900 from Cornen *et al.* (1996), and from the Braccia Gabbro (Malenco crust–mantle boundary) from Hermann (1997).

they were ultimately exposed at the sea floor. This is documented by the assemblage actinolite + albite or biotite + chlorite + albite underlining the foliation in low-temperature shear zones. The low Na^{M4} content of the actinolites suggests that they crystallized at low pressure <2 kbar (Brown 1977). The shear zones are overprinted by cataclastic deformation, implying that the gabbros were deformed also at very shallow depth or at the sea floor during their exhumation. The *P–T* path of the gabbros is summarized in Figure 15.

Basalts and dolerites

Basaltic rocks cover a large area in the Platta nappe; they occur as massive basalts, pillow lavas, pillow breccias, hyaloclastites and dolerite dykes. The extrusive rocks stratigraphically overlie the serpentinites, gabbros and the various types of breccias including ophicalcites. Sedimentary breccias underlying the submarine volcanic rocks contain clasts of serpentinite, continental basement rocks and pre-rift sedi-

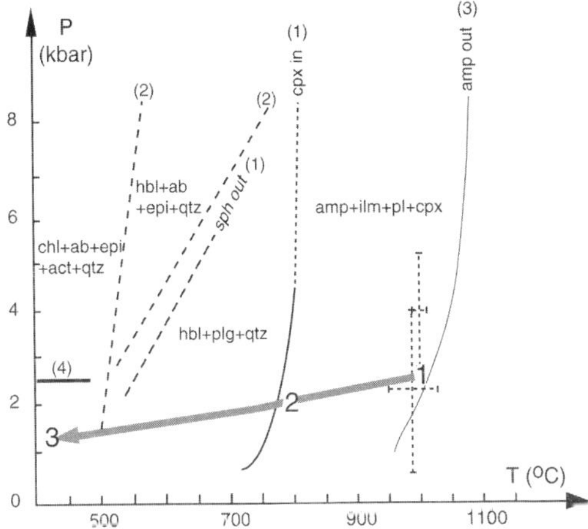

Fig. 15. P, T paths for the gabbros and the diorite of the Platta nappe. 1, magmatic crystallization; 2, oceanic hydration; 3, greenschist deformation and alteration. Pressure conditions during intrusion were determined by the cpx-geobarometer (Nimis & Ulmer 1998). Reactions (1) from Spear (1981); reactions (2) from Apted & Liou (1983); (3) Amph-in reaction under quartz–fayalite–magnetite redox condition from Heltz (1982); (4) depth of serpentinization front from Boillot (1992).

ments, and clearly show that tectonic emplacement of the extensional allochthons and exhumation of the mantle rocks predated the extrusion of the basaltic rocks. Clasts of foliated gabbro, deformed at high-temperature conditions and occurring in weakly deformed pillow breccias (Fig. 9d) show that intrusion, deformation and exhumation of the gabbros also preceded the emplacement of the volcanic rocks at these specific locations. The volcanic rocks are overlain by post-rift sediments of the Radiolarite or the Calpionella Limestone Formations.

The dolerite dykes show a typical intersertal structure between plagioclase and augite, and chilled margins along the contact with serpentinites or gabbros expressed by a grain size reduction along the edges of the dykes. Rodingitization of these dykes is limited to the first centimetre at the contact with the serpentinite, suggesting intrusion into an already cool and serpentinized mantle.

The dolerites and basalts show a major element composition typical for tholeiitic magmas (Frisch *et al.* 1994). Frisch *et al.* distinguished two groups of basalt; one including the dolerites and part of the pillow basalts shows a major and trace element composition typical for mid-ocean ridge basalt (MORB), whereas

the second group of pillow basalts shows an enrichment in incompatible elements. Frisch *et al.* (1994) interpreted this difference by a change from a subcontinental to a MOR magma source. However, all pillow basalts analysed so far have high εNd (+7.8 to 9.8) and low $^{87}Sr/^{86}Sr$ ratios (0.7028–0.7049), indicating a depleted mantle source compatible with a ridge environment (Stille *et al.* 1989; Schaltegger, pers. comm.).

Discussion

Steinmann (1905) thought that all the rocks of his trinity were intrusive; nevertheless, he recognized that the serpentinites were the oldest rocks, followed by the gabbros and finally the basalts and 'variolites' (Steinmann 1927). However, in contrast to Steinmann's interpretation that the ophiolites observed in the Alps and Apennines were parts of a consanguineous magmatic association (like the ophiolites of Troodos, Semail, etc.), the mantle and the magmatic rocks of the transitional crust appear to be related to different geodynamic processes. The serpentinites and associated ophicalcites document exhumation of mantle rocks and their exposure at the sea floor; the gabbros and related dykes intruded into already serpentinized mantle rocks during exhumation of the mantle; the basaltic pillow breccias finally contain fragments of flaser gabbro and therefore are younger than the gabbros with which they occur together today (we do not see the plutonic equivalents of the basalts and dolerites). However, as we shall discuss below, this does not exclude that the deformed gabbros and the basalts belong to the same magmatic cycle and reflect the interplay between tectonic and magmatic processes.

Mantle exhumation

Exhumation of mantle rocks and their exposure at the sea floor are well documented in various present-day tectonic settings, e.g. along oceanic transform faults (Fox *et al.* 1976; Bonatti *et al.* 1980), slow-spreading ridges (Lagabrielle & Cannat 1990), along the foot of 'non-volcanic' passive continental margins (Boillot *et al.* 1988; Whitmarsh *et al.* 1998) and along compressional ridges such as the Gorringe Bank in the central Atlantic (Lagabrielle & Auzende 1982). In many of these settings, tectonic and/or sedimentary breccias resembling the various types of ophicalcites are associated with serpentinized peridotites (e.g. Bonatti *et al.* 1974; Whitmarsh *et al.* 1998). In the Alps, the exposure of serpentinized peridotites at the ocean floor has been interpreted (1) within the context of Mesozoic transform faults (Lemoine 1980; Weissert & Bernoulli 1985), (2) by exhumation along a slow-spreading ridge (Lagabrielle & Lemoine 1997) and (3) by exhumation along a system of low-angle detachment faults along a lower-plate continental margin (Lemoine *et al.* 1987; Froitzheim & Manatschal 1996; Manatschal & Bernoulli 1999). In the case of the Platta nappe, final exhumation of mantle rocks along a system of low-angle detachments is suggested by the association of the mantle rocks with extensional continental allochthons that overlie them along gently inclined fault planes (Manatschal & Nievergelt 1997). In addition, the tectono-sedimentary breccias form a more or less continuous carapace along the surface of the serpentinites. Manatschal & Bernoulli (1999) have emphasized the numerous analogies between mantle exhumation along the south Pennine–Austroalpine transition zone and the passive continental margin west of Iberia.

As the serpentinized lherzolites preserve a high-temperature spinel foliation and are stratigraphically overlain by sediments, they must have been exhumed from the deep lithosphere to the sea floor. Hermann *et al.* (1997) showed in Val Malenco that mantle rocks, ultimately derived from the asthenosphere, were in a shallow mantle position long before rifting initiated. Therefore the structures and mineralogical changes recorded in the mantle rocks are not necessarily related to the rifting process alone, and it is not possible without further age constraints to unambiguously relate the anhydrous high-temperature shear zones showing a top-to-the-continent sense of shear to Jurassic rifting. However, we assume that most of the hydration processes within the mantle were associated with Jurassic rifting as documented for the Malenco complex by Müntener *et al.* (2000). In the Platta nappe, static high-temperature hydration is documented by the growth of Mg-hornblende at the expense of diopside, which in turn was followed by further hydration at lower temperature as testified by the crystallization of tremolite. We do not know the age of these transformations; however, Müntener *et al.* (2000) related the hydration of lower-crustal and upper-mantle rocks in the Malenco complex to early rifting during early Jurassic time (± 190 Ma), based on Ar/Ar age determinations on pargasitic hornblende (Villa *et al.* 2000). We therefore think, that the top-to-the-continent high-temperature shear zones predate the Jurassic rifting and may be related to an earlier phase of extension (see Manatschal & Bernoulli 1998).

Serpentinization preceded or was contemporaneous with the following deformation at falling temperatures. Low-temperature deformation produced serpentinite mylonites and tectonic breccias overprinting them, and was followed by the replacement of serpentinite minerals by calcite and by final exhumation and the formation of the tectono-sedimentary breccias at the sea floor (ophicalcites I). The observations that (1) the sense of shear found in the serpentinite mylonites is top-to-the-ocean as along the low-angle detachments cutting across the distal continental margin (Manatschal & Nievergelt 1997) and (2) the gabbros intruding the already serpentinized mantle show evidence of intrusion in a tectonically active setting suggest that serpentinization and intrusion were contemporaneous with the exhumation of the mantle rocks at the sea floor.

Gabbro intrusion and exhumation

Gabbros associated with rifted continental margins represent either magmatic underplating of continental crust, passively exhumed during much later rifting or, alternatively, initiation of partial melting during late rifting and incipient sea-floor spreading. Both types of occurrences have been described along present-day ocean–continent transitions (Schärer *et al.* 1995; Cornen *et al.* 1999, and references therein). In many cases the unknown stratigraphic and structural relationships and the lack of age and thermobarometric data do not allow for an unambiguous interpretation. However, deep-sea drilling and submersible observations along the non-volcanic Galicia margin and in the Iberia Abyssal Plain off western Spain and Portugal have shown that only minor volumes of synrift basaltic melts were produced along this 'non-volcanic' margin before initiation of sea-floor spreading (Manatschal *et al.* 2001). Likewise, the proportion of gabbro in the transitional crust of the Platta nappe is relatively small.

The gabbros of the Platta nappe intruded into already serpentinized mantle rocks in an active tectonic system at *c.* 161 Ma (Desmurs *et al.* 1999). A shallow level of intrusion is supported by our mineral chemistry data, and the chilled margins along the intrusions. The gabbros were subsequently intruded by basaltic dykes, and finally exposed on the sea floor, as recorded by their occurrence as clasts in pillow and sedimentary breccias stratigraphically overlying the exhumed mantle rocks. The microstructures in the gabbros indicate synmagmatic deformation, followed by deformation under greenschist and lower temperature conditions during final exhumation. On the basis of these observations and considering the short time gap between crystallization and exposure on the sea floor, we favour the interpretation that the gabbros were emplaced during or after continental break-up and cooled rapidly during their exhumation in the footwall of detachment structures active during extension.

Basalt extrusion and related dykes

Our field observations document that the massive basalts, pillow lavas and pillow breccias stratigraphically overlying the exhumed mantle rocks and the associated tectono-sedimentary breccias are younger than the gabbros that today are found in the Platta nappe: clasts of the previously deformed gabbros are found together with clasts of albitites in pillow breccias and sedimentary breccias associated with them. On the other hand, we found no basalt but only serpentinite and gabbro clasts in the tectono-sedimentary breccias overlying the exhumed mantle. Undeformed basaltic dykes show chilled margins along their contacts with the serpentinites and the gabbros. This clearly shows that the emplacement of the basalts and of the dykes related to them occurred after serpentinization of the mantle rocks, and after the intrusion, deformation, cooling and exposure of the gabbros on the sea floor. The latest basalts are thus clearly post-tectonic and are the youngest magmatic rocks emplaced at the foot of this passive margin.

Magmatic sources

The gabbros and the basalts yield the same high εNd values (Stille *et al.* 1989; Schaltegger, pers. comm.). This means that both were not contaminated by continental lithosphere and come from similar asthenospheric sources or from the same source. This geochemical signature is similar to that of the magmatic rocks of the internal Liguride ophiolites (Borsi *et al.* 1996; Rampone *et al.* 1998). Only in the slightly older magmatic rocks of the Gets nappe (166 $\pm$ 1 Ma, Bill *et al.* 1997) did pillow basalts and dykes intruding continental crustal rocks yield lower εNd values and rare earth element patterns indicating contamination of the magma by a continental component (Bill *et al.* 2000). Together, these data document an evolution similar to the evolution of the sources of the magmatic rocks recorded along the Galicia margin. There, the εNd of the syn- and post-rift gabbros and basalts suggests contamination of the liquids extracted from the asthenosphere by the overlying subcontinental mantle. However, the younger the magmatic rocks, the higher the εNd, suggesting a decrease in crustal contamination during rifting and thinning of the continental lithosphere (Charpentier *et al.* 1998). In the Platta nappe, the geochemical signatures of the magmatic rocks are compatible with the field and petrological observations that the magmatic rocks were emplaced when the continental lithosphere was extremely thinned, i.e. during or shortly after the break-up of the continental lithosphere.

Timing of emplacement of the various rock types

A relative chronology of the emplacement of the rocks of the Platta nappe at the deep-sea floor can be established that corresponds to the sequence of relative age proposed by Steinmann (1927) for his trinity. Our data show that exhumation of the mantle rocks to the sea floor was accompanied by the intrusion of the gabbros into an already cooled, serpentinized mantle. These gabbroic rocks were exhumed to the sea floor before the extrusion of still younger basalts and deposition of the post-rift sediments. Rates of processes, however, are more difficult to establish. The synrift sediments preserved in the extensional allochthons and locally onlapping onto exposed inactive segments of the low-angle detachment(s) are not directly dated but are younger than the Pliensbachian hardground along the top of the pre-rift sediments of the distal margin ($\pm$190 Ma in the time scale of Gradstein *et al.* (1995)). Ar/Ar determination of phlogopite in a pyroxenite of the Totalp peridotite of the Arosa zone to the north yielded an age of 160 $\pm$ 8 Ma (Peters & Stettler 1987; the error includes the mid-Bathonian to late Oxfordian interval in the time scale of Gradstein *et al.* (1995)). This age is interpreted as the age of mantle exhumation to 10 km depth or less and fits well with the crystallization age of the gabbros of the Platta nappe (161 $\pm$ 1 Ma, Desmurs *et al.* 1999).

Dating of the oceanic basalts by the age of the overlying oceanic sediments proves to be difficult. Both the transitional crust with its tectonically emplaced allochthons as well as the accumulations of pillow lavas present a pronounced submarine relief. The oldest sediments of slow-spreading oceans are typically ponded between the submarine highs of the volcanic

Table 3. *Summary of the geochronological data from the south Pennine–Austroalpine boundary zone, Eastern Alps*

Evolution of the sedimentary cover (Err, Platta,Totalp) 1, 2	Syn-rift (proximal margin)	Syn-rift (distal margin)	? Deposition of Radiolarite Formation
Evolution of the gabbroic rocks. (Platta) 3			Intrusion of gabbros
Evolution of the mantle rocks. (Totalp) 4		Exhumation of mantle rocks	
Evolution of the pre-rift crust-mantle boundary. (Val Malenco) 5	Initiation of rifting	Exhumation of mantle rocks	
	220	200 180	160 140 (Ma)

1, Froitzheim & Eberli (1990); 2, Weissert & Bernoulli (1985); 3, Desmurs *et al.* (1999); 4, Peters & Stettler (1987); 5, Villa *et al.* (2000).

basement on which they rest with an onlap (see, e.g. Lancelot *et al.* 1972, fig. 27); the same observation is made along distal 'non-volcanic' margins (e.g. Whitmarsh *et al.* 1998). The age of the sediments directly overlying transitional or oceanic crust may thus vary over a short distance: in the Alps and Apennines within the same tectonic unit (see Decandia & Elter 1972). In the Platta nappe, the oldest post-rift sediments unconformably overlying the exposed mantle rocks and the pillow lavas are bedded cherts and siliceous shales of the Radiolarite Formation. This formation is not directly dated in our area. In less deformed areas in the Alps, its base is dated to late Bathonian to early Callovian time (Bill *et al.* 1997; between 166.5 and 162 Ma in the time scale of Gradstein *et al.* (1995)) or younger. This age would be slightly older than or the same as our intrusion age of the gabbros. The base of the Radiolarite Formation, however, is certainly diachronous and, in our area, most probably younger than early Callovian time. Nevertheless, we suspect that only a few million years separated the intrusion of the gabbros and the (late Callovian?) onset of deposition of the radiolarites (Table 3). A short time span for this entire evolution is suggested in the Gets nappe of the French Alps, where only a few million years separated the intrusion of the gabbros (166 ± 1 Ma) and deposition of the overlying radiolarites (166.5–162 Ma, Bill *et al.* 1997). Some uncertainty, however, remains because of the still considerable errors in the calibration of the biochronol-

ogy of the mid-Jurassic time interval (±4 Ma, Gradstein *et al.* 1995).

Relationships between tectonic and magmatic processes

Along non-volcanic margins, the transition from continental to oceanic crusts occurs over a transition zone with an intermediate, transitional crust that consists of serpentinized peridotites. Along the Iberian margin (Manatschal *et al.* 2001) and along the Err–Platta margin (Froitzheim & Manatschal 1996; Manatschal & Nievergelt 1997), these exhumed mantle rocks are overlain by extensional allochthons, stranded klippen of continental basement, emplaced along low-angle detachment systems. Magmatic rocks are subordinate in this type of crust, but increase in volume towards the future ocean, and in the case of the Platta nappe, they carry a typical isotopic MORB signature without indications of continental crustal contamination (Stille *et al.* 1989). Towards the ocean, and with continuing extension and mantle exhumation, we laterally pass from a regime in which extension was accommodated, by and large, by mechanical deformation to a regime in which it was more and more compensated by the emplacement of magmatic rocks. In the context of 'non-volcanic' margins, the passage from tectonic extension to sea-floor spreading appears to be gradual, and it becomes difficult to clearly separate continental and ocean

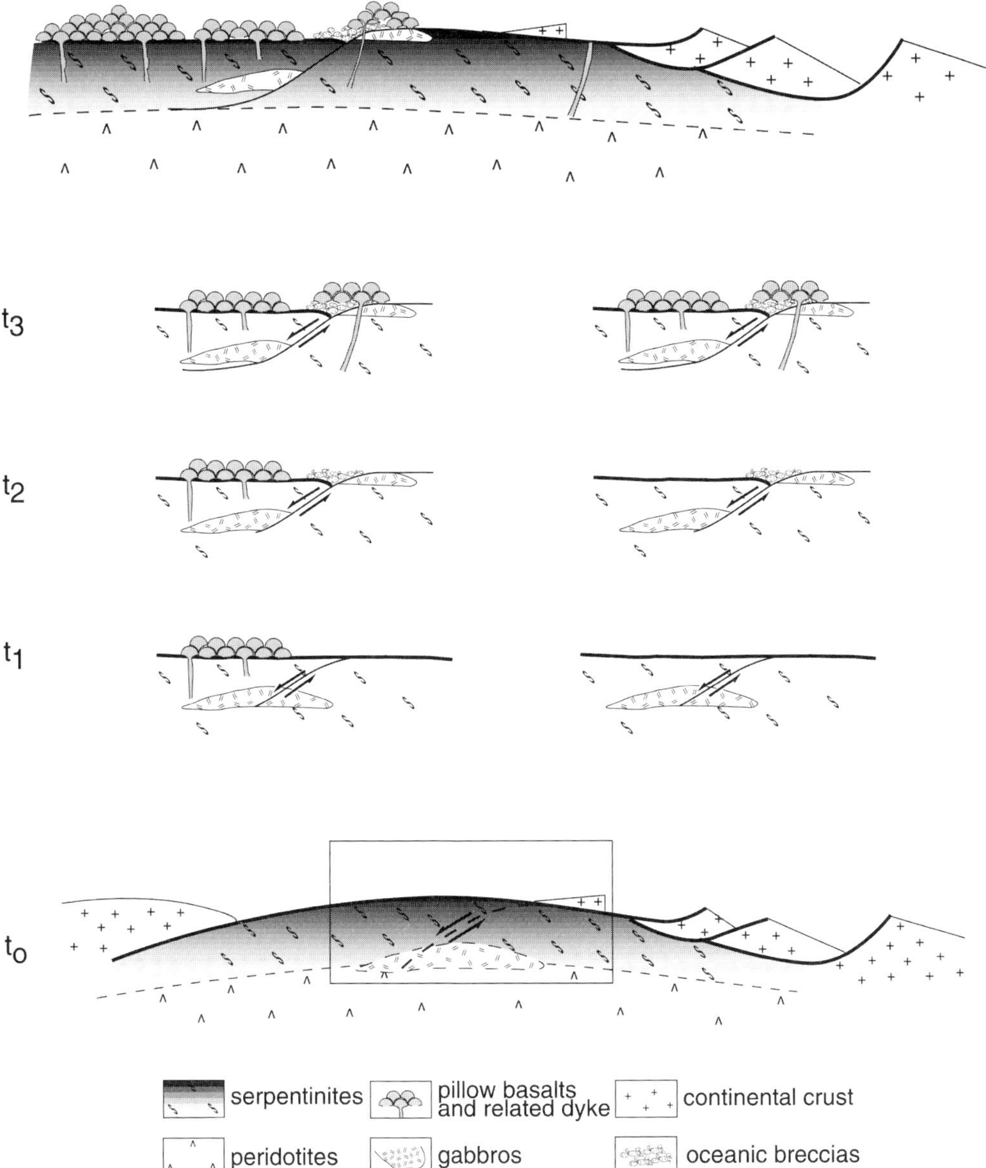

Fig. 16. Sketch illustrating the evolution of emplacement of the various basement rocks. t_0, detachment faulting leading to the exhumation of mantle rocks and intrusion of the gabbros near the serpentinization front. t_1 to t_3, exhumation of the gabbros and emplacement of the pillow basalts. Two hypotheses are shown: on the left, gabbros and basalts are part of the same magmatic cycle; on the right, all the extrusive rocks were emplaced after the exhumation of the gabbros. Both scenarios can lead to the geometries and stratigraphic relationships observed today.

lithosphere and to define continental break-up. Although it is evident that this transition records the processes finally leading to sea-floor spreading and the formation of oceanic lithosphere, we do not yet understand the precise temporal evolution.

Our chronological data do not allow us to clearly separate extensional and magmatic events (Table 3). During late rifting, extension and magmatic processes overlap in time: gabbros intruding the exhumed mantle rocks were deformed at falling temperature conditions, and were finally exhumed at the sea floor and covered in turn by pillow lavas (Fig. 16). Cross-cutting relationships on an outcrop scale show that mantle exhumation started before the first mafic melts intruded during a late stage of rifting and into previously serpentinized mantle. These relationships also show that undeformed basaltic dykes and extrusive rocks are, on a local scale, distinctly younger than the deformed gabbros we observe today at shallow depth and at the surface of the former transitional crust. As gabbros and basalts crystallized at different crustal levels (the basalts at the sea floor, the gabbros within the serpentinized mantle rocks) the gabbros could not have been the magma chambers of the basalts with which they are spatially associated today (e.g. Val da Natons). Unfortunately, we observe today only the uppermost few hundred metres of the former transitional crust. We can therefore not decide whether gabbros and basalts record two different phases of magmatism separated by a phase of extension (Fig. 16, right) or whether gabbros and basalts were more or less contemporaneous and the products of the same steady processes, i.e. of processes similar to those envisaged for slow-spreading ridges. In this case, gabbros and basalts that formed together were subsequently separated by faulting, whereas basalts directly overlying exhumed gabbros are younger (Fig. 16, left), an interpretation that we favour for the time being. These overlapping extensional and magmatic processes might record the initiation of a slow-spreading ridge.

Conclusions

In its type area, the south Pennine Platta nappe, the Steinmann Trinity is dominated by serpentinites and includes significant amounts of pillow basalts and oceanic sediments. Careful mapping of the field relationships shows that the subcontinental mantle rocks were exhumed to the sea floor along a low-angle detachment system similar to that observed along the Iberian non-volcanic margin. Gabbros intruded at shallow depth into the already serpentinized mantle rocks but were continuously deformed at declining temperatures and exposed at the sea floor during continuing extension. With the overlying locally younger basalts the gabbros share typical isotopic MORB signatures and we suspect that they are only the older intrusive members of a magmatic suite, which were deformed, brought to the surface and overlain by younger extrusive rocks of the same suite in a steady process of extension, magma gener-

ation and emplacement leading to the evolution of a slow-spreading ridge.

Our work is part of the research project 'Comparative anatomy of passive continental margins: Iberia and Eastern Alps' supported by the Swiss National Science Foundation, projects 21-049117.96/1 and 20-55284.98. We thank U. Schaltegger, O. Müntener, G. Bernasconi-Green, N. Froitzheim, G. Boillot, Y. Lagabrielle, F. Chalot-Prat, V. Gardien, J. Hermann and V. Trommsdorff for stimulating discussions and helpful suggestions, and Y. Lagabrielle, F. Chalot-Prat, O. Müntener and N. Froitzheim for critical review.

References

ALLEN, J.C., BOETTSCHER, A.L. & MARLAND, G. 1975. Amphibole in andesite and basalt: I. stability as a function of $P-T-f_{O2}$. *American Mineralogist*, **60**, 1069–1085.

ANDERSON, J.L. & SMITH, D.R. 1995. The effect of the temperature and f_{O2} on the Al in hornblende barometer. *American Mineralogist*, **80**, 549–559.

ANONYMOUS 1972. Penrose Field Conference Ophiolites. *Geotimes*, **17**, 2, 24–25.

AOKI, K. & SHIBA, I. 1973. Pyroxenes from lherzolite inclusions of Itinome-gata, Japan. *Lithos*, **6**, 41–51.

APTED, M.J. & LIOU, J.G. 1983. Phase relations among greenschist, epidote-amphibolite and amphibolite in a basaltic system. *American Journal of Science*, **283**, 328–354.

AVIGAD, D., CHOPIN, C., GOFFÉ, B. & MICHARD, A. 1993. Tectonic model for the evolution of the western Alps. *Geology*, **21**, 659–662.

BAILEY, E.B. & MCCALLIEN, W.J. 1950. The Ankara Mélange and the Anatolian Thrust. *Nature*, **166**, 938–943.

BAUMGARTNER, P.O. 1987. Age and genesis of Tethyan radiolarites. *Eclogae Geologicae Helvetiae*, **80**, 831–879.

BENDER, J.F., HODGES, F.N. & BENCE, A.E. 1978. Petrogenesis of basalts from the project FAMOUS area: experimental study from 0 to 15 kbars. *Earth and Planetary Science Letters*, **41**, 277–302.

BERNOULLI, D. & WEISSERT, H.J. 1985. Sedimentary fabrics in Alpine ophicalcites, South Pennine Arosa zone. *Geology*, **13**, 755–758.

BILL, M., BUSSY, F., COSCA, M., MASSON, H. & HUNZIKER, J. 1997. High-precision U–Pb and ^{40}Ar/^{39}Ar dating of an Alpine ophiolite (Gets nappe, French Alps. *Eclogae Geologicae Helvetiae*, **90**, 43–54.

BILL, M., NÄGLER, T.F. & MASSON, H. 2000. Geochemistry, Sm–Nd and Sr isotopes of mafic rocks from the earliest oceanic crust of Alpine Tethys. *Schweizerische Mineralogische und Petrographische Mitteilungen*, **80**, 131–145.

BOILLOT, G., BESLIER, M.O. & COMAS, M. 1992. Seismic images of undercrusted serpentinite beneath a rifted margin. *Terra Nova*, **4**, 25–33.

BOILLOT, G., COMAS, M. C., GIRARDEAU, J. & 5 others 1988. Preliminary results of the Galinaute cruise: dives of the submersible *Nautile* on the western Galicia margin, Spain. *In*: BOILLOT G., WINTERER E.L. *et al.* (eds) *Proceedings of the Ocean Drilling Program, Scientific Results, 103.* Ocean Drilling Program, College Station, TX, 37–51.

BOILLOT, G., RECQ, M. & WINTERER, E.L. 1987. & 21 others 1987. Tectonic denudation of the upper mantle along passive margin: a model based on drilling results (ODP Leg 103, Western Galicia Margin, Spain). *Tectonophysics*, **132**, 334–342.

BONATTI, E. & SEYLER, M. 1987. Crustal underplating and evolution in the Red Sea Rift: uplifted gabbro/gneiss crustal complexes on Zabargad and Brothers islands. *Journal of Geophysical Research*, **92**, 12803–12821.

BONATTI, E., CHERMAK, S. & HONNOREZ, J. 1980. Tectonic and igneous emplacement of crust in oceanic transform zones. *In*: TALWANI, M., HARRISON, C.G. & HAYES, D. (eds) *Deep Drilling Results in the Atlantic Ocean: Ocean Crust.* American Geophysical Union, Maurice Ewing Series, **2**, 239–248.

BONATTI, E., FERRARA, G., HONNOREZ, J. & RYDELL, H. 1974. Ultramafic carbonate breccias from the equatorial mid-Atlantic Ridge. *Marine Geology*, **16**, 83–102.

BORSI, L., SCHÄRER, U., GAGGERO, L. & CRISPINI, L. 1996. Age, origin and geodynamic significance of plagiogranites in lherzolites and gabbros of the Piedmont–Ligurian ocean basin. *Earth and Planetary Science Letters*, **140**, 227–241.

BOUDIER, F., NICOLAS, A., JI, S., KIENAST, J.R. & MEVEL, C. 1988. The gneisses of Zabargad Island: deep crust of a rift. *Tectonophysics*, **150**, 209–227.

BOYER, S. & ELLIOTT, D. 1982. Thrust systems. *AAPG Bulletin*, **66**, 1196–1230.

BREY, G.P. & KOHLER, T. 1990. Geothermobarometry in four-phase lherzolites, II. New thermobarometers and practical assessment of existing thermobarometers. *Journal of Petrology*, **31**, 1353–1378.

BROWN, E.H. 1977. The crossite content of Ca-amphibole as guide to pressure of metamorphism. *Journal of Petrology*, **18**, 53–72.

CARROL-WEBB, S.A. & WOOD, B.J. 1986. Spinel–pyroxene–garnet relationships and their dependence on Cr/Al ratio. *Contributions to Mineralogy and Petrology*, **92**, 471–480.

CHARPENTIER, S., KORNPROBST, J., CHAZOT, G., CORNEN, G. & BOILLOT, G. 1998. Interaction entre lithosphère et asthenosphère au cours de l'ouverture océanique: données isotopiques préliminaires sur la Marge passive de Galice (Atlantique-Nord). *Comptes Rendus de l'Académie des Sciences, Sciences de la Terre et des Planètes*, **326**, 757–762.

CORNEN, G., BESLIER, M.-O. & GIRARDEAU, J. 1996. Petrology of mafic rocks cored in the Iberia Abyssal Plain. *In*: WHITMARSH, R.B., SAWYER, D.S., KLAUS, A. & MASSON, D.G. (eds) *Proceedings of the Ocean Drilling Program, Scientific Results, 149.* Ocean Drilling Program, College Station, TX, 449–465.

CORNEN, G., GIRARDEAU, J. & MONNIER, C. 1999. Basalts, underplated gabbros and pyroxenites record the rifting process of the West Iberian margin. *Mineralogy and Petrology*, **67**, 111–142.

DECANDIA, F.A. & ELTER, P. 1972. La 'zona' ofiolitifera del Bracco nel settore compreso tra Levanto e la Val Graveglia (Appennino ligure). *Memorie della Società Geologica Italiana*, **11**, 503–530.

DESMURS, L., SCHALTEGGER, U., MANATSCHAL, G. & BERNOULLI, D. 1999. Geodynamic significance of gabbros along ancient ocean–continent transitions: Tasna and Platta nappes, Eastern Alps. *Terra Abstracts*, **10**, 379.

DEWEY, J.F. & BIRD, J. 1970. Mountain belts and the new global tectonics. *Journal of Geophysical Research*, **75**, 2625–2647.

DIETRICH, V. 1970. Die Stratigraphie der Platta-Decke: Fazielle Zusammenhänge zwischen Oberpenninikum und Unterostalpin. *Eclogae Geologicae Helvetiae*, **63**, 631–671.

DÜRR, S.B. 1992. Structural history of the Arosa Zone between Platta and Err nappes east of Marmorera (Grisons): multi-phase deformation at the Pennine–Austroalpine plate boundary. *Eclogae Geologicae Helvetiae*, **85**, 361–374.

ELTHON, D. & SCARFE, C.M. 1984. High pressure equilibria of a high-magnesian basalt and the genesis of primary oceanic basalt. *American Mineralogist*, **69**, 1–15.

EVANS, B.W. 1977. Metamorphism of alpine peridotite and serpentinite. *Annual Review of Earth and Planetary Sciences*, **5**, 397–447.

EVANS, B.W. 1982. Amphiboles in metamorphosed ultramafic rocks. *In*: VEBLEN, D.R. & RIBBE, P.H. (eds) *Amphiboles: Petrology and Experimental Phases Relations.* Mineralogical Society of America, Reviews in Mineralogy, **9B**, 98–112.

FERREIRO MÄHLMANN, R. 1995. Das Diagenese-Metamorphose-Muster von Vitrinit-Reflexion und Illit-'Kristallinität' in Mittelbünden und im Oberhalbstein, Teil 1: Bezüge zur Stockwerktektonik. *Schweizerische Mineralogische und Petrographische Mitteilungen*, **75**, 85–122.

FOX, P.J., SCHREIBER, E., ROWLETT, H. & MCCAMY, K. 1976. The geology of the Oceanographer Fracture zone: a model for fracture zones. *Journal of Geophysical Research*, **81**, 4117–4128.

FRISCH, W., RING, U., DÜRR, S., BORCHERT, S. & BIEHLER, D. 1994. The Arosa zone and Platta nappe ophiolites (Eastern Swiss Alps): geochemical characteristics and their meaning for the evolution of the Penninic Ocean. *Jahrbuch der Geologischen Bundesanstalt (Wien)*, **137**, 19–33.

FROITZHEIM, N. & EBERLI, G.P. 1990. Extensional detachment faulting in the evolution of a Tethys passive continental margin (Eastern Alps, Switzerland). *Geological Society of America Bulletin*, **102**, 1297–1308.

FROITZHEIM, N. & MANATSCHAL, G. 1996. Kinematics of Jurassic rifting, mantle exhumation, and passive-margin formation in the Austroalpine and Penninic nappes (eastern Switzerland). *Geological Society of America Bulletin*, **108**, 1120–1133.

FROITZHEIM, N., SCHMID, S.M. & CONTI, P. 1994. Repeated change from crustal shortening to orogen-parallel extension in the Austroalpine units of Graubünden. *Eclogae Geologicae Helvetiae*, **87**, 559–612.

FROITZHEIM, N., SCHMID, S.M. & FREY, M. 1996. Mesozoic paleogeography and the timing of eclogite-facies metamorphism in the Alps: a working hypothesis. *Eclogae Geologicae Helvetiae*, **89**, 81–110.

GLENNIE, K.W., BOEUF, M.G.A., HUGHES CLARKE, M.W., MOODY STUART, M., PILAAR, W.F. & REINHARDT, B.M. 1974. Geology of the Oman Mountains. *Verhandelingen van het Koninklijk Nederlands Geologisch Mijnbouwkundig Genootschap*, **31**, 1–423.

GRADSTEIN, F., AGTERBERG, F.P., OGG, J.G., HARDENBOL, J., VAN VEEN, V., THIERRY, J. & HUANG, Z. 1995. A Triassic, Jurassic and Cretaceous time scale. *In*: BERGGREN, W.A., KENT, D.V., AUBRY, M.-P. & HARDENBOL, J. (eds) *Geochronology, Time Scales and Global Stratigraphic Correlation*. Society of Economic Paleontologists and Mineralogists, Special Publication, **54**, 94–126.

GREEN, D.H. & RINGWOOD, A.E. 1967. The genesis of basaltic magma. *Contributions to Mineralogy and Petrology*, **15**, 103–190.

HANDY, M.R., HERWEGH, M. & REGLI, C. 1993. Tektonische Entwicklung der westlichen Zone von Samedan (Oberhalbstein, Graubünden, Schweiz). *Eclogae Geologicae Helvetiae*, **86**, 785–817.

HÉBERT, R., CONSTANTIN, M. & ROBINSON, P.T. 1991. Primary mineralogy of Leg 118 gabbroic rocks and their place in the spectrum of oceanic mafic igneous rocks. *In*: VON HERZEN, R.P., ROBINSON, P.T. *et al.* (eds) *Proceedings of the Ocean Drilling Program, Scientific Results, 118*. Ocean Drilling Program, College Station, TX, 3-20.

HELTZ, R.T. 1973. Phase relations of basalts in their melting range at $P_{H2O} = 5$ kb as a function of oxygene fugacity, Part I. Mafic phases. *Journal of Petrology*, **14**, 249–302.

HELTZ, R.T. 1982. Phase relations and compositions of amphiboles produced in studies of the melting behaviour of rocks. *In*: VEBLEN, D.R. & RIBBE, P.H. (eds) *Amphiboles: Petrology and Experimental Phase Relations*. Mineralogical

Society of America, Reviews in Mineralogy, **9B**, 279–346.

HERMANN, J. 1997. *The Braccia Gabbro (Malenco, Alps): Permian intrusion at the crust–mantle interface and Jurassic exhumation during rifting*. PhD thesis, ETH Zurich.

HERMANN, J., MÜNTENER, O. & GÜNTHER, D. 2001. Differentiation of mafic magma in a continental crust-to-mantle transition zone. *Journal of Petrology*, **42**, 189–206.

HERMANN, J., MÜNTENER, O., TROMMSDORFF, V., HANSMANN, W. & PICCARDO, G.B. 1997. Fossil crust to mantle transition, Val Malenco (Italian Alps). *Journal of Geophysical Research*, B9, **102**, 20123–20132.

HESS, H.H. 1955. Serpentines, orogeny, and epeirogeny. *In*: POLDERVAART, A. (ed.) *The Crust of the Earth*. Geological Society of America, Special Papers, **62**, 391–408.

LAGABRIELLE, Y. 1987. Les ophiolites: marqueurs de l'histoire tectonique des domaines océaniques. Le cas des Alpes franco-italiennes (Queyras-Piémont), comparaison avec les ophiolites d'Antalya (Turquie) et du Coast Range de Californie. Thèse d'état, Université de Brest.

LAGABRIELLE, Y. & AUZENDE, J.M. 1982. Active *in situ* disaggregation of oceanic crust and mantle on Gorringe Bank: analogy with ophiolite massives. *Nature*, **297**, 490–493.

LAGABRIELLE, Y. & CANNAT, M. 1990. Alpine Jurassic ophiolites resemble the modern central Atlantic basement. *Geology*, **18**, 319–322.

LAGABRIELLE, Y. & LEMOINE, M. 1997. Alpine, Corsican and Apennine ophiolites: the slow-spreading ridge model. *Compte Rendus de l'Académie des Sciences, Sciences de la Terre et des Planètes*, **325**, 909–920.

LAGABRIELLE, Y., POLINO, R., AUZENDE, J.-M. & 8 others 1984. Les témoins d'une tectonique intra-océanique dans le domaine téthysien: analyse des rapports entre les ophiolites et leur couverture métasédimentaire dans la zone piémontaise des Alpes franco-italiennes. *Ofioliti*, **9**, 67–88.

LAIRD, J. 1982. Amphiboles in metamorphosed basaltic rocks. *In*: VEBLEN, D.R. & RIBBE, P.H. (eds) *Amphiboles: Petrology and Experimental Phase Relations*. Mineralogical Society of America, Reviews in Mineralogy, **9B**, 113–159.

LANCELOT, Y., HATHAWAY, J.C. & HOLLISTER, C.D. 1972. Lithology of sediments from the western North Atlantic, Leg 11, Deep Sea Drilling Project. *In*: HOLLISTER, C.D., EWING, J. *et al.* (eds) *Initial Reports of the Deep Sea Drilling Project, 11*. US Government Printing Office, Washington, DC, 901–949.

LAUBSCHER, H. 1969. Mountain building. *Tectonophysics*, **7**, 551–563.

LEAKE, B.E. 1978. Nomenclature of amphiboles. *Mineralogical Magazine*, **42**, 533–563.

LEMOINE, M. 1980. Serpentinites, gabbros and ophicalcites in the Piemont–Ligurian domain of the Western Alps: possible indicators of oceanic fracture zones and of associated serpentinite protrusions in the Jurassic–Cretaceous Tethys. *Archives des Sciences (Genève)*, **33**, 103–115.

LEMOINE, M., TRICART, P. & BOILLOT, G. 1987. Ultramafic and gabbroic ocean floor of the Ligurian Tethys (Alps, Corsica, Apennines): in search of a genetic model. *Geology*, **15**, 622–625.

MANATSCHAL, G. & BERNOULLI, D. 1998. Rifting and early evolution of ancient ocean basins: the record of the Mesozoic Tethys and of the Galicia–Newfoundland margins. *Marine Geophysical Researches*, **20**, 371–381.

MANATSCHAL, G. & BERNOULLI, D. 1999. Architecture and tectonic evolution of non-volcanic margins: present-day Galicia and ancient Adria. *Tectonics*, **18**, 1099–1119.

MANATSCHAL, G. & NIEVERGELT, P. 1997. A continent–ocean transition recorded in the Err and Platta nappes (Eastern Switzerland). *Eclogae Geologicae Helvetiae*, **90**, 3–27.

MANATSCHAL, G., FROITZHEIM, N., RUBENACH, M. & TURIN, B.D. 2001. The role of detachment faulting in the formation of an ocean–continent transition; insights from the Iberia Abyssal Plain. *In*: WILSON, R.C.L., WHITMARSH, R.B., TAYLOR, B. & FROITZHEIM, N. (eds) *Non-volcanic Rifting of Continental Margins: a Comparison of Evidence from Land and Sea*. Geological Society, London, Special Publications, **187**, 405–428.

MOORES, E.M. & VINE, F.J. 1971. The Troodos Massif, Cyprus and other ophiolites as oceanic crust: evaluations and implications. *Philosophical Transactions of the Royal Society of London, Series A*, **268**, 443–466.

MÜNTENER, O., HERMANN, J. & TROMMSDORFF, V. 2000. Cooling history and exhumation of lower-crustal granulite and upper mantle (Malenco, eastern Central Alps). *Journal of Petrology*, **41**, 175–200.

NIMIS, P. & ULMER, P. 1998. Clinopyroxene geobarometry of magmatic rocks: Part I. An expanded structural geobarometer for anhydrous and hydrous, basic and ultrabasic systems. *Contributions to Mineralogy and Petrology*, **133**, 122–135.

OTTEN, M.T. 1984. The origin of brown hornblende in the Artfjället gabbro and dolerites. *Contributions to Mineralogy and Petrology*, **86**, 189–199.

PETERS, T.J. & STETTLER, A. 1987. Radiometric age, thermobarometry and mode of emplacement of the Totalp peridotite in the Eastern Swiss Alps. *Schweizerische Mineralogische und Petrographische Mitteilungen*, **67**, 285–294.

RAMPONE, E., HOFMANN, A.E. & RACZECK, I. 1998. Isotopic contrast within the Internal Liguride Ophiolite (N. Italy): the lack of a genetic mantle–crust link. *Earth and Planetary Science Letters*, **163**, 175–189.

RING, U., RATSCHBACHER, L. & FRISCH, W. 1988. Plate-boundary kinematics in the Alps: motions in the Arosa suture zone. *Geology*, **16**, 696–698.

RING, U., RATSCHBACHER, L., FRISCH, W., BIEHLER, D. & KRALIK, M. 1989. Kinematics of the Alpine plate-margin: structural styles, strain and motions along the Penninic–Austroalpine boundary in the Swiss Austrian Alps. *Journal of the Geological Society*, **146**, 835–849.

RUBATTO, D., GEBAUER, D. & FANNING, M. 1998. Jurassic formation and Eocene subduction of the Zermatt–Saas Fee ophiolites: implications for the geodynamic evolution of the Central and Western Alps. *Contributions to Mineralogy and Petrology*, **132**, 269–287.

SCHÄRER, U., KORNPROBST, J., BESLIER, M.-O., BOILLOT, G. & GIRARDEAU, J. 1995. Gabbro and related rock emplacement beneath rifting continental crust: U–Pb geochronological and geochemical constraints for the Galicia passive margin (Spain). *Earth and Planetary Science Letters*, **130**, 187–200.

SPEAR, F.S. 1981. An experimental study of hornblende stability and compositional variability in amphibolite. *American Journal of Science*, **281**, 697–734.

STAKES, D., MÉVEL, C., CANNAT, M. & CHAPUT, T. (1991). Metamorphic stratigraphy of Hole 735B. *In*: VON HERZEN, R.P., ROBINSON, P.T. *et al.* (eds) *Proceedings of the Ocean Drilling Program, Scientific Results, 118*. Ocean Drilling Program, College Station, TX, 153–180.

STEINMANN, G. 1905. Geologische Beobachtungen in den Alpen, II. Die Schardtsche Ueberfaltungstheorie und die geologische Bedeutung der Tiefseeabsätze und der ophiolitischen Massengesteine. *Berichte der Naturfoschenden Gesellschaft zu Freiburg im Breisgau*, **16**, 18–67.

STEINMANN, G. 1925. Gibt es fossile Tiefseeablagerungen von erdgeschichtlicher Bedeutung? *Geologische Rundschau*, **14**, 435–468.

STEINMANN, G. 1927. Die ophiolitischen Zonen in den mediterranen Kettengebirgen. *Compte-Rendu, XIVe Congrès Géologique International, 1926*, Madrid, Gráficas Reunidas, **2**, 637–667.

STILLE, P., CLAUER, N. & ABRECHT, J. 1989. Nd isotopic composition of Jurassic seawater and the genesis of Alpine Mn deposits: evidence from Sr–Nd isotope data. *Geochimica et Cosmochimica Acta*, **53**, 1095–1099.

TROMMSDORFF, V. 1983. Metamorphose magnesiumreicher Gesteine: Kritischer Vergleich von Natur, Experiment und thermodynamischer Datenbasis. *Fortschritte der Mineralogie*, **61**, 283–308.

TROMMSDORFF, V. & EVANS, B. 1974. Alpine metamorphism of peridotitic rocks. *Schweizerische Mineralogische und Petrographische Mitteilungen*, **54**, 333–352.

ULMER, P. 1986. Basische und ultrabasische Gesteine des Adamello. PhD thesis, ETH Zürich.

VILLA, I., HERMANN, J., MÜNTENER, O. & TROMMSDORFF, V. 2000. ^{39}Ar–^{40}Ar dating of multiply zoned amphibole generations (Malenco, Italian

Alps). *Contributions to Mineralogy and Petrology*, **140**, 363–381.

WEISSERT, H.J. & BERNOULLI, D. 1985. A transform margin in the Mesozoic Tethys: evidence from the Swiss Alps. *Geologische Rundschau*, **74**, 665–679.

WHITMARSH, R.B., BESLIER, M.-O. & WALLACE, P.J. *Proceedings of the Ocean Drilling Program, Initial Reports, 173*. Ocean Drilling Program, College Station, TX.

WITT-EICKSCHEN, G. & SECK, H.A. 1991. Solubility of Ca and Al in orthopyroxene from spinel peridotite: an improved version of an empirical geothermometer. *Contributions to Mineralogy and Petrography*, **106**, 431–439.

The role of lower crust and continental upper mantle during formation of non-volcanic passive margins: evidence from the Alps

OTHMAR MÜNTENER[1,2] & JÖRG HERMANN[1,3]

[1]*Institut für Mineralogie und Petrographie, ETH Zürich, CH 8092 Zürich, Switzerland*
[2]*Present address: Geology Institute, University of Neuchâtel, CH-2007 Neuchâtel, Switzerland (e-mail: othmar.muentener@unine.ch)*
[3]*Present address: Research School of Earth Sciences, ANU, Canberra, A.C.T. 0200, Australia*

Abstract: The remnants of a Mesozoic passive continental margin and of the Tethyan ocean floor are preserved in the Austroalpine and Upper Penninic nappes in eastern Switzerland and northern Italy. Reconstructions of the continent–ocean transition indicate that large areas of subcontinental mantle rocks, but only limited areas of lower-crustal rocks were exposed on the Tethyan sea floor. Microstructures, large shear zones, and the retrograde metamorphic evolution of peridotite and gabbro from Malenco (northern Italy) are investigated to evaluate the role of lower crust and upper mantle during formation of non-volcanic passive continental margins. The combination of petrological constraints and microstructures suggests two contrasting stages: (1) high-temperature (> 650 °C) shearing and annealing of microstructures are attributed to pre-rift tectonics; (2) localized mylonitic shear zones cut the high-temperature structures and developed during nearly isothermal decompression (*T* <600 °C), followed by cooling and hydration of the rocks. These shear zones formed during exhumation of the lower crust and upper mantle and are related to early rifting of the Adriatic passive continental margin. The microstructures of the hydrous mylonites display drastic grain-size reduction, which results from a combination of dynamic recrystallization and metamorphic hydration reactions at temperatures <650 °C. Strain softening facilitated the formation of crustal-scale shear zones along which the lower crust and upper mantle were exhumed to shallow crustal levels of *c.* 10–15 km. Such large shear zones excised 10–20 km of mostly intermediate and lower crust, and are linked to and contemporaneous with the formation of rift-related basins in the upper crust. Boudinage of the lower crust during early rifting is proposed as a major process to explain the scarcity of exposed lower crust along non-volcanic passive margins. The compilation of pressure–temperature data and rift-related structures in the deep crust and upper continental mantle from the Alps suggests that most peridotites preserve a high-temperature evolution that is not related to Mesozoic rifting. Granulite-facies rocks occur in pre-rift lower and middle continental crust. Exhumed granulites along passive continental margins preserve much of a history that is not related to the exhumation itself, but to tectonic processes predating rifting.

Much of what is currently known about the formation of non-volcanic passive continental margins stems from a combination of geophysical data and drilling results from modern oceanic basins (e.g. Boillot *et al.* 1995*b*; Reston *et al.* 1996) and stratigraphic and tectonic investigations from orogens where the remnants of former oceans and margins are exposed (e.g. Froitzheim & Eberli 1990; Froitzheim & Manatschal 1996; Hermann & Müntener 1996). Compared with upper-crustal processes the structural and metamorphic response of the deep crust and upper mantle to passive rifting is less well known. Because of its rare exposure, the role of the lower crust in particular remains one of the mysteries during the formation of passive continental margins and currently there are more models than field observations from which to infer the behaviour of the lower crust. However, the lower crust and upper mantle play a key role in distinguishing the mode of extension (i.e. simple shear v. pure shear, one- v. two-phase rifting, brittle v. ductile behaviour of lower crust and upper mantle during rifting). Structural observations and petrological analysis of lower-crustal and upper-mantle rocks from both present-day (e.g. West Iberia) and ancient (e.g. Alps) margins thus place important constraints on models of passive rifting.

Exhumation of lower crust and upper mantle during rifting is dependent on a variety of factors such as temperature, pressure, stress, strain rate, grain size, rock composition and fluid activity. Handy (1989), for example, showed that the consideration of these factors results in a rheological model of the lithosphere that is much more complex than commonly assumed in modelling studies (e.g. Hopper & Buck 1996) or passive rifting models that are based primarily on observations from the top 5 km of the crust (Froitzheim & Eberli 1990; Froitzheim & Manatschal 1996).

Two main areas in the Alps, the Ivrea zone in the Southern Alps and the Malenco region in the Eastern Central Alps, are suitable to study rift-related processes in the deep crust. Both areas contain lower-crustal rocks preserving pre-Alpine structures and metamorphism. A crucial problem in both cases is the clear identification of pre-rift and syn-rift structures in a polymetamorphic basement. To extract the rifting-relevant observations in lower crust and upper mantle, knowledge of three factors is crucial: (1) the initial temperature and pressure conditions of the lower crust–upper mantle at the beginning of rifting; (2) the availability of fluid; (3) the pre-rift structure and composition of a particular lithospheric section.

In this paper we concentrate on constraints from Malenco (Eastern Central Alps) where a

Permian fossil crust-to-mantle transition that was exhumed during Jurassic rifting has been preserved (Trommsdorff et al. 1993; Müntener & Hermann 1996; Hermann et al. 1997). These rocks permit reconstruction of the pressure–temperature–deformation–time evolution of lower crust and upper mantle during rifting. Special emphasis was placed on the microstructures in gabbros and peridotites because they are a key factor in understanding the rheology of lower crust and upper mantle during rifting. Petrological constraints indicate that the Malenco lower crust and upper mantle had a very low H_2O content before the onset of rifting, and that strain localization during rifting is closely associated with the ingress of fluids and synkinematic breakdown reactions of anhydrous minerals. Our data and those from the Ivrea zone (e.g. Brodie & Rutter 1987; Handy 1987; Handy & Zingg 1991) show that strain localization favours large, rift-related shear zones that exhumed lower crust and upper mantle. These data support the hypothesis that localized deformation and hydration play a major role in the exhumation of lower crust and upper mantle during passive rifting. We review available rift-related pressure–temperature paths from lower-crustal and upper-mantle rocks from the Alps and the West Iberia margin, emphasizing similarities and also major differences compared with the Malenco rocks. Finally, we discuss their significance for the formation of non-volcanic passive continental margins.

Alpine tectonics and palaeogeographical reconstructions

Alpine tectonics along the Penninic–Austroalpine boundary zone

The sedimentary and upper-crustal tectonic evolution of the Adriatic margin can be reconstructed along several transects in the Southern and Eastern Alps (i.e. Bernoulli et al. 1990; Bertotti et al. 1993; Manatschal & Bernoulli 1999). Here we focus on the reconstruction of the Adriatic margin (Fig. 1) as exposed along the Penninic–Lower Austroalpine zone of southeastern Switzerland and northern Italy (Fig. 2a). This former rifted passive margin was overprinted by thrusting during late Cretaceous time. Alpine metamorphism increases from prehnite–pumpellyite facies in the north (Davos–Arosa) to epidote–amphibolite facies in the south (Val Malenco) (Trommsdorff 1983). Accordingly, imbrication tectonics predominates in the northern part (Fig. 2b), whereas

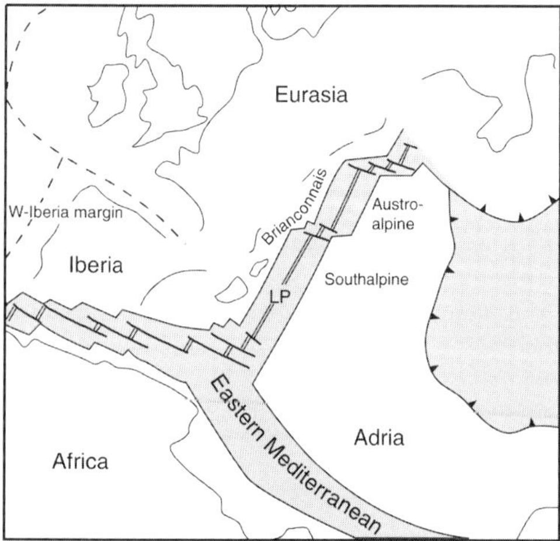

Fig. 1. Schematic palaeogeography of Europe, northern Africa and Adria for late Jurassic to early Cretaceous time. LP, Ligurian–Piemontese ocean.

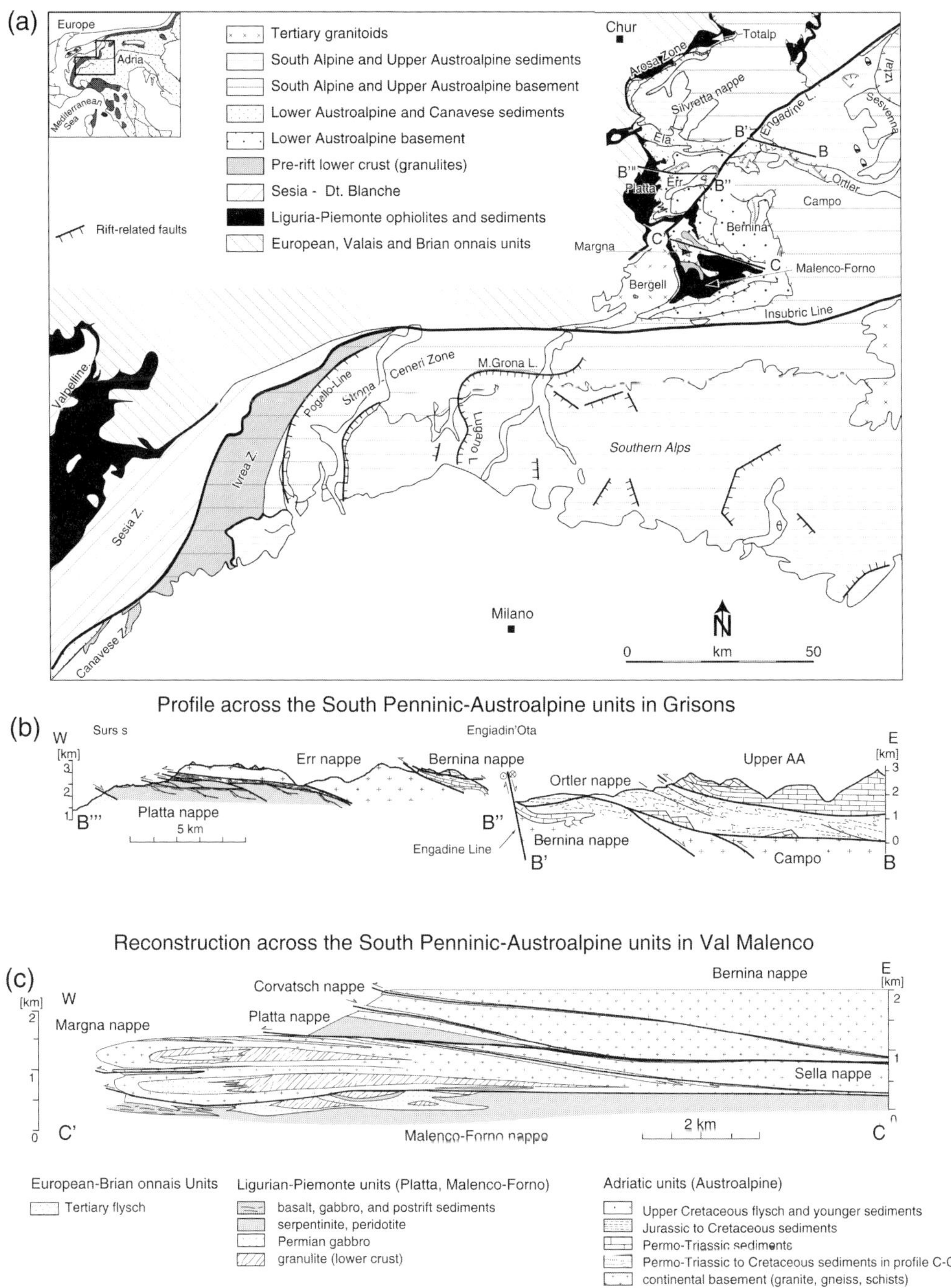

Profile across the South Penninic-Austroalpine units in Grisons

Reconstruction across the South Penninic-Austroalpine units in Val Malenco

Fig. 2. (**a**) Geological overview of the Eastern and Southern Alps, modified after Bernoulli *et al.* (1990) and Manatschal & Bernoulli (1999). (**b**) Cross-section of the south Pennine–Austroalpine units in northern Grisons after Manatschal & Nievergelt (1997). (**c**) Reconstructed nappe pile of the south Pennine–Austroalpine units in Val Malenco (northern Italy) after Spillmann (1993) and Hermann & Müntener (1996).

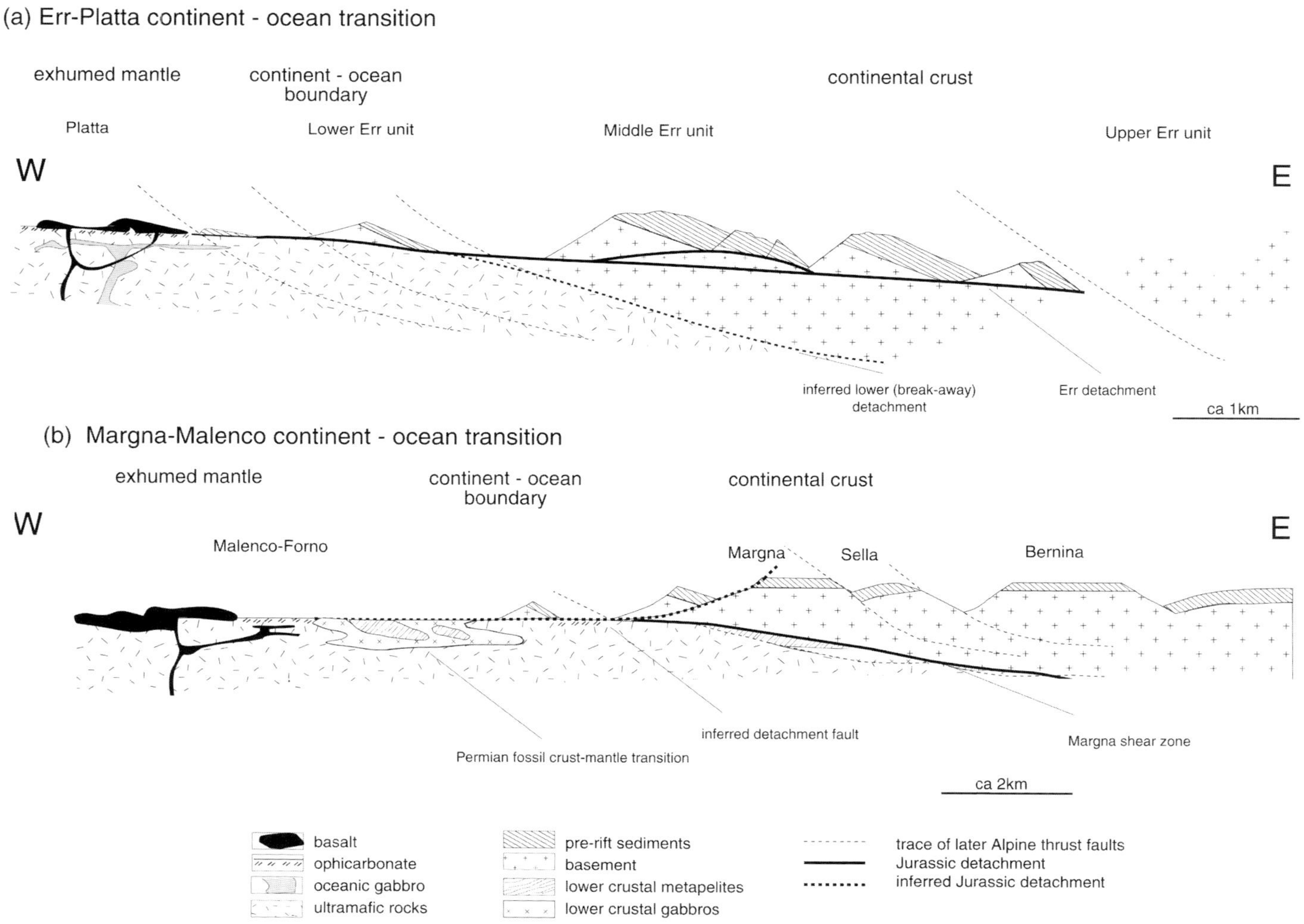

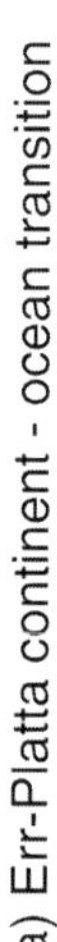

Fig. 3. Palaeotectonic reconstructions across the continent–ocean transition. (**a**) Err–Platta section, modified after Manatschal & Nievergelt (1997). (**b**) Bernina–Margna–Malenco–Forno section, modified after Hermann & Müntener (1996). Syn- and post-rift sediments are not shown.

ductile deformation resulted in large recumbent folds in the southern part (Fig. 2c). The resulting nappe edifice includes the following nappe units from top to bottom: upper Austroalpine nappes (Campo, Oetztal, Ortler, Silvretta), lower Austroalpine nappes (Bernina–Ela, Sella, Err, Margna) and South Pennine nappes (Platta, Davos–Arosa, Malenco–Forno).

Two orogenic cycles affected the former passive margin. East–west-directed nappe stacking and subsequent extension both occurred during Late Cretaceous time (Hermann & Müntener 1992; Handy *et al.* 1993, 1996; Spillmann 1993; Froitzheim *et al.* 1994; Villa *et al.* 2000). Younger north- and south-directed thrusts and east–west-trending folds related to the Tertiary continent–continent collision overprinted the Cretaceous nappe stack (e.g. Hermann & Müntener 1992; Handy *et al.* 1993, 1996; Froitzheim *et al.* 1994) but generally did not disrupt the continuity of the Cretaceous structures in cross-sections of Figure 2. Tertiary structures are therefore not considered further in the retrodeformation of the Alpine nappe stack.

Palaeogeographical reconstructions

From the palaeotectonic reconstructions presented in Figure 3, three main domains can be recognized: (1) an exhumed mantle domain (Platta, Malenco units) with serpentinized ultramafic rocks, ophicarbonate breccias and mid-ocean ridge (MOR) type basaltic intrusive and extrusive rocks covered by a Jurassic–Cretaceous deep-water sedimentary succession; (2) a continent–ocean transition with small slivers of continental crust and pre-rift sediments 'stranded' on top of subcontinental mantle as a consequence of west-directed detachment faulting; (3) a distal margin (Err, Margna units) comprising continental basement rocks with well-preserved pre-rift Permo-Mesozoic sedimentary sequences. The northern transect preserves (Fig. 3a) 'shallow features of rifting' such as the geometry of the rifted basins, the relationships of pre-, syn- and post-rift sedimentation with respect to fault activity, and major detachment faults (e.g. Froitzheim & Manatschal 1996; Manatschal & Nievergelt 1997). The transect in the southern part preserves the 'deep features of rifting'. A unique association of lower crust, Permian gabbroic intrusions and continental upper mantle represents the pre-rift crust–mantle boundary (Müntener & Hermann 1996; Hermann *et al.* 1997) and shows early rift-related structures from the deep crust and upper mantle that

formed in the footwall of a major shear zone (Hermann & Müntener 1996).

The reconstructed passive margin was characterized by a heterogeneous crustal structure with exhumed and partially serpentinized subcontinental mantle, locally associated with continental lower crust, partially overlain by ophicarbonate breccias, extensional allochthons of continental basement, and MOR-type basalts with locally preserved pillow structures and breccias. The thickness and areal extent of basaltic rocks increase oceanwards in both transects. Basaltic dykes intruded the most distal continental blocks (Puschnig 1998; Bissig & Hermann 1999). These dykes have major element chemistry and rare earth element (REE) abundances similar to those of the other basalts. Jurassic gabbroic rocks are rare and are preserved only in the Err–Platta transect. Single zircon U–Pb data from these gabbros yielded an age of 161 ± 1 Ma (Desmurs *et al.* 2001) that is close to the oldest post-rift sediments (radiolarites) overlying the basaltic rocks and the serpentinites. In contrast, the gabbroic rocks in the Malenco transect are of Permian age (Hansmann *et al.* 1996) and associated with lower-crustal rocks. The 'ophiolites' preserved along the Pennine–Austroalpine boundary are distinctly different from typical fast-spreading ridge associations and can be compared with transitional crust of non-volcanic passive margins in present-day oceans (e.g. Galicia margin: Boillot *et al.* 1995*b*). The transition to a slow-spreading system was most probably developed further oceanwards (see also Desmurs *et al.* 2001).

In the northern part of the studied area a complication arises from the position of the Platta nappe, which is sandwiched between two lower Austroalpine units (Fig. 2). This is a crucial problem in any palinspastic reconstruction and its solution depends on how one interprets the pinching-out of the Platta ultramafic rocks SE of the Engadine line (Schmid *et al.* 1990). One hypothesis, put forward originally by Trümpy (1975), is that the Margna–Sella nappes form a microcontinent. Later, Froitzheim & Manatschal (1996), Froitzheim *et al.* (1996) and Handy (1996) interpreted Margna–Sella as an extensional allochthon resulting from detachment faulting during the advanced stages of rifting, implying oceanic crust between the Margna–Sella and the Austroalpine units (e.g. Froitzheim & Manatschal 1996, fig. 10). However, studies of the Alpine tectonic evolution (Liniger 1992; Spillmann 1993) have shown that there is no ophiolitic suture between the lower Austroalpine Bernina and

Margna–Sella nappes east of the point where the Platta nappe pinches out. An alternative hypothesis therefore suggested that the lower Austroalpine crystalline units represent a continuous segment of basement rocks in the southern part of the study area (Fig. 4) and that the arrangement of continental and oceanic units reflects the original palaeogeography of a segmented passive continental margin. Consequently, Schmid *et al.* (1990), and later Spillmann (1993) proposed that the pre-Alpine extension direction and Alpine thrusting are not strictly parallel in this area and that the southernmost part of the Platta nappe became sandwiched between lower Austroalpine nappes during Cretaceous thrusting.

We suggest that the fragmentation of the continental margin was caused by rifting and that in the northern part (Fig. 4) the continent–ocean transition is formed by the Bernina–Err–Platta domain. However, there is no field evidence at all that the Margna–Sella domain represented an extensional allochthon surrounded by oceanic crust (Froitzheim & Manatschal 1996). Consequently, we propose that the Platta oceanic crust did not exist in the south and the Bernina–Margna–Malenco domain represented the Adriatic passive continental margin in Late Jurassic times (Fig. 4). Such a geometry can best explain why cross-sections along continent–ocean transitions (e.g. the Err–Platta and Margna–Malenco domain) vary laterally (see cross-sections of Froitzheim & Manatschal 1996; Handy (1996); Hermann & Müntener 1996).

Exhumation metamorphism of lower crust and upper mantle during Jurassic rifting

Rift-related metamorphic processes in the lower crust can be studied in Val Malenco (Fig. 2). There, an undisturbed Permian fossil crust–mantle transition has been preserved and is now exposed (Trommsdorff *et al.* 1993; Müntener & Hermann 1996). The Braccia gabbro intruded *c.* 278 Ma ago along the crust–mantle transition zone and caused granulite-facies metamorphism of lower-crustal and upper-mantle rocks (Hermann *et al.* 1997). Recent SHRIMP dating of zircons from the granulite-facies metapelites yielded the same age as obtained for the gabbro (Rubatto, pers. comm.). Two stages of retrograde metamorphism followed (Fig. 5). Mineral parageneses in pelitic granulites, metagabbros and metaperidotites record a first stage of cooling under anhydrous conditions from 850 °C to 650 °C with moderate decompression from 1.0 GPa to 0.8 ± 0.1 GPa, and dP/dT < *c.* 0.15 GPa/100 °C (Müntener *et al.* 2000). This period of near-isobaric cooling is most probably related to Permian transtensional tectonics. The crust was thinned by *c.* 3–6 km of the crustal section. The stabilized crust of *c.* 30 km thickness then persisted over a period of *c.* 50 Ma into Late Triassic time. Exhumation of the crust–mantle complex began with the onset of continental rifting. This

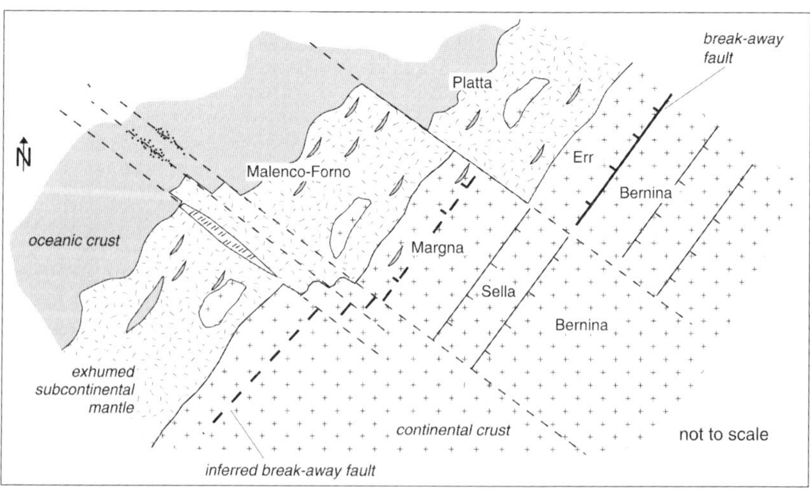

Fig. 4. Schematic palaeogeography (map view) of the continent–ocean transition along the Adriatic continental margin. Tectonic klippen of continental rocks (extensional allochthons) overlying the exhumed subcontinental mantle should be noted.

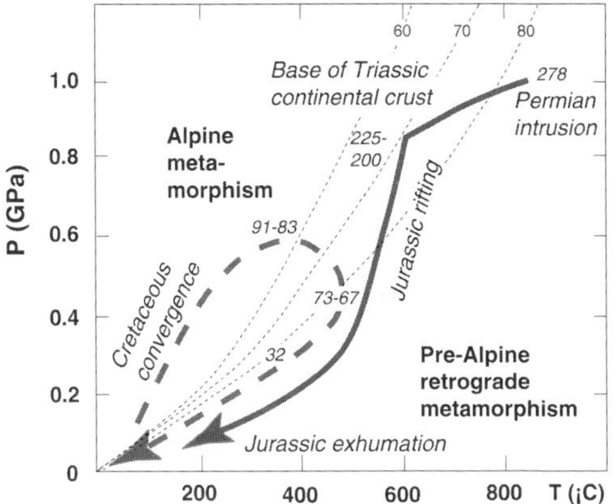

Fig. 5. Pressure–temperature–time diagram for the Malenco crust–mantle complex, after Müntener (1997), Müntener *et al.* (2000) and Villa *et al.* (2000). Black arrow indicates the Permian to Jurassic retrograde metamorphic evolution; dashed arrow indicates the burial and exhumation related to the Alpine cycle. Geotherms corresponding to different surface heat flows (in mW m^{-2}) are taken from Chapman (1986).

second stage of retrograde metamorphism is recorded by near-isothermal decompression and partial hydration of the granulitic mineral assemblages. P–T conditions at the initial stage of rifting were 0.8 ± 0.1 GPa and 600 ± 50 °C. Subsequent retrograde metamorphism involved decompression of c. 0.4 GPa and cooling to c. 500 °C. Finally, the crust–mantle complex was exhumed on the Tethyan ocean floor. The two-stage retrograde path of the Malenco granulites separated by >50 Ma suggests that Permian transtensional tectonics and Jurassic rifting are two independent tectonic processes (Müntener *et al.* 2000).

A crucial point in the petrogenesis of the Malenco rocks is that the initial stage of rifting is associated with hydration and that the average metamorphic conditions (0.8 ± 0.1 GPa and 600 ± 50 °C) correspond to the base of a continental crust about 30 km thick. This is the average depth of continental crust beneath large parts of western central Europe that were not affected by rifting and the Alpine orogeny (Ansorge *et al.* 1992). Müntener *et al.* (2000) showed that the fluids responsible for hydration had an external source and were Cl-rich brines with low XCO_2. Thus, during the initial stage of rifting, temperatures <650 °C combined with partial hydration of previously dry granulites are important boundary conditions controlling the formation of rift-related structures and shear zones.

Large-scale structures related to rifting in the deep crust

An example of a major shear zone is exposed in the Margna nappe (Fig. 2c). The Margna nappe is a composite basement unit consisting of granulitic lower crust intruded by (presumably Permian) tholeiitic gabbros identical to the Braccia gabbro (Bissig & Hermann 1999), and of Variscan upper crust with a Permo-Mesozoic sedimentary cover. These two rock associations represented respectively the lower and upper crust before rifting, and were assembled during rift-related thinning of the continental crust. The contact between upper-crustal and lower-crustal rocks as exposed today in the Margna nappe forms two large recumbent folds within the latter (Fig. 2c). The significance of this contact for pre-Alpine extension becomes clear only when the lower Austroalpine nappe stack (Fig. 2c) is retrodeformed into its pre-collisional orientation (Fig. 3b). Restoring the original geometry of this shear zone results in a low-angle shear zone dipping below the Adriatic continent (Hermann & Müntener 1996; Bissig & Hermann 1999). On the basis of the 'pressure gap' of at least 5 kbar between upper-crustal granitoid rocks and lower-crustal granulites, we conclude that rifting excised 15–20 km of mostly intermediate crust. Thus this shear zone accommodated extensional exhumation of the pre-rift lower crust to a depth of 10 km or less.

Microstructures

Microstructures in the gabbros

On the basis of field and microstructural observations we have identified two major classes of gabbro structures: (1) coarse-grained flaser gabbros; (2) fine-grained mylonites crosscutting the flaser gabbros. The flaser gabbros consist of two distinct microstructural types. A texture indicative of magmatic flow is characterized by large pyroxene crystals (0.5–10 cm) aligned in parallel bands. Concentrations of pyroxene and plagioclase led to a pronounced modal layering. The magmatic flow texture, which is sometimes cut by leucogranites (Müntener & Hermann 1996), has an orientation that is oblique to the later foliations and is preserved in lenses surrounded by a tectonic flaser structure (Fig. 6a). This flaser structure, marked by stretched pyroxenes in a plagioclase matrix, is the main structure within the gabbro. Pyroxene and plagioclase grains (1–2 mm) are nearly equant and exhibit planar grain boundaries that often form

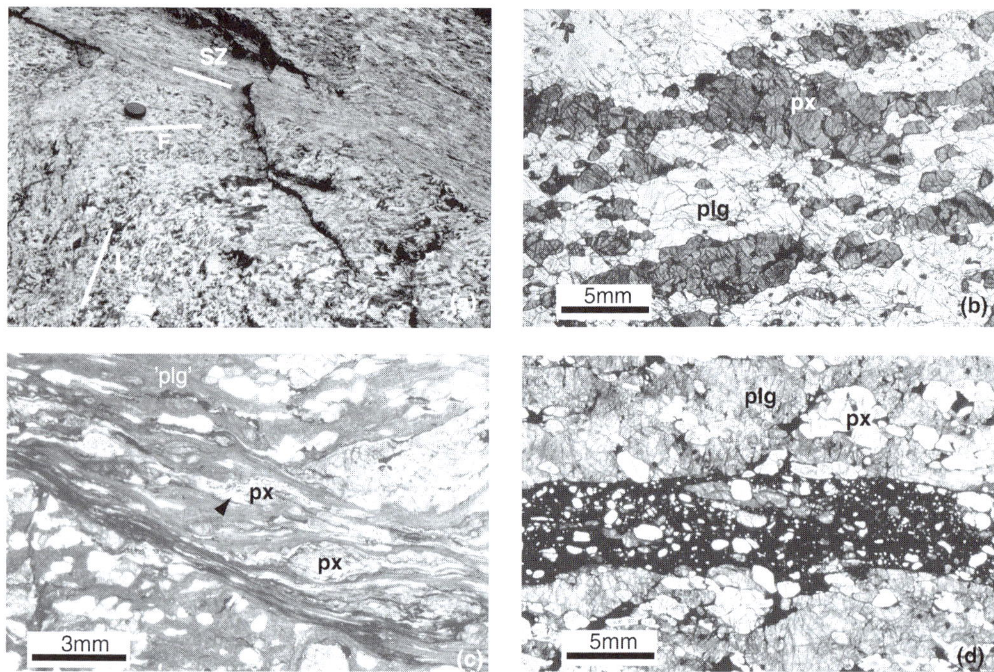

Fig. 6. Field aspects and microstructures of the Braccia gabbro. (**a**) Igneous (I) flow textures cut by tectonic flaser textures (F). These high-temperature structures are cut by shear zones (SZ). Diameter of lens cover 5 cm. (**b**) Photomicrograph of a high-temperature flaser texture. Pyroxene (px) and plagioclase (plg) show annealing. (**c**) Porphyroclastic microstructure illustrates localized reduction of grain size. The foliation in the mylonite is made up of ultra-fine-grained bands of albite, clinozoisite and amphibole ('plg') surrounding clinopyroxene porphyroclasts and pseudomorphosed 'orthopyroxene'. Pressure shadows of pyroxenes (indicated by arrow) are made of plagioclase and hornblende. (**d**) Flaser texture cut by a black shear zone possibly representing ancient pseudotachylite.

triple junctions (Fig. 6b). These microstructures indicate annealing and re-equilibration during subsequent granulite-facies metamorphism. Estimates from dynamically recrystallized ortho- and clinopyroxene indicate temperatures of c. 850 °C and pressures of c. 1.0 GPa (Müntener *et al.* 2000). Therefore the flaser texture is related to postmagmatic deformation and predates the rift-related structures.

The tectonic flaser textures are crosscut by later shear zones and mylonites (Fig. 6a). Incipient to complete reaction of pyroxene porphyroclasts to amphibole and chlorite, and dynamic recrystallization of plagioclase are associated with a drastic grain-size reduction from >1 mm outside to <50 μm inside the shear zone. Clino- and orthopyroxene porphyroclasts are boudinaged and show no evidence of dynamic recrystallization (Fig. 6c). In places, pyroxene pseudomorphs show the same microstructures as unaltered porphyroclasts but occasionally the pseudomorphs are strongly deformed with aspect ratios of >5:1. Syntectonic alteration of

pyroxene clasts to amphibole + chlorite also occurred along intragranular cracks and along cleavages. These mineral transformations indicate that shearing and hydration were coeval, at fluid pressures high enough to induce microfracturing of pyroxene and plagioclase porphyroclasts. The coexistence of amphibole and plagioclase suggests that the mylonites formed during incipient hydration under amphibolite-facies conditions, i.e. during the initial stages of rifting (Müntener *et al.* 2000).

Narrow black shear zones were occasionally found and belong to a third type of microstructure (Fig. 6d). These shear zones truncate the flaser texture and contain rounded pyroxene in an extremely fine-grained matrix (<5 μm) of pyroxenes and altered plagioclase. However, such black shear zones were never found in completely hydrated gabbros. The clast-to-matrix ratio is about 1:2. The contacts of the black zones are extremely sharp and locally the black zones seem to 'intrude' the surrounding flaser gabbros or they form a network in the surrounding flaser gabbros. These observations suggest that the black shear zones were originally pseudotachylites. However, the preferred orientation of the clasts subparallel to the shear-zone boundary (Fig. 6d) suggests that these pseudotachylites were mylonitically overprinted. The field relations demonstrate that these black shear zones overprint the Permian high-temperature deformation. We suggest that they formed at the earliest stage of rifting, before the development of weak, hydrous shear zones, but we cannot rule out that they formed during the last stage of rifting at low-grade metamorphic conditions.

Microstructures in the peridotites

There are three major classes of pre-Alpine microstructures in the Malenco peridotite: (1) coarse-grained porphyroclastic peridotites; (2) anhydrous peridotite mylonites; (3) hydrous peridotite mylonites.

On a map scale, coarse-grained tectonites are predominant (Fig. 7a), although some less deformed 'boudins' of nearly undeformed peridotite occur (Müntener 1997). In thin section, spinel and pyroxene porphyroclasts are aligned within the foliation and olivine is often flattened with subgrain walls subperpendicular to the lineation. Orthopyroxene grains may be stretched to aspect ratios of up to 5:1, whereas clinoyproxene occurs as elongate clusters of several grains (Fig. 7c), suggesting dynamic recrystallization. Similar coarse-grained tectonites have been described from xenoliths, peridotite massifs and abyssal peridotite, and are interpreted to form during high-temperature plastic flow (e.g. Nicolas 1989).

The coarse-grained spinel tectonites are overprinted to various degrees by a network of narrow shear zones with small dynamically recrystallized grains ($<50\,\mu m$) of olivine, orthopyroxene, clinopyroxene, spinel and Ti-pargasite. These narrow shear zones generally form anastomosing networks around large porphyroclastic minerals (Fig. 7b). The mineral compositions of the recrystallized grains are retrograde with respect to the porphyroclasts, and thermobarometry indicates granulite-facies conditions (Müntener 1997; Müntener *et al.* 2000). This type of shear zone is often found in dunites, indicating that monomineralic olivine layers might be weaker than polymineralic peridotites during anhydrous high-temperature deformation. Generally, these networks are rarely observed in the Malenco peridotite because many of these shear zones were reactivated and recrystallized during later defor-

mation. The coarse-grained tectonites and the fine-grained shear zones consist of anhydrous minerals and formed at high temperatures ($T > 750\,°C$), indicating that they predate hydration and most probably Jurassic rifting (see below).

Rift-related structures occur in hydrated peridotite mylonites that overprint the original peridotite tectonites. Incipient to complete reaction of ortho- and clinopyroxene to fine-grained chlorite + tremolite + olivine is apparent along grain boundaries and cleavages, and indicates progressive infiltration of H_2O-rich fluid during deformation. Pseudomorphs after stretched pyroxene clasts are recrystallized to a fine-grained mixture of chlorite + tremolitic amphibole (Fig. 7c). Pressure shadows of pyroxene porphyroclasts consist of olivine + tremolite + chlorite $\pm$ ilmenite $\pm$ ferrichromite. In some of the hydrated peridotite mylonites spinel, olivine and pyroxene porphyroclasts are fractured and show incipient recrystallization of small grains of amphibole and chlorite (Fig. 7d). Veins of tremolite and orthopyroxene occasionally occur in olivine. The synkinematic assemblage olivine + tremolite indicates upper amphibolite-facies conditions during the initial stages of deformation. Different mylonitic textures that exhibit pronounced grain-size reduction are shown in Figure 7d and e. The porphyroclasts are composed of rounded olivine or clinopyroxene with grain sizes varying from $<100\,\mu m$ to $>2\,mm$. Sometimes polycrystalline porphyroclasts of clinopyroxene are observed (Fig. 7d). Pressure shadows of olivine porphyroclasts in dunite and in pyroxene-poor zones in peridotites consist entirely of olivine and generally show a larger grain size (Fig. 7f) than recrystallized grains in pressure shadows of pyroxenes (Fig. 7d). In the latter, an extremely fine-grained matrix of recrystallized grains ($<10\,\mu m$) of olivine, tremolite, chlorite and rarely talc was observed (Fig. 7e). This suggests that grain growth may have been impeded by the presence of secondary phases (e.g. Olgaard 1990). Presuming that the dominant deformation mechanism is diffusion-accommodated granular flow or diffusion creep, we think that the polyphase recrystallized bands may be the weakest zones in recrystallized peridotites and are likely to predominate in lherzolitic (i.e. pyroxene-rich) compositions. The boudinage of dunite on the outcrop scale and the formation of olivine porphyroclasts indicates that a strength inversion occurred during retrograde metamorphism, and that the rheology of the peridotite mylonites is ultimately controlled by

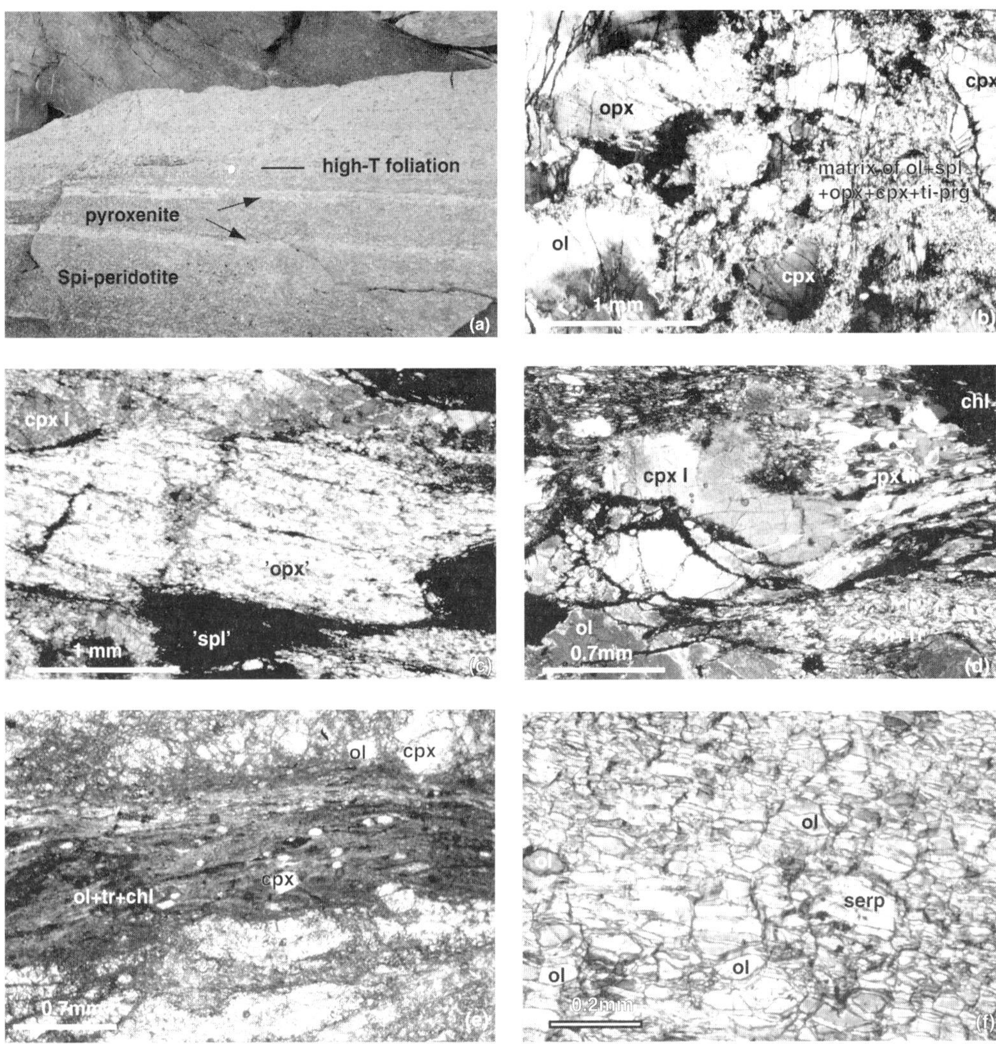

Fig. 7. Field aspects and microphotographs of the Malenco peridotite. (**a**) Field aspect of peridotite tectonite with pyroxenite layers parallel to the foliation. Diameter of coin 3 cm. (**b**) Porphyroclastic spinel peridotite with neoblasts forming a fine-grained matrix. Compositional variation of the minerals has been described by Müntener (1997). ol, olivine; cpx, clinopyroxene; opx, orthopyroxene; spl, spinel; ti-prg, titanian pargasite. (**c**) Pseudomorphs of amphibole and chlorite after stretched orthopyroxene. (**d**) Recrystallized olivine + tremolite + chlorite in pressure shadows of olivine and clinopyroxene porphyroclasts. Arrow points to fractures in clinopyroxene that are filled with tremolite + chlorite ± olivine. (**e**) Peridotite mylonite; cpx clasts occur in an extremely fine-grained (<10 μm) mylonitic matrix of olivine + tremolite + chlorite. (**f**) Dunite mylonite. (Note the difference in grain size compared with (**e**).)

the fine-grained matrix of olivine + hydrous silicates (amphibole, chlorite, talc).

Discussion

In the following, we discuss rheological and petrological aspects of the lower crust and upper mantle in the evolution of non-volcanic passive continental margins, in particular the processes controlling strain localization and the formation of detachment structures in the deep crust and uppermost mantle during the early stages of rifting. We propose that localized shear zones during the early stages of passive rifting are required to exhume lower crust and upper mantle to shallow depth of *c.* 10 km, and that during an advanced stage of rifting detachment faulting finally exhumes lower crust and mantle. The distribution of early and advanced detachment structures largely controls the spatial occurrence of lower crust and upper mantle at passive continental margins.

Strain localization in hydrous mylonites during the early stages of rifting

The rheological conditions in the lower continental crust and uppermost mantle are reflected in the textural development and the $P-T$ history of gabbro and peridotite during the rifting phase. The structures in these rocks postdate high-temperature granulite-facies shear zones and record a sequence in the deformation history of successively finer-grained microstructures overprinting each other. The mylonites illustrated in Figures 6c and 7e have porphyroclasts, which are surrounded by an extremely fine-grained matrix. Grain-size reduction is enhanced by cataclasis and synkinematic retrograde breakdown reactions of anhydrous to hydrous phases. Thermobarometric estimates on the earliest hydrated assemblages of the Malenco rocks (P 0.8 $\pm$ 0.1 GPa, T 600 $\pm$ 50 °C, Müntener et al. 2000) indicate that the metamorphic conditions corresponded to amphibolite facies during the initial stage of rifting. Therefore, the combination of fluid infiltration and syntectonic breakdown reactions of anhydrous minerals concomitant with deformation at T <650 °C was the principal mechanism for the observed strain localization.

A consequence of hydration reactions is that the strength of the rocks becomes controlled by the relatively weak and fine-grained reaction products in mylonites (e.g. Brodie & Rutter 1985). Deformation experiments on dry and wet olivine aggregates have shown that grain-size reduction coupled with low temperature could favour a change in the deformation mechanism from dislocation creep to diffusion creep (e.g. Karato et al. 1986). Deformation mechanism maps for olivine (e.g. Handy 1989; Jaroslow et al. 1996) show that diffusion creep may be the dominant mechanism at temperatures of c. 500–600 °C and grain sizes of 10 μm or less. Polymineralic rocks such as hydrated lherzolites or hydrated gabbros may be weaker than monomineralic rocks such as dunite because synkinematic grain growth is impeded (Olgaard 1990). Vissers et al. (1995) proposed the strength of polymineralic mylonites to be two to four orders of magnitude lower than the strength expected for homogeneous deformation controlled by dislocation creep of olivine alone.

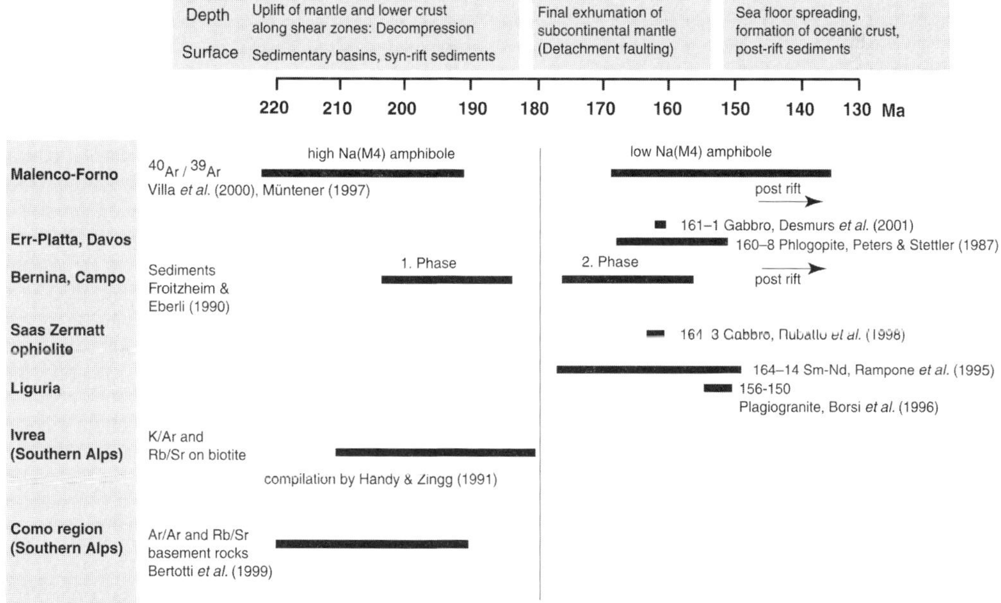

Fig. 8. Compilation of rift-related ages from the Alpine chain. Early rifting is separated from detachment faulting and the onset of Tethyan magmatism. (For discussion, see text.)

Fine-grained dynamic recrystallization of plagioclase and synkinematic hydration of pyroxene resulting in the formation of fine-grained chlorite and amphibole (Fig. 6c) drastically reduce the strength of the gabbroic rocks. Experimental evidence for considerable reaction-enhanced weakening in shear zones has been presented recently by Stünitz & Tullis (2001). They studied the retrograde hydration reaction plagioclase + H_2O ⇌ zoisite + albite + kyanite + quartz. At $900\,°C$ and $1.0\,GPa$ plagioclase deformed homogeneously by dislocation creep. In 'wet' samples at $750\,°C$ and $1.0\,GPa$ deformation localized in fine-grained shear bands and the deformation mechanism involved grain boundary sliding accommodated by diffusion. In contrast, at the same conditions, 'dry' samples are strong and deformed mainly by dislocation creep (Stünitz & Tullis 2001). The positive feedback mechanisms between hydration, grain-size reduction and strain localization ('reaction-enhanced ductility', White & Knipe 1978) seen in the experiments are consistent with observations from mafic rocks in Val Malenco (Fig. 6c) and the Ivrea zone (e.g. Brodie & Rutter 1985; Handy & Zingg 1991). We suggest that in polymineralic rocks reaction softening (White & Knipe 1978; Rubie 1983) reduced the strength of mantle and lower-crustal shear zones and facilitated a change to diffusion creep as the dominant deformation mechanism. The positive feedback mechanism between grain-size reduction and focusing of fluid flow (Wark & Watson 2000) might further increase fluid access along fine-grained hydrous mylonites. Such weak zones are likely to promote the development of large shear zones (detachments) in the lower crust and upper mantle (see next section). However, there remains the classical 'chicken-or-the-egg-first' dilemma of whether mylonite formation facilitated fluid access or vice versa. The observation that black shear zones cut the flaser texture in the gabbros but do not show any obvious relation to the hydrous mylonites suggests that they formed during the initial stage of rifting. One might speculate that at the initial stage of rifting, brittle failure ('microseismicity') led to the formation of these black shear zones (e.g. Fig. 6d). Along these zones mylonitization was probably initiated, thus permitting hydration of previously dry rocks. A close relationship between seismicity and early rifting has also been demonstrated by Curewitz & Karson (1999) for the East Greenland rifted margin. Alternatively, these black shear zones could have formed during the last stage of rifting at sub-greenschist-facies conditions.

Shear zones in the deep crust of the Adria margin related to early rifting

The field relations, deformation textures and the equilibration temperature and pressure of the lower-crustal gabbros and peridotites support the existence of localized deformation at the base of a $30\,km$ thick continental crust. Such zones of weakness are a prerequisite for the formation of large shear zones, which play an important role in the exhumation of lower crust and upper mantle at continental margins (e.g. Handy 1989). Observed hydration of lower crust at $25–30\,km$ depth requires crustal-scale shear zones (Müntener et al. 2000). Low-angle normal faulting associated with early Mesozoic rifting excised c. $15–20\,km$ of mostly intermediate (and lower) crust and exhumed subcontinental mantle and lower crust to a depth of $10\,km$ or less in the Ivrea zone and in Malenco. The available kinematic data recording deformation in the deep crust and upper mantle indicate a top-to-the-continent sense of shearing for these deep-seated fault zones (Pogallo shear zone, Handy 1987; Margna shear zone, Hermann & Müntener 1996; Bissig & Hermann 1999).

In the Malenco area the maximum amount of displacement along the Margna shear zone has been estimated to be c. $50\,km$ (Hermann & Müntener 1996) whereas the estimated maximum displacement along the Pogallo shear zone is c. $20\,km$ (Handy 1987). The Pogallo shear zone was active sometime between 180 and $210\,Ma$ as constrained by correlating the temperature range of retrograde syntectonic metamorphism with radiometric cooling curves constructed for the Ivrea–Verbano zone (Handy 1987). There are no age constraints on the activity of the Margna shear zone because of a strong Alpine metamorphic overprint. However, syntectonic hydration during Jurassic rifting is supported by $^{40}Ar/^{39}Ar$ analyses of amphibole in the footwall of the Margna shear zone, which indicate radiometric ages of c. $190–225\,Ma$ (Müntener 1997; Villa et al. 2000).

The good correlation of ages with evidence from the sedimentary record indicates that the activity of these large-scale shear zones in the deep crust coincides with the first rifting phase (Fig. 8; Froitzheim & Eberli 1990; Bertotti et al. 1993). Such localized shear zones are probably responsible for uplift of the deep crust and subcontinental mantle to depths of $10\,km$ or less. In both areas extension was accommodated by shear zones, which are discrete in the middle to lower crust and upper mantle. These observations do not support models of a hot and entirely ductile weak lower crust (e.g. Manatschal & Bernoulli 1999) during the early stages of rifting. However, localized shear zones in the deep crust of the Adriatic margin do not necessarily indicate that early rifting was asymmetric. Indeed, there may be conjugate shear zones dipping under the European continent (see Fig. 10, below).

P–T paths and their significance for exhumation of lower crust and upper mantle during rifting

Figure 9 summarizes the available P–T–t data from some examples of lower crust and upper mantle of various terranes in the Alps and of the West Iberia margin (see Figs 1 and 2 for locations). Two features of the P–T evolution are of particular interest in the context of rifting: (1) the P–T conditions of the lithosphere at the onset of rifting; (2) the P–T path during rifting. Concerning the first characteristic feature, Müntener et al. (2000) presented results from phase equilibria in the lower crust of Malenco, which indicate that the initial temperature at the onset of rifting (Late Triassic to Early Jurassic time) did not exceed 650 °C (at c. 0.8 GPa). These conditions would correspond to a surface heat flow of about 70 mW m^{-2} (thermal model of Chapman (1986)). Such a heat flow is slightly higher than the average value of 60 mW m^{-2} measured today (Pollack & Chapman 1977), but may be typical for young continental crust. The second characteristic feature, the P–T path during rifting, is expressed in near-isothermal decompression followed by cooling. This leads to an increase in the surface heat flow (Fig. 5), yet it does not require that the rocks themselves were heated during the initial stages of rifting.

It is not possible to construct a single P–T path of the entire Ivrea crustal section because

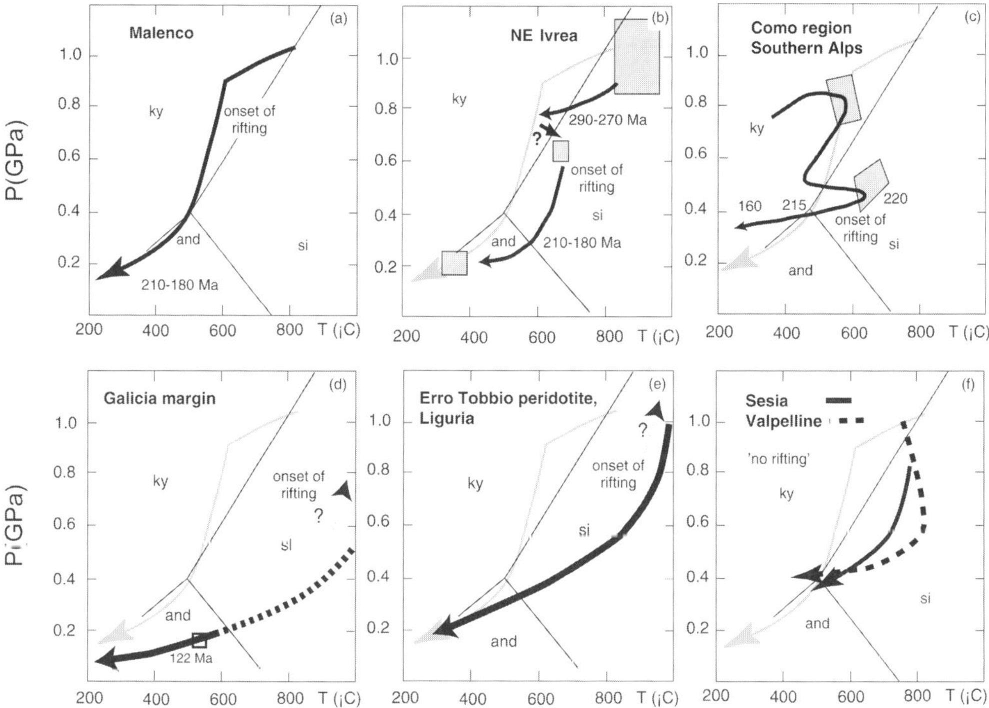

Fig. 9. P–T–t paths from lower-crustal and upper-mantle rocks from the Alps and the West Iberia margin. (**a**) Malenco (Müntener et al. 2000); (**b**) northeastern Ivrea (Handy et al. 1999); (**c**) South Alpine (Sanders et al. 1996); (**d**) Galicia margin (after Boillot et al. (1995a) and Beslier et al. (1996)); (**e**) Erro Tobbio peridotite (Hoogerduijn Strating et al. 1993); (**f**) Valpelline series (Gardien et al. 1994) and Sesia (Lardeaux & Spalla 1991). The grey arrow in (**b**)–(**f**) refers to the pre-Alpine P–T path from Malenco (Fig. 5). Aluminosilicate stability fields are from Holdaway (1971).

the Ivrea zone exposes different levels of the crust. From the numerous $P-T-t$ data published in the literature (e.g. Handy 1987; Zingg *et al.* 1990; Brodie 1995; Henk *et al.* 1997; Handy *et al.* 1999; Barboza & Bergantz 2000) we refer here to the recent $P-T-t$ compilation of Handy *et al.* (1999) that combines structural and petrological data. It must be emphasized that the $P-T-t$ path in Figure 9b is valid only for the northeastern part of the Ivrea zone where rifting effects are most pronounced and suitable for a comparison with the Malenco rocks. The compilation of the Ivrea zone reveals the following major stages: (1) intrusion of mafic rocks (type 2 and type 3 mafic rocks of Zingg *et al.* (1990)) and granulite-facies regional metamorphism (320–300 Ma); (2) decompression and intrusion of gabbro diorite (type 4 mafic rocks of Zingg *et al.* (1990)) (285 Ma) and local granulite-facies metamorphism; (3) near-isobaric cooling; (4) hydrothermal activity, alkaline magmatism (210 ± 12 Ma); (5) rift-related exhumation (210–180 Ma). The structural features found in the Ivrea zone may be correlated with this general metamorphic evolution, to distinguish between rift-related and pre-rift structures. The difference between high-temperature shear zones (T >650 °C) and later more localized deformation (T <650 °C) in the Ivrea zone was discussed by Brodie & Rutter (1987), Zingg *et al.* (1990) and Handy & Zingg (1991). The latter workers preferred the interpretation that they belong to two independent tectonic settings although the extension direction was similar and radiometric age determinations yielded ambiguous results. Most of the high-temperature shear zones probably formed under relatively'dry'conditions ('dry' in the sense of very limited external fluid input). Such anastomosing shear zones were responsible for late Carboniferous to early Permian crustal thinning before Jurassic rifting, presumably during and after underplating of mafic magmas (Brodie & Rutter 1987; Handy & Zingg 1991; Handy & Streit 1999; Snoke *et al.* 1999). Estimates of late Carboniferous to early Permian crustal thinning vary between 2 and 5 km, based on thermal modelling (Brodie & Rutter 1987) and analysis of radiometric cooling ages (Handy 1987), and 6–12 km, based on thermobarometric estimates (Brodie 1995) in the Ivrea zone.

Rift-related structures in the northeastern part of the Ivrea zone record a slightly different evolution from that of the Malenco rocks. Stünitz (1998) described two types of high-temperature shear zones, one with retrograde parageneses, and another that experienced initially prograde then retrograde conditions. Handy *et al.* (1999) proposed heating at the onset of rifting (Fig. 9a) based on anhydrous symplectitic intergrowth of plagioclase, orthopyroxene and spinel in metabasic rocks that are structurally contiguous with the Pogallo line in the narrow, northeastern part of the Ivrea zone. These data indicate upper amphibolite- to granulite-facies conditions at somewhat lower pressures (*c.* 0.6 GPa) than in Val Malenco at the onset of rifting (Fig. 9b).

Unambiguous mineral growth related to rifting is documented in synkinematic, retrograde amphibolite- to greenschist-facies parageneses in the Pogallo mylonite zone (Handy 1987). The metamorphic conditions indicate that the initial temperature of rifting did not exceed 650 °C (compilation of $P-T$ data, Handy *et al.* 1999). $^{40}Ar/^{39}Ar$ analyses of retrograde hornblende (Brodie & Rutter 1985) in mafic mylonites yielded ages of 210–215 Ma (Brodie *et al.* 1989) that are similar to $^{40}Ar/^{39}Ar$ ages of pargasite obtained from Malenco (Villa *et al.* 2000). Late Triassic to Early Jurassic ages most probably indicate the onset of the rift-related Pogallo phase (Handy *et al.* 1999). From all these data we conclude that most structures formed at high temperature (> 650 °C) in the lower crust and subcontinental upper mantle are probably older and not related to Jurassic rifting.

A different scenario with a different $P-T$ path for Early Jurassic rifting was published by Sanders *et al.* (1996) for mid-crustal rocks of the Southern Alps (Figs 1 and 9c) indicating a heating event at the onset of rifting. Bertotti *et al.* (1999) reached similar conclusions for rocks exposed along the Lugano–M. Grona line (Fig. 1). Through combined $^{40}Ar/^{39}Ar$, Rb/Sr and fission-track dating, Bertotti *et al.* (1999) argued that the thermal evolution of the South Alpine margin during rifting is mainly controlled by relaxation of isotherms that were assumed to be elevated because of inferred large mafic intrusions in the lower crust in mid- to late Triassic time (Bertotti & ter Voorde 1994). Thermal gradients decreased from >60 °C km^{-1} at the initial stage of rifting to *c.* 20 °C km^{-1} after breakup. This decrease in the thermal gradient obtained by Bertotti *et al.* (1999) is opposite to what is preserved in Malenco and the Ivrea zone (Fig. 9a,b), where the thermal gradient increases because of near-isothermal decompression during rifting. Although there is field and isotopic evidence to support limited igneous activity during mid- and late Triassic time (e.g. Ferrara & Innocenti 1974; Stähle *et al.* 1990; Lu *et al.* 1997), the modelled regional-scale heating in late Triassic time (Bertotti & ter Voorde

1994) is not supported by the work of Vavra *et al.* (1999) and Vavra & Schaltegger (1999). These workers argue that alkaline magmatism is limited and that widespread hydrothermal activity at 210 ± 12 Ma caused alteration of zircons in Ivrea zone granulites. Such hydrothermal activity may well be associated with the onset of rifting, and we propose that fluid accessibility is a major feature of early rifting in the deep crust. There may be spatial and temporal differences in how the deep crust and upper mantle respond to rifting, but we favour the hypothesis that regional-scale heating was not important at the onset of rifting. We suggest that granulite-facies conditions in the lower crust are related to Early Permian decompression and associated magmatism, which was then followed by a period of near-isobaric cooling at all crustal levels.

The $P-T$ paths and their relation to rifting are especially important for unravelling the structural record of denuded subcontinental mantle where no lower-crustal rocks are attached. Microstructural analysis of exposed peridotites at the present-day West Iberian margin (e.g. Boillot *et al.* 1995*a*; Beslier *et al.* 1996) and ancient Adria (Erro Tobbio peridotite, Vissers *et al.* 1991, 1995; Hoogerduijn Strating *et al.* 1993) led to the hypothesis that both high-temperature and low-temperature structures were associated with rifting. This led to high-temperature, low-pressure $P-T$ paths for these rocks (Fig. 9d,e) because in both cases plagioclase peridotites are preserved. A likely alternative is that only the latest ('hydrous?') part of the structural evolution is associated with rifting and that the anhydrous high-temperature evolution (i.e. $>750\,°\mathrm{C}$) predates rifting, as it is the case in Malenco and the Ivrea zone. This hypothesis would imply that plagioclase peridotites are not necessarily related to rifting and that their role has to be reassessed. It remains to be tested if the plagioclase peridotites in the Alps (e.g. Lanzo) and in the West Iberian margin indeed represent processes related to rifting, or, more likely in our opinion, that they record infiltration processes into the subcontinental mantle. The lack of a genetic link between oceanic magmatism and underlying mantle rocks has been demonstrated by Rampone *et al.* (1995) for the Ligurian ophiolites and by Müntener & Hermann (1996) for the Malenco peridotites, and is also the case in the Platta ophiolites (Müntener *et al.*, unpubl. data).

Metagabbros and pelitic granulites that were exhumed together with subcontinental mantle may help us to better understand the signifi-cance of granulite-facies rocks along passive continental margins. Granulites are often considered to represent lower crust and their presence (or absence) has been used as argument to favour different rifting models (e.g. Lemoine *et al.* 1987; Stampfli & Marthaler 1990; Froitzheim & Manatschal 1996; Froitzheim *et al.* 1996). The $P-T$ estimates for the Malenco granulites and the Ivrea zone at the onset of rifting correspond to amphibolite-facies conditions. These rocks are only granulites because of high-temperature metamorphism and mafic underplating *c.* 50 Ma before rifting. However, in areas without mafic underplating, the lower-most continental crust may consist of amphibolite facies rocks that are hard to distinguish from upper-crustal basement rocks. On the other hand, as pointed out by Harley (1989) and Rudnick & Fountain (1995), not all exhumed granulites represent the lowermost continental crust. For example, zircon fission-track ages of granulites from the continent–ocean transition of the Galicia margin indicate that these rocks were already at upper-crustal levels before rifting (Fügenschuh *et al.* 1998) and do not represent the Mesozoic pre-rift lower crust. Granulite-facies rocks from the Valpelline series and the Sesia zone in the Western Alps (Fig. 9f) show a $P-T$ path with near-isothermal decompression at much higher temperature than the Malenco granulites. For this reason, Gardien *et al.* (1994) favoured a late Variscan formation of the Valpelline granulites as a result of post-collisional thinning, and consequently exhumation of these granulites to mid-crustal levels is not related to Jurassic passive rifting. These examples demonstrate that the presence of granulite-facies rocks next to passive margin sediments is not proof of exhumation of the lowermost crust during rifting, even though the Sesia and Valpelline rocks are in an Alpine tectonic position similar to the Malenco granulites. On the other hand, the metamorphic conditions in the lower crust at the onset of rifting reach only amphibolite-facies conditions. It is thus possible that parts of the lower crust are gneisses and amphibolites, which are not *a priori* recognizable as lower crust once they are exhumed during passive rifting.

A simple model for the formation of passive continental margins

There has been a rapid shift in thinking about the formation of passive continental margins from essentially one-phase models (pure shear:

e.g. McKenzie 1978; Le Pichon & Sibuet 1981; simple shear: e.g. Wernicke 1985; Lister *et al.* 1986; first applied to the Alps by Lemoine *et al.* 1987) to more complex two-phase models, which may include two simple shear systems separated in time (Hermann & Müntener 1996), simple shear systems with spatially different dip along Tethyan passive margins (Stampfli *et al.* 1991; Favre & Stampfli 1992), or early pure shear followed by later simple shear along low-angle detachment faults with a top-to-the ocean sense of movement (Handy & Zingg 1991; Froitzheim & Manatschal 1996; Handy 1996; Manatschal & Bernoulli 1999). An important point is that kinematic inversions of seismic sections of the present-day Galicia margin (Manatschal *et al.* 2001) and reconstructions from the ancient Tethyan margins (e.g. Manatschal & Nievergelt 1997) show that the crust within the distal margin was already considerably thinned before detachment faulting became active. This indicates that detachment faulting is a late event in the formation of a passive continental margin, which can explain only a part of the thinning of the continental crust. Thus, most of the current debate considers the early rifting phase and the question of how lower crust and upper mantle are exhumed to shallow crustal levels, so that they can be denuded during late detachment faulting.

The problem in the Alps is that the lower crust is exposed only in small parts of the Adria margin and not at all in the opposing European margin. Therefore most rifting models to date employ the polarity of facies distributions and sediment thickness in asymmetrical rifts to infer the dip directions of detachment faults at depth. For example, Schmid *et al.* (1997) invoked an earlier west- and a later east-dipping detachment fault to explain the lack of lower crust in the Briançonnais (European) margin. Yet, the continentward dipping Margna and Pogallo shear zones are the only geological evidence for large extensional shear zones in the lower crust along the ancient Tethyan passive margins. On the basis of the available geological evidence we propose that the scarcity of lower crust can be explained by the following simple model (Fig. 10).

During an early rifting phase, stretching is accommodated by non-uniform large shear zones, which are probably symmetrically distributed on the scale of the whole rifted area, but are asymmetrical on the 10–100 km scale (Fig. 10a). Seismic and field evidence (e.g. Bertotti *et al.* 1993) from sedimentary basins indicates that shear zones in the upper crust have a listric geometry and flatten at mid-crustal levels (at *c.* 10–15 km depth), but are symmetrically distributed over the whole rifted area. On the other hand, shear zones in the deep crust and upper mantle uplifted lower-crustal and subcontinental mantle rocks to a depth of *c.* 10–15 km. Although such a model could reconcile observations from the top 5 km of the crust with those from the lower crust, several important consequences must be addressed. (1) The site of major shear zones in the lower crust and upper mantle is laterally displaced from major rifting at the surface. In this case, there must be a 'weak horizon' in the middle crust, where the rheology of the crust changes significantly. This indicates that the upper crust may be decoupled from the lower crust and upper mantle by a weak mid-crustal layer. Such a weak layer might correspond to quartz-rich rocks, presumably at temperatures higher than 300 °C (Handy 1987), around the brittle to plastic transition of quartz deformation. (2) The lower crust is considerably stronger than commonly assumed in many modelling studies that used quartz flow laws as an analogue for the lower crust (e.g. Dunbar & Sawyer 1989). In our model (Fig. 10a) the lower crust is 'boudinaged' if it is assumed that such large shear zones occurred on both future margins and a 'decoupling horizon' was present in the middle crust.

During an advanced stage of rifting, extension was dominated by simple shear along low-angle detachments faults with a top-to-the-ocean sense of movement as documented and discussed by Froitzheim & Manatschal (1996) and Manatschal & Nievergelt (1997). The scarcity of lower crust along non-volcanic passive margins can be explained in three different ways: (1) the lower crust was exhumed close to the surface but remains covered by later emplaced blocks of upper continental crust (Manatschal & Bernoulli 1999); (2) the lower crust was not exhumed because detachment faults generally occurred in places where the lower crust was previously excised by localized shear zones of early rifting (lithospheric boudinage; see traces of future detachments in Fig. 10b); (3) the lower crust consists of amphibolites and gneisses and is therefore not easily distinguished from the mid- or upper crust unless detailed cooling ages combined with microstructures and thermobarometry are available. Although this is highly debated we favour the second hypothesis, because the geometry depicted in Figure 10 could explain the observation that subcontinental mantle is exhumed, but exhumation of the pre-rift lower crust on the ocean floor is extremely scarce and is so far demonstrated for only the Malenco area.

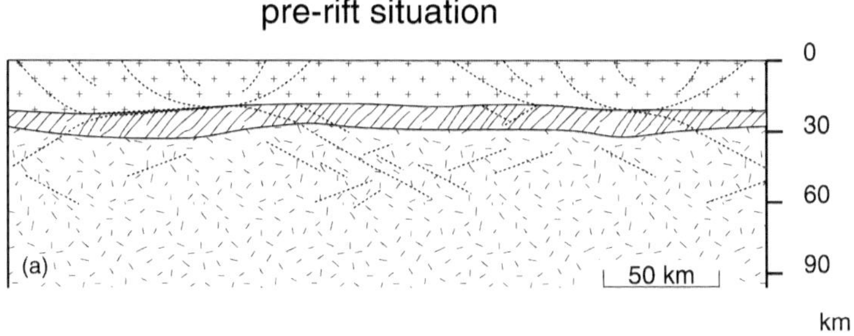

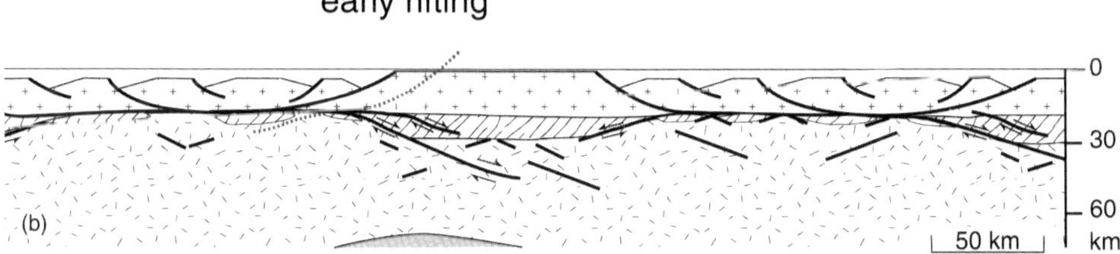

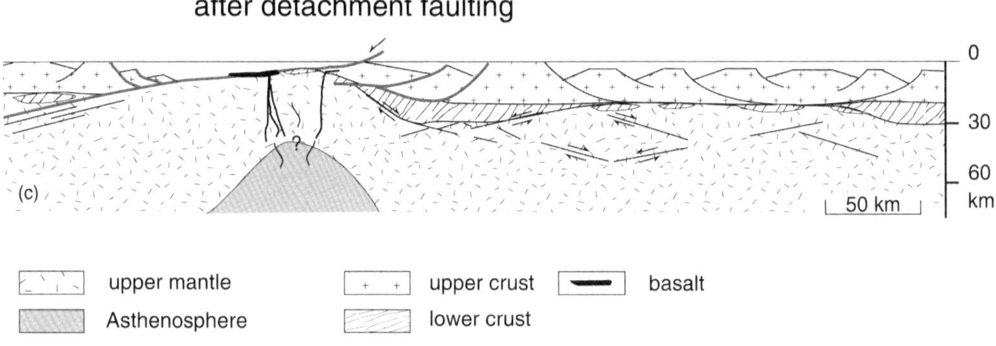

Fig. 10. A simple model for the formation of non-volcanic passive continental margins, using a refined version of a model developed by Manatschal (pers. comm.). This model combines data from the lower crust and upper mantle (e.g. Hermann & Müntener 1996) with data from the surface (e.g. Manatschal & Bernoulli 1999). (**a**) Pre-rift situation; with lateral variations in lower-crustal thickness. Future faults are indicated with dotted lines. (**b**) During early rifting, the major site of lower-crustal and upper-mantle deformation is laterally displaced from the formation of basins in the upper crust. The lower crust is boudinaged and exhumed with lithospheric mantle to shallow crustal levels of 10–15 km. (**c**) During advanced rifting, detachment faulting cuts from the upper crust into the upper mantle, which finally leads to exhumation of upper mantle and, in rare cases, of lower crust. It should be noted that the spatial distribution of detachment faults largely controls the occurrence of lower crust at the continent–ocean transition of passive margins. Early formed mafic rocks intrude distal continental rocks and subcontinental mantle, and have a chemical composition corresponding to T-MORB (Puschnig 1998).

Implications for lower crust and upper mantle during rifting

(1) The combination of field observations, microstructural studies and thermobarometric data demonstrates that deformation in the Mal-enco complex occurred in two styles: (a) high-temperature shearing ($T > 650\,°C$) with partial to complete annealing of the microstructures is related to pre-rift, Permian extension or, in the case of the mantle peridotite, to even older high-temperature events; (b) localized myloniti-zation with drastic grain-size reduction at tem-

peratures <600 °C occurred during Early Jurassic rifting. Microstructural observations in the lower-crustal and upper-mantle rocks in the Malenco area and the Ivrea zone indicate that the combination of reaction-induced softening and hydration was the principal mechanism for strain localization and grain-size reduction (see also Handy 1989). Grain-size reduction was most effective in polymineralic rocks where syntectonic breakdown of anhydrous minerals (olivine, pyroxenes, spinel and plagioclase) forms weaker, finer-grained and mostly hydrous phases (amphibole, chlorite, clinozoisite, white mica).

(2) Results from phase equilibria in lower crust of the Malenco and the Ivrea zone indicate that the initial temperature of rifting did not exceed 650 °C. The combination of external fluid input and grain-size reduction probably resulted in a change in the dominant deformation mechanism from dislocation creep to grain-size sensitive diffusion creep (e.g. Handy 1989; Vissers *et al.* 1995; Jaroslow *et al.* 1996). Consequently, we suggest that the rheology of the lower crust and upper mantle during early rifting is largely controlled by fine-grained 'hydrous' shear zones. Such shear zones may promote the development of large shear zones (Ivrea zone: Handy & Zingg 1991; Malenco: Hermann & Müntener 1996) in the lower crust and upper mantle.

(3) Hydration at lower-crustal depth and the observed microstructures indicate that noncoaxial shearing dominated in the lower crust and upper mantle during early rifting. This suggests that localized shear zones such as the Pogallo mylonite zone and the Margna shear zone accommodated substantial displacement during the early stages of rifting. These intracrustal major shear zones indicate a top-to-the Adria-continent sense of shear during the early phases of rifting, excised 15–20 km of mainly middle and lower crust, and are largely responsible for the uplift of lower crust and upper mantle to shallow depths of *c.* 10–15 km. These observations are consistent with models for passive rifting that incorporate a relatively weak, quartz-rich (and possibly hydrous) middle crust (Handy 1987) that decouples a strong lower crust–upper mantle from a brittle upper crust.

(4) To investigate the role of granulites and subcontinental mantle during rifting, it is crucial to understand which structures and which part of a *P–T* path are related to rifting. The compilation of *P–T* data and rift-related structures in the deep crust and upper continental mantle from the Alps suggests that peridotites may often preserve a high-temperature structural evolution that is not related to rifting. Amphibolite-facies rocks are stable at the base of the continental crust with an average heat flow of *c.* 60 mW m^{-2}, and granulites only partially represent pre-rift lower continental crust. In general, granulites that formed the pre-rift lower crust show near-isothermal decompression at lower temperatures than granulites that were exhumed to mid-crustal levels at the end of an orogenic cycle. Thus pre-rift lower crust is not necessarily granulitic, and exhumed granulites along passive continental margins may not represent the pre-rift lower crust.

This research was supported by Swiss NSF grant 2000-037388.93/1 to V. Trommsdorff. This paper benefited from numerous discussions with G. Manatschal, A. Puschnig, L. Desmurs, D. Bernoulli and V. Trommsdorff. N. Froitzheim, M. Seyler and (especially) M. R. Handy provided very helpful reviews. The authors are grateful to M. R. Handy for sharing his knowledge of the Ivrea zone, and for many helpful comments and ideas, which greatly improved the presentation. R.B. Whitmarsh and N. Froitzheim are thanked for their invitation to present this work at the 1999 meeting at the Geological Society, London.

References

ANSORGE, J., BLUNDELL, D. & MUELLER, S. 1992. Europe's lithosphere—seismic structure. *In*: BLUNDELL, D., FREEMAN, R. & MUELLER, S. (eds) *A Continent Revealed: The European Geotraverse*. Cambridge University Press, Cambridge.

BARBOZA, S.A. & BERGANTZ, G.W. 2000. Metamorphism and anatexis in the mafic complex contact aureole, Ivrea zone, Northern Italy. *Journal of Petrology*, **41**, 1307–1327.

BERNOULLI, D., BERTOTTI, G. & FROITZHEIM, N. 1990. Mesozoic faults and associated sediments in the Austroalpine–south Alpine continental margin. *Memorie della Società Geologica Italiana*, **45**, 25–38.

BERTOTTI, G. & TER VOORDE, M. 1994. Thermal effects of normal faulting during rifted basin formation, 2. The Lugano–Val Grande normal fault and the role of pre-existing thermal anomalies. *Tectonophysics*, **240**, 145–157.

BERTOTTI, G., PICOTTI, V., BERNOULLI, D. & CASTELLARIN, A. 1993. From rifting to drifting: tectonic evolution of the South-Alpine upper crust from the Triassic to the early Cretaceous. *Sedimentary Geology*, **86**, 53–76.

BERTOTTI, G., SEWARD, D., WIJBRANS, J., TER VOORDE, M. & HURFORD, A.J. 1999. Crustal thermal regime prior to, during, and after rifting: a geochronological and modeling study of the Mesozoic South Alpine rifted margin. *Tectonics*, **18**, 185–200.

BESLIER, M.-O., CORNEN, G. & GIRARDEAU, J. 1996. Tectono-metamorphic evolution of peridotites from the ocean/continent transition of the Iberia Abyssal Plain margin. *In*: WHITMARSH, R.B., SAWYER, D.S., KLAUS, A. & MASSON, D.G. (eds) *Proceedings of the Ocean Drilling Program, Scientific Results, 149*. Ocean Drilling Program, College Station, TX, 397–412.

BISSIG, T. & HERMANN, J. 1999. From pre-Alpine extension to Alpine convergence: the example of the southwestern margin of the Margna nappe (Val Malenco, N-Italy). *Schweizerische Mineralogische und Petrographische Mitteilungen*, **79**, 363–380.

BOILLOT, G., BESLIER, M.-O. & GIRARDEAU, J. 1995*a*. Nature, structure and evolution of the ocean–continent boundary: the lesson of the west Galicia margin (Spain). *In*: BANDA, E., TALWANI, M. & TORNE, M. (eds) *Rifted Ocean–Continent Boundaries*. NATO Advanced Study Institute Series, Series C, **463**, 219–229.

BOILLOT, G., BESLIER, M.-O., KRAWCZYK, C.M., RAPPIN, D. & RESTON, T.J. 1995*b*. The formation of passive margins: constraints from the crustal structure and segmentation of the deep Galicia margin, Spain. *In*: SCRUTTON, R.A., STOKER, M.S., SHIMMIELD, G.B. & TUDHOPE, A.W. (eds) *The Tectonics, Sedimentation and Palaeoceanography of the North Atlantic Region*. Geological Society, London, Special Publications, **90**, 71–91.

BORSI, L., SCHÄRER, U., GAGGERO, L. & CRISPINI, L. 1996. Age, origin and geodynamic significance of plagiogranites in the lherzolites and gabbros of the Piedmont–Ligurian ocean basin. *Earth and Planetary Science Letters*, **140**, 227–241.

BRODIE, K.H. 1995. The development of orientated symplectites during deformation. *Journal of Metamorphic Geology*, **13**, 499–508.

BRODIE, K.H. & RUTTER, E.H. 1985. On the relationship between deformation and metamorphism with special reference to the behaviour of basic rocks. *In*: THOMPSON, A.B. & RUBIE, D.C. (eds) *Metamorphic Reactions: Kinetics, Textures and Deformation. Advances in Physical Geochemistry 4*. Springer, Berlin, 138–179.

BRODIE, K.H. & RUTTER, E.H. 1987. Deep crustal extensional faulting in the Ivrea Zone of Northern Italy. *Tectonophysics*, **140**, 193–212.

BRODIE, K.H., REX, D. & RUTTER, E.H. 1989. On the age of deep crustal extensional faulting in the Ivrea zone, northern Italy. *In*: COWARD, M.P., DIETRICH, D. & PARK, R.G. (eds) *Alpine Tectonics*. Geological Society, London, Special Publications, **45**, 203–210.

CHAPMAN, D.S. 1986. Thermal gradients in the continental crust. *In*: DAWSON, J.B., CARSWELL, D.A., HALL, J. & WEDEPOHL, K.H. (eds) *The Nature of the Lower Continental Crust*. Geological Society, London, Special Publications, **24**, 63–70.

CUREWITZ, D. & KARSON, J.K. 1999. Ultracataclasis, sintering and frictional melting in pseudotachylites from East Greenland. *Journal of Structural Geology*, **21**, 1693–1713.

DESMURS, L., MANATSCHAL, G. & BERNOULLI, D. 2001. The Steinmann Trinity revisited: mantle exhumation and magmatism along an ocean–continent transition: the Platta nappe, eastern Switzerland. *In*: WILSON, R.C.L., WHITMARSH, R.B., TAYLOR, B. & FROITZHEIM, N. (eds) *Non-volcanic Rifting of Continental Margins: a Comparison of Evidence from Land and Sea*. Geological Society, London, Special Publications, **187**, 235–266.

DUNBAR, J.A. & SAWYER, D.S. 1989. How pre-existing weaknesses control the style of continental breakup. *Journal of Geophysical Research*, **90**, 3551–3557.

FAVRE, P. & STAMPFLI, G.M. 1992. From rifting to passive margin: the examples of the Red Sea, Central Atlantic and Alpine Tethys. *Tectonophysics*, **215**, 69–97.

FERRARA, G. & INNOCENTI, F. 1974. Radiometric age evidence of a Triassic thermal event in the Southern Alps. *Geologische Rundschau*, **63**, 572–581.

FROITZHEIM, N. & EBERLI, G.P. 1990. Extensional detachment faulting in the evolution of a Tethys passive continental margin, Eastern Alps, Switzerland. *Geological Society of America Bulletin*, **102**, 1297–1308.

FROITZHEIM, N. & MANATSCHAL, G. 1996. Kinematics of Jurassic rifting, mantle exhumation, and passive-margin formation in the Austroalpine and Penninic nappes (eastern Switzerland). *Geological Society of America Bulletin*, **108**, 1120–1133.

FROITZHEIM, N., SCHMID, S.M. & CONTI, P. 1994. Repeated change from crustal shortening to orogen-parallel extension in the Austroalpine units of Graubünden. *Eclogae Geologicae Helvetiae*, **87**, 559–612.

FROITZHEIM, N., SCHMID, S.M. & FREY, M. 1996. Mesozoic paleogeography and the timing of eclogite-facies metamorphism in the Alps: a working hypothesis. *Eclogae Geologicae Helvetiae*, **89**, 81–110.

FÜGENSCHUH, B., FROITZHEIM, N. & BOILLOT, G. 1998. Cooling history of granulite samples from the ocean–continent transition of the Galicia margin: implications for rifting. *Terra Nova*, **10**, 96–100.

GARDIEN, V., REUSSER, E. & MARQUER, D. 1994. Pre-Alpine metamorphic evolution of the gneisses from the Valpelline series (Western Alps, Italy). *Schweizerische Mineralogische und Petrographische Mitteilungen*, **74**, 489–502.

HANDY, M.R. 1987. The structure, age and kinematics of the Pogallo Fault Zone, a deep crustal dislocation in the Southern Alps, northwestern Italy. *Eclogae Geologicae Helvetiae*, **80**, 593–632.

HANDY, M.R. 1989. Deformation regimes and the rheological evolution of fault zones in the lithosphere: the effects of pressure, temperature, grain size and time. *Tectonophysics*, **163**, 119–152.

HANDY, M.R. 1996. The transition from passive to active margin tectonics: a case study from the Zone of Samedan (eastern Switzerland). *Geologische Rundschau*, **85**, 832–851.

HANDY, M.R. & STREIT, J.E. 1999. Mechanics and mechanisms of magmatic underplating: inferences from mafic veins in deep crustal mylonite. *Earth and Planetary Science Letters*, **165**, 271–286.

HANDY, M.R. & ZINGG, A. 1991. The tectonic and rheological evolution of an attenuated cross section of the continental crust: Ivrea crustal section, southern Alps, northwestern Italy and southern Switzerland. *Geological Society of America Bulletin*, **103**, 236–253.

HANDY, M.R., FRANZ, L., HELLER, F., JANOTT, B. & ZURBRIGGEN, R. 1999. Multistage accretion and exhumation of the continental crust (Ivrea crustal section, Italy and Switzerland). *Tectonics*, **18**, 1154–1177.

HANDY, M.R., HERWEGH, M., KAMBER, B., TIETZ, R. & VILLA, I. 1996. Geochronological, petrological, and kinematic constrains on the evolution of the Err–Platta boundary, part of a fossil continent–ocean suture in the Alps (eastern Switzerland). *Schweizerische Mineralogische und Petrographische Mitteilungen*, **76**, 453–473.

HANDY, M.R., HERWEGH, M. & REGLI, R. 1993. Tektonische Entwicklung der westlichen Zone von Samedan (Oberhalbstein, Graubünden, Schweiz). *Eclogae Geologicae Helvetiae*, **86**, 785–818.

HANSMANN, W., HERMANN, J. & MÜNTENER, O. 1996. U–Pb-Datierung des Fedozer Gabbros, einer Intrusion an der Krusten Mantel Grenze. *Schweizerische Mineralogische und Petrographische Mitteilungen*, **76**, 116–117.

HARLEY, S.L. 1989. The origin of granulites: a metamorphic perspective. *Geological Magazine*, **126**, 215–247.

HENK, A., FRANZ, L., TEUFEL, S. & ONCKEN, O. 1997. Magmatic underplating, extension, and crustal equilibration: insights from a cross-section through the Ivrea zone and Strona–Ccncri zone, northern Italy. *Journal of Geology*, **105**, 367–377.

HERMANN, J. & MÜNTENER, O. 1992. Strukturelle Entwicklung im Grenzbereich zwischen dem penninischen Malenco-Ultramafitit und dem Unterostalpin (Margna- und Sella-Decke). *Schweizerische Mineralogische und Petrographische Mitteilungen*, **72**, 225–240.

HERMANN, J. & MÜNTENER, O. 1996. Extension-related structures in the Malenco–Margna-system: implications for paleogeography and consequences for rifting and Alpine tectonics. *Schweizerische Mineralogische und Petrographische Mitteilungen*, **76**, 501–519.

HERMANN, J., MÜNTENER, O., TROMMSDORFF, V., HANSMANN, W. & PICCARDO, G.B. 1997. Fossil crust-to-mantle transition, Val Malenco (Italian Alps). *Journal of Geophysical Research*, **102**, 20123–20132.

HOLDAWAY, M.J. 1971. Stability of andalusite and the aluminium silicate phase diagram. *American Journal of Science*, **271**, 97–131.

HOOGERDUIJN STRATING, E.H., RAMPONE, E., PICCARDO, G.B., DRURY, M.R. & VISSERS, R.L.M. 1993. Subsolidus emplacement of mantle peridotites during incipient oceanic rifting and opening of the Mesozoic Tethys (Voltri massif, NW Italy). *Journal of Petrology*, **34**, 901–927.

HOPPER, J.H. & BUCK, W.R. 1996. The effect of lower crustal flow on continental extension and passive margin formation. *Journal of Geophysical Research*, **101**, 20175–20194.

JAROSLOW, G.E., HIRTH, G. & DICK, H.J.B. 1996. Abyssal peridotite mylonites: implications for grain-size sensitive flow and strain localization in the oceanic lithosphere. *Tectonophysics*, **256**, 17–37.

KARATO, S.I., PATERSON, M.S. & FITZGERALD, J.D. 1986. Rheology of synthetic olivine aggregates: influence of grain size and water. *Journal of Geophysical Research*, **91**, 8151–8176.

LARDEAUX, J.M. & SPALLA, M.I. 1991. From granulites to eclogites in the Sesia zone (Italian Western Alps): a record of opening and closure of the Piemont ocean. *Journal of Metamorphic Geology*, **9**, 35–59.

LEMOINE, M., TRICART, P. & BOILLOT, G. 1987. Ultramafic and gabbroic ocean floor of the Ligurian Tethys (Alps, Corsica, Apennines). In search of a genetic model. *Geology*, **15**, 622–625.

LE PICHON, X. & SIBUET, J.C. 1981. Passive margins: a model of formation. *Journal of Geophysical Research*, **86**, 3708–3720.

LINIGER, M. 1992. *Der ostalpin penninische Grenzbereich im Gebiet der nördlichen Margna-Decke (Graubünden, Schweiz).* PhD thesis, ETH Zürich.

LISTER, G.S., ETHERIDGE, M.A. & SYMONDS, P.A. 1986. Detachment faulting and the evolution of passive continental margins. *Geology*, **14**, 246–250.

LU, M., HOFMANN, A.W., MAZZUCCELLI, M. & RIVALENTI, G. 1997. The mafic–ultramafic complex near Finero (Ivrea–Verbano Zone), II. Geochronology and isotope geochemistry. *Chemical Geology*, **140**, 223–235.

MANATSCHAL, G. & BERNOULLI, D. 1999. Architecture and tectonic evolution of non-volcanic margins: present-day Galicia and ancient Adria. *Tectonics*, **18**, 1099–1119.

MANATSCHAL, G. & NIEVERGELT, P. 1997. A continent–ocean transition recorded in the Err and Platta nappes (eastern Switzerland). *Eclogae Geologicae Helvetiae*, **90**, 3–27.

MANATSCHAL, G., FROITZHEIM, N., RUBENACH, M. & TURRIN, B.D. 2001. The role of detachment faulting in the formation of an ocean–continent transition: insights from the Iberia Abyssal Plain. *In*: WILSON, R.C.L., WHITMARSH, R.B.,

TAYLOR, B. & FROITZHEIM, N. (eds) *Non-volcanic Rifting of Continental Margins: a Comparison of Evidence from Land and Sea*. Geological Society, London, Special Publications, **187**, 405–428.

MCKENZIE, D.P. 1978. Some remarks on the development of sedimentary basins. *Earth and Planetary Science Letters*, **40**, 25–32.

MÜNTENER, O. 1997. *The Malenco peridotites (Alps): petrology and geochemistry of subcontinental mantle and Jurassic exhumation during rifting*. PhD thesis, ETH-Zürich.

MÜNTENER, O. & HERMANN, J. 1996. The Val Malenco lower crust–upper mantle complex and its field relations (Italian Alps). *Schweizerische Mineralogische und Petrographische Mitteilungen*, **76**, 475–500.

MÜNTENER, O., HERMANN, J. & TROMMSDORFF, V. 2000. Cooling history and exhumation of lower-crustal granulites and upper mantle (Malenco, Eastern Central Alps). *Journal of Petrology*, **41**, 175–200.

NICOLAS, A. *Structures of Ophiolites and Dynamics of the Oceanic Lithosphere*. Kluwer, Dordrecht.

OLGAARD, D.L. 1990. The role of second phase in localizing deformation. *In*: KNIPE, R.J. & RUTTER, E.H. (eds) *Deformation Mechanisms, Rheology and Tectonics*. Geological Society, London, Special Publications, **54**, 175–181.

PETERS, Tj. & STETTLER, A. 1987. Radiometric age, thermobarometry and mode of emplacement of the Totalp peridotite in the Eastern Swiss Alps. *Schweizerische Mineralogische und Petrographische Mitteilungen*, **67**, 285–294.

POLLACK, H.N. & CHAPMAN, D.S. 1977. On the regional variation of the heat flow, geotherms and lithosphere thickness. *Tectonophysics*, **38**, 279–296.

PUSCHNIG, A.R. 1998. *The Forno unit (Rhetic Alps): evolution of an ocean floor sequence from rifting to Alpine orogeny*. PhD thesis, ETH Zürich.

RAMPONE, E., HOFMANN, A.W., PICCARDO, G.B., VANUCCI, R., BOTTAZZI, P. & OTTOLINI, L. 1995. Petrology, mineral and isotope geochemistry of the External Liguride peridotites (Northern Appennines, Italy). *Journal of Petrology*, **36**, 81–105.

RESTON, T.J., KRAWCZYK, C.M. & KLAESCHEN, D. 1996. The S reflector west of Galicia (Spain): evidence from prestack depth migration for detachment faulting during continental breakup. *Journal of Geophysical Research*, **101**, 8075–8091.

RUBATTO, D., GEBAUER, D. & FANNING, M. 1998. Jurassic formation and Eocene subduction of the Zermatt–Saas–Fee ophiolites: implications for the geodynamic evolution of the Central and Western Alps. *Contributions to Mineralogy and Petrology*, **132**, 269–287.

RUBIE, D.C. 1983. Reaction enhanced ductility: the role of solid–solid univariant reactions in deformation of the crust and mantle. *Tectonophysics*, **96**, 331–352.

RUDNICK, R.L. & FOUNTAIN, D.M. 1996. Nature and composition of the continental crust: a lower crustal perspective. *Reviews in Geophysics*, **33**, 267–309.

SANDERS, C.A.E., BERTOTTI, G., TOMMASINI, S., DAVIES, G.R. & WIJBRANS, J.R. 1996. Triassic pegmatites in the Mesozoic middle crust of the southern Alps (Italy): fluid inclusions, radiometric dating and tectonic implications. *Eclogae Geologicae Helvetiae*, **89**, 505–525.

SCHMID, S.M., PFIFFNER, O.A. & SCHREURS, G. 1997. Rifting and collision in the Penninic zone of eastern Switzerland. *In*: PFIFFNER, O.A., LEHNER, P., HEITZMANN, P., MÜLLER, S.T. & STECK, A. (eds) *Deep Structure of the Swiss Alps: Results of NFP 20*. Birkhäuser, Basel, 160–185.

SCHMID, S.M., RÜCK, P. & SCHREURS, G. 1990. The significance of the Schams nappes for the reconstruction of the paleotectonic and orogenic evolution of the Penninic zone along the NFP-20 East traverse (Grisons, eastern Switzerland). *Mémoirs de la Société Géologique de la France*, **156**, 263–287.

SNOKE, A.W., KALAKAY, T.J., QUICK, J.E. & SINIGOI, S. 1999. Development of a deep-crustal shear zone in response to syntectonic intrusion of mafic magma into the lower crust, Ivrea–Verbano zone, Italy. *Earth and Planetary Science Letters*, **166**, 31–45.

SPILLMANN, P. 1993. *Die Geologie des penninisch–ostalpinen Grenzbereichs im südlichen Berninagebirge*. PhD thesis, ETH-Zürich.

STÄHLE, V., FRENZEL, G., KOBER, B., MICHARD, A., PUCHELT, H. & SCHNEIDER, W. 1990. Zircon syenite pegmatites in the Finero peridotite (Ivrea zone)—evidence for a syenite from a mantle source. *Earth and Planetary Science Letters*, **101**, 196–205.

STAMPFLI, G. & MARTHALER, M. 1990. Divergent and convergent margins in the North-Western Alps: confrontation to actualistic models. *Geodinamica Acta*, **4**, 159–184.

STAMPFLI, G., MARCOUX, J. & BAUD, A. 1991. Tethyan margins in space and time. *Palaeogeography, Palaeoclimatology, Palaeoecology*, **87**, 373–409.

STÜNITZ, H.H. 1998. Syndeformational recrystallization—dynamic or compositionally induced? *Contributions to Mineralogy and Petrology*, **131**, 219–236.

STÜNITZ, H.H. & TULLIS, J. 2001. Weakening and strain localization produced by syn-deformational reaction of plagioclase. *International Journal of Earth Sciences*, **90**, 127–135.

TROMMSDORFF, V. 1983. Metamorphose magnesiumreicher Gesteine: kritischer Vergleich von Natur, Experiment und thermodynamischer Datenbasis. *Fortschritte der Mineralogie*, **61**, 283–308.

TROMMSDORFF, V., PICCARDO, G.B. & MONTRASIO, A. 1993. From magmatism through metamorphism to sea floor emplacement of subcontinental Adria lithosphere during pre-Alpine rifting

(Malenco, Italy). *Schweizerische Mineralogische und Petrographische Mitteilungen*, **73**, 191–203.

TRÜMPY, R. 1975. Penninic–Austroalpine boundary in the Swiss Alps: a presumed former continental margin and its problems. *American Journal of Science*, **279**, 209–238.

VAVRA, G. & SCHALTEGGER, U. 1999. Post-granulite facies monazite growth and rejuvenation during Permian to lower Jurassic thermal and fluid events in the Ivrea zone (Southern Alps). *Contributions to Mineralogy and Petrology*, **134**, 405–414.

VAVRA, G., SCHMID, R. & GEBAUER, D. 1999. Internal morphology, habit and U–Th–Pb microanalysis of amphibolite-to-granulite facies zircons: geochronology of the Ivrea zone (Southern Alps). *Contributions to Mineralogy and Petrology*, **134**, 380–404.

VILLA, I.M., HERMANN, J., MÜNTENER, O. & TROMSDORFF, V. 2000. ^{40}Ar/^{39}Ar dating of multiply zoned amphibole generations (Malenco Italy). *Contributions to Mineralogy and Petrology*, **140**, 363–381.

VISSERS, R.L.M., DRURY, M.R., HOOGERDUIJN STRATING, E.H., SPIERS, C.J. & VAN DER WAL, D. 1995. Mantle shear zones and their effect on lithosphere strength during continental breakup. *Tectonophysics*, **249**, 155–171.

VISSERS, R.L.M., DRURY, M.R., HOOGERDUIJN STRATING, E.H. & VAN DER WAL, D. 1991. Shear zones in the upper mantle: a case study in an Alpine lherzolite massif. *Geology*, **19**, 990–993.

WARK, D.A. & WATSON, E.B. 2000. Effect of grain size on the distribution and transport of deep-seated fluids and melts. *Geophysical Research Letters*, **27**, 2029–2032.

WERNICKE, B. 1985. Uniform-sense normal simple shear of the continental lithosphere. *Canadian Journal of Earth Sciences*, **22**, 108–125.

WHITE, S.H. & KNIPE, R.J. 1978. Transformation and reaction enhanced ductility in rocks. *Journal of the Geological Society, London*, **135**, 513–516.

ZINGG, A., HANDY, M.R., HUNZIKER, J.C. & SCHMID, S.M. 1990. Tectonometamoprhic history of the Ivrea zone and its relationship to the crustal evolution of the southern Alps. *Tectonophysics*, **182**, 169–192.

A tale of two kinds of normal fault: the importance of strain weakening in fault development

W. ROGER BUCK[1] & LUC L. LAVIER[2,3]

[1]*Lamont–Doherty Earth Observatory of Columbia University, Route 9W, Palisades, NY 10964, USA (e-mail: buck@ldeo.columbia.edu)*

[2]*GeoForschungsZentrum Potsdam, Albert-Einstein-Strasse, Telegrafenberg, 14473 Potsdam, Germany*

[3]*Present address: California Institute of Technology, Seismology Laboratory 252-21, Pasadena, CA 91125, USA*

Abstract: We search for a description of fault formation consistent with the main features of two very different types of extensional faults: (1) large-offset, low-angle normal faults; (2) small-offset, high-angle normal faults. We use an advanced numerical model to predict how the pattern of faulting varies as a function of the imposed magnitude and rate of weakening of an extending Mohr–Coulomb layer. We assume that fault weakening is due to reduction of cohesion with fault offset. Faults initiate and slip at high dip angles. When the fault offset is large (i.e. comparable with layer thickness) then the inactive footwall fault surface can be rotated to a flat orientation. We find two requirements for development of a large-offset fault. The magnitude of cohesion loss must be greater than *c.* 20% of the initial total extensional yield strength. Also, the rate of weakening with fault offset has to be moderate: the fault offset to lose cohesion has to be less than *c.* 2 km and more than *c.* 100 m, with the lower bound being harder to define. Using the same cohesion and rate of offset weakening, extension of a thick layer can lead to development of multiple, small-offset, high-angle faults rather than a single 'low-angle' fault. For cohesion reduction of 20 MPa a brittle lithosphere thicker than 20 km leads to multiple faults. Finally, we show that inclusion of thermal advection weakening can shift the transition to thinner layers for the same magnitude and rate of cohesion weakening.

Faults are, by definition, weaker than the rock surrounding them. If this were not so, then strain would not concentrate on faults. One of the fundamental controversies in tectonics is just how weak faults are, and specifically whether some faults have a lower than normal coefficient of friction. Geological constraints on the geometry of thrust faults indicate that they slip at levels of shear stresses much lower than expected for rocks under hydrostatic pore pressure. A viable explanation for 'weak' thrust faults is that pore pressures in them are very large, even approaching lithostatic values (Hubbert & Rubey 1959). These faults can slip at a very low dip angle, even though they have a friction coefficient in the same range as measured on laboratory samples (e.g. Byerlee 1978).

Some workers claim that observations around the San Andreas fault indicate that strike-slip faults do slip at lower than expected stress levels (e.g. Lachenbruch & Sass 1980; Mount

& Suppe 1987; Zoback & Healy 1992). However, others (e.g. Scholz 2000) show that the observations do not require an anomalously weak fault.

The debate over the strength of normal faults is driven by the observation of two distinct classes of normal faults (Fig. 1). The most commonly seen faults have high dip angles (>30°) consistent with the predictions of simple Mohr Coulomb failure theory for faults with normal coefficients of friction (Anderson 1951; Sibson 1985). These faults also have relatively small offsets, typically <10 km (e.g. Vening Meinesz 1950; Stein *et al.* 1988). In contrast, the inactive, up-dip parts of some normal faults dip between 30° and horizontal, and some appear to have accommodated offsets of as much as 50 km or more (Davis & Lister 1988; Tucholke *et al.* 1998). Some workers contend that geological evidence requires that some of these normal faults formed and slipped with a low dip (e.g. Wernicke 1985) whereas others

From: WILSON, R.C.L., WHITMARSH, R.B., TAYLOR, B. & FROITZHEIM, N. 2001. *Non-Volcanic Rifting of Continental Margins: A Comparison of Evidence from Land and Sea.* Geological Society, London, Special Publications, **187**, 289–303. 0305-8719/01/$15.00 © The Geological Society of London 2001.

(a)
Gulf of Suez

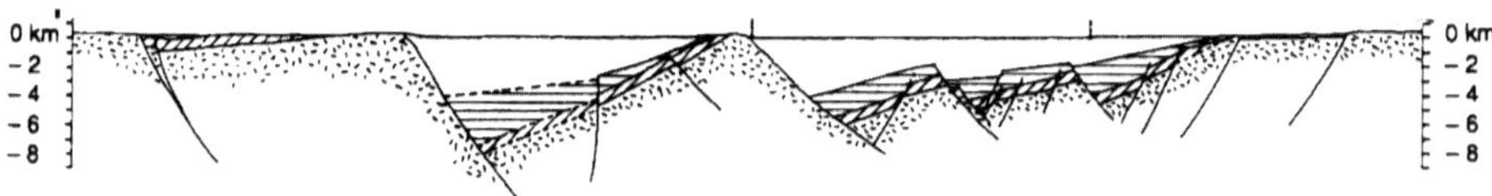

(b)
Whipple Mountains

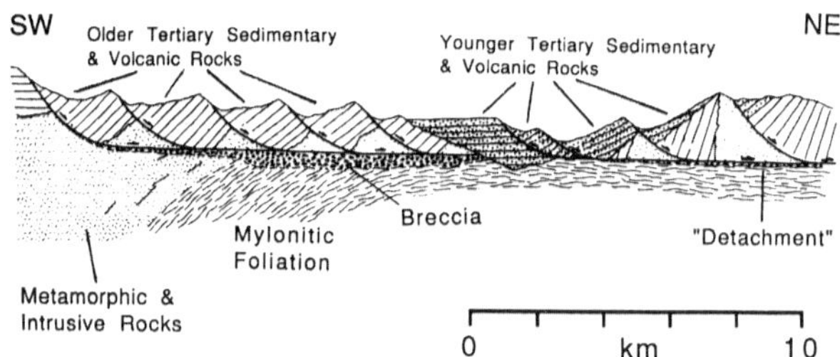

Fig. 1. (**a**) High-angle normal faults across the Gulf of Suez based on the interpretation of seismic and well data (after Patton *et al.* 1994). (**b**) Schematic cross-section of the Whipple Mountains showing the low-angle normal fault interpreted from geological mapping (Davis 1980).

contend that the geological evidence does not require such slip (e.g. Anders & Christie-Blick 1994).

It is particularly difficult to understand why low-angle normal faults should form with low dip angles. High pore pressures are not likely to be maintained consistently above hydrostatic values in an extensional environment. Even if pore pressures were initially high, extension would still tend to favour breaking of high-angle faults (Wills & Buck 1997). An alternative is that these faults have normal frictional strength, and were formed and slipped at a high dip angle. The footwalls of these faults could later be rotated to a low, or even negative, dip angle without the fault being weak (Spencer 1984; Buck 1988, 1993; Hamilton 1988; Wernicke & Axen 1988).

The question we address in this paper is what physical weakening mechanisms might produce the main features of these two distinct classes of normal faults without having to assume different fault properties for each. We would like to simulate the development of different fault patterns by changing something that we know varies between regions, such as the initial lithospheric structure.

Our approach is to use a recently developed numerical model for 2D elasto-visco-plastic deformation, which allows spontaneous, self-consistent generation of faults (Poliakov & Buck 1998). We have to choose a particular parameterization of the strength of the model lithosphere and how it weakens. We assume that the brittle strength of the lithosphere is controlled by friction and cohesion. Faults form as a result of cohesion loss with strain. This is certainly not the only and may not be the best way to describe how faults develop in the Earth, but this is a simple way to start. It is worth noting that before we started our numerical study, we did not suspect how important the rate of fault weakening with strain could be in controlling extensional fault patterns.

This paper is meant to give an overview of recent work on numerical simulation and analysis of normal faulting. More detailed descriptions of some of the models and analyses have been given by Buck & Poliakov (1998); Poliakov & Buck (1998); Lavier *et al.* (1999; 2000). We present new work on combined thermal evolution of a rift and fault formation. Some effort is made to illustrate how we think that regional force changes caused by layer bending and strain weakening of faults

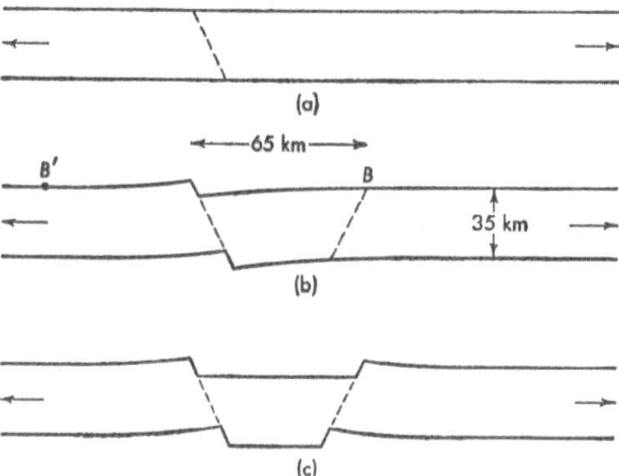

Fig. 2. Model for graben formation after Vening Meinesz (1950).

combine to explain some of the numerical fault patterns. Finally, we speculate about whether different types of fault weakening, such as strain-dependent friction reduction, are consistent with observed fault patterns and fault-related topography.

Previous work

In contrast to strike-slip faults, the offset of a dip-slip fault produces topography and so changes the stresses around the fault. For a high-angle normal fault, the topographic relief should build up quickly as the fault is offset. Vening Meinesz (1950) recognized this and was among the first to suggest that the stress changes related to normal fault offset could result in new faults being formed. Offset of an elastic layer by slip of one fault produces maximum bending stresses at about a flexural wavelength from the fault. Vening Meinesz (1950) assumed that the next fault would break where the bending stresses at the surface were maximally extensional, and result in a graben, as shown in Figure 2. This assumption is reasonable, as the yield stress (the stress needed to

break and slip on a fault) is at a minimum at the surface.

A different approach to analysing the effect of normal fault offset on stresses was suggested by Forsyth (1992). In contrast to Vening Meinesz (1950), he ignored the direct effect of bending stresses in promoting layer breaking and instead estimated the increase in the average regional tectonic stress in a layer as a result of the build-up of fault-related topography. To estimate when a new fault is formed he assumed that the yield strength of the layer is described by Mohr–Columb theory (Fig. 3). The layer is taken to have an Andersonian stress field, where principal stresses are either horizontal or vertical. The initial fault is assumed to be cohesionless but having a finite coefficient of friction. To break a new fault requires the regional stresses to build up enough to overcome the cohesion and frictional strength of the unfaulted layer (Fig. 3). This approach does not specify where a new fault forms, but how much one fault can be offset before a new fault forms.

In the Forsyth (1992) formulation the increase in tectonic force needed to continue fault slip is related to the work done to make topography. As the wavelength of the topography depends on how the layer bends, we sometimes refer to this as the tectonic force increase caused by bending. To quantify this, Forsyth made several assumptions: (1) that the layer cut by a normal fault could be treated as a thin elastic plate; (2) that the stress changes associated with topography did not affect the stresses needed to overcome friction on the initial fault; (3) that the hanging wall and footwall of the fault could be treated as floating beams oppositely loaded at a single horizontal position. These assumptions allowed an analytical description of the increase in the tectonic force for continued slip on one fault. Figure 4 shows that the tectonic force related to topographic build-up increases linearly with fault offset

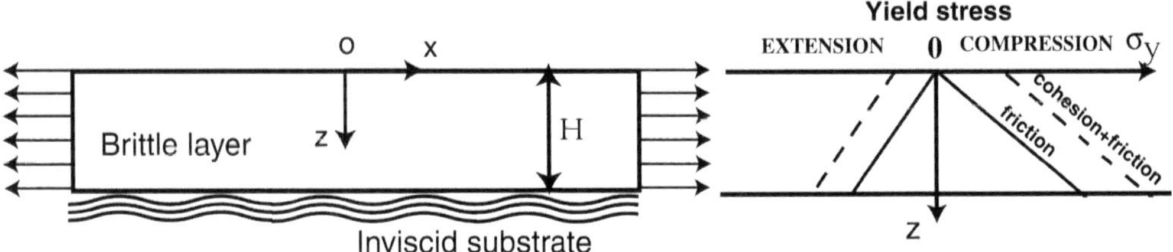

Fig. 3. Schematic model set-up for calculation showing the initial layer geometry, and, on the right, that the brittle layer strength depends on friction and cohesion. The layer floats on an inviscid substrate, as shown in all but the last set of calculations, where the temperature-dependent viscosity is included.

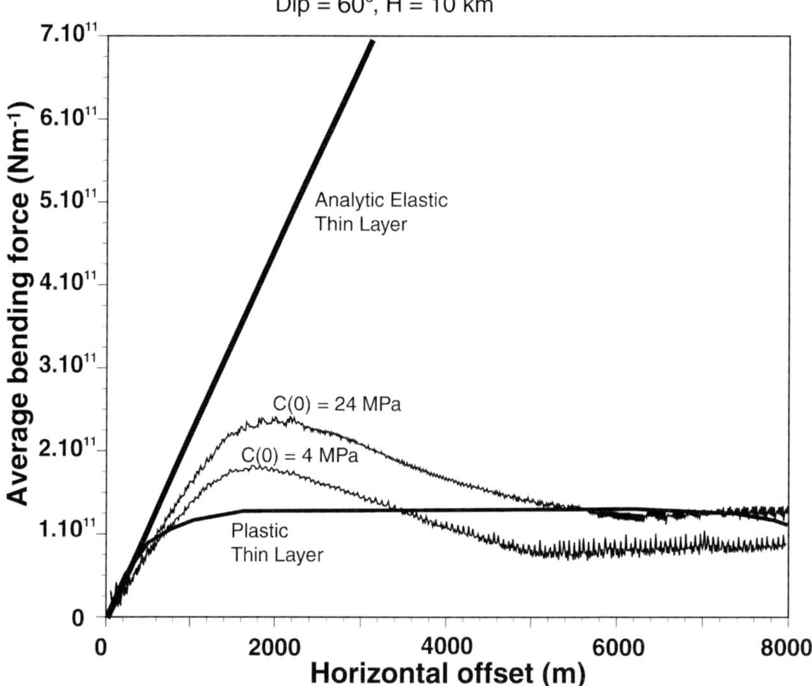

Fig. 4. Different estimates of the increase in tectonic force associated with offset of a brittle layer of 10 km thickness on a 60° dipping fault. The straight line is from the analysis of a thin, purely elastic plate by Forsyth (1992). The bold curved line is from the thin plate, elastic–plastic analysis of Buck (1993). The two fine lines are results of the fully 2D calculations described in the text. Here we inserted an initial 'fault' of three-element width in the numerical grid. These fault elements have zero cohesion and zero friction. In one case the surrounding lithosphere has normal friction ($\mu = 0.6$) and cohesion $C = 4$ MPa and does not change with strain. The other case has the same friction, but C is initially 24 MPa but strain weakens to 4 MPa.

according to the Forsyth (1992) model. Using reasonable values of cohesion, Forsyth (1992) calculated that a normal fault cutting a 10 km layer could slip only about 1 km before being replaced by another fault.

Buck (1993) used the approach of relating work building topography to tectonic forces as suggested by Forsyth (1992). Although he still used the thin plate approximation, he modified the three simplifying assumptions listed above by accounting for changes in the stress needed for slip on the initial fault as a result of plate bending stresses, and by more fully considering the effect of the fault geometry. The most significant change was the inclusion of the effect of finite yield strength on bending stresses. A Mohr–Coulomb plate will bend more easily than a purely elastic plate, as bending stresses cannot exceed the yield stress. Inclusion of finite yield stress in this model radically lowers the size of the tectonic force increase as a result of fault-related topography (as shown in Fig. 4). Using reasonable values for friction and

cohesion for a layer of 10 km thickness, the Buck (1993) model predicts possibly unlimited fault offset. For a thicker layer the fault offset may be limited to an amount smaller than the layer thickness.

Model formulation

To better understand how extension produces both low-angle and high-angle normal faults we would like to treat the regional force changes, discussed by Forsyth (1992) and Buck (1993), as well as the local bending stresses emphasized by Vening Meinesz (1950). We would like to solve for the fully 2D stress and strain fields without having to make the thin plate approximation made by those workers. Ideally, we would not have to assume a pre-existing weak fault, as have most previous numerical studies (e.g. Melosh & Williams 1989; King & Ellis 1990). We would specify the same process of weakening to develop the primary fault as we do for any secondary faults.

Brittle rheology and fault weakening

We assume that the yield stress for brittle material is given by the Mohr–Coulomb failure criterion and that when that criterion is met, flow follows a rule for non-associated plasticity (e.g. Poliakov & Buck 1998). This formulation allows localization of plastic deformation in shear zones or 'faults'. The shear stress at yield τ is given by Mohr–Coulomb theory:

$$\tau = \mu\sigma_n + C(\varepsilon_{ps}). \qquad (1)$$

In Equation (1), μ is the coefficient of friction, σ_n is the normal stress, and C is the cohesion, which depends on the total plastic strain, ε_{ps}. The plastic strain is the non-recoverable strain accumulated when the stress in the layer is locally equal to the yield stress.

The cohesion is reduced with increasing strain after yielding. Up to the point where material loses all cohesion, the reduction of cohesion with strain is linear:

$$C(\varepsilon_{ps}) = C(0)[1 - (\varepsilon_{ps}/\varepsilon_c)] \qquad (2)$$

where $C(0)$ is the initial cohesion of the layer. We define ε_c as a characteristic value of plastic strain. When the plastic strain reaches ε_c the fault is cohesionless. However, as seen in all models allowing for the localization of deformation (e.g. Cundall 1989), the width of a fault, Δw, is consistently about three times the grid size. Thus, for a given fault displacement, the strain is dependent on the grid size. As we do not want our results to depend on grid size we specify the fault offset for cohesion loss Δx_c. Then the characteristic plastic strain ε_c for fault weakening is set equal to $\Delta x_c/\Delta w$, where Δw equals three times the grid spacing. To scale the characteristic strain and the rate of cohesion weakening between two models with different grid sizes, we use characteristic offset, $\Delta w = \varepsilon_c \Delta w$ as a measure of the amount of deformation needed to form a cohesionless fault.

Numerical scheme

We use an explicit finite-element method similar to the FLAC (fast Lagrangian analysis of continua) technique of Cundall (1989), which has been used to simulate localized deformation in elastic–plastic materials in a variety of problems (Hobbs & Ord 1989; Poliakov *et al.* 1993; Poliakov & Herrmann 1994; Hassani & Chery 1996; Buck & Poliakov 1998). As this method is Lagrangian (i.e. the numerical grid follows the deformations), the simulation of very large deformation involves remeshing to overcome the problem of degradation of numerical precision when elements are distorted. The treatment of regridding was developed by Poliakov (Poliakov & Buck 1998). We trigger remeshing when any triangle in a grid element is so distorted that one of its internal angles becomes smaller than 10°. Every time remeshing occurs, strains and other properties at each grid point are interpolated between the old deformed mesh and the new undeformed mesh using a nearest-neighbour algorithm. The disadvantage of our approach is that our model calculations are very slow (taking days to weeks to run on typical workstations), and that it is difficult to isolate the parameters that affect the pattern of faulting.

The width of the model domain is taken to be 10–15 times the layer thickness, H. We take the density, ρ, of the brittle layer and the inviscid substrate equal to $2700\,\mathrm{kg}\ \mathrm{m}^{-3}$, and the acceleration of gravity, g, equal to $10\,\mathrm{m}\ \mathrm{s}^{-2}$. At the upper surface, shear and normal stresses are assumed to be zero. The right and left sides of the box are pulled steadily apart. At the bottom, we apply normal stress equal to the lithostatic pressure in the brittle layer and zero shear stress (Winkler foundation).

A small perturbation, one to three elements large, is initially placed at the centre of the model domain to induce the formation of the first fault. That area is set to have a plastic strain equal to ε_c, to be cohesionless. Without such a 'flaw' faults always first break at the boundaries, as a result of the dynamic stresses associated with finite boundary velocities, as detailed by Lavier *et al.* (2000).

Here we look at a few experiments to illustrate simple behaviour we have noted in a large number of experiments, some of which have previously been discussed (e.g. Buck & Poliakov 1998; Lavier *et al.* 1999, 2000). We consider only a simplified geometry of brittle lithosphere to avoid some of the complexity seen in some past experiments. For example, Buck & Poliakov (1998) showed that a chaotic pattern of faults could develop during extension of lithosphere in which thermally defined lithospheric thickness increased with distance from a 'spreading centre'. We now understand that faults migrate relative to fixed thermal structure in a way that causes faults moving into thicker lithosphere to be replaced by faults cutting thinner lithosphere. These geometric complications made it difficult for us to identify how parameters relating to fault weakening might affect the pattern of faulting.

In the first set of experiments we consider extension of a lithosphere of nearly constant thickness. The lithosphere floats on an inviscid substrate and we set the lithospheric base to a constant depth each time regridding occurs, following the set-up used by Lavier *et al.* (1999). This allows us to isolate the effect of changes in the amount and rate of fault weakening on the extensional fault pattern.

In the second set of experiments we start with a lithosphere of constant thickness, but allow thermal advection to modify the lithospheric thickness with time. In these experiments we define an initial thermal structure and take material strength to depend on temperature. The initial temperature distribution is laterally uniform and there is a linear increase in temperature with depth. The surface temperature is maintained at $10\,°C$ and the bottom temperature at $640\,°C$.

The effective viscosity is defined in terms of power-law creep (e.g. Kirby & Kronenberg 1987):

$$\mu = A^{-1/n} \dot{\varepsilon}_{\mathrm{eff}}^{(1/n)-1} \exp(Q/nRT) \qquad (3)$$

where A is a constant, n is the power-law exponent, $\dot{\varepsilon}_{\mathrm{eff}}$ is defined as the second invariant of the strain-rate tensor, T is temperature in Kelvin, and R is the universal gas constant. For the following example we took $n = 3$, $Q = 442\,\mathrm{kJmol}^{-1}$, and $A = 1.73 \times 10^5 (\mathrm{Pas})^{1/n}$ (plagioclase from Kirby & Kronenberg (1987)). After each time step the temperature and the viscosity are updated.

Results and discussion

In all our numerical calculations the initially forming fault or faults form with a high (*c.* $60°$) dip angle, in agreement with the internal friction coefficient assumed in the brittle layer ($\mu = 0.6$). In some cases one fault can develop very large offsets (relative to the layer thickness) with minor deformation focused on other model faults. In other cases several faults form and are significantly offset. Below, we separate out the results of varying each of several parameters. In the first sets of results the layer thickness is kept constant, and in a later section we consider advective thinning.

Effect of layer thickness or relative amount of weakening

Figure 5 compares results of the distribution of plastic strain after finite extension for cases that differ only in the constant brittle layer thickness. In Figure 5a the layer is $10\,\mathrm{km}$ thick and offset is concentrated on one shear zone or fault. The inactive part of this fault is rotated to a nearly horizontal position. In contrast, Figure 5b shows that several faults form when a layer of $20\,\mathrm{km}$ thickness is extended, even though the magnitude and the rate of cohesion reduction are the same as for the calculation shown in Figure 5a.

The numerical results confirm that the thickness of a layer can strongly affect the pattern of faulting, as long as the amount of strength loss on a fault is independent of layer thickness. This result can be understood in terms of the idea that normal fault offset is resisted by stress changes related to plate bending (Forsyth 1992). Figure 6 is an approximate representation of how force changes might evolve during extension of a brittle layer with only one fault developing. The tectonic force increase is based on the kind of analysis illustrated in Figure 4.

To define the force decrease as a result of strain (or in terms of fault offset) we multiply layer thickness by the average yield stress of the layer. The strength, S, of a layer of thickness H is the force needed to form or slip on an ideally oriented, throughgoing fault assuming an Andersonian stress state, and is given by

$$S(\varepsilon_{\mathrm{p}}) = [(\rho g H^2/2) + C(\varepsilon_{\mathrm{p}})H]K_{\mu} \qquad (4)$$

where $K_{\mu} = 1/[(1 + \mu^2)^{1/2} + \mu]$ and $C(\varepsilon_{\mathrm{p}})$ is given by Equation (2) (see Lavier *et al.* 2000). We define the fault weakening as the change in fault strength ΔS as a result of strain:

$$\Delta S = S(\varepsilon_{\mathrm{p}}) - S(0). \qquad (5)$$

This change in strength is always negative, as shown in Figure 6. Even the initial value is negative, as we impose a small flaw, or perturbation, to start the first fault.

The maximum force increase as a result of layer bending scales approximately with the square of the layer thickness (Buck 1993; Lavier *et al.* 2000), whereas the maximum reduction in strength as a result of loss of cohesion on a fault scales linearly with the layer thickness (Equations (4) and (5)). The force changes shown in Figure 6 and later plots are defined relative to the force needed to start faulting on a single fault (Equation (4)). Thus, if the sum of the bending and the weakening force changes becomes positive then there would be enough force to form a new fault.

For a relatively thin layer the weakening as a result of cohesion reduction on the first fault is greater than the force increase caused by layer

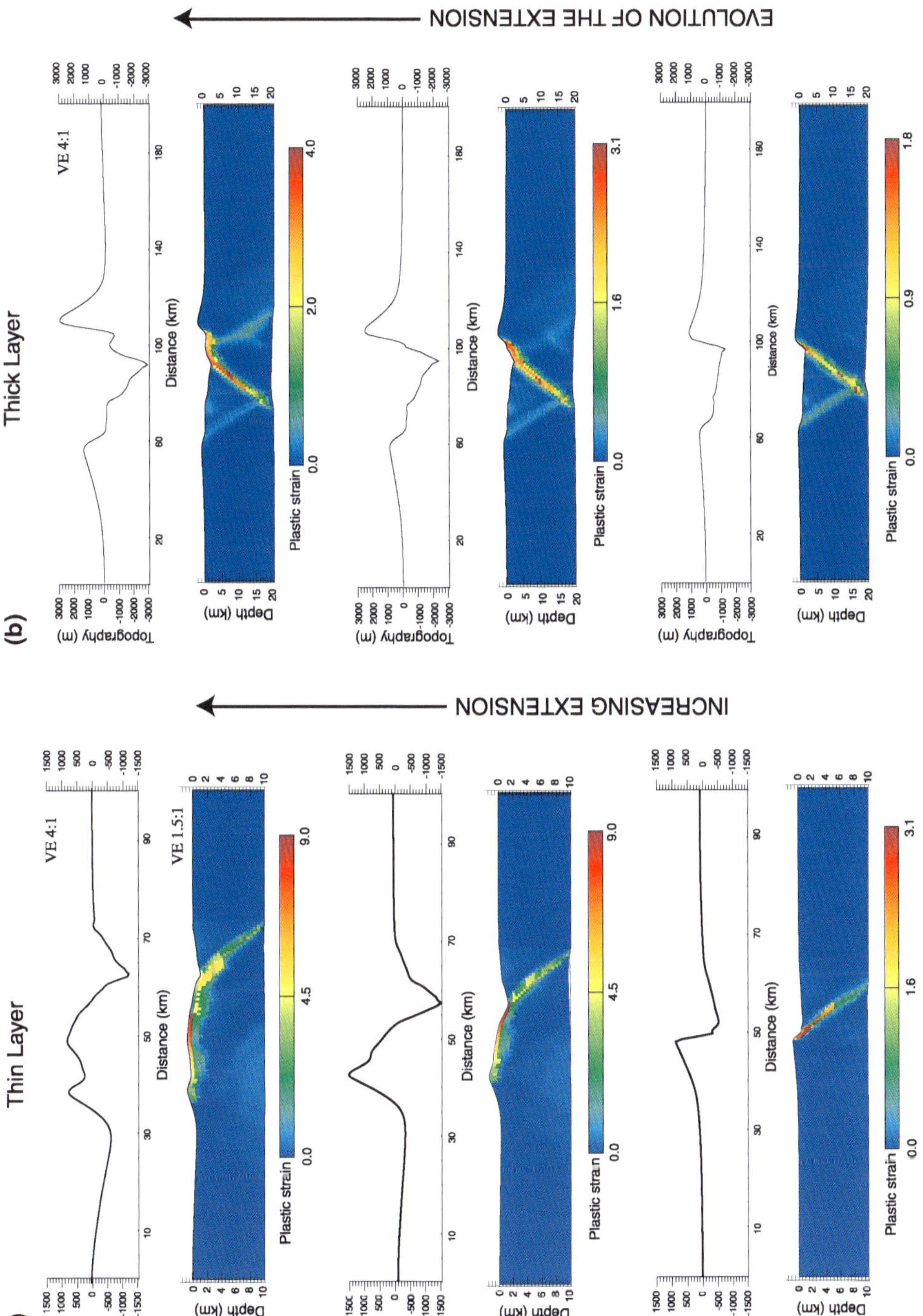

Fig. 5. Results showing the distribution of plastic strain for two model calculations that differ only in the layer thickness. Here the critical fault offset for total cohesion loss, Δx_c, is 1.5 km. Areas of large plastic strain are considered to be model faults. (**a**) The thinner (10 km) layer mainly extends on a single high-angle fault, and the inactive footwall of the fault has rotated to a quasi-flat orientation. This snapshot is after 20 km of layer extension. (**b**) The thick (20 km) layer extends on several interacting faults, after the 5 km of extension pictured here. VE, vertical exaggeration.

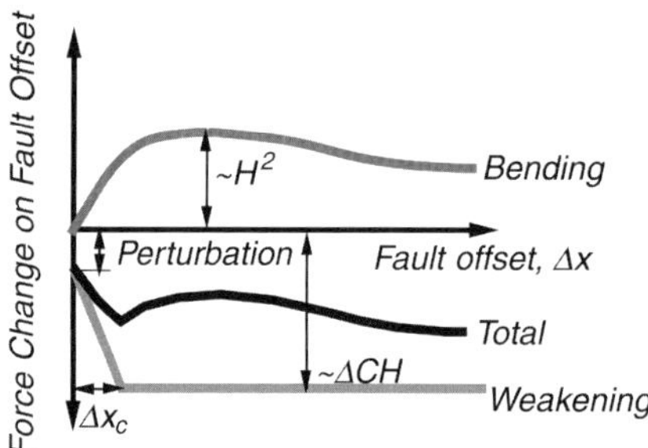

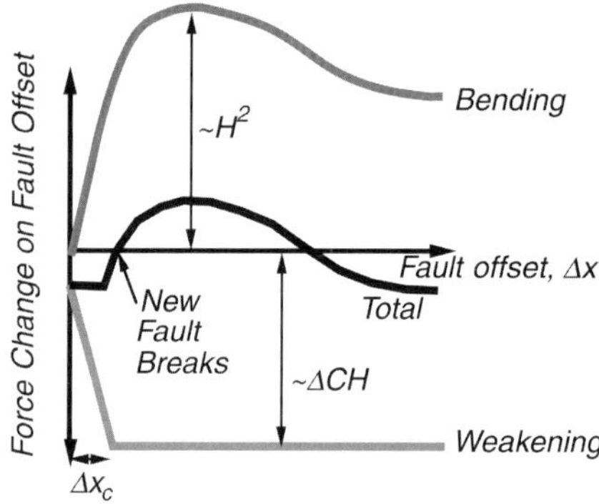

Fig. 6. Schemes showing the effect of different layer thicknesses on the average tectonic forces required to continue extension of a single fault. These curves of force increase as a result of layer bending, and are based on the numerical analysis shown in Figure 4, but are not actually calculated for any model. The fault weakening consists of the initial weak perturbation that 'seeds' the fault and the prescribed strain-dependent weakening up to the critical horizontal fault offset, Δx_c. (**a**) shows that for a thin layer the sum of force changes stays negative, so no new fault forms. (**b**) For a thick layer the total force becomes positive, indicating that a new fault could form.

bending, as shown in Figure 6a. The total force change is always negative so there is not enough regional force to form a new fault in cohesive lithosphere. This graphical representation makes it easy to see that the transition from thin-layer behaviour, where the fault offset could be unlimited, to thick-layer behaviour depends on the value of cohesion reduction. What is important here is the amount of cohesion reduction relative to the total strength of the layer. For a fixed value of cohesion this ratio depends on the layer thickness.

To obtain a large-offset fault with an inactive portion rotated to horizontal requires extension of a thin layer. For a cohesion loss of 20 MPa the brittle layer would have to be less than c. 20 km to produce a large-offset fault with a low-angle inactive part. This is consistent with the prediction of Buck (1993), based on analysis of this elastic–plastic plate theory. Although a thin layer (or a relatively large amount of cohesion reduction) is a necessary condition for large-offset fault development, it is not a sufficient condition.

Effect of fault weakening rate

Figure 7 contrasts two calculations with differing rates of cohesion loss with strain or fault offset. Other parameters, such as layer thickness and cohesion, are the same as for the case shown in Figure 5a. For Figure 7a the rate of weakening was very slow, as the amount of fault offset to lose cohesion, Δx_c, was 4.5 km. In contrast, for the case shown in Figure 5a, Δx_c was 1.5 km. For Figure 7a there was immediate breaking of a second fault centred on the flaw we introduced to seed the first fault. A symmetrical graben results. With this slow rate of cohesion loss there is no possibility of large offset on one fault. Also, it should be noted that slow cohesion reduction results in a perfectly symmetrical graben. As most real graben are asymmetric, we suspect that natural fault weakening is faster than shown in Figure 7a.

A case with very rapid cohesion loss with fault offset is shown in Figure 7b. Here Δx_c equals 0.15 km. In this case secondary faults break on the hanging-wall side of the first formed fault after a small amount of slip on the primary fault. Eventually, faults break on the footwall side of the initial fault, and significant offset of these tertiary faults develops.

This shows that the second necessary condition for large-offset normal fault to develop, given our model parameterization, is that the rate of cohesion loss with strain is 'moderate'. What we mean by 'moderate' weakening rate can be understood by considering extreme cases. Let us imagine that the rate of strain weakening were so slow that 1 km of offset caused only 1% of cohesion loss on a fault. The finite fault offset would bend the layer and produce topography, and so increase the tectonic force needed to continue offset of the primary fault. As fault weakening negligibly reduced the tectonic force then it should be sufficient to break new faults in other places (Fig.

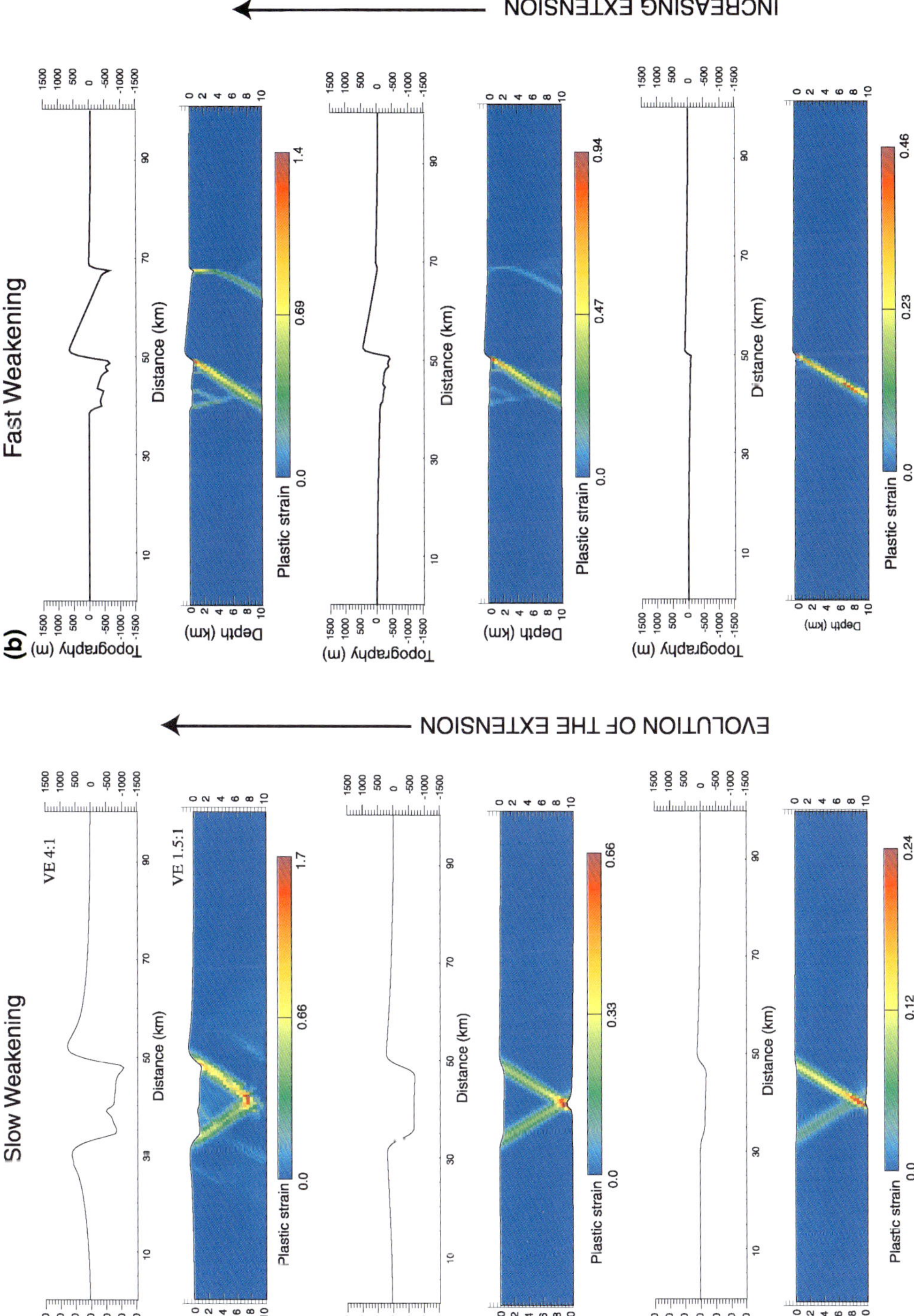

Fig. 7. Results of two models with the same parameters as in Figure 5a but with different rates of strain weakening. (**a**) For very slow strain weakening ($\Delta x_c = 4.5$ km) new faults break away from the initial fault. (**b**) For very fast weakening ($\Delta x_c = 0.15$ km) new faults develop, first in the hanging wall, and later in the footwall.

8). These faults can form pervasively through the width of an extended layer.

In the other extreme, let us imagine that almost no strain is required to completely reduce the cohesion on a fault. Small layer bending (i.e. resulting from fault offsets that are small compared with the layer thickness) produces bending stresses that are sufficient to cause yielding of the top of the hanging wall of the primary fault. As almost no strain is required for significant weakening, a new fault starts to form at the top of the layer. Cohesion loss at the top of the layer concentrates stress at the base of the newly formed break, allowing the fault to propagate downward. In a sense the layer shatters. Fast strain weakening is very hard to model numerically, but we think that the secondary faults may be able to replace the first faults.

Advective layer thinning

In the previous experiments, the brittle layer was floating on an inviscid fluid and the layer thickness was kept relatively constant (i.e. during each regridding its base was set at a fixed depth). To illustrate the possible effects of thinning of the brittle layer by thermal and mechanical necking, we set up a calculation in which the brittle layer is floating on a non-Newtonian visco-elastic layer (Maxwell visco-elastic behaviour).Thermal advection and diffusion are calculated in this model. Advection during fault offset should change the temperature structure of the lithosphere, causing local thinning under a fault.

This process of thinning is rate controlled; if the rate of extension of the layer is rapid, advection would dominate over heat diffusion and the layer could significantly weaken after a small amount of extension. If the rate of extension is slow, diffusion dominates and the layer would weaken far less for the same amount of extension. We would like to illustrate the contrasting behaviour of a system in which the brittle layer thickness is constant (Fig. 5a) with one in which the brittle layer thickness varies. Therefore we chose the same parameters for fault strength and weakening of the brittle layer as in the case with constant layer thickness, which led to the formation of a single large-offset normal fault (Fig. 5a).

Figure 9a shows the effect of layer thinning on the initial distribution of faulting in the layer. The figure shows the state of strain, temperature and viscosity after *c.* 2 km of extension. The temperature field reflects the advection of heat resulting from the extension.

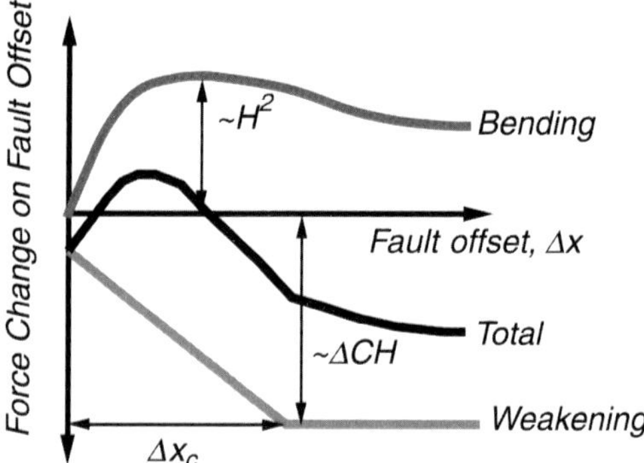

Fig. 8. Scheme showing how slow strain weakening can lead to the kind of secondary faulting shown in Figure 7a.

The viscosity field indicates the shallowing of the brittle–viscous transition. The total strain field shows that after 2 km of extension another fault formed in the hanging wall of the initial fault (as indicated by zones of highly concentrated strain).

The evolution of faulting in the brittle layer is illustrated in Figure 9b. After the formation of the first fault another fault rapidly forms. The two faults then accumulate deformation and develop into an asymmetric graben. After further extension and thinning of the layer another fault develops and faulted blocks rotate in a manner akin to tilted blocks in a narrow rift.

Advective thinning of the lithosphere tends to promote development of secondary faults near the primary fault. This occurs because the 'extra force' needed to break a new fault is roughly equal to the cohesion times the layer thickness. As the layer thickness locally decreases it takes less 'extra force' to break the new fault. The force increase as a result of layer bending is not initially affected much by the local thinning of the layer. Essentially, the first fault still has to offset a thick plate.

Figure 10 shows how local layer thinning reduces the strength difference between the initial fault and the fully cohesive but thinned layer. The strength difference scales linearly with layer thickness, so with reduced thickness it takes less stress to break a new, secondary fault. Because offset of the initial fault still has to bend the entire layer we ignore any changes in the tectonic force increase because of fault offset. Eventually, as the layer thins over a wider region this assumption should break

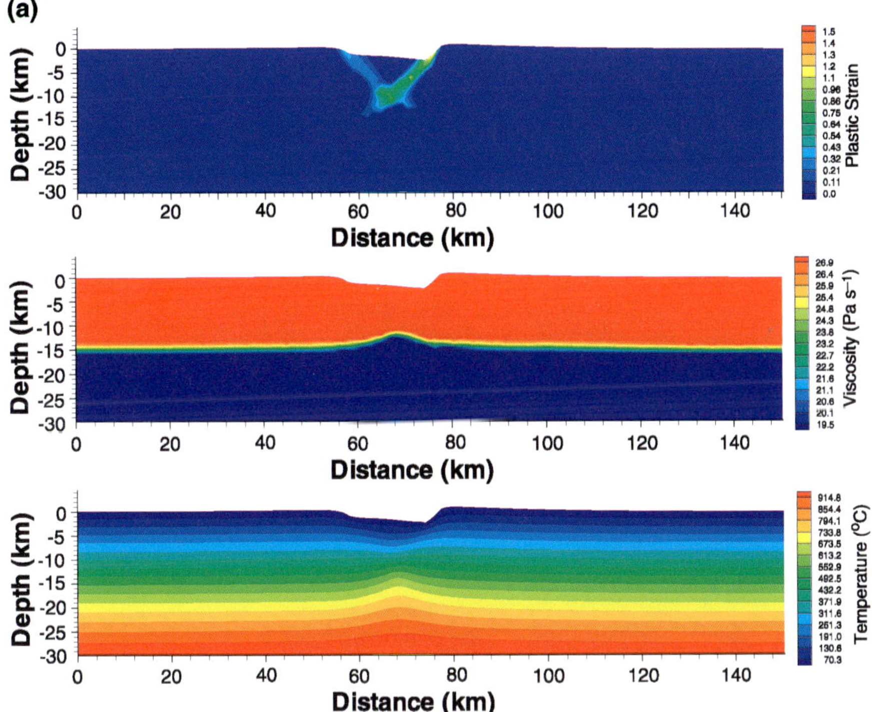

Fig. 9. Continued on next page.

down. However, this very approximate way of representing forces and strength changes does allow us to visualize a way in which layer thinning may promote secondary fault development.

Speculations

There are many other ways to formulate the problem of fault weakening and even thermal weakening of extending lithosphere. We have looked at the results of varying a few parameters of a particular model formulation. On the basis of what we have learnt from these relatively simple experiments we can speculate about what we would see if different processes were considered.

Fault width and layer thickness

One interesting possibility is that fault width might scale with brittle layer thickness. Certainly when a slab of rock of 1 cm thickness breaks it does so over a width of 1 mm or less. Fault zones in the Earth are often metres thick and various strands of a fault may be many kilometres wide.

If the strain required to reduce fault cohesion (ε_c in Equation (2)) is constant, but the width of a fault zone scales with layer thickness, then the critical fault offset for weakening (Δx_c) also scales with layer thickness. Thus, a fault in a thin layer would weaken quickly (i.e. after small fault offset), whereas a fault in a thick layer would weaken after much longer offset. This could mean that very thin layers were more prone to shatter on extension. Extension of very thick layers could result in multiple faults because the rate of strain weakening is so low.

Hydrothermal cooling

As another example of processes not yet considered, hydrothermal circulation might affect lithospheric thermal structure. This might counter some of the lithospheric thinning caused by thermal advection. Precisely this kind of cooling seems to be necessary to explain the thermal structure inferred for mid-ocean ridges (see Phipps Morgan & Chen 1993). Thus, an initially thin layer, say 10 km thick, might be

(b)

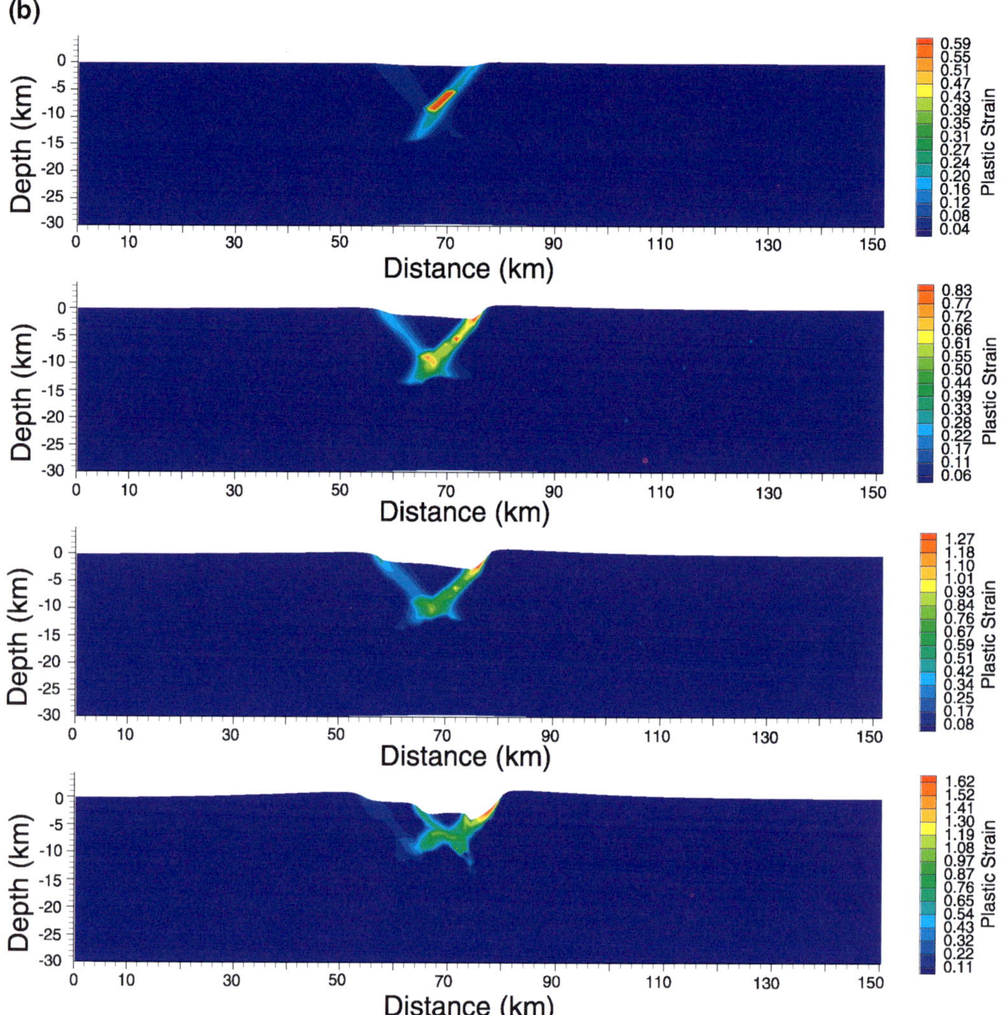

Fig. 9. Results of a model calculation where the temperatures control strength (through the viscosity) and the advection and diffusion of heat are included. (**a**) The strong layer has thinned after 2 km of extension and a second fault has formed. (**b**) The evolution of plastic strain and extension. First, one secondary fault forms antithetic to the first, and then another forms further from the first fault.

thinned little during extension as a result of active hydrothermal cooling in the vicinity of the active fault. Thus, the layer might extend in much the way that was shown for the cases with constant layer thickness in Figure 5a.

The depth of penetration of efficient hydrothermal circulation may be limited to 5–10 km where overburden is insufficient to close cracks (the pathway of fluids). If that is the case, then for extension of thick lithosphere thermal advection should still cause large lateral variations in lithospheric thickness. These vari-

ations should promote secondary faulting as they did in the model case we ran.

Strain-dependent friction

The assumption that cohesion loss with strain is the dominant mechanism for fault weakening implies that changes in the brittle layer thickness should be a major control on the pattern of faulting. The most obvious alternative to our approach of cohesion reduction is to assume that the coefficient of friction on a fault is

Advective Thinning of Initially Thin Layer with a Moderate Weakening Rate

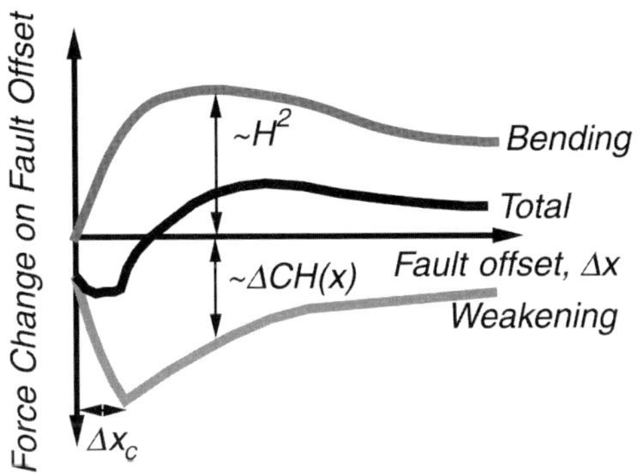

Fig. 10. Scheme indicating how layer thinning can affect average tectonic stresses.

reduced with strain. If we were to assume friction loss with strain then we would expect very different results.

Fault offset should wear down the rocks in the fault zone to a smaller grain size. This might promote hydration reactions that would transform the rocks to clays or serpentinites. Whatever the mechanism, friction might be reduced by a set fraction and this would reduce the strength of a fault by the same fraction, as long as cohesive strength was negligible (Equations (2), (4) and (5)). On the basis of our analysis of models with cohesion reduction, we expect that at least a 20% reduction in strength is needed to allow a large-offset fault to develop. Thus, to obtain a large-offset normal fault, with only friction reduction, requires a 20% drop in friction. Such a magnitude of friction reduction would allow large fault offset to develop for any layer thickness.

We base this on our calculations showing that the magnitude of the tectonic force change scales with the square of the layer thickness. The reduction in fault strength also scales with thickness squared (Equation (4)). Therefore, the force increase related to plate bending would not be able to dominate fault weakening and cause formation of secondary faults.

The maximum topographic relief produced by fault offset should roughly scale with layer thickness. We see >2 km of relief for extension of a 10 km layer (Fig. 5a). Extension of a layer of 50 km thickness could, for the friction reduction model, produce topographic relief of >10 km. Such fault-related relief is not seen on Earth. In the cohesion reduction model, fault offset and topographic relief are limited to a few kilometres even for very thick lithosphere.

There may be a way to reconcile a friction reduction model to the observation that extensional topographic relief is not gigantic and that extension of thick lithosphere seems to produce many faults. Let us imagine that the tectonic force available for extension is not greater than the 'ridge push' force level ($c.\ 2 \times 10^{12}$ N m^{-1}). This is enough to extend brittle lithosphere thinner than $c.\ 20$ km (assuming $\mu = 0.6$ and no cohesion in Equation (3)). If tectonic force is limited, then thicker lithosphere could be extended if there was injection of melt through part or all of the lithosphere. Assuming melt has about the same density as the lithosphere it is intruding, then injection of dykes through the whole lithosphere could occur with almost no tectonic force. If there was insufficient magma to accommodate all the extension, then dykes might not reach the surface, but extension could proceed at a reduced tectonic force level (Rubin & Pollard 1987).

Many repeated intrusive events could bring in massive amounts of heat, so that the lithosphere would be locally thinned. When the lithosphere becomes thinner than $c.\ 20$ km, extension could occur at moderate tectonic force levels, even without continued magmatism. From our experience with thermal variations in lithospheric thickness we expect that secondary faults could develop centred on the thermal perturbation. Thus, magmatically assisted extension could still produce a non-volcanic margin.

Summary

Our goal here was to define lithospheric and fault properties that are consistent with some first-order observed geological characteristics of development of faults. Normal faults are a useful starting point, because (1) they theoretically should not be greatly affected by high pore pressures, as thrust faults may be, and (2) there is a great variety of faulting patterns observed in continental rifts and mid-ocean ridges. So far we have focused mostly on the difference between small- and large-offset normal faults. We have found one plausible, although not unique, formulation for faulting that is based on cohesion loss with strain. We have identified several conditions needed for development of large-offset normal faults. Our results suggest that thicker layers can extend by breaking of multiple small-offset, high-angle normal faults.

One way to look at the problem is that we have found a self-consistent way for large-offset, low-angle normal faults to develop. The fact that many normal faults do not have large offset means that something stops these faults and in many cases prompts growth of other faults near an initial fault. We show that one way to explain small-offset faults is in terms of layer thickness. If the cohesion loss is moderate and the rate of weakening with fault offset is independent of layer thickness, then thick layers should be broken by multiple, high-angle faults during extension.

The results of one calculation with thermal advection do bear some resemblance to continental rift structures, in terms of showing multiple, fault-bounded graben. If this case had not included advective lithospheric thinning, then no clearly defined secondary faults would have developed. This basic result leads us to believe that the thermal conditions in which a rift develops have a major impact on the style of faulting.

We have not considered several processes that could influence the pattern of fault development during rifting. For example, extensional thinning of crust, which overlies denser mantle, would produce gravitational stresses that could limit offset of normal faults. In future models we plan to consider such effects as well as magmatic intrusion. We trust that we then may be able to make more detailed comparisons between our models and geological data.

We thank A. Poliakov for help developing the numerical models used here. J. Braun, D. Forsyth and N. Froitzheim provided very helpful reviews. Support was from NSF grants OCE-98-19866 and EAR-98-14576. L. Lavier also thanks the Geo-Forschungs Zentrum, Potsdam, for support while he completed this work. This paper is Lamont Contribution 6126.

References

ANDERS, M. & CHRISTIE-BLICK, N. 1994. Is the Sevier Desert reflection of west–central Utah a normal fault? *Geology*, **22**, 771–774.

ANDERSON, E.M. *The Dynamics of Faulting and Dyke Formation, with Applications to Britain*. Oliver and Boyd, Edinburgh.

BUCK, W.R. 1988. Flexural rotation of normal faults. *Tectonics*, **7**, 959–973.

BUCK, W.R. 1993. Effect of lithospheric thickness on the formation of high- and low-angle normal faults. *Geology*, **21**, 933–936.

BUCK, W.R. & POLIAKOV, A.N.B. 1998. Abyssal hills formed by stretching oceanic lithosphere. *Nature*, **392**, 272–275.

BYERLEE, J.D. 1978. Friction of rocks. *Pure and Applied Geophyiscs*, **116**, 615–626.

CUNDALL, P.A. 1989. Numerical experiments on localization in frictional materials. *Ingenieur-Archiv*, **58**, 148–159.

DAVIS, G.A. 1980. Problems of intraplate extensional tectonics, western United States. *In*: NATIONAL RESEARCH COUNCIL (ed.) *Continental Tectonics*. National Academy of Sciences, Washington, DC, 89–95.

DAVIS, G.A. & LISTER, G.A. 1988. Detachment faulting in continental extension; perspectives from the southwestern U.S. Cordillera. *In*: CLARK, S.P., BURCHFIELD, B.C. & SUPPE, J. (eds) *Processes in Continental Lithospheric Deformation*. Geological Society of America, Boulder, CO, 133–159.

FORSYTH, D.W. 1992. Finite extension and low-angle normal faulting. *Geology*, **20**, 27–30.

HAMILTON, W. 1988. Extensional faulting in the Death Valley region (abstract). *Geological Society of America, Abstracts with Programs*, **20**, 165–166.

HASSANI, R. & CHERY, C. 1996. Anelasticity explains topography associated with Basin and Range normal faulting. *Geology*, **24**, 1095–1098.

HOBBS, B.E. & ORD, A. 1989. Numerical simulation of shear band formation in frictional–dilatational materials. *Ingenieur-Archiv*, **59**, 209–220.

HUBBERT, M.K. & RUBEY, W. 1959. Role of fluid pressure in mechanics of over-thrust faulting, Pts. I and II. *Geological Society of America Bulletin*, **70**, 115–205.

KING, G. & ELLIS, M. 1990. The origin of large local uplift in extensional regions. *Nature*, **348**, 689–692.

KIRBY, S.H. & KRONENBERG, A.K. 1987. Rheology of the lithosphere: selected topics. *Review of Geophysics*, **25**, 1219–1244.

LACHENBRUCH, A. & SASS, J. 1980. Heat flow and energetics of the San Andreas fault zone. *Journal of Geophysical Research*, **85**, 6185–6222.

LAVIER, L., BUCK, W.R. & POLIAKOV, A.N.B. 1999. Self-consistent rolling-hinge model for the evolution of large-offset low-angle normal faults. *Geology*, **27**, 1127–1130.

LAVIER, L., BUCK, W.R. & POLIAKOV, A.N.B. 2000. Factors controlling normal fault offset in an ideal brittle layer. *Journal of Geophysical Research*, **105**, 23431–23442.

MELOSH, H.J. & WILLIAMS, C.A. 1989. Mechanics of graben formation in crustal rocks: a finite element analysis. *Journal of Geophysical Research*, **94**, 13961–13973.

MOUNT, V. & SUPPE, J. 1987. State of stress near the San Andreas fault: implications for wrench tectonics. *Geology*, **115**, 1143–1146.

PATTON, T.L., MOUSTAFA, A.R., NELSON, R.A. & ABDINE, S.A. 1994. Tectonic evolution and structural setting of the Suez Rift. *In*: LANDON, S.M. (ed.) *Interior Rift Basins*. American Association of Petroleum Geologists, Memoir, **59**, 9–55.

PHIPPS MORGAN, J. & CHEN, Y.J. 1993. The genesis of oceanic crust: magma injection, hydrothermal circulation and crustal flow. *Journal of Geophysical Research*, **98**, 6283–6297.

POLIAKOV, A. & BUCK, W.R. 1998. Mechanics of stretching elastic–plastic–viscous layers: applications to slow-spreading mid-ocean ridges. *In*: BUCK, W.R., DELANEY, P.T., KARSON, J.A. & LAGABRIELLE, Y. (eds) *Faulting and Magmatism at Mid-Ocean Ridges*. Geophysical Monograph, American Geophysical Union, **106**, 305–324.

POLIAKOV, A.N.B. & HERRMANN, H. 1994. Self-organized criticality in plastic shear bands. *Geophysical Research Letters*, **21**, 2143–2146.

POLIAKOV, A.N.B., PODLADCHIKOV, Y. & TALBOT, C. 1993. Initiation of salt diapirs with frictional overburden: numerical experiments. *Tectonophysics*, **228**, 199–210.

RUBIN, A.M. & POLLARD, D.D. 1987. Origins of Blake-like Dikes in Volcanic Rift Zones. US Geological Survey, Professional Paper, **1350**, 1449–1470.

SCHOLZ, C. 2000. Evidence for a strong San Andreas fault. *Geology*, **28**, 163–166.

SIBSON, R.H. 1985. A note on fault reactivation. *Journal of Structural Geology*, **7**, 751–754.

SPENCER, J.E. 1984. Role of tectonic denudation in warping and uplift of low-angle normal faults. *Geology*, **12**, 95–98.

STEIN, R., KING, G. & RUNDLE, J. 1988. The growth of geological structures by repeated earthquakes: 2, Field examples of continental dip-slip faults. *Journal of Geophysical Research*, **93**, 13319–13331.

TUCHOLKE, B., LIN, J. & KLEINROCK, M. 1998. Megamullions and mullion structure defining oceanic metamorphic core complexes on the Mid-Atlantic Ridge. *Journal of Geophysical Research*, **103**, 9857–9866.

VENING MEINESZ, F.A. 1950. Les graben africains résultant de compression ou de tension dans la croute terrestre? *Bulletin de l'Institut Royal Colonial Belge*, **21**, 539–552.

WERNICKE, B. 1985. Uniform-sense normal simple shear of the continental lithosphere. *Canadian Journal of Earth Science*, **22**, 108–125.

WERNICKE, B. & AXEN, G.J. 1988. On the role of isostasy in the evolution of normal fault systems. *Geology*, **16**, 848–851.

WILLS, S. & BUCK, W.R. 1997. Stress-field rotation and rooted detachment faults: a Coulomb failure analysis. *Journal of Geophysical Research*, **102**, 20503–20514.

ZOBACK, M.D. & HEALY, J.H. 1992. *In situ* stress measurements to 3.5 km depth in the Cajon Pass scientific research borehole: implications for the mechanics of crustal faulting. *Journal of Geophysical Research*, **97**, 5039–5057.

Evidence for seismogenic normal faults at shallow dips in continental rifts

GEOFFREY A. ABERS

Department of Earth Sciences, 685 Commonwealth Avenue, Boston University, Boston, MA 02215, USA (e-mail: abers@bu.edu)

Abstract: Several recent observations indicate that normal faulting earthquakes occasionally occur on faults dipping <35°, dips often considered shallow. Most of these occur in the Woodlark and Aegean rifts. These two rifts are found to generate significantly more earthquakes than others and are the most rapidly extending, and so display the widest variety of fault behaviour. Even within the Woodlark Rift system extension rates vary along strike, with the shallowest-dipping faults confined to the most rapidly rifting segment. Here, several events (M_w 6.0–6.8) feature nodal planes dipping 23–35°. These planes are subparallel to shear zones bounding nearby metamorphic core complexes, including one imaged to 8–9 km depth by seismic reflection profiling. In the western Gulf of Corinth at least one large event (M_w 6.4) occurred on a fault dipping *c.* 33°. Similarly to the Woodlark example, this rift segment exhibits a high opening rate (10–20 mm a^{-1}). Several other cases elsewhere, based on older historical data, microseismicity, or geological inference suggest seismic slip at similar or shallower dips. However, no documented large earthquake exhibits seismic slip on subhorizontal surfaces (dip <10–15°). Stress rotation may explain the 23–35° dips, but thus far no realistic mechanism has been found. More likely, these faults represent surfaces somewhat weaker than surrounding rock, through some combination of modest cohesion of the surrounding rock and slightly lower frictional coefficients on the fault. Such weakening may be a consequence of high slip rates, which rapidly generate large offsets, and of mature fault systems.

Although fault mechanics appears to preclude brittle slip on low-angle normal faults, dipping <30° (e.g. Anderson 1951), several structures found in the geological record appear to suggest that such slip is possible (e.g. Wernicke 1995). The seismic record, until recently, did not show any well-documented normal-faulting earthquakes with such dips (Jackson & White 1989), in part prompting the development of theories for generating low-angle faults without requiring brittle slip at those dips (e.g. Buck 1988; Wernicke & Axen 1988). In many of these theoretical studies, dips of the active portion of normal faults are limited to ≥45° (e.g. Lavier *et al.* 1999). However, in recent years several examples of normal faulting earthquakes have been documented for which dips <35° have been claimed, some of which may show dips <30° (Doser 1987; Abers 1991; Johnson & Loy 1992; Rietbrock *et al.* 1996; Rigo *et al.* 1996; Abers *et al.* 1997; Bernard *et al.* 1997; Miller & John 1999). This evidence has been used as demonstration that low-angle normal faulting can occur seismically.

The core of this paper is a review of these observations and the settings in which they occur, with focus on the Woodlark Rift of Papua New Guinea. A global analysis of seismicity data shows that background seismicity rates in rifts correlate well with extension rates, so it should come as no surprise that most evidence for these 'unusual' earthquakes comes from those rifts that are most rapidly extending (Woodlark and Aegean rifts). Such a correlation may reflect sampling bias (Wernicke 1995), or may reflect more favourable conditions (fault weakening, block rotation, etc.) at high extension rates. The focal mechanisms for the potential low-angle earthquakes reveal two things. First, several large earthquakes clearly rupture fault planes dipping 23–35°, so models of fault generation requiring brittle slip at dips of ≥45° require revision. The 23–35° dips are consistent with frictional properties expected for mature faults with well-developed gouge zones (e.g. Byerlee & Savage 1992), and unusual mechanics may not be needed. Second, at present, no large earthquake has been found showing slip on a fault unequivocally dipping

From: WILSON, R.C.L., WHITMARSH, R.B., TAYLOR, B. & FROITZHEIM, N. 2001. *Non-Volcanic Rifting of Continental Margins: A Comparison of Evidence from Land and Sea.* Geological Society, London, Special Publications, **187**, 305–318. 0305-8719/01/$15.00 © The Geological Society of London 2001.

<20°, although such slip can be found in microseismicity. However, the record of reliable focal mechanisms is no more than 35 years long, much less than large-earthquake repeat times, even in the few rifts where 25–35° dips are found. Thus, the presence of low-angle normal faults in the geological record remains enigmatic: we cannot yet conclude whether shallower-dipping faults are truly aseismic, or merely rupture infrequently.

The use of the term 'low-angle' varies considerably; it most often refers to normal faults dipping <30° but is sometimes applied to a wider range of dips. Although the 30° dip threshold has mechanical significance, it lies in the middle of the uncertainty range of dips for many of the earthquake-generating faults discussed here, so use of the term 'low-angle' may confuse. The events discussed here are interesting because they show faulting at dips close to or <30° or because they have been claimed to be 'low-angle'; in any case, the faults slip despite the relatively high shear stresses required for neighbouring rock. An alternative term is needed to describe these earthquakes. Rather than use the term 'low-angle', I will describe these fault geometries as 'shallow-dipping', referring to specific dip ranges wherever possible, and, it is hoped, will offend fewer readers.

Seismicity in rifts

As in other tectonic settings, seismicity in rifts is controlled by a combination of plate motion rates, seismic coupling, fault geometry, and the short duration of instrumental records of earthquakes. To illustrate this, a comparison of the seismicity rates for several of the world's rifts to known extension velocities shows good correlation (Fig. 1, Table 1). Other compilations have primarily concentrated on the largest earthquakes by cataloguing fault-plane solutions (e.g. Jackson & White 1989) or cumulative seismic moment (Jackson & McKenzie 1988), and so provide useful insights into extension rates, as the largest earthquakes constitute the majority of the extension accommodated by earthquakes. However, such studies are often hampered by the relatively short period of instrumental recording compared with the recurrence time of most faults, making the present comparison a useful complement.

Table 1 compares seismicity rates with rift opening rates for the several rifts studied; Ruppel (1995) discussed other parameters. Seismicity rates calculated here are based on seismic moment estimates derived from standard catalogues (Fig. 1), normalized by the along-strike length of the rifts. These data provide both seismicity rates as a function of moment-magnitude and estimates of detection thresholds in different rifts. Again, the seismicity rates do not provide estimates of overall extension rates in a region, which is dominated by the occurrence of infrequent, large earthquakes, but describe the background seismicity that makes up most of the earthquake record.

Rift opening rates are compiled from a number of sources, with preference given to geodetic rates in the Basin and Range (Dixon *et al.* 1995), Aegean (Le Pichon *et al.* 1995), Rio

Table 1. *Active continental rift systems*

Rift	Length L (km)	Width (km)	Opening rate V (mm a^{-1})	Maximum earthquake depth H (km)	Extrapolated number of events N with M $\geqslant 4$ per year, per	
					L ($\times 10^{-5}$ m^{-1} a^{-1})	LVH ($\times 10^{-7}$ m^{-3})
Woodlark	600	100	25–40	7–9	6.8 ± 0.9	2.6 ± 0.7
Aegean	800	500	35	10–15	12.8 ± 0.2	2.9 ± 0.5
Basin and Range	1200	600	10	15	1.7 ± 0.1	1.2 ± 0.3
Red Sea–Suez	2200	250	7–12	5–10	1.2 ± 0.1	1.7 ± 0.6
Baikal	1500	100	4.5	25	1.4 ± 0.2	1.2 ± 0.7
East Africa	3000	100	<3	35	2.0 ± 0.1	3.8 ± 3.1
Rhine	400	100	0.5–1	20	0.09	0.6 ± 0.3
Rio Grande	1000	100	<3	15	0.1 ± 0.1	0.5 ± 0.4

Descriptive data from Ruppel (1995) and other sources as discussed in text. Opening rates and lengths refer to continental segments only. Annual earthquake rates extrapolated from maximum-likelihood regressions (Fig. 1), normalized to along-strike length of rift, or to product of L, seismogenic zone depth H, and opening rate V. Uncertainties are $1-\sigma$, for N from regression; propagated uncertainties in V and H assumed uniformly distributed at 5 mm a^{-1} and 2.5 km, respectively, unless range is shown. Seismicity data for Rhine Graben from Ahorner (1983), from ISC catalogue elsewhere.

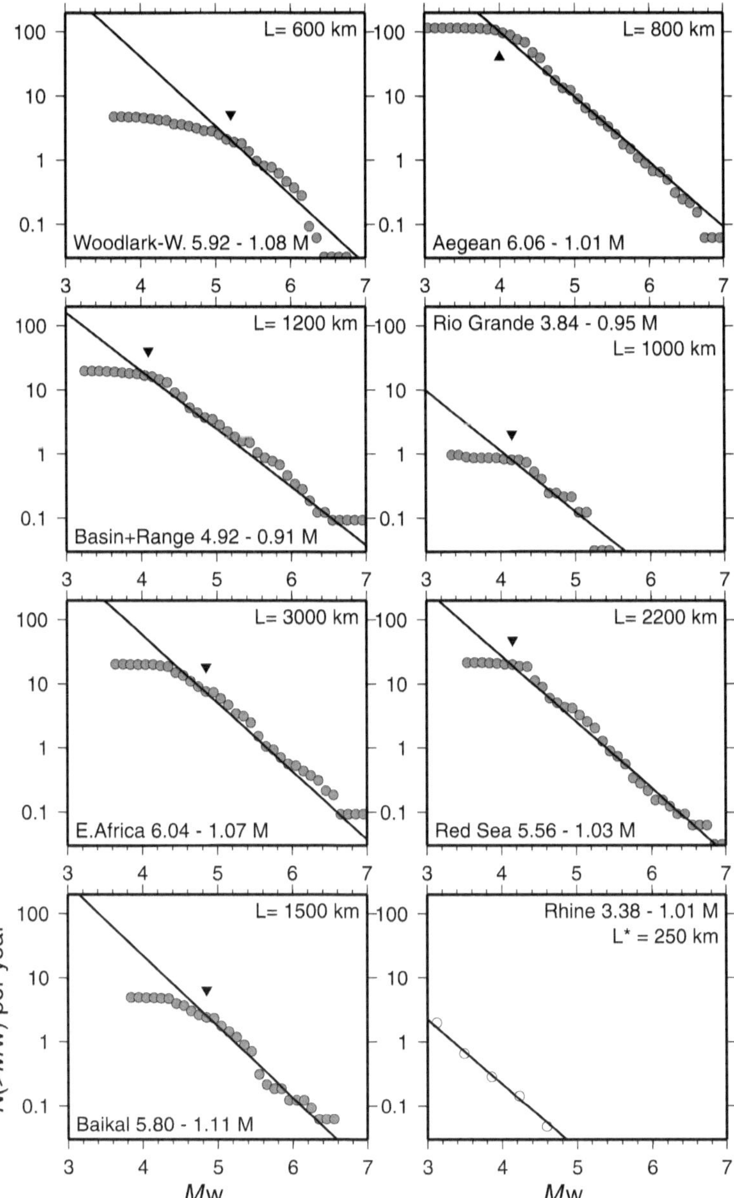

Fig. 1. Seismicity rates from the world's active continental rifts (Table 1). Plots show number of events (N) larger than a given moment magnitude (M_w), from a combination of International Seismological Centre (ISC) and Harvard centroid-moment tensor (HCMT) catalogues, spanning 1964–1995 and 1977–1995, respectively. Within each region normal faulting predominates. Magnitudes are taken (in order of preference) from the HCMT catalogue, M_s and m_b values from the ISC catalogue. M_s and m_b are converted to M_w following Ekström & Dziewonski (1986); Nuttli (1985), respectively. Lines show fits to $\log(N) = a - bM_w$ from a maximum-likelihood regression (Weichert 1980) that determines a and b. Minimum magnitudes for fits (▼) are those below which estimates of b become unstable and low. In all regions, the maximum possible magnitude is assumed to be 7.0, although variation of this parameter by one M_w unit changed results insignificantly. Data and regression result for the Rhine graben are taken from Ahorner (1983), utilizing his M_L–M_w relation, rather than the insufficient ISC catalogue; L^* gives rift length covered by Ahorner (1983). These regressions are used to predict the number of events with $M_w \geqslant 4.0$ in Table 1, after normalizing to an along-strike rift length (L). (Note the wide range in both seismicity levels and detection.)

Grande (Savage *et al.* 1985), Baikal (Calais *et al.* 1998), and the Rhine Graben (Ahorner 1983). Elsewhere, opening rates are based on magnetic lineations in the Woodlark Rift (Taylor *et al.* 1995) and the Red Sea–Gulf of Suez. Finally, an upper limit of extension rates for East Africa can be gained by closure of global plate velocity circuits, following DeMets *et al.* (1990). Uncertainties typically are in the range of a few millimetres per year, so differences between fast and slow rifts are significant.

These compilations illustrate several trends in the rift-seismicity record. First, seismicity rates correlate well with plate opening rates. When normalized to extension rates (V) and depth of seismogenic zone (H), seismicity rates show remarkable constancy (Table 1). This suggests that large variations in seismic coupling sometimes inferred from seismic moment release rate may be partly attributable to the short historical record for large earthquakes, as may be true for subduction zones (McCaffrey 1997). Second, detection of earthquakes is poor in many but not all rifts. For example, the Woodlark Rift shows evidence of incomplete sampling at M_w <5.2 and almost no earthquakes with M_w <4.8, whereas earthquakes as small as M_w 3.0 are routinely reported for the Aegean and Basin and Range.

After correcting for sampling, two rifts stand out as having high seismicity, the Woodlark and Aegean (Table 1). These are the two most likely places to see rift-related earthquakes, and they are also regions where extensional geometries that accommodate large amounts of extension are favoured.

Large earthquakes

The Woodlark Rift, Papua New Guinea

Setting. The Woodlark Rift (Fig. 2) shows a full transition from sea-floor spreading in the east to extension in quasi-continental crust farther west (e.g. Mutter *et al.* 1996; Taylor *et al.* 1999). The continental section features development of several metamorphic core complexes along the rift (Davies & Warren 1988). Extension takes place about a pole somewhere west of 147°E, with rates in the continental section ranging from 10 to 40 mm a^{-1} (Taylor *et al.* 1999). The easternmost extensional feature, Moresby 'Seamount', lies immediately west of the oceanic rift tip and also shows a north-dipping master fault zone that exposes

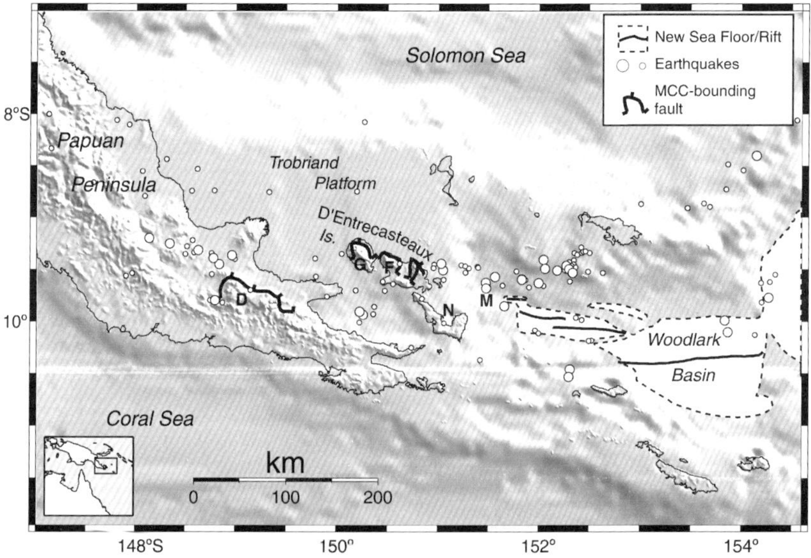

Fig. 2. Tectonic setting and seismicity of the Woodlark Rift, relocated, from Abers *et al.* (1997). ○, events for which waveform-derived focal mechanisms are available; ○, others. The D'Entrecasteaux Islands are Goodenough, Fergusson and Normanby, labelled G, F and N respectively. M, Moresby 'Seamount'; D, Dayman Dome, a core complex on the Papuan Peninsula. Thick black lines show spreading centre in eastern oceanic half of the rift (Taylor *et al.* 1999), or faults bounding metamorphic core complexes (MCC) in western half, with barbs on hanging walls (Davies & Warren 1988). Dashed line surrounding white region denotes new (<5 Ma) sea floor.

basement rocks. The westernmost identified core complex, the Dayman Dome, lies on the Papuan Peninsula near 149°E longitude (Davies & Warren 1988) and appears to be active (Ollier & Pain 1980). Although the fault systems connecting these structures are poorly known except by seismicity, all show bounding shear zones dipping to the north and appear to lie in a single extensional domain. Subordinate deformation, reflected in seismicity and subsurface faults south of the D'Entrecasteaux Islands, suggests that some deformation may occur there west of 150.5°E (e.g. Mutter *et al.* 1996); its kinematic relationship to the main deformation zone is unclear.

The large core complexes of the D'Entrecasteaux Islands (Fig. 2) appear very young, and probably are still in the process of being exhumed. Despite tropical erosion conditions, the bounding shear zones still control the shapes of land surfaces and in many places appear fresh near their base, which along with other geomorphological observations suggests continued activity (Ollier & Pain 1980). Their footwall rocks show some of the most recent rapid uplift known, with footwall rocks experiencing 700–900 °C and 5–6 kbar conditions as recently as 3–4 Ma, and 4–5 kbar at similar temperatures at 1.5–2 Ma (Baldwin *et al.* 1993; Hill & Baldwin 1993), requiring unroofing from mid-crustal depths within the last 2 Ma. Late cooling recorded by apatite fission-track ages on shear-zone rocks occurred at 0.4–1.0 Ma (Baldwin *et al.* 1993). Although such dates reveal little about the last 0.5 Ma history, the continued structural control of land surface morphology suggests that the rapid Quaternary exhumation continues to the present (e.g. Senior & Billington 1987).

The very rapid unroofing of these rocks correlates well with the rapid extension here, which ranges from 20 to 40 mm a^{-1} across the core complex belt over the *c.* 6 Ma history of extension. The dominant shear zones bounding the D'Entrecasteaux Islands core complexes dip consistently to the north at 20–25°, although some evidence of more complex faulting patterns exists (Hill *et al.* 1992).

Earthquake locations. Seismicity has been reprocessed and relocated utilizing 3D velocity models and fully nonlinear relocation procedures, as reported elsewhere (Abers *et al.* 1997). The arrival times constraining the locations largely come from teleseismic catalogues, limiting earthquakes to magnitudes greater than *c.* 4.8 (Fig. 1). The resulting epicentres (Fig. 2) show seismicity to be localized in a band of *c.* 25 km width and confined to the

north side of the rift zone near 9.5°S, at least east of 150°E. Most of these epicentres lie near the major fault zones bounding the core complexes of the Fergusson and Goodenough Islands and the Dayman Dome, and lie just north of Moresby Seamount (e.g. Mutter *et al.* 1996; Taylor *et al.* 1999). Hence, much of the seismicity appears to follow a contiguous system of faults along strike, and is generally consistent with movement on faults that lie on the north side of the core complexes. Such a pattern would be expected were the faults responsible for core complex exhumation active at present.

Earthquake fault planes. First-motion focal mechanisms (Ripper 1982; Weissel *et al.* 1982) document the extensional nature of the Woodlark–D'Entrecasteaux Islands region. These methods can have relatively large uncertainties, as the ray geometries are rarely ideal. Studies of Woodlark Rift earthquakes based on waveform inversion provide much clearer constraints on the geometry of faulting, and are discussed here (Abers 1991; Abers *et al.* 1997).

Overall, focal mechanisms reveal north–south extension (Fig. 3), consistent with magnetic lineations (Weissel *et al.* 1982; Taylor *et al.* 1995, 1999). Normal faulting extends west to 148°E, the central Papuan Peninsula, throughout the region of metamorphic core complex development; the pole of opening must lie farther to the west. Waveform modelling shows source depths consistently <8–9 km below sea level (Abers *et al.* 1997); this cut-off depth is somewhat shallower than seen at other continental rifts (e.g. Jackson & White 1989) and is consistent with elevated temperatures during rapid rifting.

The most unusual aspect of faulting here is the abundance of normal faulting mechanisms with dips of 20–35° near the oceanic rift tip (151.7°E; Fig. 3). Between 150.5°E and 153°E, the dips of the shallower nodal planes for normal-faulting earthquakes all lie between 23° and 36°. By contrast, shallow planes for earthquakes in the Papuan Peninsula rift segment (147–150.3°E) all dip between 38° and 51°. This variation correlates with extension rates: rates are 20–40 mm a^{-1} between 150.5°E and 153°E, and lower to the west. In at least two of the eastern cases, discussed below, there is good evidence that the shallower plane is the fault plane. Hence, the tendency toward shallow–dipping fault planes appears to correlate with extension rate.

1985 Gameta earthquake. The largest event (M$_w$ 6.8; 1 in Fig. 3) occurred on 29 October 1985, 10–40 km east of the eastern core com-

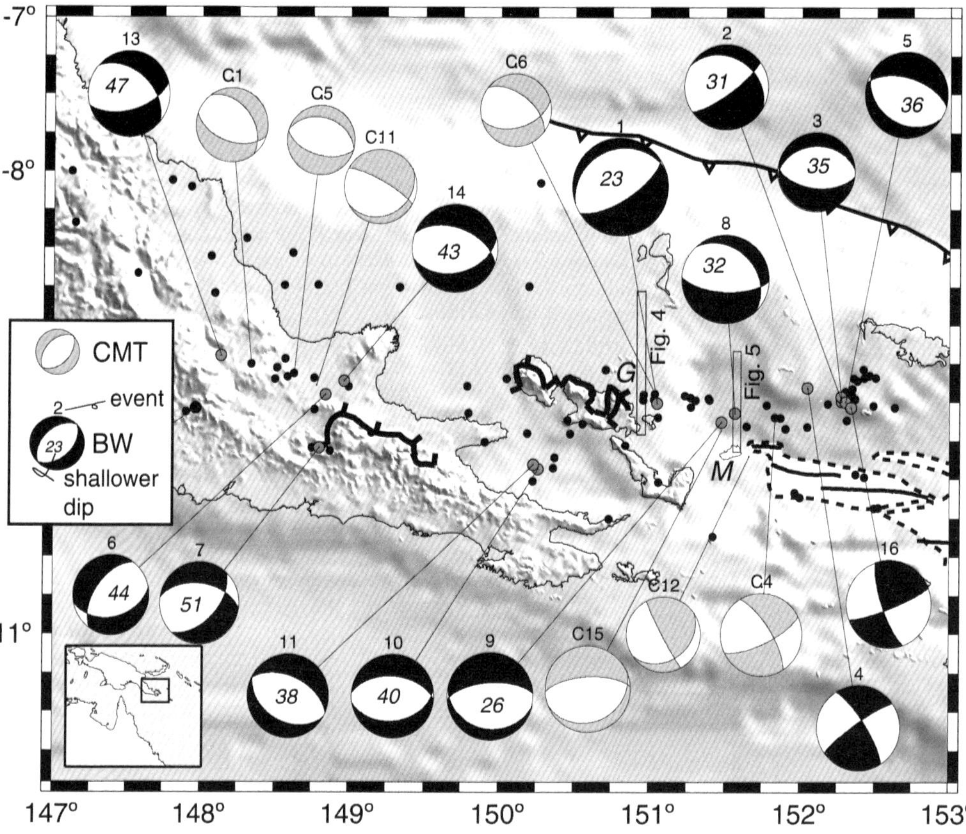

Fig. 3. Focal mechanisms of the Woodlark Rift determined from waveform inversion. Earthquake locations same as in Figure 2. Focal mechanisms are lower-hemisphere equal-area projections; black, from body-wave studies (Abers 1991; Abers *et al.* 1997); grey, from HCMT catalogue (Dziewonski *et al.* 1981). Event numbers above focal mechanisms correspond to Table 1 of Abers *et al.* (1997), as do dips of shallower planes labelled inside dilatational quadrants for normal faults. Locations of cross-sections (Figs 4 and 5) are shown. G, Gameta; M, Moresby Seamount.

plex on Fergusson Island as determined by relocation, directly along strike from the dominant shear zone that bounds the Oiatabu core complex of NE Fergusson Island (Abers 1991). The focal mechanism shows one plane dipping NNW at 23°; waveforms rule out dips for that plane as steep as 30° (Abers 1991). Unfortunately, no direct method exists to choose the fault plane from among the two nodal planes, but comparison with the nearby shear zone favours the shallow-dipping plane. First, a north-dipping plane would be expected were the shear-zone bounding faults active in their present orientation. The shear zone dips 20–25° to the north. Second, the north-dipping plane is consistent with uplift of the core complex whereas the south-dipping plane suggests that the island is submerging (Fig. 4). The former scenario is preferred, as the core complexes

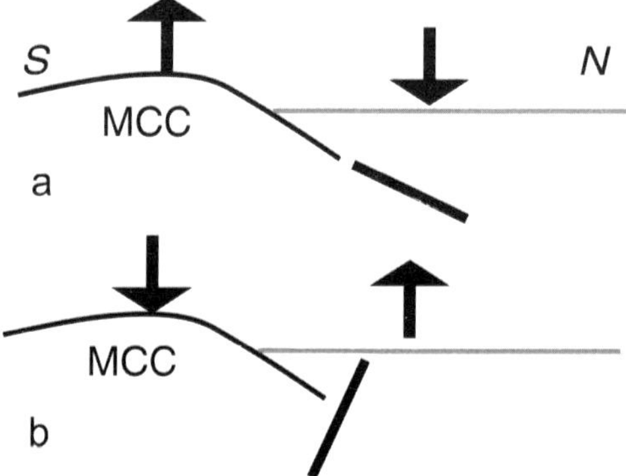

Fig. 4. Illustration of the two possible fault planes for the 1985 Gameta earthquake, and the uplift pattern predicted by each. Only the north-dipping plane is consistent with the continuing exhumation of the Oiatabu core complex, which lies immediately south of the earthquake.

appear to be actively exhuming, as discussed above. The alternative, that major faulting occurs today on south-dipping faults, suggests a radical (and *ad hoc*) termination to several million years of core complex exhumation in the last 0.4 Ma, whereas the geomorphological expression (e.g. Ollier & Pain 1980) continues to show core complex exhumation.

Moresby Seamount earthquakes. A normal fault dipping *c.* 30° seems likely along the north-dipping fault that bounds Moresby Seamount, the easternmost active fault block (Taylor *et al.* 1995; Mutter *et al.* 1996; Abers *et al.* 1997). Here, several earthquakes show normal-faulting focal mechanisms with one plane dipping near 30° (Fig. 3). Multi-channel seismic lines reveal the presence of a planar north-dipping surface here that appears to be a downdip extension of the northern flank of the seamount. From an on-axis line this feature's dip is constrained via dip-moveout techniques to lie between 25° and 35° (Abers *et al.* 1997), and parallel off-axis seismic lines have been used to infer dips closer to 25° (Taylor *et al.* 1995). One significant earthquake, the closest to the central seismic profile, shows a north-dipping fault plane dipping 25–35° or nearly coincident with the imaged surface (Fig. 5). This earthquake suggests that the feature imaged on the seismic line is a fault, and one that exhibits brittle failure (i.e. earthquakes) along its surface. This interpretation was confirmed by ODP Leg 180 drilling, which sampled the exposed fault surface at Site 1117 (arrow in Fig. 5). Undeformed rock (gabbro) at *c.* 100 m below sea floor grades upward toward the fault surface into increasingly sheared breccias and mylonites; the upper 4 m below sea

floor are interpreted as fault gouge of brittle origin (Shipboard Scientific Party 1999).

The water depth overlying the Moresby Seamount earthquake (M_w 6.2; 8 in Fig. 3) is confirmed by the presence of prominent water multiples in its seismograms, and places tight constraints on the epicentre of this event (Abers *et al.* 1997). The period of these reverberations is exceedingly sensitive to the exact water depth overlying the source, and here requires a water depth of 3.0 ± 0.2 km, placing it under the deepest part of the basin (Fig. 5). There, only the shallow north-dipping plane shows prominence in the seismic reflection record; the conjugate south-dipping fault system is overlain by only *c.* 2 km of water. Hence the choice of fault plane is clear, and it seems likely that this event occurred on a normal fault dipping near 30°.

Gulf of Corinth

The Gulf of Corinth (Fig. 6) accommodates a large fraction of extension in the Aegean, perhaps 15 mm a^{-1} of the *c.* 35 mm a^{-1} total (Clarke *et al.* 1998). Recent global positioning satellite (GPS) geodetic measurements in the western Gulf, east of Patras, have shown that *c.* 12 mm a^{-1} extension is localized across the single fault system underlying the Gulf (Clarke *et al.* 1997), accompanied by abundant microseismicity in the Patras area (Rigo *et al.* 1996). Utilizing waveform cross-correlation methods, Rietbrock *et al.* (1996) were able to reduce hypocentral uncertainties for the microseismicity to a few tens of metres, and show that most events lay in a plane dipping north at 12–20°. Although focal mechanisms of some

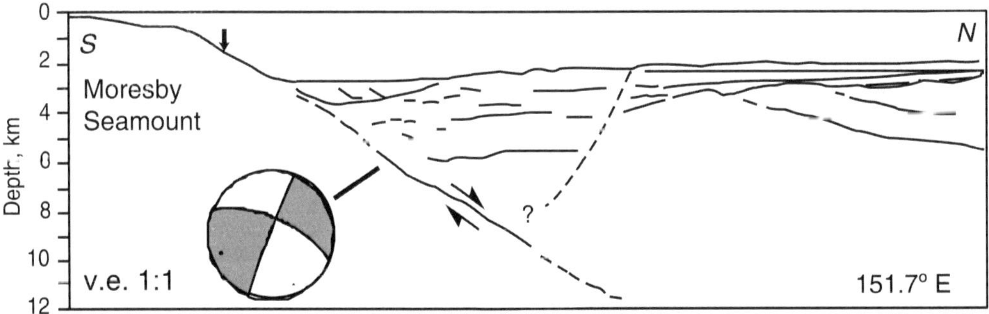

Fig. 5. Interpretation of seismic line L1218 across Moresby Seamount, and focal mechanism of nearest earthquake (6 January 1970). Data have been discussed and presented by Abers *et al.* (1997). Focal mechanism is side looking, projected into the plane of the profile, shown at the preferred depth. The near-coincidence of the north-dipping nodal plane with the inferred fault surface should be noted. Vertical arrow shows approximate position of fault gauge sampled by ODP Leg 180 (Shipboard Scientific Party 1999).

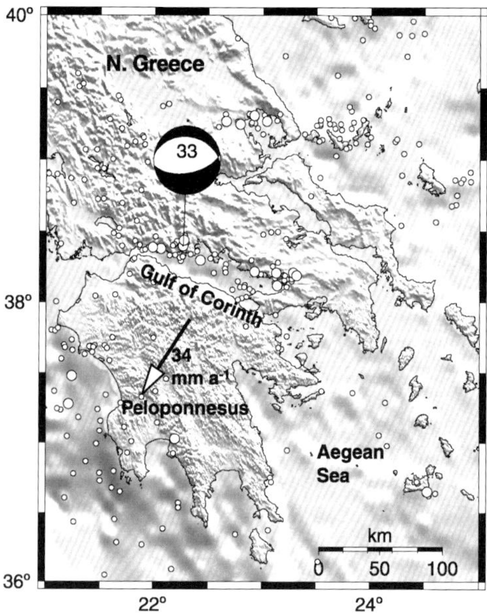

Fig. 6. Extension in the Gulf of Corinth region. ○, earthquakes in ISC catalogue with seismic moment estimated at $>10^{16}$ N m as estimated from magnitudes as in Fig. 1. ○, earthquakes in HCMT catalogue with apparent normal-faulting mechanisms (P-axis plunge $>45°$). Focal mechanism shows waveform-based result (Bernard *et al.* 1997) for the 15 June 1995 Aigion earthquake (M_s 6.2); dip of shallow plane labelled. Large arrow shows motion of Peloponnesian peninsula relative to Eurasia, averaged from vector velocities on peninsula of Clarke *et al.* 1998; 10–20 mm a^{-1} of this rate is taken up across the Gulf.

microearthquakes show a nodal plane of similar dip (Rigo *et al.* 1996), reanalysis of an expanded dataset shows that the north-dipping planes most often dip at *c.* 30° at depth in the western Gulf (Hatzfeld *et al.* 2000). Perhaps the plane of seismicity reflects the brittle–ductile transition rather than a fault plane (Hatzfeld *et al.* 2000).

On 6 June 1995, a large earthquake (M_s 6.2) ruptured throughout the centre of the micro-earthquake and GPS network, causing extensive damage at the town of Aigion (Tselentis *et al.* 1996). Detailed waveform inversion of this earthquake shows normal faulting at a depth of 7.2 km with a plane dipping north at 33° ± 5° (Bernard *et al.* 1997), determined using methods similar to those for the Woodlark Rift studies. In this case, good microseismic and geodetic data confirm that the shallow-dipping nodal plane is the fault. That fault geometry (Fig. 7) coincides with GPS and interferometric

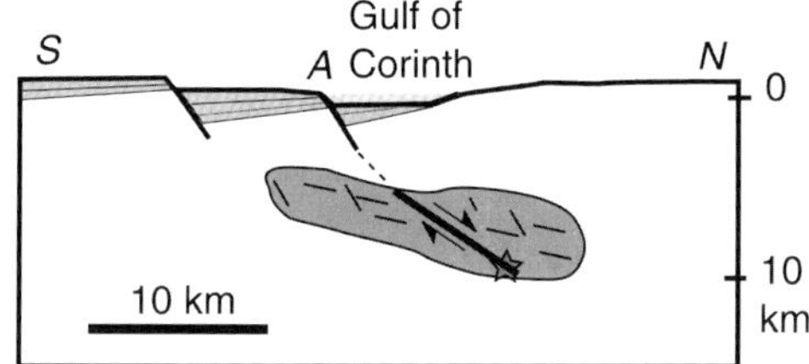

Fig. 7. Schematic diagram illustrating seismic zone in the western Gulf of Corinth. Grey oval shape shows region of background seismicity, abstracted from Rigo *et al.* (1996). This shows focal mechanisms consistent with dips of 10–25° to the north (12–20° in the relocations of Rietbrock *et al.* (1996)). Star shows focus of 1995 Aigion mainshock and heavy line shows fault plane inferred from seismic waveforms and geodetic observations (Bernard *et al.* 1997). North–south section transects western Gulf near 22.0°E longitude. A, the town of Aigion, which was heavily damaged.

estimates of surface displacement, which suggest a fault slip of 0.9 m at depths between 2.5 and 10 km (Bernard *et al.* 1997). The fault dips steeper than inferred from the microseismic studies discussed above, although depths are similar.

In summary, the western Gulf of Corinth shows an example of normal faulting at the shallow end of dips commonly observed (i.e. 33°), a geometry supported by geodetic and seismic observations. As with the Woodlark examples, extension here is concentrated along a single fault system that accommodates very rapid slip (10–20 mm a^{-1} here); such rates are possible only in the most rapidly opening rifts (Table 1). Also as with the Woodlark example, along-strike changes in opening rates correlate with changes in fault dips: in the eastern Gulf of Corinth fault plane dips increase to *c.* 45° and extension rates decrease to *c.* 6 mm a^{-1} (Clarke *et al.* 1997; Hatzfeld *et al.* 2000).

Peruvian Andes

In the Peruvian High Andes, normal faulting occurs at high elevations over an otherwise convergent plate boundary (e.g. Dalmayrac & Molnar 1981). Much of the evidence for extension comes from Quaternary fault scarps, as earthquakes are relatively rare. The largest normal-fault earthquake in the Andes occurred near Ancash, Peru, in 1946, and produced extensive surface rupture along the west-dipping Quiches fault (Bellier *et al.* 1991). First motions and waveform modelling show that the fault plane dips 30° for this event, with a source depth of 15–17 km and an M_w of 6.8 (Doser 1987); a 10° uncertainty in dip was claimed by Doser. The actual uncertainties could be larger, as instrumentation in 1946 was both poorly distributed and of uncertain calibration. The seismic moment and surface rupture give a source dimension of many tens of kilometres; a slightly larger seismic moment (M_w 7.0) was estimated from fault offset and length (Bellier et al. 1991). Trenching across the surface trace reveals 1.7–3 m of offset across a fault, which, together with one previous post-glacial rupture, implies a vertical displacement rate <0.25 mm a^{-1}. Hence, the Ancash earthquake represents a third region in which moderately low-angle normal faulting may have produced an earthquake. Unlike the previous two examples, it occurred in a setting where the long-term extension rate is fairly low; geodetic data limit extension across the Altiplano to less than a few millimetres per year (Norabuena et al. 1998).

Prehistoric examples

The geological record contains many examples of normal faults that appear to have moved at shallow dips (Axen 1992; Wernicke 1995). In the two examples below, some evidence exists for seismic slip or surface offset; low-angle faulting is claimed to occur via earthquakes.

Johnson & Loy (1992) showed that late Quaternary fault scarps along the Santa Rita fault, southern Arizona, extend to depth as planar features dipping 20° on seismic reflection sections. Although no earthquakes have been observed on this fault, geomorphological evidence suggests movement within the last 100 ka, over an area corresponding to an earthquake with M_w 6.7–7.6. Although seismic slip is not confirmed, this fault probably moved at depths normally considered within the brittle regime and at a dip near 20°.

Near the Chemehuevi–Sacramento Mountains, depositional patterns led Miller & John (1999) to infer earthquakes along a low-angle fault (dip ⩽30°), active between 23 and 12 Ma. They argued on the basis on the geometry and distribution of coarse-grained strata that sediments were deposited directly upon a detachment fault while that fault lay at shallow dip (<30°) and was active. Megabreccias represent catastrophic depositional events, which those workers argued were triggered by earthquakes with magnitude ⩾6.0, the only clear source for which is the detachment fault itself. Fault slip rates here were fairly rapid, 7–8 mm a^{-1} during peak extension.

Discussion

These observations suggest that earthquakes do occasionally occur on normal faults that dip near 30°, the low end of the range found by Jackson & White (1989). However, their occurrence is rare and perhaps tied to special geological circumstances. In most cases, such faulting is found in fault systems of high slip rate (>10 mm a^{-1}), the one exception being the Ancash, Peru event, on which dip constraints are relatively weak. Furthermore, except for microseismicity the record does not yet include an earthquake on a low-angle fault dipping <20°. Whether that absence reflects a fundamental physical limit of normal faulting or merely a longer recurrence time for such events (Wernicke 1995) remains to be seen.

Explanations for such faulting generally fall into two categories, those that rely upon basal tractions (e.g. Yin 1989; Melosh 1990) or magmatic intrusion (Parsons & Thompson 1993) to rotate stress into a favourable geometry, and those that appeal to fault weakening (e.g. Axen 1992). Although the former approach has many attractions, it has been shown that most stress-rotation models do not generate sufficient shear stresses to produce brittle sliding or (in the magmatic intrusion case) promote low-angle faulting in very small areas (Wills & Buck 1997). Some situations with extreme flow rates at the base of a brittle layer may produce the sufficient differential stresses (Westaway 1999). Such a situation may indeed be a consequence of rapid extension, particularly if extension is driven from below in some way (e.g. Abers et al. 1997). However, without constraints on the temperature, flow and rheology of the lower crust, these models are difficult to test and remain speculative.

Failure requirements

Analytical estimates. Do the observed dips actually require very weak faults? Here, we calculate the stress conditions observed on faults for the dips observed, for simple (i.e. Anderson-like) driving forces, and find that only slight deviations from Byerlee's law are necessary to explain the observed fault geometries. Figure 8 illustrates the calculation: a fault with dip δ is cohesionless with static frictional coefficient F, whereas the surrounding medium is cohesive. At the relatively shallow depths of faulting observed (Figs 4–7) the effective maximum compressive (vertical) stress σ_1' is low and the tensile strength T limits rock strength in extension (Fig. 8b). The minimum effective stress

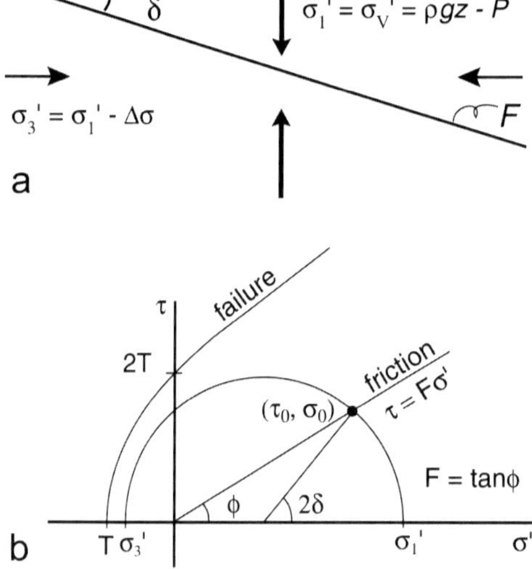

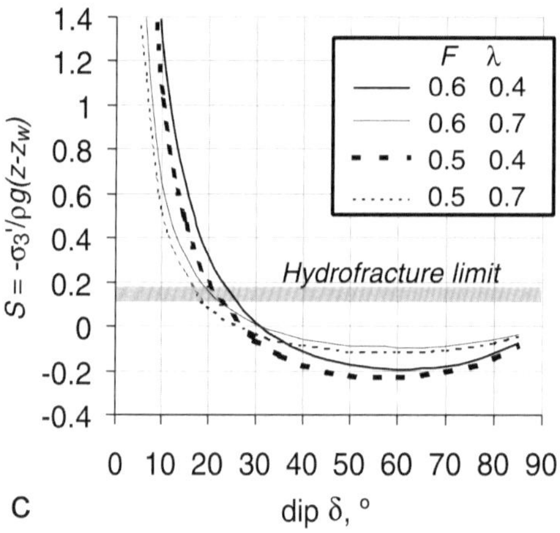

Fig. 8. Relationship between fault dip (δ), rock tensile strength (T), and effective principal stresses ($\sigma_1' > \sigma_3'$) required for low-angle normal faulting when $-T < \sigma_3' < 0$. (**a**) Assumed normal fault geometry in cross-section; fault has frictional coefficient F and no cohesion. (**b**) Mohr circle diagram showing relation between shear stresses (τ) and effective normal stresses (σ'). Stresses on the fault lie on the circle; the line labelled 'friction' is the frictional failure envelope ($F = 0.6$) and that labelled 'failure' is the assumed Coulomb–Griffith fracture envelope for intact surrounding rock with tensile strength T (after Sibson 1998). Hydraulic (tensile) fracture occurs if σ_3' becomes less than $-T$. (**c**) Calculated stress ratio S, defined in text, v. fault dip δ for various assumptions of F and λ (pore pressure ratio). Grey bar shows the range of hydrofracture limits, T', estimated in text for the base of the seismogenic zone in the Woodlark Rift.

during faulting (σ_3') may be negative in this situation: as long as σ_3' is $>-T$, tensile failure or hydraulic fracture does not occur. To estimate this limit, the parabolic Griffith–Coulomb failure criterion is assumed for the intact rock (Sibson 1998), whereas the fault is cohesionless, with stresses calculated as in Figure 8. In this scenario, faulting continues on the cohesionless fault at non-optimal angles until either Coulomb (shear) fracture or tensile (hydraulic) fracture occurs in the surrounding rock. Although the true stress state is probably affected by bending and topography (e.g. Forsyth 1992; Buck 1993), this calculation provides a first-order estimate of the physical limits to fault dip, with few assumptions of plate rheology.

Below sea floor, effective vertical stress is

$$\sigma_1' = \rho g(z - z_w)(1 - \lambda)$$

where ρ is the rock density, g is gravitational acceleration, $z-z_w$ is the depth below the sea floor, and λ is the ratio of pore pressure to lithostatic pressure ($c.$ 0.4 for hydrostatic pore pressure) (e.g. Brace & Kohlstedt 1980). If the fault slides frictionally at coefficient of friction F and dip δ, a relation exists between σ_1', σ_3', δ and F on the fault. A nondimensional horizontal stress can be defined by

$$S = \frac{-\sigma_3'}{\rho g(z - z_w)} = (1 - \lambda)\left(\frac{1 - F\cot\delta}{1 + F\tan\delta}\right)$$

(following Sibson 1985). The hydrofracture condition, $-\sigma_3' < T$, can then be recast in terms of dimensionless parameters such that hydrofracture does not occur so long as

$$S < \frac{T}{\rho g(z - z_w)} = T'.$$

Figure 8c shows values of S v. δ for several appropriate values of λ and F. For typical rocks with simple faults F is $c.$ 0.6 (e.g. Brace & Kohlstedt 1980), but in well-developed fault zones with thick gouge layers, F $c.$ 0.5 may be more appropriate (see review by Lockner (1995)). Many scenarios (and some direct evidence) call for super-hydrostatic conditions with $\lambda > 0.4$ (e.g. Axen 1992; Hickman *et al.* 1995) although that may not be necessary here. The hydrofracture (tensile failure) condition is less clear; T in crystalline rock varies from 10 to 40 MPa, with values of 25 MPa typical for mafic rocks (Lockner 1995), comparable with cohesion in intact rock assumed in more complex faulting models (Forsyth 1992; Buck 1993; Lavier *et al.* 1999). In the Woodlark Rift, earthquakes do not exceed 8 km depth and water depth varies from 0 to 3 km, so $\rho g(z - z_w)$ is c. 130–210 MPa, giving a hydrofracture threshold of $T' = 0.11-0.19$ at the base of the seismogenic zone as indicated in Figure 8c. T' could be considerably larger at shallower depths. Additional calculations show that extensional failure should be reached before shear (Coulomb) fracture as long as T' exceeds c. $0.25(1 - \lambda)$, for reasonable internal friction coefficients, over the relevant depth range, so Coulomb fracture can be ignored. Thus, the limitations set by extensional failure (hydrofracture) control the minimum allowed fault dip. A slight weakening and small cohesion can be very important in this scenario, as effective stresses are small.

Interpretation. Figure 8c shows that faults dipping as low as 25° should slip without hydraulic fracture for all conditions, and this limit may be decreased to 15–20° with modest decreases in strength or increases in pore pressure. Normal faults rapidly enter the hydraulic fracture regime as dips decrease below 10–15°, and slip on such faults would require other physical mechanisms (e.g. Axen 1992; Wernicke 1995). All observed large earthquakes occur with dips above this limit, and mechanical problems exist only if the microseismicity evidence for dips <20° (e.g. Rietbrock *et al.* 1996; Rigo *et al.* 1996) reflects large-scale stresses. The large event with the shallowest dip, the Gameta earthquake, has a centroid depth of 3 km (Abers 1991). At these depths the hydrofracture limit is not reached until $S = 0.3$, a condition easily reached on faults dipping near 20° provided cohesion remains high. Thus, we conclude that observed normal faults could be understood without requiring unusual frictional conditions.

Topographic effects. The stress calculation described above assumes simple stress geometry, in particular one with no stress rotations. Topographic variations could produce large changes in stress fields at least near the surface (e.g. McTigue & Mei 1981). One example is given in Figure 9, based on topographic effects caused by Moresby Seamount (calculation described by Abers *et al.* (1997)). Here, tectonic stresses are applied by stretching the crust until frictional failure at optimal angles ($F = 0.7$) is achieved somewhere at each depth. To isolate the topographic effect, cohesion is ignored and hydrostatic pore pressure is assumed. The faulting region is in a state of decreased shear stress owing to a stress shadowing effect, but stresses are rotated slightly to favour low-angle faulting. The net result is that failure is possible on the fault

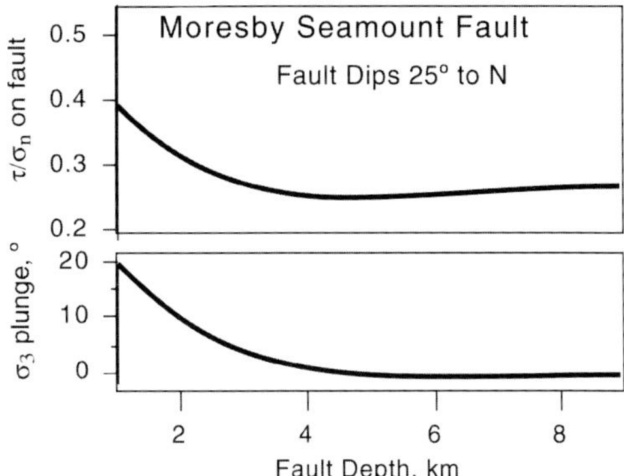

Fig. 9. Stresses predicted on the Moresby Seamount Fault, Papua New Guinea, in the presence of both topographic and extensional forces, from Abers *et al.* (1997). Fault is assumed to be planar and dip c. 25°, as in Figure 5. Extension drives the region to frictional failure beneath its weakest point (beneath the 'seamount' massif), assuming a coefficient of friction of 0.7 and hydrostatic pore pressure beneath the sea floor. Additional forces are not considered, such as basal tractions or transient effects, and material is assumed cohesionless.

dipping 25° at F near 0.3. The added effects of elevated pore pressure or cohesion should allow sliding at F closer to 0.5–0.6, as suggested in the previous section. Similar topographic stresses are expected for the Gulf of Corinth, where topography likewise defines a narrow rift basin between tall mountains.

These two sets of calculations are not exhaustive, and many alternative mechanical simulations can be found in the literature. Nevertheless, they show that reasonable assumptions about material strength and stress can account for normal-faulting earthquakes with dips of 20–35°.

Conclusions

Seismicity in rifts tends to correlate with overall rate of extension. This observation explains the tendency for unusual behaviour such as low-angle faulting to be reported most commonly in the Woodlark Rift and Aegean, the two most rapidly opening rifts. In these locales, normal faults dipping 23–35° are observed in the regions of fastest extension: 25–40 mm a^{-1} near the rift tip in the Woodlark–D'Entrecasteaux region, and 10–20 mm a^{-1} across the western Gulf of Corinth. These observations only slightly extend the range of previously reported dips for normal faults, but demonstrate that the low end of the dip range is commonly achieved; models of faulting that restrict dips to $\geq 45°$ cannot explain many earthquakes. Several other examples of potentially seismogenic low-angle normal faults have been reported in the literature but none as well constrained. These observations suggest that processes associated with high extension rates (and high overall extension) on individual fault systems may produce the conditions needed for low-angle faulting. For example, well-worn faults should achieve effective frictional coefficients near 0.5. Such coefficients do not require very unusual materials but may be expected in faults that achieve large offsets in short time spans and so can develop mature, uncemented gouge zones.

The physical conditions for such faulting remain poorly understood. Except for the Aegean case, little is known locally about the geometry of faulting, and neither the conditions within the fault zones nor the behaviour of the deeper ductile crust are well known. In particular, the role of the lower crust in these processes may be critical, yet few direct observations exist on its state. Also, direct measures of the cohesion, mechanical strength and permeability structure of these active fault zones would allow several models to be tested. Future experiments and observations aimed at resolving these ambiguities may be necessary.

This work was inspired by discussions at the Geological Society workshop on non-volcanic rifted margins. It benefited from reviews by J. Jackson and T. Reston. This research is supported by the National Science Foundation, grants OCE-9012730, OCE-9730567 and OCE-9996317.

References

ABERS, G.A. 1991. Possible seismogenic shallow-dipping normal faults in the Woodlark–D'Entrecasteaux extensional province, Papua New Guinea. *Geology*, **19**, 1205–1208.

ABERS, G.A., MUTTER, C.Z. & FANG, J. 1997. Shallow dips of normal faults during rapid extension: earthquakes in the Woodlark–D'Entrecasteaux rift system, Papua New Guinea. *Journal of Geophysical Research*, **102**, 15301–15317.

AHORNER, L. 1983. Seismicity and neotectonic structural activity of the Rhine Graben system in Central Europe. *In*: RITSEMA, A.R. & GÜRPINAR, A. (eds) *Seismicity and Seismic Risk in the Offshore North Sea Area*. Reidel, Dordrecht, 101–112.

ANDERSON, E.M. *The Dynamics of Faulting*. Oliver and Boyd, Edinburgh.

AXEN, G.J. 1992. Pore pressure, stress increase, and fault weakening in low-angle normal faulting. *Journal of Geophysical Research*, **97**, 8979–8992.

BALDWIN, S.L., LISTER, G.S., HILL, E.J., FOSTER, D.A. & McDOUGALL, I. 1993. Thermochronologic constraints on the tectonic evolution of active metamorphic core complexes, D'Entrecasteaux Islands, Papua New Guinea. *Tectonics*, **12**, 611–628.

BELLIER, O., DUMONT, J.F., SEBRIER, M. & MERCIER, J.L. 1991. Geological constraints on the kinematics and fault-plane solution of the Quiches fault zone reactivated during the 10 November 1946 Ancash earthquake, northern Peru. *Bulletin of the Seismological Society of America*, **81**, 468–490.

BERNARD, P., BRIOLE, P., MEYER, B. & 25 others 1997. The MS = 6.2 June 15, 1995 Aigion earthquake (Greece): evidence for low angle normal faulting in the Corinth rift. *Journal of Seismology*, **1**, 131–150.

BRACE, W.F. & KOHLSTEDT, D.L. 1980. Limits on lithospheric stress imposed by laboratory experiments. *Journal of Geophysical Research*, **85**, 6248–6252.

BUCK, W.R. 1988. Flexural rotation of normal faults. *Tectonics*, **7**, 959–973.

BUCK, W.R. 1993. Effect of lithospheric thickness on the formation of high- and low-angle normal faults. *Geology*, **21**, 933–936.

BYERLEE, J.D. & SAVAGE, J.C. 1992. Coulomb plasticity within the fault zone. *Geophysical Research Letters*, **19**, 2341–2344.

CALAIS, E., LESNE, O., DEVERCHERE, J. & 7 others 1998. Crustal deformation in the Baikal Rift from GPS measurements. *Geophysical Research Letters*, **25**, 4003–4006.

CLARKE, P.J., DAVIES, R.R., ENGLAND, P.C. & 11 others 1998. Crustal strain in central Greece from repeated GPS measurements in the interval 1989–1997. *Geophysical Journal International*, **135**, 195–214.

CLARKE, P.J., DAVIES, R.R., ENGLAND, P.C. & 8 others 1997. Geodetic estimate of seismic hazard in the Gulf of Korinthos. *Geophysical Research Letters*, **24**, 1303–1306.

DALMAYRAC, B. & MOLNAR, P. 1981. Parallel thrust and normal faulting in Peru and constraints on the state of stress. *Earth and Planetary Science Letters*, **55**, 473–481.

DAVIES, H.L. & WARREN, R.G. 1988. Origin of eclogite-bearing, domed, layered metamorphic complexes ('core complexes') in the D'Entrecasteaux Islands, Papua New Guinea. *Tectonics*, **7**, 1–21.

DEMETS, C., GORDON, R., ARGUS, D. & STEIN, S. 1990. Current plate motions. *Geophysical Journal International*, **101**, 425–478.

DIXON, T.H., STEFANO, R., LEE, J. & REHEIS, M.C. 1995. Constraints on present-day Basin and Range deformation from space geodesy. *Tectonics*, **14**, 755–772.

DOSER, D.I. 1987. The Ancash, Peru, earthquake of 1946 November 10; evidence for low-angle normal faulting in the high Andes of northern Peru.

Geophysical Journal of the Royal Astromical Society, **91**, 57–71.

DZIEWONSKI, A.M., CHOU, T.A. & WOODHOUSE, J.H. 1981. Determination of earthquake source parameters from waveform data for studies of global and regional seismicity. *Journal of Geophysical Research*, **86**, 2825–2852.

EKSTRÖM, G. & DZIEWONSKI, A.M. 1986. Evidence of bias in estimations of earthquake size. *Nature*, **332**, 319–323.

FORSYTH, D.W. 1992. Finite extension and low-angle faulting. *Geology*, **20**, 27–30.

HATZFELD, D., KARAKOSTAS, V., ZIAZIA, M. & 5 others 2000. Microseismicity and fault geometry in the Gulf of Corinth (Greece). *Geophysical Journal International*, **141**, 438–456.

HICKMAN, S., SIBSON, R. & BRUHN, R. 1995. Introduction to special section: mechanical involvement of fluids in faulting. *Journal of Geophysical Research*, **100**, 12831–12840.

HILL, E.J. & BALDWIN, S.L. 1993. Exhumation of high pressure metamorphic rocks during crustal extension in the D'Entrecasteaux Islands, Papua New Guinea. *Journal of Metamorphic Geology*, **11**, 261–277.

HILL, E.J., BALDWIN, S.L. & LISTER, G.S. 1992. Unroofing of active metamorphic core complexes in the D'Entrecasteaux Islands, Papua New Guinea. *Geology*, **20**, 907–910.

JACKSON, J. & MCKENZIE, D. 1988. The relationship between plate motions and seismic moment tensors, and the rates of active deformation in the Mediterranean and Middle East. *Geophysical Journal International*, **93**, 45–73.

JACKSON, J.A. & WHITE, N.J. 1989. Normal faulting in the upper continental crust: observations from regions of active extension. *Journal of Structural Geology*, **11**, 15–36.

JOHNSON, R.A. & LOY, K.L. 1992. Seismic reflection evidence for seismogenic low-angle faulting in southeastern Arizona. *Geology*, **20**, 597–600.

LAVIER, L.L., BUCK, W.R. & POLIAKOV, A.N.B. 1999. Self-consistent rolling-hinge model for the evolution of large-offset low-angle normal faults. *Geology*, **27**, 1127–1130.

LE PICHON, X., CHAMOT-ROOKE, N., LALLEMANT, S., NOOMEN, R. & VEIS, G. 1995. Geodetic determination of the kinematics of central Greece with respect to Europe; implications for eastern Mediterranean tectonics. *Journal of Geophysical Research*, **100**, 12675–12690.

LOCKNER, D.A. 1995. Rock failure. *In*: AHRENS, T.J. (ed.) *Rock Physics and Phase Relations, American Geophysical Union Reference Shelf 3*. 127–147. American Geophysical Union, Washington, DC.

MCCAFFREY, R. 1997. Statistical significance of the seismic coupling coefficient. *Bulletin of the Seismological Society of America*, **87**, 1069–1073.

MCTIGUE, D.F. & MEI, C.C. 1981. Gravity-induced stresses near topography of small slope. *Journal of Geophysical Research*, **86**, 9268–9278.

MELOSH, H.J. 1990. Mechanical basis for low-angle normal faulting in the Basin and Range province. *Nature*, **343**, 331–335.

MILLER, J.M.G. & JOHN, B.E. 1999. Sedimentation patterns support seismogenic low-angle normal faulting, southeastern California and western Arizona. *Geological Society of America Bulletin*, **111**, 1350–1370.

MUTTER, J.C., MUTTER, C.Z. & FANG, J. 1996. Analogies to oceanic behaviour in the continental breakup of the western Woodlark basin. *Nature*, **380**, 333–336.

NORABUENA, E., LEFFLER-GRIFFIN, L., MAO, A., DIXON, T., STEIN, S., SACKS, I.S., OCOLA, L. & ELLIS, M. 1998. Space geodetic observations of Nazca–South America convergence across the Central Andes. *Science*, **279**, 358–362.

NUTTLI, O.W. 1985. Average seismic source-parameter relations for plate-margin earthquakes. *Tectonophysics*, **118**, 161–174.

OLLIER, C.D. & PAIN, C.F. 1980. Active rising surficial gneiss domes in Papua New Guinea. *Journal of the Geological Society of Australia*, **27**, 33–44.

PARSONS, T. & THOMPSON, G.A. 1993. Does magmatism influence low-angle normal faulting? *Geology*, **21**, 247–250.

RIETBROCK, A., TIBERI, C., SCHERBAUM, F. & LYON-CAEN, H. 1996. Seismic slip on a low-angle normal fault in the Gulf of Corinth: evidence from high-resolution cluster analysis of microearthquakes. *Geophysical Research Letters*, **23**, 1817–1820.

RIGO, A., LYON-CAEN, H., ARMIJO, R., DESCHAMPS, A., HATZFELD, D., MAKROPOULOS, K. & PAPADIMITRIOU, P. 1996. A microseismic study in the western part of the Gulf of Corinth (Greece): implications for large scale normal faulting mechanisms. *Geophysical Journal International*, **126**, 663–688.

RIPPER, I.D. 1982. Seismicity of the Indo-Australian/Solomon Sea plate boundary in the southeast Papua region. *Tectonophysics*, **87**, 355–369.

RUPPEL, C. 1995. Extensional processes in continental lithosphere. *Journal of Geophysical Research*, **100**, 24187–24215.

SAVAGE, J.C., LISOWSKI, M. & PRESCOTT, W.H. 1985. Strain accumulation in the Rocky Mountain states. *Journal of Geophysical Research*, **90**, 10310–10320.

SENIOR, B.R. & BILLINGTON, W.G. 1987. Structural interpretation of northwest Fergusson Island, Papua New Guinea. *In*: BRENNAN, E. (convenor) *Pacific Rim Congress 87*. Australasian Institute of Mining and Metallurgy, Gold Coast, 387–392.

Shipboard Scientific Party 1999. Leg 180 summary. *In*: TAYLOR, B., HUCHON, P., KLAUS, A. *et al.* (eds) *Proceedings of the Ocean Drilling Program, Initial Reports, 180*. Ocean Drilling Program, College Station, TX, 1–77.

SIBSON, R.H. 1985. A note on fault reactivation. *Journal of Structural Geology*, **7**, 751–754.

SIBSON, R.H. 1998. Brittle failure mode plots for compressional and extensional tectonic regimes. *Journal of Structural Geology*, **20**, 655–660.

TAYLOR, B., GOODLIFFE, A., MARTINEZ, F. & HEY, R. 1995. Continental rifting and initial sea-floor spreading in the Woodlark basin. *Nature*, **374**, 534–537.

TAYLOR, B., GOODLIFFE, A.M. & MARTINEZ, F. 1999. How continents break up: insights from Papua New Guinea. *Journal of Geophysical Research*, **104**, 7497–7512.

TSELENTIS, G.A., MELIS, N.S., SOKOS, E. & PAPATSIMPA, K. 1996. The Egion June 15, 1995 (6.2 ML) earthquake, Western Greece. *Pure and Applied Geophysics*, **147**, 83–98.

WEICHERT, D.H. 1980. Estimation of the earthquake recurrence parameters for unequal observations for different magnitudes. *Bulletin of the Seismological Society of America*, **70**, 1337–1346.

WEISSEL, J.K., TAYLOR, B. & KARNER, G.D. 1982. The opening of the Woodlark Basin, subduction of the Woodlark spreading system, and the evolution of northern Melanesia since mid-Pliocene time. *Tectonophysics*, **87**, 253–277.

WERNICKE, B. 1995. Low-angle normal faults and seismicity: a review. *Journal of Geophysical Research*, **100**, 20159–20174.

WERNICKE, B. & AXEN, G.A. 1988. On the role of isostasy in the evolution of normal fault systems. *Geology*, **16**, 848–851.

WESTAWAY, R. 1999. The mechanical feasibility of low-angle normal faulting. *Tectonophysics*, **308**, 407–443.

WILLS, S. & BUCK, W.R. 1997. Stress-field rotation and rooted detachment faults; a Coulomb failure analysis. *Journal of Geophysical Research*, **102**, 20503–20514.

YIN, A. 1989. Origin of regional rooted low-angle normal faults: a mechanical model and its tectonic implications. *Tectonics*, **8**, 469–482.

Deformation fabrics of faulted rocks, and some syntectonic stress estimates from the active Woodlark Basin detachment zone

SYBILLE ROLLER[1], JAN H. BEHRMANN[1] & ACHIM KOPF[2]

[1]*Geologisches Institut, Universität Freiburg, Albertstr. 23-B, D-79104 Freiburg, Germany*
(e-mail: rollers@sun2.ruf.uni-freiburg.de)

[2]*Laboratoire de Géodynamique Sous-Marine, B.P. 48, F-06235 Villefranche-sur-Mer Cedex,*
France

Abstract: Ocean Drilling Program (ODP) Leg 180 investigated, in the western Woodlark Basin off Papua New Guinea, the nature and evolution of continental extension, eventually leading to crustal break-up and sea-floor spreading. At Moresby Seamount, the rift-related extension is localized at a recently active low-angle (30°) detachment fault, partly buried beneath a Pliocene–Pleistocene sedimentary synrift sequence. Data from three drillsites sample the detachment fault itself, secondary faults in its hanging wall and a steep normal fault cutting the footwall. The fault plane itself is manifested as a strongly altered fault gouge. Deformation of turbiditic sediments in several fault zones in the hanging wall is dominated by brittle mechanisms, and accompanied by intensive veining and pervasive diagenetic cementation. The metabasic rocks of the footwall below the detachment show an unusual transition from ductile to brittle deformation fabrics with increasing depth. Many fracture systems show evidence of repeated opening and healing during multistage hydrothermal mineralization. Syn-mylonitic microstructures and vein fill mineralogy suggest exhumation of the detachment footwall from considerable depth in the crust. Two palaeo-piezometers were applied to calcite-filled veins that show evidence of plastic deformation. Differential stress values of similar magnitude and probably close to the rock failure strength are found in both the hanging wall and footwall.

The nature and origin of large extensional detachments was first described from so-called metamorphic core complexes (e.g. Coney & Harms 1984). Later, the study of passive, rifted continental margins (e.g. Lister *et al.* 1986) showed that some of them are floored by extensional detachments, which play an important role in accommodating whole-lithosphere stretching. In both cases the extensional detachments are fossil tectonic systems that do not allow a study of deformation processes *in vivo* and *in situ*. One of the principal goals of Ocean Drilling Program (ODP) Leg 180 in the western Woodlark Basin (east of Papua New Guinea; see Taylor *et al.* 1999*b*) was to penetrate an active extensional detachment system (Abers *et al.* 1997) by deep ocean drilling, and thus increase understanding of the nature and the evolution of a low-angle normal fault in an active tectonic environment.

Investigations in a recently active rifting–spreading transition zone (e.g. Taylor *et al.* 1999*a*) can improve our understanding of mechanisms and deformation history of low-angle detachment faulting of comparable fossil tectonic settings. Our microstructural observations and data are based on faulted rocks that show structures related to very young or even recent tectonic activity, and show no signs of later static retrograde metamorphic overprinting and/ or diagenesis. Hence, these structures are expected to store a direct response to extensional tectonics and associated fluid activity on all scales. Their interpretation can lead to a better understanding of detachment faulting mechanisms.

Geological setting

The study area (around 151°35′E and 9°50′S) is located in a narrow transition zone between continental rifting in the west and oceanic spreading in the east (Fig. 1). Oceanic spreading in the Woodlark Basin has propagated to the west since 6 Ma (oldest sea-floor magnetization; Taylor *et al.* 1995) with the segmented, roughly east–west-oriented spreading axis reaching a length of 500 km and producing a maximum of

From: WILSON, R.C.L., WHITMARSH, R.B., TAYLOR, B. & FROITZHEIM, N. 2001. *Non-Volcanic Rifting of Continental Margins: A Comparison of Evidence from Land and Sea.* Geological Society, London, Special Publications, **187**, 319–334. 0305-8719/01/$15.00 © The Geological Society of London 2001.

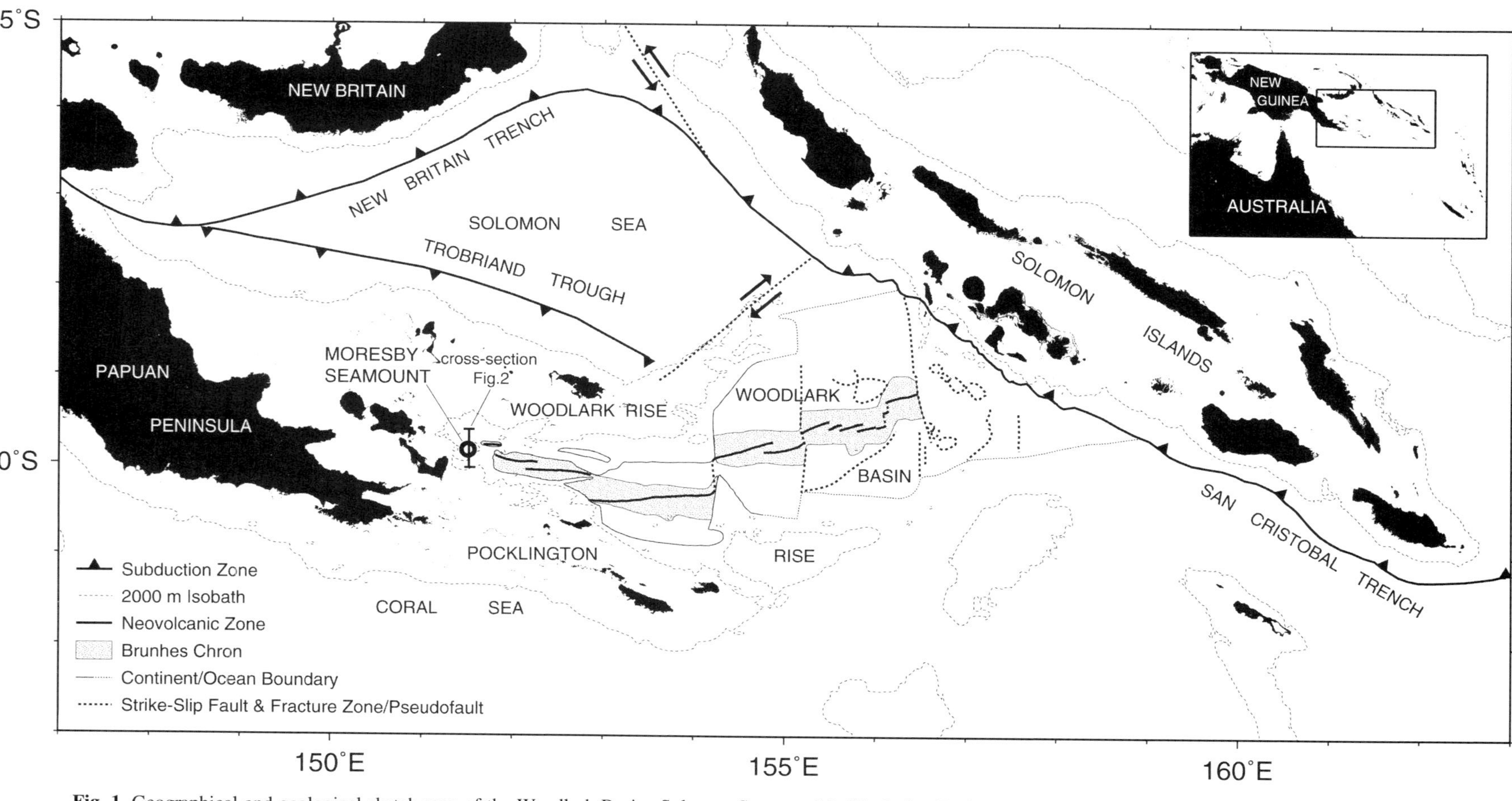

Fig. 1. Geographical and geological sketch map of the Woodlark Basin–Solomon Sea area. Modified after Taylor *et al.* (1999*b*).

350 km of new oceanic crust in the easternmost part of the basin (Taylor *et al.* 1996). In the working area, the temporal and spatial succession of a certain amount of continental rifting preceding subsequent oceanic spreading becomes evident. From the sedimentary record it can be seen that in the Woodlark Basin the two processes coexist for *c.* 1 Ma after oceanic spreading initiated (Taylor *et al.* 1995). The structurally asymmetric Papuan basement hence is divided into the Woodlark Rise and Pocklington Rise, which respectively mark the northern and southern disrupted and subsided continental margins traceable to the west (Abers & Roecker 1991). The total amount of extension at the longitude of investigations is estimated at 200 km (Taylor *et al.* 1996). Drilling was performed some tens of kilometres west of the incipient oceanic spreading tip. Seismic cross-sections reveal a strongly asymmetric east–west-oriented graben structure showing intense block faulting. A complex pattern of low- and high-angle normal faults bordering several discrete basins and rotated fault blocks can be observed (Mutter *et al.* 1996).

At 151.35°E Moresby Seamount is situated at the central graben axis between the Woodlark and Pocklington Rises. There, extensional tectonics is manifested by seismically active low-angle normal faulting with the deformation style having neither purely oceanic nor continental character (Mutter *et al.* 1996). Moresby Seamount is interpreted to be an unroofed footwall block of lower Papuan basement beneath a north-dipping low-angle detachment fault (Fig. 2). The detachment plane dips *c.* 30° to the NNE, and is traceable in refraction seismic sections to a depth of 9 km beneath the pre- and synrift sedimentary rocks and the basement rocks of the hanging wall. A published estimate of fault heave is 12 km (Taylor *et al.* 1999*b*). The graben is bordered to the north by a steep, 60° south-dipping normal fault. Water depth varies between 120 m at the summit of the seamount and 2600 m at the deepest point of the half-graben. The rocks involved in faulting are sedimentary rocks of a Pliocene–Pleistocene synrift sequence of maximum 2 km thickness unconformably overlying partly eroded Miocene prerift sedimentary rocks and metamorphic basic basement rocks (Taylor *et al.* 1999*b*). The latter show greenschist- and amphibolite-facies metamorphism and are along-strike equivalents of rocks cropping out on the Papuan Peninsula (Abers & Roecker 1991).

Fault system architecture and mesoscopic deformation fabrics

We focused our investigations on fault rocks from three drillsites of ODP Leg 180: Sites 1108B, 1114A and 1117A (Fig. 2), all situated on the northern flank of Moresby Seamount. The single core logs can be connected in offset style to form a synoptic section through the detachment fault, its footwall and its hanging wall (Fig. 3). General descriptions of drilled sections, lithologies and on-board measurements incorporated in the following have been given by Taylor *et al.* (1999*b*).

Site 1108B drills through 485 m of the synrift sequence in the hanging wall to the main detachment fault (Fig. 3) at the northern slope of Moresby Seamount (Fig. 2). Because of the turbiditic depositional regime, the rock sequence sampled is strongly variable in grain size and composition. Detritus mostly consists of material derived from the continuously uplifted seamount (Taylor *et al.* 1999*b*).

Two fault zones are encountered downhole at Site 1108B. They are characterized by populations of core-scale faults of variable spacing depending on local lithology and depth. Throughout the section, mineralized veins indicate fluid-assisted fracturing. Fault zone I extends from 160 to 200 m below sea floor (mbsf), and has a vertical offset of *c.* 200 m. Altogether, 400 m of the upper part of the sedimentary column are missing. According to porosity curves, magnetostratigraphy and sedimentation rates, one half of this amount was eroded and the other half removed by faulting (0.58 Ma age offset at 165 mbsf, Taylor *et al.* 1999*b*). Fault zone II extends from 350 mbsf to the bottom of the drillhole at 485 mbsf. The main detachment zone is expected at a depth of *c.* 900 mbsf, as inferred from seismic profiles at Site 1108. The rocks in the vicinity of the upper fault zone show extensional fractures and small faults with slickensides. Strain is concentrated in fine-grained clayey–silty layers. The deformation style in the lower fault zone depends on grain size, with two zones of scaly fabrics in clay-rich lithologies, and fragmentation and moderately dipping extensional faults in coarse-grained sandstones (Taylor *et al.* 1999*b*). Most of the brittle deformation in the fault zones is concentrated in fine-grained sediments. Neither fault zone could be imaged by the Formation Microscanner. For this reason exact dip directions and dip angles cannot be given.

Site 1114 was spudded north of the top of Moresby Seamount at 406 m below sea level (mbsl) and marks the location of a steep SSW-

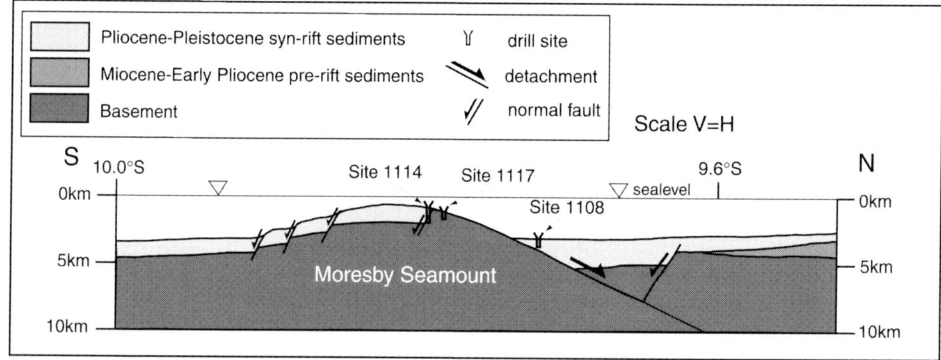

Fig. 2. Simplified true-scale cross-section through Moresby Seamount showing type of bedrock and principal structural features that can be deduced from seismic profiles. The low-angle detachment crops out at the northern flank of the seamount and builds the southern boundary of the deep graben, filled with Pliocene–Pleistocene syn-rift turbidites. A set of antithetic steep normal faults crosscuts the top of the seamount. Locations of Sites 1108, 1114 and 1117 are projected into the section plane.

dipping normal fault zone revealed by seismic cross-sections. The fault apparently offsets the metamorphic basement by >2 km and is the northernmost of a set of steep antithetic normal faults (Taylor *et al.* 1999*b*) as shown in Figure 2. Drilling provided a section through 286 m of faulted, upward-fining turbidites (Fig. 3). Porosity data and a hiatus in the microfossil record indicate an erosional removal of 200–400 m of the uppermost part of the stratigraphic column (Taylor *et al.* 1999*b*). The sedimentary rocks are separated from the underlying metamorphic basic basement rocks by a tectonic breccia of 6 m thickness, which, according to shipboard data (Taylor *et al.* 1999*b*), marks the location of the 60–65° SSW-dipping normal fault mentioned above. Its angular components consist of greenschist-grade metamorphic, highly fractured and mineralized diabase with abundant early formed quartz and epidote veins, and show a downward decrease in size.

On drillcore scale, deformation of the hanging-wall sedimentary rocks appears as intense fracturing in coarse-grained strata, and inclined bedding and scaly fabrics in fine-grained strata. The clay-rich layers become more intensely deformed with depth. Mineralization of extensional fractures occurs only subordinately. Brecciation and shearing intensity of the metadiabase of the footwall rocks (Fig. 3) decreases with increasing distance from the main fault plane. The basement rocks show a high density of multistage veins, mostly filled with quartz and minor calcite. Ultracataclasites drilled near the normal fault are the most intensely deformed rocks. With increasing depth, deformation and alteration become less penetrative

and less intense. In general, deformation is inhomogeneous and strongly localized in narrow zones of foliated cataclasite surrounded by massive wall rocks. Vein geometry is complex and affected by both ductile and brittle deformation. Often the older veins are folded, and the younger veins are disrupted and sheared.

Site 1117 is situated at the upper northern flank of Moresby Seamount, where the low-angle detachment fault plane is exposed at the sea floor (Fig. 2). The uppermost cores cut through at least 12 m of clayey fault gouge cropping out at the sea floor (Fig. 3). The fault gouge probably marks the position of the high-strain detachment plane. The dominant grain-size class is clay with varying amounts of dark, polished pebbles. Mineralogical composition is variable, with talc, chlorite, calcite, chrysotile, antigorite, lizardite, minor quartz and amphibole, all of them indicating a mafic to ultramafic protolith akin to the gabbros recovered at greater depth (Taylor *et al.* 1999*b*).

Beneath the fault gouge, mylonites and ultracataclasites probably occupy a section of 50 m thickness of very low recovery (6% at Site 1117A). Both rock types show that the main detachment has a structural record of intense, cohesive ductile (mylonites) and intermittent ductile and brittle (cataclasites) deformation. Beneath a fault breccia at *c*. 65 mbsf a succession of 50 m thickness of metagabbros and their faulted, altered and fractured equivalents was recovered (Fig. 3). Viewed altogether, the remarkable fact about fabrics at Site 1117 is that an evolution from ductile to brittle deformation is observed both with depth and with

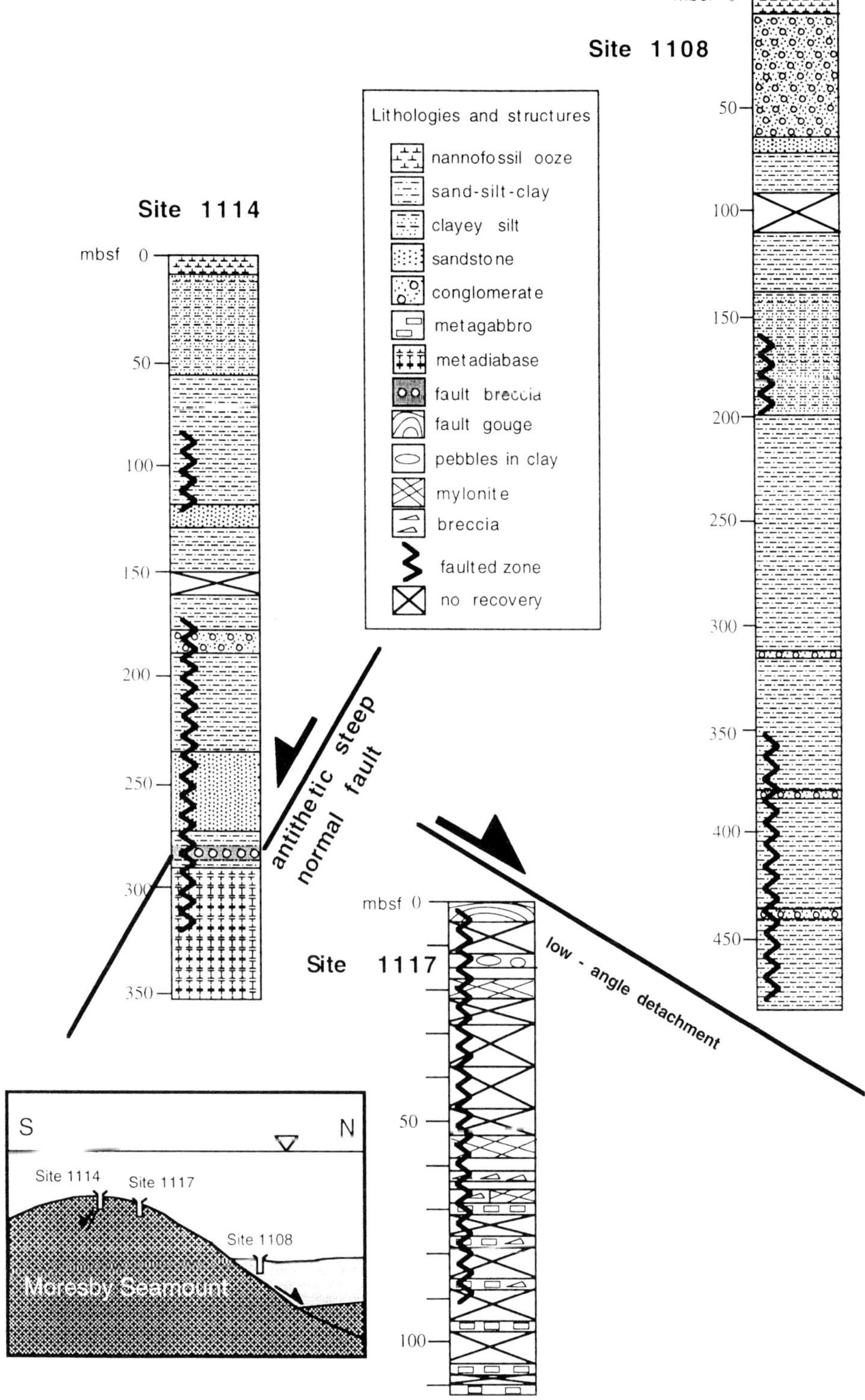

Fig. 3. Synoptic litho-log representation of ODP Sites 1108, 1114 and 1117. (See text for discussion.)

time for the upper parts of the drilled section (Fig. 3).

Ductile and brittle deformation microfabrics

On a mesoscopic scale, the sedimentary rocks in the hanging wall to the detachment at Site 1108B show brittle deformation fabrics, such as steeply dipping faults with slickensides (on a centimetre scale), scaly fabrics and mineralized veins. Deformation intensity varies locally and is mostly related to the two fault zones (Fig. 3). The abundance of fracturing and mineralization at all scales indicates that the sediments had reached some degree of induration at the onset of faulting.

On a microscopic scale, the samples provide insight into complex polyphase dewatering structures and veining. Most of the filled fractures observed in the sediment samples of the hanging wall to the main detachment at Site 1108 are purely extensional with no or only very small vein-parallel displacement. This is documented by the microscopic pull-apart structure in Figure 4a. Vein mineralization is dominated by coarse equidimensional calcite crystals (Fig. 4b). Only at greater depths fibrous calcitic veins are observed (Fig. 4c). Geometry of the veins and vein networks is dependent on depth. Generally, vein boundaries become less irregular and vein paths straighter with increasing depth. Fracturing of detrital mineral grains and rock fragments is common in deeper parts of the section, attesting to high shear strength of the cemented sediments where transgranular and intragranular fractures were formed. Crosscutting relationships between veins are ubiquitous, with angles between veins mostly between 30 and 90°. Often, disrupted wall-rock fragments are captured within the calcitic vein infill. Some veins show zonation of the infill geometry, with coarse-grained outer parts and fine-grained inner parts.

Carbonate-rich fluid migration was not restricted to single vein pathways but fluids percolated through the matrix. This is proven by pervasive calcitic cementation and by dense networks of strongly anastomosing veinlets (Fig. 4b). Often, the veinlets are subparallel to major veins. Pervasive cementation is often observed in fine-grained intervals, whereas discrete veins or geometrically regular patterns of veinlets occur in coarser-grained sediments (Fig. 4b,c). There, anastomosing patterns of both veins and veinlets around clastic components are a characteristic feature for less lithified parts of the sec-

tion (Fig. 4b). Where vein boundaries are sharp and well defined, shear parallel to the vein surface and associated structures are often developed (Fig. 4d). Fibrous calcite veins show crystal laths of constant width along a single vein (Fig. 4d). Very often, a dark median line (not always equidistant from the vein walls) consisting of small wallrock particles indicates an antitaxial growth of the calcite infill (Fig. 4b,c). In some cases, antitaxial fibres and equidimensional mineral growth laterally replace each other (Fig. 4d).

The two observed vein-infill types probably have different origins. The fibrous veins (Fig. 4c) have opened slowly and incrementally, so that mineral growth could fill the produced void simultaneously. Their genesis is thought to be related to a regime of persistent high fluid pressure (see Passchier & Trouw 1996). In contrast, the mosaic-type filled veins are interpreted to have opened during a sudden, probably microseismic event and have been filled by crystallization of minerals subsequently.

At Site 1114, sedimentary rocks of the hanging wall are pervasively sheared in two fault zones, showing scaly fabrics in fine-grained strata and predominantly steep normal faults and strike-slip faults in the intervals of sandy facies (Taylor *et al.* 1999*b*). Vein mineralization occurs only very subordinately in the faulted sedimentary rocks of Site 1114. In the fault breccia, between the fractured components and the matrix, late calcite veins occur. These veins are partly sheared parallel to the layering of the breccia (Taylor *et al.* 1999*b*).

The main detachment zone itself was sampled by drilling at Site 1117. Because of the very low recovery the belt of mylonites and cohesive ultracataclasites between about 15 and 60 mbsf is represented by only a few, non-reorientable samples. Some of the mylonites occur as tectonically reworked fragments ('pebbles') in the overlying fault gouge. A photomicrograph with typical mylonitic microstructure is shown in Figure 5a. A darkish, ultra-fine-grained foliated chlorite-rich matrix alternates with stretched layers of calcite, around asymmetrically tailed porphyroclasts consisting of polycrystalline, dynamically recrystallized quartz (centre) and calcite (lower left). The porphyroclasts are strain enclaves and preserve microstructures diagnostic of pre- or early mylonitic deformation and metamorphism. Whereas the mylonites attest to ductile flow with crystal plastic deformation and/or fluid-assisted mass transfer, the cohesive, foliated cataclasites found in the same drilled section (Fig. 5b) probably were generated by a complex interplay of mainly

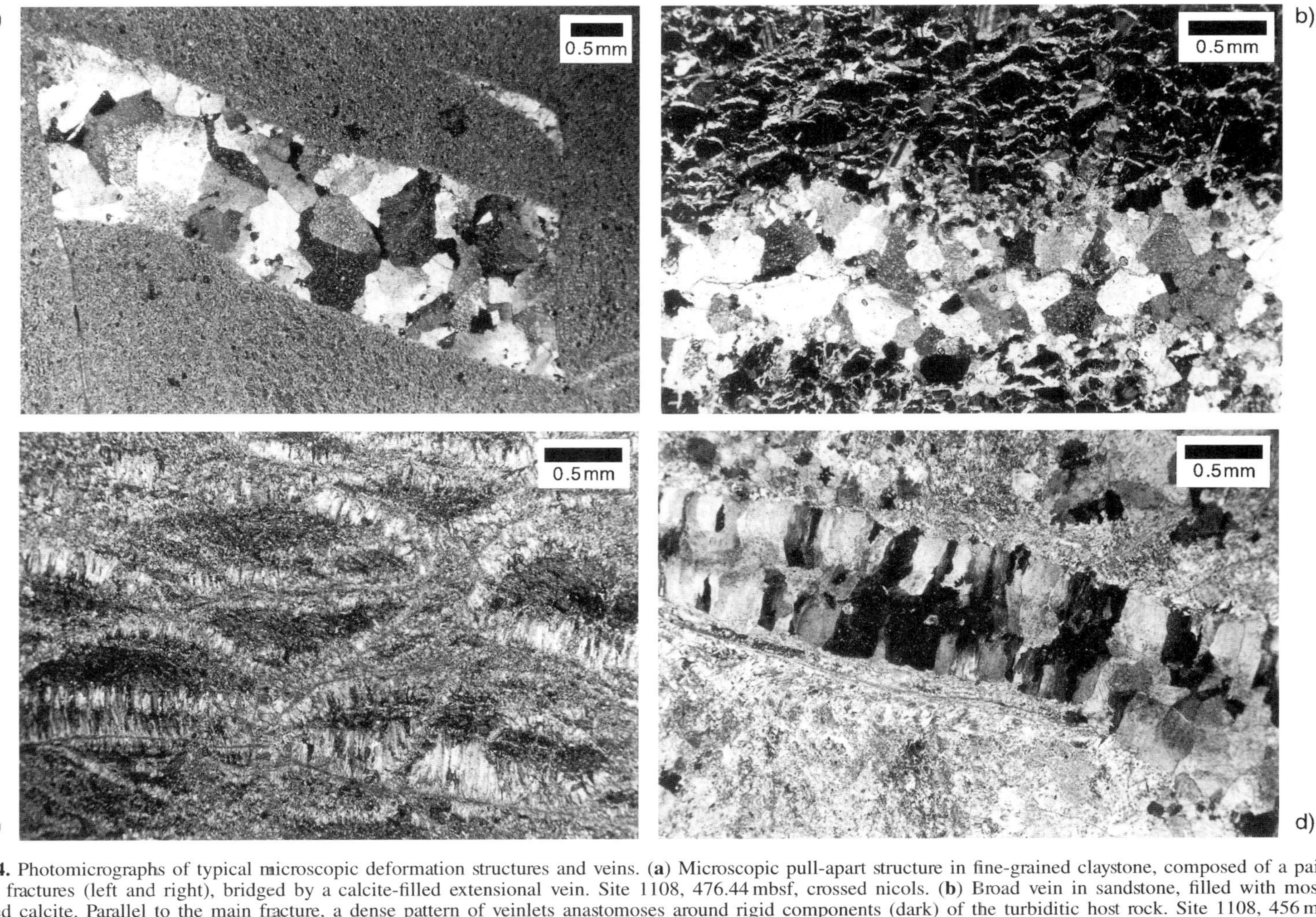

Fig. 4. Photomicrographs of typical microscopic deformation structures and veins. (**a**) Microscopic pull-apart structure in fine-grained claystone, composed of a pair of shear fractures (left and right), bridged by a calcite-filled extensional vein. Site 1108, 476.44 mbsf, crossed nicols. (**b**) Broad vein in sandstone, filled with mosaic-shaped calcite. Parallel to the main fracture, a dense pattern of veinlets anastomoses around rigid components (dark) of the turbiditic host rock. Site 1108, 456 mbsf, crossed nicols. (**c**) Anastomosing network of veins in claystone filled with antitaxially grown fibrous calcite. Most vein fills display an asymmetric position of the dark median lines. Thin-section plane is perpendicular to core axis. The fine-grained turbiditic host rock suffered pervasive carbonate cementation. Site 1108, 475 mbsf, crossed nicols. (**d**) Extensional vein filled with fibrous (left and centre) and blocky (lower right) calcite, in matrix of strongly altered metabasite. Vein walls are sheared and made up of ultracataclasite. Site 1117, 3.99 mbsf, crossed nicols.

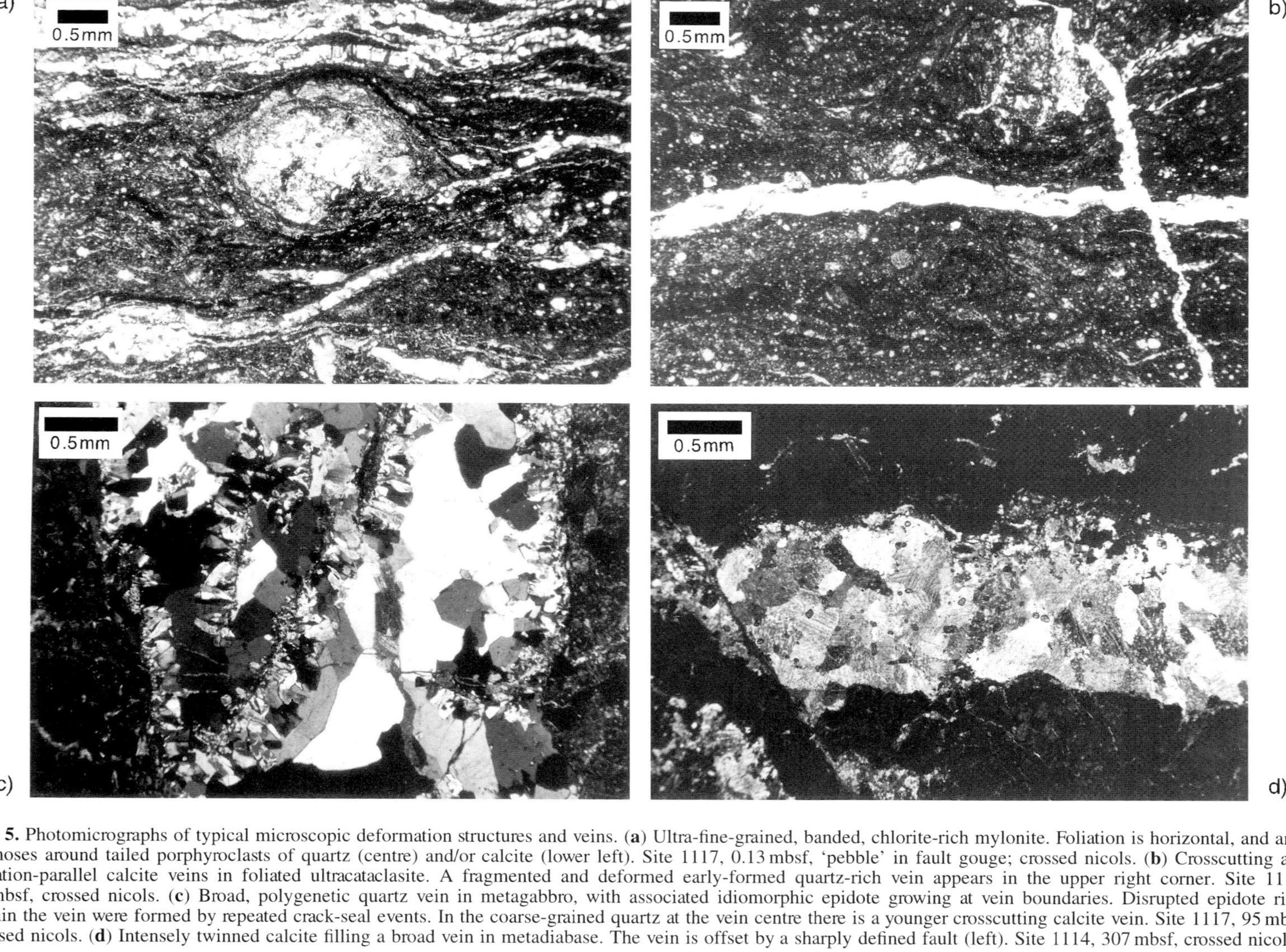

Fig. 5. Photomicrographs of typical microscopic deformation structures and veins. (**a**) Ultra-fine-grained, banded, chlorite-rich mylonite. Foliation is horizontal, and anastomoses around tailed porphyroclasts of quartz (centre) and/or calcite (lower left). Site 1117, 0.13 mbsf, 'pebble' in fault gouge; crossed nicols. (**b**) Crosscutting and foliation-parallel calcite veins in foliated ultracataclasite. A fragmented and deformed early-formed quartz-rich vein appears in the upper right corner. Site 1117, 66 mbsf, crossed nicols. (**c**) Broad, polygenetic quartz vein in metagabbro, with associated idiomorphic epidote growing at vein boundaries. Disrupted epidote rims within the vein were formed by repeated crack-seal events. In the coarse-grained quartz at the vein centre there is a younger crosscutting calcite vein. Site 1117, 95 mbsf, crossed nicols. (**d**) Intensely twinned calcite filling a broad vein in metadiabase. The vein is offset by a sharply defined fault (left). Site 1114, 307 mbsf, crossed nicols.

brittle and partly ductile deformation. Foliation is expressed by clast alignment and quartz-rich bands in an ultra-fine-grained chlorite-rich matrix. The quartz band (upper right in Fig. 5b) is folded, and crosscut by a sub-vertical calcite-filled extensional fracture. The fracture itself is sheared and openly folded.

The microstructural record in samples from the detachment zone is too fragmentary to constrain a complete history of the mechanical behaviour. Some important interpretations can be derived, however: the fact that fragments of mylonite are reworked in the fault gouge indicates that mylonites and probably ultracataclasites are products of shearing under conditions of elevated temperature and pressure early in the exhumation history of the seamount. This occurred before formation of the fault gouge under upper-crustal conditions. We now take the approach to shift scales of observation and use overprinting relationships (Fig. 6). The dynamically recrystallized quartz found as relict microfabric in porphyroclasts indicates ductile flow at

a temperature of at least 300 °C (e.g. Voll 1976). Later, deformation was exclusively taken up by the fine-grained, chlorite-rich matrix of the ultracataclasites under conditions of low-grade to very low-grade metamorphism. The latest increments of extensional deformation led to the development of the clay-rich fault gouge under near-surface conditions. The retrograde deformation history deduced from the overprinting relationships of brittle over ductile structures probably is controlled by the continuous exhumation and cooling of the detachment zone by crustal extension.

In the lower part of the drillhole at Site 1117 the deformation style changes significantly and rapidly with distance from the detachment. Uppermost mylonites and ultracataclasites pass vertically into cataclasites and these, in turn, pass into breccia (Fig. 3). The lowermost rock type recovered is a fresh, only slightly fractured gabbro. The variety of different mineralization types in extensional fractures is larger in the footwall than in the hanging wall. The crosscut-

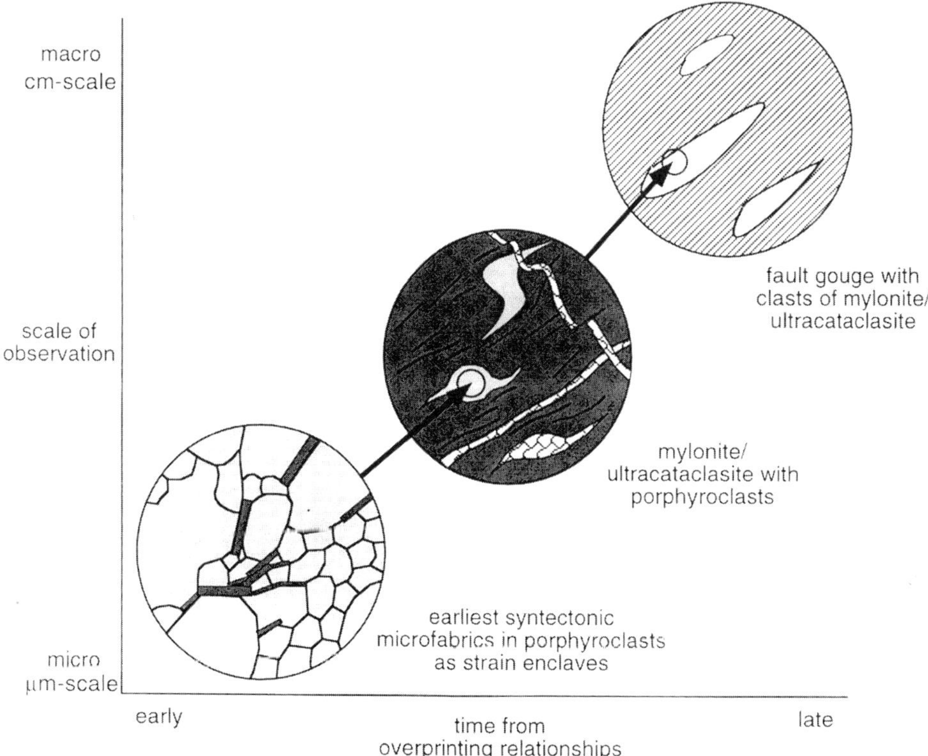

Fig. 6. Schematic representation of overprinting relationships and scales of observations in the mylonites and ultracataclasites of the Woodlark extensional detachment. Strain enclaves at various scales reveal progressively older parts of the deformational history of the detachment rocks.

ting relations and deformation fabrics of the veins and their infill locally allow the reconstruction of the relative timing of deformation and precipitation events. For example, Figure 5c shows a pervasive network of crosscutting vein generations of early quartz, or combined quartz and epidote, overprinted by later extensional fractures filled with calcite. At Site 1117 the gabbroic wall rock becomes less altered and less fractured with depth, being almost unaffected by veining in the deepest parts of the borehole.

At Site 1114 deformation is only locally developed in the metadiabase flooring the fault breccia below 295 mbsf in the drilled section (Fig. 3). Brecciation of the metadiabase with early formed quartz–epidote veins was followed by calcite precipitation in extensional fractures (Fig. 5d). Most of the calcite veins were later strongly affected by shearing and folding. Continuing deformation produced pervasive twinning of the calcite, and offsets of the vein fill along discrete brittle microshears.

Deformation structures of the clayey fault gouge recovered at Site 1117A are scaly fabrics, shear faults and drilling-induced folding. Onboard measurements revealed a high anisotropy of the material in magnetic susceptibility and static permeability (Taylor *et al.* 1999*b*), both indicating a strong planar texture as a result of shearing parallel to the fault plane.

Syntectonic stress estimations using calcite twins

Many of the extensional fissures in the faulted rocks of the hanging wall and the footwall of the detachment contain infills of calcite. A variable proportion of the calcite grains is twinned in response to tectonic stresses that were transmitted through the rocks after vein formation. Twinning on crystallographic e-directions induces a shear strain (e.g. Nicolas & Poirier 1976) and is a deformation mechanism frequently observed in polycrystalline calcite in the low-temperature plasticity regime (e.g. Schmid 1982). The occurrence of twinning is a function of a critical shear stress resolved in the twinning plane (estimates are 2–12 MPa, Passchier & Trouw 1996; 10 MPa, Jamison & Spang 1976) and is largely independent of temperature, strain and strain rate (Tullis 1980).

We have applied two empirically calibrated methods to estimate palaeostress from twinned calcite polycrystals (Jamison & Spang 1976; Rowe & Rutter 1990). The method of Rowe & Rutter (1990) for stress estimation is based on the determination of the twinning density (D). D is defined as the number of twins per millimetre, measured perpendicular to the lamellae. On the basis of deformation experiments on marbles at temperatures between 200 and 800 °C, Rowe & Rutter (1990) derived an empirical relationship between D and differential stress ($\Delta\sigma$):

$$\Delta\sigma = -52.0 + 171.1\log D$$

with a standard error of 43 MPa.

The second method we have applied is based on the percentage of grains showing one, two or three twin sets (Jamison & Spang 1976). The relationship between this number and differential stress has been tested experimentally, with the result that the method is applicable for calcite that has suffered low strain. The critical resolved shear stress (τ_r) for e-twin formation is taken as 10 MPa, and the percentages of grains with one, two or three twin sets on the number of all counted grains are related to a 'resolved shear stress coefficient' S_1 for calcite. For differential stress ($\Delta\sigma$) calculation the following relationship (Jamison & Spang 1976) is used:

$$\tau_r = S_1\Delta\sigma$$

.

The analytical procedure was as follows. We determined grain size of all grains under the microscope, counted grain numbers and proportions of twinned grains to determine S_1, counted the twin lamellae and measured the grain diameter perpendicular to the lamellae to determine D. Between 200 and 400 grains per thin section were counted. Only monomineralic calcitic veins showing unimodal grain-size distribution (for an example, see Fig. 7) and randomly oriented, mosaic-shaped, and equidimensional grains were examined (see Jamison & Spang 1976; Rowe & Rutter 1990). These conditions ensure reliable results, as inclusions of other mineral grains influence both the stress field and resulting twinning. The same is true for lattice preferred orientation or a preferred shape of the grains. As Newman (1994) stated, heterogeneity of grain sizes also influences twinning activity, as grain contacts present points of stress concentration, which enhance the probability of twinning.

Results for the samples analysed are given in Table 1. Evidently, the estimates for differential stress magnitude vary greatly between the two methods applied: 195–309 MPa by the method of Rowe & Rutter (1990), and 22–62 MPa by the method of Jamison & Spang (1976). These vast differences may be related to limitations of

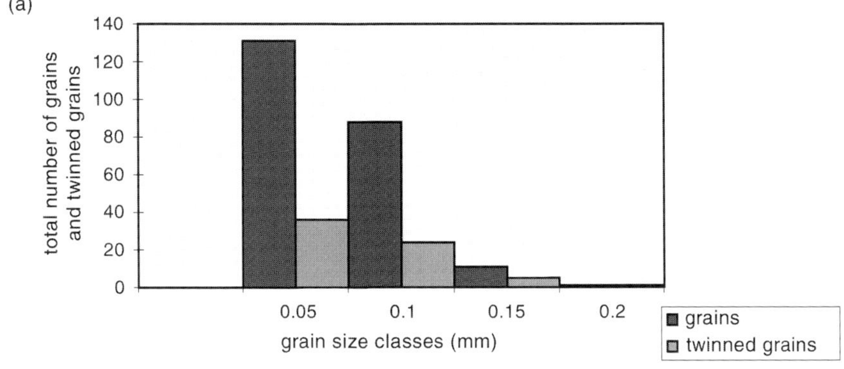

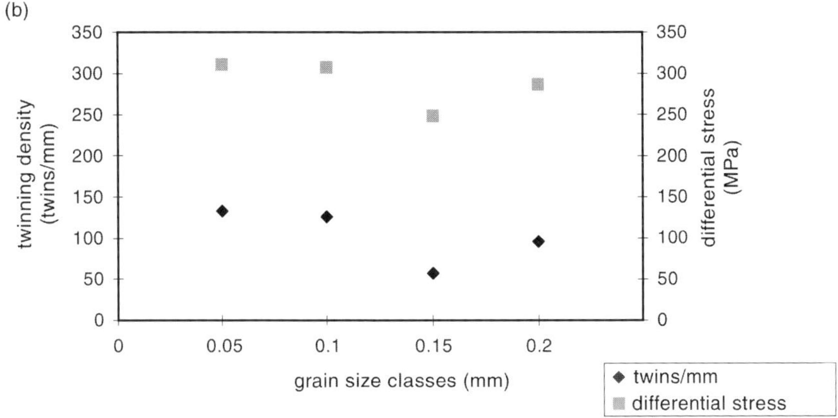

Fig. 7. An example of measured parameters for shear-induced twinning. Sample No. 8B17a, Site 1108, 456.35 mbsf. (**a**) Grain-size distribution, and numbers of twinned grains per grain-size class. (**b**) Twin density and inferred differential stress as functions of grain size.

the two methods, as pointed out in a recent discussion by Ferrill (1998).

In the Woodlark tectonic setting it is probable that during repeated seismic relative movement of the faulted blocks the calcitic vein infills were affected by a large number of seismic stick–slip cycles. At first glance, both methods seem to be appropriate for application to the naturally deformed calcite. Twinning density is supposed to be largely insensitive to repeated stress events because the number of twin nuclei at a grain boundary should not be affected by reorientation of stresses (Rowe & Rutter 1990). The percentage of twinned grains as used by Jamison & Spang (1976) is more sensitive to repeated stress events but grains already twinned normally react to further deformation by strain hardening.

Despite their postulated broad applicability, both methods have restrictions rooted in the basic assumptions made. Using the Jamison & Spang (1976) method, grain size of the calcite grains remains uninvolved although twinning activity is dependent on grain size (Rowe & Rutter 1990). Grain boundaries impede the formation and spreading of twins (Hall–Petch Law; see Nicolas & Poirier 1976), so that initiation of twinning in fine-grained rocks requires higher differential stresses than in coarse-grained rocks. Grain-size averages for the twinned and untwinned calcite grains of our samples range between 30 and 400 μm. At least for the fine-grained samples, therefore, underestimation of stresses must be considered. Conversely, the method of Rowe & Rutter (1990) can lead to overestimations by a factor of 2–9 for

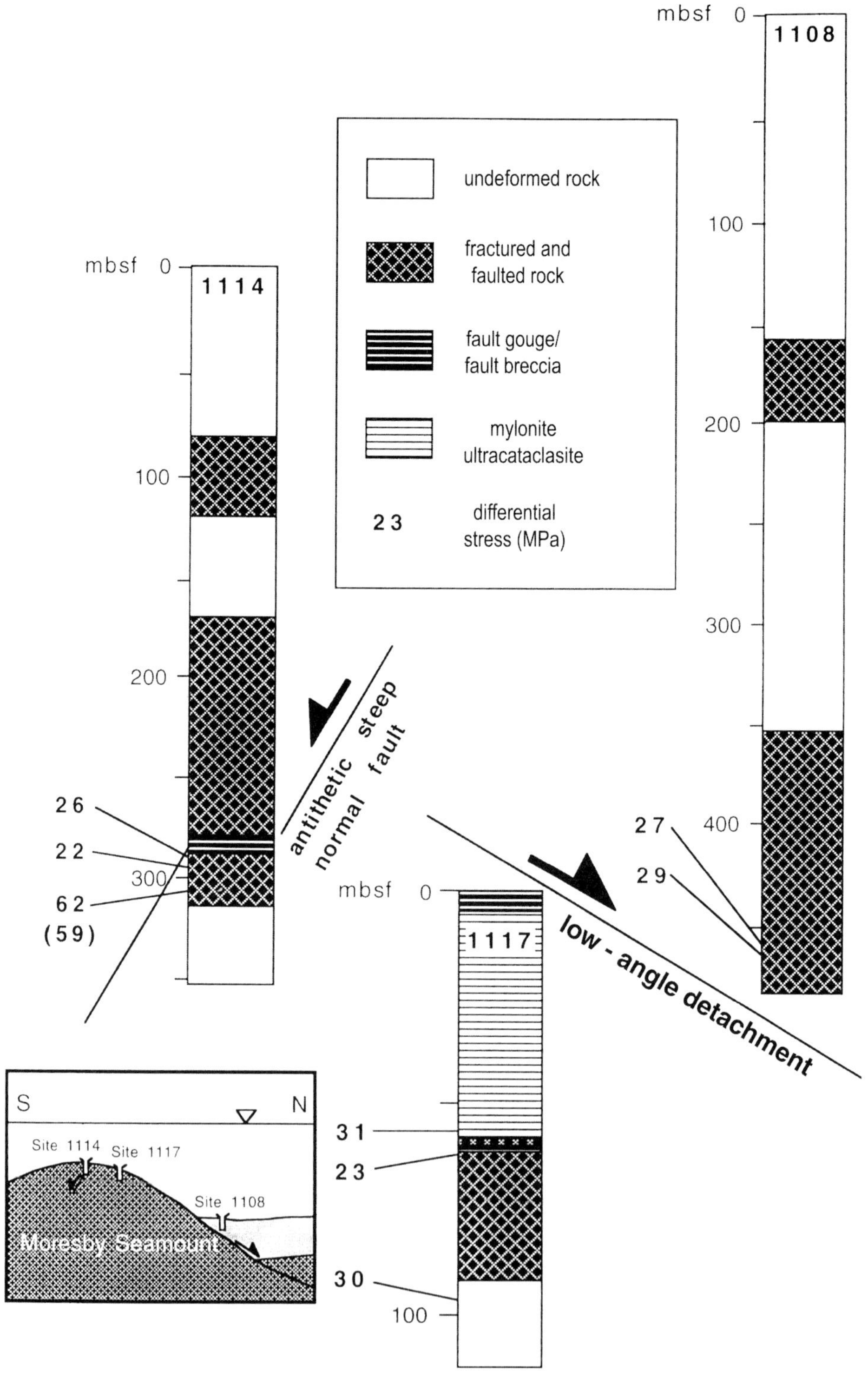
mbsf 0
1108
100
200
300
400
mbsf 0
1114
100
200
300
26
22
62
(59)
undeformed rock
fractured and
faulted rock
fault gouge/
fault breccia
mylonite
ultracataclasite
23
differential
stress (MPa)
antithetic steep
normal fault
low - angle detachment
27
29
mbsf 0
1117
31
23
30
100
S
N
Site 1114
Site 1117
Site 1108
Moresby Seamount

Table 1 *Estimations of differential stresses and comparison of the applied palaeopiezometric methods for calcite twinning of Jamison & Spang (1976); Rowe & Rutter (1990)*

Sample no./Site	Depth (mbsf)	Jamison & Spang (1976)		Rowe & Rutter (1990)	
		% of grains twinned	$\Delta\sigma$ (MPa)	Twinning density (twins per mm)	$\Delta\sigma$ (MPa)
8B17/1108	456.35	29 (S1 = 0.37)	27	110.15	297
8B18/1108	456.37	33 (S1 = 0.35)	29	57.639	248
17A4/1117	57.02	41 (S1 = 0.32)	31	57.84	249
17A9/1117	61.64	14 (S1 = 0.43)	23	30.48	205
17A7/1117	95.79	38 (S1 = 0.33)	30	50	239
14A2/1114	296.19	23 (S1 = 0.39)	26	30.48	195
14A4/1114	296.38	12 (S1 = 0.45)	22	40.9	224
14A26/1114 TSB153	307.27	72 (2nd set: 14) (S1 = 0.16; 2nd set: 0.17)	62 (2nd set: 59)	128.79	309
14A25/1114 TSB152	307.03			64.08	257

differential stresses if applied to twinning fabrics generated in the low-temperature regime (<200 °C; see Ferrill 1998). Experiments as well as field studies revealed the existence of a 'thin twin regime' and a 'thick twin regime' separated by the 200 °C boundary (Burkhard 1993; Ferrill 1998). The 'thick twin regime' is more characteristic of greenschist-grade metamorphic conditions (Schmid *et al.* 1987; Burkhard 1993). Thin, low-temperature twins are commonly <5 μm wide, and straight. If strain continues after formation of first twin lamellae, it is accomplished by increasing twin density in the crystal. In contrast, at higher (>200 °C) temperatures, existing twin lamellae widen when subjected to continuing strain (Ferrill 1998). Twins in the samples analysed by us are characteristically between 1 and 4 μm wide, and therefore qualify as thin 'low-temperature' twins, generated late in the strain history of the Woodlark Detachment at temperatures below 200 °C. From this reasoning it follows that the lower (22–62 MPa) differential stress estimates (Table 1; Fig. 8) using the method of Jamison & Spang (1976) are probably the more realistic reflection of the palaeostresses in the zone of contact strain after formation of the calcite veins.

There is one more important argument from the tectonic setting to support this inference: the twinned calcite in the sediment samples from Site 1108 was probably never subjected to a maximum compressive stress (σ_1) exceeding *c.* 38 MPa. This value is computed assuming σ_1 to be vertical and assuming 3 km of water overburden plus 650 m of sedimentary rock overburden (including 200 m of sedimentary column later removed at fault zone I) of average sediment density of 2.1 g cm^{-3} (Taylor *et al.* 1999*b*) minus 650 m of hydrostatic pore water of density 1.05 g cm^{-3}. Taking into account an estimated tensional strength of sandstone or claystone of 6 MPa (see Lockner 1995), and a Poisson ratio, ν, of 0.25–0.33 (Twiss & Moores 1992), a hyperbolic Mohr envelope can be constructed, to give an idea of maximum possible differential stresses in the rocks near the base of the drillhole (Fig. 9). These are in the range of 25–28 MPa. Assuming a reduction of σ_v and σ_h as a result of increased pore pressure, the genesis of the observed extensional veins becomes intelligible. Similar inferences for the footwall basement rocks from Sites 1114 and 1117

Fig. 8. Spatial distribution of ductile (mylonite) and brittle deformation (ultracataclasite, fault gouge, fractured rocks) in the Woodlark extensional detachment system. Numbers beside drillhole logs denote palaeostress estimates from twinning fabrics of calcite using the calibration of Jamison & Spang (1976).

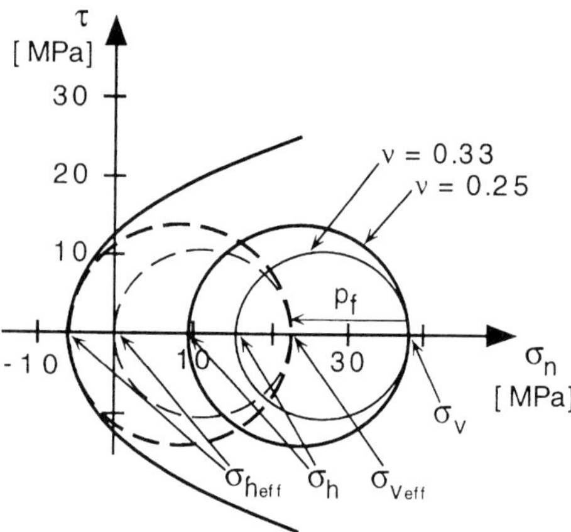

Fig. 9. Mohr circle construction for subcritical and critical stress state at likely conditions near the base of Site 1108 at 450 mbsf. Continuous-line circles, subcritical stress state at low pore fluid pressure. Vertical applied stress (σ_v), which equals maximum stress (σ_1), was calculated with ρ of 2.1 g cm^{-3} for 650 m of sediment overburden (including 200 m removed by faulting) and a 3 km seawater column with ρ_{water} of 1.05 g cm^{-3}. Poisson ratios (ν) of 0.25 and 0.33 (σ_h of 0.3σ_v and 0.49σ_v) give range of expected values for least horizontal stresses σ_h. Dashed-line circles, critical state at elevated pore fluid pressure (p_f) reducing principal stresses to lower values (σ_{veff} and σ_{heff}), and leading to tensional or transtensional failure. Mohr envelope is constructed assuming tensile strength of 6 MPa (Lockner 1995) and cohesion to be twice the tensile strength (Axen 1992).

cannot be made, as there is no clear evidence for the depth of burial at the time when the calcite in the veins was deformed and twinned.

Some tectonic implications

The microfabrics in the faulted rocks in the Woodlark extensional detachment zone lead to some interesting questions regarding the tectonic evolution of the detachment itself. If viewed as a product of continuous fabric evolution in a regime of decreasing temperature, the belt of fault gouge, mylonites and ultracataclasites depicts a history of continuous ductile or brittle–ductile flow in the footwall part of the shear zone, of c. 60 m width. Early deformation fabrics are preserved in quartz-rich porphyroclasts in the mylonite samples, and the fact that the quartz shows evidence for dynamic recrystallization suggests ductile creep under at least lowermost greenschist-grade conditions.

Therefore, the mylonites of the detachment zone must have originated from their metabasic protoliths at considerable depth in the crust with quartz-veining predating and/or accompanying the ductile deformation processes. In the rift zone north of Moresby Seamount, the present geothermal gradients are around 100 °C km^{-1} (Site 1108; Taylor *et al.* 1999*b*). However, the thermal structure of the Woodlark Basin before rifting is probably better depicted by the present thermal gradients outside the rift zone, which are 31 and 28 °C km^{-1} for Sites 1109 and 1115, respectively (Taylor *et al.* 1999*b*). Assuming a lower temperature boundary of about 300 °C for the plastic deformation in the mylonites, their depth of origin may be as much as 10 km. This is a maximum estimate, however, as it assumes the unlikely case of a linear geotherm. Transient effects, such as a heat flow pulse during rifting, or fault zone heating by advection of fluids, may have modified the thermal structure. More precise estimates can be obtained only from heat flow data from deep (>5 km) boreholes or modelling of basin thermal evolution (e.g. Cochran 1983; Brigaud & Vasseur 1989). Tectonic models to explain the kinematics and the deformation of the Woodlark Detachment and its wall rocks must account for a possible throw of this magnitude (10 km) on the low-angle normal fault. Less than half of the vertical offset can be constrained by the difference in elevation between the location of Site 1117 (1163 mbsl, Taylor *et al.* 1999*b*) and the floor to the sedimentary fill of the half-graben at c. 5000 mbsl in the vicinity of Site 1108 (see fig. F4 of Taylor *et al.* 1999*b*). A significant remainder of the fault throw, therefore, must have involved a process that neither generated strong uplift of the footwall block to the detachment fault nor resulted in strong attenuation of the hanging wall. In the case studied here, hanging-wall attenuation is unlikely to be important, as the pre-rift and syn-rift sediment sequences upward and northward from the detachment show no signs of strong fragmentation into fault blocks and graben or half-graben fills (e.g. Taylor *et al.* 1999*b*). What makes the Woodlark Basin tectonic environment special is the fact that unroofing and exhumation of the footwall fault block cannot be assisted by substantial erosional denudation because of the submerged nature of the whole tectonic system.

A second interesting observation is that predominantly ductile and predominantly brittle deformations are (1) spatially distributed in the vertical dimension and (2) reflect the uplift history of the footwall rocks (see Fig. 8). Rocks containing fabrics that attest to a history of ductile flow (mylonites, cohesive cataclasites, fault gouges) are restricted to the footwall part, of 60 m thickness, of the detachment zone that was intersected at Site 1117 (Figs 3 and 8). The abundance of mineralized veins in the mylonites, however, indicates that deformation was not continuously and exclusively ductile, but included embrittlement as a result of either local strain rate or stress build-up or reduction of the effective stresses by pulses of near-lithostatic fluid pressure. Many of the veins are oriented parallel to foliation planes or folded, because they were generated syntectonically and later deformed compatibly with the mylonite matrix. Exclusively brittle deformation histories, however, are restricted to the rocks away from the detachment's fault zone. Uplift of the footwall to progressively shallower depth probably was related to a contemporaneous widening of the fault zone with change to brittle deformation mode.

If evolution of fractures and fracture systems is genetically related to palaeo-microseismicity, then this observation implies that the microseismicity of extensional detachment systems (e.g. Wernicke 1995) such as the Woodlark system probably is not restricted to the major faults, but also involves large rock volumes in the hanging wall and footwall (compare also with data on the Rhine Graben Rift; Bonjer 1997). Conversely, it will not be easy to locate active major faults simply from earthquake microzonation studies.

Conclusions

(1) Our account of the microfabrics of deformed rocks from the Woodlark extensional detachment shows an interesting localization pattern of dominantly brittle and dominantly ductile deformation histories. The rocks in the major detachment record a history of exclusively ductile flow. The rock volumes in the hanging wall and the footwall indicate prevailing brittle behaviour.

(2) The exhumation of the basement rocks of Moresby Seamount and its northern shoulder begins deep in the crust at ambient temperatures of at least $300\,^{\circ}C$. Assuming that the present-day, far-field geothermal gradient ($28-31\,^{\circ}C$ km^{-1}) reflects the conditions at the time when exhumation began, this corresponds to a depth of $c.\ 10\,km$.

(3) ODP Leg 180 drillcores helped to assess the applicability of two empirically calibrated calcite twin palaeo-piezometers. We are able to show that the calibration of Jamison & Spang (1976) is probably preferable for this situation. We also infer that shear twinning in the calcite veins occurred at differential stresses in the range of $22-62\,MPa$ at temperatures lower than $200\,^{\circ}C$. This is the likely magnitude of stress build-up in and around the Woodlark extensional detachment in the later part of the exhumation history of Moresby Seamount.

We wish to thank B. Taylor, P. Huchon, A. Klaus and all ODP Leg 180 shipboard scientific and technical personnel for discussion and feedback at various stages of our work. Discussions with R. Buck gave us valuable new insights. Incisive reviews by D. Hindle and G. Manatschal were most helpful and sharpened our thinking. This study was funded by Deutsche Forschungsgemeinschaft Grant Be 1041/13 to J.H.B. Deutscher Akademischer Austauschdienst (DAAD) is acknowledged for making funds available for scientific interchange between the University of Freiburg and Laboratoire de Géodynamique Sous-Marine, Villefranche-sur-Mer, under the 'Procope' agreement. Thanks go to J. Mascle for actively supporting these contacts.

References

ABERS, G.A. & ROECKER, S.W. 1991. Deep structure of an arc–continent collision: earthquake relocation and inversion for upper mantle P and S wave velocities beneath Papua New Guinea. *Journal of Geophysical Research*, **96**, 6379–6401.

ABERS, G.A., MUTTER, C.Z. & FANG, J. 1997. Earthquakes and normal faults in the Woodlark D'Entrecasteaux rift system, Papua New Guinea. *Journal of Geophysical Research*, **102**, 15301–15317.

AXEN, G.J. 1992. Pore pressure, stress increase, and fault zone weakening in low-angle normal faulting. *Journal of Geophysical Research*, **97**, 8979–8991.

BONJER, K.P. 1997. Seismicity pattern and style of seismic faulting at the eastern border fault of the Southern Rhine Graben. *Tectonophysics*, **275**, 41–69.

BRIGAUD, F. & VASSEUR, G. 1989. Mineralogy, porosity, and fluid control on thermal conductivity of sedimentary rocks. *Geophysical Journal*, **98**, 525–542.

BURKHARD, M. 1993. Calcite twins, their geometry, appearance and significance as stress–strain markers and indicators of tectonic regime: a review. *Journal of Structural Geology*, **15**, 351–368.

COCHRAN, J.R. 1983. Effects of finite rifting times on the development of sedimentary basins. *Earth and Planetary Science Letters*, **66**, 289–302.

CONEY, P.J. & HARMS, T.A. 1984. Cordilleran metamorphic core complexes: Cenozoic extensional relics of Mesozoic compression. *Geology*, **12**, 550–554.

FERRILL, D.A. 1998. Critical reevaluation of differential stress estimates from calcite twins in coarse-grained limestone. *Tectonophysics*, **285**, 77–86.

JAMISON, W.R. & SPANG, J.H. 1976. Use of calcite twin lamellae to infer differential stress. *Geo-*

logical Society of America Bulletin, **87**, 868–872.

LISTER, G.S., ETHERIDGE, M.A. & SYMONDS, P.A. 1986. Detachment faulting and the evolution of passive continental margins. *Geology*, **14**, 246–250.

LOCKNER, D.A. 1995. Rock failure. *In*: AHRENS, T.J. (ed.) *Rock Physics and Phase Relations: a Handbook of Physical Constants*. American Geophysical Union, Washington, DC, 127–145.

MUTTER, J.C., MUTTER, C.Z. & FANG, J. 1996. Analogies to oceanic behaviour in the continental breakup of the western Woodlark Basin. *Nature*, **380**, 333–336.

NEWMAN, J. 1994. The influence of grain size and grain size distribution on methods for estimating paleostresses from twinning in carbonates. *Journal of Structural Geology*, **16**(12), 1589–1601.

NICOLAS, A. & POIRIER, J.P. *Crystalline Plasticity and Solid State Flow in Metamorphic Rocks*. Wiley, London.

PASSCHIER, C.W. & TROUW, R.A.J. *Microtectonics*. Springer, Berlin.

ROWE, K.J. & RUTTER, E.H. 1990. Paleostress estimations using calcite twinning: experimental calibration and application to nature. *Journal of Structural Geology*, **12**, 1–17.

SCHMID, S.M. 1982. Microfabric studies as indicators of deformation mechanisms and flow laws operative in mountain building. *In*: HSÜ, K.J. (ed.) *Mountain Building Processes*. Academic Press, London, 95–110.

SCHMID, S.M., PANOZZO, R. & BAUER, S. 1987. Simple shear experiments on calcite rocks: rheology and microfabric. *Journal of Structural Geology*, **9**, 747–778.

TAYLOR, B., GOODLIFFE, A.M. & MARTINEZ, F. 1999a. How continents break up: insights from Papua New Guinea. *Journal of Geophysical Research*, **104**, 7497–7512.

TAYLOR, B., GOODLIFFE, A., MARTINEZ, F. & HEY, R. 1995. Continental rifting and initial sea-floor spreading in the Woodlark Basin. *Nature*, **374**, 534–537.

TAYLOR, B., HUCHON, P., KLAUS, A., *et al.* (eds) 1999b. *Proceedings of the Ocean Drilling Program, Initial Reports, 180*. Ocean Drilling Program, College Station, TX.

TAYLOR, B., MUTTER, C., GOODLIFFE, A. & FANG, J. 1996. Active continental extension: the Woodlark Basin. *JOI/USSAC Newsletter*, **9**, 1–4.

TULLIS, T.E. 1980. The use of mechanical twinning in minerals as a measure of shear stress magnitudes. *Journal of Geophysical Research*, **85**, 6263–6268.

TWISS, R.J. & MOORES, E.M. *Structural Geology*. W.H. Freeman, New York.

VOLL, G. 1976. Recrystallization of quartz, biotite and feldspars from Erstfeld to the Leventina Nappe, Swiss Alps, and its geological significance. *Schweizerische Mineralogische und Petrographische Mitteilungen*, **56**, 641–647.

WERNICKE, B. 1995. Low-angle normal faults and seismicity: a review. *Journal of Geophysical Research*, **100**, 20159–20174.

Evolution of the Miocene–Recent Woodlark Rift Basin, SW Pacific, inferred from sediments drilled during Ocean Drilling Program Leg 180

A.H.F. ROBERTSON[1], S.A.M. AWADALLAH[2], S. GERBAUDO[3], K.S. LACKSCHEWITZ[4], B.D. MONTELEONE[5], T.R. SHARP[6] & other members of The Shipboard Scientific Party:

C.P. HUCHON & B. TAYLOR (Co-chief Scientists), A. KLAUS (Staff Scientist), C.K. BROOKS, B. CÉLÉRIER, E.H. DECARLO, J. FLOYD, G.M. FROST, V. GARDIEN, A.M. GOODLIFFE, J.K. HAUMU, N. ISHIKAWA, G. KARNER, P.M. KIA, A. KOPF, R. LARONGA, B. LE GALL, I.D. MATHER, R.C.D. PEREMBU, J.M. RESIG, E.J. SCREATON, W.G. SIESSER, S.C. STOVER, K. TAKAHASHI, P. WELLSBURY

[1]*Department of Geology and Geophysics, West Mains Road, University of Edinburgh, Edinburgh EH9 3JW, UK (e-mail: Alastair.Robertson@glg.ed.ac.uk)*
[2]*Earth Sciences Department, Memorial University of Newfoundland, St. John's, NF A1B 3X5, Canada*
[3]*Dipartimento di Scienze della Terra, Università degli Studi di Genova, Corso Europa, 26, Genova 16132, Italy*
[4]*Institut für Geowissenschaften, Johannes Gutenberg-Universtät Mainz, 55099 Mainz, Germany*
[5]*Department of Geosciences, Gould–Simpson Building 210, University of Arizona, Tuscon, AZ 85721, USA*
[6]*Department of Environmental Sciences, University of Technology, Sydney, PO Box 123, Broadway, Sydney 2007, N.S.W., Australia*

Abstract: The results of drilling during Ocean Drilling Program Leg 180 provide insights into fundamental processes of continental break-up, because rifting can be related to westward propagation of a spreading centre into continental crust. A generally north–south transect of holes was drilled across the Woodlark Rift on the uplifted northern rift margin on the Moresby Seamount (Sites 1114 and 1116), on the hanging wall of the low-angle (25–30°) extensional Moresby Detachment Fault (Sites 1108, 1110–1113 and 1117) and across the downflexed northern rift margin (Sites 1118, 1109 and 1115). The results, when placed in the regional tectonic context, document a history of Palaeogene ophiolite emplacement, followed by Miocene arc-related sedimentation. Regional uplift and emergence of the forearc area took place in Late Miocene time. Submergence to form the Woodlark Rift began in latest Miocene time, marked by widespread marine transgression and shallow-water deposition, accompanied by input of air-fall tephra and volcaniclastic sediments. During Pliocene time, deposition within the rift basin was dominated by deep-water turbidites, including high-density turbidites in the south. Strong extension along the north-dipping Moresby Detachment Fault was active during Pleistocene time, associated with uplift of the Moresby Seamount and shedding of fault-derived talus, mainly of meta-ophiolitic origin. During Pleistocene time, a carbonate platform was constructed to the NW, trapping clastic sediment and resulting in a switch to slower, more pelagic and hemipelagic deposition within the Woodlark Rift Basin. The marked change in rift basin configuration during Pleistocene time may relate to westward propagation of the Woodlark oceanic spreading centre at *c.* 2 Ma.

From: WILSON, R.C.L., WHITMARSH, R.B., TAYLOR, B. & FROITZHEIM, N. 2001. *Non-Volcanic Rifting of Continental Margins: A Comparison of Evidence from Land and Sea.* Geological Society, London, Special Publications, **187**, 335–372. 0305-8719/01/$15.00 © The Geological Society of London 2001.

A generally north–south transect of holes, drilled across the Woodlark Rift directly west of the tip of the spreading oceanic Woodlark Basin, has elucidated the history and processes of rifting from Late Miocene to Recent time (Taylor *et al.* 1999*a*). The Woodlark Rift is one of the few areas where rifting, coeval with sea-floor spreading, has been studied in detail by scientific drilling. When the primary objective of Ocean Drilling Program (ODP) Leg 180, to drill and instrument an active extensional detachment fault, proved impossible to achieve, time was devoted to completing a transect of holes on the northern and southern margins of the Neogene–Recent Woodlark Rift. This was outstandingly successful, and resulted in the most detailed record of a young active rift system yet achieved by either academic or industry drilling. Here, we discuss the palaeo-environmental and sedimentary evolution of the Woodlark Rift as a series of time slices and present an integrated interpretation within the regional geological context.

Regional setting

The Woodlark Rift is located along the northern periphery of the Papuan Peninsula and the offshore D'Entrecasteaux Islands (Fig. 1). The rift basin deepens eastwards into the contemporaneous oceanic Woodlark spreading centre, of *c.* 500 km length (Weissel *et al.* 1982; Binns & Whitford 1987; Goodliffe *et al.* 1993, 1997; Benes *et al.* 1994; Taylor *et al.* 1999*b*; Martinez, pers. comm.). The spreading tip, located at 9.8°N, 171.7°E, reaches within 1 nautical mile of the most easterly drill site (Site 1108).

The Woodlark Rift is divisible into a relatively narrow (*c.* 15 km), deep (*c.* 3000 m) fault-bounded rift basin in the south and a wider (*c.* 50 km), shallower (*c.* 1000–2000 m) northern margin (Fig. 2). The Pliocene–Pleistocene succession is marked by a shallow discordance between lower, southward-dipping reflectors, and overlying more horizontal reflectors (Taylor *et al.* 1999*a*; Fig. 3a,b). Beneath this, the southerly part of the downflexed northerly rift margin is unconformably underlain by

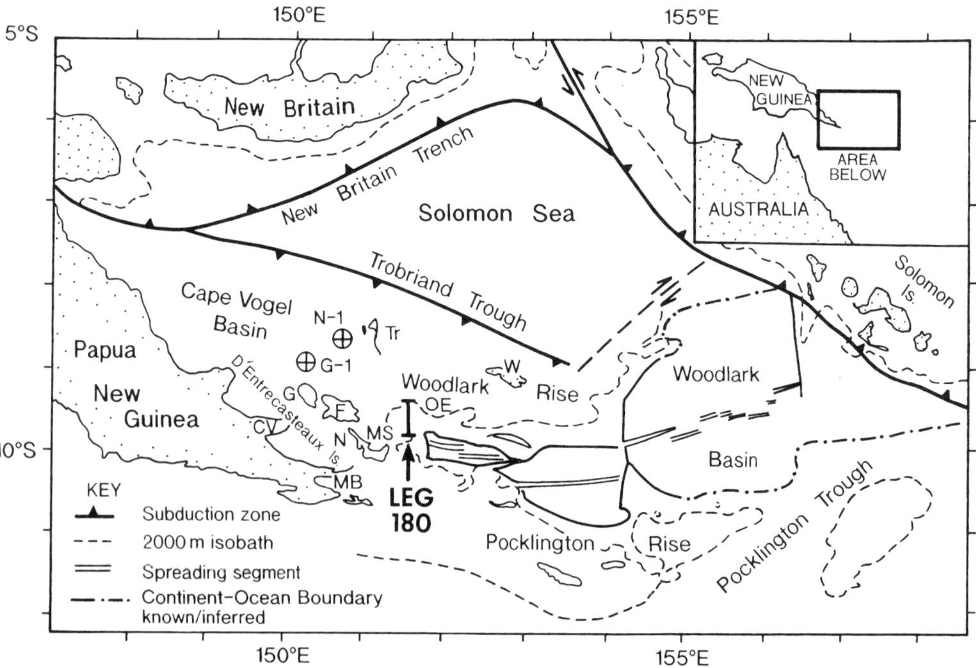

Fig. 1. Regional tectonic setting of the Woodlark Rift adjacent to the Papuan Peninsula; inset shows wider regional setting. G-1, Goodenough-1 well; N-1, Nubiam-1 well; CV, Cape Vogel Basin (onshore); E, Egum Island; F, Fergusson Island; G, Goodenough Island; MB, Milne Bay; MS, Moresby Seamount; N, Normanby Island; Tr, Trobriand Island; W, Woodlark Island. Modified from Taylor *et al.* (1999*a*).

an inferred Palaeogene ophiolitic unit, whereas the outer rift margin is underlain by Miocene sediments correlated with the Trobriand fore-arc, based on drilling during Leg 180 (Taylor *et al.* 1999*a*). The southern margin of the Woodlark Rift is delineated by the Moresby Detachment Fault, which dips at 25–30° to the NNE, away from the southern margin of the rift, rep-

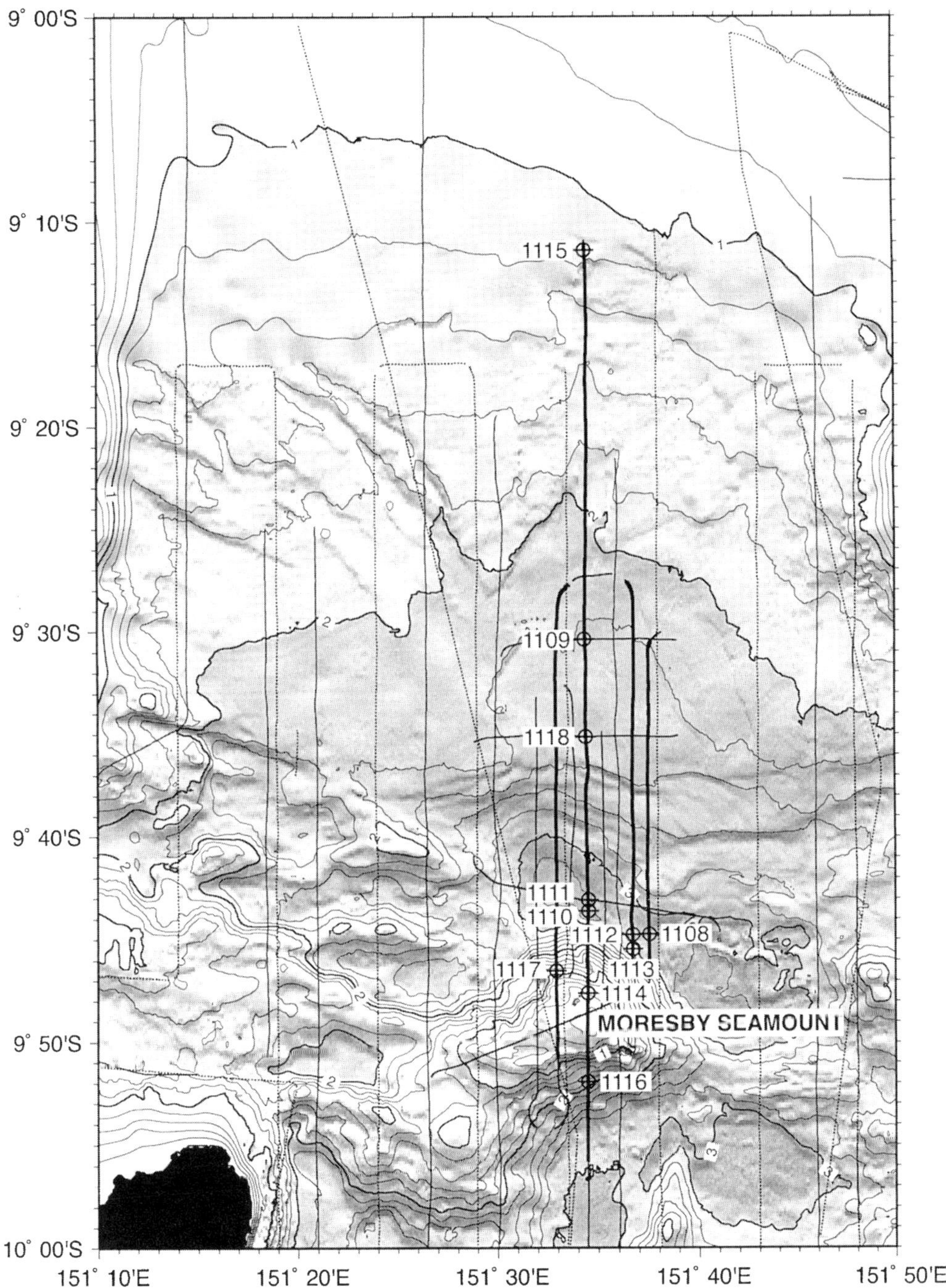

Fig. 2. Locations of sites drilled during Leg 180. Modified from Taylor *et al.* (1999*a*).

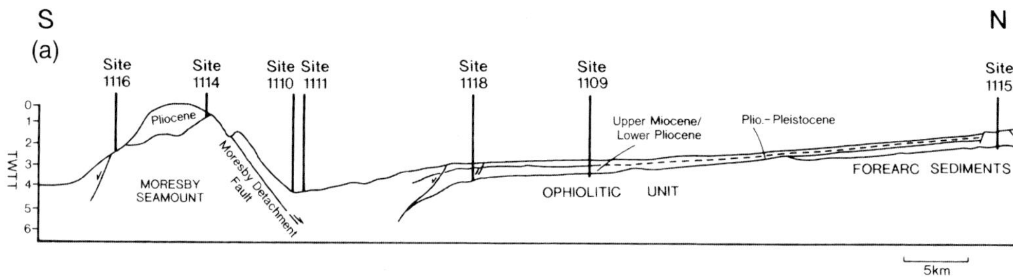

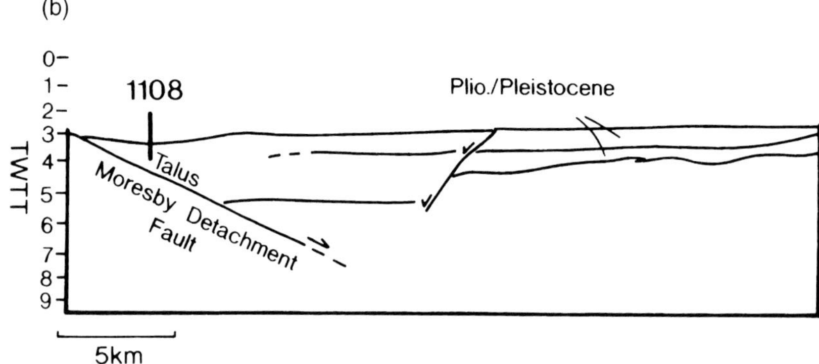

Fig. 3. Line diagrams showing interpretations of north–south seismic profiles along which the Leg 180 sites were drilled (see Taylor *et al.* 1995, 1999*a*). (**a**) westerly profile; (**b**) easterly profile. See Figure 2 for locations of the sites drilled on these two lines. Seismic profiles on which these figures are based are given in the Leg 180 Initial Results (Taylor *et al.* 1999*a*). TWTT, two-way travel time in seconds.

resented by the Moresby Seamount. Numerous north–south, normal and oblique faults imaged on the Moresby Seamount are consistent with regional north–south extension, as deduced from earthquake fault-plane solutions (Abers 1991, 2001; Hegler *et al.* 1995; Mutter *et al.* 1996; Abers *et al.* 1997), and global positioning system (GPS) measurements (Tregoning *et al.* 1998). Drilling during ODP Leg 180 confirmed that the Moresby Seamount is composed of continental crust, rather than being a typically igneous seamount, confirming earlier dredge results from the northeastern flank of this structure, which produced clasts including psammite, phyllite, pelite, greenschist, metagabbro and microgranite (Binns *et al.* 1990).

North of the Woodlark Rift, the Woodlark Rise, part of the Trobriand forearc, is interspersed with several volcanic islands (e.g. Woodlark Island) and seamounts (e.g. Egum atoll), beyond which lies the Trobriand Trough, interpreted as an active south-dipping subduction zone related to the Trobriand Arc; further north again is the Solomon Sea, which is floored by oceanic crust (Davies *et al.* 1987; Honza *et al.* 1987; Fig. 1).

The Woodlark Rift passes northwestwards into a large region, here termed the Cape Vogel Basin as a whole. In detail, the Cape Vogel Basin includes onshore Neogene basins in Papua New Guinea and an adjacent offshore area dominated by terrigenous clastic sediments derived from onshore. Further offshore within the Cape Vogel Basin as a whole there is a large carbonate platform ('Trobriand Platform'), which is studded with coral reefs, atolls and islands and interspersed with submarine channels (Fig. 1).

Regional tectonic history

Before Miocene time the Papua New Guinea area was the site of northeastward subduction of Cretaceous oceanic crust. This subduction gave rise to a Paleogene volcanic arc and related oceanic crust, now obducted on a regional scale as the NW–SE-trending Papuan Ultramafic Belt (Davies *et al.* 1997; Davies 1980*a*; see Figs 4 and 5), exposed over a distance of 400 km in the Owen Stanley Range. Other related units include the Milne Basic Complex (Smith 1982), located closest to the

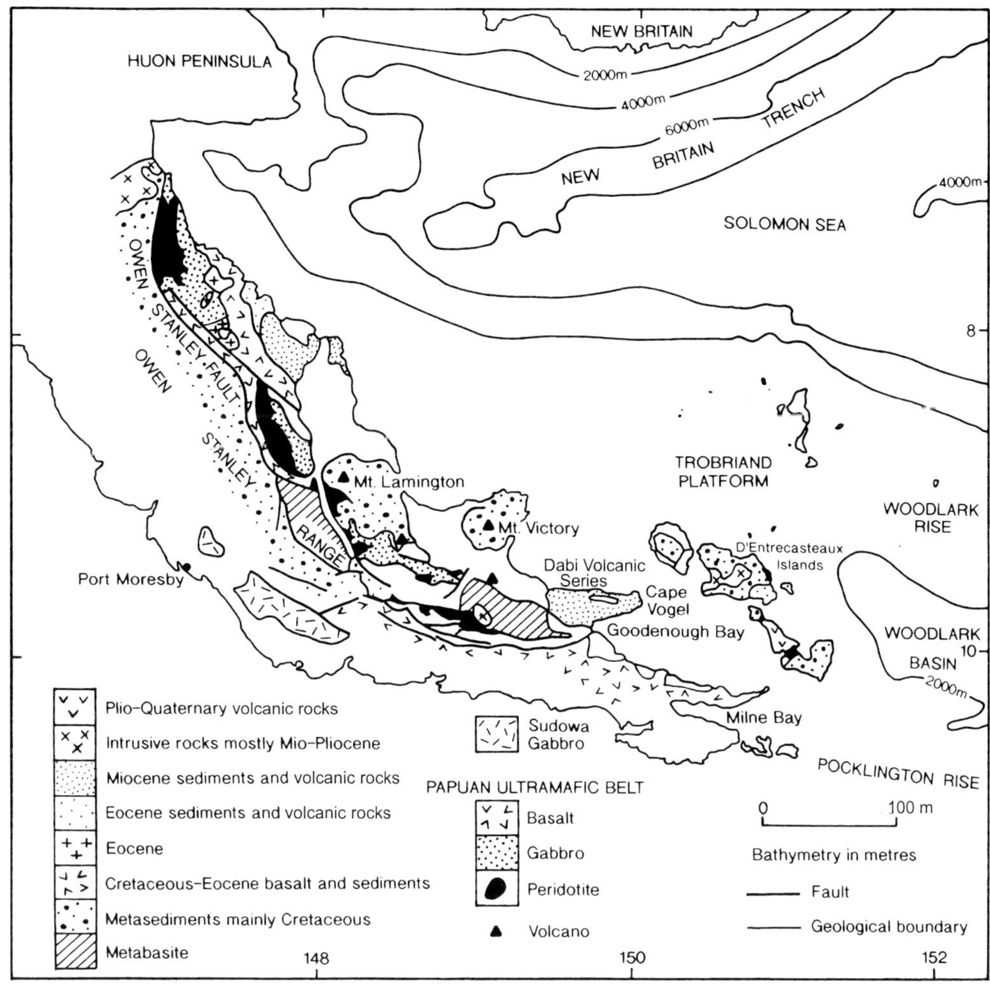

Fig. 4. Geological map of Papua New Guinea showing the main tectonic units. Modified after Davies & Jaques (1984). Submarine features simplified from Taylor *et al.* (1999*a*).

Leg 180 sites, the Cape Vogel Boninites (Francis 1985) and the Dabi volcanic series (Walker & McDougall 1982); also, more easterly extensions of these rocks on the Pocklington Rise and the Woodlark Rise (Davies & Smith 1971; Davies *et al.* 1984, 1997) (Fig. 1). In places, a complete ophiolite sequence is present, an interpretation supported by regional geophysical studies (Finlayson *et al.* 1976). The ophiolite is overlain by rarely preserved Palaeocene foraminiferal pelagic carbonates. The ophiolite is also locally cut by intermediate–acid extrusive rocks, which have been radiometrically dated as Palaeogene in age (Rogerson *et al.* 1993). The Cape Vogel Boninites were also assigned to a Palaeogene age by the $^{40}Ar/^{39}Ar$ method (Walker & McDougall 1982), as was a

gabbro from the Papuan Ultramafic Belt (Duncan, pers. com., cited by Taylor *et al.* 1999*a*). It is unclear if all these ophiolitic units are of the same age and whether some of the published ages may record an obduction-related isotopic resetting rather than primary magmatic ages of ocean crust genesis. At present, we believe the Papuan ophiolitic rocks were created and emplaced between Late Cretaceous and Early Eocene time, but more dating work is needed.

In the central Papuan Peninsula ophiolitic thrust sheets are locally underlain by a preserved metamorphic sole including granulites, and this, in turn, is underlain by felsic schists and gneiss (Owen Stanley Metamorphic Series; Davies 1980*a*), interpreted as a metamorphosed accretionary complex (Fig. 5).

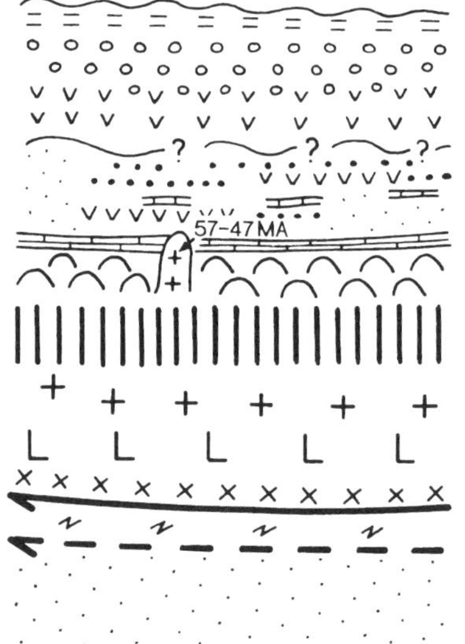

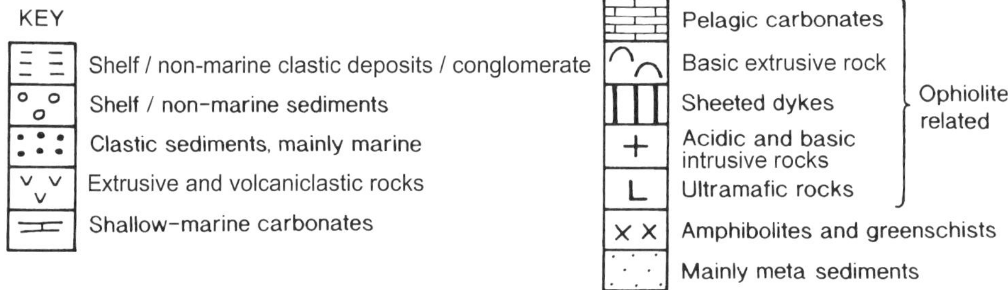

Fig. 5. Simplified rock-relations diagram of major lithostratigraphic units exposed in the Papuan Peninsula and the D'Entrecasteaux Islands. Not to scale. (See text for explanation and literature sources.)

Also, underlying metabasic rocks (Emo Metamorphic Series) are interpreted as accreted Cretaceous oceanic crust (Worthing 1988). In addition, beneath the Cape Vogel Basin of the SE Papuan Peninsula (Fig. 1), ophiolitic rocks are underlain by high-pressure blueschists, including lawsonite and blue amphibole (Kagi Metamorphic Series). Meta-ophiolites and metasediments (e.g. schist, gneiss) are also exposed in the D'Entrecasteaux Islands (Davies 1980*b*, 1990; Dow 1997).

In the Cape Vogel area, ophiolite-related extrusive rocks (Lokanu Volcanic Series) are overlain by Upper Paleocene hemipelagic carbonates, then by Lower Eocene subduction-related calc-alkaline volcanic rocks, volcaniclastic sediments and conglomerates (Eia Beds;

Rogerson *et al.* 1987). These lithologies record the construction of a volcanic arc above an ophiolitic basement. Northward subduction culminated in collision of the North Australian passive continental margin, activating a southward-propagating thrust belt that included obducted ophiolites and underlying accretionary unit (Rogerson *et al.* 1993). After collision, subduction flipped to southwestward from Miocene time onwards (Cooper & Taylor 1987), setting the scene for development of the Miocene Trobriand Arc and subsequent formation of the Woodlark Rift.

The Miocene–Recent Trobriand Arc relates to subduction of Palaeogene oceanic crust of the West Solomon Sea, which had an original width of up to 1000 km (Davies *et al.* 1987;

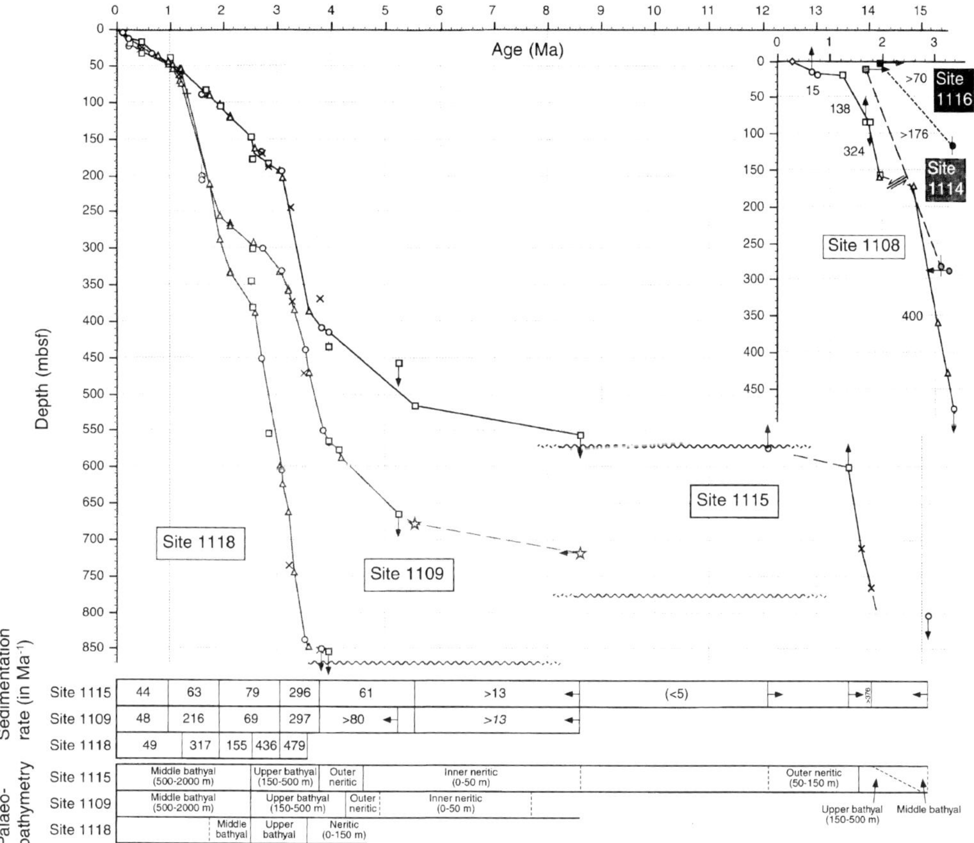

Fig. 6. Age–depth chart for the Leg 180 sites, including a summary of inferred sedimentation rates and depths (below). Complied by the Leg 180 shipboard biostatigraphers and magnetostratigraphers. Modified after Taylor *et al.* (1999*b*). Key: sedimentation curves at Sites 1115 (continuous line), 1109 (dashed line), 1118 (dotted line), 1108 (continuous line upper right), 1114 (dashed line upper right), and 1116 (dotted line upper right), based on nannofossil (□) and planktonic foraminifer (○) datum events, magnetic chron and subchron boundaries (△), and lithostratigraphical correlation (☆), shaded to differentiate sites. Symbols with arrows denote that actual datum point can be above or below and older or younger than indicated by the symbols. Wavy lines denote unconformities. Shown below are average sedimentation rates, calculated for intervals separated by vertical lines and palaeobathymetry, based on benthic foraminifers, at Sites 1115, 1109 and 1118. Broken lines indicate uncertainty in the placement of palaeodepth boundaries.

Honza *et al.* 1987). Subduction of *c.* 700 km of oceanic crust is inferred, both southwestwards beneath Papua New Guinea, forming the Miocene–Recent Trobriand Arc, and also northwards beneath New Britain and the Bismarck Sea (Davies *et al.* 1987; Honza *et al.* 1987; Taylor 1987). The Trobriand Arc comprises a backbone of granitic batholiths in the SE Papuan Peninsula. Arc volcanic rocks are exposed in the Papuan Peninsula, on the offshore D'Entrecasteaux Islands, and as a small number of volcanic centres further north (e.g. Woodlark Island). Initial high-K volcanism of Mid-Miocene age was succeeded by Upper Miocene–Recent, dominantly calc-alkaline

volcanic rocks. Local Middle Pliocene–Recent peralkaline–shoshonitic extrusive rocks are interpreted to reflect volcanism from subduction-influenced mantle in an extensional setting (Smith *et al.* 1977; Smith & Milsom 1984; Stolz *et al.* 1993).

Sedimentary evolution of the Woodlark Rift Basin

The northern rift margin sites are discussed first (Figs 7–10), as they are intact, well dated and can be correlated using seismic reflection data. By contrast, the Moresby Seamount sites are

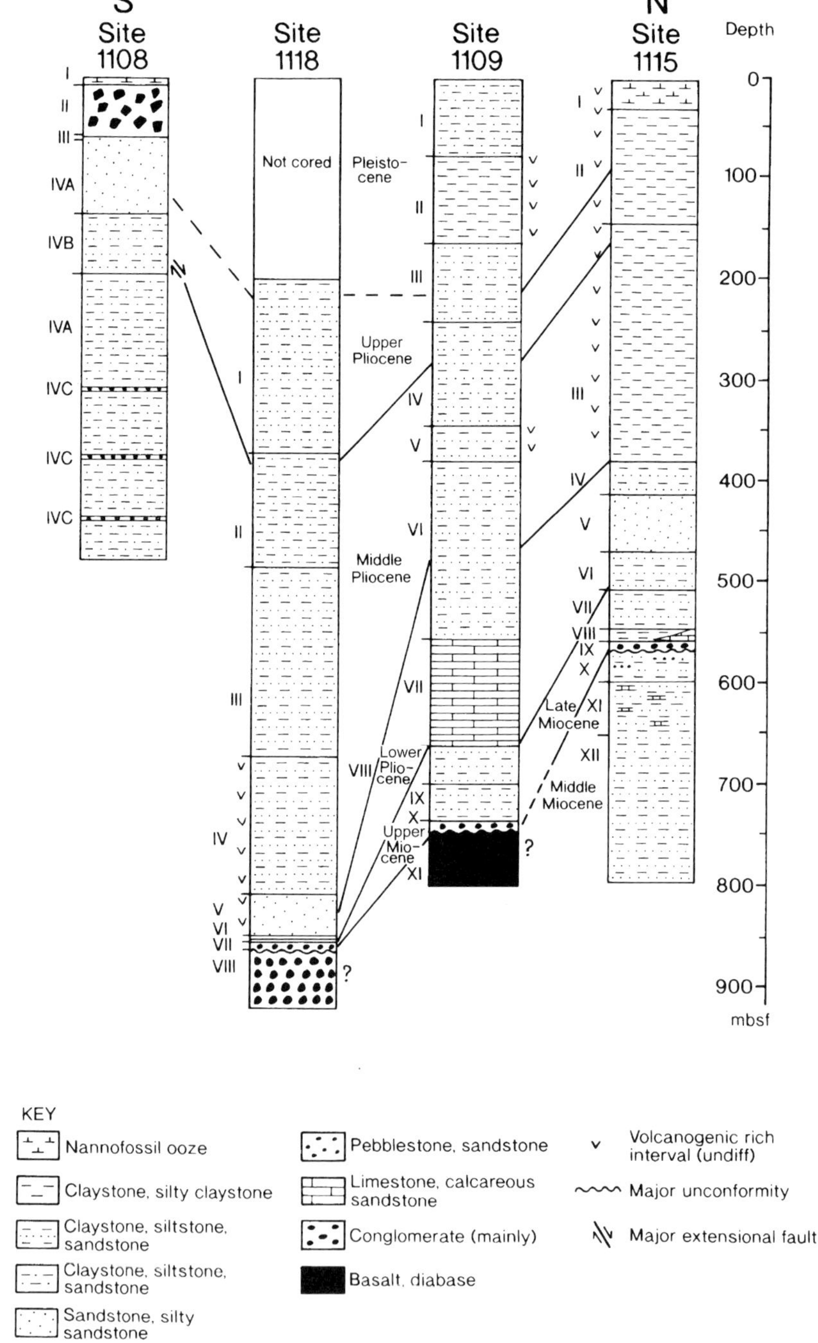

Fig. 7. Summary of the sedimentary successions drilled on near axial and northern margin sites of the Woodlark Rift. Modified after Taylor *et al.* (1999*a*).

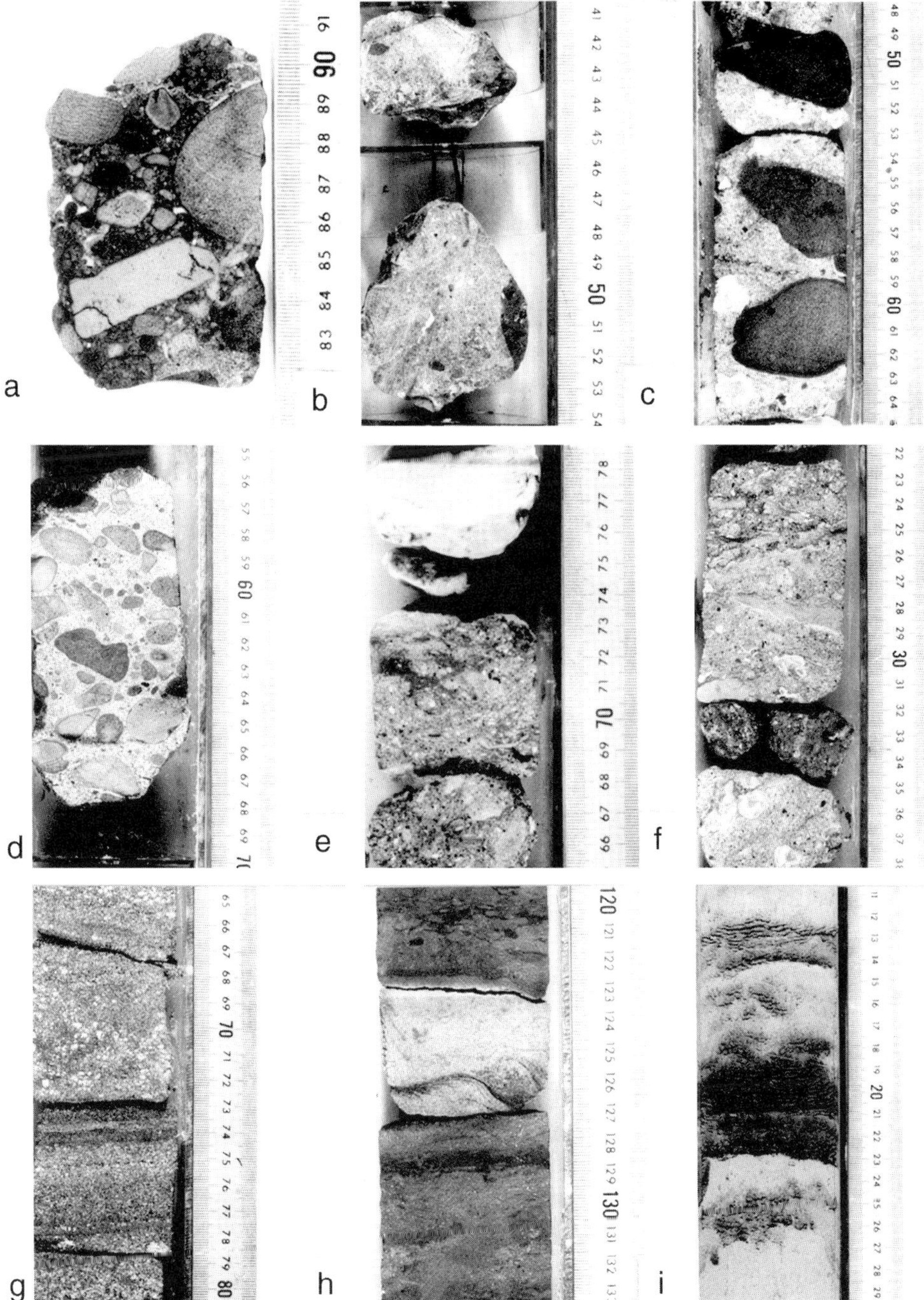

Fig. 8. Photographs of cores from the near axial and northern rift margin sites. (**a**) Conglomerate with clasts of basalt–diabase, Site 1109D-43R-1. (**b**) Basalt clasts in neritic carbonate matrix; Site 1118A-70R-2. (**c**) Well-rounded basalt clasts in shallow-water carbonate; Site 1118A-69R-3. (**d**) Small well-rounded igneous clasts in shallow-water carbonate matrix; Site 1115C-30R-5. (**e**) Basalt-derived sandstone overlain by algal limestone (contact not recovered); Site 1118A-70R-1. (**f**) Algal-rich neritic limestones; Site 1118A-70R-1. (**g**) Volcaniclastic sandstone, Site 1118A-68R-2. (**h**) Typical turbiditic volcaniclastic siltstone–fine sandstone; Site 1115C-48R-1. (**i**) Tephra-rich silt interbedded with volcanogenic mud, Site 1115B-10H-3.

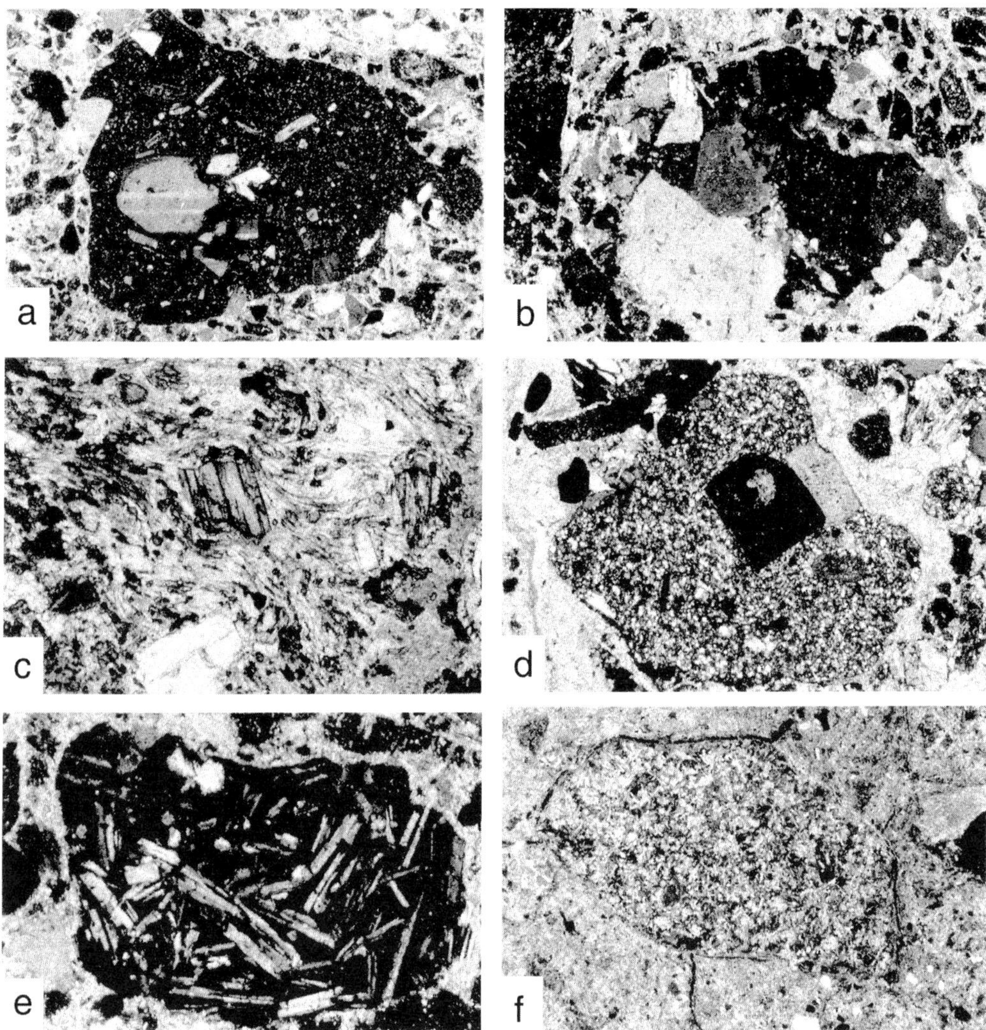

Fig. 9. Photomicrographs illustrating: (**a**)– (**f**) volcanic-derived constituents; (**g**)– (**l**) sediments of the Woodlark Rift. (**a**) Plagioclase-, hornblende- and biotite-phyric acidic volcanic detrital grain surrounded by mafic volcanic and feldspar grains set in a calcite-cemented coarse-grained lithic sandstone. Surrounding detrital grains include acidic and mafic volcanic and metamorphic rocks (Site 1108, 14.5 mbsf, field of view 4 mm, cross-polarized light). (**b**) Detrital grain composed of intergrown feldspar, biotite and hornblende (granite fragment) surrounded by mafic volcanic and feldspar detrital grains in calcite-cemented coarse-grained lithic arenite (Site 1108, 33.86 mbsf, field of view 1.55 mm, cross-polarized light). (**c**) Hornblende- and feldspar-phyric vitric volcanic detrital grain with an internal trachytic texture (Site 1115, 394 mbsf, field of view 1.55 mm, cross-polarized light). (**d**) Feldspar- and biotite-phyric acidic volcanic detrital grain surrounded by acidic volcanic, mafic volcanic and feldspar detrital grains in a calcite-cemented coarse-grained lithic arenite (Site 1116, 0.58 mbsf, cross-polarized light). (**e**) Detrital basaltic grain with intersertal glass and intergranular clinopyroxene (Site 1108, 14.5 mbsf, field of view 1.0 mm, cross-polarized light). (**f**) A rounded dolerite granule–pebble within a matrix-supported conglomerate with a highly altered siltstone matrix (Site 1118, 879 mbsf, field of view 4 mm, cross-polarized light).

more tectonically deformed, less well dated, and difficult to correlate. The age constraints are based on shipboard biostratigraphical and magnetostratigraphical analysis (Taylor *et al.* 1999*a*), slightly modified by post-cruise studies (Resig *et al.* 2001; Takahashi pers. comm.). The time scale is that of Berggren *et al.* (1995). The water depth estimates were obtained

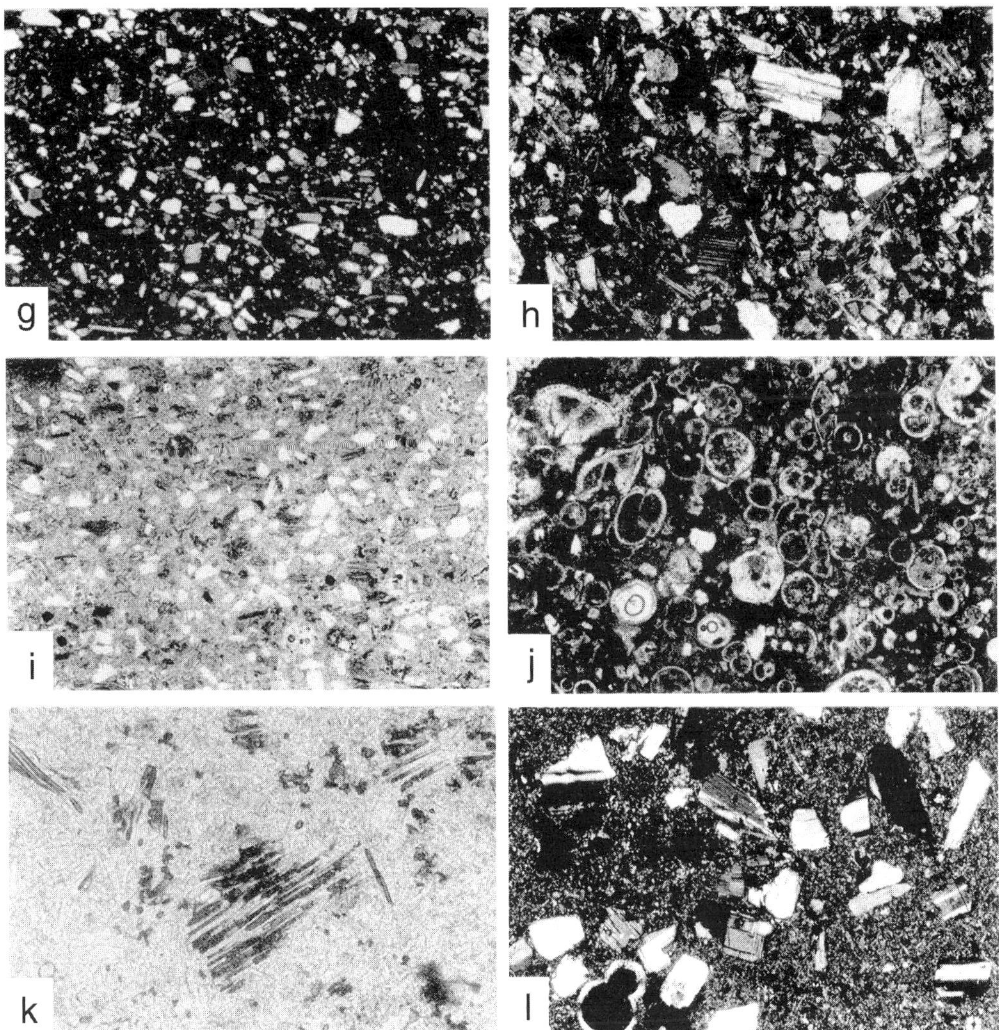

(**g**) Moderately sorted fine-grained feldspathic sandstone present at the base of normally graded sandstone laminae. Rare detrital biotite, amphibole, pyroxene and common vitric volcanic grains (Site 1115, 698.7 mbsf, field of view 4 mm, cross-polarized light). (**h**) Poorly sorted medium-grained feldspathic sandstone with common detrital hornblende and biotite grains (Site 1108, 276.39 mbsf, field of view 4 mm, cross-polarized light). (**i**) Fine-grained crystal vitric ash with detrital grains of feldspar, biotite, hornblende and clear glass, in a glassy matrix (Site 1109, 371.68 mbsf, field of view 1.55 mm, plane-polarized light). (**j**) Foraminiferal packstone (Site 1109, 590.05 mbsf; field of view 4 mm, plane polarized light). (**k**) Fresh pipe vesicle glass fragments in a glassy matrix of a vitric ash (Site 1118, 564.4 mbsf, field of view 0.5 mm, plane-polarized light). (**l**) Sandy siltstone containing fresh angular detrital grains of plagioclase (commonly zoned), biotite, hornblende and foraminiferal tests (Site 1115, 369.55 mbsf, field of view 1.55 mm, cross-polarized light).

mainly using benthic foraminifers (Taylor *et al.* 1999*a*). The reader is referred to Taylor *et al.* (1999*a*) for further discussion.

Northern rift margin sites

Middle Miocene forearc turbiditic succession. The Middle Miocene interval, recovered only at the most northerly site (1115), begins with homogeneous, weakly calcareous (5–10% $CaCO_3$) sandy siltstone, alternating with silty sandstone and rare silty claystone (657.8–802.5 m below sea floor (mbsf); Unit XII; Fig. 7). The sandstones contain abundant feldspar, common biotite, and small amounts of amphibole and pyroxene (Fig. 9g). In addition,

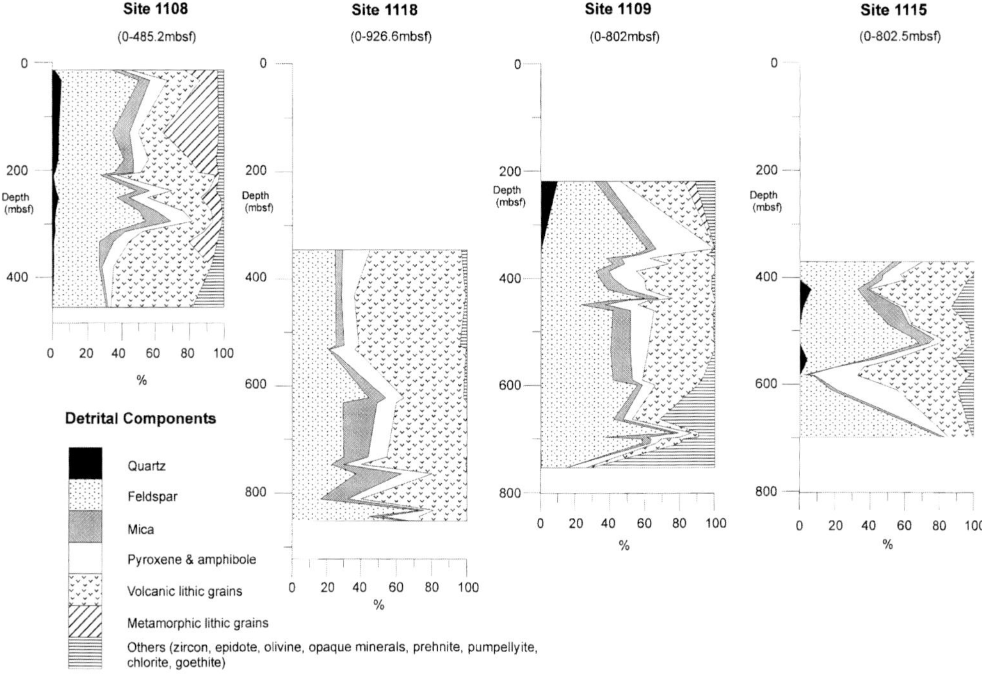

Fig. 10. Plots of downhole variation in the modal composition of sandstone; from the near axial and northerly margin of the Woodlark Rift, calculated from point count data from 66 sandstone thin sections by T. Sharp.

pyroxene becomes abundant towards the top of the Miocene succession. Lithic detritus includes grains of basalt (commonly chloritized), variolitic glass, palagonite, acidic extrusive rocks, and distinctive pyroxene-phyric basic extrusive rocks, again towards the top of the succession. The source rocks were mainly basaltic andesite, of inferred volcanic-arc origin. In addition, Unit XII accumulated at upper bathyal depths (150–500 m) at an average sedimentation rate of >135 m Ma^{-1}. Sandy siltstone and silty sandstone, with considerable input of terrestrial organic matter, underwent extensive bioturbation, indicating well-oxidized bottom conditions. The siltstones and sandstones are interpreted as deposits from turbidity currents (Fig. 8h).

The succession at Site 1115 passes upwards into sandstone and carbonate packstone (with coral, calcareous algae, bivalve shell fragments and bryozoans), interbedded with siltstone and claystone and minor conglomerate (603.8–657.8 mbsf; Unit XI). Formation MicroScanner images reveal a homogeneous succession, with occasional thin (<5 cm), inferred sandstone beds (relatively resistant) within muddy sediment. Possible cross bedding is imaged between 677 and 679 mbsf. This interval continued to

accumulate in an upper bathyal setting (150–500 m) at >135 m Ma^{-1}. Packstone was redeposited from a shallow-water setting, possibly by high-density turbidity currents. A slope setting is inferred from sedimentary structures (e.g. small-scale convolute lamination, synsedimentary faults), indicative of slumping, sliding and dewatering.

The mainly turbiditic Middle Miocene succession (Units XI and XII) accumulated in a relatively deep-water (150–500 m) forearc basin, adjacent to the front of the Trobriand Arc. Lithologies included distinctive pyroxene-phyric basic extrusive rocks. Comparison with seismic stratigraphic records (Taylor *et al.* 1999*a*) suggests that the cored interval represents only the top of a kilometres-thick forearc succession (Fig. 3a). Upwards, the increased influx of redeposited carbonate (Unit XI) could reflect derivation from carbonate build-ups bordering the Trobriand Arc, or a related outer arc high.

The Middle Miocene succession generally coarsens upwards into weakly calcareous ($<2\%$), well-graded, mainly volcaniclastic, sandstone, siltstone and minor granule conglomerate (571.9–603.8 mbsf; Unit X) that accumulated in an outer neritic setting (50–

150 m), based on the evidence of neritic large fossils and planktonic microfossils. Sedimentation rate is poorly defined, because of limited biostratigraphical control. The FMS logs and limited core recovery indicate the presence of thin sandstones, <15 cm thick, spaced every 0.5–1 m on average. Neritic shell material was derived from a carbonate-influenced coastline, or shoal. A common reddish colour suggests a well-oxidized *in situ* depositional setting. The granule conglomerates are interpreted as high-density turbidity current deposits. Shallowing of the uppermost part of the Middle Miocene succession at Site 1115 may reflect infilling of the forearc basin by sedimentary processes, or possibly tectonic uplift (see below).

Late Miocene forearc emergence and paralic deposition

The shallowing-upward succession at Site 1115 (>5.4 Ma) culminated in emergence and an unconformity. This unconformity is overlain, at this site, by an emergent, shallow-marine succession, dated to Late Miocene time (see below). Counterparts of these shallow-water deposits were also cored at Sites 1109 and 1118 further south.

At Site 1115, an Upper Miocene succession (565.7–571.9 mbsf; Unit IX) is characterised by poorly recovered sandstone, siltstone, conglomerate and rare calcareous claystone. The well-rounded nature of conglomerate clasts (mainly basalt, with some siltstone and bioclastic limestone) is suggestive of sediment transport in a high-energy shallow marine (e.g. coastal or beach), or fluvial setting (Fig. 8d). The geophysical logs, especially FMS, are interpreted to indicate that conglomerate is more abundant than the minimal recovery would suggest (574–567 mbsf).

Around 27.5 km further south, at Site 1109, the lowest sedimentary material recovered (737.2–772.9 mbsf; Unit X) comprises minor (undated) conglomerate, *c.* 3 m thick (Fig. 8a), rare sandstone and claystone (which can be lithologically correlated with the igneous-derived conglomerate at Site 1115). FMS data reveal a thick interval (*c.* 30 m) dominated by conglomerate, overlying a basement of diabase rocks, of assumed ophiolitic origin. The irregular contact (located at *c.* 772 mbsf) may represent an unconformity. Above this, the FMS data are suggestive of alternations of conglomerate and finer-grained sediment. The clasts are mainly basalt and diabase, with subordinate epiclastic (reworked) sandstone, together with rare clasts and grains of acidic extrusive rock and possible intrusive rock. All the recovered material is highly altered, with abundant smectite.

The recovery in the overlying 30 m of the succession (705.3–737.2 mbsf; Unit IX) is dominated by altered clay-rich, greenish grey, to locally orange brown, colour-mottled siltstone, silty claystone and fine-grained sandstone, with a few basalt or diabase clasts. The geophysical logs (e.g. FMS) indicate that this interval again includes minor conglomerate. The sediments include rare goethite concretions, interpreted as bog iron ore (e.g. Blatt *et al.* 1980). Colour mottling reflects a variable oxidation state, typical of such non-marine (paralic) sediments. Occasional carbonate concretions resemble caliche, some of which is reworked as small carbonate clasts, consistent with a penecontemporaneous origin. Abundant smectite is inferred to be of pedogenic origin (e.g. Bjørlykke 1993). Also, freshwater charophytes are locally present (Taylor *et al.* 1999a). Overall, the recovered sediments are suggestive of deposition in a non-marine, paralic setting, possibly a mangrove swamp.

In addition, at Site 1118, *c.* 3.5 km further south, the lowest (undated) interval recovered (873.1–929.6 mbsf; Unit VIII) comprises *c.* 60 m of diabase with rare granule conglomerate composed of diabase, again of inferred ophiolitic origin, together with sandstone and silty claystone intercalations. Recovery in the lowest 40 m of the well (below 890 mbsf) (not imaged by FMS) was restricted to minor diabase and sedimentary rock. Above this (873.5–890 mbsf), FMS images indicate the presence of poorly sorted mainly clast-supported conglomerate composed of rounded clasts of basalts, dolerite and rare gabbro (up to 0.5 m thick), surrounded by smaller (<5 cm) angular to sub-angular clasts, and finer-grained material. Sediment between the diabase clasts originated as a primary sedimentary infill of well-oxidized palaeosol. Subsequently, texturally immature, black (non-oxidized) breccia and sand filtered down cracks between the larger clasts.

Comparisons. The Late Miocene conglomerates at Sites 1115, 1109 and 1108 were eroded from a tropically weathered landmass, probably a Palaeogene ophiolite, and deposited by fluvial processes, based on the following evidence: (1) marine fossils are absent; (2) chemical alteration of the clasts is suggestive of subaerial weathering; (3) a probable root trace was observed at Site 1115; (4) carbonate nodules are interpreted as incipient caliche at Site 1115;

(5) goethite nodules are identified as bog iron ore at Site 1109; (6) freshwater charophytes are present at Site 1109; (7) inferred palaeosols occur at Site 1118; (8) abundant smectite is probably of pedogenic origin. The conglomerates thin from south to north, from *c.* 56 m at Site 1118, to *c.* 30 m at Site 1109, and to *c.* 3 m at Site 1115, suggesting a source area towards the present Papuan Peninsula area.

Late Miocene–Early Pliocene initial transgression

At Sites 1109 and 1115 the paralic deposits, discussed above, are overlain by Upper Miocene–Lower Pliocene lagoonal to shallow-marine facies.

At Site 1109, the paralic interval was followed by fine-grained, largely unbioturbated, sediments with abundant clay and woody debris suggesting derivation from a nearby landmass in quiet water (671.8–705.3 mbsf; Unit VIII). Post-cruise studies indicate that the woody material was derived from eucalyptus trees that were presumably swept downstream during floods (Karner, pers. comm.). Geophysical logs (e.g. FMS) suggest the presence of sand–mud intercalations (although hole conditions were poor). Abundant shelly debris is mainly composed of bivalves (rarely articulated) and some gastropods, with little sign of current reworking. The local presence of small planktonic foraminifers (globigerines) indicates a periodical marine influence. A brackish-water biota was identified in Cores 1109D-35R to -38R (Taylor *et al.* 1999*a*). The scarcity of bioturbation in the dark, organic-rich sediments is suggestive of low-oxygen conditions. The existence of dolomite might indicate periods of hypersalinity. The fine-grained sediments are interbedded with volcaniclastic sand, altered glassy basalt and rare acidic volcanic pebbles derived from basic extrusive igneous rocks, similar to those of the underlying conglomerates.

Further north, at Site 1115, only a minor amount of dark organic-rich sediment was recovered as thin coal-rich laminae and grains within siltstone, fine sandstone and rare bioclastic limestone (551.8–565.7 mbsf; Unit VIII). The gamma-ray and FMS logs are suggestive of the presence of coal-rich layers, up to 2 m thick. In addition, more carbonate-rich layers are present, containing benthic foraminifers and shell fragments in a micritic matrix. A marine, inner neritic setting (<50 m) is inferred, with a sedimentation rate of >13 m Ma^{-1}. The succes-

sion passes into a coarse, but still mainly muddy succession (513.4–551.8 mbsf; Unit VII), with rare thin (*c.* 19 cm) sandstones, as inferred from the FMS logs. The recovered core reveals abundant aragonitic shelly bioclasts and volcaniclastic material. Some shell-rich beds may represent bioclastic lags that resulted from wave or tidal activity in a marine lagoonal setting. Volcaniclastic sediment includes grains of volcanic quartz, plagioclase (fresh and altered), plagioclase (zoned), biotite, hornblende, basalt (altered), acidic volcanic rocks, rare palagonite, perthite and muscovite. A continuing relatively low sedimentation rate is inferred (>13 m Ma^{-1}), still at inner neritic water depths (<50 m).

In the south, at Site 1118, a contrasting Early Pliocene, or older, carbonate-rich paraconglomerate (*c.* 14 m thick) was recovered (859.0–873.1 mbsf; Unit VII), directly overlying inferred fluvial conglomerates (see above). The paraconglomerate is poorly sorted, with clasts of basalt and dolerite set in a matrix of shallow-water-derived bioclastic carbonate. FMS and other logs (e.g. photoelectric effect) show that the lower contact of this interval is sharp, with a discrete carbonate-rich unit near the base, followed by a section dominated by conglomerate. Seismic profiles (Taylor *et al.* 1999*a*) suggest that Site 1118 then formed a palaeo-high, presumably an island, until it was transgressed in latest Early Pliocene time (see below).

Comparisons. Only the inferred lagoonal deposits at Site 1115 (Unit VIII) in the north are well dated to Late Miocene time, based on calcareous microfossils. Facies of the two northerly sites (1115, 1109), record very similar, semi-enclosed, lagoonal conditions. For a time, isolated islands and banks remained in the vicinity of Sites 1115 and 1109, promoting semi-enclosed lagoons, until these, too, were submerged. By contrast, the southerly site (1118) shows an apparently abrupt marine transgression directly above fluvial conglomerates.

Latest Miocene–Early Pliocene time: rift-related subsidence

Within the limits of magneto-biostratigraphic resolution, Sites 1115 and 1109 underwent effectively synchronous subsidence from shallow-marine (neritic), to deeper-marine (upper bathyal) conditions during latest Miocene–Mid-Pliocene time (8.6–4 Ma; Fig. 6). By contrast, at Site 1118 marine transgression was delayed until latest Early Pliocene time (*c.* 3.6 Ma).

At Site 1109 in the north geophysical logs and recovered sediments chart a transition, during Early–Mid-Pliocene time (*c.* 3.8–5.6 Ma) (570.4–671.8 mbsf; Unit VII), from bioclastic limestone (packstone to grainstone), with 40–80% $CaCO_3$ (643–672 mbsf), to slightly calcareous sandstone with *c.* 20% $CaCO_3$ (599–643 mbsf), then to calcareous, bioclastic sandstone with 45–50% $CaCO_3$ (570–599 mbsf). The FMS images of the carbonate-rich interval reveal a highly resistive 'blotchy' effect, interpreted as concentrations of bioclastic material and isolated carbonate clasts (yielding resistive bright patches). The carbonate component, most abundant near the base, is dominated by shell fragments, micritic grains, minor calcite spar, planktonic (and rare) benthic foraminifers, rare echinoderm plates, bryozoans and nannofossils. The packstone or grainstone is well sorted and cemented by calcite spar (Fig. 9j). Pyrite occurs as scattered cubes in the matrix and infills planktonic foraminifer tests. Both planktonic and agglutinating benthic foraminifers are present. The upper part of this interval (599–643 mbsf) includes abundant volcaniclastic sandstone, mainly derived from basic–intermediate composition extrusive igneous rocks, as confirmed by the presence of small lithoclasts of basalt or andesite, and abundant grains of feldspar, quartz, biotite, hornblende and pyroxene (in decreasing order of abundance). The well-rounded shape of many of the extrusive lithoclasts suggests that they were reworked before final deposition (i.e. by epiclastic processes). Overall, this Early–Mid-Pliocene interval accumulated in a shallow-marine setting (inner to outer neritic, <150 m). The sands show evidence of reworking in a well-oxygenated, current-influenced setting. Carbonate grains were presumably derived from coral and algal build-ups. Rapid sedimentation rates of >78 m Ma^{-1} are inferred. The composition of the volcaniclastic sediments is unlike that of the underlying basaltic conglomerates, and a calc-alkaline volcanic-arc source is inferred.

At Site 1115, further north, fine- to medium-grained silty sandstone of mainly volcaniclastic origin (417.3–474.9 mbsf; Unit V) accumulated during Early to Mid?-Pliocene time (3.9–3.6 Ma), at a sedimentation rate, initially, of 284 m Ma^{-1}. Acidic (i.e. volcanic rock fragments, quartz), as well as basic volcanic material (basalt, hornblende, biotite, altered palagonite, clinopyroxene, olivine) is common, along with a small number of plutonic igneous-derived grains (e.g. coarse plagioclase). The geophysical log response is indicative of clay-rich sediments, passing upwards into a mixed sandy and silty succession, including a small number of discrete sandstone layers (<15 cm thick). The limited core recovery (<20%) indicates a transition from poorly sorted, highly bioturbated sandstone in the lower part of the interval, to planar laminated, wavy laminated and cross-laminated sandstone in the mid-part of the interval. These sedimentary structures are suggestive of reworking under the influence of traction currents (wave or tidal), consistent with accumulation at outer neritic depths (50–150 m).

At Site 1118, further south, the entire sequence above the paraconglomerate is post-Miocene in age, being dated as ≤3.6 Ma above 860 mbsf (Fig. 6). The paraconglomerate is overlain by up to 3 m of shallow-water packstone–grainstone (857.1–859.0 mbsf; Unit VI), as confirmed by the photoelectric log (from 860 to 857 mbsf). This interval directly overlies an unusual sandstone with a Th/K ratio of up to ten on the geophysical logs. The limestone contains coralline algae, coral, echinoderms, benthic foraminifers and other bioclasts (Fig. 8b). Shallow-water carbonate debris (coral, algae, rhodoliths, bivalves, shell fragments; Figures 8b–e and 9f) is mixed with rounded pebbles and granules of basalt and dolerite. Rounding of the igneous clasts took place before mixing with carbonate sediment, as shown by the preservation of non-abraded, calcareous algal coatings on some individual pebbles. The conglomeratic material is interspersed with micritic carbonate and algal deposits interpreted as algal mats (Fig. 8f). Rounded clasts were probably supplied fluvially from a neighbouring landmass, or were reworked from the subjacent paraconglomerate. An open-marine lagoonal setting is inferred, affected by waves and currents adjacent to a submerging landmass.

The succession at Site 1118 continues upwards into an interval, dated to latest Early Pliocene–Mid-Pliocene time (810.8–857.1 mbsf; Unit V), comprising alternations of sandstones, siltstones and volcaniclastic sandstone (Fig. 8g). Coarse-grained sandstone rich in carbonate bioclasts is present in the lowermost 10 m of the unit. Above this, the succession is dominated by poorly sorted, mixed sandy and silty claystones, siltstones and cemented sandstones, with local lithic fragments (Fig. 10). $CaCO_3$ values are high, ranging from 10 to 22%. The geophysical logs also reveal major spikes in the gamma-ray, thorium and potassium logs from 860.2 to 816.7 mbsf, which correlates with an increase in feldpar, as revealed by petrographic study. The occurrence of such 'radiogenic' spikes is tentatively correlated with input of volcaniclastic sands. These mixed sediments accumulated in upper bathyal depths (150–500 m), at estimated high sedimentation rates of 485 m Ma^{-1}. Proximity to a carbonate margin is indicated by the abundance of shallow-water bioclasts, including coral, coralline algae and benthic foraminifers. The sediments were mainly deposited by turbidity currents, but with some influence by traction currents, as indicated by the presence of sharp tops to many beds. A slope setting adjacent to a landmass is inferred from the sedimentary structures and the presence of sporadic wood fragments.

Comparisons. The Early–Mid-Pliocene interval of each of the three sites (latest Early Pliocene–Mid-Pliocene at Site 1118) records a change from a relatively shallow-water, current-influenced, neritic setting, to a deeper-water setting, dominated by turbidity current deposition. Shallow-water carbonate input was initially abundant at two Sites (1109 and 1118), but waned upsection, suggestive of submergence of the source of the shallow-water carbonate sounce. The volcaniclastic turbidites, dominantly at Site 1118, are suggestive of contemporaneous volcanism in the source region (see below). Sites 1109 and 1115 subsided synchronously through 150 m water depth (from outer neritic to upper bathyal), at *c.* 3.8 Ma. By contrast, Site 1118, located on a palaeo-island, underwent a delayed, but then abrupt subsidence. As Site 1118 is inferred to have been the most proximal (landward) during the preceding Late Miocene period of non-marine deposition, very strong subsidence of the (future) axial rift zone south of Site 1118 is implied, compared with the more northerly sites (Sites 1109 and 1115). Site 1108 eventually subsided rapidly at *c.* 3.8 Ma. All three sites then subsided through 500 mbsf (from upper to middle bathyal) at *c.* 2.6 Ma.

Mid–Late Pliocene time: continuing rift subsidence

At all three sites (1118, 1109 and 1118), successions of Mid–Late Pliocene age (3.5–1.8 Ma) are dominated by similar turbiditic sediments, although subtle differences in sedimentary structures and composition exist (Fig. 10). The geophysical logs, especially FMS, show an upward continuation of similar sedi-

mentation from the underlying Early–Mid-Pliocene interval.

At Site 1115, the lower part of the Mid–Late Pliocene interval (388.0–417.3 mbsf; Unit IV) is dominated by greenish grey claystone, admixed with siltstone and claystone and numerous shell fragments. Hornblende- and feldspar-phyric volcaniclastic grains to granules are locally concentrated in siltstone (Fig. 9c,l). In addition, sandy and calcareous siltstones at Site 1115 record pervasive intermixing of volcaniclastic-derived sandstone, volcanic ash and calcareous siltstone. The mineral grains are fresh and eudehral. Only minor discrete (unbioturbated) fine-grained volcaniclastic sandstone interbeds are present. The FMS logs show that this interval is relatively uniform and mainly muddy (non-resistive). Primary sedimentary structures are strongly modified, or obliterated, by very extensive bioturbation. The lowermost 5 m at this site (1115, Unit IV) of this interval accumulated at outer neritic depths (0–150 m), whereas, above this, sediments accumulated at upper bathyal depths (150–500 m) at rates of >284 m Ma^{-1}.

The succession at Site 1115 continues up into a thick uniform interval (Unit III), cored in two holes (Hole 1115B, 149.7–293.1 mbsf; Hole 1115C, 283.2–388.5 mbsf), which accumulated in upper bathyal depths (150–500 m), at rapid sedimentation rates of initially 284 m Ma^{-1}, decreasing to 79 m Ma^{-1} above 190 mbsf (Taylor *et al.* 1999*a*; Fig. 6). The above interval is characterized by an increased input of thin, graded beds of volcaniclastic sand (particularly above 206.7 mbsf) within silty clay (greatly affected by bioturbation), together with rare beds of volcanic ash. CaCO$_3$ values within the fine-grained sediments show a general increase from around 20% to around 30% above 250 mbsf, reflecting a relative increase in the abundance of calcareous micro-organisms. The FMS logs indicate a muddy succession, with sandy intervals. Occasional resistive (bright) layers (<5 cm thick), are interpreted as volcaniclastic ash layers, mainly deposited by turbidity currents. A thin dolomite layer is present from 291 to 292 mbsf, as confirmed by the photoelectric log and X-ray diffraction analysis. The graded volcaniclastic sands and silts observed in the cores are interpreted as turbidity current deposits (Fig. 8i). Volcanic glass of mainly rhyolitic composition (Lackschewitz *et al.* 2001) is locally abundant, but mainly reworked, along with lithic and biogenic material; air-fall tephra is, however, rare in this interval. X-ray diffraction of fine-grained sediments reveals that illite occurs commonly below 169 mbsf, but is rare, or absent, higher in the succession, which may reflect a regional change in sediment provenance. One curious, but unexplained feature, is a marked increase in magnetic susceptibility around 210 mbsf, which is apparently not reflected in any systematic change in facies or composition at this level.

A very similar Middle–Upper Pliocene succession was recovered at Site 1109 (387.6–570.4 mbsf; Unit VI), dominated by clayey siltstone and silty claystone, with subordinate thin beds of fine- to medium-grained sandstone and siltstone. Geophysical logs (e.g. gamma log) are interpreted to indicate the presence of sandy and clay-rich layers. $CaCO_3$ decreases above 500 mbsf, based on photoelectic log data, in keeping with the disappearance of shallow-water-derived carbonate, seen lower in the succession. The FMS logs further reveal a well-layered, rather uniform succession with numerous more resistive turbiditic layers, 5–10 cm thick, exhibiting sharp bases and gradational tops, within a mainly muddy succession. Minor bottom current reworking is suggested by the presence of repetitive lamination and small ripples. Overall, these sediments accumulated at an inferred high sedimentation rate of 266 m Ma^{-1}, in an upper bathyal setting (150–500 m). The volcaniclastic sandstones include abundant relatively unaltered material of basaltic–andesitic origin (i.e. quartz, plagioclase, biotite, common green hornblende) and generally devitrified fragments of basic and rare acidic extrusive rocks, plus minor amounts of neritic carbonate (e.g. gastropods).Occasional tephra layers (e.g. Fig. 9l) are mostly composed of colourless acidic glass (thin platy and bubble wall), with minor brown glass shards.

The succession at Site 1109 continues upwards as a mainly clay-rich deposit (Hole 1109C, 362.2–375.7 mbsf; Hole 1109D, 352.8–387.6 mbsf; Unit V), as confirmed by the geophysical logs. There is also increased volcanic glass as colourless platy shards and vesicular grains, together with lithic fragments (basalt, andesite, acidic extrusive rocks) and individual crystals (e.g. amphibole, plagioclase, quartz, biotite, clay minerals). This interval (Unit V) contains four volcanic ash layers (1–7 cm thick), comprising relatively unaltered, colourless volcanic glass, mainly as platy and bubble-wall shards. Occasional peaks in the radioactivity logs (U, Th) are correlated with volcaniclastic intervals. The interval accumulated in an upper bathyal setting (150–500 m), at continuing sedimentation rate of 266 m Ma^{-1}. Bioturbated silty clay and clayey silt are interpreted as background calcareous hemi-

pelagic sediments that accumulated on a well-oxygenated sea floor. Graded glass-rich sand, plus a few ash layers, were deposited by turbidity currents, continuing a well-established pattern. The volcaniclastic sediment input may relate to a discrete volcanic event, dated to Mid-Pliocene time, by $^{39}Ar/^{40}Ar$ dating of volcaniclastic mineral grains (Lackschewitz et al. 2001).

The succession at Site 1109 passes upwards into mainly clayey siltstone and silty claystone (246.7–362.2 mbsf; Unit IV). Sedimentation rates decreased markedly to around 68 m Ma^{-1} and water depths increased from upper bathyal (150–500 m), to middle bathyal depths (500–2000 m). The geophysical logs, including FMS (Taylor et al. 1999a), suggest the presence of alternating muddy and sandy layers, which are particularly sandy around 341 mbsf. Around 305–320 mbsf dips of 5–10° to the SSE and SSW are inferred from the FMS data, which could reflect a tectonic tilt towards the rift depocentre, as dips are too great to be simply depositional. Volcanic ash is rare is this interval and, where present, is disseminated. Minor metamorphic lithic grains appear around 350 mbsf (Fig. 10). In addition, a thin interval, estimated as 40 m thick, is inferred to have accumulated in response to repeated sediment redeposition related to slumping (Resig et al. 2001).

Comparable sediments continued to accumulate above this level at Site 1109 (169.7–246.7 mbsf; Unit III), although study was hampered by low recovery. Interpretation of the geophysical logs suggests that an important, but poorly recovered, interval of resistive sand is present from 219 to 233 mbsf (log unit L2; Taylor et al. 1999a). Minor core recovery included sand, silt and mud intraclasts, volcanic rock fragments and wood particles. The sands were derived from a volcanic arc, with a range of basic to acidic extrusive lithologies, including mainly colourless volcanic glass, quartz, plagioclase, biotite, hornblende and olivine grains. Otherwise conditions remained little changed, with accumulation at middle bathyal depths (500–2000 m), at high sedimentation rates reaching 225 m Ma^{-1}.

At Site 1118, a very similar Mid–Late Pliocene succession accumulated, although seismic interpretation indicates that this interval is more expanded than at Sites 1109 and 1115 further north, reflecting more rapid accumulation towards the rift depocentre. The lowest part of this interval, of Mid-Pliocene age (679.3–810.8 mbsf), is characterized by variably bedded volcaniclastic sandstone (rich in biotite and hornblende phenocrysts), siltstone, mixed sediments (e.g. sandy silty claystone) and common tephra layers (with common colourless glass shards). Both the ash-rich and associated volcaniclastic sandstones are compositionally similar, suggesting that both may relate to an episode of calc-alkaline volcanism. The typical background hemipelagic sediment contains about 20–30% $CaCO_3$. Although most beds exhibit sharp bases and graded tops (partial Bouma sequences), typical of turbidity current deposits, a few exhibit sharp tops suggestive of reworking by bottom currents. Rare packstones composed of planktonic foraminiferal shells are interpreted as calciturbidites. Overall, this part of the succession accumulated at *c.* 485 m Ma^{-1} at upper bathyal depths (150–500 m).

The continuation of the Mid–Late Pliocene succession at Site 1118 (492.3–679.3 mbsf; Unit III) accumulated very rapidly at an average rate of 485 m Ma^{-1}, at upper bathyal depths (150–500 m). The geophysical logs indicate that this is a relatively sandy interval, correlating with the presence of cored volcaniclastic turbidites. Some of the volcaniclastic sandstones are very rich in lithic clasts derived from basic extrusive igneous rocks; thin tephra layers are also present (Fig. 9k).

Higher in the succession at Site 1118 (377.8–492.3 mbsf; Unit II) sediments, still of Mid–Late Pliocene age (>2.5 Ma), accumulated at high sedimentation rates of 435 m Ma^{-1}, also at upper bathyal depths (150–500 m). Distinctive reddish, finely laminated, mainly fine-grained intervals show only minimal bioturbation, which is an unusual feature. Pyrite is effectively absent, in contrast to the typically greenish grey sediments above and below. One possibility is that this red interval was originally relatively rich in fine-grained iron oxide, possibly reflecting input from a tropically weathered landmass. Alternatively, these sediments were oxidized on, or beneath, the sea floor in contact with oxidizing waters. Similar red sediments were not recorded elsewhere during Leg 180 drilling.

Comparisons. The relatively expanded Mid–Late Pliocene successions at Sites 1115, 1109 and 1118 provide insights into rift evolution. Sedimentation was fastest nearest the rift depocentre (Site 1118; 435–485 m Ma^{-1}). All sites subsided through 500 mbsl at about 2.6 Ma (Fig. 6). Site 1109 continued to deepen to middle bathyal depths (500–2000 m). Northerly rift margin sediments (Sites 1115 and 1109) are strongly bioturbated, possibly related to oxidizing bottom conditions and/or reflecting somewhat lower sedimentation rates. Deposition was dominantly by turbidity currents, with localized reworking by bottom currents. Where present, primary air-fall ash layers are characterized by rhyolitic bubble-wall shards, very vesicular pipe-shaped shards and minor phenocrysts. Volcaniclastic turbidites commonly contain brown and colourless glass shards, plagioclase crystals (often zoned), quartz, biotite, amphibole, pyroxene, opaque minerals and common, variably altered, lithic fragments. Phenocrysts may comprise up to 50% by volume of the sediment. Tephra layers are particularly abundant (*c.* 100 layers) during the Early–Mid-Pliocene transition (3.5–3.8 Ma) at Site 1118 (Fig. 11). Then, during Mid-Pliocene time, tephra input was generally subordinate to volcaniclastic turbidites. During Mid–Late Pliocene time, abundant volcaniclastic turbidites were deposited at Site 1118, whereas at Sites 1109 and 1115 only minor volcaniclastic input is represented by mainly acidic air-fall tephra (Fig. 11). In addition, poorly consolidated sandy intervals (rarely well recovered) include minor continentally derived material (wood, muscovite, serpentinite, mixed-layer clay minerals).

Late Pliocene–Pleistocene time: deepening rift basin

At Site 1115, the Late Pliocene–Pleistocene interval from 35.7 to 149.7 mbsf (Unit II) is mainly characterized by clay, with subordinate volcaniclastic sand layers, passing up into nannofossil-rich silty clay. Deposition took place at middle bathyal depths (500–2000 m), at an average sedimentation rate of 59–63 m Ma^{-1}. The volcaniclastic sand and silt are interpreted as volcaniclastic turbidites and subordinate airfall ash. The Pleistocene interval (e.g. 0–4.4 mbsf in Hole 1115A; Unit I) accumulated at middle bathyal depths at rates of 44 m Ma^{-1}. Background pelagic ooze is composed of nannofossils, planktonic foraminifers, radiolarians and shell fragments. Calcite, aragonite and quartz dominate the mineralogy, as shown by X-ray diffraction analysis (Taylor *et al.* 1999*a*). In addition, mainly siliceous volcanic glass is interpreted as tephra (Fig. 11). Electron microprobe analysis shows that the volcanic glass in the tephra layers studied is of high-K calc-alkaline to calc-alkaline composition (Lackschewitz, pers. comm.).

Further south, at Site 1109, the Pleistocene interval from 83.4 to 167.7 mbsf (Unit II) is dominated by repeated (normally) graded beds of volcaniclastic sands, silty clay, clayey silt

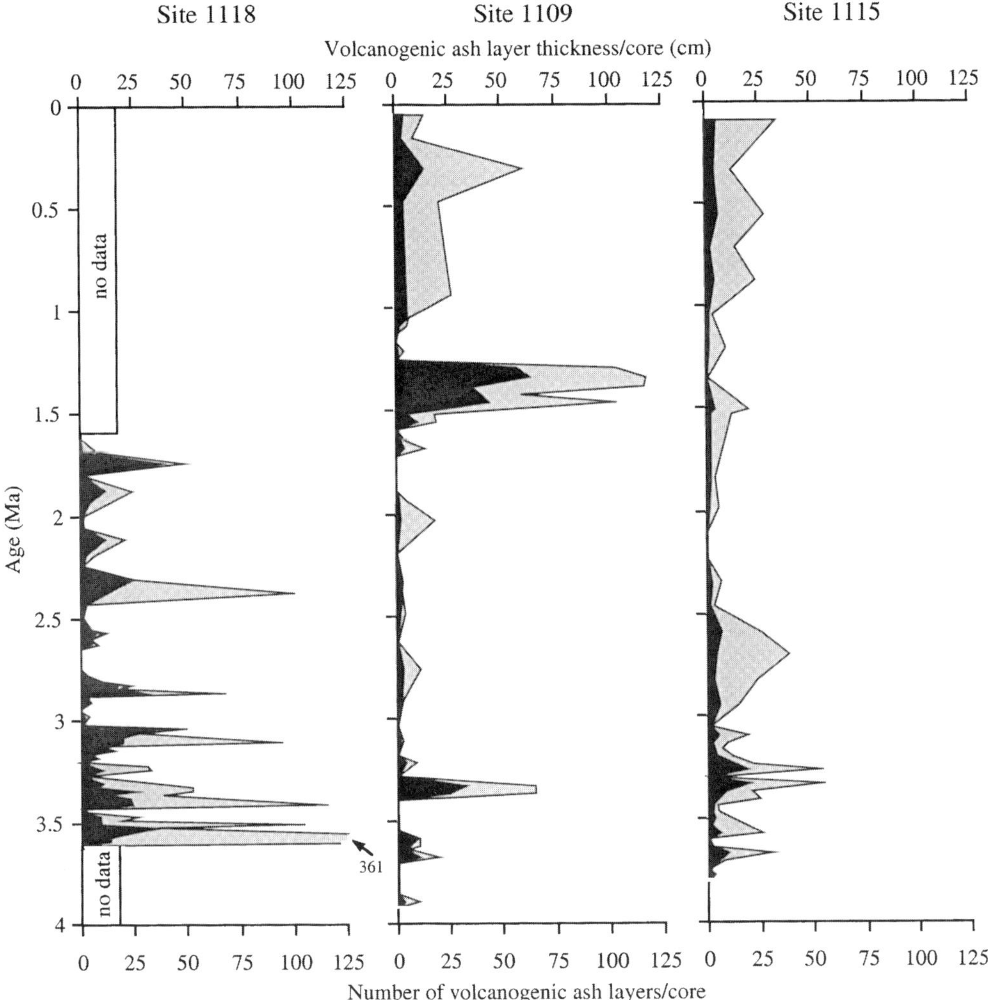

Fig. 11. Summary of the occurrence of volcanogenic ash layers at the near axial and northern rift margin sites, shown in terms of thickness (grey) and number of volcanic ash layers (black) per million years from Pliocene–Pleistocene time. It should be noted that the 'spiky' pattern is partly an artefact of core recovery. After Taylor *et al.* (1999*a*).

and clay, together with rare carbonate packstone (interpreted as calciturbidites) and volcanic ash. The FMS logs confirm a succession of alternating clay- and sand-rich layers. A carbonate-rich layer, inferred from the photoelectric effect (*c.* 161 mbsf) may correspond to one of several thin calciturbidites recovered in cores. Deposition took place at middle bathyal depths (500–2000 m), at high sedimentation rates, estimated at 225 m Ma^{-1}. A single interval of silt, sand and granules (Core 1109C-18X) is interpreted as a rare high-density turbidity current deposit. In addition, a few packstones (calciturbidites), containing calcar-

eous algae, shell fragments and planktonic foraminifers, were derived from a shallow-water setting. Volcanic glass is interpreted as air-fall tephra, commonly interbedded with volcaniclastic turbidites (Fig. 11). Above, the Late Pliocene–Pleistocene interval (Unit I, cored in three adjacent holes; Taylor *et al.* 1999*a*) is interpreted as deep-water calcareous volcanogenic clays interspersed with volcaniclastic sediment. Sedimentation rates decreased to *c.* 44 m Ma^{-1} above 56 mbsf. Accumulation was at middle bathyal water depths (500–2000 mbsf). Hemipelagic clays are interbedded with volcanogenic clays and numerous well-sorted volcanic tephra layers, interpreted as coeval air-fall ash, within additional, rare, white carbonate sands, interpreted as calciturbidites derived from a shallow-water setting (mainly algal in origin).

Comparisons. The Late Pliocene–Pleistocene sediments at Sites 1115 and 1109 show some significant differences compared with the underlying successions. Coeval volcanic ash layers are relatively abundant, suggesting a period of enhanced, largely acidic, volcanism in the source region. Especially at Site 1115, the background Pleistocene sediments are relatively rich in calcareous (and minor siliceous) biogenic material, suggesting relatively high plankton productivity. The presence of rare shallow-water derived calciturbidites points to enhanced input from shallow-water carbonate reefs fringing the basin. Also, sedimentation rates decreased, which could reflect a reduction in detrital sediment input caused by a changed palaeogeography of the source area (see below). A prominent episode of mainly volcaniclastic turbidites of early Pleistocene age is recorded at Site 1109, but not further north at Site 1115 (Fig. 11). Also, at both Sites 1109 and 1115 abundant acidic ash of mainly platy and bubble-wall type is found within the Pleistocene succession, indicating a phase of relatively recent explosive activity.

Southern margin of the Woodlark Rift Basin: deposition on the Moresby Seamount

Contrasting sedimentation was recorded on the southern margin of the Woodlark Rift on the Moresby Seamount at Sites 1114 and 1116, although limited by poor recovery, limited log data, as well as imprecise dating (Figs 12–14).

Crest of the Moresby Seamount

Site 1114 is located near the crest of the Moresby Seamount, just south of the Moresby Detachment Fault (Figs 2 and 3). The sedimentary succession there is terminated by a thin tectonic breccia that separates rift sediments, above, from metadiabase of inferred ophiolitic origin, below (Taylor *et al.* 1999*a*). Geophysical logs, especially FMS, allow the succession at Site 1114 to be reconstructed despite poor core recovery, whereas no logs were obtained from Site 1116, where recovery was again very limited (mainly <20%).

The lowest (undated) sediments (<16% recovery) at Site 1114 (237.6–276.1 mbsf; Unit IV) are dominated by variably lithified fine- to medium-grained, and rarely coarse-grained dusky red sandstone, with scattered granules (up to 4 mm in size). Petrographic study indicates derivation from a calc-alkaline-type volcanic arc, including relatively unaltered volcanic rocks.

The overlying interval at Site 1114 (55.0–237.6 mbsf; Unit III) is tentatively dated to ?Mid–Late Pliocene time. Although the recovery was poor (<25%), missing intervals were reconstructed using geophysical logs (e.g. FMS), revealing much evidence of tectonic faulting and fracturing in addition to primary sedimentary features (Taylor *et al.* 1999*a*). The geophysical logs reveal a distinctive resistive interval (179–181.5 mbsf) that could correlate with an unusually coarse interval in the limited core recovery (180.1–180.7 mbsf; Subunit IIIB). The logged Th/U ratios reach a maximum at this depth, which might indicate a high input of relatively radiogenic volcaniclastic sand. Around 120–140 mbsf the gamma-ray and resistivity log peaks are interpreted to suggest the presence of relatively sandy beds, possibly correlating with a sandy interval in the cores (123.2–141.6 mbsf; Unit IIIA). Otherwise, the FMS images reveal a well-layered succession, interpreted as mainly mudstone with resistive sandstone interbeds, *c.* 10 cm thick.

A distinctive granule conglomerate that was observed in the lower part of the succession (180.1–180.7 mbsf) contains clasts, up to 0.9 cm in size, set in a matrix of coarse, poorly sorted sandstone. Individual clasts show a vague alignment. Reverse-to-normal grading was rarely observed. The clasts are mainly volcanic derived (including abundant basalt), plus a few carbonate bioclasts. Sandstones below and above the granule conglomerate are compositionally similar. In addition, a distinctive sandy interval higher in the succession (123.2–141.6 mbsf) consists of texturally mature calcite-cemented sandstone and rare bioclastic limestone. Siliciclastic grains are mainly quartz, feldspar and sub-angular, altered volcanic rock fragments (<2 mm). Carbonate grains include coral, calcareous algae and shell fragments.

The above interval was deposited by turbidity currents (rarely of high-density type) at middle bathyal depths at rates estimated to be ≥176 m Ma^{-1}. The paucity of burrowing is suggestive of deposition under sub-oxic sea-floor conditions. However, fully reducing conditions probably existed beneath the sea floor, as suggested by the subdued colour and abundance of woody organic matter and pyrite. The

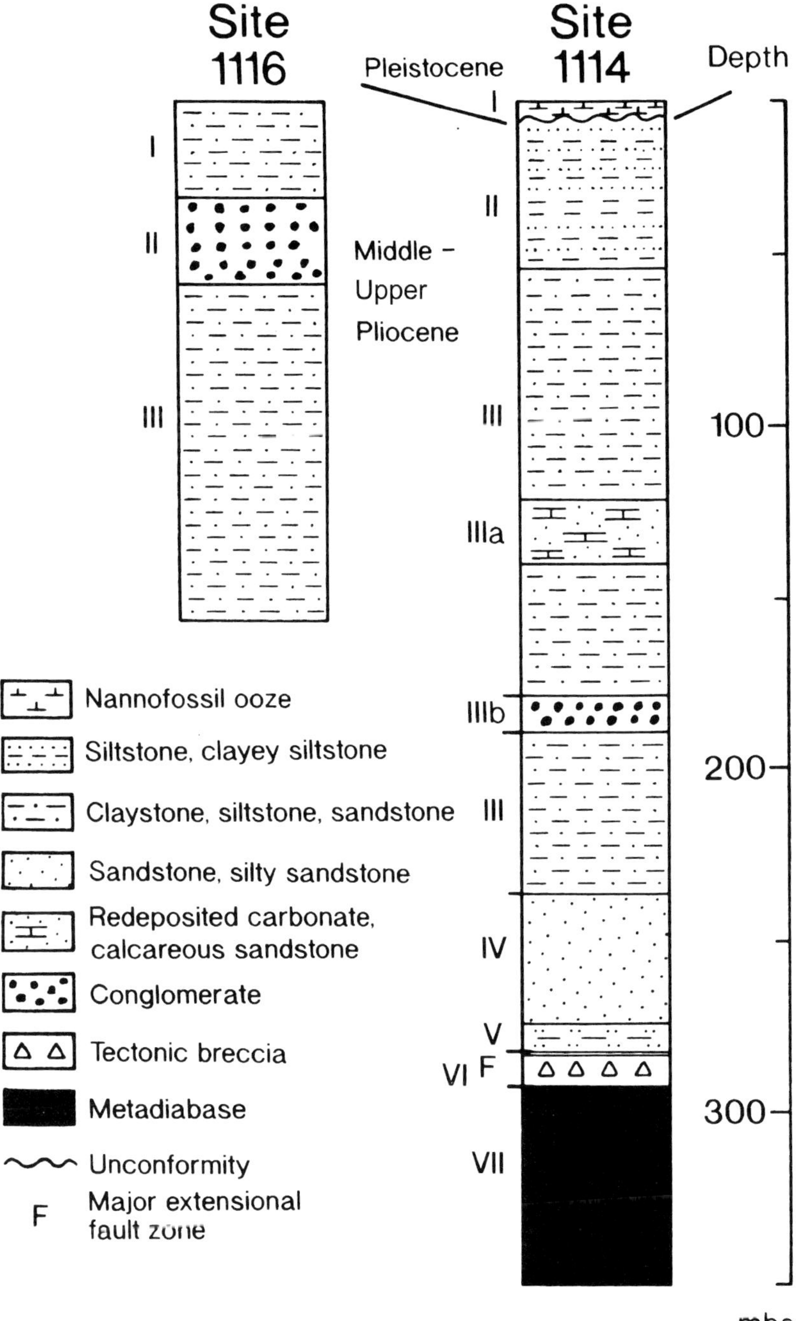

Fig. 12. Summary of the sedimentary succession drilled on the southern margin of the Woodlark Rift. Modified after Taylor *et al.* (1999*a*).

provenance was from a volcanic arc, including a range of basic, andesitic and acidic extrusive rocks. The presence of clasts of variolitic basalt points to sub-aqueous eruption. Volcanic glass is, however, relatively minor and, where present, is typically altered. Relatively unaltered

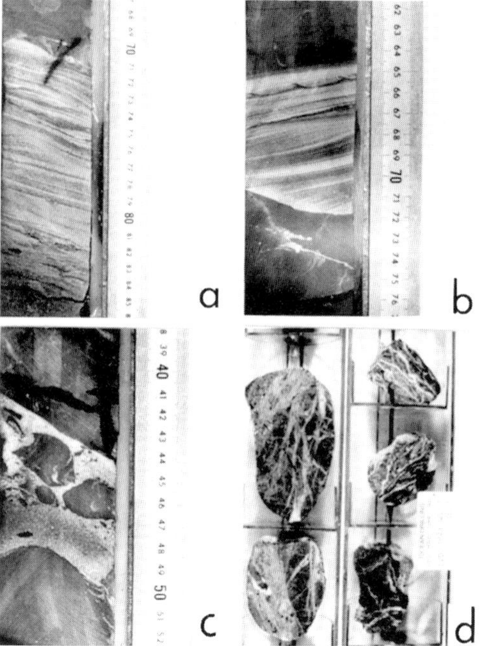

Fig. 13. Core photographs, footwall sites. (**a**) and (**b**) turbiditic sandstone showing partial Bouma sequence; Site 1116A-16R-2 and Site 1116A-17R-2, respectively. (**c**) Debris flows with mud rip-up clasts in a sandy matrix; Site 1116A-16R-2. (**d**) Clasts of veined metadiabase; Site 114, indicative of the basement of part of the Moresby Seamount, Site 1112A-4R and -5R.

clasts of glassy basalt and olivine are locally present, especially near the base of this succession, suggesting that some of the volcanic material escaped weathering before final deposition.

The succession passes upwards into a generally finer-grained interval (6.6–55.0 mbsf; Unit II), estimated as Late Pliocene–Pleistocene in age. This interval comprises lithified clay-rich hemipelagic sediments interspersed with siltstone and fine- to medium-grained sandstone, interpreted as turbidites. The benthic foraminiferal assemblage indicates acummulation at middle bathyal (500–2000 m) water depths. Bottom-water conditions were oxidizing, as inferred from the abundance of *Chondrites* burrows. Sandstone provenance continued to be from a volcanic-arc terrane, but includes a minor input of metamorphic lithoclasts (Fig. 14).

The uppermost 0.37 m of the succession (Unit 1) differs strongly from the underlying well-lithified succession, discussed above, as it

consists of unconsolidated Pleistocene foraminifer-bearing nannofossil ooze and clay, with subordinate silt and sand. This thin, superficial interval is interpreted as calcareous pelagic and hemipelagic sediment that accumulated at middle bathyal water depths (500–2000 m). Planktonic foraminiferal sands may record episodes of relatively high productivity, or winnowing of the clay-rich fraction. The abrupt change to well-lithified sediments beneath these soft sediments is interpreted as an unconformity. Sediments beneath the Pleistocene interval are dated as >1.67 to 3.09–3.25 Ma, suggesting that a significant hiatus exists. The probable explanation of the strong difference in consolidation state above and below the unconformity is that several hundred metres of sediments (estimated from porosity–depth relationships at 224–>400 m; Taylor *et al.* 1999*a*) were additionally deposited, compacted and then removed. The possible cause was subaerial erosion of the Moresby Seamount, although current erosion or slumping could also have removed some sediment. The unconformably overlying hemipelagic and pelagic sediments accumulated in a setting starved of gravity sediment input, probably after uplift of the Moresby Seamount to near its present position.

Southern flank of the Moresby Seamount

Site 1116 is positioned on the southern flank of the Moresby Seamount, 8 km south of Site 1114, within a tilted fault block bounded by normal faults that each offset the basement by >1 km to the SSW (Figs 2 and 3; Taylor *et al.* 1999*a*). The whole of the recovered succession accumulated during Pliocene time at middle bathyal depths (500–2000 m), at an average sedimentation rate of >70 m Ma^{-1}. The lower part of the recovered succession (62.6–158.9 mbsf; Unit III) comprises sandstones and siltstone, with only very limited recovery. Occasional thick-bedded sandstones, exhibiting inverse grading, are interpreted as high-density turbidites. Parallel lamination and ripple lamination in the upper part of these beds is suggestive of reworking within the tail of a turbidity current (Allen 1982). Other beds (Fig. 13a,b) are attributed to the Ta, Tb, Tc and Td Bouma divisions of classical turbidites (Bouma 1962). The presence of climbing-ripple cross-lamination is suggestive of high rates of sediment fallout from suspension. However, complete Bouma Ta–Te divisions are not developed. Rare thick-bedded mudstones with graded sand–silt–clay couplets are seen as low-density

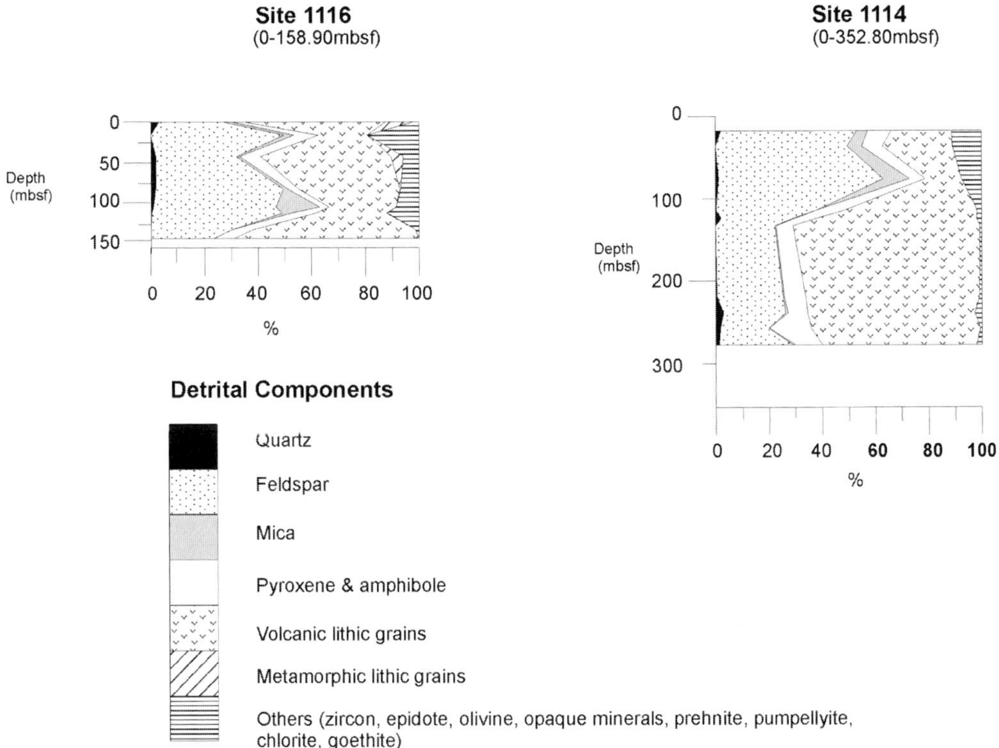

Fig. 14. Plots of downhole variation in the modal composition of sandstone from the footwall of the Woodlark Rift, based on points of 18 sandstones in thin sections by T. Sharp.

turbidity current deposits (Piper 1978). Units that show alternations of ripple lamination and wavy lamination are interpreted as reflecting deposition from pulsating currents, possibly bottom currents (Pickering *et al.* 1989). Well-developed structures resulting from injection of coarse sand into finer-grained sediments (Lowe 1975) reflect fluid escape, possibly triggered by rift-related earthquakes (Fig. 13c). The sands and silts were mainly derived from ubiquitous basic to acidic extrusive rocks of calc-alkaline type, including fresh basalt, glassy basalt, chloritized basalt, palagonite, felsic volcanic rocks (Fig. 9d), zoned plagioclase, biotite, clinopyroxene, rare basalt and ophiolite-derived rocks (serpentinite, rare chromite, diabase, gabbro). There are also spasmodic occurrences of shallow-water carbonate grains (bivalve shell fragments, echinoderm plates, benthic foraminifers, bryozoans). Minor metamorphic lithic grains appear above 111 mbsf (Fig. 14).

The succession continues upwards into a distinctive paraconglomerate with subordinate sandstone and siltstone (33.8–62.6 mbsf; Unit II), interpreted as a debris flow deposit. Well-rounded pebbles and granules underwent rounding before final deposition, probably related to wave or current action in a shallow-water setting. Shallow-water-derived benthic foraminifers of inner neritic origin (0–50 m) are present within some clasts. In addition to a calc-alkaline provenance, a subordinate ophiolitic source is indicated by rare grains of mafic cumulate and pyroxene (both fresh and altered). The uppermost part of the succession, also of Pliocene age (0–33.8 mbsf; Unit I), is recorded only as a number of isolated clasts and rare coherent intervals of mainly lithified sandstone, siltstone and claystone, of typical hemipelagic and turbiditic origin. The clasts possibly reflect parts of the subjacent succession that were reworked as talus on the steeply sloping southern flank of the Moresby Seamount.

Hanging wall of the Moresby Detachment Fault

Sites 1108 and Sites 1110–1113 (Figs 2 and 3) shed light on the evolution of Woodlark Rift and the Moresby Detachment Fault. A thick (>400 m) succession was recovered at Site 1108, whereas recovery at Sites 1110–1113 was fragmentary to minimal.

Site 1108 is located near the base of slope of the Moresby Detachment Fault. The succession begins with turbiditic clastic sediments (72.3–485.2 mbsf) of mainly Mid–Late Pliocene age (1.67–1.75 Ma at 82.8 mbsf to <3.35 Ma at the base). Deposition was at middle bathyal depths (500–2000 m), increasing to lower bathyal depths (>2000 m) above 410 mbsf. Sedimentation rates decreased up-section from 400 m Ma^{-1} at 2.6–3.2 Ma, to 324 m Ma^{-1} at 1.7–2 Ma (Taylor *et al.* 1999*a*). Unusually, during Leg 180 the lower part of the succession (72.3–485.2 mbsf; Unit IV) was subdivided according to the dominant sediment facies (Subunit IVA), with two specific subunits (IVA and IVB) being recognized.

Near the base (200.2–485.2 mbsf; Subunit IVA) the dominant sediments comprise interbedded dark greenish grey and dark grey, coarse- to medium-grained, graded sandstone, claystone and siltstone. The sandstones mainly exhibit normal grading, with some reverse grading. Classical turbidites exhibit partial Bouma sequences (Bouma 1962), including parallel lamination and cross-lamination, but more commonly sandstones are massive, or show irregular cross-lamination or reverse-to-normal grading. Similar turbiditic sediments occur near the top of Unit IV (72.30–139.4 mbsf; Subunit IVA). Rare amalgamated beds may record deposition from successive turbiditic currents, perhaps triggered by repeated seismic shocks. Occasional sharp-topped planar laminae are suggestive of scouring by bottom currents.

Of the two specifically defined intervals, Subunit IVB (139.4–200.2 mbsf) is dominated by greenish grey to dark grey and black foraminifer-bearing, clay-rich silt, interbedded with thin- to medium-bedded, mainly fine- to medium-grained, turbiditic sandstone of Mid-Late Pliocene age. X-ray diffraction analysis indicates the presence of chlorite, illite and probably smectite (Taylor *et al.* 1999*a*). The fine-grained sediments are calcareous, with values from 27 to 41% CaCO$_3$. Subdued grey–green, to grey and black hues, with minor disseminated pyrite, are suggestive of relatively low-oxygen conditions. This part of the succession is truncated by an important inclined dip-slip fault zone, *c.* 40 m thick, concentrated around 160–173 mbsf, along which an estimated up to 200 m of the succession has probably been removed, based on biostratigraphic, physical

property and core structural evidence (Taylor *et al.* 1999*a*). The second distinctive lithofacies (Subunit lVC) forms relatively coarse intercalations of conglomerate (313.34–314.04, 379.7–380.80 and 437.39–438.06 mbsf), interpreted as deposits from high-density turbidity currents.

The provenance of the Mid-Pliocene–Pleistocene succession indicates erosion from mainly extrusive igneous rocks and (rarely) intrusive igneous and metamorphic rocks. A source area probably included ophiolitic rocks, in view of the presence of rare grains of chromite and serpentinite. Coarse-grained feldspathic sandstone with common lithic grains occurs throughout much of Site 1108 (Fig. 9h). Metamorphic rock fragments become more abundant above *c.* 380 mbsf at Site 1108 (Fig. 14). Ubiquitous reddened lithic grains could represent land-derived laterite. Occasional calcareous bioclastic fragments were derived from shallow water (i.e. corals, algae, benthic foraminifers).

The overlying interval at Site 1108 (62.7–72.3 mbsf; Unit III) is characterized by a thin (undated) interval of unconsolidated coarse sand, interpreted as having been deposited by high-density turbidity currents. These unlithified sands contrast strongly with the underlying well-lithified turbiditic succession. The probable explanation of this abrupt change is that a substantial interval of the original succession was removed, creating an unconformity, as a result of subaerial erosion, current erosion or slumping. The inferred age of sediment removal is constrained only by the late Pleistocene age of overlying surficial sediments (1.25 Ma) and the Late Pliocene–early Pleistocene age (1.25–1.71 Ma), beneath (upper part of Unit IV).

The unconsolidated sands (Unit III) are overlain by an undated interval (14.5–62.8 mbsf; Unit II) that is characterized by clasts of plutonic and extrusive igneous rocks, metamorphic rocks and sedimentary rocks, collectively termed talus, and inferred to have been shed from upslope on the Moresby Detachment Fault by debris flow and/or rock-fall processes. The plutonic rocks are mainly greenschist-facies metadiabase, often epidote rich, of inferred ophiolitic origin, although other lithologies (e.g. microgranite and gneiss) were also identified at Site 1108. Many of the clasts exhibit mylonitic or cataclastic fabrics, again suggesting that the Moresby Detachment Fault was the source. Occasional sandstone clasts were derived from a turbiditic succession, similar to that cored near the crest of the Moresby Seamount at Site 1114. In addition, a number of clasts of fresh basalt, basalt breccia and rare feldspar porphyry were recovered from the upper part of the section of Unit II (Fig. 9a,b). Quench textures and a glassy groundmass imply a young age and submarine eruption. The fresh basaltic rock clasts were possibly derived from thin lava flows, or hyaloclastic deposits, related to the Woodlark spreading centre, located only one nautical mile to the east. Fresh basalts would be difficult to drill and could have disintegrated into the abraded clasts, as recovered by drilling.

The talus at Site 1108 is overlain by thin upper Pleistocene sediments (0–8.6 mbsf; Unit I), dominated by nannofossil clay, with a high content of volcanogenic clay and sparse volcaniclastic turbidites. A relatively low sedimentation rate of $15 \, \text{m} \, \text{Ma}^{-1}$ is estimated, reflecting a relatively low detrital input.

Minor amounts of sediments, sedimentary rocks and lithoclasts of meta-igneous rocks were also recovered at Sites 1110–1112 (Fig. 13d), but only lithoclasts at Site 1113. The well-lithified nature and mainly mafic–intermediate composition of the detrital material in the sedimentary clasts of Site 1112 suggests that this material was derived from erosion of an Early–Mid-Pliocene turbiditic succession similar to that cored at Sites 1114 and 1116. The minor recovery of dated sediments suggests that a full Pliocene sedimentary succession was originally present at Sites 1110–1112, extending from NN21 to NN19 (nannofossils) and N22 to N23 (planktonic foraminifers). The talus includes clasts of greenschist, dark pelitic schist and variably altered and deformed diabase–gabbro. In addition, rare clasts of acid–intermediate porphry, quartz 'trachyte' and 'lamprophyre' (with high K, Sr and Ba) were identified at Sites 1110–1113 (Taylor et al. 1999a). These could relate to the existence of minor intrusions within the Moresby Seamount, and represent part of a widespread 'high-K' (shoshonitic) suite known in the Papuan Peninsula and the D'Entrecasteaux Islands, and attributed to the Miocene Trobriand Arc (Ashley & Flood 1981). Unaltered basalts were not recorded at Sites 1110–1113, west of Site 1108, perhaps as these were further from the Woodlark spreading centre.

Direct sampling of the preserved up-dip limit of the Moresby Detachment Fault was achieved at Site 1117, where, despite very poor recovery, the base of the cored section was identified as quartz-bearing gabbro, of inferred ophiolitic origin, overlain by brecciated and mylonitized equivalents of this rock, and finally by unconsolidated fault gouge that crops out directly on the sea floor (Taylor et al. 1999a). It is probable that the mylonitic rocks relate to extension along the Woodlark Detachment Fault, although a compressional origin related to previous (Paleogene) ophiolite emplacement cannot at present be ruled out (Karner pers. comm.).

Sedimentary–tectonic evolution of the Woodlark Rift

This is now presented as a series of time slices, as summarized in Figures 15 and 21.

Palaeogene time

U/Pb SHRIMP dating yielded an age of $66.4 \pm 1.5 \, \text{Ma}$ on zircon for a gabbro at Site 1117, while the $^{40}\text{Ar}/^{39}\text{Ar}$ method gave $58.9 \pm 5.8 \, \text{Ma}$ using Plagioclase separates from a diabase at Site 1109. Whole-rock Ar/Ar dating also yielded Palaeogene ages from Sites 1109, 1117 and 1118 (Brooks pers. comm.). Shipboard and shore-based petrological and limited geochemical studies of unmetamorphosed fresh basalt and dolerite from the northern margin sites (1109, 1115 and 1118), of metadiabase from a southern margin site (1114) and of metabasic rocks forming talus from the Moresby Detachment Fault (Sites 1108, 1110–1113 and 1117) suggest that all of these basic intrusive and extrusive igneous rocks can be correlated with the Papuan Ophiolite Belt. In addition, the gneiss and micaschist clasts from the talus could be correlated with the inferred subduction–accretion complex that regionally underlies the Palaeogene Papuan ophiolite.

The maximum northerly extent of the emplaced Papuan Ophiolite Belt is unknown, but may have reached Woodlark Island, where a basement of pre-Miocene (?Cretaceous–Eocene) low-K tholeiitic basalts, dolerites and minor sediments is unconformably overlain by Miocene volcanic rocks and sediments of the Trobriand Arc (Ashley & Flood 1981). Thus, the Woodlark Rift and both its southern and northern margins are probably underlain by Palaeogene ophiolitic rocks. To the north of Site 1109 the existence of a Miocene forearc sedimentary succession was confirmed by drilling at Site 1115.

One apparent problem with correlating the basalt, dolerite and gabbro cored during Leg 180 with the Palaeogene Papuan ophiolite is that preliminary shipboard and shore-based chemical analysis of 'immobile' elements (e.g. Ti, Nb) suggests that these exhibit a relatively

WOODLARK BASIN

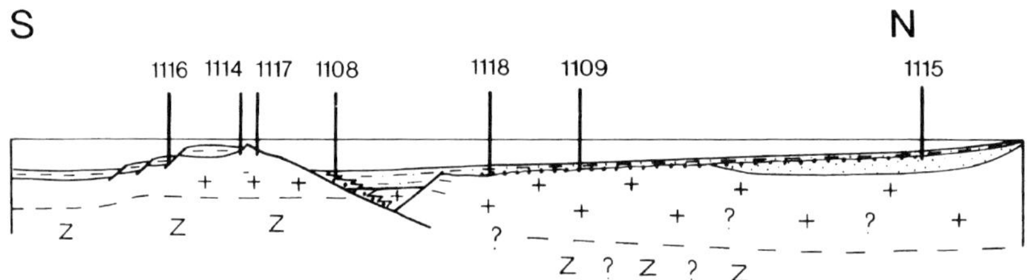

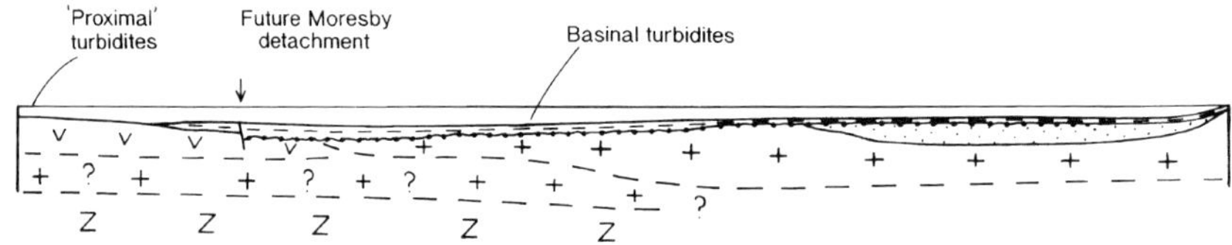

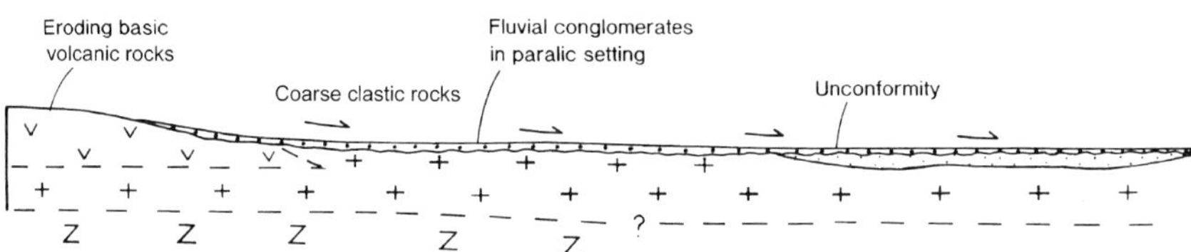

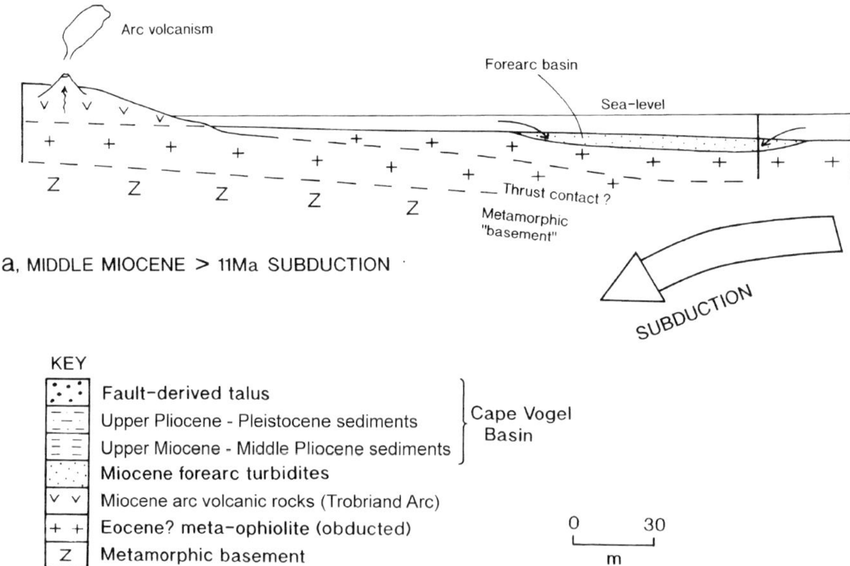

Fig. 15. Inferred major stages in tectonic development of the Woodlark Basin: (**a**) Middle Miocene pre-rift; (**b**) Late Miocene phase of regional uplift; (**c**) Early–Mid-Pliocene initial rifting; (**d**) Late Pliocene–Pleistocene later rift stage, with drill site locations of Leg 180.

'enriched' signature similar to mid-ocean ridge-type basalts (Taylor *et al.* 1999*a*; Brooks pers. comm.), rather than a 'depleted' signature typical of ophiolites interpreted as having formed above a subduction zone (e.g. Troodos ophiolite, Cyprus; Pearce *et al.* 1984; Robertson & Xenophontos 1993). The Papuan ophiolites were in the past inferred to have formed in a supra-subduction zone (forearc) setting (Davies & Jaques 1984). The Palaeogene ophiolitic rocks on land range from the highly depleted Cape Vogel Boninites (Francis 1985), to the relatively 'enriched' Milne Basin Complex (Brooks pers. comm.). Possibly, the Palaeogene Papuan ophiolitic rocks could represent a range of tectonically emplaced units formed in different tectonic settings (e.g. forearc; mid-ocean ridge, seamount), and it is also possible that the origin of the Woodlark basic igneous rocks could be distinct from all of the ophiolitic land exposures. On the other hand, dredging of the Tonga forearc area, thought to have formed by supra-subduction zone spreading in Eocene time (Bloomer *et al.* 1995), has recently revealed a range of both 'depleted' and 'less depleted' basaltic rocks (Bloomer pers. comm.), suggesting that considerable compositional variation can exist within a single supra-subduction zone-type ophiolite. The inferred ophiolitic rocks cored during Leg 180 may, thus, be compatible with an origin as parts of the regionally emplaced Palaeogene Papuan ophiolite, which exhibited local variation in chemical composition. In any scenario, ophiolitic extrusive rocks originally emplaced above the dolerite–gabbro on the Moresby Seamount (Site 1114) were possibly faulted out or previously eroded, as meta-extrusive ophiolitic clasts were not identified in the talus derived from the Moresby Detachment Fault.

Miocene time

Southwestward subduction of oceanic crust of the Solomon Sea initiated the Miocene–Recent Trobriand volcanic arc (Davies *et al.* 1984; Lock *et al.* 1987), which is well developed onshore in the Papuan Peninsula and also cored in offshore industry wells (Fig. 16). The front of the Trobriand Arc was located in the vicinity of the Moresby Seamount and the D'Entrecasteaux Islands. This is supported by the presence of Middle Miocene forearc basin turbidites, cored at Site 1115, and similar volcaniclastic deposits in the Goodenough-1 and Nubiam-1 wells (Figs 16 and 17). Unusually, arc-type magmas were erupted in the Trobriand forearc region, as shown by volcanic outcrops on outly-

ing islands. Specifically, on Woodlark Island, Early Miocene limestones and volcaniclastic sediments are overlain by later Miocene high-K volcanic rocks; co-magmatic intrusions and younger volcanic rocks are also present (Ashley & Flood 1981). This unusual distribution of arc volcanism complicates interpretation of the provenance of volcaniclastic sediments recovered during Leg 180. The outer high of the Trobriand Arc is inferred to be located about 100 km to the NE of Site 1109 (Davies *et al.* 1984; Smith & Milsom 1984; Lock *et al.* 1987) and represents the most northerly possible source of gravity-derived sediment to the Woodlark Rift.

Late Miocene time (<5.4–9.93 Ma)

In the classical model of arc rifting (Karig & Moore 1975), as subduction proceeded the Trobriand Arc would split, with a frontal portion remaining active while a back-arc rift opened, bordered by a remnant arc. The former forearc basin would remain submerged. However, the Woodlark area deviated strongly from this model, as there is evidence of regional emergence of the future Woodlark Rift area, which also affected a larger area extending northwestwards into the Cape Vogel Basin region. During Late Miocene time the forearc succession, as recorded at Site 1115 in the north, shallowed, culminating in emergence and an unconformity (Fig. 17). Similar emergence is recorded at Sites 1109 and 1118 further south, as fluvial conglomerates and overlying paralic deposits. Emergence and an unconformity were also recorded in the Nubiam-1 well, 100 km NW of the Woodlark Rift (Francis *et al.* 1987) (Fig. 16), where a similar facies is dated to <9.3 Ma (Harris *et al.* 1985). By contrast, no marked stratigraphical unconformity was observed in the more southerly Goodenough-1 well, or onshore in Papua New Guinea (Fig. 16). However, both of the industry wells penetrated an interval of Upper Miocene deltaic shallow-marine to paralic argillaceous and clastic sediments, including lignite and conglomerate (Ruaba unit) (Francis 1985; Stewart *et al.* 1986). A comparable Upper Miocene succession is also present in the Cape Vogel Basin onshore (Fig. 16). The unconformity in Numiam-1 well can be dated as younger than 9.63 Ma (NN9–NN10 boundary), as the upper part of the underlying Nubiam shale is dated at NN10–NN11 and N-16 in the Nubiam-1 well. Also, the top of the deltaic clastic unit in Goodenough-1 is similarly dated, as it lies above NN10–NN11 (Harris *et al.* 1985; Fig. 16).

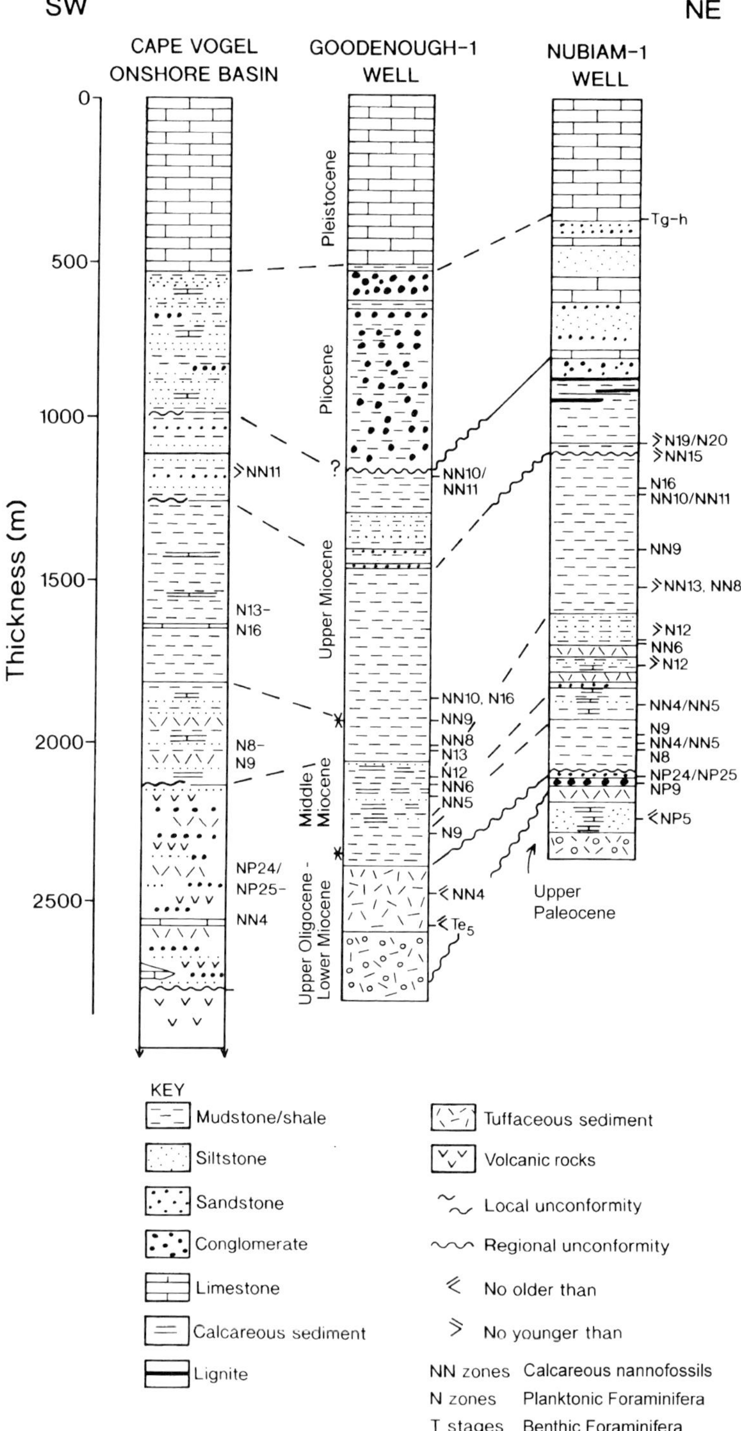

Fig. 16. Summary of the complete Paleocene–Recent successions cored at industry wells located in the shelf area to the northwest of the Leg 180 sites, and the onshore succession for comparison. Modified from Tjhin (1976); Stewart *et al.* (1986). (See Fig. 1 for locations.)

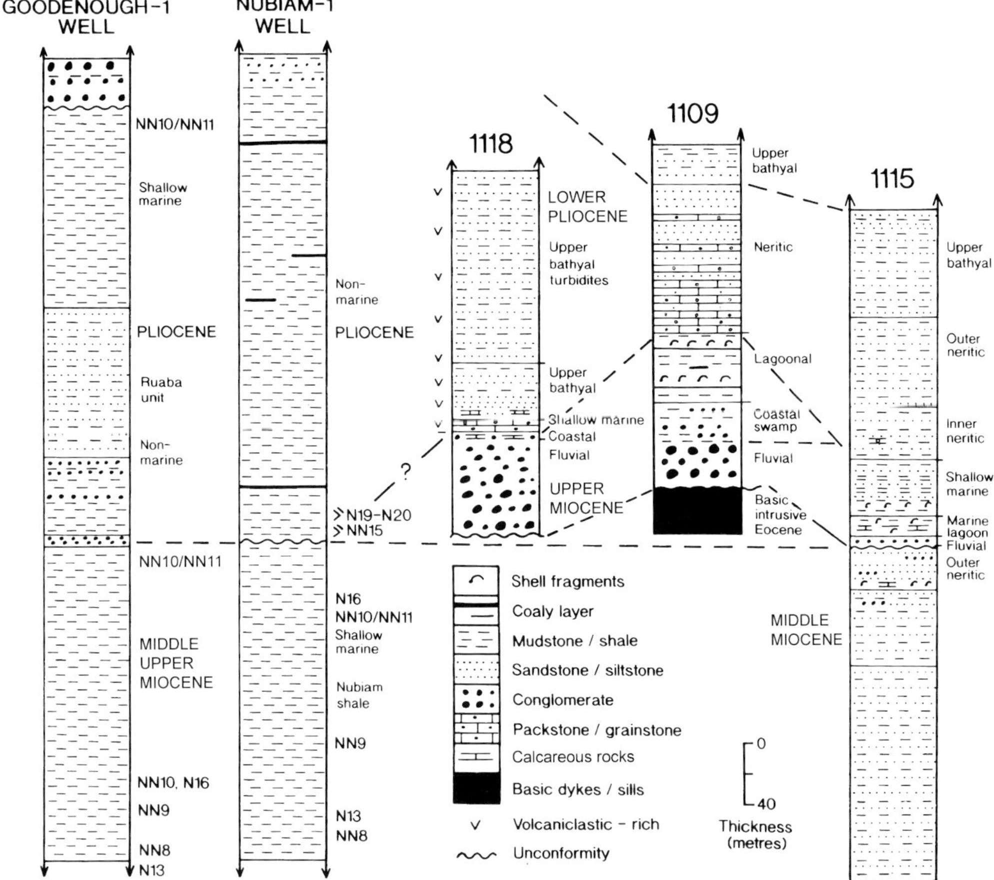

Fig. 17. Summary of Upper Miocene–Lower Pliocene successions drilled at ODP Sites 1109, 1115 and 1118, compared with well logs of the Nubiam-1 and Goodenough-1 wells, industry data from Tjhin (1976); Stewart *et al.* (1986).

Combining this with the biostratigraphic results from Site 1115 on the assumption that the locally observed unconformities are regionally synchronous, the inferred age of a regional discontinuity in sedimentation is inferred to be younger than 9.63 Ma, but older than 5.54 Ma (i.e. of Tortonian–Messinian age).

Study of extensive seismic profiles (2250 km) of the offshore Cape Vogel Basin (NW of the Woodlark Rift) reveals evidence of reportedly pre-Pliocene-aged, high-angle thrusts, reverse faults and possible oblique-slip faults (Tjhin 1976; Pinchin & Bembrick 1985; Stewart *et al.* 1986; Francis *et al.* 1987; Home *et al.* 1990). This evidence suggests that the Late Miocene emergence of the Leg 180 Sites (1109, 1115 and 1118) was not the direct result of, for example, simple infilling of the Trobriand forearc basin, or rift-related flexural uplift, but was

instead caused by a regional compressional event. One possibility would be the collision and underthrusting of a crustal unit (seamount, microcontinent or oceanic plateau) beneath the Trobriand forearc. Whatever the cause of emergence, during Late Miocene time the Trobriand outer-arc–forearc area, mainly composed of exposed Palaeogene ophiolitic rocks, was sub-aerially exposed, tropically weathered and eroded, yielding fluvial conglomerates, which prograded mainly northwards, over the forearc sedimentary succession, as at Site 1115.

Late Miocene–Early Pliocene time (8.6–3.5 Ma)

In the north, at Sites 1109 and 1115 a transition from terrestrial, to marginal (coastal–lagoonal),

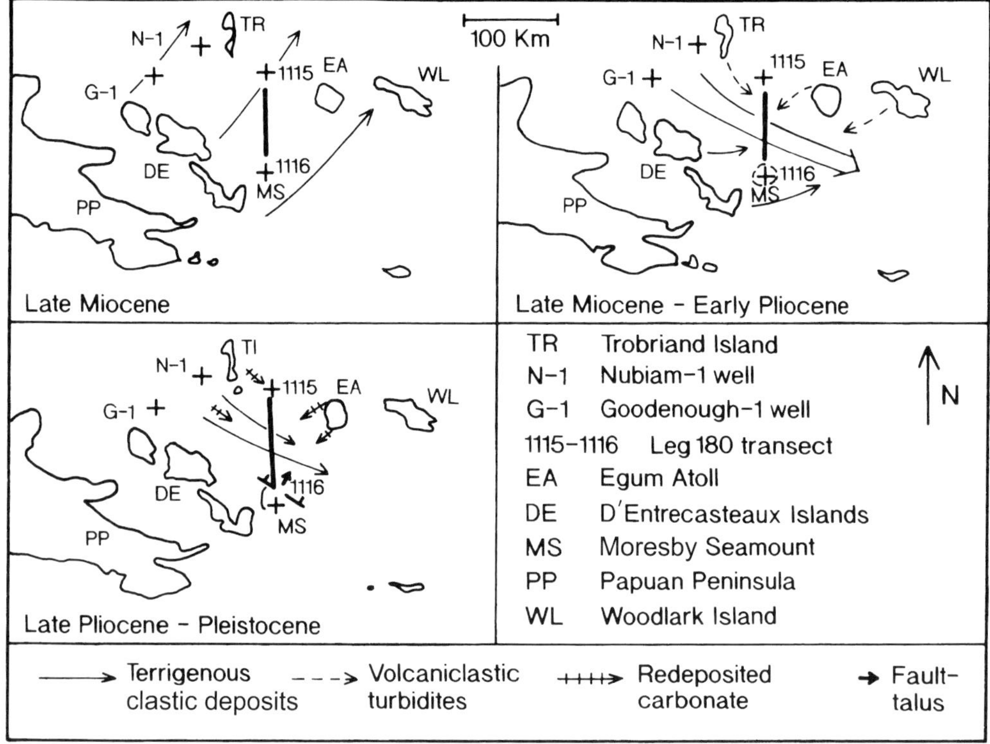

Fig. 18. Inferred palaeocurrent directions at different stages in the regional tectonic evolution. Not palinspastically restored. (See text for further explanation and discussion.)

to shallow-marine deposition took place during Late Miocene–Early Pliocene time (Fig. 17). This marine transgression could reflect submergence (e.g. tectonic subsidence of the Trobriand forearc area, or eustatic sea-level change). However, regional evidence shows that during Late Miocene–Early Pliocene time the Woodlark Rift formed part of a wider zone of crustal extension continuing as far as south Papua New Guinea (e.g. Milne Bay area) (Binns *et al.* 1989, 1990; Taylor *et al.* 1999*b*). A rift-related cause of the Late Miocene–Early Pliocene transgression is thus likely. The rift depocentre was located in the south, between Sites 1108 and 1118 (Fig. 3), where deeply buried synrift sediments were not penetrated during Leg 180.

Early–Late Pliocene time (5.3–1.8 Ma)

Despite being located relatively close to the rift depocentre, Site 1118 remained as a palaeo-island until Late Early Pliocene time (*c.* 3.6 Ma). There was then an abrupt transition from open-marine neritic carbonates of latest

Early Pliocene age, to upper bathyal (150–500 m), mainly turbiditic sediments. The northern margin of the Woodlark Rift deepened markedly after 4 Ma, accompanied by an increase in sedimentation rate (Sites 1109, 1115 and 1118). Further south, similar turbiditic Pliocene sediments accumulated, both at the Moresby Seamount sites (1114 and 1116), and at the proximal rift hanging-wall site (1108); that is, both on the now uplifted rift flanks and in the present rift basin. These sediments include inferred high-density turbidites of relatively proximal origin that accumulated in deep water (upper bathyal, 150–500 m). Provenance was mainly from calc-alkaline volcanic rocks, metamorphic rocks and rare plutonic ophiolitic rocks, for which the D'Entrecasteaux Islands, Papuan Peninsula or any submerged (formally subaerial) equivalents are the obvious source areas (Fig. 18). Early–Late Pliocene rifting was marked by explosive volcanism, recorded by common air-fall tuff and volcaniclastic turbidites that are most abundant nearest the rift depocentre (Site 1118). Radiometric dating (Ar/Ar) of volcanic ash from Sites 1109, 1115 and

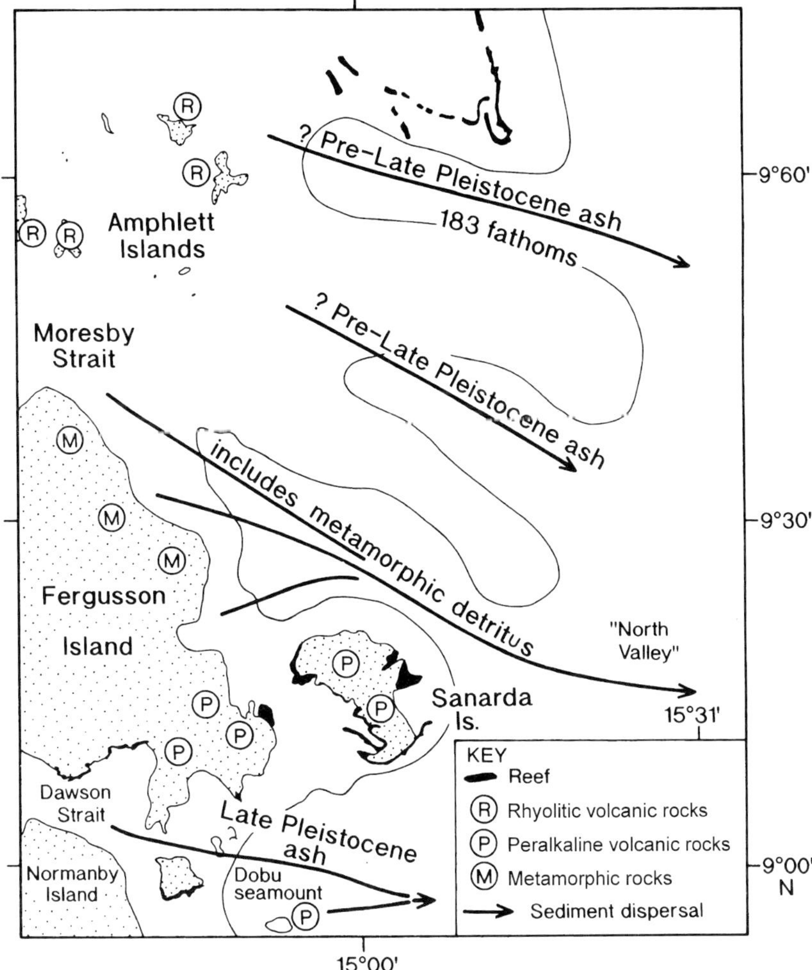

Fig. 19. Inferred present-day depositional pathways into the depocentre of the Woodlark Rift Basin from the northwest. (See text for further explanation.)

1118 shows that the age determined is generally similar to the independently determined sediment age (using microfossils and magnetostratigraphy), indicating that arc volcanism and the deposition of related volcaniclastic turbidites were generally coeval (Lackschewitz *et al.* pers. comm.). However, older volcaniclastic material is present at Site 1118, indicating that some reworking of volcanic material has taken place. Downhole trends of sandstone composition indicate that metamorphic grains (e.g. muscovite and schist) first appeared in the Woodlark Rift Basin (Sites 1108, 1109, 1115 and 1118) at *c.* 2.5–3.3 Ma (Sharp, pers. comm.). This probably reflects regional exhumation of metamorphic rocks in the D'Entre-

casteaux Islands and adjacent areas. There is a general tendency for calc-alkaline volcaniclastic sediments to become more potassic after 2.5 Ma, which may reflect volcanism in an increasingly extensional regional setting.

British Admiralty charts show the presence of NW–SE-oriented submarine channels linking the Woodlark Rift depocentre in the southeast with a shallow basinal area to the northwest, offshore from the D'Entrecasteaux Islands (part of the Cape Vogel Basin; Fig. 19). The most southerly of these channels extends to near the northerly end of Fergusson Island. Assuming sediment pathways similar to those of today, the probable source of the Pliocene turbidites lay to the northwest (Fig. 18). The main source of ash and volcaniclastic turbidites was possibly the Amphlett Islands and surrounding areas, where Plio-Pleistocene volcanic rocks occur (e.g. Dobu Seamount; Smith & Milsom 1984; Binns *et al.* 1987). These volcanic rocks include distinctive peralkaline rhyolites (comendites), interpreted to reflect rifting of subduction-influenced mantle lithosphere (Smith 1976; Smith *et al.* 1977; Smith & Johnson 1981; Stolz *et al.* 1993). An additional source of detrital sediment, including volcanic and metamorphic debris, is the D'Entrecasteaux Islands, specifically the eastern flank of Fergusson Island and adjacent area of detachment faulting (Fig. 4), which were probable sources of metamorphic lithoclasts in the Woodlark Rift after *c.* 3 Ma. Large volumes of eroded clastic sediment were shed into the neighbouring shallow seas of the Cape Vogel Basin, as cored in the Goodenough-1 well (Fig. 17), and then underwent rapid gravity transport southeastwards (axially) into the deep Woodlark Rift Basin.

The terrigenous sediment supply can be related to extensional exhumation of the southern margin of the Woodlark Rift, particularly from the D'Entrecastreaux Islands. Extensional detachments there are known to have been active in Plio-Pleistocene time (Senior & Billington 1987; Davies & Warren 1988; Hill *et al.* 1992; Baldwin & Ireland 1995), creating a topography up to 2.5 km high. In addition, some exhumation took place earlier, as metamorphic rocks including eclogites are unconformably overlain by Late Oligocene–Early Miocene shallow-marine sediments (Davies & Warren 1988).

Relatively thin Early–Late Pliocene successions were cored at Sites 1109 and 1115 north of the thick sediments within the Woodlark Rift depocentre. These sediments include abundant unaltered volcanic debris, of typical calc-alkaline character. Site 1109 is now located within a submarine channel carrying sediment southwards from the northern margin of the rift basin. Assuming a similar sediment pattern in Pliocene time, the probable sediment source was active arc volcanoes within the Woodlark Rise, including the Lusancay Islands, the Trobriand Islands and possibly Egum atoll (Smith & Milsom 1984).

Latest Pliocene–Pleistocene time (<2 Ma)

During latest Pliocene–Pleistocene time deepwater sedimentation continued on the northern margin of the Woodlark Rift (Sites 1109, 1115 and 1118), whereas the southern margin was uplifted to the present-day relatively shallow water depths (406.5 m at Site 1114). Both Sites 1114 and 1108 exhibit an unconformity above a lithified mainly Pliocene turbiditic succession. The inferred unconformities may record uplift and emergence of the Moresby Seamount that can be only approximately dated to Early Pleistocene time (>0.46 Ma, but <1.67 Ma), mainly based on evidence from Site 1108. The similarities in Pliocene turbiditic sedimentation at Sites 1114 and 1116 on the southern margin with Site 1108 near the rift depocentre further north suggest that the intervening Moresby Detachment Fault was not then in existence as a major topographic feature. For example, the well-rounded granules in the Late Pliocene–early Pleistocene conglomerate intercalations at Site 1108 do not show signs of derivation from the present nearby fault scarp, in contrast to the overlying Pleistocene angular talus. The probable cause of emergence of the Moresby Seamount was flexural uplift of the footwall of the Moresby Detachment Fault related to regional crustal extension (during a time of periodically lowered sea level). Interpretation of seismic profiles shows that the Seamount was bounded by high-angle normal faulting along its southern margin (Taylor *et al.* 1999*a*), which must also postdate deposition of the Pliocene turbiditic sediments at Site 1116.

After emergence of the Moresby Detachment Fault on the sea floor as a major inclined fault scarp (25–30°), talus began to be shed, as observed at Sites 1108 and 1110–1112. The original Pliocene sedimentary cover of the Moresby Seamount was reworked as clasts at Sites 1108 and 1110–1113, and as slumps and talus at Site 1116. Extensional faulting unroofed meta-ophiolitic rocks (dolerite, gabbro, quartz gabbro, metagabbro) that were then reworked downslope as talus. As extensional faulting proceeded, deeper structural levels were potentially unroofed, exhuming micaschists and gneiss from structurally beneath the ophiolite, now observed as clasts generally higher in the talus at Sites 1111–1113 (Taylor *et al.* 1999*a*).

Marine surveys show that the Moresby Detachment Fault is a regionally important structure, which can be traced along the northwestern margin of the Moresby Seamount (Binns *et al.* 1989, 1990; Taylor *et al.* 1999*b*). Further northwest, pervasive extensional unroofing is documented by core complexes in the D'Entrecasteaux Islands, as noted above (e.g. Suckling–Dayman Massif; Davies 1980*b*; Davies & Warren 1988, 1992; Hill *et al.* 1992; Baldwin *et al.* 1993; Hill & Baldwin 1993; <?tjlHill 1994). To the southwest the Moresby

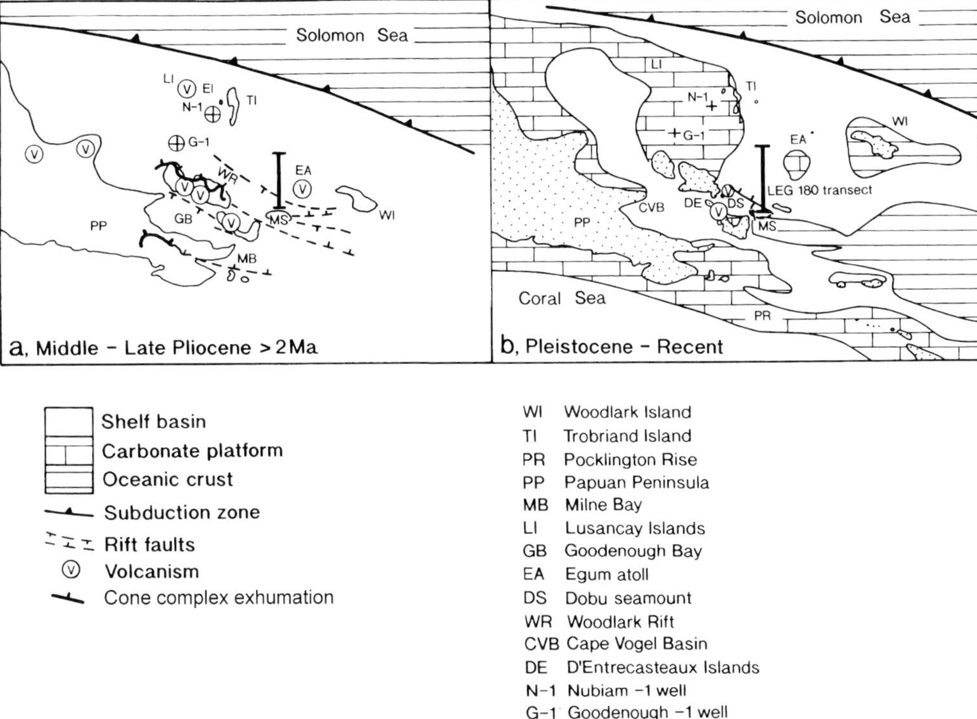

Fig. 20. Regional tectonic interpretation of the Woodlark Rift incorporating results from Leg 180. (**a**) Mid–Late Pliocene time (with approximate palinspastic restoration before spreading); (**b**) Pleistocene–Recent time. (See text for further explanation.)

Detachment Fault it is reported to be truncated by a transverse fault zone (Binns *et al.* 1987, 1989). In addition, a poorly explored extensional lineament, confirmed by dredging (Binns *et al.* 1987, 1989) also borders young (<2 Ma) oceanic crust of the oceanic Woodlark Basin further southeast, adjacent to the Pocklington Rise (Fig. 1; Taylor *et al.* 1999*b*).

Clastic input from the D'Entrecasteaux Islands and adjacent Papua New Guinea to the Cape Vogel Basin as a whole waned during Pleistocene time, as suggested by evidence from Goodenough-1 well, where terrigenous sediments are overlain by neritic carbonates (Fig. 16). There was a marked corresponding overall reduction in deposition rate in the Woodlark Rift to the south (Sites 1108, 1109 and 1115). At Site 1108 there was a switch to ooze and hemipelagic sediment with reduced turbiditic input, as confirmed by regional coring and dredging (Binns *et al.* 1987, 1989). A probable cause of the decrease in clastic sediment input was the growth of a carbonate platform on the outer part of the Cape Vogel Basin ('Trobriand Platform') (Tjhin 1976), which trapped clastic sediments

within inshore basins and changed palaeoslopes, in effect limiting gravity supply to the Woodlark Rift Basin. The growth of neritic carbonate build-ups was also reflected in spasmodic input of shallow-water derived calciturbidites to the Woodlark Rift Basin (e.g. Site 1109).

The latest Pliocene–Pleistocene glass-rich tephra layers record explosive acidic volcanism. Possible source areas of calc-alkaline volcanic rocks include Woodlark Island, the eastern Papuan Peninsula or the D'Entrecasteaux Islands (e.g. Fergusson, Goodenough and Normanby Islands) (Smith 1976, 1982; Smith *et al.* 1977; Davies *et al.* 1984; Smith & Milsom 1984; Hegner & Smith 1992; Lackschewitz *et al.* 2001) (Figs 18 and 19). Geochemical analysis of *c.* 50 ash layers is suggestive of sources in the Dawson Strait and Moresby Strait areas, and rarely from Dobu Seamount (Lackschewitz pers. comm.).

Implications for rift processes

Although back-arc rift sediments were previously drilled by ODP, notably in the Izu–

Bonin (Leg 126; Taylor *et al.* 1990) and Lau basins (Leg 135, Parson *et al.* 1992; SW Pacific) the Woodlark Basin is the first 'back-arc basin' in the Pacific region to be documented in detail from its pre- to syn-rift history. Only the Tyrrhenian back-arc basin in the central Mediterranean (Leg 107; Kastens & Mascle 1990; Robertson *et al.* 1990) has been studied in comparable detail, although its tectonic setting as a semi-enclosed syn-collisional basin differs markedly from that of the Woodlark Basin.

The results of Leg 180 are consistent with a two-stage model for extension of the Woodlark Rift (Fig. 20), when combined with regional information, although these may be local stages in the continuing extension of a wider area (Taylor *et al.* 1999*b*).

Stage 1. Extension, distributed over a broad zone (*c.* 100–150 km), formed a series of sub-parallel rift basins during Late Miocene to Mid-Pliocene time (earlier than *c.* 4 Ma), probably related to regional back-arc extension behind the Trobriand Arc. This first stage is reflected in regional Late Miocene–Pliocene extensional basin formation, including the Cape Vogel Basin (onshore and offshore), the Woodlark Rift, and also the Goodenough Bay, Mullins Harbour and Milne Bay rifts (Davies & Smith 1971; Smith & Simpson 1972). At this stage the northern and southern margins of the present Woodlark Rift along the Leg 180 transect formed part of a single broad extensional basin, accommodating deepening Pliocene turbiditic deposition.

Stage 2. During Pleistocene time the present Moresby Seamount was uplifted, splitting the pre-existing rift basin into the present-day southern and northern rift margins separated by a deep fault-bounded rift depocentre. Uplift of the southern margin to produce the Moresby Seamount was accommodated by extension along the low-angle (20–30°) Moresby Detachment Fault, and numerous smaller higher-angle faults mainly along the southern margin of the seamount. The northern rift margin was correspondingly downflexed. The regional palaeogeography of the Cape Vogel Basin to the NW was also reorganized during Pleistocene time. The regional cause of these changes was possibly the westward passage of the Woodlark spreading tip beyond the Moresby Transform, slightly after 2 Ma (Taylor *et al.* 1999*b*), an event that may well have triggered strong activity on the Moresby Detachment Fault and exhumation of the Moresby Seamount. A comparable pulsed extension history is apparently recorded in other areas including the Western

Alps and the related Iberia margin (Wilson *et al.* 2001).

Conclusions

(1) Leg 180 results document the tectonic–sedimentary evolution of a Late Miocene–Pleistocene rift, behind the Miocene Trobriand volcanic arc (Fig. 21). A two-stage rift history is envisaged, with initial Late Miocene–Pliocene rifting of a wide area including both the present northern and southern margins of the Woodlark Rift. This was followed in Pleistocene time by a second stage of more focused rifting, resulting in uplift of the southern rift margin as the Moresby Seamount and other regional palaeogeographical changes. A possible cause was a pulse of westward advance of the Woodlark oceanic spreading tip soon after 2 Ma, as indicated by regional geophysical studies.

(2) Drilling during Leg 180 established that the basement of both the southern and northern margins of the Woodlark Rift includes basic igneous rocks (basalt, dolerite and gabbro), which are regionally correlated with Palaeogene ophiolites exposed on land in Papua New Guinea.

(3) The northernmost Woodlark Rift margin (Site 1115) is underlain by Upper Miocene volcaniclastic sedimentary rocks that are correlated with the Miocene forearc basin of the Trobriand volcanic arc.

(4) The frontal arc–forearc region of the Trobriand Arc underwent uplift, subaerial exposure and weathering in a non-marine, paralic environment during Late Miocene time, as confirmed by drilling at Sites 1109, 1115 and 1118, at industry wells to the NW, and land exposures.

(5) Subsidence, probably related to rifting, began in latest Miocene time on the northern rift margin (Sites 1109 and 1115), with a transition through swamp, coastal and lagoonal settings. By contrast, Site 1118, nearer the rift depocentre, remained as a palaeo-island to shallow-marine high during latest Miocene–Early Pliocene time, followed by abrupt subsidence after 3.6 Ma.

(6) Turbidity current deposition dominated during Pliocene time on both the (present) southern and northern margins of the Woodlark Rift. Pliocene deposition on the southern margin (Sites 1114 and 1116) includes rare high-density-type turbidites, with volcaniclastic, terrigenous and shallow-water carbonate material. Northern margin deposition included abundant volcaniclastic turbidites and localized

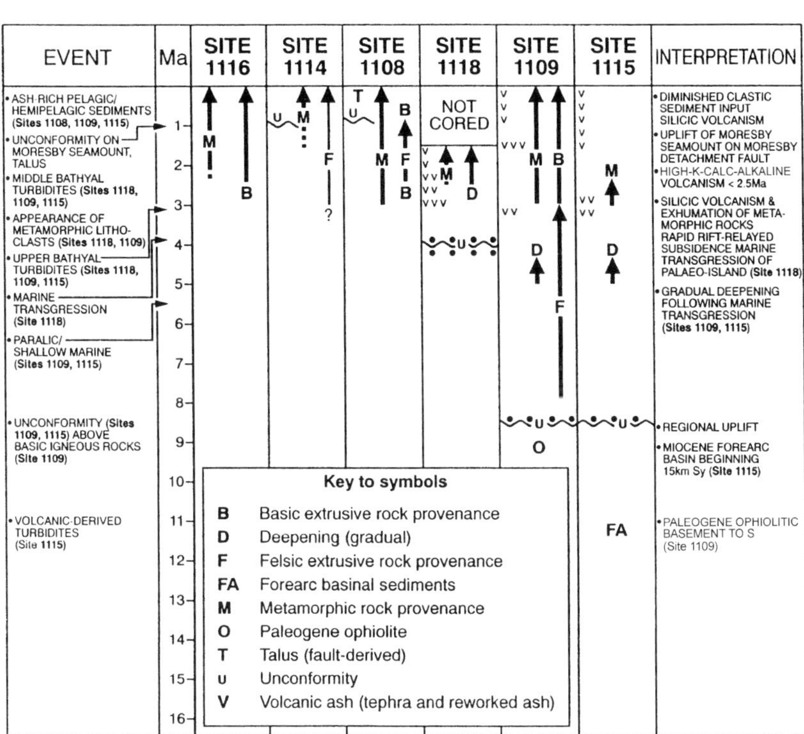

Fig. 21. Time–activity plot showing the main depositional events at each site drilled during Leg 180 and their interpretation. (See text for discussion.)

air-fall tephra, which were particularly abundant around 3–3.6 Ma and south of the northern margin sites (1118).

(7) The main source of Pliocene volcaniclastic sediment of calc-alkaline composition at the southerly sites (1114 and 1116) was probably the D'Entrecasteaux Islands, Papua New Guinea and now submerged adjacent areas. However, volcanic islands (unusually) located within the forearc area (e.g. Trobriand Islands) are seen as additional sources of volcaniclastic turbidites and tephra, especially at the most northerly rift margin site (Site 1115).

(8) Metamorphic-derived material is a widespread, but volumetrically minor component of the Plio-Pleistocene southern and northern rift margin successions, appearing at c. 3 Ma. The probable source is metamorphic rocks within the D'Entrecasteaux Islands and Papua New Guinea that were exhumed by coeval extensional faulting (e.g. core complex of Fergusson Island).

(9) The Moresby Seamount, forming the present southern margin of the Woodlark Rift, was uplifted and partially eroded in Pleistocene time (>0.46 Ma, but < 1.67 Ma), probably owing to extension along the low-angle Moresby Detachment Fault. Emergence of this fault on the sea floor as scarps triggered shedding of coarse angular talus

(10) During Pleistocene time, the construction of a carbonate platform to the NW of the Woodlark Rift resulted in a decreased input of terrigenous turbidites, and a switch to less rapidly accumulating pelagic and hemipelagic sediments in the Woodlark Rift (e.g. Site 1108).

References

ABERS, G.A. 1991. Possible seismogenic shallow-dipping normal faults in the Woodlark D'Entrecasteaux extensional province, Papua New Guinea. *Geology*, **19**, 1205–1208.

ABERS, G.A. 2001. Evidence for seismogenic normal faults at shallow dips in continental rifts. *In*: WILSON, R.C.L., WHITMARSH, R.B., TAYLOR, B. & FROITZHEIM, N. (eds) *Non-volcanic Rifting of Continental Margins: a Comparison of Evi-*

dence from Land and Sea. Geological Society, London, Special Publications, **187**, 305–318.

ABERS, G.A., MUTTER, C.Z. & FANG, J. 1997. Shallow dips of normal faults during rapid extension: earthquakes and normal faults in the Woodlark D'Entrecasteaux rift system, Papua New Guinea. *Journal of Geophysical Research,* **102**, 15301–15317.

ALLEN, J.R.L. *Sedimentary Structures. Their Character and Physical Basis.* Elsevier, Amsterdam.

ASHLEY, P.M. & FLOOD, R.H. 1981. Low-K tholeiites and high-K igneous rocks from Woodlark Island, Papua New Guinea. *Journal of the Geological Society of Australia,* **28**, 227–240.

BALDWIN, S.L. & IRELAND, T.R. 1995. A tale of two eras: Plio-Pleistocene unroofing of Cenozoic and late Archean zircons from active metamorphic core complexes, Solomon Sea, Papua New Guinea. *Geology,* **23**, 1023–1026.

BALDWIN, S.L., LISTER, G.S., HILL, E.J., FOSTER, D.A. & MCDOUGALL, I. 1993. Thermochronologic constraints on the tectonic evolution of active metamorphic core complexes, D'Entrecasteaux Islands, Papua New Guinea. *Tectonics,* **12**, 611–628.

BENES, V., SCOTT, S.D. & BINNS, R.A. 1994. Tectonics of rift propagation into a continental margin: western Woodlark Basin, Papua New Guinea. *Journal of Geophysical Research,* **99**, 4439–4455.

BERGGREN, W.A., KENT, D.V., SWISHER, C.C. III & AUBRY, M.-P. 1995. A revised Cenozoic geochronology and chronostratigraphy. *In*: BERGGREN, W.A., KENT, D.V., AUBRY, M.-P. & HARDENBOL, J. (eds) *Geochronology, Time Scales and Global Stratigraphic Correlation.* Society of Economic Paleontologists and Mineralogists, Special Publications, **54**, 129–212.

BINNS, R.A. & WHITFORD, D.J. 1987. Volcanic rocks from the western Woodlark Basin, Papua New Guinea. *Proceedings of the Pacific Rim Congress,* **87**, 525–534.

BINNS, R.A., SCOTT, S.D. & PACLARK Participants 1987. Western Woodlark Basin: potential analogue setting for volcanogenic massive sulphide deposit. *Proceedings of the Pacific Rim Congress,* **87**, 531–535.

BINNS, R.A., SCOTT, S.D. & PACLARK Participants 1989. Propagation of sea-floor spreading into continental crust, western Woodlark Basin, Papua New Guinea. *CCOP/SOPAC Miscellaneous Report,* **79**, 14–16.

BINNS, R.A., SCOTT, S.D., WHELLER, G.E., & BENES, V. 1990. *Report on the SUPACLARK cruise, Woodlark Basin, Papua New Guinea, April 8–28 1990, RV Akademik Mstislav Keldysh.* Commonwealth Science and Research Organisation Report **176R**.

BJØRLYKKE, K. 1993. Diagenetic reactions in sandstones. *In*: PARKER, A.B. & SELLWOOD, B.W. (eds) *Sediment Diagenesis.* North Atlantic Treaty Organization, Advanced Research Institute Series C115. Reidel, Dordrecht, 169–213.

BLATT, H., MIDDLETON, G.V. & MURRAY, R. *Origin of Sedimentary Rocks* (2nd). Prentice–Hall, Englewood Cliffs, NJ.

BLOOMER, J.H., TAYLOR, B., MACLEOD, C.T., HERN, R.J., FRYER, P., HAWKINS, T.W. & JOHNSON, L. 1995. Early arc volcanism and the ophiolite problem: a perspective from drilling in the West Pacific. *In*: TAYLOR, B. & NATLAND, J.J. (eds) *Active Margins and Marginal Basins of the Western Pacific.* Geophysical Monograph, American Geophysical Union, **88**, 1–3.

BOUMA, A. *Sedimentology of some Flysch Deposits; a Graphic Approach to Facies Interpretation.* Elsevier, Amsterdam.

COOPER, P. & TAYLOR, B. 1987. Seismotectonics of New Guinea: a model for arc reversal following arc–continent collision. *Tectonics,* **6**, 53–67.

DAVIES, H.L. 1980*a*. A crustal structure and emplacement of ophiolite in southeastern Papua New Guinea. *Colloques Internationale du CNRS,* **272**, 17–33.

DAVIES, H.L. 1980*b*. A folded thrust and associated metamorphics in the Suckling–Dayman massif, Papua New Guinea. *American Journal of Science,* **280**, 171–191.

DAVIES, H.L. 1990. Structure and evolution of the border region of New Guinea. *In*: CARMAN, G.J. & CARMAN, Z. (eds) *Petroleum Exploration in Papua New Guinea: Proceedings of the 1st Papua New Guinea Petroleum Convention, Port Moresby,* 245–250.

DAVIES, H.L. & JAQUES, A.L. 1984. Emplacement of ophiolite in Papua New Guinea. *In*: GASS, I.G., LIPPARD, S.J. & SHELTON, A.W. (eds) *Ophiolites and Oceanic Lithosphere.* Geological Society, London, Special Publications, **13**, 341–350.

DAVIES, H.L. & SMITH, I.E. 1971. Geology of eastern Papua. *Geological Society of American Bulletin,* **82**, 3299–3312.

DAVIES, H.L. & WARREN, R.G. 1988. Origin of eclogite-bearing, domed, layered metamorphic complexes ('core complexes') in the D'Entrecasteaux Islands, Papua New Guinea. *Tectonics,* **7**, 1–21.

DAVIES, H.L. & WARREN, R.G. 1992. Eclogites of the D'Entrecasteaux Islands. *Contributions to Mineralogy and Petrology,* **112**, 463–474.

DAVIES, H.L., HONZA, E., TIFFIN, D.L. & 5 OTHERS 1987. Regional setting and structure of the western Solomon Sea. *Geo-Marine Letters,* **7**, 153–160..

DAVIES, H.L., SYMONDS, P.A. & RIPPER, I.D. 1984. Structure and evolution of the southern Solomon Sea region. *Bureau of Mineral Resources, Journal of Australian Geology and Geophysics,* **9**, 49–68.

DAVIES, H.L., WINN, R.D., KENGEMAR, P. & PEREMBO, R.C.S. 1997. Evolution of the New Guinea orogen. *PNG Geology.* Department of Geology, University of Papua New Guinea, Port Moresby, 247–266.

Dow, D.B. 1997. Geology of Papua New Guinea. *PNG Geology*. Department of Geology, University of Papua New Guinea, Port Moresby, 1–16.

Finlayson, D.M., Muirhead, K.J., Webb, J.B., Gibson, G., Furumoto, A.S., Cooke, R.J.S. & Russel, A.J. 1976. Seismic investigation of the Papuan Ultramafic Belt. *Geophysical Journal of the Royal Astronomical Society*, **29**, 245–253.

Francis, G. 1985. *Stratigraphy of the Cape Vogel Basin*. Geological Survey of Papua New Guinea Report **85/4**.

Francis, G., Lock, J. & Okuda, Y. 1987. Seismic stratigraphy and structure of the area to the southeast of the Trobriand Platform. *Geo-Marine Letters*, **7**, 121–128.

Goodliffe, A., Taylor, B., Hey, R. & Martinez, F. 1993. Seismic images of continental breakup in the Woodlark Basin, PNG *EOS Transactions, American Geophysical Union*, **74**, 606.

Goodliffe, A., Taylor, B., Martinez, F., Hey, R., Maeda, K. & Ohno, K. 1997. Synchronous reorientation of the Woodlark Basin spreading center. *Earth and Planetary Science Letters*, **146**, 233–242.

Harris, C.S., Varol, O. & Mortimer, C.P. 1985. *Biostratigraphy of the Goodenough-1 and Nubiam-1 wells from the Cape Vogel Basin and the L'étoile-1 well from the Bougainville Basin*. Robertson Research Report **1486**.

Hegler, G., Das, S. & Woodhouse, J.H. 1995. A seismological study of the eastern New Guinea and the western Solomon Sea regions and its tectonic implications. *Geophysics Journal International*, **122**, 961–981.

Hegner, E. & Smith, I.E.M. 1992. Isotopic compositions of late Cenozoic volcanics from southeast Papua New Guinea: evidence for multi-component sources in arc and rift environments. *Chemical Geology*, **97**, 233–249.

Hill, E.J. 1994. Geometry and kinematics of shear zones formed during continental extension in eastern Papua New Guinea. *Journal of Metamorphic Geology*, **16**, 1093–1105.

Hill, E.J. & Baldwin, S.L. 1993. Exhumation of high-pressure metamorphic rocks during crustal extension in the D'Entrecasteaux region: Papua New Guinea. *Journal of Metamorphic Geology*, **11**, 261–277.

Hill, E.J., Baldwin, S.L. & Lister, G.S. 1992. Unroofing of active metamorphic core complexes in the D'Entrecasteaux Islands, Papua New Guinea. *Geology*, **20**, 907–910.

Home, P.C., Dalton, D.G. & Brannan, J. 1990. Geological evolution of the Western Papuan basin. *In*: Carman, G.J. & Carman, Z. (eds) *Petroleum Exploration in Papua New Guinea: Proceedings of the 1st Papua New Guinea Petroleum Convention*. Papual New Guinea Petroleum Society, Port Moresby, 107–117.

Honza, E., Davies, H.L., Keene, J.B. & Tiffin, D.L. 1987. Plate boundaries and evolution of the Solomon Sea region. *Geo-Marine Letters*, **7**, 161–168.

Karig, D.E. & Moore, G.F. 1975. Tectonically controlled sedimentation on marginal basins. *Earth and Planetary Science Letters*, **26**, 233–238.

Kastens, K.A., Mascle, J. & Shipboard Scientific Party 1990. Ocean Drilling Program Leg 107 in the Tyrrhenian Sea: insights into passive marginal back-arc basin evolution. *Geological Journal of America Bulletin*, **100**, 1140–1156.

Lackschewitz, K.S., Bogaard, P.v.D. & Mertz, D.F. 2001. ^{40}Ar/^{39}Ar ages of fallout tephra layers and volcaniclastic deposits in the sedimentary succession of the western Woodlark Basin, Papua New Guinea: the marine record of Miocene–Pleistocene volcanism. *In*: Wilson, R.C.L., Whitmarsh, R.B., Taylor, B. & Froitzheim, N. (eds) *Non-volcanic Rifting of Continental Margins: a Comparison of Evidence from Land and Sea*. Geological Society, London, Special Publications, **187**, 373–388.

Lock, J., Davies, H.L., Tiffin, D.L., Murakami, F. & Kisomoto, K. 1987. The Trobriand subduction system in the Western Solomon Sea. *Geo-Marine Letters*, **7**, 129–134.

Lowe, D.R. 1975. Water escape structures in coarse grained sediments. *Sedimentology*, **22**, 157–204.

Monteleone, B., Baldwin, S.L., Ireland, T.R. & Fitzgerald, P.G. 2002. *Thermochronologic constraints for the tectonic evolution of the Moresby Seamount, Woodlark Basin, Papua New Guinea*. ODP Leg 180 Scientific Results, in press.

Mutter, J.C., Mutter, C.Z. & Fang, J. 1996. Analogies to oceanic behaviour in the continental breakup of the western Woodlark Basin. *Nature*, **380**, 333–336.

Parson, L., Hawkins, J., Allan, J., *et al.* (eds) 1992. *Proceedings of the Ocean Drilling Program, Scientific Results, 135*. Ocean Drilling Program, College Station, TX.

Pearce, J.A., Lippard, J.J. & Roberts, S. 1984. Characteristics and tectonic significance of supra-subduction zone ophiolites. *In*: Kokelaar, B.P. & Howells, M.F. (eds) *Marginal Basin Geology*. Geological Society, London, Special Publications, **16**, 77–89.

Pickering, K.T., Hiscott, R.N. & Hein, F.J. *Deep Marine Environments. Clastic Sedimentation and Tectonics*. Unwin Hyman, London.

Pinchin, J. & Bembrick, C. 1985. Cape Vogel Basin, PNG—tectonics and petroleum potential. *14th Bureau of Mineral Resources Symposium Extended Abstracts: Petroleum Geology of South Pacific Island Countries*. Bureau of Mineral Resources of Australia, Records, **32**, 31–37.

Piper, D.W.J. 1978. Turbidite muds and silts on abyssal plains. *In*: Stanley, D.J. & Kelling, G. (eds) *Sedimentation in Submarine Fans, Canyons and Trenches*. Dowden, Hutchinson & Ross, Stroudsburg, PA, **1**, 63–176.

Resig, J.M., Frost, G.M., Ishikawa, N. & Perembo, R.C.B. 2001. Micropalaeontological and palaeomagnetic approaches to stratigraphic anomalies in rift basins, ODP Site 1109, Woodlark Basin. *In*: Wilson, R.C.L., Whitmarsh,

R.B., TAYLOR, B. & FROITZHEIM, N. (eds) *Non-volcanic Rifting of Continental Margins: a Comparison of Evidence from Land and Sea*. Geological Society, London, Special Publications, **187**, 389–404.

ROBERTSON, A.H.F. & XENOPHONTOS, C. 1993. Development of concepts concerning the Troodos ophiolite and adjacent basins in Cyprus. *In*: PRICHARD, H.M., ALABASTER, T., HARRIS, N.B. & NEARY, C.R. (eds) *Magmatic Process and Plate Tectonics*. Geological Society, London, Special Publications, **70**, 85–120.

ROBERTSON, A.H.F., HEIKE, W., MASCLE, G., MCCOY, F., MCKENZIE, J., REHAULT, J.P. & SARTORI, R. 1990. Summary and synthesis of the Tyrrhenian Sea, Western Mediterranean: Leg 107 of the Ocean Drilling Program. *In*: KASTENS, K. & MASCLE, J. (eds) *Proceedings of the Ocean Drilling Program, Scientific Results, 107*. Ocean Drilling Program, College Station, TX, 639–688.

ROGERSON, R., HILYARD, D., FRANCIS, G. & FINLAYSON, E. 1987. The foreland thrust belt of Papua New Guinea. *Proceedings of the Pacific Rim Congress*, **87**, 579–583.

ROGERSON, R., QUEEN, L. & FRANCIS, G. 1993. The Papuan Ultramafic Belt Arc Complex. *In*: WHELLER, G.E. (ed.) *Islands and Basins: Correlation and Comparison of Onshore and Offshore Geology*. CCOP/SOPAC Miscellaneous Report, **159**, 28–29.

SENIOR, B.R. & BILLINGTON, W.G. 1987. Structural interpretation of Northwest Fergusson Island, Papua New Guinea. *Proceedings of the Pacific Rim Congress*, **87**, 387–392.

SMITH, I.E.M. 1976. Peralkaline rhyolites from the D'Entrecasteaux Islands, Papua New Guinea. *In*: JOHNSON, R.W. (ed.) *Volcanism in Australasia*. Elsevier, Amsterdam, 275–285.

SMITH, I.E.M. 1982. Volcanic evolution in eastern Papua. *Tectonophysics*, **87**, 315–333.

SMITH, I.E.M. & JOHNSON, R.W. 1981. Contrasting rhyolite suites in the late Cenozoic of Papua New Guinea. *Journal of Geophysical Research*, **86**, 10257–10272.

SMITH, I.E.M. & MILSOM, J.S. 1984. Late Cenozoic volcanism and extension in eastern Papua. *In*: KOKELAAR, B.P. & HOWELLS, M.F. (eds) *Marginal Basin Geology*. Geological Society, London, Special Publications, **16**, 163–171.

SMITH, I.E. & SIMPSON, C.J. 1972. Late Cenozoic uplift in the Milne Bay area, eastern Papua New Guinea. *Bureau of Mineral Resources of Australia, Bulletin*, **125**, 29–35.

SMITH, I.E.M., CHAPPELL, B.W., WARD, G.K. & FREEMAN, R.S. 1977. Peralkaline rhyolites associated with andesitic arcs of the southwest Pacific. *Earth and Planetary Science Letters*, **37**, 230–236.

STEWART, W.P., FRANCIS, G. & DEIBERT, D.H. 1986. *Hydrocarbon potential of the Cape Vogel Basin, Papua New Guinea*. Geological Survey of Papua New Guinea Report **86/10**.

STOLZ, A.J., DAVIES, G.R., CRAWFORD, A.J. & SMITH, I.E.M. Sr 1993. Nd and Pb isotopic compositions of calc-alkaline and peralkaline silicic volcanics from the D'Entrecasteaux Islands, Papua New Guinea, and their tectonic significance. *Mineralogy and Petrology*, **47**, 103–126.

TAYLOR, B. 1987. A geophysical survey of the Woodlark–Solomons region. *In*: TAYLOR, B. & EXON, N.F. (eds) *Marine Geology, Geophysics and Geochemistry of the Woodlark Basin–Solomon Islands, Earth Science Series, 7*. Circum-Pacific Council for Energy and Mineral Resources, Houston, TX, 25–48.

TAYLOR, B., FUIJOKA, K., KLAUS, A., *et al.* (eds) 1990. *Proceeding of the Ocean Drilling Program, Scientific Results, 126*. National Science Foundation, Joint Oceanographic Institutions, Inc.

TAYLOR, B., GOODLIFFE, A. & MARTINEZ, F. 1999*b*. How continents break up: insights from Papua New Guinea. *Journal of Geophysical Research*, B4, **104**, 7437–7512.

TAYLOR, B., GOODLIFFE, A., MARTINEZ, F. & HEY, R. 1995. Continental rifting and initial seafloor spreading in the Woodlark basin. *Nature*, **374**, 534–537.

TAYLOR, B. HUCHON, P. KLAUS, A., *et al.*, 1999*a*. Active continental extension in the Western Woodlark Basin, Papua New Guinea, Sites 1108–1118, 7 June–11 August 1998. *In*: TAYLOR, B. HUCHON, P. KLAUS, A., *et al.* (eds) *Proceedings of the Ocean Drilling Program, Initial Reports, 180*. National Science Foundation, Joint Oceanographic Institutions, Inc.

TJHIN, K.T. 1976. Trobriand Basin Exploration, Papua New Guinea. *The APEA Journal*, **1976**, 81–90.

TREGONING, P., LAMBECK, K., STOLZ, A. & 8 OTHERS 1998. Estimation of current plate motions in Papua New Guinea from Global Positioning Systems observations. *Journal of Geophysical Research*, **103**(B6), 12181–12204.

WALKER, D.A. & MCDOUGALL, I. 1982. ^{40}Ar/^{39}Ar and K–Ar dating of altered glassy volcanic rocks: the Dabi Volcanics P.N.G. *Geochimica et Cosmochimica Acta*, **46**, 2181–2190.

WEISSEL, J.K., TAYLOR, B. & KARNER, G.D. 1982. The opening of the Woodlark Basin, subduction of the Woodlark spreading system and the evolution of northern Melanesia since mid-Pliocene time. *Tectonophysics*, **87**, 253–277.

WILSON, R.C.L., MANATCHAL, G. & WISE, S. 2001. Rifting along non-volcanic passive margins: stratigraphic and seismic evidence from the Mesozoic successions of the Alps and western Iberia. *In*: WILSON, R.C.L., WHITMARSH, R.B., TAYLOR, B. & FROITZHEIM, N. (eds) *Non-volcanic Rifting of Continental Margins: a Comparison of Evidence from Land and Sea*. Geological Society, London, Special Publications, **187**, 429–452.

WORTHING, M.A. 1988. Petrology and tectonic setting of blueschist facies metabasites from the Emo Metamorphics of Papua New Guinea. *Australian Journal of Earth Science*, **35**, 159–168.

^{40}Ar/^{39}Ar ages of fallout tephra layers and volcaniclastic deposits in the sedimentary succession of the western Woodlark Basin, Papua New Guinea: the marine record of Miocene–Pleistocene volcanism

K.S. Lackschewitz[1,2], P.V.D. Bogaard[3] & D.F. Mertz[1]

[1]*Johannes Gutenberg-Universität Mainz, Institut für Geowissenschaften, 55099 Mainz, Germany*

[2]*Present address: Universität Bremen, Fachbereich Geowissenschaften, 28334 Bremen, Klagenfurter Str. 2, Germany (e-mail: klacksch@uni-bremen.de)*

[3]*GEOMAR Forschungszentrum, Wischhofstrasse 1–3, 24148 Kiel, Germany*

Abstract: Five fallout tephra layers and 13 heterolithological volcaniclastic deposits drilled at Holes 1115A, 1115B, 1115C, 1109C, 1109D and 1118A, during Leg 180 on the downflexed northern margin of the western Woodlark Basin, have been dated by single-crystal laser ^{40}Ar/^{39}Ar analyses. The fallout tephra layers range in age from 0.135 ± 0.008 Ma to 2.84 ± 0.03 Ma. Sedimentation ages determined for the volcaniclastic deposits range from 1.75 ± 0.29 Ma to 3.79 ± 0.01 Ma, closely matching the nannofossil, planktonic foraminifer and palaeomagnetic chronostratigraphies of the holes. However, two volcaniclastic deposits from 516.91 m below seafloor (mbsf) and 632.5 mbsf in Hole 1118A are significantly older than indicated by biostratigraphic and palaeomagnetic data, probably because of the presence of older reworked volcanic crystals. The youngest ash layer is derived from explosive eruptions in the Dawson Strait area of the D'Entrecasteaux Islands, whereas the four older tephra layers are attributed to explosive eruptions in the Moresby Strait area of the D'Entrecasteaux Islands. The ^{40}Ar/^{39}Ar ages of volcaniclastic sand layers in Holes 1115C and 1118A indicate a transition from a shallow-water succession (<150 m) to a deeper-water succession ($150–500$ m) with rapid deposition of volcaniclastic sands, mainly by turbiditic currents, at 3.8 Ma. This transition is related to the subsidence of the margin during rifting of the Woodlark Basin. Two volcaniclastic deposits with ages of 13.84 ± 0.02 Ma and 14.04 ± 0.03 Ma, respectively, provide important time markers in the middle Miocene sedimentary sequence at Hole 1115C, where biostratigraphic ages are scarce. Our ^{40}Ar/^{39}Ar ages represent the first marine record of Miocene to Pleistocene volcanism in the area of eastern Papua.

Miocene to Pleistocene submarine tephra layers and volcaniclastic deposits recovered during Leg 180 (Western Woodlark Basin; Fig. 1) reflect major Cenozoic volcanic episodes in the western Woodlark Basin–Papuan Peninsula region of Papua New Guinea and in the sedimentary evolution of the island's submarine volcaniclastic wedges, respectively. This paper presents the results of ^{40}Ar/^{39}Ar geochronological analyses of mid-Miocene to Pleistocene fallout tephra layers and volcaniclastic sediments recovered at Sites 1109, 1115 and 1118 (Fig. 2). The aim is to date chronostratigraphic markers that permit direct correlation of subaerial ocean-island volcanic phases with the marine volcanic record of the sedimentary succession and to evaluate the chronostratigraphic significance of volcaniclastic deposits in the sedimentary record through ^{40}Ar/^{39}Ar laser dating of their volcanic mineral components.

Fallout tephra deposits, and various types of subaerial and submarine pyroclastic flow deposits, are directly derived from explosive volcanic eruptions. The time between cooling and closure of the Ar isotopic system of juvenile crystals in primary tephra deposits differs only insignificantly from the actual time of deposition. Contaminant older crystals are rare in most fallout tephra layers and can be distinguished from juvenile crystal components by their mineral and chemical composition. Thus, isotopically homogeneous crystal populations

From: WILSON, R.C.L., WHITMARSH, R.B., TAYLOR, B. & FROITZHEIM, N. 2001. *Non-Volcanic Rifting of Continental Margins: A Comparison of Evidence from Land and Sea.* Geological Society, London, Special Publications, **187**, 373–388. 0305-8719/01/$15.00 © The Geological Society of London 2001.

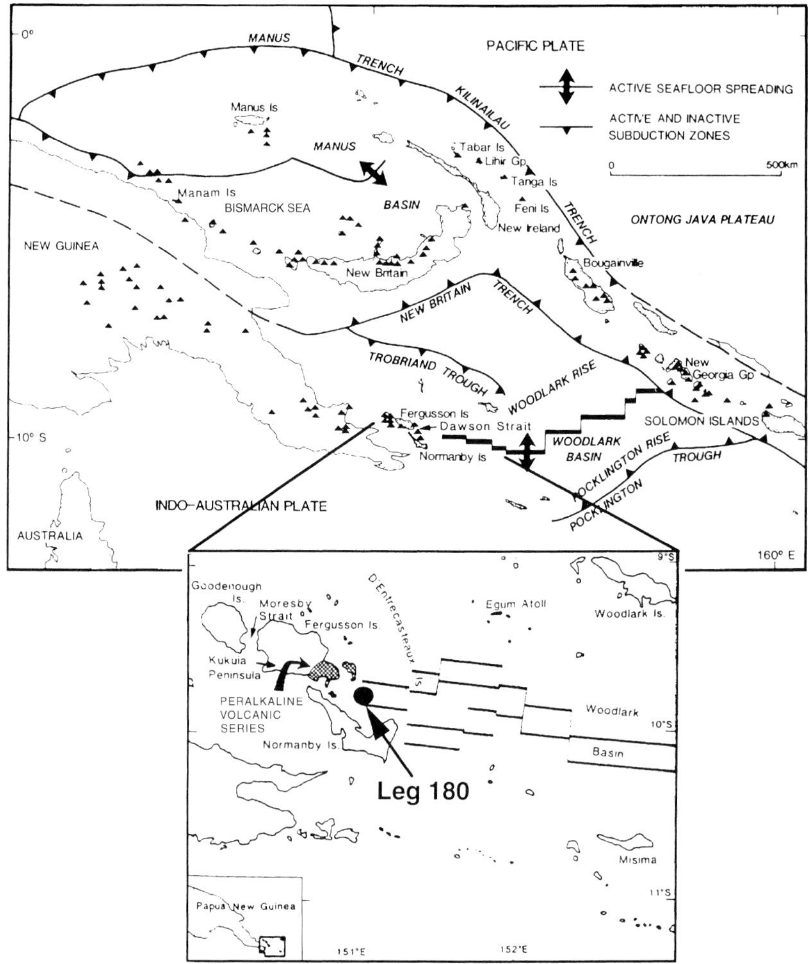

Fig. 1. Map of the Woodlark Basin region showing the location of Leg 180 and possible source areas for fallout tephra layers and volcaniclastic sediments (modified after Stolz *et al.* 1993). ▲, late Cenozoic volcanoes.

make fallout tephra layers useful chronostratigraphic markers that can be traced over long distances and in different environments, and whose ages can be determined with high precision by K–Ar and ^{40}Ar/^{39}Ar dating.

Volcaniclastic deposits range from tephra layers that were remobilized and transported over short distances during or shortly after depostion with little admixing of non-volcanic components, to epiclastic deposits such as lahars, debris flows and turbidity currents that mix lithic and biogenic sediment material from many different sources generally varying in age and composition. The age of deposition of volcaniclastic sediments provides important constraints on the rates and processes of mass transfer from the volcanic sources into the sur-

rounding sedimentary basins, and erosional intervals and major tectonic events of the volcanically active regions themselves (e.g. Schmincke & Bogaard 1990). In contrast to primary tephra layers, however, volcaniclastic sediments are more difficult to date. If there are crystals of different age and provenance, which cannot be distinguished by means of their macroscopic or microscopic characteristics, single-crystal ^{40}Ar/^{39}Ar analysis is a suitable method to differentiate age populations (e.g. Chen *et al.* 1996; Bogaard 1998). The youngest volcanic crystals in a volcaniclastic sediment can give a maximum age of the formation of the deposits, in some cases even the accurate age.

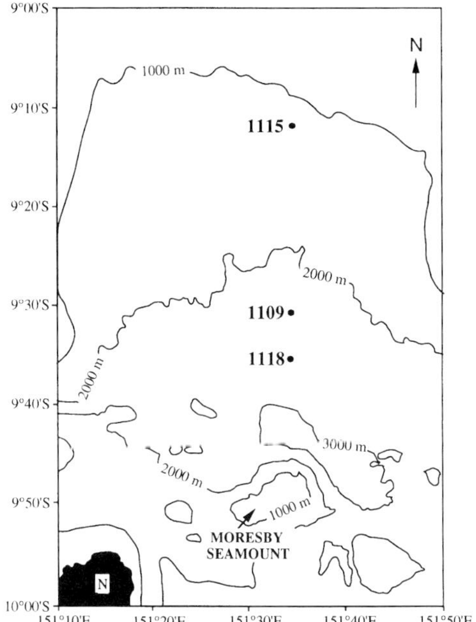

Fig. 2. Locations of Sites 1109, 1115 and 1118.

Geological setting of Sites 1109, 1115 and 1118

During Leg 180, three sites (1109, 1115 and 1118) were drilled in a sedimentary basin of the western Woodlark Basin in the hanging wall of the north-dipping Moresby Detachment Fault (Fig. 3). Location and sampling data for each hole have been given by Taylor *et al.* (1999). Robertson *et al.* (2001) provide an overview of the sedimentary fill of the western Woodlark Basin. The volcanic deposits studied here are from the mid-Miocene to Pleistocene lithostratigraphic Units I, V and VI of Holes 1109C and 1109D; Units I, II, III and XII of Holes 1115A, 1115B and 1115C; and Units III, IV and V of Hole 1118A (Fig. 4).

At Site 1109, Unit I is composed of calcareous sand, silt and clay, interbedded with a small number of volcaniclastic sand beds and fallout tephra layers. Unit V is characterized by relatively uniform, greenish grey, clayey siltstones and silty claystones interbedded with volcaniclastic silt and sand layers, rare tephra layers and quartz-rich sand layers. Unit VI is distinguished from Unit V by clayey siltstone and silty claystone, with subordinate thin beds of fine- to medium-grained sandstone–siltstone.

At Site 1115, Unit I consists of nannofossil ooze, nannofossil-rich silty clay, calcareous silty clay–claystone, volcaniclastic silt and sand beds, and tephra layers. Tephra layers occur throughout the entire Unit I. Unit II is marked by a transition from mainly ooze to clay as the background fine-grained sediment. Fallout tephra layers are less abundant than in Unit I, but numerous volcaniclastic layers of epiclastic origin occur. Lithostratigraphic Unit III is characterized by the appearance of discrete thin graded beds of volcaniclastic sand with calcareous silty clay. The final Unit XII at Site 1115 is marked by a change from heterogeneous, variably calcareous sedimentary rocks

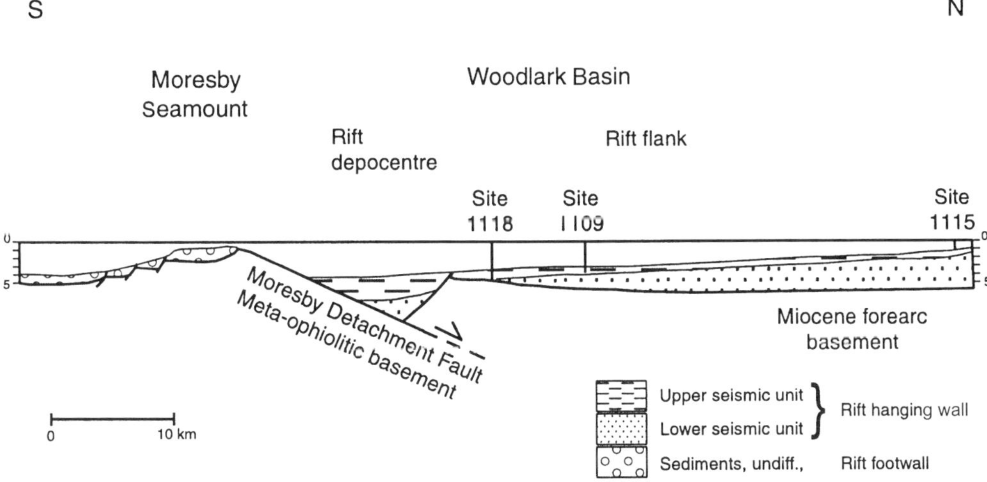

Fig. 3. Line diagram showing interpretations of north–south profiles along which the Leg 180 sites were drilled (modified after Robertson *et al.* 2001).

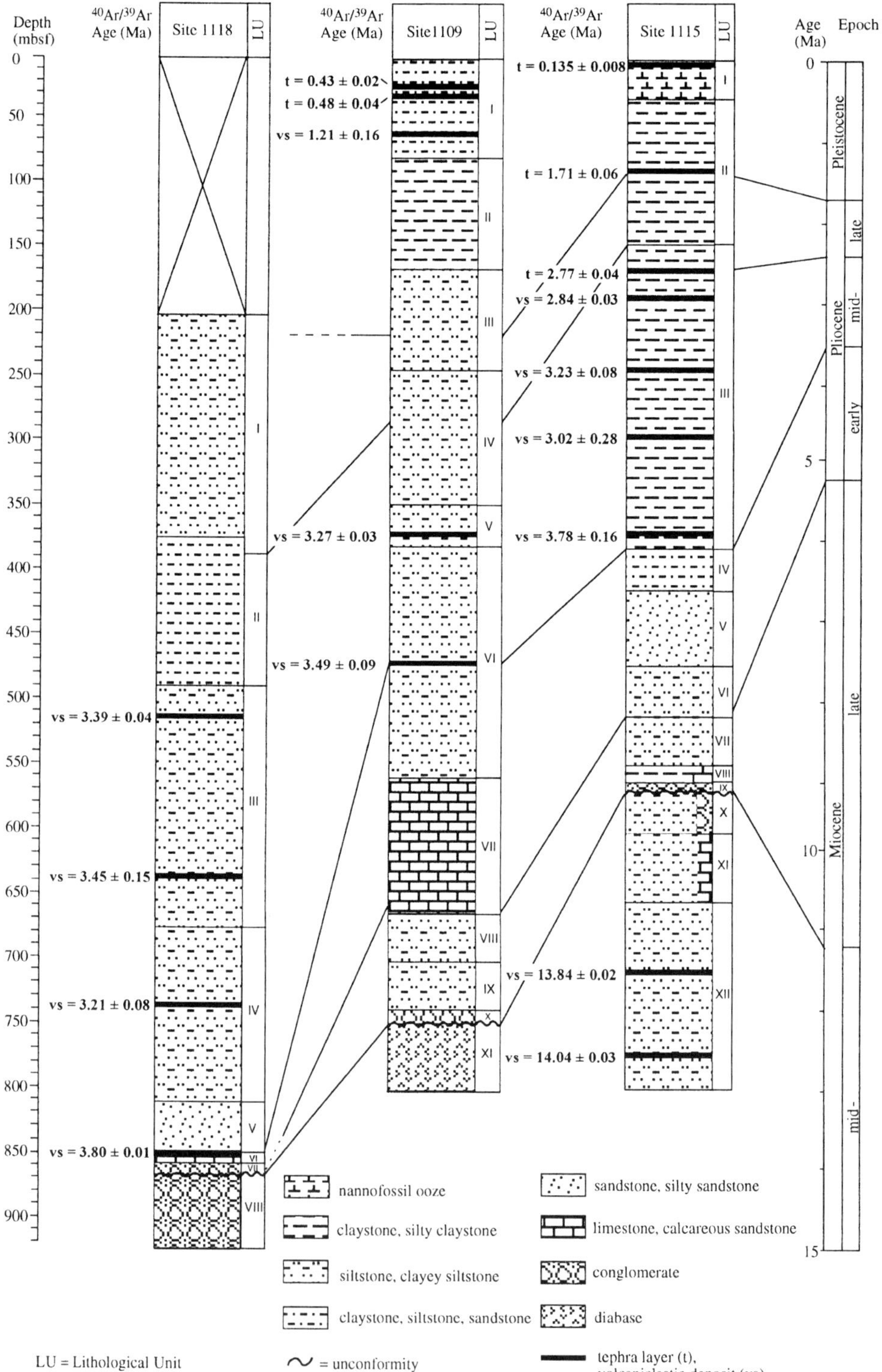

Fig. 4. Lithostratigraphy and stratigraphic position of dated tephra layers and volcaniclastic deposits recovered at Sites 1109, 1115 and 1118 (modified after Taylor *et al.* 1999).

of the overlying unit to a much more homogeneous finer-grained, less calcareous succession of sandy siltstone alternating with silty sandstone and rare silty claystone. In this unit, only two medium- to coarse-grained volcaniclastic layer with sharp upper and lower contacts were observed.

At Site 1118, Unit III is characterized by a mainly greyish green succession of sandstones, siltstones and hemipelagic sediments. Thin beds of volcaniclastic sandstone occur sporadically. Lithostratigraphic Unit IV is dominated by silty claystones and clayey siltstones interbedded with volcaniclastic sandstones and siltstones. Mixed sandstone, siltstones and volcaniclastic sandstone are present in Unit V. Volcaniclastic sandstone is restricted to near the base of the unit.

Analytical methods

All ash layers and volcanicastic deposits studied here have been identified by the Shipboard Scientific Party (Taylor *et al.* 1999) and were sampled on the *JOIDES Resolution*. The samples comprised five distinctive, well-sorted, homogeneous fallout ash layers and 13 volcaniclastic deposits. Samples were washed and wet sieved, and sanidine, plagioclase and biotite crystals were hand-picked under a binocular microscope using a vacuum tweezer. Separated crystals were then etched in 5% hydrofluoric acid and cleaned with an ultrasonic disintegrator.

^{40}Ar/^{39}Ar laser probe analyses were carried out in the geochronology laboratory of the GEOMAR research centre. The crystals were irradiated in Al sample-holders in the 5 MW reactor at the GKSS (Gesellschaft für Kernforschung und Strahlenschutz) Research Centre (Geesthacht, Germany) in two capsules. The first capsule, containing single-crystal samples, was irradiated for 12 h (reactor position 36) at an average fast neutron flux of 1.12 $\times$ 10^{13} N cm^{-2} s^{-1}. The second capsule, containing laser step-heating samples, was irradiated for 14 days (reactor position 44) at an average fast neutron flux of 3.90 $\times$ 10^{11} N cm^{-2} s^{-1}, using Cd shielding of 0.5 mm thickness to reduce ^{40}Ar production by neutron interactions with potassium. The neutron flux was monitored using TCR sanidine (27.92 Ma; Dalrymple & Duffield 1988; Duffield & Dalrymple 1990) in both cases. Vertical variations in the neutron flux were quantified by cosine function fit.

For the single-crystal analyses, individual sanidine, plagioclase and biotite crystals of >100 µm grain size were fused, and basically evaporated, in a single heating step with a focused 15–20 W beam of a 25 W Spectra Physics argon ion laser. Step-heating analyses of sanidine, plagioclase and biotite mineral separates of <100 µm grain size were carried out with a defocused laser beam, incrementally increasing the nominal laser power output from 50 or 100 mW to 20 W in 25–30 steps. Complete gas release of step-heating samples was assured by additional final fusions with an incrementally focused laser beam.

The Ar isotope compositions were measured using a MAP 216 series gas source mass spectrometer. Raw mass spectrometer peaks were corrected for mass discrimination and system blanks, measured before and after every fifth or sixth analysis. Interfering neutron reactions were quantified using pure K_2SO_4 crystals and optical-grade CaF_2 crystals that were irradiated together with the feldspar and biotite samples. The single-crystal dating results are evaluated by means of isotope correlation (York 1969) and their isochron and mean apparent ages. Step-heating analyses are evaluated by apparent age spectra (assuming an initial 'atmospheric' ^{40}Ar/^{36}Ar ratio of 295.5 and excluding monitor uncertainties) and isotope correlations. Analytical uncertainties of plateau, isochron and mean apparent ages include uncertainty in the ^{40}Ar*/^{39}Ar$_K$ ratio of the neutron flux monitor but exclude uncertainty in the age of the monitor, and are given as 1σ errors. Decay constants and isotope ratios used for age calculation are those given by Steiger & Jäger (1977).

Age determinations

The ^{40}Ar/^{39}Ar dating results of single-crystal and step-heating analyses are summarized in Tables 1 and 2. The full analytical data can be obtained from the Society Library or the British Library Document Supply Centre, Boston Spa, Wetherby, West Yorkshire LS23 7BQ, UK as Supplementary Publication No. SUP 18163 (5 pages).

Fallout ash layers

Sample 1115A-1H-4, 96–103 cm represents a single, platy and bubble-wall shard-rich, crystal-bearing fallout ash layer of 7 cm thickness. Nineteen sanidine crystals (0.158–1.080 mg) from this sample range in apparent age from 0.04 ± 0.006 to 0.23 ± 0.06 Ma. Inverse isochron age (0.16 ± 0.05 Ma) and initial ^{40}Ar/^{36}Ar ratio (235 ± 108) are poorly constrained (Table 1; Fig. 5a); however, the mean

Table 1.

Leg, core, section, interval (cm)	Depth (mbsf)	Mineral	N	Mean apparent age (Ma) $\pm 1\sigma$	MSWD	Inverse isochron age (Ma) $\pm 1\sigma$	Initial $^{40}Ar/^{36}Ar \pm 1\sigma$	MSWD	$J \pm 1\sigma$	Lab. no.
180-1115A-1H-4, 96–103	3.98	S	19	0.135 ± 0.008	1.37	0.16 ± 0.05	235 ± 108	1.38	$3.5004E-03 \pm 4.7E-06$	9005
180-1115B-10H-4, 60–62	88.3	P	8	1.65 ± 0.17	0.16	2.16 ± 0.73	163 ± 183	0.13	$3.5399E-03 \pm 3.9E-06$	9015
		B	1	1.72 ± 0.06	n.c.					
		P + B	9	1.71 ± 0.06	0.16	1.61 ± 0.35	309 ± 45	0.16		
180-1115B-26X-6, 52–54	243.42	P	8	4.1 ± 0.21	0.5	4.17 ± 0.21	290 ± 5	0.34	$3.5004E-03 \pm 4.7E-06$	9004
		B	5	3.18 ± 0.09	2.69	3.25 ± 0.08	290 ± 4	3.03		
		P + B	13	3.23 ± 0.08	2.74	3.3 ± 0.07	290 ± 3	2.71		
180-1115B-31X-5, 27–29	289.77	P	9	3.02 ± 0.28	1.4	4.5 ± 0.45	84 ± 63	0.86	$3.5004E-03 \pm 4.7E-06$	9018
180-1115C-9R-7, 7–9	367.94	P	6	3.78 ± 0.16	0.64	3.73 ± 0.71	302 ± 76	0.79	$3.5004E-03 \pm 4.7E-06$	9006
180-1115C-45R-7, 107–108	709.68	S	10	13.84 ± 0.02	0.53	13.9 ± 0.08	274 ± 15	0.42	$3.5004E-03 \pm 4.7E-06$	9008
180-1115C-51R-2, 18–20	765.4	S	10	14.03 ± 0.04	1.3	13.98 ± 0.1	308 ± 17	1.45	$3.5004E-03 \pm 4.7E-06$	9007
		B	1	14.15 ± 0.15	n.c.					
		S + B	11	14.04 ± 0.03	1.23	13.97 ± 0.1	311 ± 15	1.3		
180-1109C-3H-7, 36–38	26.26	S	7	0.43 ± 0.02	1.47	0.46 ± 0.03	287 ± 7	1.48	$3.5399E-03 \pm 3.9E-06$	9011
180-1109C-3H-7, 61.5–63.5	26.51	S	8	0.48 ± 0.04	0.68	0.53 ± 0.07	281 ± 15	0.62	$3.5399E-03 \pm 3.9E-06$	9009
180-1109D-3R-2, 141–145	371.61	P	10	3.19 ± 0.07	1.16	3.18 ± 0.08	296 ± 5	1.3	$3.5004E-03 \pm 4.7E-06$	9012
		B	4	3.29 ± 0.04	1.57	3.18 ± 0.18	316 ± 32	2.13		
		P + B	14	3.27 ± 0.03	1.31	3.27 ± 0.04	295 ± 5	1.42		
180-1109D-13R-6, 87–89	471.51	P	9	3.56 ± 0.11	0.18	3.6 ± 0.15	294 ± 4	0.17	$3.5004E-03 \pm 4.7E-06$	9013
		B	1	3.37 ± 0.14	n.c.					
		P + B	10	3.49 ± 0.09	0.28	3.49 ± 0.11	295 ± 4	0.31		
180-1118A-45R-4, 110–112	632.54	P	9	3.45 ± 0.15	0.44	3.38 ± 0.31	306 ± 41	0.49	$3.5004E-03 \pm 4.7E-06$	9016
180-1118A-56R-1, 107–111	734.87	P	6	3.15 ± 0.10	0.24	3.23 ± 0.23	285 ± 25	0.26	$3.5004E-03 \pm 4.7E-06$	9014
		B	1	3.35 ± 0.15	n.c.					
		P + B	7	3.21 ± 0.08	0.42	3.34 ± 0.18	277 ± 22	0.38		
180-1118A-68R-1, 142–144	850.72	P	10	3.64 ± 0.06	1.22	3.63 ± 0.11	297 ± 12	1.37	$3.5399E-03 \pm 3.9E-06$	9010
		B	7	3.8 ± 0.13	0.19	3.81 ± 0.04	294 ± 4	0.21		
		P + B	17	3.79 ± 0.01	1.23	3.78 ± 0.04	297 ± 4	1.3		

S, sanidine; P, plagioclase; B, biotite; N, number of crystals analysed; MSWD, mean square weighted deviation; J, dimensionless irradiation parameter.

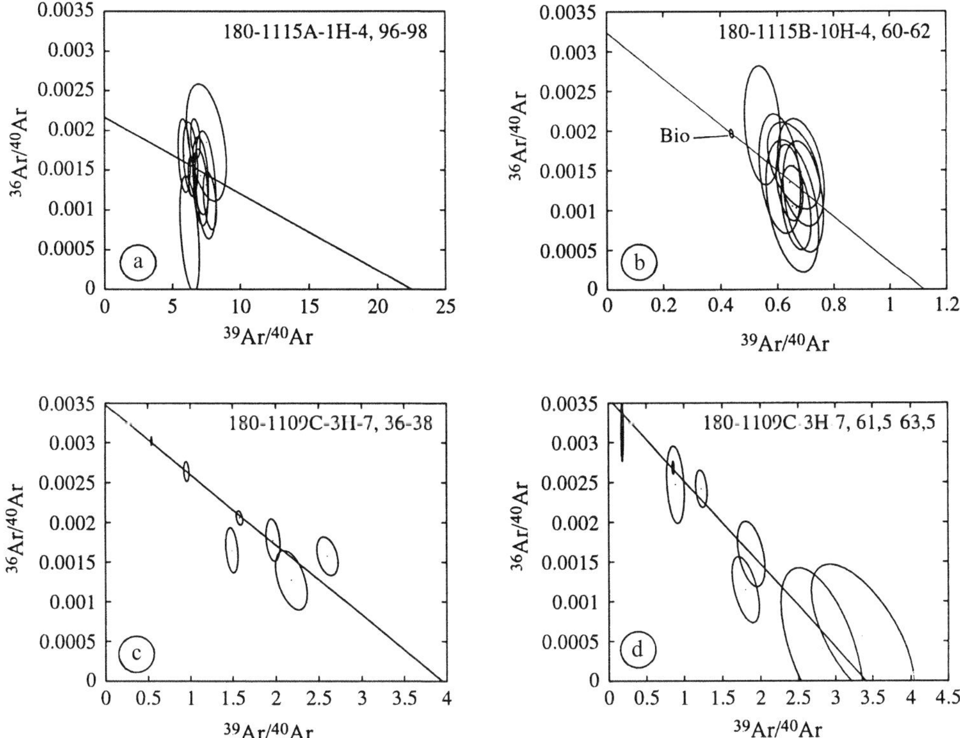

Fig. 5. Isotope correlation diagrams showing the argon isotope composition of feldspar and biotite single-crystal fusions from fallout tephra layers in Holes 1115A, 1115B and 1109C. Analytical uncertainties are indicated by error ellipses. Numerical results of the inverse isochrons are given in Table 1. Bio, biotite.

apparent age appears to be geological significant (0.135 ± 0.008 Ma; mean square weighted deviation (MSWD) = 1.37) and is interpreted as the eruption age.

Sample 1115B-10H-4, 60–62 cm represents a dark green, fine-grained, well-sorted, vitric ash layer. Ten crystals, comprising nine plagioclase and one biotite crystal, yield apparent ages from 1.28 to 1.90 Ma. Isotope correlation returns an inverse isochron age of 1.61 ± 0.35 Ma with an atmospheric initial $^{40}\text{Ar}/^{36}\text{Ar}$ ratio of 309 ± 45 (MSWD = 0.16) (Fig. 5b). The weighted mean apparent age of the entire population of 1.71 ± 0.06 Ma (MSWD = 0.16) is indistinguishable from the isochron age considering analytical uncertainties, and interpreted to reflect the eruption age.

Sample 1115B-19H-1, 60–62 cm is a greyish green, well-sorted, pumice-rich ash layer with some plagioclase and biotite crystals. Feldspar step-heating yields a poorly resolved age spectrum with a seven-step plateau representing c. 91% of the total ^{39}Ar

released (Fig. 6a), and a plateau age of 2.32 ± 0.06 Ma (Table 2). The biotite step-heating analysis, in turn, indicates a higher apparent age of 2.73 ± 0.02 Ma, based on a 16-step well-resolved plateau that represents c. 96% of the total ^{39}Ar released from this sample (Fig. 6b). Feldspar and biotite plateau step isotope ratios, however, are isochronous at an initial $^{40}\text{Ar}/^{36}\text{Ar}$ ratio of 293.6 ± 2.4, yielding a multiphase isochron age of 2.77 ± 0.04 Ma (MSWD = 1.11) (Table 2).

Sample 1109C-3H-7, 36–38 cm represents the bottom part of a coarse-grained, vitric fallout ash layer of 10 cm thickness. Seven sanidine crystals (0.244–0.577 mg) were dated with apparent ages varying from 0.36 to 0.62 Ma with a mean apparent age of 0.43 ± 0.02 Ma (MSWD = 1.47). This population defines an inverse isochron with an age of 0.46 ± 0.03 Ma, with an initial $^{40}\text{Ar}/^{36}\text{Ar}$ ratio of 287 ± 7 (Fig. 5c). Isochron and mean apparent ages are concordant and 0.43 ± 0.02 Ma

Table 2.

Leg, core, section, interval (cm)	Hole and sample data			Total gas data				Plateau/weighted mean			Isochron data			
	Depth (mbsf)	Mineral	Mass (mg)	Steps	Vol. ^{39}Ar (CCSTP/mg)	Age Ma) $\pm 1\sigma$	% ^{39}Ar	Age (Ma) $\pm 1\sigma$	MSWD	Age (Ma) $\pm 1\sigma$	MSWD	Initial $\pm$ 1σ	Lab no.
180-1115B-19H-1, 60–62	169.3	Fsp	4.782	30	1.69E−10	2.35 ± 0.12	90.80	2.32 ± 0.06	0.53	2.30 ± 0.11	0.64	301.4 ± 22.8	8526
		B	1.531	32	2.33E−09	2.67 ± 0.03	95.80	2.73 ± 0.02	1.28	2.74 ± 0.04	1.36	294.4 ± 2.9	
		Fsp + B						2.69 ± 0.03	3.23	2.77 ± 0.04	1.11	293.6 ± 2.4	
180-1115B-21H-2W, 42.5–45	189.62	Fsp	7.934	24	3.68E−10	2.78 ± 0.03	97.50	2.81 ± 0.02	0.65	2.82 ± 0.03	0.70	283.5 ± 19.6	8525
		B	1.739	32	2.48E−09	2.96 ± 0.03	96.20	2.97 ± 0.02	0.84	2.94 ± 0.03	0.66	298.7 ± 1.7	
		Fsp + B						2.89 ± 0.02	2.92	2.84 ± 0.03	1.73	303.7 ± 2.0	
180-1109C-7H-6W, 50–52	62.9	Fsp	2.041	29	1.54E−10	1.75 ± 0.29	96.90	1.21 ± 0.16	0.49	1.02 ± 0.46	0.59	320.8 ± 61.6	8524
180-1118A-33R-4, 63–65	516.91	Fsp	5.089	27	3.58E−10	3.59 ± 0.05	68.10	3.38 ± 0.04	1.67	3.36 ± 0.05	0.67	302.7 ± 9.5	8523
		B	0.794	31	2.15E−09	3.22 ± 0.06	95.60	3.37 ± 0.05	1.43	3.41 ± 0.07	1.45	293.5 ± 2.4	
		Fsp + B						3.37 ± 0.03	1.41	3.39 ± 0.04	1.44	294.2 ± 1.9	

Plateau criteria: consecutive steps with identical age within 2σ and representing >50% of total ^{39}Ar released. Plateau age calculation: weighted mean of plateau steps, excluding steps with ^{39}Ar <1% and ^{40}Ar$_{atm}$ >100%. CCSTP, cm^3 at standard temperature and pressure; Fsp, feldspar; B, biotite; MSWD, mean square weighted deviation.

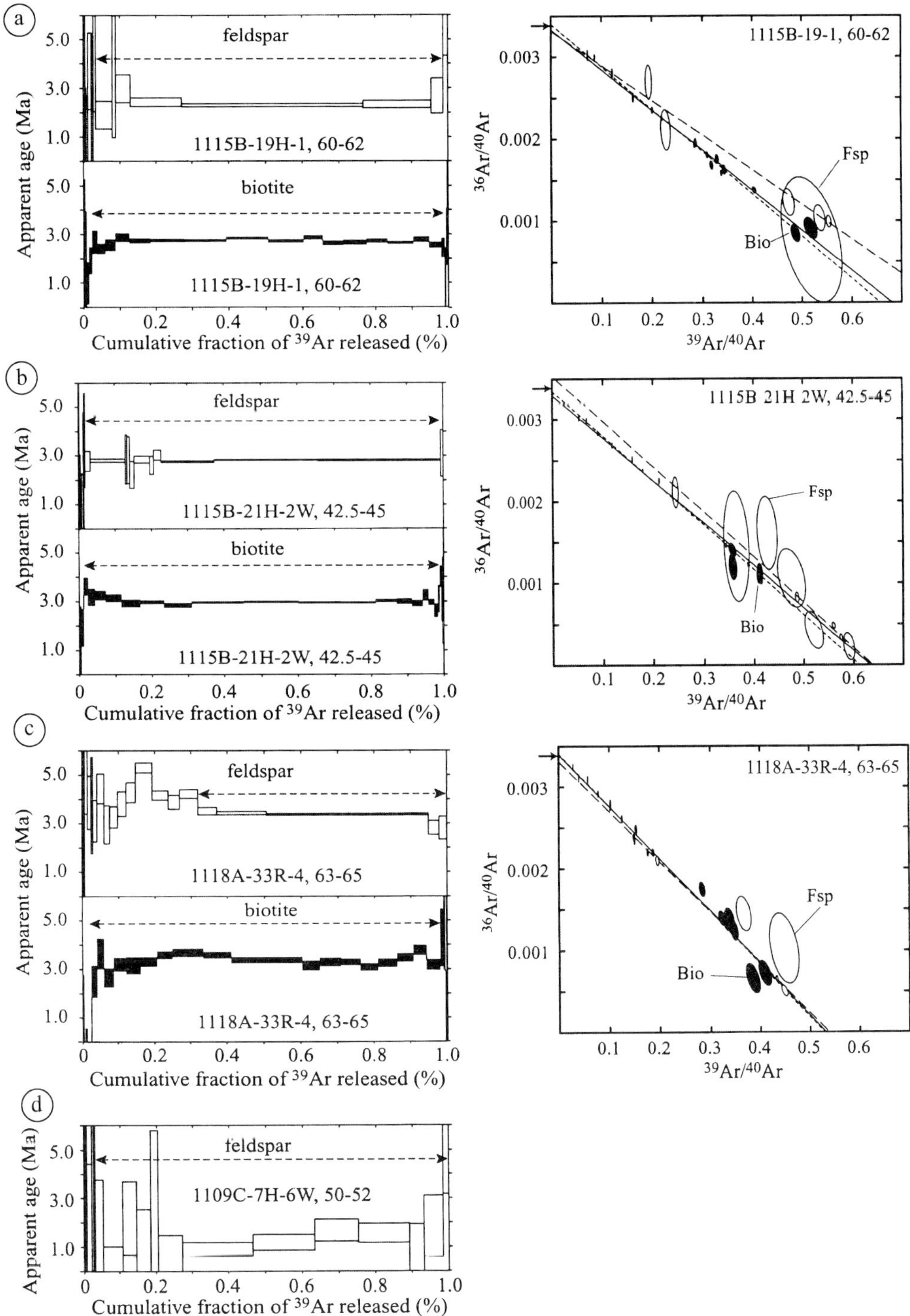

a
feldspar
1115B-19H-1, 60-62
biotite
1115B-19H-1, 60-62
Fsp
Bio
1115B-19-1, 60-62
b
feldspar
1115B-21H-2W, 42.5-45
biotite
1115B-21H-2W, 42.5-45
Fsp
Bio
1115B 21H 2W, 42.5-45
c
feldspar
1118A-33R-4, 63-65
biotite
1118A-33R-4, 63-65
Fsp
Bio
1118A-33R-4, 63-65
d
feldspar
1109C-7H-6W, 50-52

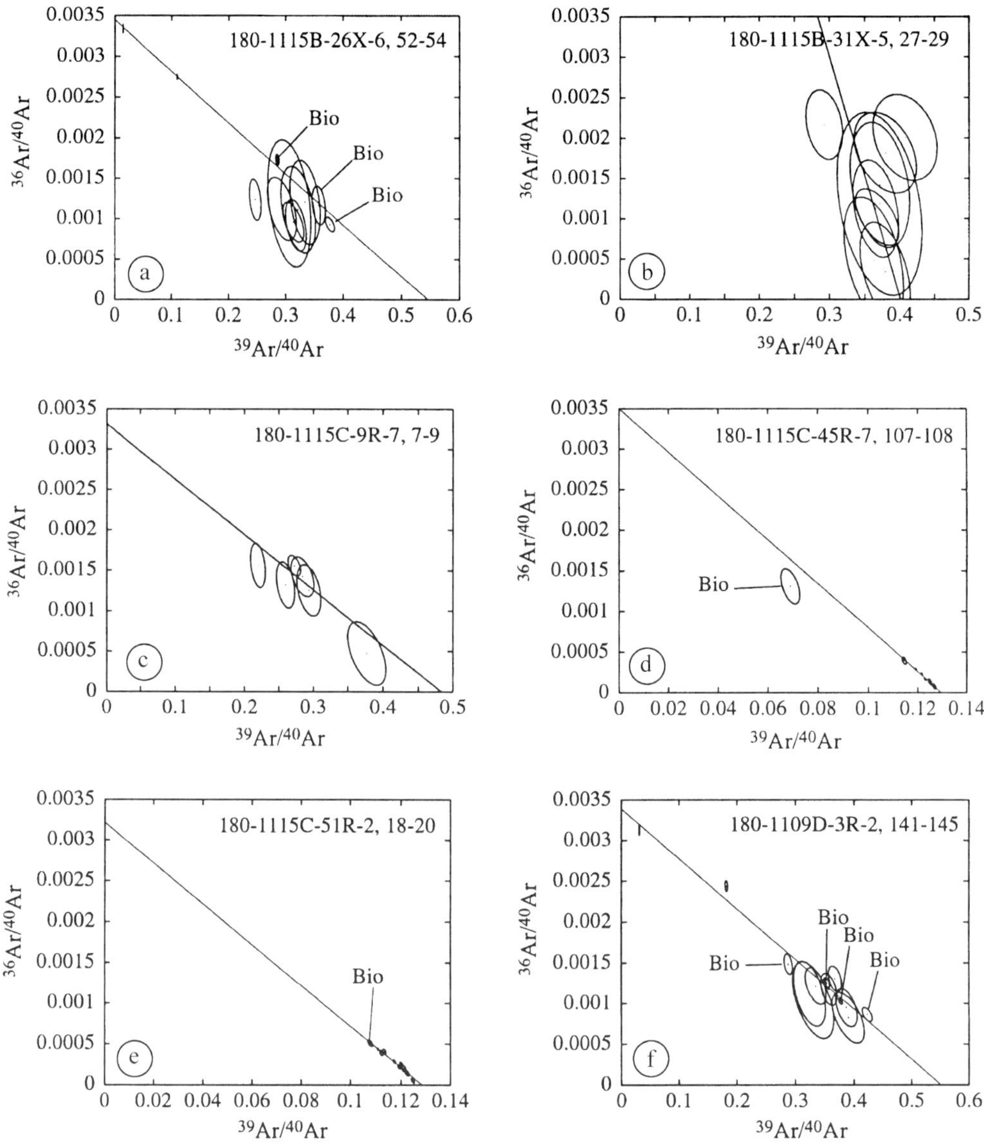

Fig. 7. Isotope correlation diagrams showing the argon isotope composition of feldspar and biotite single-crystal fusions from volcaniclastic deposits in Holes 1115B, 1115C, 1109C and 1109D. Analytical uncertainties are indicated by error ellipses. Numerical results of the inverse isochrons are given in Table 1. Bio, biotite.

Fig. 6. Laser step-heating analyses from a tephra layer and volcaniclastic deposits in Holes 1115B, 1109C and 1118A. (**a**) Apparent-age spectrum and inverse isochron of laser step-heating of feldspar and biotite crystals from tephra layer 1115B-19H-1, 60–62 cm. (**b**) Apparent-age spectrum and inverse isochron of laser step-heating of feldspar and biotite crystals from volcaniclastic layer 1115B-21H-2W, 42.5–45 cm. (c) Apparent-age spectrum and inverse isochron of laser step-heating of feldspar and biotite crystals from volcaniclastic layer 1118A-33R-4, 63–65 cm. Bio, biotite, Fsp, feldspar. (**d**) Apparent-age spectrum of laser step-heating of feldspar crystals from volcaniclastic layer 1109C-7H-6W, 50–52 cm.

(MWSD = 1.47) is interpreted as the eruption age.

Sample 1109C-3H-7, 61.5–63.5 cm is from a fine-grained, well-sorted, vitric fallout ash layer of 12 cm thickness. Eight sanidine crystals show apparent ages from 0.42 to 0.72 Ma. Isotope correlation gives an isochron with an age of 0.53 ± 0.07 Ma (initial $^{40}Ar/^{36}Ar$ ratio of 281 ± 15; MWSD = 0.62), largely overlapping with the mean apparent age of 0.48 ± 0.04 Ma (MWSD = 0.68) within error limits (Fig. 5d). The mean apparent age of the population is interpreted as the eruption age.

Volcaniclastic deposits

Sample 1115B-21H-2, 42.5–45.0 cm is a greenish grey, medium-grained volcaniclastic sandstone. This layer is rich in feldspar and biotite crystals (125–250 μm grain size), which were separated for step-heating analysis. Both step-heating analyses yield age spectra with well-defined plateaux (Fig. 6b), representing *c.* 98% of the total ^{39}Ar release (10 steps) in case of the feldspar, and *c.* 96% in case of the biotite analyses (16 steps). Plateau ages, however, differ significantly at the 1σ or 2σ confidence level, indicating isotopic closure of the feldspars at 2.81 ± 0.02 Ma, compared with 2.97 ± 0.02 Ma for the biotite crystals (Table 2). If combined into a single multiphase isochron (Fig. 6b), however, both feldspar and biotite plateau isotope ratios are concordant at a slightly elevated initial $^{40}Ar/^{36}Ar$ ratio (303.7 ± 2.0), and an isochron age of 2.84 ± 0.03 Ma (MSWD = 1.73).

Sample 1115B-26X, 52–54 cm represents a dark grey volcaniclastic sand, rich in fresh volcanic glass and crystals. Thirteen crystals (0.04–0.581 mg) were analysed, comprising eight plagioclase crystals with a mean apparent age of 4.1 ± 0.2 Ma, and five biotite crystals whose individual ages (3.07–3.47 Ma) are compatible with a mean apparent age of 3.23 ± 0.08 Ma (MSWD = 2.74) (Table 1, Fig. 7a).

Sample 1115B-31X-5, 27–29 cm is a light grey, normally graded, volcaniclastic sand deposit rich in crystals and glass. Nine plagioclase crystals were analysed, their single grain ages ranging from 1.8 to 4.26 Ma. Isotope correlation gives an isochron age of 4.5 ± 0.45 Ma (MSWD = 0.86), but a poor estimate of the initial $^{40}Ar/^{36}Ar$ ratio (84 ± 63) (Fig. 7b).

Sample 1115C-9R-7, 7–9 cm represents a fine-grained volcaniclastic sand rich in volcanic glass shards and crystals. Six plagioclase crystals yield ages from 3.57 to 4.42 Ma, and an isochron age of 3.73 ± 0.71 Ma (initial

$^{40}Ar/^{36}Ar$ ratio of 302 ± 76); their mean apparent age is 3.78 ± 0.16 Ma (MWSD = 0.64) (Fig. 7c).

Sample 1115C-45R-7, 107–108 cm is a medium- to coarse-grained ash-rich volcaniclastic layer. Ten sanidine crystals show apparent ages from 13.8 to 13.9 Ma. Isotope correlation gives an isochron with an age of 13.87 ± 0.08 Ma (Fig. 7d; initial $^{40}Ar/^{36}Ar$ ratio of 386 ± 14 and a mean apparent age of 13.84 ± 0.02 Ma (MWSD = 0.7).

Sample 1115C-51R-2, 18–20 cm is from a poorly sorted, crystal-rich volcaniclastic sand. Eleven crystals (0.063–0.703 mg) were analysed, comprising ten sanidine crystals with ages of 13.9–14.2 Ma and one biotite crystal individual age of 14.1 Ma compatible with a mean apparent age of 14.04 ± 0.03 Ma (MSWD = 1.23) (Fig. 7e).

Sample 1109C-7H-6, 50–52 cm is a greyish green, very fine-grained volcaniclastic sandstone comprising predominantly plagioclase. Laser step-heating of a plagioclase separate from this unit gives a low-precision age spectrum dominated by a 12-step plateau that represents *c.* 97% of the total ^{39}Ar release (Fig. 6d). Total gas age (1.75 ± 0.29 Ma) and plateau age (1.21 ± 0.16 Ma) are identical within their 1σ uncertainties, indicating isotopic closure of the plateau fraction at 1.21 ± 0.16 Ma (Table 2).

Sample 1109D-3R-2, 141–145 cm represents a volcaniclastic sand containing colourless glass and sand-sized phenocrysts of amphibole, pyroxene, plagioclase, quartz and biotite. Fourteen crystals were dated. Four biotite crystals (0.053–0.227 mg) give apparent ages from 3.18 to 3.52 Ma. The ages of ten plagioclase crystals (0.363–2.682 mg) range from 2.77 to 4.43 Ma. The mean apparent age is 3.27 ± 0.03 Ma (MSWD = 1.31) (Fig. 7f).

Sample 1109D-13R-6, 87–89 cm is from a volcaniclastic sand layer rich in volcanic glass. Ten crystals were analysed, comprising one biotite with an apparent age of 3.37 Ma and nine plagioclase crystals whose individual ages (ranging from 3.42 to 3.71 Ma) are compatible with a mean apparent age of 3.49 ± 0.09 Ma (MSWD = 0.28) (Fig. 8a).

Sample 1118A-33R-4, 63–65 cm is a dark grey, fine-grained volcaniclastic sandstone, rich in feldspar, biotite, amphibole and pyroxene crystals with grain sizes <250 μm. Laser probe $^{40}Ar/^{39}Ar$ step-heating analyses were carried out on feldspar and biotite separates. Feldspar step-heating yields an age spectrum slightly disturbed in the low-temperature fraction, but a well-defined plateau in the high-temperature

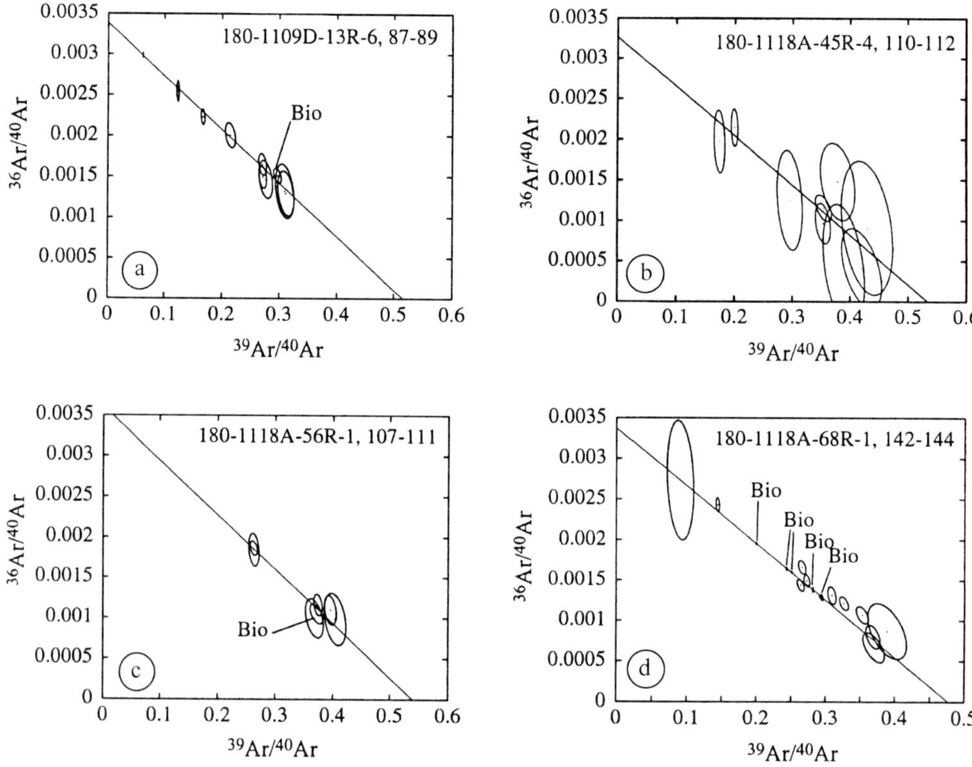

Fig. 8. Isotope correlation diagrams showing the argon isotope composition of feldspar and biotite single-crystal fusions from volcaniclastic deposits in Holes 1109D and 1118A. Analytical uncertainties are indicated by error ellipses. Numerical results of the inverse isochrons are given in Table 1. Bio, biotite.

range, comprising five heating steps overlapping in age within 2σ error and representing $c.$ 68% of the total ^{39}Ar release (Fig. 6c). Isotope correlation yields an age of 3.36 ± 0.05 Ma and an initial ^{40}Ar/^{36}Ar ratio estimate of 302.7 ± 9.5 (Table 2), identical to the present-day atmospheric ratio of 295.5 within error limits. The weighted mean plateau age calculated from these steps is 3.38 ± 0.04 Ma. Biotite step-heating analysis yields widely concordant apparent step ages representing $c.$ 96% (19 steps) of the total ^{39}Ar release with a plateau age of 3.37 ± 0.05 Ma (Fig. 6c; Table 2). Plateau and isochron ages are concordant, and the initial ^{40}Ar/^{36}Ar ratio of 293.5 ± 2.4 is atmospheric. Combining feldspar and biotite plateau step analyses a weighted mean age of 3.39 ± 0.04 Ma (MSWD = 1.41) is derived (Table 2).

Sample 1118A-45R-4, 110–112 cm is from the base of a fine-grained volcaniclastic turbidite of 4 cm thickness consisting predominantly of euhedral to subhedral plagioclase and biotite crystals. Apparent ages of nine plagioclase crystals range from 2.69 to 4.34 Ma. Isotope corre-

lation gives an isochron with an age of of 3.38 ± 0.31 Ma (Fig. 8b; ^{40}Ar/^{36}Ar initial ratio of 306 ± 41), overlapping with the mean apparent age of 3.45 ± 0.15 Ma (MSWD = 0.44) within error limits.

Sample 1118A-56R-1, 107–111 cm represents a volcaniclastic sandstone composed of plagioclase, biotite and amphibole. Six plagioclase crystals (0.36–0.96 mg) show apparent ages from 3.08 to 3.46 Ma. A single biotite crystal was dated and gives an apparent age of 3.35 ± 0.15 Ma. Regression of ^{40}Ar/^{36}Ar v. ^{40}Ar/^{39}Ar yields a well-defined isochron with an age of 3.23 ± 0.23 Ma (Fig. 8c; ^{40}Ar/^{36}Ar initial ratio of 285 ± 25) identical to the mean apparent age of 3.21 ± 0.08 Ma (MSWD = 0.42) within error limits.

The crystal-rich medium-grained volcaniclastic sandstone of Sample 1118A-68R-1, 142–144 cm was obtained from the upper part of a volcaniclastic sandstone sequence of $c.$ 4 m thickness, which formed the entire sections 1 and 2, and most of section 3 of Core 180-1118A-68R. Ten plagioclase crystals give

Site 1115

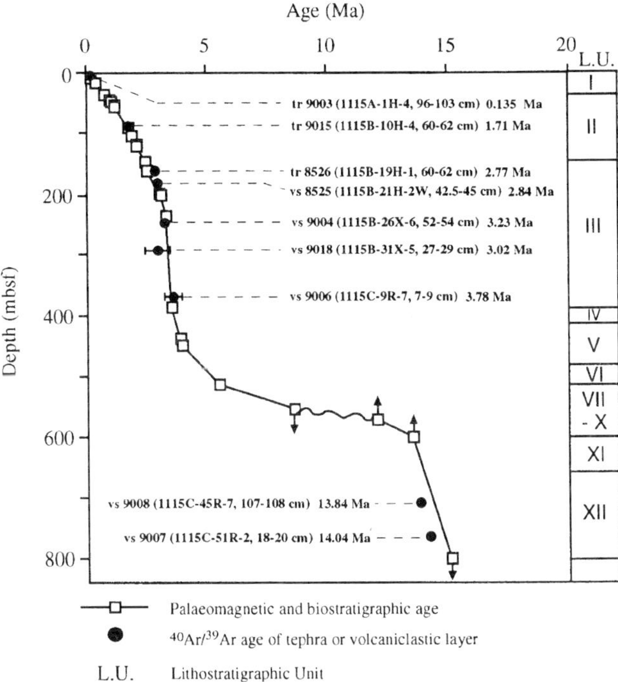

Fig. 9. Age v. depth diagram showing the stratigraphic position and inferred ages of tephra layers and volcaniclastic sediments from Site 1115. Age–depth curve is based on palaeomagnetic and biostratigraphic data points (Taylor *et al.* 1999). Symbols with arrows denote actual datum point can be above or below and older or younger than indicated by the symbols.

apparent ages from 3.44 to 3.93 Ma. The ages of seven biotite crystals range from 3.78 to 3.81 Ma. Isotope correlation gives an isochron with an age of 3.78 ± 0.04 Ma and an initial ^{40}Ar/^{36}Ar ratio of 297 ± 4 (Fig. 8d). Their mean apparent age of 3.79 ± 0.01 Ma (MSWD = 1.23) constrains the maximum age of Sample 68R-1, 142–144 cm.

Discussion and conclusions

On the basis of Miocene to Pleistocene biostratigraphic tiepoints (Taylor *et al.* 1999) and the ^{40}Ar/^{39}Ar ages of tephra layers and volcaniclastic deposits of Holes 1115A, 1115B, 1115C, 1109C and 1118A, age v. depth diagrams allow a reasonable estimate of the depositional ages of stratigraphic units (Figs 9–11).

Fallout tephra layers from the upper 27 mbsf at Hole 1109C and 170 mbsf at Hole 1115B erupted between 0.135 and 2.84 Ma. The ash beds form chronostratigraphic markers at 3.91 mbsf (0.135 ± 0.008 Ma) in lithological Unit I, 88.3 mbsf (1.71 ± 0.06) in lithological Unit II and 169.3 mbsf (2.84 ± 0.03 Ma) in lithological Unit III of Hole 1115B, and 26.26 mbsf (0.43 ± 0.02 Ma) and 26.51 mbsf (0.48 ± 0.04 Ma) in lithological Unit I of Hole 1109C. The ^{40}Ar/^{39}Ar ages of all fallout ash layers are compatible with the age and sedimentation rate estimates from nannofossil, planktonic foraminifer and palaeomagnetic chronostratigraphies (Figs 9 and 10).

The Pliocene to Pleistocene fallout layers record explosive silicic volcanism, inferred to have been sited in the Trobriand arc to the south, or eastern Papua New Guinea including the D'Entrecasteaux Islands (Johnson *et al.* 1978; Davies *et al.* 1984). The glass originates either from calc-alkaline volcanism of the Trobriand arc (Davies *et al.* 1984) or from calc-alkaline volcanism of eastern Papua and peralkaline volcanism of the Dawson Strait area (Smith 1976, 1981; Johnson *et al.* 1978; Stolz *et al.* 1993). The Trobriand arc volcanism is related to southward subduction of oceanic

Site 1109

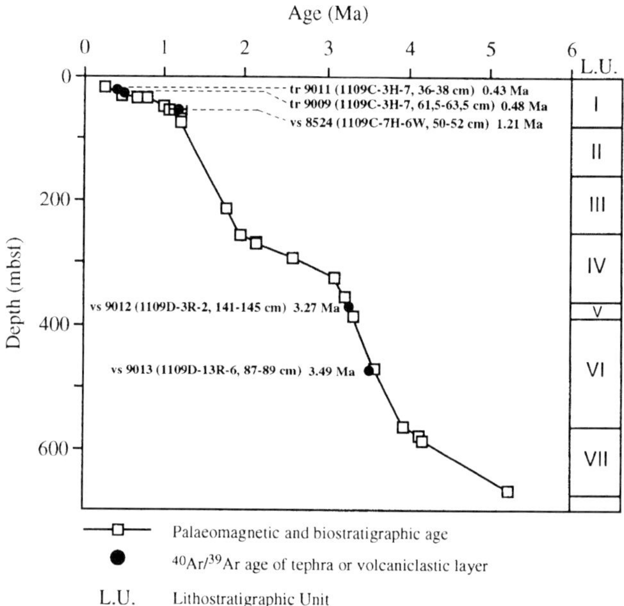

Fig. 10. Age v. depth diagram showing the stratigraphic position and inferred ages of tephra layers and volcaniclastic sediments from Site 1109. Age–depth curve is based on palaeomagnetic and biostratigraphic data points (Taylor *et al.* 1999).

crust from the Solomon Sea, whereas Dawson Strait area volcanism is attributed to rifting of the Woodlark Basin. The Eastern Papua peninsula has been volcanically active from mid-Miocene to recent time, and there have been eruptions on many of the surrounding islands (Smith & Milsom 1984). Because the various volcanic source areas and magmatic provinces are related to different tectonic environments, they have developed characteristic magma compositions. Considering the petrogenesis and radiometric ages of erupted material from the potential sources (Smith 1976; 1981, 1982; Smith *et al.* 1977; Smith & Johnson 1981; Smith & Milsom 1984), the mineral compositions, and ^{40}Ar/^{39}Ar ages of the Pliocene to Pleistocene fallout tephra layers in Holes 1115A, 1115B and 1109C, the fallout layers are interpreted to be derived from explosive eruptions of the D'Entrecasteaux Islands (e.g. Goodenough Island, Fergusson Island and Normanby Island, including Dawson Strait and Moresby Strait).

The youngest fallout ash bed dated from Hole 1115A (Sample 180-1115A-1H-4, 96–103 cm) is composed of peralkaline rhyolitic glass (Lackschewitz, unpubl. data). The most important source is the Dawson Strait, whose volcanic centres are characterized by peralkaline rhyolitic magmas and by volcanic activity during most of the Pleistocene to recent period (Smith 1976, 1981). The peralkaline magmatism in the Dawson Strait area appears to relate to extensional tectonics resulting from the westward propagation of the Woodlark spreading ridge into eastern Papua (Stolz *et al.* 1993).

The older tephra layers from Core 1115B and Hole 1109C are of calc-alkaline rhyolitic composition (Lackschewitz, unpubl. data). Calc-alkaline rhyolites of Late Cenozoic age in Papua New Guinea are known only from the Moresby Strait area (Smith 1976, 1981; Smith & Johnson 1981), which is probably related to deactivated subduction along the Trobriand Trough (Stolz *et al.* 1993). Thus, calc-alkaline rhyolitic eruptions in the Moresby area are probably the major source for the four tephra layers of Sample 1115B-10H-4, 60–62 cm; Sample 1115B-19H-1, 60–62 cm; Sample 1109C-3H-7, 36–38 cm; and Sample 1109C-3H-7, 61.5–63.5 cm.

The ^{40}Ar/^{39}Ar ages of volcaniclastic deposits in Holes 1115B, 1115C, 1109C and 1118A provide important time markers in the lower to

Site 1118

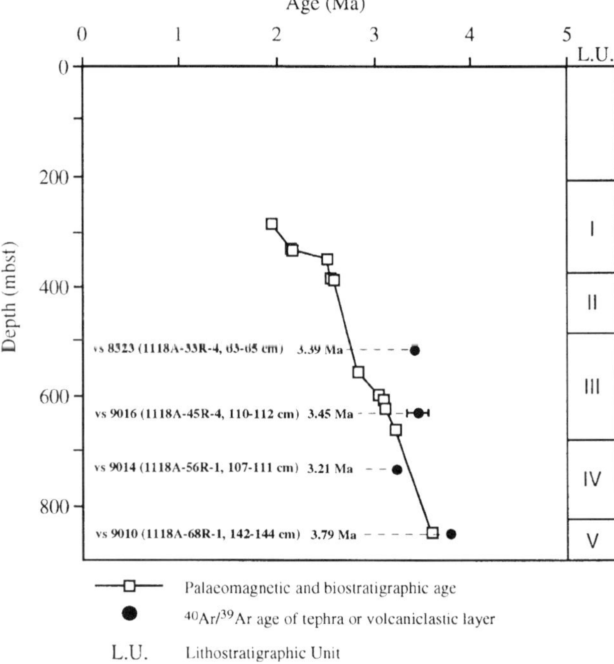

Fig. 11. Age v. depth diagram showing the stratigraphic position and inferred ages of tephra layers and volcaniclastic sediments from Site 1118. Age–depth curve is based on palaeomagnetic and biostratigraphic data points (Taylor *et al.* 1999).

middle Pliocene sedimentary sequence, where biostratigraphic ages are scarce (Taylor *et al.* 1999). However, the ^{40}Ar/^{39}Ar ages obtained from feldspars and biotites in the volcaniclastic sand layer of Samples 1118A-33R-4, 63–65 cm and 1118A-45R-4, 110–112 cm are older than the depositional ages inferred from palaeomagnetic and biostratigraphic evidence (Fig. 11). These feldspars and biotites are interpreted as epiclastic components in the volcaniclastic deposits, derived from the erosion of coastal and/or shelf areas.

The ^{40}Ar/^{39}Ar ages of the volcaniclastic sand layers 1115C-9R-7, 7–9 cm and 1118A-68R-1, 142–144 cm date a transition from a shallow-water succession (<150 m) to a deeper-water succession (150–500 m) with rapid deposition (284 m Ma^{-1}) of muds and volcaniclastic sands at 3.8 Ma. This appears to be related to the subsidence of the margin during the rifting of the Woodlark Basin.

Two volcaniclastic layers, derived from Miocene volcanic rocks, are present only in the northernmost Site 1115. The ^{40}Ar/^{39}Ar ages (13.84 Ma and 14.04 Ma) are significantly younger than the stratigraphic ages estimated from the foraminifer and nannofossil time scale (Fig. 10). The mid-Miocene nannofossil assemblage indicates an age of 13.6 Ma or younger whereas an assemblage of planktonic foraminifers points to an age of 15.1 Ma or older (Taylor *et al.* 1999). On the basis of poor Miocene biostratigraphic tiepoints and isotopically homogeneous crystal populations, the volcaniclastic layers form stratigraphic markers in Unit XII of Site 1115.

During mid- to late Miocene time, Site 1115 is inferred to have formed part of a submerged forearc *c.* 70 km NE of the front of the Trobriand arc (Taylor *et al.* 1999). Considering the isotopically homogeneous crystal populations, the ^{40}Ar/^{39}Ar ages of these two volcaniclastic layers are interpreted to be derived from eruptive phases of the Trobriand arc, including former volcanoes in the Amphlett Islands and Equm Atoll.

The results of our ^{40}Ar/^{39}Ar geochronological studies provide the first marine record of

Miocene to Pleistocene volcanic events in the area of eastern Papua, which may or may not be preserved on the continental volcanic source areas. Most of the volcanic and volcaniclastic ash layers are related to the Late Cenozoic tectonic history of eastern Papua, which is characterized by continental rifting and spreading of the Woodlark Basin.

We are indebted to M. Kummetz for mineral separation work and J. Sticklus for assistance with argon laser dating. Thanks also go to the Leg 180 Shipboard Scientific Party and the crew of *JOIDES Resolution* for constructive collaboration. A. Robertson and B. Turrin provided thorough and helpful reviews, which led to significant improvements in the paper. This project was supported by the Deutsche Forschungsgemeinschaft via grant Me 1155/3.

References

BOGAARD, P.V.D. 1998. ^{40}Ar/^{39}Ar ages of Pliocene–Pleistocene fallout tephra layers and volcaniclastic deposits in the sedimentary aprons of Gran Canaria and Tenerife (Sites 953, 954, and 956). *In*: WEAVER, P.P.E., SCHMINCKE, H.-U., FIRTH, J.V. & DUFFIELD, W. (eds) *Proceedings of the Ocean Drilling Program, Scientific Results, 157.* Ocean Drilling Program, College Station, TX, 329–341.

CHEN, Y., SMITH, P.E., EVENSEN, N.M. & YORK, D. 1996. The edge of time: dating young volcanic ash layers with the ^{40}Ar–^{39}Ar laser probe. *Science*, **274**, 1176–1178.

DALRYMPLE, G.B. & DUFFIELD, W.A. 1988. High precision ^{40}Ar/^{39}Ar dating of Oligocene from the Mogollon–Datil volcanic field using a continuous laser system. *Geophysical Research Letters*, **15**, 463–466.

DAVIES, H.L., SYMONDS, P. & RIPPER, I.D. 1984. Structure and evolution of the southern Solomon Sea region. *Bureau of Mineral Resources Journal of Australian Geology and Geophysics*, **9**, 49–68.

DUFFIELD, W.A. & DALRYMPLE, G.B. 1990. The Taylor Creek Rhyolite of New Mexico: a rapidly emplaced field of lava domes and flows. *Bulletin of Volcanology*, **52**, 475–487.

JOHNSON, R.W., MCKENZIE, D.E. & SMITH, I.E. 1978. Delayed partial melting of subduction modified mantle in Papua New Guinea. *Tectonophysics*, **46**, 197–216.

ROBERTSON, A.H.F., AWADALLAH, S.A.M., GERBAUDO, S., LACKSCHEWITZ, K.S., MONTELEONE, B.D. & other members of the Shipboard Scientific Party 2001. Evolution of the Miocene–Recent Woodlark Rift Basin, SW Pacific: inferred from sediments drilled during ODP Leg 180. *In*: WILSON, R.C.L., WHITMARSH, R.B., TAYLOR, B. & FROITZHEIM, N. (eds) *Non-volcanic Rifting of Continental Margins: a Comparison of Evidence from Land and Sea.* Geological Society, London, Special Publications, **187**, 335–372.

SCHMINCKE, H.-U. & BOGAARD, P.v.d. 1990. Tephra layers and tephra events. *In*: EINSELE, G., RICKEN, W. & SEILACHER, A. (eds) *Cycles and Events in Stratigraphy.* Springer, Berlin, 392–429.

SMITH, I.E.M. 1976. Peralkaline rhyolites from the D'Entrecasteaux Islands, Papua New Guinea. *In*: JOHNSON, R.W. (ed.) *Volcanism in Australasia.* Elsevier, Amsterdam, 275–285.

SMITH, I.E.M. 1981. Young volcanoes in eastern Papua. *Geological Survey of Papua New Guinea Memoir*, **10**, 257–265.

SMITH, I.E.M. 1982. Volcanic evolution in eastern Papua. *Tectonophysics*, **87**, 315–333.

SMITH, I.E.M. & JOHNSON, R.W. 1981. Contrasting rhyolite suites in the Late Cenozoic of Papua New Guinea. *Journal of Geophysical Research*, B11, **86**, 10257–10272.

SMITH, I.E.M. & MILSOM, J.S. 1984. Late Cenozoic volcanism and extension in eastern Papua. *In*: KOKELAAR, B.P. & HOWELLS, M.F. (eds) *Marginal Basin Geology.* Geological Society, London, Special Publications, **16**, 163–171.

SMITH, I.E.M., CHAPPELL, B.W., WARD, G.K. & FREEMAN, R.S. 1977. Peralkaline rhyolites associated with andesitic arcs of the southwestern Pacific. *Earth and Planetary Science Letters*, **37**, 230–236.

STEIGER, R.H. & JÄGER, E. 1977. Subcommission on geochronology: convention on the use of decay constants in geo- and cosmochronology. *Earth and Planetary Science Letters*, **36**, 359–362.

STOLZ, A.J., DAVIES, G.R., CRAWFORD, A.J. & SMITH, I.E.M. Sr 1993. Nd and Pb isotopic composition of calc-alkaline silicic volcanics from D'Entrecasteaux Islands, Papua New Guinea, and their tectonic significance. *Mineralogy and Petrology*, **47**, 103–126.

TAYLOR, B., HUCHON, P., KLAUS, A. *et al.* (eds) 1999. *Proceedings of the Ocean Drilling Program, Initial Reports, 180.* Ocean Drilling Program, College Station, TX.

YORK, D. 1969. Least-squares fitting of a straight line with correlated errors. *Earth and Planetary Science Letters*, **5**, 320–324.

Micropalaeontological and palaeomagnetic approaches to stratigraphic anomalies in rift basins: ODP Site 1109, Woodlark Basin

JOHANNA M. RESIG[1], GINA M. FROST[2], NAOTO ISHIKAWA[3] & RUSSELL C.B. PEREMBO[4]

[1]*Department of Geology and Geophysics, University of Hawaii, Honolulu, HI 96822, USA*
(e-mail: jresig@soest.hawaii.edu)
[2]*Hawaii Institute of Geophysics and Planetology, University of Hawaii, Honolulu, HI 96822, USA*
[3]*School of Earth Sciences, Kyoto University, Kyoto 606-8501, Japan*
[4]*Department of Geology, University of Papua New Guinea, University PO, Box 414, Papua New Guinea*

Abstract: The stratigraphic succession of the western Woodlark Basin is examined in detail relative to ODP Site 1109, where *c.* 55 m of Pliocene sediment lies anomalously shallow owing to slumping between *c.* 1.10 and 0.65 Ma in the tectonically active rift basin. Palaeomagnetic reversals and varying percentages of the characteristic microfossils, *Globigerinoides fistulosus* and discoasters, within the deformed sediment defining the slump indicate that the Pliocene sediment was periodically introduced by a number of slumps rather than being emplaced by a single slump event. Palaeomagnetic and biostratigraphic events in the remainder of the hemipelagic section of Site 1109 to *c.* 4 Ma appear undisturbed and consistent with those at adjoining sites. *Globorotalia truncatulinoides* first appeared between 2.65 and 2.71 Ma at these low-latitude sites, which extends the previously reported area of evolution of the species in the southwestern Pacific toward the Equator. Seven of the nine coiling changes in *Pulleniatina* through time are recognized and dated using sedimentation rates. Along with palaeomagnetic events and microfossil species datum levels, these coiling changes can be used to correlate between sites in the area.

Westward extension of the Woodlark Basin spreading centre has resulted in rifting and subsidence of the eastern extremity of the Papuan peninsula over the past 6 Ma (Taylor *et al.* 1995, 1999*a*). Sediment cores retrieved from this newly formed western basin during Ocean Drilling Program (ODP) Leg 180 yielded biostratigraphically and palaeomagnetically determined age–depth profiles for a sedimentary record in which pyroclastic layers and turbidites are prominent components (Taylor *et al.* 1999*b*). This dynamic tectonic setting is stratigraphically complex. Immediately to the south of the area cored during ODP Leg 180, Barash & Kuptsov (1997) reported slumps detected through radiocarbon dating of microfossils, as well as turbidite basin-fill. Shafik & Belford (1987) found reworked nannofossils off the spreading centre to the east of the Leg 180 sites.

Slumping of Pliocene sediments at ODP Site 1109 was initially detected from seismic and palaeomagnetic stratigraphies. These slumped deposits contain diagnostic Pliocene microfossils, particularly *Globigerinoides fistulosus* and rare *Discoaster brouweri* to the exclusion of species definitive of a younger age. Although these Pliocene species were present only sporadically for *c.* 125 m below the slumped sediment, the Pliocene–Pleistocene boundary was placed at the top of the slump until it was noted that Pliocene sediment occurred lower in the section at other sites and lateral tracing of those layers did not intersect the Site 1109 section at the level palaeontologically designated. Furthermore, palaeomagnetic reversal sequences favoured the stratigraphically deeper Pliocene.

In this paper biostratigraphic calibration of the section at Site 1109 is established through the use of first and last appearance events of planktonic foraminiferal species and through the correlation of rarely used coiling changes in *Pulleniatina* coupled with palaeomagnetic

From: WILSON, R.C.L., WHITMARSH, R.B., TAYLOR, B. & FROITZHEIM, N. 2001. *Non-Volcanic Rifting of Continental Margins: A Comparison of Evidence from Land and Sea*. Geological Society, London, Special Publications, **187**, 389–404. 0305-8719/01/$15.00 © The Geological Society of London 2001.

dating. These correlations are verified with reference to other ODP sites in the area.

Pulleniatina coiling and palaeomagnetic polarity

In the modern ocean, the planktonic foraminifer, *Pulleniatina,* is limited to tropical waters, where it coils almost exclusively clockwise (dextrally) when observed from the spiral side of the test. Although its areal distribution and faunal association in upper Miocene to Pleistocene strata indicate that the lineage was restricted to the tropics historically, temporal changes in its coiling direction have occurred over the past 4 Ma (Saito 1976).

Coiling changes in various planktonic foraminiferal species were discussed at length by Boltovskoy & Wright (1976). Coiling directions have been linked to present and past temperature regimes in *Neogloboquadrina pachyderma* (Ericson 1959; Bandy 1960; Kent *et al.* 1971), *Globorotalia truncatulinoides* (Ericson *et al.* 1954) and *Globorotalia menardii* (Ericson & Wollin 1964) among others, in which predominantly left-coiled or sinistral populations are associated with cooler waters. Because *Pulleniatina* is right-coiled today, even at the extremes of its areal distribution, and has shown no coiling changes during the numerous late Pleistocene glacial events, other factors besides temperature must play a role. Bolli (1950, 1971) suggested that randomly coiled early representatives of a species or lineage assume preferred coiling directions through time. *Pulleniatina* coiling is at odds with this scenario because the longest interval of unidirectional coiling occurred early in the lineage. The stimulus for these coiling changes in the *Pulleniatina* lineage is at present unknown.

Hallock & Larsen (1979) observed that coiling in benthic, reef-associated *Amphistegina lobifera* is predominantly sinistral in the western tropical Pacific and dextral progressively eastward to Hawaii. Those workers found that, during asexual reproduction (schizogony), both dextrally and sinistrally coiled parent individuals produced young that were coiled both dextrally and sinistrally within a single clone and in the proportions shown by the local populations. Thus, the minority coiling direction is maintained irrespective of the conditions stimulating change.

Saito (1976) attempted to increase biostratigraphic resolution through the correlation of the coiling changes in *Pulleniatina* with palaeomagnetic chronology in cores from the tropical Atlantic, Indian and Pacific Oceans. He designated intervals of left coiling down section as follows: L1, shortly before the Brunhes epoch; L2, near the Jaramillo event; L3, between the Jaramillo and the Olduvai events; L4–L8, from the early Matuyama to the late Gauss epoch; L9, from the Gilbert epoch and earlier.

Stage L9 was found in three oceans, after which *Pulleniatina* disappeared from the Atlantic only to reappear in early Matuyama time with coiling ratios no longer in phase with those of the Indo-Pacific. Saito's database for the establishment of the L1–L9 events for the Indo-Pacific was small (three piston-cored sites for each ocean) and other researchers have encountered difficulties in using these left-coiling events for stratigraphic purposes (e.g. Chaproniere *et al.* 1994; Chaisson 1995). The widely recognized L9 change, however, has been dated at 3.95 Ma (Berggren *et al.* 1995).

Woodlark Basin Site 1109 (Fig. 1) is well suited for a test of Saito's correlations not only because *Pulleniatina* occurs there in sufficient abundance and stratigraphic continuity but also because a large non-carbonate component of the sediment retains the magnetic signature. Other western equatorial Pacific drill sites such as Deep Sea Drilling Project (DSDP) Site 62.1 on the Eauripik Rise and ODP Site 806 on the Ontong Java Plateau lack adequate palaeomagnetic stratigraphy, although the pattern of coiling changes at Site 62.1 (Brönnimann & Resig 1971) is similar to that in the central equatorial Pacific (Hays *et al.* 1969; Saito 1976).

Analytical methods

The ODP sites studied (Site 1109 for its anomalous stratigraphy and Sites 1118 and 1115 for comparative purposes) are located in the western Woodlark Basin north of Moresby Seamount and on the hanging wall of a normal fault along which basin extension has occurred (Fig. 1). Their coordinates (latitude, longitude) and water depths are as follows: Site 1118: 9°35.110′S, 151°34.421′E, 2304 m; Site 1109: 9°30.392′S, 151°34.390′E, 2222 m; Site 1115: 9°11.382′S, 151°34.437′E, 1149 m.

Palaeomagnetic reversal sequences for the studied ODP cores were determined aboard ship by N.I. and G.M.F. using an automated pass-through cryogenic DC-SQUID magnetometer, which permitted the measurement of cores with remanent intensities as weak as *c.* 10^{-5} A m^{-1}. Oriented discrete samples were taken from most cores, generally at two samples per core, and their remanences were measured on ship using a cryogenic magnet-

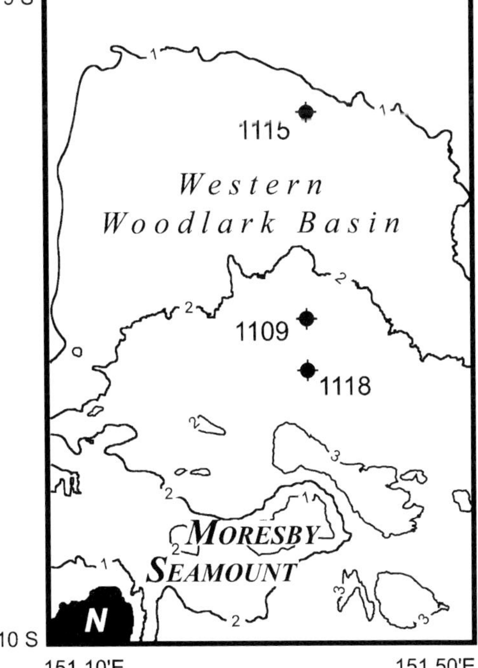

Fig. 1. Location of the western Woodlark Basin in the tropical western Pacific. ODP Sites 1115, 1109 and 1118 form a north–south basinal traverse on the hanging wall of a low-angle normal fault that has rifted the continental crust of Papua New Guinea. Bathymetric contour interval 1 km. N, Normanby Island. (Modified from Taylor *et al.* 1999*b*)

ometer. Details of sampling and analysis have been presented in the Explanatory Notes section of the Initial Reports Volume 180 (Shipboard Scientific Party 1999).

Samples for foraminiferal study were taken as plugs of 5 or 10 cm^3, generally at one sample per core section (*c.* 1.5 m sampling interval), and as core-catcher samples of variable volume. Dry weight of each sectional sample was determined (for later studies involving species abundance) before wet sieving the sample through a

0.62 mm mesh screen. For examination of the dried sand fraction, which was also weighed for determination of the percent sand, a portion was distributed evenly over a tray and coiling directions of 50 specimens of *Pulleniatina* were systematically counted microscopically. Viewed from the spiral side of the test, clockwise coiling is dextral and counterclockwise is sinistral. Occasionally, owing to detrital input and especially near the lowest-in-section occurrence of *Pulleniatina* where biofacies are shallow, only a small number of specimens were present in the available samples. These low counts are so identified in the data tables available from the Society Library or the British Library Document Supply Centre, Boston Spa, Wetherby, West Yorkshire LS23 7BQ, UK, as Supplementary Publication No. SUP 18160 (13 pages).

For each sample, a search was made for species with previously dated (Berggren *et al.* 1995) evolution and extinction events. In most cases, these dated events corresponded closely to the dates derived from the calculation of sedimentation rates between palaeomagnetic reversals. In a few instances, *in situ* ages were determined from these sedimentation rate calculations.

Results

Site 1109

Palaeomagnetic and foraminiferal biostratigraphic chronologies and species datum events as well as the sinistral or left-coiling (L) intervals of *Pulleniatina* are shown in Figure 2 for Holes 1109C and 1109D. The overall palaeobathymetry at the site shows subsidence from a doleritic basement and overlying subaerial conglomerate, sandstone and siltstone between 700 and 746 m below the sea floor (mbsf), through neritic to upper, middle and lower bathyal depths of 2211 m, the present water depth (Taylor *et al.* 1999*b*). At 591 mbsf the ancient water depth was sufficient to permit the first influx of *Pulleniatina*, after which its distribution was essentially continuous to the top of the section.

The stratigraphy of the section that includes *Pulleniatina* has been affected to some degree by three disturbances: (1) the oldest, a fault zone at 360–362 mbsf, may have displaced by up to 10 m the N21–N20 planktonic foraminiferal zone boundary, as defined by the first appearance datum (FAD) of *Globorotalia tosaensis*; (2) an influx of shallow detritus at 219–233 mbsf in the top part of the interval defining the Olduvai event has expanded the relative proportion of that part of the section;

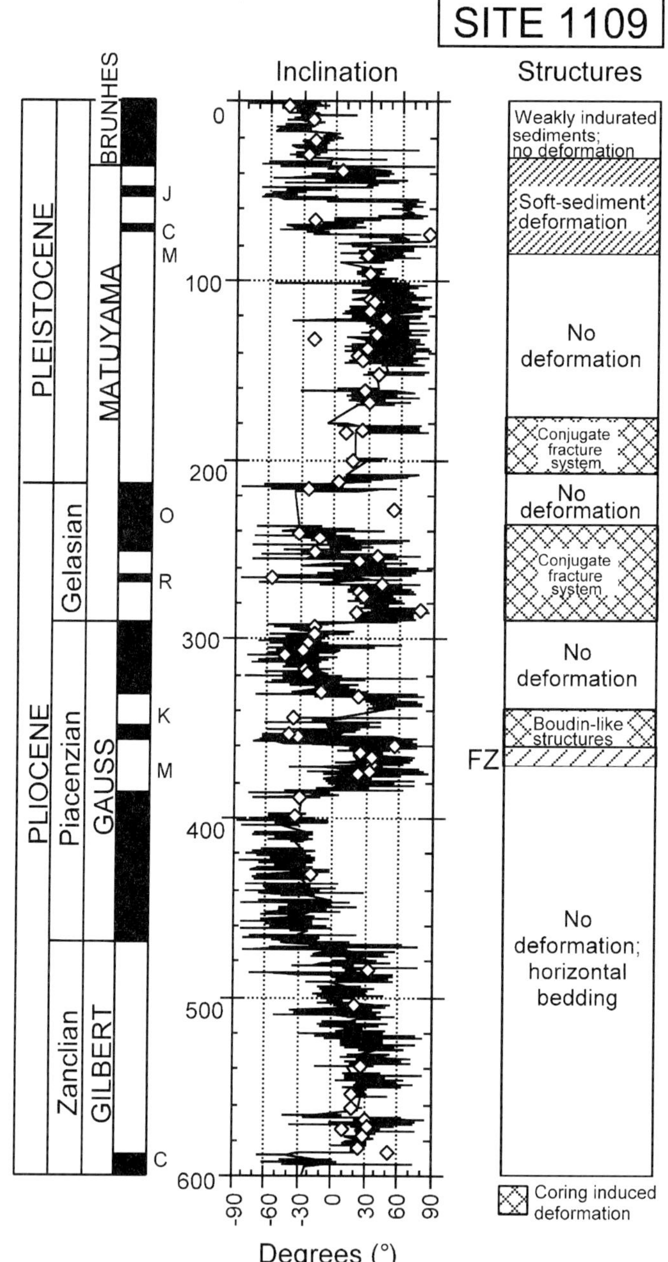

SITE 1109
Inclination
Structures
BRUNHES
PLEISTOCENE
MATUYAMA
J
C
M
Gelasian
O
R
PLIOCENE
Piacenzian
GAUSS
K
M
Zanclian
GILBERT
C
FZ
0
100
200
300
400
500
600
-90
-60
-30
0
30
60
90
Degrees (°)
Weakly indurated sediments; no deformation
Soft-sediment deformation
No deformation
Conjugate fracture system
No deformation
Conjugate fracture system
No deformation
Boudin-like structures
No deformation; horizontal bedding
Coring induced deformation

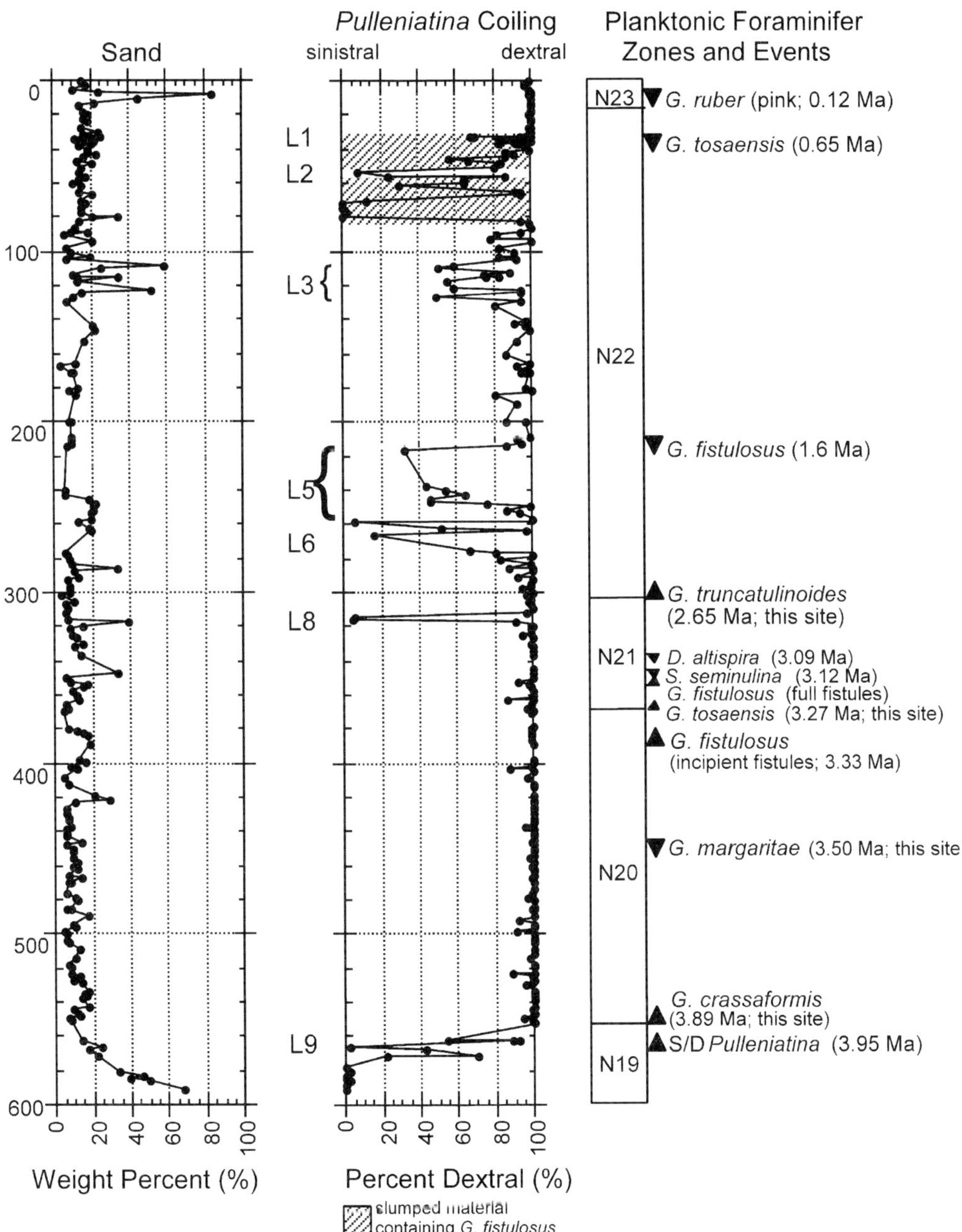
Sand
Pulleniatina Coiling
sinistral
dextral
Planktonic Foraminifer
Zones and Events
0
100
200
300
400
500
600
L1
L2
L3
L5
L6
L8
L9
N23
N22
N21
N20
N19
G. ruber (pink; 0.12 Ma)
G. tosaensis (0.65 Ma)
G. fistulosus (1.6 Ma)
G. truncatulinoides
(2.65 Ma; this site)
D. altispira (3.09 Ma)
S. seminulina (3.12 Ma)
G. fistulosus (full fistules)
G. tosaensis (3.27 Ma; this site)
G. fistulosus
(incipient fistules; 3.33 Ma)
G. margaritae (3.50 Ma; this site)
G. crassaformis
(3.89 Ma; this site)
S/D Pulleniatina (3.95 Ma)
0
20
40
60
80
100
Weight Percent (%)
0
20
40
60
80
100
Percent Dextral (%)
slumped material
containing G. fistulosus

(3) *c*. 55 m of upper Matuyama–lower Brunhes section exhibiting soft sediment deformation (29–84 mbsf) contains microfossils indicative of Pliocene age. This deformed section, originally considered by shipboard palaeontologists to represent the top of the Pliocene succession, has been shown to consist of sediment slumped over a long period of time (Taylor *et al.* 1999*b*) during which the magnetic grains were reset in later magnetic fields. Examination of foraminifera from *c*. 1.5 m intervals of the deformed section has revealed the presence of fossil *Globigerinoides fistulosus* in many of the samples, but the fistule-like chamber extensions are not well developed in some of the samples and other samples have no specimens of *G. fistulosus* at all; thus these appear to be *in situ* deposits, as indicated in Figure 2.

The polarity of the remanent magnetization was determined primarily from the inclinations. Trends within the inclination data corroborated by discrete sample analysis and intensity data facilitated the polarity interpretation at this site. The polarity reversal between 35.8 and 36 mbsf represents the Brunhes–Matuyama boundary, within the interval of soft sediment deformation mentioned above; the Jaramillo and Cobb Mountain events also occur in this deformed interval. However, correction of the palaeomagnetic data for what were perceived to be tilted and overturned beds changed the polarity and removed limited transition data indicated by the inclinations. Because of uncertainties in structural interpretation, these corrections were viewed with caution and are not used here.

The Olduvai and Reunion events are identified between *c*. 213–255 and 266–269 mbsf, respectively, in spite of poor core recovery and a scattered dataset (Taylor *et al.* 1999*b*). This interpretation is supported by seismic stratigraphic correlation between Sites 1109 and 1118 (Taylor *et al.* 1999*b*). Likewise, the base of the Kaena event is not well defined but occurs between *c*. 342 and 353 mbsf. All other polarity transitions are well defined (Fig. 2).

The stratigraphic position of first (FAD) and last (LAD) appearance datum events of planktonic foraminiferal species indicated in Figure 2 are shown with the age assignments of Berggren *et al.* (1995), except for four species events that show different ages relative to palaeomagnetic chronology. These events and their calculated ages at Site 1109, assuming uniform sedimentation rates between dated intervals, are as follows: FAD *Globorotalia truncatulinoides* 2.65 Ma; FAD *Globorotalia tosaensis* 3.27 Ma; LAD *Globorotalia margaritae* 3.50 Ma; FAD *Globorotalia crassaformis* 3.89 Ma.

Dowsett (1988) reported that *Globorotalia truncatulinoides* appeared earlier in the southwestern Pacific (Lord Howe Rise) than in equatorial Pacific and Atlantic locations. Hills & Thierstein (1989) found its age in the Indo-Pacific between 20°S and 35°S to be between 2.6 and 2.7 Ma. The Lau Basin sites between 18°S and 19°S also include the early appearance of *G. truncatulinoides* within the later part of the Gauss chron (Chaproniere *et al.* 1994). The FAD of *G. truncatulinoides* at Site 1109 is calculated to be 2.65 Ma based on sedimentation rate, but Site 1109 is at lower latitude (within 10° of the Equator) than the previously reported sites.

Of the three other planktonic foraminiferal events mentioned above, both the LAD of *Globorotalia margaritae* and the FAD of *G. crassaformis* agree with the results of Chaproniere *et al.* (1994); the FAD of *G. tosaensis*, however, occurs within the Mammoth event rather than between the Kaena and the Mammoth events, as reported by those workers.

Left-coiling intervals of *Pulleniatina* relative to the palaeomagnetic chronology of Hole 1109 are shown in Fig. 2, as follows: L1, at the start of the Brunhes chron; L2, probably at the start of the Jaramillo event (within the interval of slump deposits); L3, between the Cobb Mountain and the Olduvai events; L5, within the Olduvai event and just before; L6, within the Reunion event; L8, within the later Gauss normal chron; L9, just after the Cochiti event.

Other Woodlark Basin sites

Sections showing the commencement of rifting were penetrated at two additional Woodlark

---

Fig. 2. Palaeomagnetic and planktonic foraminiferal stratigraphy and events at ODP Site 1109 to *c*. 4 Ma. Increased sand percentage near the base of the section studied represents a shallow-water facies in which planktonic foraminifera are rare or absent. Polarity: black, normal; white, reversed. J, Jaramillo; CM, Cobb Mountain; O, Olduvai; R, Reunion; K, Kaena; M, Mammoth; C, Cochiti; FZ, fault zone. Left-coiling intervals: L1, L2, etc. Foraminifer zones: N23, N22, etc.

Basin sites, Sites 1115 and 1118 (Taylor *et al.* 1999*b*). At Site 1115, *c.* 35 km north of Site 1109, drilling deformation and poor recovery of the lower section studied here created stratigraphic problems. Site 1118, 9 km south of 1109, was drilled to 205 mbsf without coring. The proximity of the site to Moresby Seamount resulted in the foraminiferal components being diluted by detrital sediment. Between the two sections, however, most of the stratigraphic events recorded at Site 1109 are discernible.

Site 1115. Approximately 230 m of upward shallowing middle Miocene Trobriand forearc basin sediments were penetrated at this site. The sea floor had emerged, as indicated by an unconformity at 574 mbsf, and then subsequently subsided to its present depth of 1150 m. The site was sufficiently submerged at 456 mbsf to contain the first populations of *Pulleniatina*, which are thereafter continuously present to the top of the section.

Palaeomagnetic and biostratigraphic events for Holes 1115C and 1115D are shown in Figure 3. The palaeomagnetic reversal sequence is well displayed in the holes; however, the Gauss chron is of unusually long vertical extent and the Mammoth subchron does not show up in the palaeomagnetic data. Drilling deformation is present between *c.* 264 and 370 mbsf and in the lower part of the Gauss chron, between *c.* 312 and 331 mbsf, core recovery was low. The LAD of *Globorotalia margaritae* and the FAD of *Globigerinoides fistulosus* occur in reverse of the accepted order in this low recovery section and below it. With the exception of these two foraminiferal datum levels, foraminiferal events and palaeomagnetic chronologies are in agreement.

The following intervals of left-coiling *Pulleniatina* relative to palaeomagnetic chronology were recognized at Site 1115: L1, just before the Brunhes chron; L3, between the Cobb Mountain and the Olduvai events; L5, within the Olduvai event and just before; L6, just before the Reunion event; L8, within the later Gauss normal chron; L9, within the later Gilbert chron.

Site 1118. This site, located 9 km south of Site 1109, lies relatively closer to Moresby Seamount and has higher sedimentation rates because of the input of detrital sand. At the base of the site at 873 mbsf, a doleritic con-

glomerate indicative of subaerially exposed basement rocks is succeeded by a sequence of marine sediments deposited in progressively deeper water leading to the present water depth of 2304 m (Taylor *et al.* 1999*b*). The top 250 m at this site was not cored and recovery below 660 m was, for the most part, relatively poor. The paucity of *Pulleniatina* in many samples at this site, particularly in the lower section, may be attributed to dilution by a high influx of sand derived from a volcanic terrane. Soft sediment deformation was noted between *c.* 272 and 584 mbsf (Taylor *et al.* 1999*b*).

Palaeomagnetic chronology and foraminiferal event and coiling data for Site 1118 are shown in Fig. 4. Coring began here near the end of the Olduvai event in deposits that contain *Globigerinoides fistulosus* and dextral *Pulleniatina*. Left-coiling intervals of *Pulleniatina* relative to palaeomagnetic chronology are as follows: L5, within the Olduvai event and just before; L6, later than and in the Reunion event; L8, within the later Gauss normal chron.

The FAD of *Globorotalia truncatulinoides* at 449.9 mbsf marks the base of planktonic foraminiferal zone N22, calculated to be 2.71 Ma at this site. At 456.85 mbsf, a single isolated sample with predominantly left-coiling *Pulleniatina* is grouped here with L8. Although L8 shows multiple left-coiling intervals at Site 1115, none of them stand far removed from the others. These isolated left-coiled individuals may be reworked or they may be separated from the majority of L8 by turbidite sedimentation. Below this interval at 609 mbsf lies the LAD of *D. altispira*.

Discussion

A summary of palaeomagnetic and planktonic foraminiferal events for Site 1109 relative to Sites 1118 and 1115 is shown in Table 1. At Site 1115, the Cochiti subchron (Berggren *et al.* 1995: C3n.1n; 4.18–4.29 Ma), although not identified from the palaeomagnetic data, probably occurred between *c.* 460 and 480 mbsf in an interval of low sediment recovery, as indicated by the L9 (3.95 Ma) coiling event just above that depth. Also at Site 1115, the Mammoth subchron is not defined from the palaeomagnetic data and the appearances of *Globigerinoides fistulosus* and *Globorotalia*

Fig. 3. Palaeomagnetic and planktonic foraminiferal stratigraphy and events at ODP Site 1115 to *c.* 4 Ma. Lithification and poor recovery have resulted in loss of data in the lower portion of the section. Abbreviations as in Figure 2.

J.M. RESIG *ET AL.*

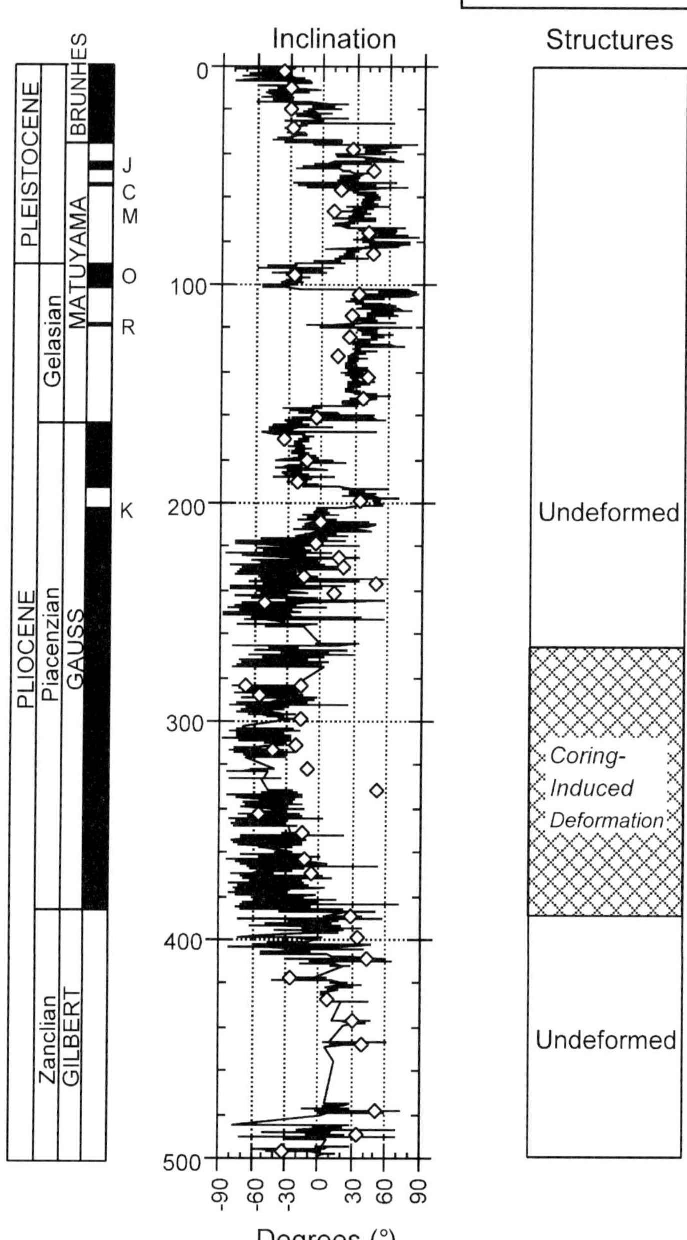

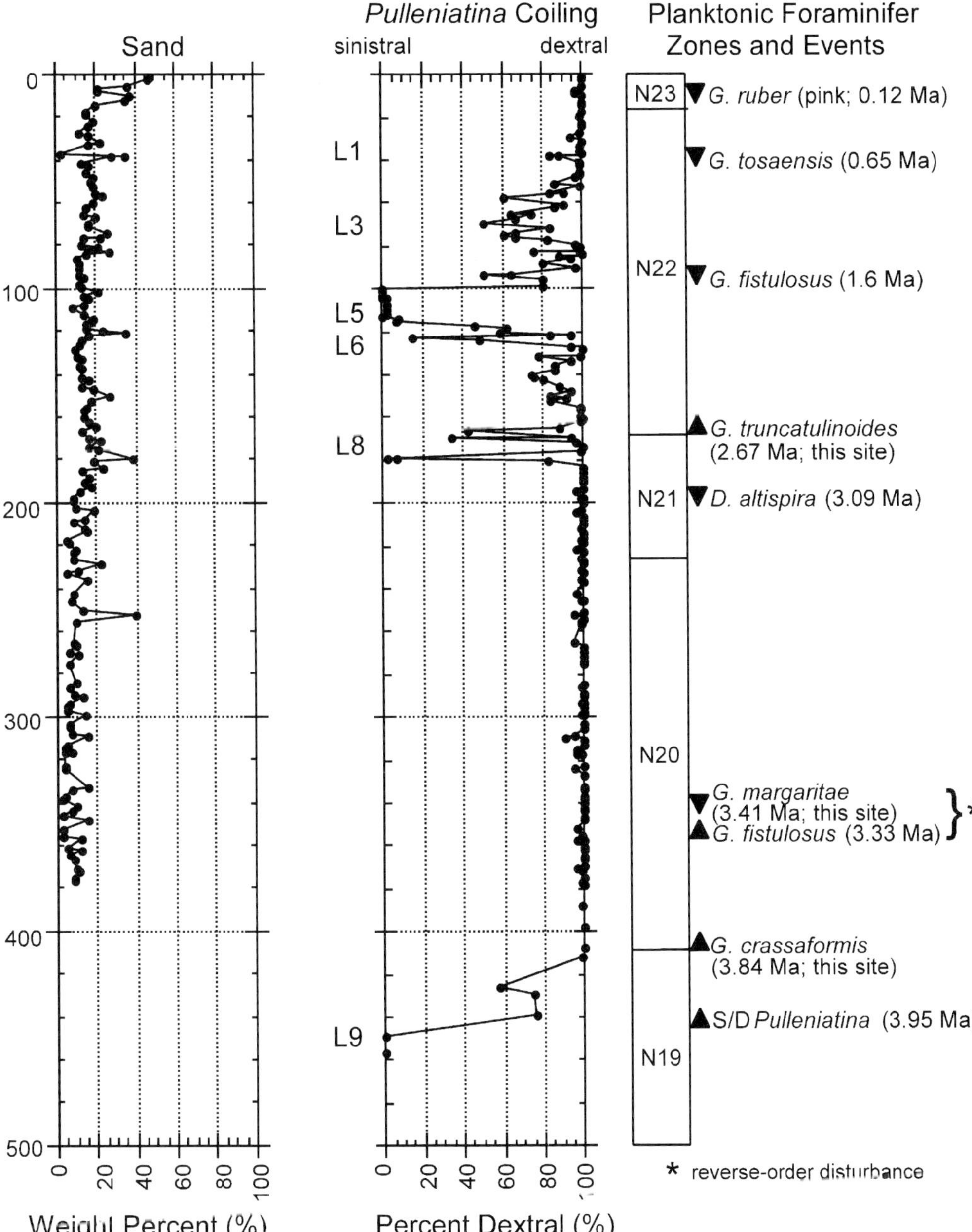
Sand
Pulleniatina Coiling
sinistral
dextral
Planktonic Foraminifer
Zones and Events
0
L1
L3
100
L5
L6
L8
200
300
400
L9
500
0
20
40
60
80
100
Weight Percent (%)
0
20
40
60
80
100
Percent Dextral (%)
N23
G. ruber (pink; 0.12 Ma)
G. tosaensis (0.65 Ma)
N22
G. fistulosus (1.6 Ma)
G. truncatulinoides
(2.67 Ma; this site)
N21
D. altispira (3.09 Ma)
N20
G. margaritae
(3.41 Ma; this site)
G. fistulosus (3.33 Ma)
}*
G. crassaformis
(3.84 Ma; this site)
S/D Pulleniatina (3.95 Ma)
N19
* reverse-order disturbance

Table 1. Depth in hole (mbsf) and age (Ma) of events at Sites 1118, 1109 and 1115

Event	Site 1118, 9°35'S Depth (mbsf) Low	High	Av.	Age (Ma)	Site 1109, 9°30'S Depth (mbsf) Low	High	Av.	Age (Ma)	Site 1115, 9°11'S Depth (mbsf) Low	High	Av.	Age (Ma)
LAD G. ruber (pink)		Not recovered		–	5.39	2.43	3.91	0.12	5.44	2.45	3.95	0.12
LAD G. tosaensis		Not recovered		–	32.62	32.50	32.56	0.65	33.15	31.65	32.4	0.65
Brunhes–Matuyama		Not recovered		–	36	35.8	35.9	0.78	36	33.5	34.75	0.78
L1, Top		Not recovered			32.62	32.50	32.56	0.65	37.78	36.65	37.22	0.84
L1, Bottom		Not recovered		–	32.95	32.80	32.87	0.66	38.15	37.78	37.96	0.86
Jaramillo, Top		Not recovered		–	48	47.5	47.8	0.99	43.5	43.5	43.5	0.99
Jaramillo, Bottom		Not recovered		–	54	53.5	53.8	1.07	47	47	47	1.07
L2, Top		Not recovered		–	53.85	50.85	52.35	1.06		Not identified		–
L2, Bottom		Not recovered		–	55.85	55.49	55.67	1.10		Not identified		–
Cobb Mountain, Top		Not recovered		–	70.5	70.5	70.5	1.201	53	53	53	1.201
Cobb Mountain, Bottom		Not recovered		–	74	74	74	1.211	54.5	54.5	54.5	1.211
L3, Top		Not recovered		–	109.36	107.78	108.57	1.35	57.15	55.65	56.40	1.24
L3, Bottom		Not recovered		–	122.54	121.66	122.1	1.40	79.15	77.65	78.40	1.59
LAD G. fistulosus		Not recovered		–	208.51	199.87	204.19	1.6	88.35	85.65	87.0	1.6
Olduvai, Top		Not recovered/ not identified		–	213	213	213	1.77	90.5	90.5	90.5	1.77
L5, Top	214.43	205.00	209.43	1.78[a]	216.27	213.70	214.99	1.78	92.62	90.14	91.53	1.78
L5, Bottom	306.05	301.92	303.99	2.01	263.25	261.74	262.49	2.08[b]	120.99	119.97	120.48	2.16
Olduvai, Bottom	288	288	288	1.95	255	255	255	1.95	103.5	102.5	103	1.95
Reunion, Top	332	332	332	2.14	266	266	266	2.14	118	118	118	2.14
Reunion, Bottom	334.5	334.5	334.5	2.15	269	269	269	2.15	119.5	119.5	119.5	2.15
L6, Top	322.78	321.25	324.32	2.11[b]	266.01	263.25	264.63	2.11[b]	122.14	120.99	121.57	2.17

Table 1. *Continued*

Event	Site 1118, 9°35'S Depth (mbsf) Low	High	Av.	Age (Ma)	Site 1109, 9°30'S Depth (mbsf) Low	High	Av.	Age (Ma)	Site 1115, 9°11'S Depth (mbsf) Low	High	Av.	Age (Ma)
L6, Bottom	348.37	345.39	346.88	2.24	276.54	275.50	276.02	2.29[c]	126.64	123.65	125.15	2.21
Matuyama–Gauss	387.5	387.5	387.5	2.58	291.5	290	290.75	2.58	162	162	162	2.58
FAD *G. truncatulinoides*	456.85	449.93	453.39	2.71	297.23	295.75	296.49	2.65	169.16	166.15	167.66	2.67
L8, Top	456.85	455.35	456.10	2.72	314.23	311.32	312.78	2.82	166.15	164.65	165.4	2.63
L8, Bottom	551.74	549.75	550.75	2.91	316.65	315.15	315.9	2.86	180.65	179.15	179.9	2.85
Kaena, Top	599	599	599	3.04	331	331	331	3.04	192.5	192.5	192.5	3.04
LAD *D. altispira*	608.2	595.72	601.96	3.09	333.83	330.85	332.34	3.09	193.15	190.15	191.65	3.09
Kaena, Bottom	623	623	623	3.11	353	342	347.5	3.11	202	202	202	3.11
Mammoth, Top	662.5	662.5	662.5	3.22	358	358	358	3.22		Not identified		–
Mammoth, Bottom	744.5	744.5	744.5	3.33	384.5	384.5	384.5	3.33		Not identified		–
FAD *G. tosaensis*	695.27	688.26	691.77	3.24	369.14	367.83	368.49	3.27	236.35	232.76	234.56	3.19
LAD *G. margaritae*	840.54	835.32	837.93	3.56	444.49	442.08	443.29	3.50	323.90	322.52	323.21	3.41
Gauss–Gilbert	849.5	846	847.8	3.58	472	470	471	3.58	387	386	386.5	3.58
FAD *G. crassaformis*	Not identified, facies change			–	551.00	548.80	549.9	3.89	412.07	407.31	409.69	3.84
L9 = S/D *Pulleniatina*	Not identified, facies change			–	566.55	562.71	564.63	3.95	425.81	412.07	418.94	3.95
Cochiti, Top	Not identified, facies change			–	587	587	587	4.18		Not identified		–

[a]Based on age at Sites 1109 and 1115.
[b]Based on sedimentation rate between Bottom Olduvai and Top Reunion.
[c]Based on sedimentation rate between Bottom Reunion and Matuyama–Gauss boundary.
mbsf, metres below sea floor, given as the low and high depth, one of which is the sample depth in which the event was first or last recorded, and the other, the sample depth above or below which lacks the event, the average depth between them was used to represent the depth of the event in Figs 2–4; Av., average; LAD, last appearance datum; FAD, first appearance datum; S/D, sinistral/dextral; L1, L2, etc., left-coiling *Pulleniatina* event.

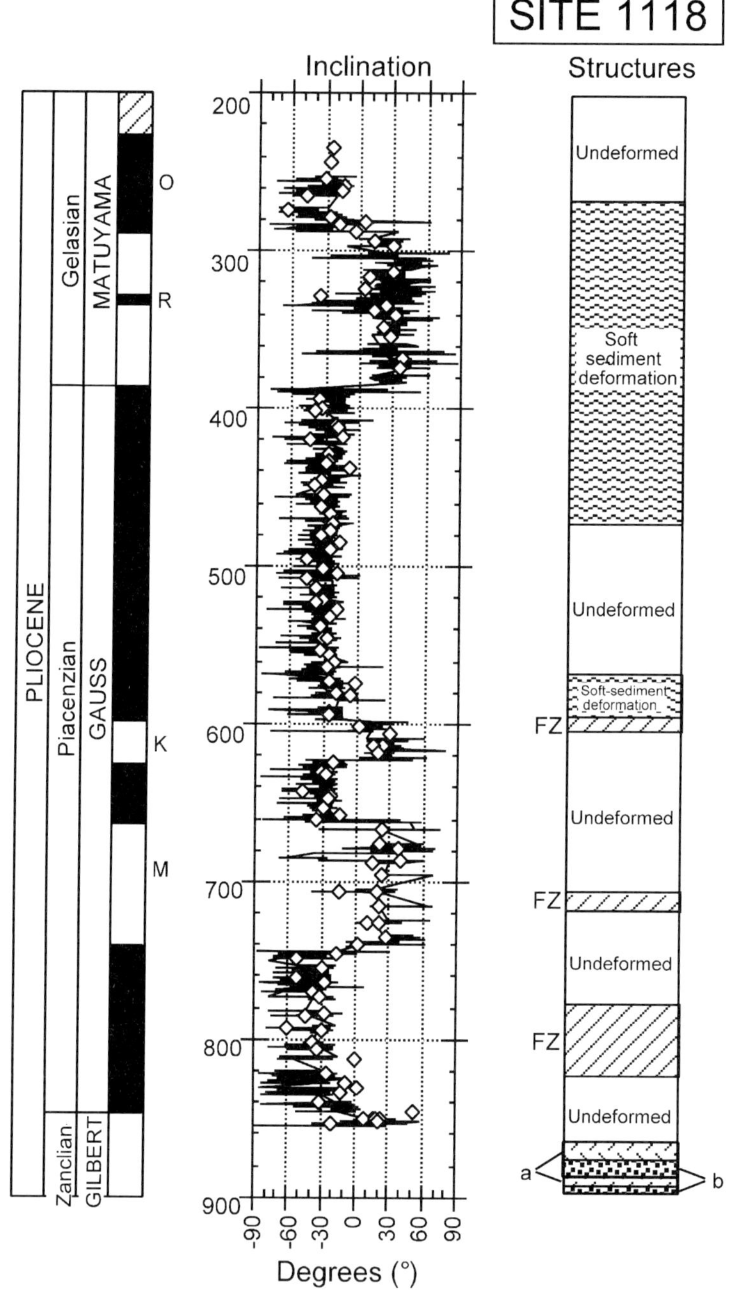
SITE 1118
Inclination
Structures
Undeformed
Soft
sediment
deformation
Undeformed
Soft-sediment
deformation
FZ
Undeformed
FZ
Undeformed
FZ
Undeformed
a
b
PLIOCENE
Gelasian
Piacenzian
Zanclian
MATUYAMA
GAUSS
GILBERT
O
R
K
M
200
300
400
500
600
700
800
900
-90
-60
-30
0
30
60
90
Degrees (°)

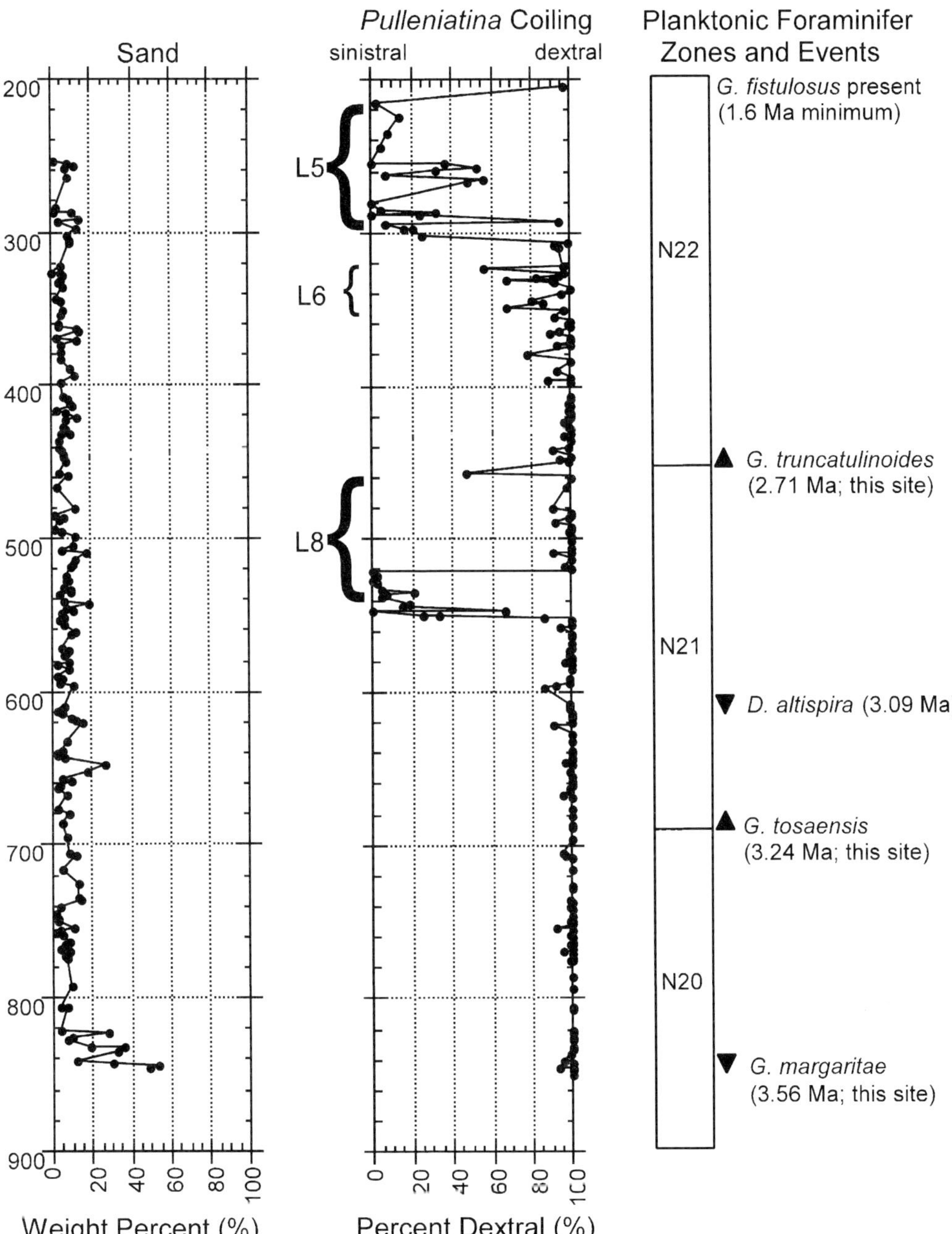
Sand
Pulleniatina Coiling
sinistral
dextral
Planktonic Foraminifer
Zones and Events
G. fistulosus present
(1.6 Ma minimum)
L5
L6
L8
N22
N21
N20
G. truncatulinoides
(2.71 Ma; this site)
D. altispira (3.09 Ma)
G. tosaensis
(3.24 Ma; this site)
G. margaritae
(3.56 Ma; this site)
200
300
400
500
600
700
800
900
0
20
40
60
80
100
Weight Percent (%)
0
20
40
60
80
100
Percent Dextral (%)

margaritae are in reverse of their known stratigraphic order, within an interval of coring-induced deformation. Among the foraminiferal evolutionary events, the FAD of *Globorotalia truncatulinoides* ranges from 2.65 to 2.71 Ma at the three sites, where the difference in age may have resulted from the averaging of sedimentation rates between the bounding palaeomagnetic reversals. In any case, these ages are consistently older than the 2.00 Ma age (Berggren *et al.* 1995) attributed to the evolution of the species outside of the southwestern tropical Pacific.

These data indicate a stratigraphically ordered sequence of deposition at Site 1109, except for the anomalous Pliocene deposits high in the section. Table 1 also shows a fair amount of correspondence among the left-coiling *Pulleniatina* intervals at the three sites. However, L1 at Site 1109 occurs above the Brunhes–Matuyama boundary rather than below, which is its position at Site 1115 and as described by Saito (1976). This younger occurrence is at the top of the deformed sediment that marks the slump deposits where prolonged sediment settlement and reworking may have been involved.

Table 2 compares the calculated ages of left-coiling *Pulleniatina* at the three Woodlark Basin sites with those calculated for the Pacific sites of Saito (1976), based on the graphic data of Hays *et al.* (1969), Saito *et al.* (1975), and Saito (1976), converted to the time scale of Berggren *et al.* (1995). These results, as well as the downcore coiling patterns, should be viewed with several caveats.

Saito (1976) described the nine left-coiling intervals in *Pulleniatina* from sections of 11–28 m length of central equatorial Pacific piston cores in which the *c.* 10 cm sampling interval converts to about 80–110 ka resolution during the Brunhes chron. In contrast, the *Pulleniatina*-bearing section at the ODP sites under study is 460–600 m long, but the *c.* 1.5 m interval between samples converts to about 50– 60 ka resolution during the Brunhes chron. Thus, differences in detail of left-coiling intervals between this study and Saito's may be expected owing to the greatly expanded section and higher stratigraphic resolution at the ODP sites. Short-term fluctuations in coiling would tend to

be averaged out in the more condensed sections. Also of note are some differences in age range of left-coiling events among the cores of Saito's Pacific sites, as calculated from sedimentation rates between palaeomagnetic events. L4 is weakly defined in one core only; L1 and L7 are short events.

Ages of the left-coiling intervals of the Woodlark Basin sites generally fall within the range exhibited by Saito's cores; however, L4 (weakly defined in Saito's dataset) and L7 (a short interval in Saito's dataset) were not identified, and the calculated age of L1 is somewhat younger at Site 1109 than at Sites 1115 and 1118. Whereas Berggren *et al.* (1995) indicated an age of 3.95 Ma for the initial left to right coiling change of *Pulleniatina*, calculations based on sedimentation rate indicate an age somewhat greater than 4.0 Ma for Saito's cores and for Site 1109.

The close comparison of ages of the coiling changes in *Pulleniatina* between the central equatorial Pacific and Woodlark Basin cores indicate that most of the originally designated intervals are useful stratigraphic markers for correlation purposes. Correlation between the Woodlark Basin sites using these markers as well as palaeomagnetic events and foraminiferal species datum levels is in progress and will be presented elsewhere.

Conclusions

Tectonically active rift basins are prone to stratigraphic problems, exemplified by the anomalously shallow occurrence of Pliocene sediment at Woodlark Basin ODP Site 1109. The tripartite stratigraphic approach followed here, involving palaeomagnetic reversals, first and last appearance datum levels of planktonic foraminifera, and coiling changes in *Pulleniatina*, has established a precise chronology of Site 1109, defined the portion of section affected by Pliocene slump material, and shown the remainder of the section down to *c.* 4 Ma to be stratigraphically consistent with other sites in the basin.

The FAD of *Globorotalia truncatulinoides* is between 2.65 and 2.71 Ma at the basin sites, extending northward the previously reported area of origin of this species in the southwestern Pacific to within 10° of the Equator. Other

---

Fig. 4. Palaeomagnetic and planktonic foraminiferal stratigraphy and events at ODP Site 1118. The uppermost *c.* 200 m was not recovered at this site and a facies change to shallow-water deposition occurred before the initial coiling change in *Pulleniatina*. Abbreviations as in Figure 2. Polarity: diagonal lines indicate no data. Structures: a, unbrecciated; b, brecciated.

Table 2. *Comparison of ages (Ma) of Pulleniatina left-coiling events for equatorial Pacific sites*

Coiling event	Central equatorial Pacific sites[a]			Woodlark Basin sites[b]		
	V 24-58	V 24-59	RC 12-66	1118	1109	1115
L1	0.85	c. 0.85	c. 0.78	Not cored	0.65–0.66	0.84–0.86
L2	0.96–1.15	0.99–1.28	1.06–1.20	Not cored	1.06–1.10	Not identified
L3	1.46–1.58	c. 1.49	1.32–1.48	Not cored	1.35–1.40	1.24–1.59
L4	Not identified	Not identified	c. 1.70	Not identified	Not identified	Not identified
L5	1.77–2.09	1.77–2.05	1.77–2.09	1.78–2.01	1.78–2.08	1.78–2.16
L6	Not cored	2.15–2.36	2.16–2.44	2.11–2.24	2.11–2.29	2.17–2.21
L7	Not cored	c. 2.5	c. 2.5	Not identified	Not identified	Not identified
L8	Not cored	2.58–2.72	2.58–2.81[c]	2.72–2.91	2.82–2.86	2.63–2.85
L9	Not cored	c. 4.03	c. 4.18	Facies change	3.95[d]	3.95[d]
					4.05[e]	

[a]Data from Hays *et al.* (1969), Saito *et al.* (1975) and Saito (1976).
[b]Data from this study (refer to Table 1).
[c]Lower limit based on Saito *et al.* (1975).
[d]Based on Berggren *et al.* (1995).

species appearance events in the Woodlark Basin that appear to be shared with the southwestern Pacific pool include the LAD of *Globorotalia margaritae* (3.5 Ma) and the FAD of *G. crassaformis* (3.9 Ma).

The chronology of coiling changes in *Pulleniatina* as reported by Saito (1976) is recognized in the Woodlark Basin except for events L4 and L7. These changes provide additional means of confirming the stratigraphy of the basin sections and correlating between them. Owing to higher sedimentation rates, these coiling fluctuations are more detailed in the Woodlark Basin than in the more condensed sections in which they were originally described. The high detrital sediment contribution to the rift basin has preserved the palaeomagnetic reversal record at this low-latitude location and permitted microfossil events to be dated more precisely than in areas of slow deposition.

Part of the data for this study resulted from the shipboard collaboration of our disciplinary colleagues W. Siesser and K. Takahashi, as well as the other ODP Leg 180 scientific participants and crew of the R.V. *JOIDES Resolution.* The reviewers, M. Leckie, R.C. L. Wilson and W. Wise, provided advice on clarifying parts of the manuscript. This research was funded through the National Science Foundation's Ocean Drilling Program. The aid of these individuals and institutions is gratefully acknowledged. This is University of Hawaii SOEST Contribution 5261.

References

BANDY, O.L. 1960. The geologic significance of coiling ratios in the foraminifera *Globigerina pachyderma* (Ehrenberg). *Journal of Palaeontology*, **34**, 671–681.

BARASH, M.S. & KUPTSOV, V.M. 1997. Late Quaternary palaeoceanography of the western Woodlark Basin (Solomon Sea) and Manus Basin (Bismarck Sea), Papua New Guinea, from planktic foraminifera and radiocarbon dating. *Marine Geology*, **142**, 171–187.

BERGGREN, W.A., HILGEN, F.J., LANGEREIS, C.G. & 5 others 1995. Late Neogene chronology: new perspectives in high-resolution stratigraphy. *Geological Society of America Bulletin*, **107**, 1272–1287.

BOLLI, H.M. 1950. The direction of coiling in the evolution of some Globorotaliidae. *Cushman Foundation for Foraminiferal Research, Contributions*, **1**, 82–89.

BOLLI, H.M. 1971. The direction of coiling in planktonic foraminifera. *In*: FUNNELL, B.M. & RIEDEL, W.R. (eds) *The Micropalaeontology of the Oceans*. Cambridge University Press, London, 105–149.

BOLTOVSKOY, E. & WRIGHT, R. 1976. *Recent Foraminifera*. W. Junk, The Hague, 358–365.

BRÖNNIMANN, P. & RESIG, J. M. 1971. A Neogene globigerinacean biochronologic time-scale of the southwestern Pacific. *In*: WINTERER, E. L. *et al.* (eds) *Initial Results of the Deep Sea Drilling Project, 7*. US Government Printing Office, Washington, DC, 1235–1469.

CHAISSON, W. 1995. Planktonic foraminiferal assemblages and palaeoceanographic change in the trans-tropical Pacific Ocean: a comparison of west (Leg 130) and east (Leg 138), latest Miocene to Pleistocene. *In*: PISIAS, N.G., MAYER, L.A., JANACEK, T.R., PALMER-JULSON, A. & VAN ANDEL, T.H. (eds) *Proceedings of the Ocean Drilling Program, Scientific Results, 138*. Ocean Drilling Program, College Station, TX, 555–597.

CHAPRONIERE, G. C. H., STYZEN, M. J., SAGER, W. W., NISHI, H., QUINTERNO, P. J. & ABRAHAM-

SEN, N. 1994. Late Neogene biostratigraphic and magnetostratigraphic synthesis, Leg 135. *In:* HAWKINS, J., PARSON, L., ALLAN, J. *et al.* (eds) *Proceedings of the Ocean Drilling Program, Scientific Results, 135*. Ocean Drilling Program, College Station, TX, 857–877.

DOWSETT, H.J. 1988. Diachrony of late Neogene microfossils in the southwest Pacific Ocean: application of the graphic correlation method. *Palaeoceanography*, **3**, 209–222.

ERICSON, D.B. 1959. Coiling direction of *Globigerina pachyderma* as a climatic index. *Science*, **130**, 219–220.

ERICSON, D.B. & WOLLIN, G. 1964. *The Deep and the Past*. Knopf, New York.

ERICSON, D.B., WOLLIN, G. & WOLLIN, J. 1954. Coiling direction of *Globorotalia truncatulinoides* in deep-sea cores. *Deep-Sea Research*, **2**, 152–158.

HALLOCK, P. & LARSEN, A.R. 1979. Coiling direction in *Amphistegina. Marine Micropalaeontology*, **4**, 33–44.

HAYS, J.D., SAITO, T., OPDYKE, N.D. & BURCKLE, L.H. 1969. Pliocene–Pleistocene sediments of the equatorial Pacific: their palaeomagnetic, biostratigraphic, and climatic record. *Geological Society of America Bulletin*, **80**, 1481–1514.

HILLS, S.J. & THIERSTEIN, H.R. 1989. Plio-Pleistocene calcareous plankton biochronology. *Marine Micropalaeontology*, **14**, 67–96.

KENT, D., OPDYKE, N.D. & EWING, M. 1971. Climatic change in the North Pacific using ice-rafted detritus as a climatic indicator. *Geological Society of America Bulletin*, **82**, 2741–2754.

SAITO, T. 1976. Geologic significance of coiling direction in the planktonic foraminifera *Pulleniatina. Geology*, **4**, 305–309.

SAITO, T., BURKLE, L.H. & HAYS, J.D. 1975. Late Miocene to Pleistocene biostratigraphy of equatorial Pacific sediments. *In:* SAITO, T. & BURKLE, L.H. (eds) *Late Neogene Epoch Boundaries*. Micropalaeontology Press, New York, 226–244.

SHAFIK, S. & BELFORD, D.J. 1987. Quaternary calcareous microfossils in deep-sea samples from the Woodlark–Solomons region, tropical Pacific. *In:* TAYLOR, B. & EXON, N.F. (eds) *Marine Geology, Geophysics, and Geochemistry of the Woodlark Basin–Solomon Islands*. Circum-Pacific Council for Energy and Mineral Resources Earth Science Series, **7**, 347–363.

SHIPBOARD SCIENTIFIC PARTY 1999. Explanatory notes. *In:* TAYLOR, B., HUCHON, P., KLAUS, A. *et al.* (eds) *Proceedings of the Ocean Drilling Program, Initial Reports, 180*. Ocean Drilling Program, College Station, TX, 1–75.

TAYLOR, B., GOODLIFFE, A.M. & MARTINEZ, F. 1999*a*. How continents break up: insights from Papua New Guinea. *Journal of Geophysical Research*, **104**, 7497–7512.

TAYLOR, B., GOODLIFFE, A., MARTINEZ, F. & HEY, R. 1995. Continental rifting and initial sea-floor spreading in the Woodlark Basin. *Nature*, **374**, 534–537.

TAYLOR, B., HUCHON, P., KLAUS, A. *et al.* (eds) 1999*b*. *Proceedings of the Ocean Drilling Program, Initial Reports, 180*. Ocean Drilling Program, College Station, TX, 1–77.

The role of detachment faulting in the formation of an ocean–continent transition: insights from the Iberia Abyssal Plain

G. MANATSCHAL[1,2], N. FROITZHEIM[3], M. RUBENACH[4] & B.D. TURRIN[5]

[1]*Geologisches Institut, ETH Zürich, CH-8092 Zürich, Switzerland*
[2]*Present address: EOST-CGS (UMR 7517) CNRS, Université Louis Pasteur, 1 rue Blessig, F-67064 Strasbourg Cedex, France (e-mail: manatschal@illite.u-strasbg.fr)*
[3]*Geologisches Institut, Universität Bonn, Nussallee 8, D-53115 Bonn, Germany*
[4]*Department of Earth Sciences, James Cook University, Townsville, 4811 Qld., Australia*
[5]*Lamont–Doherty Earth Observatory, Route 9W, Palisades, NY 10964, USA*

Abstract: The Iberia Abyssal Plain segment of the West Iberia margin was drilled during Ocean Drilling Program Legs 149 and 173 and has been extensively studied geophysically. We present new microstructural investigations and new age data. These, together with observed distribution of upper- and lower-crustal and mantle rocks along the ocean–continent transition suggest the existence of three detachment faults, one of which was previously unrecognized. This information, together with a simple kinematic inversion of the reinterpreted seismic section Lusigal 12, allows discussion of the kinematic evolution of detachment faulting in terms of the temporal sequence of faulting, offset along individual faults, and thinning of the crust during faulting. Our study shows that the detachment structures recognized in the seismic profile became active only during a final stage of rifting when the crust was already considerably thinned to *c.* 12 km. The total amount of extension accommodated by the detachment faults is of the order of 32.6 km corresponding to a β factor of about two. During rifting, the mode of deformation changed oceanwards. Initial listric faulting led to asymmetric basins, accommodating low amounts of extension, and was followed by a situation in which the footwall was pulled out from underneath a relatively stable hanging wall accommodating high amounts of extension. Deformation along the latter faults resulted in a conveyor-belt type sediment accumulation in which the exhumed footwall rocks were exposed, eroded and redeposited along the same active fault system.

Low-angle detachment systems have been recognized to play an important role in continental extensional tectonics, such as in the metamorphic core complexes in the SW United States (e.g. Wernicke 1981; Davis & Lister 1988). Lister *et al.* (1986), Boillot *et al.* (1987) and Lemoine *et al.* (1987) proposed that low-angle detachment faults occur also in rifted continental margins. The main arguments for such an inference were the occurrence of prominent subhorizontal reflections within the basement of the distal margin and ocean–continent transition, the exposure of mantle rocks at the sea floor, and the asymmetry of conjugate pairs of margins. However, because no directly observable examples of rift-related low-angle detachment faults were known from continental margins, the validity of these models remains in dispute.

In recent years, several examples of exposed rift-related detachment faults have been described from the ancient Tethyan margins in the Alps (Froitzheim & Eberli 1990; Florineth & Froitzheim 1994; Froitzheim & Manatschal 1996; Manatschal & Nievergelt 1997) and from the Red Sea margin (Talbot & Ghebreab 1997). These studies yielded important information about the kinematic, sedimentary and metamorphic evolution of rift-related detachment systems. In the Tethyan margins the detachment faults are shallow crustal structures that formed during a late stage of rifting, followed by the onset of sea-floor spreading. Along these faults extensional allochthons of upper continental crust overlie exhumed subcontinental mantle rocks forming the ocean–continent transition. Displacements of up to 10 km have been determined and shown to have occurred along thin gouge zones (Manatschal 1999). The dynamic evolution of these faults, however, is not yet fully understood.

In the active Woodlark–D'Entrecasteaux rift system in the Western Pacific, Abers *et al.* (1997) demonstrated that earthquakes associated with normal faulting occurred at depths <10 km in the brittle regime in an area with a seismically well-imaged reflection dipping 25–30° (Abers 2001). This reflection has been drilled recently during Ocean Drilling Program (ODP) Leg 180 (Taylor *et al.* 1999). The drilling results show that it represents a fault marked by a gouge zone that is exposed at the sea floor. This shows that, notwithstanding Andersonian mechanics (Anderson 1942), normal faults can be active at angles of 25–30° within the brittle upper crust. Global positioning system (GPS) measurements between two islands located on the footwall and hanging wall, respectively, indicate very fast extension across the system (*c.* 30 mm a^{-1}; Tregoning *et al.* 1998). Compared with many fossil detachment faults observed on land, which have dip angles of <10°, the normal fault in the Woodlark–D'Entrecasteaux rift is still relatively steep. Evidence for earthquakes along subhorizontal detachment faults (<10–15° dip) is still to be found (Abers 2001).

High-resolution multichannel seismic reflection experiments along present-day passive continental margins, ideally combined with deep-sea drilling, can reveal the architecture of rifting in great detail. However, even in margins where both types of data are available, interpretations of kinematic evolution remain controversial, as is the case of the southern Iberia Abyssal Plain (Fig. 1). An ocean–continent transition zone of *c.* 100 km width exists in this area, between unequivocal continental crust in the east and oceanic crust in the west. For the formation of this ocean–continent transition, four models were proposed after ODP Leg 149. Krawczyk *et al.* (1996) proposed exhumation of subcontinental mantle accommodated by at least two detachment faults dipping oceanwards, away from Iberia. Brun & Beslier (1996) assumed mantle exhumation along conjugate low-angle shear zones with a lower shear zone within the mantle being rooted beneath Iberia. Whitmarsh & Sawyer (1996) presented two models, one assuming that the ocean–continent transition was formed by thinned continental crust intruded by magmatic rocks, and the other suggesting that it was formed by very slow-spreading oceanic crust.

There are now ten ODP sites that sampled a transect perpendicular to the strike of the margin across the ocean–continent transition (Fig. 1). Furthermore, there is a wealth of seismic reflection, magnetic and gravity data from this area. These data impose strong constraints on the kinematics of continental break-up and the beginning of sea-floor spreading. In addition, insights gained from the study of former Tethyan margins in the Alps and of active rifting in the Woodlark Basin have furthered the understanding of the detachment faulting process. This increased body of knowledge allows us to reconsider the tectonic evolution of the Iberia Abyssal Plain.

In this paper we present new geological results from ODP Leg 173, which lead to a reinterpretation of the depth-migrated seismic profile Lusigal 12 across the Iberia Abyssal Plain. We propose the existence of a previously unrecognized detachment fault, the Hobby High detachment fault (HHD). We use a simple kinematic inversion of our reinterpreted Lusigal 12 profile to reconstruct the kinematic evolution of detachment faulting in terms of temporal sequence of faulting, offset along individual faults and pre-faulting crustal thickness.

Tectonic setting

The Iberia margin resulted from rifting and continental break-up between the North American and Iberian plates during Early Cretaceous time. It is a typical example of a non-volcanic, sediment-starved margin. The margin may be divided into several segments (from north to south): the Galicia segment (also known as the Galicia margin), the Iberia Abyssal Plain segment (the subject of this paper), and the Tagus Abyssal Plain segment (Fig. 1).

In its northern (Galicia margin) and middle segments (Iberia Abyssal Plain) the margin is characterized by strong shallow-dipping reflections underlying tilted blocks of continental crust: the S reflection of the Galicia margin and the H reflection of the Iberia Abyssal Plain. Researchers have interpreted these reflections in different ways, but Reston *et al.* (1996) and Krawczyk *et al.* (1996) provided strong geophysical evidence that they represent detachment faults.

Geophysical investigations and the results of three ODP Legs (103, 149 and 173), and dredging demonstrated that oceanwards of the thinned continental crust, exhumed and serpentinized mantle is exposed at the sea floor or at the base of the post-rift sedimentary cover. In the southern Iberia Abyssal Plain these mantle rocks form at least part of the >100 km wide ocean–continent transition (Fig. 1). Unequivocal oceanic crust begins only west of Site 1070.

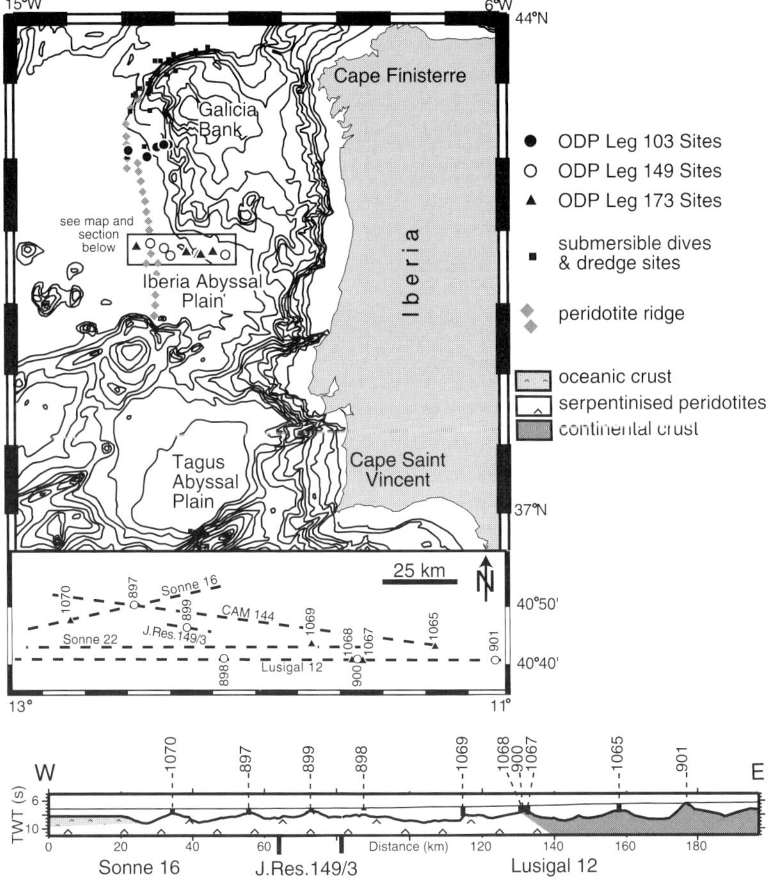

Fig. 1. Overview map of the West Iberia margin showing the location of ODP Sites, and of submersible dive and dredge sites (modified after Whitmarsh *et al.* 1998). Map below shows locations of drill sites in the Iberia Abyssal Plain relative to seismic reflection profiles discussed in the text and used to draw the composite section shown below and in Figure 3. Composite west-to-east cross-section at the base illustrates the composition of the crust based on the results of ODP Legs 149 and 173. TWT, two-way travel time.

The magnetic anomaly pattern indicates northward-propagating continental break-up along the Iberian margin. The magnetic anomaly chart of West Iberia shows the J anomaly to die out northward at about 41°30' in the Iberia Abyssal Plain. This anomaly is thought to be slightly older than chron M0 (Whitmarsh & Miles 1995). Hence sea-floor spreading in the southern Iberia Abyssal Plain started shortly before the formation of the J anomaly whereas the break-up off Galicia, farther north, postdates this anomaly (Pinheiro *et al.* 1996). Detailed modelling of one deep-towed magnetic profile across the Iberia Abyssal Plain by Whitmarsh & Miles (1995) suggests that the onset of sea-floor spreading within this segment of the Iberia Abyssal Plain

took place at *c.* 126 Ma (during Barremian time according to the time scale of Gradstein & Ogg (1996)), at a spreading rate of 10 mm a^{-1}.

Seismic data

A network of multichannel seismic reflection profiles has been collected over the southern Iberia Abyssal Plain. (For an overview of the data, see Discovery 215 Working Group (1998) and Whitmarsh *et al.* (2000)). East–west-directed profiles reveal (from east to west) tilted fault blocks of continental crust, an ocean–continent transition zone of relatively low basement relief merging oceanwards into one or more peridotite ridges, and basement

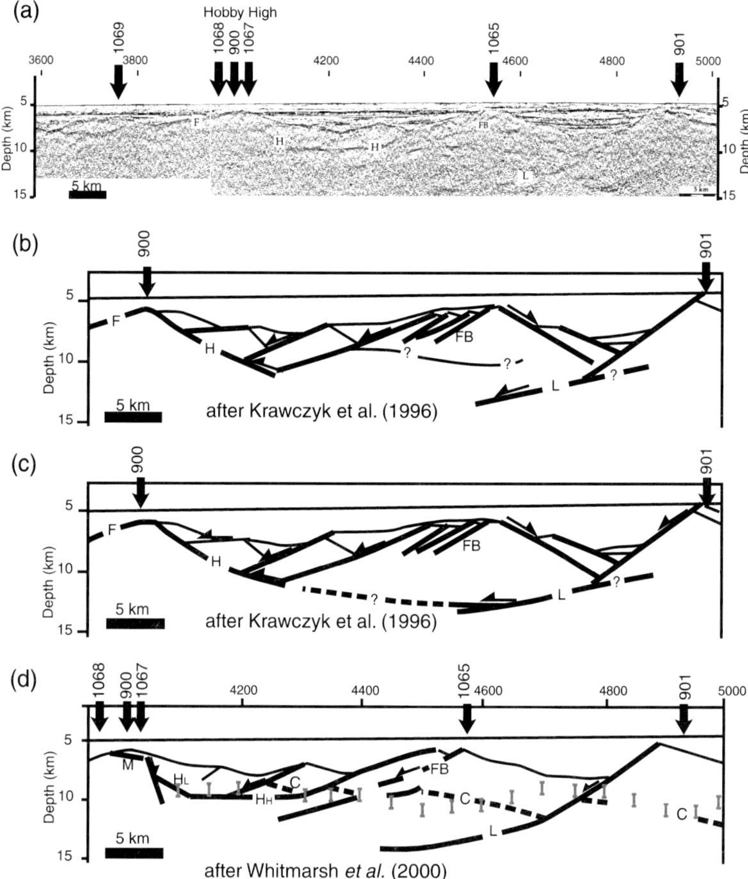

Fig. 2. (**a**) East–west-directed depth-migrated section of profile Lusigal 12 across the Iberia Abyssal Plain (taken from Krawczyk *et al.* 1996). (For location of the profile, see Fig. 1.) Numbers at the top of the section refer to sites drilled during ODP Legs 149 and 173; strong characteristic reflections in the basement are labelled L, FB, H and F. Shot spacing is 50 m, and there is no vertical exaggeration. (**b**) and (**c**) two interpretations of the depth-migrated section of Lusigal 12 between Sites 901 and 900 by Krawczyk *et al.* (1996); in (**b**) H and L are interpreted as independent faults whereas in (**c**) L and H belong to the same large-scale fault zone. (**d**) Alternative interpretation of the same section proposed by Whitmarsh *et al.* (2000). Reflections are labelled C, L, FB, H_H, H_L and M. Vertical bars represent the possible ranges of crust–mantle boundaries computed from isostasy (for further explanation, see Whitmarsh *et al.* 2000).

with the geophysical signature of oceanic crust (Pickup *et al.* 1996) (Fig. 1).

The tilted fault blocks are bounded by strong reflectors (Beslier *et al.* 1993; Krawczyk *et al.* 1996). These mostly seaward-dipping reflectors are partly listric on time sections and are observed to penetrate to depths of several kilometres. On the east–west profile Lusigal 12, oriented roughly perpendicular to the strike of the margin, a number of strong intrabasement reflections is imaged, which dip both landward and oceanward (Fig. 2a). Krawczyk *et al.* (1996) described several reflections (L, H and F

in Fig. 2b,c), which were partly reinterpreted by Whitmarsh *et al.* (2000) (L, H_H, H_L and M in Fig. 2d).

Reflection L, a bright reflection between SP 4900 and 4400 in Figure 2a, appears listric on a depth section. It bounds the western edge of a tilted fault block of continental crust at Site 901 (Fig. 2). Reflection H was interpreted by Krawczyk *et al.* (1996) to curve up towards the west in the direction of an isolated, slightly north–south elongated basement high called the Hobby High (ODP Leg 173 Shipboard Scientific Party 1998) located near SP 4000.

Whitmarsh *et al.* (2000) found that this curving-up is not continuous, but that Hobby High is bounded continentwards by a high-angle fault, which offsets reflection H_L (Fig. 2d). The equivalent of the H_L reflection on the oceanward side of the high-angle fault is called the M reflection. Reflections H_L and H_H, the continentward-dipping and oceanward-dipping parts, respectively, of reflection H, were interpreted by Whitmarsh *et al.* (2000) to lie either entirely within continental crust or at the crust–mantle boundary. Reflection M, in contrast, was interpreted to form the interface between continental crust above and serpentinized mantle rocks below. Reflection F bounds the western flank of Hobby High and is not clearly imaged seismically below the top of the acoustic basement on this profile. On nearby profile SONNE 22 (Fig. 1) it can be traced further to the west to a depth of 9.6 s. The F reflection appears to dip at no more than 17° and in places it appears to coincide with the top of the acoustic basement.

In the ocean–continent transition zone, a normal incidence Moho reflection is absent (Pickup *et al.* 1996) and so is a clear wide-angle Moho reflection; instead there appears to be a gradual downward increase to upper-mantle velocities (Chian *et al.* 1999). Continentwards, a reflection called C is seen east of SP 4250 on Lusigal 12 (Fig. 2d) (Whitmarsh *et al.* 2000). At the eastern end of seismic line CAM 144 (Fig. 1) only a few kilometres north of Lusigal 12, Chian *et al.* (1999) (see their fig. 9a) found a 6.5 ± 2 to $7.3\,km\,s^{-1}$ interface at a similar depth to reflection C on Lusigal 12. Whitmarsh *et al.* (2000) interpreted this reflection as a contact between continental crust above and serpentinized peridotite below. Despite the high quality of the seismic data it is difficult to evaluate the geological nature of this reflection (e.g. tectonic v. primary contact), and to distinguish between lower-crustal rocks and serpentinized mantle rocks (e.g. underplated gabbro v. serpentinized peridotite).

Evidence from ODP Sites

The deep-sea drilling sites of ODP Legs 149 and 173 are located along an east–west-directed transect (Fig. 1). Because drilling focused only on basement highs, sampling is biased and the drilled sections did not sample the basinal record in which a more complete sedimentary record of rifting can be expected. Nevertheless, the existing cores sampled the pre-rift lithosphere and so constrain reconstructions of the break-up of the continental crust

and the exhumation of mantle rocks in the Iberia Abyssal Plain. The rocks from ODP Leg 149 have been described by Sawyer *et al.* (1994) and those from Leg 173 by Whitmarsh *et al.* (1998). The drilling results are illustrated in Figure 3. In the following sections, we summarize some of the results that we think are important for the understanding of the tectonic evolution of the area.

Syn- and post-rift sediments

An overview of the seismic stratigraphy recognized in the Iberia Abyssal Plain and its link to lithostratigraphic and biostratigraphic data obtained from ODP Leg 149 was given by Wilson *et al.* (1996). ODP Leg 173 did not add much new data regarding post-rift evolution. The drilling sites are located over basement highs and so a complete sedimentary record for the Early to Late Cretaceous time interval was not recovered. It is not possible, therefore, to quantify tectonically and thermally driven syn-rift and early post-rift subsidence from the sedimentary record. The data suggest that the infill of depressions between basement highs was completed by Eocene time as indicated by burial of the basement highs by sediments of this age. Applying accepted criteria for the recognition of syn-rift packages on seismic sections, only few equivocal examples of syn-rift sequences were recognized by Wilson *et al.* (2001). Thus, the rift episode may have been very brief, and/or deposition rates were very low, so that syn-rift sedimentary sequences are not sufficiently thick to be seismically resolved.

In the Iberia Abyssal Plain, breccias were recovered at Sites 1070, 897, 899 and 1068 (Fig. 3) and it appears that they are characteristic of the ocean–continent transition zone. The breccias were found in various tectonic positions suggesting different mechanisms of formation.

Mass-flow deposits recovered at Sites 897 and 899 are composed of reworked serpentinized peridotite and a minor proportion of basalt clasts within a matrix of late Barremian to early Aptian age (de Kaenel & Bergen 1996; Fig. 3). These mass-flow deposits were described by Comas *et al.* (1996) and Gibson *et al.* (1996), and were interpreted as generated by submarine slope failure on a large fault scarp. Because the matrix of the breccias is of Aptian age, they are younger than the onset of sea-floor spreading, which took place around 126 Ma (Barremian) along this transect (Whitmarsh & Miles 1995). The occurrence of these post-rift mass-flow deposits on basement highs,

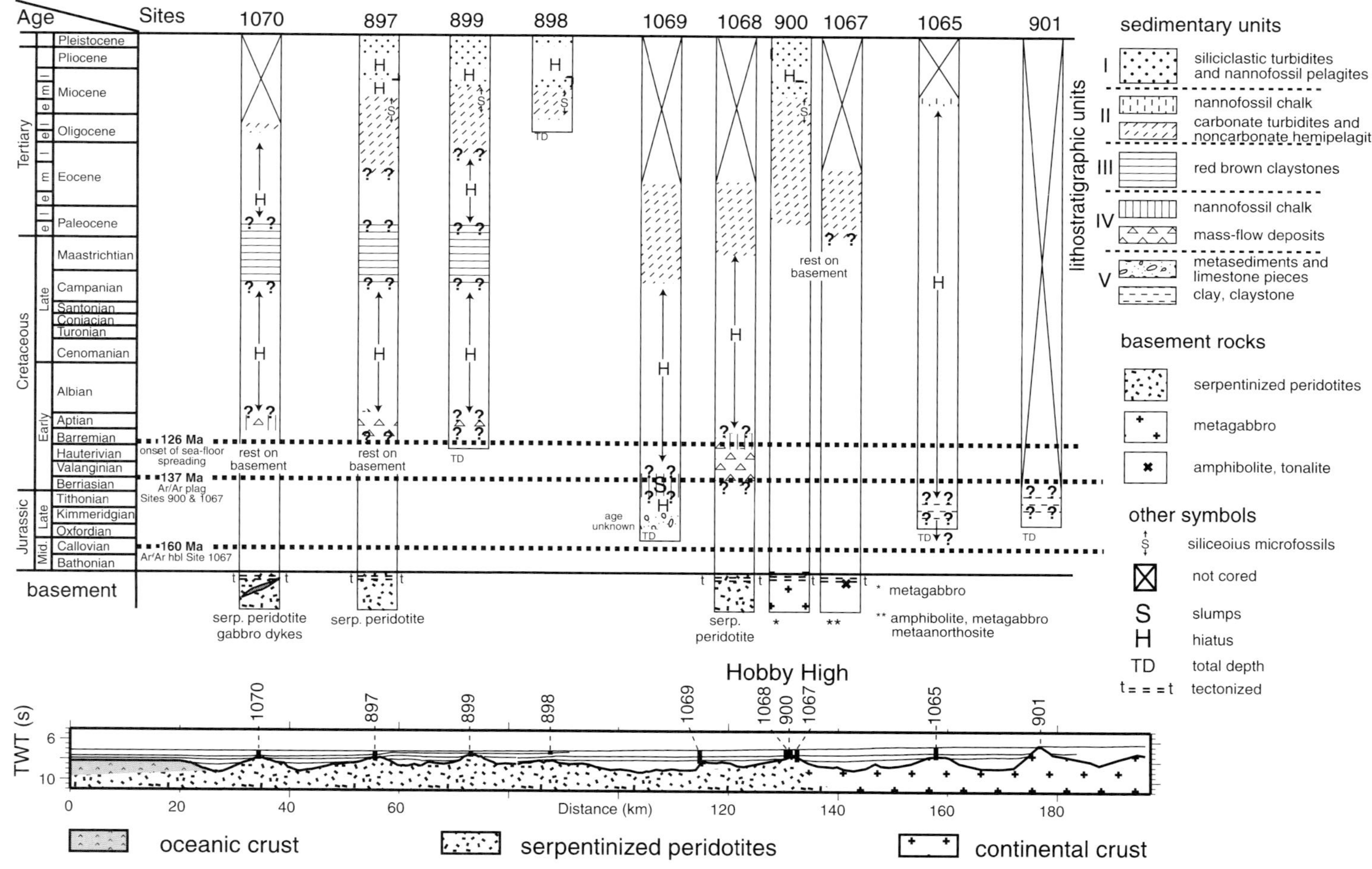

Fig. 3. Summary of drilling results from ODP Legs 149 and 173 in the Iberia Abyssal Plain (modified after Whitmarsh *et al.* 1998).

however, suggests that the basement topography within the ocean–continent transition has been modified after their deposition in Aptian time; that is, tectonic movements continued after break-up.

Sedimentary breccias recovered at Site 1068 are clast supported to matrix supported and consist of clasts of metagabbro, meta-anorthosite, amphibolite, minor metatonalite, and arkosic wacke in a calcareous matrix containing, in the upper part, nannofossils indicating an imprecise Valanginian to Barremian age (Wilson *et al.* 2001). These breccias formed as mass flows, rockfalls and talus deposits (Whitmarsh *et al.* 1998). Down-hole, the breccias become tectonized and their matrix recrystallized suggesting hydrothermal circulation associated with tectonic activity along a basal fault zone (Skelton & Valley 2000). At the base, the preferred orientation of elongated clasts and the occurrence of gouge zones characterized by high matrix-to-clast ratios, rounded clasts and a strong foliation indicate a transition to tectonic breccias. These tectonized breccias are juxtaposed against serpentinized peridotite. The gradual tectonic overprint of the breccia down-hole, the lack of serpentinite and basalt clasts, as well as the Valanginian to Barremian age of the sedimentary matrix, distinguish these breccias from those recovered at Sites 897 and 899.

Breccias recovered at Site 1070 are cut by gouge horizons and grade downward into massive serpentinized peridotite. These breccias consist only of serpentinite and rare gabbro clasts, poorly sorted and embedded in calcite cement. The fabric of these breccias is very similar to those of alpine ophicalcites described by Bernoulli & Weissert (1985) and Lemoine *et al.* (1987). Replacement structures of calcite after serpentine occur; sedimentary structures, however, were not observed. We assume that the Site 1070 breccias are tectonic rather than sedimentary in origin and belong to the tectonized basement discussed below.

Pre-rift sediments

Pre-rift sediments were recovered at Sites 901, 1065 and 1069 (Fig. 3). At Sites 901 and 1065 drilling penetrated into variably cemented clay and claystone with thin sandstone and conglomerate layers of Tithonian age. These sediments are broadly comparable with the onshore upper Jurassic Abadia Formation, and so are interpreted to have been deposited in a shelf setting (see also Whitmarsh *et al.* 1998, p. 77). These neritic sediments were deposited at a depth of

c. 200 m (Wilson *et al.* 2001). Although the underlying basement rocks have not been recovered, the relatively shallow-water character indicates that they were deposited on equilibrated (i.e. *c.* 30 km thick) continental crust.

At Site 1069, Tertiary and Upper Cretaceous post-rift sediments unconformably overlie thin Lower Cretaceous nannofossil chalk with slumped intervals and gravel layers, underlain by a thin layer of Tithonian clay (Wilson *et al.* 2001, fig. 13). Below this, shallow-water limestone clasts and partly rounded clasts of weakly metamorphosed Palaeozoic metasediments were drilled. The matrix to these clasts was not recovered, but we interpret these clasts to document probably a Tithonian or older conglomerate. The occurrence of Tithonian pre-rift sediment at Site 1069 indicates continental basement at this locality.

Lower-crustal rocks

Lithology. Lower-crustal rocks were drilled at Sites 900 and 1067 and were found as clasts within breccias at Site 1068. All these sites are located on Hobby High, a basement high draped by Upper Cretaceous to Paleocene post-rift sediments.

At Site 900 drilling penetrated foliated and brecciated metagabbro, which Seifert *et al.* (1996; 1997) interpreted as metamorphosed cumulate gabbro formed from a mid-ocean ridge basalt (MORB) magma during sea-floor spreading. In contrast, Cornen *et al.* (1996, 1999) interpreted this metagabbro, based on evidence for high-pressure (>0.8 GPa) conditions during deformation, as underplated during rifting at the base of continental crust.

At Site 1067, which is only 800 m east of Site 900, a series of variably deformed and brecciated amphibolite, metatonalite and meta-anorthosite was drilled. A transition downward occurs from strongly foliated amphibolite and metatonalite to less strongly foliated amphibolite and meta-anorthosite, some of which retain their original igneous textures. The amphibolite contains euhedral plagioclase included in poikilitic amphibole and regions of subhedral equigranular plagioclase indicating a gabbroic origin. It is unclear whether the amphibole crystallized from a melt (hornblende gabbro) or formed by replacement of magmatic pyroxene under later amphibolite-facies conditions. The meta-anorthosite also preserved relict igneous textures. A comparison of the amphibolite of Site 1067 with the metagabbro of Site 900 shows that the latter is more primitive (higher *Mg*-numbers, higher abundance of compatible elements, and lower abundance of incompatible elements; Whitmarsh *et al.* 1998).

Foliated amphibolite, microamphibolite, meta-anorthosite and metagabbro occur as clasts in the breccia at Site 1068. The petrological, mineralogical and structural characteristics of these clasts are very similar to those of the basement rocks recovered at Sites 900 and 1067. Therefore, and because of the typical proximal facies of this breccia, we assume that the clasts in the breccia were derived from Hobby High.

Granulite- to amphibolite-facies deformation. Microstructures of the basement rocks from Hobby High and of the breccia clasts at Site 1068 record a deformation history evolving from upper amphibolite or lower granulite facies to sea-floor conditions. Evidence for deformation under granulite-facies conditions was observed only in a few clasts of the breccia recovered at Site 1068. There, pyroxenes were affected by crystal-plastic deformation leading to the formation of subgrains along the margins of elongated clinopyroxene crystals. The flaser structure preserved in a few metagabbros at Site 900 formed under amphibolite or higher metamorphic conditions, as indicated by dynamic recrystallization of amphibole and plagioclase. The present-day paragenesis found in these basic rocks equilibrated under lower amphibolite-facies conditions, as indicated by the typical mineral assemblage consisting of blue–green to green–brown amphibole plus andesine.

Evidence for greenschist-facies mylonitization in Site 1067 metatonalite. Deformation under greenschist-facies conditions is recorded in the metatonalite lenses or veins found in the amphibolite at Site 1067 (Fig. 4b). Such metatonalite veins are restricted to the uppermost part of the cored amphibolite section at Site 1067 (Whitmarsh *et al.* 1998). Contacts between metatonalite and amphibolite are parallel to the foliation in both rocks. A micro-augen-structure is defined in the metatonalite by porphyroclasts of plagioclase with undulatory extinction within a foliated matrix of recrystallized quartz and phyllosilicates including chlorite (Fig. 5a). Chlorite occurs also in pressure shadows of plagioclase porphyroclasts and along shear bands, suggesting that deformation occurred, or at least continued, under greenschist-facies conditions. Quartz layers are deflected around the plagioclase porphyroclasts. Recrystallized quartz grains exhibit preferred grain-shape and crystallographic orientations (Fig. 5b). In sections parallel to the stretching lineation and perpendicular to the foliation, consistent shear-sense indicators provide ample evidence for strongly non-coaxial deformation. These shear-sense indicators include σ-type porphyroclasts (Fig. 5a), shear bands, s–c fabrics and oblique grain shapes in recrystallized quartz layers (Fig. 5b). Figure 5c shows an example of a quartz *c*-axis pattern. In spite of the considerable deflection of the quartz layers around plagioclase porphyroclasts, the *c*-axes show a strong preferred orientation. The pattern is a single girdle with only a slight tendency towards a cross girdle. This is again indicative of strongly non-coaxial flow close to simple shear (Schmid & Casey 1986). The absence of a *c*-axis maximum at a high angle to the foliation suggests that basal $\langle a \rangle$ glide did not occur or was less important than rhomb $\langle a \rangle$ and prism $\langle a \rangle$ glide. This is not a measuring artefact because the pole figure is combined from measurements in two perpendicular planes.

Although, unfortunately, the original orientation of the mylonitized metatonalite veins could not be determined, the evidence for simple-shear deformation is important. It strongly suggests that at Site 1067 the uppermost part of the basement, to which the metatonalite veins are restricted, represents a shear zone active under greenschist-facies conditions.

Brittle deformation. Probably contemporaneously with ductile deformation in the metatonalite, the amphibolite, metagabbro and meta-anorthosite deformed brittly as indicated by greenschist-facies paragenesis in veins and along brittle faults. This first generation of veins (epidote, chlorite, plagioclase, iron oxide and rare calcite veins) is overprinted by pervasive calcite veining, which is late syn- to post-kinematic and occurred near or at the sea floor (Skelton & Valley 2000).

The occurrence of breccia, cataclasite, fault gouge, veins and pseudotachylite (Fig. 4a) in the cores from Hobby High as well as the recovery of small localized normal faults indicate that the basement at Hobby High was strongly affected by brittle deformation. At Site 1067, the extent of brittle deformation clearly decreases from the top of the basement downward, as reflected by the core recovery: 9.4% in the uppermost, 92.1% in the lowermost basement core. The distribution of brecciated zones, in particular at the top of the basement at Sites 900 and 1067, suggests that brecciation is linked to fault zones. Within the brecciated zones at Sites 900 and 1067, three types of textures can be distinguished. (1) Clast-supported breccias are dominated by millimetre- to decimetre-sized angular clasts, showing a 'jigsaw' arrangement in places. These

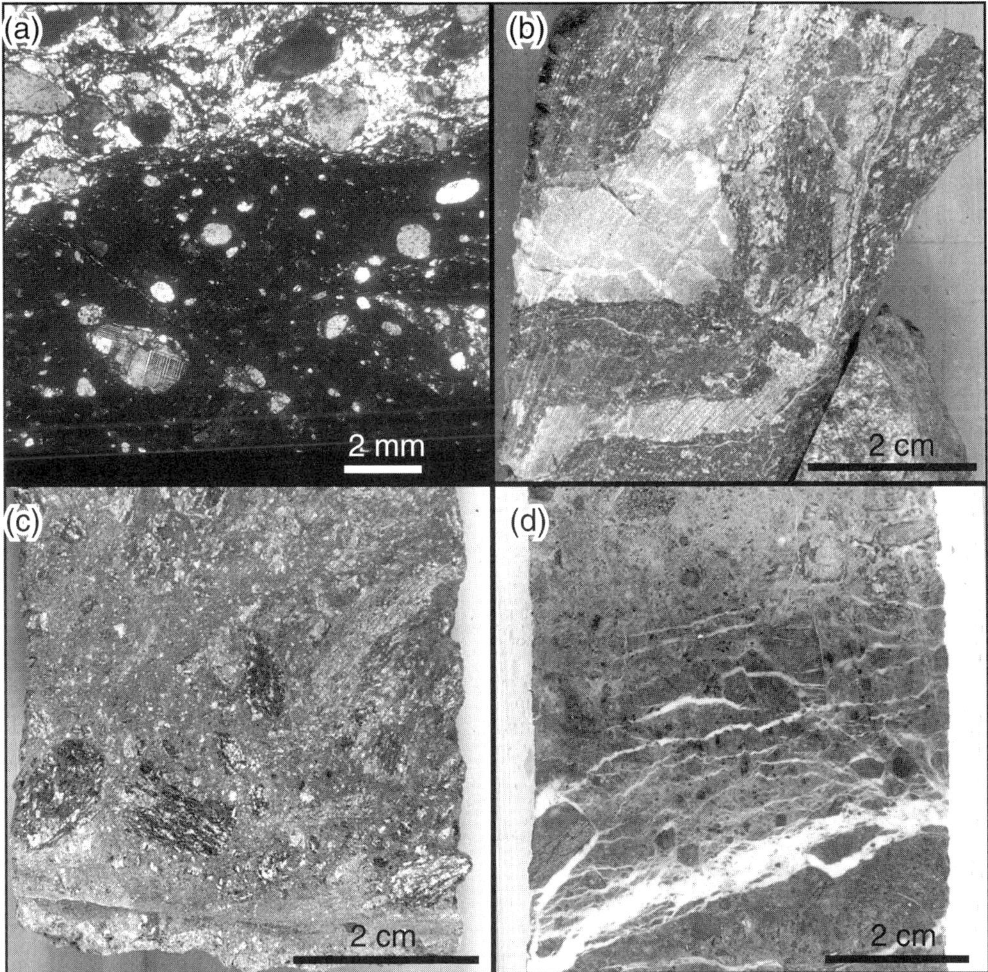

Fig. 4. (a) Thin-section photomicrograph of a pseudotachylite cutting the foliation of a mylonitized tonalite. Sample 173-1067A-14R-1 (Piece 3, 8–15 cm). (b) Folded tonalite lenses within foliated amphibolite. Interval 173-1067A-17R-1, 60–70 cm. (c) Transition from clast-supported to matrix-supported breccia with rounded amphibolite clasts. At its base the breccia is cut by a localized, subhorizontal gouge zone of *c*. 0.5 cm thickness. Interval 173-1067A-17R-1, 60–70 cm. (d) Serpentine gouge with serpentine veins occurring at the top of the exhumed mantle at Site 1068. Interval 173-1068A-20R-7, 52–60 cm.

breccias consist of angular foliated amphibolite clasts set in a network of fractures filled with clasts of the wall rock and milled material plus, in places, newly crystallized epidote, amphibole and plagioclase. (2) Matrix-supported breccias are usually polygenetic and consist of amphibolite, metatonalite and epidosite (lower part of Fig. 4c). Their matrix consists of very fine-grained epidote and milled host-rock minerals. In general, higher matrix/clast ratios correlate with rounder and smaller clasts. A weak foliation within these breccias is indicated by the alignment of elongated clasts. (3) Fault gouges

occur in zones of 1–20 cm width with a high matrix/clast ratio and rounded or elongated clasts. These zones form moderately dipping normal faults as well as subhorizontal localized fault zones (lowermost part of Fig. 4c). The minerals of the fault gouge, identified by X-ray diffraction (XRD) analysis (Whitmarsh *et al.* 1998), comprise plagioclase, amphibole, chlorite, calcite, Fe-oxyhydroxides and magnetite. Clay minerals were not found in the fault gouge or the matrix-supported breccia, which suggests that syntectonic, retrograde mineral reactions occurred mostly under greenschist-facies conditions and became less important when the temperatures dropped below 300 °C.

(a)

(b)

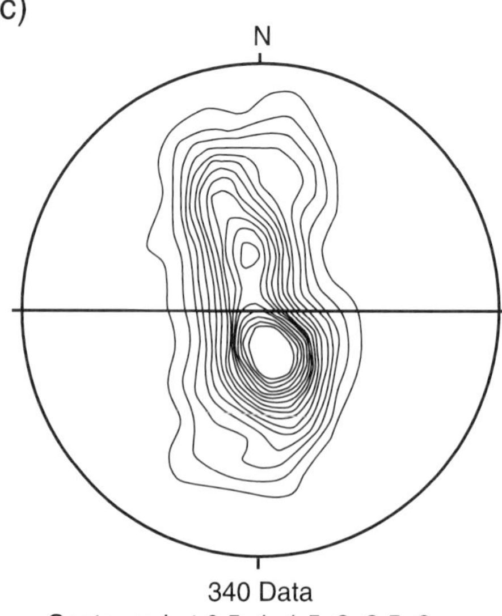

(c)

Study by scanning electron microscopy (SEM) shows that gouge zones are localized preferentially within former cataclastic shear zones and that fluid–rock interaction occurred before, during and after localization of displacement within these zones. Hydration reactions appear to be responsible for the weakening process. Localization of displacement occurred in zones with a high amount of fine-grained minerals forming a matrix of <5 μm grain size. In scanning electron micrographs, these minerals show idiomorphic or irregular, angular outlines indicating that they formed by both cataclasis and recrystallization. In general, the higher the matrix/clast ratio, the stronger the preferred orientation of the matrix minerals and the better rounded the clasts. Commonly, the deformation microstructures are masked by later post-kinematically grown calcite, indicating that crystallization of new grains outlasted deformation. In summary, we conclude that the brittle fault rocks recovered from Hobby High are associated with a large-scale shear zone that formed under greenschist-facies and lower metamorphic conditions and was finally exposed at the sea floor. We attribute the mylonitization of the tonalite veins and lenses to the same shear zone and assume that the contrasting behaviour (mylonitization v. cataclasis) reflects different protolith composition (tonalite v. amphibolite).

Age data. U–Pb SHRIMP dating on zircons from one metatonalite from Site 1067 revealed two age populations, 340 ± 10 Ma for the rims, and 595 Ma and older for the cores.

Fig. 5. (**a**) Thin-section micrograph of mylonitized tonalite from Site 1067 (sample 1067A-14R-1). Thin section cut perpendicular to foliation and parallel to stretching lineation (horizontal). Dark material is phyllosilicate, mostly chlorite; light grey porphyroclasts are plagioclase; light layers are recrystallized quartz. Note σ-type porphyroclast in the centre, indicating dextral shear, and chlorite in pressure shadows indicating deformation under greenschist-facies conditions. Plane-polarized light, field of view is 2 mm wide. (**b**) Thin-section micrograph of mylonitized tonalite from same core section as (**a**) but different sample, showing dynamically recrystallized quartz layer (middle) between plagioclase porphyroclasts (upper and lower parts). Thin section cut perpendicular to foliation and parallel to stretching lineation (horizontal). Fine-grained, recrystallized quartz exhibits oblique grain shapes indicating dextral shear sense. Crossed polars, field of view is 0.5 mm wide. (**c**) Quartz *c*-axis pole figure for sample shown in (**b**). Orientation as in (**b**), foliation horizontal, lineation east–west. Simple-shear flow is indicated by single-girdle pattern, with a weak tendency towards a cross girdle. (See text for discussion.)

However, zircons with typical magmatic morphologies from the surrounding amphibolite yielded 270 ± 3 Ma.

Ages similar to those found in the metatonalites were reported by Ordóñez Casado (1998) from the Ossa Morena Zone of onshore Iberia, the likely equivalent to the continental basement of the Iberia Abyssal Plain margin (Capdevila & Mougenot 1988). The ages correspond to Cadomian and Variscan regional events, respectively. The question, however, remains how veins of Variscan metatonalite can occur within a post-Variscan amphibolite that recrystallized at 270 ± 3 Ma. The most likely interpretation is that the Variscan and Cadomian zircons in the tonalite are inherited. Although the data are at the moment still ambiguous in this respect, the zircon ages clearly show that the rocks recovered at Site 1067 formed at 270 ± 3 Ma or earlier, and did not form during the rifting event.

$^{40}Ar/^{39}Ar$ step-heating plateau ages were obtained from a hornblende and plagioclase mineral pair from an amphibolite from Site 1067. The hornblende yielded a plateau age of 161 ± 1 Ma and the plagioclase one of 137.2 ± 0.5 Ma. A similar $^{40}Ar/^{39}Ar$ plateau age of 136.4 ± 0.3 Ma was obtained for a plagioclase from a Site 900 metagabbro (Féraud et al. 1996). The $^{40}Ar/^{39}Ar$ ages obtained from the Site 1067 hornblende and plagioclase represent cooling ages dating the time when the rocks cooled below the blocking temperatures for hornblende (c. 500 °C) and plagioclase (c. 150 °C). This supports our interpretation of the U−Pb zircon ages, that the amphibolites of Site 1067 were part of the pre-rift continental crust.

Serpentinized mantle peridotites

Lithology. Serpentinized peridotites were drilled at Sites 897, 899, 1068 and 1070. They comprise spinel dunite and harzburgite, spinel–plagioclase harzburgite, lherzolite and pyroxenite. The peridotites underwent limited partial melting and some impregnation by undersaturated alkali magmas leading to plagioclase crystallization (Cornen et al. 1996). The serpentinized peridotites recovered from the Iberia Abyssal Plain show a variable petrological and tectono-metamorphic evolution.

At Site 897 the entire ultramafic section is extensively serpentinized and exhibits a transition from carbonate-rich serpentinite associated with serpentinite breccias, at the top, to massive serpentinized peridotite, locally displaying plagioclase- and clinopyroxene-enriched zones, at the base.

At Site 899 ultramafic rocks were found in breccias and as olistoliths of serpentinized peridotite several tens of metres across, intercalated in sediments of late Barremian to early Aptian age (de Kaenel & Bergen 1996). The ultramafic clasts and olistoliths have a large compositional range from plagioclase-bearing peridotites to pyroxene- and olivine-rich peridotites. Up to 30% plagioclase occurs in patches or veinlets in the peridotites or rims the spinel.

At Site 1068 serpentinized peridotites occur below tectonized breccias. At the top, the serpentinized peridotites are strongly brecciated and deformed, as indicated by the occurrence of serpentine gouges with a strong foliation. Down-hole this fabric disappears and is replaced by brittly deformed and veined serpentinites preserving a high-T foliation marked by spinel grains. Primary minerals such as pyroxene are preserved only to a minor extent.

At Site 1070 massive serpentinized mantle rocks containing gabbro pegmatites are overlain by a poorly sorted, clast-supported breccia, which consists of serpentinized peridotite and gabbro clasts and is cemented by blocky calcite. At the contact between these 'ophicalcites' and the massive serpentinized mantle, serpentine gouge was observed. Clasts of the gouge occur also in the ophicalcites. The pegmatitic gabbro is composed of altered plagioclase, clinopyroxene, amphibole and ilmenite. It appears to form an intrusion into the mantle peridotite. The underlying serpentinized peridotite varies in composition between pyroxene-rich and -poor layers and has a weak, shallowly dipping high-T foliation.

The composition of the Site 897 and 899 peridotites was interpreted in terms of a low degree of depletion and partial melting (probably <10%) of the whole ultramafic unit (Cornen et al. 1996). Thus, the mantle exposed in the Iberia Abyssal Plain has a subcontinental origin, or, alternatively, it formed below a very slow-spreading ridge (c. 20 mm a^{-1}). Compositional variations, such as the presence or absence of plagioclase (plagioclase-bearing lherzolite occurs at Sites 897 and 899 but is absent at Sites 1068 and 1070), indicate that the mantle exposed in the Iberia Abyssal Plain is rather heterogeneous.

Deformation. The deformation history recorded by the Site 897 and 899 mantle rocks was described and subdivided into the following five distinct stages by Beslier et al. (1996): (1) high-temperature shear deformation at 1000–900 °C; (2) limited partial melting; (3) subsolidus re-equilibration in the plagioclase stability field; (4) poorly developed mylonitic shear deformation at *c.* 700 °C under high deviatoric stress and low pressure (i.e. lithospheric conditions); (5) hydration and brittle deformation under greenschist-facies and lower metamorphic conditions. The peridotites recovered at Sites 1068 and 1070 lack most of the high-*T* deformation except for a spinel foliation. Thus, most of the deformation stages described by Beslier *et al.* (1996) are not homogeneously distributed within the mantle rocks underlying the Iberia Abyssal Plain, suggesting that most of the deformation was localized in shear zones. It has to be kept in mind, however, that during serpentinization much of the pre-existing microstructure might have been erased. Thus, the high-*T* deformation history is in many cases difficult to evaluate.

Late deformation occurring at shallow depth is well recorded by gouge zones at the top of the mantle rocks (Fig. 4d). This is clearly documented by the drilling results. Wherever mantle rocks were encountered, their top is formed by brecciated serpentinites containing zones of foliated serpentine gouge. Microstructural studies of these rocks show that they represent high-displacement fault zones formed during a late stage of mantle exhumation and were impregnated with or replaced by calcite at or near the sea floor.

The mantle rocks from the Iberia margin, record, like those from the Alps, an exhumation path that can be traced back to asthenospheric conditions. This, however, does not mean that the complete exhumation path is necessarily related to rifting. For the Malenco peridotite exposed in the Eastern Alps, Müntener *et al.* (2000) documented an exhumation history with many similarities to that of the mantle rocks from the Iberia margin. In the Malenco, however, only hydration and related deformation, i.e. the last phase determined by Beslier *et al.* (1996), was linked to the rifting event that preceded continental break-up. All the other stages predated rifting by >60 Ma and are related to Variscan and pre-Variscan tectonic events. The lack of radiometric data for the high-*T* deformation in the Iberia Abyssal Plain margin makes it difficult to determine whether all the deformation stages determined by Beslier *et al.* (1996) are related to rifting, or only the late stages.

Syn-rift magmatic rocks

Syn-rift magmatism appears to be of minor importance in the Iberia Abyssal Plain (Pinheiro *et al.* 1996; ODP Leg 173 Shipboard Scientific Party 1998). The only candidates are the metagabbro at Site 900, the pegmatitic gabbro at Site 1070 and basalts so far only recovered as clasts from mass-flow deposits at Site 899.

Although Cornen *et al.* (1996) interpreted the metagabbro of Site 900 as being underplated at *c.* 24 km depth below slightly thinned continental crust, the available data do not indicate when and in what circumstances the gabbro crystallized. So far, the data show only that the gabbro cooled below the closure temperature of Ar in plagioclase, i.e. below *c.* 150 °C at 136.4 ± 0.3 Ma (Féraud *et al.* 1996). Therefore a pre-rift age for these rocks cannot be excluded and is even probable given the 161 ± 1 Ma ^{40}Ar/^{39}Ar amphibole cooling age from nearby Site 1067.

For the hornblende–gabbro pegmatites intruded into peridotite at Site 1070, ^{40}Ar/^{39}Ar step-heating on hornblende gave a plateau age of 119 ± 0.7 Ma. The lack of crystallization ages makes it impossible to decide whether these gabbro pegmatites are syn- or post-rift.

Basalt has not been drilled in the southern Iberia Abyssal Plain, except for basalt clasts recovered at Site 899 in the mass flows. Thus, the age of the oldest basalt is not yet known.

Tectonic evolution of the Iberia Abyssal Plain

A geological interpretation of the Lusigal 12 profile

Parts of the Lusigal 12 profile were published by Beslier *et al.* (1993), Krawczyk *et al.* (1996) and Whitmarsh *et al.* (2000). Details concerning the recovery, processing and analysis of the Lusigal 12 seismic data, as well as an interpretation of the profile in terms of the tectonic evolution of the West Iberian margin using the results of ODP Leg 149, were given by Krawczyk *et al.* (1996). Whitmarsh *et al.* (2000) reviewed the seismic data and used also drilling results of ODP Leg 173. Their tectonic interpretation is strongly based on the geophysical data and is in some respects different from that presented here. Our interpretation is mainly based on the results of ODP Legs 149 and 173, but was also constrained in an iterative way by our attempts to make a kinematic inversion. Where the seismic data allowed alternative interpretations, as for instance in the area of Hobby High (Fig. 6), often one of these could be excluded because it led to impossible situations in the kinematic inversion (see interpretations 1 and 2 and their inverted profiles in

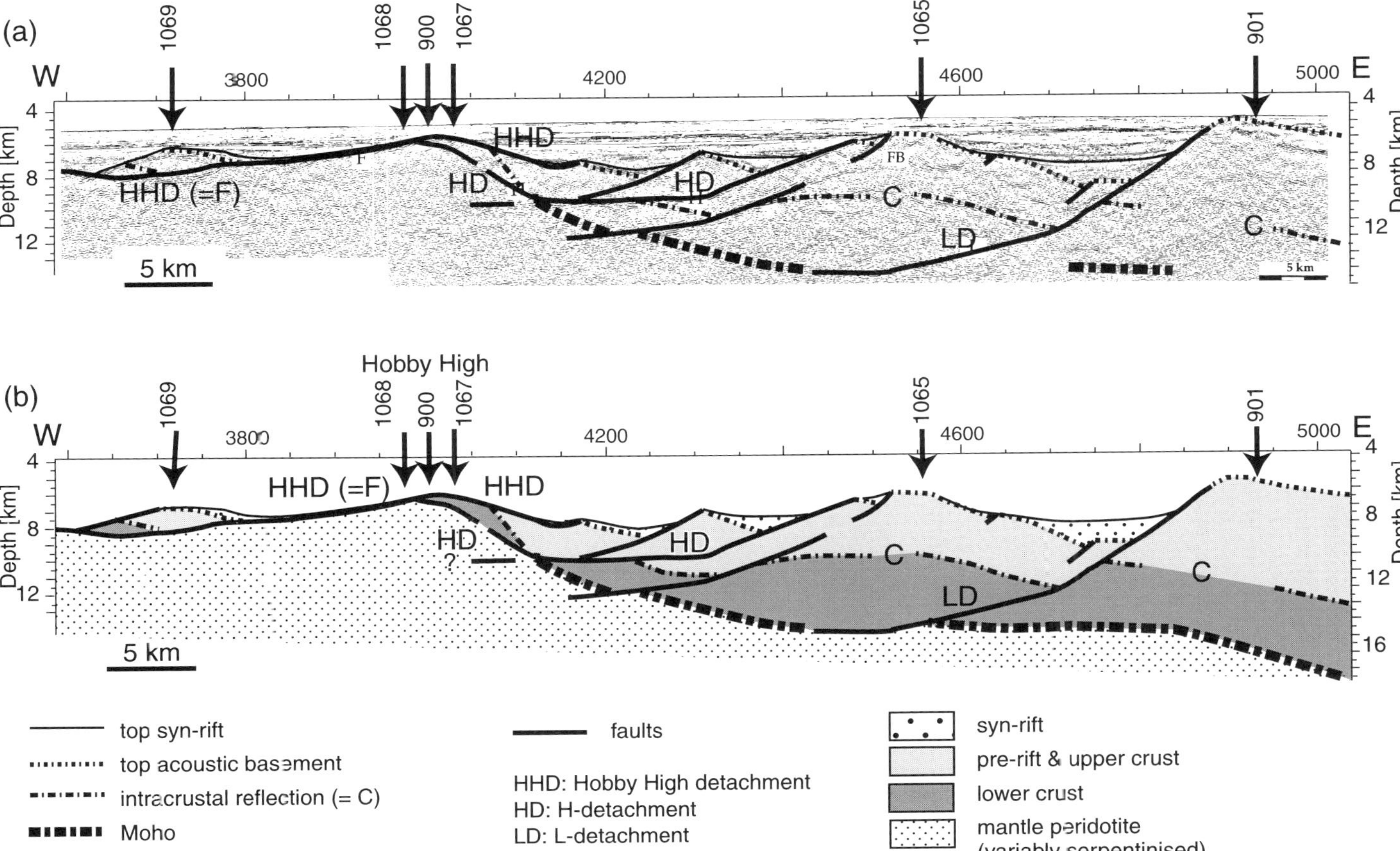

Fig. 6. (a) Depth-migrated section Lusigal 12 (from Krawczyk *et al.* 1996) showing the reflections discussed in the text. **(b)** Geological interpretation of the depth-migrated section of Lusigal 12 showing the distribution of upper- and lower-crustal and mantle rocks.

Fig. 7). We use the depth-migrated seismic profile Lusigal 12 and the seismic reflections described by Krawczyk *et al.* (1996).

The Hobby High detachment fault (HHD). The various rock types recovered from Hobby High not only demonstrate the geological complexity of this high, but also that of the entire ocean–continent transition. We will discuss the relationship between deformation, exhumation and sedimentation, which led to our interpretation of the evolution of the Iberia Abyssal Plain.

Site 1067, on the continent side of the Hobby High, penetrated amphibolite with metatonalite lenses (Fig. 4b). These rocks record a deformation history that evolved from upper amphibolite-facies ductile conditions to near sea-floor brittle deformation. The existing pre-rift (Palaeozoic) U–Pb zircon ages and the 161 Ma $^{40}Ar/^{39}Ar$ amphibole age demonstrate that the rocks from Hobby High are pre-rift continental material. The $^{40}Ar/^{39}Ar$ plagioclase ages (137 Ma at Site 1067 and 136 Ma at Site 900) indicate not only a common late cooling history for both sites but also that the basement rocks at Sites 1067 and 900 had not reached the sea floor at this time (i.e. at the Berriasian–Valanginian boundary). Cooling, exhumation and exposure at the sea floor of these rocks as well as their retrograde deformation history can be best explained assuming that the rocks were in the footwall of a detachment fault, the Hobby High detachment fault (HHD) (Fig. 6a). Although this detachment had been partly eroded over Hobby High, as indicated by footwall clasts in the breccias at Site 1068, the small volume of syn-rift sediments (Wilson *et al.* 2001) precludes very extensive submarine erosion. The low recovery at the top of the basement, the occurrence of gouges and pseudotachylites, and the occurrence of small-scale subhorizontal low-temperature shear zones within the basement cores at Sites 900 and 1067 support the idea that the present-day contact between the acoustic basement and the post-rift sediments represents an exhumed detachment fault surface.

Oceanwards, the HHD can be traced to Site 1068, where it has been drilled. At this site the HHD is a shear zone marked by serpentine gouge (Fig. 4d) separating sedimentary breccias from massive, serpentinized peridotites. Thus, the HHD is underlain by continental crust at Sites 1067 and 900 and by serpentinized peridotite at Site 1068, indicating that the HHD cut from the crust into the mantle (Fig. 6b). On the basis of the cooling ages and the pervasive brittle deformation associated with the HHD, we assume that the mantle was already at a shallow level and probably already serpentinized before it was exhumed by the HHD.

Further oceanwards, the HHD appears to coincide with the F reflection (Fig. 6a). This reflection can be traced into a reflection underlying the basement high that was drilled at Site 1069 and that consists of upper continental crust overlain by post-rift sediments. The continental crust of this high appears on the profile as a wedge-shaped body, about 1.5 km thick and 9 km long, defined by the basement–sediment contact above and the prolongation of reflection F below. That this body represents an olistolith derived from Galicia Bank to the north appears very unlikely to us because of its size and its shape (Discovery 215 Working Group 1998). It forms a north–south-striking ridge, which is consistent with a fault block origin rather than an olistolith. We interpret this block, therefore, as an extensional allochthon of continental crust overlying mantle along the HHD (Fig. 6b). This interpretation requires a large displacement for this block and an origin continentwards of Hobby High. On the continentward side of Hobby High, the HHD may coincide approximately with the top of the acoustic basement. We assume the breakaway of the detachment fault to be in the vicinity of SP 4150. This interpretation implies that the HHD was exposed over long distances at the sea floor. Similar situations are known from the Woodlark Basin (Taylor *et al.* 1999) as well as from the Alps, where rift-related detachment faults are directly overlain by syn- and post-rift sediments (Florineth & Froitzheim 1994; Manatschal & Nievergelt 1997). We therefore interpret the HHD as a major fault that (1) formed at shallow crustal levels, (2) accommodated a displacement on the order of 20 km, and (3) cut from the continental crust directly into a mantle that was already at a shallow depth when the detachment became active.

Interpretation of the H, L and C reflections. The strong reflections L and H were discussed by Krawczyk *et al.* (1996) and Whitmarsh *et al.* (2000) (Fig. 2). For the interpretation of the L reflection, there is general agreement that it represents a listric fault cutting down into the mantle (Whitmarsh *et al.* 2000) or at least reaching the base of the continental crust (Krawczyk *et al.* 1996). We follow these interpretations and call this reflection LD (L detachment) (Fig. 6a).

More controversial is the H reflection (Fig. 2). In all interpretations H has one or several breakaway(s) on the continentward side and can be traced towards Hobby High, towards

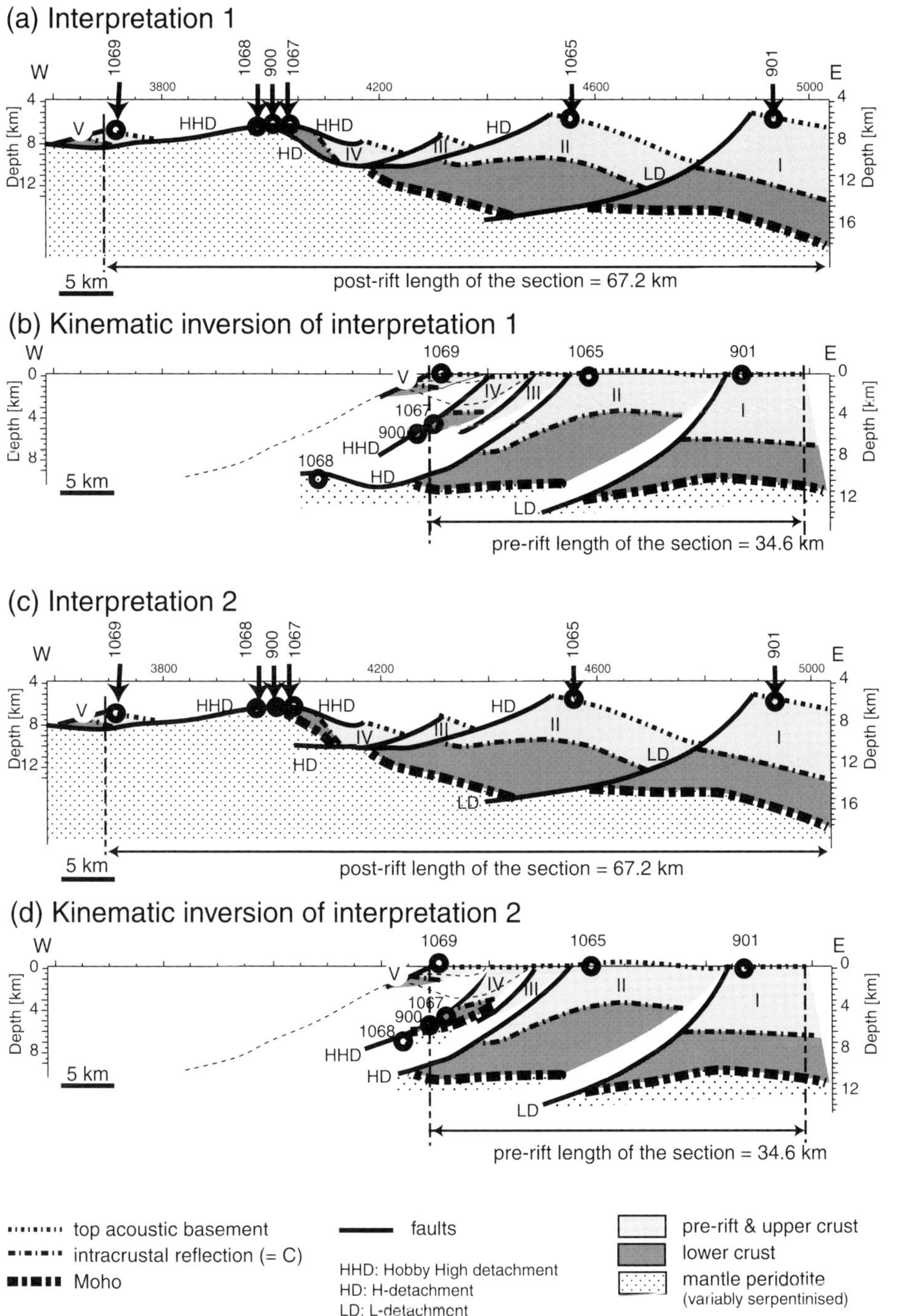

Fig. 7. Kinematic inversion of two alternative interpretations of the depth-migrated section of Lusigal 12. (**a**) Seismic interpretation 1. (**b**) Kinematically inverted section of interpretation 1 shown in (**a**). (**c**) Seismic interpretation 2. (**d**) Kinematically inverted section of interpretation 2 shown in (**c**). The different initial position of Site 1068 in the inverted profiles (**b**) and (**d**) relative to the HD should be noted (in (**a**) Site 1068 lies in the footwall of the HD, in (**c**) Site 1068 lies in the hanging wall of the HD). (For further explanation and discussion, see text.)

which controversy arises. Krawczyk *et al.* (1996) interpreted H to curve up towards Hobby High and assumed that Site 900 represents the footwall to H (Fig. 2b and c). In contrast, Whitmarsh *et al.* (2000) distinguished between two reflections: one (H_H) that is flat and appears to die out east of Hobby High, and a second (H_L) that dips continentwards, rises towards Hobby High and is offset by a high-angle fault, on the western side of which it may be represented by the reflection M (Fig. 2d). M underlies the rocks drilled at Sites 900 and 1067, so that the lower-crustal rocks recovered at these sites sampled the hanging wall, whereas the mantle rocks at Site 1068 sampled the footwall of the M reflection. Thus, M separates mantle rocks from crustal rocks. Our interpretation is very similar to that of Whitmarsh *et al.* (2000). We interpret H as a detachment fault (HD: H detachment) with a breakaway just oceanwards of Site 1065 (Fig. 6a), which can be followed toward SP 4100, where the reflection appears to branch into one short reflection that remains at constant depth and another curving up towards Hobby High. This allows for two alternative interpretations (see interpretations 1 and 2 in Fig. 7a and c). In interpretation 1 (Fig. 7a), HD separates upper crust from upper crust near the breakaway, upper crust from lower crust and mantle further oceanwards, and lower crust from mantle at Hobby High and HD is truncated by HHD between Sites 900 and 1068. In interpretation 2 (Fig. 7c), HD cuts near SP 4100 directly into the mantle. In this case, the reflection curving up towards the Hobby High represents a real Moho rather than a detachment and the mantle rocks recovered at Site 1068 represent a mantle section bounded by HHD above and a flat HD below.

The C reflection has been described by Whitmarsh *et al.* (2000) east of SP 4250 on Lusigal 12 (Fig. 2d) as a weak, discontinuous reflection within the continent side of the margin (Fig. 6) and has been interpreted by those workers as a crust–mantle boundary separating continental basement from serpentinized mantle. Only a few kilometres north of Lusigal 12, Chian *et al.* (1999) found a 6.5 ± 2 to 7.3 km s^{-1} interface at the eastern end of seismic line CAM 144 (Fig. 1) at a similar depth to that of reflection C in the Lusigal 12 (see their fig. 9a). Thus, an alternative interpretation is that the rocks below C represent lower crust that has acoustic properties similar to those of serpentinized mantle. What makes us favour interpreting these rocks as lower crust is the fact that LD and HD cut across C and flatten only at a deeper level. If the material below C was serpentinized peridotite, which is a relatively weak material, one might expect the faults to flatten or even sole out at the C reflection. We tentatively assume, therefore, that C separates continental upper-crustal rocks from lower-crustal rocks, with both exhibiting similar strength under such shallow crustal conditions. The crust–mantle boundary probably occurs at a deeper level, at about 12–14 km, where LD and HD do indeed flatten out (Fig. 6b). This is compatible with the velocity model for seismic line CAM 144 proposed by Chian *et al.* (1999). In their velocity model they showed that velocities higher than 7.6 km s^{-1}, compatible with a weakly serpentinized mantle with <10% of serpentine (Christensen 1966, 1972), occur at a depth of *c.* 12 km.

Kinematic inversion of the seismic section

To constrain qualitatively the displacement accommodated by LD, HD and HHD, and to reconstruct the pre-detachment situation, we made a kinematic inversion of two alternative interpretations of the depth-migrated seismic profile Lusigal 12 presented in Figure 6. The kinematic inversion was kept as simple as possible, considers only rotational and translational deformation of seismically well-defined blocks, and conserves the cross-sectional areas within the profile, assuming that the blocks moved parallel to the profile.

To perform the kinematic inversion, we defined blocks consisting of continental crust. These blocks were labelled I to V (Fig. 7). Minor faults within block II were restored to zero offset. At the top of each block we defined the top of the pre-rift sediments, which we assumed to coincide with the top of the acoustic basement. For blocks I and II we defined the base of the crust by assuming that it coincides with the horizon where the reflections flatten. LD, two branches of HD, and HHD separate the blocks from each other. The reconstruction was performed in a graphics program by the following procedure: the blocks were identified as explained above, were cut, rotated until the top of the pre-rift sediments of each block was in a subhorizontal position, and then aligned so that the pre-rift top of the upper crust became continuous. The misfit between the single blocks, which probably results from the fact that internal deformation within the blocks was not considered, increases oceanwards. Internal deformation may have consisted of an oceanward increasing amount of vertical shearing (west-side-up).

The major results obtained by reversing the two profiles are that: (1) the total extension accommodated by LD, HD and HHD is of the order of 32.6 km (initial, pre-detachment length is 34.6 km, final post-detachment length is 67.2 km), which results in a β factor of about two; (2) the crust was thinned to 12 km already before LD, HD and HHD became active; (3) assuming that the C reflection separates lower- from upper-crustal rocks, the rocks on both sides of the C reflection do not appear to deform in a different way during faulting along LD, HD and HHD, indicating that the crust behaved in a rheologically homogeneous way during detachment faulting. If the interpretation of Krawczyk *et al.* (1996) is inverted using the same procedure, the total extension is only of the order of 9.3 km, resulting in a β factor of 1.14. Interpreting the C reflection as a crust–mantle boundary as proposed by Whitmarsh *et al.* (2000), the pre-detachment thickness of the crust becomes about 6–7 km, resulting in even higher values for pre-detachment thinning. The different interpretations of the H reflection (Fig. 7a,c) do not change the amount of accommodated extension, but result in a different distribution of the crustal thickness before onset of detachment faulting. For interpretation 1 the crustal thickness remains constant throughout the zone affected by later detachment faulting, whereas for interpretation 2 the crust thins towards the future ocean–continent transition (Fig. 7b,d). However, kinematic inversion of interpretation 2 results in a duplication of the Moho in the pre-detachment configuration (Fig. 7d), a situation that is unlikely to occur. Therefore, we reject interpretation 2 and will continue to discuss only interpretation 1.

Our results show that the crust was thinned already before detachment faulting became active. Except for the C reflection, which we interpret as a reflection separating upper- from lower-crustal rocks, no other pre-detachment intrabasement structure could be identified in seismic sections from the Iberia Abyssal Plain. Reflection C might have acted as a décollement structure separating upper-crustal from lower-crustal deformation during pre-detachment pure-shear thinning of the crust. The lack of further structures associated with pre-detachment thinning can be explained in two ways. Either they exist but did not produce impedance contrast so that they are not imaged by seismic reflection profiles, or deformation was homogeneous in large parts of the continental crust and did not produce localized structures. It appears, however, that seismic profiles along rifted margins image predominantly structures associated with the latest evolution of the margins.

The β factor obtained for the extension accommodated by the detachment faults differs from the β factor obtained from the total thinning of the crust, which is much higher. There are two reasons for this: (1) the simple-shear geometry of the detachment faulting (the assumption that vertical thinning is inversely proportional to horizontal extension is valid only for pure shear deformation); (2) the presence of pre-detachment thinning as mentioned above. Our study shows that the structures that have been recognized and described so far by other workers interpreting the Lusigal 12 profile can account for only a fraction of the total thinning of the crust. Our interpretation of the profile leads to a higher β value, which, however, still explains only about half of the thinning. Thus, β factors calculated on the basis of seismically imaged structures may not necessarily reflect the total stretching and thinning of the crust.

Discussion

Kinematic evolution of detachment faulting

In Figure 7a and b the interpreted present-day geometry and the pre-detachment crustal thickness are shown. Anti-clockwise rotation of the blocks (looking north) leads from the present-day to the initial geometry. This, as well as the occurrence of breakaways on the continentward end of each fault, clearly indicates a top-to-the-ocean sense of shear for the detachment faults. To understand the kinematic evolution, the sequence of faulting has to be known. The observation that the crustal thickness between breakaway and mantle along LD, HD and HHD decreases oceanwards is compatible with a stepping of detachment faulting towards the future ocean, i.e. the younger a fault, the thinner the crust, and the shorter the distance from its breakaway to the crust–mantle boundary. Thus, we infer that the temporal sequence of faulting is LD, HD then HHD, which is also consistent with the interpretation that HD is cut by HHD on the oceanward side of Hobby High (Fig. 6).

After establishing the final and initial architecture, the kinematics and the age relationships between the different faults, we are able to describe the evolution of detachment faulting, its relation to exhumation of crustal and mantle rocks, and the formation of sedimentary basins. In Figure 8 we present the stages of detachment

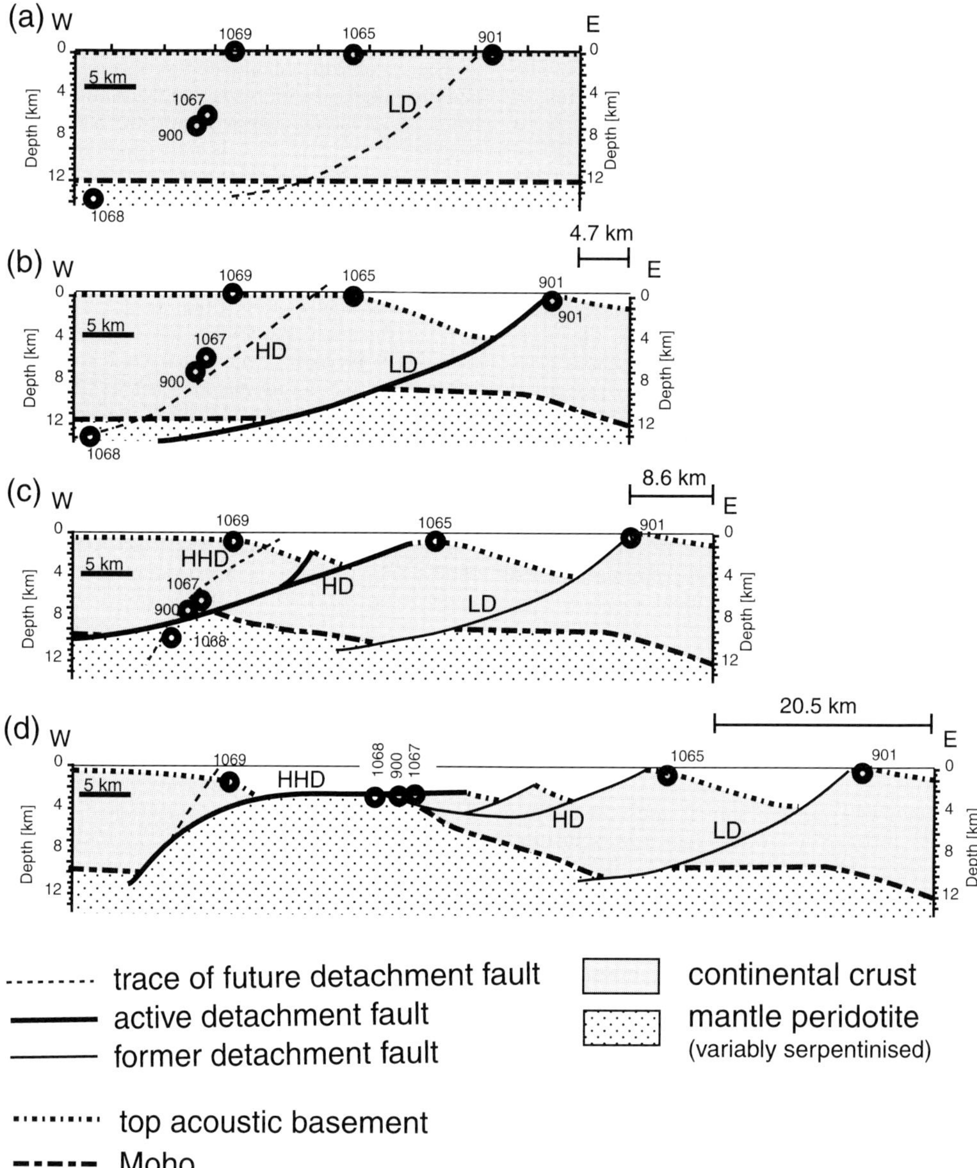

Fig. 8. Model showing the temporal and spatial evolution of faulting along the Iberia Abyssal Plain: (**a**) pre-'detachment' stage; (**b**) LD stage; (**c**) HD stage; (**d**) HHD stage. (For discussion, see text.)

faulting and we estimate and discuss the displacement, thinning and style of faulting.

Pre-'detachment' stage

The latest Jurassic–earliest Cretaceous history is recorded in the microfaunas and nannofloras of the sediments recovered at Sites 1069, 1065 and 901. Wilson *et al.* (2001) demonstrated that these show a transition from a Jurassic clastic interior basin with restricted circulation to an open marine, shelf-slope carbonate margin. The early Valanginian relatively outer shelf to upper slope nannofossil chalks are directly overlain by a Campanian deep-water turbidite–hemipelagite sequence, so that no continuous record is provided of the post-Valanginian subsidence to abyssal depths. Thus, the margin remained at shallow-water to slope conditions up to Valanginian time (140–135 Ma). Assuming that thinning of the crust and subsidence are intimately related, we can infer that substantial thinning of the crust (i.e. β >1.2) did not occur before Valanginian time.

The thickness of the pre-detachment crust and the horizontal distribution of the sites shown in Figure 8a are taken from the reconstructed profile presented in Figure 7b. From the kinematic inversion we obtain a crust of 12 km thickness for the stage at which extension initiated along LD. This stage does not correspond to the pre-rift stage, for which an equilibrated crust (i.e. 30 km thick) is assumed (Cordoba *et al.* 1987). The tectonic processes that led to a thinning of the crust from a pre-rift stage (*c.* 30 km thickness) to a pre-detachment stage (*c.* 12 km thickness) will not be discussed in this paper. A model to explain the pre-detachment thinning has been proposed by Manatschal & Bernoulli (1999). They proposed a conceptual model on the basis of data from Iberia and the Tethyan margins in which initial thinning of the crust is accommodated by pure-shear extension on a lithospheric scale and only the latest rifting stage is controlled by detachment faulting leading to the exhumation of subcontinental mantle at the sea floor.

The circles labelled with the Site numbers of ODP Legs 149 and 173 shown in Figure 8a correspond to the assumed pre-detachment position of the Sites before onset of faulting. The occurrence of pre-rift sediments at Sites 901, 1065 and 1069 shows that these sites were located on the top surfaces of tilted blocks. On the basis of the data presented above, the basement rocks recovered at Sites 1067 and 900 are derived from the lower continental crust. It is difficult to decide, however, on the depth of these rocks when extension along LD began. The 5–7 km depths shown in Figure 8a are based on assumptions discussed below. Also, the pre-LD position of the mantle rocks recovered at Site 1068 before onset of detachment faulting is not well constrained. In Figure 8a, we place these rocks near the top of the subcontinental mantle.

LD stage

LD is listric on a depth section (Fig. 8b). It bounds a basin up to 4 km deep and 15 km wide. The horizontal displacement accommodated along LD is of the order of 4.7 km. The occurrence of Tithonian pre-rift sediments in the footwall block near the breakaway at Site 901, as well as the shape of the high, suggests that this block remained below sea level and

was not eroded significantly during rifting. Thus, the footwall block was not significantly uplifted whereas the hanging wall was rotated and subsided, resulting in the formation of a basin showing the classical asymmetry of rift basins.

A remarkable implication of our reconstruction is that LD cuts across what is interpreted to represent the continental crust without changing its dip significantly on the depth-migrated section (Fig. 2). This suggests that the crust was rheologically homogeneous. Only at the base of the crust at about 12 km does the fault change its dip to become subhorizontal within a layer with a velocity higher than 7.3 km s^{-1}. The flattening of LD at 12–14 km suggests the occurrence of weak material at the crust–mantle boundary during activity along LD. There are four possible interpretations that can account for this: (1) the lowermost crust contains quartz-rich and therefore weak rocks such as tonalites; (2) the lowermost crust was hot enough for ductile deformation of feldspar; (3) the uppermost mantle was already serpentinized and was therefore weak; (4) the weak material was formed by hot, underplated gabbro. The available geophysical data do not allow us to distinguish between these possibilities, despite the fact that this assumed weak layer can be traced towards Hobby High where it has been sampled.

HD stage

HD is in many regards similar to LD, but was affected more strongly by post-detachment uplift during fault activity along HHD. We interpret HD to cut from its breakaway down to the base of the crust without changing significantly its dip and to flatten, like LD, at the crust–mantle boundary (Fig. 8c). The combined horizontal displacement along HD and the associated smaller fault in its hanging wall, which we interpret as kinematically linked to HD, is 8.6 km. Despite the higher amount of extension, the basin bounded by HD is much smaller and shallower compared with the basin bounded by LD. On the footwall high bounded by the breakaway of HD, evidence for erosion and subaerial exposure has not been found in the seismic image, nor in the cores drilled at Site 1065.

We assume that during activity along HD the Site 1068 mantle rocks were juxtaposed against the lower-crustal rocks of Sites 900 and 1067, leading to the close proximity of these rocks today at Hobby High. Whether HD penetrated into the mantle or soled out within the weakly serpentinized top of the mantle as indicated in Figure 8c cannot be resolved with the available data.

HHD stage

In contrast to LD and HD, which represent prominent intrabasement reflections in the Lusigal 12 profile, HHD coincides over a large distance with the top of the acoustic basement. The tectonic nature of HHD has been demonstrated by drilling at Sites 900, 1067 and 1068. On seismic profiles, however, the tectonic origin of Hobby High becomes clear only where it is overlain by a hanging-wall block such as at Site 1069. The oceanward transition from continental rocks at Sites 1067 and 900 to mantle rocks at Site 1068, occurring in the footwall of the HHD, as well as the occurrence of a block of upper continental crust in the hanging wall of the HHD (Site 1069), oceanwards of a tectonic window of exhumed mantle, strongly suggests a top-to-the-ocean sense of shear for this detachment structure. Assuming that the pre-rift sediments recovered in the hanging wall at Site 1069 have their equivalent at the breakaway of HHD in the footwall, the displacement along this fault amounts to 20.5 km.

A high amount of displacement along the HHD is also supported by the exhumation of lower-crustal rocks at Sites 900 and 1067, indicated by structural and mineralogical overprinting relationships documenting a retrograde tectono-metamorphic history extending from amphibolite-facies to sea-floor conditions. However, the ^{40}Ar/^{39}Ar hornblende cooling age of 161 Ma dating cooling to *c.* 500 °C is older than the shallow-water Tithonian sediments recovered at Sites 901, 1065 and 1069, which are interpreted as pre-rift (Wilson *et al.* 2001). Thus, the amphibolite at Site 1067 and probably also the gabbro at Site 900 are pre-rift in age. Therefore their deformation history as well as the retrograde metamorphic evolution of these rocks need not necessarily be related to exhumation along HHD only. If we assume a simple planar geometry for the HHD and take into account that the rocks of Hobby High cooled to *c.* 150 °C during exhumation along this fault, 5 km is a realistic initial depth. This results in an initial dip of 30° for the HHD. Although the HHD is regionally corrugated parallel to the margin as a result of late- and possibly also post-kinematic uplift, it dips at present nowhere more than 17° (Fig. 6). This low inclination, and the fact that the angle between the syn-rift breccias recovered at Site 1068 and the detachment fault is <10° indicates that the fault

was at a low angle when it was exposed at the sea floor. Major tilting after its activity can be excluded, because the overlying strata are subhorizontal. Hence, the tilting of the fault from 30° to 17° must have occurred during fault activity. Moreover, if a displacement of 20.5 km, as proposed above, had been accommodated by a fault dipping 30° with hanging wall and footwall remaining rigid, either the hanging wall, i.e. Site 1069, would have subsided by 10 km, or the footwall would have been uplifted above sea level. There is, however, no evidence for subaerial exposure during rifting of the footwall, nor for subsidence of the hanging wall of the HHD. The lower Valanginian rocks recovered at Site 1069 in the hanging wall of the HHD are yellow chalks dominated by nannoconids and micrantholiths representing a nannoflora assemblage thought to be indicative of relatively shallow-water, open-marine, outer shelf or slope environments (Wilson *et al.* 2001). Thus, we assume that along the HHD the footwall was pulled out from underneath a hanging wall that remained relatively stable, a situation described by the rolling-hinge model (Buck 1988; Wernicke & Axen 1988). The different tectonic style of the HHD as compared with the LD is also reflected in the architecture of the basins bounding HHD and LD. HHD is not associated with an asymmetric basin, and characteristic synsedimentary packages showing thickening towards the fault and onlapping onto the hanging wall are not observed, features that may be present in the basin bounded by the LD (the 'enigmatic wedges' of Wilson *et al.* 2001).

Although onset and duration of fault activity beneath the Iberia Abyssal Plain have not yet been dated, the available data permit some constraints to be put on HHD. The ^{40}Ar/^{39}Ar plagioclase ages of about 136 Ma (Site 900) and 137 Ma (Site 1067) date the activity of HHD, under the assumption that cooling to *c.* 150 °C was related to exhumation along this fault, which is supported by the structural data presented above. Consequently, HHD was active around 137 Ma (i.e. the Berriasian–Valanginian boundary according to the time scale of Gradstein & Ogg (1996)). The progressive tectonic overprint within the sedimentary breccias at Site 1068 towards the base, as well as the occurrence of clasts preserving the deformation history observed in the footwall rocks at Hobby High, are consistent with deposition of these breccias during activity along HHD. Cataclastic deformation along the fault may have facilitated the formation and erosion of clasts, and seismic failure indicated by the occurrence of pseudotachylite (Fig. 4a) may have triggered mass flows.

The clast stratigraphy in the sedimentary breccias of Site 1068 also supports the idea that tectonic exhumation of the Hobby High and deposition of the breccias are simultaneous. Above a zone of strongly tectonized and altered breccias composed of serpentinite, metagabbro and meta-anorthosite, the lowermost unequivocal sedimentary breccias (Core 19R-2) consist exclusively of amphibolite clasts identical with the rocks recovered at Site 1067. Metagabbro clasts identical with those found at Site 900 occur for the first time in Core 18R-1 and become dominant upwards in the section. This suggests that the first basement rocks that became exposed and available to erosion were the amphibolites from Site 1067 whereas the gabbros from Site 900 became exhumed and eroded only later. This observation not only confirms the sense of shear determined for HHD, but also supports the idea that the exhumation of lower-crustal and even mantle rocks was accommodated by a low-angle detachment fault that was active at the sea floor during sedimentation. Unfortunately, the occurrence of Valanginian to Barremian nannofossil assemblages within the matrix of the breccia recovered at Site 1068 does not throw much light on the age of the HHD and therefore also on the exhumation of Hobby High. The plagioclase cooling ages obtained from the Site 900 gabbros and Site 1067 amphibolites and their occurrence as clasts in the sedimentary breccias recovered at Site 1068 indicate, however, that tectonic activity along HHD might have occurred during Valanginian to Barremian time. Assuming that detachment faulting predates the onset of sea-floor spreading determined to occur in the Iberia Abyssal Plain close to 126 Ma (early Barremian) (Whitmarsh & Miles 1995), detachment faulting occurred around 137 Ma and had to terminate before 126 Ma, suggesting a short duration of detachment activity.

Clasts of metagreywacke are very rare, but occur throughout the breccia section at Site 1068. Composition of these clasts is identical to that of the clasts recovered at Site 1069 in the hanging wall of the HHD. A possible interpretation is that these clasts were deposited on the footwall (= Hobby High) when it was pulled from underneath its hanging wall, i.e. the upper-crustal block underlying Site 1069.

The interpretation proposed here for HHD is new for the Iberia Abyssal Plain. In the Eastern Alps, however, structures similar to that of HHD are spectacularly exposed and can be observed directly in the field (Manatschal & Bernoulli 1999). The occurrence of blocks of upper continental crust overlying exhumed subcontinental mantle (analogous to the block underlying Site 1069) has been described by Manatschal & Nievergelt (1997) from the Platta nappe, the geometry of low-angle detachment faults within the ocean–continent transition has been described by Florineth & Froitzheim (1994) from the Tasna nappe, and detachment faults cutting across the continental crust have been described by Froitzheim & Eberli (1990) and Manatschal & Nievergelt (1997) from the Err nappe.

Conclusions

The importance of detachment faulting in the extension of the NE Atlantic margins from the Armorican margin of the northern Bay of Biscay to the southern Iberia Abyssal Plain has been emphasized by many workers (e.g. Boillot *et al.* 1987; Le Pichon & Barbier 1987; Sibuet 1992; Krawczyk *et al.* 1996; Reston *et al.* 1996). In this paper, we have evaluated the role of detachment faulting based on a reinterpretation of the Lusigal 12 profile in the light of the results of ODP Legs 149 and 173. Our study led to the following results.

(1) The crust within the distal margin was already considerably thinned before the detachment structures recognized in the seismic profile became active. Hence, detachment faulting is a late event, which can explain only part of the thinning of the continental crust. This is in line with the numerical modelling results of Lavier *et al.* (1999) suggesting that detachment faulting is favoured when the brittle layer is thin.

(2) The detachment faults cut across the thinned crust without significant changes in their dip. They flatten only at the crust–mantle boundary, suggesting that the crust was rheologically homogeneous during the final stage of rifting, and that decoupling took place at the crust–mantle boundary. This occurred because the mantle was serpentinized at the top, quartz-rich rocks occurred at the base of the lower crust, the temperatures were sufficiently high that feldspar deformed by crystal-plastic processes, or magmatic underplating took place during detachment faulting.

(3) Faulting propagated towards the future ocean, which resulted in a temporal and spatial evolution of tectonic activity and sedimentation within the ocean–continent transition that is reflected in the architecture of the sedimentary basins.

(4) The amount of displacement along individual faults increased oceanward from 4.7 km along LD to 20.5 km along the younger HHD. This change in displacement is accompanied by a change in the geometry from downward flattening, listric, in the case of LD, to upward flattening in the case of HHD. The upward flattening geometry, compatible with a rolling-hinge model, allowed exhumation of deeper crustal levels and simultaneous sedimentation on an active subhorizontal fault plane, both of which are observed along HHD. The across-strike changes in geometry of the detachment faults were accompanied by a change in the mode of deformation. Hanging-wall subsidence along LD leading to a large asymmetric basin was followed by the footwall being pulled from beneath a stable hanging wall along HHD, the latter leading to a sort of conveyor belt along which stratal relationships are completely different from that within classical rift basin (Wilson *et al.* 2001). Because HD shows a transitional stage between LD and HHD we assume that the evolution of faulting from listric to flattening upward with a rolling hinge, or anti-listric, is a gradual process accompanied by an increase in displacement (possibly also rate of displacement) and rate of exhumation, and a change in the overall rheology.

(5) The 161 Ma $^{40}Ar/^{39}Ar$ amphibole age from the Site 1067 amphibolite suggests that only the part of the deformation history that evolved below *c.* 500 °C is related to detachment faulting along the ocean–continent transition. Thus, most of the extension leading to the mantle exhumation is accommodated within the brittle field as far as mafic rocks are concerned, and is localized within shear zones.

(6) The kinematic reconstruction forces us to assume initial dips of at least 30° for the detachment faults that were rotated to shallower dips during their activity (Fig. 8).

We thank D. Bernoulli, B. Whitmarsh, C. Wilson, B. Taylor, G. Molli and G. Axen for comments, suggestions and very helpful reviews. We would also like to thank the Shipboard Scientific Party, the ODP technicians and the SEDCO drilling crew of Leg 173 for their help. The work of G.M. and N.F. has been supported by the Swiss National Science Foundation Project 21-049117.96/1.

References

ABERS, G.A. 2001. Evidence for seismogenic normal faults at shallow dips in continental rifts. *In*: WILSON, R.C.L., WHITMARSH, R.B., TAYLOR, B. & FROITZHEIM, N. (eds) *Non-volcanic Rifting of Continental Margins: a Comparison of Evidence from Land and Sea.* Geological Society, London, Special Publications, **187**, 305–318.

ABERS, G.A., MUTTER, C.Z. & FANG, J. 1997. Shallow dips of normal faults during rapid extension: earthquakes in the Woodlark–D'Entrescasteaux rift system, Papua New Guinea. *Journal of Geophysical Research*, B7, **102**, 15301–15317.

ANDERSON, E.M. *The Dynamics of Faulting* (1st). Oliver and Boyd, Edinburgh, 1–183.

BERNOULLI, D. & WEISSERT, H. 1985. Sedimentary fabrics in Alpine ophicalcites, South Pennine Arosa zone, Switzerland. *Geology*, **13**, 755–758.

BESLIER, M.-O., ASK, M. & BOILLOT, G. 1993. Ocean–continent boundary in the Iberia Abyssal Plain from multichannel seismic data. *Tectonophysics*, **218**, 383–393.

BESLIER, M.-O., CORNEN, G. & GIRARDEAU, J. 1996. Tectono-metamorphic evolution of peridotites from the ocean/continent transition of the Iberia Abyssal Plain margin. *In*: WHITMARSH, R.B., SAWYER, D.S., KLAUS, A. & MASSON, D.G. (eds) *Proceedings of the Ocean Drilling Program, Scientific Results, 149.* Ocean Drilling Program, College Station, TX, 397–412.

BOILLOT, G., RECQ, M., WINTERER, E.L. & 21 OTHERS 1987. Tectonic denudation of the upper mantle along passive margins: a model based on drilling results (ODP Leg 103, western Galicia margin, Spain). *Tectonophysics*, **132**, 335–342..

BRUN, J.P. & BESLIER, M.-O. 1996. Mantle exhumation at passive margins. *Earth and Planetary Science Letters*, **142**, 161–173.

BUCK, W.R. 1988. Flexural rotation of normal faults. *Tectonics*, **7**, 959–973.

CAPDEVILA, R. & MOUGENOT, D. 1988. Pre-Mesozoic basement of the western Iberian continental margin and its place in the Variscan belt. *In*: BOILLOT, G. & WINTERER, E.L. (eds) *Proceedings of the Ocean Drilling Program, Scientific Results, 103.* Ocean Drilling Program, College Station, TX, 3–12.

CHIAN, D., LOUDEN, K.E., MINSHULL, T.A. & WHITMARSH, R.B. 1999. Deep structure of the ocean–continent transition in the southern Iberia Abyssal Plain from seismic refraction profiles: Ocean Drilling Program (Legs 149 and 173) transect. *Journal of Geophysical Research*, B4, **104**, 7443–7462.

CHRISTENSEN, N.I. 1966. Elasticity of ultramafic rocks. *Journal of Geophysical Research*, **71**, 5921–5931.

CHRISTENSEN, N.I. 1972. The abundance of serpentinites in the oceanic crust. *Journal of Geology*, **80**, 709–719.

COMAS, M.C., SÁNCHEZ-GÓMEZ, M., CORNEN, G. & de Kaenel, E. 1996. Serpentinised peridotite breccia and olistostrome on basement highs of the Iberia Abyssal Plain: implications for tectonic margin evolution. *In*: WHITMARSH, R.B., SAWYER, D.S., KLAUS, A. & MASSON, D.G. (eds) *Proceedings of the Ocean Drilling

Program, Scientific Results, 149. Ocean Drilling Program, College Station, TX, 577–591.

CORDOBA, D., BANDA, E. & ANSORGE, J. 1987. The Hercynian crust in northwestern Spain: a seismic survey. *Tectonophysics*, **132**, 321–333.

CORNEN, G., BESLIER, M.-O. & GIRARDEAU, J. 1996. Petrology of the mafic rocks cored in the Iberia Abyssal Plain. *In*: WHITMARSH, R.B., SAWYER, D.S., KLAUS, A. & MASSON, D.G. (eds) *Proceedings of the Ocean Drilling Program, Scientific Results, 149*. Ocean Drilling Program, College Station, TX, 449–469.

CORNEN, G., GIRARDEAU, J. & MONNIER, C. 1999. Basalts, underplated gabbros and pyroxenites record the rifting process of the West Iberian margin. *Mineralogy and Petrology*, **67**, 111–142.

DAVIS, G.A. & LISTER, G.S. 1988. Detachment faulting in continental extension: perspectives from the Southwestern US Cordillera. *In*: CLARK, S.P. Jr, BURCHFIEL, B.C. & SUPPE, J. (eds) *Processes in Continental Lithospheric Deformation*. Geological Society of America, Special Papers, **218**, 133–159.

DE KAENEL, E. & BERGEN, J.A. 1996. Mesozoic calcareous nannofossil biostratigraphy from Sites 897, 899, and 901, Iberia Abyssal Plain: new biostratigraphic evidence. *In*: WHITMARSH, R.B., SAWYER, D.S., KLAUS, A. & MASSON, D.G. (eds) *Proceedings of the Ocean Drilling Program, Scientific Results, 149*. Ocean Drilling Program, College Station, TX, 27–59.

Discovery 215 Working Group 1998. Deep structure in the vicinity of the ocean–continent transition zone under the southern Iberia Abyssal Plain. *Geology*, **26**, 743–746.

FÉRAUD, G., BESLIER, M.-O. & CORNEN, G. 1996. $^{40}Ar/^{39}Ar$ dating of gabbros from the ocean/continent transition of the western Iberia margin: preliminary results. *In*: WHITMARSH, R.B., SAWYER, D.S., KLAUS, A. & MASSON, D.G. (eds) *Proceedings of the Ocean Drilling Program, Scientific Results, 149*. Ocean Drilling Program, College Station, TX, 489–495.

FLORINETH, D. & FROITZHEIM, N. 1994. Transition from continental to oceanic basement in the Tasna nappe (Engadine window, Graubünden, Switzerland): evidence for Early Cretaceous opening of the Valais ocean. *Schweizerische Mineralogische und Petrographische Mitteilungen*, **74**, 437–448.

FROITZHEIM, N. & EBERLI, G.P. 1990. Extensional detachment faulting in the evolution of a Tethys passive continental margin, Eastern Alps, Switzerland. *Geological Society of America Bulletin*, **102**, 1297–1308.

FROITZHEIM, N. & MANATSCHAL, G. 1996. Kinematics of Jurassic rifting, mantle exhumation, and passive-margin formation in the Austroalpine and Penninic nappes (eastern Switzerland). *Geological Society of America Bulletin*, **108**, 1120–1133.

GIBSON, I.L., MILLIKEN, K.L. & MORGAN, J.K. 1996. Serpentinite-breccia landslide deposits generated during crustal extension at the Iberia margin. *In*: WHITMARSH, R.B., SAWYER, D.S., KLAUS, A. & MASSON, D.G. (eds) *Proceedings of the Ocean Drilling Program, Scientific Results 149*. Ocean Drilling Program, College Station, TX, 571–575.

GRADSTEIN, F.M. & OGG, J.G. 1996. A Phanerozoic Time Scale. *Episodes*, **19**, 3–5.

KRAWCZYK, C.M., RESTON, T.J., BESLIER, M.-O. & BOILLOT, G. 1996. Evidence for detachment tectonics on the Iberia Abyssal Plain rifted margin. *In*: WHITMARSH, R.B., SAWYER, D.S., KLAUS, A. & MASSON, D.G. (eds) *Proceedings of the Ocean Drilling Program, Scientific Results, 149*. Ocean Drilling Program, College Station, TX, 603–615.

LAVIER, L.L., BUCK, W.R. & POLIAKOV, A N.B. 1999. Self-consistent rolling-hinge model for the evolution of large-offset low-angle normal faults. *Geology*, **27**, 1127–1130.

LEMOINE, M., TRICART, P. & BOILLOT, G. 1987. Ultramafic and gabbroic ocean floor of the Ligurian Tethys (Alps, Corsica, Apennines): in search of a genetic model. *Geology*, **15**, 622–625.

LE PICHON, X. & BARBIER, R. 1987. Passive margin formation by low-angle faulting within the upper crust: the northern Bay of Biscay margin. *Tectonics*, **6**, 133–150.

LISTER, G.S., ETHERIDGE, M.A. & SYMONDS, P.A. 1986. Detachment faulting and the evolution of passive continental margins. *Geology*, **14**, 246–250.

MANATSCHAL, G. 1999. Fluid- and reaction-assisted low-angle normal faulting: evidence from rift-related brittle fault rocks in the Alps (Err nappe, Switzerland). *Journal of Structural Geology*, **21**, 777–793.

MANATSCHAL, G. & BERNOULLI, D. 1999. Architecture and tectonic evolution of non-volcanic margins: present-day Galicia and ancient Adria. *Tectonics*, **18**, 1099–1119.

MANATSCHAL, G. & NIEVERGELT, P. 1997. A continent–ocean transition recorded in the Err and Platta nappes (Eastern Switzerland). *Eclogae Geologicae Helvetiae*, **90**, 3–27.

MÜNTENER, O., HERMANN, J. & TROMMSDORFF, V. 2000. Cooling history and exhumation of lower crustal granulites and upper mantle (Malenco, Eastern Central Alps). *Journal of Petrology*, **41**, 175–200.

ODP Leg 173 Shipboard Scientific Party 1998. Drilling reveals transition from continental breakup to early magmatic crust. *EOS Transactions, American Geophysical Union*, **79**, 173–181.

ORDÓÑEZ CASADO, B. 1998. *Geochronological studies of the pre-Mesozoic basement of the Iberian Massif: the Ossa Morena zone and the Allochthonous Complexes within the Central Iberian zone*. PhD thesis, ETH Zürich.

PICKUP, S.L.B., WHITMARSH, R.B., FOWLER, C.M.R. & RESTON, T.J. 1996. Insight into the nature of the ocean–continent transition off West Iberia from a deep multichannel seismic reflection profile. *Geology*, **24**, 1079–1082.

PINHEIRO, L.M., WILSON, R.C.L., PENA DOS REIS, R., WHITMARSH, R.B. & RIBEIRO, A. 1996. The western Iberia margin: a geophysical and geological overview. *In*: WHITMARSH, R.B., SAWYER, D.S., KLAUS, A. & MASSON, D.G. (eds) *Proceedings of the Ocean Drilling Program, Scientific Results, 149*. Ocean Drilling Program, College Station, TX, 3–23.

RESTON, T.J., KRAWCZYK, C.M. & KLAESCHEN, D. 1996. The S reflector west of Galicia (Spain): evidence from prestack depth migration for detachment faulting during continental breakup. *Journal of Geophysical Research*, B4, **101**, 8075–8091.

SAWYER, D.S., WHITMARSH, R.B., KLAUS, A. *et al.* (eds) 1994. *Proceedings of the Ocean Drilling Program, Initial Reports, 149*. Ocean Drilling Program, College Station, TX.

SCHMID, S.M. & CASEY, M. 1986. Complete fabric analysis of some commonly observed quartz *c*-axis patterns. *In*: HEARD, H.C. & HOBBS, B.E. (eds) *Mineral and Rock Deformation: Laboratory Studies. The Paterson Volume*. Geophysical Monograph, American Geophysical Union, **36**, 263–286.

SEIFERT, K.E., CHENG-WEN, C. & BRUNOTTE, D.A. 1997. Evidence from ODP Leg 149 mafic igneous rocks for oceanic crust in the Iberia Abyssal Plain ocean–continent transition zone. *Journal of Geophysical Research*, **102**, 7915–7928.

SEIFERT, K.E., GIBSON, I., WEIS, D. & BRUNOTTE, D. 1996. Geochemistry of metamorphosed cumulate gabbros from Hole 900A, Iberia Abyssal Plain. *In*: WHITMARSH, R.B., SAWYER, D.S., KLAUS, A. & MASSON, D.G. (eds) *Proceedings of the Ocean Drilling Program, Scientific Results, 149*. Ocean Drilling Program, College Station, TX, 471–488.

SIBUET, J.C. 1992. New constraints on the formation of the non-volcanic continental Galicia–Flemish Cap conjugate margins. *Journal of the Geological Society, London*, **149**, 829–840.

SKELTON, A.D.L. & VALLEY, J.W. 2000. The relative timing of serpentinization and mantle exhumation at the ocean–continent transition, Iberia: constraints from oxygen isotopes. *Earth and Planetary Science Letters*, **178**, 327–338.

TALBOT, C.J. & GHEBREAB, W. 1997. Red Sea detachment and basement core complexes in Eritrea. *Geology*, **25**, 655–658.

TAYLOR, B., HUCHON, P., KLAUS, A. *et al.* (eds) 1999. *Proceedings of the Ocean Drilling Program, Initial Reports, 180*. College Station, TX. Available at: http://www-odp.tamu.edu/publications/180_IR/180ir.htm.

TREGONING, P., LAMBECK, K., STOLZ, A. & 7 OTHERS 1998. Estimation of current plate motions in Papua, Global Positioning System observations. *Journal of Geophysical Research*, **103**, 12181–12203.

WERNICKE, B. 1981. Low-angle normal faults in the Basin and Range Province: nappe tectonics in an extending orogen. *Nature*, **291**, 645–648.

WERNICKE, B. & AXEN, G.J. 1988. On the role of isostasy in the evolution of normal fault systems. *Geology*, **16**, 848–851.

WHITMARSH, R.B. & MILES, R.P. 1995. Models of the development of the West Iberia rifted continental margin at 40°30′N deduced from surface and deep-tow magnetic anomalies. *Journal of Geophysical Research*, **100**, 3789–3806.

WHITMARSH, R.B. & SAWYER, D.S. 1996. The ocean/continent transition beneath the Iberia Abyssal Plain and continental-rifting to seafloor-spreading processes. *In*: WHITMARSH, R.B., SAWYER, D.S., KLAUS, A. & MASSON, D.G. (eds) *Proceedings of the Ocean Drilling Program, Scientific Results, 149*. Ocean Drilling Program, College Station, TX, 713–733.

WHITMARSH, R.B., BESLIER, M.-O., WALLACE, P.J. *et al.* (eds) 1998. *Proceedings of the Ocean Drilling Program, Initial Reports, 173*. Ocean Drilling Program, College Station, TX.

WHITMARSH, R.B., DEAN, S.M., MINSHULL, T.A. & TOMPKINS, M. 2000. Tectonic implications of exposure of lower continental crust beneath the Iberia Abyssal Plain, Northeast Atlantic Ocean: geophysical evidence. *Tectonics*, **19** (5), 919–942.

WILSON, R.C.L., MANATSCHAL, G. & WISE, S. 2001. Rifting along non-volcanic passive margins: stratigraphic and seismic evidence from the Mesozoic successions of the Alps and Western Iberia. *In*: WILSON, R.C.L., WHITMARSH, R.B., TAYLOR, B. & FROITZHEIM, N. (eds) *Non-volcanic Rifting of Continental Margins: a Comparison of Evidence from Land and Sea*. Geological Society, London, Special Publications, **187**, 429–452.

WILSON, R.C.L., SAWYER, D.S., WHITMARSH, R.B., ZERONG, J. & CARBONELL, J. 1996. Seismic stratigraphy and tectonic history of the Iberia Abyssal Plain. *In*: WHITMARSH, R.B., SAWYER, D.S., KLAUS, A. & MASSON, D.G. (eds) *Proceedings of the Ocean Drilling Program, Scientific Results, 149*. Ocean Drilling Program, College Station, TX, 617–633.

Rifting along non-volcanic passive margins: stratigraphic and seismic evidence from the Mesozoic successions of the Alps and western Iberia

R.C.L. WILSON[1], G. MANATSCHAL[2,3] & S. WISE[4]

[1]*Department of Earth Sciences, The Open University, Milton Keynes MK7 6AA, UK*
(e-mail: r.c.l.wilson@open.ac.uk)
[2]*Geologisches Institut, ETH, Zurich, Switzerland*
[3]*Present address: EOST-UMR 7517 CRNS, Université Louis Pasteur, 1 rue Blessig,*
F-67064 Strasbourg Cedex, France
[4]*Department of Geology, Florida State University, Tallahassee, FL 32305-4100, USA*

Abstract: The paper examines aspects of the sedimentology and stratigraphy of rift basins that evolved in deep marine settings near the ocean–continent transition. It focuses on the applicability of a low-angle extensional detachment model developed in the Alps to the West Iberian margin, and on difficulties of objectively identifying syn-rift stratigraphic intervals in both areas. The paper examines evidence obtained from Ocean Drilling Program holes drilled in the Iberia Abyssal Plain. Despite the fact that all the holes were sited above highs in the acoustic basement and so did not penetrate a complete sedimentary record of rifting, they do provide some constraints on the age and mechanism of rifting. We suggest that published identifications of syn-rift intervals in distal basins off West Iberia and in the Southern Alps have not demonstrated, using objective criteria, the occurrence of syn-rift stratigraphic intervals. They have, therefore, probably overestimated the duration of rifting by as much as 20 Ma. The absence of syn-rift related stratal divergence towards fault footwalls may be due to resedimentation of syn-rift sediments towards basin centres, lack of significant hanging-wall rotation along flat detachment faults, or the syn-rift interval being too thin to resolve on seismic data. The syn-rift episode beneath the deep Galicia margin postdates Tithonian–Berriasian shallow-water carbonates, and predates Late Valanginian turbiditic sediments. Drilling results from the Iberia Abyssal Plain suggest a similar age because Tithonian siliciclastic mudrocks are overlain by Berriasian pelagic chalks. It seems likely that in both regions rifting lasted for <5 Ma, probably from late Berriasian to early Valanginian. At Site 1068 in the Iberia Abyssal Plain, the interpretation from seismic reflection data, of a low-angle detachment dipping about 10° west, was confirmed by drilling, which revealed sedimentary and tectonic breccias containing clasts of lower-crustal rocks overlying a fault zone below which occurs serpentinized peridotite showing a downward decrease in deformation. At least 20.5 km of displacement is interpreted to have occurred along this fault, but it is not accompanied by large-amplitude, rift-related topography. This paradox is resolved if the detachment developed as a deepening-downwards, rolling-hinge fault.

This paper examines aspects of the sedimentology and stratigraphy of basins developed during large extension (i.e. β >2) and in very deep marine settings. The features of such basins have received scant attention in the voluminous literature concerning rift basins, which focuses on those formed by relatively low extension, and filled by continental, shallow marine or submarine fan depositional systems.

Comparing evidence from land and sea is a crucial part of our approach. This is not straightforward, for different datasets collected across a range of spatial scales have to be evaluated: cores, outcrops, seismic reflection profiles and other geophysical data (Fig. 1). The outcrop scale is crucial because it allows the spatial gulf to be bridged between Deep Sea Drilling Project (DSDP) and Ocean Drilling Program (ODP) cores, and marine seismic data. There is also the problem that basins on land and beneath the sea inevitably have had different post-rift histories, resulting in their

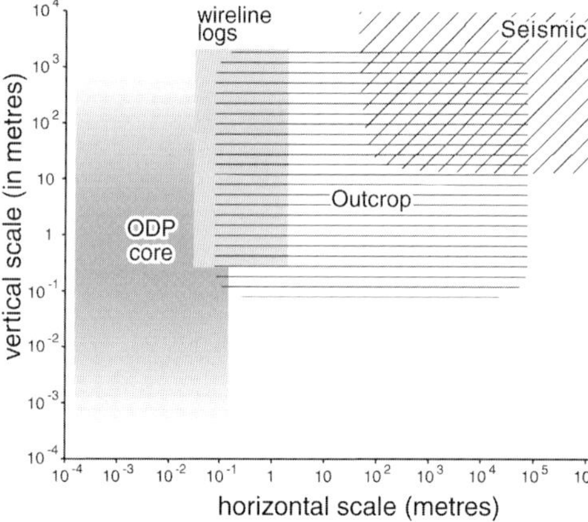

Fig. 1. Contrasting scales, shown as a log–log plot, of datasets obtained from studies of continental margins. (Note the crucial position of the outcrop scale in bridging the spatial gulf between ODP cores and seismic reflection profiles (modified from Wilson 1998).)

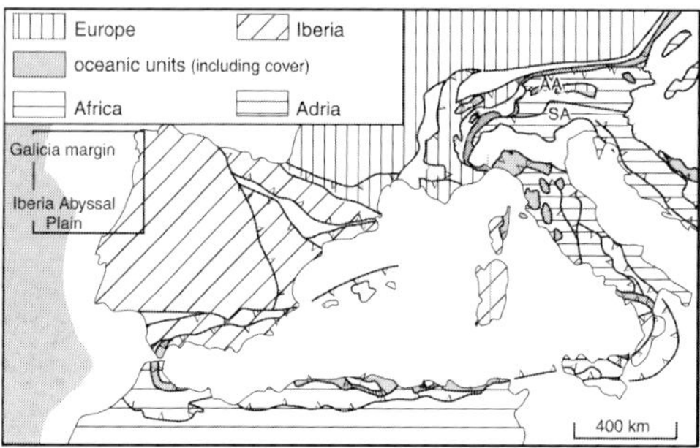

Fig. 2. Simplified tectonic map of the Alps, western Mediterranean and Iberia, showing the distribution of Mesozoic continental and oceanic crust.

contrasting present-day elevations (Fig. 2). In the Alps, portions of continental margins and oceanic crust are superbly exposed, but dismembered by subsequent compressional events. Off Iberia, extensional features have been only slightly deformed by compressional events, but are buried beneath significant thicknesses of post-rift sediments.

There are two underlying themes to the paper. One concerns the application of a low-angle detachment model developed in the Alps (Froitzheim & Manatschal 1996; Manatschal & Nievergelt 1997) to the interpretation of the seismic reflection and core data obtained off West Iberia. The other is related to the problem of objectively defining syn-rift stratigraphic units at the outcrop and seismic scales.

Researchers working in the field in the Alps, and on seismic reflection sections shot off the Galicia margin, have identified sedimentary wedges that accumulated between tilted fault blocks as syn-rift units. As discussed by Prosser (1993), a considerable part of such wedges are likely to be post-rift, rather than syn-rift, in origin. We, like Prosser, and Driscoll *et al.* (1995), believe that the only reliable means of identifying syn-rift intervals is to observe features indicative of fault block rotation. These are divergence of seismic reflectors or lithostratigraphic units towards fault footwalls, and their associated rotation. In addition, 'each phase of differential subsidence is recorded by an onlapping stratigraphic package' (Driscoll *et al.* 1995). Dating this basal onlap or unconformity, therefore, provides a means of establishing the age of rift onset. There may be several rift onset unconformities within a basin, indicating multiple phases of rifting, e.g. the Jeanne d'Arc Basin off Newfoundland (Driscoll *et al.* 1995).

In the absence of direct, or indirect (by modelling oceanic magnetic anomalies) evidence concerning the age of onset of sea-floor spreading, the break-up unconformity (Falvey 1974) is used as a proxy for estimating when this occurred (Driscoll *et al.* 1995). The break-up unconformity also signals when the rifting phase of evolution of passive margins came to an end. Driscoll *et al.* (1995) used the following three stratigraphic criteria to identify the break-up unconformity: (1) in contrast to thickening towards fault footwalls 'sediments overlying the unconformity typically have greater spatial persistence and more uniform thickness reflecting regional subsidence associated with the cooling and contraction of the lithosphere'; (2) 'growth faults associated with expanded (syn-rift) sedimentary sections on the downthrown block normally occur beneath the unconformity'; (3) 'faulting and offset should diminish markedly across the break-up unconformity'. We show that these objective criteria have not been rigorously applied, or were not reported by previous workers who studied the Galicia margin or the Southern Alps. We question, therefore, whether some of the criteria are applicable to the late-stage, high-extension tectonic regimes that led to continental separation in these areas.

The paper begins by reviewing briefly the geological setting of the Mesozoic rift basins discussed in this paper, focusing on the role of simple (i.e. high β) and pure shear (low β) in the development of the proximal and distal portions of continental margins. We then examine the stratigraphic features of the distal parts of these margins, focusing on evidence that allows the timing and duration of rifting episodes to be constrained; we conclude that rifting off Iberia that heralded ocean opening lasted for a much shorter period of time than reported by previous workers. In the light of the review, the likely generic features of sediments filling deep marine basins developed in low and high extensional regimes are discussed.

Geological setting

Figure 3 shows that whereas rifting and ocean opening events along the Alpine margins and off Iberia were not synchronous, they were part of a linked system of movements. The Ligurian–Piemonte Ocean, which was kinematically linked to the Central Atlantic, began to open during Mid-Jurassic time, whereas ocean opening between Iberia and Newfoundland began in Early Cretaceous time.

Manatschal & Bernoulli (1998, 1999) have shown that the undeformed Cretaceous Iberia–Newfoundland margins and the relics of Jurassic Tethyan margins preserved in the Alps show similar rift histories. An early phase of rifting occurred across a wide zone, which became the proximal parts of the future margins (Bernina–Ortler regions of the Austroalpine zone (Fig. 4a); Gozzano high, Lombardian Basin of the Southern Alps (Fig. 4b); Galicia Bank, Interior Basin, Porto Basin of the Iberian margin (Fig. 5c)). This phase is characterized by β values of <1.3 to 1.6 and half-grabens up to 20–40 km across. More localized rifting occurred 20–40 Ma later in the zones that became the distal margins adjacent to newly formed oceanic crust (Err region of the Austroalpine zone (Fig. 4a); Deep Galicia Margin and Iberia Abyssal Plain of the Iberian margin (Fig. 5b,d)). The latter tectonic episode resulted in an asymmetry between the opposing margins indicative of the development of upper- and lower-plate margins. The distal parts of the West Iberia margin, and Adriatic margins preserved in the Alps, are characterized by high β values (>2), low-angle detachment faults, tilted fault blocks <10 km wide, and exposure of serpentinized mantle rocks at the sea floor. All these features are consistent with a lower-plate setting, but major post-rift subsidence that occurred in both areas is a feature of an upper-plate setting. Driscoll & Karner (1998) suggested that large amounts of post-rift subsidence may be generated by lower-crustal extension with little attendant upper-crustal brittle

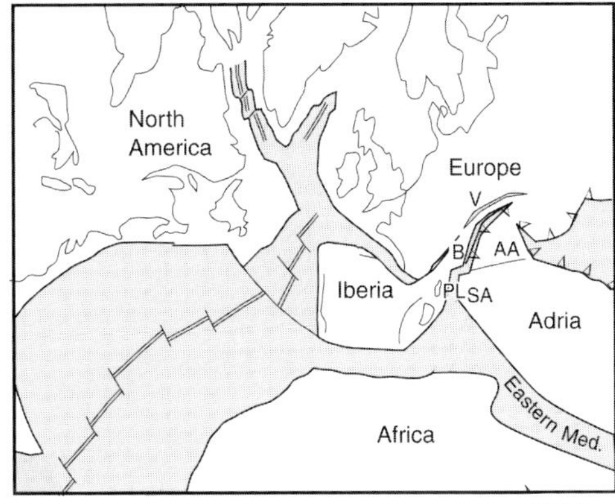

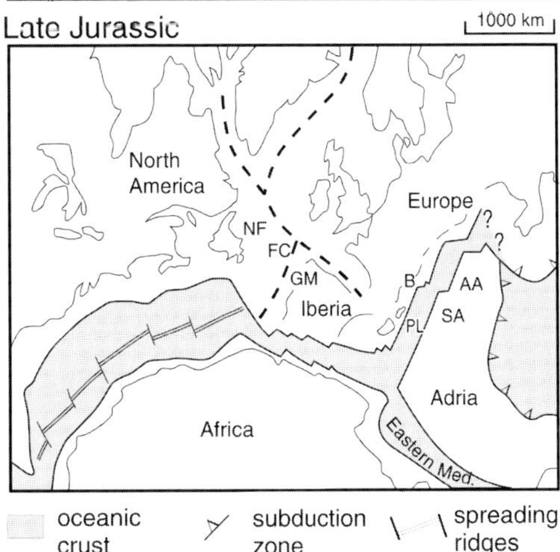

Fig. 3. Two main stages in the Mesozoic development of the North Atlantic Ocean. AA, Austroalpine zone; B, Briançonnais; FC, Flemish Cap; GM, Galicia margin; NF, Newfoundland; PL, Piemonte–Ligurian Ocean; SA, Southern Alps; V, Valais ocean. From Manatschal & Bernoulli (1999).

deformation during the late stages of rifting just before continental break-up. This led them to conclude that the terms upper and lower plate 'are only applicable when describing the morphology of brittle deformation observed on conjugate margins'. Furthermore, they believe that: 'in terms of describing the distribution and style of subsidence observed on conjugate margins the use of upper and lower plate is misleading because both margins may display subsidence patterns that are characteristic of the upper plate (e.g. Newfoundland and Iberia)'.

It should be remembered, however, that the original definition by Lister *et al.* (1988) of a lower-plate margin states that it comprises 'the deeper crystalline rocks of the lower plate

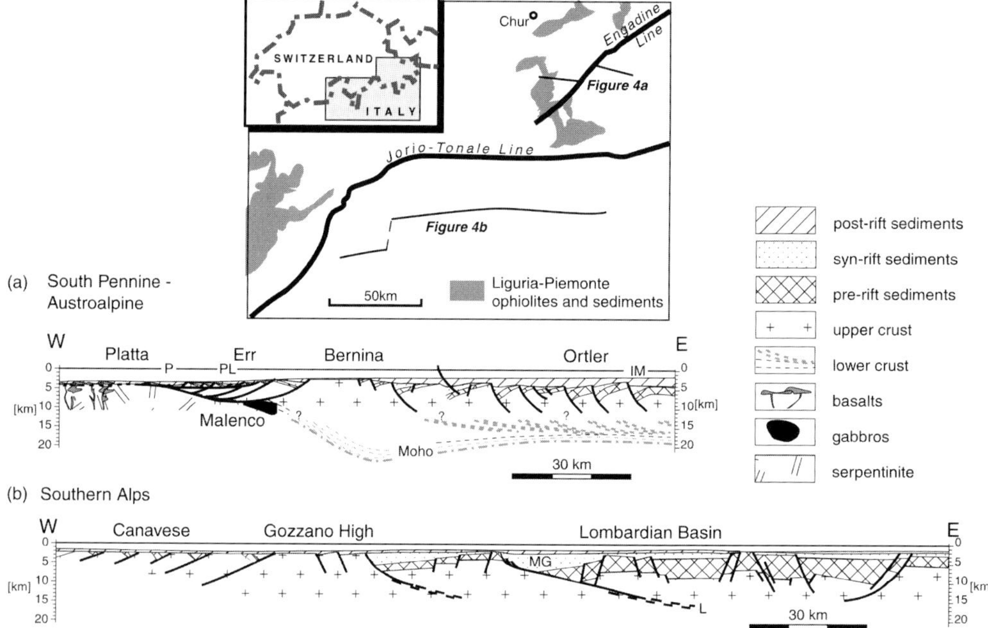

Fig. 4. Palinspastic Mid-Jurassic cross-sections across the Adriatic margin of the Alps. For location, see Figure 3a. From Manatschal & Bernoulli (1999).

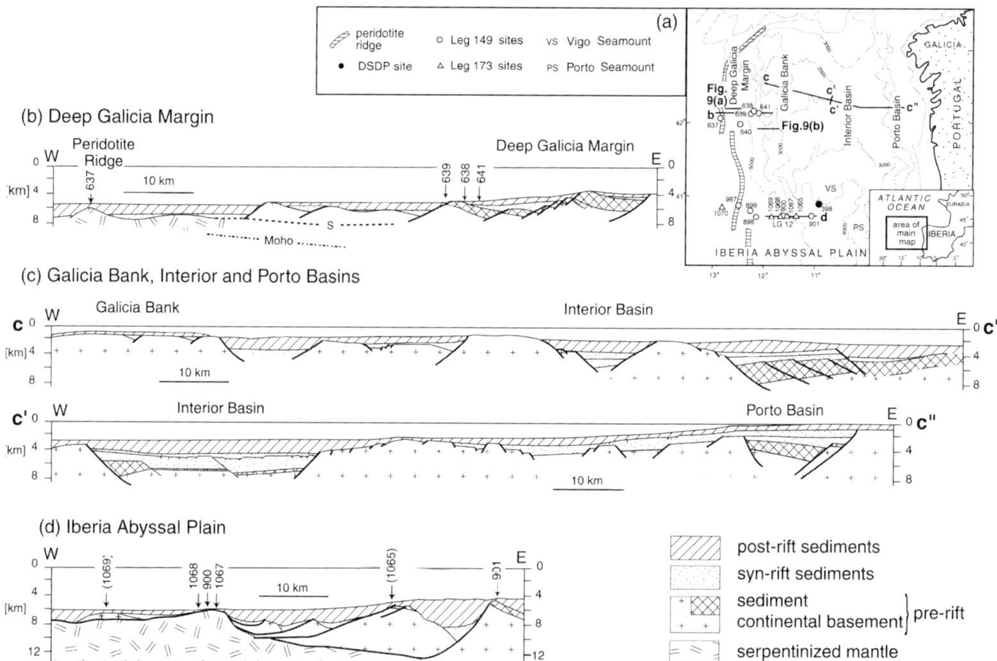

Fig. 5. The NW Iberia margin. Map showing location of main structural features and deep-sea drilling sites (**a**), and cross-sections showing structural interpretations of seismic reflection lines (**b**) from Boillot & Winterer (1988), (**c**) modified from Murillas *et al.* (1990) and (**d**) from Manatschal *et al.* (2001).

commonly overlain by highly faulted remnants of the upper plate'. In this paper, therefore, we use the terms lower and upper plate to describe the detachment architecture and syn-rift subsidence. In the examples we discuss, we believe that the lower- and upper-plate scenario is applicable to only the very latest rifting history, which is localized (i.e. only distal) and short-lived. The subsidence of most of the future continental margins (i.e. their proximal parts as defined by Manatschal & Bernoulli 1998, 1999) is typical of extension by pure shear.

According to Froitzheim & Eberli (1990, Bertotti *et al.* (1993), Driscoll & Karner (1998) and Manatschal & Bernoulli (1998), the change from broadly distributed and largely symmetrical deformation over future continental margins (pure shear) to localized deformation near the end of the rifting phase 'must reflect a change in the rheology of the lithosphere'. Manatschal & Bernoulli (1998) suggested that: 'Cooling of exhumed upper mantle and lower crust led to strengthening of the previously extended lithosphere in the proximal margins and may be responsible for the shifting of the site of rifting to the previously weakly extended area of the future distal margins. Because of cooling of the lower crust, upper crust and upper mantle were no longer decoupled, favouring fault planes penetrating and finally unroofing the upper mantle at the sea floor. At this stage, large-scale deformation resembles simple shear.'

The next two sections of the paper review and discuss published identifications of syn-rift stratigraphic units in distal margin settings. They are followed by a more detailed description and discussion of data obtained from seismic reflection profiles and ODP drilling in the Iberia Abyssal Plain.

Southern and Eastern Alps

In these regions, rifting that preceded, and appears related to future ocean opening, ceased towards the end of the late Early Jurassic period. This rifting episode now characterizes the proximal margins of the Piemonte–Ligurian Ocean (the Bernina and Ortler domains shown in Fig. 4a, and the Gozzano High and Lombardian Basin shown in Fig. 4b). Rifting along the distal margin (the Err–Platta domain shown in Fig. 4a, and possibly the Canavese Zone in Fig. 4b) commenced later, during Toarcian or earliest Mid-Jurassic time (Manatschal & Bernoulli 1998, 1999).

Figure 6 is a palinspastic reconstruction of the Err–Platta domain. The dashed lines in this figure show the traces of Alpine thrust faults that dismembered the former margin. The reconstruction shows a Jurassic detachment that is displayed wonderfully in the field. The reconstruction shows the situation after rifting had stopped and post-rift sediments had begun to accumulate, at which time upwarping resulting from the removal of the upper plate (Lister *et al.* 1988) caused the detachment to dip gently away from oceanic crust. Tilted blocks of continental crystalline rocks and pre-rift sediments occur above the detachment. In places the syn-rift Saluver Formation rests directly on the detachment. To the west mantle rocks occur in the footwall of the detachment and are overlain by allochthons consisting of upper continental crust and pre-rift sediments. Further oceanwards (i.e. further to the west) mantle rocks are overlain by pillow basalts.

A simplified composite stratigraphic column for the Err domain is shown in Figure 7a. The schematic section in Figure 7b shows the estab-

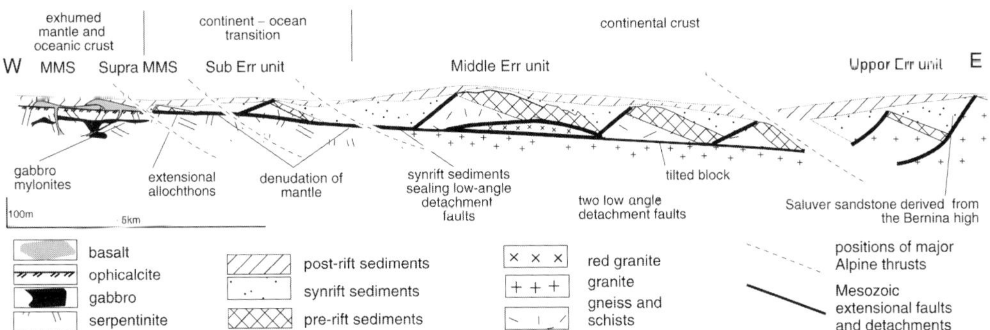

Fig. 6. Palinspastic reconstruction of the Err–Platta domain of the Eastern Alps showing low-angle detachment and related features characteristic of this distal margin. MMS, Mazzaspitz Marmorea Serpentinite.

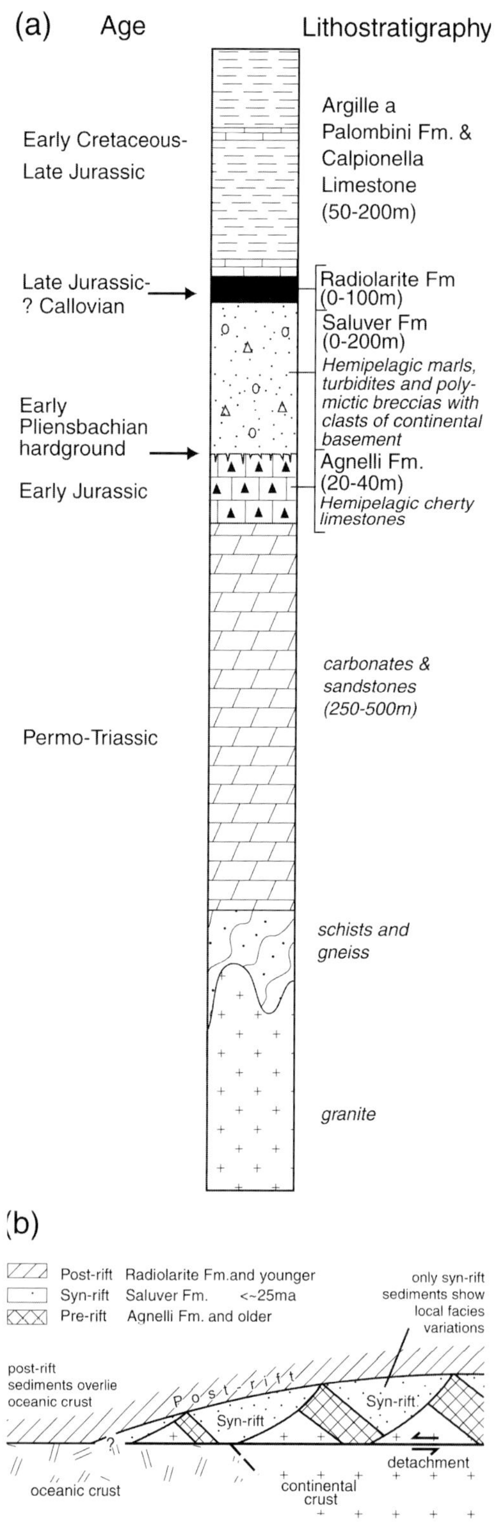

lished interpretation of pre-, syn- and post-rift stratigraphic units (Manatschal & Nievergelt 1997). The Saluver Formation is identified as the syn-rift interval for the following reasons. First, it underlies the formations interpreted as post-rift. The base of the post-rift interval is defined by the Radiolarite Formation because it rests on oceanic crust to the west. The age of the base of the Radiolarite Formation is almost the same as the crystallization age of gabbros in the oceanic crust on which the Formation rests (Desmurs *et al.* 2001).

The interpretation of the timing of the ending of rifting using the fact that the Radiolarite Formation rests on oceanic crust assumes that break-up immediately follows rifting, and that the segments of ocean crust exposed today sample the oldest products of sea-floor spreading. This interpretation also ignores the fact that the initiation of sea-floor spreading propagates along the strike of margins, resulting in continental break-up being time transgressive (Taylor *et al.* 1999). This means that sediments deposited on oceanic crust in one place may be contemporaneous with syn-rift sediments in one direction, and with post-rift sediments in the opposite direction. However, the time differences involved are likely to be relatively small (e.g. *c.* 3 Ma along *c.* 400 km for the Woodlark Basin example discussed by Taylor *et al.* (1999)) and so fall within the error ranges of current dating methods applicable to Mesozoic margins. However, significant age differences are likely to occur in the initiation of sea-floor spreading between sectors of continental margins separated by major transfer faults (e.g. the Tagus Abyssal Plain and Iberia Abyssal Plain off Iberia (Pinheiro *et al.* 1996). Second, the syn-rift interval shows local facies variations, unlike the formations between which it is sandwiched. Third, the underlying Agnelli Formation does not show abrupt changes of thickness across faults, indicating that it was not deposited during faulting.

What is lacking is any direct evidence as to exactly when deepening due to extension began and the fault blocks were rotating. This is because the Saluver Formation contains no palaeodepth indicators: it was deposited some-

Fig. 7. Lithostratigraphy and tectonostratigraphy of the Err–Platta domain (from Manatschal & Nievergelt 1997). (**a**) Simplified composite lithostratigraphic column, showing the constraints available to determine the ages of pre-, syn- and post-rift units. (**b**) Schematic summary of the published rationale for identifying the Saluver Formation as a syn-rift unit.

where below wave base and above the CCD. Neither is it possible to determine whether any sedimentary packages thicken into footwalls. So the definition of the entire Saluver Formation as entirely syn-rift is problematic: all or part of it could be early post-rift rather than syn-rift. In addition, the absence of diagnostic fossils within the Formation makes it impossible to determine its precise age, save to say that it is younger than the hardground capping the underlying Agnelli Formation (of early Pliensbachian age), and older than the overlying Radiolarite Formation (of Callovian age). This does not mean that the Formation was deposited over this period of time; it could have been deposited in a period of a few million years or less. The rifting episode in the Err–Platta domain, therefore, could have lasted for a much shorter period than 25 Ma

Galicia margin

Published studies of the Galicia distal margin state that rifting in this area lasted for c. 25 Ma, from Valanginian to almost the end of Albian time (Groupe Galice 1979; Mauffret & Montadert 1987, 1988; Boillot & Winterer 1988). This conclusion was based on a comparison of stratigraphic results from DSDP Site 398 and ODP Leg 103 sites, and seismostratigraphic studies (Fig. 8). Mauffret & Montadert (1987) stated that: 'The overall tectonic style of the Lower [sic] Cretaceous main rifting event is characterized by a series of tilted blocks bounded by westerly facing listric faults. These blocks delineate half grabens infilled by Hauterivian [Leg 103 drilled older sediments of Valanginian age] to Aptian syn-rift sediments (Formation 4). Reflectors within this formation converge towards the tilted fault block crests demonstrating a 'trapdoor' style of faulting (Groupe Galice 1979).'

Referring to the figure reproduced as Fig. 9a in this paper, Mauffret & Montadert (1988) stated that 'a divergent configuration is characteristic' of Formation 4 being syn-rift in origin. Neither the descriptions given by those workers nor any of their illustrations show clearly any of the stratigraphic criteria summarized earlier in this paper that characterize syn-rift intervals and the break-up unconformity. The reflection convergence they described is typical of deposition in a previously existing basin and compactional drape of sediments over basement topography produced by half-graben formation. We have seen no convincing examples of thickening of reflection packages towards faults on published Galicia margin seismic lines, nor

of growth faults terminating at the same seismic reflection that can be mapped over a large area. We conclude, therefore, that Formation 4 is not syn-rift in origin, but early post-rift. This reduces the duration of rifting from >20 Ma to <5 Ma (i.e. the period during which Seismostratigraphic Unit 5A was deposited. This means that the boundary between Units 4 and 3 is not the break-up unconformity (Fig. 8a,c). The fact that the exact age of the onset of sea-floor spreading off the Galicia margin is not known (Pinheiro et al. 1996) also calls into question the recognition of the boundary between the syn- and post-rift intervals defined by previous workers as the break-up unconformity. This means this boundary cannot be used to date the commencement of sea-floor spreading off the distal part of the Galicia margin as suggested by Boillot et al. (1989). If the eastern boundary of true oceanic crust could be located and dated off Galicia, then sediments immediately overlying it could be traced eastwards to define the top of the syn-rift package as has been done in the Alps (Fig. 7b). This is not possible for two reasons. First, the first westward occurrence of normal oceanic crust has not been determined by dredging or drilling. Second, the precise age of the onset of sea-floor spreading has not been determined by modelling magnetic anomalies. The problem is complicated further because of the difficulty in distinguishing 'true' oceanic lithosphere derived from the asthenosphere from deep rifted continental lithosphere exposed along distal margins (Boillot & Froitzheim 2001; Bernoulli, pers. comm.).

Mauffret & Montadert (1988) stated that Seismostratigraphic Sub-unit 5A is syn-rift in origin, but offered no detailed rationale for their interpretation. However, in the same volume Boillot & Winterer (1988) stated that: 'The 250 m thick upper Valanginian sandy turbidite section sampled at Sites 638 and 639 is part of a wedge that thickens eastward into a half-graben typical of those on the Galica margin [Fig. 8b in this paper]. Active tectonism is indicated not only by the great volume of these coarse sediments in the region seaward of Galicia Bank ... but also by the geometry seen on seismic profiles.'

In Figure 8b–d there is no evidence of eastward thickening of Seismostratigraphic Sub-unit 5A; indeed, its significantly steeper dip compared with younger units indicates fault block rotation before later sediments were deposited. However, one seismic line published by Mauffret & Montadert (1988), reproduced here as Figure 9b, does show eastward thickening into a half-graben, although the poor quality of the data makes this interpretation difficult to verify. Their reconstructed depth section of this line shows no eastward thickening, merely reflection convergence to the east and west characteristic of deposition in a previously existing basin and compactional drape over basement topography.

So there seem to be conflicting lines of seismic evidence concerning the tectonic setting of Seismostratigraphic Sub-unit 5A. It shows fault-block rotation indicative of a pre-rift origin and post-rift features consistent with the infilling of depressions formed by earlier faulting. Does the evidence obtained from the cores obtained during ODP Leg 103 throw any light on this problem? There is no doubt that the shallow-water Tithonian (and possibly earliest Berriasian) carbonates equivalent to Seismostratigraphic Sub-unit 5B are pre-rift in origin. They are overlain by *c.* 36 m of late Valanginian nannofossil and calpionellid marlstone and marl. The presence of such fossils 'together with the lack of any traces of benthic fauna of larger invertebrates in the lower part of the marlstone sequence, suggests an accumulation site below the photic zone, in somewhat pelagic conditions' (Shipboard Scientific Party 1987*a*). There is, therefore, clear evidence of deepening from Tithonian to Valanginian time, and, of course, a hiatus, with the absence of most, if not all, of a sedimentary record of Berriasian and Early Valanginian time. The 'somewhat pelagic' sediments are overlain by the turbiditic succession described above (Fig. 8a).

It is now accepted that the influx of clastic sediments is more likely to characterize post-rift fills of sedimentary basins because of the time it takes for sediment supply systems to adjust to the new topography formed during rifting (Prosser 1993; Driscoll *et al.* 1995; Ravnås & Steel 1998). Unless carbonate platforms become established (e.g. the Upper Triassic Hauptdolomite Formation of the Southern Alps (Bertotti *et al.* 1993)) sediment starvation is, therefore, more likely to be characteristic of rifting episodes, particularly in distal basins. We suggest, therefore, on the basis of the sedimentary record, that rifting along the Galicia margin could have occurred during Berriasian and Early Valanginian time (a time interval of *c.* 10 Ma), and that the entire Lower Cretaceous section drilled by Leg 103 is post-rift in origin. However, as stated above, the seismic evidence concerning the tectonic setting of deposition during Valanginian time is equivocal: whereas it shows no clear features characteristic of a syn-rift origin, it does show both pre-rift and

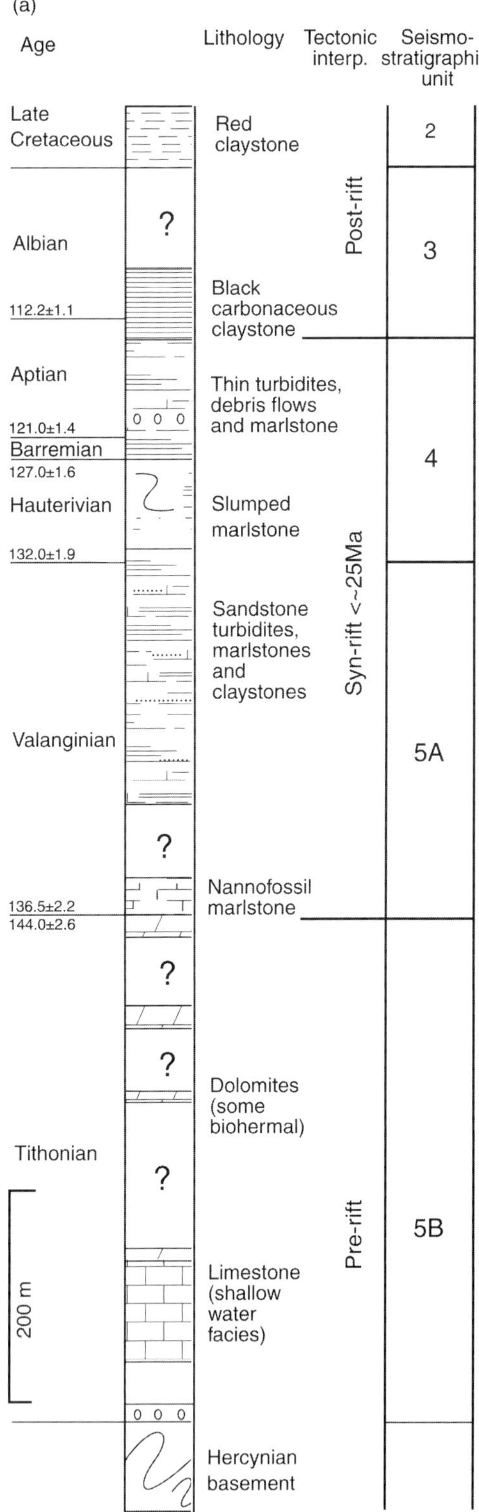

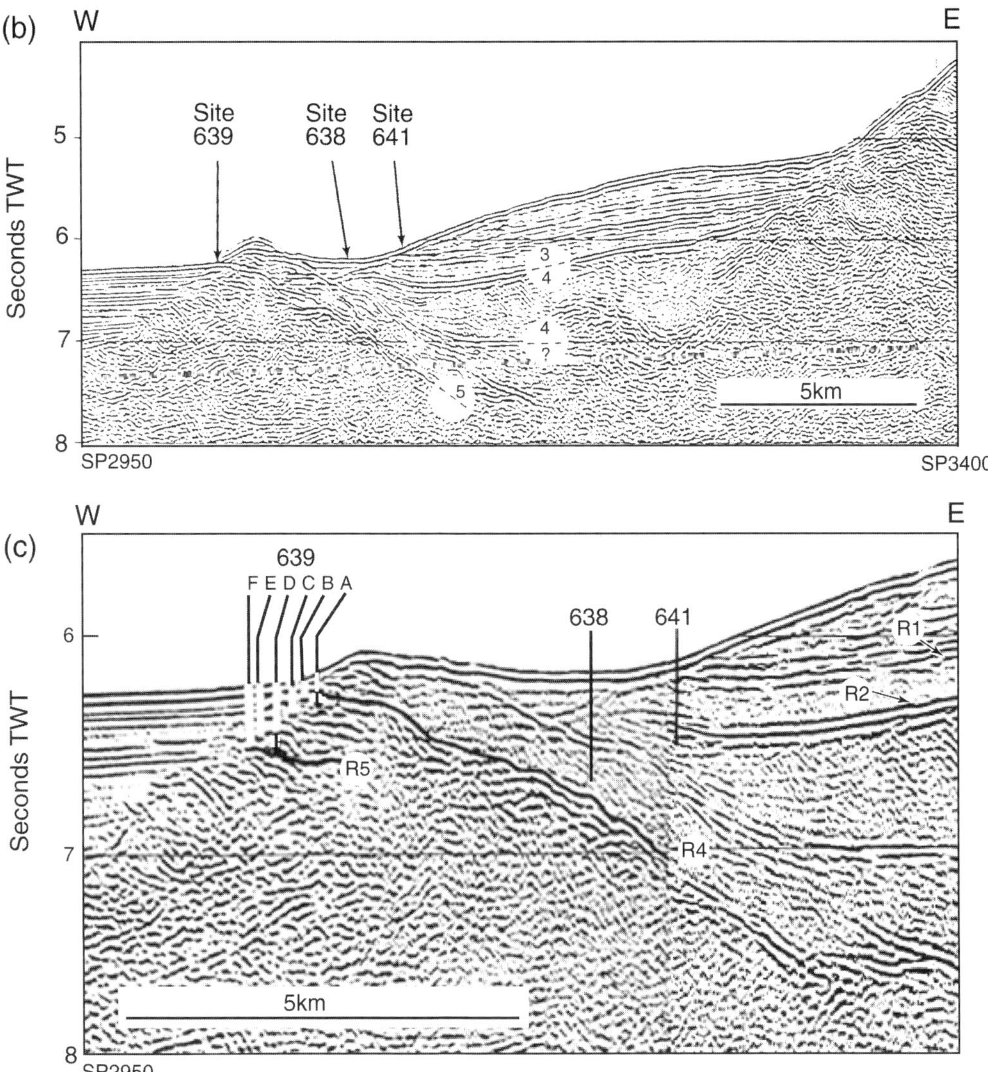

Fig. 8. Lithostratigraphy and seismostratigraphy of the Galicia Margin adjacent to ODP Sites 638, 639 and 641. (**a**) Composite lithostratigraphy based on core recovery from the three sites (Boillot & Winterer 1988). Ages of stage boundaries shown are from Gradstein & Ogg (1996). It should be noted that Jansa *et al.* (1988) suggested that the carbonate succession might extend into the Berriasian. (**b**) Part of seismic line GP 101 showing seismostratigraphic units and location of the three sites (Boillot & Winterer 1988). (**c**) Part of seismic line GP 101 showing the positions of the holes drilled at Site 638, 639 and 641 (from Shipboard Scientific Party 1987*a*, 1987*b*). For each hole at Site 639, the Neogene section is shown in white, and Mesozoic rocks in black. Reflections R4 and R5 bound seismostratigraphic Sub-unit 5B of Mauffret & Montadert (1987; 1988). R4 is the seismic reflection correlated with the top of the Tithonian (possibly Early Berriasian) carbonates, and R5 the base of these carbonates, which may rest on Hercynian basement. Reflection R1 lies immediately above a transparent layer that is correlated with the Late Aptian–Albian black shales cored at Site 641. Reflection R2 is correlated with the break-up unconformity identified by Mauffret & Montadert (1987, 1988) and Boillot & Winterer (1988).

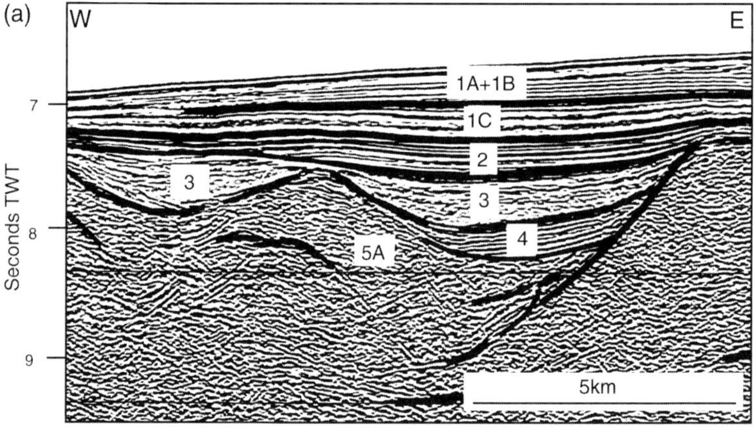

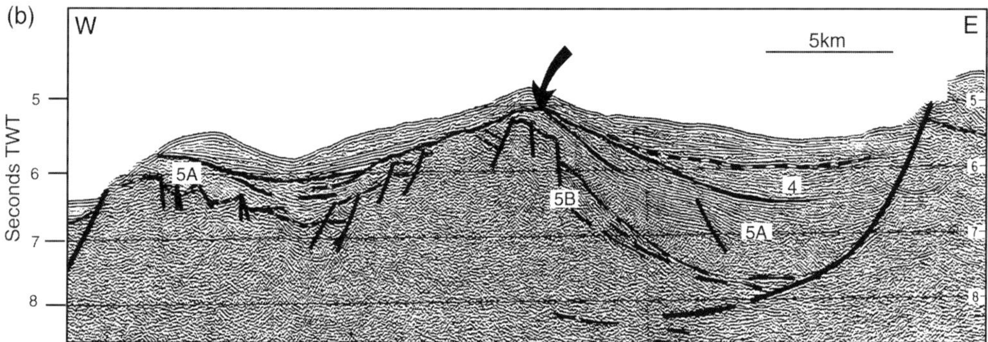

Fig. 9. Seismostratigraphic and tectonostratigraphy of the Galicia margin: exemplar seismic lines from Mauffret & Montadert (1987, 1988, their figs 10 and 7, respectively), showing seismostratigraphic units (their lithostratigraphic equivalents are shown in Figure 8a of this paper). See text for discussion, and Figure 8a for lithostratigraphic equivalents of the seismostratigraphic units. (**a**) Part of line GP 11. Note that Seismostratigraphic Units 1–4 show contrasting features, the data quality elsewhere on the section is poor, and it is difficult to discern how the base of Sub-unit 5A was identified. (**b**) A thick Valanginian fill (Sub-unit 5A) to the eastern half-graben is interpreted on this section. Unfortunately, the poor quality data beneath Sub-unit 5A and near the eastern boundary fault, plus the thick interpretive lines, make it difficult to judge the evidence for the interpretation shown.

post-rift features. Clearly, modern high-quality seismic data are needed to resolve this problem.

Iberia Abyssal Plain

Seismic evidence and tectonic history

Seismic reflection lines shot across the Iberia Abyssal Plain show a layered cover unit, the reflections of which drape the topography of acoustic basement (Fig. 10a). The reflection geometry of the cover unit indicates compactional drape over basement highs, or the effects of localized Miocene compressional folding over peridotite ridges (Masson *et al.* 1994) that

is visible on larger-scale seismic lines than that shown in Figure 10a.

Some low-angle, seaward-dipping reflections within the acoustic basement are clearly linked to faults imaged at shallow levels (e.g. H and L on Fig. 10). The relationship of these faults to the shape of the top of acoustic basement suggests the occurrence of a series of tilted fault blocks, especially between Sites 901 and 1068. Krawczyk *et al.* (1996) presented two alternative interpretations of these structures as low-angle detachment systems occurring to the east of Site 900 (see fig. 2b and c of Manatschal *et al.* 2001) and Manatschal *et al.* (2001) have suggested that the detachment system extends further west beneath Site 1069

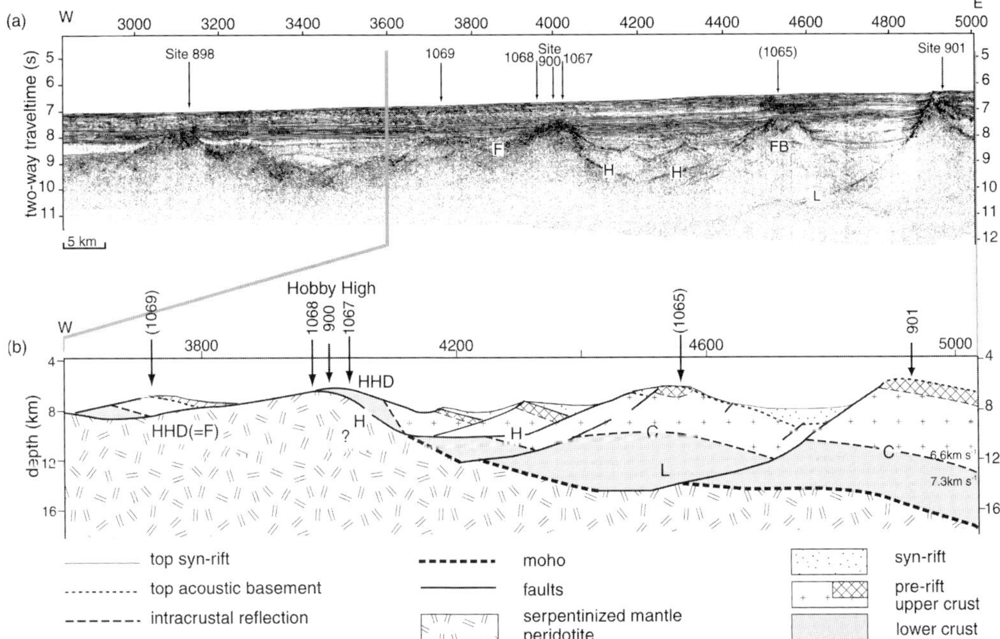

Fig. 10. Part of seismic reflection line Lusigal 12. (**a**) Migrated time section showing the location of ODP sites and detachment faults. (**b**) Detachment fault interpretation of Manatschal *et al.* (2001) of part of (**a**). HHD, Hobby High detachment; H and L, detachments interpreted by Krawczyk *et al.* (1996). F and B are fault structures described by Krawczyk *et al.* (1996); C is an intracrustal reflection shown in fig. 6 of Manatschal *et al.* (2001).

(Fig. 10b). Manatschal *et al.* (2001) kinematically inverted their interpretation of seismic line Lusigal 12 (Fig. 10b), and concluded that: (1) faulting propagated westwards towards the future Atlantic Ocean (even if syn-rift sediments had been reliably identified and drilled, the resolution of dating methods would not be precise enough to detect the likely age differences involved); (2) the amounts of displacement along individual faults increases oceanwards (4.7 km along detachment L, 8.6 km along detachment H, and 20.5 km along the Hobby High detachment (HHD)); (3) the change in displacement amount was accompanied by a change in geometry from upward steepening, listric (L, H) to downward steepening (HHD).

On the assumption that cooling through 150 °C was related to tectonic exhumation along the HHD, ^{40}Ar/ ^{39}Ar plagioclase ages from basement rocks at Site 900 (*c.* 136 Ma, Féraud *et al.* 1996) and Site 1068 (*c.* 137 Ma, Manatschal *et al.* 2001) indicate that displacement began before the Valanginian (using the time scale of Gradstein & Ogg (1996)).

Seismostratigraphy

Krawczyk *et al.* (1996) and Manatschal *et al.* (2001) identified syn-rift intervals on their interpretations, despite the fact that on seismic lines these exhibit no reflection divergence towards fault footwalls. On a few seismic lines, however, there are some wedge-shaped seismostratigraphic units that show rather enigmatic features that could indicate the presence of syn-rift sediments that occur below intervals identified by these workers as syn-rift. Wilson *et al.* (1996) suggested that one of these enigmatic wedges, situated to the west of the basement high beneath Site 1069, was deposited as an eastward prograding submarine fan or talus deposit because its faint internal reflection pattern shows westward onlap and eastward downlap onto acoustic basement (Fig. 11b). They defined this occurrence as their Seismostratigraphic Sub-unit 6C, and identified this unit on line Lusigal 12 at 9 s two-way travel time beneath shot point 3400 (Fig. 10a). This wedge-shaped unit is imaged on more recently shot seismic lines, and can be mapped almost

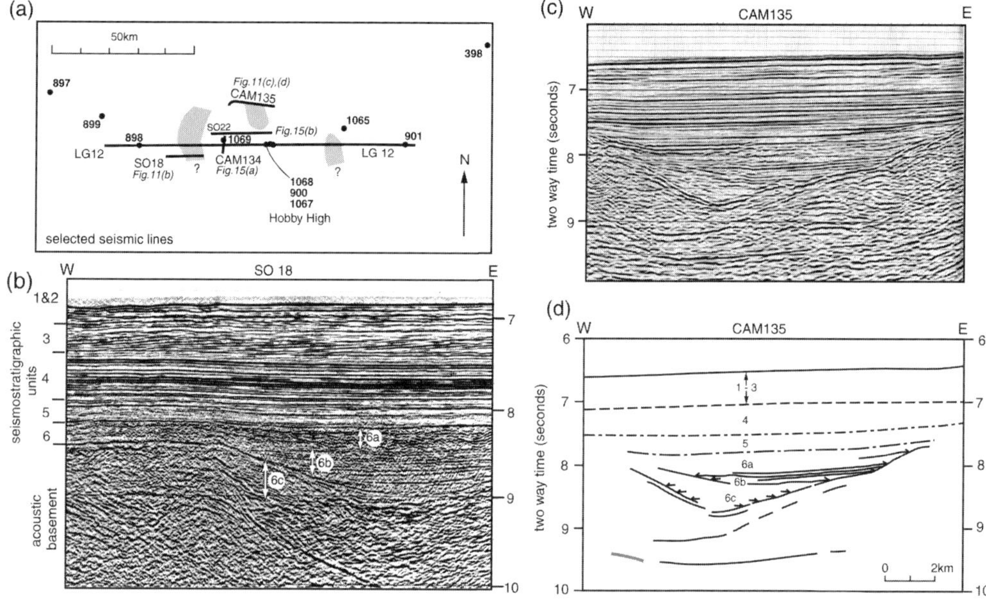

Fig. 11. Seismostratigraphic Sub-unit 6C of Wilson *et al.* (1996). (**a**) Map showing location of enigmatic wedge-shaped seismic reflection packages. (**c**) and (**d**) Seismic line CAM135 and interpretation. The seismostratigraphic units of Wilson *et al.* (1996) are shown. It should be noted that the reflection terminations at the base of this sub-unit show an onlap to the west and downlap to the east. Within the unit, reflection divergence to the east seems to be present but when viewed at a larger scale, a more complex relationship between reflections is evident. (**b**) Part of line Sonne 22, showing wedge-shaped Seismostratigraphic Sub-unit 6C interpreted by Wilson *et al.* (1996) as a possible eastward prograding submarine fan or talus deposit.

20 km to the north of line Lusigal 12 (Fig. 11a). The onlapping and downlapping relationships described above are present on all seismic lines that cross the wedge, and all show low- to moderate-amplitude reflections. One line shows an undulose geometry with some suggestion of eastward reflection divergence (Fig. 11c). The possible divergence may indicate syn-rift fault block rotation resulting in thickening of sedimentary units towards the HHD. Seismostratigraphic Sub-unit 6C, may, therefore, be a prograding depositional feature, or result from rotation of a hanging-wall block. If it is a syn-rift unit, it would have been deposited against a steeper part of this fault and subsequently rotated to its present orientation. This would account for the eastward dip of the top of Seismostratigraphic Sub-unit 6C, and its onlap by Sub-unit 6B (Fig. 11d). Reprocessing of the seismic lines that cross these possible syn-rift packages might allow origin to be confirmed with more confidence using the criteria for identifying rift onset and break-up unconformities, and syn-rift reflection divergence.

To the east of Hobby High another enigmatic wedge occurs (Fig. 11a), but it does not show clear downlapping and onlapping relationships with basement and therefore does not have the same origin as the wedges situated adjacent to Hobby High. If these wedges are tectonic in origin, can seismostratigraphic evidence throw any light on the possible age of the displacement that produced them? The ages of the seismostratigraphic units identified by Wilson *et al.* (1996) are summarized in Table 1. The location of ODP sites over the crests of basement highs means that none of the enigmatic seismostratigraphic wedges has been drilled, and that information about the possible ages and lithology of the units from the lower part of Seismostratigraphic Unit 4 downwards has to be extrapolated from DSDP Site 398. It is possible, however, that the feather edges of the lithological equivalents of these units may have been drilled at Sites 1069 and 1068 (see below). Wilson *et al.* (1996) concluded that the top of Seismostratigraphic Unit 6 is near the Aptian–Albian boundary, and that its base may be of Valanginian age or older.

Seismostratigraphic Unit 5, which is of latest Aptian to Albian age (Wilson *et al.* 1996), extends across most of that part of the Iberia

Table 1. *Summary of the principal features of seismostratigraphic units identified beneath the Iberia Abyssal Plain (modified from Wilson et al. 1996)*

Unit	Age of base (Units 4–5 by comparison with DSDP Site 398)	Thickness (ms TWT)	Geometry, onlap, etc.	Probable lithology
1	Late Pliocene	100–200	Sheet, but thickens significantly on west sides of folds; base onlaps or truncates Unit 2, and erodes into Unit 3 in crestal locations	Siliciclastic turbidites and pelagic nannofossil oozes
2	Mid-Miocene	0–460	Wedge, thinning and onlapping toward crests of folds, and absent over crests	As above, with bioturbated hemipelagites and pelagic nannofossil-rich sediments at base
3	Mid-Eocene	150–520	Lower boundary defined by base of inclined reflection interval and its correlative reflector; sheet-like geometry with gradual westward thinning	Carbonate turbidites (with minor siliciclastic deposits at base) siliciclastic hemipelagites, and contourites
4	?Santonian–Coniacian	240–600	Sheet, with high-amplitude continuous reflections but basal part onlaps highs in acoustic basement	As above, but becomes more siliciclastic sand-rich in the lower part of the Eocene succession at Sites 900 and 1068, and in the Paleocene succession at Site 1069
5	?Albian	0–200	Onlaps basement highs, occupying basinal positions between them, and showing overall westward thickening (Fig. 11); low-amplitude reflections (almost transparent)	Comparison with Site 398 suggests nannofossil chalks and claystones and dark grey to black claystones
6	Not known	0–600	Sub-unit C: discontinuous, slightly undulose to chaotic reflections; transitional boundary with Sub-unit B, onlaps Sub-unit A and acoustic basement. Sub-unit B: continuous to discontinuous parallel reflections; onlaps Sub-unit A and acoustic basement	Comparison with Site 398 suggests siliciclastic turbidites and debris flows with carbonate clasts for Sub-unit 6C

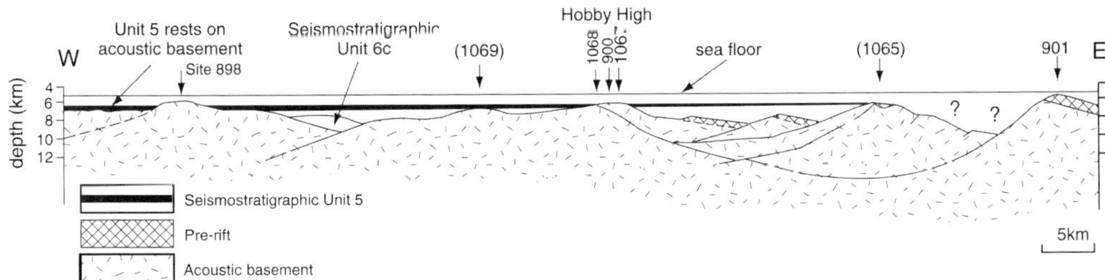

Fig. 12. The distribution of the Wilson *et al.* (1996) Seismostratigraphic Units 5 and 6C on seismic line Lusigal 12. The locations of ODP sites are shown, with numbers in parentheses indicating sites projected onto the line of section. (For location, see Fig. 10a.)

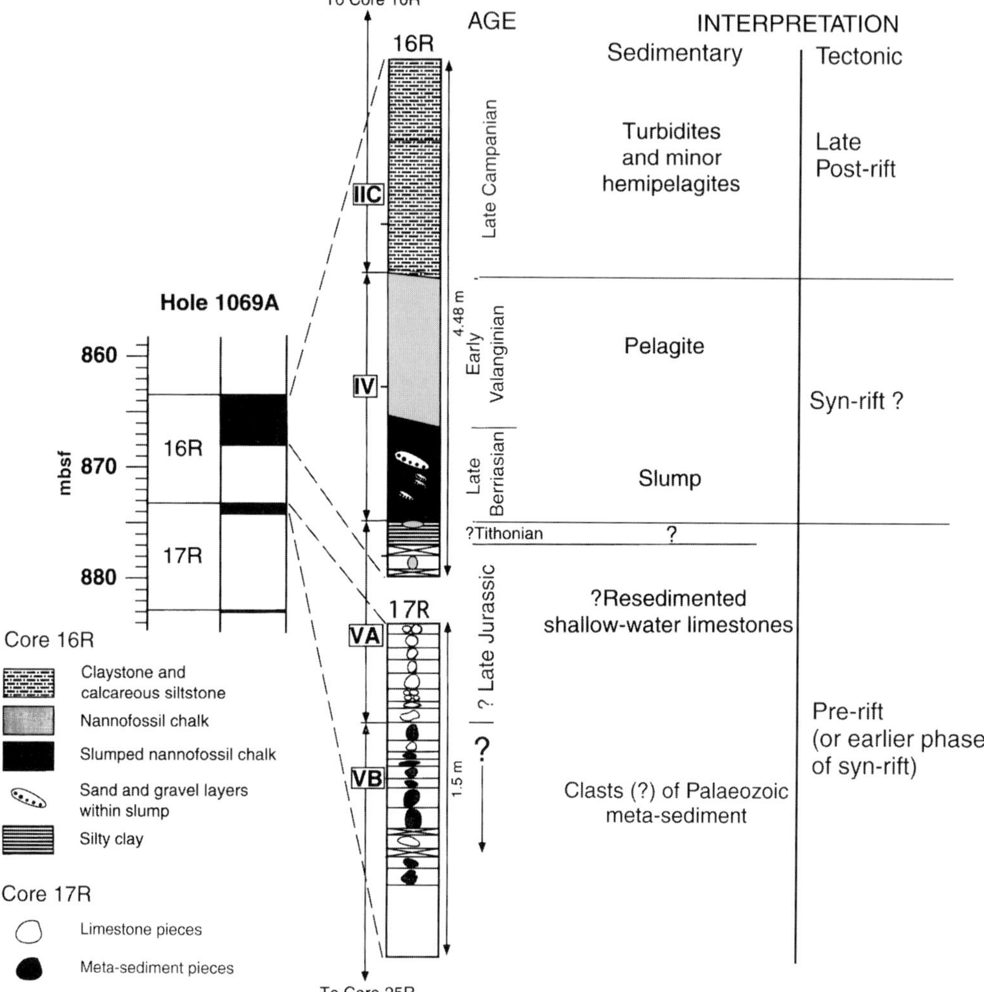

Fig. 13. Log and interpretation of Cores 16R and 17R at ODP Site 1069 (modified from Shipboard Scientific Party 1998*b*).

Abyssal Plain shown in Figure 12. It thickens to the west, and comes to rest on flat-topped acoustic basement to the east of the presumed peridotite ridge beneath Site 898. The maximum thickness of sediments situated in basement hollows beneath Unit 5 decreases to the west if the thickness of the enigmatic Sub-unit 6C is ignored. The distribution of Unit 5 and the sediments beneath it is reminiscent of the conventionally defined syn- and post-rift units in the Alps shown in Figure 7b and discussed earlier in this paper. However, it is clear that it is not the earliest post-rift unit, for, as Sub-units 6A and B drape acoustic basement and sub-unit 6C, they must also be post-rift.

Evidence from ODP Sites

Site 1069

Two cores at Site 1069 appear to have sampled the pre-rift to post-rift interval preserved on top of an extensional allochthon (Fig. 10b). The nature and origin of the pieces of metasediment in Core 17R (Fig. 13) was the subject of much shipboard speculation during Leg 173. This was because of the paradoxical situation in which only very poor recovery (<3%) of hard metasediments was achieved, yet each of the last nine rotary cores was drilled unusually quickly in 25–80 min, suggesting that a high proportion of

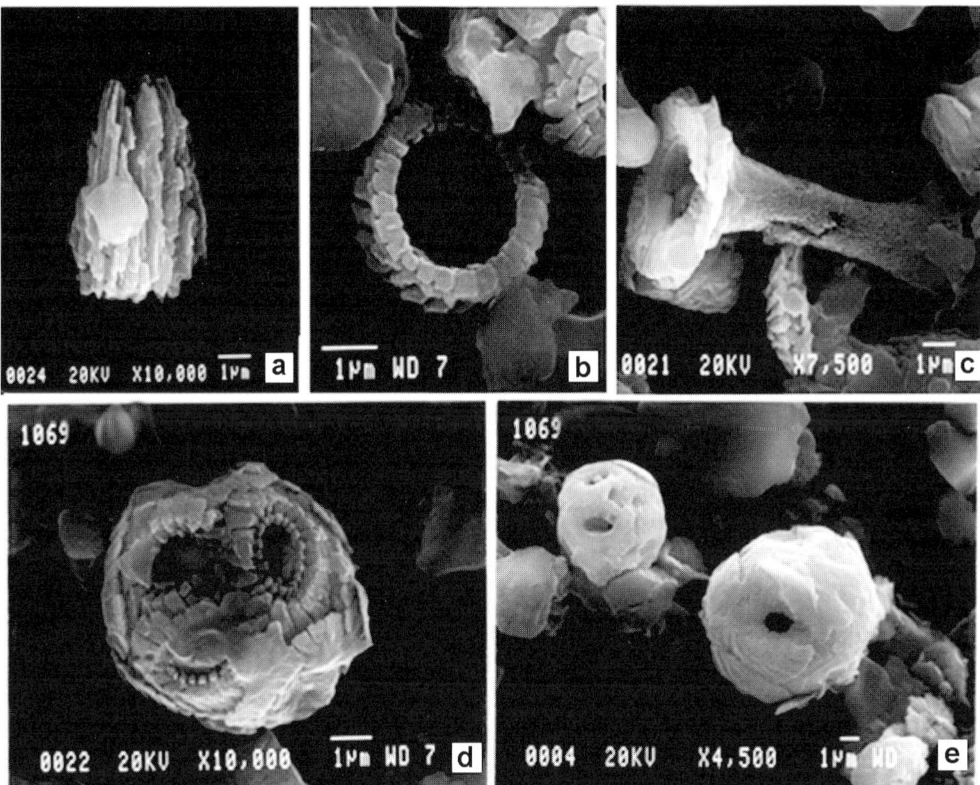

Fig. 14. Scanning electron micrographs of Tithonian calcareous nannofossils from ODP Sample 1069A-16R-3, 129 cm. (**a**) Inner core of *Conusphaera mexicana* Trejo, (**b**) *Diazomatholithus galicianus* de Kaenel and Bergen. (**c**) *Discorhabdus patulus* (Deflandre in Deflandre and Fert) Noël. (**d**) and (**e**) Watznaueriacid coccospheres; the prevalence of intact coccospheres such as these denotes the restricted nature of the sedimentary basin, including intervals of dysaerobic or anoxic bottom-water conditions.

the rock column consists of friable material; maybe unlithified sand, fissile shale, or pelites shattered by cataclasis. However, the presence of Palaeozoic metasediments and pieces of shallow-water limestones of probable Late Jurassic age does point to the existence of a near-normal thickness crust at the end of the Jurassic. This conclusion is reinforced by some of the features of the sediments recovered in Core 16R.

Features observed in the Tithonian to lower Valanginian sediments in Core 16R suggest the following environmental settings:

(1) Tithonian silty clays are environmentally restricted, as indicated by the high organic contents (up to 0.8 wt % of organic carbon (Shipboard Scientific Party 1998*a*, table 7) and the prevalence of coccospheres among the sparse nannofossil assemblages (Fig. 14; see also Concheryo & Wise (2001) for details of the assemblages).

(2) The slumped upper Berriasian chalks with pebbles of limestone and metasediment show a more open connection to the sea, based on their low organic carbon contents (below the limits of detection (Shipboard Scientific Party 1998*b*, table 9)) and pervasive abundance of nannofossils. The absence of diagnostic Foraminifera precludes any water depth interpretation. The Late Berriasian age of this assemblage is constrained by the co-occurrence of the nannofossil taxa *Umbria granulosa granulose*, *Tuibodiscus jurapelagicus*, *T. verenae*, *Rucinolithus wisei* and *Rhagodiscus nebulosus* according to the calibration of Bergren (1994).

(3) The lower Valanginian chalks were deposited in relatively shallow water, equivalent to outer shelf–upper slope, open marine environment based on the paleobathymetry of the benthic foraminiferal assemblages coupled with the prevalence of micrantholiths and nannoconids among the nannofossils.

The change from environmentally restricted Tithonian silty clays to Berriasian and Valanginian pelagic chalks indicates a shutdown in the supply of terrigenous sediments. This could have been caused by the creation of rift-related topographic features that ponded back turbidites in proximal areas of the margin. Rifting, therefore, could have occurred for a short period from latest Tithonian to Early Berriasian (<5 Ma). However, the amount of deepening (down to outer shelf–upper slope depths during the Early Valanginian) is not as much as would be expected to be associated with the 20.5 km displacement along the HHD estimated by Manatschal *et al.* (2001). We will return to this paradox after discussing results from Site 1068.

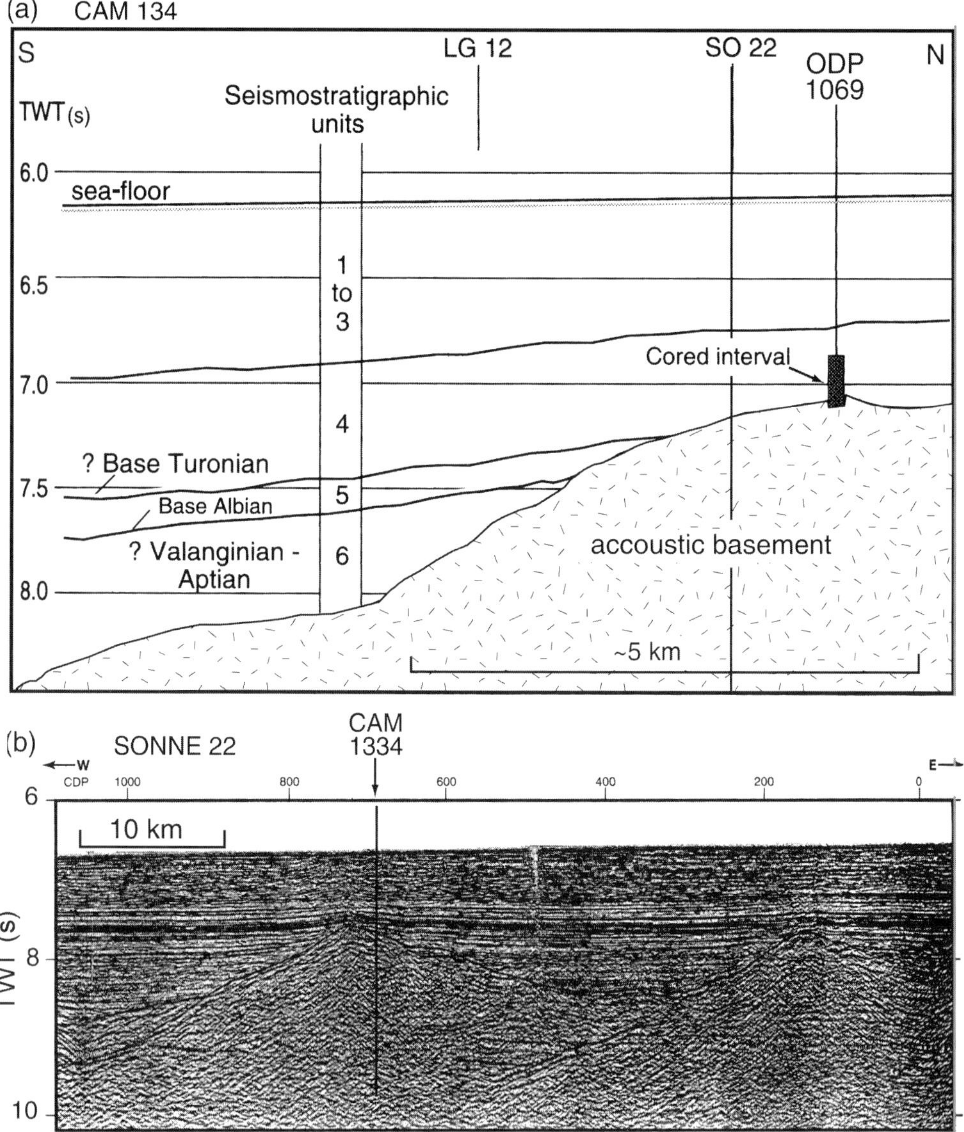

Fig. 15. The structural and stratigraphic setting of ODP Site 1069 (from Shipboard Scientific Party 1998*b*). (**a**) Interpretation of part of seismic line CAM 134. (**b**) Part of seismic line Sonne 22, showing two tilted fault blocks in acoustic basement.

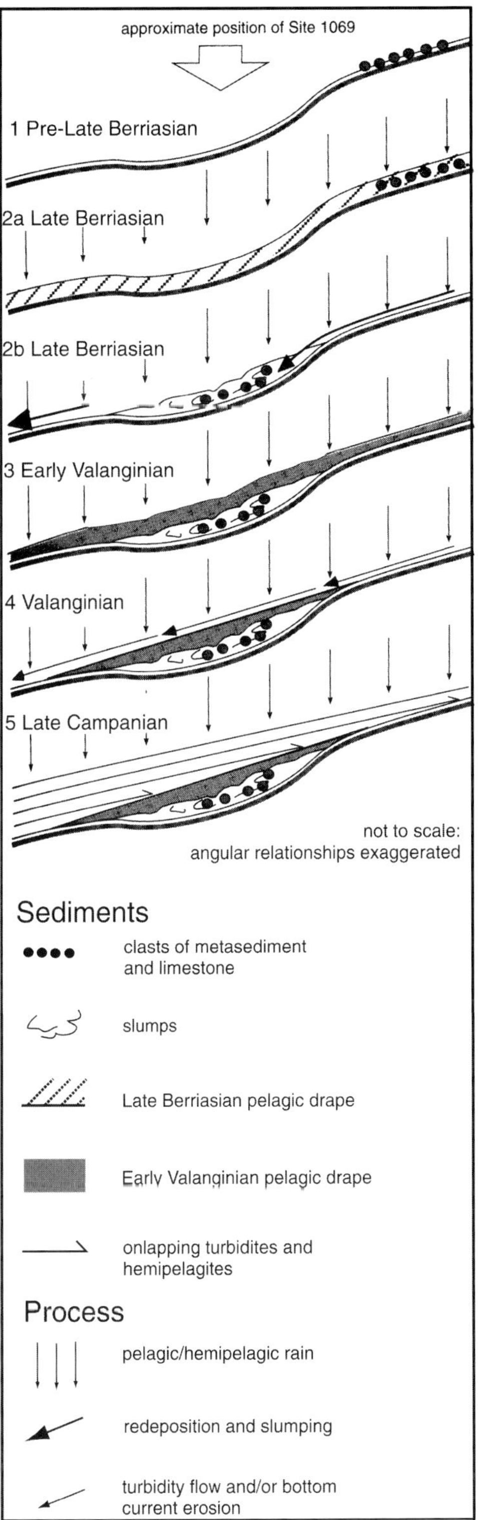

Fig. 16. Possible events in the burial of the basement high beneath ODP Site 1069 (modified from Shipboard Scientific Party 1998*b*). (See text for explanation.)

The nannofossil chalks occur on a slightly inclined slope near the summit of the high formed by the presumed extensional allochthon. Why does this thin lower Cretaceous pelagic drape occur at Site 1069 when seismic evidence suggests that possible Valanginian to Turonian age sediments accumulated at lower levels on the flanks of the high (Fig. 15)? Why were no pelagites, or other sediment types, preserved at this site until Late Campanian time? The answer may be that the Early Cretaceous pelagic–hemipelagio drape was preserved only in a small localized depression (Fig. 16). Sediments deposited away from such depressions would have been transported down on the flank of the basement high by gravity-driven transport processes. Deposition would have resumed at Site 1069 only when sediments lapped onto the summit area of the high by the Late Campanian. The core evidence therefore suggests that the basement topography formed before or during the early part of Early Berriasian. This is consistent with the conclusion reached by Wilson *et al.* (1996) that rifting occurred between Tithonian and Early Valanginian, and the ^{40}Ar/^{39}Ar plagioclase ages referred to earlier (136 or 137 Ma, within the error bar of the Gradstein & Ogg (1996) age of the Valanginian–Berriasian boundary).

Site 1068

Hole 1068A penetrated serpentinized peridotite, above which occur tectonic and sedimentary breccias containing clasts of serpentinized peridotite, meta-anorthosite, metagabbro and foliated and micro-amphibolites (Fig. 17). The upper part of the sedimentary breccias (Unit IVA of Shipboard Scientific Party 1998*b*) contains a matrix of nannofossil chalk which yield an age range from Valanginian to Barremian time. The proportion of the matrix ranges from 70% to 30%, indicating that both matrix- and grain-supported varieties are present. The angular clasts are poorly sorted, and seldom exceed 10 cm in length. In places sand- to granule-sized clasts are concentrated on the upper surfaces of larger clasts, suggesting that some settling through the matrix occurred after emplacement of the breccias by debris flows.

The near-horizontal tops to these accumulations of smaller clasts indicates that there has been no tectonic rotation.

Variations of the colour of the chalk matrix suggest that at least three debris flow units may be present. In addition to clasts composed of meta-igneous rocks a few small clasts of fine-grained Tithonian–Lower Cretaceous calpionellid limestone, and biotite schist, occur.

The lower part of the sedimentary breccias (Unit IVB of Shipboard Scientific Party 1998*a*) are entirely clast supported, set in a multicoloured matrix of clay- and silt-sized carbonate crystals (microsparite) containing silt- to sand-sized metamorphic–igneous rock fragments. They are interpreted to have been formed as rock fall or scree deposits.

In Sections 3–6 of Core 19R, a mixture of clast type occurs: micro-amphibolite, foliated amphibolite, metagabbro and minor amounts of metamorphic anorthosite. In Sections 1–3 of this core, and Sections 2 and 3 of Core 18R, only amphibolites occur, but from Section 1 of Core 18R upwards, the proportion of metagabbro increases, as does the proportion of metamorphic anorthosite from Section 6 of Core 16 upwards. The upward change in clast composition (shown in Shipboard Scientific Party 1998*a*, fig. 23) suggests progressive unroofing of Hobby High, with the anorthosites drilled at Site 1067 exposed first by the detachment, and then the gabbros of Site 900. The occurrence of a few clasts of calpionellid limestone and biotite schist in the uppermost part of the sedimentary breccias indicates an input from a different source, possibly the hanging wall of the HHD.

Down-hole, the first evidence for a tectonic overprint in the lower part of the sedimentary breccias is indicated by the occurrence of jigsaw style disintegrated clasts in Core 18R (869.2–875.2 m below sea floor (mbsf)) as well as by single faults transecting clasts, and the occurrence of a shape preferred orientation indicated by elongated clasts. Cataclastic bands in which strain is localized are observed for the first time in Core 19R (875.2 mbsf) and become more frequent downwards. At the base of the breccias (Unit IVC of Shipboard Scientific Party (1998*a*) in Core 173-1068A-20R), the tectonic overprint makes it impossible to recognize features indicating a primary sedimentary origin. The fabric is characterized by a high matrix to clast ratio, rounded clasts, a weak foliation wrapping around the clasts, and anastomosing serpentine veins that are sub-parallel to the foliation. The underlying serpentinized peridotite is tectonized at the top; however, in Core 173-1068A-23R remnants of a former high-*T* spinel foliation occur, which are overgrown statically by serpentine minerals. Thus, late deformation in mantle rocks was strongly localized. The downward increase in deformation is accompanied by penetrative mineral reactions and veining, suggesting that deformation activity along the basal fault zone was assisted by fluid infiltration leading to alteration processes. Thus there is ample evidence to show that the lower part of the sedimentary breccias were overprinted by brittle deformation after deposition, and that the base of the breccia succession represents a tectonic contact that juxtaposes serpentized mantle against sedimentary breccias.

The association of the sedimentary and tectonic breccias, the fault zone contact with mantle rocks, and the 10° west inclination of the top of acoustic basement (Fig. 10b) are consistent with sedimentation and deformation along an active low-angle fault. The upward changes in clast composition described above probably reflect the stratigraphy (in reverse order) exposed in the footwall of the HHD. This indicates a top-to-the-ocean sense of shear.

The age of the HHD can be poorly constrained in three ways. First, the age of the uppermost breccias with a chalk matrix in Cores 15-17R give an upper age limit because they show no evidence of fault-related deformation. However, movement could have continued at the base of the breccias without disturbing the most recently deposited material. Unfortunately, nannofossils in the chalk matrices yield only an imprecise age between Barremian and Valanginian. The breccias could be younger than this age range if the chalk matrix was derived from older pelagic drapes that became entrained in the debris flows. They could have been emplaced after movement on the HHD ceased, as the gradient on this feature would have been sufficient to cause sediment instability. Second, the plagioclase cooling age of *c*. 137 Ma obtained from basement rocks at Site 1067 indicates that tectonic exhumation must have begun before this date. Third, the *c*. 137 Ma cooling age also indicates that the basement rocks at Site 1067 had not reached the sea floor at this time (i.e. the Berriasian–Valanginian boundary). However, the seismostratigraphic evidence suggests that the basement topography was formed before the deposition of Sub-units 6A and B, which indicates at least a pre-Hauterivian age, and possibly a Valanginian or even later age (Table 1).

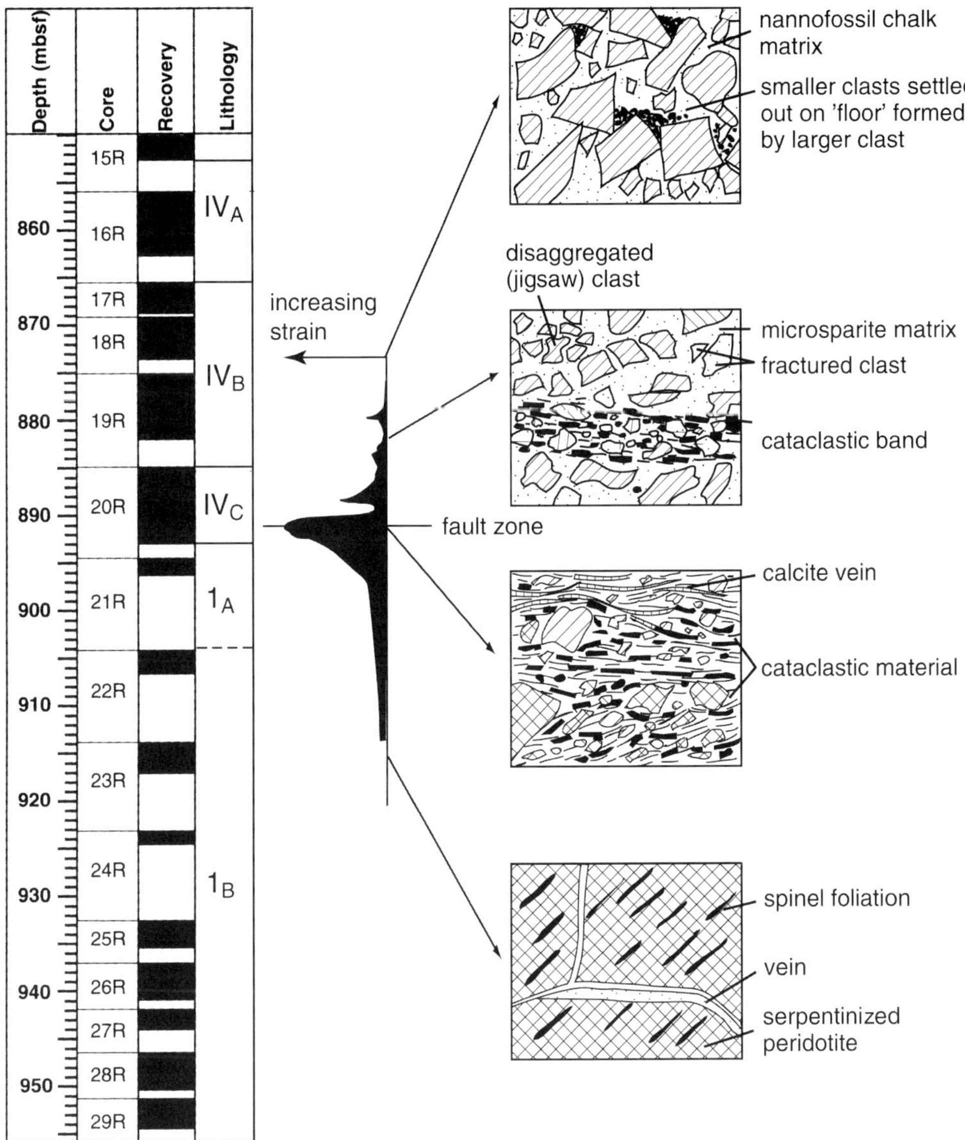

Fig. 17. The breccia succession at ODP Site 1068. It should be noted that Units IVA–C are sedimentary, and 1A and 1B are serpentinized peridotite; Units IVC and 1A are tectonic breccias. (For explanation, see text.)

Discussion

The absence of syn-rift intervals on seismic sections

How might we explain the absence of thickening of reflection packages into fault footwalls off Galicia and beneath the Iberia Abyssal Plain? There are three explanations for the absence of such syn-rift intervals. The first is that they are too thin to resolve on seismic sections because sedimentation rates were very low during rifting. This, as discussed by Wilson *et al.* (1996), is because the creation of a series of long hanging-wall slopes and footwall scarps will baffle turbidity and debris flows, which would then be ponded back in more landward sub-basins. Therefore, only the more dilute

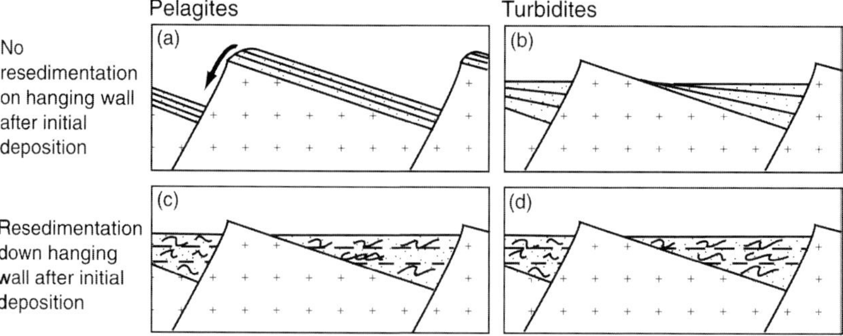

Fig. 18. Cartoons showing likely features of synrift pelagites and turbidites in deep-water settings. (For explanation, see text.)

portions of turbidity currents are likely to make it through the maze of ridges and hollows created by rifting. During the rift-climax phase, the deeper parts of the rift are starved of sediment, so that sedimentation is most likely to be dominated by pelagic and hemipelagic settling. In such a situation, rifting and contemporaneous deposition would have to continue for several million years to produce seismically resolvable reflection packages showing divergence.

The second explanation for the absence of reflection divergence towards footwalls is that syn-rift sediments are too unstable to retain their original syn-rift geometry and so are resedimented into the half-grabens. The cartoons in Figure 18 show sediment accumulation above hanging walls situated at depths well below storm wave base. Here there would be no current-related processes except for those linked to deep thermohaline currents; these are ignored in Figure 18. At such depths there is no erosion of footwall crests, save by mass wasting, and so only deposition of pelagites–hemipelagites and turbidites is shown.

Figure 18a and b illustrates the accumulation of sediments that remain cohesive after deposition, and so are not redeposited by gravity-driven processes into the central parts of half-grabens. In this cohesive sediment scenario, pelagic drape of hanging walls results in parallel sheets of sediments inclined towards the footwall, which could easily be mistaken for pre-rift sediments (Fig. 18a). Successive packages of turbiditic sediments would be expected to thicken towards footwalls (Fig. 18b), and the thinning-up and offlapping geometry shown would result from a progressive reduction of turbiditic influx as the new rift-related topography grew and ponded turbidites

in more proximal half-grabens. The 'pure' pelagite and turbidite scenarios shown are theoretical end members of a spectrum of possibilities; both processes are likely to have contributed to syn-rift deposition.

In Figure 18c and d, the sediments are resedimented to the centre of the half-graben because they remain uncohesive for some time after deposition. This is certainly the case for Recent to upper Pliocene sediments deposited on the Iberia Abyssal Plain. These siliciclastic turbiditic and hemipelagic sediments, and pelagic oozes (defined as Lithostratigraphic Unit 1 by Sawyer *et al.* (1994)) equivalent to Seismostratigraphic Unit 1 (Table 1) were sampled entirely by piston coring, and suffered significant deformation during coring down to depths between 60 and 110 m. Although the lithology of Unit 1 may not be the same as that of any syn-rift sediments occurring beneath the Iberia Abyssal Plain, their susceptibility to coring deformation does suggest that resedimentation down hanging-wall slopes as they rotate is a likely process. This could obliterate the characteristic syn-rift signature of reflection divergence towards footwalls. Such sedimentary packages are likely to be imaged as zones of very discontinuous or chaotic reflections. Such features, however, are not a common feature in the deepest reflection packages that occur off Galicia or beneath the Iberia Abyssal Plain.

The third explanation for the lack of divergence of reflections during fault movement is that there was no hanging-wall rotation because the faults were almost flat. This explanation could apply to the HHD west of Sites 1068, 900 and 1067, but not to fault-bounded basins to the east of it, nor to the half-graben basins of the Galicia margin.

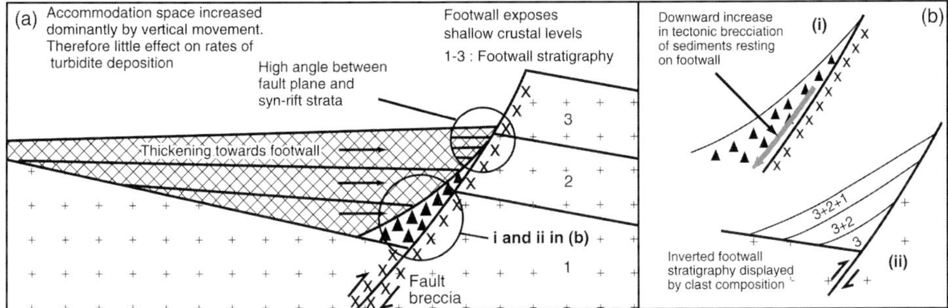

Fig. 19. Schematic summary of the principal features of synrift sediments accumulating in deep water during in low-β settings described and discussed in this paper.

The origin of the Hobby High detachment

As stated above, Manatschal *et al.* (2001) have suggested that the horizontal displacement along the low-angle HHD is 20.5 km, but what is the vertical displacement? The ^{40}Ar/^{39}Ar cooling ages indicate that the footwall rocks of Hobby High were at 150 °C at 137 Ma, so they must have been at a depth of at least 3–5 km. How was this vertical displacement accommodated along the fault? There is no evidence of a deep basin associated with it (i.e. between Hobby High and Site 1069), no evidence of subaerial exposure of the footwall, and no evidence of a dramatic deepening during Berriasian–Valanginian time at Site 1069. There is only one geometrical solution that permits the exhumation of deep crustal levels along a fault with a large displacement without creating very large topographic features. This requires the fault to deepen downwards (the rolling-hinge model of Buck (1988)) and for the footwall to

be pulled from beneath the hanging wall. This model explains all the features described from the Iberia Abyssal Plain: the virtual lack of reflection divergence into footwalls, strata being sub-parallel or parallel to the footwall, and erosion of deep crustal levels exposed in the footwall.

The rolling-hinge model of Lavier *et al.* (1999) produced a topographic amplitude of 2100 m after 27 km of extension. This amplitude is comparable with that observed along the HHD (Fig. 10b) along which a displacement of at least 20.5 km occurred. This amplitude is almost half that developed along detachment L, which has an estimated displacement of only 4–7 km. This difference may be explained by a change from extension along a listric fault in the case of detachment L, to the development of a rolling-hinge fault that formed the HHD to the west. The rolling hinge model does not explain why the Iberia Abyssal Plain subsided to its present-day depth of *c.* 5 km after the

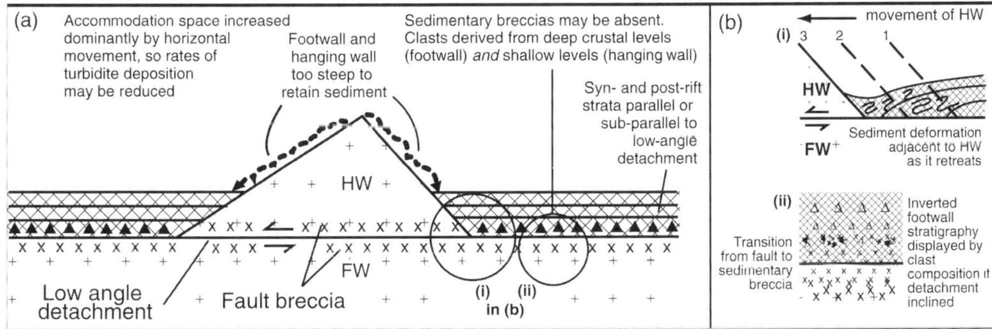

Fig. 20. Schematic summary of the principal features of synrift sediments accumulating in deep water in distal high-β settings described in this paper. It should be noted that if the hanging-wall allochthon is transported along a rolling-hinge fault, it would be carried passively by movement of the footwall, and so the arrow showing displacement to the left is not appropriate. Similarly, in (**b**) (i) there would be no sediment deformation, as the hanging wall would not retreat to the left, but would be carried passively on top of the footwall.

Valanginian. However, as we remarked in the introductory part of the paper, Driscoll & Karner (1998) suggested that large amounts of post-rift subsidence may result from lower-crustal extension that is not accompanied by brittle deformation in the upper crust.

Features of low- and high-β fault basins

Based on our observations of Alpine and Iberian margins, Figures 19 and 20 contrast the key features that might be expected in relatively low- and high-extension rifts in distal deep-water settings.

In relatively low-β systems (Fig. 19) we would expect to see the following: (1) thickening of sedimentary packages towards footwalls if sedimentary rates were high enough, and the deposits were cohesive immediately after deposition; (2) high angles between fault planes and syn- and post-rift strata; (3) acccumulation of sediments dominantly above the hanging wall; (4) inverted stratigraphy displayed by clast composition in footwall-related talus slopes; (5) downward increase in tectonic brecciation of sediments resting on the footwall; (6) accommodation space increased dominantly by vertical movement. There would, therefore, be little effect on rates of turbidite deposition, although developing rift topography would reduce density flow input.

A flat detachment is shown in the high-β setting depicted in Figure 20, but mass wasting could have occurred along it if it was inclined. The features that might be formed in this setting include the following: (1) syn- and post-rift strata are parallel to sub-parallel to the detachment, but high-angle relationships are possible on both footwalls and hanging walls of extensional allochthons; (2) sediments accumulate dominantly above the footwall; (3) both hanging walls and footwalls of extensional allochthons are too steep to retain sediments, save for small amounts trapped by local topographic features; (4) clasts are derived from deep crustal levels below the detachment and shallower levels in extensional allochthons above; (5) inverted footwall stratigraphy is displayed by breccias above the detachment if it is inclined, or was inclined but later rotated to become horizontal; (6) sediment deformation occurs adjacent to the hanging wall above the detachment as it retreats; this would not occur if the allochthons were part of a rolling-hinge fault system, as the detachment would be inactive; (7) large extension increases the surface area of the basins, and so, combined with the ponding effects of developing rift topography, turbidite deposition may be reduced.

Conclusions

Published identifications of syn-rift units in distal continental margins in the Alps and Iberia are not based on evidence for thickening of sedimentary units towards fault footwalls caused by hanging-wall rotation. This probably means that early post-rift sediments were interpreted to be syn-rift in origin, resulting in the duration of rifting to be grossly overestimated. Instead of lasting for *c.* 25 Ma, rifting may have occurred for <5 Ma.

The absence of syn-rift related reflection divergence towards footwalls on seismic lines along the Galicia margin and beneath the Iberia Abyssal Plain (with a few areally restricted exceptions) may be explained by one or more of the following reasons: (1) shutdown of sediment supply and consequent sediment starvation of distal rift sub-basins as a result of the development of inboard rift-related topographic features ponding back turbidity flows, resulting in thin syn-rift deposits not resolvable on seismic sections; (2) resedimentation of uncohesive syn-rift sediments towards the deepest parts of half-graben basins; (3) no hanging-wall rotation because faults were almost flat.

Seismostratigraphic evidence from the Galicia margin and the Iberia Abyssal Plain (i.e. rotation of pre-rift units followed by post-rift drape of rift-related topography) suggests that rifting occurred over a period of <5 Ma during Berriasian to early Valanginian time. This is consistent with evidence from ODP cores recovered in both areas. Along the Galicia margin beneath Site 639, Tithonian–Berriasian shallow-water carbonates are overlain by a thin interval of nannofossil mudstones followed by thick turbidites (both of Late Valanginian age), and beneath the Iberia Abyssal Plain, Tithonian siliciclastic mud-dominated sediments are overlain by Berriasian pelagic chalks at Site 1069.

At Site 1068 on the western flank of the Hobby High beneath the Iberia Abyssal Plain, the association of sedimentary and tectonic breccias, the upward change in clast stratigraphy, the downward decrease in deformation of serpentinized peridotite occurring beneath the breccias and the *c.* 10° westward dip of the top of acoustic basement are consistent with the development of a low-angle detachment (the HHD). Kinematic inversion of the detachment system interpreted from a seismic reflection line suggests an oceanward increase in displacement along separate faults, with at least 20.5 km occurring along the HHD. The paradox of such a large displacement not being accompanied by a large-amplitude, rift-related, topography, but none the less exposing deep crustal levels, is resolved by postulating the development of a deepening downwards, rolling-hinge fault.

Beneath the Iberia Abyssal Plain, we would expect that features shown in Figure 19 would be likely to occur to the east of Hobby High, and those depicted in Figure 20 to occur to the west. This is because of the change in mode of deformation 'from initial listric faulting leading to asymmetric basins, accommodating low amounts of stretching, to a situation where the footwall was pulled from underneath a relatively stable hanging wall accommodating high amounts of stretching' (Manatschal *et al.* 2001).

The work of G.M. was supported by the Swiss National Science Foundation Project 21-049117.96/1. We wish to thank D. Bernoulli, G, Karner, R. Steel and B. Taylor for their constructive reviews, J. Taylor for preparing some of the diagrams, and J. Matela for patiently word-processing successive drafts of the paper.

References

BERGREN, J.A. 1994. Berriasian to early Aptian calcareous nannofossils from the Vocontian Trough (SE France) and Deep Sea Drilling Site 534: new nannofossil taxa and a summary of low latitude biostratigraphic events. *Journal of Nannoplankton Research*, **16**, 59–69.

BERTOTTI, G., PICOTTI, V., BERNOULLI, D. & CASTELLARIN, A. 1993. From rifting to drifting: tectonic evolution of the South-Alpine upper crust from the Triassic to the Early Cretaceous. *Sedimentary Geology*, **86**, 53–76.

BOILLOT, G. & FROITZHEIM, N. 2001. Non-volcanic rifted margins, continental break-up and the onset of sea-floor spreading: some outstanding questions. *In*: WILSON, R.C.L., WHITMARSH, R.B., TAYLOR, B. & FROITZHEIM, N. (eds) *Non-volcanic Rifting of Continental Margins: a Comparison of Evidence from Land and Sea*. Geological Society, London, Special Publications, **187**, 9–30.

BOILLOT, G. & WINTERER, E.L. 1988. Drilling on the Galicia Margin: retrospect and prospect. *In*: BOILLOT, G., WINTERER, E.L. *et al.* (eds) *Proceedings of the Ocean Drilling Program, Scientific Results, 103*. Ocean Drilling Program, College Station, TX, 809–828.

BOILLOT, G., GIRARDEAU, J. & WINTERER, E.L. 1989. Rifting processes of the west Galicia margin, Spain. *In*: TANKARD, A.J. & BALKWILL, H.R. (eds) *Extensional Tectonics and Stratigraphy of the North Atlantic Margins*. American Association of Petroleum Geologists, Memoirs, **40**, 363–377.

BUCK, W.R. 1988. Flexural rotation of normal faults. *Tectonics*, **7**, 959–973.

CONCHERYO, A. & WISE, S.J. 2001. Jurassic calcareous nannofossils from prerift sediments drilled during ODP Leg 173, Iberia Abyssal Plain and their implications for rift tectonics. *In*: BESLIER, M.-O., WHITMARSH, R.B., WALLACE, P.J. & GIRADEAU, J. (eds) *Proceedings of the Ocean Drilling Program, Scientific Results, 173*. Ocean Drilling Program, College Station, TX, 1–24 (online).

DESMURS, L., MANATSCHAL, G. & BERNOULLI, D. 2001. The Steinmann Trinity revisited: mantle exhumation and magmatism along an ocean–continent transition: the Platta nappe. *In*: WILSON, R.C.L., WHITMARSH, R.B., TAYLOR, B. & FROITZHEIM, N. (eds) *Non-volcanic Rifting of Continental Margins: a Comparison of Evidence from Land and Sea*. Geological Society, London, Special Publications, **187**, 235–266.

DRISCOLL, N.W. & KARNER, G.D. 1998. Lower crustal extension acxross the Northern Caernarvon basin, Australia: evidence for an eastward dipping detachment. *Journal of Geophysical Research*, **103**, 4975–4991.

DRISCOLL, N.W., HOGG, J.R., CHRISTIE-BLICK, N. & KARNER, G.D. 1995. Extensional tectonics in the Jeanne d'Arc Basin, offshore Newfoundland: implications for the timing of break-up between Grand Banks and Iberia. *In*: SCRUTTON, R.A., STOKER, M.A., SHIMMIELD, G.B. & TUDHOPE, A.W. (eds) *The Tectonics, Sedimentation and Palaeoceanography of the North Atlantic Region*. Geological Society, London, Special Publications, **90**, 1–28.

FALVEY, D.A. 1974. The development of continental margins in plate tectonic theory. *Australian Exploration Journal*, **14**, 95–106.

FÉRAUD, G., BESLIER, M.-O. & CORNEN, G. 1996. ^{40}Ar/^{39}Ar dating of gabbros from the ocean/continent transition of the western Iberia margin: preliminary results. *In*: WHITMARSH, R.B., SAWYER, D.S., KLAUS, A. & MASSON, D.G. (eds) *Proceedings of the Ocean Drilling Program, Scientific Results, 149*. Ocean Drilling Program, College Station, TX, 489–495.

FROITZHEIM, N. & EBERLI, G.P. 1990. Extensional detachment faulting in the evolution of a Tethys passive continental margin, Eastern Alps, Switzerland. *Geological Society of America Bulletin*, **102**, 1297–1308.

FROITZHEIM, N. & MANATSCHAL, G. 1996. Kinematics of Jurassic rifting, mantle exhumation, and passive-margin formation in the Austroalpine and Penninic nappes (eastern Switzerland). *Geological Society of America Bulletin*, **108**, 1120–1133.

GRADSTEIN, F.M. & OGG, J.G. 1996. A Phanerozoic time scale. *Episodes*, **19**, 3–5.

GROUPE GALICE, 1979. The continental margin off Galicia and Portugal: acoustical stratigraphy, dredge stratigraphy and structural evolution. *In*: SIBUET, J.-C., RYAN, W.B.F. *et al.* (eds) *Initial Reports, Deep Sea Drilling Program, 47 (Part 2)*. US Government Printing Office, Washington, DC, 633–662.

JANSA, L.F., COMAS, M.C., SART, M. & HAGGARTY, J.A. 1988. Late Jurassic. Carbonate platform of the Galicia margin. *In*: BOILLOT, G., WINTERER, E.L. *et al.* (eds) *Proceedings of the Ocean Drilling Program, Scientific Results, 103*. Ocean Drilling Program, College Station, TX, 171–192.

KRAWCZYK, C.M., RESTON, T.J., BESLIER, M.-O. & BOILLOT, G. 1996. Evidence for detachment tectonics on the Iberia Abyssal Plain rifted margin. *In*: WHITMARSH, R.B., SAWYER, D.S., KLAUS, A. & MASSON, D.G. (eds) *Proceedings of the Ocean Drilling Program, Scientific Results, 149*. Ocean Drilling Program, College Station, TX, 603–615.

LAVIER, L.L., BUCK, W.R. & POLIAKOV, A.N.B. 1999. Self consistent rolling-hinge model for the evolution of large-offset low angle faults. *Geology*, **27**, 1127–1130.

LISTER, G.S., ETHERIDGE, M.A. & SYMONDS, P.A. 1988. Detachment faulting and the evolution of passive continental margins. *Geology*, **14**, 246–250.

MANATSCHAL, G. & BERNOULLI, D. 1998. Rifting and early evolution of ancient ocean basins: the record of the Mesozoic Tethys and of the Galicia–Newfoundland margins. *Marine Geophysical Researches*, **20**, 371–381.

MANATSCHAL, G. & BERNOULLI, D. 1999. Architecture and tectonic evolution of non-volcanic margins: present day Galicia and ancient Adria. *Tectonics*, **18**, 1099–1119.

MANATSCHAL, G. & NIEVERGELT, P. 1997. A continent–ocean transition recorded in the Err and Platta nappes (Eastern Switzerland). *Eclogae Geologicae Helvetiae*, **90**, 3–27.

MANATSCHAL, G., FROITZHEIM, N., RUBENACH, M. & TURRIN, B.D. 2001. The role of detachment faulting in the formation of an ocean–continent transition: insights from the Iberia Abyssal Plain. *In*: WILSON, R.C.L., WHITMARSH, R.B., TAYLOR, B. & FROITZHEIM, N. (eds) *Non-volcanic Rifting of Continental Margins: a Comparison of Evidence from Land and Sea*. Geological Society, London, Special Publications, **187**, 405–426.

MASSON, D.G., CARTWRIGHT, J.A., PINHEIRO, L.M., WHITMARSH, R.B., BESLIER, M.-O. & ROESER, H. 1994. Localised deformation at the ocean–continent transition in the NE Atlantic. *Journal of the Geological Society, London*, **151**, 603–613.

MAUFFRET, A. & MONTADERT, L. 1987. Rift tectonics on the passive continental margin off Galicia (Spain). *Marine and Petroleum Geology*, **4**, 49–70.

MAUFFRET, A. & MONTADERT, L., 1988. Seismic stratigraphy off Galicia. *In*: BOILLOT, G., WINTERER, E.L. *et al.* (eds) *Proceedings of the Ocean Drilling Program, Scientific Results,*

103. Ocean Drilling Program, College Station, TX, 13–30.

MURILLAS, J., MOUGENOT, D., BOILLOT, G., COMAS, M.C., BANDA, E. & MAUFFRET, A. 1990. Structure and evolution of the Galicia Interior basin (Atlantic western Iberian continental margin). *Tectonophysics*, **184**, 297–319.

PINHEIRO, L.M., WILSON, R.C.L., PENA DOS REIS, R., WHITMARSH, R.B. & RIBEIRO, A. 1996. The western Iberia margin: a geophysical and geological overview. *In*: WHITMARSH, R.B., SAWYER, D.S., KLAUS, A. & MASSON, D.G. (eds) *Proceedings of the Ocean Drilling Program, Scientific Results, 149*. Ocean Drilling Program, College Station, TX, 3–23.

PROSSER, S. 1993. Rift-related linked depositional systems and their seismic expression. *In*: WILLIAMS, G.D. & DOBB, A. (eds) *Tectonics and Seismic Stratigraphy*. Geological Society, London, Special Publications, **71**, 35–66.

RAVNÅS, R. & STEEL, R.J. 1998. Architecture of marine rift-basin successors. *AAPG Bulletin*, **82**, 110–146.

SAWYER, D.S., WHITMARSH, R.B., KLAUS, A. *et al.* (eds) 1994. *Proceedings of the Ocean Drilling Project, Initial Reports, 149*. Ocean Drilling Program, College Station, TX.

SHIPBOARD SCIENTIFIC PARTY 1987a. Site 639. *In*: Boillot, G., Winterer, E.L., Meyer, A.W. *et al.* (eds) *Proceedings of the Ocean Drilling Program, Initial Reports, 103*. Ocean Drilling Program, College Station, TX, 409–532.

SHIPBOARD SCIENTIFIC PARTY 1987b. Site 641. *In*: Boillot, G., Winterer, E.L., Meyer, A.W. *et al.* (eds) *Proceedings of the Ocean Drilling Program, Initial Reports, 103*. Ocean Drilling Program, College Station, TX, 571–660.

SHIPBOARD SCIENTIFIC PARTY 1998a. Site 1068. *In*: Whitmarsh, R.B., Beslier, M.-O., Wallace, P.J. *et al.* (eds) *Proceedings of the Ocean Drilling Program, Initial Reports, 173*. Ocean Drilling Program, College Station, TX, 163–218.

SHIPBOARD SCIENTIFIC PARTY 1998b. Site 1069. *In*: Whitmarsh, R.B., Beslier, M.-O., Wallace, P.J. *et al.* (eds) *Proceedings of the Ocean Drilling Program, Initial Reports, 173*. Ocean Drilling Program, College Station, TX, 219–263.

TAYLOR, B., GOODLIFFE, A.M. & MARTINEZ, F. 1999. How continents break up: insights from Papua New Guinea. *Journal of Geophysical Research*, **104**, 7497–7512.

WILSON, R.C.L. 1998. Sequence stratigraphy: a revolution without a cause? *In*: BLUNDELL, D.J. & SCOTT, A.C. (eds) *Lyell: the Past is the Key to the Present*. Geological Society, London, Special Publications, **143**, 303–314.

WILSON, R.C.L., SAWYER, D.S., WHITMARSH, R.B., ZERONG, J. & CARBONELL, J. 1996. Seismic stratigraphy and tectonic history of the Iberia Abyssal Plain. *In*: WHITMARSH, R.B., SAWYER, D.S., KLAUS, A. & MASSON, D.G. (eds) *Proceedings of the Ocean Drilling Program, Scientific Results, 149*. Ocean Drilling Program, College Station, TX, 617–630.

Tectonic evolution of the NW Red Sea–Gulf of Suez rift system

S.M. KHALIL & K.R. MCCLAY

Fault Dynamics Research Group, Geology Department, Royal Holloway University of London, Egham TW20 0EX, UK (e-mail: ken@gl.rhul.ac.uk)

Abstract: The NW Red Sea–Gulf of Suez rift system was initiated during Late Oligocene time and underwent extension in a N65°E direction, almost orthogonal to pre-existing WNW-trending Pan African shear-zone fabrics in the crystalline basement of the Sinai–African plate. Earliest syn-rift sediments are Upper Oligocene continental clastic deposits with minor syn-rift basalts. Early Miocene sedimentation was dominated by shallow marine clastic deposits, which developed variable stratigraphic architectures as a response to the interaction of extensional faulting, sea-level changes, sediment supply and dispersal. Analysis of fault geometries, fault kinematics and sedimentation patterns indicates that rift-normal extension predominated throughout the Late Oligocene–Early Mid-Miocene evolution of the rift. Reactivation of the Precambrian basement fabrics was the main factor controlling the fault architecture, fault linkage and evolution of the NW Red Sea–Gulf of Suez rift. Individual faults were initially strongly segmented and offset across 'soft-linked' relay structures. With increased extension these faults became linked by breaking down relay structures with the development of local 'hard-linked' transfer faults, thus giving rise to the rhomboidal fault pattern of the rift system. In Mid-Miocene time, the Levant–Gulf of Aqaba transform boundary was established, linking the Red Sea rift plate boundary to the convergent Bitlis–Zagros plate boundary. This resulted in a dramatic decrease in extension rates within the Gulf of Suez whereas the northern Red Sea continued to extend, with significant syn-rift sediments deposited in Late Miocene–Pliocene time in offshore fault-bounded basins.

The Late Cenozoic NW Red Sea–Gulf of Suez rift system is one of the best exposed and best studied examples of a continental rift environment. The Gulf of Suez has long been recognized as one of the best examples of along-axis segmentation into sub-basins with different dip polarities (Moustafa 1976, 1993; Bosworth 1994; Patton *et al.* 1994; Colletta *et al.* 1988; McClay *et al.* 1998). It also displays superb examples of the interaction between extensional tectonics and sedimentation (Gawthorpe *et al.* 1997; Gupta *et al.* 1999; Sharp *et al.* 2000). The Gulf of Suez and NW Red Sea rift systems are remarkably non-volcanic with only a few, volumetrically insignificant, late pre-rift to early syn-rift basic dykes and isolated basaltic flows (see Bosworth & McClay 2001).

Recent studies have evaluated the relative roles of 'hard-linked' and 'soft-linked' displacement transfer in intra-basin fault linkages, and the significance of pre-rift structures in controlling the style of linkage (McClay & Khalil 1998; McClay *et al.* 1998; Younes & McClay 1998). This paper summarizes recent research on the tectonic evolution of the central eastern section of the Gulf of Suez rift and the NW Red Sea, focusing on the interaction of pre-existing basement fabrics with the Cenozoic extension. The structural styles exhibited in the NW Red Sea–Gulf of Suez are compared with those of with other non-volcanic rift systems.

Plate tectonic setting of the NW Red Sea–Gulf of Suez rift system

The Late Oligocene–Early Miocene Gulf of Suez and NW Red Sea (Fig. 1) is the northern termination of the Gulf of Aden–Red Sea rift system. This developed as a result of the northeastward separation of the Arabian plate from the African plate (McKenzie *et al.* 1970; Coleman 1974, 1993; Cochran 1983; Girdler & Southren 1987; Hempton 1987; Joffe & Garfunkel 1987). Two contrasting models have been postulated for the early (Late Oligocene–Early Miocene) evolution of the Red Sea–Gulf of Suez rift system. Model 1 invokes regional northeastward extension, approximately normal to the present NW trend of the rift, whereas model 2 invokes early strike-slip faulting and formation of pull-apart basins (Jarrige *et al.*

From: WILSON, R.C.L., WHITMARSH, R.B., TAYLOR, B. & FROITZHEIM, N. 2001. *Non-Volcanic Rifting of Continental Margins: A Comparison of Evidence from Land and Sea*. Geological Society, London, Special Publications, **187**, 453–473. 0305-8719/01/$15.00 © The Geological Society of London 2001.

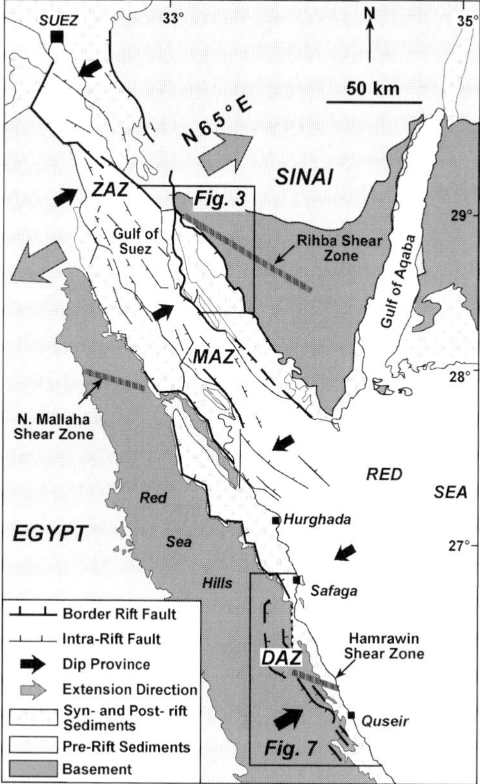

Fig. 1. Tectonic map of the Gulf of Suez and the NW Red Sea. ZAZ, MAZ and DAZ indicate the Zaafarana, Morgan and Duwi accommodation zones, respectively.

1986, 1990; Montenat *et al.* 1986, 1988; Makris & Rihm 1991; Rihm & Henke 1998).

In the first model, the Red Sea–Gulf of Suez rift system formed by anticlockwise rotation of Arabia away from Africa about a pole of rotation located in the central or south–central Mediterranean Sea (Freund 1970; McKenzie *et al.* 1970; Le Pichon & Francheteau 1978; Meshref 1990; Morgan 1990). Extension decreases westwards along the Gulf of Aden and northward along the Red Sea and into the Gulf of Suez, consistent with, and constraining, the Arabian pole of rotation. There is general agreement that magnetic anomalies in the central and southern Red Sea axis (below latitude 24°N) were generated by sea-floor spreading that began in Early Pliocene time (*c.* 4.5 Ma) (Cochran 1983; Girdler & Southren 1987; Coleman 1993). The Gulf of Suez attained stretching factors of about two (Bosworth 1994), but this and the northern Red Sea remained floored by continental crust (Freund 1970; Joffe & Gar-

funkel 1987). In late Mid-Miocene time, extension decreased in the Gulf of Suez and the opening of the Red Sea was linked to sinistral strike-slip displacement along the Gulf of Aqaba–Dead Sea transform fault system (Quennell 1958; Freund 1970; Ben-Menahem *et al.* 1976; Steckler *et al.* 1988; Abdel Khalek *et al.* 1993). The Gulf of Aqaba–Dead Sea fault system accumulated *c.* 65 km of horizontal displacement in Mid-Miocene to Pliocene time and a further 40 km in Pliocene to Quaternary time (Freund 1970; Bartov *et al.* 1980; Eyal *et al.* 1981; Quennell 1984; Hempton 1987; Joffe & Garfunkel 1987; Eyal 1996).

In contrast to this simple model of northeastward movement of the Arabian plate away from the African plate, the strike-slip models for the early opening of the Red Sea–Gulf of Suez rift require more complicated fault and plate movements. Jarrige *et al.* (1986, 1990) and Montenat *et al.* (1986, 1988) invoked a NW–SE compression along the proto-rift axis such that reactivation of NNE–SSW ('Aqaba' trend) and ESE–WNW ('Duwi' trend) pre-existing basement fabrics produced rhombic, weakly subsiding sub-basins. Makris & Rihm (1991) and Rihm & Henke (1998) proposed early sinistral strike-slip motion parallel to the axis of the rift, producing highly extended pull-apart basins with oceanic crust along a transform plate boundary. Structural analysis of both field and seismic data carried out during the course of the research summarized in this paper did not show any evidence of large-scale strike-slip tectonics and hence model 1, as discussed above, is favoured.

Structural and stratigraphic framework of the NW Red Sea–Gulf of Suez rift system

The NW Red Sea–Gulf of Suez rift system is exposed for *c.* 400 km along the NW Red Sea coast from near Quseir to the tip of the Gulf of Suez at Suez City (Fig. 1). The Suez rift and the NW Red Sea rift are characterized by a zigzag fault pattern, composed of NW–SE- and north–south- to NNE–SSW-striking extensional fault systems both at the rift borders and within the rift basins (Garfunkel & Bartov 1977; Jarrige *et al.* 1986; Moretti & Chenet 1987; Colletta *et al.* 1988; Meshref 1990; Moustafa 1993; Patton *et al.* 1994; Schutz 1994; Bosworth 1995; McClay *et al.* 1998; Montenat *et al.* 1988) (Fig. 1). Other subsidiary fault trends (WNW–ESE and NE–SW) are also exist in the Gulf of Suez and NW Red Sea. Four distinct depocentres (sub-basins) separated by complex accommodation zones occur within the Gulf of Suez and the NW Red Sea (Fig. 1). Each sub-basin is asymmetric, bounded on one side by a major NW-trending border fault system with large throws (3–6 km in general) with a dominant stratal dip direction toward the border fault system. The accommodation zones are oblique to the rift trend (see Moustafa 1976; Bosworth 1985; Coffield & Schamel 1989) (Fig. 1). Within the sub-basins, second-order depocentres are formed by individual fault blocks, each of which has its own characteristic syn-rift stratigraphy (see Montenat *et al.* 1986, 1988; Darwish & El-Azabi 1993; Bosworth 1995; Gawthorpe *et al.* 1997; Bos-

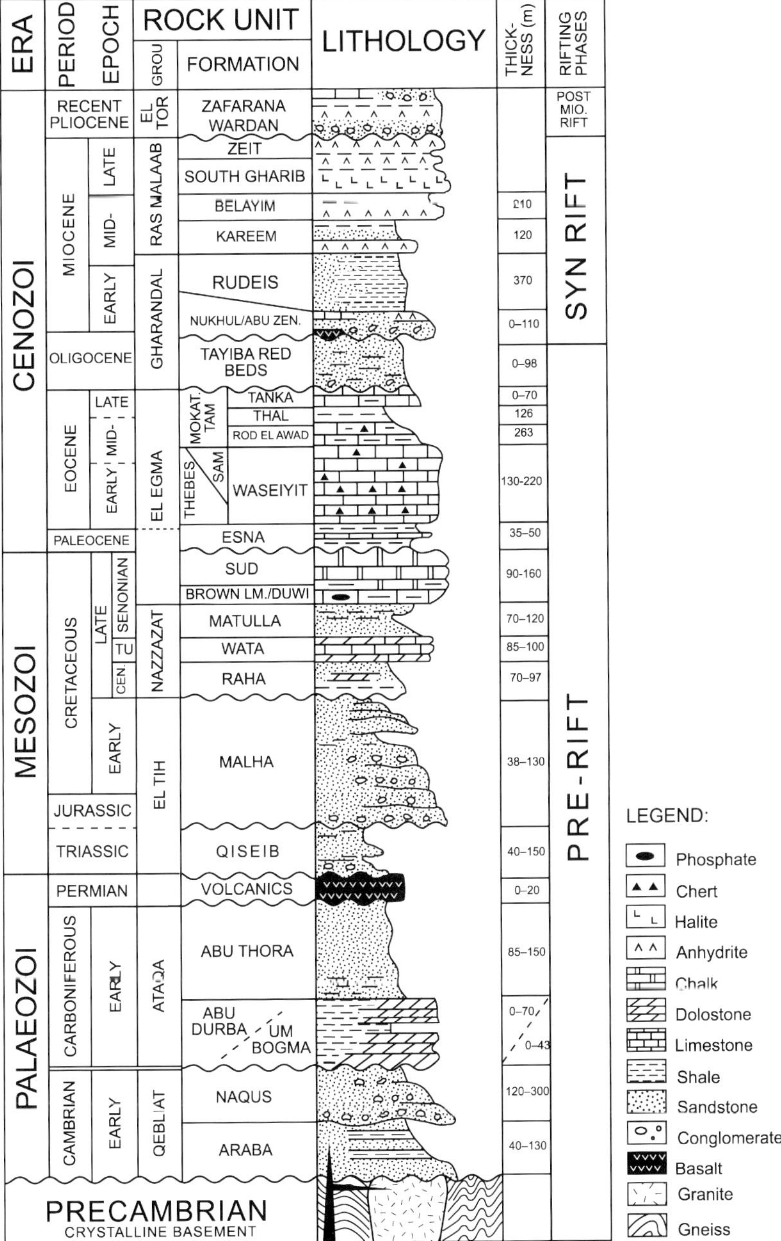

Fig. 2. Regional stratigraphy of the Gulf of Suez basin showing the basement and the pre-rift sedimentary section of 1.7 km thickness above the basement (after Darwish & El-Azabi 1993).

worth *et al.* 1998; Khalil 1998; McClay *et al.* 1998; Gupta *et al.* 1999).

Eastern margin, Gulf of Suez

The central eastern margin of the Gulf of Suez contains some of the best exposed tilted fault blocks and syn-rift stratigraphy of the entire Gulf. Figure 2 summarizes the stratigraphy of this area, where the most complete stratigraphic section is exposed (Khalil 1998; McClay *et al.* 1998). This is typical for most of the Gulf of Suez (see Patton *et al.* 1994; Bosworth *et al.* 1998; Montenat *et al.* 1988), although thick-

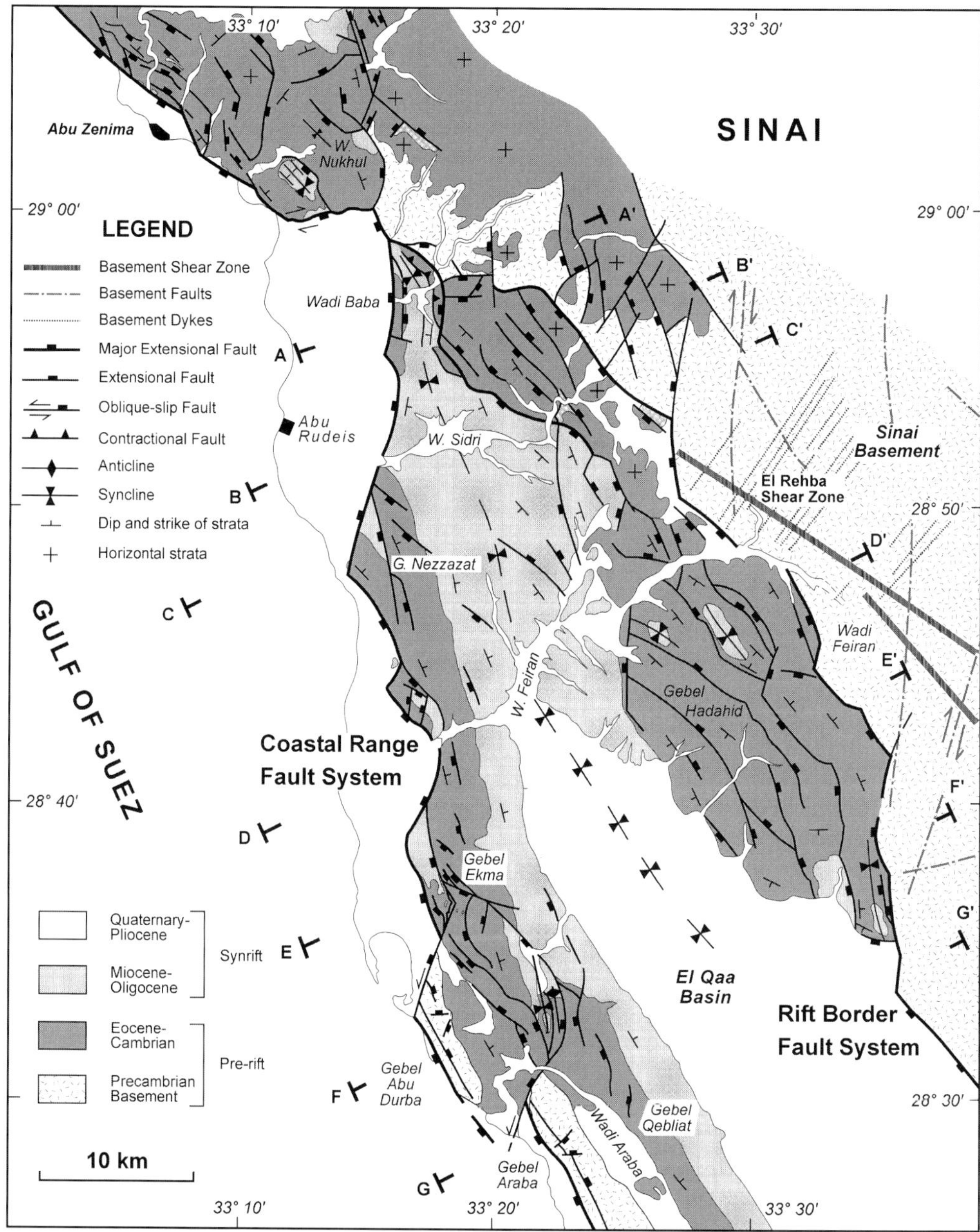

Fig. 3. Surface geological map of the central eastern margin of the Gulf of Suez rift (after Khalil 1998; McClay *et al.* 1998). Locations of cross-sections shown in Figure 4 are labelled A–A′ to G–G′.

nesses, facies and local formation names may vary. Stratigraphic and structural relationships indicate that rifting along the northern Red Sea and the Gulf of Suez appears to have commenced during Late Oligocene (Chattian) time (see Patton *et al.* 1994; McClay *et al.* 1998; Bosworth & McClay 2001).

Stratigraphy. The pre-rift stratigraphy (Fig. 2), above the Precambrian crystalline basement is about 1.7 km thick on the eastern margin of the Gulf of Suez. Syn-rift strata (Upper Oligocene to Pliocene units) vary in thickness from a few hundred metres in places along the rift flanks to as much as 6 km in basins in the centre of the rift. Detailed descriptions of the pre rift and syn-rift stratigraphy have been given by Ghorab *et al.* (1964), Darwish (1992), Patton *et al.* (1994), Khalil (1998), McClay *et al.* (1998), Plaziat *et al.* (1998*a,b*) and Bos-

worth & McClay (2001), and only a brief summary is presented here. Figure 3 is a simplified geological map of this area, and representative cross-sections are shown in Figure 4.

Basement. The crystalline basement of the eastern Gulf of Suez is formed by Precambrian gneisses, metavolcanic rocks and metasediments deformed and metamorphosed during the Pan-African orogeny, which terminated at *c.* 550 Ma (e.g. Stern 1981, 1994; Kröner 1984, 1993; Stoeser & Camp 1985). These are intruded by syn- and post-tectonic granites and granodiorites that range in age from 700 to 500 Ma (Greenberg 1981; Hassan & Hashad 1990). The Sinai basement contains a number of strong pre-rift fabrics: faults, joints, dykes and shear zones (Fig. 1).

Pre-rift strata. The lowermost section of pre-rift strata on the eastern margin of the Gulf of

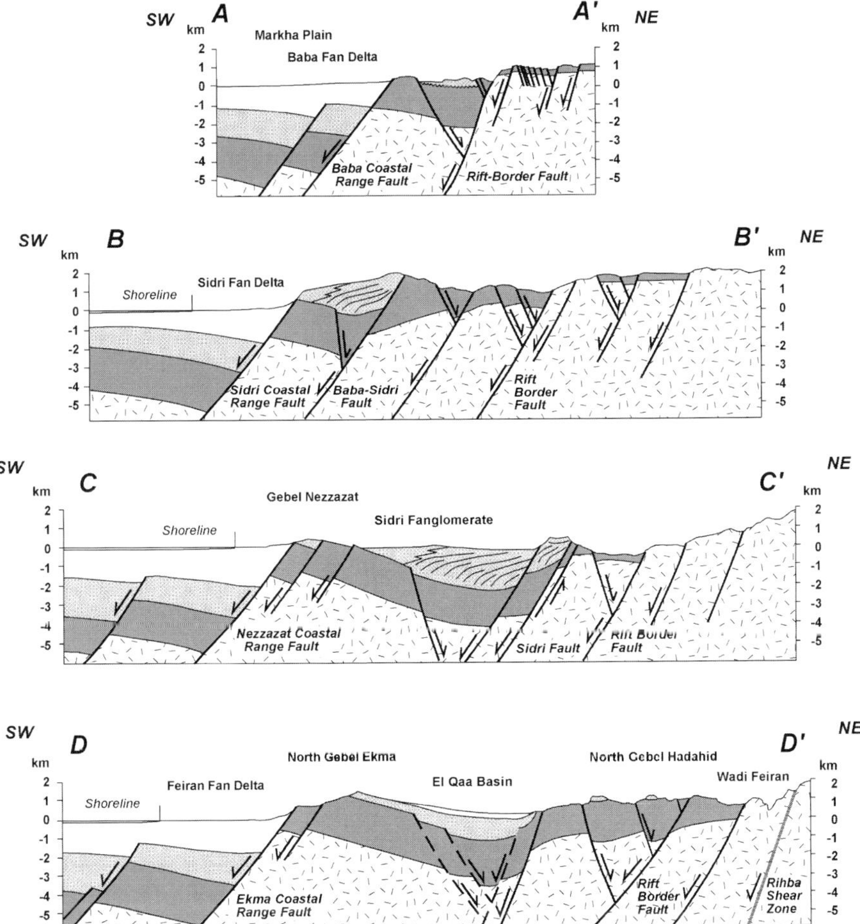

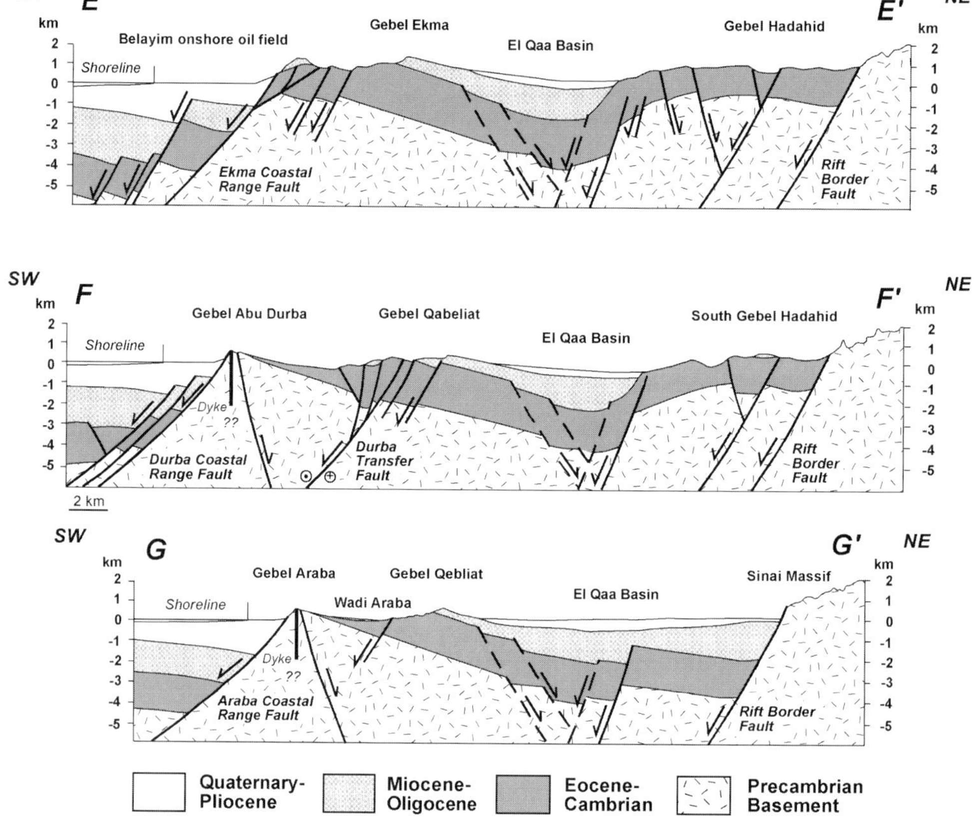

Fig. 4. Cross-sections through the central eastern margin of the Gulf of Suez. (See Fig. 3 for locations.)

Suez consists of a 650 m siliciclastic-dominated succession of shallow marine to fluvial sandstones with subordinate carbonates and shales (of Carboniferous age) and minor basalts (of Permian age) (Fig. 2). These range from Cambrian to Early Cretaceous age and are unconformably overlain by a *c.* 300 m Upper Cretaceous sequence of mixed siliclastic rocks and carbonates (Nezzazat Group, Ghorab 1961) (Fig. 2). Above this there are *c.* 750 m of chalks, chalky limestones, cherty limestones and limestones of Late Cretaceous Sennonian to Late Eocene age. The uppermost pre-rift strata consist of patchily developed (preserved) continental red beds of the Tayiba Formation (Hume *et al.* 1920) (Fig. 2).

Syn-rift strata. The Tayiba Formation is locally unconformably overlain by similar facies red beds of the Abu Zenima Formation, which locally contains subaqueous basalt flows as well as conglomerates with basalt cobbles. The Abu Zenima Formation, of Late Oligocene

(Chattian) age, is generally taken as the transition between pre-rift and syn-rift strata (see Hantar 1965; Sellwood & Netherwood 1984; Patton *et al.* 1994; Plaziat *et al.* 1998*b*; Bosworth & McClay 2001) and it marks the onset of extension in the Suez rift. It is unconformably overlain by shallow marine clastic deposits of the Miocene syn-rift Nukhul and Rudeis formations (Fig. 2). At the rift margins, near active faults, coarse-grained clastic sediments predominate, whereas finer-grained clastic rocks occur in the depocentres that were distal to the active fault systems. During the early stages of rifting, individual extensional fault blocks along the rift margins developed their own subbasins, each with a distinct stratigraphy. The earliest syn-rift deposits, therefore, vary in age in different fault blocks and overlie different pre-rift strata as a result of varying amounts and timing of extension, rotation, uplift, erosion and subsidence below the erosional base-level. In any particular sub-basin, the lowermost syn-

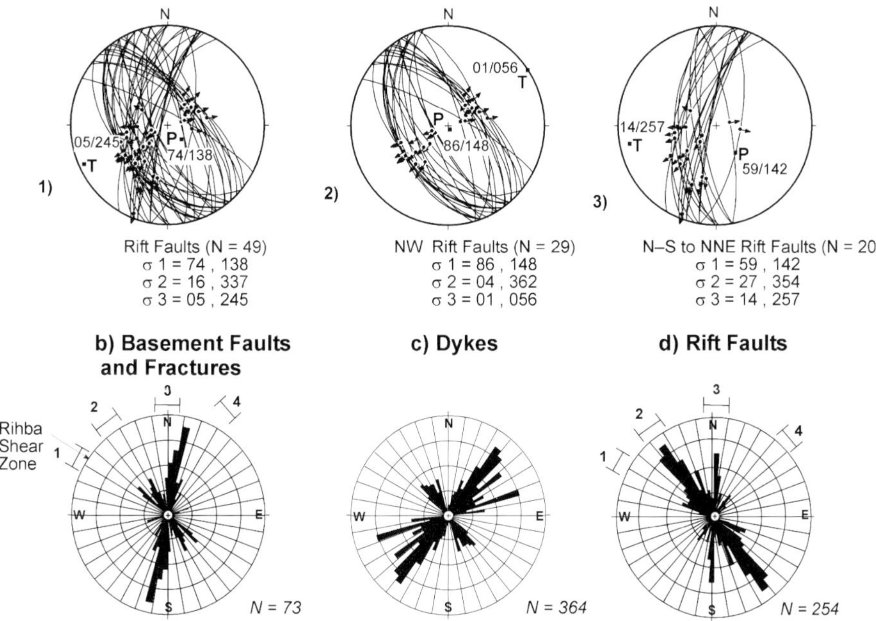

Fig. 5. Analysis of fault trends and basement fabrics in the Gulf of Suez, determined both from field mapping and analysis of Landsat TM images. (**a**) Lower-hemisphere equal-area stereograms of faults and striae. (1) NW and north–south to NNE rift faults; (2) NW faults; (3) north–south to NNE faults. Data in (2) and (3) are sorted according to the orientation of faults shown in (1). T and P indicate the extension and shortening axes, respectively. (**b**) Basement faults and fractures. (c) Dykes. (d) Rift faults: 1, trend of Pan-African shear zones: 2, trend of NNW rift fault systems: 3, trend of north–south to NNE basement fracture zones: 4, trend of NE cross-faults.

rift deposits are commonly coarse clastic deposits, which, based on lithostratigraphic correlations, are generally assigned to the Nukhul Formation.

On the eastern margin of the Gulf of Suez the shallow marine Nukhul Formation is of Aquitanian age (Ghorab *et al.* 1964; McClay *et al.* 1998; Plaziat *et al.* 1998*b*) and it unconformably overlies the Abu Zenima Formation and older Late Eocene strata. The Nukhul Formation consists dominantly of carbonate clasts and grains together with chert pebbles derived from the underlying Eocene limestone successions. The Burgidalian Rudeis Formation overlies Nukhul Formation. It is dominantly fine-grained mudstones and marls, which were deposited during an episode of basin deepening (Richardson & Arthur 1988; McClay *et al.* 1998). The Rudeis Formation locally may also be coarse grained (Abu Alaqa conglomerate, Garfunkel & Bartov 1977) in fan-delta systems developed adjacent to active faults (Gawthorpe *et al.* 1997; Khalil 1998; McClay *et al.* 1998; Gupta *et al.* 1999; Sharp *et al.* 2000). In the fan

deltas the bed forms show typical onlap patterns and syn-depositional unconformities, which indicate renewed and continued displacement on the Border Fault system during the deposition of the Abu Alaqa conglomerate. Moreover, the basement clasts occurring in the upper Rudeis Formation and Abu Alaqa conglomerates suggest significant uplift of the rift flanks during Late Burdigalian time. Above the Rudeis Formation the strata become increasingly evaporitic with the development of the Kareem, Belayim, South Gharib and Zeit formations (Fig. 2). This reflects the increased isolation of both individual fault-bounded basins as well as of the rift in general (Bosworth & McClay 2001).

Pliocene–Recent post-rift strata. The Miocene syn-rift strata of the eastern margin of the Gulf of Suez are unconformably overlain by a post-rift sequence of highly variable thickness of gravels and sands of the Wardan and Zafarana formations (Fig. 2). The Pliocene–Quaternary sediments may locally exceed 1 km in thickness. These are dominantly clastic deposits with thin evaporites and subordinate carbonates. In the eastern margin of the Gulf of Suez alluvial fans and fan deltas are developed at the mouth of main wadis (e.g. Wadi Feiran, Wadi Sidri and Wadi Baba, Fig. 3) whereas carbonate reef platform developed in the offshore southern Gulf of Suez (Bosworth & McClay 2001).

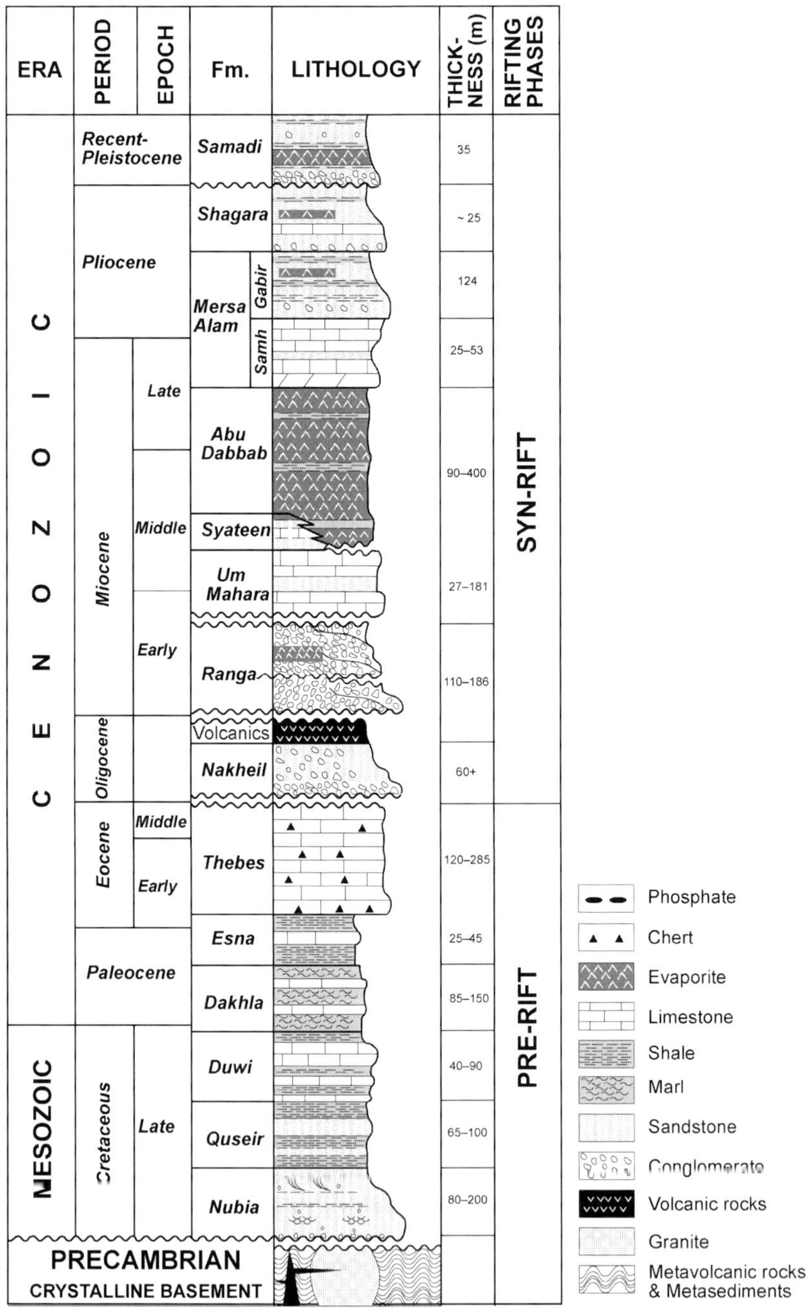

Fig. 6. Regional stratigraphy of the NW Red Sea showing the Precambrian basement and the pre- and syn-rift sedimentary sections. Data from Said (1990), Purser & Bosence (1998) and present study.

Structure. The structure of the eastern margin of the Gulf of Suez is characterized by two dominant fault systems: the Rift Border Fault system and the Coastal Range Fault system (Fig. 3). Both of these fault systems consist of two main fault trends: a dominant NW trend and a NNE trend that, where linked together, form a characteristic rhomboidal or zig-zag fault pattern. In cross-section the dominant fault geometry is that of a series of SW-dipping, planar to near-planar domino-style extensional faults that form fault blocks of 2–8 km width (Fig. 4). The Rift Border Fault system typically has throws of the order of 1–3.5 km (Khalil 1998) and separates the Precambrian basement to the east from tilted fault blocks of pre-rift strata and syn-rift strata to the west (Fig. 3). Exposed Lower Miocene syn-rift units occur mainly in hanging-wall synclines within tilted fault blocks between the Rift Border Fault system and the Coastal Range Fault system (Figs 3 and 4). The Coastal Range Fault system typically has throws of the order of 3.5–5.8 km (Khalil 1998) and bounds the westernmost fault blocks from the low-lying Quaternary gravels and sands of the coastal plains (Figs 3 and 4).

Two basement-cored tilted fault blocks, Gebel Araba and Gebel Abu Durba, occur along the southern part of the Coastal Range Fault system (Figs 3 and 4). These fault blocks are offset by NNE-trending, 'hard-linked', oblique-slip transfer faults (McClay & Khalil 1998). To the east of the Rift Border Fault system (Fig. 3) the Sinai crystalline basement consists of Precambrian gneisses and granites exposed in mountains up to 2.6 km above sea level (see cross-sections in Fig. 4). This reflects significant rift flank uplift and exhumation as a result of flexural and isostatic uplift in response to rifting. Fission-track dating of apatites separated from the basement rocks of Sinai revealed that uplift of the basement is *c.* 5 km (Kohn & Eyal 1981) and this amount probably equals or exceeds that in other areas bordering the Red Sea. The fault blocks in the footwall to the Coastal Range Fault system were also uplifted, reaching elevations in excess of 700 m above sea level. This resulted from local isostatic uplift of individual fault blocks as rifting focused inwards towards the centre of the rift and thus produced exposure of the syn-rift sediments between the Coastal Range Fault and Rift Border Fault systems (see Fig. 3).

Measurement of slickensides on exposed fault surfaces throughout the eastern margin of the Gulf of Suez indicates that the dominant extension direction during rifting was *c.* N65°E (Fig. 5a) (see Khalil 1998). Classification of the fault-slip data according to the fault trends shows that the NW-trending faults are dominantly pure dip-slip extensional faults whereas the north–south- to NNE-trending faults include pure dip-slip and sinistral oblique-slip faults (Fig. 5a). The analysis indicates extension (T) and shortening (P) axes oriented 1°/56° and 86°/148°, respectively, for the NW-trending faults and 14°/257° and 59°/142°, respectively, for the north–south- to NNE-trending faults. This indicates that the NW- and north–south- to NNE-trending faults are formed in response to the same regional extensional stresses (N65°E S65°W ± 10°). Furthermore, the analysis shows that the plunge of slickensides on the majority of the north–south- to NNE-trending oblique-slip faults has values (45–65°) that fall within the range of the dips of the NW-trending dip-slip extensional faults (Fig. 5a). This apparently indicates that the NW- and north–south- to NNE-trending faults are kinematicaly interdependent. The sinistral oblique slip on these faults forms an integral part of the extensional deformation that affected the eastern margin of the Gulf of Suez, with no need for NW–SE compression or an early stage of strike-slip faulting (as proposed by Jarrige *et al.* (1986, 1990) and Montenat *et al.* (1988)) to account for these oblique-slip faults.

Basement fabrics. The crystalline basement on the eastern flank of the Gulf of Suez is cut by a number of fabrics that include NE-trending intermediate to basic Late Pan-African basic dykes, north–south- to NNE-trending faults and fractures that offset Pan-African granitoid bodies, and, in particular, a WNW-trending, Pan-African shear zone, the Rhiba shear zone (Figs 3 and 5b–d). The Rihba shear zone is a zone of about 50 km length and 200 m width of intensely fractured and brecciated granites and foliated gneisses. It juxtaposes the Feiran gneisses in the west against the crystalline basement in the east (dated at 632 ± 3 Ma and 780 Ma, respectively; Stern & Hedge 1985). The shear zone is cut by the NE-trending dykes (590 ± 9 Ma; Stern & Manton 1987) and is probably of Precambrian age because it does not affect the Phanerozoic sediments. These fabrics produced a strong fracture anisotropy in the pre-rift basement and controlled the fault architecture and evolution of the rift system as discussed below.

Fault linkages. The fault systems on the eastern margin of the Gulf of Suez are characterized by different styles of linkages that resulted in a strong rhomboidal pattern. The Rift Border Fault system consists of a series of fault segments of 20–25 km length, dominantly trending NW and partially linked by north–south- to NNE-trending segments. Overlapping segments produced complexly faulted relay ramps; for example, between Wadi Sidri and Wadi Feiran (Fig. 3). These relay ramps may have been regions where early syn-rift sediment input points were developed during early Miocene time (see Moustafa 1992; Gawthorpe & Hurst 1993; Gupta *et al.* 1999).

In contrast, the Coastal Range Fault system is more strongly 'hard-linked' with a continuous fault trace throughout the map area (Fig. 3). This reflects the overall greater displacements on this fault system, which are typically between 3 and 5.7 km (Khalil 1998). The southern sectors of the Coastal Range Fault system at Gebel Araba and Gebel Abu Durba display two well-developed NNE-trending hard-linked transfer faults in the Coastal Range Fault system. McClay & Khalil (1998) showed that these faults were dominantly oblique-slip extensional faults that resulted from the fracturing of original, 'soft-linked' relay ramps between overlapping separate fault segments and that their orientation was probably controlled by reactivated NNE-trending basement fabrics.

Northwestern Red Sea margin

The NW Red Sea margin between the towns of Safarga and Quseir (Fig. 7) has a similar basement, pre-rift and syn-rift stratigraphy to the eastern margin of the Gulf of Suez and exhibits similar structural styles. The stratigraphy of this area is shown in Figure 6. A summary structural map and cross-sections are presented in Figures 7 and 8, respectively.

Stratigraphy. The pre-rift stratigraphy (Fig. 6), above the Precambrian crystalline basement is 500–700 m thick on the NW Red Sea margin. Syn-rift (Upper Oligocene to Recent) strata vary in thickness from <100 m in isolated outcrops within the westernmost fault blocks (Fig. 7), to hundreds of metres along the coastal sections, and can be as much as 5 km thick in the offshore basins (Heath *et al.* 1998). Detailed descriptions of the pre-rift and syn-rift stratigraphy have been given by Youssef (1957), Abd El-Razik (1967), Issawi *et al.* (1969), Said (1990), Bosworth *et al.* (1998), Montenat *et al.* (1986) and Plaziat *et al.* (1998*a,b*), and only a brief summary is presented here.

Basement. The crystalline basement of the NW Red Sea is formed largely by Precambrian metavolcanic rocks and metasediments intruded by syn- and post-tectonic granites and grano-diorites, probably similar in age to those in Sinai. The basement forms the flanks to the NW Red Sea margin (Fig. 9), and, as in Sinai, contains a number of strong pre-rift fabrics including faults, joints, dykes and, in particular, the Hamrawin and El Quwyh shear zones (Fig. 9).

Pre-rift strata. In contrast to the Sinai margin of the Gulf of Suez, the pre-rift strata range in age from Late Cretaceous to Mid-Eocene time (Fig. 6). The lowermost unit is shallow marine to fluvial sandstones of the Nubia Formation. This is overlain by a succession of interbedded shales, sandstones and limestones of the Quesir, Duwi, Dakhla and Esna formations that range in age from Late Cretaceous to Paleocene time (Fig. 6). Overlying these units is a thick succession (*c.* 200 m) of limestones and cherty limestones of the Lower to Middle Eocene Thebes Formation.

Syn-rift strata. The transition from pre-rift to syn-rift in the NW Red Sea is marked by the sandstones and conglomerates of the Oligocene Nakheil Formation, which unconformably overlies the Eocene Thebes Formation. The conglomerates of the Nakheil Formation consist of chert and limestone clasts derived from the underlying Thebes Formation, indicating minor uplift and erosion during the early stage of rifting. The sandstones and conglomerates of the Nakheil Formation are locally overlain by thin basaltic flows. The first distinct syn-rift sediments are coarse-grained sandstones and conglomerates of the Lower Miocene Ranga Formation (Fig. 6). These form submarine fans controlled by NW-trending extensional faults along the coastal region of the NW Red Sea margin (Fig. 7). The pebble- and boulder-size conglomerate occurring adjacent to the faults indicates considerable footwall uplift during Late Early Miocene time. Overlying these basal syn-rift clastic units are Middle Miocene reefal limestones and clastic deposits followed by a distinctive Middle–Upper Miocene evaporite sequence, the Sayateen and Abu Dabbab formations (Fig. 6). Upper Miocene limestones and Pliocene to Recent syn-rift clastic deposits overlie these evaporites. In contrast to the Gulf of Suez, extension continued from Pliocene time to the present day (indicated by extensional faults affecting the Pliocene and Quaternary sediments) and no post-rift strata have yet developed.

Structure. The structure of the NW Red Sea is characterized by two main fault systems, the Border Fault system to the west within the basement of the rift margin and the Coastal Fault system bordering the Red Sea (Fig. 7). Basement outcrops dominate west of the Coastal Fault system except for the immediate hanging walls of the Border Fault system, where pre-rift strata crop out together with isolated outcrops of syn-rift Nakheil Formation (Fig. 7). The Border Fault system consists of a series of WNW-trending fault segments. From south to north these are the Hamadat, Nakheil, Kallahin and Rabah faults (Fig. 7). The Hamadat and Nakheil faults dip southwest whereas the Kallahin and Rabah fault systems dip northeast. This change in polarity, as shown in Fig. 8, occurs across the Duwi Accommodation Zone (Fig. 1), which is centred on a structural high region between and including the Hamrawin and Quwyh shear zones (Fig. 7). In contrast, the Coastal Fault system trends northwest and separates the basement outcrops from pre-rift, syn-rift and Quaternary sediments along the coastal margin of the Red Sea. To the north of Quseir the Coastal Range faults dip northeast whereas south of Quseir they dip southwest (Fig. 7). Detailed fault displacements are difficult to calculate because erosion has removed the pre-rift strata from the footwalls of most fault blocks. Estimates of fault displacements range from 1.8 to 3.5 km along the Border Fault system and from 0.5 to 2 km along the Coastal Fault system (estimates of fault displacements based on cross-sections of Fig. 8).

As in the Gulf of Suez, the Coastal Fault system appears to have been active later than the Border Fault system in that younger syn-rift sediments are found in the hanging-wall basins, and the footwalls to the Border Fault system are more strongly uplifted and denuded. This indicates that rift faulting focused inwards such that the more recently active extensional fault systems now occur offshore in the Red Sea, as shown in seismic sections (see Heath *et al.* 1998). In contrast to the Gulf of Suez, both the Coastal Fault system and fault systems offshore (see Heath *et al.* 1998) continued to extend during Late Miocene to Pleistocene time.

In cross-section the rift extensional faults are dominantly planar domino in style (Fig. 8). The immediate hanging walls to many of the large-displacement faults (e.g. Nakheil Fault, Fig. 7) show well-developed hanging-wall synclines as a result of extensional fault-propagation folding (see Moustafa 1987; Hardy & McClay 1999; Khalil & McClay 2001; Sharp *et al.* 2000).

Measurement of slickensides on exposed fault surfaces in the northwestrn margin of the Red Sea indicates that the dominant extension direction during rifting was N65°E (Fig. 9a). As in the Gulf of Suez, sorting of the fault-slip data according to the fault trends shows that the NW- and north–south- to NNE-trending faults are dominantly pure dip-slip extensional faults.

Only few north–south-trending faults show sinistral oblique-slip component of movement (Fig. 9a). Both fault trends are formed in response to the same ENE–WSW-oriented extension, similar to that in the Gulf of Suez.

Basement fabrics

The NW Red Sea margin displays basement fabrics similar to those found in Sinai (Fig. 9b–d). The dominant fabrics are WNW-trending fractures and Pan-African shear zones: the Hamrawin and El Quwyh shears (Figs 7 and 9b). These shear zones clearly control the change in polarity of the rift fault systems along this NW Red Sea margin. In addition, NW- and north–south- to NNE-trending faults and fractures and NE-trending dykes are also well developed (Fig. 9b–d). The basement fabric orientations for both the eastern Gulf of Suez and the NW Red Sea (Figs 5 and 9) show similar fault or fracture distributions with three distinct groupings: (1) WNW-trending Pan-African shear zones; (2) NW-trending major rift faults; (3) north–south- to NNE-trending faults. A minor set of NE-trending faults (group 4, Fig. 9) orthogonal or at high angles to the main NW-trending rift fault system has also been identified. The shear zones in the NW Red Sea and Gulf of Suez are probably related to the Late Precambrian NW-trending Najd shear system, which extends from the eastern margin of the Red Sea to the Eastern Desert and south Sinai (Abdel Gawad 1969; Stern 1985; Husseini 1988; Sultan *et al.* 1988).

Fault linkages

Both the Border Fault and Coastal Fault systems of the NW Red Sea are strongly segmented with NW-, WNW- and NNE-trending elements. The Border Fault system dominantly consists of the Hamadat fault (HF, Fig. 7) together with the Nakheil and Kallahin faults (NF and KF, Fig. 7). Here the dominant trend is controlled by the Pan-African shear zones whereas the Coastal Fault system is dominantly NW-trending, orthogonal to the general N65°E extension direction. The Coastal Fault system clearly cuts the reactivated Pan-African shear zones and the NW-trending fault systems, indicating that rifting focused inwards through time. Along the Border Fault system many fault traces show distinct kinks where original relay ramps have been breached by transfer faults (e.g. the Nakheil fault (NF), in Wadi Nakheil, Fig. 7). Individual segments along this fault system are also delineated by individual doubly

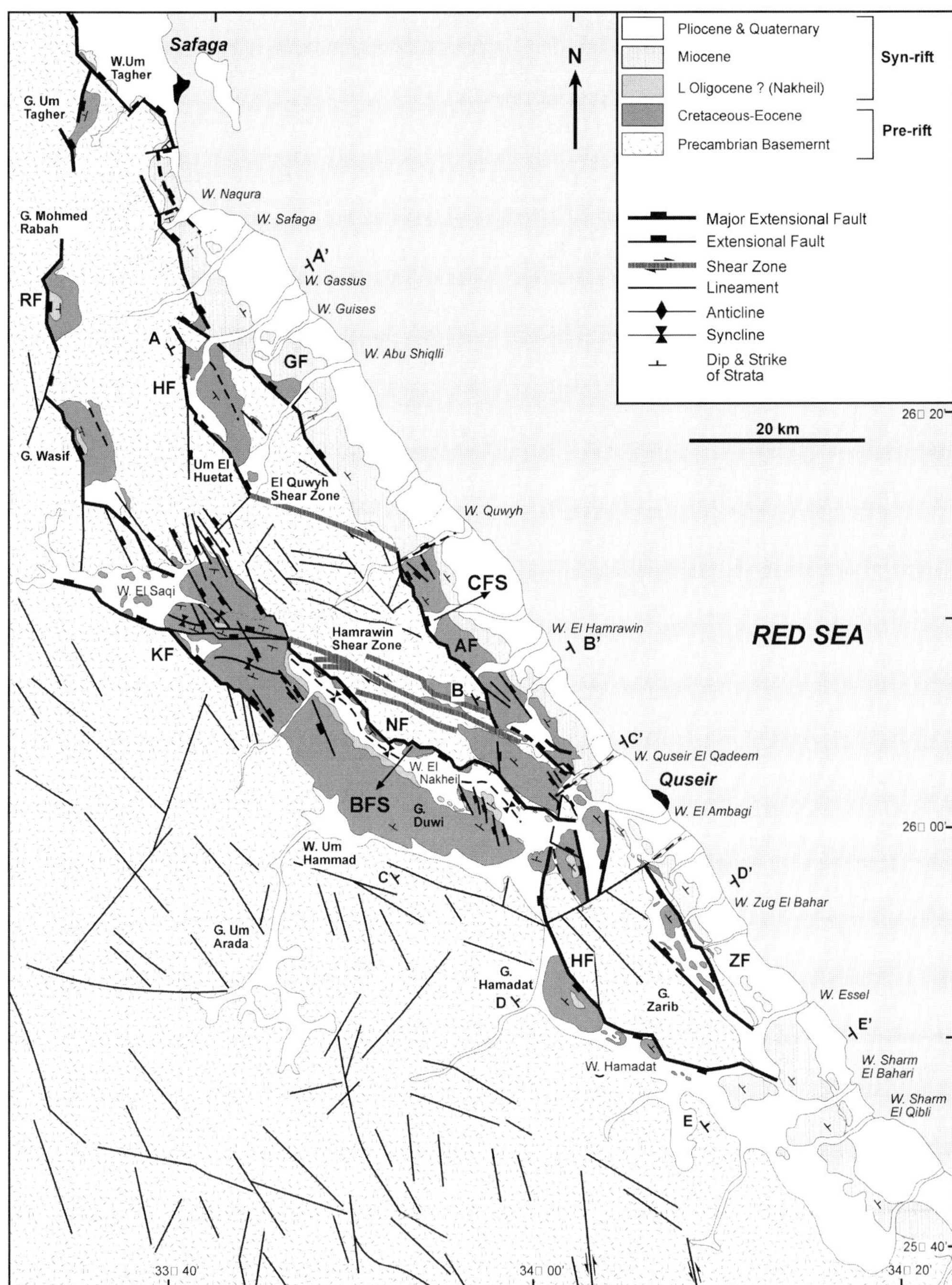

Fig. 7. Surface geological map of the NW Red Sea margin, Egypt. Locations of cross-sections shown in Figure 7 are labelled A–A' to E–E'. RF, KF, NF and HF indicate the Rabah, Kallahin, Nakheil and Hamadat segments of the Border Fault system, respectively; GF, AF and ZF indicate the Gerfan, Anz and Zug El Bahar segments of the Coastal Fault system, respectively.

plunging hanging-wall synclines as seen, for example, along the Hamadat fault (HF, Fig. 7).

The segmentation of the Coastal Range fault system has resulted in the development of relay ramps between overlapping fault tips (e.g. at Wadi Gassus, south of Safaga, Fig. 7). These relay structures appear to have controlled the sediment input sites along this part of the NW Red Sea margin.

Discussion

The non-volcanic NW Red Sea and Gulf of Suez rift systems are both characterized by a similar structural style of domino planar extensional faults, most of which dip inwards towards the central part of the rift. Along-strike rift segmentation occurs where both the border and interior fault systems flip polarity across poorly defined, rather wide (up to 20 km) accommodation zones (Fig. 1). These accommodation zones are oblique to both the rift axis and to the overall N65°E regional extension direction (Fig. 1) and appear to be strongly controlled by the loci of penetrative Pan-African shear zones (Younes & McClay 2001). In the Gulf of Suez these accommodation zones are formed by overlapping half-graben systems (Fig. 1) with major border faults on opposite sides of the rift and dipping in different directions. Although these zones are structurally very complex with many tilted fault blocks, the

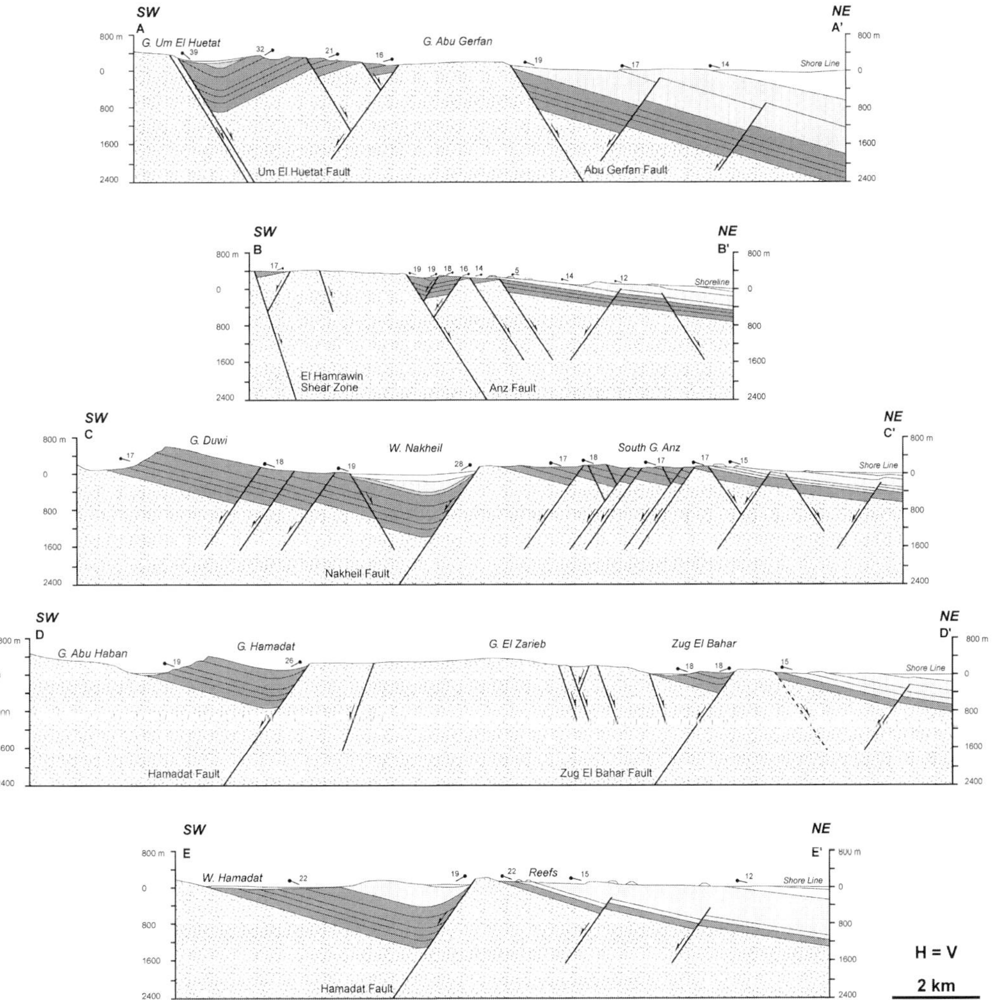

Fig. 8. Cross-sections through the NW Red Sea margin, Egypt. (See Fig. 7 for locations.)

     S. M. KHALIL & K. R. McCLAY

dominant deformation is extensional. There are no through-going strike-slip fault systems accommodating the changes in displacement and fault polarities along the strike of the rift system. Similar style accommodation zones are found in the East African rift system (Bosworth 1985; Rosendahl *et al.* 1986), Lake Tanganyika, Kenya (Morley 1988, 1999), in the Rio Grande rift, New Mexico (Russell & Snelson 1994) and in the northern Viking Graben, North Sea (Scott & Rosendahl 1989).

Basement fabrics

The widespread pre-existing zones of weakness in the crystalline basement of the Gulf of Suez and NW Red Sea areas are at high angles to the regional ENE–WSW-directed extension. This produced rift-parallel faults that were initially strongly segmented (see Khalil 1998). With increased displacement and rifting the fault segments became hard-linked by oblique and cross-faults, some of which were controlled by the reactivation of the underlying basement fabrics. This produced the overall rhomboidal pattern of faulting within the rift system. Particular well-exposed examples of this transition from 'soft linkage' to hard linkage (see terminology of Walsh & Watterson (1991)) are seen in the SE Gulf of Suez at Gebel Araba–Gebel Abu Durba (Fig. 3; McClay & Khalil 1998). Clear evidence of reactivation of pre-existing basement fabrics in the Gulf of Suez has also been presented by Younes *et al.* (1998). They demonstrated that the NNW-trending rift faults also resulted from the reactivation of basement fractures.

Rift models

Structural analysis of both the NW Red Sea and the Gulf of Suez rift systems has shown that N65°E-directed extension operated throughout most of the rift history. Analysis of fault slip data (Khalil (1998) and the present study) and fault patterns has not found any strong evidence for an early phase of strike-slip faulting. Fault-slip data have shown that the NW- and north–south- to NNE-trending faults exhibit uniform orientation and direction of

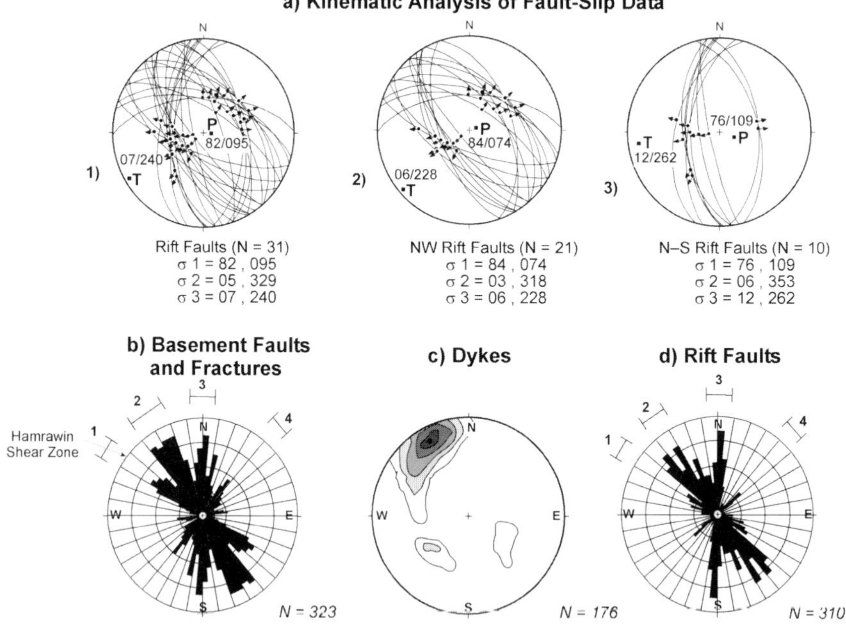

Fig. 9. Analysis of fault trends and basement fabrics in the NW Red Sea, determined both from field mapping and analysis of Landsat TM images. (**a**) Lower-hemisphere equal-area stereograms of faults and striae. (1) NW and north–south to NNE rift faults; (2) NW faults; (3) north–south to NNE faults. Data in (2) and (3) are sorted according to the orientation of faults shown in (1). T and P indicate the extension and shortening axes, respectively. (**b**) Basement faults and fractures. (**c**) Dykes (plotted as 2σ contours). (**d**) Rift faults: 1, trend of Pan-African shear zones; 2, trend of NNW rift fault systems; 3: trend of north–south to NNE basement fracture zones; 4, trend of NE cross-faults.

slip. Both of these fault trends were formed at the same time in response to the extensional stresses during the Miocene rifting (Khalil 1998). Although some of the north–south to NNE faults may have sinistral oblique-slip movement (a maximum of 500 m; Khalil 1998; McClay & Khalil 1998), these faults do not cut across the entire rift and the oblique slip on these faults is due to their obliquity to the extension direction. In addition, seismic data did not show any evidence of pull-apart basins, 'pop-up', 'flower' structures or Riedel shears, which are normally associated with strike-slip tectonics. The strike-slip model for the early evolution of the Gulf of Suez and NW Red Sea as postulated by Jarrige *et al.* (1986, 1990),

Montenat *et al.* (1986, 1988) and Makris & Rihm (1991); Rihm & Henke (1998) is therefore not considered to be valid.

Flexural modelling by Willacy (1996) has shown that the Gulf of Suez basin geometry and the rift-flank uplift and exhumation is best explained by rifting of cold, crystalline crust of ≥30 km thickness. This is in general agreement with the limited number of crustal thickness estimates from the Gulf of Suez and NW Red Sea (Gaulier *et al.* 1988; Makris & Rihm 1991; Rihm & Henke 1998). In addition, there is no firm evidence of oceanic crust in the Gulf of Suez–NW Red Sea (see Bosworth & McClay 2001). The dominant style of faulting is that of planar, domino extensional faults, and no low-

(a) **CONTINENTAL RIFT INITIATION**

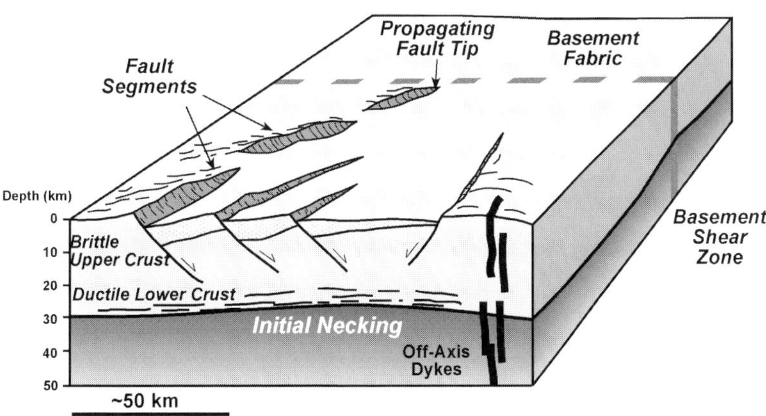

(b) **LINKED RIFT SUB-BASINS**

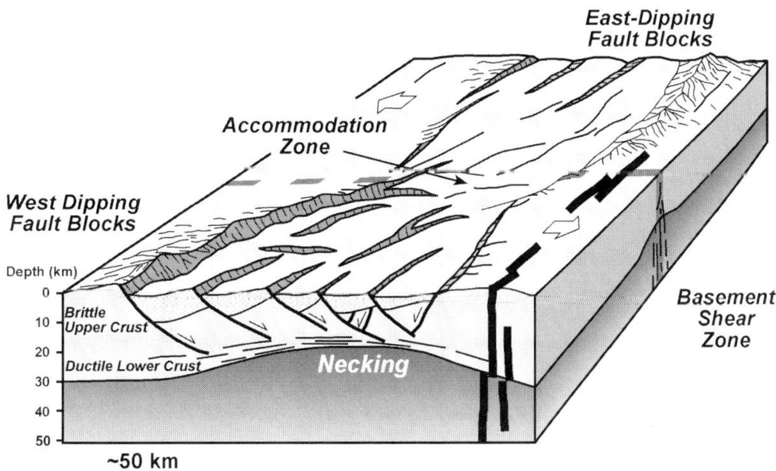

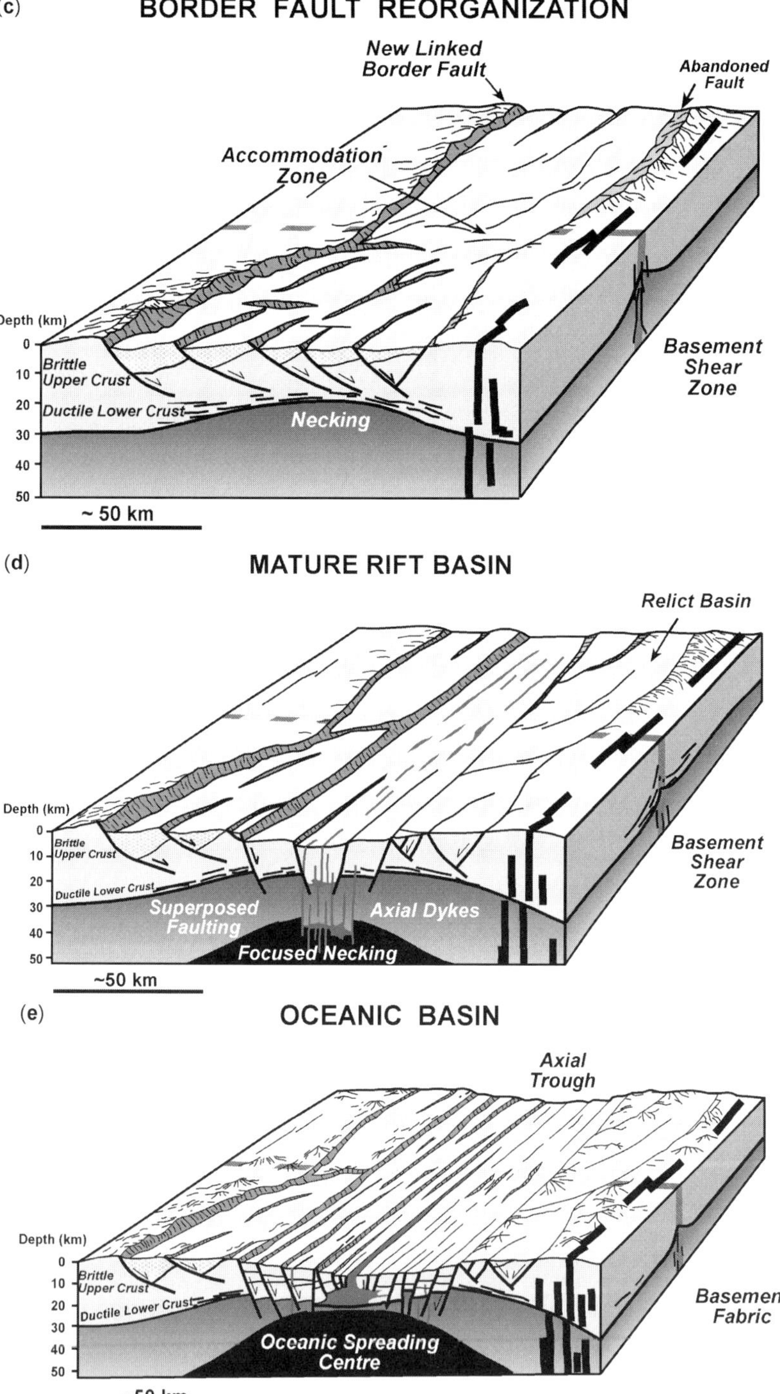

Fig. 10. Evolutionary model for the development of the Gulf of Suez and NW Red Sea rift systems (modified from Bosworth 1994). (**a**) Initial rift basin development; (**b**) linked sub-basin development; (**c**) border fault reorganization; (**d**) mature rift basin; (**e**) oceanic basin.

angle detachment faults (see Lister *et al.* 1986) have been found. Through time, the rifting focused inwards towards the centre of the rift (as evidenced by both cross-cutting fault relationships as well as younger ages of syn-rift strata in fault-bounded basins). Therefore a crustal necking model (see Cloetingh *et al.* 1995) is favoured for the development of the Gulf of Suez–NW Red Sea rift system and is summarized in Figure 10 (adapted after Bosworth 1994). In this model early rift sub-basins are influenced by reactivated basement fabrics, which in particular control the propagation of the main border fault systems (Fig. 10a). As extension develops, linked sub-basins form where the accommodation zones are focused on major basement discontinuities (Fig. 10b). With increased extension, border faults become linked (Fig. 10c) and rifting focuses inwards towards the centre of the basin with new faults cutting though older faults (Fig. 10d). If extension proceeds to crustal rupture (as in the southern Red Sea) then an oceanic basin forms. Such a model explains in general terms the evolution of the NW Red Sea–Gulf of Suez rift systems, where the Border Fault systems are strongly controlled by the basement fabrics and the younger, inner faults are near-orthogonal to the regional extension direction (see Figs 3 and 7).

The rift to passive margin model involving largely domino-style planar to gently listric upper-crustal faults as proposed in this paper (Fig. 10) contrasts with low-angle detachment models as proposed by Lister *et al.* (1991) and Manatschal & Bernoulli (1999). Manatschal & Bernoulli (1999) proposed that initial rifting occurred by upper-crustal listric faults and that advanced rifting and crustal separation was dominated by simple shear along low-angle detachment faults with a top-to-the-ocean sense of shear. In this study, no evidence of low-angle detachment faults was found in the strongly extended southern Gulf of Suez (β >2.0) or in the NW Red Sea margin.

During Mid-Miocene time, the Levant–Gulf of Aqaba transform boundary was established, linking the Red Sea rift plate boundary to the convergent Bitlis–Zagros plate boundary (Quennell 1958; Freund 1970; Ben-Menahem *et al.* 1976; Steckler *et al.* 1988; Abdel Khalek *et al.* 1993). This resulted in a dramatic drop in extension rates across the Gulf of Suez (see McClay *et al.* 1998; Bosworth & McClay 2001) such that rifting focused in the NW Red Sea continuing through Late Miocene and Pliocene time to the present day. As a result, the NW Red Sea rift system is more highly evolved compared with the Gulf of Suez, with greater faulting offshore together with more uplift and exhumation of the rift flanks.

Conclusions

Analysis of fault geometries, fault kinematics and sedimentation patterns indicates that rifting in the Gulf of Suez and NW Red Sea began in Oligocene to earliest Miocene time and was caused by ENE–WSW-directed extension. Dominant extensional fault trends are NW and WNW, in part reflecting pre-existing basement fabrics and reactivation of older Pan-African shear zones. In addition, however, reactivation of pre-existing NNE-trending basement structures resulted in the development of local N10–20°E transfer faults and rhombic basin geometries. There is no convincing evidence for significant Oligocene-age strike-slip faulting in the Gulf of Suez or in the NW Red Sea. Initial rift faults were controlled by reactivation of strong basement fabrics under the thin pre-rift sedimentary section. Individual faults were initially strongly segmented and offset across 'soft-linked' relay structures. With increased extension these faults became linked by breaking down relay structures with the development of local 'hard-linked' transfer faults, thus giving rise to the strong rhomboidal fault pattern of the rift system. As rifting progressed, N10–20°E transfer faults were locally abandoned and younger rift-normal transfer faults predominated. With greater extension, rifting focused into the centre of the basins. Along-strike segmentation of the rift system occurred across complex accommodation zones that were controlled by strong, pre-existing basement fabrics. As a result of sinistral displacement along the Dead Sea transform system, during Mid-Miocene time, extension rates decreased in the Gulf of Suez and rifting focused in the NW Red Sea. The NW Red Sea–Gulf of Suez rift systems are, apart from minor intrusions and basalt flows, non-volcanic and are thus type areas for comparison with other non-volcanic rifts and margins.

The research presented in this paper was supported by the Natural Environment Research Council (NERC) ROPA Grant GR3/R9529, with additional support from the Fault Dynamics Project (sponsored by ARCO British Limited, PETROBRAS UK Ltd., BP Exploration, Conoco (UK) Limited, Mobil North Sea Limited and Sun Oil Britain). K.R.M. also gratefully acknowledges support from ARCO British Limited. B. Bosworth and Marathon Petroleum Egypt are thanked for logistical support and for many fruitful discussions on Gulf of Suez and NW Red Sea geology. S.M.K. was supported by a research grant from BG International. BG Egypt and M. Ayyad kindly assisted with additional fieldwork in the NW Red Sea. This paper is Fault Dynamics Publication No. 103.

References

ABDEL GAWAD, M. 1969. New evidence of transcurrent movements in Red Sea area and petroleum implications. *AAPG Bulletin*, **53**, 1466–1479.

ABDEL KHALEK, M.L., ABDEL WAHED, M. & SEHIM, A. 1993. *Wrenching deformation and tectonic setting of the northwestern part of the Gulf of Aqaba. Geodynamics and sedimentation of the Red Sea–Gulf of Aden Rift system.* Geological Survey of Egypt, Special Publication, **1**, 409–445.

ABD EL-RAZIK, T.M. 1967. Stratigraphy of the sedimentary cover of the Anz–Atshan–south Duwi district. *Bulletin of the Faculty of Science, Cairo University*, **431**, 135–179.

BARTOV, Y., STEINITZ, G., EYAL, M. & EYAL, U. 1980. Sinistral movement along the Gulf of Aqaba—its age and relation to the opening of the Red Sea. *Nature*, **285**, 220–221.

BEN-MENAHEM, A., NUR, A. & VERED, M. 1976. Tectonics, seismicity and structure of the Afro-Eurasian junction—the breaking of an incoherent plate. *Physics of the Earth and Planetary Interiors*, **12**, 1–50.

BOSWORTH, W. 1985. Geometry of propagating continental rifts. *Nature*, **316**, 625–627.

BOSWORTH, W. 1994. A model for the three-dimensional evolution of continental rift basins, northeast Africa. *Geologische Rundschau*, **83**, 671–688.

BOSWORTH, W. 1995. A high-strain rift model for the southern Gulf of Suez (Egypt). *In*: LAMBIASE, J.J. (ed.) *Hydrocarbon Habitat in Rift Basins*. Geological Society, London, Special Publications, **80**, 75–112.

BOSWORTH, W. & McCLAY, K.R. 2001. Structural and stratigraphic evolution of the Neogene Gulf of Suez, Egypt: a synthesis. *In*: CAVAZZA, W., ROBERTSON, A.H.F. & ZIEGLER, P. (eds) *Peritethyan Rift/Wrench Basins and Passive Margins*. Mémoires du Muséum National d'Histoire Naturelle de Paris: Peritethys Programme (PTP) and IGCP 369, Special Publication, in press.

BOSWORTH, W., CREVELLO, P., WINN, R.D. Jr & STEINMETZ, J. 1998. Structure, sedimentation, and basin dynamics during rifting of the Gulf of Suez and northwestern Red Sea. *In*: PURSER, B.H. & BOSENCE, D.W.J. (eds) *Sedimentation and Tectonics of Rift Basins: Red Sea–Gulf of Aden*. Chapman and Hall, London, 77–96.

CLOETINGH, S., VAN WEES, J.D., VAN DER BEEK, P.A. & SPADINI, G. 1995. Role of pre-rift rheology in kinematics of extensional basin formation: constraints from thermomechanical models of Mediterranean and intracratonic basins. *Marine and Petroleum Geology*, **12**, 793–807.

COCHRAN, J.R. 1983. A model for development of the Red Sea. *AAPG Bulletin*, **67**, 41–69.

COFFIELD, D.Q. & SCHAMEL, S. 1989. Surface expression of an accommodation zone within the Gulf of Suez rift, Egypt. *Geology*, **17**, 76–79.

COLEMAN, R.G. 1974. Geologic background of the Red Sea. *In*: BURKE, C.A. & DRAKE, C.L. (eds) *The Geology of Continental Margins*. Springer, Berlin, 743–751.

COLEMAN, R.G. *Geologic Evolution of the Red Sea.* Oxford Monographs on Geology and Geophysics 24. Oxford University Press, Oxford.

COLLETTA, B., LE QUELLEC, P., LETOUZY, J. & MORETTI, I. 1988. Longitudinal evolution of the Suez rift structure (Egypt). *Tectonophysics*, **153**, 221–233.

DARWISH, M. 1992. Facies development of the Upper Paleozoic–Lower Cretaceous sequences in the Northern Galala Plateau and evidences for their hydrocarbon reservoir potentiality, Northern Gulf of Suez, Egypt. *In: Proceedings of 1st International Conference on Geology of the Arab World*. Cairo University, Cairo **1**, 75–214.

DARWISH, M. & EL-AZABI, M. 1993. Contributions to the Miocene sequences along the western coast of the Gulf of Suez, Egypt. *Egyptian Journal of Geology*, **37**, 21–47.

EYAL, Y. 1996. Stress field fluctuations along the Dead Sea rift since the Middle Miocene. *Tectonics*, **15**, 157–170.

EYAL, M., EYAL, Y., BARTOV, Y. & STEINITZ, G. 1981. Tectonic development of the western margin of the Gulf of Elat (Aqaba) rift. *Tectonophysics*, **80**, 39–66.

FREUND, R. 1970. Plate tectonics of the Red Sea and Africa. *Nature*, **228**, 453.

GARFUNKEL, Z. & BARTOV, Y. *The Tectonics of the Suez Rift*. Geological Survey of Israel Bulletin, **71**.

GAULIER, J.M., LE PICHON, X., LYBERIS, N., AVEDIK, F., GELI, L., MORETTI, I., DESCHAMPS, A. & HAFEZ, S. 1988. Seismic study of the crust of the northern Red Sea and Gulf of Suez. *Tectonophysics*, **153**, 55–88.

GAWTHORPE, R.L. & HURST, J.M. 1993. Transfer zones in extensional basins: their structural style and influence on draining development and stratigraphy. *Journal of the Geological Society, London*, **150**, 1137–1152.

GAWTHORPE, R.L., SHARP, I., UNDERHILL, J.R. & GUPTA, S. 1997. Linked sequence stratigraphic and structural evolution of propagating normal faults. *Geology*, **25**, 795–798.

GHORAB, M.A. *Abnormal Stratigraphic Features in Ras Gharib Oil Field*. Bulletin, III Arab Petroleum Congress, Alexandria.

GHORAB, M.A. et al. *Oligocene and Miocene Rock Stratigraphy of the Gulf of Suez Region*. Egyptian General Petroleum Corporation Consultative Stratigraphical Committee, Cairo.

GIRDLER, R.W. & SOUTHREN, T.C. 1987. Structure and evolution of the northern Red Sea. *Nature*, **330**, 716–721.

GREENBERG, J.K. 1981. Characteristics and origin of Egyptian younger granites. *Geological Society of America Bulletin*, **92**, 224–232.

GUPTA, S., UNDERHILL, J.R., SHARP, I.R. & GAWTHORPE, R.L. 1999. Role of fault interactions in controlling syn-rift sediment dispersal patterns: Miocene Abu Alaqa Group, Suez Rift, Sinai, Egypt. *Basin Research*, **11**, 167–189.

HANTAR, G. 1965. *Remarks on the distribution of Miocene sediments in the Gulf of Suez region*. 5th Arabian Petroleum Congress, Cairo.

HARDY, S. & MCCLAY, K.R. 1999. Kinematic modelling of extensional forced folding. *Journal of Structural Geology*, **21**, 695–702.

HASSAN, M.A. & HASHAD, A.H. 1990. Precambrian of Egypt. *In*: SAID, R. (ed.) *The Geology of Egypt*. Balkema, Rotterdam, 201–245.

HEATH, R., VANSTONE, S., SWALLOW, J. & 7 others 1998. Renewed exploration in the offshore north Red Sea Region—Egypt. *In*: *Proceedings of the 14th Petroleum Conference*. Egyptian General Petroleum Corporation, Cairo, 16–34.

HEMPTON, M. 1987. Constraints on Arabian plate motion and extensional history of the Red Sea. *Tectonics*, **6**, 687–705.

HUME, W.F., MADGWICK, T.G., MOON, F.W. & SADEK, H. *Preliminary Geological Report on the Gebel Tanka Area*. Petroleum Research Bulletin, **4**.

HUSSEINI, M.I. 1988. The Arabian Infracambrian extensional system. *Tectonophysics*, **148**, 93–103.

ISSAWI, B., FRANCIS, M., EL-HINNAWI, M. & MEHANNA, A. *Contribution to the Structure and Phosphate Deposits of Quseir Area*. Geological Survey of Egypt Paper, **50**.

JARRIGE, J.J., OTT D'ESTEVOU, P., BUROLLET, P.F. & 6 OTHERS 1986. Inherited discontinuities and Neogene structure: the Gulf of Suez and northwestern edge of the Red Sea. *Philosophical Transactions of the Royal Society of London, Series A*, **317**, 129–139.

JARRIGE, J.-J., OTT D'ESTEVOU, P., BUROLLET, P.F., MONTENAT, C., PRAT, P., RICHERT, J.-P. & THIRIET, J.-P. 1990. The multistage tectonic evolution of the Gulf of Suez and northern Red Sea continental rift from field observations. *Tectonics*, **9**, 441–465.

JOFFE, S. & GARFUNKEL, Z. 1987. Plate kinematics of the circum Red Sea—a re-evaluation. *Tectonophysics*, **141**, 5–22.

KHALIL, S.M. 1998. *Tectonic evolution of the eastern margin of the Gulf of Suez, Egypt*. PhD thesis, Royal Holloway, University of London.

KHALIL, S. & MCCLAY, K. 1998. Structural architecture of the eastern margin of the Gulf of Suez: field studies and analogue modelling results. *In: Proceedings of the 14th Exploration Conference*. Egyptian General Petroleum Corporation, Cairo 1, 201–211.

KHALIL, S. & MCCLAY, K. 2001. Extensional fault-related folding, northwestern Red Sea, Egypt. *Journal of Structural Geology, Special Issue*, in press.

KOHN, B.P. & EYAL, M. 1981. History of uplift of the crystalline basement of Sinai and its relation to opening of the Red Sea as revealed by fission track dating of apatites. *Earth and Planetary Science Letters*, **52**, 129–141.

KRÖNER, A. 1984. Late Precambrian plate tectonics and orogeny: a need to redefine the term Pan-African. *In*: KLERKX, J. & MICHOT, J. (eds) *African Geology*. Musée Royal l'Afrique Centrale, Tervuren, 23–28.

KRÖNER, A. 1993. The Pan African belt of northeastern and Eastern Africa, Madagascar, southern India, Sri Lanka and East Antarctica: terrane amalgamation during the formation of the Gondwana supercontinent. *In*: THORWEIHE, U. & SCHANDELMEIER, H. (eds) *Geoscientific Research in Northeast Africa*. Balkema, Rotterdam, 3–9.

LE PICHON, X. & FRANCHETEAU, J. 1978. A plate tectonic analysis of the Red Sea–Gulf of Aden area. *Tectonophysics*, **46**, 369–406.

LISTER, G.S., ETHERIDGE, M.A. & SYMONDS, P.A. 1986. Detachment faulting and the evolution of passive continental margins. *Geology*, **14**, 246–250.

LISTER, G.S., ETHERIDGE, M.A. & SYMONDS, P.A. 1991. Detachment models for the formation of passive continental margins. *Tectonics*, **10**, 1038–1064.

MAKRIS, J. & RIHM, R. 1991. Shear controlled evolution of the Red Sea: pull apart model. *Tectonophysics*, **198**, 441–466.

MANATSCHAL, G. & BERNOULLI, D. 1999. Architecture and tectonic evolution of nonvolcanic margins: present-day Galicia and ancient Adria. *Tectonics*, **18**, 1099–1119.

MCCLAY, K. & KHALIL, S. 1998. Extensional hard linkages, eastern Gulf of Suez, Egypt. *Geology*, **26**, 563–566.

MCCLAY, K.R., NICOLS, G.J., KHALIL, S.M., DARWISH, M. & BOSWORTH, W. 1998. Extensional tectonics and sedimentation, eastern Gulf of Suez, Egypt. *In*: PURSER, B.H. & BOSENCE, D.W.J. (eds) *Sedimentation and Tectonics of Rift Basins: Red Sea–Gulf of Aden*. Chapman and Hall, London, 223–238.

MCKENZIE, D.P., DAVIES, D. & MOLNAR, P. 1970. Plate tectonics of the Red Sea and east Africa. *Nature*, **226**, 243–248.

MESHREF, W.M. 1990. Tectonic framework. *In*: SAID, R. (ed.) *The Geology of Egypt*. Balkema, Rotterdam, 113–155.

MONTENAT, C., OTT D'ESTEVOU, P., JARRIGE, J.-J. & RICHERT, J.-P. 1998. Rift development in the Gulf of Suez and the north-western Red Sea: structural aspects and related sedimentary processes. *In*: PURSER, B.H. & BOSENCE, D.W.J. (eds) *Sedimentation and Tectonics of Rift Basins: Red Sea–Gulf of Aden*. Chapman and Hall, London, 97–116.

MONTENAT, C., OTT D'ESTEVOU, P. & PURSER, B. 1986. Tectonic and sedimentary evolution of the Gulf of Suez and northwestern Red Sea: a review. *In*: MONTENAT, C. (ed.) *Ecological Studies on the Gulf of Suez, the Northwestern Red Sea Coasts, Tectonic and Sedimentary Evolution of a Neogene Rift*. Documents et Travaux, Institut Géologique Albert de Lapparent, **10**, 7–18.

MONTENAT, C., OTT D'ESTEVOU, P., PURSER, B. & 9 OTHERS 1988. Tectonic and sedimentary evolution of the Gulf of Suez and the northern western Red Sea. *Tectonophysics*, **153**, 166–177.

MORETTI, I. & CHENET, P.Y. 1987. The evolution of the Suez rift: a combination of stretching and secondary convection. *Tectonophysics*, **133**, 229–234.

MORGAN, P. 1990. Egypt in the tectonic framework of global tectonics. *In*: SAID, R. (ed.) *The Geology of Egypt*. Balkema, Rotterdam, 91–111.

MORLEY, C.K. 1988. Variable extension in Lake Tanganyika. *Tectonics*, **7**, 785–801.

MORLEY, C.K. *Geoscience of Rift Systems: Evolution of East Africa*. American Association of Petroleum Geologists, Studies in Geology, **44**.

MOUSTAFA, A.M. 1976. Block faulting of the Gulf of Suez. Presented at 5th Exploration Seminar, Egyptian General Petroleum Company, Cairo.

MOUSTAFA, A.R. 1987. Drape folding in the Baba–Sidri area, eastern side of the Suez rift. *Egypt Journal of Geology*, **31**, 15–27.

MOUSTAFA, A.R. 1992. Structural setting of the Sidri–Feiran area, eastern side of the Suez rift. *Ain Shams University, Middle East Research Centre, Earth Science Series*, **6**, 44–54.

MOUSTAFA, A.R. 1993. Structural characteristics and tectonic evolution of the east-margin blocks of the Suez rift. *Tectonophysics*, **223**, 381–399.

PATTON, T.L., MOUSTAFA, A.R., NELSON, R.A. & ABDINE, S.A. 1994. Tectonic evolution and structural setting of the Suez Rift. *In*: LANDON, S.M. (ed.) *Interior Rift Basins*. American Association of Petroleum Geologists, Memoirs, **59**, 7–55.

PLAZIAT, J.-C., BALTZER, F., CHOUKRI, A. & 5 OTHERS 1998a. Quaternary marine and continental sedimentation in the northern Red Sea and Gulf of Suez (Egyptian coast): influences of rift tectonics, climatic changes and sea-level fluctuations. *In*: PURSER B.H. & BOSENCE D.W.J. (eds) *Sedimentation and Tectonics of Rift Basins: Red Sea–Gulf of Aden*. Chapman and Hall, London, 537–573.

PLAZIAT, J.-C., MONTENAT, C., BARRIER, P., JANIN, M.-C., ORSZAG-SPERBER, F. & PHILOBBOS, E. 1998b. Stratigraphy of the Egyptian syn-rift deposits: correlations between axial and peripheral sequences of the north-western Red Sea and Gulf of Suez and their relations with tectonics and eustacy. *In*: PURSER, B.H. & BOSENCE, D.W.J. (eds) *Sedimentation and Tectonics of Rift Basins, Gulf of Aden*. Chapman and Hall, London, 211–222.

QUENNELL, A.M. 1958. The structural and geomorphic evolution of the Dead Sea Rift. *Journal of the Geological Society, London*, **114**, 1–24.

QUENNELL, A.M. 1984. The western Arabian rift system. *In*: DIXON, J.E. & ROBERTSON, A.H.F. (eds) *The Geological Evolution of the Eastern Mediterranean*. Blackwell Scientific, Oxford, 775–788.

RICHARDSON, M. & ARTHUR, A.M. 1988. The Gulf of Suez–northern Red Sea Neogene rift. A quantitative basin analysis. *Marine and Petroleum Geology*, **5**, 247–270.

RIHM, R. & HENKE, C.H. 1998. Geophysical studies on early tectonic controls on Red Sea rifting, opening and segmentation. *In*: PURSER, B.H. & BOSENCE, D.W.J. (eds) *Sedimentation and Tectonics of Rift Basins: Red Sea–Gulf of Aden*. Chapman and Hall, London, 29–49.

ROSENDAHL, B.R. *et al.* 1986. Structural expressions in rifting: Lake Tanganyika, Africa. *In*: FRASLICK, L.E., RENAULT, R.W., REID, I., & TIETCELIN, J.J. (eds) *Sedimentation in the African Rifts*. Geological Society, London, Special Publications, **25**, 29–43.

RUSSELL, L.R. & SNELSON, S. 1994. Structure and tectonics of the Albuquerque Basin segment of the Rio Grande Rift: insights from reflection seismic data. *In*: KELLER, G.R. & CATHER, S.M. (eds) *Basins of the Rio Grande Rift: Structure, Stratigraphy and Tectonic Setting*. Geological Society of America, Special Papers, **291**, 83–112.

SAID, R. *The Geology of Egypt*. Balkema, Rotterdam.

SCHUTZ, K. 1994. Structure and stratigraphy of the Gulf of Suez, Egypt. *In*: LANDON, S.M. (ed.) *Interior Rift Basins*. American Association of Petroleum Geologists, Memoirs, **59**, 57–95.

SCOTT, D.L. & ROSENDAHL, B.R. 1989. North Viking Graben: an East African perspective. *AAPG Bulletin*, **73**, 155–165.

SELLWOOD, B.W. & NETHERWOOD, R.E. 1984. Facies evolution in the Gulf of Suez area: sedimentation history as an indicator of rift initiation and development. *Modern Geology*, **9**, 43–69.

SHARP, I.R., GAWTHORPE, R.I., UNDERHILL, J.R. & GUPTA, S. 2000. Fault-propagation folding in extensional settings: example of structural style and syn-rift sedimentary response from the Suez rift, Sinai, Egypt. *Geological Society of America Bulletin*, **112**, 1877–1899.

STECKLER, M.S., BERTHELOT, F., LYBERIS, N. & LE PICHON, X. 1988. Subsidence in the Gulf of Suez: implications for rifting and plate kinematics. *Tectonophysics*, **153**, 249–270.

STERN, R.J. 1981. Petrogenesis and tectonic setting of Late Precambrian ensimatic volcanic rocks, Central Eastern Desert of Egypt. *Precambrian Research*, **16**, 195–230.

STERN, R.J. 1985. The Najd fault system, Saudi Arabia and Egypt: a Late Precambrian rift-related transform system? *Tectonics*, **4**, 497–511.

STERN, R.J. 1994. Arc assembly and continental collision in the Neoproterozoic East African orogen: implications for the consolidation of Gondwanaland. *Annual Review of Earth and Planetary Sciences*, **22**, 319–351.

STERN, R.J. & HEDGE, C.E. 1985. Geochronological and isotopic constraints on Late Precambrian crustal evolution in the Eastern Deser of Egypt. *American Journal of Science*, **285**, 79–127.

STERN, R.J. & MANTON, W.I. 1987. Age of Feiran basement rocks, Sinai: amplications for late Precambrian crustal evolution in the northern Arabian–Nubian shield. *Journal of the Geological Society, London*, **144**, 569–575.

STOESER, D.B. & CAMP, V.E. 1985. Pan African microplate accretion of the Arabian shield. *Geological Society of America Bulletin*, **96**, 817–826.

SULTAN, M., ARVIDSON, E.R. & DUNCAN, I.J. 1988. Extension of the Najd shear system from Saudi Arabia to the Central Eastern Desert of Egypt based on integrated field and Landsat observations. *Tectonics*, **7**, 1291–1306.

WALSH, J.J. & WATTERSON, J. 1991. Geometric and kinematic coherence and scale effects in normal fault systems. *In*: ROBERTS, A.M., YIELDING, G. & FREEMAN, B. (eds) *The Geometry of Normal Faults*. Geological Society, London, Special Publications, **56**, 129–203.

WILLACY, C. 1996. *Thermo-mechanical modelling of extensional fault systems*. PhD thesis, Royal Holloway, University of London.

YOUNES, A.I. & MCCLAY, K.R. 1998. Role of basement fabric on Miocene rifting in the Gulf of Suez–Red Sea. *In: Proceedings of 14th Petroleum Conference*. Egyptian General Petroleum Corporation, Cairo **1**, 35–50.

YOUNES, A.I., ENGELDER, T. & BOSWORTH, W. 1998. Fracture distribution in faulted basement blocks: Gulf of Suez, Egypt. *In*: COWARD, M.P., DALTABAN, T.S. & JOHNSON, H. (eds) *Structural Geology in Reservoir Characterization*. Geological Society, London, Special Publications, **127**, 167–190.

YOUNES, A. & MCCLAY, K.R. 2001. Development of accommodation zones in the Gulf of Suez–Red Sea Rift, Egypt. *AAPG Bulletin, Special Issue*, in press.

YOUSSEF, M.I. 1957. Upper Cretaceous rocks in Kosseir area. *Bulletin de l'Institut du Desert d'Egypt*, **7**, 35–53.

Transfer zones normal and oblique to rift trend: examples from the Perth Basin, Western Australia

Tingguang Song, Peter A. Cawood & Mike Middleton

Tectonics Special Research Centre, School of Applied Geology, Curtin University, GPO Box U1987, Perth, W.A. 6845, Australia (e-mail: tsongt@lithos.curtin.edu.au)

Abstract: The Perth Basin is a major tectonic province along the western margin of the Australian continent. Basin morphology is controlled by north-striking faults formed during Permian rifting and reactivated during later tectonic events, notably during continental break-up in Late Jurassic–Early Cretaceous time. Transfer structures, including those normal and oblique to the major faults, compartmentalized the basin into segments of distinctive character. East–west transfer faults, perpendicular to the basin trend, were active throughout the rift stage of basin development and are recognized only in the northernmost onshore part of the Perth Basin, corresponding to the depocentre for Permian sediment accumulation. Northerly trending normal faults change in character and/or terminate at these east–west structures. The NW-striking transfer zones influenced deformational features formed during the Late Jurassic–Early Cretaceous break-up. No continuous fault plane has been identified with these zones in the sedimentary sequences. They are characterized by the termination and/or swing of major normal faults at the transfer zones. Sinistral strike-slip movement of at least 16 km is recognized across the Abrolhos Transfer Zone on the basis of offset in the trend of the Beagle Fault system. The orientation, age of activation, and position of these zones are similar to those of transform faults in the adjoining Indian Ocean, suggesting that the two structures are contiguous.

Extensional basins are often compartmentalized by fractures oriented oblique or perpendicular to the overall basin trend, which may transfer displacement among normal fault systems (Gibbs 1984; Lister *et al.* 1986*a*; Gawthorpe & Hurst 1993; Morley 1995). The linkages include regional and local transfer zones. The former include transfer faults and shear zones, which control the deposition and geometry of major elements within the basin (Nelson *et al.* 1992; Moustafa 1997; McClay & Khalil 1998). The latter may be localized to a single fault belt and include features such as relay ramps, small transfer faults and accommodation zones (Gibbs 1990; Morley *et al.* 1990; Destro 1995). Transfer linkages may develop synchronously with basin extension or reactivate pre-existing structures (Moustafa 1997; McClay & Khalil 1998).

Regional transfer zones can be orthogonal or oblique to rift-related normal faults (Gibbs 1984; Bosworth 1986; Lister *et al.* 1986*a*; McClay & Khalil 1998). Lister *et al.* (1986*b*) maintained that transfer faults with a high dip-angle are parallel to the extension direction, whereas Bosworth (1986), on the basis of studies along the East African rift system, argued that such faults are seldom parallel to extension. In nature, the pre-existing structural and basement architecture can exert a strong influence on the geometry and orientation of transfer zones. Transfer zones oblique to the rift trend may result from either a reactivation of pre-existing faults or weak zones (McClay & Khalil 1998) or a change of regional extension direction (Lister *et al.* 1986*a,b*; Dauteuil & Brun 1993). Thus, multiphase extension in variable directions may lead to regional transfer zones with different orientations in a single basin. The Perth Basin provides such an example, where basin segment linkages, both perpendicular and oblique to major normal faults, develop because of changes in extension direction through the basin history.

Geological setting

The Perth Basin is a major tectonic element, and an established petroleum province, along the western margin of the Australian continent (Fig. 1). It covers an area of *c.* 45 000 km^2 onshore and 55 000 km^2 offshore

From: Wilson, R.C.L., Whitmarsh, R.B., Taylor, B. & Froitzheim, N. 2001. *Non-Volcanic Rifting of Continental Margins: A Comparison of Evidence from Land and Sea*. Geological Society, London, Special Publications, **187**, 475–488. 0305-8719/01/$15.00 © The Geological Society of London 2001.

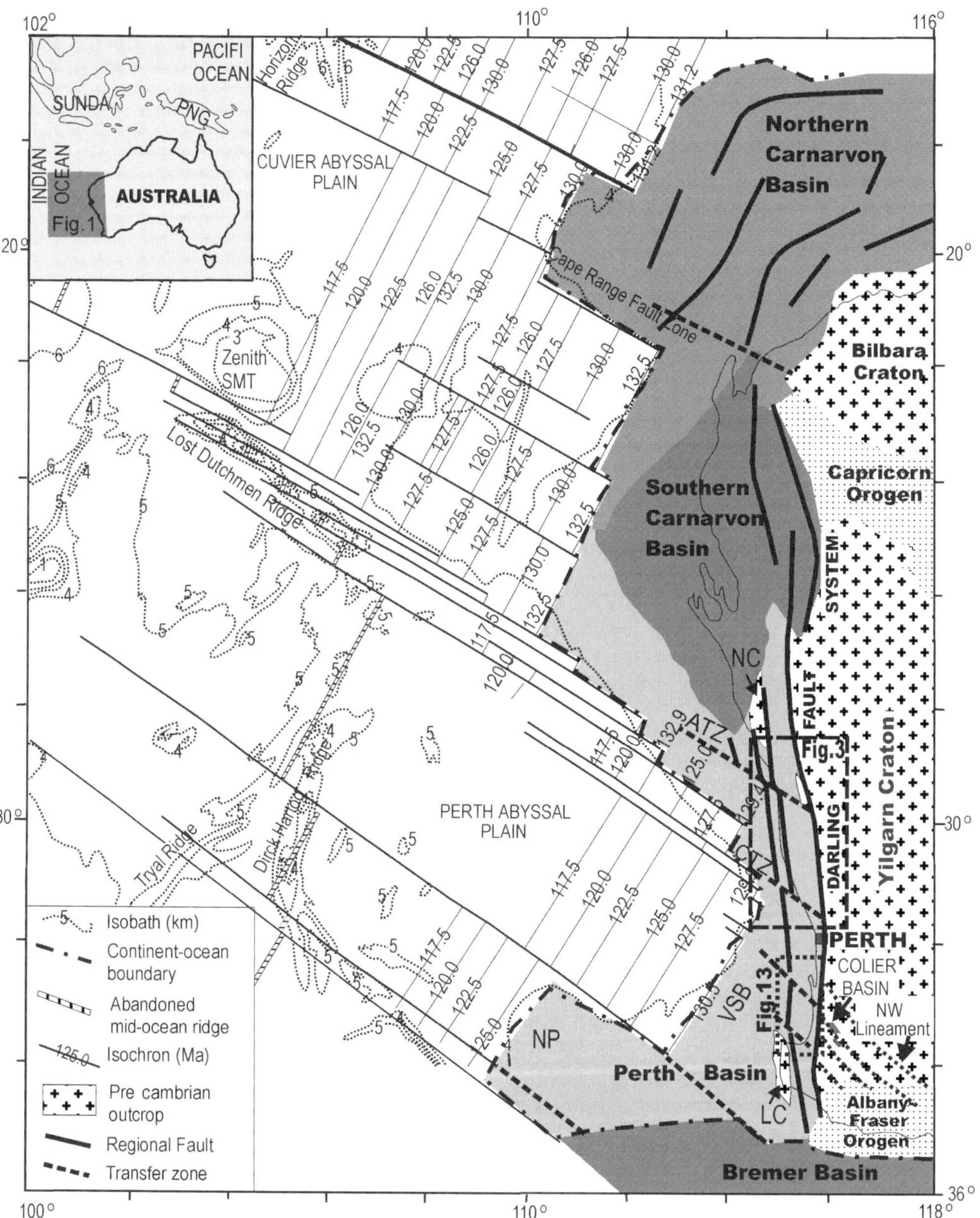

Fig. 1. Tectonic setting of the Perth Basin showing generalized sea-floor spreading isochrons and transform faults in the adjoining Indian Ocean; modified after Powell *et al.* (1988) and Muller *et al.* (1998). ATZ, Abrolhos Transfer Zone; CTZ, Cervantes Transfer Zone; NP, Naturaliste Plateau; SMT, Sea Mountain Plateau; VSB, Vlaming Sub-basin.

(Playford *et al.* 1976), and consists of a series of normal fault controlled northerly striking sub-basins, troughs and ridges. The basin is separated from the Yilgarn Craton in the east by the Darling Fault and extends west to the continent–Indian Ocean boundary. The southern Carnarvon Basin and the Bremer Basin lie at the northern and southern boundaries of the Perth Basin, respectively (Fig. 1). Sedimentary cover within the basin is largely of Permian to Cretaceous age (Fig. 2) and locally reaches 15 000 m in thickness in major depocentres such as the Dandaragan Trough and Vlaming Sub-basin. Basement to the Perth Basin con-

sists of the Proterozoic rocks of the Pinjarra Orogen (Myers 1990; Dentith *et al.* 1994).

The Perth Basin developed through the interplay of phases of extension and transtension, resulting in a complex history of faulting and synsedimentary fault block movement (Etheridge & O'Brien 1994; Harris 1994; Mory & Iasky 1994). Two main phases of basin tectonism, in Permian and Jurassic to Early Cretaceous time, have been recognized in both the offshore (Marshall *et al.* 1989; Quaife *et al.* 1994) and onshore (Iasky *et al.* 1993; Mory & Iasky 1996) segments of the basin and account for the overall structural complexity of the region. The younger event corresponds to the final rifting and break-up of Gondwana lithosphere between Australia and Greater India, and entailed dextral strike-slip deformation along major faults and basin inversion as well as extension (Song & Cawood 2000). Most hydrocarbon accumulations are in structures associated with the strike-slip movement during

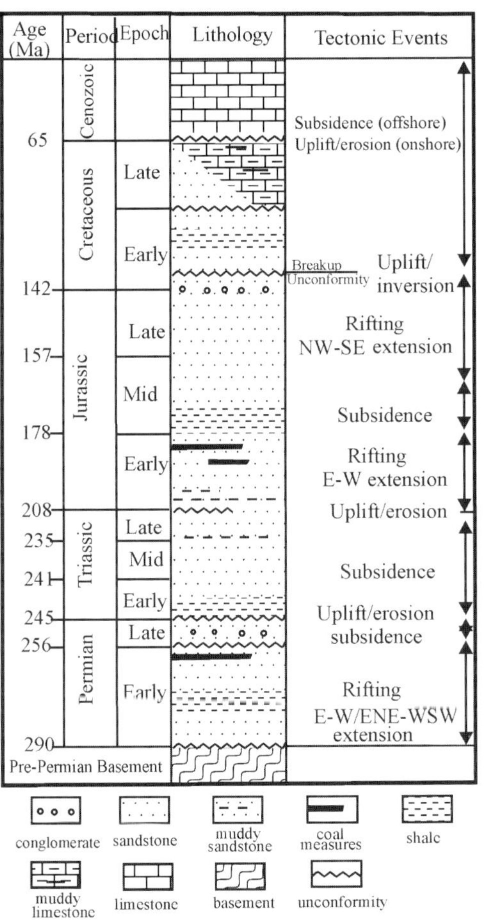

Fig. 2. Generalized stratigraphic column for the Perth Basin outlining corresponding tectonic events (integrated from data of Playford *et al.* (1976) and Mory & Iasky (1996)).

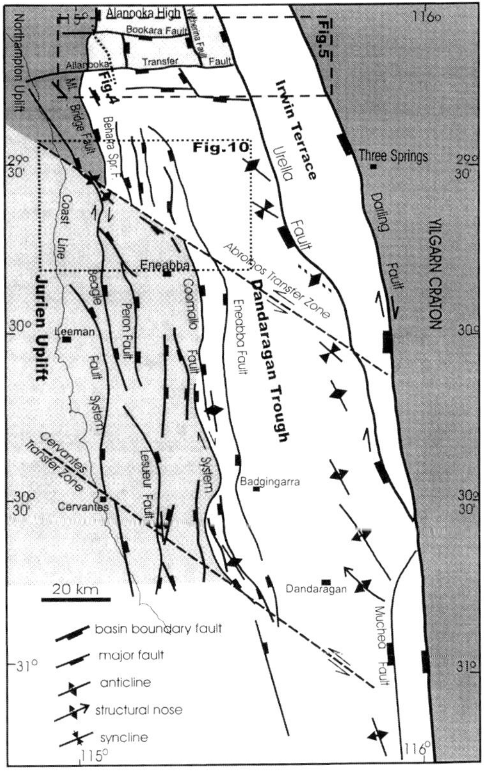

Fig. 3. Generalized structural features in the onshore northern Perth Basin.

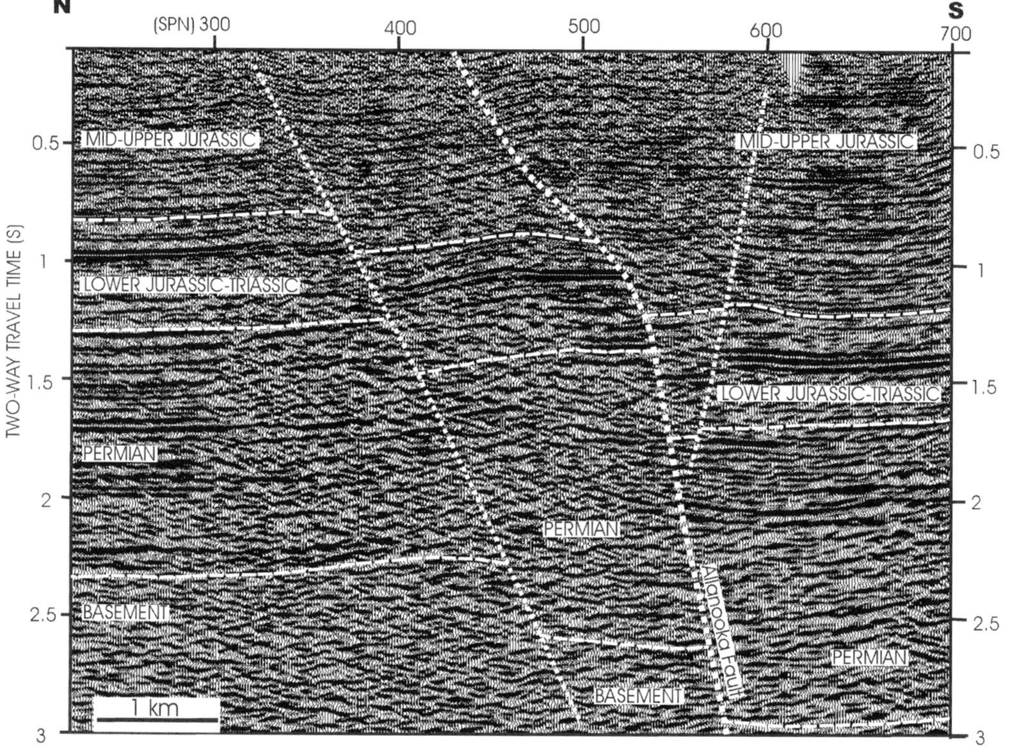

Fig. 4. An interpreted seismic reflection profile crossing the Allanooka and Bookara transfer faults. It should be noted that the Allanooka Fault has a steeper dip in lower levels than in upper levels. (See Fig. 3 for location.)

continental break-up (Crostella 1995), but minor accumulations also occur in extension-related rollover anticlines (Song & Cawood 1999). Basement structures were reactivated during basin formation and control the linear, north-striking, structural grain of the basin (Byrne & Harris 1992; Dentith *et al.* 1994). The basin is compartmentalized by a series of NW-striking transfer zones, which formed during break-up (e.g. Abrolhos and Cervantes transfer zones, Mory & Iasky 1996), but which are thought to be localized along pre-existing basement structures (Harris *et al.* 1994).

East–west transfer faults

The east–west-striking transfer faults are recognized only in the northernmost onshore part of the Perth Basin. They are approximately perpendicular to the major northerly striking normal faults and are limited to the west of the Urella Fault (Fig. 3). The Allanooka Transfer Fault (ATF), the most significant east–west-oriented transfer zone in this region, bounds the

Dandaragan Trough to the south and Allanooka High to the north. It extends west of the north-striking Mountain Bridge Fault (MBF), at least into the basement block of the Northampton

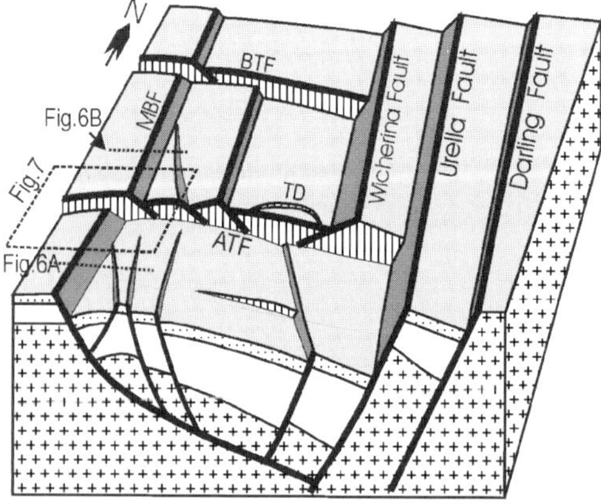

Fig. 5. Block diagram indicating geometric relationship between easterly striking transfer faults and northerly striking normal faults in northernmost onshore Perth Basin. TD, Transfer Duplex.

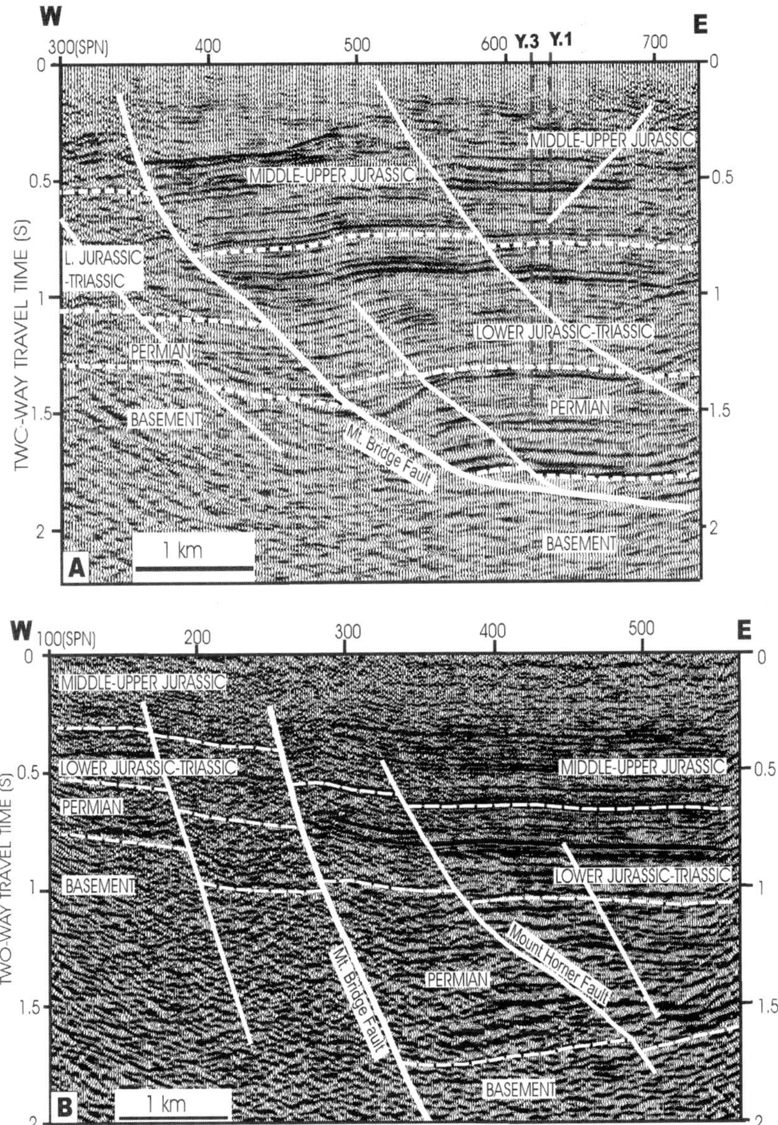

Fig. 6. Interpreted seismic cross-sections showing contrasting geometry of the Mountain Bridge Fault to the south (upper section) and north (lower section) of the Allanooka Transfer Fault.

Uplift (Fig. 3). It dips south towards the Dandaragan Trough at an angle of about 65–70° (Fig. 4). The eastern part of the Allanooka Fault has a steeper dip in lower horizons than at upper levels (Fig. 4; Song & Cawood 1999), suggesting a wrench origin for the structure (see Lowell 1985). Displacement varies significantly along the length of the fault, ranging from 100 m to 2200 m (Song & Cawood 1999), increasing where major normal faults, such as

the Mountain Bridge and Wicherina, are offset or terminate against the fault. Another east–west transfer fault, the Bookara Transfer Fault (BTF), lies north of the Allanooka Fault. It intersects the Mountain Bridge Fault but is restricted to the downthrown western side of the Wicherina Fault (Figs 3 and 4). Its influence on deposits of Permian–Jurassic age is relatively insignificant (Fig. 4), possibly because of its location within the structural high.

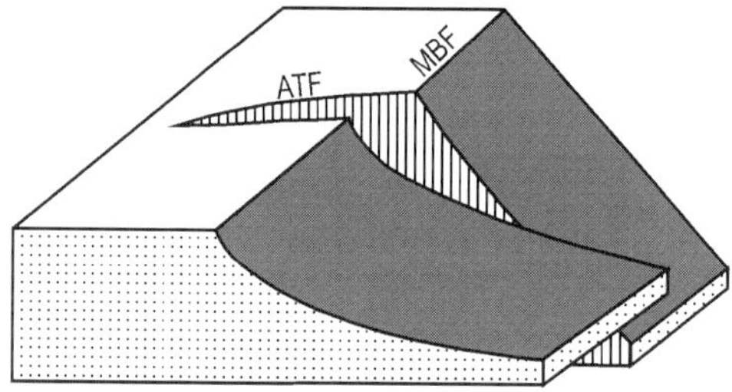

Fig. 7. Schematic block diagram showing geometric variation along the Mountain Bridge Fault (MBF) where it crosses the Allanooka Transfer Fault (ATF).

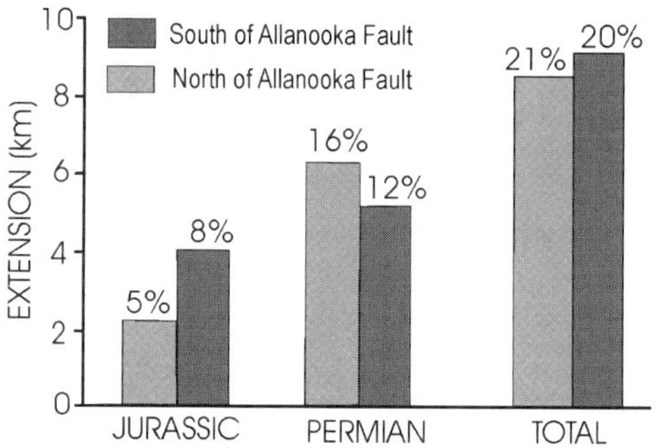

Fig. 8. Bar diagram indicating variation of extension across the Allanooka Transfer Fault. Although the total extensional ratio is similar across the transfer, the Jurassic extension is larger in the south than in the north of the fault, whereas the Permian extension is larger in the north. The percentage amount is the extension ratio of extension by the original length of the cross-section of relevant stage. Seismic reflection profiles of 850 km length has been used to calculate the extension, which is a sum of horizontal displacement by individual faults with modification of block-tilting.

Figure 5 postulates the relationship between the northerly oriented, extensional fault system and their hard linkages, the transfer faults. Characteristics of the extensional faults, including their geometry and displacement as well as stratigraphic thickness in bounding units, vary significantly as they cross the transfer structures (Figs 6 and 7). The Mountain Bridge Fault changes from a listric profile south of the Allanooka Transfer Fault to a planar profile to the north. In addition, there is significant temporal variation in the amount of extension that occurred north and south of the transfer fault, although total bulk extension across the structure is similar. South of the Allanooka Transfer Fault, total horizontal extension for the Permian to Jurassic sequence is estimated to amount of 8900 m, representing 20% extension, whereas to the north of the transfer it is around 8000 m, representing 21% extension (Fig. 8). This is a minimum estimate of extension, as basin-wide uplift and erosion, and inversion associated with continental break-up may have masked the original maximum fault displacement. Analysis of extension during the Permian and Jurassic intervals indicates marked variations in displacement amounts. Extension in the northern block was 16% in Permian time and only 5% in Jurassic time, whereas the southern block underwent 12% extension in Permian time and 8% in Jurassic time. This, together with evidence that the Permian sequences thicken to the north (Mory & Iasky 1996; Iasky *et al.* 1998; Mory *et al.* 1998), suggests that the sedimentary depocentre during Permian time was located in the northern Perth and the southern Carnarvon basins. In contrast, the depocentre during Triassic and Jurassic time was situated within the southern Perth Basin in the Vlaming Sub-basin and Dandaragan Trough (Playford *et al.* 1976).

The east–west transfer faults were initiated under a regime of east–west or ENE–WSW Permian extension (Harris 1994). During Early Jurassic time, basin extension was again oriented east–west, which resulted in a reactivation of these pre-existing hard linkages (Song & Cawood 1999). The variation in Jurassic extension ratio across the Allanooka Transfer Fault (Fig. 8) reflects the effects of this reactivation event. But there is no evidence that this second phase of extension resulted in the generation of any new east–west-oriented transfer faults.

NW transfer zones

NW–SE-oriented extension during Late Jurassic–Early Cretaceous continental break-up (Harris 1994) generated transfer zones or reactivated basement weak zones striking NW, and also resulted in oblique slip movement along the pre-existing normal faults (Song & Cawood 1999). These regional transfer zones oblique to the basin trend exerted a pronounced effect on basin development immediately before break-up (Mory & Iasky 1996). Two NW-striking transfer zones, the Abrolhos and Cervantes, are well defined in the onshore Perth Basin (Fig. 3). Their trend is *c.* 300° and they lie at an angle of about 65° to the major northerly striking extensional faults in the Perth Basin (Fig. 3). No continuous fault has been identified along these trends within the sedimentary cover of the basin (Fig. 9, B–B′), but a few discon-

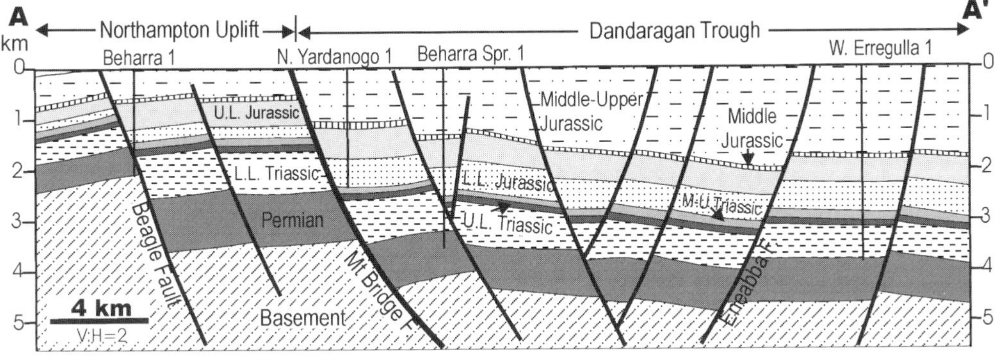

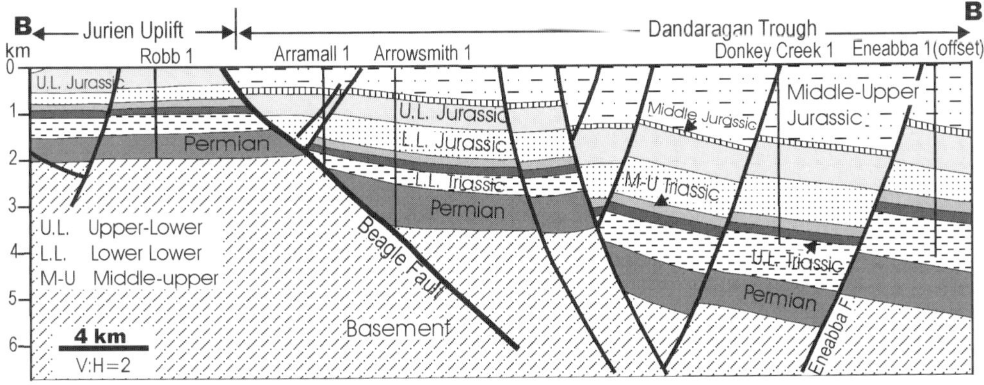

Fig. 9. Cross-sections showing variation in structural and stratigraphic geometry on both sides of the Abrolhos Transfer Zone. The Beagle Fault south of the transfer acts as the east boundary of the uplift (B–B'), whereas the Mountain Bridge Fault to the north of the transfer separates the uplift and the trough (A–A'). There is no discontinuity, at the transfer zone, offsetting sedimentary sequences (B–B'). (See Fig. 10 for location.) (After Mory & Iasky 1996.)

tinuous minor fractures occur within, or adjacent to, the zones. The transfer zones are defined and recognized by deflections or terminations of normal faults (Figs 3 and 10). Within the Abrolhos Transfer Zone the trend of the Beagle Fault swings to the northwest from its overall NNW orientation (Fig. 10) suggesting a sinistral strike-slip offset of at least 16 km along the transfer zone. Dextral strike-slip, as well as normal dip-slip (Fig. 9), along the northerly oriented major faults would have resulted from NW–SE extension during Late Jurassic to Early Cretaceous break-up. Under the oblique extension, the deflection of the Beagle Fault defines a constraining bend (Fig. 10) resulting in formation of a compressive anticline in the hanging wall to the fault (Fig. 11). The Eneabba Fault demonstrates a similar feature where it crosses the Abrolhos Transfer Zone (Fig. 3).

Some regional north-striking faults terminate at the transfer zones. The Mountain Bridge and Beharra Spring faults occur only north of the Abrolhos Transfer Zone (Figs 9 and 10) whereas the Coomallo Fault runs only between the Abrolhos and Cervantes transfer zones (Fig. 3). The faults either terminate directly at the transfer zone or, in the case of the Coomallo Fault, diverge into a series of splay faults as they approach the transfer zone (Fig. 12). Displacement on the Mountain Bridge Fault, which constitutes the east boundary of the Northampton Uplift (Fig. 9, A–A'), may be accommodated, and transferred, south of the Abrolhos structure by the Beagle Fault, which bounds the Jurien Uplift to the west (Fig. 9, B–B') and which is considered the southern extension of the Northampton Uplift (Fig. 3).

In the offshore Vlaming Sub-basin, NW–SE transfer zones are locally interpreted from seismic profiles. Several such transfer zones can be

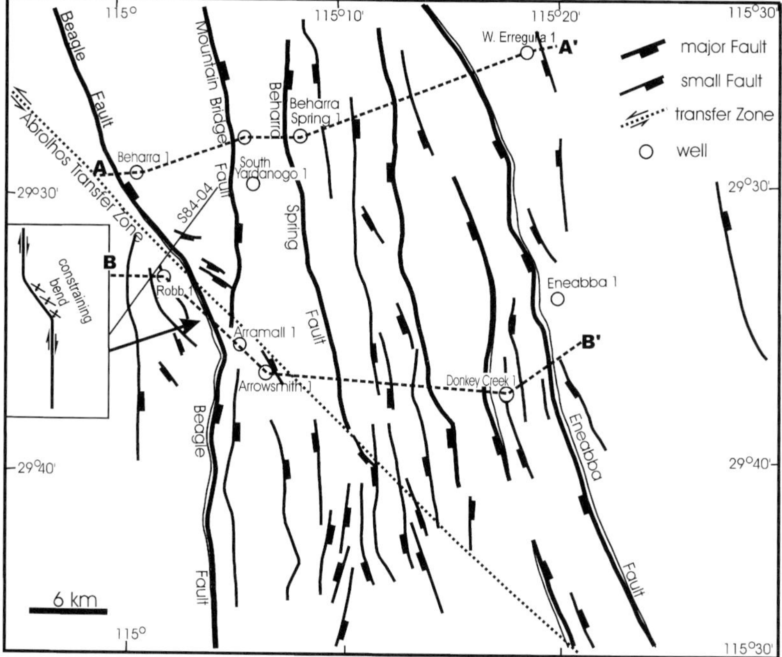

Fig. 10. A generalized fault map demonstrating influence of the NW Abrolhos Transfer Zone on northerly striking structures, which may either diminish toward the transfer zone (Mt. Bridge and Beharra Spring faults) or swing to a NW trend (Beagle Fault) when crossing the transfer. Small NW faults also occur within the transfer zone. (See Fig. 3 for location.)

delineated on the basis of changes in the sedimentary sequence and in the orientation of northerly trending structures across the inferred transfer structures (Fig. 13). Figure 14 illustrates a transfer zone characterized by faults with high dip and complex deformation in the vicinity of the zone. Such a flower structure suggests strike-slip movement along the NW transfer zone. The inferred northern extension of the Dunsborough Fault mapped from seismic data appears to be offset by the series of inferred transfer structures traversing the Vlaming Sub-basin. Marshall *et al.* (1993) postulated that a series of NW-oriented transfer faults offset and link the northerly striking structures. Such a model is supported by our structural interpretation of the Vlaming Sub-basin based on accessible seismic data (Fig. 13).

In addition, some northwest transfer zones may be characterized by a structural high, such as the Harvey Ridge, which separates the Dandaragan Trough to the north from the Bunbury Trough to the south (Dentith *et al.* 1994). Aeromagnetic data suggest that the Harvey Ridge also marks a change in the nature of the eastern boundary of the basin (Dentith *et al.* 1994). South of the ridge, the Darling Fault is the main basin-bounding structure. However, to the north, another fault, which lies a few kilometres to the west of the Darling Fault, shared bulk displacement during Jurassic extension with the Darling Fault (Dentith *et al.* 1994). Northwestern structural highs have also been interpreted and considered as transfer zones in the Vlaming Sub-basin (Spring & Newell 1993).

The above data demonstrate that the northwest transfer zones have exerted strong effects on basin deformation associated with continental break-up. However, interpretation from available data shows no northwest regional discontinuity through sedimentary sequences (Fig. 9, B–B'), apart from small-scale NW-striking faults (Figs 3 and 10) within the northwest transfer zones. In the offshore Vlaming Sub-basin, however, the basement transfer zones may locally extend up into the sedimentary cover (Figs 13 and 14). The general absence of through-going faults within the Perth Basin sedimentary sequence suggests that the oblique transfer zones may represent reactivated basement fractures in which rheological contrasts between the basement and the cover prevented the development of discrete discontinuities in the latter.

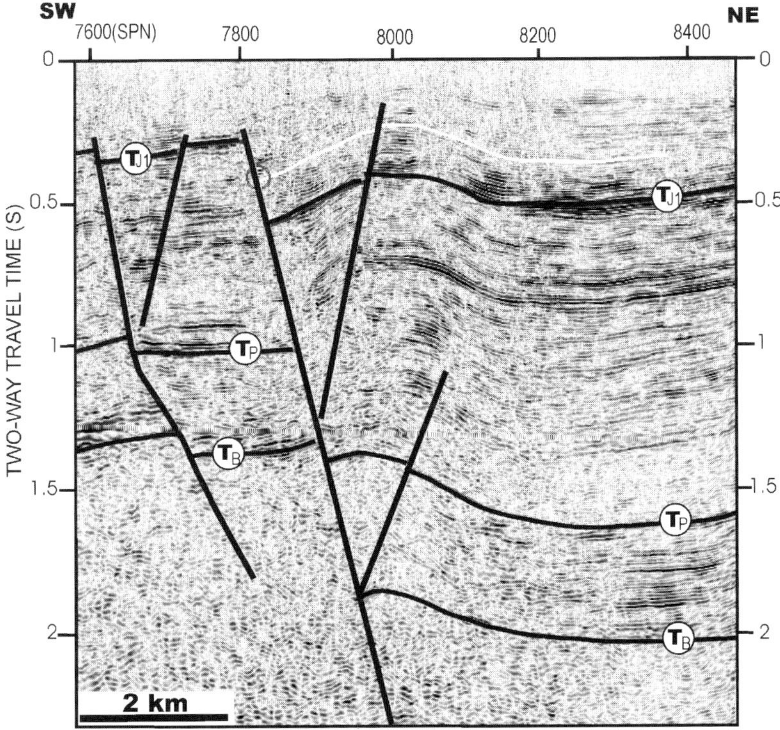

Fig. 11. Interpreted seismic profile (S84-04) showing a compressive fold developed at a constraining bend of the Beagle Fault where it crosses the Abrolhos Transfer Zone. T_{J1}, uppermost Lower Jurassic units; T_P, uppermost Permian units; T_B, uppermost basement units. (See Fig. 10 for location.)

Correlation between NW transfer zones and oceanic transform faults

Mory & Iasky (1994) proposed that the northwestern accommodation zones in the Perth

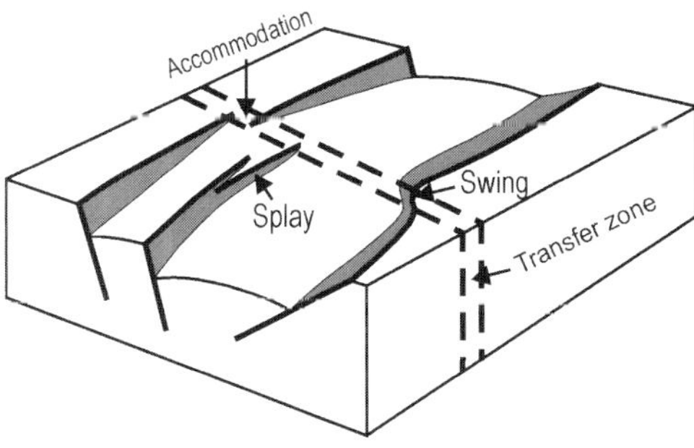

Fig. 12. Block diagram illustrating the influence of transfer zones on major normal faults. Accommodation zones, fault swings and splay faults occur either within or in the vicinity of the transfer zone.

Basin formed during late Jurassic to Early Cretaceous continental break-up. Dentith *et al.* (1994) correlated the transfer zones with similarly oriented lineaments located in the Yilgarn Craton to the east, which are believed to have originated in Archaean time and subsequently reactivated during development of the Neoproterozoic Pinjarra Orogen, which forms the basement to the Perth Basin, and again in Phanerozoic time. The transfer zones do not, however, influence or displace the Darling Fault (Figs 1, 3 and 14). Thus, although the northwest transfer zones may have been influenced by, or even originated during the Proterozoic Pinjarra Orogeny, their reactivation during basin evolution must be limited by the Darling Fault. The northwest lineaments are also proposed to represent a continuation of transform faults in oceanic crust (Harris 1994), and the location and trend of the two shows a strong (Fig. 1), but not a one-by-one, correlation (see Lister *et al.* 1986a).

The oceanic transform faults strike *c.* 300° NW and are perpendicular to, and displace,

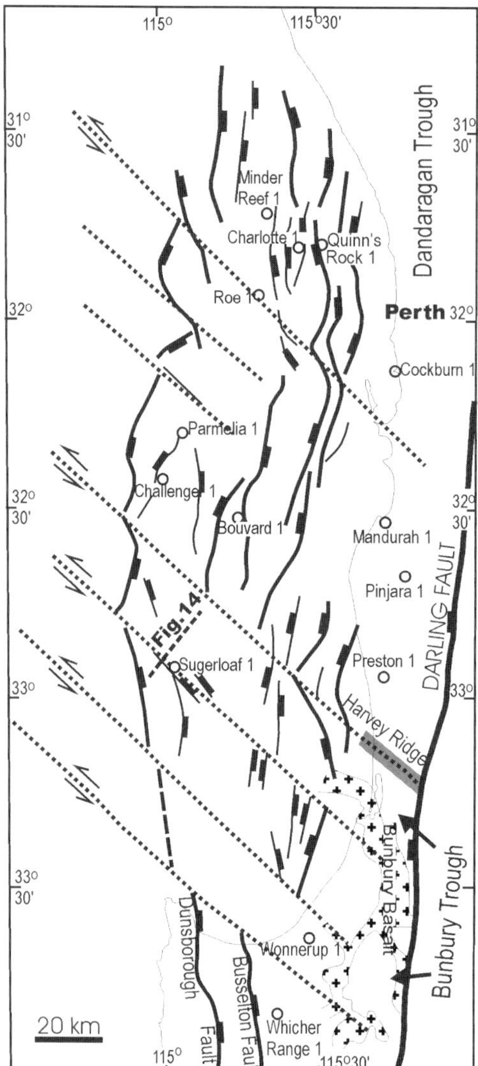

Fig. 13. Generalized structures of the Vlaming Sub-basin based on seismic interpretation indicate that northerly striking faults are linked by northwest transfer zones or faults. Data integrated from Marshall *et al.* (1993).

Indian Ocean magnetic anomalies (Fig. 1) (Powell *et al.* 1988; Muller *et al.* 1998). These transform faults may extend southeast into continental lithosphere as shear or weak zones. Some of these shear zones may be limited to the deep part of the continental crust and thus have negligible influence on deformation of the younger sedimentary sequences. On the other hand, some may have exerted strong control on characteristics of structures formed during Late

Jurassic–Early Cretaceous break-up (Fig. 15) such as the Abrolhos and Cervantes transfer zones in the onshore northern Perth Basin (Fig. 3) and those in the Vlaming Sub-basin (Figs 13 and 14). These structures acted as accommodation zones and undertook sinistral strike-slip movement during the break-up stage. The shear or weak zones may influence deformation to a width of a few kilometres (Fig. 10). During the Late Jurassic–Early Cretaceous NW–SE extension, most pre-existing primary faults had been reactivated and their deformation was controlled by the northwest transfer zones. The Abrolhos and Cervantes transfer zones lie along-strike from a series of oceanic transforms that run through the Lost Dutchmen Ridge, with both sets of structures showing a sinistral sense of offset. There are no major transform faults offsetting the continental margin adjacent to the Vlaming Sub-basin (Fig. 1; Muller *et al.* 1998). A few minor transform faults are recognized offshore of the Vlaming Sub-basin and show a similar sense of sinistral strike-slip movement to those observed in the basin (Fig. 13).

The western boundary of the Australian continent, including the Perth and Carnarvon basins, consists of a series of NE-trending segments offset by the perpendicular transform faults (Figs 1 and 15). The most southwestern element of the Perth Basin is the Naturaliste Plateau, which is bounded by two NW-trending transform faults (Fig. 1). The transform along the northeast boundary of the plateau offsets the magnetic anomalies in a dextral sense. The Perth and Curvier abyssal plains and their adjacent continental crust elements are also delineated by the transform faults (Fig. 1) (Muller *et al.* 1998). On the Northwest Shelf of the Australian margin, the margins of the Exmouth Plateau are also controlled by continental transfer faults that can be traced oceanward into transform faults (Lister *et al.* 1986*a*).

Conclusions

Transfer zones oriented both normal and oblique to primary NNW-trending extensional faults have been recognized in the Perth Basin. The transfer faults perpendicular to the basin trend were initiated during Early Permian rifting and reactivated during later tectonic events, in particular during the Jurassic extension before the continental break-up. These east–west structures have been recognized only in the northernmost onshore part of the Perth Basin, which corresponds to the basin depocentre in Permian time. They separate northerly

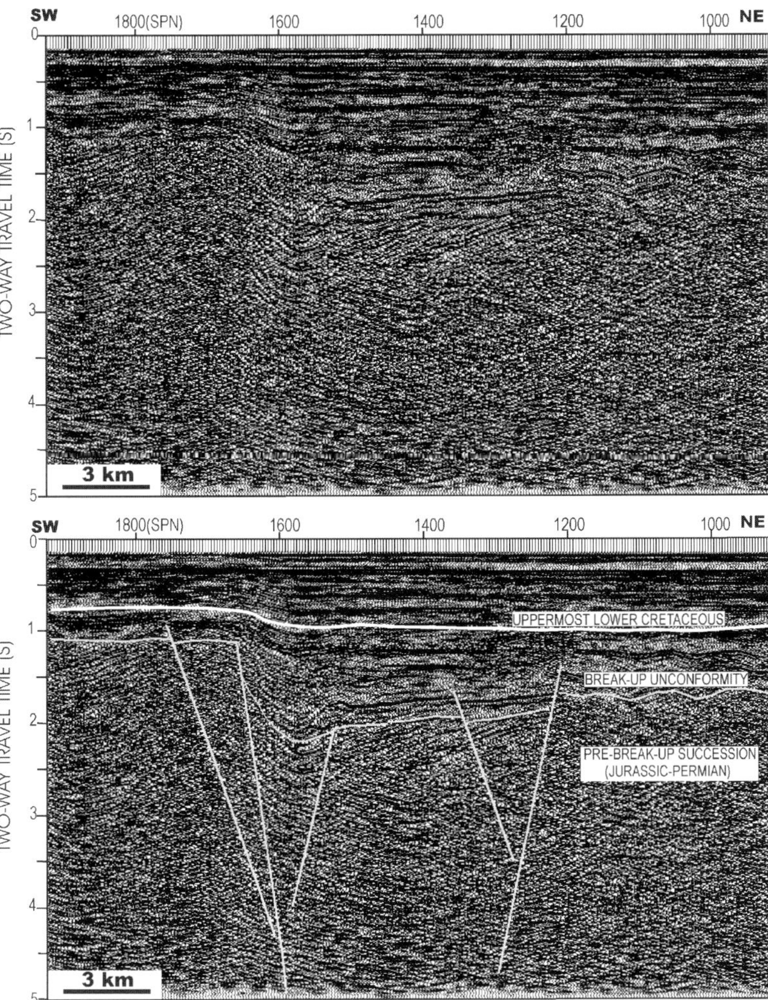

Fig. 14. Seismic section crossing a northwest transfer zone in the Vlaming Sub-basin. Faults of steep dip and their adjacent complex deformation comprise a wrench feature–flower structure. (See Fig. 13 for location.)

trending normal faults into segments of distinguishable characteristics, limit the along-strike extension of some normal faults, and transfer displacement among the normal faults. Additionally, the east–west faults are restricted to west of the Urella Fault (Figs 3 and 5), a major branch of the Darling Fault system.

The NW-striking transfer zones have influenced deformation during Jurassic time, although no continuous fault plane can be identified within the sedimentary sequences. Termination of, and a swing in the orientation of major normal faults within the transfer zones

(Figs 3, 10 and 12) provide evidence for the presence of these structures in the onshore northern Perth Basin. The northwest transfer zones may locally extend up through the pre-break-up sequences in the offshore Vlaming Sub-basin (Figs 13 and 14). Sinistral strike-slip movement, no less than 16 km along the Abrolhos Transfer Zone (Fig. 10), can also be observed on the basis of deflection of the Beagle Fault system. The NW-striking transfer faults have a similar orientation to oceanic transform faults in the adjoining Indian Ocean and were activated at the same time, during

                    TINGGUANG SONG *ET AL.*

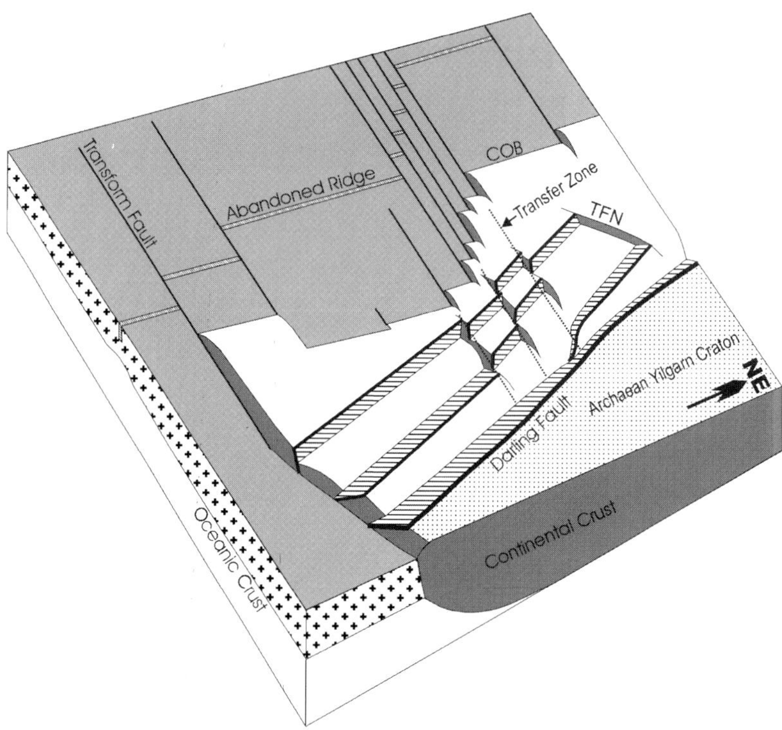

Fig. 15. Block diagram illustrating relationship between transform faults in oceanic crust and transfer zones in a continental margin. It should be noted that the transfer zones are oblique to pre-existing normal faults. The northwest transfer zones normally remain within basement but are exaggerated here as through-going structures. COB, continental–oceanic crust boundary; TFN, transfer fault normal to basin trend.

Late Jurassic–earliest Cretaceous time, according to the oceanic magnetic anomaly pattern (Powell *et al.* 1988) and analysis of deformation in the basin (Mory & Iasky 1996). Therefore, it is suggested that the transfer zones in the continental margin and the transform faults in the ocean floor are related structures (Fig. 15).

This paper is Tectonics Special Research Centre Publication 116.

References

BOSWORTH, W. 1986. Comment and reply on 'Detachment faulting and the evolution of passive continental margins': Comment. *Geology,* **14**, 890–891.

BYRNE, D.R. & HARRIS, L.B. 1992. Fault patterns during normal and oblique rifting and the influence of basement discontinuities: application to models for the tectonic evolution of the Perth Basin, Western Australia. *In*: RICKARD, M.J., HARRINGTON, H.J. & WILLIAMS, P.R. (eds) *Basement Tectonics 9: Australia and Other Regions.* Kluwer, Dordrecht, **9**, 23–42.

CROSTELLA, A. 1995. *An Evaluation of the Hydrocarbon Potential of the Onshore Northern Perth Basin, Western Australia.* Western Australian Geological Survey, Report **43**.

DAUTEUIL, O. & BRUN, J.-P. 1993. Oblique rifting in a slow-spreading ridge. *Nature,* **361**, 145–148.

DENTITH, M.C., LONG, A., SCOTT, J., HARRIS, L.B. & WILDE, S.A. 1994. The influence of basement on faulting within the Perth Basin, Western Australia. *In*: PURCELL, P.G. & PURCELL, R.R. (eds) *The Sedimentary Basins of Western Australia 1: Proceedings of Petroleum Exploration Society of Australia Symposium.* Petroleum Exploration Society of Australia, Perth, 791–799.

DESTRO, N. 1995. Release fault: a variety of cross fault in linked extensional fault systems, in the Sergipe–Alagoas Basin, NE Brazil. *Journal of Structural Geology,* **17**, 615–629.

ETHERIDGE, M.A. & O'BRIEN, G.W. 1994. Structural and tectonic evolution of the Western Australian margin basin system. *Journal of the Petroleum Exploration Society of Australia*, **22**, 45–64.

GAWTHORPE, R.L. & HURST, J.M. 1993. Transfer zones in extensional basins; their structural styles and influence in drainage development and stratigraphy. *Journal of the Geological Society, London*, **150**, 1137–1152.

GIBBS, A.D. 1984. Structural evolution of extensional basin margins. *Journal of the Geological Society, London*, **141**, 609–620.

GIBBS, A.D. 1990. Linked fault families in basin formation. *Journal of Structural Geology*, **12**, 795–803.

HARRIS, L.B. 1994. Structural and tectonic synthesis for the Perth Basin, Western Australia. *Journal of Petroleum Geology*, **17**, 129–156.

HARRIS, L.B., HIGGINS, R.I., DENTITH, M.C. & MIDDLETON, M.F. 1994. Transtensional analogue modelling applied to the Perth Basin, Western Australia. *In*: PURCELL, P.G. & PURCELL, R.R. (eds) *The Sedimentary Basins of Western Australia 1: Proceedings of Petroleum Exploration Society of Australia Symposium*. Petroleum Exploration Society of Australia, Perth, 801–809.

IASKY, R.P., MORY, A.J. & SHEVCHENKO, S.I. 1998. A structural interpretation of the Gascoyne Plateform, Southern Carnarvon Basin, WA. *In*: PURCELL, P.G. & PURCELL, R.R. (eds) *The Sedimentary Basins of Western Australia 2: Proceedings of Petroleum Exploration Society of Australia Symposium*. Petroleum Exploration Society of Australia, Perth, 589–598.

IASKY, R.P., YOUNG, R.A. & MIDDLETON, M.F. 1993. Structural study of the southern Perth Basin by geophysical methods. *Australian Society of Exploration Geophysicists Journal*, **22**, 199–206.

LISTER, G.S., ETHERIDGE, M.A. & SYMONDS, P.A. 1986*a*. Detachement faulting and the evolution of passive continental margins. *Geology*, **14**, 246–250.

LISTER, G.S., ETHERIDGE, M.A. & SYMONDS, P.A. 1986*b*. Comment and reply on 'Detachment faulting and the evolution of passive continental margins': Reply. *Geology*, **14**, 891–892.

LOWELL, J.D. *Structural Styles in Petroleum Exploration*. Oil & Gas Consultants International, Inc., Tulsa, OK.

MARSHALL, J.F., LEE, C.S., RAMSAY, D.C. & MOORE, A.M.G. 1989. Tectonic controls on sedimentation and maturation in the offshore north Perth Basin. *Australian Society of Exploration Geophysicists Journal*, **29**, 450–465.

MARSHALL, J.F., RAMSAY, D.C., MOORE, A.M.G., SHAFIK, S., GRAHAM, T.G. & NEEDHAM, J. 1993. *The Vlaming Sub-basin, Offshore Southern Perth Basin*. Australian Geological Survey, Continental Margin Program, Folio **7**.

MCCLAY, K. & KHALIL, S. 1998. Extensional hard linkages, eastern Gulf of Suez, Egypt. *Geology*, **26**, 563–566.

MORLEY, C.K. 1995. Developments in the structural geology of rifts over the last decade and their impact on hydrocarbon exploration. *In*: LAMBIASE, J.J. (ed.) *Hydrocarbon Habitat in Rift Basins*. Geological Society, London, Special Publications, **80**, 1–32.

MORLEY, C.K., NELSON, R.A., PATTON, T.L. & MUNN, S.G. 1990. Transfer zones in the East African Rift System and their relevance to hydrocarbon exploration in rifts. *AAPG Bulletin*, **74**, 1234–1253.

MORY, A.J. & IASKY, R.P. 1994. Structural evolution of the onshore northern Perth Basin, Western Australia. *In*: PURCELL, P.G. & PURCELL, R.R. (eds) *The Sedimentary Basins of Western Australia 1: Proceedings of Petroleum Exploration Society of Australia Symposium*. Petroleum Exploration Society of Australia, Perth, 781–790.

MORY, A.J. & IASKY, R.P. 1996. *Stratigraphy and Structure of the Onshore Northern Perth Basin, Western Australia*. Geological Survey of Western Australia, Report **46**.

MORY, A.J., IASKY, R.P. & SHEVCHENKO, S.I. 1998. The Coolcalalaya Sub-basin: a forgotten frontier 'between' the Perth and Carnarvon basins. *In*: PURCELL, P.G. & PURCELL, R.R. (eds) *The Sedimentary Basins of Western Australia 2: Proceedings of Petroleum Exploration Society of Australia Symposium*. Petroleum Exploration Society of Australia, Perth, 613–622.

MOUSTAFA, A.R. 1997. Controls on the development and evolution of transfer zones: the influence of basement structure and sedimentary thickness in the Suez rift and Red Sea. *Journal of Structural Geology*, **19**, 755–768.

MULLER, R.D., MIHUD, D. & BALDWIN, S. 1998. A new kinematic model for the formation and evolution of the west and northwest Australian margin. *In*: PURCELL, P.G. & PURCELL, R.R. (eds) *The Sedimentary Basins of Western Australia 2: Proceedings of Petroleum Exploration Society of Australia Symposium*. Petroleum Exploration Society of Australia, Perth, 55–72.

MYERS, J.S. *Pinjarra Orogen, Geology and Mineral Resources of Western Australia*. Western Australia Geological Survey, Perth, 265–274.

NELSON, R.A., PATTON, T.L. & MORLEY, C.K. 1992. Rift-segment interaction and its relation to hydrocarbon exploration in continental rift systems. *AAPG Bulletin*, **76**, 1153–1169.

PLAYFORD, P.E., COCKBAIN, A.E. & LOW, G.H. 1976. Geology of the Perth Basin Western Australia. *Geological Survey of Western Australia, Bulletin*, **124**.

POWELL, C.M., ROOTS, S.R. & VEEVERS, J.J. 1988. Pre-breakup continental extension in East Gondwanaland and the early opening of the eastern Indian Ocean. *Tectonophysics*, **155**, 261–283.

QUAIFE, P., ROSSER, J. & PAGNOZZI, S. 1994. The structural architecture and stratigraphy of the offshore northern Perth Basin, Western Australia. *In*: PURCELL, P.G. & PURCELL, R.R. (eds) *The Sedimentary Basins of Western Australia 1: Proceedings of Petroleum Exploration Society of Australia Symposium*. Petroleum Exploration Society of Australia, Perth, 811–822.

SONG, T. & CAWOOD, P.A. 1999. Multistage deformation of linked fault systems in extensional regions: an example from the northern Perth Basin, Western Australia. *Australian Journal of Earth Sciences*, **46**, 897–903.

SONG, T. & CAWOOD, P.A. 2000. Structural styles in the Perth Basin associated with Mesozoic break-up of Greater India and Australia. *Tectonophysics*, **317**, 55–72.

SPRING, D.E. & NEWELL, N.A. 1993. Depositional systems and sequence stratigraphy of the Cretaceous Warnbro Group, Vlaming Sub-basin, Western Australia. *Australian Petroleum Exploration Association Journal*, **33**, 190–204.

Patterns of extension and magmatism along the continent–ocean boundary, South China margin

PETER D. CLIFT, JIAN LIN & ODP LEG 184 SCIENTIFIC PARTY

Department of Geology and Geophysics, Woods Hole Oceanographic Institution, Woods Hole, MA 02543, USA (e-mail: pclift@whoi.edu)

Abstract: Early Oligocene sea-floor spreading in the South China Sea was preceded by at least two episodes (in Maastrichtian and Mid-Eocene time) of continental extension that generated a series of rift basins on the South China margin, which are separated from the continent–ocean boundary (COB) by an outer structural high. Regional multichannel seismic profiles showing faulting of the pre-rift basement allow the amount of extension in the upper crust to be measured. The total subsidence across the South China margin is far in excess of that predicted using a forward flexural-cantilever model of extension and the degree of faulting measured seismically in the upper crust. This mismatch suggests preferential extension of the lower crust, increasing towards the COB to account for the subsidence. The same feature is seen in the Nam Con Som Basin, which is located close to the southwest end of an extinct propagating spreading ridge offshore from Vietnam. However, in the Beibu Gulf Basin, which is not adjacent to the COB, subsidence is approximately compatible with uniform extension in the upper and lower crust across the entire basin, if not at all locations. We predict that extension of the lower crust exceeds that in the lithospheric mantle along the COB. Heat-flow measurements at Ocean Drilling Program (ODP) sites on the Chinese continental slope and on the conjugate Dangerous Grounds margin yield values consistent with, or slightly higher than, those predicted by models of uniform extension in the lithosphere. Although there is no magmatism comparable with the seaward-dipping volcanic rocks of rifted volcanic margins, there is seismic evidence of rift-related volcanic rocks spanning a width of *c.* 25 km landward of the COB. Simple adiabatic melting models do not predict magmatism, and we suggest that the presence of water in the mantle lithosphere, together with residual pre-rift heat, may instead be responsible for increasing melting here. Deep-water syn-rift sediments recovered by the ODP near the COB indicate that volcanism was submarine and that rifting culminated in a mass wasting event that marks a break-up unconformity. The average extension in South China Sea is much less than that seen in the extreme 'non-volcanic' Iberian margin. The South China margin may represent an intermediary form of continental extension between the end member extremes of the Iberia-type non-volcanic and the Greenland-type volcanic margin.

Extension during continental break-up is a fundamental part of the plate tectonic cycle, and is typically considered to occur in one of two discrete modes, volcanic or non-volcanic. Modern understanding of non-volcanic margins is principally the result of surveys of the Iberia–Galicia margin and to a lesser extent its conjugate Newfoundland margin (e.g. Welsink *et al.* 1989; Boillot *et al.* 1995; Reston *et al.* 1995). In this type of margin extension produces rotated basement fault blocks exposed over >100 km landward from the continent–ocean boundary (COB), as well as mantle peridotite exposures adjacent to the oldest oceanic crust. The Goban Spur margin of the North Atlantic is also seen to be of this type, but does not clearly manifest the peridotite ridge along the COB (Horsefield *et al.* 1994). This style of extension contrasts with the sharp COBs, voluminous subaerially erupted basalts, and thick bodies of gabbro underplated to the base of the original crust seen in the volcanic margins (e.g. White *et al.* 1987; Eldholm *et al.* 1995; Kelemen & Holbrook 1995). In this study we describe the nature of the South China margin, which appears to have some similarities to both these end members, and suggest that rifted margins cannot be simply divided into the existing two classifications. We further describe how the extension is partitioned between different parts of the lithospheric plate and note the differences in the mode of extension seen in

From: WILSON, R.C.L., WHITMARSH, R.B., TAYLOR, B. & FROITZHEIM, N. 2001. *Non-Volcanic Rifting of Continental Margins: A Comparison of Evidence from Land and Sea*. Geological Society, London, Special Publications, **187**, 489–510. 0305-8719/01/$15.00 © The Geological Society of London 2001.

intracontinental rift basins and passive margins that culminate in sea-floor spreading.

We choose the South China Sea as being ideal for this study (Fig. 1) because it has several advantages over the classic Atlantic margins in examining continental break-up processes. Unlike the Atlantic, where the onset of spreading can be difficult to pinpoint during the Cretaceous magnetic quiet period, the South China Sea shows well-developed anomalies in marine magnetic patterns (Taylor & Hayes 1983; Briais *et al.* 1993). The Miocene to recent time frame is ideal for interpreting the timing of break-up deformation, as the frequent magnetic reversals provide a well-constrained spreading pattern with a clearly discernible ENE to WSW propagation pattern. The Neogene rift and post-rift cover is also relatively

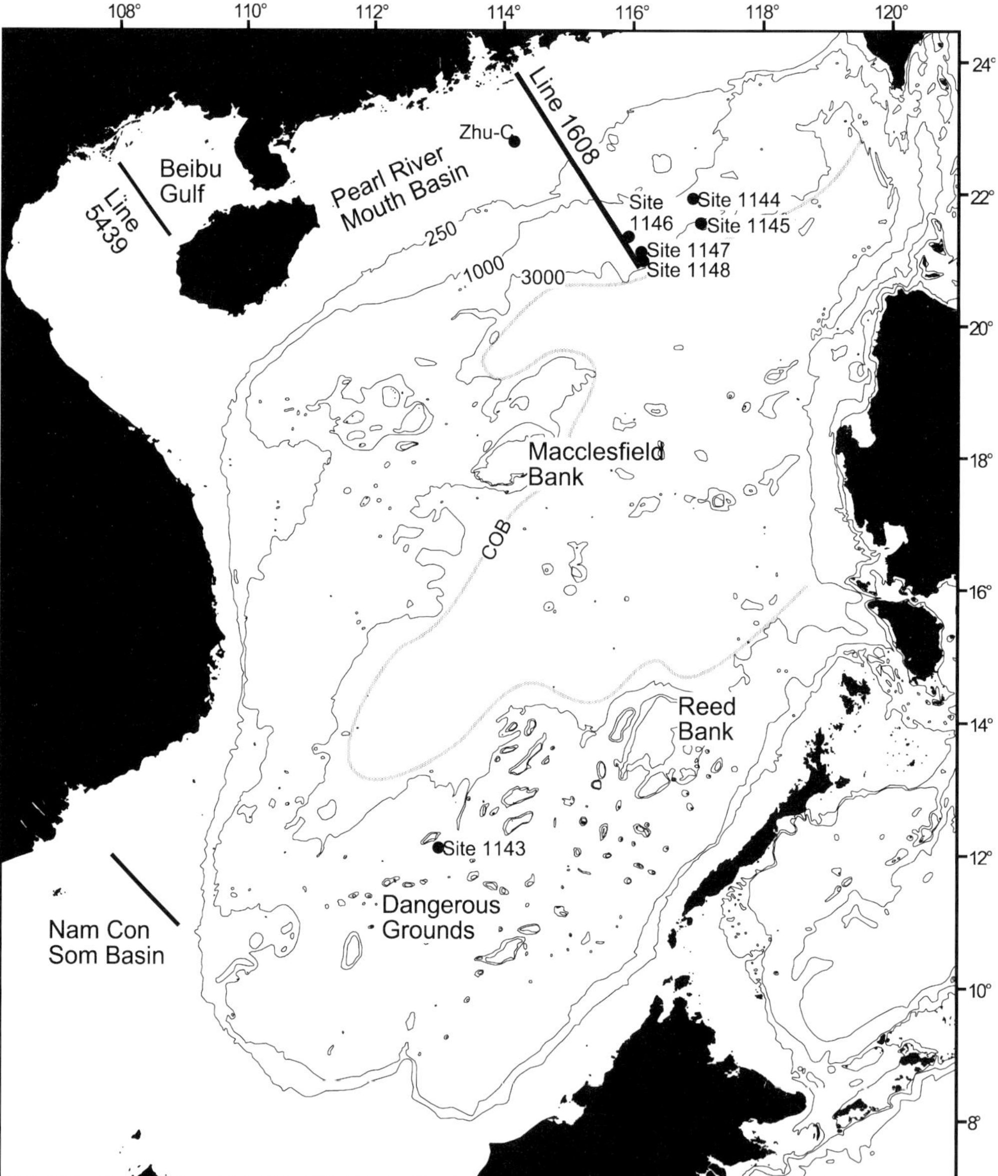

Fig. 1. Bathymetric map of the South China Shelf showing the location of the drill sites, together with the trace of the interpreted multichannel seismic data considered in this study. Water depths in metres.

easy to date accurately because of the high resolution of the biostratigraphic time scale compared with that of the Mesozoic period (e.g. Berggren *et al.* 1995). The short duration of the nannofossil and foraminiferal zones allows tectonic events identified from bore-holes or in interpreted seismic sections to be pinpointed to within *c.* 0.5 millions of years. The modest thickness of the sedimentary cover further allows drilling to readily sample the pre-rift basement, something that is rare in the heavily sedimented margins of the Atlantic.

In this work we consider three parts of the extensional system in an attempt to see how extensional deformation changes during break-up. We chose to compare interpreted seismic lines from the South China margin, the Beibu Gulf Basin and the Nam Con Som Basin because they were extended at similar times and because the South China margin and Beibu Gulf Basin (and possibly the Nam Con Som Basin) are composed of the same pre-existing continental crust of the South China block. The aim is to examine differences in the style of extension between that seen in intracontinental 'failed' rifts, in rifts on the verge of break-up, and in those that have culminated in sea-floor spreading.

Geological setting and previous work

Although extension in the South China Sea is believed to have started in Late Cretaceous–Early Paleocene time (*c.* 75–60 Ma; Schlüter *et al.* 1996), this did not advance into sea-floor spreading until Oligo-Miocene time (anomaly 11, *c.* 30 Ma; Taylor & Hayes 1980; Lu *et al.* 1987; Briais *et al.* 1993). Extension was focused in the location of a pre-existing continental arc along the south coast of China (Jahn *et al.* 1976; Hamilton 1979; Sewell & Campbell 1997), which ceased activity after 140 Ma (Davis *et al.* 1997). However, new ^{40}Ar/^{39}Ar ages of granites cored from the basement of the Pearl River Mouth Basin now suggest that some magmatism continued into Late Cretaceous–Paleocene time (Lee *et al.* 1999) It seems unlikely therefore that the lithosphere had reached thermal equilibrium before the start of extension (compare McKenzie 1978). The Nam Con Som Basin lies in the SW extreme of the South China Sea immediately ahead of the propagating sea-floor spreading ridge that marks the centre of the deep basin. Extension in this basin is dominantly of Eocene–Oligocene age, but the modest Miocene extension that has been noted on the South China margin at *c.* 14 Ma (Clift & Lin

1998) is more strongly present here (Matthews *et al.* 1997). The Nam Con Som Basin has been frozen at the stage of extension at which sea-floor spreading was about to occur. The Beibu Gulf Basin represents an earlier phase of extension, and is isolated from the COB by an intervening block of less extended crust, which is partly exposed on Hainan (Fig. 1). The Beibu Gulf Basin began to extend in Maastrichtian and Paleogene time (*c.* 75–60 Ma), but like the Pearl River Mouth Basin its greatest extension was in Eocene–Oligocene time, before the onset of sea-floor spreading (Webb 1992). The influence of the nearby Red River Fault system on the development of the Beibu Gulf Basin is not apparent away from the immediate vicinity of the major faults.

On the Chinese margin, extension before sea-floor spreading resulted in the generation not only of tilted fault blocks adjacent to the continent–ocean transition, but also of a series of sedimentary basins. These are separated from the main rift axis by blocks of less extended crust, most notably in the Pearl River Mouth Basin (Fig. 1; Edwards 1992). The driving mechanism for the regional extension is contentious. Rival theories advocate either shear as a result of the southward motion of Indochina relative to China (e.g. Tapponier *et al.* 1986; Peltzer & Tapponnier 1988) or northward subduction beneath the Philippines, driving a back-arc type of extension in the overriding plate (e.g. Taylor & Hayes 1980; Packham 1996). The resolution of this debate is outside the scope of this work and does not affect the conclusions presented.

Rift-related passive margin sediments have been studied in the Pearl River Mouth Basin, where they have been sampled by numerous petroleum exploration wells. The syn-rift sediments of the Pearl River Mouth Basin are typically continental facies and show a deepening upward into shelf siliciclastic sediments. Thin Maastrichtian sediments are present locally, succeeded by continental alluvial Paleocene to Early Eocene siltstones and sandstones (Su *et al.* 1989). From Early to Mid-Eocene time, dark lacustrine shales were deposited, followed by Early Oligocene coal-bearing swamp and littoral plain sediments. After the major extensional episode of Eocene–Oligocene time, sedimentation became submarine. Dating of the sediments is well constrained following marine transgression in Early Oligocene time, but is limited before that time as a result of the continental facies of the syn-rift deposits. Palynological methods provide a poorly defined Eocene–Oligocene age for these rocks in the

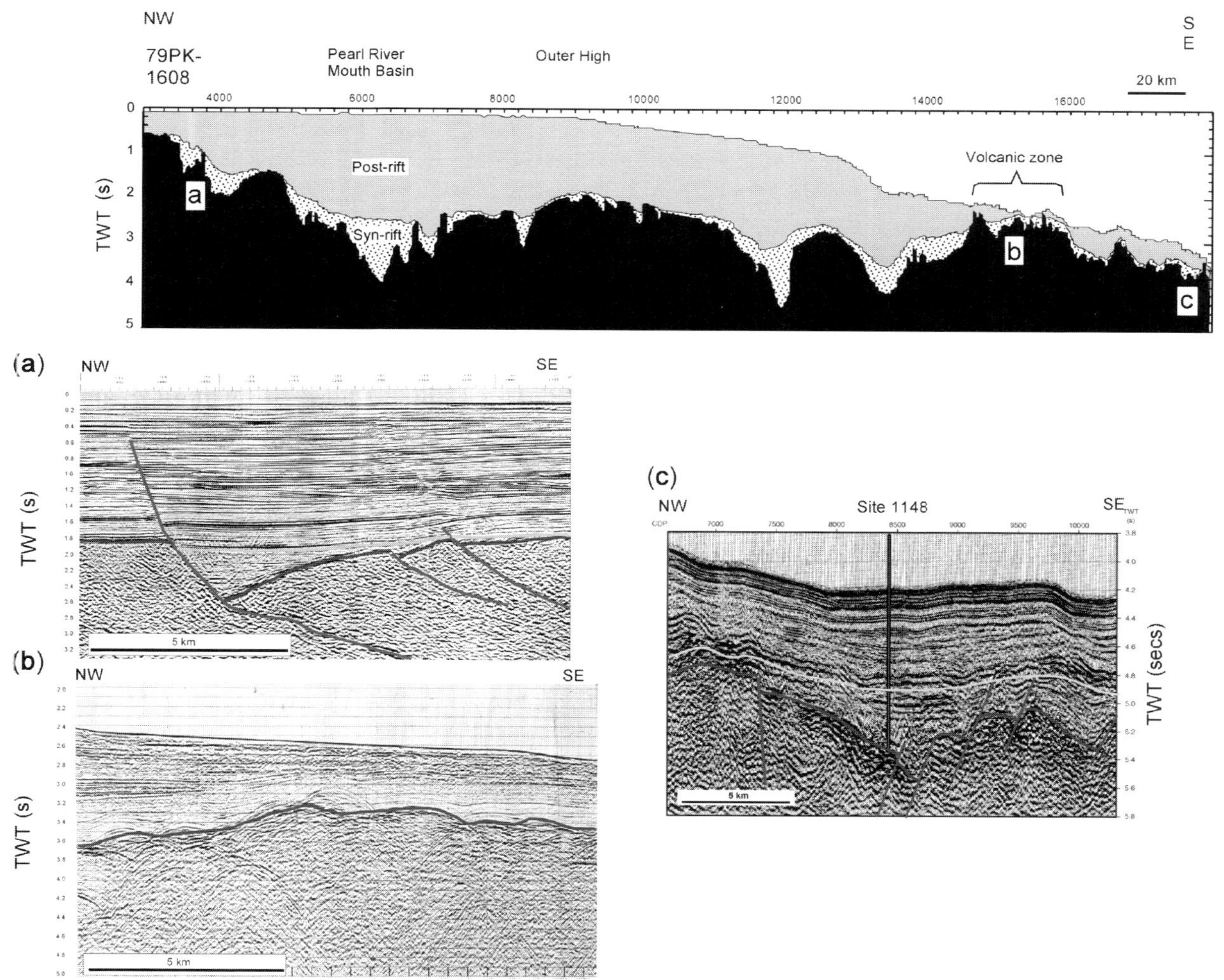

NW
79PK-1608
Pearl River Mouth Basin
Outer High
S
E
20 km
TWT (s)
Post-rift
Syn-rift
Volcanic zone
a
b
c
(a)
NW
SE
TWT (s)
5 km
(b)
NW
SE
TWT (s)
5 km
(c)
NW
Site 1148
SE
TWT (secs)
5 km

Pearl River Mouth Basin. Consequently, the start and duration of the syn-rift episode is not tightly controlled, although the end of rifting can be limited to earlier than *c.* 28 Ma in the Pearl River Mouth Basin based on nannofossil determinations from petroleum industry wells. The best constrained date for the end of active extension comes from Ocean Drilling Program (ODP) Site 1148 located close to the COB (Fig. 1), which indicates that active extension was complete by 28 Ma (Wang *et al.* 2000).

Earlier seismic studies of the margins to the South China Sea demonstrate that tilted fault blocks dominate the structure of the northern and southern conjugate margins (Edwards 1992; Feng *et al.* 1992; Hayes *et al.* 1995; Schlüter *et al.* 1996). The existence of several deep refraction seismic transects makes the South China margin a good place to examine the nature of strain accommodation during break-up (Hayes *et al.* 1995; Nissen *et al.* 1995*a*,*b*) because there are independent estimates of the total amount of crustal thinning across the margin. The refraction data in the east of the South China Shelf appear to show lower crust with a higher velocity ($>7.0 \text{ km s}^{-1}$) than that in the west. Along volcanic margins such velocities have been interpreted as gabbro underplated to the base of the crust during break-up (e.g. Hatton Bank, White *et al.* 1987; US East Coast, Holbrook & Kelemen 1993). However, locally such velocity anomalies have been attributed to serpentinized upper mantle (Hauser *et al.* 1995; Pickup *et al.* 1996; Minshull *et al.* 1998). The lack of the seaward-dipping reflectors that characterize volcanic margins (Hinz 1981) means that South China may not be a classic volcanic margin, but the suggestion of underplating does raise the possibility of it being intermediate in terms of magmatism between the end-member examples described to date. In this paper we present information suggesting that although it is not magmatic in the style of the NE Atlantic, South China may have more syn-rift magmatism than the Iberia-type non-volcanic margin.

Structure of the continent–ocean boundary

The COB in the South China Sea spans a broad zone of rift basins across the shelf before passing rapidly into the well-ordered fabric of the magmatic oceanic crust. A representative cross-section is shown in Figure 2, a line drawing of a regional petroleum seismic profile, which runs out to close to the COB in the vicinity of the ODP Leg 184 sites. This figure shows the broad-scale sub-division into syn-rift, post-rift and acoustic pre-rift basement. On the shelf the structure is dominated by relatively low-angle, sometimes listric, normal faults forming half-graben within the overall Pearl River Mouth Basin (Fig. 2a). These structures show the normal, classic, rotated syn-rift geometries, which almost fill the available syn-rift accommodation space. The Pearl River Mouth Basin is separated from the slope by an outer structural high in the basement (Fig. 2). Although much of the slope is also composed of rotated blocks, there is a zone *c.* 25 km across, just inboard of the COB, where the basement has a rubbly seismic character and no major faulting. This basement, an example of which is shown in Figure 2b, seems to take the form of a volcanic construct, albeit usually buried under post-rift strata. If these features are volcanic rocks, they clearly penetrate the continental crust because they are separated from the oceanic crust by another zone of tilted continental blocks (Fig. 2c), which were drilled into the syn-rift sediment fill at ODP Site 1148. Clear magnetic anomalies are not seen in the crust until south of the tilted block morphology. The volcanic zone is seen in many of the industrial seismic profiles examined that show that part of the margin, although it is not seen further west, in the vicinity of Hainan, where the rifted margin is also narrower (Clift & Lin, unpublished data). As such, there may be a correlation between the volcanic features and the high-velocity crust of Hayes *et al.* (1995). There is no evidence for any structure that looks analogous to the peridotite ridge of the Iberia margin in any profiles. Therefore, although there is much more similarity between South China and Iberia than the North Atlantic volcanic margins, there are important differences that point to a magmatic component to the break-up, at least in the central and eastern parts of the basin.

Methods

In this study we examine the 2D subsidence histories of the South China margin, the Beibu

---

Fig. 2. Line drawing of regional multichannel seismic line PK79-1608. Insets show the seismic character of the extension in different parts of the margin. (**a**) Listric half-graben within the Pearl River Mouth Basin; (**b**) rubbly volcanic basement landward underlying the middle slope; (**c**) high-angle normal faulted continental basement along the continent–ocean boundary.

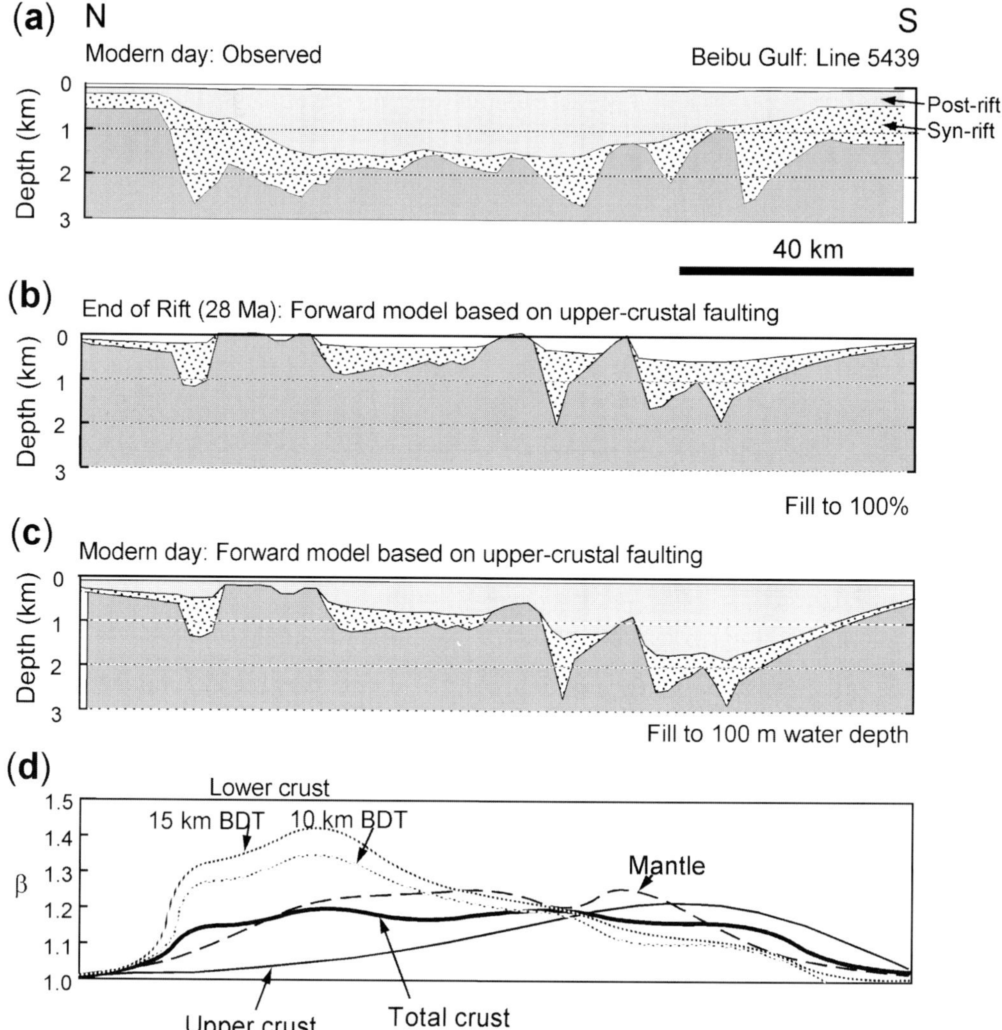

Fig. 3. (a) Line drawing of regional multichannel seismic line 5439 of the Beibu Gulf Basin. (b) Forward extensional model of the basin at the end of active extension, assuming extension in the lower crust equals that seen in the brittle faulting. Sedimentation is set to fill the basin to 100 m of water; Te is 3 km; erosion is 80% of subaerial topography. (c) Forward model showing predicted post-rift sedimentation after 32 Ma. (d) Estimates of the lateral variation in the degree of extension at different levels in the continental lithosphere across the basin. BDT, brittle–ductile transition. Two curves are shown for the lower crust, indicating the effect of depth of faulting on the degree of lower-crustal strain.

Gulf Basin and the Nam Con Som Basin, to investigate how the extensional strain is distributed within the lithosphere. In particular, we focus on the relative amounts of extension in the mantle lithosphere, and the lower and upper continental crust.

Figures 2–4a show interpreted seismic profiles cross-cutting the structural grain in each basin. Depth conversion from the seismic travel-time section was performed using averaged stacking velocities from the multichannel seismic lines, cross-checked with well data and the sonobuoy data of Nissen *et al.* (1995*b*). The stratigraphy is simply divided into syn- and post-rift sequences. The sections were then forward modelled using the flexural-cantilever model (Kusznir & Egan 1989; Kusznir *et al.* 1991). In this process the amount of horizontal extension across each seismically recognizable fault was measured and a synthetic forward

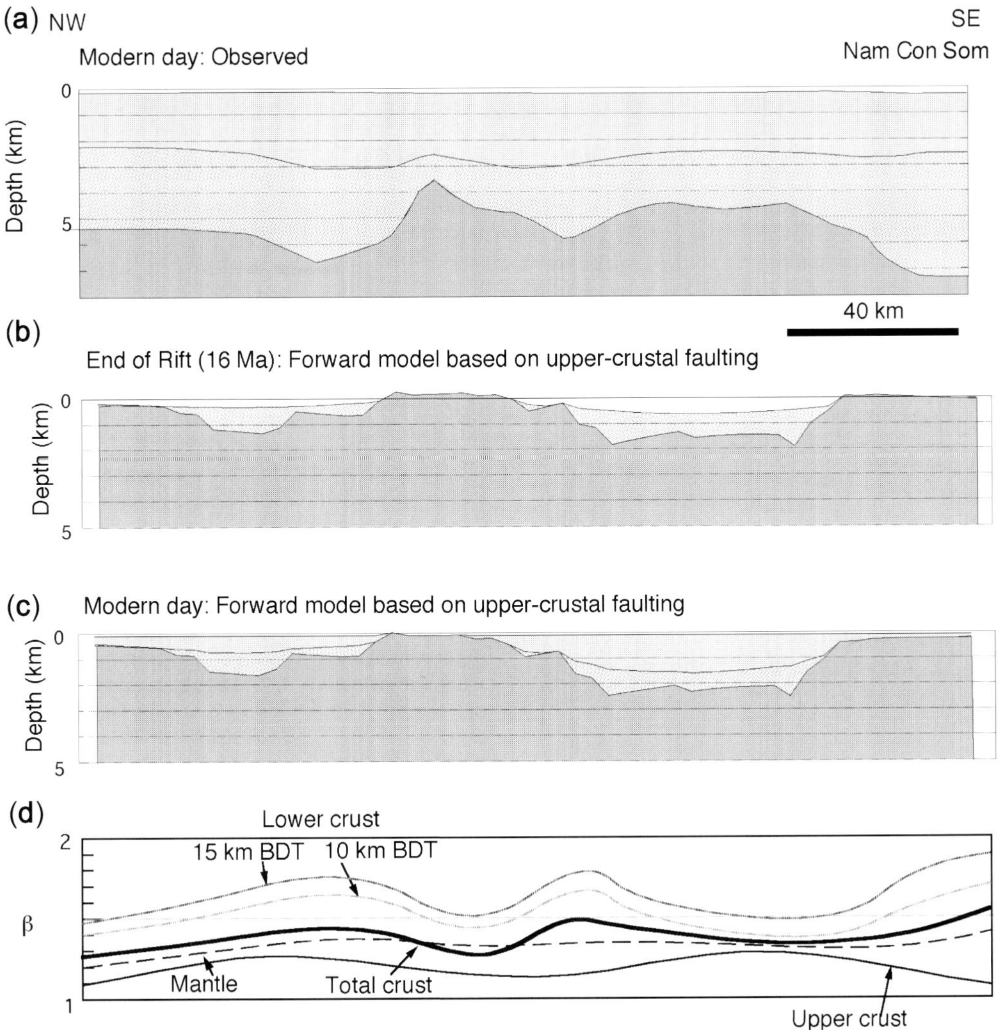

Fig. 4. (a) Line drawing of a multichannel seismic line through the Nam Con Som Basin (Matthews *et al.* 1997). **(b)** Forward extensional model of the basin at the end of active extension, assuming extension in the lower crust equals that seen in the brittle faulting. Sedimentation is set to fill the basin to 100 m of water; *Te* is 3 km; erosion is 80% of subaerial topography. **(c)** Forward model showing predicted post-rift sedimentation after 32 Ma. **(d)** Estimates of the lateral variation in the degree of extension at different levels in the continental lithosphere across the basin. BDT, brittle–ductile transition. Two curves are shown for the lower crust, indicating the effect of depth of brittle faulting on the degree of lower-crustal strain.

model of a rift basin was produced. This model assumes a pre-rift lithosphere in thermal equilibrium, which, as noted above, may not be the case in the South China Sea. Thus greater observed post-rift subsidence in the profiles compared with the model will partly reflect pre-rift heat and partly additional mantle extension beyond that predicted by the flexural-cantilever model. The modelling does not account for the effect of regional subduction on subsidence pattern (Lithgow-Bertelloni & Gurnis

1997). Wheeler & White (1997) suggested that because the crustal thicknesses derived from subsidence modelling of well data and from gravity modelling were similar then basin formation was basically controlled by rift tectonics and that dynamically driven subsidence caused by subduction is not a factor in the South China Sea.

The flexural–cantilever model assumes that strain is accommodated in brittle upper crust by simple shear along discrete faults, and that the ductile lower crust and mantle deform by pure shear, with deformation distributed over a wavelength of *c.* 100 km (Fig. 5). The model assumes that the total extension in the upper crust, lower crust and mantle lithosphere is equal. In this study upper-crustal extension is estimated from faults interpreted from the seismic lines. This is clearly a significant simplification, as Walsh *et al.* (1991) noted that as much as 40% of faulting is below seismic resolution. The challenge with correcting for this underestimate is exactly where to add the missing 40% to produce a more realistic model. It is likely that more extension is missing from areas of stronger deformation than in areas of no apparent deformation. Consequently, if the seismic profiles and models underpredict extension then a corrected version would involve greater faulting in areas of resolved faulting. The result of this correction would be to deepen existing modelled graben and increase footwall uplift of horsts. Such a correction would not cause subsidence in areas when none is predicted by the resolved faults.

The flexural–cantilever model has been applied to rift systems, including the North Sea (Roberts *et al.* 1993) and East Africa (Kusznir *et al.* 1995), as well as passive margin basins (e.g. Jeanne D'Arc basin, Kusznir & Egan 1989). In these cases reasonably good matches of the rift architecture and sedimentary fill have allowed the lateral distribution of extension to be studied (e.g. Kusznir *et al.* 1995). However, the flexural–cantilever model has been criticized for the fact that it only produces realistic models using lower elastic thickness (*Te*) of the continental lithosphere than are normally estimated by gravity and seismic data. For a given *Te* the model overestimates the amount of rift flank uplift, most notably in the Baikal Rift (van der Beek 1997). However, recent studies of global *Te* and seismicity patterns, by Maggi *et al.* (2000), have now suggested that *Te* has often been overestimated in traditional studies. Those workers suggested that in continental settings the strong part of the plate is limited to upper, seismogenic crust, and does not include the upper mantle, as is the case in the oceanic lithosphere. Roberts & Kusznir (1998) pointed out that alternative lithospheric necking models do not incorporate brittle faulting of the upper crust and, although useful on a large scale, necking models are inappropriate when considering the tilting and stratigraphic development. Because necking models of lithospheric deformation (e.g. Watts & Stewart 1998) cannot account for the footwall uplift of individual

faults they tend to underpredict rift flank topography.

The flexural–cantilever model does not account for upper-mantle strength, and so is least applicable to rifts in cold cratonic areas (e.g. Baikal, East Africa) and works best in areas where the crust and mantle are more decoupled, i.e. in a relatively 'hot' environment (e.g. Aegean Sea, Western USA). In the case of the South China Sea we are modelling the flexural unloading of a Paleogene rifted margin in crust that was in a supra-subduction zone setting during Mesozoic time, and consequently we anticipate a hotter than normal lithosphere that might be suitable for the flexural–cantilever model.

Despite its limitations, the ability of the model to allow for the flexural strength of the lithosphere when calculating the depth to basement makes it superior to simple 1D models that assume local isostatic compensation. We used an effective elastic thickness (*Te*) of 3 km (compare studies of the Jurassic North Sea;- Kusznir *et al.* 1991). A detailed gravity study of the margin is required to accurately determine the effective *Te*, and such work is beyond the scope of this present study. However, only low *Te* values (0–10 km) provide forward models that are reasonable matches to the observed basin architecture. Because the model is not very sensitive to the exact value of *Te* the results presented will be substantially correct, provided *Te* remains in this low range. Low *Te* is anticipated because gravity work on the South China margin where it is in collision with Taiwan does indicate low *Te* (10–13 km, Lin & Watts, pers. comm.). In addition, seismicity under the South China Shelf is shallower than *c.* 15 km (Harvard Catalogue), again suggestive of low strength (ductile rheology) in the lower crust and mantle lithosphere (Maggi *et al.* 2000).

Sedimentation was modelled to match that seen in the interpreted sections, and typically involved filling the available accommodation space to close to sea level. This approach is in accord with the shallow marine palaeobathymetry interpreted from post-rift sedimentary rocks cored in the Pearl River Mouth Basin. Erosion was set at 80% of the initial subaerial topography, also to provide the best possible fit to the observed topography of the basement along each profile.

Estimating extension within the lithosphere

Once a model section has been generated it can then be compared with the modern observed section, allowing mismatches to be identified and interpreted in terms of departures of the extension patterns from that input and assumed by the model. Estimates of changing extension with depth in the plate can be made by considering different aspects of the subsidence history. Extension in the upper crust is estimated by summing the extension of the seismically observable faults, which forms the input to the flexural–cantilever forward model. Extension of the mantle lithosphere is estimated through the amount of post-rift thermal subsidence, after correcting for the loading effects of the sediments and water, using the compaction parameters of Sclater & Christie (1980). Post-rift subsidence is normally considered to be purely thermally driven, i.e. caused by conductive cooling and thickening of the lithosphere alone (McKenzie 1978). Cooling of the crust represents a small fraction of this total effect, so that post-rift subsidence can be used as a proxy for mantle cooling and extension. As water

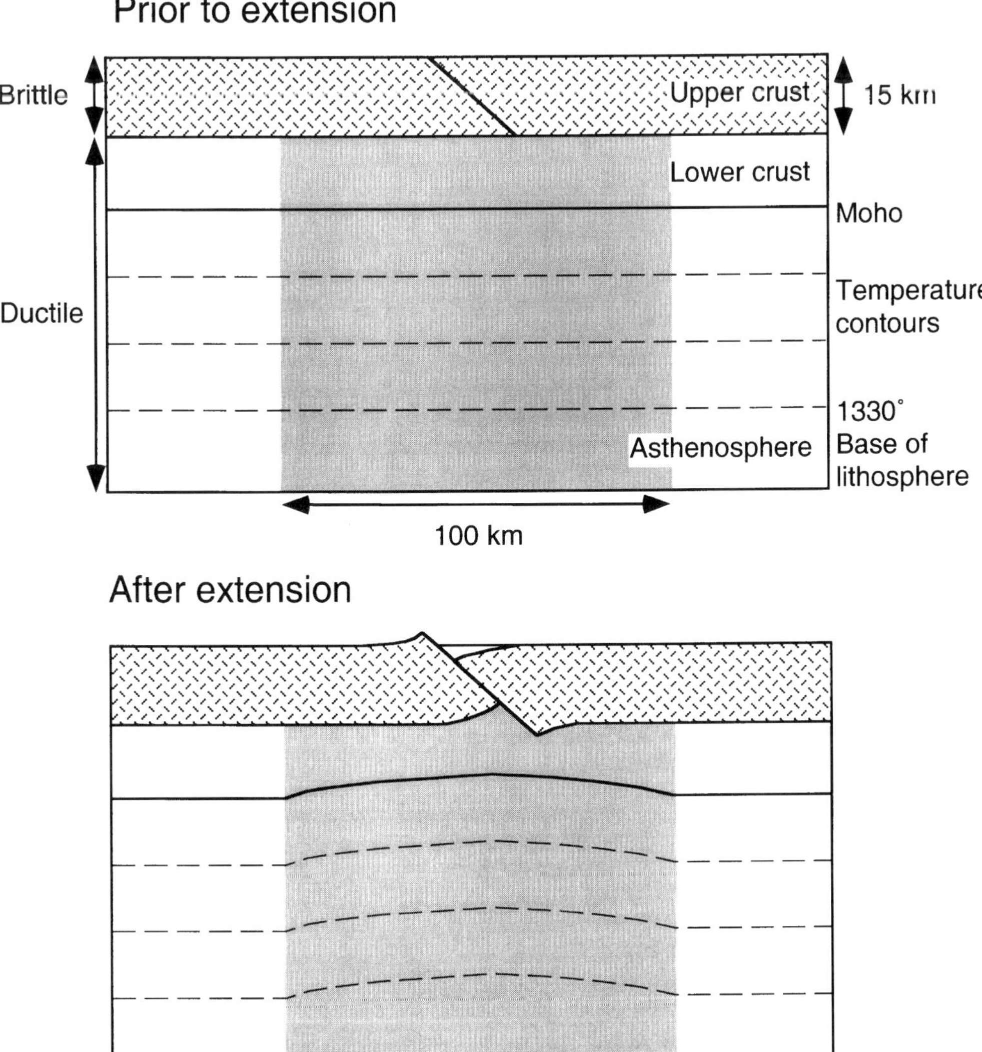

Fig. 5. A schematic representation of the flexural-cantilever model for lithospheric deformation showing assumptions of simple shear faulting in the brittle upper crust and pure shear in the lower crust and mantle (redrawn from Kusznir *et al.* 1991).

depth has been in the shallow shelf range since the end of rifting we use a value of 100 m as an average water depth. Uncertainty on the water depths will not seriously affect the mantle extension estimate because the sediment thicknesses are significantly greater than this value.

Total crustal extension is estimated from the amount of syn-rift subsidence, again after correcting for the sediment and water loading effect, as well as uplift driven by mantle thinning during rifting. Given the predicted background of mantle-driven uplift (calculated from the amount of post-rift subsidence) it is possible to calculate how much total crustal extension is need to produce the net amount of subsidence observed. The continental–shallow marine palaeobathymetries of the syn-rift strata in the basin provide confidence that the syn-rift extension estimates have low uncertainties.

Once total crustal extension has been derived from the total degree of accommodation space produced by rifting, the degree of lower-crustal extension may be calculated by subtracting the upper-crustal extension from the total crustal extension. The precise amount of lower-crustal extension will be determined by the original crustal thickness, assumed to be 32 km. This value is taken from the crustal thicknesses seen closest to the coast on the seismic refraction profiles of Hayes *et al.* (1995), as well as estimates for crustal thicknesses under southern China (Li & Mooney 1997). The base of the brittle upper crust is derived from the depth to which brittle faulting extends. For continental lithosphere with a quartz-dominated rheology and a geothermal gradient of $15-18\,^{\circ}\mathrm{C\,km^{-1}}$, the ductile–brittle transition is usually estimated as lying between 10 and 15 km (e.g. Zuber *et al.* 1986). The observed depth of global seismic activity, and especially in the modern South China area (e.g. Bufe *et al.* 1977), supports this range. Although seismicity is noted at >25 km in rifted cratons, such in the East African and Baikal rifts (e.g. Jackson & Blenkinsop 1993), in the case of the South China lithosphere its history as an active margin during Mesozoic time makes comparison with the cold, thick lithosphere of a craton inappropriate. We calculate extension in the lower crust for both 10 and 15 km depth to the ductile–brittle transition, to assess the range of reasonable extension values at this level.

Beibu Gulf Basin

Figure 3 shows the application of the forward model to BP/CNOOC line 5439 across the Beibu Gulf Basin. The rift section was forward modelled to account for 28 Ma of thermal subsidence (Fig. 3c), compatible with data from wells in the Beibu Gulf Basin (Su *et al.* 1989; Webb 1992). During the post-rift period the basin was filled to 100 m water depth to match the inferred palaeobathymetry for that time. The predicted basin architecture is then compared with the modern structure interpreted from the seismic profiles (Fig. 3a). This analysis shows that although there is a broad similarity there are important misfits between model and observation, most notably insufficient post-rift subsidence in the model in the northern half of the basin and excessive post-rift subsidence in the south.

The results of the extension calculations are shown in Figure 3d. It is clear that extension is not uniform within the basin: lower-crustal extension is highest in the north and least in the south, whereas the upper crust shows the opposite pattern. However, all depths of the plate show extension decreasing to the north and south ends of the profile. This suggests that the extension is operating in a self-contained, closed system, i.e. uniform but distributed differently across the basin. At any given point there are differences in the degree of extension with depth; however, the total amount of extension integrated across the basin is the same at all levels within the lithosphere.

Certainly the total crustal extension is very close to the mantle extension, so that preferential extension of the lower crust in the north of the basin is compensated by lower extension in the upper crust, whereas the reverse is true in the south of the basin. This observation is important to understanding the thermal state of the pre-rift lithosphere. Because the Beibu Gulf, like the other studied profiles, involves extension of thermally disturbed continental arc lithosphere, it is noteworthy that the crust and mantle extension is of similar magnitude. If the lithosphere had been much thinner than equilibrium before rifting then post-rift subsidence, and the mantle extension calculated from this quantity, would have been much higher than the crustal extension. That this is not so suggests that not much of the post-rift thermal subsidence is driven by inherited heat, in this basin or elsewhere in the South China Sea region. In practice, this means that subsidence discrepancies between model and observation are driven by departure of the rifting process from the model and are not primarily caused by inherited pre-rift heat.

Nam Con Som Basin

Figure 4a illustrates the simplified basin stratigraphy for the Nam Con Som Basin. The total stratigraphic thicknesses are very large here because of the large sediment flux from the Mekong River. The forward model is derived from an interpreted profile from Matthews *et al.* (1997), who noted that there is a significant, albeit small, extensional event in Miocene time. Unfortunately, the flexural-cantilever model employed here cannot model two or more events, so for simplicity we consider all the extension to be of Oligocene age, i.e. pre-28 Ma. The result of this approach is that there will be an overprediction of the amount of post-rift subsidence because there has been less time for post-rift subsidence to occur. The for-

ward model must then be considered to be a maximum estimate of the amount of subsidence that could have taken place given the amount of extension seen.

What is clear in the forward model is that the modelled basin that results from the observed brittle faulting and its ductile equivalent at depth (Fig. 4d) is much shallower than interpreted from the observed reflection data. This is despite the fact that the model assumes the older rift age and thus overpredicts subsidence. We particularly draw attention to the lack of subsidence of the central structural high and the margins of the basin predicted by the cantilever model. These predictions are derived from the simple observation that all faults dip away from the high and there is little extension

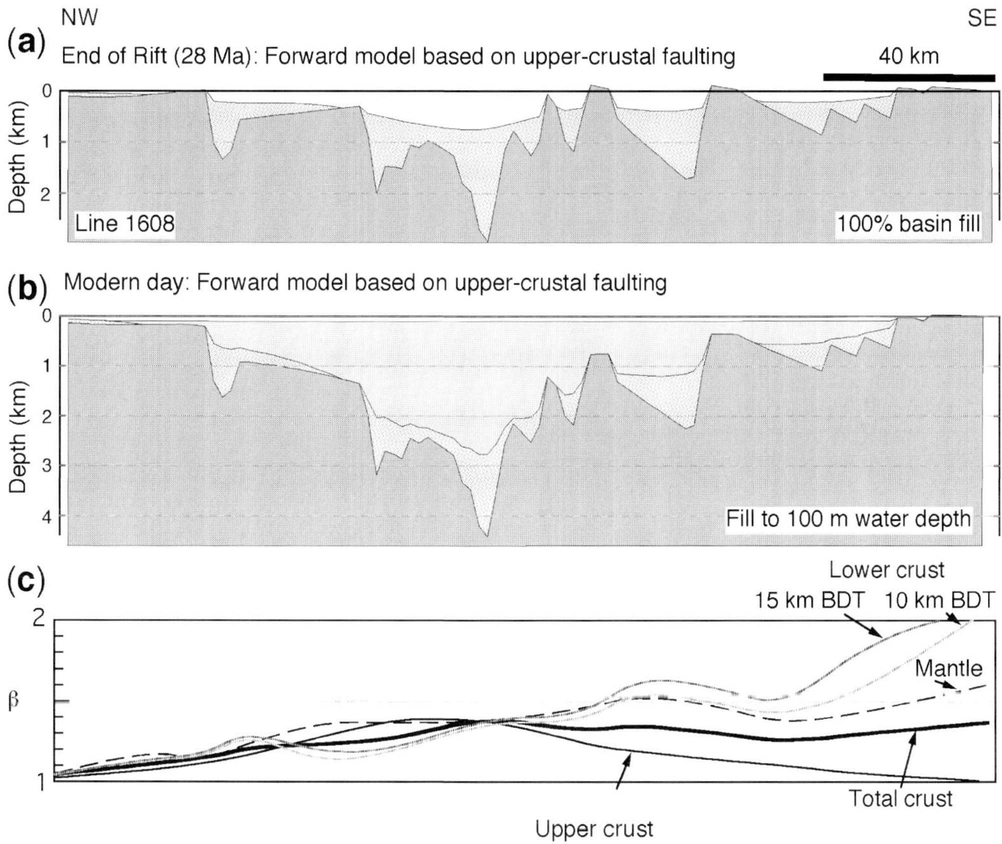

Fig. 6. (**a**) Forward extensional model of the Pearl River Mouth Basin along PK79-1608 showing rift structure predicted by the flexural–cantilever model of Kusznir *et al.* (1991). Syn-rift sedimentation is set to fill the basin before isostatic readjustment; *Te* is 3 km; erosion is 80% of subaerial topography. (**b**) Forward model showing predicted post-rift sedimentation after 32 Ma. (**c**) Estimates of the lateral variation in the degree of extension at different levels in the continental lithosphere across the basin. (Note very high lower-crustal extension at the seaward edge of the profile.)

noted in faults towards the edge of the profile. There is no clear mechanism for crustal thinning and producing the accommodation space seen without preferential thinning of the crust below the depth of faulting. Estimates of the extension at different levels show that whereas the crust and mantle have very similar values of extension the bulk of the crustal extension is in the lower crust. Unlike the Beibu Gulf Basin, mass is not conserved in this profile because the lower crust has been preferentially lost compared with the upper crust.

South China Slope

In the Pearl River Mouth Basin we chose to model the more landward part of the profile seen in Figure 2, because the palaeobathymetry is well constrained. On the continental slope water depth assumptions are not applicable, and even where well data are available, reliable palaeobathymetric data for deep-water sediments are notoriously difficult to derive and come with large uncertainties. Figure 6 shows that in the north, closest to the shore, the pattern of extension in the Pearl River Mouth Basin is such that the crust and mantle experience similar extension. The crustal strain is variably focused in the upper or lower crust, similar to the situation inferred for the Beibu Gulf Basin. However, further south a significant pattern emerges of upper-crustal extension approaching zero over the outer structural high, which in turn requires large lower-crustal extension to account for the observed subsidence. Mantle extension alone generates uplift because it is assumed that the cold, dense mantle root of a lithospheric plate is removed and replaced by warmer, less dense mantle. Thus subsidence of the outer high must indicate crustal extension. What is significant is that whatever depth of the brittle–ductile transition is chosen the amount of lower-crustal extension quickly exceeds that of the mantle lithosphere towards the southern edge. This profile does not share the mass conservancy seen in the Beibu Gulf Basin. However, it also differs from the Nam Con Som profile because total crustal extension is significantly less than that in the mantle lithosphere towards the southern end of the profile. Although some of this could represent additional subsidence as a result of heat inherited from the pre-rift period, the data from Beibu Gulf argue against this being important. If a correction could be applied for inherited heat then this would reduce the estimated mantle extension and only accentuate the preferential extension of the lower crust over other levels of the plate.

In the case of this profile, the proximity of the line to the seismic refraction profile of Nissen *et al.* (1995*a*) allows the crustal thickness estimate from the subsidence to be cross-checked against this independent method. Because Profile 1608 and the profile of Nissen *et al.* (1995*a*) do not perfectly overlie one another, a deep basin imaged by Nissen *et al.* (1995*a*), but not by Profile 1608, leads to a mismatch in the centre of the Pearl River Mouth Basin. However, elsewhere across the transect both subsidence and seismic refraction techniques provide a total crustal β of *c.* 1.4 over the outer structural high. Both of these estimates are more than the upper-crustal extension derived from shallow seismic reflection imaging of faulting. The correspondence between subsidence and seismic refraction-derived estimates of whole-crustal extension provides confidence to the modelling technique. Lower-crustal extension is much greater than that of the whole crust or mantle in the same region. We infer that the style of deformation associated with continental break-up is not simply an extreme version of that seen in the Beibu Gulf intracontinental rift.

Rift-to-drift transition

The nature of the change in extension from rifting to sea-floor spreading has been illuminated by the penetration of syn-rift sediments at ODP Site 1148. Unfortunately, the pre-rift basement was not recovered, so that the timing and rates of extension at the start of rifting cannot be determined. However, a rapid change in the rate of sediment accumulation at *c.* 28 Ma (nannofossil zone NP24) followed by a brief hiatus (lowermost part of NP25 and NN2 are missing) coincides with the top of the rotated reflectors that fan into the faults seen in Figure 2c (Fig. 7). This allows us to constrain active rifting close to the COB as being pre-28 Ma. The exact duration of the hiatus is unclear but does not seem to exceed 4 Ma (Wang *et al.* 2000). Given the setting, this hiatus surface can be termed the break-up unconformity within the scheme of Falvey (1974). Before the hiatus the sedimentation is bathyal, characterized by deep-water (>1500 m), calcareous mudstones with typical *Zoophycos* ichnofacies assemblages (Fig. 8a; Ekdale *et al.* 1984). The uncertainty in the water depths makes the conversion of the accumulation rates into tectonic subsidence data impossible in any meaningful way. The end of rifting is determined only from the

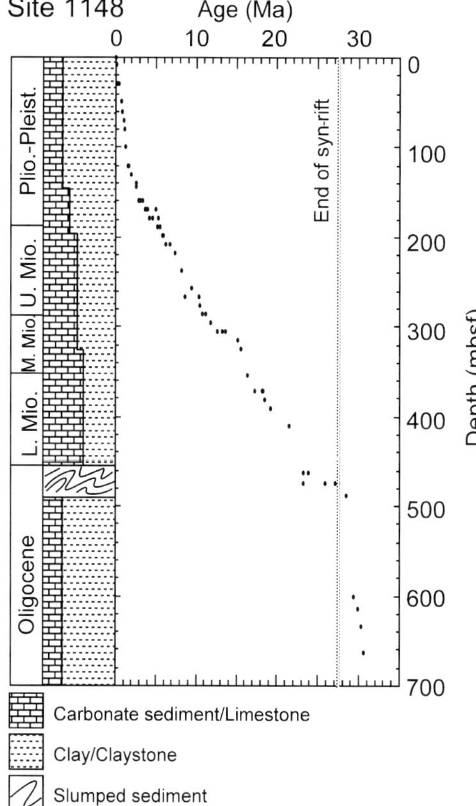

Fig. 7. Depth–age plot for ODP Site 1148 showing the rapid sedimentation related to active extension before and immediately following the onset of sea-floor spreading.

top of fanning reflectors in the seismic profile. Although rapid subsidence would be predicted during the syn-rift period, it is not possible to isolate this here because there is no trend to deepening water up-section. Rapid sedimentation by itself does not indicate active rifting. This implies that either the earliest, shallowest part of the section was not cored, or that the region of ODP Site 1148 was already in deep water at the onset of extension. It is noteworthy that although the Pearl River Mouth Basin lay close to sea level before rifting, there is evidence of a deep-water trough south of there. Deep-water Middle to Lower Eocene sediments have been identified from the Reed Bank on the south side of the South China Sea (Sampaguita-1 well; Taylor & Hayes 1980). If ODP Site 1148 also lay in this trough then continuous deep-water sedimentation would be predicted. Deepening during the active rift period would then not be

resolvable using sediment facies analysis or benthic foraminifer data.

Although the sedimentation during the syn-rift period is gradual the same cannot be said for that following the supposed break-up hiatus. The break-up is marked by the deposition of a slump deposit of *c.* 40 m thickness, showing a series of sedimentary features typical of soft sediment deformation, i.e. slump folding, water escape structures and intraformational breccias (e.g. Fig. 8b). Although the facies indicate marked downslope mass movement, it is noteworthy that the sediments are fine grained throughout the section. None of the sandstone that characterizes the syn-rift deposits in the Pearl River Mouth Basin is noted here, implying that this basin must be a very effective sediment trap. Up-section at ODP Site 1148 the deep-water carbonate and clay facies continue to the present day, but no further mass wasting is seen. The end of redeposition may reflect the cessation of faulting on the margin, and the loss of the potential for earthquake-triggered slumping events.

Heat flow

The nature of continental extension can further be examined through consideration of the regional present-day heat flow. New ODP sites provide important new heat-flow constraints on the thinned continental crust. At each well, three to five downhole temperature measurements were made *in situ* with the piston core temperature tool. In addition, a bottom-water temperature measurement was taken before coring. Original temperature records were analysed to establish the equilibrium temperature at depth. The estimated errors in equilibrium temperature vary from 0.2 to 0.4 °C, reflecting subjectivity in the specific section of the data selected for analysis and uncertainties in thermal conductivity. Depth errors are of the order of ±0.5 m. The observed heat flow is shown in Table 1.

At each site where a heat-flow measurement was made we have calculated the predicted heat flow assuming the 1D, uniform extension model of McKenzie (1978). We calculated the sediment-unloaded depth to basement at each site, using the seismic profiles collected over each site (Wang *et al.* 2000), to estimate the thickness of the sedimentary cover. Time sections were converted to depth using the interval velocities derived during processing of the multichannel seismic data (Table 1). Assuming that rifting ended at 28 Ma, this allows us to estimate the degree of extension and the pre-

Fig. 8. Core photograph showing (**a**) deep-water clays of the Oligocene syn-rift sediments at ODP Site 1148, and (**b**) an example of the overlying slump deposits. These latter deposits represent a break-up unconformity along the South China margin.

dicted modern heat flow using the equations of McKenzie (1978; Fig. 9a). The predicted and measured heat-flow values are compared in Figure 9b. It may be seen that several of the sites plot close to the predicted values, with one, ODP Site 1145, being marginally warmer. Conversely, ODP Site 1144 is much cooler than predicted. Deviations from the model may be in part due to the non-uniform extension revealed by the subsidence modelling in the previous sections. Preferential extension of the lower compared with the upper crust, for example, will result in higher long-term heat flow, as the most radioactive and more thermogenic elements are preferentially concentrated in the upper crust. Unlike uniform extension these will not be significantly lost if extension is focused in the lower crust. Nissen *et al.* (1995a) reported higher than expected heat flow on the South China margin, which these data only weakly support. Those workers suggested a thin lithosphere, a highly radioactive crust or magmatic underplating to account for their high values.

Rapid sedimentation may be an important method of reducing heat flow, as the blanketing of the margin by a cool, low-conductivity layer of sediment will reduce heat flow. It is noteworthy that the lowest heat-flow measurement was taken at ODP Site 1144, where sedimentation was most rapid, causing a lack of thermal equilibrium locally. ODP Site 1146, which had lower than predicted heat flow, also experienced high Pleistocene sedimentation rates. However, looking at average rates of sedimentation to the depth of the deepest heat-flow measurement, it may be seen that ODP Site 1146 is exceeded by ODP Site 1145. In contrast to ODP Site 1146, heat flow at ODP Site 1145 exceeds the prediction from the uniform extension model. This may indicate that ODP Site 1145 would have even higher heat flow were it not for the insulating effects of the sediment cover, similar to the conclusion of Nissen *et al.* (1995a). The uniform extension model would thus be inappropriate to describe the thermal evolution at this site and other parts of the South China margin.

Discussion

The data presented above provide an image of a rifted margin somewhat different in its devel-

Table 1. *Heat-flow data for the South China Sea taken during ODP Leg 184*

Site	Water depth (m)	Depth of deepest measurement (mbsf)	Rate of Pleistocene sedimentation (m Ma^{-1})	Average sedimentation rate to depth (m Ma^{-1})	Observed temperature gradient (deg km^{-1})	Calculated heat flow (mW m^{-2})	(hfu)	Predicted heat flow (mW m^{-2})
1143	2830	145	54	47	86	68.8	1.64	68.5
1144	2050	149	500	900	24	19.2	0.46	56.8
1145	3190	96	88	125	90	72.0	1.72	68.5
1146	2093	150	106	115	59	47.2	1.13	59.4
1148	3232	162	79	58	83	66.4	1.58	69.0

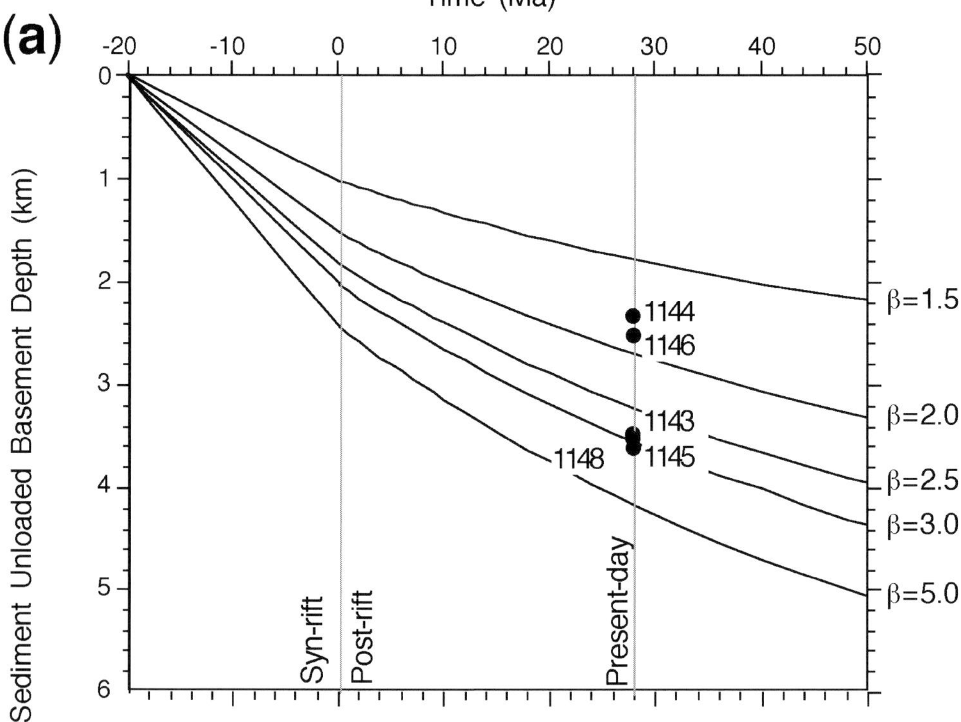

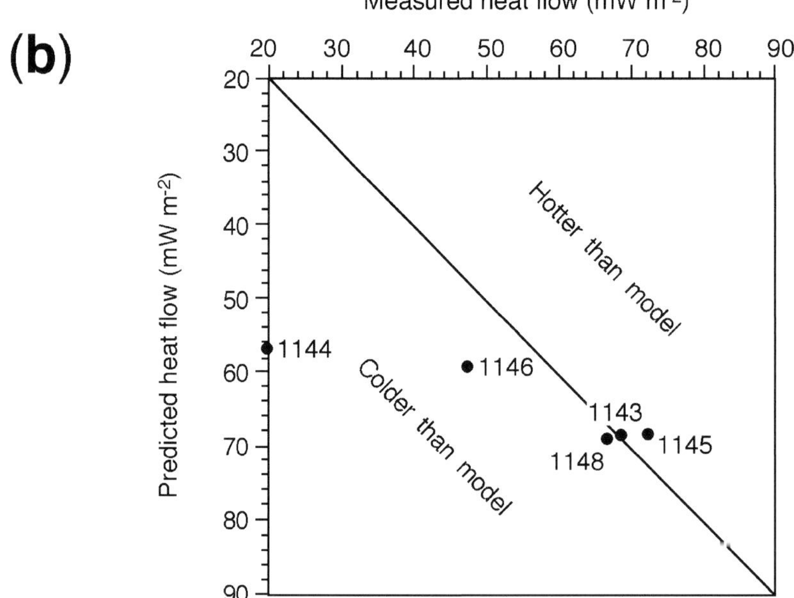

Fig. 9. (**a**) Diagram showing the sediment unloaded depths to basement at each of the ODP drill sites and the predicted uniform extension of the crust and mantle lithosphere. (**b**) Predicted heat-flow values v. measured values at each drill site. (Note lower values correspond to sites of high Pleistocene sedimentation.)

opment from the classic examples of the Central and North Atlantic. Break-up is seen to generate an unconformity surface, despite the lack of subaerial exposure, which is related to mass wasting at the end of rifting. It is not at present clear what tectonic process would trigger such mass wasting. If this redeposition is simply a slope stability effect caused by sediment accumulation on the slope, oversteepening and finally gravitational collapse, then there is no reason that it should not reoccur higher in the section. Its temporal association with the break-up is noteworthy.

Rift magmatism

The occurrence of rift-related magmatism penetrating the rotated fault block structure of the continental slope is very different from the normal non-volcanic margin structure. However, the great width of the extensional deformation and the lack of regionally developed seaward-dipping, subaerially erupted volcanic rocks (compare East Greenland; Larsen & Jakobsdottir 1988) makes South China distinct from the typical volcanic margin. The presence of deep-water (bathyal) sediments within the syn-rift section at ODP Site 1148 clearly demonstrates that the volcanic rocks, which are mantled by the post-rift cover, must have been erupted in deep water and cannot be associated with major regional thermally driven uplift. Indeed, the seismic character of the volcanic feature in Figure 2b is reminiscent of submarine volcanism, and does not have the ordered layers of subaerial activity. Although ordered seismic reflectors may be observed in deep submarine volcanic flood lavas (e.g. Planke & Alvestad 1999), it is rare to find subaerial volcanic sequences with a totally chaotic seismic character. The evidence for syn-rift volcanism is supported by the sampling of volcaniclastic sediments and tephras at the base of the syn-rift section within the Pearl River Mouth Basin itself (e.g. Well Zhu-C, Fig. 1).

Comparison of the South China Sea data with the predictions of the Bown & White (1995) model for volcanism in finite duration rifts yields a surprising conclusion. We consider two locations on the continental margin: the Pearl River Mouth Basin, with its modest volcanism, and the lower continental slope, with the apparent large volcanic features. Extension at each location is conservatively estimated at $\beta = 1.0–1.8$ and $\beta = 2.5–3.5$, respectively, based on the data presented above. The duration of rifting is taken to last from Mid-Eocene time (*c.* 37–49 Ma) to the break-up at

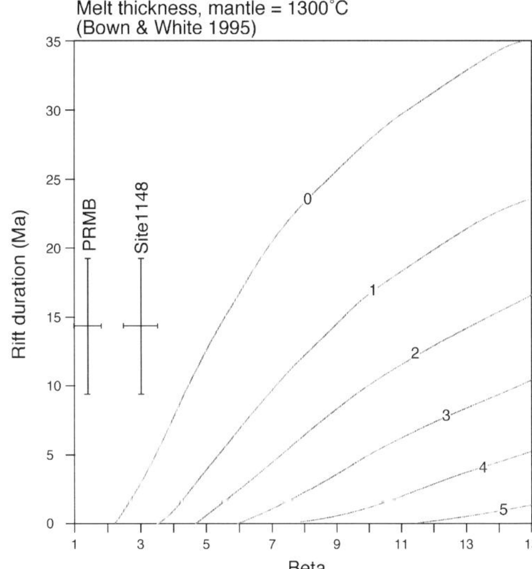

Fig. 10. Diagram from Bown & White (1995) showing the relationship between rift duration, extension and melt generation. Numbered contours indicate melt thickness in kilometres. Despite the uncertainties in the rift duration the model predicts no melting within the South China margin for normal upper-mantle temperatures. PRMB, Pearl River Mouth Basin.

28 Ma. This implies a rift duration of 9–19 millions of years. Figure 10 shows that at normal, ambient asthenosphere temperatures (1300 °C), no melting would be predicted in either location, which implies that the Chinese margin was hotter than thermally mature continental lithosphere. The recent dating of basement granites in the Pearl River Mouth Basin to Late Cretaceous to latest Paleocene time (Lee *et al.* 1999) indicates that residual heat from the earlier subduction system may have been important in the final break-up process. In addition, water, remnant from the earlier Mesozoic subduction episode and frozen into the base of the mantle lithosphere at the end of that time, may have been an important influence in increasing melt beyond that predicted by models such as that of Bown & White (1995). Stretching, adiabatic melting and remobilization of these fluids would act to increase the degree of melting by lowering the solidus without a rise in the ambient mantle asthenospheric temperature.

Rift architecture

The apparent evolution of the South China margin is from a wide rift into a narrow rift during the rift-to-drift transition. Figure 11 shows how

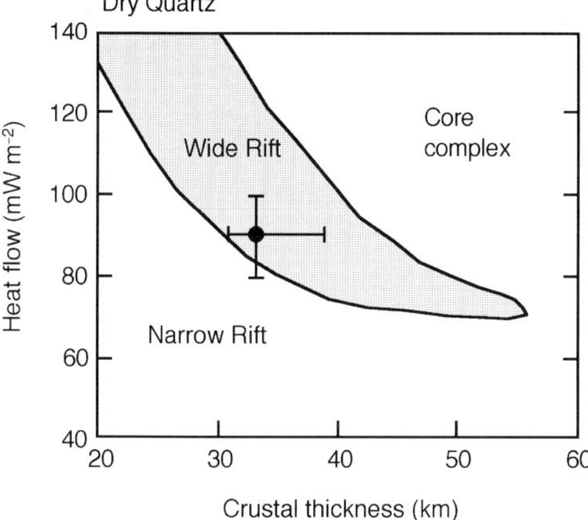

Fig. 11. Diagram from Buck (1991), which successfully predicts the wide-rift morphology of the South China Sea during the early syn-rift phase.

the South China Sea compares with the model of Buck (1991). Crustal thickness under the original arc crust was at least 32 km, as that is the seismically determined crustal thickness at the coast close to Hong Kong (Nissen *et al.* 1995*b*). In practice, the crust could have been thicker towards the centre of the arc, maybe as much as 40 km. The heat flow during break-up can be estimated from the modern heat flow at ODP Site 1148 (i.e. 69 mW m^{-2}), which would have been *c.* 90 mW m^{-2} at 28 Ma. This estimate assumes uniform extension (McKenzie 1978) and a β value of three calculated from the sediment unloaded modern depth to basement at that time. Within the measurement uncertainties these parameters place the South China margin within the wide-rift field of the Buck (1991) model, assuming a dry quartz-dominated crustal rheology. The results from South China thus support this model of continental extension. As the rift evolved, crust would become thinner, driving the extension into the narrow-rift field. This narrow rift presumably reflects the final break-up axis, along which the spreading centre propagated. It is noteworthy that at ODP Site 1148, located within a few kilometres of the COB, extension progressed only to $\beta = 3$. Although this is normal for volcanic rifted margins, much higher values typify non-volcanic margins, ranging from 4.4 (Goban Spur) to 12 (Iberia Abyssal Plain; Bowling & Harry 2001). Thus, as with the rift volcanism, the South China Sea bears some of the characteristics of a volcanic rifted margin, despite its lack of seaward-dipping volcanic rocks.

Strain accommodation

The patterns of strain accommodation across the Pearl River Mouth, Nam Con Som and Beibu Gulf Basins represent an important test of extensional models at passive margins. The similar crustal composition and ages, as well as timing of rifting, allow greater understanding of the processes that generate the differences in extension style between the intracontinental case, the incipient break-up and the fully rifted margin. In all profiles the estimated extension is rarely uniform with depth in any one place. In the Beibu Gulf and Nam Con Som examples, the total extension in the crust and mantle are close to being equal. In the Beibu Gulf the total extension at all levels of the plate decreases to zero towards the edge of the basin, demonstrating that this is a mass conserving system. The same cannot be said of the Nam Con Som Basin, where subsidence is significant at the edges and centre of the profile, but upper-crustal extension is small, implying significant lower-crustal and mantle lithospheric extension across much of the section. The lower-crustal volume is not conserved during rifting in the Nam Con Som Basin. This observation has recently been independently confirmed by modelling of original seismic profiles in the same region by Roberts & Kusznir (pers. comm.).

Like the Nam Con Som Basin, the Pearl River Mouth Basin does not appear to show mass conservation within the section. In this case, the landward side of the section shows low variable values of extension, with upper-crustal extension locally exceeding that in the lower crust. What is most significant here is that towards the continental slope the degree of lower-crustal extension accelerates, exceeding both mantle and upper crust. Lower crust has been preferentially removed from under the outer slope, in a manner similar to that predicted by finite-element modelling of the Baltimore Canyon Trough of the eastern US margin (Sawyer & Harry 1991). However, this style of extension is in contrast to the uniform mode proposed for the Pearl River Mouth Basin by Su *et al.* (1989) and Wheeler & White (1997), based on 1D subsidence analysis of well data. Westaway (1994) also suggested that extension here was not uniform with depth, but proposed that the lower-crustal material that was removed from the shelf now lies under southern China. Given that the lower crust is less dense than the mantle, we believe it is more likely that the lower crust was displaced in an oceanward direction. We therefore suggest that the lower-crustal material removed from along the

COB of the South China margin is now either under the conjugate Palawan margin or lies in an oceanward position, i.e. close to or beyond the COB.

An important question to answer is when this preferential extension began relative to the start of rifting and the onset of sea-floor spreading. Our work suggests that, as mass is not conserved in the Nam Con Som and Pearl River Mouth Basins, lower crust is preferentially thinned adjacent to, and just in advance of, a propagating spreading ridge. The loss of the lower crust therefore seems to be linked to the break-up process and distinguishes it from normal intracontinental extension. The flow of lower-crustal material during break-up may be considered a consequence of its quartz-dominated lithologies, which result in a weaker medium at typical lower-crustal temperatures and pressures compared with the olivine-dominated upper mantle (e.g. Kirby & Kronenberg 1987; Sonder & England 1989). In the thermally juvenile crust of the rifting South China Sea it is this unit that is most easily able to be preferentially extended and flow oceanward.

Preferential extension of the lower part of the South China margin would be consistent with simple shear. However, the preferential extension of the lower crust compared with the mantle there, and under both incipient conjugate margins in the Nan Con Som Basin, does not support the application of simple shear models to the passive margins (e.g. Wernicke 1985). Instead, our results are more in accord with the recent model of Driscoll & Karner (1998), who noted that many margins tend to have low degrees of upper-crustal extension compared with the total subsidence. Studies of the conjugate margin are now required to test whether the lower crust missing from under the South China margin is present there or if the two conjugate margins are symmetrical.

Conclusions

From the data discussed above we conclude that the South China margin represents an intermediate style of margin between the non-volcanic and volcanic end members known from the Atlantic. Although there is no evidence for hotspot influence, the existence of submarine volcanic-like features close to the COB appears to suggest a significant magmatic influence during the break-up process. This may reflect the influence of residual, pre-rift heat and water in the lithosphere boosting melting under the rift axis. Continental break-up occurs in deep water, and culminated in a hiatus and mass-wasting period

at 28 Ma, which correlates with the break-up unconformity of Falvey (1974). Extension close to the COB does not far exceed $\beta = 3$, before passing into oceanic crust, in contrast to the high β values measured in the 'non-volcanic' Iberia margin. The width and rate of stretching are consistent with the 'wide-rift' mode of extension defined by Buck (1991). Studies of faulting and subsidence show that extension is not uniform with depth, but also that simple shear is not readily applicable either. Preferential lower-crustal extension is interpreted to have preceded sea-floor spreading along the South China margin and in the Nam Con Som Basin ahead of the propagating sea-floor spreading axis, but was not seen in the intracontinental Beibu Gulf Basin because it does not juxtapose oceanic crust. Heat-flow values on the South China slope appear to be close to those predicted by uniform extension models, but after correction for sedimentation rates exceed those values. This is consistent with preferential extension in the lower crust and mantle lithosphere.

The multichannel seismic data used in this study were released by an agreement with the Chinese National Offshore Oil Company (CNOOC) and BP Exploration Operating Company Ltd. We are grateful to R. Miller and D. Roberts at BP Exploration for their help in providing data for this study, and R. J. Stern of UT Dallas for help in accessing the MSc thesis of T. Webb. A. Roberts, M. Davis and N. Kusznir are thanked for their help with the STRETCH and FLEX-DECOMP software packages. The paper was improved thanks to comments from M. Davis, B. Whitmarsh and an anonymous reviewer. This work was supported by the US Naval Oceanographic Office, Office of Naval Research, the National Science Foundation, JOI/USSAC and the Woods Hole Oceanographic Institution. This paper is WHOI Contribution 10274.

References

BERGGREN, W.A., KENT, D.V., SWISHER, C.C. & AUBRY, M.P 1995. A revised Cenozoic geochronology and chronostratigraphy. *In*: BERGGREN, W.A., KENT, D.V., AUBRY, M.P. & HARDENBOL, J. (eds) *Geochronology, Time Scales and Global Stratigraphic Correlation.* Society of Economic Paleontologists and Mineralogists, **54**, 129–212.

BOILLOT, G., BESLIER, M.-O., KRAWCZYK, C.M., RAPPIN, D. & RESTON, T.J. 1995. The formation of passive margins: constraints from the crustal structure and segmentation of the deep Galicia margin, Spain. *In*: SCRUTTON, R.A., STOKER, M.S., SHIMMIELD, G.B. & TUDHOPE, A.W. (eds) *The Tectonics, Sedimentation and Palaeoceanography of the North Atlantic Region.* Geological Society, London, Special Publications, **90**, 71–91.

BOWLING, J.C. & HARRY, D.L. 2001. Geodynamic models of continental extension and the formation of non-volcanic rifted continental margins. *In*: WILSON, R.C.L., WHITMARSH, R.B., TAYLOR, B. & FROITZHEIM, N. (eds) *Non-volcanic Rifting of Continental Margins: a Comparison of Evidence from Land and Sea*. Geological Society, London, Special Publications, **187**, 511–536.

BOWN, J.W. & WHITE, R.S. 1995. Finite duration rifting, melting and subsidence at continental margins. *In*: BANDA, E., TORNÉ, M. & TALWANI, M. (eds) *Rifted Ocean–Continent Boundaries*. Kluwer, Dordrecht, 31–54.

BRIAIS, A., PATRIAT, P. & TAPPONNIER, P. 1993. Updated interpretation of magnetic anomalies and seafloor spreading stages in the South China Sea: implication for the Tertiary tectonics of southeast Asia. *Journal of Geophysical Research*, **98**, 6299–6328.

BUCK, W.R. 1991. Modes of continental lithospheric extension. *Journal of Geophysical Research*, **96**, 20161–20178.

BUFE, C.G., HORSH, P.W. & BURFORD, R.O. 1977. Steady-state seismic slip—precise recurrence model. *Geophysical Research Letters*, **4**, 91–94.

CLIFT, P.D. & LIN, J. 1998. Timing and distribution of extension prior to seafloor spreading on the South China margin. *EOS Transactions, American Geophysical Union*, **79**, 337.

DAVIS, D.W., SEWELL, R.J. & CAMPBELL, S.D.G. 1997. U–Pb dating of Mesozoic igneous rocks from Hong Kong. *Journal of the Geological Society, London*, **154**, 1067–1076.

DRISCOLL, N.W. & KARNER, G.D. 1998. Lower crustal extension across the Northern Carnarvon basin, Australia: evidence for an eastward dipping detachment. *Journal of Geophysical Research*, **103**, 4975–4991.

EDWARDS, P.B. 1992. Structural evolution of the western Pearl River Mouth Basin. *In*: WATKINS, J.S., FENG, Zhiqiang & MCMILLEN, K.J. (eds) *Geology and Geophysics of Continental Margins*. American Association of Petroleum Geologists, Memoirs, **53**, 43–52.

EKDALE, A.A., BROMLEY, R.G. & PEMBERTON, S.G. *Ichnology. Trace Fossils in Sedimentation and Stratigraphy*. Society of Economic Paleontologists and Mineralogists, Short Course, **15**.

ELDHOLM, O., SKOGSEID, J., PLANKE, S. & GLADCZENKO, T. 1995. Volcanic margin concepts. *In*: BANDA, E., TORNÉ, M. & TALWANI, M. (eds) *Rifted Ocean–Continent Boundaries*. Kluwer, Dordrecht, 1–16.

FALVEY, D.A. 1974. The development of continental margins in plate tectonic theory. *Journal of the Australian Petroleum Exploration Association*, **14**, 95–106.

FENG, Zhiqiang, MIAO, Wancen, ZHENG, Weijun & CHEN, Shengyuan 1992. Structure and hydrocarbon potential of the para-passive continental margin of the northern South China Sea. *In*:

WATKINS, J.S., FENG, Zhiqiang & MCMILLEN, K.J. (eds) *Geology and Geophysics of Continental Margins*. American Association of Petroleum Geologists, Memoirs, **53**, 27–41.

HAMILTON, W. *Tectonics of the Indonesian Region*. US Geological Survey, Professional Papers, **1087**.

HAUSER, F., O'REILLY, B.M., JACOB, A.W., SHANNON, P.M., MAKRIS, J. & VOGT, U. 1995. The crustal structure of the Rockall Trough; differential stretching without underplating. *Journal of Geophysical Research*, **100**, 3, 4097–4116.

HAYES, D.E.S., NISSEN, S., BUHL, P., DIEBOLD, J., YAO, Bochu, ZHENG, Weijun & CHEN, Yongqin 1995. Through-going crustal faults along the northern margin of the South China Sea and their role in crustal expansion. *Journal of Geophysical Research*, **100**, 22435–22446.

HINZ, K. 1981. A hypothesis on terrestrial catastrophes: wedges of very thick oceanward dipping layers beneath passive continental margins, their origin and paleoenvironmental significance. *Geologisches Jahrbuch*, **E22**, 3–28.

HOLBROOK, W.S. & KELEMEN, P.B. 1993. Large igneous province on the US Atlantic margin and implications for magmatism during continental break-up. *Nature*, **364**, 433–436.

HORSEFIELD, S.J., WHITMARSH, R.B., WHITE, R.S. & SIBUET, J.-C. 1994. Crustal structure of the Goban Spur rifted continental margin, NE Atlantic. *Geophysical Journal International*, **119**, 1–19.

JACKSON, J. & BLENKINSOP, T. 1993. The Malawi earthquake of March 10, 1989; deep faulting within the East African Rift system. *Tectonics*, **12**(5), 1131–1139.

JAHN, B., CHEN, P.Y. & YEN, T.P. 1976. Rb–Sr ages of granitic rocks from southeastern China and their tectonic significance. *Geological Society of America Bulletin*, **87**, 763–776.

KELEMEN, P.B. & HOLBROOK, W.S. 1995. Origin of thick high-velocity igneous crust along the U.S. east coast margin. *Journal of Geophysical Research*, **100**, 10077–10094.

KIRBY, S.H. & KRONENBERG, A.K. 1987. Rheology of the lithosphere: selected topics. *Reviews in Geophysics*, **25**, 1219–1244.

KUSZNIR, N.J. & EGAN, S.S. 1989. Simple-shear and pure-shear models of extensional sedimentary basin formation: application to the Jean d'Arc basin, Grand Banks of Newfoundland. *In*: TANKARD, A.J. & BALKWILL, H.R. (eds) *Extensional Tectonics and Stratigraphy of the North Atlantic Margins*. American Association of Petroleum Geologists, Memoirs, **46**, 305–322.

KUSZNIR, N.J., MARSDEN, G. & EGAN, S.S. 1991. A flexural cantilever simple shear/pure shear model of continental extension. *In*: ROBERTS, A.M., YIELDING, G. & FREEMAN, B. (eds) *The Geometry of Normal Faults*. Geological Society, London, Special Publications, **56**, 41–61.

KUSZNIR, N.J., ROBERTS, A.M. & MORLEY, C.K. 1995. Forward and reverse modelling of rift basin formation. *In*: LAMBIASE, J.J. (ed.) *Hydro-*

carbon Habitat in Rift Basins. Geological Society, London, Special Publications, **80**, 33–56.

LARSEN, H.C. & JAKOBSDOTTIR, S. 1988. Distribution, crustal properties and significance of seaward-dipping sub-basement reflectors off East Greenland. *In*: MORTON, A.C. & PARSONS, L.M. (eds) *Early Tertiary Volcanics and the Opening of the NE Atlantic*. Geological Society, London, Special Publications, **39**, 95–114.

LEE, T.Y., LO, C.H., CHUNG, S.L., LAN, C.Y., WANG, P.L. & LEE, J.C. 1999. Cenozoic tectonics of the South China continental margins and the opening of the South China Sea. *EOS Transactions, American Geophysical Union*, **80**, 1042.

LI, S. & MOONEY, W.D. 1997. Crustal structure of China from deep seismic sounding profiles. *EOS Transactions, American Geophysical Union*, **78**, 470.

LITHGOW-BERTELLONI, C. & GURNIS, M. 1997. Cenozoic subsidence and uplift of continents from time-varying dynamic topography. *Geology*, **25**, 735–738.

LU, W., KE, C., WU, J., LIU, J. & LIN, C. 1987. Characteristics of magnetic lineations and tectonic evolution of the South China Sea basin. *Acta Oceanologica Sinica*, **6**, 577–588.

MAGGI, A., JACKSON, J.A., MCKENZIE, D. & PRIESTLEY, K. 2000. Earthquake focal depths, effective elastic thickness, and the strength of the continental lithosphere. *Geology*, **28**, 495–498.

MATTHEWS, S.J., FRASER, A.J., LOWE, S., TODD, S.P. & PEEL, F.J. 1997. Structure, stratigraphy and petroleum geology of the SE Nam Con Som Basin, offshore Vietnam. *In*: FRASER, A.J., MATTHEWS, S.J. & MURPHY, R.W. (eds) *Petroleum Geology of Southeast Asia*. Geological Society, London, Special Publications, **126**, 89–106.

MCKENZIE, D.P. 1978. Some remarks on the development of sedimentary basins. *Earth and Planetary Science Letters*, **40**, 25–32.

MINSHULL, T.A., MULLER, M.R., ROBINSON, C.J., WHITE, R.S. & BICKLE, M.J. 1998. Is the oceanic Moho a serpentinization front? *In*: MILLS, R.A. & HARRISON, K. (eds) *Modern Ocean Floor Processes and the Geological Record*. Geological Society, London, Special Publications, **148**, 71–80.

NISSEN, S.S., HAYES, D.E., YAO, Bochu, ZENG, Weijun, CHEN, Yongqin & NU, Ziaupin 1995*a*. Gravity, heat flow, and seismic constraints on the processes of crustal extension: northern margin of the South China sea. *Journal of Geophysical Research*, **100**, 22447–22483.

NISSEN, S.S., HAYES, D.E., BUHL, P., DIEBOLD, J., YAO, Bochu, ZENG, Weijun & CHEN, Yongqin 1995*b*. Deep penetration seismic soundings across the northern margin of the South China Sea. *Journal of Geophysical Research*, **100**, 22407–22433.

PACKHAM, G. 1996. Cenozoic SE Asia: reconstructing its aggregation and reorganization. *In*:

HALL, R. & BLUNDELL, D. (eds) *Tectonic Evolution of Southeast Asia*. Geological Society, London, Special Publications, **106**, 123–152.

PELTZER, G. & TAPPONNIER, P. 1988. Formation and evolution of strike-slip faults, rifts, and basins during the India–Asia collision: an experimental approach. *Journal of Geophysical Research*, **93**, 15085–15117.

PICKUP, S.L.B., WHITMARSH, R.B., FOWLER, C.M.R. & RESTON, T.J. 1996. Insight into the nature of the ocean–continent transition off West Iberia from a deep multichannel seismic reflection profile. *Geology*, **24**, 1079–1082.

PLANKE, S. & ALVESTAD, E. 1999. Seismic volcanostratigraphy of the extrusive breakup complexes in the northeast Atlantic: implications from ODP/DSDP drilling. *In*: LARSEN, H.C., DUNCAN, R.A., ALLAN, J.F., BROOKS, K.et al. (eds) *Proceedings of the Ocean Drilling Program, Scientific Results, 163*. Ocean Drilling Program, College Station, TX, 3–16.

RESTON, T.J., KRAWCZYK, C.M. & HOFFMAN, H.-J. 1995. Detachment tectonics during Atlantic rifting: analysis and interpretation of the S reflection, the west Galicia margin. *In*: SCRUTTON, R.A., STOKER, M.S., SHIMMIELD, G.B. & TUDHOPE, A.W. (eds) *The Tectonics, Sedimentation and Palaeoceanography of the North Atlantic Region*. Geological Society, London, Special Publications, **90**, 93–109.

ROBERTS, A.M. & KUSZNIR, N.J. 1998. Comments on 'Flank uplift and topography at the central Baikal Rift (SE Siberia): a test of kinematic models for continental extension' by Peter van der Beek. *Tectonics*, **17**, 322–323.

ROBERTS, A.M., YIELDING, G., KUSZNIR, N.J., WALKER, I. & DORN-LOPEZ, D. 1993. Mesozoic extension in the North Sea: constraints from flexural backstripping, forward modelling and fault populations. *In*: PARKER, J.R. (ed.) *Petroleum Geology of NW Europe; Proceedings of the 4th Conference*. Geological Society, London, 1123–1136.

SAWYER, D.S. & HARRY, D.L. 1991. Dynamic modeling of divergent margin formation; application to the U.S. Atlantic margin. *Marine Geology*, **102**, 1-4, 29–42.

SCHLÜTER, H.U., HINZ, K. & BLOCK, M. 1996. Tectono-stratigraphic terranes and detachment faulting of the South China Sea and Sulu Sea. *Marine Geology*, **130**, 39–78.

SCLATER, J.G. & CHRISTIE, P.A.F. 1980. Continental stretching: an explanation of the post Mid-Cretaceous subsidence of the central North Sea basin. *Journal of Geophysical Research*, **85**, 3711–3739.

SEWELL, R.J. & CAMPBELL, S.D.G. 1997. Geochemistry of coeval Mesozoic plutonic and volcanic suites in Hong Kong. *Journal of the Geological Society, London*, **154**, 1053–1066.

SONDER, L.J. & ENGLAND, P.C. 1989. Effects of a temperature-dependent rheology on large-scale continental extension. *Journal of Geophysical Research*, **94**, 7603–7619.

SU, D., WHITE, N. & MCKENZIE, D. 1989. Extension and subsidence of the Pearl River mouth basin, northern South China Sea. *Basin Research*, **2**, 205–222.

TAPPONIER, P., PELTZER, G. & ARMIJO, R. 1986. On the mechanics of the collision between India and Asia. *In*: COWARD, M.P. & RIES, A.C. (eds) *Collision Tectonics*. Geological Society, London, Special Publications, **19**, 115–157.

TAYLOR, B. & HAYES, D.E. 1980. The tectonic evolution of the South China Basin. *In*: HAYES, D.E. (ed.) *The Tectonic and Geologic Evolution of Southeast Asian Seas and Islands*. Geophysical Monograph, American Geophysical Union, **23**, 89–104.

TAYLOR, B. & HAYES, D.E. 1983. Origin and history of the South China Sea basin. *In*: HAYES, D.E. (ed.) *The Tectonic and Geologic Evolution of Southeast Asian Seas and Islands; Part 2*. Geophysical Monograph, American Geophysical Union, **27**, 23–56.

VAN DER BEEK, P. 1997. Flank uplift and topography at the central Baikal Rift (SE Siberia): a test of kinematic models for continental extension. *Tectonics*, **16**, 122–136.

WALSH, J., WATTERSON, J. & YIELDING, G. 1991. The importance of small-scale faulting in regional extension. *Nature*, **351**, 391–393.

WANG, P., PRELL, W. & BLUM, P. *Proceedings of the Ocean Drilling Program, Initial Reports, 184*. Ocean Drilling Program, College Station, TX.

WATTS, A.B. & STEWART, J. 1998. Gravity anomalies and segmentation of the continental margin offshore West Africa. *Earth and Planetary Science Letters*, **156**, 239–252.

WEBB, T. 1992. *Extension and subsidence in the Beibu Gulf basin, South China Sea*. MSc thesis, University of Texas at Dallas..

WELSINK, H.J., SRIVASTAVA, S.P. & TANKARD, A.J. 1989. Basin architecture of the Newfoundland continental margin and its relationship to ocean crust fabric during extension. *In*: TANKARD, A.J. & BALKWILL, H.R. (eds) *Extensional Tectonics and Stratigraphy of the North Atlantic Margins*. American Association of Petroleum Geologists, Memoirs, **46**, 197–213.

WERNICKE, B. 1985. Uniform sense of normal simple shear of the continental lithosphere. *Canadian Journal of Earth Sciences*, **22**, 108–125.

WESTAWAY, R. 1994. Re-evaluation of extension across the Pearl River Mouth Basin South China Sea: implications for continental lithosphere deformation mechanisms. *Journal of Structural Geology*, **16**, 823–838.

WHEELER, P.J. & WHITE, N.J. 1997. Basin kinematics in South East Asia—tectonics and dynamic implications. *EOS Transactions, American Geophysical Union, Fall Meeting Supplement*, **78**, F658.

WHITE, R.S., WESTBROOK, G.K., BOWEN, A.N. & 7 OTHERS 1987. Hatton Bank (northwest UK) continental margin structure. *Geophysical Journal of the Royal Astronomical Society*, **89**, 265–272..

ZUBER, M.T., PARMENTIER, E.M. & FLETCHER, R.C. 1986. Extension of continental lithosphere: a model for two scales of Basin and Range deformation. *Journal of Geophysical Research*, **91**, 4826–4838.

Geodynamic models of continental extension and the formation of non-volcanic rifted continental margins

JERRY C. BOWLING & DENNIS L. HARRY

Department of Geological Sciences, University of Alabama, Box 870338, Tuscaloosa, AL 35487-0338, USA (e-mail: dharry@wgs.geo.ua.edu)

Abstract: Finite-element models of continental rifting show that formation of non-volcanic rifted margins may be the result of extension of a rheologically homogeneous crust. In such circumstances lithosphere necking does not become well developed until late in the rift history, delaying the onset of decompression melting in the asthenosphere until the last 10% of the rifting episode. This result is robust over a broad range of mantle temperatures, margin geometries, and extension rates. A cool mantle is not required, so the models are able to account for the production of oceanic crust at the end of amagmatic rifting episodes. The duration of the syn-rift melting episode is most sensitive to changes in extension rate, with higher extension rates leading to shorter periods of melt production. The duration of the rifting episode is controlled by extension rate and initial crustal thickness, and the geometry of the margin after continental break-up is controlled by initial crustal thickness and the distribution of pre-existing rheological heterogeneity in the crust. The model results are generally compatible with the dimensions and extension rates of rifted continental margins across the globe, and provide a particularly good fit to the evolution of the Iberia Abyssal Plain margin.

Continental extension is typically accompanied by widespread and long-lived magmatic episodes. The primary basaltic melts are usually thought to result from decompression melting of the asthenosphere (e.g. Foucher *et al.* 1982; McKenzie 1984; Keen 1985), with production of more evolved melts being attributed to fractionation of the primary mafic magmas or to crustal anatexis. Decompression melting of the lithospheric mantle may also contribute to melt production during the early stages of extension if the lithosphere has undergone a previous melt and/or fluid metasomatic episode (e.g. Daley & DePaolo 1992; Gallagher & Hawkesworth 1992; Leeman & Harry 1993). Magmatism usually begins within a few million years after the onset of extension and progressively increases in volume and becomes more localized until sea-floor spreading is established at the time of continental break-up (break-up is here considered to be the time at which new oceanic lithosphere, including both crust and mantle, begins to form). Variations in the volume and timing of melt production on most rifted margins and continental rifts probably result from differences in mantle potential temperature, the abundance of volatiles in the mantle, or in the dimensions and vigour of mantle convection beneath the rift axis.

The duration of the syn-rift magmatic episode typically varies between 5 and 35 Ma, and usually represents a significant portion of the rift history in both long-lived and short-lived rift systems (White *et al.* 1987; White & McKenzie 1989; White 1992). Widespread, prolonged, and often voluminous magmatic episodes are common on the vast majority of rifted continental margins, leading to their classification as volcanic rifted margins (VRMs). Characteristics of VRMs include the common presence of extrusive and intrusive rocks distributed throughout the region of extended continental crust, anomalously high seismic velocity indicative of underplated mafic rocks at the base of the extended continental crust, the presence of seaward-dipping seismic reflections that indicate thick sequences of subaerial lava flows in the ocean–continent transition zone, and unusually thick oceanic crust adjacent to the margin. Most rifted continental margins, both modern and ancient, are volcanic.

In contrast, on non-volcanic rifted margins (NVRMs) seaward-dipping seismic reflector sequences and high-velocity layers are lacking, syn-rift magmatic rocks are rare and of comparatively small volume, and the oldest oceanic crust adjacent to the margin is often anomalously thin (*c.* 5 km thick). Some NVRMs are

From: WILSON, R.C.L., WHITMARSH, R.B., TAYLOR, B. & FROITZHEIM, N. 2001. *Non-Volcanic Rifting of Continental Margins: A Comparison of Evidence from Land and Sea.* Geological Society, London, Special Publications, **187**, 511–536. 0305-8719/01/$15.00 © The Geological Society of London 2001.

also characterized by an ocean–continent transition zone underlain by a broad region of exhumed subcontinental mantle (Beslier *et al.* 1993; Chian *et al.* 1995*b*, 1999). NVRMs are relatively rare in comparison with VRMs, and the lack of melt production on these margins remains enigmatic. Well-documented examples include the Iberian margin, the Bay of Biscay, the Goban Spur, the southern Australia Otway margin, the Exmouth Plateau of NW Australia, and the southern Labrador Sea (Mutter *et al.* 1989; Beslier *et al.* 1993; Horsefield *et al.* 1994; Chian *et al.* 1995*a*; Garcia-Mondejar 1996; Finlayson *et al.* 1998).

The difference between VRMs and NVRMs is often attributed to differences in the temperature of the mantle (Bown & White 1995), differences in structural style (in particular, the presence or absence of lithosphere-scale detachment faults) (Lister *et al.* 1986), or differences in the vigour and scale of mantle convection beneath the rift axis (Hopper *et al.* 1992). However, several observations suggest that alternative hypotheses be considered. In particular, the presence of new oceanic crust adjacent to NVRMs requires ascent of relatively warm mantle to shallow depths at the end of the rifting episode. This is difficult to reconcile with models that invoke a cool mantle to inhibit synrift magmatism, and would appear to require a prolonged period of widespread syn-rift magmatism, regardless of whether extension is accommodated by large-scale ductile necking or by slip on a lithosphere-scale detachment fault. Models that invoke subdued mantle convection or conductive cooling of the mantle during rifting to inhibit magmatism predict that the formation of NVRMs should be correlated with either extension rate or margin geometry. However, the dimensions and extension rates on NVRMs and VRMs are rather variable, and have considerable overlap. Furthermore, some margins change abruptly from VRM to NVRM without a corresponding change in geometry (e.g. the SW Greenland margin) (Chian *et al.* 1995*a*), and the geometry of many margins changes abruptly without corresponding changes in magmatic character (e.g. the Iberian margin) (Pinheiro *et al.* 1996). This paper examines an alternative hypothesis in which the difference between NVRMs and VRMs is related to intraplate processes controlled by the pre-extensional rheological structure of the lithosphere. This model accounts for both volcanic and non-volcanic rifting without requiring major temporal or spatial changes in mantle temperature and is robust over the range of dimensions, duration, and rates of extension on various VRMs and NVRMs. Additionally, the model accounts for the transition from amagmatic rifting to the onset of sea-floor spreading on NVRMs and the relative rarity of NVRMs in comparison with VRMs.

Previous models for volcanic and non-volcanic rifting

As noted above, variation in the temperature of the mantle provides one explanation for differing amounts of melt production during rifting. This is consistent with the association of hotspot tracks with many large igneous provinces (LIPs) that are located near continental rifts and rifted margins (White & McKenzie 1989; Coffin & Eldholm 1992). Smaller amounts of melt produced on VRMs that are not associated with LIPs can be explained by greater distance between the rift axis and the mantle plume, which results in a lower mantle temperature that is similar to that beneath most oceanic rifts. This concept has been extended to explain low melt production on NVRMs through the proposition of anomalously cool mantle beneath the rift axis before rifting and/or by conductive cooling of the mantle during rifting. Both slow extension rates and episodic extension have been proposed as mechanisms to promote cooling of the mantle during rifting (Keen *et al.* 1994; Bown & White 1995; Tett & Sawyer 1996). These mechanisms are generally consistent with a lack of syn-rift melt production, but they do not explain the production of oceanic crust at the time of break-up. The creation of oceanic crust requires that relatively warm (*c.* 1300 °C) mantle ascend to shallow depths during the late stages of extension. Furthermore, rare earth element concentrations in alkaline basalts on NVRMs in the North Atlantic require mantle temperatures of *c.* 1450 °C, suggesting an anomalously hot mantle during rifting (Williamson *et al.* 1995). Ascent of mantle that is sufficiently warm to form new oceanic crust at the end of the rifting episode appears to require a preceding transitional period of syn-rift magmatism that increases in volume as the asthenosphere ascends (Harry & Bowling 1999). This is difficult to reconcile with the amagmatic character of the transition zone adjacent to most NVRMs (Whitmarsh *et al.* 1996). Another problem is presented by the temperature dependence of rock strength (Kusznir & Park 1984). Cooling of the mantle during rifting should create a strong region in the lithosphere, resulting in a shift in the locus of extension (Sonder & England 1989; Harry *et al.* 1993). Inhibition of melt production by cooling of the mantle implies that non-volcanic rifting is localized in regions where the lithosphere is strongest. Finally, geodynamic models suggest that conductive cooling of the mantle should be most pronounced when the amount of extension is small and the duration of rifting large (Bown & White 1995). However, many NVRMs have smaller combined amounts of extension and rift duration than most volcanic rifts and rifted margins (Table 1 and Fig. 1). For example, volcanic rifting on the US east coast and in western North America lasted about 50 Ma, and, in the case of western North America, had a relatively low extension rate of c. 15 mm a^{-1} (Klitgord *et al.* 1988; Ward 1991). In contrast, the final stage of rifting of the Iberian–Grand Banks margins probably lasted <25 Ma and had an extension rate of 22–26 mm a^{-1} (Srivastava & Tapscott 1986; Whitmarsh & Miles 1995; Whitmarsh *et al.* 1996).

An alternative model accounts for the lack of magmatism on NVRMs by slip along a lithosphere-scale detachment fault (Lister *et al.* 1986; Buck *et al.* 1988). This can result in large amounts of crustal thinning and even exhumation of the subcontinental mantle without requiring large amounts of thinning of the lithospheric mantle. As a result, ascent of asthenospheric mantle is minimized and syn-exten-

sional melt production is inhibited. The detachment model is supported by seismic reflection profiles that have imaged low-angle syn-rift faults on several NVRMs, including the Galicia Bank and Iberia Abyssal Plain (IAP) (Beslier *et al.* 1993; Krawczyk *et al.* 1996; Pickup *et al.* 1996; Reston 1996*a,b*). On the IAP, the detachment exhumed subcontinental mantle over a region of c. 100 km width (Whitmarsh & Sawyer 1993; Discovery Working Group 1998). Similar structures exposed in the Italian Alps document a nearly horizontal detachment slightly below the base of the continental crust (Froitzheim & Manatschal 1996; Hermann *et al.* 1997). Thus, it seems clear that detachment faulting is a common feature on NVRMs. However, several lines of evidence suggest that the formation of detachment faults is not a sufficient condition for the formation of NVRMs. First, continental break-up does not occur until the lithosphere has separated into two distinct tectonic plates, with new oceanic lithosphere forming between. In the classical lithosphere-scale detachment model, this requires an extended period of syn-rift magmatism that begins sometime after exhumation of the subcontinental mantle. The volume of melt increases with time until sufficient melt is produced to create new oceanic crust. The region of asthenospheric upwelling is broad, resulting in magmatism across much of the ocean–conti-

Table 1. *Duration and amount of extension on rifted margins (numbers in parentheses are references)*

Margin	Duration of rifting (Ma)	β at rift axis at break-up
Non-volcanic margins		
Iberia Abyssal Plain	21 ± 8 (1,2)	12.0 ± 3.0 (3)
Tagus Abyssal Plain	25 ± 5 (3)	5.4 ± 1.4 (3)
Galicia Bank	24.4 ± 4.5 (4)	8.5 ± 2.7 (3)
Biscay	20.5 ± 6.5 (5)	9.5 ± 2.5 (3)
Goban Spur	21.5 ± 5.5 (6)	4.4 ± 1.2 (3)
Otway	52.0 ± 5.0 (7)	5.0 ± 1.5 (7)
SW Greenland	44.5 ± 8.5 (8)	11.3 ± 3.6 (9)
Labrador	44.5 ± 8.5 (8)	11.3 ± 3.6 (9)
Volcanic margins		
Barents Sea	11 ± 3 (10)	2.0 ± 0.6 (10)
Lofoten	18 ± 2 (11)	2.2 ± 0.1 (12)
More	18 ± 2 (11)	1.7 ± 0.1 (12)
Voring Plateau	18 ± 2 (11)	2.1 ± 0.6 (12)
North Sea	60 ± 18 (10)	1.5 ± 0.5 (10)
Edoras Bank	57 ± 1 (13)	3.4 ± 0.6 (13)
Western Black Sea	30 ± 9 (10)	3.0 ± 0.9 (10)
Eastern Black Sea	8 ± 2 (10)	2.2 ± 0.7 (10)
Saudi Red Sea	5 ± 3 (10)	2.5 ± 1.0 (10)

1, Pinheiro *et al.* (1996); 2, Whitmarsh *et al.* (1990); 3, Bown & White (1995) and references therein; 4, Boillot & Winterer (1987); 5, Montadert *et al.* (1979); 6, Masson *et al.* (1985); 7, Finlayson *et al.* (1998); 8, Balkwill (1987); 9, Chian & Louden (1994); 10, Cloetingh *et al.* (1995) and references therein; 11, Skogseid *et al.* (1992); 12, Skogseid (1994); 13, Barton & White (1997).

                    J.C. BOWLING & D.L. HARRY

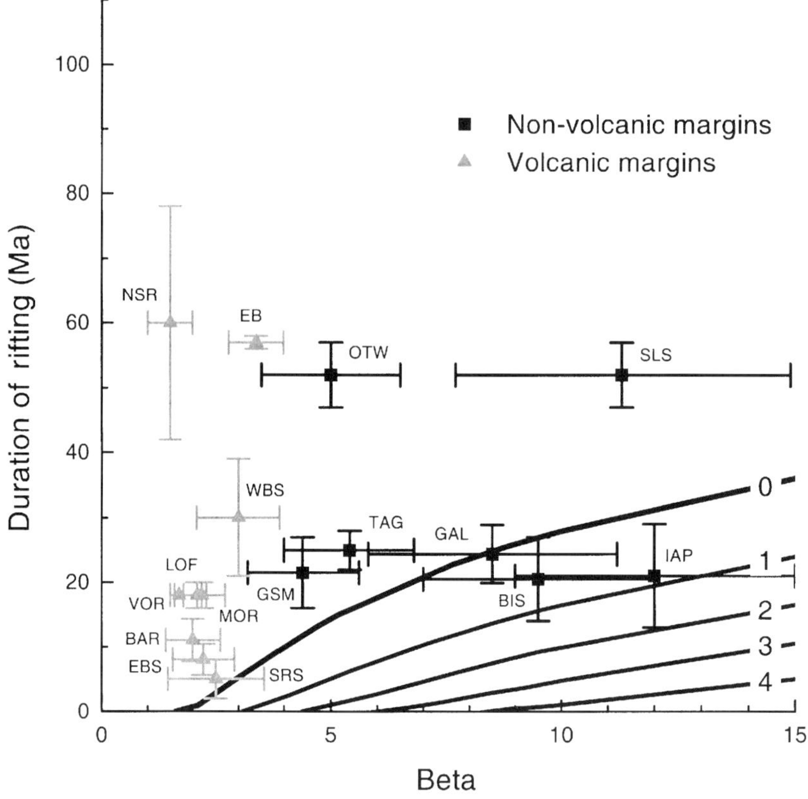

Fig. 1. Duration of rifting v. β on rifted continental margins. β is the ratio of the crustal thickness before and after extension. The duration of rifting is the time elapsed between the onset of the most recent stage of extension and initial sea-floor spreading (or, in the case of the North Sea, the cessation of extensional tectonism). See Table 1 for data sources. Continuous lines indicate thickness of melt (in km) produced at the end of the rifting episode according to the model of Bown & White (1995), assuming a mantle potential temperature of 1300 °C. This model predicts non-volcanic rifting in the upper left portion of the diagram and volcanic rifting in the lower right portion of the diagram, which is contrary to the data shown. BAR, Barents Sea; BIS, Biscay margin; EB, Edoras Bank; EBS, eastern Black Sea; GAL, Galicia margin; GSM, Goban Spur margin; IAP, Iberia Abyssal Plain; LOF, Lofoten; MOR, More; NSR, North Sea rift; OTW, Otway margin; SLS, Southern Labrador Sea; SRS, Saudi Red Sea; TAG, Tagus Abyssal Plain; VOR, Voring margin; WBS, western Black Sea.

nent transition zone (Fig. 2). Second, seismic imaging of detachment faults on rifted continental margins and in continental metamorphic core complexes, the metamorphic grade of rocks exposed in core complexes and in mantle rocks exhumed on extant and relict continental margins, and analogue laboratory models in layered brittle–ductile materials all suggest that detachment faults sole at or near the base of the crust (Allmendinger *et al.* 1983; McCarthy *et al.* 1991; Wilson *et al.* 1991; Applegate *et al.* 1992; Foster *et al.* 1992; Johnson & Loy 1992; Beslier *et al.* 1993; Brun & Beslier 1996; Pickup *et al.* 1996; Reston 1996*a,b*). In the absence of lithosphere-scale detachment faults, ductile necking remains the most likely form of deformation in the deeper lithosphere. Consequently, the timing and geometry of asthenospheric upwelling is not likely to differ greatly from that predicted by ductile stretching models, which generally predict formation of volcanic rifts unless the asthenosphere is anomalously cool. Finally, cooling histories, structural relationships, and the lack of thick sequences of syn-rift sediments deposited on exhumed mantle on the Iberian margin suggest that detachment faulting may have developed late during the extensional episode (Scharer *et al.*

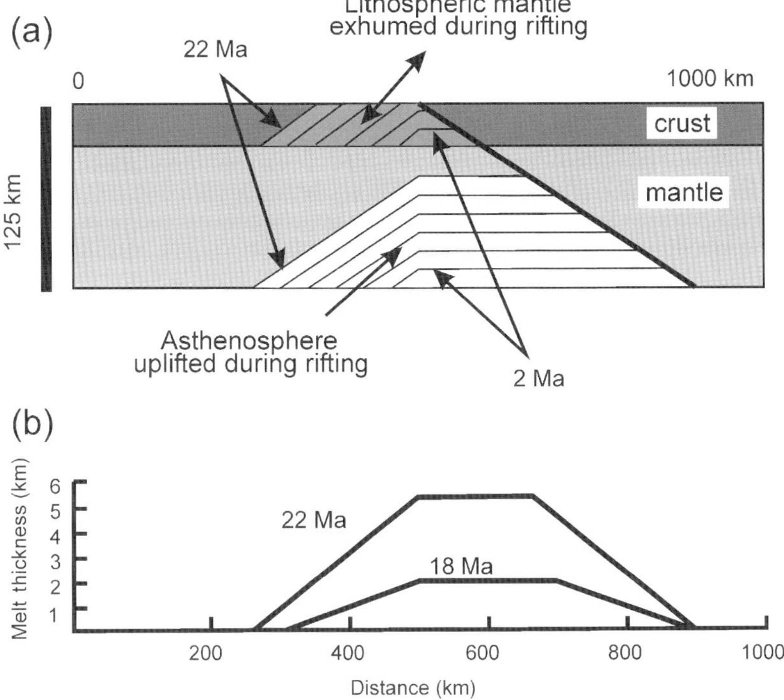

Fig. 2. Whole-lithosphere simple shear rifting. (**a**) Geometry of lithosphere during rifting. Bold line indicates detachment fault dipping 30°. Fine lines indicate the top of the lithospheric mantle and the top of the astheno-sphere 0, 2, 6, 10, 14, 18 and 22 Ma after the onset of extension. Footwall structure is determined kinemati-cally assuming 60° oblique simple shear. Modified from Buck *et al.* (1988). (**b**) Distribution of melt produced by decompression melting of the asthenosphere. Melt thickness is determined by integrating the percentage of melt produced in the asthenosphere over the thickness of the melt column at the times indicated. Initial melt-ing begins *c.* 15 Ma after the onset of extension. If break-up occurs when enough melt is produced to create ocean crust of 5 km thickness, then *c.* 7 Ma of syn-rift magmatism is predicted. This represents *c.* 30% of the rift history, and large thicknesses (>3 km) of syn-rift magmatic rocks are distributed over a region of 400 km width.

1995; Fuegenschuh *et al.* 1998). If so, ascent of the asthenosphere would not have been influ-enced by the presence of the detachment fault throughout most of the rift history. Thus, although detachment faults appear to have been important in determining the formation of rift-related structures within the crust on NVRMs, those that have been documented so far appear not to involve subcrustal lithosphere and may have developed late in the rift episode. The detachment faults therefore probably had little influence on the timing, distribution, and volume of melt generated during rifting.

A third explanation for differences in the amount of magmatism on continental margins focuses on convection patterns within the asthe-nosphere. If the rift is narrow or the transition from unextended to highly extended lithosphere is abrupt, small-scale convection cells should develop in the asthenosphere beneath the rift axis (Mutter *et al.* 1988). Small-scale convec-tion will not affect the timing of the onset of magmatism or the duration of the magmatic episode, but it may have a profound influence on the volume and distribution of melt that is produced during rifting. For narrow rifts or rifts with abrupt changes in lithospheric thickness, vigorous convection may cycle large volumes of fresh mantle material (i.e. mantle that has not previously undergone partial melting during the rifting episode) through the melt zone. This allows for melt extraction from a larger volume of mantle than would occur if the asthenosphere ascends vertically. Broad rifts or rifts with gra-dational changes in lithospheric thickness should have more subdued convection systems and therefore a lesser abundance of magmatic rocks. The apparent abrupt transition from magmatic rifting on the Cuvier segment of the NW Australian continental margin south of the Cape Range Fracture Zone (CRFZ) to amagmatic rifting on the Exmouth Plateau north of the CRFZ strongly supports this model (Lorenzo *et al.* 1991; Hopper *et al.* 1992). If the difference in melt production on the two margin segments is due to differences in asthenosphere temperature, then a similarly abrupt change in mantle temperature is required, and it must have persisted for the duration of rifting. Furthermore, the transition from hot to cooler mantle must have maintained its position relative to the overlying lithosphere as the Indian and Australian plates separated, requiring that either the lithospheric plate system was moving in a direction parallel to asthenosphere isotherms or that the asthenosphere was moving with the lithosphere. Small-scale convection is a more likely explanation. The seaward transition from unextended to highly extended continental crust is markedly more abrupt on the Cuvier margin than on the Exmouth Plateau, supporting the thesis that small-scale convection induced by narrow rifting resulted in enhanced melt production (Hopper *et al.* 1992). However, it is unclear if the model applies as well to other rifted margins. Except for the abundance of magmatic rocks, the dimensions and crustal structure of many VRMs and NVRMs are similar (White 1992; Skogseid 1994). Furthermore, some margins change from VRM to NVRM along-strike without major changes in the geometry of the margin (e.g. the SW Greenland margin) (Chian *et al.* 1995*a*). Conversely, the width of the Iberian margin changes abruptly along-strike, from a broad margin on the Iberia Abyssal Plain to a narrow margin off Galicia Bank, but both segments of the margin are non-volcanic.

In summary, although variations in mantle temperature are recognized to have a significant role in determining the amount of melt produced during extension, it is difficult to account for the transition from non-volcanic rifting to sea-floor spreading on NVRMs without requiring unusual temporal changes in mantle temperature. It is also difficult to account for abrupt changes from VRM to NVRM along-strike of many rifted margins without requiring unusual, similarly abrupt, spatial variations in mantle temperature. An abundance of evidence supports the existence of detachment faults on many NVRMs, but the transition from non-volcanic rifting to sea-floor spreading would require a broad region of crust subjected to syn-extensional magmatism if detachment faulting progresses to sea-floor spreading. Small-scale convection provides an attractive explanation for differences in melt production on many VRMs and NVRMs, particularly when abrupt along-strike changes in the magmatic nature of the rift correspond to similar changes in width of the rift zone. However, the widths of some NVRMs changes abruptly along-strike without obvious changes in magmatic character, and some rifted margins change from VRM to NVRM without pronounced changes in rift geometry. It seems clear, then, that the difference between volcanic rifting and non-volcanic rifting may not be solely, or even primarily, due to any of the explanations considered above.

Method

One commonality between the latter two models discussed above is the recognition that the pattern and rate of asthenospheric upwelling is controlled by the lithosphere-scale geometry of the rift. Many geodynamic models have shown that rift geometry is closely related to the pre-existing rheological structure of the lithosphere. Accordingly, a fourth model of rifting is investigated, which focuses primarily on the processes occurring within the lithosphere and how they affect the timing and distribution of melt production within the asthenosphere. Finite-element (FE) simulations of rifting are used to examine a range of initial and boundary conditions to identify the parameter space that results in formation of NVRMs without requiring anomalously low mantle temperatures. Successful models are required to restrict melt production to a narrow region near the site of initial sea-floor spreading, restrict the timing of melt production to the latest stages of extension, and result in a maximum melt volume that is consistent with creation of thin (*c.* 5 km thick) oceanic crust at the time of break-up. These characteristics of NVRMs should be produced without requiring anomalous asthenospheric temperatures and should be independent of the mantle convection pattern beneath the rift axis.

The models do not consider asthenosphere processes in detail: the partial melting process is evaluated using simple decompression melting relations and vertical ascent paths. Instead, the models focus on lithospheric processes, particularly extension rate, the degree of rheological heterogeneity in the lithosphere, and the pre-rift crustal thickness. Two classes of thermal boundary conditions are considered; a hot end-member that uses a constant temperature at the base of the lithosphere and a cool end-member that uses a constant heat flux boundary condition. The latter boundary condition allows the base of the lithosphere to cool by conduction during rifting.

The models do not explicitly include lateral conductive heat loss in the ascending asthenosphere, loss of latent heat through fusion, development of detachment faults, or small-scale convection in the mantle beneath the rift zone. Loss of heat from the rising asthenosphere through fusion and lateral heat conduction would lower the asthenosphere temperature during rifting, reducing the volume of melt. The models presented here therefore probably overestimate melt production, so any that are successful in simulating formation of NVRMs can be considered to be robust. Small-scale convection would increase the volume of melt, but this would not affect the timing of melt production. In the models presented here, all melting occurs near the end of the rifting episode, so although the models may underestimate melt production at the time of continental break-up, the prediction of amagmatic rifting throughout most of the rift episode is robust. Although detachment faults undoubtedly play a role during rifting on many continental margins, the evidence presented previously suggests that they are crustal rather than lithospheric in scale and that they develop late during the rifting episode. They therefore probably have little influence on the pattern of deformation in the mantle or, consequently, on the timing and distribution of melt production. This is discussed further in the last section of this paper.

The FE computer program solves the Navier–Stokes equations governing incompressible flow in a 2D viscoplastic medium to determine the strain rate and temperature throughout the FE mesh (Dunbar 1988). The strain rates determine the velocity vectors at each node in the mesh, and the mesh geometry is stepped forward in time using Euler's finite difference approximation. A Crank–Nicholson scheme is used to update the thermal structure of the mesh at the new time. The procedure is repeated to continue stepping forward in time until continental break-up occurs. The size of the time step and the mesh geometry are determined by starting with a coarse mesh and large time step and repeatedly halving the element and time-step sizes until the results of successive computations on a test model agree to within a specified precision.

The physical properties assigned to each element in the mesh correspond to granodiorite for the crust (Hansen & Carter 1982) and dunite for the mantle (Chopra & Paterson 1981) (Table 2). Power-law flow describes the rock rheology in the ductile regime and Byerlee's relationship (Byerlee 1978) is used to describe the failure strength of the rock in the brittle regime. The choice between brittle and ductile behaviour is determined for each element at each time step according to which of the two modes of deformation gives the lowest yield strength. This results in a layered strength profile, with strong regions in the middle crust and uppermost mantle (Fig. 3). Radiogenic heat production is included in the crust, with the abundance of heat-producing elements assumed to decrease exponentially with depth (Sclater *et al.* 1980) (Table 2).

Table 2. *Physical properties used in the models*

		Crust	Mantle
Ductile flow $\epsilon = A\sigma^n\exp(Q_c/nRT)$	A (Pa^{-n} s^{-1}) n Q_c (kJ mol^{-1})	5×10^{-18} 2.4 219	4×10^{-25} 4.5 498
Brittle failure $\tau = \tau_0 + \lambda P_L$	τ_0 (MPa) λ	60 0.6	60 0.6
Heat production $H(z) = H_0\exp(-z/D)$	H_0 (W m^{-3}) D (km)	3×10^{-6} 10	0 –
Thermal conductivity Density Thermal expansion coefficient	K (W m^{-1} °C^{-1}) ρ_0 (kg m^{-3}) α (°C^{-1})	2.5 2850 3.1×10^{-5}	3.4 3300 3.1×10^{-5}
Specific heat	C (J kg^{-1} °C^{-1})	875	1250

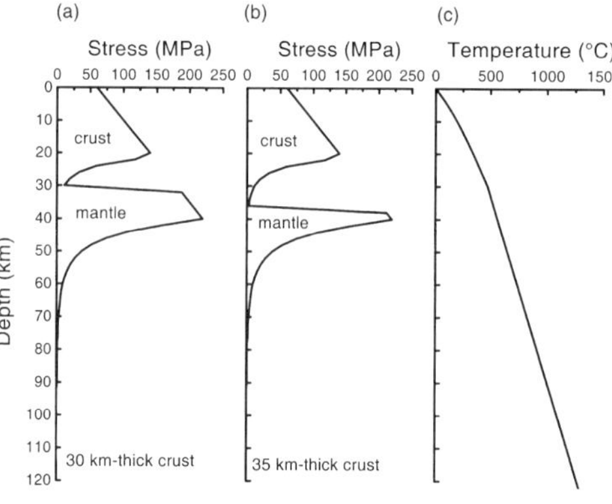

Fig. 3. Rheological structure of the lithosphere in the FE models. (**a**) Brittle strength increases linearly with depth as a result of increasing lithostatic pressure, ductile strength decreases as a result of rising temperature, and an abrupt increase in strength occurs at the transition from intermediate to mafic compositions at the crust–mantle boundary. The result is a layered rheology, with strong layers in the middle crust and uppermost mantle. (**b**) Increasing the thickness of the crust replaces strong mantle material with weaker crust, resulting in a net decrease in the strength of the lithosphere. (**c**) Geotherm used in the FE models. (See Table 2 for physical properties.)

The degree of partial melting is estimated using the method described by McKenzie & Bickle (1988). Ascent of the mantle is assumed to be adiabatic in the models that use a constant temperature basal boundary condition. This assumption is not strictly consistent with the thermal behaviour of the lithosphere in the FE model, as adiabatic ascent of the base of the lithosphere should result in a minor amount of cooling. However, the constant temperature basal boundary condition is only slightly super-adiabatic, overestimating the temperature at the base of the lithosphere by no more than 30 °C in any of the models. This small temperature difference has a negligible effect on the rift geometry, which is the important factor controlling ascent of the asthenosphere. The assumption of adiabatic ascent in the asthenosphere maximizes melt production for a given mantle potential temperature, and so provides an upper bound on the amount of melt generated during rifting. In the models using constant heat flux boundary conditions the geotherm beneath the rift axis is determined by assuming that the mantle temperature below some reference depth is not affected by rifting. Temperatures between the base of the model lithosphere and the reference depth are determined by linear interpolation. In the models shown here, the reference depth is chosen to be the depth to the base of the lithosphere before rifting. This results in an extremely subadiabatic ascent path.

To localize strain and induce lithospheric necking, a triangular-shaped region of over-thickened crust is included in the centre of the

Table 3. *Characteristics of non-volcanic rifted margins (numbers in parentheses are references)*

Margin	Amount of extension (km)	Average extension rate (mm a^{-1})	Initial sea-floor spreading rate (mm a^{-1})	Minimum thickness of extended crust (km)	Thickness of oldest oceanic crust (km)
Iberia Abyssal Plain	200 ± 50 (1)	12.2 ± 7	11 (2)	2.5 ± 0.4 (2)	4.9 ± 0.8 (2)
Tagus Abyssal Plain	200 ± 50 (1)	8.8 ± 3.8	10 (3)	6.55 (3)	1.8 (3)
Galicia Bank	200 ± 50 (1)	8.9 ± 3.7	6.5 (5)	3.3 ± 0.5 (5)	3.6 ± 0.6 (5)
Biscay	120 ± 15 (4)	7 ± 3		3.5 (5)	6.4 ± 0.6 (6)
Goban Spur	44 ± 3 (4)	22 ± 7	7.5 (6)	7.6 (7)	5.7 (7)
SW Greenland	25 ± 10 (8)	0.63 ± 0.34	4.5 ± 1.5 (9)	<7 (10)	6.9 (10)
Labrador	100 ± 25 (8)	2.47 ± 1	4.5 ± 1.5 (9)	<6 (9)	5.5 (9)

Amount of extension, average extension rate and sea-floor spreading rate do not include extension on the conjugate margin. Extension estimates are based on total tectonic subsidence for the Iberia and Tagus Abyssal Plains, Galicia Bank, SW Greenland, and Labrador margins. Extension on Biscay and Goban Spur margins is estimated from seismic constraints on the amount of crustal thinning, with crustal thickness estimates corrected for magmatic underplating. Average extension rate is estimated by dividing the amount of extension by the duration of rifting (Table 1). 1, Tett & Sawyer (1996); 2, Whitmarsh *et al.* (1990); 3, Pinheiro *et al.* (1992); 4, Skogseid (1994); 5, Whitmarsh *et al.* (1996); 6, White (1992); 7, Horsefield *et al.* (1994); 8, Dunbar & Sawyer (1989); 9, Chian *et al.* (1995*b*); 10, Masson *et al.* (1985).

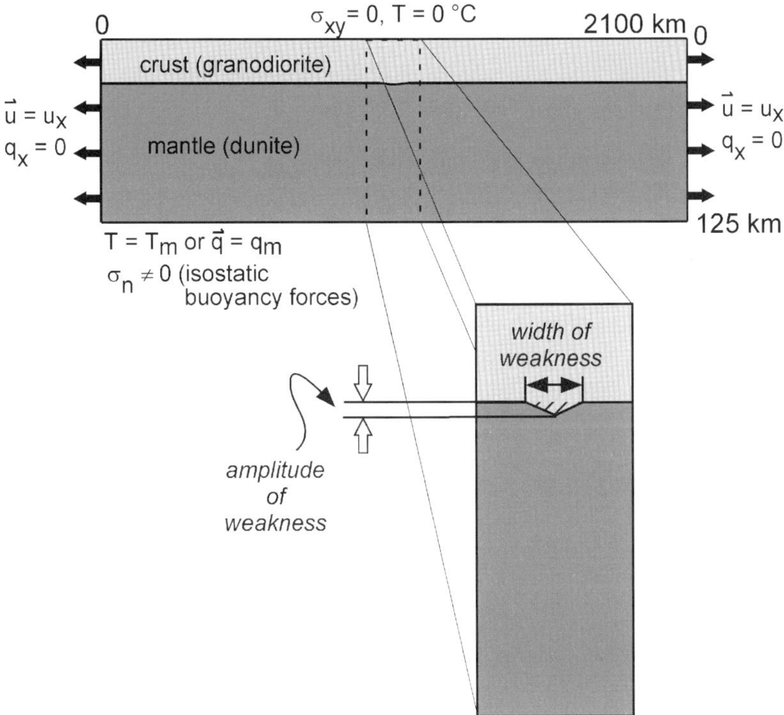

Fig. 4. Boundary and initial conditions in the FE models. Constant extension rate and zero heat flux conditions are applied to the sides of the model. The surface is held at 0 °C with no shear stress. The base of the model has either a constant heat flux or constant temperature boundary condition, with a normal stress dictated by isostatic balance. The centre of the model has a triangular-shaped region of overthickened crust, which weakens the uppermost mantle (see Fig. 3).

model. In effect, strong mantle material is replaced with weaker crustal rocks (Fig. 4). Other than this region of weakened mantle in the centre of the model, the lithosphere has a laterally homogeneous rheology. Several models focus on the impact that the geometry (width and amplitude) of the mantle weakness has on rifting and associated melt production. The models are evaluated to determine the duration of syn-extensional melting episodes, the width of the zone of melt emplacement, the total amount of extension that occurs before break-up, and the thickness of the continental crust at the rift axis at the time of break-up. The continuum nature of the FE technique prevents the models from explicitly breaking into separate tectonic plates, so the time of continental break-up must be assessed by other means. We identify the time of break-up according to crustal thickness and the amount of melt produced at the centre of the model. On most NVRMs the continental crust has been attenuated to less than 10 km thick (Table 3).

The most landward oceanic crust is also unusually thin, typically *c.* 5 km thick. Accordingly, break-up in the models is defined to occur when the continental crust has thinned to <10 km and sufficient melt is generated to create oceanic crust of 5 km thickness. For comparison, we also consider the case in which break-up is defined to occur when sufficient melt is produced to create oceanic crust of 7 km thickness, which is the approximate global average thickness of oceanic crust created during steady-state sea-floor spreading (Mooney *et al.* 1998).

Results

For purposes of comparison, a reference model is defined with a crust of 30 km thickness before extension and a mantle of 95 km thickness. The thickness of the crust in a region of 100 km width in the centre of the model is increased to a maximum of 31 km (Fig. 4). Extension rate in the reference model is 10 mm

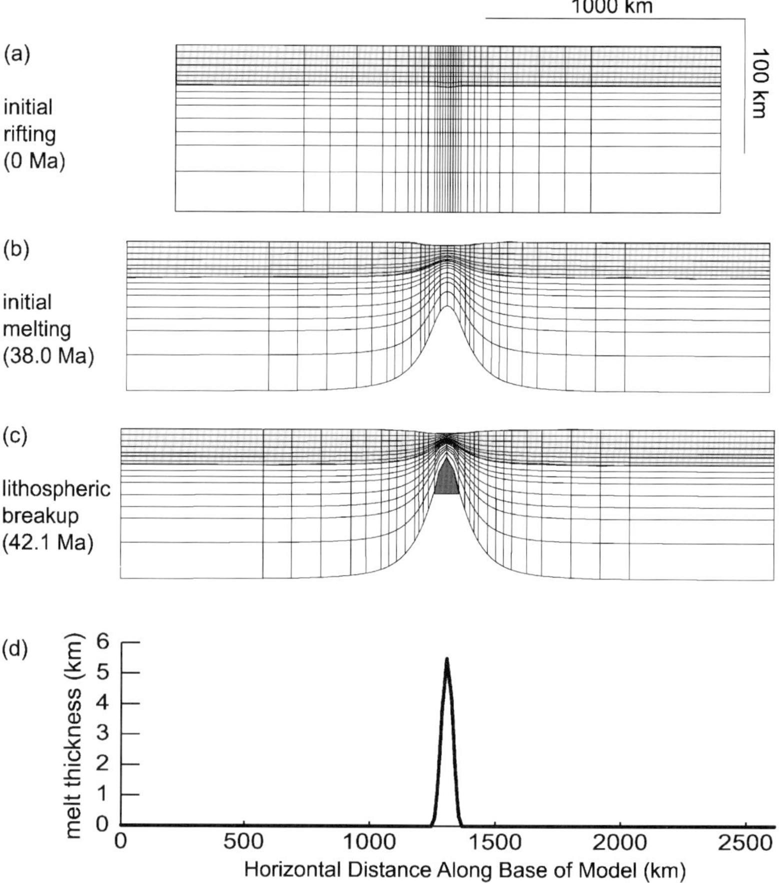

Fig. 5. Finite element mesh and melt production during rifting in the reference model. Melting does not occur until the last *c.* 4 Ma of extension and is restricted to a narrow region near the rift axis. (**a**) FE mesh before rifting. Grey is crust; white is mantle. (**b**) At 38 Ma after the onset of extension, necking is beginning to develop and the asthenosphere has just risen to shallow enough depths for decompression melting to begin. (**c**) Continental break-up, 42.1 Ma after the onset of extension. Sufficient melt (dark grey) is being generated to produce oceanic crust of 5 km thickness. (**d**) Total thickness of melt produced at the time of break-up. Melt production is restricted to a narrow region near the rift axis. Extension rate in this model is 10 mm a^{-1}, width and amplitude of the mantle weakness are 100 km and 1 km, respectively, and a constant temperature (1300 °C) basal thermal boundary condition is used.

a^{-1}, and a constant temperature (1300 °C) basal boundary condition is used. The weakness used to localize strain in the model is symmetric, resulting in a symmetric pattern of rifting (Fig. 5). Melt production is sufficient to create an oceanic crust of 5 km thickness after 42.1 Ma, which is taken to be the time of break-up. The total amount of extension in the model is 421 km and the continental crust has been thinned to 5 km in the centre of the model at the time of break-up, in general agreement with the structure of most NVRMs (Table 3). The most notable aspect of the model is the timing of melt generation. The asthenosphere does not ascend to shallow enough depths for partial melting to begin until 38 Ma after the onset of rifting. By this time, strain in the model is highly focused in the centre of the rift zone and substantial thermal weakening of the lithosphere has occurred. As a result, rifting progresses rapidly to break-up, limiting melt production to only the last 4.1 Ma of the rift history. The rift is non-volcanic over 90% of its history. Melting does not begin until necking is well established, so the zone of melting is narrow in comparison with the width of the rift

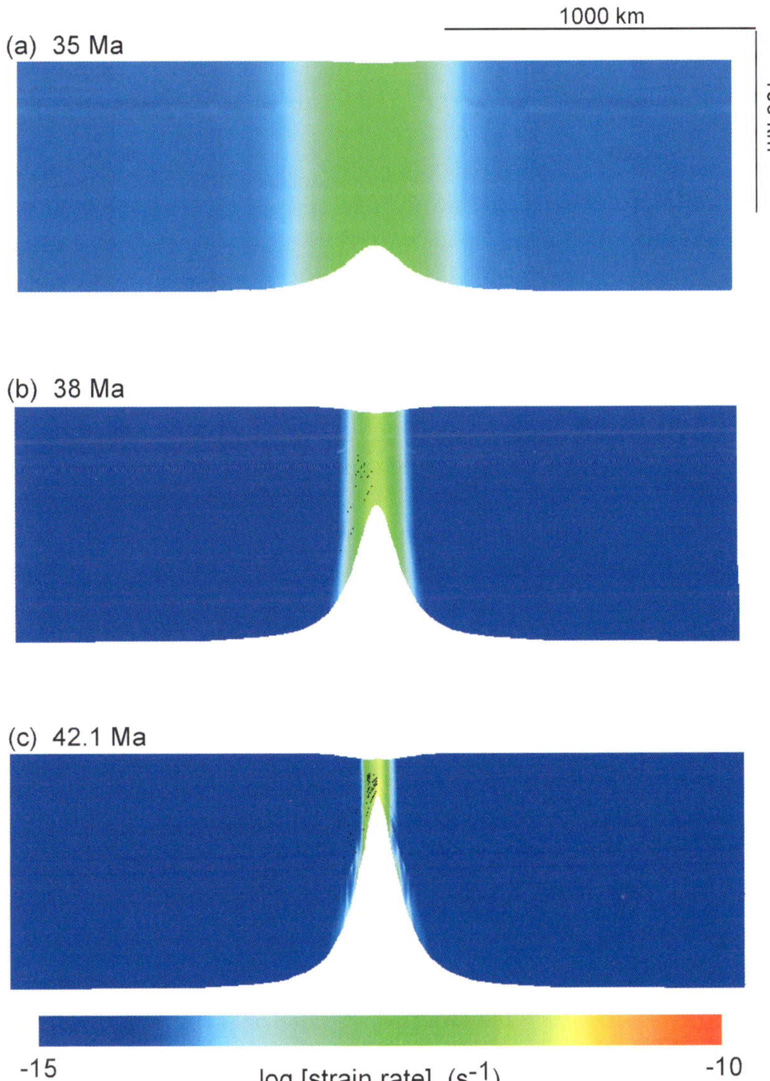

Fig. 6. Second invariant of strain rate in the reference model. (Compare with Fig. 5.) (**a**) Ductile necking of the lithosphere does not begin to develop until 35 Ma after the onset of extension. At this time, strain in the lithosphere is just beginning to become focused near the rift axis. (**b**) At 38 Ma after the onset of extension, strain has become highly focused and necking has progressed sufficiently to produce magma by decompression melting of the asthenosphere. (**c**) At 42.1 Ma after the onset of extension, strain is localized at the incipient sea-floor spreading centre and sufficient melt is being produced to create new oceanic lithosphere. The time elapsed between establishment of the lithosphere neck and break-up constitutes 16% of the rift history, with melt being produced only in the last 10%.

(the term 'rift' as used here encompasses the broad region of attenuated continental crust). The half-width of the melt zone is only 27 km, so the model predicts that most of the melt is produced in a relatively narrow region near the landward edge of the oceanic crust. The model successfully simulates non-volcanic rifting, producing broad rift margins underlain by highly attenuated continental crust and little syn-extensional melt. The model also predicts creation of new oceanic crust, but all melting occurs during the latest stages of extension and is restricted to

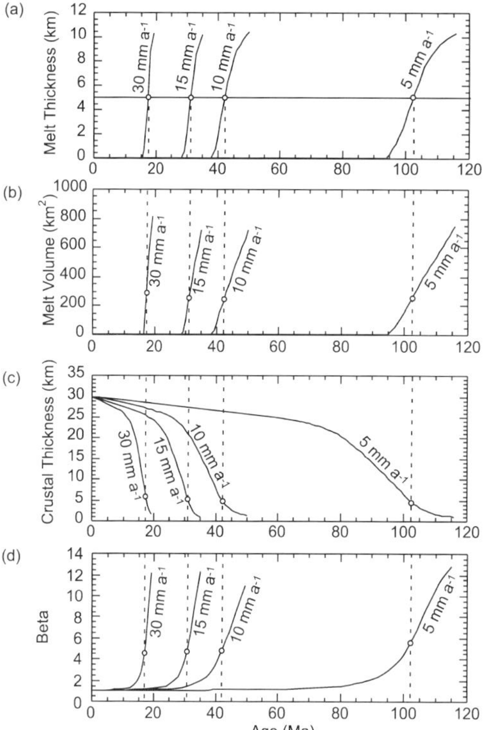

Fig. 7. The effect of extension rate on the FE model results illustrated as a function of time elapsed since the onset of extension. (**a**) Total thickness of basaltic magma produced at the rift axis. (**b**) Total volume of melt generated throughout the rift zone. (**c**) Thickness of the continental crust at the rift axis. (**d**) Extension factor (β) at the rift axis. The horizontal line in the top figure indicates the amount of melt required to produce oceanic crust of 5 km thickness, which is taken to mark the time of break-up. This criterion defines a time of break-up in each of the models that is marked by the dashed vertical lines.

within a few tens of kilometres of the location of initial sea-floor spreading. The key to this behaviour is the laterally homogeneous crust. In such circumstances, strain in the lithosphere is distributed over a broad region for most of the rift history, preventing the formation of a well-developed zone of necking and consequently inhibiting decompression melting (Fig. 6a). The width of the region undergoing extension becomes progressively more focused as extension proceeds. At the same time, temperatures in the lithosphere above the ascending asthenosphere rise and deviatoric stress in the rift axis increases because extensional forces are distributed through a thinner lithosphere. Because strain rate is strongly dependent on

both temperature and stress (Carter & Tsenn 1987; Ranalli & Murphy 1987), the focusing of strain and lithospheric thinning accelerate. By the time the asthenosphere rises to depths shallow enough to begin melting, rifting is highly focused and the net strength of the lithosphere is dramatically reduced (Fig. 6b). As a result, the lithosphere rapidly proceeds to failure and continental break-up (Fig. 6c), leading to a period of melt production that is short in comparison with the duration of rifting.

The reference model was modified to consider a range of extension rates (5, 10, 15 and 30 mm a^{-1}) that bound those suggested for most NVRMs (Table 3). Melt production is restricted to the last 10% of the rift history in all of these models, but the duration of the rifting episode varies substantially, from 102.5 Ma at the lowest extension rate to 17.1 Ma at the highest (Fig. 7). However, the slope of the melt production v. time curve increases as the rate of extension increases (Fig. 7a,b), leading to a decrease in the duration of the melting episode (i.e. the time elapsed between the onset of melt production and creation of new oceanic crust). Magmatism lasts <2.5 Ma at extension rates >15 mm a^{-1}. Thus, extension rate has a significant impact on the duration of melt production in these models, but in an opposite sense from that in models that appeal to cooling of the mantle during rifting to inhibit magmatism (e.g. Bown & White 1995; Tett & Sawyer 1996). If the lithosphere is rheologically homogeneous, increasing extension rate reduces the duration of syn-extensional melt production. Variations in extension rate have little influence on the thickness of the continental crust at the time of break-up, which is <5 km at the rift axis in all cases. This is consistent with the thickness of continental crust on most NVRMs (Table 3).

Varying the thickness of the continental crust before extension has little effect on the rift history if the crust is initially thinner than c. 35 km (Fig. 8). Models with initial crustal thicknesses of 20 km and 30 km have virtually identical rift duration, melt production histories, amounts of extension, and continental crust thicknesses at the time of break-up. Larger initial crustal thickness prolongs the rift episode and increases the duration of syn-rift melt production. The abrupt change in model behaviour for crustal thickness >30 km (Fig. 8) can be understood by considering the rheological structure of the lithosphere. At strain rates of c. 10^{-15} s^{-1} and temperatures below c. 500 °C (typical of the upper mantle in these models) the mantle is in the semi-brittle deformation regime, and is much stronger than the crust

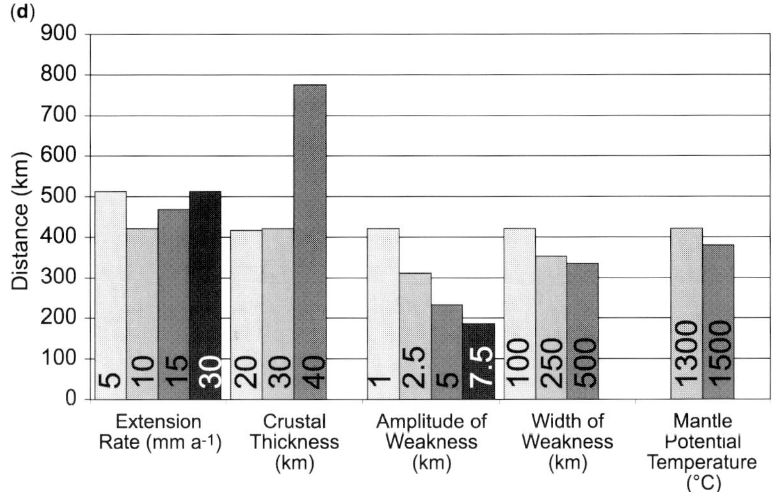
(d)
Distance (km)
900
800
700
600
500
400
300
200
100
0
5
10
15
30
20
30
40
1
2.5
5
7.5
100
250
500
1300
1500
Extension Rate (mm a-1)
Crustal Thickness (km)
Amplitude of Weakness (km)
Width of Weakness (km)
Mantle Potential Temperature (°C)

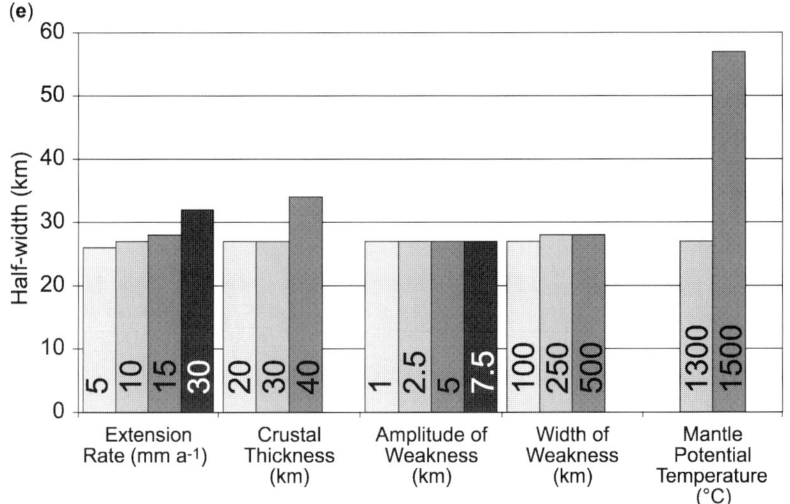
(e)
Half-width (km)
60
50
40
30
20
10
0
5
10
15
30
20
30
40
1
2.5
5
7.5
100
250
500
1300
1500
Extension Rate (mm a-1)
Crustal Thickness (km)
Amplitude of Weakness (km)
Width of Weakness (km)
Mantle Potential Temperature (°C)

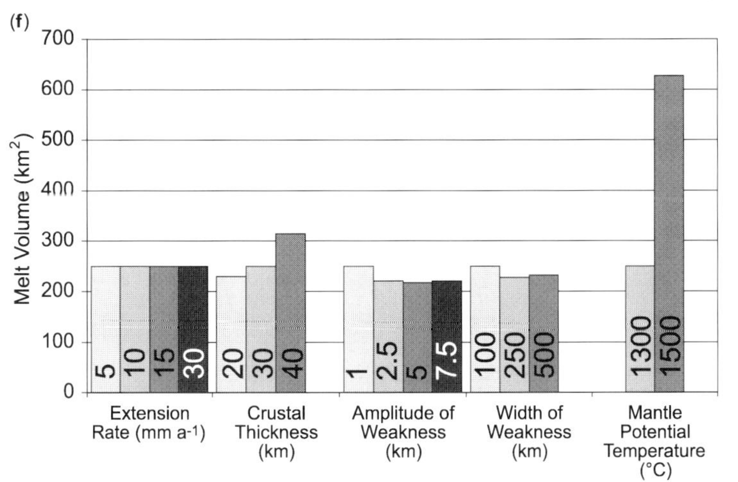
(f)
Melt Volume (km2)
700
600
500
400
300
200
100
0
5
10
15
30
20
30
40
1
2.5
5
7.5
100
250
500
1300
1500
Extension Rate (mm a-1)
Crustal Thickness (km)
Amplitude of Weakness (km)
Width of Weakness (km)
Mantle Potential Temperature (°C)

the asthenosphere beneath the rift axis is sufficiently rapid that the top of the asthenosphere reaches the depth required for the onset of melting in 1300 °C mantle only 7 Ma later than it reaches the depth required for melting to begin in 1500 °C mantle.

There is a trade-off in the sensitivities of the time elapsed until the onset of melt production and the duration of the rifting episode to the amplitude and width of the mantle weakness. As a result, the total duration of the melting episode has little dependence on the geometry of the mantle weakness (Fig. 11c). There is a weak dependence on crustal thickness, with thicker crust leading to longer periods of melt production, but this effect is minor, and changes the melt production episode by only c. 2 Ma. Higher mantle temperatures prolong the melt production episode, but even at 1500 °C syn-extensional melting lasts only 7 Ma in the reference model. Extension rate has the strongest influence on the duration of melt production, with slower extension rates favouring longer melting episodes. Extension rate has little effect on the total amount of extension (Fig. 11d), which is controlled most strongly by the geometry of the mantle weakness (particularly the amplitude) unless the crust is thicker than c. 30 km. Neither the extension rate, crustal thickness, nor geometry of the weakness has a significant effect on the width of the zone of melting or the total volume of melt generated per kilometre of rift length (Fig. 11e and f). Higher mantle temperature increases the width of the melt zone and the total volume of melt generated (Fig. 11e and f), but in all of these models the duration of melt production is short in comparison with the rift duration and most magmatism is restricted to within a few kilometres of the locus of initial sea-floor spreading.

Discussion

The models presented here are in many ways similar to FE models published elsewhere that simulate the formation of volcanic margins (Harry & Sawyer 1992*a,b*). The key difference is the absence of lateral variations in crustal rheology. Heterogeneity in the crust prolongs the period of time required for strain to become focused into a narrow rift zone, but may allow for asthenospheric necking to develop relatively early during the rifting episode. This generally leads to a two-stage rift evolution. During the first stage, extension in the crust is centred in the area where the crust is weakest, whereas mantle necking is centred beneath the region where the mantle is weakest. If these two positions do not coincide, the locus of strain in the crust and the apex of the asthenosphere diapir are offset (Fig. 12a–c). Eventually, the mantle weakness takes control because the influx of heat from the rising asthenosphere drastically reduces the strength of the lithosphere above the mantle neck. The locus of extension in the crust then shifts to become coincident with the locus of extension in the mantle, and necking progresses rapidly to break-up (Fig. 12d). If the crust is heterogeneous, sufficient mantle thinning can occur in the first stage to produce large amounts of melt relatively early during the rift history, leading to formation of a VRM. It should be noted that the model shown in Figure 12 requires a mantle potential temperature of 1400 °C (c. 100 °C higher than the typical mantle potential temperature) to match the melt production history of the Baltimore Canyon Trough margin. Calculations based on this model that use a 'normal' mantle potential temperature of 1300 °C produce smaller volumes of melt, and the onset of melt production is slightly delayed. However, unless the mantle potential temperature is lower than c. 1100 °C, melt production occurs throughout a large portion of the rift history and is distributed over a wide region. Hence, the 1400 °C mantle potential temperature used in the model shown in Figure 12 is not the sole cause of the magmatic character of the rifting episode.

The FE models presented here show that the presence of a laterally homogeneous crust before rifting promotes formation of non-volcanic rifts and rifted margins. In such a case, lithospheric necking does not become well established until strain in the lithosphere is highly focused. By this time, lithospheric stress has become so concentrated that the rift pro-

Fig. 11. Sensitivity of rift behaviour to model parameters. (**a**) Duration of rifting. (**b**) Time elapsed until the onset of melting. (**c**) Duration of the melt production episode. (**d**) Total amount of extension at the time of break-up. (**e**) Half-width of the zone of melting at the time of break-up. (**f**) Total volume of melt produced per unit length of rift at the time of break-up. The model with a mantle potential temperature of 1100 °C did not generate any melt, so rift duration (time elapsed between the onset of extension and the generation of new oceanic crust of 5 km thickness) and measures of melt production history are undefined.

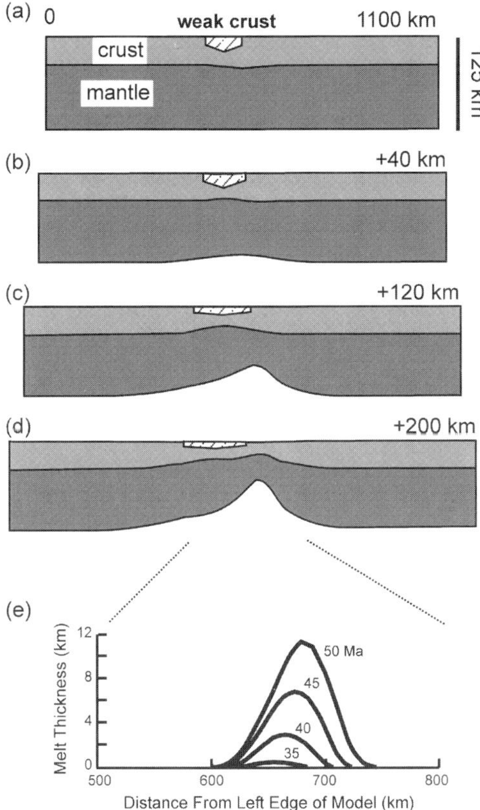

Fig. 12. FE model of extension on the Baltimore Canyon Trough margin of eastern North America. (**a**) Pre-rift configuration of the model includes two weaknesses: a region of overthickened crust, which produces a mantle weakness, and a region of reduced strength in the crust (striped pattern), which produces heterogeneity in the crust. (**b**) Extensional strain at shallow levels (indicated by crustal thinning) is initially localized where the crust is weakest, whereas mantle necking localizes beneath the weakest part of the upper mantle. (**c**) The offset between the pre-existing crust and mantle weaknesses results in an offset between the position of the rift axis at the surface and the locus of asthenospheric upwelling. This allows decompression melting of the asthenosphere to begin well before the crust above the mantle neck has been sufficiently thinned to allow for continental break-up. (**d**) Heat advection and progressive concentration of strain above the asthenosphere diapir eventually overcome the pre-existing weakness in the crust. This leads to a shift in the locus of strain in the crust and eventual continental break-up where the mantle was weakest before the onset of extension. (**e**) Because mantle necking is established before strain in the crust becomes sufficiently localized to allow for break-up, decompression melting in the asthenosphere begins relatively early during the rift history, leading to a prolonged period of magmatism and the

gresses rapidly to break-up, resulting in a very short (*c.* 2 Ma) period of magmatism in a narrow region near the rift axis that occurs during the last stages of extension. The FE models demonstrate that this scenario for non-volcanic rifting is robust over a broad range of rift duration, extension rate, and rift margin geometry. In most of the models presented here, magmatism is restricted to the last *c.* 10% of the rift history, and the half-width of the melt zone is less than *c.* 25 km. The models do not require abrupt spatial or temporal variations in mantle temperature and are consistent with the wide range of rift dimensions and duration on NVRMs. Most importantly, the models account for the transition from amagmatic rifting to production of new oceanic crust at the time of continental break-up.

Most rifts localize in regions where the lithosphere has been weakened by prior tectonic events, and generally lie parallel or subparallel to older orogenic belts. Rifting in regions where the crust is relatively homogeneous is probably an uncommon event, accounting for the rarity of NVRMs in comparison with VRMs. It is important to recognize that, in the absence of major differences in mantle temperature, it is the heterogeneity of the crust rather than the presence or absence of pre-existing weaknesses in the lithosphere that dictates whether a rift is volcanic or non-volcanic. For example, VRMs on the US and African Atlantic margins, in the Norwegian and Greenland Seas, and off western Britain formed by rifting along the trend of the Appalachian and Caledonide orogenic belts (Fig. 13). Here, the mantle weakness that ultimately controls the location of continental break-up can be attributed to a thickened crust in the core of the orogen. Crustal heterogeneity can be attributed to the contrast between the strength of the rocks within the Palaeozoic orogenic belts and the adjacent Precambrian basement (e.g. Harry & Sawyer 1992*b*). The NVRMs in the southern Labrador Sea developed within relatively homogeneous Precambrian crust (the northward transition to a VRM on this margin can be explained by closer proximity to the Iceland plume at the time of break-up). Interestingly, the NVRMs adjacent to Iberia, the Grand Banks, and southern Britain all lie within the Variscan, Caledonide or Aca-

formation of a volcanic rifted margin. Extension rate in this model is 15 mm a^{-1}, and melt production is calculated assuming a mantle potential temperature of 1400 °C. After Harry & Sawyer (1992*b*).

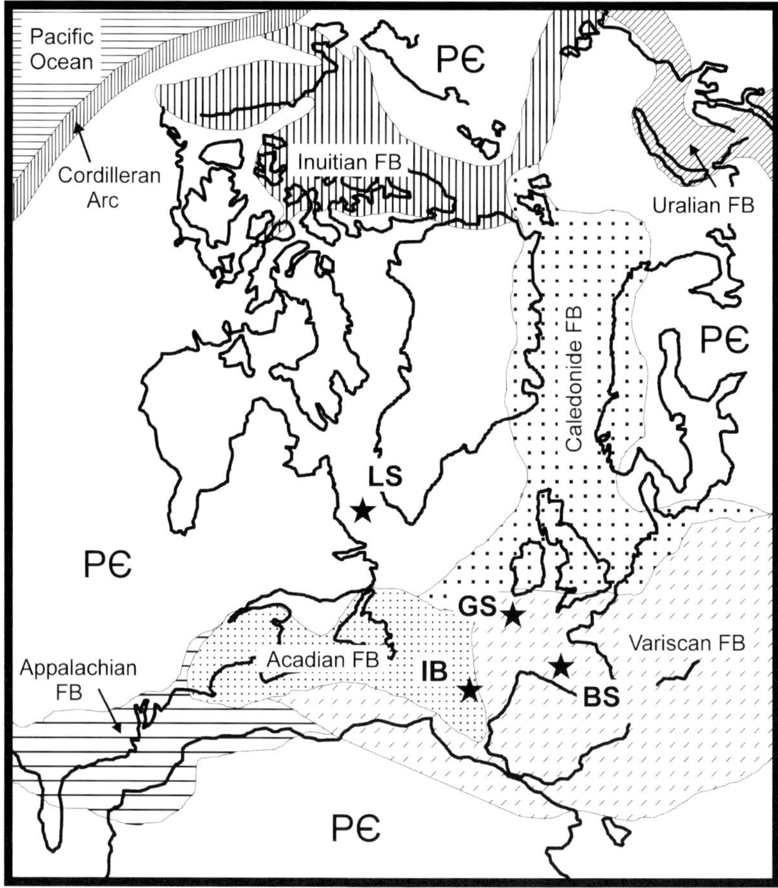

Fig. 13. Late Permian reconstruction of the North Atlantic continents. Norwegian–Greenland Sea and US–African VRMs developed in regions of heterogeneous crust, where Palaeozoic orogenic belts are sandwiched between Precambrian crust. NVRMs of the North Atlantic typically developed in regions involving either uniformly old (Precambrian) crust or uniformly young (Palaeozoic) crust within the interiors of the orogenic belts. Stars indicate non-volcanic margins. BS, Biscay; GB, Goban Spur; IB, Iberia; LS, Labrador Sea. After Ziegler (1989).

dian orogenic belts, or near one of the boundaries between these orogenic belts. However, none of the NVRMs lies near a boundary separating Phanerozoic and Precambrian basement. Within the context of the FE models presented here, this is interpreted to indicate that thickened crust in the orogens provides the mantle weakness necessary to localize rifting, but there is little difference in the strength of the crust within and between orogenic belts. In summary, comparison of the spatial distribution of VRMs and NVRMs in the North Atlantic with basement age suggests that the contrast between Precambrian and Phanerozoic crust provides sufficient heterogeneity to promote

volcanic rifting, even if the mantle temperature is not unusually high. If the suture between Precambrian and Phanerozoic crust lies far from the rift axis the root of the orogen may weaken the mantle enough to localize rifting, but the crust is sufficiently homogeneous to promote non-volcanic rifting.

Application to the Iberia Abyssal Plain non-volcanic margin

It is relatively easy to envision how the FE models presented here would relate to NVRMs such as the Goban Spur margin (Fig. 14a). In such a case, the overall geometry of the margin

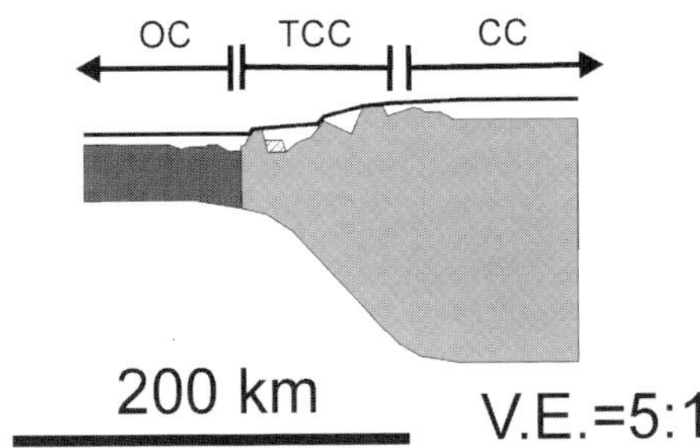

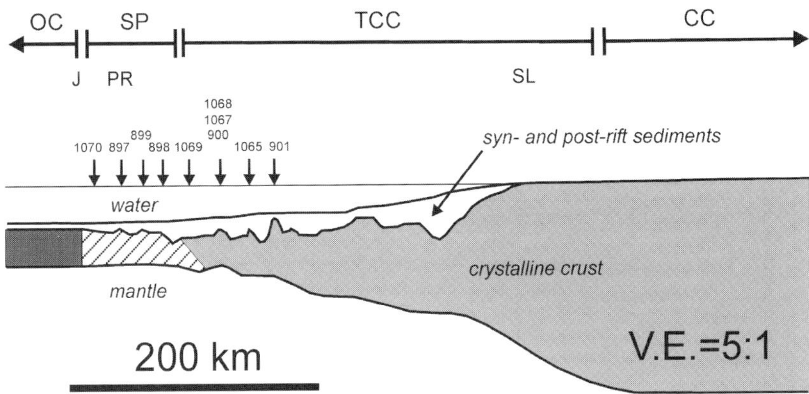

Fig. 14. Crustal structure of non-volcanic rifted margins. (**a**) Goban Spur (after Horsefield *et al.* 1994). (**b**) Iberia Abyssal Plain continental margin. OC indicates oceanic crust (dark grey), SP indicates serpentinized peridotite in the region of exhumed mantle (ruled), and light grey indicates unextended continental crust (CC) and continental crust that was thinned during the rifting episode (TCC). White indicates post-rift sediments. The location of the J-anomaly at the landward edge of oceanic crust (J), the peridotite ridge sampled on ODP Leg 149 (PR), and the shoreline (SL) are indicated. Vertical arrows indicate the locations of Ocean Drilling Program sites (Sawyer *et al.* 1994; Whitmarsh *et al.* 1998). Other aspects of the IAP cross-section are constrained by seismic reflection, refraction, and tomographic studies (Whitmarsh *et al.* 1990; Corchette *et al.* 1995; Chian *et al.* 1999).

is not greatly dissimilar to that of the FE models, with the major difference being that shallow crustal extension and attenuation is accommodated by high-angle normal faults rather than by ductile stretching. However, the structure of the IAP differs from the models presented here in that it contains an expanse of *c*. 100 km width of exhumed mantle in the transition zone separating oceanic and thinned continental crust (Fig. 14b). Ocean Drilling Program (ODP) drilling and seismic reflection profiling indicate that the mantle was exhumed along a low-angle detachment fault during rifting (Krawczyk *et al.* 1996; Pickup *et al.* 1996; Reston 1996*a*). A similar geometry has been suggested for the SW Greenland NVRM and for a Jurassic NVRM exposed in the Swiss and Italian Alps (Froitzheim & Eberli 1990; Chian *et al.* 1995*a*,*b*; Froitzheim & Manatschal 1996). On the Iberian margin and in the Alpine exposure the detachment fault soles at the base of the crust or in the uppermost mantle, and it

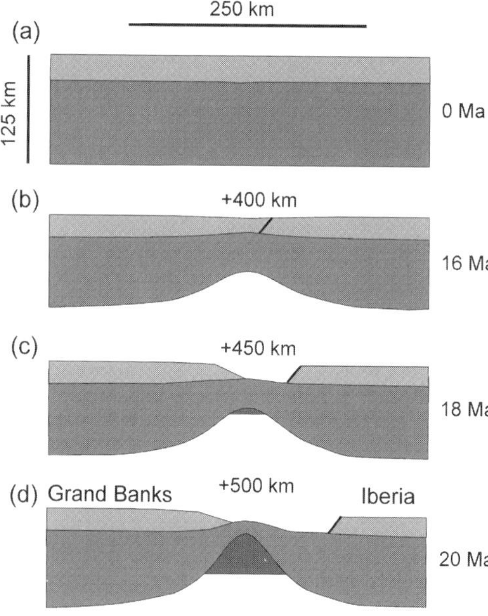

Fig. 15. Schematic illustration of exhumation of the Iberia Abyssal Plain interpreted within the context of the FE models. Extension rate is 25 mm a^{-1}. Ages indicate time elapsed since the onset of extension. Evolution of the mantle (medium grey) and melt production (dark grey) is determined from the FE model. Evolution of the crust (light grey) is identical to the FE model before 16 Ma. At that time, formation of a low-angle detachment fault that soles at the base of the crust is postulated according to the rolling-hinge model of Buck (1988). Subsequent evolution of the hanging wall is governed by 60° oblique simple-shear.

probably developed late during the rifting episode (Scharer *et al.* 1995; Froitzheim & Manatschal 1996; Krawczyk *et al.* 1996;

Fuegenschuh *et al.* 1998). It is argued here that the detachment faults serve primarily to control the geometry of rift-related structures in the crust but have little influence on deformation in the deeper lithosphere, and therefore little impact on the magmatic history of the rift. As such, the formation and width of the transition zone on NVRMs merely reflect the response of the crust to extension in the lithosphere as a whole. This bulk pure shear behaviour of the lithosphere with simple shear deformation in the crust is observed in laboratory analogue models of extension (Brun & Beslier 1996), which produce an overall lithosphere-scale evolution similar to that observed in the FE models. In this scenario, rifting is initially characterized by lithosphere-scale ductile stretching (Fig. 15a and b). As the crust thins and temperatures at the rift axis rise the rheological structure of the lithosphere becomes similar to a two-layer model, with a ductile mantle overlain by a thin but predominantly brittle crust. Eventually the crust becomes thin enough to break, creating the detachment fault and exhuming the upper mantle (Fig. 15b and c). By this time, however, necking of the lithosphere mantle is well established and rifting quickly proceeds to break-up (Fig. 15d). Development of the detachment fault during the late stages of rifting is consistent with the cooling histories of lower-crustal rocks on the Iberian margin and with structural relations on both the Iberian and Alpine rifted margins (Scharer *et al.* 1995; Froitzheim & Manatschal 1996; Fuegenschuh *et al.* 1998). Furthermore, this interpretation is consistent with ascent of relatively warm asthenosphere during rifting, which is required to account for the composition of the earliest magmatic rocks generated during rifting on the NE Canadian margin and for the creation of oceanic crust at the time of break-

Table 5. *Comparison of FE model (best-fitting FE model for IAP uses 25 mm a^{-1} extension rate) with Iberia Abyssal Plain (numbers in parentheses are references)*

Parameter	Observed	Modelled
Onset of rifting	147 Ma (1)	–
Onset of melting	135–130 Ma (2)	18 Ma after extension begins
Time of break-up	126 Ma (2)	20 Ma after extension begins
Duration of rifting event	21 Ma	20 Ma
Duration of magmatic episode	4–9 Ma	2 Ma
Continental crust thickness	2.1–3.3 km (3)	5 km
Oceanic crust thickness	4.9 ± 1.5 km (3)	5 km
β	5.7 (1)	5
Total amount of extension	400 km (4)	500 km

Modelled ages are time elapsed since the onset of extension. 1, Wilson *et al.* (1996); 2, Pinheiro *et al.* (1996); 3, Whitmarsh *et al.* (1990); Chian *et al.* (1999); 4, Tett & Sawyer (1996).

up (Williamson *et al.* 1995; Harry & Bowling 1999). A FE model with an extension rate of $25\,mm\,a^{-1}$ provides a particularly good fit to the geometry and duration of rifting on the IAP (Table 5), and is reasonably consistent with the $22-26\,mm\,a^{-1}$ sea-floor spreading rate that is estimated at the time of break-up (Srivistava *et al.* 1990; Whitmarsh & Miles 1995). Differences in the width of the transition zone on other NVRMs can be attributed to differences in the time at which the detachment fault nucleates, with earlier nucleation producing a wider region of exhumed mantle

Summary

The FE models demonstrate that a rheologically homogeneous crust and moderately rapid extension rate promote formation of non-volcanic rifted margins, even at normal to slightly elevated mantle temperatures. The key to this behaviour is delaying the ascent of asthenosphere to sufficiently shallow depths to begin melting until after lithosphere necking is well established. Rifting then proceeds rapidly to break-up, minimizing both the duration of syn-extensional melting and the spatial distribution of syn-rift magmatic rocks. Differences in extension rate allow for a broad variation in rift duration, as appears to be required by the geological record, and differences in the geometry of the mantle weakness account for differences in the total amount of extension and geometry of the rifted margins that ultimately develop. The ranges in margin widths, crustal thickness, rift duration, and extension rate in the models compare favourably with those estimated on various NVRMs, and are a particularly good fit to the Iberian Abyssal Plain margin. The combination of homogeneous crust and moderately rapid extension rates is likely to be an uncommon occurrence, accounting for the relative rarity of NVRMs in comparison to VRMs. The models do not require temporal or spatial changes in mantle temperature to account for the difference between volcanic and non-volcanic rifting, the abrupt changes in margin geometry and magmatic character that are observed on many rifted margins, or the transition from amagmatic rifting to the generation of new oceanic crust at the time of break-up.

This work was supported by the US National Science Foundation under Grant OCE-9906889 and was conducted as shore-based research associated with Ocean Drilling Program Leg 173. We are grateful for insightful reviews from T.A. Minshull and D.S. Sawyer.

References

ALLMENDINGER, R.W., SHARP, J.W., TISH, D.V., SERPA, L., BROWN, L., KAUFMAN, S. & OLIVER, J. 1983. Cenozoic and Mesozoic structure of the eastern Basin and Range province, Utah, from COCORP seismic-reflection data. *Geology*, **11**, 532–536.

APPLEGATE, J.D.R., WALKER, J.D. & HODGES, K.V. 1992. Late Cretaceous extensional unroofing in the Funeral Mountains metamorphic core complex, California. *Geology*, **20**, 519–522.

BALKWILL, H.R. 1987. Labrador basin: structural and stratigraphic style. *In*: TANKARD, A.J. (ed.) *Sedimentary Basins and Basin-Forming Mechanisms*. Canadian Society of Petroleum Geologists, Memoir, **12**, 17–43.

BARTON, A.J. & WHITE, R.S. 1997. Crustal structure of Edoras Bank continental margin and mantle thermal anomalies beneath the North Atlantic. *Journal of Geophysical Research*, **102**, 3109–3129.

BESLIER, M.-O., ASK, M. & BOILLOT, G. 1993. Ocean–continent boundary in the Iberia abyssal plain from multichannel seismic data. *Tectonophysics*, **218**, 383–393.

BOILLOT, G. & WINTERER, E.L. *Proceedings of the Ocean Drilling Program, Initial Reports, 103.* Ocean Drilling Program, College Station, TX.

BOWN, J.W. & WHITE, R.S. 1995. Effect of finite extension rate on melt generation at rifted continental margins. *Journal of Geophysical Research*, **100**, 18011–18030.

BRUN, J.P. & BESLIER, M.O. 1996. Mantle exhumation at passive margins. *Earth and Planetary Science Letters*, **142**, 161–173.

BUCK, W.R. 1988. Flexural rotation of normal faults. *Tectonics*, **7**, 959–973.

BUCK, W.R., MARTINEZ, F., STECKLER, M.S. & COCHRAN, J.R. 1988. Thermal consequences of lithospheric simple shear: pure and simple. *Tectonics*, **7**, 213–234.

BYERLEE, J.D. 1978. Friction in rocks. *Pure and Applied Geophysics*, **116**, 615–626.

CARTER, N.L. & TSENN, M.C. 1987. Flow properties of continental lithosphere. *Tectonophysics*, **136**, 27–64.

CHIAN, D. & LOUDEN, K.E. 1994. The continent–ocean crustal transition across the southwest Greenland margin. *Journal of Geophysical Research*, **99**, 9117–9136.

CHIAN, D., KEEN, C., REID, I. & LOUDEN, K.E. 1995*a*. Evolution of nonvolcanic rifted margins: new results from the conjugate margins of the Labrador Sea. *Geology*, **23**, 589–592.

CHIAN, D., LOUDEN, K.E., MINSHULL, T.A. & WHITMARSH, R.B. 1999. Deep structure of the ocean–continent transition in the southern Iberia Abyssal Plain from seismic refraction profiles, Ocean Drilling Program (Legs 149 and

173) transect. *Journal of Geophysical Research*, **104**, 7443–7462.

CHIAN, D., LOUDEN, K.E. & REID, I. 1995*b*. Crustal structure of the Labrador Sea conjugate margin and implications for the formation of nonvolcanic continental margins. *Journal of Geophysical Research*, **100**, 24239–24254.

CHOPRA, P.N. & PATERSON, M.S. 1981. The experimental deformation of dunite. *Tectonophysics*, **78**, 453–473.

CLOETINGH, S., van Wees, J.D., van der Beek, P.A. & SPADANI, G. 1995. Role of pre-rift rheology in kinematics of extensional basin formation: constraints from thermomechanical models of Mediterranean and intracratonic basins. *Marine and Petroleum Geology*, **12**, 793–807.

COFFIN, M.F. & ELDHOLM, O. 1992. Volcanism and continental breakup: a global compilation of large igneous provinces. *In*: STOREY, B.C., ALABASTER, T. & PANKHURST, R.J. (eds) *Magmatism and the Causes of Continental Breakup*. Geological Society, London, Special Publications, **68**, 21–34.

CORCHETTE, V., BADAL, J., SERON, F.J. & SORIA, A. 1995. Tomographic images of the Iberian subcrustal lithosphere and asthenosphere. *Journal of Geophysical Research*, **100**, 24133–24146.

DALEY, E.E. & DEPAOLO, D.J. 1992. Isotopic evidence for lithospheric thinning during extension: southeastern Great Basin. *Geology*, **20**, 104–108.

DISCOVERY WORKING GROUP 1998. Deep structure in the vicinity of the ocean–continent transition zone under the southern Iberia abyssal plain. *Geology*, **26**, 743–746.

DUNBAR, J. A., 1988. *Kinematics and dynamics of continental breakup*. PhD thesis, University of Texas at Austin.

DUNBAR, J.A. & SAWYER, D.S. 1989. Patterns of continental extension along the conjugate margins of the central and north Atlantic oceans and Labrador Sea. *Tectonics*, **8**, 1059–1077.

FINLAYSON, D.M., COLLINS, C.D.N., LUKASZYK, I. & CHUDYK, E.D. 1998. A transect across Australia's southern margin in the Otway Basin region: crustal architecture and the nature of rifting from wide-angle seismic profiling. *Tectonophysics*, **288**, 177–189.

FOSTER, D.A., MILLER, C.F., HARRISON, T.M. & HOISCH, T.D. 1992. 40Ar/39Ar thermochronology and thermobarometry of metamorphism, plutonism, and tectonic denudation in the Old Woman Mountains area, California. *Geological Society of America Bulletin*, **104**, 176–191.

FOUCHER, J.P., LE PICHON, X. & SIBUET, J.C. 1982. The ocean–continent transition in the uniform stretching model: role of partial melting in the mantle. *Philosophical Transactions of the Royal Society of London*, **305**, 27–43.

FROITZHEIM, N. & EBERLI, G.P. 1990. Extensional detachment faulting in the evolution of a Tethys pasive continental margin, Eastern Alps, Switzerland. *Geological Society of America Bulletin*, **102**, 1297–1308.

FROITZHEIM, N. & MANATSCHAL, G. 1996. Kinematics of Jurassic rifting, mantle exhumation, and passive-margin formation in the Austroalpine and Penninic nappes (eastern Switzerland). *Geological Society of America Bulletin*, **108**, 1120–1133.

FUEGENSCHUH, B., FROITZHEIM, N. & BOILLOT, G. 1998. Cooling history of granulite samples from the ocean–continent transition of the Galicia margin: implications for rifting. *Terra Nova*, **10**, 96–100.

GALLAGHER, K. & HAWKESWORTH, C. 1992. Dehydration melting and the generation of continental flood basalts. *Nature*, **358**, 57–59.

GARCIA-MONDEJAR, J. 1996. Plate reconstruction of the Bay of Biscay. *Geology*, **24**, 635–639.

HANSEN, F.D. & CARTER, N.L. 1982. Creep of selected crustal rocks at 1000 MPa. *EOS Transactions, American Geophysical Union*, **63**, 437.

HARRY, D.L. & BOWLING, J.C. 1999. Inhibiting magmatism on nonvolcanic rifted margins. *Geology*, **27**, 895–898.

HARRY, D.L. & SAWYER, D.S. 1992*a*. Basaltic volcanism, mantle plumes, and the mechanics of rifting: the Parana flood basalt province of South America. *Geology*, **20**, 207–210.

HARRY, D.L. & SAWYER, D.S. 1992*b*. A dynamic model of extension in the Baltimore Canyon trough region. *Tectonics*, **11**, 420–436.

HARRY, D.L., SAWYER, D.S. & LEEMAN, W.P. 1993. The mechanics of continental extension in western North America: implications for the magmatic and structural evolution of the Great Basin. *Earth and Planetary Science Letters*, **117**, 59–71.

HERMANN, J., MUNTENER, O., TROMMSDORFF, V., HANSMANN, W. & PICCARDO, G.B. 1997. Fossil crust-to-mantle transition, Val Malenco (Italian Alps). *Journal of Geophysical Research*, **102**, 20123–20132.

HOPPER, J.R., MUTTER, J.C., LARSON, R.L., MUTTER, C.Z. & Northwest Australia Study Group 1992. Magmatism and rift margin evolution: evidence from Northwest Australia. *Geology*, **20**, 853–857.

HORSEFIELD, S.J., WHITMARSH, R.B., WHITE, R.S. & SIBUET, J.-C. 1994. Crustal structure of the Goban Spur rifted continental margin, NE Atlantic. *Geophysical Journal International*, **119**, 1–19.

JOHNSON, R.A. & LOY, K.L. 1992. Seismic reflection evidence of seismogenic low-angle faulting in southeastern Arizona. *Geology*, **20**, 597–600.

KEEN, C.E. 1985. The dynamics of rifting: deformation of the lithosphere by active and passive driving forces. *Geophysical Journal of the Royal Astronomical Society*, **80**, 95–120.

KEEN, C.E., COURTNEY, R.C., DEHLER, S.A. & WILLIAMSON, M.-C. 1994. Decompression melting at rifted margins: comparison of model predictions with the distribution of igneous rocks on the eastern Canadian margin. *Earth and Planetary Science Letters*, **121**, 403–416.

KLITGORD, K.D., HUTCHINSON, D.R. & SCHOUTEN, H. 1988. U.S. Atlantic continental margin; structural and tectonic framework. *In*: SHERIDAN, R.E. & GROW, J.A. (eds) *The Atlantic Continental Margin*. Geological Society of America, Geology of North America, **I-2**, 19–55.

KRAWCZYK, C.M., RESTON, T.J., BESLIER, M.-O. & BOILLOT, G. 1996. Evidence for detachment tectonics on the Iberia Abyssal Plain rifted margin. *In*: WHITMARSH, R.B., SAWYER, D.S. & KLAUS, A. (eds) *Proceedings of the Ocean Drilling Program, Scientific Results, 149*. Ocean Drilling Program, College Station, TX, 603–615.

KUSZNIR, N.J. & PARK, R.G. 1984. Intraplate lithosphere deformation and the strength of the lithosphere. *Geophysical Journal of the Royal Astronomical Society*, **79**, 513–538.

LEEMAN, W.P. & HARRY, D.L. 1993. A binary source model for extension-related magmatism in the Great Basin, western North America. *Science*, **262**, 1550–1554.

LISTER, G.S., ETHERIDGE, M.A. & SYMONDS, P.A. 1986. Detachment faulting and the evolution of passive continental margins. *Geology*, **14**, 246–250.

LORENZO, J.M., MUTTER, J.C., LARSON, R.L. & NW Australia Study Group 1991. Development of the continent–ocean transform boundary of the southern Exmouth Plateau. *Geology*, **19**, 843–846.

MASSON, D.G., MONTEDERT, L. & SCRUTTON, R.A. 1985. Regional geology of the Goban Spur continental margin. *In*: DE GRACIANSKY, P.C., POAG, C.W. & CUNNINGHAM, R. (eds) *Initial Reports of the Deep Sea Drilling Project, 60*. 1115–1139. US Government Printing Office, Washington, DC.

MCCARTHY, J., LARKIN, S.P., FUIS, G.S., SIMPSON, R.W. & HOWARD, K.A. 1991. Anatomy of a metamorphic core complex: seismic refraction/wide-angle reflection profiling in southeastern California and western Arizona. *Journal of Geophysical Research*, **96**, 12259–12292.

MCKENZIE, D. 1984. The generation and compaction of partially molten rock. *Journal of Petrology*, **25**, 713–765.

MCKENZIE, D. & BICKLE, M.J. 1988. The volume and composition of melt generated by extension of the lithosphere. *Journal of Petrology*, **29**, 625–679.

MONTADERT, L., ROBERTS, D.G., CHARPEL, O.D. & GUENNOC, P. 1979. Rifting and subsidence of the northern continental margin of the Bay of Biscay. *In*: MONTADERT, L., ROBERTS, D.G. & AUFFRET, G.A. (eds) *Initial Reports of the Deep Sea Drilling Project, 48*. US Government Printing Office, Washington, DC, 1025–1060.

MOONEY, W.D., LASKE, G. & MASTERS, T.G. 1998. CRUST 5.1: a global crustal model at $5° \times 5°$. *Journal of Geophysical Research*, **103**, 727–747.

MUTTER, J.C., BUCK, W.R. & ZEHNDER, C.M. 1988. Convective partial melting, 1, A model for the formation of thick basaltic sequences during the initiation of spreading. *Journal of Geophysical Research*, **93**, 1031–1948.

MUTTER, J.C., LARSON, R.L. & NORTHWEST AUSTRALIA STUDY GROUP 1989. Extension of the Exmouth Plateau, offshore northwestern Australia: deep seismic reflection/refraction evidence for simple and pure shear mechanisms. *Geology*, **17**, 15–18.

PICKUP, S.L.B., WHITMARSH, R.B., FOWLER, C.M.R. & RESTON, T.J. 1996. Insight into the nature of the ocean–continent transition off West Iberia from a deep multichannel seismic reflection profile. *Geology*, **24**, 1079–1082.

PINHEIRO, L., WHITMARSH, R. & MILES, P. 1992. The ocean–continent boundary off the western continental margin of Iberia—II. Crustal structure in the Tagus abyssal plain. *Geophysical Journal International*, **109**, 106–124.

PINHEIRO, L.M., WILSON, R.C.L., PENA DOS REIS, R., WHITMARSH, R.B. & RIBEIRO, A. 1996. The western Iberia margins: a geophysical and geological overview. *In*: WHITMARSH, R.B., SAWYER, D.S., KLAUS, A. & MASSON, D.G. (eds) *Proceedings of the Ocean Drilling Program, Scientific Results, 149*. Ocean Drilling Program, College Station, TX, 3–23.

PURDY, G.M. 1975. The eastern end of the Azores–Gibraltar plate boundary. *Geophysical Journal of the Royal Astronomical Society*, **43**, 973–1000.

RANALLI, G. & MURPHY, D.C. 1987. Rheological stratification of the lithosphere. *Tectonophysics*, **132**, 281–295.

RESTON, T.J. 1996a. The S reflector off Galicia: the seismic signature of a detachment fault. *Geophysical Journal International*, **127**, 230–244.

RESTON, T.J. 1996b. The S reflector west of Galicia (Spain): evidence from prestack migration for detachment faulting during continental breakup. *Journal of Geophysical Research*, **101**, 8075–8091.

SAWYER, D.S., WHITMARSH, R.B. & KLAUS, A. *Proceedings of the Ocean Drilling Program, Initial Reports. 149*. Ocean Drilling Program, College Station, TX.

SCHARER, U., KORNPROBST, J., BESLIER, M.-O., BOILLOT, G. & GIRARDEAU, J. 1995. Gabbro and related rock emplacement beneath rifting continental crust: U–Pb geochronological and geochemical constraints for the Galicia passive margin (Spain). *Earth and Planetary Science Letters*, **130**, 187–200.

SCLATER, J.G., JAUPART, C. & GALSON, D. 1980. The heat flow through oceanic and continental crust and the heat loss of the Earth. *Reviews of Geophysics and Space Physics*, **18**, 269–311.

SKOGSEID, J. 1994. Dimensions of the Late Cretaceous–Paleocene northeast Atlantic rift derived from Cenozoic subsidence. *Tectonophysics*, **240**, 225–247.

SKOGSEID, J., PEDERSON, T., ELDHOLM, O. & LARSEN, B.T. 1992. Tectonism and magmatism during NE Atlantic continental break-up: the Voring Margin. *In*: STOREY, B.C., ALABASTER, T. & PANKHURST, R.J. (eds) *Volcanism and Continental Break-up: a Global Compilation of Large Igneous Provinces*. Geological Society, London, Special Publications, **68**, 305–320.

SONDER, L.J. & ENGLAND, P.C. 1989. Effects of temperature-dependent rheology on large-scale continental extension. *Journal of Geophysical Research*, **94**, 7603–7619.

SRIVASTAVA, S.P. & TAPSCOTT, C.R. 1986. Plate kinematics of the North Atlantic. *In*: VOGT, P.R. & TUCHOLKE, B.E. (eds) *The Western North Atlantic Region. Geological Society of America*. Geological Society of America, The Geology of North America, **M**, 379–404.

SRIVISTAVA, S.P., ROEST, W.R., KOVACS, L.C., LEVESQUE, S., VERHOEF, J. & MACNAB, R. 1990. Motion of Iberia since the Late Jurassic: results from detailed aeromagnetic measurements in the Newfoundland basin. *Tectonophysics*, **184**, 229–260.

TETT, D.L. & SAWYER, D.S. 1996. Dynamic models of multiphase continental rifting and their implications for the Newfoundland and Iberia conjugate margins. *In*: WHITMARSH, R.B., SAWYER, D.S., KLAUS, A. & MASSON, D.G. (eds) *Proceedings of the Ocean Drilling Program, Scientific Results, 149*. Ocean Drilling Program, College Station, TX, 635–648.

WARD, P.L. 1991. On plate tectonics and the geologic evolution of southwestern North America. *Journal of Geophysical Research*, **96**, 12479–12496.

WHITE, R.S. 1992. Crustal structure and magmatism of North Atlantic continental margins. *Journal of the Geological Society, London*, **149**, 841–854.

WHITE, R. & MCKENZIE, D. 1989. Magmatism at rift zones: the generation of volcanic continental margins and flood basalts. *Journal of Geophysical Research*, **94**, 7685–7729.

WHITE, R.S., SPENCE, G.D., FOWLER, S.R., MCKENZIE, D.P., WESTBROOK, G.K. & BOWEN, A.N. 1987. Magmatism at rifted continental margins and flood basalts. *Nature*, **330**, 439–444.

WHITMARSH, R.B. & MILES, P.R. 1995. Models of the development of the west Iberia rifted continental margin at 40°30′N deduced from surface and deep-tow magnetic anomalies. *Journal of Geophysical Research*, **100**, 3789–3806.

WHITMARSH, R.B. & SAWYER, D.S. 1993. The ocean/continent transition beneath the Iberia abyssal plain and continental-rifting to seafloor-spreading processes. *In*: WHITMARSH, R.B., SAWYER, D.S., KLAUS, A. & MASSON, D.G. (eds) *Proceedings of the Ocean Drilling Program, Scientific Results, 149*. Ocean Drilling Program, College Station, TX, 713–733.

WHITMARSH, R.B., BESLIER, M.-O. & WALLACE, P.J. *Proceedings of the Ocean Drilling Program, Initial Reports, 149*. Ocean Drilling Program, College Station, TX.

WHITMARSH, R.B., MILES, P.R. & MAUFFRET, A. 1990. The ocean–continent boundary off the western continental margin of Iberia—I. Crustal structure at 40°30′N. *Geophysical Journal International*, **103**, 509–531.

WHITMARSH, R.B., WHITE, R.S., HORSEFIELD, S.J., SIBUET, J.-C., RECQ, M. & LOUVEL, V. 1996. The ocean–continent boundary off the western continental margin of Iberia: crustal structure west of Galicia Bank. *Journal of Geophysical Research*, **101**, 28291–28314.

WILLIAMSON, M.-C., COURTNEY, R.C., KEEN, C.E. & DEHLER, S.A. 1995. The volume and rare earth concentrations of magmas generated during finite stretching of the lithosphere. *Journal of Petrology*, **36**, 1433–1453.

WILSON, J.M., MCCARTHY, J., JOHNSON, R.A. & HOWARD, F.A. 1991. An axial view of a metamorphic core complex: crustal structure of the Whipple and Chenehuevi Mountains, southeastern California. *Journal of Geophysical Research*, **96**, 12293–12312.

WILSON, R.C.L., SAWYER, D.S., WHITMARSH, R.B., ZERONG, J. & CARBONELL, J. 1996. Seismic stratigraphy and tectonic history of the Iberia Abyssal Plain rifted margin. *In*: WHITMARSH, R.B., SAWYER, D.S., KLAUS, A. & MASSON, D.G. (eds) *Proceedings of the Ocean Drilling Program, Scientific Results, 149*. Ocean Drilling Program, College Station, TX, 617–633.

ZIEGLER, P.A. 1989. Evolution of the North Atlantic—an overview. *In*: TANKARD, A.J. & BALKWILL, H.R. (eds) *Extensional Tectonics and Stratigraphy of the North Atlantic Margins*. American Association of Petroleum Geologists, Memoirs, **46**, 111–129.

Anomalous melt production after continental break-up in the southern Iberia Abyssal Plain

T.A. MINSHULL[1], S.M. DEAN[1], R.S. WHITE[2] & R.B. WHITMARSH[3]

[1]*School of Ocean and Earth Science, University of Southampton, Southampton Oceanography Centre, European Way, Southampton SO14 3ZH, UK (e-mail: tmin@soc.soton.ac.uk)*

[2]*Bullard Laboratories, Department of Earth Sciences, University of Cambridge, Madingley Road, Cambridge CB3 0EZ, UK*

[3]*Challenger Division, Southampton Oceanography Centre, European Way, Southampton SO14 3ZH, UK*

Abstract: Recent geophysical work and Ocean Drilling Program drilling in the southern Iberia Abyssal Plain have indicated that, in a transition zone up to 170 km wide between thinned continental crust and oceanic crust, the basement consists of serpentinized peridotite mantle with sparse mafic intrusive or extrusive rocks. There is no evidence for the addition of significant magmatic material to the stretched continental crust landward of this zone during the last phase of rifting, whereas seaward of this zone, where the half-spreading rate is about $10\,\mathrm{mm\,a^{-1}}$, the crust rapidly reaches a thickness of $c.\ 6\,\mathrm{km}$, which is normal for Atlantic oceanic crust. Models of melt generation during pure shear, finite-duration continental rifting can successfully reproduce the observed absence of significant syn-rift magmatism on, within and beneath the thinned continental crust if the rifting episode is longer than 10–20 Ma. However, for normal mantle potential temperatures, such models predict significant melt generation in the transition zone seaward of the thinned continental crust even for rift durations longer than 20 Ma. Restricted melting beneath the transition zone might be explained partly by lateral heat loss to the adjacent continental lithosphere, by anomalously low mantle potential temperatures at break-up time, or by depth-dependent stretching such that the observed infinite stretching factor for the crust is not representative of the lithosphere as a whole. An additional mechanism for restricted melt production involves a transitional state between the end of continental extension and the onset of steady-state sea-floor spreading, during which mantle upwelling is less focused than at normal oceanic spreading centres.

Seismic observations of rifted continental margins over the past two decades have led to the definition of 'non-volcanic' and 'volcanic' end-members. Non-volcanic margins preserve the clearer record of lithospheric deformation during rifting because the tectonic signature of continental break-up is largely obscured at volcanic margins by magmatism and subaerial erosion. The West Iberia margin in the southern Iberia Abyssal Plain (Fig. 1) has become the best-studied section of non-volcanic rifted margin in the world, forming the target for extensive geophysical studies (e.g. Pickup *et al.* 1996; Whitmarsh & Miles 1995; Discovery 215 Working Group 1998) and two Ocean Drilling Program (ODP) legs (Sawyer *et al.* 1994; Whitmarsh *et al.* 1998). The most recent geophysical results are summarized in the accompanying paper by Whitmarsh *et al.* (2001).

Here we focus on numerical models of melt generation and their application to the southern Iberia Abyssal Plain. The main observation that needs to be satisfied by these models is that very little melt appears to have been generated during or immediately after continental break-up (Whitmarsh *et al.* 2001), either beneath the stretched continental crust or beneath the transition zone, which reaches a width of $c.\ 170\,\mathrm{km}$ on the IAM-9 seismic profile (Fig. 1; Pickup *et al.* 1996), although oceanic crust of normal thickness was generated within $c.\ 30\,\mathrm{km}$ seaward of the transition zone (Dean *et al.* 2000). The transition zone is characterized by relatively low seismic velocities (typically 4–$5\,\mathrm{km}$ $\mathrm{s^{-1}}$) at the top of the crystalline basement, a steep velocity gradient in the upper $2\,\mathrm{km}$ of basement to reach velocities of 7.0–$7.5\,\mathrm{km\ s^{-1}}$, and then a smooth low-gradient transition to

From: WILSON, R.C.L., WHITMARSH, R.B., TAYLOR, B. & FROITZHEIM, N. 2001. *Non-Volcanic Rifting of Continental Margins: A Comparison of Evidence from Land and Sea*. Geological Society, London, Special Publications, **187**, 537–550. 0305-8719/01/$15.00 © The Geological Society of London 2001.

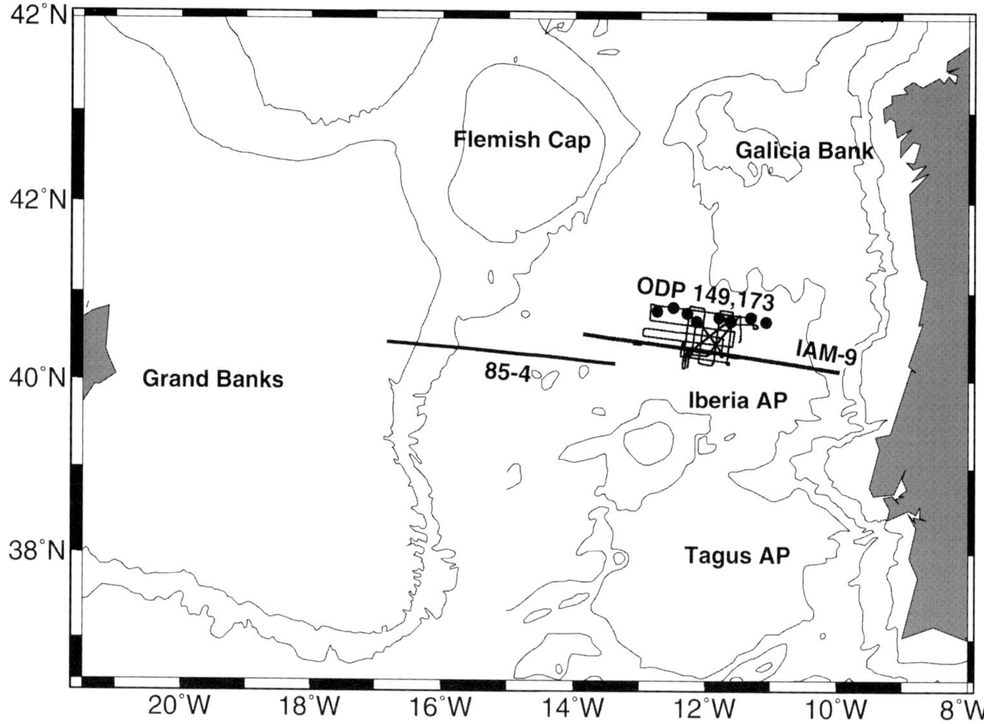

Fig. 1. Reconstruction of the West Iberia margin to the time of magnetic anomaly M0, using the pole of Srivastava *et al.* (1990), with Iberia fixed. Fine lines mark 1 km, 2 km and 4 km bathymetric contours on the Iberia side and 1 km and 2 km contours on the Newfoundland side. Bold continuous lines mark tracks of multichannel seismic profiles IAM-9 and 85-4 (Fig. 3), which overlap in the reconstruction because both extend across the M0 isochron; lines of intermediate thickness mark other profiles with wide-angle seismic coverage in the southern Iberia Abyssal Plain (Discovery 215 Working Group 1998). ●, borehole sites from ODP Legs 149 and 173.

normal upper-mantle velocities of 7.9–8.0 km s^{-1} over a depth interval of at least 2 km, with no major velocity discontinuity marking the Moho (Fig. 2). ODP drilling on basement highs at Sites 897, 899 and 1068 in the transition zone and at its borders has recovered significant quantities of peridotite mantle but no *in situ* syn-rift melt products; from the drilling and geophysical evidence we intepret the transition zone basement as consisting primarily of upper-mantle material that was exhumed to the sea bed between the end of continental rifting and the onset of sea-floor spreading, and intruded by small volumes of magmatic material (Whitmarsh *et al.* 2001).

Zones adjacent to continental margins with the geophysical characteristics of the transition zone in the southern Iberia Abyssal Plain appear also to exist elsewhere on the West Iberia margin, on its conjugate, and on several other non-volcanic margins in the North Atlantic (Fig. 2), although no other margin is so well characterized, and in several cases the velocity structures have not previously been interpreted as belonging to transition zones. A common characteristic of the lithosphere beneath the abyssal plain immediately adjacent to these margins is the presence of a region around 2–6 km into the basement with velocities in the range 7.2–7.6 km s^{-1}, which is never found in normal oceanic crust. Such velocities, although typically toward the lower end of this range, are commonly found at volcanic rifted margins, where they are attributed to magmatic underplating or lower-crustal magmatic intrusion (e.g. White & McKenzie 1989), but on volcanic margins the high-velocity material occurs at greater depths (typically >5 km) below basement than in the transition zones of non-volcanic margins and also extends landward beneath the continental slope.

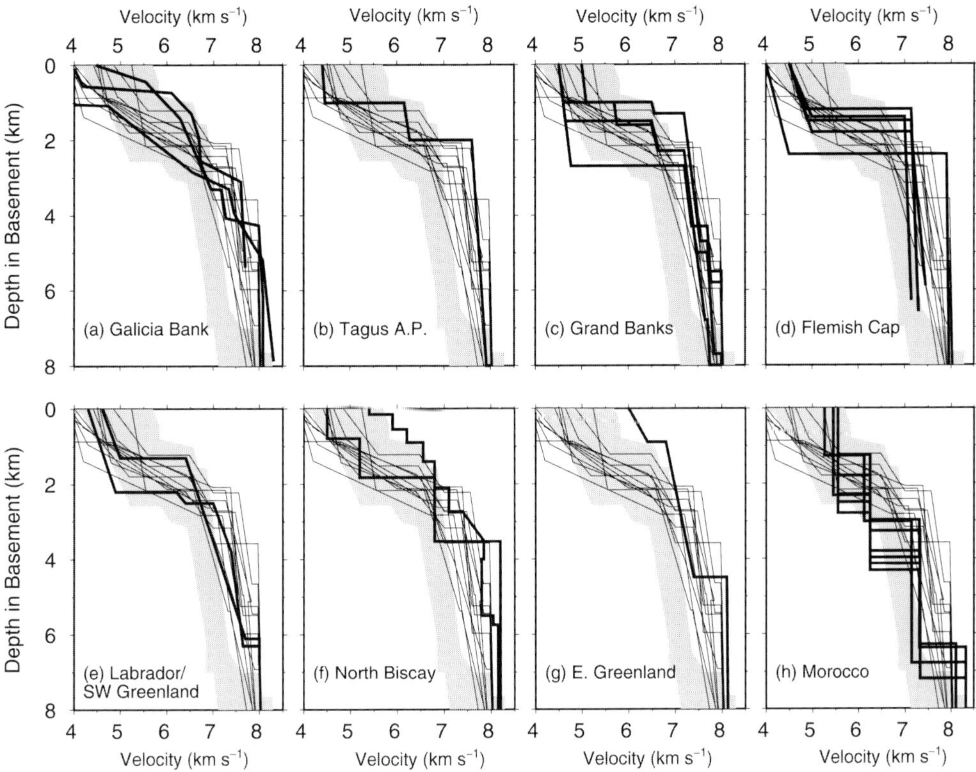

Fig. 2. Seismic velocity structure of acoustic basement in possible ocean–continent transition zones on the West Iberia margin, its conjugate, and other non-volcanic margins in the North Atlantic. In each panel, bold lines mark seismic velocities for the margin indicated, finer lines are velocity structures for various locations in the transition zone in the southern Iberia Abyssal Plain (as compiled by Whitmarsh *et al.* 2001, fig. 5a), and stippled area marks the envelope for Atlantic oceanic crust aged 59–170 Ma (White *et al.* 1992). (**a**) Galicia Bank; structures at the crossing points between Line 6 and Lines 4, 7E and 7W of Whitmarsh *et al.* (1996). (**b**) Tagus Abyssal Plain; velocity structure is from the centre of Line 5 of Pinheiro *et al.* (1992). (**c**) Grand Banks; velocity structures at OBS B, D, K and R of Reid (1994). (**d**) Flemish Cap; 1D velocity models from lines HU-1, HU-2, HU-6 and HU-18 of Todd & Reid (1989). (**e**) Conjugate margins of Labrador and SW Greenland; velocity structures from the centre of the transition zone in the model of Chian & Louden (1994) and at OBS A in the model of Chian *et al.* (1995). (**f**) Northern bay of Biscay; 1D velocity models from Limond *et al.* (1972); Whitmarsh *et al.* (1986), who interpreted the structure as representing thinned continental crust. (**g**) East Greenland margin at the Liverpool Land shelf; velocity structure at OBS 6 of Weigel *et al.* (1995). This portion of the margin was formed in Eocene time by a ridge jump, remote from the Iceland plume, in contrast to the remainder of the East Greenland margin. (**h**) Moroccan margin; models for sonobuoys 25, 26, 27, 40, 41, 47 and 86 of Holik *et al.* (1992), who interpreted velocities of 7.0–7.2 km s⁻¹ as due to magmatic underplating.

Margin cross-sectional shape and rift duration

Melt generation before continental break-up may be inhibited by both vertical and lateral heat conduction (Pederson & Ro 1992; Bown & White 1995); the relative importance of these two processes depends on the geometry of the seaward-thinning lithospheric wedge formed at the continental margin and on the duration of the preceding rifting episodes. In the calculations that follow, we define the location of continental break-up as the point between the seaward limit of stretched continental crust and the landward limit of the transition zone. This is the point of break-up of continental crust, which is readily defined by geophysical data, but may or may not correspond to the point of break-up of continental lithospheric mantle; the definition is unambiguous on IAM-9, but would have to be qualified further for the ODP Legs 149 and 173 transect to the north, where Site 1069 sampled continental material seaward of serpentinized peridotite transition zone material at Site 1068 (ODP Leg 173 Shipboard Scientific Party 1998).

We define the onset of sea-floor spreading to occur at the seaward limit of the transition zone. We also assume that formation (i.e. exhumation) of the transition zone followed continental break-up, so that there was no significant continuing extension of the continental crust contemporaneous with the formation of the transition zone. The sparsity in seismic reflection profiles of syn-rift sediment packages over continental fault blocks suggests that these blocks, which presumably represent the final stages of continental break-up, may not have not had a protracted history of rotation, and therefore that this final stage was relatively short-lived (Wilson *et al.* 2001).

The amount of continental extension may be assessed from the cross-sectional shape of a margin (e.g. Sawyer 1985). It is clear that for a full analysis, data from both conjugate margins are required. Unfortunately, break-up between Iberia and North America appears to have occurred towards the end of the M-series of sea-floor spreading magnetic anomalies and shortly before the Cretaceous normal polarity superchron (Pinheiro *et al.* 1996), so there are few sea-floor spreading anomalies immediately seaward of the transition zone, and there is no clear set of gravity lineations such as might define fracture zones that establish the precise direction of extension. Consequently, plate reconstructions have considerable uncertainty and it is difficult to know unambiguously the pre-break-up locations and direction of extension of the conjugate margins. Here we use the M0 pole of Srivastava *et al.* (1990) to define the configuration shortly after break-up in the southern Iberia Abyssal Plain (Fig. 1). In this reconstruction, the Lithoprobe profile 85-4 of Keen & de Voogd (1988) on the North American margin is only 30–40 km offset from IAM-9 on the Iberia margin.

On the Iberia side, the amount of continental extension may be estimated from the crustal thickness variations defined by wide-angle seismic data (Dean *et al.* 2000). There are still approximations involved because the prerift configuration is not known precisely and the crust–mantle boundary at the continental end of Dean *et al.*'s model is constrained only by gravity modelling. On the Newfoundland side, there are no published refraction data on line 85-4, but the shape of the continental crustal

wedge defining the margin can be estimated (following, e.g. Dunbar & Sawyer 1989) by assuming that the margin is in local isostatic equilibrium. Sea-bed and basement two-way travel times were picked from the seismic reflection profile, and basement depths were computed using the sediment velocity function of Tucholke & Fry (1985). Sediment densities were computed using an exponential porosity function (Sclater & Christie 1980) with a sea-bed porosity of 60%, a compaction length of 1.82 km and a sediment grain density of $2762\,\mathrm{kg\ m^{-3}}$; isostatic calculations assumed densities of 1030, 2800 and $3330\,\mathrm{kg\ m^{-3}}$ for the sea water, crust and mantle, respectively. The resulting crustal structure (Fig. 3) shows that at the latitude of IAM-9, the margin is broadly symmetric, and that on both sides of the rift the crust thins abruptly from a prerift thickness to *c.* 5–7 km in a horizontal distance of only *c.* 100 km. There is some degree of asymmetry in overall basement depth; the origin of this asymmetry is unknown.

Once the margin cross-sectional shape has been defined, the amount of crustal extension, and hence of lithospheric plate separation, required to form this shape may be estimated (assuming large-scale pure shear) by restoring the thinning wedge to a rectangular block with the prerift crustal thickness. Here we consider only the final stage of extension that led to break-up. On IAM-9, the crustal thickness immediately before this stage is taken to be 29 km, which is the thickness at the landward end of the model of Dean *et al.* (2000). On the Newfoundland side, the seismic experiment of Reid (1994), *c.* 200 km to the south of profile 85-4, gives a thickness of *c.* 27 km for the crystalline continental crust; significantly larger values would imply unusually large values for the oceanic crustal thickness in the isostatically balanced section of Figure 3. Uncertainties of a few kilometres in the prerift crustal thickness make little difference to the estimate of total crustal extension across the margin, which is determined as *c.* 40 km on IAM-9 and 56 ± 4 km (for a prerift thickness of 27 ± 2 km) on profile 85-4, to give a total extension of *c.* 100 km.

The main uncertainty in the above calculation arises from the presence of marginal basins on the Grand Banks, and onshore and on the continental shelf of Portugal, in which contemporaneous extension is well documented (e.g. Tankard & Welsink 1989; Pinheiro *et al.* 1996). Although this extension is well determined in many locations by subsidence data, the corresponding integrated plate separation is

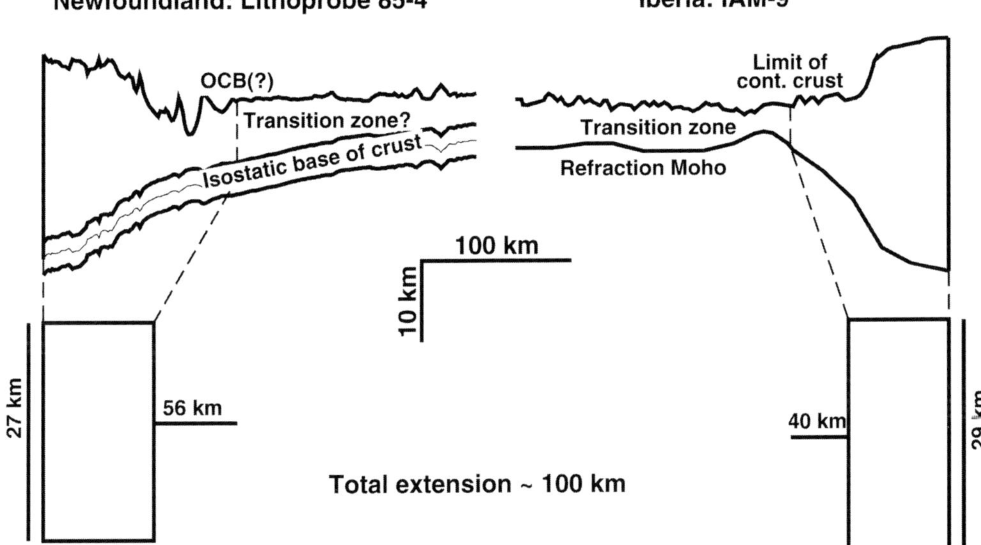

Fig. 3. Cross-sections of the West Iberia and Newfoundland margins. Basement, Moho and the seaward limit of continental crust on the Iberia margin are taken from the model of Dean *et al.* (2000). On the Newfoundland margin, basement depths come from the time-migrated seismic line, depth of the crust–mantle boundary is estimated (within bounds) by the approach described in the text, and the ocean–continent boundary (OCB) of Keen & de Voogd (1988) is taken to mark the seaward limit of continental crust. Rectangular blocks have the same area as the corresponding blocks of stretched continental crust.

difficult to quantify. Here we merely assume that the cumulative amount of extension further landward on both margins in the final break-up stage, as calculated above, is not more than the extension concentrated at the margin, i.e. not more than, and probably considerably less than, 100 km. The total accumulated extension during this stage was then in the range 100–200 km. Other uncertainties arise from a lack of knowledge of the true crustal thickness before the final stage of rifting, uncertainty in the isostatic crust–mantle boundary determined for profile 85-4, and uncertainty in the location of the seaward limit of continental crust, particularly on the Newfoundland margin, but these uncertainties are likely to be proportionally smaller.

The total extension, the rift duration and the extension rate are all closely related, and the extension rate immediately before continental break-up, the extension rate during formation of the transition zone, and the full sea-floor spreading rate immediately after formation of the transition zone probably varied as a continuous function of time. The extension rate may have varied with time in a complex manner, as has been inferred for many continental basins from the analysis of subsidence data (e.g. Newman & White 1999). Unfortu-

nately, this time variation is not usefully constrained by subsidence data on the margin in the southern Iberia Abyssal Plain because of the sparsity of syn-rift sediments and the lack of recovery, by drilling, of sediments deposited at depths intermediate between shelf and abyssal depths. Indeed, the large uncertainty of palaeodepth estimates in this intermediate range (e.g. Murray 1991, pp. 288–290) means that the time variation could only be recovered with any confidence on a margin where the sedimentation was sufficiently rapid to keep the sea bed at shelf depths right up to the time of break-up, and drilling to basement on such a margin, even if one were identified, is not feasible with current technology.

The simplest assumption that can be made is that rifting started instantaneously at its final rate; such a scenario is often used in numerical models (e.g. Dunbar & Sawyer 1989; Bassi 1995) and is consistent with data from the Woodlark Basin (Taylor *et al.* 1999). The next simplest assumption is that the extension rate increases linearly from zero to its final value at the time of continental break-up (Fig. 4), and then remains constant during formation of the transition zone and subsequent sea-floor spreading. Such a variation would result from rift propagation at a constant rate, with a constant angular velocity about a pole moving at constant velocity. In this scenario, the rift duration along a small circle about the pole is given by the total continental extension divided by half the final extension rate. The initial sea-floor spreading half-rate from modelling of magnetic anomalies could be as high as $17.5 \, \text{mm a}^{-1}$, but a value of *c.* $10 \, \text{mm a}^{-1}$ gives the best match to deep-towed magnetometer data (Whitmarsh & Miles 1995). For symmetric spreading, assuming no rate changes during the formation of the transition zone, and for a total continental extension of $100-200 \, \text{km}$ the rift duration on IAM-9/85-4 is therefore *c.* $10-20 \, \text{Ma}$ (Fig. 5), with values toward the lower end of this range preferred. These values would be halved if extension started instantaneously at its final rate, and also reduced if the final rate was significantly higher.

Given an age of $127 \, \text{Ma}$ for the beginning of magnetic anomaly M3 (Gradstein *et al.* 1994), the first clearly identified sea-floor spreading anomaly at the latitude of IAM-9 (Whitmarsh & Miles 1995), and that *c.* $17 \, \text{Ma}$ would be required to form the $170 \, \text{km}$ wide transition zone on IAM-9, this estimate would imply an age of *c.* $154-164 \, \text{Ma}$ for the initiation of the final stage of continental rifting on IAM-9. If rifting propagated from south to north, as suggested by magnetic anomalies (Whitmarsh & Miles 1995), and propagated rather slowly ($10 \, \text{mm a}^{-1}$), the younger limit of these ages is consistent with evidence for rift initiation no earlier than $149 \, \text{Ma}$ on the drilling transect $50 \, \text{km}$ to the north (ODP Leg 173 Shipboard Scientific Party 1998; Dean *et al.* 2000). For a third scenario in which the extension rate increases linearly with time from the initiation of rifting to the initiation of sea-floor spreading, the rift duration would be $21-33 \, \text{Ma}$ with the initiation of rifting at $171-181 \, \text{Ma}$ (Fig. 4). The discrepancy in the inferred onset of rifting with the ages derived from the drilling transect appears then to be unreasonably large. The discrepancy is also large if the mean extension half-rate in the transition zone was significantly less than $10 \, \text{mm a}^{-1}$, as suggested by Srivastava *et al.* (2000), as none of the three scenarios considered above allows the duration of continental extension to be $<5 \, \text{Ma}$, and only $22 \, \text{Ma}$ are available between rift initiation on the drilling transect and anomaly M3 time. From these observations we conclude that the extension half-rate in the transition zone is likely to be close to $10 \, \text{mm a}^{-1}$.

Models of melt generation at continental margins

With the above constraints on rift duration established, we now apply the numerical models of melt production during finite-duration rifting of Bown & White (1995) to the specific case of the West Iberia margin at IAM-9. More recent models have addressed the deformation of the lithosphere in response to extensional stresses in addition to the melt generation (e.g. Harry & Bowling 1999); the advantage of the Bown & White (1995) model is that the deformation field and its evolution over time can be specified by the user to match geological and geophysical observations. Previous application of this model to stretched continental crust on the West Iberia margin has shown that the observed lack of significant syn-rift melt is successfully reproduced (Bown & White 1995; Whitmarsh *et al.* 1996); here we focus on melt production at the time of continental break-up.

We first investigate the effects of lateral heat conduction, which may be expected to be significant because on this profile the crust thins abruptly over a distance of only *c.* $100 \, \text{km}$ (Fig. 3). For 2D modelling we assume that the rift is grossly symmetric, which appears to be a reasonable assumption (Fig. 3). Finite-duration rifting is described in the 2D model by the thinning of a series of rectangular blocks of equal initial width; blocks of $1 \, \text{km}$ width were used to give a reasonably smooth final structure. The model is run to an extension factor of 50, which reproduces adequately the melt thickness for infinite extension, as indicated by the generation of $6 \, \text{km}$ of melt, a normal oceanic crustal thickness, when the rift duration is zero (Fig. 6a; Whitmarsh *et al.* 1996). The results of this modelling show that, for a rift duration of $10-20 \, \text{Ma}$ and a normal mantle potential temperature of $1300 \, ^\circ\text{C}$, the 2D model predicts the generation of $2.5-4 \, \text{km}$ of melt at infinite extension, and a difference in melt thickness between the 1D and 2D models of $<1 \, \text{km}$ for this range of rift durations. For the shorter rift durations that would result if rifting started at its final rate, the model predicts even larger melt thicknesses. The total melt thickness for a given final degree of extension is relatively insensitive to the way the strain rate varies over time (Bown & White 1995). These results apply immediately seaward of the seaward limit of continental crust; further seaward, one would expect the melt thickness to increase

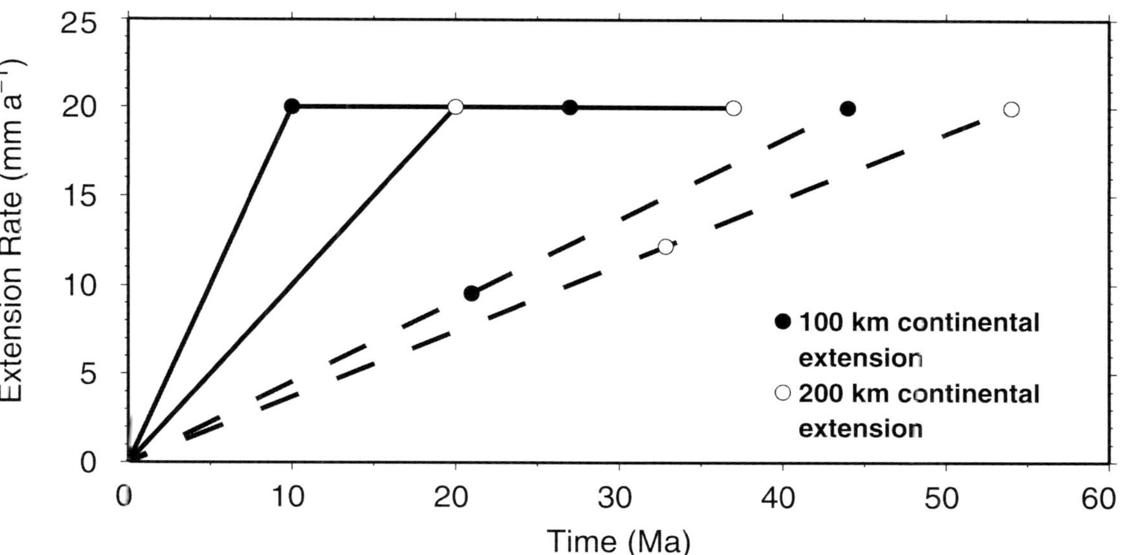

Fig. 4. Possible variation over time of the rate of extension on the West Iberia–Newfoundland margin at the latitude of IAM-9, as discussed in the text. Continuous lines mark scenario in which extension rate increases linearly with time during continental extension and remains constant during formation of the transition zone (preferred model). Dashed lines mark scenario where extension rate increases linearly with time from the onset of rifting to the onset of sea-floor spreading. ●, ○, end of continental extension and onset of sea-floor spreading for 100 km and 200 km continental extension, respectively. Total transition zone width is assumed to be twice the width observed on IAM-9.

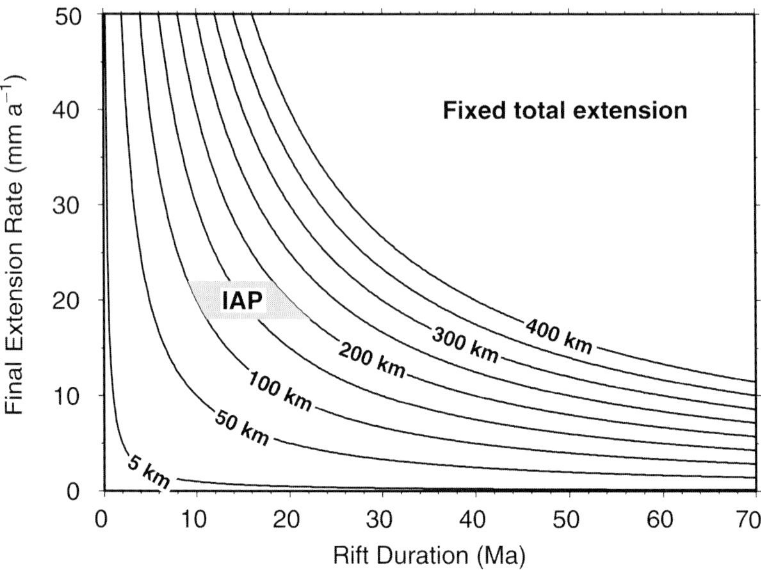

Fig. 5. Final full extension rate as a function of rift duration for a variety of values of total extension of the continental crust on the margin, for an assumed linear increase of extension rate between the initiation of rifting and break-up. For a total extension of 100–200 km on IAM-9 and 85-4, and a final extension rate equal to the full sea-floor spreading rate of *c.* 20 mm a^{-1}, the inferred rift duration is 10–20 Ma (see text).

further from the above values toward a normal oceanic crustal thickness.

In contrast to the above result, geophysical data indicate that very little solidified melt is present in the 170 km wide transition zone on IAM-9 (Whitmarsh *et al.* 2001). On the basis of seismic data alone, Dean *et al.* (2000) inferred a maximum possible melt thickness of 4.5 km in the transition zone, using the extreme assumption that the upper *c.* 3 km of basement, where velocities fall within the envelope for oceanic crust (Fig. 2), consists entirely of melt products, and the next 4 km, where velocities are around 7.5 km s^{-1}, consists of *c.* 40% melt products. However, magnetic data indicate that very little magmatic material is present in the upper basement layer, and that melt products are likely to be confined mainly to sparse intrusions in the lower layer (Whitmarsh *et al.* 2001). A precise upper limit on mean melt thickness that takes account of all the geophysical observations is hard to define; for the purpose of illustration we choose an upper limit of 2 km, corresponding to *c.* 40% of the lower layer (as above) plus *c.* 10% of the upper layer, to define the stippled area of Figure 6a. However, it should be borne in mind that our preferred interpretation is to the lower left of this area. It is clear that lateral heat conduction

alone cannot explain the lack of significant melt products in the transition zone.

A second possibility is that melting was suppressed by an anomalously low mantle potential temperature at the time of continental break-up. Although regions of anomalously high mantle potential temperature appear to be rather common on the Earth (e.g. White & McKenzie 1989), mantle 'cold' spots are rarer. At one such cold spot, the Australian–Antarctic discordance (AAD), the anomalously deep mid-ocean ridge axis implies that the oceanic crust is thinned significantly (e.g. Cochran *et al.* 1997), and this cold spot has a length scale of *c.* 1000–2000 km, somewhat larger than that of the West Iberia margin. One-dimensional modelling shows that a temperature anomaly of 50–100 °C would be required to suppress the melting sufficiently on IAM-9 (Fig. 6b), with a smaller anomaly required if 2D effects are also important. The temperature anomaly required is comparable with anomalies of 80–250 °C (Kuo 1993) or 55–100 °C (Cochran *et al.* 1997) inferred for the AAD. However, the apparent ubiquity at non-volcanic margins of transition zones with similar structures (Fig. 2) makes this an awkward interpretation, as the implication is that volcanic margins were underlain by anomalously hot mantle, non-volcanic margins were underlain by anomalously cold man-

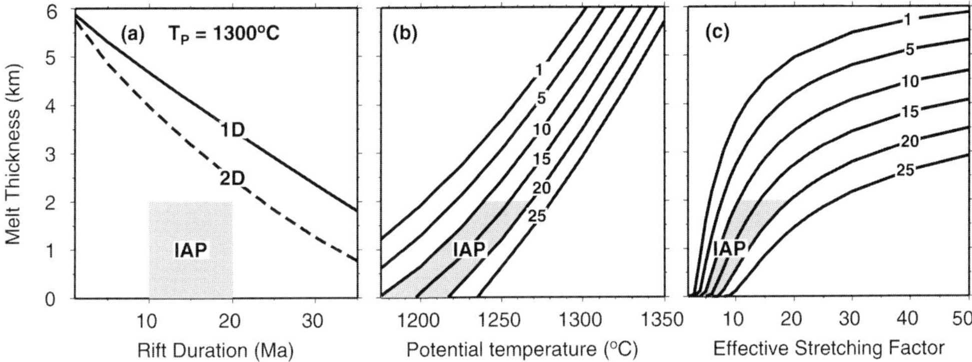

Fig. 6. (**a**) Melt thickness resulting from adiabatic decompression of the mantle according to the model of Bown & White (1995). The 1D solution (continuous line) is for a stretching factor of 50, which is effectively infinite, and is as shown by Whitmarsh *et al.* (1996); the 2D solution comes from a model in which the stretching factor increases linearly over time from one to a value determined at each location by the thickness of stretched continental crust on IAM-9 (Fig. 3). In both cases the mantle potential temperature is 1300 °C, the initial lithospheric thickness is 125 km, and other parameters are as given by Bown & White (1995). Stippled region marks inferred rift duration and estimated average melt thickness (see text) in the transition zone on IAM-9. (**b**) Curves mark melt thickness as a function of mantle potential temperature for a range of rift durations (marked in Ma on the curves). All other parameters are as in (**a**). (**c**) Curves mark melt thickness as a function of lithospheric stretching factor, again for marked rift durations in Ma. All other parameters are as in (**a**).

tle, and few if any margins were underlain by mantle with the same temperature as the mantle underlying most of the ocean basins at the time of their formation. Further research is required to determine whether transition zones of the kind found on the West Iberia margin are indeed ubiquitous at non-volcanic margins.

A third possibility is that the effective infinite stretching factor inferred from the lack of continental crust in the transition zone is not in fact representative of the whole lithosphere. One way this can be achieved is by lithospheric simple shear, which always leads to the generation of less melt for a given extension factor than the corresponding pure shear case (Latin & White 1990). A completely simple shear mechanism for the formation of the margin at the latitude of IAM-9 seems unlikely given the apparent symmetry of the conjugate margins (Fig. 3) and the lack of seismic imaging of a corresponding lithospheric detachment (Pickup *et al.* 1996). Detachments mostly within continental crust have been imaged around the drilling transect (Krawczyk *et al.* 1996; Whitmarsh *et al.* 2000), but these detachments appear to have been formed toward the end of continental deformation. Low-angle detachments could exhume large expanses of mantle lithosphere by the 'rolling-hinge' mechanism of Lavier *et al.* (1999), and might be difficult to detect in seismic reflection profiles if they follow the basement surface and then plunge steeply into basement. However, mantle exhumation by this mechanism requires a relatively thin brittle layer and is likely only during late stages of rifting, when the asthenosphere is already shallow; transport of deeper mantle material is required to fill space. For a broad zone of exhumation, it is not clear that decompression melting would be reduced significantly compared with that resulting from pure shear deformation.

However, a similar effect may be achieved by depth-dependent stretching, as has been inferred from analysis of syn-rift and post-rift sediment thicknesses on several North Atlantic margins (Keen & Dehler 1993). In this case, the result can be symmetric conjugate margins where on both sides the continental lithospheric mantle has been pulled out from beneath the continental crust during the final stages of rifting. We are not able to explicitly model this process, but we are able to estimate the effective lithospheric stretching factor that would be required (Fig. 6c). A factor in the range *c.* 5–15 is required, which for infinite crustal stretching and initial crustal and lithospheric thicknesses of 29 km and 125 km, respectively, corresponds to a stretching factor of 4–12 for the mantle lithosphere. Such a process requires some simple shear deformation, but if that deformation were sufficiently distributed it might not be readily detected in seismic reflection profiles. However, although this process could explain a narrow transition zone of 20–30 km width, it is difficult to see how melt generation could be suppressed by such a process for sufficiently long to generate a zone of 170 km width as on IAM-9, because when no crust is left, all the extension must be taken up in the mantle.

A conceptual model for the transition zone

The broad ocean–continent transition zone in the southern Iberia Abyssal Plain, and the limited seismic evidence for such transition zones on many other non-volcanic margins, poses a problem for models of melting at rifted margins. Although the effects investigated in the previous section may all contribute to suppressing melt production to some degree in some cases, they do not appear to provide a complete explanation. One major consideration is missing from the above analysis: neither the thermal structure nor the chemistry of the mantle beneath the margin at break-up time are the same as that of a steady-state mid-ocean ridge with a spreading rate the same as the extension rate at break-up time (Fig. 7). The effect of horizontal heat conduction has been shown to be relatively small, but the thermal structure also affects mantle flow; as a result, ocean–continent transition zones may result from a transitional stage on the way to steady-state sea-floor spreading in which the mantle flow pattern is different. Here we consider qualitatively such effects by reference to models of melt generation at mid-ocean ridges, which have been developed much further than the equivalent models for margins.

In addition to compositional factors, such as the relationship between melt fraction and temperature, a major control on mantle melting at mid-ocean ridges is the rate of upwelling of the asthenosphere at the ridge axis (e.g. Forsyth 1992). In a passive flow model and for slow spreading rates, the main control on this upwelling rate is the shape of the upwelling region, or equivalently the shape of the base of the lithosphere of the two diverging plates. A narrow upwelling region can lead to a large upwelling rate, and hence efficient melt production, even at very slow spreading rates. In contrast, if the upwelling region remains broad at slow spreading rates, the upwelling rate and hence the predicted melt production and crustal thickness decrease rapidly as the spreading rate decreases (e.g. Shen & Forsyth 1992). The melt generation model used in Fig. 6 does not incorporate the physics determining the shape of the upwelling region. The predicted melt thickness

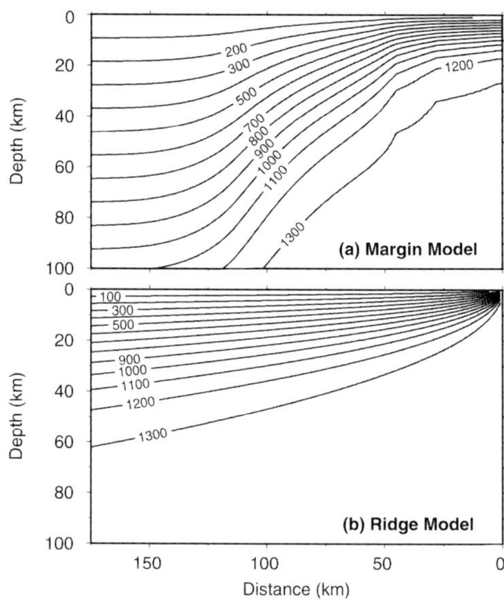

Fig. 7. (**a**) Thermal structure of the lithosphere at break-up time for the 2D model of Fig. 6a with a rift duration of 10 Ma (equivalent to a full extension rate of 20 mm a^{-1} for a total extension of 100 km). (**b**) Thermal structure of the lithosphere for a mid-ocean ridge with a full spreading rate of 20 mm a^{-1} for the plate cooling model of Parsons & Sclater (1977).

immediately seaward of continental crust does decrease as the final extension rate decreases, but this decrease appears only through the relationship between this extension rate, the rift duration, and the total extension (Fig. 8a). In this model, if the extension rate is low then the rift duration is long and the upwelling rate is small, unless the rift is very narrow. Figure 8a is generated using the 1D melting model. The effects of lateral heat conduction will be greatest for a narrow rift and a low value for the total extension, whereas for a broad rift these effects will be negligible; the results for a 2D model would be to move the curves of Figure 8a slightly closer together.

In the case of mid-ocean ridge models, the effect on melt production of the viscosity structure, which controls the shape of the upwelling zone, is illustrated in Figure 8b. For the extreme case of a constant viscosity melting model with plate-driven flow and passive upwelling, the predicted melt thickness drops off rapidly at slow spreading rates as a result of vertical heat conduction over a broad upwelling zone (Figure 8b, continuous curve). For a passive upwelling model in which the thickening lithosphere is represented by a temperature-dependent viscosity structure, the melt thickness also

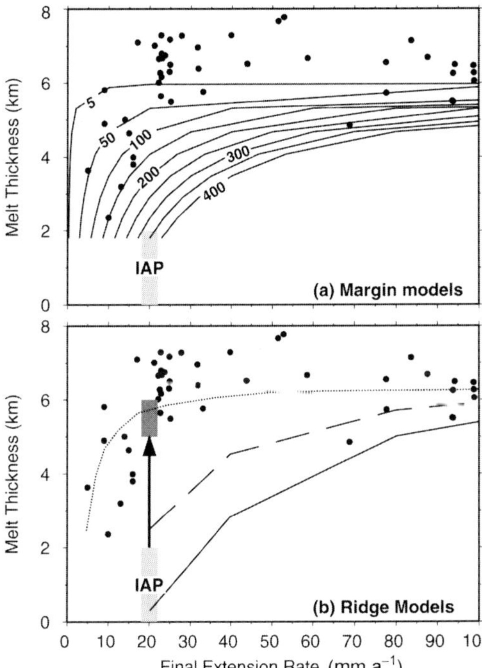

Fig. 8. (**a**) ●, dataset of Bown & White (1994) for oceanic crustal thickness as a function of full spreading rate, with the addition of two along-segment average thicknesses for the Southwest Indian Ridge (Muller *et al.* 1999, 2000) and an average thickness for the Mohns Ridge (Klingelhoefer *et al.* 2000*a*). Curves mark melt thickness as a function of final extension rate for various values of total extension in kilometres of the continental lithosphere; these curves are derived from those of Figure 5 by relating melt thickness to rift duration with the 1D melting model of Figure 6a. Stippled region marks the ocean–continent transition in the southern Iberia Abyssal Plain (see text). (**b**) ●, as in (**a**). Curves mark melt thickness as a function of spreading rate derived from models of melt generation at mid-ocean ridges. Continuous curve marks result of Shen & Forsyth (1992) for a constant mantle viscosity. Dashed curve marks result of Shen & Forsyth (1992) for a viscosity proportional to the third power of temperature (to represent flow by dislocation creep), which generates a lithospheric wedge conformal to isotherms. Dotted curve marks the solution of Bown & White (1994), which has the slope of the base of the lithosphere expressed as an analytical function of the spreading rate. Paler stippled region as in (**a**); darker stippled region marks the first formed oceanic crust on the IAM-9 transect (Dean *et al.* 2000).

decreases at slow spreading rates, but somewhat less rapidly (Fig. 8b, dashed curve) because of the narrowing of the upwelling zone controlled by the model isotherms. The observed variation of crustal thickness with spreading rate can only be achieved in a passive upwelling model if the upwelling zone narrows more rapidly, as implemented in the model of Bown & White (1994) (Fig. 8b, dotted line). In this model the base of the lithosphere is parameterized as a plane of constant slope that is an analytical function of spreading rate.

Although it is clear that part of the geophysically defined crust at slow-spreading ridges consists of serpentinized peridotite (e.g. Cannat 1993), the fact that geophysically determined crustal thicknesses are not normally significantly greater than geochemically determined melt thicknesses for oceanic crust, even at very slow spreading rates (White *et al.* 1992; Klingelhoefer *et al.* 2000*b*), suggests that the proportion of peridotite is usually small. Therefore, we use seismic crustal thicknesses to validate melting models for oceanic crust. The Bown & White (1994) model matches well the observed thicknesses (Fig. 8b), but does not explain the physical origin of the narrowing of the upwelling zone beyond that suggested by the shape of the isotherms. One possible explanation arises from the recent observation that the viscosity of the mantle is strongly dependent on the water content (Karato 1986; Hirth & Kohlstedt 1996). If the water, which is strongly incompatible, is removed by melting in a narrow zone beneath the ridge axis, then upwelling may be confined to a very narrow zone defined by the large viscosity contrasts bounding a 'compositional lithosphere' (Hirth & Kohlstedt 1996; Phipps Morgan 1997).

If the above analysis is correct, then the flow regime required to produce normal thickness oceanic crust at full spreading rates of *c.* 20 mm a⁻¹ can be generated only in the presence of melting, i.e. it is maintained by a positive feedback effect. During most of the continental rifting phase at non-volcanic margins, no melting occurs because the mantle has not risen sufficiently shallow and rapidly enough to cross its solidus (Bown & White 1995). At the time of continental break-up the compositional lithosphere (as defined above) is not present, mantle upwelling is spread over a broad region, and the isotherms close to the rift axis are close to horizontal (Fig. 7a). Under these conditions, the melt generation is similar to that predicted by a model with a constant viscosity mantle (Shen & Forsyth 1992; Fig. 8b, continuous curve). We speculate that during the formation of the transition zone, the thermal and viscosity structures evolve slowly towards those of a mid-ocean ridge, but the final stage of this process proceeds rapidly once sufficient melting has been initiated to focus the flow as a result of the large viscosity increase that occurs as the water in the mantle is removed. The instability involved may explain the observed large variation in transition zone width, even within the southern Iberia Abyssal Plain (Chian *et al.* 1999; Dean *et al.* 2000). Hence, on IAM-9, a normal 6 km thickness oceanic crust lies within *c.* 30 km (representing 2.5–3 Ma of spreading) of a transition zone of 170 km width with little lateral variation in structure that represents *c.* 17 Ma of evolution of the margin.

Conclusions

From our detailed analysis of the basement structure along the IAM-9 profile in the southern Iberia Abyssal Plain and a conjugate profile on the Newfoundland margin, and the application of melting models to this region, we reach the following conclusions.

(1) The deformation of the conjugate Iberia and Newfoundland margins is broadly symmetrical at the latitude of IAM-9.

(2) The duration of continental rifting at this latitude was relatively short, probably *c.* 10 Ma.

(3) Melt production at the time of continental break-up may have been reduced by the effects of lateral heat conduction, depth-dependent stretching or reduced mantle temperatures, but none of these factors on its own can readily explain the observed broad region of low melt production on IAM-9 and at other non-volcanic rifted margins.

(4) Melt generation after the break-up of the continental crust remained limited for a period of up to *c.* 17 Ma during which time the ocean–continent transition zone was formed as the thermal structure evolved to that of a spreading ridge axis and asthenospheric flow became increasingly focused at the site of the new ridge axis.

Our geophysical work on the West Iberia margin was supported by NERC grant GR3/9354. T.A.M. was supported by a Royal Society University Research Fellowship. We thank D. Harry, B. Taylor and T. Reston for thoughtful reviews.

References

BASSI, G. 1995. Relative importance of strain rate and rheology for the mode of continental extension. *Geophysical Journal International*, **122**, 195–210.

BOWN, J.W. & WHITE, R.S. 1994. Variation with spreading rate of oceanic crustal thickness and geochemistry. *Earth and Planetary Science Letters*, **121**, 435–449.

BOWN, J.W. & WHITE, R.S. 1995. The effect of finite extension rate on melt generation at continental rifts. *Journal of Geophysical Research*, **100**, 18011–18030.

CANNAT, M. 1993. Emplacement of mantle rocks in the seafloor at mid-ocean ridges. *Journal of Geophysical Research*, **98**, 4163–4173.

CHIAN, D. & LOUDEN, K.E. 1994. The continent–ocean transition across the southwest Greenland margin. *Journal of Geophysical Research*, **99**, 9117–9135.

CHIAN, D., LOUDEN, K.E., MINSHULL, T.A. & WHITMARSH, R.B. 1999. Deep structure of the ocean–continent transition in the southern Iberia Abyssal Plain from seismic refraction profiles: 1. Ocean Drilling Program (Legs 149 and 173) transect. *Journal of Geophysical Research*, **104**, 7443–7462.

CHIAN, D., LOUDEN, K.E. & REID, I. 1995. Crustal structure of the Labrador Sea conjugate margins and implications for the formation of nonvolcanic continental margins. *Journal of Geophysical Research*, **100**, 24239–24253.

COCHRAN, J.R., SEMPERÉ, J.-C. & SEIR Scientific Team 1997. The Southeast Indian Ridge between 88°E and 118°E: gravity anomalies and crustal accretion at intermediate spreading rates. *Journal of Geophysical Research*, **102**, 15463–15487..

DEAN, S.M., MINSHULL, T.A., WHITMARSH, R.B. & LOUDEN, K.E. 2000. Deep structure of the ocean–continent transition in the southern Iberia Abyssal Plain from seismic refraction profiles: II. The IAM-9 transect at 40°20'N. *Journal of Geophysical Research*, **105**, 5859–5885.

DISCOVERY 215 WORKING GROUP 1998. Deep structure in the vicinity of the ocean–continent transition zone under the southern Iberia Abyssal Plain. *Geology*, **26**, 743–746.

DUNBAR, J.A. & SAWYER, D.S. 1989. Patterns of continental extension along the conjugate margins of the Central and North Atlantic Oceans and Labrador Sea. *Tectonics*, **8**, 1059–1077.

FORSYTH, D.W. 1992. Geophysical constraints on mantle flow and melt generation beneath mid-ocean ridges. *In*: MORGAN, J.P., BLACKMAN, D.K. & SINTON, J.M. (eds) *Mantle Flow and Melt Generation at Mid-Ocean Ridges*. Geophysical Monograph, American Geophysical Union, **71**, 1–65.

GRADSTEIN, F.M., AGTERBERG, F.P., OGG, J.G., HARDENBOL, J., van Veen, P., THIERRY, J. & HUANG, Z. 1994. A Mesozoic time scale. *Journal of Geophysical Research*, **99**, 24015–24074.

HARRY, D.L. & BOWLING, J.C. 1999. Inhibiting magmatism on nonvolcanic rifted margins. *Geology*, **27**, 895–898.

HIRTH, G. & KOHLSTEDT, D.L. 1996. Water in the oceanic upper mantle: implications for rheology, melt extraction and the evolution of the lithosphere. *Earth and Planetary Science Letters*, **144**, 93–108.

HOLIK, J.S., RABINOWITZ, P.D. & AUSTIN, J.A. 1992. Effects of the Canary hotspot volcanism on structure of oceanic crust off Morocco. *Journal of Geophysical Research*, **96**, 12039–12067.

KARATO, S. 1986. Does partial melting reduce the creep strength of the upper mantle. *Nature*, **319**, 309–310.

KEEN, C.E. & DEHLER, S.A. 1993. Stretching and subsidence: rifting of conjugate margins in the North Atlantic region. *Tectonics*, **12**, 1209–1229.

KEEN, C.E. & DE VOOGD, B. 1988. The continent–ocean boundary at the rifted margins off eastern Canada: new results from seismic reflection studies. *Tectonics*, **7**, 107–124.

KLINGELHOEFER, F., GÉLI, L. & MATHIAS, L. 2000a. Crustal structure of a super-slow spreading centre: a seismic refraction study of Mohns Ridge. *72°N. Geophysical Journal International*, **141**, 509–526.

KLINGELHOEFER, F., GÉLI, L. & WHITE, R.S. 2000b. Geophysical and geochemical constraints on crustal accretion at the very slow-spreading Mohns Ridge. *Geophysical Research Letters*, **27**, 1547–1550.

KRAWCZYK, C.M., RESTON, T.J., BESLIER, M.-O. & BOILLOT, G. 1996. Evidence for detachment tectonics on the Iberia Abyssal Plain rifted margin. *In*: SAWYER, D.S., WHITMARSH, R.B & KLAUS, A. (eds) *Proceedings of the Ocean Drilling Program, Scientific Results, 149*. Ocean Drilling Program, College Station, TX, 603–616.

KUO, B.-Y. 1993. Thermal anomalies beneath the Australian–Antarctic discordance. *Earth and Planetary Science Letters*, **119**, 349–364.

LATIN, D.M. & WHITE, N.J. 1990. Generating melt during lithospheric extension: pure shear vs simple shear. *Geology*, **18**, 327–331.

LAVIER, L.L., BUCK, W.R. & POLIAKOV, A.N.B. 1999. Self-consistent rolling-hinge model for the evolution of large-offset low-angle normal faults. *Geology*, **27**, 1127–1130.

LIMOND, W.Q., GRAY, F., GRAU, G. & PATRIAT, PH. 1972. Mantle reflections in the Bay of Biscay. *Earth and Planetary Science Letters*, **15**, 361–366.

MULLER, M.R., MINSHULL, T.A. & WHITE, R.S. 1999. Segmentation and melt supply at the Southwest Indian Ridge. *Geology*, **27**, 867–879.

MULLER, M.R., MINSHULL, T.A. & WHITE, R.S. 2000. Crustal structure of the Southwest Indian Ridge at the Atlantis II Fracture Zone. *Journal of Geophysical Research*, **105**, 25809–25828.

MURRAY, J.W. *Ecology and Palaeoecology of Benthic Foraminifera*. Longman, Harlow.

NEWMAN, R. & WHITE, N. 1999. The dynamics of extensional sedimentary basins: constraints from subsidence inversions. *Philosophical Transactions of the Royal Society of London, Series A*, **357**, 805–834.

ODP Leg 173 SHIPBOARD SCIENTIFIC PARTY 1998. Drilling reveals transition from continental breakup to early magmatic crust. *EOS Transactions, American Geophysical Union*, **79**, 173–181.

PARSONS, B. & SCLATER, J.G. 1977. An analysis of the variation of ocean floor bathymetry and heat flow with age. *Journal of Geophysical Research*, **82**, 803–827.

PEDERSON, T. & RO, H.E. 1992. Finite duration extension and decompression melting. *Earth and Planetary Science Letters*, **113**, 15–22.

PHIPPS MORGAN, J. 1997. The generation of a compositional lithosphere by mid-ocean ridge melting and its effect on subsequent off-axis hotspot upwelling and melting. *Earth and Planetary Science Letters*, **146**, 213–232.

PICKUP, S.L.B., WHITMARSH, R.B., FOWLER, C.M.R. & RESTON, T.J. 1996. Insight into the nature of the ocean–continent transition off West Iberia from a deep multichannel seismic reflection profile. *Geology*, **24**, 1079–1082.

PINHEIRO, L.M., WHITMARSH, R.B. & MILES, P.R. 1992. The ocean–continent boundary off the western continental margin of Iberia—II. Crustal structure in the Tagus Abyssal Plain. *Geophysical Journal International*, **109**, 106–124.

PINHEIRO, L.M., WILSON, R.C.L., PENA DOS REIS, R., WHITMARSH, R.B. & RIBEIRO, A. 1996. The west Iberia margin: a geophysical and geological overview. *In*: SAWYER, D.S., WHITMARSH, R.B & KLAUS, A. (eds) *Proceedings of the Ocean Drilling Program, Scientific Results, 149*. Ocean Drilling Program, College Station, TX, 3–23.

REID, I.D. 1994. Crustal structure of a nonvolcanic rifted margin east of Newfoundland. *Journal of Geophysical Research*, **99**, 15161–15180.

SAWYER, D.S. 1985. Total tectonic subsidence: a parameter for distinguishing crust type at the U.S. Atlantic continental margin. *Journal of Geophysical Research*, **90**, 7751–7769.

SAWYER, D. S., WHITMARSH, R. B., KLAUS, A. & SHIPBOARD SCIENTIFIC PARTY 1994. *Proceedings of the Ocean Drilling Program, Initial Reports, 149*. College Station, TX, Ocean Drilling Program.

SCLATER, J.G. & CHRISTIE, P.A.F. 1980. Continental stretching: an explanation of the post-mid-Cretaceous subsidence of the central North Sea basin. *Journal of Geophysical Research*, **85**, 3711–3739.

SHEN, Y. & FORSYTH, D.W. 1992. The effects of temperature- and pressure-dependent viscosity on three-dimensional passive flow of the mantle beneath a ridge-transform system. *Journal of Geophysical Research*, **97**, 19717–19728.

SRIVASTAVA, S.P., ROEST, W.R., KOVACS, L.C., OAKEY, G., LÉVESQUE, S., VERHOEF, J. & MACNAB, R. 1990. Motion of Iberia since the Late Jurassic: results from detailed aeromagnetic measurements in the Newfoundland Basin. *Tectonophysics*, **184**, 229–260.

SRIVASTAVA, S.P., SIBUET, J.-C., CANDE, S., ROEST, W.R. & REID, I.D. 2000. Magnetic evidence for slow seafloor spreading during the formation of the Newfoundland and Iberian margins. *Earth and Planetary Science Letters*, **182**, 61–76.

TANKARD, A.J. & WELSINK, H.-J. 1989. Mesozoic extension and styles of basin formation in Atlantic Canada. *In*: TANKARD, A.J. & BALKWILL, H.R. (eds) *Extensional Tectonics and Stratigraphy of the North Atlantic Margins*. American Association of Petroleum Geologists, Memoirs, **46**, 175–195.

TAYLOR, B., GOODLIFFE, A.M. & MARTINEZ, F. 1999. How continents break up: insights from Papua New Guinea. *Journal of Geophysical Research*, **104**, 7497–7512.

TODD, B.J. & REID, I. 1989. The continent–ocean boundary south of Flemish Cap: constraints from seismic refraction and gravity. *Canadian Journal of Earth Sciences*, **26**, 1392–1407.

TUCHOLKE, B.E. & FRY, V.A. 1985. Basement structure and sediment distribution in the Northwest Atlantic Ocean. *AAPG Bulletin*, **69**, 2077–2097.

WEIGEL, W., FLÜH, E.R., MILLER, H. & 10 OTHERS 1995. Investigations of the East Greenland continental margin between 70° and 72°N by deep seismic sounding and gravity studies. *Marine Geophysical Researches*, **17**, 167–199.

WHITE, R. & MCKENZIE, D. 1989. Magmatism at rift zones: the generation of volcanic continental margins and flood basalts. *Journal of Geophysical Research*, **94**, 7685–7730.

WHITE, R.S., MCKENZIE, D. & O'NIONS, R.K. 1992. Oceanic crustal thickness from seismic measurements and rare earth element inversions. *Journal of Geophysical Research*, **97**, 19683–19715.

WHITMARSH, R.B. & MILES, P.R. 1995. Models of the development of the west Iberia rifted continental margin at 40°30′N deduced from surface and deep-tow magnetic anomalies. *Journal of Geophysical Research*, **100**, 3789–3806.

WHITMARSH, R.B., AVEDIK, F. & SAUNDERS, M.R. 1986. The seismic structure of thinned continental crust in the Bay of Biscay. *Geophysical Journal of the Royal Astronomical Society*, **86**, 589–602.

WHITMARSH, R.B., BESLIER, M.-O. & WALLACE, P.J. (eds) *Proceedings of the Ocean Drilling Program, Initial Reports, 173*. Ocean Drilling Program, College Station, TX.

WHITMARSH, R.B., DEAN, S.M., MINSHULL, T.A. & TOMPKINS, M. 2000. Tectonic implications of exposure of lower continental crust beneath the Iberia Abyssal Plain, Northeast Atlantic Ocean: geophysical evidence. *Tectonics*, **19**, 919–942.

WHITMARSH, R.B., MINSHULL, T.A., RUSSELL, S.M., DEAN, S.M., LOUDEN, K.E. & CHIAN, D. 2001. The role of syn-rift magmatism in the rift-to-drift evolution of the West Iberia continental margin: geophysical observations. *In*: WILSON, R.C.L., WHITMARSH, R.B., TAYLOR, B. & FROITZHEIM, N. (eds) *Non-volcanic Rifting of Continental Margins: a Comparison of Evidence from Land and Sea*. Geological Society, London, Special Publications, **187**, 107–124.

WHITMARSH, R.B., WHITE, R.S., HORSEFIELD, S.J., SIBUET, J.-C., RECQ, M. & LOUVEL, V. 1996. The ocean–continent boundary off the western continental margin of Iberia: crustal structure west of Galicia Bank. *Journal of Geophysical Research*, **101**, 28291–28314.

WILSON, R.C.L., MANATSCHAL, G. & WISE, S. 2001. Rifting along non-volcanic passive margins: stratigraphic and seismic evidence from the Mesozoic successions of the Alps and Western Iberia. *In*: WILSON, R.C.L., WHITMARSH, R.B., TAYLOR, B. & FROITZHEIM, N. (eds) *Non-volcanic Rifting of Continental Margins: a Comparison of Evidence from Land and Sea*. Geological Society, London, Special Publications, **187**, 429–452.

Serpentinization and magmatism during extension at non-volcanic margins: the effect of initial lithospheric structure

M. PÉREZ-GUSSINYÉ, T.J. RESTON & J. PHIPPS MORGAN

Geomar FZ, Wischhofstr. 1–3, D24148 Kiel, Germany (e-mail: mperez@geomar.de)

Abstract: At several non-volcanic margins serpentinized peridotites occur within a wide continent–ocean transition (COT) and beneath the edge of the thinned continental crust. However, other margins such as the Woodlark Basin appear to have a sharp COT and no reported serpentinites. We investigate the thermal, magmatic and rheological evolution of margins during extension as a function of initial lithospheric structure, rift duration and stretching factor. For cratonic and old orogen models, the entire crust should become brittle at stretching factors *c.* 3–4. The resultant crust-cutting faults allow water to reach and serpentinize the mantle, leading to the development of serpentinite décollements at the crust–mantle boundary and exhumation of mantle at the COT. Our predictions are consistent with the spatial limit and thickness of serpentinites at the SW Greenland and West Iberia margins, and the Rockall Trough. They also explain the absence of a broad zone of unroofed, serpentinized mantle at the COT of the Woodlark Basin: here the crust was too thick and hot for serpentinites to form before break-up. Larger melt production than in the West Iberia type margins and concentration of the lithospheric strength in the crust leads to synchronous crustal separation and lithospheric failure, yielding a sharp COT.

Although many rifted margins are characterized by voluminous magmatic activity at or close to break-up, others exhibit little or no magmatic activity, no seaward-dipping reflector sequences (SDRS) and are termed non-volcanic (e.g. papers in Banda *et al.* 1995). Well-known examples of such margins include the West Iberia–Newfoundland conjugate margins (Boillot & Winterer 1988; Reid 1994; Whitmarsh & Sawyer 1996; ODP Leg 173 Shipboard Scientific Party 1998) (Fig. 1), the south Australian margin (Nicholls *et al.* 1981; Finlayson *et al.* 1998), and portions of the West Greenland–Labrador conjugate margins (Keen *et al.* 1994; Chian *et al.* 1995; Srivastava & Roest 1995; Chalmers 1997) (Figs 1 and 2). All these margins formed by the rifting of cool, normal thickness post-orogenic or cratonic crust, and appear in places to be characterized by the presence of serpentinized peridotites within a broad continent–ocean transition (COT; Fig. 2). Of these, the West Iberia margin has become the 'type example' with intensive geochronological, petrological, magnetic and seismic data collected in the last few years (Boillot & Winterer 1988; Krawczyk *et al.* 1996; Reston *et al.* 1996; Whitmarsh & Sawyer 1996; Whitmarsh *et al.* 1996; Discovery 215 Working Group 1998; ODP Leg 173 Shipboard Scientific

Party 1998; Chian *et al.* 1999; Dean *et al.* 2000). At this margin the last stages of continental extension seem to have occurred along detachment faults, the 'S reflector', in the Deep Galicia Margin (DGM, Reston *et al.* 1996), and the 'H reflector', at the Iberia Abyssal Plain (IAP, Krawczyk *et al.* 1996) (Fig. 3). Parts of these detachments appear to be décollements following the crust–mantle boundary (CMB) and separating very thin continental crust (*c.* 3 km thick, implying a crustal stretching factor of 10) from underlying partially serpentinized mantle (Boillot *et al.* 1989; Chian *et al.* 1999; Figs 2 and 3). Landward of the detachment faults the basement is continental in origin, with the crust thickening landward. Oceanward, the basement consists of exhumed and serpentinized mantle (Boillot & Winterer 1988; Discovery 215 Working Group 1998; ODP Leg 173 Shipboard Scientific Party 1998; Chian *et al.* 1999; Dean *et al.* 2000) in a zone up to 170 km wide (Figs 2a and 3; Dean *et al.* 2000).

In contrast, the Woodlark Basin, another margin described as 'non-volcanic' (Taylor *et al.* 1995, 1999; Mutter *et al.* 1996) (Fig. 1) as SDRS are absent, exhibits a sharp continent–ocean boundary where no mantle serpentinites have been reported, and formed relatively quickly in thick orogenic crust and hot, weak

From: WILSON, R.C.L., WHITMARSH, R.B., TAYLOR, B. & FROITZHEIM, N. 2001. *Non-Volcanic Rifting of Continental Margins: A Comparison of Evidence from Land and Sea.* Geological Society, London, Special Publications, **187**, 551–576. 0305-8719/01/$15.00 © The Geological Society of London 2001.

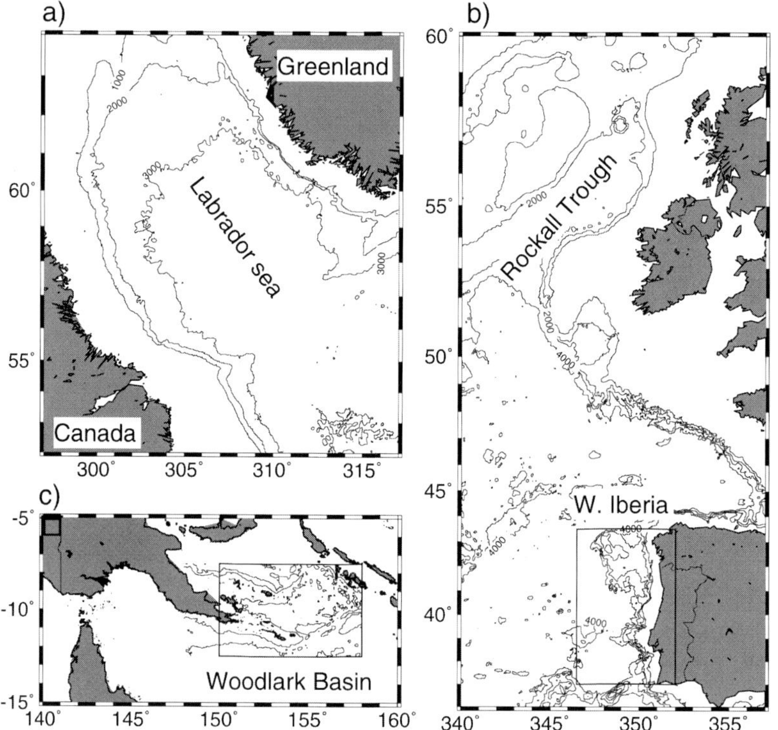

Fig. 1. Location of the four margins discussed here: (**a**) SW Greenland margin; (**b**) Rockall Trough and West Iberia Margin; (**c**) Woodlark Basin.

lithosphere (Mutter 1995; Taylor *et al.* 1999). At the continental margins of this basin numerous small volcanoes, a few kilometres in diameter and often erupted along faults, have been described (Taylor *et al.* 1995), indicating that the degree of magmatism during extension was somewhat higher than at non-volcanic margins of the 'West Iberia type'. The region of active continental extension west of the propagating spreading centre is characterized by numerous full and half-grabens and large metamorphic core complexes on the D'Entrecasteaux islands and the Papuan peninsula (Davies 1980; Davies & Warren 1988).

In this paper we investigate whether the differences between the break-up styles of these margins can be related to differences in their rheological and magmatic evolution. These differences, in turn, arise from the different initial lithospheric structure of each margin, and from variations in the duration and amount of the margin's extension before break-up.

Effect of progressive extension

As extension at finite extension rates proceeds two main processes take place, as follows.

First, the deep crust is gradually brought to shallower depths and cools (Jarvis & McKenzie 1980). The reduction in pressure reduces the brittle yield stress; the decline in temperature increases the required plastic yield stress. Both effects tend to move rocks that were originally in the plastic field into the brittle deformation field. (We follow Rutter (1986) in preferring the term 'plastic' rather than 'ductile' to describe the rheology of rocks deforming by pressure-insensitive creep.) As a result, the originally plastic portions of the lower crust will eventually move into the brittle field and the entire crust will become brittle. When this happens, faults can cut across the entire crust and allow hydrous fluids to penetrate to the mantle. Under appropriate $P-T$ conditions serpentinization will begin (Pérez-Gussinyé & Reston 2001).

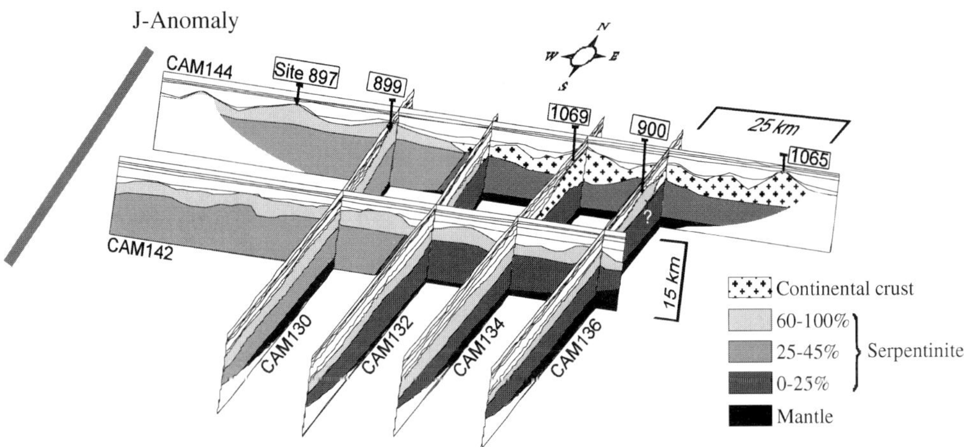

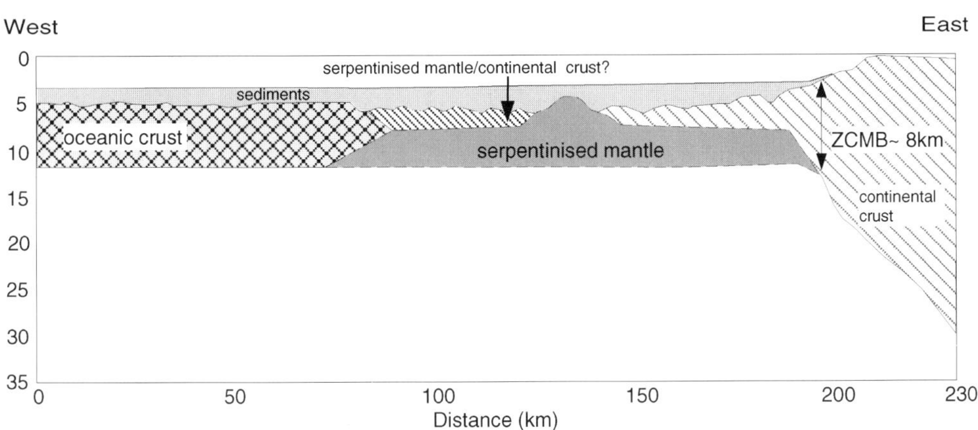

Fig. 2. (a) Three-dimensional interpretation, constrained by results of ODP drilling (Leg 149, 173; see Fig. 3), of wide-angle velocity profiles in the southern Iberia Abyssal Plain (IAP) of the West Iberia margin (Chian *et al.* 1999). **(b)** Cross-section through the SW Greenland margin (after Chian & Louden 1994), showing occurrence of serpentinized peridotites within the continent–ocean transition (COT). Serpentinites occur both beneath thinned continental crust and as top basement where total crustal separation has occurred. The crustal thickness, z_{CMB}, at the landward limit of serpentinization is measured, at the West Iberia margin, at the location shown in Figure 3, and **(b)** at the SW Greenland margin, at the location indicated by the arrow. From here seawards, the entire crust might just have entered the brittle regime, allowing faults to cut into the mantle and bring sea water to serpentinize it.

Second, lithospheric extension induces upwelling of the underlying hot asthenosphere. During upwelling the mantle starts to partially melt when the geotherm intersects the mantle solidus (e.g. McKenzie & Bickle 1988). Bown & White (1995) showed that the stretching factor at which melting starts and the amount of melt produced are both controlled by the rate of extension. At most non-volcanic margins such as West Iberia, little evidence for synrift magmatism has been found, from which Bown & White (1995) inferred that stretching was slow enough for little melt to be formed.

In this paper, we consider the rheological evolution of non-volcanic margins with different initial thermal structures. In particular, we

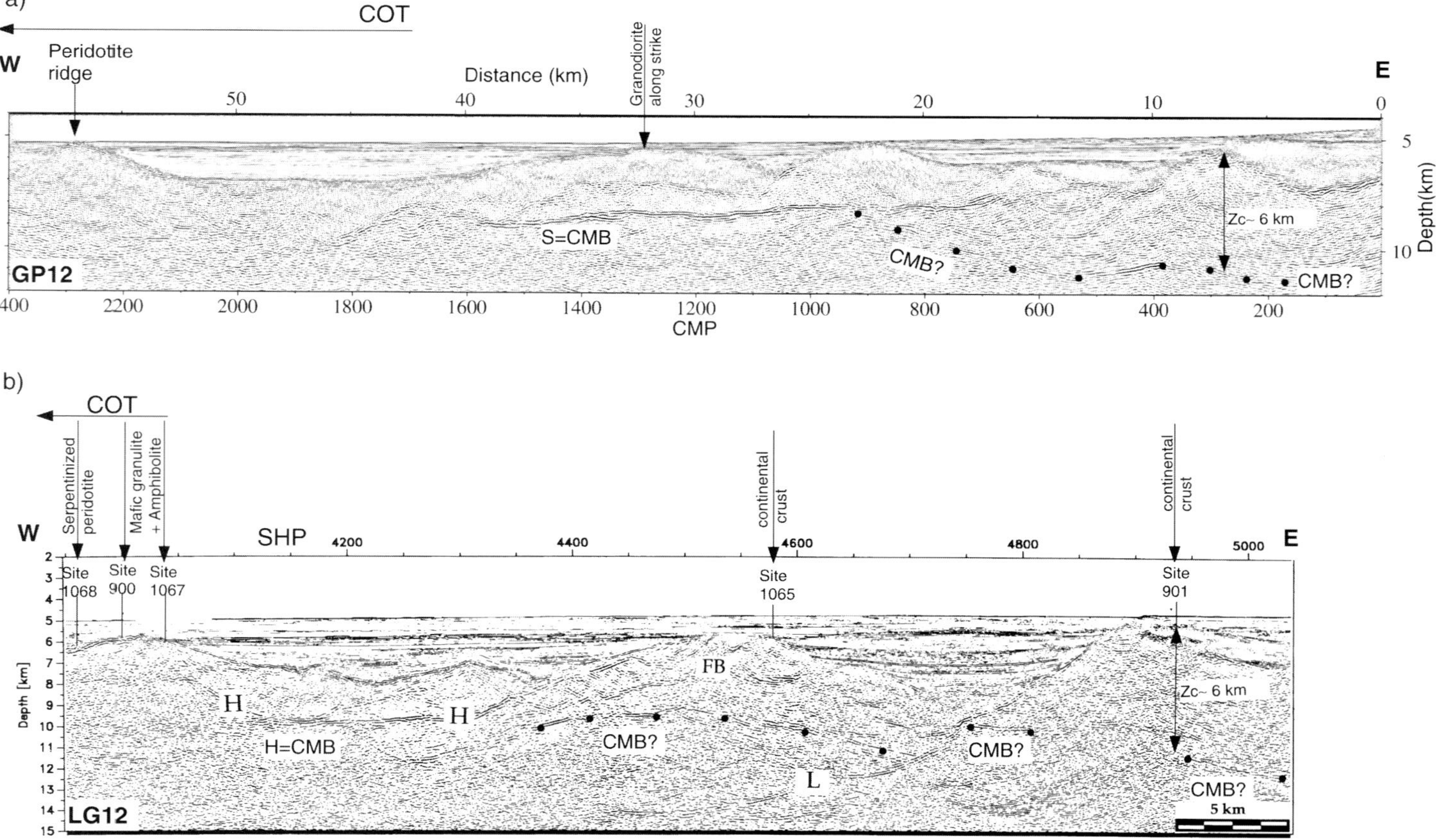
a)
COT
W
E
Peridotite ridge
Distance (km)
Granodiorite along strike
50
40
30
20
10
0
Depth(km)
5
10
Zc~ 6 km
S=CMB
CMB?
CMB?
GP12
2400
2200
2000
1800
1600
1400
1200
1000
800
600
400
200
CMP
b)
COT
W
E
Serpentinized peridotite
Mafic granulite + Amphibolite
continental crust
continental crust
SHP
4200
4400
4600
4800
5000
Site 1068
Site 900
Site 1067
Site 1065
Site 901
Depth [km]
H
H
H=CMB
FB
CMB?
L
CMB?
Zc~ 6 km
CMB?
CMB?
LG12
5 km

Table 1. *Initial lithospheric models*

Symbol	Name	Value
	Craton (e.g. SW Greenland)	
z_{CMB}	Depth to crust–mantle boundary	35 km
z_L	Depth to lithosphere's base	155 km
T_{CMB}	Temperature at the crust–mantle boundary	475 °C
T_L	Temperature at the lithosphere's base	1300 °C
H	Radiogenic heat production	4.3 μW m^{-3}
q_{So}	Initial heat flux at surface	65 mW m^{-2}
	Old orogen (e.g. Iberia)	
z_{CMB}	Depth to crust–mantle boundary	32 km
z_L	Depth to lithosphere's base	125 km
T_{CMB}	Temperature at the crust–mantle boundary	515 °C
T_L	Temperature at the lithosphere's base	1300 °C
H	Radiogenic heat production	4.45 μW m^{-3}
q_{So}	Initial heat flux at surface	71.4 mW m^{-2}
	Young orogen (e.g. Woodlark)	
z_{CMB}	Depth to crust–mantle boundary	50 km
z_L	Depth to lithosphere's base	95.6 km
T_{CMB}	Temperature at the crust–mantle boundary	850 °C
T_L	Temperature at the lithosphere's base	1300 °C
H	radiogenic heat production	4.68 μW m^{-3}
q_{So}	initial heat flux at surface	80 mW m^{-2}

investigate at which stretching factors, β_b, the entire crust begins to deform in the brittle deformation field, so that significant amounts of water can penetrate the mantle and large-scale serpentinization can begin. In addition, we predict the thickness of the potential serpentinite layer and of the melt produced during extension, to compare these predictions with observations made at several margins.

We consider a 1D model in which the lithosphere undergoes extension by uniform pure shear at a constant strain rate. Uniform pure shear seems to be a reasonable approximation at moderate amounts of extension (e.g. McKenzie 1978); the effects of variations in strain rate during rifting have been discussed by Pérez-Gussinyé & Reston (2001). However, as discussed below, at high degrees of extension the serpentinization of the mantle may lead to the development of a décollement at the CMB and hence to a departure from uniform pure shear. The effect this has on the calculations of the potential thickness of the serpentinizing zone has been considered by Pérez-Gussinyé & Reston (2001) and is discussed briefly below. Additionally, we calculate the amount of melt produced during rifting. The treatment of melting takes into account the increase in the solidus temperature with increasing degree of melting. A detailed description of this model is given in the Appendix.

Fig. 3. Prestack depth migrations of (**a**) GP12 (after Reston *et al.* 1996) in the deep Galicia Margin (DGM), and (**b**) LG12 (after Krawczyk *et al.* 1996) in the Iberia Abyssal Plain (IAP). Sampling results are shown with an arrow marking each site. In the continent–ocean transition (COT) serpentinized peridotite was exposed during rifting (Boillot & Winterer 1988; ODP Leg 173 Shipboard Scientific Party 1998; Whitmarsh & Sawyer 1996). Both profiles image detachment faults (S and H) that cut across deeper (CMB) reflections (●). We suggest that CMB reflections are reflections from the crust–mantle boundary. The portions of these detachment faults oceanward of the intersection with the CMB (GP12 (**a**), Common Mid Point gather (CMP) 900–1600 and LG12 (**b**), Shot Point gather (SHP) 4100–4250) are seismically very bright, mark the boundary between crust and partially serpentinized mantle rocks (Chian *et al.* 1999; Zelt pers. comm.) and may represent serpentinite décollements (GP12: CMP 900–1600; LG12: SHP 4100–4250). We suggest that the detachment faults (or décollements) developed where hydrous fluids could pass down crustal penetrating faults to serpentinize the mantle. Just landward (arrows) of the intersection of the detachment fault with the CMB, the entire crust was on the point of entering the brittle regime during rifting: here the crustal thickness (z_c) is *c.* 6 km. FB, fault block; L, a fault reflection discussed in the text.

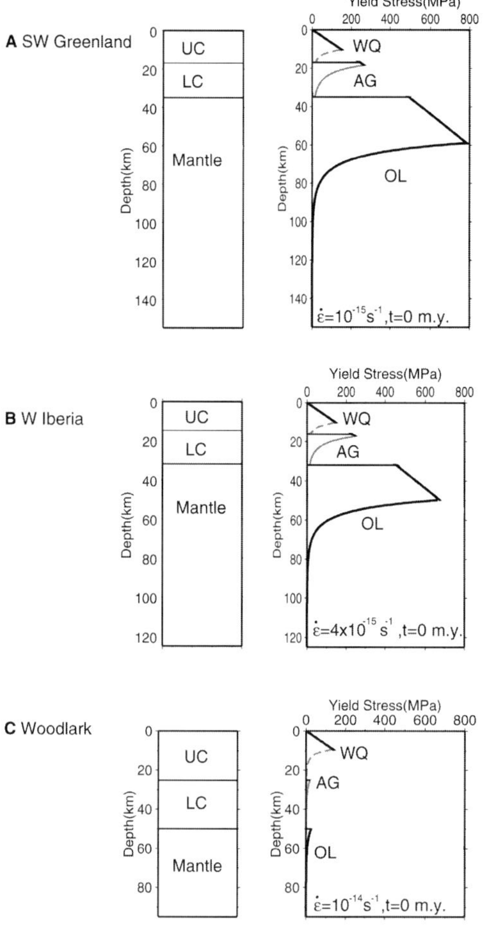

Fig. 4. Initial lithospheric models and rheological profiles at the onset of rifting for (**a**) SW Greenland, (**b**) West Iberia, and (**c**) Woodlark Basin. In each case the strain rate shown is one of a broad range we have considered. UC, upper crust; LC, lower crust; OL, olivine rheology assumed for mantle; WQ, wet quartz rheology assumed for the upper crust; AG, aggregate rheology (50% dry quartz, 50% anorthosite) assumed for the lower crust (see Table 2 for parameters).

Initial lithospheric models

We consider three initial lithospheric models to represent the rifting of a craton, of an old collapsed orogen, and of a young orogen (see Table 1 for key parameters, and Fig. 4). To calculate the initial thermal models for each margin we used the published data on heat flow, crustal thickness, radiogenic heat production and pressure–temperature–time ($P–T–t$) paths. When these data are not directly available, we

assumed standard values appropriate for the tectonic setting.

We have assumed that the present-day onshore crustal thickness at each margin represents the crust before rifting. This gives a lower bound for this value, as it does not take erosion into account. Increasing the initial crustal thickness by a few kilometres will have the effect of increasing the stretching factor at which the entire crust moves into the brittle field. However, when the initial crustal thickness is increased, the observed stretching factors have also a higher value and hence the agreement between the predicted and observed stretching factors varies only within the error bars of this analysis (for a discussion on this topic, see Pérez-Gussinyé & Reston (2001)).

The initial heat flow values have been estimated using global compilations of average heat flow values in various tectonic settings. These compilations show that the heat flow decreases with the age to the last orogenic event and hence young orogenic belts have typically high heat flow values and cratons lower ones (Sclater *et al.* 1980). Where available, present-day onshore values have been assumed to represent the initial heat flow (West Iberia: Fernández *et al.* 1998). After assuming a radioactive heat production (see Table 2) the resulting temperature at the CMB and the thickness of the lithosphere are in agreement with indirect observations (where they exist; e.g. $P–T–t$ curves of exhumed rocks near the Woodlark Basin (Hill *et al.* 1992)) and are similar to other thermal structures used to model similar tectonic settings (Buck 1991; Bassi 1995). Variations of about 50 °C on the chosen temperatures at the CMB only change the results within the errors of the analysis (see Pérez-Gussinyé & Reston 2001).

The cratonic model is based on the structure of Labrador and West Greenland before the opening of the Labrador Sea, and consists of a crust of 35 km thickness (e.g. Chian *et al.* 1995; Chalmers 1997) and a lithosphere of 155 km thickness (Fig. 4a). From these, and assuming heat flow and heat production parameters (see Table 1), we obtain an initial temperature at the Moho of 475 °C. The old orogen model is based on the structure of West Iberia immediately before rifting as deduced from the current structure of onshore Galicia (Pérez-Gussinyé & Reston 2001 and references therein). The initial model consists of a crust of 32 km thickness and a lithosphere of 125 km thickness, giving a Moho temperature of 515 °C. This is similar to many collapsed orogens (e.g. the Caledonides of the UK; Klemperer & Hobbs 1991). Finally,

Table 2. *Rheological parameters,* $\sigma_{zz} - \sigma_{xx} = (\dot{\varepsilon}/A)^{1/n}exp(E/nRT)$ *(numbers in parentheses are references)*

Material	n	A (Pa^{-n} s^{-1})	E (KJ mol^{-1})
Wet quartzite (1)	2.4	1.3×10^{-20}	134
Dry quartzite (2)	2.9	5×10^{-25}	149
Aggregate (3)	3	4.9×10^{-24}	192.4
Anorthosite (4)	3.2	5.6×10^{-23}	238
Olivine (5)	3	10^{-13}	500

1, Kronenberg & Tullis (1984); 2, Koch (1983); 3, Tullis *et al.* (1991); 4, Buck (1991), and references therein; 5, Newman & White (1997).

our young lithospheric model is based on the structure of Papua New Guinea immediately adjacent to the Woodlark Basin and on the *P–T–t* evolution of rocks exhumed from depth within the core complexes of the D'Entrecasteaux Islands (Hill *et al.* 1992). In this model, the crust is initially 50 km thick (Finlayson *et al.* 1976; Taylor *et al.* 1999), with a Moho temperature of 850 °C (Hill *et al.* 1992), and the lithosphere is in total 95 km thick.

Initial strength curves

The strength of the lithosphere at constant strain rate is determined by the interplay between rheology, temperature and depth. Thick hot crust leads to relative weakening; thin cold crust or thick cold lithosphere leads to relative strengthening. In each of the above models, a two-layer crust is assumed, with the top half of the crust being rheologically modelled as wet quartz, and the lower half of the crust being modelled as an aggregate of 50% dry quartz and 50% plagioclase (Fig. 4). Although we consider that dry quartz probably underestimates the strength of the lower crust, whereas anorthosites probably are too strong (Pérez-Gussinyé & Reston 2001), for completeness we also show results obtained assuming these two end-member rheologies (see Table 2 for rheological parameters).

Although we calculate the evolution of each model at a wide range of strain rates (and show the results in later figures), we first illustrate our approach using appropriate strain rates at which the different margins might have developed (e.g. the initial strength curves in Fig. 4). On the basis of the duration of rifting at each margin (discussed further below), these strain rates are 10^{-14} s^{-1} for the Woodlark Basin (Taylor *et al.* 1999), 4×10^{-15} s^{-1} for West Iberia and 10^{-15} s^{-1} for Labrador–West Greenland. Because the resistance to plastic rock deformation increases with strain rate, the higher strain rate for the Woodlark Basin

should increase its initial strength profile. However, this effect is minor compared with the weakening effects of thick, hot crust: the integrated lithospheric strength (the total area to the left of the strength curve) is far less for the Woodlark Basin starting model than for the other two cases.

Both the West Iberia and the Labrador–West Greenland lithospheres initially exhibit marked weak zones at the base of the upper and lower crust and towards the base of the lithosphere (Fig. 4a,b). The upper crust, the top of the lower crust and in particular the uppermost mantle are, in contrast, strong and initially brittle. The Woodlark Basin starting model is very different. Despite higher instantaneous strain rates, the strength profile (Fig. 4c) shows that beneath the brittle upper crust, the lithosphere has little strength. This model thus resembles models proposed for core complex formation in which the effective elastic thickness is small (Buck 1988) and the lower crust can flow in response to buoyancy forces (Buck 1991; Kruse *et al.* 1991).

Rheological evolution during progressive extension

As the lithosphere is extended, the crust thins and the depth of burial of deep crustal rocks is reduced. At the same time these rocks cool (see, for instance, the change in the temperature of the CMB in Fig. 5). The reduction in the overburden pressure reduces the resistance to brittle faulting; the falling temperature results in increased resistance to plastic deformation. Thus during progressive extension the deformation mechanism of rocks changes; rocks that were initially behaving plastically will eventually switch to brittle behaviour. The stretching factor at which the entire crust becomes brittle will be called β_b. We consider that this amount of extension may be a key stage in the development of widespread serpentinites beneath the crust at non-volcanic margins. Ser-

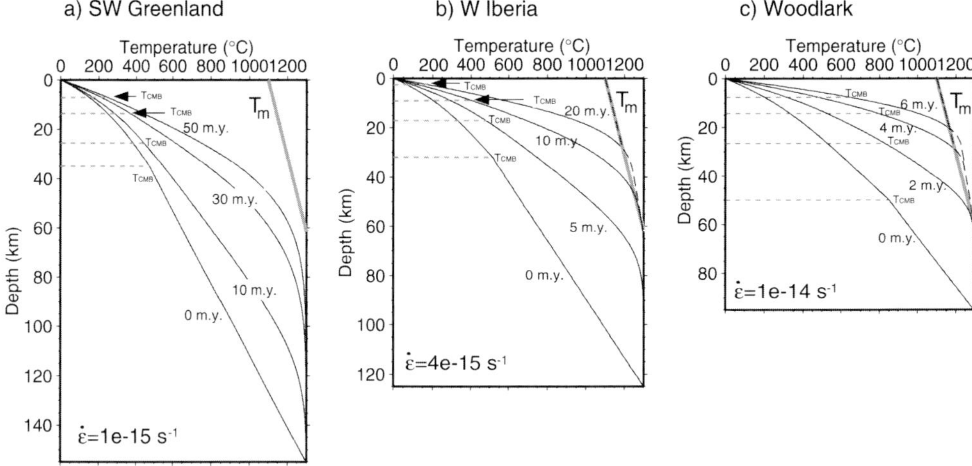

Fig. 5. Evolution of the geotherm (continuous black line) during rifting with an average strain rate appropriate for (**a**) SW Greenland, (**b**) West Iberia, and (**c**) Woodlark Basin (see Fig. 4). T_{CMB}, temperature at the CMB. Geotherm intersects solidus (grey line) after *c.* 2 Ma rifting for Woodlark Basin and after *c.* 10 Ma rifting for West Iberia, but no melting is predicted during rifting at SW Greenland margin. Dashed black line indicates the solidus temperature with increasing melt extraction (see Appendix and Figure 12 for a detailed description of the evolution of the solidus temperature, T_{m}, with increasing depletion).

pentinization requires aqueous fluids in considerable volumes (e.g. $>0.4\,\mathrm{m}^3$ of water for each cubic metre of peridotite to be fully serpentinized). The most likely source of sufficient fluid to serpentinize the uppermost mantle is from the surface along deeply penetrating brittle faults (e.g. O'Reilly *et al.* 1996). As the depth extent of such faults is probably controlled by the thickness of the brittle layer, one condition for mantle serpentinization is that the entire overlying crust has become brittle, which occurs at β_b.

Figure 6 illustrates the rheological evolution of the three models, each evolving at a geologically appropriate strain rate. Both the West Iberia and the Labrador–West Greenland cases show that at moderate stretching factors (2.7 and 2.3, respectively) the base of the upper crust becomes brittle whereas the lower crust remains plastic until slightly higher stretching factors. However, at β values of 4.4 and 3.2, respectively, the entire crust has become brittle. In contrast, the thick, hot crust of the Woodlark Basin model does not become entirely brittle until stretched by more than a factor of 10. This reflects the intial lithospheric structure (thick, hot crust) and also the rapid strain rate (which means that deep crustal rocks do not have time to cool substantially by conduction during the rifting process).

For our three models we have generalized the results of Figure 6 by further considering for each model a wide range of possible rift durations (and hence a wide variety of possible strain rates). This also allows our results to be compared with those for other margins and basins (Fig. 7). We have also modelled two other rheologies for the lower crust: the end-member rheologies of dry quartz and anorthosite, which probably under- and overestimate the strength of the lower crust, respectively (see Table 2 for parameters and Pérez-Gussinyé & Reston (2001) for discussion). As expected, the entire crust becomes brittle at lower stretching factors when a strong lower-crustal rheology (anorthosite) is used, and at higher stretching factors if a weak lower crust is used (dry quartz). Furthermore, we find that if all other parameters are kept constant, β_b decreases with increasing rift duration (decreasing strain rate) because the lithosphere is given more time to cool during rifting.

A second condition for mantle serpentinization to occur is that the temperature of the uppermost brittle mantle lies within the stability field of serpentinites. The serpentinite stability field varies with the various serpentine minerals and with the partial pressure of H_2O (O'Hanley 1996). Under retrograde conditions, before complete serpentinization, a low $P(H_2O)$ may be maintained, as inflowing water is consumed

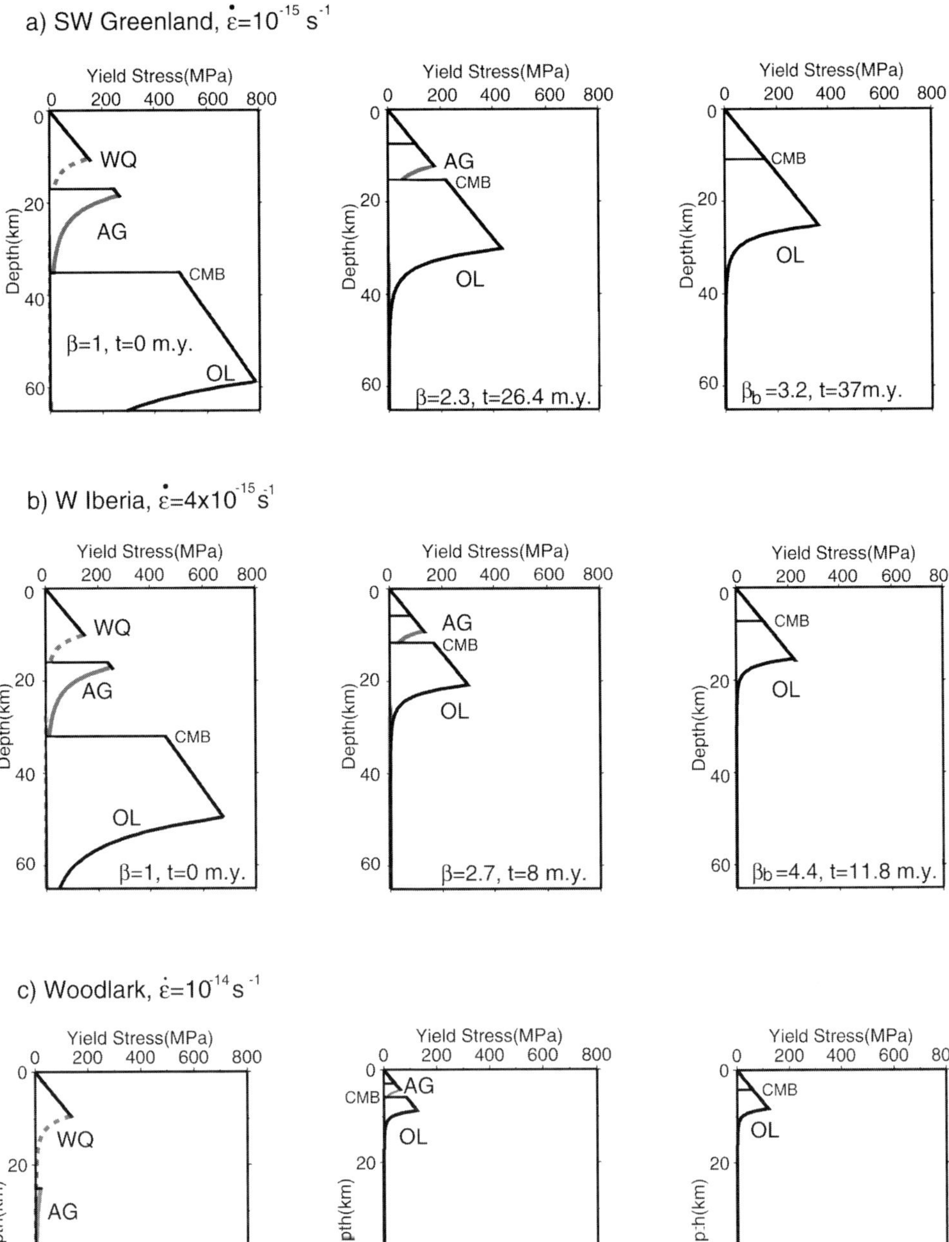

Fig. 6. Rheological evolution of model lithospheres during progressive extension at appropriate strain rates for each margin (see Fig. 4). Rheological profiles are shown at $t = 0$ (the onset of rifting), at the stretching factor when the upper crust becomes entirely brittle and at β_b, where the entire crust becomes brittle. (**a**) SW Greenland: upper crust becomes brittle after 26.4 Ma rifting at a stretching factor of *c.* 2.3, entire crust becomes brittle at $\beta_b = 3.2$ after 37 Ma rifting. (**b**) West Iberia: upper-crustal weak zone disappears at β *c.* 2.7, and entire crust becomes brittle at β_b *c.* 4.4. (**c**) Woodlark Basin: upper-crustal weak zone disappears only at β *c.* 8.5 and entire crust becomes brittle only at β_b *c.* 12.

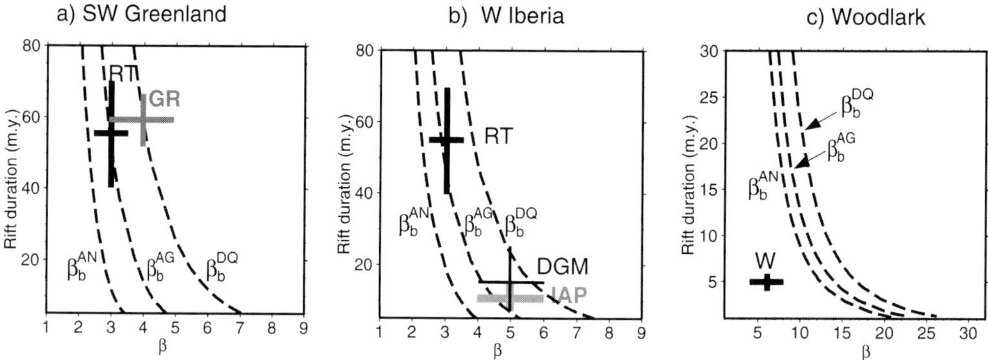

Fig. 7. Plot of the stretching factors at which the entire crust enters the brittle regime during rifting for a broad range of rift durations and lower-crustal rheologies; for anorthosite (β_b^{AN}), aggregate (β_b^{AG}) and dry quartz (β_b^{DQ}). These stretching factors are calculated using an initial thermal profile appropriate for (**a**) SW Greenland, (**b**) West Iberia and (**c**) Woodlark Basin. Crosses give error bars on estimated rift durations and observed stretching factors at the landward limit of the serpentinite zone of SW Greenland (GR), Deep Galicia Margin (DGM), Iberia Abyssal Plain (IAP) and Rockall Trough (RT). At the Woodlark Basin (W) no serpentinized and exposed mantle has been reported and the cross corresponds to error bars in the estimates of rift duration and crustal stretching factors observed at the oceanward limit of the continental crust. For Iberia, SW Greenland and Rockall Trough there is a good agreement between the predicted stretching factors at which the crust becomes brittle, β_b, and those observed at the landward limit of the serpentinite zone, indicating that serpentinization may commence when the entire crust is brittle (**a, b**). However, for the Woodlark Basin the stretching factors at which break-up occurred (at the oceanward limit of the continental crust) are far smaller than those required for whole crustal embrittlement, β_b.

by the hydration reaction (O'Hanley 1996). The Evans (1977) phase diagrams show that at low $P(H_2O)$ the maximum temperature for serpentinite formation is 400–500 °C. Thus, we consider that the 400–500 °C isotherms mark the upper limit for serpentinite formation. As we calculate the temperature structure of the lithosphere during extension (Fig. 5), we can determine whether this condition is met. Furthermore, by determining the depth beneath the CMB of the isotherm marking the upper limit of the serpentinite stability during rifting, we can estimate the thickness of the mantle layer where serpentinization may take place. Several factors such as the increase in volume and heat released during serpentinization and the heat released by the melt when it ponds and freezes, which might hinder serpentinization, have not been included in our calculations. (Although as discussed below, at those margins where serpentinites do occur, little melt production is predicted.) As a result, we emphasize that this aspect of our results should be regarded as only estimates of the potential thickness of the serpentinizing zone. In Figure 8, we contour the β_b curve (for the aggregate lower-crustal rheology) as a function of rift duration and stretching factor, together with the

thickness between the CMB and the 400 and 500 °C isotherms during rifting. The latter values are thus contours of the potential thickness of the serpentinizing zone once the entire crust has become brittle.

We consider the thickness of this zone only during rifting, as we believe that active faulting (e.g. Sibson *et al.* 1975) is required to pump sufficient water into the mantle to cause serpentinization at depth (faulting is needed to keep the cracks open in face of the sealing tendency as a result of the volume increase associated with serpentinization). Results from Ocean Drilling Program (ODP) drilling (Legs 103, 149 and 173) west of Iberia indicated that serpentinites sampled at basement highs 637, 897C, 897D, 899 and 1070 formed at low temperatures (*c.* 200 °C) probably during the postrift phase (Agrinier *et al.* 1988, 1996; Alt & Shanks 1998). However, these are unlikely to be representative of the deeper zone of anomalous velocity generally interpreted as partially serpentinized peridotite (Agrinier *et al.* 1996) but rather represent continued serpentinization and perhaps conversion of one serpentinite mineral to another during exposure at the sea floor. It should be borne in mind that these highs are onlapped by sediments as young as Eocene

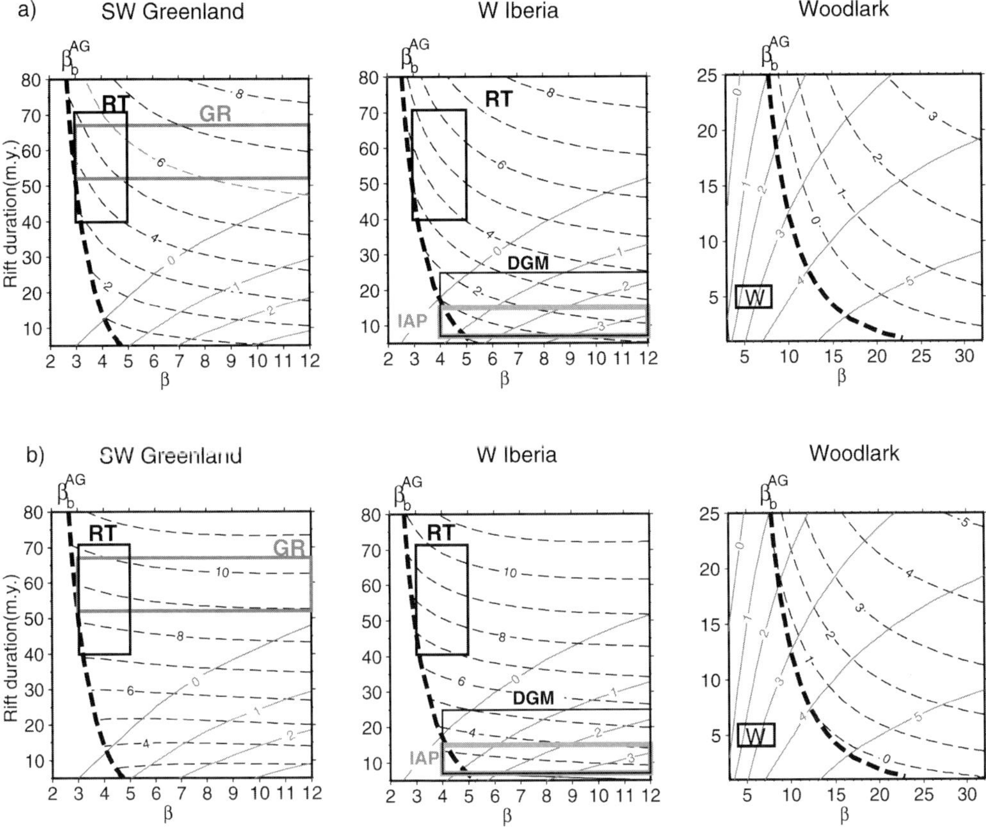

Fig. 8. Plots of predicted thicknesses (in km) of serpentinites (thin dashed contours) and of basaltic melt generated by pressure release melting (thin continuous lines) with increasing rift durations and stretching factors. β_b^{AG}, stretching factor at which the entire crust becomes brittle for an aggregate lower-crust rheology (bold dashed line); after β_b^{AG} is reached serpentinization can commence. Boxes represent range of rift durations and stretching factors observed within the COT for margins (abbreviations as in Fig. 7). Two possible upper temperature limits for serpentinite stability are considered: (**a**) 400 °C; (**b**) 500 °C. Serpentinites are predicted to occur at the West Iberia and Greenland margins and beneath the Rockall Trough, However, at Woodlark Basin melting and no serpentinites are predicted; here the crust was too thick and hot for serpentinites to be formed at a wide COT before break-up.

time and thus that the serpentinites were exposed at the sea floor for at least 60 Ma (from M0 to the beginning of Eocene time).

For all rift durations that were considered, after stretching factors at and beyond β_b, the 400 °C isotherm is below the CMB for the SW Greenland and West Iberia models, but not for the Woodlark Basin model (Fig. 8). Indeed, at the short rift durations (4–6 Ma) that are most appropriate for the Woodlark Basin, the CMB temperature is above 500 °C until stretching factors somewhat above β_b. Thus whereas for West Iberia and SW Greenland, serpentinization should commence immediately after the

entire crust becomes brittle, this is not predicted to be the case for the Woodlark Basin.

Thus along the West Iberia margin or SW Greenland margins, the calculated value of β_b gives the predicted crustal thickness at which the crust should become brittle and serpentinization might have started.

Assuming uniform pure shear throughout, the thickness of the potential serpentinizing zone increases with rift duration and stretching factor beyond β_b. Whereas both the West Iberia and the Labrador–West Greenland models predict similar thickness of serpentinites for any given rift duration and stretching factor beyond β_b, the model predicts far thinner serpentinites for the young orogen model (Woodlark), where high temperatures inhibit serpentinization, beyond a far higher value of β_b.

We compare the predicted potential thickness of the serpentinizing zone with that of the basaltic magma produced by pressure release melting during rifting. In the old orogen and cratonic models for West Iberia and Labrador–West Greenland, the thickness of total melt produced exceeds that of the serpentinized zone only at very high stretching factors and/or short rift durations (the bottom right-hand corner of the plots; e.g. a β_b value of eight, rift duration of *c.* 10 Ma). In contrast, in the range of rift durations investigated, the amount of melt produced during rifting of the young orogen model is generally greater than the thickness of the serpentinizing zone.

The seismic signature of basaltic 'underplate' and serpentinized uppermost mantle can be similar, as both can be characterized by velocities in the range $7-7.4 \text{ km s}^{-1}$. For the margins studied here, basement sampling provides a further constraint: where basement consists of serpentinized peridotites and no volcanic rocks have been reported, velocities $>7 \text{ km s}^{-1}$ probably represent partially serpentinized peridotites; where volcanic rocks but no serpentinites are present, mafic underplate may be a better explanation. However, it is also clear from our results that voluminous basaltic intrusion and mantle serpentinization are likely to occur under very different conditions. In particular, knowledge of the initial lithospheric model and the duration of rifting may thus help distinguish between serpentinites and mafic underplate at margins where no samples have been obtained.

Comparison with observations

In our 1D model, mantle-penetrating faults are a precondition for serpentinization, and thus the stretching factor at which the entire crust became brittle during rifting (β_b) marks the temporal onset of serpentinization. This stretching factor should match that at the current spatial limit of partially serpentinized mantle beneath the crust: where the crust is thicker (e.g. further landward), the faults never reached the mantle as the lower crust remained plastic during rifting; where the crust is thinner (e.g. further oceanward), faults should have penetrated into the mantle leading to its serpentinization. In this section we compare the predicted values of β_b, at which the entire crust enters the brittle regime, with the stretching factors observed at the spatial (landward, except for the Rockall Trough) limit of the serpentinite

zone beneath the crust at the West Iberia and Labrador–West Greenland margins and at the Rockall Trough (Fig. 7). No exposures of mantle serpentinites are known from Woodlark Basin: here we compare β_b with the stretching factors observed at the oceanward limit of the thinned continental crust (Fig. 7).

Furthermore, we compare our predictions of the variation in the potential thickness of the serpentinizing zone and of the amount of melt produced during rifting across these margins, with the observations of the thickness of the $7-7.8 \text{ km s}^{-1}$ layer, thought to represent partially serpentinized mantle (Fig. 8). To obtain these estimates, we consider the stretching factors (deduced from crustal thickness) from the landward limit of serpentinites to the most oceanward blocks of continental crust that are interpreted to be underlain by serpentinites. Where no mantle serpentinites have been reported or inferred (Woodlark Basin) we compare our model predictions with observations of the stretched continental crust adjacent to oceanic crust (Fig. 8) to see if our model explains the observed lack of serpentinites at this margin.

In principle, stretching factors for rifted margins can be derived from subsidence measurements (assuming a model for lithospheric extension), from crustal thickness measurements, and from fault geometries. In practice, the resolution of subsidence is limited by palaeo-waterdepth measurements (Moullade *et al.* 1988) and is inappropriate when there has been serpentinization of the underlying mantle (O'Reilly *et al.* 1996).

Stretching factors derived from seismically imaged fault geometry commonly are far smaller than values determined from crustal thickness or subsidence (e.g. Sibuet 1992). This discrepancy may represent problems in imaging all generations of faults (Reston *et al.* 1996). Multiple sets of faults are indeed to be expected at a rifted margin: where stretching factors exceed about two, faults that originally developed at *c.* 60° should have rotated to *c.* 35° and thus, according to Andersonian fault mechanics (e.g. Jackson 1987), should lock up to be cut by a second generation of steeper faults. At the DGM for instance, where stretching factors are much greater than four, we would expect at least three generations of faulting within the final rift phase, and indeed here there is evidence for three phases of faulting associated with Early Cretaceous rifting (Reston *et al.* 1996). ODP Leg 103 found deep-water Valanginian turbidites occurring as pre-tilting sediment within the fault blocks imaged seismically: these have been interpreted as representing early synrift sediments, developed before the formation of block-bounding faults imaged seismically (Moullade *et al.* 1988). A second phase of faulting is represented by the faults imaged seismically: these dip at an angle of *c.* 25–40°, detach onto the S reflector (Reston *et al.* 1996) and truncate the Valanginian turbidites. However, submersible diving (Boillot *et al.* 1988) reveals that the fault blocks imaged are cut by another, steeper set of faults, dipping *c.* 60° to the west; we interpret these, which are shown by Boillot *et al.* (1989) in their cross-section of the margin, as only the last generation of faults. The problem is that commonly only one of the phases of faulting is imaged seismically (e.g. at the DGM only the middle of the above three phases). Thus fault geometries cannot be used to estimate stretching factors at rifted margins.

For these reasons, we use crustal thickness measurements to determine β, using wide-angle data supplemented by reflection data where appropriate. Furthermore, crustal thickness gives us the information crucial to our study; namely, the depth of the CMB.

Woodlark Basin

Bathymetry and gravity data show that the crustal structure of the continental margins that surround the Woodlark Basin changes along strike: the crust is thicker west of the Moresby Transform (*c.* 154.2°) (Martinez *et al.* 1999). This is believed to reflect the different prerift histories of the margins of the western and eastern basin (west and east of the Moresby Transform, respectively). Although the crust of the basin originated as a volcanic arc (because of north Coral Sea subduction) the western margin additionally experienced collision and underthrusting by continental plateaux (Martinez *et al.* 1999). The lithospheric structure of our young orogen is meant to simulate that of the western basin, and thus the rift durations and stretching factors we use for comparison are those of this basin.

Rift onset at the Woodlark Basin is believed to have begun synchronously at 6 Ma along the length of its protomargins (Taylor *et al.* 1999). However, sea-floor spreading initiated in a time transgressive fashion from east to west (Taylor *et al.* 1999), reaching Moresby Transform at 2 Ma and 151.8°E at 0.7 Ma (Taylor *et al.* 1999). Thus the range of rift durations for the western basin spans 4–5.3 Ma: we use values from 4 to 6 Ma to allow for this variation.

Taylor *et al.* (1999) estimated average strains along the margins of the basin and obtained a maximal value corresponding to a stretching factor of four. However, we have estimated the stretching factors at the edge of the continental crust by comparing crustal thickness estimates obtained from gravity data (7–10 km, Martinez *et al.* 1999) with the initial crustal thickness from our young orogen model (50 km). This leads to a range of stretching factors at the continental edge of *c.* 5–7. We again plot a larger span of stretching factors, 4–8, to allow for variations along the margin and errors in the estimation of crustal thickness. At all these stretching factors, the crust is still partly (largely) plastic in our young orogen model when continental break-up occurs (Figs 6c, 7c and 8), so we predict no mantle serpentinization. This is compatible with the observation that the transition from continental to oceanic crust is sharp (Taylor *et al.* 1999) and with the fact that no exposures of unroofed and serpentinized mantle (mantle serpentinites) have been reported. However, the rapid rift duration combined with the warmer initial thermal structure implies that partial pressure release melting of the upwelling asthenosphere occurred during the synrift phase, producing several kilometres of new basaltic melt (Fig. 8). Although this margin lacks the SDRS and other characteristics of volcanic rifted margins, the Woodlark Basin may be a margin where neither the volcanic nor the non-volcanic label seems to apply (Mutter 1993): Taylor *et al.* (1995) reported several small volcanoes (a few kilometres in diameter) erupted along the continental margins. Basalts recovered from the continental margins have geochemical characteristics indicative of low degree of partial melting (Taylor *et al.* 1995). Thus our predictions are entirely compatible with the observations.

West Iberia

The rift duration of the West Iberia margin is somewhat controversial. Drilling results were interpreted as showing that at the DGM, where break-up occurred during the Cretaceous quiet zone, rifting lasted 25 Ma (Boillot & Winterer 1988). At the IAP drilling results combined with magnetic data were used to infer that it lasted around 15 Ma (Whitmarsh & Miles 1995). However, Wilson *et al.* (1996) questioned whether all the sequences previously interpreted as synrift truly are so. On the basis of their reinterpretation of many of these units as postrift sequences, they proposed that the rift duration on both the DGM and the southern

IAP may have been as short as 7 Ma. We compare our modelling results with these ranges of rift durations: from 7 to 15 Ma for the IAP, and from 7 to 25 Ma for the DGM.

At the DGM, the region of the breakaway to S (Fig. 3) corresponds to the landward limit of serpentinites identified beneath the crust from wide-angle velocities (e.g. Zelt pers. comm.). Further landward the CMB deepens (crustal thickness of *c.* 6 km) and is not intersected by crustal penetrating faults. On the southern Iberia Abyssal Plain, it is less clear where the landward limit of serpentinites occurs: seismic velocities of $7-7.8$ km s^{-1} are still present beneath the crust at the landward limit of the CAM wide-angle data near ODP Site 1065 (Chian *et al.* 1999; compare Figs 2a and 3). However, to the east of these profiles, a major fault L appears on profile LG12 to cut across and slightly offset a reflection at a depth of $10-12$ km (Fig. 3). We interpret this reflection as the CMB, which has been offset by late movement along the fault once the entire crust had become brittle. The crustal thickness (6 km) just landward of L (SHP 4900) thus provides the constraint that the crust is brittle when thinner than this. The same crustal thickness is found just landward of the serpentinite zone defined using seismic wide-angle data along line IAM 9, which is located 40 km south of LG12 (Dean *et al.* 2000). This crustal thickness leads to a stretching factor of 5.3 when compared with the initial crustal thickness for West Iberia (32 km). To allow for variations along and across the IAP and the DGM and for errors in our estimation of crustal thickness at the point where the entire crust became brittle during rifting, we plot (Fig. 7) a range of stretching factors $(4-6)$ to represent those at the spatial limit of the serpentinite zone, just landward of S and H.

Our model predicts that serpentinization and the formation of detachments such as S and H started when the entire crust had just entered the brittle regime (Fig. 7b). The stretching factor observed at the landward limit of partially serpentinized peridotites and one predicted (β_b) compare well for the range of rift durations that have been proposed (Fig. 7b). We note that moving oceanward the crust thins further above a region of serpentinized mantle before mantle is exposed, as interpreted from the wide-angle data (Chian *et al.* 1999; Figs 2 and 3), supporting our suggestion that serpentinization occurred beneath the crust here before total crustal separation took place.

The predicted thickness of the potential serpentinizing zone can also be compared with the

thickness of partially serpentinized mantle at these margins. Here, we consider the stretching factors observed from the landward limit of the serpentinite zone to the most oceanward continental blocks (blocks on top of S and H, Fig. 3) that are interpreted to be underlain by serpentinized peridotite (Chian *et al.* 1999; Fig. 2a; see Pérez-Gussinyé & Reston (2001) for a more detailed comparison with the Chian *et al.* (1999) results). The range of rift durations is the same as that used for Figure 7.

The thickness of the partially serpentinized mantle beneath the oceanward-thinning crust of the IAP can be estimated from the Chian *et al.* (1999) seismic velocity data to increase oceanward from *c.* 4 km to *c.* 7 km. This is somewhat greater than the thickness of the potential serpentinizing zone predicted by the uniform pure shear model presented here, where for the observed crustal stretching factors values of $1-4$ km are predicted, depending on rift duration and on the upper temperature limit of serpentinite stability. It is possible that some of the $7-7.8$ km s^{-1} zone corresponds to mantle intruded by melt, as our model predicts that melt thickness of $0-3$ km may be generated (*c.* 1 km more than the results of Bown & White (1995)). Mixtures of serpentinized mantle and gabbroic intrusions may, for instance, form much of the lower oceanic crust formed at slow-spreading centres (e.g. Cannat *et al.* 1995). However, this melt would have to be emplaced mostly beneath the serpentinizing zone if the two together can be summed to match the observed thickness of the $7-7.8$ km s^{-1} zone. We consider that a better explanation for the slight discrepancy is that once mantle serpentinization has begun, the development of a décollement at the CMB will lead to deviation from uniform pure shear in the final phases of crustal break-up (Pérez-Gussinyé & Reston 2001). Specifically, crustal extension is likely to focus over the region of serpentinization, above a broader region of less focused subcrustal extension. As a result, the thickness of the serpentinizing zone increases more rapidly during rifting as the overlying crust thins and the mantle cools (Pérez-Gussinyé & Reston 2001).

The development of a décollement and the consequent focusing of crustal extension will have the opposite effect on the amount of melt from that expected. As local excess crustal extension occurs, overall stretching factors can no longer be estimated from crustal thickness. Thus the upwelling and resulting partial melting of the asthenosphere should be less than predicted from the amount of crustal thinning above the serpentinizing zone. As almost no melt is produced before β_b is reached (Fig. 8), the subsequent development of a serpentinite décollement and consequent heterogeneous lithospheric stretching may provide an explanation for the lack of synrift magmatism reported from the West Iberia margins. It should be noted also that the lack of significant melt production before β_b for West Iberia and SW Greenland means that the value for β_b here is not influenced by the heat released when melt freezes at crustal levels.

Labrador–West Greenland

The rift duration of these margins is estimated to have been between 52 and 67 Ma (e.g. Srivastava & Roest 1995; Chalmers 1997), depending on the interpretation of the magnetic anomalies and hence of the position of the COT. The thickness of the crust at the landward limit of the partially serpentinized peridotites, *c.* 8 km, has been estimated from the Chian & Louden (1994) interpretation of a 2D velocity model at the southern West Greenland margin (Fig. 2b). For a crust with an initial 35 km thickness, this implies stretching factors of about 4.3; as previously (Fig. 7) we plot a range of stretching factors (3–5) to allow for variations along the margin and errors in the estimation of crustal thickness. On the Labrador margin, the crust appears thinner above the landward edge of the serpentinized zone, but here crustal delamination following the onset of serpentinization may have removed the upper crust (Chian *et al.* 1995; see below for further discussion of the consequences of crustal delamination). Thus we concentrate on the West Greenland side: β_b predicted by our modelling agrees well with the stretching factors at the landward edge of the serpentinites, suggesting that also here the crust might have just entered the brittle regime when serpentinization started (Fig. 7a).

Moving further oceanwards, the crust thins and the serpentinite thickness increases (e.g. Chian & Louden 1994, Fig. 2b), reaching *c.* 5 km. This is of the order of the predictions of the model (Fig. 8); at high stretching factors, the depth of the 400 and 500 °C isotherms beneath the base of the crust increases to about 7.5 and 10 km, respectively, for the maximum rift duration. With decreasing rift duration the match between observations and predictions improves. Despite these uncertainties in the rift duration at this margin, the match between the predicted potential thickness of the serpentiniz-

ing zone and the inferred thickness of serpentinized mantle is good.

Rockall Trough

The analysis developed above can also be applied to deep basins such as the Rockall Trough (Figs 1 and 9), a failed rift situated near the Atlantic margin. Although in the past thought to be floored by oceanic crust (e.g. Roberts *et al.* 1981), it is now generally accepted that the Rockall Trough is underlain by continental crust (e.g. Hauser *et al.* 1995; O'Reilly *et al.* 1996), although the evidence is somewhat equivocal (Joppen & White 1990).

The age of rifting in the Rockall Trough has also been the subject of much controversy, with some workers suggesting that it initiated in Late Palaeozoic time (e.g. Smythe 1989), although the consensus now appears that it is dominantly a Cretaceous rift (e.g. Musgrove & Michener 1996) that perhaps overprinted an earlier Mesozoic structure (e.g. Cole & Peachey 1999; Nadin *et al.* 1999). The rift duration is thus not fully clear but probably was at least 40 Ma and perhaps somewhat longer (see above references for discussion).

The Rockall Trough is situated just southeast of the well-developed seaward-dipping reflectors of the Rockall–Hatton Bank region. It is heavily intruded by Paleocene sills, which may hinder the imaging of the Pre-Cretaceous structure, and is marked in several places by major igneous centres of Late Cretaceous to Early Tertiary age (Hitchen & Ritchie 1993). However, this magmatic activity all took place after rifting, and wide-angle data have been interpreted as showing no evidence for synrift magmatism (e.g. Hauser *et al.* 1995); the Rockall Trough appears to be a fundamentally non-volcanic rift, developed in crust *c.* 30 km thick. As the Rockall Trough developed in the immediate foreland to the Caledonides, the lithospheric structure before rifting was probably most similar to the cratonic model (O'Reilly *et al.* 1996) of West Greenland. However, given its proximity to the Caledonides, we also compare observations from the Rockall Trough with the predictions of the West Iberia model.

The Rockall Trough is crossed by several deep penetration seismic profiles (e.g. Joppen & White 1990), such as the BIRPS Westline profile (England & Hobbs 1997). None of these profiles clearly images the Moho, so that constraints on the crustal structure and thickness come almost exclusively from wide-angle data. The best quality wide-angle data show that as the crust thins towards the centre of the basin,

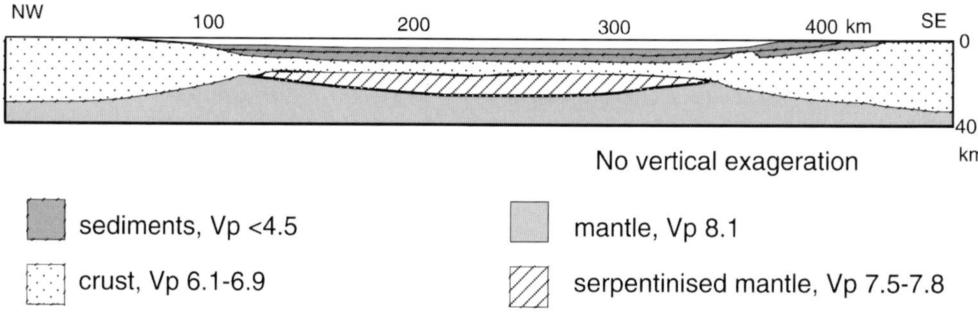

Fig. 9. Interpretation of a wide-angle velocity profile across the Rockall Trough shown at no vertical exaggeration (based on results of O'Reilly *et al.* (1996)), showing the presence of partially serpentinized peridotites beneath a broad thin region of interpreted continental crust. The crustal thickness at the spatial limit of the serpentinized zone implies stretching factors of *c.* 2.5–3.5, in agreement with the predictions of the model (see Fig. 7). The thickness of the partially serpentinized layer (5–10 km) is consistent with the predicted thickness of the serpentinizing zone (Fig. 8). It should be noted that the present model of O'Reilly *et al.* (1996) implies that the continental crust stretched to a thickness of 6 km over 200 km without breaking up.

it is underlain by a layer of 5–10 km thickness with velocity 7.7–7.8 km s^{-1}. This has been interpreted as representing partially serpentinized mantle (O'Reilly *et al.* 1996). The crustal thickness at the edge of the serpentinized mantle is *c.* 10 km (Fig. 9), implying a stretching factor of about three. Allowing for errors in the estimation of crustal thickness and for variations along the margin of the basin, we plot, in Figure 7, a range of 2.5–4 for the edge of the serpentinized zone beneath the Rockall Trough. This range of stretching factors matches well the range of β_b at which the entire Rockall crust should have become brittle during rifting, as predicted by our calculations for either the West Greenland (cratonic) or West Iberia (old orogen) model. Thus our model explains the presence of serpentinized mantle beneath the Rockall Trough.

At the central part of the Rockall Trough the crust is *c.* 6 km thick (implying a β value of five, we plot a range of stretching factors of 3–5 in Figure 8 to cover the variation in crustal thickness above the serpentinized zone). The underlying partially serpentinized mantle layer (O'Reilly *et al.* 1996) is 5–10 km thick (Fig. 9). At these stretching factors, our model predicts that between 4 and 10 km of mantle should have been partially serpentinized, but that no melt should have been produced, in good agreement with the observations. The Rockall Trough appears to be similar to the West Iberia margin except that rifting failed before the onset of sea-floor spreading and it has been overprinted by postrift (plume-related) magmatism.

The Rockall Trough does, however, pose another problem: the crustal thickness as determined from wide-angle data is *c.* 6 km over a distance across the basin of 200 km. Bassi (1995) pointed out that, once the entire crust becomes brittle and faults penetrate into the mantle, it should not be possible to stretch continental crust to such an extent. This raises the intriguing possibility that much of the 'crust' within the Rockall Trough is actually serpentinized mantle. Although the velocity structure of the top 6 km of basement has been interpreted as continental crust (e.g. O'Reilly *et al.* 1996), it may also be explained by a suitable mixture of continental crustal blocks floating in a sea of partially serpentinized mantle. This is clearly a subject for further investigation, as the nature of the basement here has implications for, among other things, heat production and source rock maturity in the Trough.

Detachments and detachments

The terms detachment and detachment fault commonly raise images of western US style core complexes and/or the Wernicke (1981) model for lithospheric extension by simple shear. To avoid confusion, we follow the following definitions of these terms: a detachment is a low-angle surface onto which other structures detach; a décollement is a form of or part of a detachment that follows a stratigraphic or structural level (e.g. the top of serpentinized peridotites); a detachment fault is an apparently low-angle brittle normal fault onto which steeper normal faults detach (e.g. see Lister &

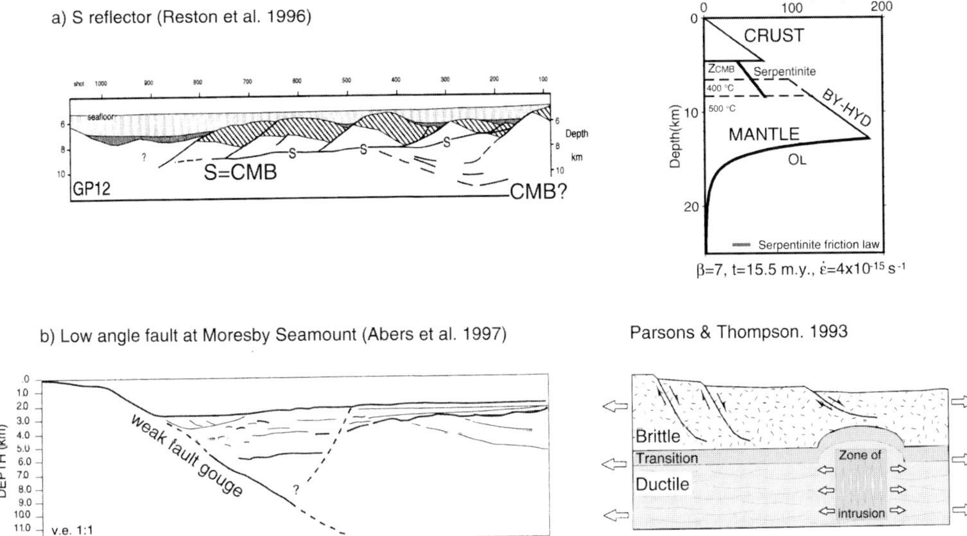

Fig. 10. Two types of detachment faults may form at rifted margins. (**a**) Weak serpentinite décollements may develop at cool margins such as West Iberia and SW Greenland: here the low friction coefficient of serpentinites together with high local pore pressures generated (Escartín *et al.* 1997) allow faults to move at very low angles. BY-HYD is Byerlee's law for hydrostatic pore pressure. (**b**) At the hotter Woodlark Basin, slight distortion of the stress field (as a result of igneous intrusion; Parsons & Thompson 1993) or decreased coefficient of friction (as a result of weak material along the fault gouge) allows faults to remain active to angles *c.* 10° lower than predicted by classical fault mechanics. Rotation of the footwall accompanying flow of hot ductile lower crust may help the development of core complexes.

Davis 1989). Thus a part of a detachment fault that follows a structural level can be a décollement. In this section we discuss how detachment faults such as S (Fig. 3), which appears to have been active at angles as low as 20°, differ from structures described from the western USA.

The formation of serpentinites may be one way in which detachment faults can develop during rifting (Fig. 10a) after the entire crust has entered the brittle regime (around 6 km thin crust) and serpentinization has started (temperatures at the CMB below 400–500 °C). The low strength of serpentinite, together with the high pore pressures that it may internally develop (Escartín *et al.* 1997), may create a weak zone at the CMB, allowing it to act as a décollement surface and thus decoupling deformation in the crust from the deformation in the mantle (Fig. 10a). Such décollements separating crustal rocks from partially serpentinized peridotites form part of the S reflector west of the Galicia Bank (Reston 1996; Reston *et al.* 1996) and the H reflector in the southern IAP (Krawczyk *et al.* 1996). As S cuts up to a breakaway to the east (Reston *et al.* 1996), clearly only one part of this structure is a décollement; overall S is a

detachment fault, as it is probably brittle. Similar structures have also been inferred beneath the West Greenland and Labrador margins (reflections G and D of Keen *et al.* (1994)).

The development of serpentinite décollement as part of a detachment fault system may provide a mechanism for the exposure of large expanses of mantle rocks within the COT of non-volcanic margins. Subhorizontal movement of crustal blocks along such a décollement, together with normal faulting along the faults between these blocks, could allow complete crustal separation and mantle exhumation (Pérez-Gussinyé & Reston 2001). Thus the embrittlement of the entire crust may be a necessary condition for total crustal separation at some margins.

One extreme model for mantle exhumation is the possibility of crustal delamination (Chian *et al.* 1995), in which deformation after the development of a serpentinite layer approximates simple shear along a detachment fault that follows the base of the crust for many tens of kilometres. More generally, it appears likely that after the onset of serpentinization, extension may become more heterogeneous, leading to isolated blocks of crust floating in a sea of serpentinites. This can have the effect of increasing the thickness of the serpentinizing zone by up to *c.* 5 km if the upper plate is locally extended far more than the lower plate (see Pérez-Gussinyé & Reston (2001) for a complete discussion of this subject).

However, the detachment faults associated with the core complexes in the Basin and Range and the Aegean Sea are very different in terms of both structure and tectonic setting. These faults juxtapose lower-grade hanging-wall rocks against a footwall comprising high-grade metamorphic rocks from the middle to lower crust (Coney 1980; Lister *et al.* 1984). The heat flux in both areas is greater than *c.* 90 mW m^{-2} and the estimated crustal thickness at the time of core complex formation is estimated to have been >50 km (Buck 1991). Present-day crustal thickness at the core complexes of the Basin and Range is 26–28 km (McCarthy *et al.* 1991), considerably greater than in the region of S. Core complex formation is associated with synextensional magmatism (Basin and Range: Parsons & Thompson 1993 and references therein; Aegean Sea: Lister *et al.* 1984) and high strain rate (Buck (1991) and references therein). Given the amount of displacement on some of these faults, the lack of either a deep basin or an uplift of the seismically imaged Moho (McCarthy *et al.* 1991) has led to the suggestion that the lower crust is extremely weak and fluid in the regions of core complexes (Buck (1991) and references therein). No such flow could have taken place in the region of the S reflector, where the entire crust is brittle. Thus, the characteristics that accompany formation of low-angle faults at core complexes are very different from those inferred for the formation of detachments at 'serpentinite' margins. Whereas in the first the crust is thick and hot, the lower crust is likely to flow, and moderate magmatism is likely to occur during extension, in the second the crust is thin and cold, the lower crust is within the brittle regime and magmatism is practically absent.

Several explanations have been proposed to explain the occurrence of apparently low-angle faults at core complexes: a decreased coefficient of friction (weak materials and fluids along fault gouges), deviation of the principal stress from the vertical (e.g. because of fluids: Axen 1992; magmatic intrusions: Parsons & Thompson 1993), or rotation of active steeply dipping active normal faults to shallowly dipping inactive faults (Buck 1988). The first two mechanisms are independent of the thermal state of the crust; however, the flexural rotation

model proposed by Buck (1988) assumes the thermal conditions found at core complexes, hence it might not be adequate to explain low-angle detachments such as those observed at 'serpentinite' margins.

The active Moresby Seamount fault is located in the Woodlark Basin near the tip of the oceanic propagator and dips at a shallow angle of *c.* 25–35° (Abers *et al.* 1997). This is only slightly lower than predicted by Andersonian fault mechanics (e.g. Abers 2001) and is thus not a very low-angle fault. The activity of this fault might be explained simply by the presence of weak material, as talc and serpentinite have been recovered from the fault gouge during ODP Leg 180, thought to represent altered ophiolitic rocks rather than unroofed mantle (Taylor *et al.* 2000; Goodliffe pers. comm.). Alternatively, the activity of this fault at an angle as low as 25° may reflect the rotation of the stress field by igneous intrusions (Fig. 10b; Parsons & Thompson 1993); synrift igneous intrusions are widespread in the D'Entrecasteaux Islands (e.g. Hill *et al.* 1992) and are predicted by our model (Fig. 10b).

In future, the Moresby fault may rotate to a lower angle while remaining active (as a genuine low-angle fault) or when inactive (following the mechanism proposed in the flexural rotation model (Buck 1988)). However, we consider it unlikely that the Moresby fault will develop into a detachment fault quite like S. Décollements at non-volcanic margins of the Iberia type occur at the edge of a broad zone of mantle exhumed before sea-floor spreading occurs. Such a zone has not been observed along the margins of the Woodlark Basin, where the boundary between continental and oceanic crust is sharp. Nor have genuinely low-angle detachments like S been imaged seismically anywhere in the Woodlark Basin. From our modelling results we suggest that the Woodlark Basin crust is too thick and warm to meet, before break-up, the low-temperature conditions present during the development of décollements at 'serpentinite' margins. The characteristics accompanying continental rifting at the Woodlark Basin are more akin to those found at the core complexes of the Aegean Sea and the Basin and Range; this is supported by the occurrence of core complexes at the D'Entrecasteaux islands (located further to the west of Moresby Seamount) and the Papuan peninsula (Davies 1980; Davies & Warren 1988). Furthermore, if lower-crustal flow accompanies extension at the Woodlark Basin (as is suggested for the Basin and Range) the crust would behave

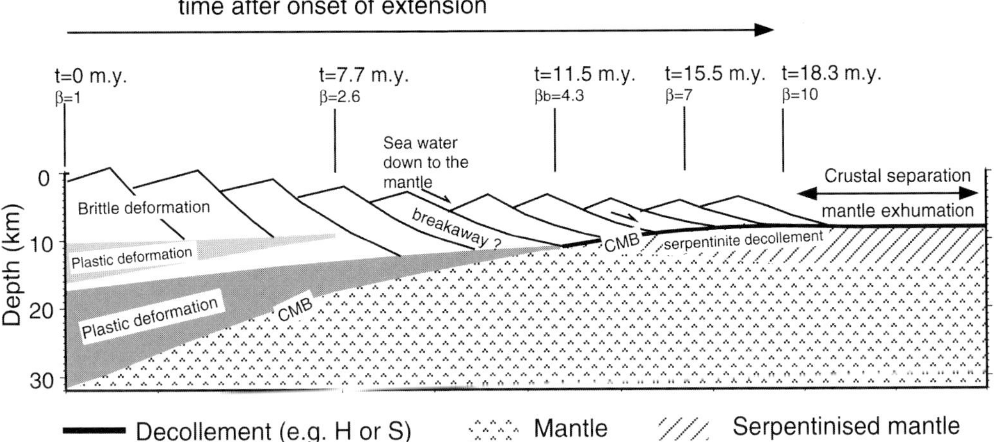
a) West Iberia type
time after onset of extension
t=0 m.y.
β=1
t=7.7 m.y.
β=2.6
t=11.5 m.y.
βb=4.3
t=15.5 m.y.
β=7
t=18.3 m.y.
β=10
Sea water down to the mantle
Crustal separation
mantle exhumation
Brittle deformation
breakaway ?
Plastic deformation
CMB
serpentinite decollement
Plastic deformation
CMB
Depth (km)
0
10
20
30
Decollement (e.g. H or S)
Mantle
Serpentinised mantle

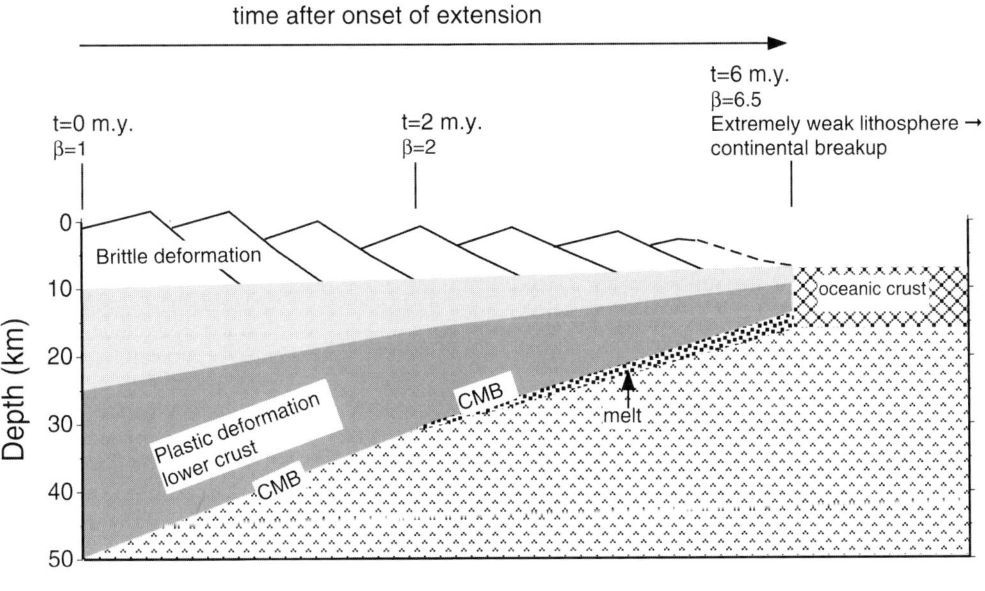
b) Woodlark type
time after onset of extension
t=6 m.y.
β=6.5
Extremely weak lithosphere →
continental breakup
t=0 m.y.
β=1
t=2 m.y.
β=2
Brittle deformation
oceanic crust
Depth (km)
0
10
20
30
40
50
Plastic deformation lower crust
CMB
CMB
melt
Core complex detachment?
Mantle
Melt generated

plastically until higher extension factors than predicted here.

Temporal evolution of non-volcanic margins

In this paper we have described two fundamentally different types of 'non-volcanic' margins: the SW Greenland and West Iberia type, where slow to moderate rifting of a cool crust leads to the formation of mantle serpentinites and the development of a décollement at the CMB; and the Woodlark Basin, where more rapid stretching of hot, thick crust does not lead to mantle serpentinization but instead to the partial melting of the mantle and the subsequent intrusion of melts into the base of the extending crust.

We can summarize the temporal evolution of such margins that is expected from our sequence of 1D 'snapshots' of rheological and thermal evolution (Fig. 11). These 'sections' illustrate the simplified temporal evolution of the centre of the rift rather than the spatial variation from the unstretched continent to the COT, but do provide some insight into how such margins develop. Faults shown are schematic and simply illustrate the level of detachment, not the evolution of the fault systems.

In the West Iberia type (Fig. 11a), the extending crust originally exhibited two plastic zones corresponding to the mid-crust and the lower crust. During progressive extension, the upper plastic zone disappeared and the upper and mid–lower-crustal levels became tightly coupled above a weak plastic zone at the base of the crust. Eventually, at stretching factors of about 4–5, the entire crust became brittle and faults cut deep into the mantle. Continuing rifting and movement along these faults pumped large volumes of water through the brittle crust and into the mantle, initiating mantle serpentinization and leading to the development of a décollement at the base of the crust. This weak zone localized crustal extension, leading to complete crustal separation and mantle exhumation.

In contrast, at the onset of extension the Woodlark Basin crust had only a thin brittle carapace over a largely plastic crust. Although these plastic zones thinned during progressive extension, at no time did the entire crust become brittle (Fig. 11b). Instead, at stretching factors greater than c. 3–4, the upwelling asthenosphere underwent partial melting, leading to the intrusion of the hot, ductile lower crust by dense basaltic magmas that were 'trapped' in the deep crust, and to the observed volcanism. We suggest that either the occurrence of weak materials along the fault gouge or local modification of the stress field by intrusion of the footwall allowed normal faults to remain active at angles as low as c. 25°. In either case, as virtually all of the strength of the Woodlark Basin lithosphere was within the crust (e.g. Fig. 6), crustal separation is synchronous with total lithospheric failure, resulting in no exposure of the subcrustal lithosphere.

Conclusions

This study suggests that a fundamental influence on the evolution and tectonic architecture of a 'non-volcanic' (i.e. not hotspot-dominated) margin is likely to be the relative importance of mantle serpentinization, controlled by the embrittlement of the overlying crust, and of asthenospheric melting, controlled by the rate and amount of mantle upwelling that accompanies rifting. These in turn are controlled by the initial lithospheric structure, and by the rate and total amount of extension.

Fig. 11. Schematic representation of the temporal evolution of the centre of the rift for the margins of the 'West Iberia type' (**a**) and the Woodlark Basin (**b**). During rifting of cool and strong lithosphere of the West Iberia and SW Greenland margins, uplift (pressure reduction) and cooling moved progressively more of the crust into the brittle regime until the entire crust became brittle at stretching factors of about four. At this point, faults cut down through the crust, allowing hydrous fluids to reach and serpentinize the upper mantle, leading to the development of a décollement at the CMB (e.g. S and H). It should be noted that faults shown are schematic and illustrate only the level of detachment, not the evolution of the fault system. Separation of crustal blocks by movement along this décollement led to the exposure of a broad zone of serpentinized mantle in the COT before the onset of sea-floor spreading. In contrast, during rifting of hot and thick crust of the Woodlark Basin (**b**), the weak lithosphere broke up before the entire crust entered the brittle regime and serpentinized mantle was not exposed at a broad COT. Instead, pressure release melting of the upwelling mantle and mafic intrusion (shown as underplate but also likely to occur as crustal intrusions + volcanism) preceded lithospheric separation and may have led to core complex formation and further lithospheric weakening. Crustal and lithospheric separation occur synchronously when crust fails, leading to a sharp COT.

Table 3. *Defined parameters*

Symbol	Name	Value
K_c	Thermal conductivity in the crust	$2.5\,\text{W m}^{-1}\,\text{K}^{-1}$
K_m	Thermal conductivity in the mantle	$3.4\,\text{W m}^{-1}\,\text{K}^{-1}$
k	Thermal diffusivity	$8 \times 10^{-7}\,\text{m}^2\,\text{s}^{-1}$
r_c	Crustal density	$2800\,\text{kg m}^{-3}$
λ	Hydrostatic pore pressure factor	0.35
g	Acceleration of gravity	$10\,\text{m s}^{-2}$
h_r	Radiogenic heat production length scale	$10^4\,\text{m}$
C_p	Specific heat	$1250\,\text{J kg}^{-1}\,\text{K}^{-1}$
R	Universal gas constant	$8.3143\,\text{J mol}^{-1}\,\text{K}^{-1}$
L'	Effective superheat	$600\,\text{K}$
dT^m/dF	Rise in solidus temperature as a result of melting	$300\,\text{K}$
dT^m/dz	Solidus gradient	$3.25\,\text{K km}^{-1}$

We have used a 1D model for the thermal evolution of the lithosphere during rifting to predict the amount of melt produced and the rheological evolution of the lithosphere during rifting. Our model predicts when the crust should become brittle, allowing mantle serpentinization to start. We have compared the predictions of our model with observed crustal thickness at the landward limit of partially serpentinized mantle and the observed or inferred serpentinized mantle thickness within the COT of the West Iberia and SW Greenland margins. In both cases, the observations match the predictions well for the range of published rift durations. The thin crust and cool, strong lithosphere of West Iberia and SW Greenland became completely brittle at moderate stretching factors, at which point fluids reaching the cool uppermost mantle could initiate serpentinization. The weak serpentinites subsequently formed a weak layer, which acted as a décollement allowing complete crustal separation well before the lithosphere failed and sea-floor spreading started. The result was the exposure of a broad expanse of serpentinized subcontinental mantle lithosphere within the continent–ocean transition at these margins.

The predictions of our model (onset of serpentinization, thickness of the serpentinized zone) also agree with the structure inferred for the Rockall Trough. However, here, although the crust is generally thought to be continental in nature, we speculate that much of the 'crust' beneath the Trough may correspond to partially serpentinized mantle.

In contrast, our model predicts that during rifting of the thick crust and hot, weak lithosphere of the Woodlark Basin, partial melting

of the upwelling asthenosphere was more pronounced than in the 'Iberia type' margins. The weak lithosphere failed at observed stretching factors of *c.* 4–5, long before the crust could become entirely brittle, thus we predict that no serpentinites were formed, and no mantle peridotites were exposed in the continent–ocean transition (none have been observed or inferred). As virtually all the lithospheric strength was concentrated in the crust, crustal separation and lithospheric failure occurred synchronously, yielding a sharp continent–ocean transition or boundary. The type of detachment fault observed at the Woodlark Basin is akin to that formed within the thick, hot, weak crust of the western USA and differs markedly from those developed within and beneath very thin, brittle crust at the West Iberia margin (S, H).

Appendix: Model calculations

We need to know the thermal structure in order to follow the rheological evolution of the lithosphere and to calculate the amount of melt produced during lithospheric extension. This is calculated using a 1D heat transport equation that takes into account vertical advection and conduction as well as the heat of fusion consumed by decompression melting whenever this occurs during extension (Turcotte & Schubert 1982):

$$\frac{\partial T^c}{\partial t} = \kappa \frac{\partial^2 T^c}{\partial z^2} + v \frac{\partial T^c}{\partial z} + \frac{H(z)}{\rho} C_p - L' \frac{dF}{dt}.$$

T^c is the temperature corrected for adiabatic effects: $T^c = T^{\text{true}} - \gamma z$, where T^{true} is the real temperature and γ is the adiabatic gradient

($\gamma \equiv \alpha g T/C_\mathrm{p}$). t is time, v is the vertical velocity, $H(z)$ is the radiogenic heat production, ρ the density, C_p the specific heat, dF/dt is the melt production rate and L' is the latent heat of melting (or heat of fusion), which we convert here into an effective 'superheat' of 600 K (Hess 1992). The values used for these constants are shown in Table 3. Because the strain rate is constant, the upwelling velocity is related to strain rate by

$$v = -\dot{\varepsilon}z$$

where $\dot{\varepsilon}$ is the strain rate and z is depth (defined to be positive downwards). The heat released by the freezing melt is not added in these calculations in large part because it is unclear at which depth melt will pond.

We assume a distribution of radioactive elements that decreases exponentially downward within the crust and is zero in the mantle:

$$H(z) = H\exp(-z/h_\mathrm{r}), \quad z < z_\mathrm{CMB}$$

$$H(z) = 0, \quad z > z_\mathrm{CMB}$$

where z_CMB is the depth of the crustal–mantle boundary (see Tables 1 and 3 for the values of the parameters).

The development of the equations for the production of melt has been described by Phipps Morgan (2001) but will be repeated here for clarity. During ascent, pressure-release melting is likely to occur because the mantle solidus temperature falls much more steeply than the adiabat of upwelling material (Fig.

12). Here we model melt production taking into account the rise in the solidus temperature, T_m, that is associated with progressive melt extraction or depletion of the ascending mantle. Peridotites have a solid-solution compositional change related to melt extraction. After melt extraction, the residual material has a higher solidus (for a given pressure) (Fig. 12b). At any depth, the heat available to drive melting, δq, during an ascent dz is proportional to the temperature difference between the solidus and the adiabat at that depth:

$$\delta q = \rho C_p \left(\frac{dT_\mathrm{m}}{dz} - \gamma\right) dz = \rho C_p \left(\frac{dT_\mathrm{m}^\mathrm{c}}{dz}\right) dz$$

where the first term within parentheses is the solidus gradient and the second the adiabatic gradient. (T_m^c is defined as the solidus temperature corrected for adiabatic effects.)

The heat consumed by the generation of an increment dF of partial melt (at a given depth and time) is

$$\delta q = -\rho \left(L + C_\mathrm{p} \frac{dT_\mathrm{m}}{dF}\right) dF$$

where the first term on the right-hand side is the heat consumed by the latent heat of melting L, and the second term is the additional heat that can be stored in solid state as a result of the melting-induced rise in the solidus temperature dT_m/dF (see Asimov *et al.* 1997; Phipps Morgan 2001). The change in solidus temperature gradient as a result of increasing depletion,

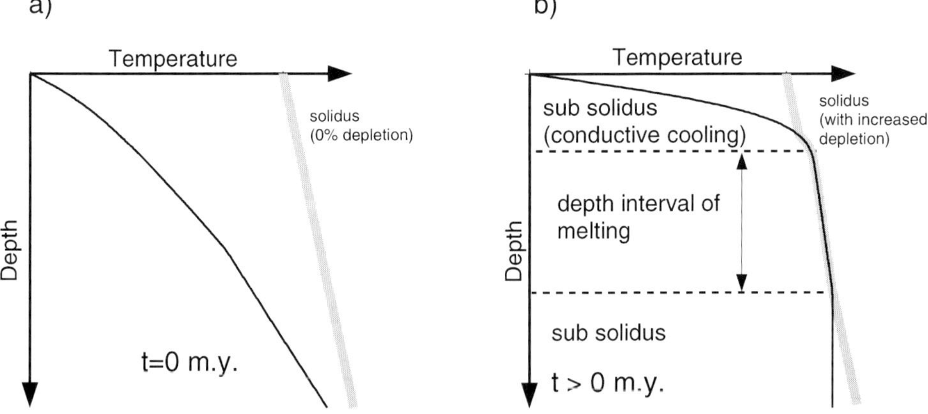

Fig. 12. Model for pressure release melting accompanying mantle upwelling during lithospheric extension. (**a**) At the onset of extension, the geotherm (black line) and the solidus (thick grey line) are shown. With continuing extension, melting occurs whenever the geotherm crosses the solidus (**b**). After melt extraction, the residual material has a raised solidus temperature. As the heat available to drive melting is equal to the heat consumed by melting, the solidus temperature is equal to the geotherm along the melting region (see text).

dT_m/dF, is assumed to be constant (see Table 3). Equating the heat available to drive melting with the heat consumed by melting gives the melt productivity (at a given time during extension):

$$\frac{dF}{dz} = -\left(\frac{dT_m}{dz} - \gamma\right)\bigg/\left(\frac{L}{C_p} + \frac{dT_m}{dF}\right)$$

$$= -\left(\frac{dT_m^c}{dz}\right)\bigg/\left(\frac{L}{C_p} + \frac{dT_m}{dF}\right)$$

which is a constant value wherever melting occurs.

The rate of melt production dF/dt (at a given depth and time) is given by the product of the depth-dependent melt productivity and the upwelling velocity v:

$$\frac{dF}{dt} = v\frac{dF}{dz} = \dot{\varepsilon}z\left(\frac{dT_m^c}{dz}\right)\bigg/\left(\frac{L}{C_p} + \frac{dT_m}{dF}\right).$$

Because all the heat available for melting is consumed during melt generation, after partial melting the temperature in the melting region coincides with the solidus temperature (Fig. 12). The change in temperature within the melting region is governed by the pressure dependence and depletion dependence of the solidus:

$$\frac{dT}{dz} = \frac{dT_m}{dz} - \left(\frac{dT_m}{dF}\right)\frac{dF}{dz}$$

where the first term on the right-hand side of the equation is the slope of the solidus for zero depletion and the second term is the rise in the solidus temperature as a result of depletion, multiplied by the melt productivity. The temperature within the melting region is found by integrating the previous equation over z:

$$T(z) = T_m(z) = \left(\frac{dT_m}{dz}\right)z - \left(\frac{dT_m}{dF}\right)\int_0^z\left(\frac{dF}{dz}\right)dz$$

$$= \left(\frac{dT_m}{dz}\right)z - \left(\frac{dT_m}{dF}\right)p$$

where

$$p = \int_0^z\left(\frac{dF}{dz}\right)dz$$

is the depletion, p.

The heat transport equation is solved using an iterative Crank–Nicholson finite difference scheme (Smith 1985). As boundary conditions, the temperature at the base and top of the region are kept constant throughout the calculations. The initial thermal state is obtained by solving the steady-state heat transport equation

setting the advective term to zero. At every time step the temperature and the solidus are first calculated using the melt production rate and depletion of the previous time step. Once the temperature is known, the new melt production rate and depletion for the actual time step are calculated and these are used to re-evaluate the temperature field and solidus. This is done iteratively until a stable solution for the melt production rate is found.

The melt produced at a time t is obtained integrating over depth the rate of melt production:

$$F(t) = \int_0^z\left(\frac{dF}{dt}\right)dz'.$$

Finally, the total thickness of melt, t_m, produced by a time t since the onset of rifting is (Bown & White 1995)

$$t_m = \frac{1}{\beta_t}\int_0^t F(t')\beta(t')dt'$$

where β_t is the stretching factor at time t.

C. R. Ranero suggested the comparison of the West Iberia margin with Woodlark Basin and contributed with numerous discussions. S. Rouzo helped with the programming of the codes. Discussion with A. Goodliffe contributed to better understanding of the Woodlark Basin. Formal review by R. Buck and G. Manatschal improved the presentation of the ideas in this paper. M.P.G. was supported by a Marie Curie Grant of the EU (ERBFMBICT961467).

References

ABERS, G.A. 2001. Evidence for seismogenic normal faults at shallow dips in continental rifts. *In*: WILSON, R.C.L., WHITMARSH, R.B., TAYLOR, B. & FROITZHEIM, N. (eds) *Non-volcanic Rifting of Continental Margins: a Comparison of Evidence from Land and Sea*. Geological Society, London, Special Publications, **187**, 305–318.

ABERS, G., MUTTER, C. & FANG, J. 1997. Shallow dips of normal faults during rapid extension: earthquakes in the Woodlark–D'Entrecasteaux rift system, Papua New Guinea. *Journal of Geophysical Research*, **102**, 15301–15317.

AGRINIER, P., CORNEN, G. & BESLIER, M.-O. 1996. Mineralogic and oxygen isotopic features of serpentinites recovered from the ocean/continent transition in the Iberia Abyssal Plain. *In*: WHITMARSH, R.B., SAWYER, D.S., KLAUS, A. & MASSON, D.G. (eds) *Proceedings of the Ocean Drilling Program, Scientific Results, 149*. Ocean Drilling Program, College Station, TX, 541–552.

AGRINIER, P., MEVEL, C. & GIRARDEAU, J. 1988. Hydrothermal alteration of the peridotites cored at the ocean/continent boundary of the Iberia margin: petrologic and stable isotope evidence. *In*: BOILLOT, G. & WINTERER, E.L. (eds) *Proceedings of the Ocean Drilling Program, Scientific Results, 103*. Ocean Drilling Program, College Station, TX, 225–234.

ALT, J.C. & SHANKS, W.C. 1998. Sulfur in serpentinized oceanic peridotite: serpentinization processes and microbial sulfate reduction. *Journal of Geophysical Research*, **103**, 9917–9929.

ASIMOV, P.D., HIRSCHMANN, M.M. & STOLPER, E.M. 1997. An analysis of variation in isentropic melt productivity. *Philosophical Transactions of the Royal Society of London, Series A*, **355**, 255–281.

AXEN, G. 1992. Pore pressure, stress increase and fault weakening in low-angle normal faults. *Journal of Geophysical Research*, **97**, 8979–8991.

BANDA, E., TORNE, M. & TALWANI, M. *Rifted Ocean–Continent Boundaries*. NATO ASI Series C, **463**.

BASSI, G. 1995. Relative importance of strain rate and rheology for the mode of continental extension. *Geophysical Journal International*, **122**, 195–210.

BOILLOT, G., WINTERER, E.L. *et al.* 1988. Drilling on the Galicia margin: retrospect and prospect. *In*: BOILLOT, G., WINTERER, E.L. *et al.* (eds) *Proceedings of the Ocean Drilling Program, Scientific Results, 103*. Ocean Drilling Program, College Station, TX, 809–828.

BOILLOT, G., COMAS, M., GIRARDEAU, J. & 5 OTHERS 1988. Preliminary results of the Galinaute cruise: dives of the submersible Nautile on the western Galicia margin, Spain. *In*: BOILLOT, G., WINTERER, E. L. *et al.* (eds) *Proceedings of the Ocean Drilling Program, Scientific Results, 103*. Ocean Drilling Program, College Station, TX, 37–51.

BOILLOT, G., FERAUD, G., RECQ, M. & GIRARDEAU, J. 1989. 'Undercrusting' by serpentinite beneath rifted margins: the examples of the west Galicia margin (Spain). *Nature*, **341**, 523–525.

BOWN, J.W. & WHITE, R.S. 1995. Effect of finite extension rate on melt generation at rifted continental margins. *Journal of Geophysical Research*, **100**, 18011–18029.

BUCK, W.R. 1988. Flexural rotation of normal faults. *Tectonics*, **5**, 959–973.

BUCK, W.R. 1991. Modes of continental extension. *Journal of Geophysical Research*, **96**, 20161–20178.

CANNAT, M., MEVEL, C., MAIA, M. & 8 OTHERS 1995. Thin crust, ultramafic exposures, and rugged faulting patterns at the Mid-Atlantic Ridge (22°–24°N). *Geology*, **23**, 49–52.

CHALMERS, J.A. 1997. The continental margin off southern Greenland: along-strike transition from an amagmatic to a volcanic margin. *Journal of the Geological Society, London*, **154**, 571–576.

CHIAN, C. & LOUDEN, K.E. 1994. The continent–ocean transition across the southwest Greenland margin. *Journal of Geophysical Research*, **99**, 9117–9135.

CHIAN, C., KEEN, C., REID, I. & LOUDEN, K. 1995. Evolution of nonvolcanic rifted margins of the Labrador Sea. *Geology*, **23**, 589–592.

CHIAN, C., LOUDEN, K.E., MINSHULL, T.A. & WHITMARSH, R.B. 1999. Deep structure of the ocean–continent transition in the southern Iberia Abyssal Plain from seismic refraction profiles: Ocean Drilling Program (Legs 149 and 173) transect. *Journal of Geophysical Research*, **104**, 7443–7462.

COLE, J.E. & PEACHEY, J. 1999. Evidence for pre-Cretaceous rifting in the Rockall Trough: an analysis using quantitative plate tectonic modelling. *In*: FLEET, A.J. & BOLDY, S.A.R. (eds) *Petroleum Geology of Northwest Europe: Proceedings of the 5th Conference*. Geological Society, London, 359–370.

CONEY, P.J. 1980. Cordilleran metamorphic core complexes: an overview. *In*: CRITTENDEN, M.D. Jr, CONEY, P.J. & DAVIS, G.H. (eds) *Cordilleran Metamorphic Core Complexes*. Geological Society of America, Memoirs, **153**, 7–31.

DAVIES, H.L. 1980. Folded thrust fault and associated metamorphics in the Suckling–Dayman Massif, Papua New Guinea. *American Journal of Science*, **280**, 171–191.

DAVIES, H.L. & WARREN, R.G. 1988. Origin of eclogite-bearing, domed, layered metamorphic complexes ('core complexes') in the D'Entrecasteaux Islands, Papua New Guinea. *Tectonics*, **7**, 1–21.

DEAN, S.M., MINSHULL, T.A., WHITMARSH, R.B. & LOUDEN, K.E. 2000. Deep structure of the ocean–continent transition in the southern Iberia Abyssal Plain from seismic refraction profiles: the IAM-9 transect at 40°20′N. *Journal of Geophysical Research*, **105**, 5859–5885.

DISCOVERY 215 WORKING GROUP 1998. Deep structure in the vicinity of the ocean–continent transition zone under the southern Iberia Abyssal Plain. *Geology*, **26**, 743–746.

ENGLAND, R.W. & HOBBS, R.W. 1997. The structure of the Rockall Trough imaged by deep seismic reflection profiling. *Journal of the Geological Society, London*, **154**, 497–502.

ESCARTÍN, J., HIRTH, G. & EVANS, B. 1997. Nondilatant brittle deformation of serpentinites: implications for Mohr–Coulomb theory and the strength of faults. *Journal of Geophysical Research*, **102**, 2897–2913.

EVANS, B.W. 1977. Metamorphism of Alpine peridotite and serpentinite. *Annual Review of Earth and Planetary Sciences*, **5**, 397–447.

FERNÁNDEZ, M., MARZÁN, I., CORREIA, A. & RAMALHO, E. 1998. Heat flow, heat production, and lithospheric thermal regime in the Iberian Peninsula. *Tectonophysics*, **291**, 29–53.

FINLAYSON, D.M., COLLINS, C.D.N., LUKASZYK, I. & CHUDYK, E.C. 1998. A transect across Australia's southern margin in the Otway basin region: crustal architecture and the nature of rifting from wide-angle seismic profiling. *Tectonophysics*, **288**, 177–189.

FINLAYSON, D.M., MUIRHEAD, J., WEBB, J.P., GIBSON, G., FURUMOTO, A.S. & COOKE, R.J. 1976. Seismic investigation of the Papuan ultramafic belt. *Geophysical Journal of the Royal Astronomical Society*, **29**, 245–253.

HAUSER, F., O'REILLY, B., BRIAN JACOB, A.W., SHANNON, P., MAKRIS, J. & VOGT, U. 1995. The crustal structure of the Rockall Trough: differential stretching without underplating. *Journal of Geophysical Research*, **100**, 4097–4116.

HESS, P.C. 1992. Phase equilibria constraints on the origin of ocean floor basalts. *In*: PHIPPS MORGAN, J, BLACKMAN, D. & SINTON, J.M. (eds) *Mantle Flow and Melt Generation at Mid-ocean Ridges*. American Geophysical Union, Washington, DC, 67–102.

HILL, E.J., BALDWIN, S.L. & LISTER, G.S. 1992. Unroofing of active metamorphic core complexes in the D'Entrecasteaux Islands, Papua New Guinea. *Geology*, **20**, 907–910.

HITCHEN, K. & RITCHIE, J.D. 1993. New K–Ar ages and a provisional chronology for the offshore part of the British Tertiary Igneous Province. *Scottish Journal of Geology*, **29**, 73–85.

JACKSON, J. 1987. Active normal faulting and crustal extension. *In*: COWARD, M.P., DEWEY, J.F. & HANCOCK, P.L. (eds) *Continental Extensional Tectonics*. Geological Society, London, Special Publications, **28**, 3–17.

JARVIS, G.T. & MCKENZIE, D.P. 1980. Sedimentary basin formation with finite extension rates. *Earth and Planetary Science Letters*, **48**, 42–52.

JOPPEN, M. & WHITE, R.S. 1990. The structure and subsidence of Rockall Trough from two ship seismic experiments. *Journal of Geophysical Research*, **95**, 19821–19837.

KEEN, C.E., POTTER, P. & SRIVASTAVA, S.P. 1994. Deep seismic reflection data across the conjugate margins of the Labrador Sea. *Canadian Journal of Earth Sciences*, **31**, 192–205.

KLEMPERER, S.L. & HOBBS, R.H. *The BIRPS Atlas, Deep Seismic Reflection Profiles around the British Isles*. Cambridge University Press, Cambridge.

KOCH, P. S. 1983. *Rheology and microstructures of experimentally deformed quartz aggregates*. PhD Thesis, University of California, Los Angeles.

KRAWCZYK, C.M., RESTON, T.J., BESLIER, M.-O. & BOILLOT, G. 1996. Evidence for detachment tectonics on the Iberia Abyssal Plain rifted margin. *In*: WHITMARSH, R.B., SAWYER, D.S., KLAUS, A. & MASSON, D.G. (eds) *Proceedings of the Ocean Drilling Program, Scientific Results, 149*. Ocean Drilling Program, College Station, TX, 603–615.

KRONENBERG, A.K. & TULLIS, J. 1984. Flow strength of quartz aggregates: grain size and pressure effects due to hydrolitic weakening. *Journal of Geophysical Research*, **89**, 4281–4297.

KRUSE, S., MCNUTT, M., PHIPPS MORGAN, J., ROYDEN, L. & WERNICKE, B. 1991. Lithospheric extension near Lake Mead, Nevada: a model for ductile flow in the lower crust. *Journal of Geophysical Research*, **96**, 4435–4456.

LISTER, G.S. & DAVIS, G.A. 1989. A model for the formation of metamorphic core complexes and mylonitic detachment terranes. *Journal of Structural Geology*, **11**, 65–94.

LISTER, G.S., BANGA, G. & FEENSTRA, A. 1984. Metamorphic core complexes of Cordilleran type in the Cyclades, Aegean Sea, Greece. *Geology*, **12**, 221–225.

MARTINEZ, F., TAYLOR, B. & GOODLIFFE, A. 1999. Contrasting styles of seafloor spreading in the Woodlark Basin: indications of rift-induced secondary mantle convection. *Journal of Geophysical Research*, **104**, 12909–12926.

MCCARTHY, J., LARKIN, S.P., FUIS, G.S., SIMPSON, R.W. & HOWARD, K.A. 1991. Anatomy of a metamorphic core complex: seismic refraction/wide-angle reflection profiling in southeastern California and western Arizona. *Journal of Geophysical Research*, **96**, 12259–12291.

MCKENZIE, D. 1978. Some remarks on the development of sedimentary basins. *Earth and Planetary Science Letters*, **40**, 25–32.

MCKENZIE, D.P. & BICKLE, M.J. 1988. The volume and composition of melt generated by extension of the lithosphere. *Journal of Petrology*, **29**, 625–679.

MOULLADE, M., BRUNET, M.-F., BOILLOT, G. *et al.* 1988. Subsidence and deepening of the Galicia Margin: the palaeoenvironmental control. *In*: BOILLOT, G., WINTERER, E.L. *et al.* (eds) *Proceedings of the Ocean Drilling Program, Scientific Results, 103*. Ocean Drilling Program, College Station, TX, 733–740.

MUSGROVE, F.W. & MICHENER, B. 1996. Analysis of pre-Tertiary rifting history of the Rockall Trough. *Petroleum Geoscience*, **2**, 353–360.

MUTTER, J.C. 1993. Margins declassified. *Nature*, **364**, 393–394.

MUTTER, J.C. 1995. Hot, fat and falling apart. *Nature*, **374**, 499–500.

MUTTER, J.C., MUTTER, C.Z. & FANG, J. 1996. Analogies to oceanic behaviour in the continental breakup of the western Woodlark Basin. *Nature*, **380**, 333–337.

NADIN, P.A., HOUCHEN, M.A. & KUSZNIR, N.J. 1999. Evidence for pre-Cretaceous rifting in the Rockall Trough: an analysis using quantitative 2D structural/stratigraphic modelling. *In*: FLEET, A.J. & BOLDY, S.A.R. (eds) *Petroleum Geology of Northwest Europe: Proceedings of the 5th Conference*. Geological Society, London, 371–378.

NEWMAN, W. & WHITE, N. 1997. Rheology of the continental lithosphere inferred from sedimentary basins. *Nature*, **385**, 621–624.

NICHOLLS, I.A., FERGUSON, J., JONES, H., MARKS, G.P. & MUTTER, J.C. 1981. Ultramafic blocks from the ocean floor southwest of Australia. *Earth and Planetary Science Letters*, **56**, 362–374.

ODP LEG 173 SHIPBOARD SCIENTIFIC PARTY 1998. Drilling reveals transition from continental breakup to early magmatic crust. *EOS Transactions, American Geophysical Union*, **79**, 180–181.

O'HANLEY, D.S. *Serpentinites: Records of Tectonic and Petrological History*. Oxford Monographs in Geology and Geophysics, 34. Oxford University Press, Oxford.

O'REILLY, B.M., HAUSER, F., JACOB, A.W.B. & SHANNON, P.M. 1996. The lithosphere below the Rockall Trough: wide-angle seismic evidence for extensive serpentinization. *Tectonophysics*, **255**, 1–23.

PARSONS, T. & THOMPSON, G. 1993. Does magmatism influence low-angle normal faulting? *Geology*, **21**, 247–250.

PÉREZ-GUSSINYÉ, M. & RESTON, T.J. 2001. Rheological evolution during extension at non-volcanic rifted margins: onset of serpentinization and development of detachments leading to continental break-up. *Journal of Geophysical Research*, **106**, 3961–3975.

PHIPPS MORGAN, J. (2001). The thermodynamics of pressure-release melting of a veined plum-pudding mantle. *G-cubed,* April (on-line journal).

REID, I. 1994. Crustal structure of a nonvolcanic rifted margin east of Newfoundland. *Journal of Geophysical Research*, **99**, 15161–15180.

RESTON, T.J. 1996. The S reflector west of Galicia: the seismic signature of a detachment fault. *Geophysical Journal International*, **127**, 230–244.

RESTON, T.J., KRAWCZYK, C.M. & KLAESCHEN, D. 1996. The S reflector west of Galicia (Spain): evidence from prestack depth migration for detachment faulting during continental breakup. *Journal of Geophysical Research*, **101**, 8075–8091.

ROBERTS, D.G., MASSON, D.G. & MILES, P.R. 1981. Age and structure of the southern Rockall Trough: new evidence. *Earth and Planetary Science Letters*, **52**, 115–128.

RUTTER, E. 1986. On the nomenclature of modes of failure transitions in rocks. *Tectonophysics*, **122**, 381–387.

SCLATER, J.G., JAUPART, C. & GALSON, D. 1980. The heat flow through oceanic and continental crust and the heat loss of the Earth. *Journal of Geophysical Research*, **18**, 269–311.

SIBSON, R.H., MCMOORE, J. & RANKIN, R.H. 1975. Seismic pumping—a hydrothermal fluid transport mechanism. *Journal of the Geological Society, London*, **131**, 653–659.

SIBUET, J.C. 1992. New constraints on the formation of the non-volcanic continental Galicia–Flemish Cap conjugate margins. *Journal of the Geological Society, London*, **149**, 829–840.

SMITH, G.D. *Numerical Solution of Partial Differential Equations: Finite Difference Methods*. Oxford Applied Mathematics and Computing Science Series. Oxford University Press, Oxford.

SMYTHE, D.K. 1989. Rockall Trough—Cretaceous or Late Paleozoic? *Scottish Journal of Geology*, **25**, 5–43.

SRIVASTAVA, S.P. & ROEST, W.R. 1995. Nature of thin crust across the SW Greenland margin and its bearing on the location of the continent–ocean boundary. *In*: BANDA, E., TORNE, M. & TALWANI, M. (eds) *Rifted Ocean–Continent Boundaries*. NATO ASI Series C, **463**, 95–119.

TAYLOR, B., GOODLIFFE, A. & MARTINEZ, F. 1999. How continents break up: insights from Papua New Guinea. *Journal of Geophysical Research*, **104**, 7494–7512.

TAYLOR, B., GOODLIFFE, A., MARTINEZ, F. & HEY, R. 1995. Continental rifting and initial seafloor spreading in the Woodlark Basin. *Nature*, **374**, 534–537.

TAYLOR, B., HUCHON, P., KLAUS, A. *et al.* (eds) (2000). *Proceedings of the Ocean Drilling Program, Initial Reports, 180*. Available from http://www-odp.tamu.edu/publications/180_IR/180TOC.HTM.

TULLIS, T.E., HOROWITZ, F.G. & TULLIS, J. 1991. Flow law of polyphase aggregates from end-member flow laws. *Journal of Geophysical Research*, **96**, 8081–8096.

TURCOTTE, D.L. & SCHUBERT, G. *Geodynamics Applications of Continuum Physics to Geological Problems*. John Wiley, New York.

WERNICKE, B. 1981. Low-angle normal faults in the Basin and Range province: nappe tectonics in an extending orogen. *Nature*, **291**, 645–648.

WHITMARSH, R.B. & MILES, P. 1995. Models of the development of the West Iberia rifted continental margin at 40°30′N deduced from surface and deep tow magnetic anomalies. *Journal of Geophysical Research*, **100**, 3789–3806.

WHITMARSH, R.B. & SAWYER, D.S. 1996. The ocean/continent transition beneath the Iberia Abyssal Plain and continental-rifting to seafloor-spreading processes. *In*: WHITMARSH, R.B., SAWYER, D.S., KLAUS, A. & MASSON, D.G. (eds) *Proceedings of the Ocean Drilling Program, Scientific Results, 149*. Ocean Drilling Program, College Station, TX, 713–733.

WHITMARSH, R.B., WHITE, R.S., HORSEFIELD, S.J., SIBUET, J., RECQ, M. & LOUVEL, V. 1996. The ocean–continent boundary off the western continental margin of Iberia: crustal structure west of Galicia Bank. *Journal of Geophysical Research*, **101**, 28291–28314.

WILSON, R.C.L., SAWYER, D.S., WHITMARSH, R.B., ZERONG, J. & CARBONELL, J. 1996. Seismic stratigraphy and tectonic history of the Iberia Abyssal Plain. *In*: WHITMARSH, R.B., SAWYER, D.S., KLAUS, A. & MASSON, D.G. (eds) *Proceedings of the Ocean Drilling Program, Scientific Results 149*. Ocean Drilling Program, College Station, TX, 617–633.

Index

Note: Page numbers in **bold** type refer to tables, those in *italic* type to illustrations